Agricultural Practices and Policies for Carbon Sequestration in Soil

Agricultural Practices and Policies for Carbon Sequestration in Soil

J.M. Kimble • R. Lal • R.F. Follett

CRC Press
Taylor & Francis Group
Boca Raton London New York

CRC Press is an imprint of the
Taylor & Francis Group, an **informa** business

CRC Press
Taylor & Francis Group
6000 Broken Sound Parkway NW, Suite 300
Boca Raton, FL 33487-2742

First issued in paperback 2020

© 2002 by Taylor & Francis Group, LLC
CRC Press is an imprint of Taylor & Francis Group, an Informa business

No claim to original U.S. Government works

ISBN 13: 978-0-367-57865-7 (pbk)
ISBN 13: 978-1-56670-581-3 (hbk)

Visit the Taylor & Francis Web site at
http://www.taylorandfrancis.com

and the CRC Press Web site at
http://www.crcpress.com

Library of Congress Cataloging-in-Publication Data

Agriculture practices and policies for carbon sequestration in soil / edited by John M. Kimble, Rattan Lal, and Ronald F. Follett.
    p. cm.
Papers from a symposium held July 1999 at Ohio State University, Columbus, Ohio.
Includes bibliographical references and index.
    ISBN 1-56670-581-9 (alk. paper)
    1. Soils—Carbon content—Congresses. 2. Carbon sequestration—Congresses. I. Kimble,
J. M. (John M.) II. Rattan Lal, 1918- III. Follett. R. F. (Ronald F.), 1939-

S592.6.C36 A44 2002
631.4'1—dc21
    2001050315
    CIP

Library of Congress Card Number 2001050315

Preface

In July 1999, the International Symposium, Agricultural Practices and Policies for Carbon (c) Sequestration in Soil, was held at Ohio State University, in Columbus, for about 75 participants from 11 countries. This was the third symposium (eighth in a series of conferences, symposiums, and workshops) organized by the Soils and Global C Cycle Program. Five of these were held at Ohio State University, one in Tunis, Tunisia, one in Belem, Brazil, and one at Las Cruces, NM. This symposium differed from the others in that the goal was to go beyond the scientific understanding of soil's roles in the global change arena; it was designed to bring together scientists, policymakers, economists, industrial representatives, and members of the farming community. The meeting was designed around several different themes with a generous amount of time set aside for discussions from all participants. Two panel sessions were used for discussion on the themes of practical framing perspective and on policy issues, although scientists often do not take part in such debates. The presentations of 43 papers and 19 posters generated discussion and set the stage for the closing panel discussions.

The overall idea of the conference was to create a dialogue between the different groups with the idea of developing common understanding and ideas. This was accomplished; the outcome is in this book, which will enable the ideas and information generated there to reach a wider audience.

There are 48 chapters under nine themed sections to this book, as well as the foreword, give the political perspective of the importance of the symposium and this volume. The chapters in this book are organized under sections concerning 1) historical perspectives, 2) conservation tillage and residue management, 3) monitoring and assessment, 4) soil management, 5) soil structure and C sequestration, 6) economics of C sequestration, 7) policy issues and industrial and farmer viewpoints, 8) regional pools, and 9) the summary. The content in these chapters, when linked to the previous works, will provide information needed for development of policy and options that will allow soil C sequestration to be considered as a serious option in the raging debate on global climate change. Soils change levels of greenhouse gases through sequestration; at the same time this sequestration will have many other positive farm benefits, such as improved yields, reduced erosion, and lower need for external inputs. Environmental and societal benefits will occur as well. The classic win–win scenario is possible, with even more wins.

Many in society look at agriculture as a problem. Although there are issues of nonpoint source pollution, perceived environmental degradation, odor, and unsafe food, at the same time the general population of the world has cheap food and, in most places, adequate supplies. Many times deficiencies are more a result of distribution than supply. This is not to say that agriculture does not contribute to problems; it does, and one of the major contributions has been the loss of soil organic carbon (SOC), which has led to reduced soil fertility, increased erosion, etc. The point is that SOC is a resource, as are other carbon compounds such as oil and gas, but SOC is a renewable resource. We can gain many environmental benefits by increasing its levels. Society needs to look at agriculture as part of the solution to an increasing concern: global climate change.

This symposium was organized with the support of USDA's Natural Resources Service, the Agricultural Research Service, and Ohio State University. The editors thank all of the authors for their outstanding contributions to this volume. Their efforts will allow others to gain from their work and will, we hope, lead to development of new polices to help to mitigate the greenhouse effect while providing many other benefits to agriculture and society. These efforts have led to a merging of science and policy.

Thanks are due to the staff of Lewis Publishers/CRC Press for their timely efforts in publishing this information and making it available to the scientific and the policy communities. In addition, numerous colleagues, graduate students, and staff at Ohio State University made valuable contributions. We especially thank Lynn Everett for her efforts in organizing the conference and in handling all of the papers included here from the first draft through the peer review process to providing the information to the publisher. We are sure all of the participants and contributors felt that she held their feet to the fire when needed. We also offer special thanks to Brenda Swank for help in preparing this material and assistance in all aspects of the symposium. We are indebted to Dr. Debbie Reed for providing the foreword to this volume, thus putting the book into a context of political debate and showing why the marriage of science and policy is needed if we are going to make progress in addressing the concerns of global climate change. The efforts of many others also were very important in publishing this relevant and important work.

The Editors

The Editors

Dr. John M. Kimble is a research soil scientist at the United States Department of Agriculture (USDA) Natural Resources Conservation Service National Soil Survey Center in Lincoln, NE, where he has worked for the past 20 years. Previously he was a field soil scientist in Wyoming for 3 years and an area soil scientist in California for 3 years. Dr. Kimble has received the International Soil Science Award from the Soil Science Society of America. While in Lincoln, he worked on a U.S. Agency for International Development Project for 15 years helping developing countries with their soil resources and still maintains an active role in international activities. For the last 10 years he has focused more on global climate change and the role soils can play in this area. His scientific publications deal with topics related to soil classification, soil management, global climate change, and sustainable development. Dr. Kimble has worked in many different ecoregions, from the Antarctic to the Arctic and all points in between. With the other editors of this book, he has led efforts to increase the overall knowledge of soils and their relationship to global climate change. In collaboration with Dr. Lal, Dr. Follett, and others, he has produced 11 books related to the topic of global climate change and the role of soils in climate change.

Dr. Rattan Lal is a professor of soil science in the School of Natural Resources at Ohio State University. Prior to joining Ohio State in 1987, he served as a soil scientist for 18 years at the International Institute of Tropical Agriculture, Ibadan, Nigeria. In Africa, Professor Lal conducted long-term experiments on soil erosion processes as influenced by rainfall characteristics, soil properties, methods of deforestation, soil tillage and crop residue management, cropping systems (including cover crops and agroforestry), and mixed and relay cropping methods. He also assessed the impact of soil erosion on crop yield and related erosion-induced changes in soil properties to crop growth and yield. Since joining Ohio State University in 1987, he has continued research on erosion-induced changes in soil quality and developed a new project on soils and global warming. Dr. Lal has demonstrated that accelerated soil erosion is a major factor affecting emission of carbon from soil to the atmosphere. Soil erosion control and adoption of conservation-effective measures can lead to carbon sequestration and mitigation of the greenhouse effect. Professor Lal is a fellow of the Soil Science Society of America, American Society of Agronomy, Third World Academy of Sciences, American Association for the Advancement of Sciences, Soil and Water Conservation Society, and Indian Academy of Agricultural Sciences. He is the recipient of the International Soil Science Award, the Soil Science Applied Research Award of the Soil Science Society of America, the International Agronomy Award of the American Society of Agronomy, and the Hugh Hammond Bennett Award of the Soil and Water Conservation Society. He is the recipient of an honorary degree of Doctor of Science from Punjab Agricultural University, India. Dr. Lal is past president of the World Association of the Soil and Water Conservation and the International Soil Tillage Research Organization. He is a member of the U.S. National Committee on Soil Science of the National Academy of Sciences. He has served on the Panel on Sustainable Agriculture and the Environment in the Humid Tropics of the National Academy of Sciences.

Dr. Ronald F. Follett is supervisory soil scientist with the Agricultural Research Service (ARS) of the USDA with 35 years of research experience. For the past 15 years he has been research leader with the ARS Soil–Plant–Nutrient Research Unit in Fort Collins, CO. He previously served for 10 years as a national program leader with ARS headquarters in Beltsville, MD, and has also been a research soil scientist with ARS in Mandan, ND, and Ithaca, NY. Dr. Follett is a Fellow of the Soil Science Society of America, American Society of Agronomy, and the Soil and Water Conservation Society. He has twice received USDA's highest award, the USDA Distinguished Service Award, and in June of 2000 he received an Individual USDA Superior Service Award "for promoting sensible management of natural, soil, and water resources for an environmentally friendly and sustainable agriculture." Dr. Follett organized and wrote the ARS strategic plans for Ground-Water Quality Protection — Nitrates and Global Climate Change — Biogeochemical Dynamics.

He has edited several books, most recently, *The Potential of U.S. Grazing Lands to Sequester Carbon and Mitigate the Greenhouse Effect*. Dr. Follett has co-authored *The Potential of U.S. Cropland to Sequester Carbon and Mitigate the Greenhouse Effect*, and served as a guest editor for the *Journal of Contaminant Hydrology*. His scientific publications include topics about nutrient management for forage production, soil-N and -C cycling, groundwater quality protection, global climate change, agroecosystems, soil and crop management systems, soil erosion and crop productivity, plant mineral nutrition, animal nutrition, irrigation, and drainage.

Contributors

J. Adams
Department of Natural Resources Science
University of Rhode Island
Kingston, Rhode Island

J. M. Adams
Department of Geographical and
 Environmental Studies
University of Adelaide
Adelaide, South Australia
Australia

V. Akala
Formerly Ohio State University
Columbus, Ohio

Stephan L. Albrecht
Columbia Plateau Conservation Research
 Center
USDA-ARS
Pendleton, Oregon

Bruno A. Alesii
Monsanto Company
Cordillera Ranch
Boerne, Texas

B. J. R. Alves
Embrapa-Agrobiologia
Seropédica
Rio de Janeiro, Brazil

John M. Antle
Department of Agricultural Economics and
 Economics
Montana State University
Bozeman, Montana

Asa L. Aradottir
Soil Conservation Service
Gunnarsholt, Iceland

Olafur Arnalds
Agricultural Research Institute
Reykjavik, Iceland

M. Banov
N. Poushkarov Institute of Soil Science and
 Agroecology
Sofia, Bulgaria

John Bennett
Saskatchewan Soil Conservation Association
Indian Head, Saskatchewan
Canada

Jeri L. Berc
USDA-NRCS
Washington, D.C.

Herby Bloodworth
USDA-NRCS
Washington, D.C.

George Bluhm
USDA-NRCS
Land, Air, Water Resources
University of California-Davis
Davis, California

R. M. Boddey
Embrapa-Agrobiologia
Seropédica
Rio de Janeiro, Brazil

Sven U. Bóhm
Department of Agronomy
Kansas State University
Manhattan, Kansas

John Brenner
USDA-NRCS and Natural Resource Ecology
 Laboratory
Colorado State University
Fort Collins, Colorado

C. A. Campbell
Agriculture and Agri-Food Canada
Eastern Cereal and Oilseed Research Center
Central Experimental Farm
Ottawa, Ontario
Canada

K. Chakalov
Niproruda
Mineralagroengineering Laboratory
Sofia, Bulgaria

T. Z. Chernogorova
Executive Agency for Soil Resources
Ministry of Agriculture and Forestry
Sofia, Bulgaria

L. J. Cihacek
Soil Science Department
North Dakota State University
Fargo, North Dakota

Jan Cipra
Natural Resource Ecology Laboratory
Colorado State University
Fort Collins, Colorado

K. Coleman
Soil Science Department
IACR–Rothamsted
Harpenden, Hertfordshire
United Kingdom

R. T. Conant
Natural Resource Ecology Laboratory
Colorado State University
Fort Collins, Colorado

Jason Cordes
Division of Science and Mathematics
University of Minnesota–Morris
Morris, Minnesota

O. C. Devevre
Department of Land, Air, and Water Resources
University of California–Davis
Davis, California

Raina Dilkova
N. Poushkarov Institute of Soil Science and
  Agroecology
Sofia, Bulgaria

T. A. Doane
Department of Land, Air, and Water Resources
University of California–Davis
Davis, California

Clyde L. Douglas, Jr.
Columbia Plateau Conservation Research
  Center
USDA–ARS
Pendleton, Oregon

Brian Dudek
Center for Environmental Management of
  Military Lands
Colorado State University, CEMML
Fort Collins, Colorado

S. W. Duiker
Department of Agronomy
Pennsylvania State University
University Park, Pennsylvania

Edward T. Elliott
School of Natural Resource Sciences
University of Nebraska–Lincoln
Lincoln, Nebraska

Marlen D. Eve
Natural Resource Ecology Laboratory
Colorado State University
Fort Collins, Colorado

L. R. Everett
Ohio State University
Byrd Polar Research Center
Columbus, Ohio

P. D. Falloon
Soil Science Department
IACR–Rothamsted
Harpenden, Hertfordshire
United Kingdom

Ekaterina Filcheva
N. Poushkarov Institute of Soil Science and
  Agroecology
Sofia, Bulgaria

M. Fisher
Centro Internacional de Agricultura CIAT
Cali, Colombia

Ronald R. Follett
Soil Plant Nutrient Research
U.S. Dept. of Agriculture
Ft. Collins, Colorado

A. J. Franzluebbers
USDA
Agricultural Research Service
J. Phillip Campbell Sr. Natural Resource
  Conservation Center
Watkinsville, Georgia

S. Gameda
Agriculture and Agri-Food Canada
Eastern Cereal and Oilseed Research Center
Central Experimental Farm
Ottawa, Ontario
Canada

Gretar Gudbergsson
Agricultural Research Institute
Reykjavik, Iceland

Kenneth M. Hinkel
Department of Geography
University of Cincinnati
Cincinnati, Ohio

Jeffrey Hopkins
USDA
Economic Research Service
Resource Economics Division
Economic Research Service
Washington, D.C.

W. R. Horwath
Department of Land, Air, and Water Resources
University of California–Davis
Davis, California

R. M. House
USDA
Economic Research Service
Resource Economics Division
Economic Research Service
Washington, D.C.

R. C. Izaurralde
Pacific Northwest Laboratory
Washington, D.C.

A. M. Johnston
Potash and Phosphate Institute of Canada
Saskatoon, Saskatchewan
Canada

Cliff T. Johnston
Department of Agronomy
Purdue University
W. Lafayette, Indiana

M. Jokova
N. Poushkarov Institute of Soil Science and
  Agroecology
Sofia, Bulgaria

Timothy Kautza
Des Moines, Iowa

George Kerchev
N. Poushkarov Institute of Soil Science and
  Agroecology
Sofia, Bulgaria

Milena Kercheva
N. Poushkarov Institute of Soil Science and
  Agroecology
Sofia, Bulgaria

Kendrick Killian
Natural Resource Ecology Laboratory
  Colorado State University
Fort Collins, Colorado

John M. Kimble
Natural Resources Conservation Service
National Soil Survey Center
Lincoln, Nebraska

Jim Kinsella
Farmer
Lexington, Illinois

O. Kostov
N. Poushkarov Institute of Soil Science and
  Agroecology
Sofia, Bulgaria

A. W. Kramer
Department of Agronomy and Range Science
University of California–Davis
Davis, California

A. Kulikov
Executive Agency for Soil Resources
Ministry of Agriculture and Forestry
Sofia, Bulgaria

Rattan Lal
School of Natural Resources
Ohio State University
Columbus, Ohio

A. Lantz
School of Natural Resources
Ohio State University
Columbus, Ohio

B. C. Liang
Pollution Data Branch
Environment Canada
Hull, Quebec, Canada

E. Lioubimtseva
Department of Geography and Planning
Grand Valley State University
Allendale, Michigan

Dian Lopez
Division of Science and Mathematics
University of Minnesota
Morris, Minnesota

Beata Madari
Department of Soil Science and Agricultural
  Chemistry
Szent Istvan University
Gödöllö, Hungary

Andrew P. Manale
Senior Policy Analyst
Henry A. Wallace Center of Winrock
  International
Arlington, Virginia

S. J. Marshall
School of Biological Sciences
University of Nottingham
Sutton Bonnington Campus
Loughborough, Leicestershire
United Kingdom

Neil Mattson
Department of Horticulture
University of Minnesota
St. Paul, Minnesota

B. G. McConkey
Agriculture and Agri-Food Canada
Semiarid Prairie Agricultural Research Center
Swift Current, Saskatchewan
Canada

D. W. Meyer
Plant Sciences Department
North Dakota State University
Fargo, North Dakota

G. J. Michaelson
University of Alaska–Fairbanks
Palmer Research Center
Palmer, Alaska

Erika Michéli
Department of Soil Science and Agricultural
  Chemistry
Szent Istvan University
Gödöllö, Hungary

F. P. Miller
Emeritus Professor of Soil Science
School of Natural Resources
Ohio State University
Columbus, Ohio

Dave Mitchell
Saskatchewan Soil Conservation Association
Indian Head, Saskatchewan
Canada

T. Mitova
N. Poushkarov Institute of Soil Science
Sofia, Bulgaria

Siân Mooney
Department of Agricultural Economics and
  Economics
Montana State University
Bozeman, Montana

A. P. Moulin
Brandon Research Center
Agriculture and Agri-Food Canada
Brandon, Manitoba
Canada

S. Nedyalkov
Executive Agency for Soil Resources
Ministry of Agriculture and Forestry
Sofia, Bulgaria

Tamás Németh
Research Institute for Soil Science and
  Agricultural Chemistry
Hungarian Academy of Sciences
Budapest, Hungary

Alan Olness
USDA
Agricultural Research Service
North Central Soil Conservation Research
  Laboratory
Morris, Minnesota

László Pásztor
GIS Lab
Research Institute for Soil Science and
  Agricultural Chemistry
Hungarian Academy of Sciences
Budapest, Hungary

Keith Paustian
Department of Soil and Crop Sciences and
  Natural Resource Ecology Laboratory
Colorado State University
Fort Collins, Colorado

M. Peters
USDA
Economic Research Service
Resource Economics Division
Economic Research Service
Washington, D.C.

L. Petrova
N. Poushkarov Institute of Soil Science and
  Agroecology
Sofia, Bulgaria

C. L. Ping
University of Alaska-Fairbanks
Palmer Research Center
Palmer, Alaska

T. Popova
Niproruda
Mineralagroengineering Laboratory
Sofia, Bulgaria

Debbie Reed
National Environmental Trust
Washington, D.C.

D. C. Reicosky
USDA
Agricultural Research Service
North Central Soil Conservation Research
  Laboratory
Morris, Minnesota

Charles W. Rice
Department of Agronomy
Kansas State University
Manhattan, Kansas

William Richards
Circleville, Ohio

Ron W. Rickman
Columbia Plateau Conservation Research
  Center
USDA–ARS
Pendleton, Oregon

S. V. Rousseva
N. Poushkarov Institute of Soil Science
Sofia, Bulgaria

D. Sabourin
Agriculture and Agri-Food Canada
Eastern Cereal and Oilseed Research Center
Central Experimental Farm
Ottawa, Ontario
Canada

Alan J. Schlegel
Southwest Research Extension Center
Tribune, Kansas

W. J. Slattery
Agriculture Victoria Rutherglen
Rutherglen Research Institute (RRI)
Rutherglen, Victoria
Australia

Gordon R. Smith
Environmental Resources Trust
Seattle, Washington

J. U. Smith
Soil Science Department
IACR-Rothamsted
Harpenden, Hertfordshire
United Kingdom

P. Smith
Soil Science Department
IACR-Rothamsted
Harpenden, Hertfordshire
United Kingdom

V. Somlev
N. Poushkarov Institute of Soil Science and
  Agroecology
Sofia, Bulgaria

M. Sperow
Natural Resource Ecology Laboratory
Colorado State University
Fort Collins, Colorado

J. L. Steiner
USDA–ARS
J. Phillip Campbell Sr. Natural Resource
  Conservation Center
Watkinsville, Georgia

A. Surapaneni
Agriculture Victoria Tatura, Institute of
  Sustainable Irrigated Agriculture (ISIA)
Tatura, Victoria
Australia

Colin Sweeney
Department of Medicine
University of Minnesota
Minneapolis, Minnesota

J. Szabó
GIS Lab
Research Institute for Soil Science and
  Agricultural Chemistry
Hungarian Academy of Sciences
Budapest, Hungary

Etelka Tombácz
Department of Colloid Chemistry
Szeged University
Szeged, Hungary

M. G. Ulmer
USDA-NRCS
Bismarck, North Dakota

Noel D. Uri
Competitive Pricing Division
Common Carrier Bureau
Federal Communication Commission
Washington, D.C.

S. Urquiaga
Embrapa-Agrobiologia
Seropédica
Rio de Janeiro, Brazil

C. van Kessel
Department of Agronomy and Range
  Science
University of California-Davis
Davis, California

W. B. Voorhees
USDA–ARS
North Central Soil Conservation Research
  Laboratory
Morris, Minnesota

Michael J. Walsh
Environmental Financial Products LLC
Chicago, Illinois

Steve Williams
Natural Resource Ecology Laboratory
Colorado State University
Fort Collins, Colorado

R. P. Zentner
Agriculture and Agri-Food Canada
Semiarid Prairie Agricultural Research
  Center
Swift Current, Saskatchewan
Canada

Contents

Foreword

During his 12-year tenure in the U.S. Senate, former Senator J. Robert Kerrey of Nebraska served on the Senate Committee on Agriculture, Nutrition, and Forestry. Representing a state with a largely rural, agricultural constituency, Senator Kerrey was an active and vocal member of the committee. He was also one of the first committee members to appeal to the agricultural sector to take heed of growing scientific evidence that global climate change was occurring, and that the potential impacts of unchecked climate change could be devastating to the sector.

Senator Kerrey's admonitions often began with the observation that agriculture, as a business conducted outdoors, faced considerable risk and adversity if regional heat waves, droughts, severe storms, or flooding from climate change were to occur. Over time, and joined by a small bipartisan group of his colleagues on the committee, Kerrey succeeded in jumpstarting a dialogue among the U.S. Department of Agriculture, representatives of farmer and commodity groups, scientists, environmental groups, and others. The dialogue was unique not because it was the first occasion of a dialogue focused solely on climate change and agriculture — numerous multidisciplinary workshops and conferences had been held around the country on the issue — but rather because it was the first occasion whereby Washington policymakers, joined by climate and environmental scientists and other interested stakeholders, engaged in a sustained discussion with farmers, farm group representatives, and agriculture sector interests to address the multifaceted issue of climate change and its potential impacts for the sector. The discussion was educational and instructive for all parties, and focused not only on potential risks to agriculture, but also potential opportunities for the sector to contribute to domestic (and, by extension, international) practices to reduce unchecked emissions of greenhouse gases into the atmosphere. Significantly, the dialogue is slated to continue through the upcoming year.

The sustained interaction afforded by this dialogue among policymakers, scientists, practitioners, and affected stakeholders to deal with a complex societal (and global) problem has been invaluable to the legislative and policy process. The interaction of scientists and policymakers, in particular, often does not occur in as regular or as orderly a manner as would benefit policymaking, in general. The crafting of legislation, in particular, is often a cumbersome, unruly process governed by rules and procedures known in theory around the country, but in practice only "inside the beltway" — the rather insular political world that is Washington, D.C. The more complex a problem requiring legislative solution is, the longer and more deliberative is the process to craft the legislation. In the case of issues with a complex scientific underpinning, such as climate change, the deliberative process demands a (more) thorough, thoughtful interface between scientists, who tend to speak in terms of "uncertainties," and policymakers, who seek bottom-line "certainties" necessary to create good policy and prudent laws.

The science and policy interface is often an uncomfortable, unlikely exchange that takes time to mature and develop, from both a process and an interpersonal perspective. Scientists, by temperament and training, are whetted to a logical, orderly method of discovering the laws of nature that order the world and our surroundings, and to presenting those discoveries in peer-reviewed scientific journals. Often it is the media that play the intermediary role of interpreting to the public — and the policymakers among them — scientific findings, and the implications of those findings to people and to good policy. Frequently, however, relevant scientific details are simplified, misinterpreted, or even omitted in the interpretation and presentation by the media. Such omissions or explanation often occur not deliberately, but as part of the process of making news relevant and newsworthy. Too often, however, the missing details are relevant to policy.

Most scientists have neither the training nor the opportunity to interact in the arena where policies are crafted and laws debated and passed. But it is imperative that this interface exist and be nurtured, particularly where problems are complex and the outcome of policies and legislation imperative. Climate change is one such area.

Scientists and scientific organizations must make it a priority to interact with policymakers and legislators to share their findings and to ensure that their materials are accessible and interpreted for policy and legislative use. Policymakers and legislators must consistently ensure that scientists as well as practitioners are engaged and consulted as part of the legislative and policy development process. In practice, this is a far more complicated and opportunistic process than it would appear. Developing the relationships and interactions with Congressional and federal offices takes time and often a devoted effort, but previous interactions or relationships become key when programs and policies or legislation are considered. Congressional and federal agency staff typically seek the engagement and input of relevant organizations when developing programs, policies, and legislation, but should consider whenever possible the value of sustained, multidisciplinary dialogues or working groups to policy or legislative development.

Global climate change as a phenomenon has been much studied and, at least in this country, much debated. Although all too often the terms of the debate are not scientific, the scientific consensus is solid: climate change is real, and human activities have contributed to the increased accumulation of gases in the atmosphere causing the Earth's surface to warm at an alarming rate — with potentially devastating consequences to ecosystems, weather patterns, and life as we know it. Land use, land-use change, and forestry, among other activities, greatly impact the exchange of greenhouse gases, particularly carbon dioxide, between terrestrial ecological systems and the atmosphere. Land use is thus an important factor in controlling or affecting global climate change.

Agricultural practices that sequester carbon typically represent good policy options and have many ancillary environmental and natural resources benefits beyond helping to reduce greenhouse gas emissions and mitigate global climate change. Soil carbon sequestration, for instance, can be enhanced via farming practices that reduce the loss of soil organic carbon (SOC), and at the same time increase the humification rate, leading to increased organic content in the soil. The addition of organic matter and thus carbon to soil improves soil fertility, productivity, and water retention capacity; increases pesticide retention and breakdown; and reduces soil erosion and the runoff and leaching of nutrients or pesticides. These results clearly benefit farmers and land managers, but additional societal benefits are realized from good land stewardship that leads to cleaner air and water.

The conversion of lands for agricultural use in this country in the last half of the 19th and the first half of the 20th centuries saw the loss of significant amounts of carbon — approximately 5000 million metric tons (MMT) of SOC, according to the editors' 1998 book, *The Potential of U.S. Cropland to Sequester Carbon and Mitigate the Greenhouse Effect* (Lal et al., 1999). By applying "best management practices" on agricultural lands, and with the use of new technologies, the authors surmise that U.S. cropland potentially can sequester 4000 to 6000 MMT of SOC in cropland soils, thus helping to reduce emissions of the greenhouse gases causing global climate change.

Public and private rangelands in the U.S. represent significant additional sequestration opportunities, as also reported by the editors in their companion text, *The Potential of U.S. Grazing Lands to Sequester Carbon and Mitigate the Greenhouse Effect* (Lal et al., 2001). Public and private grazing lands account for more than twice the area of U.S. croplands; carbon sequestration can help to restore degraded grazing land resources while also increasing biomass production and mitigating global warming. The ability of rangelands to sequester both organic and inorganic soil carbon, coupled with the vast land resources, leads the authors to estimate that net sequestration of 17.5 to 90.5 MMT (mean 54.0 MMT) carbon occurs per year on managed and unmanaged U.S. rangelands.

Forests also represent significant sequestration opportunities, as has been documented and discussed more widely in the literature than perhaps these other aspects of agriculture, but opportunities within agroforestry and silvopasture (timber and grazing systems managed on the same area of land) should not be overlooked, either. The USDA National Agroforestry Center in Lincoln, NE, indicates that tree windbreaks in the Great and Central Plains help not only to reduce the need for crop fertilizer and improve water use efficiency, but also to sequester significant additional carbon dioxide — up to 21 MMT carbon dioxide per million acres of planted windbreak (200 million trees) at 20 years. Trees planted as forested riparian buffers act to filter excess nutrients, pesticides, animal wastes, and sediments; they also sequester carbon.

In all instances for agricultural activities, adequate research and modeling and the process of teasing out, in theory and in practice, the best management processes to achieve multiple environmental goals, i.e., greenhouse gas reductions in concert with good stewardship and resource conservation, is needed. With hearings now underway in the U.S. Congress on reauthorization of Farm Bill policies and programs, and significant, growing bipartisan interest in the ability of U.S. agriculture to help mitigate and stem the unabated emissions of greenhouse gases, the timing of this book (and its predecessors) is fortuitous. Continuation of the agriculture and climate change dialogue is also critical to ensure that good legislation and good policy result from the scientific evidence presented here and elsewhere, and contribute to the good stewardship of today's farmers in preserving our natural resources and helping to prevent or mitigate the threats of climate change. Good policy and good legislation, in this case as much as in any other, are dependent upon the adoption of those practices that will sequester carbon if the world is to benefit from agricultural sequestration and its many ancillary environmental benefits, and farmers are to benefit from the stewardship practices as well as the potential sale of carbon credits under a cap-and-trade system.

Finally, even in the event that carbon is not regulated in the near- or medium-term, as some may argue, the practices and effects of the policies discussed in this book and others represent the types of "no-regrets" policies that we should advocate and strive for based solely upon good land stewardship and beneficial results to landowners, land managers, and farmers. In the event that carbon is regulated in the future, the agricultural sector will be poised to play yet another invaluable societal role, by working now to determine how best to achieve common beneficial environmental and conservation results as advocated here, and ensuring that we can adequately measure, monitor, and verify the beneficial results and the actual sequestration of greenhouse gases. U.S. agriculture produces the safest, most abundant, most economical food and fiber supply in the world. Why not also add climate change mitigation to that list?

Debbie Reed
Legislative Director
National Environmental Trust

REFERENCES

Lal et al., eds., *The Potential of U.S. Cropland to Sequester Carbon and Mitigate the Greenhouse Effect*, CRC Press, Boca Raton, FL, 1999.

Lal et al., eds., *The Potential of U.S. Grazing Lands to Sequester Carbon and Mitigate the Greenhouse Effect*, CRC Press, Boca Raton, FL, 2001.

PART I

Historical Perspectives

CHAPTER 1

Carbon Sequestration and the Integration of Science, Farming, and Policy

John M. Kimble, L. R. Everett, Ronald Follett, and Rattan Lal

CONTENTS

INTRODUCTION

Over the past 9 years National Resources Conservation Service (NRCS), Agricultural Research Service (ARS), and Ohio State University jointly organized a series of international symposiums and workshops whose purpose was to look at the relationship among soils, climate change, and potential carbon sequestration. These meetings resulted in a series of books that have had a major impact on the science of carbon sequestration (Lal et al., 2000a, b, c; 1995a, b, c; 1997; 1998a, b; 2001, and Follett et al., 2001). The underlying objective of these meetings was to develop a *scientific base* for understanding the relationship among soils, climate change, greenhouse gas fluxes, and, particularly, the role of carbon sequestration.

These meetings scientifically established that accurate point measurements of soil carbon can be made and scaled to larger areas, i.e., from the point sample to the field to regional and national levels. It was shown that different management practices can reduce the rate of enrichment of CO_2 in the atmosphere. As the concepts of the Kyoto Protocols are considered, the need to better understand how these point measurements and sequestration rates for different practices can be scaled to larger areas becomes crucial. This is particularly significant if there is movement towards the trading carbon credits where a buyer may want to purchase many million metric tons of carbon (MMTC).

In July 1999, a meeting brought together participants from all areas that have an interest in the carbon sequestration question (scientists, economists, policymakers, farmers, land managers, and industrial representatives). This meeting provided an opportunity to create a dialogue among all of the stakeholders relative to agriculture practices, processes that enhance sequestration, and rates and retention time of soil carbon. Meetings that create such a dialogue are necessary because scientists often work in their own world, as do producers and policymakers, so never shall the twain meet. Although in past meetings the science was covered, if it is not converted into workable policies it will just end as an esoteric exercise, sitting on a shelf or just used when scientists talk with each other.

Scientists know that adopting recommended management practices (RMPs) for intensification of agriculture on prime farm land, taking marginal agricultural land out of production agriculture, and restoring degraded soils and ecosystems can lead to increases in the soil C reserves at the expense of the atmospheric pool (Lal et al., 1999; Follett et al., 2001). As a result, management of soil carbon sequestration has become a priority issue and needs to be addressed not just in the scientific realm, but also within the policy arena. The development of policy options requires input from farmers, land managers, and other groups using a holistic approach. This process cannot be conducted in a vacuum but must be done in an atmosphere of cooperation involving many diverse groups working towards the benefit of all. Enrichment of atmospheric CO_2 concentration by about 80 ppmv (from 280 ppmv ca. 1850 to 365 ppmv in 1997) is due partly to land use changes and partly to fossil fuel combustion. The present rate of annual enrichment of atmospheric concentration of CO_2 is 3.3 Pg (1 Pg = petagram = 10^{15} g = 1 billion t). Until the 1950s, as much as 75% of the annual increase in atmospheric CO_2 concentration came from changes in land use, agricultural expansion, and soil cultivation. From then until the 1970s, more than 50% of the annual increase was the result of changes in land use, agricultural expansion into natural ecosystems, and mineralization of soil organic carbon. In the 1990s, at least 20% of CO_2 emissions came from soil cultivation and agricultural practices. With knowledge that this increase in atmospheric CO_2 is affecting the global climate, the need to guide policies regarding agricultural practices becomes vital.

World soils have contributed as much as 55 to 78 Pg of carbon to the total atmospheric CO_2. This is a conservative estimate, possibly even an underestimate as no reliable estimates of the total C pool in world soils are available. The total soil C reserve includes the soil organic carbon (SOC) pool, estimated at 1500 Pg to 1-m depths, and the soil inorganic carbon (SIC) pool, estimated at 750 Pg. Land-use conversion along with improper farming practices may release C from both soil pools into the atmosphere.

Although C emissions from agricultural activities caused much of the historic enrichment of atmospheric CO_2, sequestering C in world soils can reverse the trend. RMPs' intensification of agriculture on prime farm land, removal of marginal agriculture land from agricultural production, and restoration of degraded soils and ecosystems can increase the soil C reserve at the expense of the atmospheric pool. Thus, managing C sequestration in soil is a priority issue that needs to be addressed.

Increasing SOC content enhances soil quality, reduces soil erosion, improves water quality, and increases biomass and agronomic productivity. It also improves the environment by absorbing pollutants from natural waters and reducing the rate of enrichment of atmospheric CO_2 concentration. Thus, in addition to meeting basic needs (food, feed, and fiber), scientific agriculture offers a solution to several environmental issues in general and to the greenhouse effect in particular. In some cases, managed ecosystems can produce more biomass than a natural ecosystem (because natural ecosystems may be subject to more drought, sub-soil acidity, P and N deficiency, and nutrient imbalance). Therefore, adopting appropriate land use and judicious soil–nutrient–water–plant management may sequester more C than was lost historically through agricultural land use. Converting from plow-based to no-till or conservation tillage, eliminating summer fallow, incorporating cover crops into the rotation cycle, judiciously using fertilizers, integrating nutrient management by using biosolids and manure, and appropriately managing water can enhance yield, return biomass to the soil, and increase the SOC pool. Thus the sequestration of C in soil is a win–win strategy.

Using available technical information to develop an action program to sequester C in soil and terrestrial ecosystems requires a holistic approach, one that brings the scientific community together with policymakers, farmers, and land managers and that works from the bottom up, as a grass roots movement. Farmers and land managers must work with researchers and policymakers to identify strategies that can lead to widespread adoption of management practices to enhance productivity and increase the size of the soil carbon pool.

With this in mind, a 5-day conference was organized and held at Ohio State University in July 1999 with the following objectives to:

1. Assess historic C loss from world, regional, and U.S. soils
2. Evaluate the potential of adopting RMPs on C sequestration in principal ecoregions and major soils
3. Determine the economic cost of the unit quantity of SOC in terms of productivity and environmental regulatory capacity
4. Identify policy issues that enhance C sequestration in soils
5. Identify criteria and indicators of farmers' adoption of C sequestration practices
6. Develop policies to implement RMPs

The major themes that were discussed during the conference included:

1. Historic loss of soil carbon from various pools
2. Impacts of RMPs on soil carbon dynamics on regional and national scales
3. Regional and global trends in the adoption of RMPs
4. Economics of C sequestration
5. Policy considerations
6. Farmer participation and inputs

PERCEPTIONS

Several concerns and myths have stymied the ability to deal with the problem of soil and environmental degradation in general, and of the accelerated greenhouse effect in particular. These perceptions lead to a poor understanding, which results in the proliferation of soil degradation and contamination, pollution of natural waters, depletion of the SOC pool, emissions of radiatively active gases from soil to the atmosphere, and the threat of an accelerated greenhouse effect. They underscore the need for a public forum of land managers, researchers, and policymakers. More concerns and myths exist than can be addressed in this chapter, but a few that will be discussed include:

- Land (soil) is a resilient and inexhaustible commodity.
- Nature and wilderness are to be conquered and subdued.
- Soil and dirt are synonymous.
- Agricultural science as it pertains to soil carbon is a work in progress.
- Engineering technologies can fix anything.
- Experimental data are too site-specific and difficult to extrapolate or scale up.
- Soil C cannot be measured as accurately as can C stored in aboveground biomass (trees).
- Computer models are infallible and can provide all of the answers.

Such perceptions must be changed in order to effectively replace them with workable solutions. It was with this in mind that the symposium was organized, the objectives of which were to (1) present scientific facts about the attributes of soil and natural resources, (2) discuss strategies for sustainable use of these finite, nonrenewable, fragile resources, and (3) highlight advances in soil science related to sequestering C.

The real concern is why these perceptions propagate and remain in the minds of people. One obvious reason is the lack of any tangible dialogue among various stakeholders. Scientists know soil is not dirt and is not indestructible, yet they are unable to properly convey this information to policymakers; they do not really work with farmers to show them that soil should be treated as a valued commodity or important asset. Farmers have been shown how to apply more fertilizers and use improved crop varieties to overcome the slow, insidious reduction in soil quality. The result is that production has increased, but at what cost to the overall environment? More natural gas is used to produce more nitrogen fertilizers. Soil structure has been destroyed, thus causing increased soil erosion and loss of water-holding capacity, but yield is maintained by adding more and more inputs. One does not ask or perhaps chooses to ignore the question: is such a system sustainable? The answer is that it is not sustainable, but for some reason this information seems unimportant and is ignored by the general population, including policymakers.

The U.S. is fortunate because much of the grain belt region has very fertile soils and, despite poor management, a reasonable level of production has been maintained. However, it has not been possible to stop erosion in many regions and it may continue to accelerate as the soil's structure becomes more damaged, which also leads to a continued loss of SOC. Table 1.1 shows data from soils in Africa and what can happen over time if the integrity of the soil and soil carbon is not maintained. The yield differential occurs despite continued external inputs.

Table 1.1   SOC Changes over Time and the Effect on Maize Yield with Complete Fertilizer

	Egveda Series		Iwo Series		Cambari Series	
	OM (%)	Yield (kg/ha)	OM (%)	Yield (kg/ha)	OM (%)	Yield (kg/ha)
1st crop	2.8	3100	3.0	3800	2.1	2500
3rd crop	1.8	2000	2.4	2600	0.9	1000
7th crop	0.8	1000	1.8	1800	0.6	800
10th crop	0.6	700	1.0	1000	0.3	200

Source: Adapted from Agboola, A.A., in *Organic-Matter Management and Tillage in Humid and Subhumid Africa*, IBSRAM Proceedings, no. 10, 1990.

The above data demonstrate that, even with inputs, there is a significant loss in productivity. If many of these perceptions as well as current ways of farming are not changed, the desertification and massive soil degradation that exist in many other parts of the world are risked. These soils are different from the soils found in the U.S. grain belt, but not so different from the red clays of Georgia and other parts of the Southeastern U.S. Poor farming practices degraded these soils and led to much of the early migration west to new farmland. This practice cannot continue; sustainable farming systems should become a universal practice.

Many questions need to be addressed by scientists, farmers, land managers, and policymakers, including determining why these perceptions exist and how to change the ways people look at soils. There is a need to address reasons for the reduction in the adoption of no-till farming, as well as the idea that more chemicals (herbicides) are needed in no-till farming (yet many long-term, no-till farmers say that they need fewer chemicals). As the dialogue develops, these perceptions can be addressed and how people think about the soil and relevant farming practices changed.

PRACTICAL LIMITATIONS OF ADOPTING NO-TILL FARMING

Many farmers, from very different parts of the globe, emphasize numerous practical problems of and several constraints to widespread adoption of no-till or conservation tillage. In fact, the rigorous no-till practice in Brazil and Argentina is not widely followed in the U.S. specifically because of these constraints. Some perceived limitations of adopting no-till farming given by farmers include:

1. Lack of effective weed control measures
2. High risks (initially) of crop failure and low returns
3. Susceptibility to soil compaction
4. Problems of drought
5. Poor seedling establishment
6. Timeframe before change is visible to farm manager
7. Cultural resistance to change

There is a strong need to overcome these constraints or perceptions, as they are now preventing adoption of new technologies that would provide many environmental as well as societal benefits. The real concern to be addressed is determining if these are really constraints or just perceptions. Is the real problem nothing more than a cultural resistance to change? How does societal pressure correlate with benefit to the farm owner or manager? What about the fact that, even with all these perceptions and constraints, many farmers say these constraints do not really exist? One farmer commented that he had tried no-till farming several times for a year or two each time, thus illustrating a real problem, as change does not occur overnight. Science has repeatedly shown that farmers cannot expect to increase their SOC in the first year or increase the beneficial soil fauna that helps to rebuild the soil structure. How then does one convince the farmer or farm manager that waiting several years to see the improvement will, over the long term, provide a greater benefit, monetarily as well as environmentally? This then becomes a policy-related concern, helping farmers make the necessary changes to their farming practices with the knowledge that for a few years there will be some increased risk of crop failure, a possibility of lower yields and an initial need for more chemical inputs. Constraints, perceptions, and concerns often are simply a lack of understanding and a poor knowledge base.

However, the real problem may just be an inherent resistance to change. Farming has been done in a similar manner for 50 years, with better yields observed over this time, so why change? This is not taking into consideration how much of the yield increase resulted from improved varieties or greater chemical inputs. Also, many times all that is looked at is yield per acre without any consideration to the net profit and other benefits obtained from conservation tillage. A paradigm shift is needed in which only yield and monetary profit are looked at, but also net profit, i.e., the overall environmental benefit of a sustainable farming system. High yields with negative environmental benefits may not be sustainable over the long term and, in actuality, may prove to have irreversible negative impacts on the environment.

More holistic management is needed. This includes looking at other parameters that influence the whole system such as watershed processes and the effect of an upstream process on downstream users. Society will benefit from more holistic management, particularly if water quality, air quality, and other off site conditions are improved. But when this is considered the economic gain of such management, it needs to be passed back somehow to the farmer in a tangible reward system. Providing a reward for proper land stewardship encourages and propagates the use of beneficial management systems. Thus the long term goal must be to have a more sustainable farming system with a reward system in place to provide sufficient incentives to farmers for the inherent risk of changing their agricultural practices.

LINKAGES OF SCIENCE AND POLICY

Many books and journals have addressed the link between science and policy (Watson et al., 2000; Rosenberg et al., 1999; Reilly and Anderson, 1992; Kaiser and Drennen, 1993; and Cline,

1992). For example, the journal *Environmental Science and Policy* published a special issue in 1999 addressing the idea of land use, land-use change, and forestry in the Kyoto Protocol (Wisniewski, 1999). The major trouble is that, as in other such works, most are written by scientists and in a scientific format that does not present concepts and ideas in a simple readable format that most policymakers would find useful. Scientists present esoteric arguments and presume that, since they are clear to them, then they obviously are clear to everyone. Armed with this egocentric impression scientists wonder why policymakers, farmers, and land managers do not use, understand, or even identify with their arguments. To paraphrase the comic strip *Pogo*, "We have met the problem and it is us." Scientists need to develop their interactions with policymakers and put them in a context that can be understood. A book on the potential of U.S. cropland to sequester carbon (Lal et al., 1999) was written in a manner aimed specifically towards policymakers and nonscientific persons. Its popularity among policymakers, farm managers, and scientists provides evidence that science and policy can be brought together in a common forum useful to all.

Articles and books dealing with global climate change (Reilly and Anderson, 1992; Cline, 1992), the adaptation of agriculture to climate change (Schimmelpfenning et al., 1996), or the agricultural dimensions of global climate change (Kaiser and Drennen, 1993) all deal with issues, economic and otherwise, related to agriculture and climate change, but are written mostly by scientists and specifically for the scientific community. All of these books and articles contain needed and useful information, but many times are in a format that farmers and policymakers do not comprehend. In a sense an information overload has been created again, limiting the ability to create an open dialogue among all stakeholders and thus putting up a roadblock to the development of workable solutions. Many times politicians enter into the dialogue, as well they should. One example is the nonbinding vote of the U.S. Senate of 95 to 0 in opposition to the Kyoto Protocols. Once ideas are translated into the political process, they need to be worked and developed so that useful solutions are the outcome.

This workshop demonstrated that what scientists have learned from carbon sequestration can create a precedent that provides a unique opportunity for science to determine policy. As stated earlier, the combined knowledge of scientist, policymaker, and farm manager not only is good policy but will also lead to the development of useful, workable policies that will benefit all stakeholders economically as well as environmentally. Policies put into place today not only will affect the lands of the U.S., but also will have a global impact

The United Nations Framework Convention on Climate Change (UNFCCC) was established to address many issues related to global climate change. This is a necessary process, but all too often the work of this group gets bogged down in the idea of whether or not global warming is occurring. Scientists dealing with atmospheric process models make predictions as to possible future change; however, one group of scientists says there is change while another looks at the same data and says there is no change. The International Panel on Climate Change (IPCC) has issued many reports related to this topic and the general conclusion is that there is an anthropogenic effect, but no one can say for certain to what extent. The basic problem is that uncertainty exists so scientists then say, "Give us more money so we can do more studies," and the big picture gets blurred. The IPCC has done summaries for policymakers (IPCC, 1995; Watson et al., 1996) to facilitate moving the dialogue forward. The IPCC also produced a special report entitled *Land Use, Land-Use Change, and Forestry* (Watson et al., 2000). It was done at the request of the UNFCCC to provide information to the Conference of Parties (COP). This report was written by a large group of eminent scientists and the findings and information contained in it are well documented and based on good science. Many have read the report, but when it is considered in the context of policy, there is a "disconnect" among scientists, negotiators, and policy leaders. Scientists working together wrote the report and it was considered by the politicians; however, there is little or no real interaction between the two groups.

The COP looks at the report line by line, pulling it apart word by word to ensure that no group or country gains an advantage over the other. The science upon which the material was based gets lost in the political repertoire of "what do *I get* from this" or "what do *I not get* from this." At some point one needs to go beyond the hard line views taken by many people (both positive and negative) and work as a team and not as polar opposites. It is necessary to reflect on the possible effects of the rapid and measured increase in greenhouse gases (CO_2 from 280 ppmv to 365 ppmv in just over 150 years). This increase in the gases mirrors the increase in the world's population when plotted. If populations continue to rise as predicted, can one expect CO_2 to keep rising? Or will it increase even faster as the less developed world becomes more developed? The concern at the heart of many of Kyoto Protocols' political considerations is whether or not limits are placed on developing countries.

One can bury one's head in the sand or create a positive dialogue among scientists, farmer, land managers, environmental groups, and policy leaders. All should have a common goal of achieving sustainable agriculture that is both environmentally and economically beneficial. Scientists are in a win–win situation and need to move forward. The win–win concept was very evident in one of the U.S. Senate Agriculture Subcommittee hearings as the senators came up with more win–win options. The material contained in this book will help move this dialogue forward and towards a future where all stakeholders will benefit and the win–win strategy wil become global in nature.

The bottom line is that everyone lives on the same planet and, in reality, works for a common goal. How to get there will, by the nature of the problem, be different for different regions but everyone can and will reach a common goal. It is always said that the devil is in the details. Meeting together has shown that everyone can work together and differences can be overcome. In a way this meeting was the culmination of the overall goals of the meetings organized by the soils and global carbon cycle program. The first goal of this program was to establish and understand the basic science behind soils and the carbon cycle, which was accomplished in the early meetings. In this meeting the goal of the program was to present some science and at the same time integrate it with policy and on-farm management; this goal was also achieved.

Ismail Serageldin talks about Promethean science (Serageldin and Persley, 2000) and the creative and daringly original ideas of much of the Consultative Group for International Agricultural Research (CGIAR) were developed from the Promethean idea of taking for the good. However, recent generations are doing the opposite of what Prometheus promoted by releasing harmful gases into the atmosphere. Prometheus took fire from the gods and gave it to humans; now humans are putting gases into the atmosphere in ever increasing amounts that will lead to fire on the earth.

CONCLUSION

The idea that policies are needed to manage soil carbon is not new, nor is the effect that it has on land values. More than 60 years ago, Albrecht (1938) pointed out that the decrease of organic carbon led to decreased land values and lower crop yields. A quote from his paper is telling and, sadly, still true. " Up to the present, the policy — if it can be called a policy — has been to exhaust the supply, rather than to maintain it by regular additions according to the demands of the crops produced or the soil fertility removed. To continue very long with this practice will mean a further sharp decline in crop yields." In many ways that is still the case. An attempt has been made to restore fertility with inorganic fertilizers and, in many cases, the yields have been maintained with ever increasing inputs, but with a continued decline in the SOC.

The sad part is that policies are still not in place to really encourage the increase and or even the maintenance of SOC. The question now to address is that of why, after almost 70 years, this issue is still on the back burner. SOC needs to be built up in soils because the advice presented in 1938 by Albrecht was not followed. If there had been a continued supply of fresh organic matter added to the soils, the levels would not now be as low as they are. It should be noted that, in many of the soils in the U.S. grain belt, the native levels of SOC were very high and remained highly productive even with years of poor management practices. However, in the tropics and warmer areas, this has not been the case. In these regions almost all of the SOC has been lost with devastating results, i.e., sharp and continuing drops in crop yields leading to failed agriculture systems. The effect of temperature on virgin organic matter content of different soils has been shown to decrease from north to south (Jenny, 1930). For each 10°C decrease in annual temperature the average organic matter content increased two to three times under the same precipitation–evaporation ratio. Moisture also has a major effect, but the effect of temperature is very dramatic.

What will happen to SOC if the climate warms? Is enough now known to promote sustainability in the world's agricultural systems? The question becomes how to create a dialogue to address all the issues and define a workable solution. All of the parts are in place to create the win–win situation described earlier; now the dialogue must move forward so that productivity does not decline and benefits rise, globally, among all of the stakeholders.

REFERENCES

Agboola, A.A., Organic matter and soil fertility management in the humid tropics of Africa, in *Organic-Matter Management and Tillage in Humid and Subhumid Africa*, Bangkok. Thailand: International Board for Soil Research and Management, 1990. IBSRAM Proceedings no. 10.

Albrecht, W.A., Loss of soil organic matter and its restoration, in Soils and Men, Yearbook of Agriculture. USDA, U.S. Government Printing Office, 348–360, 1938.

Cline, W.R., *The Economics of Global Warming*, Institute for International Economics, Washington, D.C., 1992.

Follett, R.F., J.M. Kimble, and R. Lal (eds.), *The Potential of U.S. Grazing Lands to Sequester Carbon and Mitigate the Greenhouse Effect*, Lewis Publishers, Boca Raton, FL, 2001, 442 pp.

IPCC, *Climate Change 1995; Impacts, Adaptations, and Mitigation. Summary for Policy Makers*, IPCC, Washington, D.C., 1995.

Jenny, H., A study on the influence of climate upon the nitrogen and organic matter content of soil, *Mo. Agr. Expt. Sta. Res. Bull.*, 152, 66, 1930.

Kaiser, H.M. and T.E. Drennen (eds.), *Agriculture Dimensions of Global Climate Change*, St. Lucie Press, Delray Beach, FL, 1993, 311 pp.

Lal, R., J.M. Kimble, and B.A. Stewart (eds.), *Global Climate Change and Tropical Ecosystems*, CRC Press, Boca Raton, FL, 2000a.

Lal, R. et al. (eds.), *Global Climate Change and Pedogenic Carbonates*, CRC Press, Boca Raton, FL, 2000c.

Lal, R., J.M. Kimble, and B.A. Stewart (eds.), *Global Climate Change and Cold Regions Ecosystems*, Lewis Publishers, Boca Raton, FL, 2000b, 265 pp.

Lal, R., J. Kimble, and E. Levine (eds.), *Soil Processes and Greenhouse Gas Emissions*, USDA-SCS-NSSC, Lincoln, NE, 1995a.

Lal, R., J. Kimble, and R. Follett (eds.), *Soil Properties and Their Management for Carbon Sequestration*, USDA-NRCS-NSSC, Lincoln, NE, 1997.

Lal, R. et al. (eds.), *Soil and Global Change*, Advances in Soil Science, CRC Press, Boca Raton, FL, 1995b.

Lal, R. et al. (eds.), *Soil Management and Greenhouse Effect*, Advances in Soil Science, CRC Press, Boca Raton, FL, 1995c.

Lal, R. et al., *The Potential of U.S. Cropland to Sequester Carbon and Mitigate the Greenhouse Effect*, Ann Arbor Press, Chelsea, MI, 1999.

Lal, R. et al. (eds.), *Soil Processes and C Cycles*, Advances in Soil Science, CRC Press, Boca Raton, FL, 1998a.

Lal, R. et al. (eds.), Management of Carbon Sequestration, Advances in Soil Science, CRC Press, Boca Raton, FL, 1998b.

Lal, R. et al. (eds.), *Assessment Methods for Soil Carbon*, Lewis Publishers, Boca Raton, FL, 676, 2001.

Reilly, J.M. and M. Anderson (eds.), Economic issues in global climate change, in *Agriculture, Forestry, and Natural Resource*, Westview Press, Boulder, CO, 1992.

Rosenberg, N.J., R.C. Izaurralde, and E.L. Malone, Carbon sequestration in soils: science, monitoring, and beyond, *Proc. St. Michaels Workshop*, December 1998, Battelle Press, Columbus, OH, 1999.

Schimmelpfennig, D. et al., Agricultural Adaptation to Climate Change: Issues of Longrun Sustainability, Natural Resources and Environment Division, ERS, USDA, AER-740, 1996.

Serageldin, I. and G.J. Persley, *Promethean Science — Agriculture Biotechnology, the Environment, and the Poor*, Consultative Group on International Agricultural Research, Washington D.C., 2000.

Watson, R.T., M.C. Zinyowera, and R.H. Moss, *Technologies, Policies and Measures for Mitigating Climate Change*, IPCC, Washington, D.C., 1996.

Watson, R.T. et al., *Land Use, Land-use Change, and Forestry. A Special Report of the IPCC*, Cambridge University Press, Cambridge, England, 2000.

Wisniewski. J., Land use, land-use change and forestry in the Kyoto protocol, *Environ. Sci. Policy* (Special Issue) 2, Elsevier, New York, 216 pp.

CHAPTER 2

Trend in Use of Conservation Practices in U.S. Agriculture and Its Implication for Global Climate Change

Herby Bloodworth and Noel D. Uri

CONTENTS

INTRODUCTION

Organic carbon in soils plays a key role in the carbon cycle and has a potentially large impact on the greenhouse effect (Lal et al., 1998). Soils in the world contain an estimated 1.5×10^{18} g of carbon, or twice as much as the atmosphere and three times the level held in terrestrial vegetation (Post, 1990). Annual net release of carbon from agriculture has been estimated at 2.5×10^{15} g, or about 15% of current fossil fuel emissions globally (Smith, 1999). In addition to the influence that soil carbon has on global warming, it also plays a key role in determining long-term soil quality. Ability to sequester carbon in soils by proper tillage and erosion management provides long-term justification for soil conservation programs (Hunt, 1996 and Weinhold and Halvorson, 1998). There is, however, a paucity of information on the changes in soil organic carbon (SOC) that accrue from key soil conservation programs and policies (Donigian et al., 1994).

After recounting the trend in the use of conservation practices in U.S. agriculture, this chapter provides some estimates of the impact that the increase in the use of three conservation practices has on changes in SOC and, hence, the potential to sequester carbon. The three conservation practices considered are conservation tillage, the conservation reserve program (CRP), and conservation buffer strips.

THE TREND IN THE USE OF CONSERVATION PRACTICES

Conservation Tillage

Conservation tillage is probably the best known conservation practice, evolving from practices that range from reducing the number of trips over the field to raising crops without primary or secondary tillage. Current emphasis is on leaving crop residues on the surface after planting rather than merely reducing the number of trips across the field, although the two are closely related.

In 1963, the Soil Conservation Service of the USDA began recording the area of cropland planted by minimum tillage, which was then used on about 1.6 million planted hectares (about 1% of the total). By 1967, the area had doubled (Mannering et al., 1987).

Use of conservation tillage in the U.S. had an identifiable upward trend until the last few years, which show no discernible change. A longer term perspective can be obtained from Figure 2.1. Use of conservation tillage increased from 1% of planted hectarage in 1963 to approximately 37% of planted hectarage in 1998 (Schertz, 1986; CTIC, 1998). A disaggregated view of the use of conservation tillage can be obtained by considering specific crops and states or regions. The percentage of U.S. cropland that was conservation tilled increased from 26 to 37% over a 10-year period (1989 to 1998), but the increase differs by crop. That is, conservation tillage is used mostly on soybean, corn, and small grains. More than 45% of total corn- and soybean-planted hectarage in 1998 was conservation tilled. Where double-cropping was used, nearly 74% of soybean hectarage, 45% of corn hectarage, and 49% of sorghum hectarage employed conservation tillage systems.

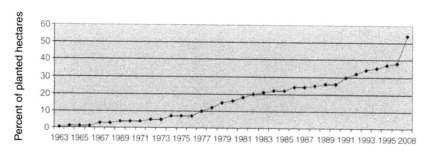

Source: CTIC, 1998

Figure 2.1   Percent of planted hectares on which conservation tillage is adopted in the U.S. and projected percent 2008 hectarage.

Full-season corn is the most extensively grown crop in the U.S., accounting for 27% of total planted hectarage in 1998. Currently, nearly 40% of planted hectarage on which corn is grown is conservation tilled.

Conservation Reserve Program

The policy to take land out of production and place it into conservation uses was first used in the soil bank program of the 1950s, and has been significantly increased in the current CRP. This program was initiated by Congress in Title XII of the Food Security Act of 1985. As a voluntary long term program, the CRP provides participants (farm owners and operators) with an annual per-ha rent and half the cost of establishing a permanent land cover (usually grass or trees) in exchange for retiring highly erodible or environmentally sensitive cropland from production for 10 to 15 years. Although the enrollment mandate established in the 1985 Act was 16.2 to 18.2 million ha by the end of the 1990 crop year, by that point 13.7 million ha had been enrolled.

Under the Food, Agriculture, Conservation, and Trade Act of 1990, an additional 1 million ha were enrolled, bringing total enrollment to 14.7 million ha by 1993. Subsequent appropriations legislation capped CRP enrollment at 15.4 million ha. The most recent farm legislation, as manifest in the Federal Agricultural Improvement and Reform Act of 1996, continues the CRP through 2002.

The establishment of perennial grass cover was an original objective of the CRP. These new grasslands, however, have the potential to reduce concomitantly atmospheric carbon emission levels due to accumulation and incorporation of litter into surface soils and the relatively large amounts of net primary production allocated toward root growth in grasslands (CAST, 1992).

Conservation Buffer Strips

Conservation buffer strips are small areas or strips of land in permanent vegetation designed to intercept pollutants and manage other environmental concerns. As with CRP ha planted to grass, conservation buffer strips have the potential to reduce atmospheric carbon emission levels by the accumulation and incorporation of litter into surface soils and the relatively large amounts of net primary production dispersed to root growth in grasslands (CAST, 1992).

Objective data on the installation of conservation buffer strips are scarce. The national resources inventory (NRI), however, is one available source based on a scientifically selected random sample (Soil Conservation Service, 1992). Figure 2.2 gives information on the number of ha of installed conservation buffer strips in 1997 and 1998. These data were taken from the 1998 Special NRI and represent the most current data available. Number of ha of grassed waterways is significantly greater than that of the other conservation practices.

CONSERVATION PRACTICES AND THE POTENTIAL TO SEQUESTER CARBON

There are different interpretations of what constitutes carbon sequestration. Among the alternatives are (1) soil organic matter (excluding litter), (2) soil organic matter plus soil litter and roots, and (3) below-ground carbon plus the minimum standing stocks of above-ground litter and biomass. The fact that different measures exist leads to some uncertainty in precisely defining the carbon sequestration potential of the various conservation practices.

The objective in what follows is to estimate the likely amount of SOC contained by cropland in the U.S. on which conservation practices are used. This will provide an estimate of the carbon sequestration potential of these practices. Two separate time periods will be considered: 1998, the period for which data on the extent of conservation practices are available, and 2008. For the latter period, forecasts of conservation tillage, CRP ha, and conservation buffer strip ha will be needed.

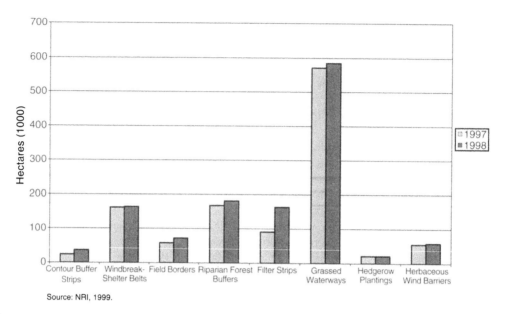

Source: NRI, 1999.

Figure 2.2   Conservation buffer strips installed, 1997 to 1998.

The Year 1998

Conservation Tillage

In 1998, 37.2% of cropland in the U.S. was conservation tilled. Of this total, 43.8% was not tilled. That is, no tillage was practiced on 19.3 million ha in 1998 (CTIC, 1998). Based on climatic conditions, SOC potential for conservation tillage will be approximately 0.27 t per ha per year depending on the climatic conditions, crops grown, soil characteristics, etc. (Lal, 1995). Thus, cropland on which no tillage was used in 1998 sequestered 5.25 million t of carbon. This value is consistent with the results of Kern and Johnson (1993), but somewhat less than the value obtained by Lal et al. (1998).

With 0.27 t per ha per year, cropland under no tillage sequesters an amount of carbon equal to 1.0% of total carbon emissions in the U.S. from burning fossil fuels (Energy Information Administration, 1995). This relative amount is consistent with the value reported by Kern and Johnson (1993).

Conservation Reserve Program

In 1998, 12.5 million ha were enrolled in the CRP. Most of the hectarage, as noted previously, was planted in grass. With CRP hectarage in grass, the amount of SOC that will be contained in the soil ranges is approximately 0.5 t per ha per year (Paustien et al., 1999). Thus, the amount of SOC annually sequestered in CRP hectarage was 6.3 metric million t. This is consistent with the results of Follett (1993), but significantly less than the value reported by Paustien et al. (1999). Using 0.5 t per ha per year, the carbon in the soil associated with the CRP equals about 1.2% of carbon emissions in the U.S. from burning fossil fuels (Energy Information Administration, 1995).

Conservation Buffer Strips

In 1998, there were 1.24 million ha of conservation buffer strips installed in the U.S. Most of the hectarage was planted in grass. With the hectarage in grass, the amount of SOC that will be contained in the soil will be comparable to what one finds for CRP hectarage in grass. Thus, the amount of SOC annually sequestered in conservation buffer strip hectarage was 6.5 million metric tons. Carbon in the soil associated with the conservation buffer strips equals about 0.1% of carbon emissions in the U.S. from burning fossil fuels (Energy Information Administration, 1995).

The Year 2008

The year 2008 was not capriciously chosen to calculate changes in the SOC associated with an increase in the use of conservation practices in the U.S. Rather, 2008 is the final year of the forecast horizon of the current USDA baseline forecasts (World Agricultural Outlook Board, 1999). Forecasts provided include total planted cropland hectarage and CRP hectarage. Not provided is a forecast of hectarage that is conservation tilled; that will be estimated here.

Conservation Tillage

There is some uncertainty about the extent to which conservation tillage — no tillage — will be used in 2008. Consequently, the forecast of its adoption and use must be provided in the context of a confidence interval. To forecast the use of no tillage in 2008, a forecast of the proportion of total planted conservation-tilled hectarage (recall that the historical data are pictured in Figure 2.1) is combined with the baseline forecast of total planted hectarage. This yields the number of ha that is conservation tilled. Assuming that the historical relationship between the proportion of ha that are mulch tilled, ridge tilled, and no tilled holds, the number of ha not to be tilled in 2008 will be obtained.

To estimate the proportion of total planted ha to be conservation tilled in 2008, a simple logistic model is fit to the historical data (Davidson and MacKinnon, 1993). The model is estimated as

$$ct_t = (1 / (1 + a \exp(-bt))) + u_t \qquad (2.1)$$

where ct denotes the proportion of ha conservation tilled, t denotes the time period, u is the stochastic term, and a and b are coefficients to be estimated.

Using nonlinear least squares with an adjustment for first order serial correlation (which is present and which is the highest order empirically observed), the estimated model is

$$ct = (1 / (1 + 5.5571 \exp(-0.1295\ t)$$

$$(0.8453)\ (0.0161) \qquad (2.2)$$

with SSE = 0.5881, log of the likelihood function = 21.7547, and the standard errors of the estimates in parentheses.

Using this functional relationship, the proportion of total planted hectarage expected to be conservation tilled in 2008 is 0.54. The associated 95% confidence interval is 0.46 to 0.62.

Combining this with the baseline forecast and the historical relationship among the three different conservation tillage types means that between 17.7 and 23.1 million ha will be no tilled in 2008. For the next step in the computation, assume that none of the previously no-tilled hectarage is tilled. That is, the net increase in no-tilled hectarage comes only from cropland that was previously conventionally tilled. Also assume, consistent with the literature, that the increase in the SOC of cropland converted from conventional tillage to no tillage is relatively rapid. If the forecast of no-tilled hectarage with its attendant 95% confidence interval is combined with the estimate of the organic carbon content of soil that is conservation tilled, then between 0.9 and 11.6 million t of carbon will be contained in soil conservation tilled in 2008. This represents about a 19% increase over 1998.

Conservation Reserve Program

The baseline forecast predicts that CRP hectarage will increase to its mandated limit of 14.7 million ha by 2002 and remain constant thereafter. Assuming that all of the CRP hectarage is planted in grass, then CRP hectarage will contain between 4.4 and 10.3 million t of carbon in 2008. This is a 17% increase in the carbon in CRP hectarage over 1998.

Conservation Buffer Strips

For conservation buffers strips, it is assumed that the target of 3.2 million ha will be reached by 2008, if not by 2002, which is the original USDA target year. These conservation buffer strips will contain between 1 and 2.4 million t of carbon, which translates into a 162% increase in SOC over the level in 1998. Combining the forecasts for each of the different conservation practices and absent any further policy initiatives, SOC associated with the use of these practices is expected to increase by approximately 25% between 1998 and 2008.

Clearly, the conservation practices considered here contribute greatly to organic carbon in the soil. An increase in their use will further increase SOC and hence enhance the potential to sequester carbon in the U.S. Policy options available to promote the adoption of conservation practices are discussed in the next section.

USE OF PUBLIC POLICIES TO PROMOTE THE USE OF CONSERVATION PRACTICES

Education and Technical Assistance

If a preferred conservation practice would be profitable for a farmer but the farmer is unaware of its benefits, education efforts can lead to its voluntary use. Educational activities generally take the form of demonstration projects and information campaigns in print and electronic media, newsletters, and meetings. Education and technical assistance can be provided by public or private sources; both will induce adoption by farmers for whom the conservation practice would be more profitable than the one previously used.

Financial Assistance

Financial assistance can be offered to overcome either short- or long-term impediments or barriers to adoption. If the conservation practice would be profitable once installed, but involves initial investment or transition and adjustment expenses, a single cost-share payment can be used to encourage switching to the preferred conservation practice. Another form of financial incentive could be granting a tax credit for investment in a particular conservation practice. Cost-share and incentive payment policies are based on the fact that targeted farmers would not voluntarily adopt the preferred conservation practice, but the public interest calls for conservation practices to be used more widely. Financial assistance is not a substitute for education and technical assistance. Even with financial assistance, a farmer will not adopt a technology if he or she is unfamiliar with it.

Research and Development

Research and development (R&D) policies can be used to enhance the benefits of a given conservation practice. The objective of research would be either to improve performance or to reduce costs of the conservation practice. R&D is a long-term policy strategy with an uncertain probability of success, but it may also reap the greatest gains in encouraging the voluntary adoption of a preferred technology because it can increase the profitability of the conservation practice for a wider range of potential adopters.

Land Retirement

The policy with the largest impact on a farmer's choice of conservation practices is land retirement. The underlying premise is that large public benefits can be gained by radically changing agricultural conservation practices on particular parcels of land, and that changes in individual conservation practices would not provide sufficient social benefits. Payment mechanisms that can be used to implement land retirement strategies are lump sum payments or annual "rental fees." The former are often referred to as easements whereby the farmer's right to engage in nonconserving uses is purchased by the public sector for a specific period. Payment to an individual to retire land would result in a voluntary change in conservation practices.

Regulation, Taxes, and Incentives

If voluntary measures prove insufficient to produce the changes in conservation practices necessary to achieve public goals, regulation can be used. Use of certain conservation practices could be prohibited, taxed, or made a basis for withholding other benefits. Preferred conservation practices could be required or tax incentives offered to promote their use thereby offsetting some of the cost of the requisite new equipment. Point sources of pollution have been subject to command-and-control policies for many years. There are recognized inefficiencies associated with technology-based regulations because the least costly technology combination to meet an environmental goal for an individual may not be permitted. It has been assumed that such loss in efficiency is made up for by ease of implementation.

REFERENCES

CAST, *Preparing U.S. Agriculture for Global Climate Change*, Council for Agriculture and Technology, Report 119, Washington, D.C., 1992.

Conservation Technology Information Center (CTIC), *National Crop Residue Management Survey*. CTIC, West Lafayette, IN, 1998.

Davidson, R. and J. Mackinnon, *Estimation and Inference in Econometrics*, Oxford University Press, Oxford, 1993.

Donigian, A. et al., *Alternative Management Practices Affecting Soil Carbon in Agroecosystems of the Central U.S.*, U.S. Environmental Protection Agency, Washington, D.C., 1994.

Energy Information Administration, *Emissions of Greenhouse Gases in the United States 1987–1994*, U.S. Department of Energy, Washington, D.C., 1995.

Follett, R., Global climate change, *J. Prod. Agric.*, 6, 181–190, 1993.

Hunt, P.G., Changes in carbon content of a norfolk loamy sand after 14 years of conservation or conventional tillage, *J. Soil Water Conserv.*, 51, 255–258, 1996.

Kern, J. and M. Johnson, Conservation tillage impacts on national soil and atmospheric carbon levels, *Soil Sci. Soc. Am. J.*, 57, 200–210, 1993.

Lal, R., Conservation tillage for carbon sequestration, paper presented at the Int. Symp. on Soil-Source and Sink of Greenhouse Gases, Nanjing, China, September 1995.

Lal, R. et al., *The Potential of U.S. Cropland to Sequester Carbon and Mitigate the Greenhouse Effect*, Ann Arbor Press, Chelsea, MI, 1998.

Mannering, J., D. Schertz, and B. Julian, Overview of Conservation Tillage. in *Effects of Conservation Tillage on Groundwater Quality*, Lewis Publishers, Chelsea, MI, 1987.

Paustien, K. et al., The contribution of grassland CRP to C sequestration and CO_2 mitigation, unpublished manuscript, Natural Resource Ecology Laboratory, Colorado State University, Ft. Collins, 1999.

Post, W. et al., The global carbon cycle, *Am. Sci.*, 78, 310–326, 1990.

Schertz, D., Conservation tillage: an analysis of acreage projections in the United States, *J. Soil Water Conserv.*, 33, 256–258, 1986.

Smith, K., After the Kyoto protocol: can soil scientists make a useful contribution? *Soil Use Manage.*, 15, 71–75, 1999.

U.S. Department of Agriculture, Natural Resources Conservation Service, *Summary of the 1998 National Resources Inventory*, Washington, D.C., 1999.

U.S. Department of Agriculture, Soil Conservation Service, *Instructions for Collecting 1992 National Resources Inventory Sample Data*, Washington, D.C., 1992.

Weinhold, B. and A. Halvorson, Cropping system influences on several soil quality attributes in the northern Great Plains, *J. Soil Water Conserv.*, 53, 254–258, 1998.

World Agricultural Outlook Board, *USDA Agricultural Baseline Projections to 2008*, U.S. Department of Agriculture, Washington, D.C., 1999.

CHAPTER 3

Why Carbon Sequestration in Agricultural Soils

Rattan Lal

CONTENTS

INTRODUCTION

It is generally agreed that, if the world emission of greenhouse gases (GHGs) continue unabated, atmospheric concentration of carbon dioxide (CO_2) may double by the end of the 21st century. Responding to this concern, world communities signed the Kyoto Protocol in December 1997 under the auspices of the United Nations Framework Convention on Climate Change (UNFCCC, 1998). The Kyoto Protocol (Baker and Barrett, 1999; Faeth and Greenhalgh, 2000) legally binds the developed nations (Annex-1 countries) to reduce emission of GHGs by 5% below the 1990 level. Under this treaty (not ratified by the U.S. Senate), the U.S. would be required to a cumulative reduction in its GHGs over the commitment period 2008 to 2012 to 7% below the 1990 level. In addition to reduction in industrial emissions, the Kyoto Protocol also proposes sequestering carbon (C) in terrestrial sinks. However, the sixth meeting of the Conference of Parties (COP-6) held in

The Hague in November 2000 did not reach any agreement with regard to allowing countries to receive credits for terrestrial sinks, especially for C sequestration in agricultural soils.

The importance of soil organic carbon (SOC) to soil quality and agronomic productivity has long been recognized. Albrecht (1938) stated that "soil organic matter is one of our most important national resources, its unwise exploitation has been devastating, and it must be given proper rank in any conservation policy as one of the major factors affecting the level of crop production in the future." The importance of a changing climate to agriculture has been known for decades, if not centuries (Russell, 1941). Yet, the concept of soil C sequestration is not widely appreciated because of lack of understanding of the role soil processes play in the global C cycle. Further, agriculture has been generally regarded as a source of GHGs and other environmental pollutants.

The idea of using soils to sequester C and mitigate risks of an accelerated greenhouse effect is rather novel. The objectives of this chapter are to (1) explain soil processes and properties relevant to reducing emissions of GHGs and sequestration of C, (2) identify practices leading to an increased SOC pool, and (3) discuss policy options that facilitate adoption of recommended agricultural practices (RAPs). There are at least ten reasons why C sequestration in agricultural soils ought to be included for credits in the Kyoto Protocol.

TEN REASONS FOR SOIL CARBON SEQUESTRATION

Soil Carbon as a Major Global Pool

The soil C pool constitutes a major global reservoir comprising both SOC and soil inorganic carbon (SIC) components. The SOC pool consists of "a mixture of plant and animal residues at various stages of decomposition, of substances synthesized microbiologically and/or chemically from the breakdown products, and of the bodies of live microorganisms and small animals and their decomposing products" (Schnitzer, 1991). The SIC includes elemental C and carbonate minerals of primary and secondary origin. Primary carbonates are derived from the parent material and secondary carbonates are formed through the reaction of atmospheric CO_2 with Ca^{+2} and Mg^{+2} (Lal and Kimble, 2000). The SOC pool is estimated at about 1500 Pg to 1-m depth, 2000 Pg to 2-m depth, and 2340 Pg to 3-m depth (Table 3.1). In comparison, the SIC pool is estimated at 750 to 950 Pg to 1-m depth (Table 3.2). The available data on the SIC pool are highly variable, probably because of the weak database and lack of standard methods.

Table 3.1 Global Estimates of the Soil Organic Carbon Pool

Reference	0–1 m (Pg)	0–2 m (Pg)	0–3 m (Pg)
Jobbagy and Jackson (2000)	1502	1993	2344
Batjes (1996)	1642–1548	2376–2456	—
Eswaran et al. (1993; 2000)	1526–1555	—	—

Table 3.2 Global Estimates of the Soil Inorganic Carbon Pool to 1-m Depth

Reference	SIC pool (Pg)
Batjes (1996)	695–748
Eswaran et al. (1995)	1738
Eswaran et al. (2000)	940
Schlesinger (1982)	780–930
Sombroek et al. (1993)	720

Table 3.3 Soil C Pool to 1-m Depth in Major Soil Orders or Land Types

Order	Area (Mha)	SOC pool (Pg)	SIC pool (Pg)
Alfisols	1262.0	158	43
Andisols	91.2	20	0
Aridisols	1569.9	59	456
Entisols	2113.7	90	263
Gelisols	1126.0	316	7
Histosols	152.6	179	0
Inceptisols	1286.3	190	34
Mollisols	900.5	121	116
Oxisols	981.0	126	0
Spodosols	335.3	64	0
Ultisols	1105.2	137	0
Vertisols	316.0	42	21
Shifting sand	532.2	2	5
Rocky land	1307.6	22	0
Total	13079.5	1526	940

Source: Adapted from Eswaran et al., 2000.

Most currently used methods cannot differentiate between primary and secondary carbonates. Principal soil orders with a large reservoir of SOC include gelisols, inceptisols, histosols, alfisols and ultisols (Table 3.3). Principal soil orders with a large reservoir of SIC pool include aridisols, entisols, mollisols, alfisols and vertisols (Table 3.3). Therefore, total soil C including both SOC and SIC pools in the active soil layer of 1-m depth constitutes about 2300 Pg. The soil C pool of 2300 Pg is three times the atmospheric pool, estimated at 760 Pg, and 3.7 times the biotic pool estimated at 620 Pg. Such a large and active C pool cannot be ignored.

Direct Link between Soil and the Atmospheric Carbon Pools

The soil C pool is highly reactive and dynamic. Cultivation of virgin soils or conversion of natural to agricultural ecosystems leads to depletion of the soil C pool with attendant emission of GHGs (e.g., CO_2, CH_4) into the atmosphere. The reduction in the soil C pool by 1 Pg is equivalent to an atmospheric enrichment of CO_2 by 0.47 ppmv. The reduction in soil C occurs because of the depletion of the SOC pool, which is caused by reduced inputs of biomass carbon (in the form of roots, residues, etc.) and an increase in the output of C from the soil–plant continuum. The C output from the system is accentuated by deforestation, biomass burning, drainage of wetlands, tillage, and removal of crop residue and biomass. Even if the amount of residues or biomass returned is the same as that under natural ecosystems, the rate of C output from agricultural ecosystems is more because of changes in temperature and moisture regimes. Soils of agricultural ecosystems are drier and warmer than those under natural ecosystems, and thus have high rates of oxidation or mineralization (Jenny, 1981).

Misused Agricultural Soils Can Be a Major Source of Greenhouse Gases

Cultivation of agricultural soils is not as obvious a source of GHGs as is direct fossil fuel combustion. However, soil misuse can lead to emissions of GHGs from soil to the atmosphere (Lal, 1999; Batjes and Bridges, 1994). The cumulative historic loss of C by cultivation since the dawn of settled agriculture is enormous (Lal, 2000). From 1850 to 1998, 270 ± 30 Pg C were emitted from fossil fuel combustion and cement production (Marland et al., 1999). During the same period, net cumulative CO_2 emissions from land use change are estimated to have been 136 ± 55 Pg (Houghton, 1997, 1999; Houghton et al., 1999, 2000). The emission from soil cultivation is estimated at 78 ± 17 Pg (Lal, 1999). Therefore, land use and soil management have been a major source of atmospheric enrichment of CO_2. In fact, until 1970, more C came from soil cultivation and deforestation than from fossil fuel combustion. During the 1990s, about 25% of global emissions came from land-use change and soil cultivation (IPCC, 1996); therefore, conversion to appropriate land use and adoption of RAPs is a prudent option to reduce the rate of atmospheric enrichment of CO_2.

Soil Carbon Sequestration Improves Soil Quality

SOC content is an important factor affecting soil quality (Doran et al., 1994; Lal, 1993), defined as soil's biomass productivity and environment moderating capacity (Figure 3.1). The SOC affects soil physical quality through change in soil structure, aggregation, total- and macroporosity, susceptibility to crusting and compaction, and ease of root system development. It has long been recognized that soil structure is the key to high yield and to erosion control (Yoder, 1946). The impacts of the SOC pool on soil fertility result because of SOC capacity to hold and slowly release plant nutrients upon decomposition or mineralization. The SOC also plays a critical role in cycling of principal elements including N, P, S, and Zn. Soil chemical quality depends on its ability to maintain a favorable balance among macro- and microelements, and on processes governing elemental cycling moderated by the SOC pool. Soil biological quality involves population and species diversity of macro- and microfauna, microbial biomass C, microbial processes leading to transformation of biomass, and denaturing and detoxification of applied chemicals and other pollutants.

Soil is a habitat for macro- and microfauna, within and on soil, which transform living and dead organic matter. Biological processes depend on the SOC pool and the characteristics of humic substances in soils. Soil hydrological properties are crucial to plant growth and the environment; the hydrological cycle and partitioning of renewable water into different components (e.g., runoff,

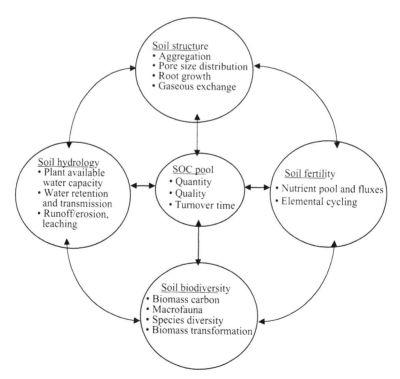

Figure 3.1   The SOC pool and its dynamic play an important role in all components of soil quality and biodiversity.

deep seepage, evapotranspiration, soil water reserve) depend on the SOC pool and processes moderated through it. The dynamics of different components of soil quality govern biomass productivity, and use efficiency of input (e.g., fertilizers, amendments, and irrigation water). Therefore, the SOC pool is as vital for soils as blood is for the animal body (Martius et al., 1999).

The Soc Pool Improves Quality of Natural Waters

Soil as a "geomembrane" is a vast reactor. It filters, buffers, and moderates quality of natural waters (Yaalon and Arnold, 2000). Soil's capacity to absorb and store and release water at variable rates and its capacity to regulate and buffer the hydrological cycle depend on the SOC pool. Through its microbial and chemical processes, soil transforms, denatures, and filters pollutants and purifies natural water.

The quality of surface waters depends on the magnitude of dissolved and suspended loads. Surface runoff from agricultural watersheds contains sediments, sediment-borne pollutants, and dissolved chemicals. The sediment load, as influenced by soil erodibility, is moderated by the SOC pool. Similarly, the nature and concentration of sediment-borne and dissolved chemicals depends on the characteristics and concentrations of humic substances. RAPs that enhance the SOC pool are an important strategy to reduce risks of soil erosion and decrease sediment load in streams and rivers. Similar to the quality of surface water, that of ground water also depends on the SOC pool. Because of the charge properties of SOC, cations and anions are absorbed and desorbed on the exchange complex.

Reduction in Silt Load and Sedimentation

Improvement in soil structure, decrease in soil erodibility, increase in infiltration capacity, and reduction in runoff rate and amount by improvements in the SOC pool lead to reductions in sediment load transported into waterways and reservoirs. The life of reservoirs is shortened by excessive siltation and can be prolonged by reduction in sediment load. The economic benefits of reduced sediment load include less cost from dredging silt, reduction in adverse impacts on fisheries and aquaculture, and low cost of water purification for use by municipalities and urban centers.

Air Quality

There are direct and indirect effects of the SOC pool on air quality. Directly, it affects the nature and rate of emission of GHGs (e.g., CO_2, CH_4, N_2O, NO_x) from soil into the atmosphere. There is a strong interaction between biotransformations of crop residues and biomass and the emission of trace gases. Aerobic transformations lead to emission of CO_2, while anaerobic processes cause the release of CH_4 and N_2O from SOM in the soil into the atmosphere. The rate of biotic processes and the gaseous emissions also depends on soil temperature and moisture regimes.

Indirectly, the SOC pool affects air quality through moderating the rate of emission of particulate material from soil to the atmosphere by wind erosional processes. Similar to erosion by water, the susceptibility to wind erosion also depends on the SOC pool and its interaction with soil texture and clay minerals. Wind-blown dust, often carried over hundreds and thousands of miles from the site of origin, can have severely adverse impacts on environmental quality and human health.

Aesthetic and Economic Value of Land

Enhancement in the SOC pool also improves economic and aesthetic values of the land. High-quality soils have a vast capacity to produce large amounts of biomass of rich nutritional value. Because of low risks of soil erosion, especially of gully erosion that dissects terrain and dismembers landscape, the aesthetic value of land with high-quality soil is high. Water resources are clean and valuable for agricultural use and human and animal consumption. Increases in the SOC pool lead to an increase in the economic worth of the land.

Biodiversity Enhancement

Biodiversity of flora and fauna depends on attributes of the land among which soil is an integral component. Soil quality enhances biodiversity directly and indirectly: directly, by increasing the soil's capacity to produce biomass and enhance use efficiency of water and nutrients, leading to an increased biotic population on and within the soil; indirectly, by improvement in soil quality and its agronomic productivity, leading to a reduction in the degradation of land. Therefore, agriculturally marginal lands can be converted to nonagricultural or restorative land use. Conversion to conservation reserve program (CRP) and afforestation of degraded lands have beneficial impacts on biodiversity. A close link exists between soil quality and soil biodiversity; in fact, there can be no life without soil and no soil without life because both have evolved together (Kellogg, 1938). The SOC pool plays an important role in maintaining the vital link between soil and all life.

Land Stewardship

Soil is the most basic among all natural resources and an integral component of land. Developing a societal understanding of "land stewardship" is important to identifying strategies for sustainable management of natural resources. Soil resources are finite and nonrenewable in terms of the human time frame, unequally distributed among ecological and geographical regions, and prone to degradation because of land misuse and soil mismanagement. As Aldo Leopold wrote, "We abuse land because we regard it as a commodity belonging to us. When we see land as a community to which we belong, we may begin to use it with love and respect" (Udall, 1963). Involving land managers and policymakers in improving soil quality through enhancement of the SOC pool strengthens societal value of the concept of land stewardship. It is important that each generation hands over to the next generation soil and land with a better quality than that received from the previous generation.

The conceptual outline in Figure 3.2 shows numerous ancillary benefits of SOC sequestration. In fact, SOC sequestration is an ancillary benefit of adopting RAPs on agricultural soils for enhancing food security. An important objective of adopting RAPs (e.g., integrated nutrient management, precision farming, conservation tillage, water table management, growing cover crops) is to enhance productivity. While doing so, this strategy also sequesters C in soil and terrestrial ecosystems, and decreases the risks of an accelerated greenhouse effect.

PRACTICES OF SOIL CARBON SEQUESTRATION

A vast body of literature exists documenting the effectiveness of land use and soil management practices for SOC sequestration in relation to soil and ecoregional characteristics (Lal et al., 1998; Batjes, 1999; IPCC, 2000; Lal, 2000). Important among these is conversion from a conventional to a conservation or no-till system of seedbed preparation based on liberal use of crop residue mulch and frequent inclusion of cover crops in the rotation cycle.

There is a long history of conservation tillage. Prior to the invention of motorized tillage equipment, reduced or shallow tillage was the traditional mode of seedbed preparation and still is in several regions of the tropics and subtropics. The modern concepts of conservation tillage have evolved since World War II because of the realization that accelerated erosion has severely adverse impacts on soil and environment qualities (Faulkner, 1942). Thus, conservation tillage has been promoted for erosion control since the 1950s (Buchele, 1956; Meyer and Mannering, 1961). The

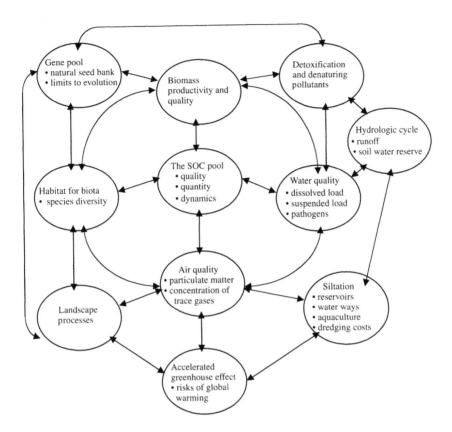

Figure 3.2 Ten reasons for soil C sequestration.

effectiveness of conservation tillage in erosion control is greatly enhanced by use of crop residue mulch, and weed control by economic availability of appropriate chemicals or growth regulators.

In the 1990s, the usefulness of conservation tillage was recognized for C sequestration to reduce the risks of an accelerated greenhouse effect (Lal et al., 1998; Lal, 2000). It is in this context that a strong need exists to extend the ecological limits of application of conservation tillage through adaptive and soil-specific research, especially in soils and crops of the tropics and subtropics. While erosion control leads to maintenance of the SOC pool, the need for C sequestration in soil strengthens development of innovative technologies for widespread adoption of conservation tillage systems.

There are several impediments adopting a conservation tillage system in developing countries. Crop residue, an important input for erosion control and enhancing the SOC pool, is needed for multifarious purposes, important among which is use of residue as fodder, fuel, and construction material. There is a need to develop farming systems that produce enough biomass to cater to alternative needs. Economic and timely availability of herbicides for effective weed control and of appropriate seeding equipment are other issues to be addressed through institutional arrangements and practical research.

STRATEGIES OF CARBON SEQUESTRATION

Carbon sequestration implies removing atmospheric carbon and storing it in natural reservoirs for extended periods. There are numerous strategies of C sequestration (DOE, 1999), most of which can be grouped under two categories: biotic and abiotic (Figure 3.3). The underlying principal for

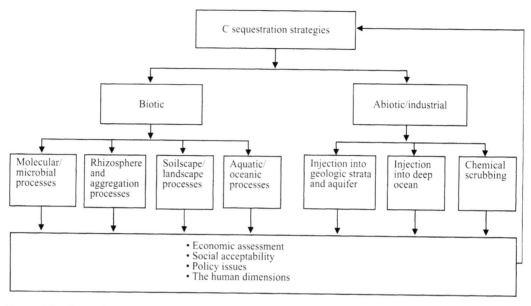

Figure 3.3   Strategies of carbon sequestration to reduce the rate of enrichment of atmospheric carbon dioxide.

biotic options is converting atmospheric CO_2 into biomass and transforming a fraction (10 to 20%) into the SOC pool. There is a wide array of biotic processes ranging from molecular or microbial reactions to those occurring in the rhizosphere, on clay surfaces leading to formation of organo–mineral complexes, within the profile involving humification and leaching of dissolved organic carbon (DOC), and over the soil and landscape involving production of biomass.

Three abiotic processes exist, two of which involve compression of CO_2 emitted by industrial complexes and injection into geologic strata and deep ocean. The third is a soil process that involves dissolution of atmospheric CO_2 leading to formation of carbonates following reaction with Ca^{+2} or Mg^{+2} cations brought in from outside the ecosystem. Leaching of carbonates into ground water is an important mechanism of C sequestration (Nordt et al., 2000). Over and above leaching, annual sequestration of C in soil carbonates and caliches can be 10 to 11 Tg C/yr (Schlesinger, 1997). A report prepared by DOE (Battelle, 2000) concluded that, among all options, SOC sequestration is the most cost-effective strategy available within the next 25 to 50 years. Further, it is a win–win strategy because of numerous ancillary benefits (Lal et al., 1998).

CONCLUSIONS

The historic loss of SOC pool due to conversion from natural to agricultural ecosystems and subsequent soil cultivation by plowing and excessive tillage has contributed a large amount of anthropogenic emissions of CO_2 to the atmosphere. In addition to enhancing the risks of accelerated greenhouse effect and reducing quality of natural waters, depletion of the SOC pool has adversely affected soil quality. Soils with depleted SOC pools have low biomass productivity, especially in farming systems with low external input. In commercial agricultural systems, the use efficiency of water and agricultural chemicals is vastly enhanced through restoration of the SOC pool to an optimum level. Among biotic and abiotic options, SOC sequestration is a win–win strategy, with numerous economic, environmental, and aesthetic benefits. In addition, it is the most cost-effective option. SOC sequestration is a bridge to the future, buying time until alternative energy options, based on noncarbon fuels, can take effect.

For the strategy of the SOC sequestration to take effect, it is important to adopt a holistic approach. This implies an objective assessment of biophysical, socio-economic policy and the human dimensions of adopting RAPs. Extension education is an important aspect of behavioral change. Such a strategy would require a multidisciplinary and integrative approach to sustainable management of soil, water, and vegetation resources.

REFERENCES

Albrecht, W.A., Loss of soil organic matter and its restoration, in *Soils and Men*, USDA Yearbook of Agriculture, Washington, D.C., 347–360, 1938.

Baker, D. and J. Barrett, Cleaning up the Kyoto protocol, Economic Policy Institute, Washington, D.C., 1999.

Batjes, N.H., Total carbon and nitrogen in soils of the world, *Eur. J. Soil Sci.*, 47: 151–163, 1996.

Batjes, N.H., Management options for reducing CO_2-concentrations in the atmosphere by increasing carbon sequestration in the soil, ISRIC Technical Paper 30, ISRIC, Wageningen, Holland, 1999.

Batjes, N.H. and E.M. Bridges, Potential emission of radiatively active gases from soil to atmosphere with special reference to methane: development of a global database (WISE), *J. Geophys. Res.*, 99: 16479–16489, 1994.

Battelle, J., Global energy technology strategy: addressing climate change, *Bulletin*, Washington, D.C., 60pp, 2000.

Buchele, W.F., Ridge farming for erosion control, *Soil Conserv.*, 12: 269–273, 1956.

DOE, *Carbon Sequestration: Research and Development*, U.S. Department of Energy, Office of Scientific and Technical Information, Oakridge, TN, 1999.

Doran, J.W. et al. (eds.), *Defining Soil Quality for Sustainable Environment*, SSSA Spec. Publ. 35, Madison, WI, 1994.

Eswaran, H.E., E. Van den Berg, and P. Reich, Organic carbon in soils of the world, *Soil Sci. Soc. Am. J.*, 57: 192–194, 1993.

Eswaran, H. et al., Global soil carbon resources, in R. Lal et al. (eds.), *Soils and Global Change*, CRC Press, Boca Raton, FL, 27–44, 1995.

Eswaran, H. et al., Global carbon stocks, in R. Lal et al. (eds.), *Global Climate Change and Pedogenic Carbonates*, CRC/Lewis Publishers, Boca Raton, FL, 15–25, 2000.

Faeth, P. and S. Greenhalgh, A climate and environment strategy for U.S. agriculture, World Resources Institute, Climate Protection Initiative, Washington, D.C., 2000.

Faulkner, E.H., *Plowman's Folly*, University of Oklahoma Press, Norma, 1942.

Houghton, J., *Global Warming: The Complete Briefing*, Cambridge University Press, Cambridge, U.K., 1997.

Houghton, R.A., The annual net flux of carbon to the atmosphere from changes in land use 1850–1990, *Tellus*, 50B: 298–313, 1999.

Houghton, R.A., J.L. Hackler, and K.T. Lawrence, The U.S. carbon budget: contributions from land use change, *Science*, 285: 574–578, 1999.

Houghton, R.A. et al., Annual fluxes of carbon from deforestation and regrowth in the Brazilian Amazon, *Nature*, 403: 301–304, 2000.

IPCC, *Climate Change 1995. Impacts, Adaptations and Mitigation of Climate Change: Scientific-Technical Analyses*, Working Group 1, IPCC, Cambridge University Press, Cambridge, U.K., 1996.

IPCC, *Land Use, Land Use Change and Forestry. Inter-Government Panel on Climate Change, Special Report*, Cambridge University Press, Cambridge, U.K., 2000.

Jabbagy, E.G. and R.B. Jackson, Below-ground processes and global change, *Ecol. Appl.*, 10: 423–436, 2000.

Jenny, H., *The Soil Resource*, Springer-Verlag, New York, 1981.

Kellogg, C.E., Soil and society, in *Soils and Men. Yearbook of Agriculture*, USDA, Washington, D.C., 863–886, 1938.

Lal, R., Tillage effects on soil degradation, soil resilience, soil quality and sustainability, *Soil Tillage Res.*, 29: 1–8, 1993.

Lal, R., Soil management and restoration for carbon sequestration to mitigate the accelerated greenhouse effect, *Prog. Env. Sci.*, 1: 307–326, 1999.

Lal, R., World cropland soils as a source or sink for atmospheric carbon, *Adv. Agron.*, 71: 145–191, 2000.

Lal, R. et al., *The Potential of U.S. Cropland to Sequester Carbon and Mitigate the Greenhouse Effect*, Ann Arbor Press, Chelsea, MI, 1998.

Lal, R. and J.M. Kimble, Pedogenic carbonates and the global carbon cycle, in R. Lal et al. (eds.), *Global Climate Change and Pedogenic Carbonates*, CRC/Lewis Publishers, Boca Raton, FL, 1–14, 2000.

Marland, G. et al., Global, regional and national CO_2 emission estimates from fossil fuel burning, cement production and gas flaring 1751–1996, Report NDP-030, Carbon Dioxide Information Analyses Center, ORNL, Oakridge, TN, 1999.

Martius, C., H. Tiessen, and P. Vlek, The challenge of managing soil organic matter in the tropics, ZEF News No. 2 (Sept. 1999), University of Bonn, Germany, 3–4, 1999.

Meyer, L.D. and J.V. Mannering, Minimum tillage for corn, *Agric. Eng.*, 41, 72–75, 1961.

Nordt, L.C., L.P. Wilding, and L.R. Drees, Pedogenic carbonate transformations in leaching soil systems: implications for the global carbon cycle, in R. Lal et al. (eds.), *Global Climate Change and Pedogenic Carbonates*, CRC/Lewis Publishers, Boca Raton, FL, 43–63, 2000.

Russell, R.J., Climatic change through the ages, in *Climate and Man, USDA Yearbook of Agriculture*, Washington, D.C., 67–97, 1941.

Schlesinger, W.H., *Biogeochemistry: An Analysis of Global Change*, Academic Press, San Diego, CA, 1997.

Schlesinger, W.H., Carbon storage in the caliche of the arid world: a case study from Arizona, *Soil Sci.*, 133: 247–255, 1982.

Schnitzer, M., Soil organic matter — the next 75 years, *Soil Sci.*, 151: 41–58, 1991.

Sombroek, W.G., F.O. Nachtergale, and A. Hebel, Amounts, dynamics and sequestration of carbon in tropical and sub-tropical soils, *Ambio*, 22: 417–426, 1993.

UNFCCC, Report of the Conference of the Parties on its Third Session held at the Kyoto from 1–11 December 1997, United Nations, New York, 1998.

Udall, S., *The Quiet Crisis*, Holt, Rinehart, and Winston, New York, 1963.

Yaalon, D.H. and R.W. Arnold, Attitudes toward soils and their societal relevance: then and now, *Soil Sci.*, 165: 5–12, 2000.

Yoder, R.E., Soil structure is key to yield, *Successful Farming*, 44, November 1946.

CHAPTER 4

Historical Perspective and Implications of Human Dimensions of Ecosystem Manipulations: Sustaining Soil Resources and Carbon Dynamics

F. P. Miller

CONTENTS

1-56670-581-9/02/$0.00+$1.50
© 2002 by CRC Press LLC

PERSPECTIVE AND CONTEXT

Land and its complements of soil resources, climates, water resources, flora and fauna, and other renewable and nonrenewable natural resources are inextricably linked to the sustenance of humankind and its level of well-being. This linkage of humans to the earth's natural resource base and its ecological processes (Table 4.1) was as applicable to the earliest stages of anthropogenic occupancy and influence of the earth's ecosystems as it is to contemporary nations and their economies. The human dimensions of ecology act as major manipulators of ecosystems and drive ecosystem changes. As the dominant occupiers of many ecosystems, human beings and their social conventions and values must be central to protocols and policies to mitigate environmental impacts and sustain those ecological processes (Table 4.1) that undergird life and human activity.

As Simpson (1999) has stated:

... humans have always been agents of profound landscape change — we are a disturbance species, especially those societies based on agriculture and industry. We modify our surroundings to better suit us in our quest to acquire the food and shelter necessary for our survival. We also change the world around us to create the comforts and communities we desire. People have always manipulated the landscape to satisfy these common physical and psychological requirements and, in so doing, have created fundamental environmental disturbances that have exhibited striking similarities throughout history and across cultures.

Land and its natural resource components have been the primary common capital from which people's sustenance and well-being have been extracted. The conversion of natural landscapes to people-dominated landscapes has left both environmental and cultural "footprints" or impacts. These anthropogenic footprints range from various forms of land degradation to changes in biological diversity and landscape patterns. Among these ecological disturbances, the natural dynamics of carbon storage and flux in land ecosystems have been significantly altered as levels of anthropogenic impacts across landscapes have increased over time (Lal et al., 1998a; Lal et al., 1998b).

Indeed, natural landscapes have given way to cultural landscapes in many regions of the earth occupied by humans. As population densities have increased across these ecosystems and as levels of affluence increase in the developed and developing nations, the degree of humankind's environmental impact and footprints likewise increases (Figures 4.1 and 4.2). As Paul Kennedy (1993) noted, the overall consensus — except for a few revisionists — is that the projected growth in world population cannot be sustained with the current levels and patterns of resource consumption. Thus, it does make a difference, as Kennedy points out, whether the planet contains 4 billion people, as it did in 1975, engaged in consuming forests, draining wetlands, burning fossil fuels, polluting rivers and oceans (and air), and ransacking the earth for ores, oil, and other raw materials vs. 8 to 9 billion, as is likely in 2025 and beyond.

Table 4.1   Ecosystem Processes, Goods, and Services

Ecosystem "processes" include:

Sustaining solar energy flux, heat dissipation
Climate modulation
Sustaining hydrological flux, hydrological cycle, filtration and water quality
Biological diversity and productivity, plant pollination, sustaining food webs, habitats
Sustaining biogeochemical flux, mineral and gaseous cycling, storage
Decomposition, weathering, soil development-stability, carving-generating landscapes
Sustaining biological diversity
Absorbing, buffering, diluting, detoxifying pollutants–xenobiotics

Ecosystem "goods" include:

Climatic zones conducive to biological diversity and production
Food, wood, fiber, forage, wildlife
Soil and water resources, precipitation, fresh water
Natural resource base for human well-being, life-styles; building materials, energy, ores
Medicinal plants, sources of pharmaceuticals
Biological diversity, large gene pool for enhancing productivity, traits, stress resistance
Natural beauty and ammenities that support tourism and recreation

Ecosystem "services" include:

Maintenance of hydrological cycles, groundwater recharge, water balance, filtration, quality
Modulation of run-off and flooding
Regulating climate, maintaining the ecological niche for humans, flora, fauna
Biological dynamics; O_2 generation, CO_2 sequestration, atmospheric composition, quality
Maintenance of biological diversity
Pollinating crops and other plants
Generating and maintaining soils
Storing and cycling essential nutrients
Absorbing and detoxifying pollutants
Providing beauty, inspiration, and recreation

Note: Healthy ecosystems carry out a diverse array of processes that provide goods and services to humanity. Here, goods refer to items given monetary value in the marketplace, whereas services from ecosystems are valued, but rarely bought or sold.

Source: Modified by F.P. Miller (1999) and others from Ehrlich and Ehrlich 1991, Lubchenco et al., 1993, and Richardson, 1994.

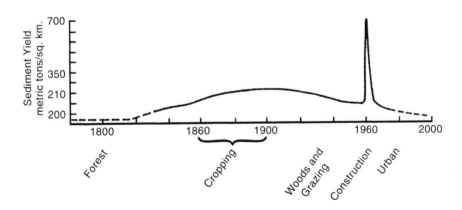

Figure 4.1   Schematic sequence: land-use and sediment yield from a fixed mid-Atlantic area landscape (Wolman, 1967).

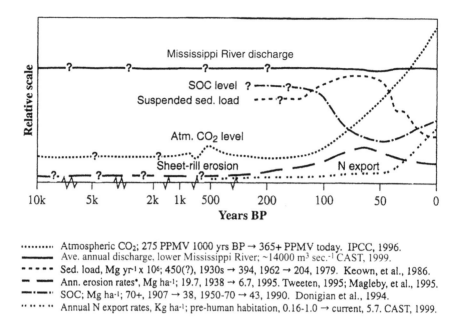

.... Atmospheric CO_2; 275 PPMV 1000 yrs BP → 365+ PPMV today. IPCC, 1996.
——— Ave. annual discharge, lower Mississippi River; ~14000 m³ sec.⁻¹ CAST, 1999.
- - - - - Sed. load, Mg yr⁻¹ x 10⁶; 450(?), 1930s → 394, 1962 → 204, 1979. Keown, et al., 1986.
— — Ann. erosion rates*, Mg ha⁻¹; 19.7, 1938 → 6.7, 1995. Tweeten, 1995; Magleby, et al., 1995.
—·—·· SOC; Mg ha⁻¹; 70+, 1907 → 38, 1950-70 → 43, 1990. Donigian et al., 1994.
·· ·· ·· Annual N export rates, Kg ha⁻¹; pre-human habitation, 0.16-1.0 → current, 5.7. CAST, 1999.

Figure 4.2   Mississippi River ecosystem responses (schematic) to anthropogenic impacts — "footprints."

Changes in land use in the last few centuries have been phenomenal. Richards (1990) notes that, in the past 300 years:

> ... the world's forests and woodlands diminished by 1.2 billion ha, or 19% of the [year] 1700. Grasslands and pastures have declined by 580 million ha, or 8% of the [year] 1700 estimate. Croplands brought into cultivation show a net increase of 1.2 billion ha, for a 466% increase in less than three centuries. ... agricultural expansion and depletion of forests and grasslands were greater in absolute terms over the 30 years between 1950 and 1985 than in the 150 years between 1700 and 1850.

And yet, with all this land use reallocation and with global food production at its highest levels within the last decade, it seems incongruous that discussions are increasingly focused on such looming topics as food security, environmental refugees, the sustainability of agriculture, global change, carbon sequestration, and sustainable development (Miller and Wali, 1995).

What went wrong? Quite simply, the uses of global landscapes and consumption of natural resources have not been commensurate with the sustainability and resilience of those ecosystem processes (Table 4.1) that undergird the renewable natural resource base and its capacity to produce. Ecosystems and natural resources have not been managed sustainably, i.e., within the conceptual framework and principles of basic ecology.

The Ecological Society of America (Christensen et al., 1996) articulated this concept as follows: "... strategies to provide ecosystem goods and services cannot take as their starting points mandated yield goals, timber supply, water demand, or arbitrarily set harvests of fish — instead, sustainability must be the primary objective, levels of commodity and amenity must be adjusted to meet that goal."

Using what is now the U.S. portion of the North American continent as a model, this chapter will briefly trace its conversion from a largely unaltered (anthropogenically) natural ecosystem to a highly developed nation. The implications of human domination of this vast ecosystem will be viewed from the perspective of 1) the ecological footprints resulting from its biological (human) invasion, 2) whether its soil resources (and by association its SOC resources) are or can be sustained, and 3) the lessons learned for establishing an agenda to mitigate these and future impacts.

HISTORICAL SETTLEMENT AND DEVELOPMENT OF THE U.S. PORTION
OF NORTH AMERICA

Hunter–Gatherer Societies

For more than 99% of the nearly 2 million years that humans existed on earth, they lived as hunter–gatherers without agriculture. Buringh (1989), Pimentel (1989), and other authors (e.g., Clark and Haswell, 1970) estimate that the productivity of flora and fauna across various landscapes needed to sustain a hunter–gatherer with a 2500 kcal/day food energy requirement would call for as much as 50,000 ha per person in subarctic lands to as low as 40 ha per person for highly productive ecosystems with ideal biodiversity.

The land requirement to sustain one hunter–gatherer under favorable to moderately favorable environmental conditions was estimated (Clark and Haswell, 1970) to require 150 ha to 250 ha per person. Buringh (1989) estimated that, in general, about 80 ha of land were needed to feed one person in a hunter–gatherer society. Based on these estimates at maximum capacity in the eastern portion of the U.S. Corn Belt, 10 million ha could sustain between 40,000 to 125,000 hunter–gatherers in what is now the State of Ohio (ca. 10.5 million ha). But such numbers over large areas were most likely never realized. Even as late as 1790, the Ohio area numbered only 5000 Indians compared to about 3000 settlers (Simpson, 1999).

Origins of Agriculture

Agriculture began about 10,000 years ago in the Middle East with cultivation of crops and animal confinement, and subsequent domestication of both plants and animals. It eventually spread across the globe under various configurations and levels of intensity ranging from shifting cultivation and bush-fallow (a few years of crop cultivation followed by 10 to 20 or more years of regrowth) systems to more traditional production systems consisting of 2-year crop-fallow systems to the more advanced continuous cropping of cereals in rotation with legumes, root crops, and grasses (Buringh, 1989). Crude implements and rudimentary management techniques were eventually devised by native peoples to increase productivity.

Native American Peoples

The North American continent was populated by a variety of native American Indian nations, tribes, and subtribes. These people and their social structures often had their own languages, customs, cultures, and art forms. Even after the origins of agriculture, most of these Indian tribes were still nomadic, moving with the animals they depended on for their sustenance, which often correlated with the seasons. These peoples tended to have relatively defined geographic areas within which they moved, with perhaps winter and summer camps where they lived.

Later the domestication and cultivation of plants during the summer months and the capacity to store food lessened the need for seasonal or periodic movement. They still fanned out from their compact villages to hunt and fish as well as harvest their natural resource needs such as wood and wood fuel, fiber, and native seasonal berries and other foods. In total, the vastness of the continent and the low density of population resulted in relatively little large-scale disturbance of the landscape or significant impacts upon the environment, although concentrations of people did leave local scars on the landscape. Also their use of fire as a tool for ecosystem management was an important factor in the subsequent biological character of woodlands and prairies affected by this recurring practice.

During the past 1500 years, some of these previously nomadic Indian nations developed very sophisticated social, political, religious, and trade systems and organizations. In some cases, large cities and concentrated communities rose from the landscape, such as the Aztec empire and their monument-centered cities in central and southern Mexico (ca. 1325–1520 AD), the Pueblo Indian-related Anasazi cliff dweller culture in the Mesa Verde area of southwest Colorado (1100–1300 AD), and the city of Cahokia built by the Cahokia subtribe of the Illini Indians (1000–1300 AD) along the American bottoms and terraces on the east side of the Mississippi River in view of what is now the city of St. Louis. Cahokia, built around a woodhenge, is the largest prehistoric Indian city north of Mexico (ca. 20,000 people, estimated) and contains the largest prehistoric earthen mound structure in the western hemisphere. It is estimated that the Cahokia Indians moved more than 1.4 million cubic meters of earth for several mounds constructed in Cahokia.

Except for the Aztec empire captured by Cortez and his Spanish armada in 1519–1520 AD, the Anasazi and Cahokia cities were eventually abandoned. Evidence suggests that the concentration of people in and around these cities exceeded the natural resource base to sustain them (Martin, 1993; Cahokia Mounds Museum Society, 1997). At the Cahokia site, there is very clear local evidence that severe erosion and increased flooding resulted from cutting trees in the local watersheds. Much of the downstream cropland became buried with sediment.

Thus, in these early quasi-urban ecosystems, the combination of overgrazing and overharvesting of trees in the watersheds and subsequent soil erosion, SOC loss, and loss of wildlife habitat eventually contributed to the collapse of these cultures. The human propensity to extract ecosystem goods and services without sustainable management has been and still is repeated often throughout human history since the dawn of civilization (Hillel, 1991; Lowdermilk, 1953; Butzer, 1976).

EUROPEAN SETTLEMENT OF NORTH AMERICA

As the North American continent and its native peoples gave way to European migration into and settlement of this vast continent, the repercussions of unsustainable ecosystem exploitation were visited upon its perpetrators once again. This occupation of North America and subsequent development of a new nation are unmatched in the annals of human history for the sheer magnitude of human impact on a natural resource base of a continent and the speed with which it occurred (Carter and Dale, 1974). As Simpson (1999) noted, nation building is a messy process.

The relatively small numbers and few concentrations of native Americans and the local impacts of those who had been farming, at least to some extent, for several centuries, coupled with the vastness of the land and their lifestyles' requirement of few natural resources, left the North American continent's natural resources virtually intact at the time of the first European settlements. These early settlements became the catalyst for mass migrations of Europeans to this new land. And come they did.

What They Inherited: The Beginnings

As a testimonial to the resources these settlers inherited, the dense forests were considered an obstacle to agriculture even though they were utilized for fuel and timber. This timber resource was so large that the center of the commercial lumbering industry did not move beyond western New York until after 1850 (Sears, 1980). Between 1850 and 1910, American farmers cleared more forest than had been cleared in the previous 250 years — about 77 million ha. This deforestation rate amounted to an average of more than 3500 ha or 35 square kilometers every day and was sustained for 60 years (Powell et al., 1993).

These forest and grassland resources covering hundreds of millions of ha of the area now within U.S. borders, coupled with temperate climates and productive soils, were a fabulous prize, to say nothing of the minerals, wildlife, and other riches that awaited their taking. It took nearly a century after the early American settlements for European nations to become convinced of the importance of this new land and the resource bounty its political control would insure.

By the time of the American Revolution, a 160 km-wide strip of land from southern Maine to Georgia had been settled, and nearly half to three quarters of this area had been cleared. The population was less than 4 million within the entire 13 colonies (USDA, 1981). Even before the Revolution, however, erosion was recognized as a problem (McDonald, 1941) — a disease of the land, including its associated SOC loss, to be carried westward with the European migrations.

"Westward Ho" into the Mississippi River Basin

While the eastern seaboard and piedmont were taking more than a century (1700 to 1830) to settle (Trimble, 1974), the westward migration had already begun to spill over the Alleghenies. By 1830, most of the better, well-drained lands east of the Mississippi River were occupied (Bennett, 1939). The push westward, however, was not uniform. Sears (1980) outlined two settlement patterns: the movement of European farmers through the northern states into the Northern Great Plains and prairies and the southern migration and settlement, which differed because of the culture of its settlers and the character of the land.

The northern farmer moved westward with unrestrained destruction of native Indian culture. Until the lands now bounded by the grain-belt states were reached and self-interest dictated better husbandry practices, farming areas were systematically impoverished and the lands exhausted. Most of these settlers were not first-class farmers, and for the most part, lacked an agricultural tradition honed through the experience of previous generations. However, their considerable practical and mechanical sense, combined with their work ethic, resulted in a brilliant record of commercial and industrial growth (Sears, 1980).

The southern migration of Europeans west of the Alleghenies resulted in a culture quite different from that of their northern counterparts (Sears, 1980). The gentlefolk of old England were granted large estates, which were worked by the impoverished and the dregs of society who flooded this new land. Also, the indentured, including many Scotch–Irish, were numerous among those who worked these landholdings and eventually settled in the Southeast. Later, black slaves, who proved unprofitable in the North, were found to be well-suited to the tobacco, cotton, indigo, and sugarcane plantations of the warmer South (Sears, 1980).

A Social Tragedy

Sears (1980) also points out another circumstance destined to exact a staggering toll on ensuing generations. The wealthy landed gentry viewed education as a luxury to be dispensed only to those with the means to pay for it. Not only blacks, because of their status as slaves, but also poorer whites were denied educational opportunities. Because the better lands in the South lay along the coastal plain and the broad valleys running inland from it, these lands were occupied and controlled by the plantation landlords. The poorer whites were relegated to lower quality lands, resulting in their literally taking to the hills and eventually developing their own culture in the backwoods and hollows of the Appalachians and the Ozarks (Sears, 1980).

The Frontier as a Talisman and the Impermanence Syndrome

On the southern plantations there was some sense of land stewardship, as exemplified by George Washington and Thomas Jefferson (McDonald, 1941; Sears, 1980). But the continued export of each harvest eventually took its toll on the soil resource and its SOC-related fertility. During westward migrations, however, the vastness of the resource base and the open and cheap land areas still available to the west became the settler's talisman and worked against a psychology of permanence (Sears, 1980). The knowledge that these western lands were still available tended to salve the anxiety of failure and was not conducive to fostering a conservation ethic or to promoting a sense of stability.

To illustrate this impermanence syndrome, Gray (1933) pointed out the tendency to deplete land and then migrate west by stating that, "over the upland soils from Virginia to Texas the wave of migration passed like a devastating scourge. Especially in the rolling piedmont lands the planting of corn and cotton in hill and drill hastened erosion, leaving the hillsides gullied and bare" (p. 446).

An Ecological Tragedy and End of the Frontier

As the northern and southern migrations into the semiarid portion of the new nation continued, the exploitation of the soil resources behind these migrations was supplemented by yet another chapter of soil erosion: the Great Dust Bowl era. The European background of the encroaching settlers and any experience that may have been acquired in the humid East left those who first entered the tallgrass prairie and then the semi-arid shortgrass region unprepared. The latter ecosystem proved the undoing of those who broke the sod. Tempted by their tradition, lured by the bait of immediate profits, and encouraged by financial and industrial interests (Sears, 1980), the settlers' breaking of the sod in this region, coupled with the vagaries of its climate, triggered one of the great ecological disasters of human history in the form of many dust storms. It was a tragic event in American history that scarred both the land and its people. The impacts of this Dust Bowl were felt from Texas to the Dakotas, damaging some 60 to 80 million ha of land (Bennett, 1939).

The westward trek of humanity across the U.S. is a story of human courage, tragedies, successes, failures, and a willingness to reach for new horizons, regardless of the motives. In the process, a great nation was built, but not without significant deterioration and scarring of its land resources. By 1890, nearly all of the better lands had been settled and the American frontier no longer existed. And by the 1920s and early 1930s, American agricultural land use reached its zenith of spatial extent.

SOIL EROSION AND SOC LOSS: A LESSON IN PHILOSOPHY, PEOPLE, AND POLICY

From the beginning of the country's feeble settlements, soil erosion and the loss of SOC were voluminous by-products of U.S. agriculture. It was an ecological price extracted for an exploitative land settlement philosophy developed from basic human instincts as modified by cultural inheritance. The lessons of such behavior had been etched in the earth by civilizations past, but like most historical precedents, humankind has proved to be a slow learner.

As Madden (1974) pointed out, the settlement and development philosophy of the young nation had two serious flaws. First, most Americans treated land as if it were indestructible and inexhaustible. Second, land was treated as a commodity; its biological role as a crucial link in the web of life was ignored. For 100 years after the U.S. gained its independence, the dominant political philosophy was to settle the nation's interior by encouraging the disposition of land to settlers. This process of land disposal enabled most Americans to gain wealth and live better and dominated early American history. Heedless and headlong land transfer to private ownership was a major social and political force throughout the 19th century. The policies derived from these forces moved two thirds of the land in the 48 conterminous states out of the public domain, regardless of its capability to sustain a farm family or community.

THE SOIL CONSERVATION MOVEMENT

While attempts were made to call the abuse of land to the attention of farmers and officials, little headway resulted until after the first conservation movement from 1890 to 1920. The chief reformers during this period were Gifford Pinchot and John Muir. Although Pinchot was aware of the soil erosion problem, his primary interest as a forester was to establish the national forests (Held and Clawson, 1965). During this period, soil conservation was neglected and soil conservation programs were left out of U.S. Department of Agriculture (USDA) programs. John Muir was a preservationist promoting a national park system.

Then, in 1928, an increasingly prominent crusader for soil conservation, Hugh Hammond Bennett, pointed out in a USDA circular entitled *Soil Erosion A National Menace* (Bennett and Chapline, 1928) that the problem affected everyone, not just farmers. Bennett's continued emphasis on this theme (e.g., Bennett, 1928) made him an effective proponent for conservation. Bennett was to soil conservation what Pinchot and Muir were to the national forests and parks, respectively. It is difficult to overestimate his influence in bringing about a national soil-conservation consciousness and program.

Eventually, through Bennett's leadership, the Soil Erosion Service was formed in 1933 to carry out conservation programs financed by the Public Works Administration in the Great Depression era. This temporary agency was housed in the Department of Interior under the direction of Bennett and later transferred to USDA. Subsequently, the visual acuity of and the national alarm over the Dust Bowl, along with Bennett's continued prodding, set the stage for Congress to pass the Soil Conservation Act in 1935. This act declared soil erosion to be a national menace and established the soil conservation program and an agency (Soil Conservation Service) on a permanent basis. After nearly 200 years, the people of the U.S., through their Congress, took responsibility for addressing the issue of protecting the nation's soil and land resources.

SOIL EROSION AND SOC LOSS IN THE U.S.: A SUBJECTIVE ASSESSMENT

The magnitude of environmental deterioration, including soil erosion and SOC loss, on the U.S. portion of the North American continent prior to the 1930s is not well documented. Evidence of this environmental impact is largely subjective, soil-focused, and given in historic accounts, often detailed and graphically eloquent, but seldom quantitatively comprehensive in geographic extent, time frame, or degree of damage.

Early Assessments: 1850–1930

In the 1800s and early 1900s, numerous state and federal bulletins, reports, and other publications, along with many private accounts, detailed the impacts of soil erosion on various segments of land. The commentary of various witnesses to the erosive demise of many landscapes provides a unique description of ecological degradation as seen through the eyes of those who experienced it. Such accounts expressed disgust and the realization that man, by his own hand, had unleashed a pillaging genie that could not be restrained and was still running amok over once fair lands. Recorded in these accounts is an intensity of feeling and despair reserved only to those personally affected by the immediacy of a tragedy; herein such shock has not been tempered by either time healing the landscape or by the more muffled accounts of historians (Miller et al., 1985).

An 1853 appraisal of Laurens County, South Carolina (Trimble, 1974), was written in apocalyptic prose:

> The destroying angel has visited these once fair forests and limpid
> streams ... The farms, the fields ... are washed and worn into unsightly
> gullies and barren slopes — everything everywhere betrays improvident
> and reckless management ... (p. 54).

Prior to the Civil War, C.C. Clay, in a speech before Congress (Gray, 1933), described soil exhaustion in Madison County, Alabama:

> In traversing that county one will ... observe fields, once fertile, now unfenced, abandoned and covered with those evil harbingers, foxtail and broomsedge... Indeed, a country in its infancy, where fifty years ago, scarce a forest tree had been felled by the axe of the pioneer, is already exhibiting the painful signs of senility and decay, apparent in Virginia and the Carolinas (p. 446).

Even U.S. government documents of the day waxed eloquent on the erosion situation, providing the reader with awesome analogies but few data. An example is the 1852 agricultural volume of the Report of the U.S. Commissioner of Patents,* which reads:

> Twice the quantity of rain falls in the Southern States in the course of a year than falls in England, and it falls in one third the time ... (p. 12). Cotton has destroyed more land than earthquakes, eruptions of burning volcanic mountains or anything else. Witness the red hills of Georgia and South Carolina that have produced cotton till the last dying gasp of the soil forbids any further attempt at cultivation and the land turned out to nature reminding the traveler, as he views the dilapidated country, of the ruins of Greece ... (p. 72). And these evils to the community and to posterity, greater than could be effected by the most powerful and malignant foreign enemies of any country, are the regular and deliberate work of benevolent and intelligent men, of worthy citizens and true lovers of the country (p. 386).

The Dust Bowl Era: From Dreams to "The Grapes of Wrath"

Sears (1980) quotes the owner of a large tract in the shortgrass country that had been plowed for wheat in the 1930s:

> We're through. It's worse than the papers say. Our fences are buried, the house is hidden to the eaves, and our pasture which was kept from blowing by the grass, has been buried and is worthless now. We see what a mistake it was to plow up all that land, but it's too late to do anything about it (p. 158).

Others were not so articulate about the problem. They simply packed their few belongings and headed west, much as they and their predecessors had done when the water-induced erosion to the east had rendered the land scarred and unyielding. These hardy but tragic people became the human pulp for Steinbeck's *The Grapes of Wrath*. They left behind a damaged ecosystem totaling many tens of millions of hectares.

SOIL EROSION IN THE U.S.: A QUANTITATIVE ASSESSMENT

Historical Assessment: Pre-1950

Figure 4.2 illustrates in a schematic way the historical trends for several human-induced environmental footprints that are data-based, including the historical sediment load carried by the Mississippi River and SOC changes in the Corn Belt portion of this basin. The first recorded sediment samples in this basin were taken in 1838 (Keown et al., 1986).

---

* The following quotations were taken from three separate sources within the same reference (Report of the Commissioner of Patents, 1853). The USDA had its origin in the U.S. Patent Office in 1837 through Henry Ellsworth's program of distributing seeds from abroad to farmers. In 1839, Congress appropriated $1000 to Ellsworth for gathering agricultural statistics, conducting experiments, and distributing seeds. He eventually established an Agricultural Division within the Patent Office which expanded its operations until 1862, when Congress established an independent USDA (Petulla, 1977).

The most comprehensive historical assessment of erosion in the U.S. is probably best documented in Bennett's 1939 accounts. He was a soil scientist by training and traveled widely throughout the U.S. Until the national erosion survey was conducted under his direction in the mid 1930s, most erosion damage assessments had been based on subjective accounts, regional erosion surveys, and soil surveys. Despite an educational and erosion investigation program begun in the late 1920s by Bennett and others (Bennett and Chapline, 1928; Bennett, 1931), Bennett stated in his 1939 book that the soil erosion impact on the land area of the U.S. may never be known.

Nevertheless, the earliest and most comprehensive assessment of soil erosion in the U.S. was the 1934 national reconnaissance erosion survey. The survey covered every county in the nation and was carried out by Soil Erosion Service soil technologists over a 2-month period in the late summer and early fall of 1934 (Bennett, 1935; Lowdermilk, 1935). The maps from this reconnaissance survey vary in scale and detail, but indicate the predominant erosion conditions; they are preserved in the National Archives. From this survey and previous survey information and data, it was possible to estimate the real extent and seriousness of erosion damage to the U.S. soil resource base.

These data showed that more than half of the U.S. land area had been affected by erosion before and during the 1930s. Excluding the 40 million ha of essentially ruined cropland, the process of soil erosion had already damaged or was threatening more than one third of arable land (Bennett, 1939). Approximately three fourths of U.S. cropland in the 1930s was susceptible to some degree of soil erosion.

Bennett's "Ruined Land"

Although Bennett and others attempted to quantify the impact of erosion on soil productivity and physical deterioration, their terminology for the soil damage classes implies a degree of destruction and permanency of injury inappropriate to many of these lands today. Much of the "ruined" land described by Bennett produces timber and forage today; some of it is still cultivated. Thus the term "ruined land," so frequently quoted from Bennett and used as an erosion classification unit (Bennett, 1939), must be understood as a designation that, while appropriate in Bennett's day, has lost its harshness and meaning in today's lexicon of land capability classification. Some lands that were severely eroded and classed as ruined in the mid 1930s remain irreparably harmed; however, most of the land that Bennett (1939) described as ruined is or can be productive under various levels of management (Miller et al., 1985).

Post-WWII Assessments

The next reconnaissance study after the 1934 soil erosion inventory (Bennett, 1935) was the 1945 soil and water conservation inventory (USDA, 1945). Subsequently, inventories of the nation's soil and water resources were conducted using more scientific and statistically reliable protocols. A series of conservation needs inventories was initiated in the late 1950s. These soil erosion and conservation needs assessments resulted in the conservation needs inventories (USDA, 1962; USDA, 1971; USDA, 1978).

Subsequent inventories of the condition of the nation's land and water resources, known as the national resource inventories (NRI), were carried out in response to the requirements of the Soil and Water Resources Conservation Act of 1977 (RCA). The first RCA appraisal (USDA, 1981) was conducted for the year 1977. The second and more comprehensive RCA–NRI appraisal (USDA, 1989) reported conditions for the nation's land as described in 1982. The 1992 and 1997 NRI updates have been completed and released (USDA, 1994; USDA, 1999). Data collectors updated the 1982 and 1987 databases to 1992 technology standards, enabling the Soil Conservation Service and Natural Resources Conservation Service to produce a 10-year trend line for the nation's natural resource uses, conditions, and trends. The 1997 NRI report and data are available via the internet (USDA, 1999).

It is difficult to compare more recent inventories with the erosion estimates of the 1930s because different survey techniques were used. To complicate the issue, even for the more recent inventories, land uses change from one survey period to the next. Furthermore, the severity of erosion and its effect on productivity are not uniform from one soil region to another. Agriculture was still in transition to mechanical power and other technologies from the 1930s to the 1950s and 1960s, thereby affecting land use and erosion potential. An indicator of the impact of the transition during this period is the decrease in the amount of hay and forage land as the number of draft animals declined.

Soil erosion (with its associated SOC loss) in the U.S. in 1992 averaged over 8 Mg (tons) per ha for a total of more than 15.5 billion Mg (tons) annually (Magleby et al., 1995). The nation's 155 million ha of cropland and 161 million ha of rangeland together account for over 60% of total estimated erosion. The rate of water and wind erosion on U.S. cropland has declined from 20 Mg ha^{-1} in 1938 to about 12 Mg ha^{-1} in 1992. Sheet and rill erosion, although not separated from wind erosion in the 1938 database, has declined from 19.7 Mg ha^{-1} in 1938 (estimated by Tweeten, 1995) to 6.9 Mg ha^{-1} in 1992 (Magleby et al., 1995) and 6.7 Mg ha^{-1} in 1995 (Tweeten, 1995), a decline of 66% over this 57-year period.

However, despite 60 years of improvements in soil erosion control and 30+ years of increased SOC sequestration in agricultural soils, ecosystem degradation via soil erosion and other processes is still not ecologically sustainable for many agricultural lands. The SOC content of the nation's cropland soils, as epitomized by simulated models for the central corn belt, is still only about 60% of the SOC content at the beginning of the 20th century (Donigian et al., 1994; Figure 4.2).

Sheet-rill and wind erosion originating from U.S. cropland in 1992 (Magleby et al., 1995) amounted to nearly 2 billion tons with an average annual erosion rate of 12.5 Mg ha^{-1}. U.S. cropland that is classified as highly erodible makes up only 27% of total cropland, yet accounts for nearly 55% of the nation's total sheet, rill, and wind erosion. Pasture, range, and forest land contribute another 2.4 billion tons of soil erosion. U.S. cropland and rangeland account for about 40% of the contiguous 48-state land area, but generate over 60% of the total estimated erosion from this land base (Magleby et al., 1995).

THE COUNTRY'S ENVIRONMENTAL FOOTPRINTS: MISSISSIPPI BASIN AS AN ECOLOGICAL MIRROR

The Basin's Character and Stock of Soil Resources and SOC

The Mississippi River Basin can be characterized as follows (CAST, 1999):

- Covers 41% of contiguous U.S.
- Accounts for 36% of total U.S. runoff
- Accommodates 52% of American farms
- Includes 55% of U.S. agricultural land
- Holds 47% of the nation's rural population
- Generates annual agricultural gross receipts approaching $100 billion
- Accounts for 52% of U.S. farm receipts

The U.S. Corn Belt is located almost entirely within the Mississippi River Basin. This highly productive ecosystem contains much of the nation's best agricultural soils with thick "A" horizons and relatively high SOC levels. However, their continuous cultivation over the last century, with many soils cultivated far longer, has exacted a toll in the form of soil erosion, lost SOC, and other impacts. These human-induced environmental "exhausts" illustrate the human impact on the nation's resource base and serves as a reminder that these natural resources cannot be liquidated indefinitely.

Anthropogenic Impacts, Ecosystem Responses, and Mitigation Options

Prior to European settlement in the Mississippi Basin, the Mississippi River, particularly below its confluence with the Missouri River, was a naturally heavy sediment-carrying waterway due to high rates of bank caving and some surface soil loss. As land within this huge watershed was cleared and cultivated during the 19th and early 20th centuries, the sediment load increased (Figure 4.2) as more sediment washed from the landscape due to lack of proper soil conservation measures (Copeland and Thomas, 1992; Keown et al., 1986). At the height of soil erosion in the 1920s and 1930s, the Mississippi River sediment load reflected this ecological impact, as did SOC decreases in the basin (Figure 4.2).

As U.S. government programs to mitigate the soil erosion problem and, by default the loss of SOC, came on-line, the basin's erosion rates and the river's sediment load began to level off and then decline due to conservation practices put on the land, construction of small sediment-retention dams on higher-order streams, and stabilization of streambanks. Also during this time, a number of multipurpose dams were constructed on major streams in the basin, which altered the character and flow dynamics of the Mississippi River and further reduced its sediment load. Subsequently, a series of navigation and flood control dams was constructed in the Missouri River Basin (1953–1967) and on the Arkansas River (1963–1970), further reducing the sediment load discharge to the Mississippi River as depicted in the second phase of sediment load reduction in Figure 4.2 (Copeland and Thomas, 1992; Keown et al., 1986). Thus, ecosystem interventions and manipulations by man can initiate environmental impacts and then mitigate them, as reflected in Figure 4.2.

Another example of ecosystem response to human impacts is illustrated in Figure 4.1, which shows how a Mid-Atlantic watershed north of Baltimore, Maryland, responded to changes in land use over time using sediment discharge as the response indicator. Soil organic carbon losses most likely mimic the sediment discharge loads in this figure. Mitigation of these impacts has taken many forms, from educational appeals for voluntarily changing land use practices to technical assistance, financial incentives for cost-sharing solutions, new technologies, and, ultimately, penalties and mandatory compliance for meeting certain standards of land stewardship.

MANIPULATING ECOSYSTEMS, MITIGATING IMPACTS: A MULTIDISCIPLINARY DOMAIN

Human and Ecological Dimensions of Ecosystem Changes — Preferences

While ecosystems respond in various ways to their manipulations, it must be remembered that nature has no preference for its equilibrium points. Yet, many environmental responses to intended and unintended ecosystem manipulations turn out to be negative from the human perspective. From the perspective of ecological forces and principles that drive nature's reactions and responses to human-induced ecosystem manipulations (if one could personify nature), it matters not at all to nature whether such manipulations result in soil and SOC losses, eutrophic and septic lakes, or hypoxia conditions beyond the mouths of many large river basins. Nature adapts without any hierarchy of what is good or bad. It is mankind who imposes these values on environmental reactions induced by its own hand. Therefore, if these negative environmental impacts as judged by humankind are to be mitigated successfully, then it must address technical solutions as well as social incentives and institutions.

The Institutional–Political Dimensions of Ecosystem Manipulations

Figure 4.3 illustrates the variety of political and individual players involved in land use decision-making and ecosystem manipulations. For each individual land owner or manager, his or her goals in using land and his or her associated value system and degree of stewardship may or may not coincide with the interests of society or the sustainability of the environment. As population density increases and land becomes more subdivided, problems and conflicts arise as some lands are committed to uses commensurate with their capabilities. Furthermore, incompatible land uses occur more frequently as people with different goals and value systems clash over how "their" lands should be used.

In the U.S. political system, these clashes are often resolved through the court system, resulting in much case law. Subsequently, case law is used to establish the legal thresholds and framework for protecting both individuals' and society's interests in land and ecosystems. In Figure 4.3, as the density of land ownership increases, the courts tend to weigh more heavily the interests of society vs. the interests and land ownership rights of the individual.

Mitigating Impacts of Ecosystem Manipulations Are Not the Sole Domain of Science

Soil erosion, SOC losses, and other negative environmental impacts are not intentionally caused by humans. They are, however, the repercussions of social behavior driven by a spectrum of individual objectives, needs, values, and desires playing out over a matrix of ecosystems with variable capacities for accommodating human incentives and cultural timescales not commensurate with nature's temporal cadence. For example, the productivity capacity of an ecosystem is governed by its inherent natural components (e.g., soils, climate, topography, plant and animal species) as modified by responses to technical inputs such as genetic improvements, fertilizers, tillage, drainage, etc. But the manipulation of these ecosystems is also modified by lending institutions' credit policies and short-term repayment schedules, the farmer's goals, ethic, and level of stewardship, and a host of other factors.

The mitigation of soil erosion, SOC losses, and other environmental problems must be the domain not only of the natural and physical sciences but also of the social sciences and humanities ranging from economics, political science, and education to psychology, philosophy, history, and religion. While scientists may come up with various strategies to address the negative repercussions of environmental impacts as well as technological fixes, it is people, their cultural heritages and value systems, as well as their social institutions (e.g, government hierarchy, lending institutions, market structure), that will ultimately control the success or failure of adopting such strategies and technologies.

Science has, indeed, wrought wonders for agriculture. But there is the danger of relying too heavily on science. The talisman of the early settlers was the knowledge of more and better land to the west; the talisman of many modern farmers and others is the continued scientific development of better production technologies coupled with a calculus of serious soil and SOC losses that differs greatly from the sustainability thresholds proposed by soil scientists and conservationists. Thus, carbon dynamics, like many natural processes, is as much a sociological phenomenon as a scientific and technical problem. Without both parties at the table, mitigation may be very hard to deliver through scientific solutions alone.

SOIL EROSION, SOC MITIGATION: BUILDING ON THE PAST

Trending Toward Rational and Sustainable Ecosystem Management

Nations are built on people's dreams and values as much as on facts and information. Thus, land use decisions and ecosystem manipulations are not necessarily rational processes. As societies

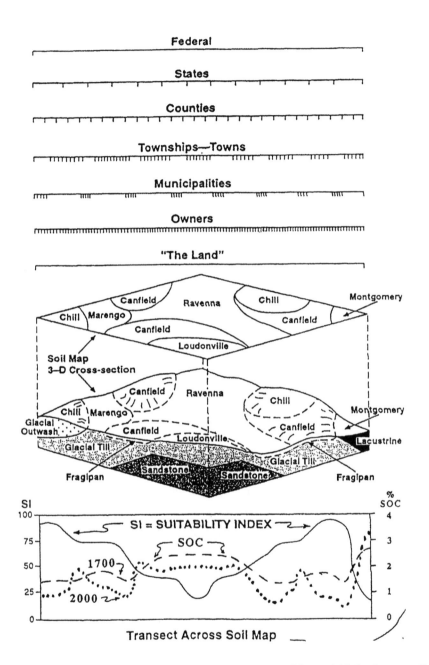

Figure 4.3   Hierarchy of land-use decision-makers and managers overlying variable landscapes with differing land-use suitability indices and hypothetical responses of SOC dynamics to land management, 1700 to 2000. (Modified and expanded from Platt, 1996.)

become more complex, population density increases, individuals begin to encroach on each other's freedoms and self-interests in uses of land, and environmental and life-style impacts become acute, there comes the motivating realization that something must change. Political compromises are forged between individuals' rights and interests in land and the greater good for society. Evidence of these compromises are concepts and words such as smart growth, sustainable management, ecology-based management, etc. that have entered our lexicon. Government agencies and laws have been established to assist in mitigating environmental problems and regulating ecological manipulations.

Where The Country Has Come from

The history of soil resource management and, by default SOC management in the U.S., offers a poignant lesson in incremental mitigation of the repercussions of unsustainable ecosystem manipulations. In the early phases of this nation's development, little heed was paid to soil erosion, SOC dynamics, land degradation, and natural resource depletion. Although there were individual calls of alarm during this early period, they were drowned out by the human stampede across this continent to grab the brass rings in this natural resource-rich land and build a new nation. As has been noted, environmental scars began to occur with increasing frequency and extent in the post-1850 period.

Bennett and Chapline (1928) and others began to lift their voices, using persuasion coupled with the visual acuity of land degradation to establish finally an institutional entity (SES/SCS) in 1933–1935 from which soil–land conservation programs were launched. These initial programs relied on educational appeals for voluntary individual stewardship coupled with various cropland acreage reduction programs starting in 1933. Although somewhat successful, it became obvious that appeals to this "disturbance species" and its human nature for voluntary conservation implementation would meet with limited success.

In the post-WWII period, USDA ratcheted up its financial cost-sharing for various practices to complement its arsenal of soil conservation tools. USDA also launched a new wave of cropland reduction programs starting with the soil bank program in 1956; the current CRP is the latest version of this concept. As the environmental era dawned and progressed, conservation programs began to have stronger teeth and focused more on critical areas as exemplified by conservation compliance provisions and CRP initiative enacted in the 1985 Food Security Act. Furthermore, farms today are subject to various state and local restrictions and penalty-based regulations.

However, as noted previously, despite much progress in reducing erosion and increasing SOC levels (Figure 4.2), there is a long way to go to achieve ecological and environmental sustainability. Many areas of highly erodible and moderately erodible croplands and rangelands still are not sustainably managed. Likewise, many farm and rangeland practices and management systems are not ecologically sustainable with respect to soil loss, SOC dynamics, and other environmental indicators such as hydrologic flux, water quality, and soil compaction.

WHERE TO FROM HERE?

Reaching beyond the "Low Hanging Fruit" toward Sustainability

The magnitude of soil erosion and the reduced SOC pool on U.S. cropland and rangeland, despite much improvement, still begs the question of how to mitigate further these and other kinds of land degradation. The nation appears to be in a near "flat-line" situation with respect to controlling soil erosion and increasing SOC sequestration. Even the rate of agricultural productivity is decelerating because the "low hanging fruit" has already been plucked with respect to solving production problems and improving plant and animal genetics. With the necessity to be competitive globally, farmers and ranchers will continue to put more pressure on the land. The country simply must reach beyond the low hanging fruit if the specter of ecosystem deterioration is to be stemmed, focusing efforts toward the goal of ecosystem sustainability.

A Suggested Agenda: Heeding Lessons from History

Although all the answers required to mitigate complex issues such as soil-land degradation and global change through such initiatives as soil erosion control and SOC sequestration strategies and protocols are not known, it does seem clear that successful mitigation strategies oriented toward sustainable ecosystem management will most likely be driven more by socio-political disciplines than by the natural sciences community. One can only lean so far on the good will of human nature to adopt science-based technologies and practices. Voluntary appeals for individuals to adhere to an ecological ethic, regardless of the incentives initiated to date, will simply not succeed. The forces of economic realities, pragmatism, multiple demands upon one's time and energy resulting in a "tyranny of the imminent," human limitations, and the forces of nature will simply derail the quest for sustainability if voluntary programs are the main line of defense. Sustainability, therefore, must involve changes in behavior, which is not the domain of the natural sciences.

As a take-home message, the following observations and suggestions are offered in the interest of striving for ecological sustainability:

- Science-based information and mitigation strategies must be the foundation undergirding the campaign to reach the goal of sustainability.
- The scientific community must be more unified and networked to deal with such a complex undertaking. Whether by design or by default within the broader science community, signs still exist of a culture of isolation, issue ownership and domination, and even discipline arrogance regarding mitigation issues spun out of man's unsustainable manipulations of ecosystems. Such a culture tends to foster exclusivity, science cliques, talking only among one's group, and competition for limited grant funds, thus precluding broader interdisciplinarity.
- Governmental agencies must consider consolidating or networking their component specialties and programs to address natural resource-ecosystem sustainability. Not only are such core specialties and programs fragmented among many agencies and divisions, but they are also isolated from each other. The long-incubated culture of bureaucratic survival and stability often stymies such partnering and networking. Neither are academic and other institutions exempt from such behavior. As the business community has learned, survival is not an entitlement; thus, teaming among the fragmented components enhances more visibility and leverage.
- Ecosystem degradation is driven by a complex of human behaviors and responses to external forces. Likewise, these same human behaviors will determine whether or not the concept of sustainability can become operational. This arena is not the domain of natural and physical scientists, but of the social sciences and humanities. Thus, if they are not at the table when problems are discussed, strategies are devised, and plans are proposed for mitigating problems associated with human-induced ecosystem manipulations, the goal of sustainability is doomed to failure.
- There must be strong agency and political leadership and a will to act if mitigation strategies are to succeed. People do not become excited or passionate about a proposed new sewage treatment plant or issues such as soil erosion and carbon sequestration. Yes, the latter two are on the national agenda, but so too are many other issues and priorities on the nation's radar screen. Unless there is an ecological catastrophe (e.g., Dust Bowl) or a mandated program, most well-meaning politicians and leaders are acting rationally in not stirring a pot without a fire under it. There is no Endangered Soil Act, Clean Soil Act, or Ecosystem Sustainability Act to bolster visibility and legitimacy of, and funding for the cause. Yet, one cannot afford to be left at the margins of the scientific and political communities. Thus, a compelling case must be made and agency heads must be empowered to step up to the plate and give strong leadership to these issues. Political leaders who can perhaps tie these initiatives to more visible and well-funded environmentally-related issues should be supported.
- The community of scientists and their social science counterparts must be empowered with both scientific and political legitimacy through such mechanisms and structures as a committee or forum under one of the boards or divisions of the National Research Council (NRC), National Academy of Science (NAS), an IPCC subpanel or task force, President's Science Advisory Council task force, Congressional advisory panel, or other options.
- The public and some political leaders are coming to understand the connection between human sustenance and well-being and the natural resource base and ecological processes that undergird all of agriculture, forestry, fisheries, wildlife, and the quality of life. Scientific and other communities must "stay on message" and encourage articulate, established writers and commentators to address this concept upon which all life is based.

- This message must be institutionalized through the curricula of the K–16 education system, not necessarily by courses devoted to these specific issues, but by including ecology or ecological concepts within core curriculum subject matter such as biology, chemistry, physics, math, history, civics, and the humanities. High school and college graduates who do not appreciate or understand the relationship of natural resources and ecosystem processes to their sustenance and well-being are no longer acceptable. While noble and well-intentioned, imparting knowledge and understanding about agriculture and sustainable natural resource and ecosystem management through exhibits and demonstrations at shopping malls and county fairs are not enough.

Again, mitigation of ecosystem problems, while scientifically based, will be largely driven by the socio-political process. Unless science can provide readily adaptable technologies that, like no-tillage and conservation tillage, fit into agricultural and natural resource management systems, the human dimensions of ecosystem management will play the major role in mitigation strategies.

A Wild Card?

As people become better educated and more discriminating in the market place, the consumer may ultimately drive the adoption of ecological sustainability and mitigation of carbon dynamics and other environmental problems. Eco-labeling, certification programs for various food and fiber production systems, and consumer rejection of certain technologies (e.g., genetically modified foods, artificial hormones) have already begun in the developed world. The momentum for these movements continues to increase. Producers are beginning to respond in the market place by certifying sustainable and environmentally friendly production systems, organic foods, and other environmentally compatible products and services. Therefore, the consumer may ultimately drive the trend toward sustainable ecosystem management, which has already begun.

EPILOGUE

A common theme throughout this chapter has been that mitigation of environmental impacts and sustainable management of ecosystems can only be achieved by changing people's behavior and value systems. This challenge, therefore, is more a social enterprise than a scientific one. Unlike native American Indians whose culture and spirituality were intimately connected to the land, European settlers of this land were driven by their dreams to stake out a new life carved from a landscape that they had dominion over. Simpson (1999) notes that "ours is a landscape made by a society devoted to individual 'life, liberty, and property.'" It was from this philosophy that the concept of land, its natural resource base, and ecological processes became disconnected from people's dreams for quality life, sustenance, and physical well-being. Any success in mitigating ecological footprints and using this cornucopia-like endowment of natural resources and ecosystems in a sustainable manner can only come about by reconnecting people to the land and those ecological processes that undergird desired lifestyles.

Simpson (1999) captures the essence of this concept in the last lines of his book:

Reliance on rules to modify ... behavior by requiring recycling or establishing quality standards for air and water, while important, will remain limited in effectiveness until our landscape values change. Then those rules become formalities. Nature preserves and bioreserves, preservation and conservation, even enlightened urban and suburban landscape planning, treat the symptoms not the cause. Only with changed values will we be able to focus on enhancing the landscape health rather than on repairing landscape injury. Only then will we become more fully connected to the land and better able to realize [our own dreams and] the wonder of our world.

REFERENCES

Bennett, H.H., The geographical relation of soil erosion to land productivity, *Geogr. Rev.*, 18(4): 583, 1928.

Bennett, H.H., National program of soil and water conservation, *J. Am. Soc. Agron.*, 23: 357–371, 1931.

Bennett, H.H., Report of the chief of the Soil Conservation Service (to the Secretary of Agriculture), Mimeograph, USDA-SCS, Washington, D.C., 1935.

Bennett, H.H., *Soil Conservation*, McGraw-Hill Book Co., New York, 1939.

Bennett, H.H. and W.R. Chapline, *Soil Erosion A National Menace*, USDA Agric. Circ. 33. U.S. Government Printing Office, Washington, D.C., 1928.

Buringh, P., Availability of agricultural land for crop and livestock production, in D. Pimentel and C.W. Hall (ed.), *Food and Natural Resources*, Academic Press, Inc. San Diego, 69–83, 1989.

Butzer, K.W., *Early Hydraulic Civilization in Egypt: A Study in Cultural Ecology*, University of Chicago Press, Chicago, 1976.

Cahokia Mounds Museum Society 1997. Cahokia Mounds. Collinsville, IL. 62234.

Carter, V.G. and T. Dale, *Topsoil and Civilization*, University of Oklahoma Press, Norman, 1974.

Christensen, N.L. et al., The report of the Ecological Society of America Committee on the Scientific Basis for Ecosystem Management, *Ecol. Appl.*, 6(3): 665–691, 1996.

Clark, C. and M. Haswell, *The Economics of Subsistence Agriculture*, Macmillan, New York, 1970.

Copeland, R.R. and W.A. Thomas, Lower Mississippi River Tarbert Landing to East Jetty sedimentation study, Technical Rpt. HL-92-6. Dept. of the Army, Waterways Experiment Station, Corps of Engineers, Vicksburg, MS, 1992.

Council for Agricultural Science and Technology (CAST) 1999, Task Force Report 134, Gulf of Mexico Hypoxia: Land and Sea Interactions. Ames, IA, 1999.

Donigian, A.S., Jr. et al., Assessment of alternative management practices and policies affecting soil carbon in agroecosystems of the central United States, Publ. No. EPA/600/R-94/067, U.S.-EPA, Athens, GA, 1994.

Ehrlich, A.H. and P.R. Ehrlich, *Healing the Planet: Strategies for Resolving the Environmental Crisis*, Addison-Wesley, Reading, MA, 1991.

Gray, L.C., *History of Agriculture in the Southern United States to 1860*, Publication 430. Carnegie Institution of Washington, D.C. Waverly Press, Inc., Baltimore, MD, 1933.

Held, R.B. and M. Clawson, Soil conservation in perspective, The Johns Hopkins University Press, Baltimore, MD, 1965.

Hillel, D.J., *Out of the Earth*, The Free Press, New York, 1991.

Intergovernmental Panel on Climate Change (IPCC), *Climate Change 1995 — Impacts, Adaptations and Mitigation of Climate Change: Scientific-Technical Analysis*, IPCC Working Group II, Cambridge University Press, Cambridge, 1996.

Kennedy, P., *Preparing for the Twenty-First Century*, Random House, New York, 1993.

Keown, M.P., E.A. Dardeau, Jr., and E.M. Causey, Historic trends in the sediment flow regime of the Mississippi River, *Water Resour. Res.*, 22(11): 1555–1564, 1986.

Lal, R. et al. (eds.), *The Potential of U.S. Cropland to Sequester Carbon and Mitigate the Greenhouse Effect*, Sleeping Bear Press, Inc., Chelsea, MI, 1998a.

Lal, R. et al. (eds.), *Soil Processes and the Carbon Cycle*, CRC Press, Boca Raton, FL, 1998b.

Lowdermilk, W.C., Soil erosion and its control in the United States, SCS-MP-3, USDA-SCS. U.S. Government Printing Office, Washington, D.C., 1935.

Lowdermilk, W.C., Conquest of the Land through Seven Thousand Years, Agricultural Information Bulletin No. 99, U.S. Department of Agriculture, Soil Conservation Service, Washington, D.C., 1953.

Lubchenco, J. et al., Priorities for an environmental science agenda for the Clinton-Gore Administration: recommendations for transition planning, *Bull. Ecol. Soc. Am.*, 74: 4–8, 1993.

McDonald, A., Early American soil conservationists, Misc. Publ. No. 449, USDA-SCS. U.S. Government Printing Office, Washington, D.C., 1941.

Madden, C.H., Land as a national resource, in C.L. Harriss (ed.), *The Good Earth of America, Planning Our Land Use*, Prentice-Hall, Englewood Cliffs, NJ, 1974, 6–30.

Magleby, R. et al., Soil erosion and conservation in the United States — an overview, Natural Resources and Environ. Div., Economic Research Service, USDA, *Agric. Info. Bull.* No. 718, 1995.

Martin, L., *The Anasazi Legacy. Mesa Verde — the Story Behind the Scenery*, U.S. National Park Service, KC Publications, Inc., Las Vegas, NV, 1993.

Miller, F.P., W.D. Rasmussen, and L.D. Meyer, Historical perspective of soil erosion in the United States, in R.F. Follett and B.A. Stewart (eds.), *Soil Erosion and Crop Productivity*, American Society of Agronomy, Madison, WI, 1985, 23–48.

Miller, F.P. and M.K. Wali, Soils, land use and sustainable agriculture: a review, *Can. J. Soil Sci.*, 75(4): 413–422, 1995.

Petulla, J.M., *American Environmental History: the Exploitation and Conservation of Natural Resources*, Boyd & Fraser Publishing Co., San Francisco, 1977.

Pimentel, D., Ecological systems, natural resources, and food supplies, in D. Pimentel and C.W. Hall (eds.), *Food and Natural Resources*, Academic Press, Inc., San Diego, 1989, 1–29.

Platt, R.H., *Land Use and Society: Geography, Law, and Public Policy*, Island Press, Washington, D.C., 1996.

Powell, D.S. et al., Forest Resources of the United States, 1992, General Technical Rept. RM-234, USDA, Forest Service, 1993.

Report of the Commissioner of Patents, Part II Agriculture for the year 1852. 32nd Congress, 2nd Session, House of Representatives Exec. Doc. No. 65, Washington, D.C., 1853.

Richards, J.F., Land transformation, in W.C. Clark and R.E. Munn (eds.), *Sustainable Development of the Biosphere*, Cambridge University Press, NY, 1990.

Richardson, C.J., Ecological functions and human values in wetlands: a framework for assessing forestry impacts, *Wetlands*, 14: 1–9, 1994.

Sears, P.B., *Deserts on the March*, 4th ed., Oklahoma University Press, Norman, 1980.

Simpson, J.W., *Visions of Paradise — Glimpses of Our Landscape Legacy*, University of California Press, Berkeley, 1999.

Trimble, S.W., *Man-Induced Soil Erosion on the Southern Piedmont, 1700–1970*, Soil Conservation Society of America, Ankeny, IA, 1974.

Tweeten, L., Farm structure graphics, ESO 2256, Department of Agricultural Economics, Ohio State University, Columbus, 1995.

USDA, Soil and Water Conservation Needs Estimates for the US. by States, Soil Conservation Service, Washington, D.C., 1945.

USDA, Basic Statistics of the national inventory of soil and water conservation needs, *Agric. Stat. Bull.* No. 317, Washington, D.C., 1962.

USDA, National inventory of soil and water conservation needs, 1967, *Stat. Bull.* No. 461. Washington, D.C., 1971.

USDA, 1977 National Resource Inventories, Soil Conservation Service, Washington, D.C., 1978.

USDA, Soil and Water Resources Conservation Act, 1980 RCA Appraisal, Parts I, II, and Summary, U.S. Government Printing Office, Washington, D.C., 1981.

USDA, The Second RCA Appraisal — Soil, Water, and Related Resources on Nonfederal Land in the U.S., U.S. Government Printing Office, Washington, D.C., 1989.

USDA, Summary Report, 1992 National Resources Inventory. Soil Conservation Service, Washington, D.C., 1994.

USDA, 1997 National Resources Inventory, www.nhq.nrcs.usda.gov/NRI/, 1999.

Wolman, M.G., A cycle of sedimentation and erosion in urban river channels, *Geogr. Ann.*, 49A: 385–395, 1967.

PART II

Conservation Tillage and Residue Management

CHAPTER 5

Mulch Rate and Tillage Effects on Carbon Sequestration and CO_2 Flux in an Alfisol in Central Ohio

S. W. Duiker and Rattan Lal

CONTENTS

ABSTRACT

The amount of carbon sequestered in an arable soil is affected, among other things by the tillage system and amount of crop residue applied. This study was conducted to determine the long-term effect of mulch application rates on the carbon pool with three different tillage systems. Zero, 2, 4, 8, and 16 Mg ha^{-1} yr^{-1} wheat straw was applied for 8 years in a no-till, plow-till and ridge-till system on a Crosby silt loam in central Ohio. Only the mulch rate had a significant effect on the SOC content on a weight basis (g/g), in the top 10 cm of the soil. However, bulk density in the no-till system was higher, resulting in a higher SOC pool in the no-till system than in the other tillage systems. Analysis of the SOC pool in the 0 to 1, 1 to 3, 3 to 5, and 5 to 10-cm depths showed that the amount of SOC sequestered per Mg of mulch applied was highest in the no-till system, followed by the plow-till system. The absence of an effect of mulch application on carbon sequestration with the ridge-till system was probably due to sampling method. Most carbon sequestered was concentrated at the surface of the soil in both no-till and plow-till systems. No statistically significant differences in CO_2 flux were measured between mulch rate treatments during a growing season. The results show a higher potential of carbon sequestration in a no-till system than in a plow-till system (10% of carbon applied sequestered per year vs. 8%) and the need to avoid burning or removing some or all crop residue.

INTRODUCTION

The interest in carbon sequestration in soils has increased, especially since the sixth meeting of the United Nations Framework for Climate Change in The Hague in November 2000. A major point of debate, which eventually led to suspension of the talks, was whether countries can claim carbon credits for carbon stored in soils and forests. No-till has been recognized as an important possibility to increase carbon sequestration in soils (Lal et al., 1999; Smith et al., 2000a, 2000b). The area under conservation tillage in the U.S. increased from 29 million ha in 1989 (26% of planted area) to 44 million ha in 1998 (37%; CTIC, 1998). This change of soil management has important implications for carbon sequestration because more crop residue is returned to the soil, decomposition rates are lower, and, therefore, more carbon is sequestered with conservation tillage compared to conventional tillage (Dick et al., 1986a, 1986b; Havlin et al., 1990).

The amount of crop residue converted into soil organic carbon (SOC) is affected by ecological factors (e.g., temperature, moisture regime, soil type) and by management (e.g., quantity and quality of residue returned to the land, soil tillage technique, fertilizer applied, irrigation, crops grown). Several field experiments have documented changes in SOC content under different crops and soil management systems (Blevins et al., 1983; Janzen, 1987; Angers et al., 1997; Paustian et al., 1997). In most of these experiments, the total amount of carbon added to the soil in crop residue was estimated because the below-ground biomass was not measured. Land managers and policymakers need to know how much crop residue is to be returned to the soil in order to maintain or increase the SOC pool to a desired level. Additionally, knowledge of the factors influencing dynamics of SOC decomposition and sequestration during the year would indicate which changes in soil management maximize carbon sequestration.

This experiment was thus initiated to determine the effect of long-term mulch application with three tillage techniques on SOC content of an Alfisol in central Ohio. Knowledge of the amount of mulch added to the soil allowed quantification of the increase in the SOC pool with mulch rate and calculation of the efficiency of conversion of residue C into SOC. Additionally, the effect of rates of mulch application on decomposition dynamics under no-till was determined during a growing season.

MATERIALS AND METHODS

Experimental Site and Statistical Design

The experiment (Figures 5.1 and 5.2) was located on the Waterman Farm of Ohio State University (40°00' N latitude and 83°01' W longitude). The average annual temperature is 11°C; the average annual precipitation is 932 mm. The soil of the experimental site is a Crosby silt loam (fine, mixed, mesic Aeric Ochraqualf). The experiment was initiated in the summer of 1989 as a split-plot design with three replicates. Tillage was the main plot and mulch rate was the subplot (2×2 m). The three tillage treatments were ridge-till, plow-till, and no-till. Five mulch treatments (based on air-dry weight) were 0, 2, 4, 8, and 16 Mg ha^{-1} yr^{-1} wheat straw. The carbon content of the air-dry wheat straw was 0.413 ± 0.004 g g^{-1}, determined with dry combustion at 600°C (Nelson and Sommers, 1986). The typical nutrient content of wheat straw is 0.006 g g^{-1} N, 0.002 g g^{-1} S, 0.006 g g^{-1} K, and 0.001 g g^{-1} P (Russell, 1973).

Figure 5.1   General view of experimental site and layout.

Figure 5.2   Detailed view of one plot and CO_2 sampling chamber.

Soil tillage was performed each spring after which mulch was applied. Plowing and ridging were done with a multiple and single moldboard plow, respectively, to a depth of approximately 20 cm, after which the soil was not tilled anymore during the year. It was observed that mulch rapidly compacted after a rainstorm (very frequent around the time of its application) and therefore no special measures were taken to keep it on the plots. No crop was planted and no fertilizer was applied. Weed control was achieved through application of Glyphosate (phosphonomethyl glycine) as and when needed.

Measurements and Analyses

Bulk soil samples for SOC analyses were taken in the autumn of 1996 at 0 to 10, 10 to 20, and 20 to 30-cm depths. Results revealed the need for more detailed sampling of SOC content in the surface layer of the soil. Therefore, soil samples were taken again, but at 0 to 1, 1 to 3, 3 to 5, and 5 to 10-cm depths in the summer of 1997. A total of three to four samples were taken per plot at each depth with a 5-cm diameter auger pushed into the soil to each subsequent required depth (using the same hole) and then emptied. In the case of ridge-till treatments samples were taken halfway up the ridge in order to obtain representative samples. The SOC content was determined on air-dried, finely ground, and sieved (<0.149 μm) composite samples for each depth from the individual plots using dry combustion at 600°C (Nelson and Sommers, 1986). Soil bulk density was determined using a Troxler density probe (Blake and Hartge, 1986). The alkali-absorption method (Anderson, 1982) was used to measure daily CO_2 flux.

White, unshaded PVC cylinders were permanently installed in the plots in the spring of 1997. Measurements were taken approximately every 2 weeks from April to October, 1997, always more than 2 days after a rainfall event. Glass jars filled with 0.02 L of 1.0 M NaOH were placed on a perforated PVC ring in a cylinder which was then closed with a white PVC cap; the rim was filled with water to minimize gas exchange with the atmosphere outside the cylinder. The jars were left for 24 hours in the closed cylinders, after which they were removed, closed with a parafin-paper sealed metal lid, and transported to the laboratory for analysis. Excess NaOH was titrated to pH 8.2 in the presence of excess $BaCl_2$ using 0.85 M HCl and phenolphtalein as indicator.

Calculations and Statistical Analysis

The SOC pool in Mg ha^{-1} was computed using the SOC content of each sampling layer multiplied by its depth (m) and the bulk density (Mg m^{-3}). The analysis of variance for F-ratio was computed to determine the effects of mulch and tillage treatments and their interactive effects on the SOC pool (Steel et al., 1997). Polynomial regression equations were computed to detect linear, quadratic, or cubic relationships between mulch rates and the SOC pool.

RESULTS AND DISCUSSION

Tillage and Mulch Rate Effects on Soil Bulk Density and Organic Carbon Content

The analysis of variance for SOC content is shown in Table 5.1. A highly significant effect of mulch rate on SOC content was observed in the 0- to 10-cm layer, but not for other depths. (A mass balance of SOC for the total 10-cm depth is presented in Duiker and Lal, 1999.) Tillage treatments had no effect on SOC content at any of the three depths sampled. Dick et al. (1986b) and Havlin et al. (1990) also reported no changes of SOC contents below 10-cm depth with different conventional and conservation tillage practices.

Table 5.1 Analysis of Variance of SOC (wt %) at Different Depths (1996 samples)

Source of Variance	Degr. of Freedom	0–10 cm Depth		10–20 cm Depth		20-30 cm Depth	
		F Ratio	p Value	F Ratio	p Value	F Ratio	p Value
Tillage	2	1.57	0.3139	1.74	0.2863	1.34	0.3581
Replicate	2	0.20	0.8283	5.00	0.0816	1.24	0.3803
Mulch	4	16.56	0.0000	0.71	0.5955	0.40	0.8090
Tillage × Mulch	8	0.99	0.4707	1.15	0.3694	0.64	0.7348

In contrast, some researchers have observed higher SOC contents at deeper depths under plow-till compared to no-till (Blevins et al., 1983; Dick et al., 1986a; Angers et al., 1997). Differences in crops grown, climate, drainage, texture, and type and depth of soil tillage may be responsible for differences in response of SOC to tillage for different locations. The lack of a tillage effect on SOC content in the surface layer observed in this first year of sampling is probably because samples represented an average of 10-cm depth. Therefore, more detailed sampling at smaller depth increments was started in 1997, 8 years after initiation of the experiment.

Soil bulk density was significantly higher in the no-till treatment than in the ridge- and plow-till treatments (Figure 5.3). Tillage method and mulch rate had significant effects on the SOC pool in the 0- to 1-, 1- to 3-, and 3- to 5-cm depths (Figure 5.4); significant and linear increases of SOC content in the 0 to 5-cm layer with mulch rate were observed for no-till and plow-till treatments. The data indicate that the increase in SOC content is highest near the surface, and that the effect decreases with depth, in both no-till and plow-till treatments. The increase of SOC content in the surface 3 cm is, however, much higher in no-till than in plow-till treatment.

No increase in SOC content with mulch rate was observed in the ridge-till treatment. This lack of response of SOC content to application of mulch with ridge-till may be due to inadvertent sampling of subsoil brought to the surface when the ridges were formed. Regression equations relating SOC content in the 0 to 10-cm depth to mulch rate indicate that the rate of SOC sequestration was slightly more for no-till than plow-till (Table 5.2). The higher decomposition rate with plow-till than with no-till is expected because of increased aeration and higher soil temperatures, especially shortly after tillage (Reicosky and Lindstrom, 1993; Reicosky et al., 1997).

Similar results of higher SOC content in the surface layer of no-till compared with plow-till have been reported by Blevins et al. (1983) for a Typic Paleudalf (silt loam) in Kentucky, by Dick et al. (1986a) for a Mollic Ochraqualf (silty clay loam) in northwest Ohio, by Dick et al. (1986b) and Bajracharya et al. (1998) for Typic Fragiudalfs (silt loam) in northeast Ohio, and for an Aeric Aqualf in central Ohio (Bajracharya et al., 1998). The results of the present study did not indicate a significant relationship between mulch rate and SOC content with ridge-till; only the intercept was highly significant.

Carbon conversion efficiency was calculated by dividing the slope of the regression line in Table 5.2 by 0.4 (carbon content of mulch) and the number of years of mulch application (8) to

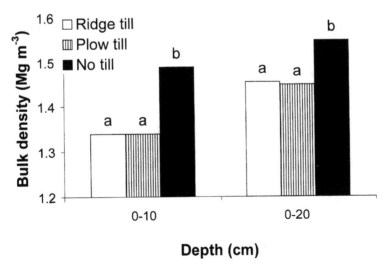

Figure 5.3 Tillage effects on bulk density (Mg m^{-3}) at two depths. (Different letters on bars indicate significant differences in bulk density [per depth].)

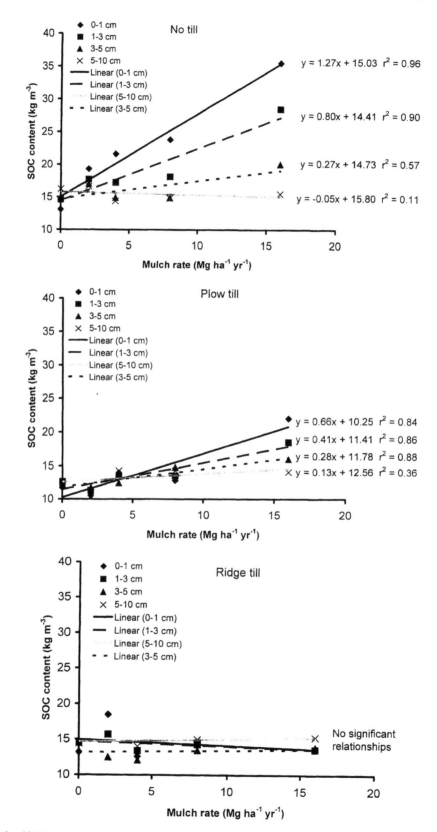

Figure 5.4   Mulch rate and tillage effects on SOC content (kg m⁻³) at four depths.

Table 5.2 SOC Content after 8 Years of Mulch Application with Different Tillage Practices

Tillage Practice	Mulch Application Rate (Mg ha⁻¹ yr⁻¹)					Linear Regression Equation*	r**
	0	2	4	8	16		
No till	15.3	17.2	15.8	16.4	21.0	SOC = 15.21 + 0.32 × M	0.68
Plow till	12.5	11.3	13.7	14.1	16.2	SOC = 11.95 + 0.27 × M	0.72
Ridge till	14.3	15.3	13.4	14.5	14.5	SOC = 14.42	0.00

* M stands for mulch application rate; ** r = 0.88 is significant (p = 0.05).

calculate the efficiency on an annual basis. The conversion efficiency was 0% with ridge-till, 8% with plow-till, and 10% with no-till. Some studies have reported conversion efficiencies in the range of 14 to 21% (Rasmussen and Collins, 1991). Low conversion efficiencies in this experiment may be due to lack of essential nutrients necessary for decomposition, since no fertilizer was applied (Nyborg et al., 1995; Himes, 1998).

Mulch Rate Effects on CO_2 Flux with No-Till

CO_2 flux was highest from the 16 Mg ha⁻¹ yr⁻¹ mulch treatment in 7 out of 13 sampling dates, and the lowest from the 0 Mg ha⁻¹ yr⁻¹ treatment in 8 out of 13 sampling dates (Figure 5.5). However, differences between treatments were generally not significant. The CO_2 flux increased with the increase in air and soil temperatures during spring. The magnitude of temporal changes in CO_2 flux measured in this experiment over the year is comparable to that reported in other studies (Buyanovsky et al., 1986; Hendrix et al., 1988; Alvarez et al., 1995; Fortin et al., 1996) and indicates a relationship between soil temperature and CO_2 flux. The best regression equation relating CO_2 flux from unmulched bare plots and the average daily soil temperature measured at 5-cm depth was CO_2 flux $= -0.22 + 0.02 \times$ (soil temperature)$^{1.49}$ ($r^2 = 0.60$). Franzenluebbers et al. (1994, 1995) reported high correlation between CO_2 flux and soil temperature and moisture content.

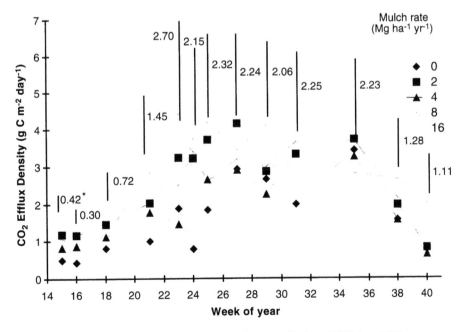

Figure 5.5   Mulch rate effect on CO_2 flux (g C m⁻² day⁻¹) on no-till plots. * LSD (p = 0.05).

Low CO_2 flux from the unmulched treatment until June is probably due to lack of organic matter substrate and low soil moisture contents in the soil surface. Consequently, even high surface soil temperatures (see Duiker and Lal, 2000) in spring did not result in high CO_2 flux. A relative increase in CO_2 flux from the end of June to the end of August in the unmulched control may be due to mineralization of SOC from subsoil layers.

The highest CO_2 flux was observed from the 16 Mg ha^{-1} yr^{-1} mulch treatment from mid May to the end of July. In early October, the CO_2 fluxes from the 0, 2, 4, and 8 Mg ha^{-1} yr^{-1} mulch treatments were similar, while that of the plots receiving 16 Mg ha^{-1} yr^{-1} was significantly higher than that of others. The thick mulch layer of the 16 Mg ha^{-1} yr^{-1} treatment may have resulted in higher nighttime temperatures compared to other treatments in which the mulch layer was thin in autumn due to compaction and decomposition. Soil temperature measurements in the early morning in autumn (data not shown) confirmed that soil temperatures were 1 to 3°C higher under the 16 Mg ha^{-1} yr^{-1} plots compared to other treatments. No significant difference in CO_2 flux between the 0 Mg ha^{-1} yr^{-1} and other treatments may be attributed to the presence of undecomposed crop residue lying on the soil surface and perhaps incorporated into the soil by faunal activity; this was not determined as SOC content with the analytical method employed here.

CONCLUSIONS

The influence of mulch and tillage treatments on SOC content in this Crosby silt loam was limited only to the surface 10-cm layer of the soil. SOC content increased with increases in mulch rates for no-till and plow-till treatments, but not for ridge-till. The most signficant regression equations relating the rate of mulch application (Mg ha^{-1} yr^{-1}) to SOC pool (Mg ha^{-1}) were SOC = 15.21 + 0.32 × Mulch (r = 0.68) for no-till, and SOC = 11.95 + 0.27 × Mulch (r = 0.72) for plow-till. Calculated carbon conversion efficiencies (% carbon applied in mulch converted into SOC pool per year) were 10% for no-till, 8% for plow-till, and 0% for ridge-till. The CO_2 flux measured on no-till plots was highly variable. No significant differences were observed in CO_2 flux in relation to mulch rate. It is possible that a large proportion of the mulch applied remained on the surface of the soil or was incorporated into the soil without decomposition. Low rates of decomposition may be due to lack of nutrients, since no fertilizer was applied. The nutrient constraint may also explain the low carbon conversion efficiencies observed in this experiment compared to those reported in the literature.

REFERENCES

Alvarez, R., O.J. Santanatoglia, and R. Garcia, Soil respiration and carbon inputs from crops in a wheat-soybean rotation under different tillage systems, *Soil Use Manage.*, 11: 45–50, 1995.

Angers, D.A. et al., Impact of tillage practices on organic carbon and nitrogen storage in cool, humid soils of eastern Canada, *Soil Tillage Res.*, 41: 191–201, 1997.

Anderson, J.P.E., Soil respiration, in Page, A.L., R.H. Miller, and D.R. Keeney (eds.), *Methods of Soil Analysis. Part 2. Chemical and Microbiological Properties*, 2nd ed., Agronomy Monograph No. 9, American Society of Agronomy, Madison, WI, 1982, 831–872.

Bajracharya, R.M., R. Lal, and J.M. Kimble, Long-term tillage effects on soil organic carbon distribution in aggregates and primary particle fractions of two Ohio soils, in Lal, R. et al. (eds.), *Management of C Sequestration in Soils*, CRC Press, Boca Raton, FL, 1998, 113–123.

Blake, G.R. and K.H. Hartge, Bulk density, in Klute, A. (ed.), *Methods of Soil Analysis. Part I. Physical and Mineralogical Methods*, 2nd ed., Agronomy Monograph No 9, American Society of Agronomy, Madison, WI, 1986, 363–382.

Blevins, R.L. et al., Influence of conservation tillage on soil properties, *J. Soil Water Conserv.*, 38: 301–305, 1983.

Buyanovsky, G.A., G.H. Wagner, and C.J. Gantzer, Soil respiration in a winter wheat ecosystem, *Soil Sci. Soc. Am. J.*, 50: 338–344, 1986.

CTIC, National Crop Residue Management Survey, 1998. Conservation Tillage Information Center, West Lafayette, IN, 1998.

Dick, W.A. et al., Influence of long-term tillage and rotation combinations on crop yields and selected soil parameters. I. Results obtained for a Mollic Ochraqualf soil, *Res. Bull.* 1180, OARDC, Wooster, OH, 1986a.

Dick, W.A. et al., Influence of long-term tillage and rotation combinations on crop yields and selected soil parameters. II. Results obtained for a Typic Fragiudalf soil, *Res. Bull.* 1181, OARDC, Wooster, OH, 1986b.

Duiker, S.W. and R. Lal, Crop residue and tillage effects on carbon sequestration in a Luvisol in central Ohio, *Soil Tillage Res.*, 52: 73–81, 1999.

Duiker, S.W. and R. Lal, Carbon budget study using CO$_2$ flux measurements from a no till system in central Ohio, *Soil Tillage Res.*, 54: 21–30, 2000.

Fortin, M.-C., P. Rochette, and E. Pattey, Soil carbon dioxide fluxes from conventional and no tillage small-grain cropping systems, *Soil Sci. Soc. Am. J.*, 60: 1541–1547, 1996.

Franzenluebbers, K., A.S.R. Juo, and A. Manu, Decomposition of cowpea and millet amendments to sandy Alfisol in Niger, *Plant Soil*, 167: 255–265, 1994.

Franzenluebbers, A.J., F.M. Hons, and D.A. Zuberer, Tillage-induced seasonal changes in soil physical properties affecting soil CO$_2$ evolution under intensive cropping, *Soil Tillage Res.*, 34: 41–60, 1995.

Havlin, J.L. et al., Crop rotation and tillage effects on soil organic carbon and nitrogen, *Soil Sci. Soc. Am. J.*, 54: 448–452, 1990.

Hendrix, P.F., Chun-Ru Han, and P.M. Groffman, Soil respiration in conventional and no tillage agroecosystems under different winter cover crop rotations, *Soil Tillage Res.*, 12: 135–148, 1988.

Himes, F.L., Nitrogen, sulfur and phosphorus and the sequestering of carbon, in Lal, R. et al. (eds.), *Soil Processes and the C Cycle*, CRC Press, Boca Raton, FL, 1998, 315–320.

Janzen, H.H., Soil organic matter characteristics after long-term cropping to various spring wheat rotations, *Can. J. Soil Sci.*, 67: 845–856, 1987.

Lal, R. et al., Managing U.S. cropland to sequester carbon in soil, *J. Soil Water Conserv.*, 54: 374–381, 1999.

Nelson, D.W. and L.E. Sommers, Total carbon, organic carbon, and organic matter, in Page, A.L., R.H. Miller, and D.R. Keeney (eds.), *Methods of Soil Analysis, Part 2. Chemical and Microbiological Properties*, 2nd ed., Agronomy Series No 9, American Society of Agronomy, Madison, WI, 1986, 539–579.

Nyborg, M. et al., Fertilizer N, crop residue, and tillage alter soil C and N content in a decade., in Lal, R. et al. (eds.), *Soil Management and Greenhouse Effect*, Advances in Soil Science, CRC Press, Boca Raton, FL, 1995, 93–99.

Paustian, K., H.P. Collins, and E.A. Paul, Management controls on soil carbon, Paul, E.A. et al. (eds.), *Soil Organic Matter in Temperate Agroecosystems, Long-Term Experiments in North America*, CRC Press, Boca Raton, FL, 1997, 15–49.

Rasmussen, P.E. and H.P. Collins, Long-term impacts of tillage, fertilizer, and crop residue on soil organic matter in temperate semiarid regions, *Adv. Agron.*, 45: 93–134, 1991.

Reicosky, D.C. and M.J. Lindstrom, Fall tillage method: effect on short-term carbon dioxide flux from soil, *Agron. J.*, 85: 1237–1243, 1993.

Reicosky, D.C., W.A. Dugas, and H.A. Torbert, Tillage-induced soil carbon dioxide losses from different cropping systems, *Soil Tillage Res.*, 41: 105–118, 1997.

Russell, E.W., *Soil Conditions and Plant Growth*, 10th ed., Longman, London, 1973.

Smith, P. et al., Meeting Europe's climate change commitments: quantitative estimates of the potential for carbon mitigation by agriculture, *Global Change Biol.*, 6: 525–539, 2000a.

Smith, W.N., R.L. Desjardins, and E. Pattey, The net flux of carbon from agricultural soils in Canada 1970–2010, *Global Change Biol.*, 6: 557–568, 2000b.

Steel, R.G.D., J.H. Torrie, and D.A. Dick, *Principles and Procedures of Statistics: A Biometrical Approach*, 3rd ed., McGraw-Hill, New York, 1997.

CHAPTER 6

Effects of Tillage on Inorganic Carbon Storage in Soils of the Northern Great Plains of the U.S.

L. J. Cihacek and M. G. Ulmer

CONTENTS

INTRODUCTION

Research on carbon (C) sequestration by soils during the past decade has focused on two principle themes: (1) carbon sequestration by soil as organic carbon and (2) carbon sequestration in the surface 10 to 30 cm of the soil profile. Soils have the potential to sequester significant quantities of carbon deposited as biological residues from plants and animals; restricting C release back to the atmosphere as CO_2 contributes to the greenhouse effect. We have examined C cycling near the soil surface in a semiarid region of the U.S. northern Great Plains under native grass and crop-fallow cultures (Cihacek and Ulmer, 1995), examined, as well as differences in organic C storage in the soil profile in these cultures (Cihacek and Ulmer, 1997). However, little work has been done on examining the role of inorganic C on the overall function of soils for C sequestration and storage.

Since the early 1900s, crop-fallow rotations have regularly been practiced in the semiarid cultivated lands of the U.S. Great Plains west of the 100° meridian. This reduced the chance of crop failure by conserving water and providing mineral nitrogen (N) for subsequent crops. Typically, wheat (*Triticum aestivium* L.) or other small grains are seeded in alternate years in a crop-fallow system. During the intervening fallow period, vegetative growth is restricted by shallow cultivation or herbicides. Characteristically, the northern Great Plains experience cool temperatures and irregular rainfall patterns generally deficient for optimum plant growth. This has encouraged an agriculture based on hard red spring wheat or spring small grain production in this region (Norum et al., 1957).

In the crop-fallow cycle, carbon is sequestered in the crop residue during the period when the crop grows on the soil. After the crop is harvested, tillage incorporates the residue back into the soil where the carbon is released by microbial activity. The carbon released from the residue then enters the soil organic matter pool precipitates as inorganic carbonates and bicarbonates or is released to the atmospheric pool as CO_2 (Alexander, 1961).

Inorganic carbon precipitated as carbonates or bicarbonates originates from CO_2 supplied to the soil by several pathways, including: (1) CO_2 dissolved in rainfall as a weak solution of carbonic acid (H_2CO_3), (2) CO_2 evolved in the soil by microbial breakdown of soil organic matter and dissolved in the soil solution, (3) CO_2, $CO_3^=$, or HCO_3^- evolved by plant root respiration or nutrient transport processes and dissolved in the soil solution, and (4) dissolution of soil carbonate minerals by the acidic nature of precipitation or soil solution containing dissolved CO_2 (carbonic acid). A source of Ca^{2+} ion such as gypsum ($CaSO_4$) is also required. McFadden et al. (1991) and Chadwick et al. (1994) have discussed the dynamics of CO_2 relationships in the formation and transformation of soil minerals and soil carbonates.

Cihacek and Ulmer (1995) recently estimated that, in Major Land Resource Area (MLRA) 54, annual C loss in a wheat-fallow cropping system averaged 0.93 Mg C ha^{-1} in the surface 15 cm of the soil. This loss was due to carbon cycling within the cropping system, but, in part, may contribute to carbonate formation and inorganic C storage in these soils. Cihacek and Ulmer (1997) have also shown that organic carbon may accumulate at depths below 50 cm in the soil profile due to increased water infiltration in a crop-fallow cropping system.

This study continues previous work to examine the long-term effects of crop-fallow agriculture on carbon distribution and storage in major cropland soil types in MLRA 54, which is part of the Northern Plains spring wheat region of southwestern North Dakota. Objectives were to: (1) compare profile inorganic carbon distribution in grassland and cultivated soils and (2) evaluate effects of comparable cultivation on inorganic carbon storage within the profiles of selected soil series.

METHODS

Twenty-one selected paired cultivated and native rangeland sites were sampled throughout MLRA 54, the location of which has been previously reported by Cihacek and Ulmer (1995, 1997). The paired sites were within 75 m of each other and had similar slope and parent material parameters. Soils from all sites were classified as either coarse- or fine-loamy, mixed Typic Haploborolls located on Pleistocene and Tertiary age deposits. Classification and physiographical information of these soils are shown in Table 6.1. Cropland sites have been under cultivation for over 60 years; most cropland soils in this MLRA were brought under cultivation by 1910. The predominant grass species on the native grassland sites included blue grama (*Bouteloua gracilis* (H.B.K.) Lag. ex Griffiths), green needlegrass (*Stipa viridula* Trin.), needle-and-thread (*Stipa comata* Trin. and Reysr.), and, occasionally, western wheatgrass (*Agropyron smithii* Rydo.).

Three 6-cm diameter cores were collected to a depth of 1.2 m or to contact with residuum from an area representative of the cultivated or rangeland soils at each site. The cores were composited by horizon, then mixed, air-dried, and crushed to pass a 2-mm screen. Inorganic C (IC) was determined by the titrimetric procedure of Bundy and Bremner (1972) and soil bulk density was determined by the core method described by Blake and Hartge (1986).

Inorganic C content was calculated by adjusting soil mass for bulk density to correct for changes in bulk density due to cultivation (Cihacek and Ulmer, 1995). Profile data were combined for each

Table 6.1   Classification and Physiography of Soils

Soil Series	Number of Paired Sites	Classification	Landscape Position	Slope
Amor	4	Fine-loamy, mixed Typic Haploborolls	Upland sideslope	1–6%
Shambo	10	Fine-silty, mixed Typic Haploborolls	Terrace/fan	1–6%
Stady	4	Fine-loamy, mixed Typic Haploborolls	Terrace	1–3%
Tally	1	Coarse-loamy, mixed Typic Haploborolls	Terrace/fan	1–6%
Vebar	2	Coarse-loamy, mixed Typic Haploborolls	Upland sideslope	1–6%

Source: Adapted from Patterson and Heidt, 1987.

soil series by horizon within each management system and averaged for horizon thickness. The data are presented as the density of C ($Kg\ m^{-2}$) by horizon per/m depth or total profile depth if less than 1 m. Statistical differences were determined by using the ANOVA routine of the statistical analysis system SAS Institute (1985).

RESULTS AND DISCUSSION

Profile of Inorganic Carbon Distribution

Soil inorganic C density distributions for the five selected soils series are shown in Figure 6.1. The IC distributions are shown as plots for each 5-cm depth increment to smooth out differences between the horizons. Nearly all of the profiles showed evidence of near contact or actual contact with underlying bedrock or depositional materials. However, the relationship of the profile to the underlying material did not appear to influence the IC in the lowest 10 to 20 cm of the profiles. All soils, both cultivated and grassland, showed very low IC levels in the upper 20 cm of the profile; this low IC extended to 45 cm in the Shambo and Stady soils.

With the exception of the Tally soils, all soils showed increases in IC below a depth of 50 cm. The Amor and Shambo soils showed increasing IC with depth under grassland management. Under

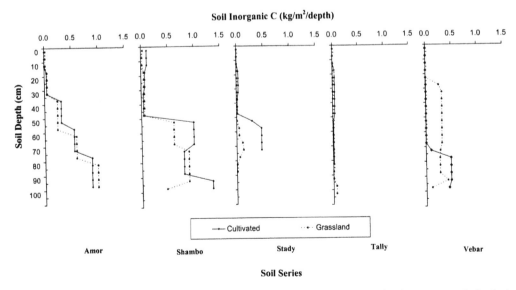

Figure 6.1   Soil inorganic carbon density distribution under cultivated and grassland management of selected soil series from MLRA 54. Profile depths less than 100 cm are due to contact with gravel, shale, or sandstone.

cultivation, the Amor showed a slightly higher IC density but similar distribution throughout the profile. On the other hand, the Shambo soil showed a marked increase in IC between 50-cm to 70-cm depth and again from 90-cm to 100-cm depth. The Stady soil showed very little accumulation of IC in the grassland profile, but showed a marked increase in IC between 50 and 70 cm, at which depth a lithologic discontinuity occurred. The coarser-textured Vebar soil showed a nearly uniform distribution of IC in the profile below 20 cm in the grassland sites. Little IC was observed in the cultivated Vebar soils up to a depth of 75 cm, at which point the IC density increased to a level nearly double that found in the grassland soils.

Two divergent philosophies predominate on how increases in carbonates occur in soil. The first involves the concept that materials, containing carbonate materials such as aeolian materials and aerosols, are deposited on the soil surface and then are mineralized and leached into the soil (Gile et al., 1966). The second involves the concept of weathering of minerals in place due to dissolved CO_2 in rain water and soil solution (McFadden et al., 1991; Chadwick et al., 1994).

Cerling (1984) has proposed that soil carbonate formation is a function of the proportion of biomass using the C_4 photosynthetic pathway and CO_2 respiration rate of the soil and that, at low soil respiration rates, a significant contribution of atmospheric CO_2 can occur. He also suggests that soils that freeze at depth may also precipitate carbonates due to an increase in ion concentrations resulting from ion exclusion during ice formation. In frozen soils, atmospheric CO_2 can contribute to a significant portion of the soil carbonates formed by this process.

The water storing effect of crop-fallow management will increase water movement to a greater depth in cultivated soils, thereby providing opportunities to move IC deeper into the soil profile and causing it to accumulate in the vicinity of the wetting front. Muir et al. (1976) noted that, in western Nebraska, the water content of soil was significantly higher in loessial soils under wheat-fallow cropping than under native range to a depth of 8 to 9 m. This occurred in a region of 400 to 500 mm annual rainfall. Annual precipitation in MLRA 54 can be similar, but with a wider range of 250 to 550 mm of annual precipitation.

The increases in IC between 50 and 80 cm in the cultivated finer textured soils (Amor, Shambo, and Stady) and below 80 cm in the coarser-textured Vebar soils represent the average depth of water infiltration into cultivated soils from individual precipitation events. These soils also regularly freeze to a depth of 1 m or more, thereby providing ample opportunity for carbonates to precipitate even during periods of slow soil respiration.

Although C_4 species are dominant in the uncultivated grasslands in this study, deep accumulation of IC does not occur because the grass cover intercepts precipitation and utilizes it before it infiltrates to any significant depth in the soil. The grass cover also insulates the soil during cold periods by trapping snow and thus reducing the depth and intensity of freezing compared to that occurring on cultivated fallow soils.

Changes in Soil Inorganic Carbon Distribution

Figure 6.2, shows changes in inorganic C density distribution in cropland soils relative to grassland soils. These changes in distribution reflect the data shown in Figure 6.1, but are based on values obtained by subtracting cropland soil IC from grassland soil IC. Most changes range $\leq \pm 0.5$ kg m^{-2} within a 5-cm profile segment, indicating an absolute change of 1 kg m^{-2} or 10,000 kg ha^{-1}. Compare this to the long-term loss of 14,344 kg organic C ha^{-1} in the surface horizon of cultivated soils in MLRA 54 (Cihacek and Ulmer, 1995) due to cultivation losses from residue cycling, erosion, and natural biological oxidation. The absolute magnitude of change in the surface 1 m in these soils due to cultivation is 0.7 times the carbon loss from the surface horizon above. This can be an important offset for the C loss from the surface horizon.

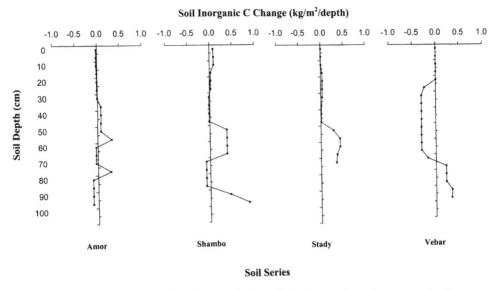

Figure 6.2   Changes in soil inorganic carbon density in cultivated soils relative to grassland soils.

Inorganic Carbon Storage

Inorganic C storage in the 1-m profile of each soil series examined in this study is shown in Figure 6.3. The finer-textured soil series (Amor, Shambo, Stady) showed a trend toward higher IC storage in the cultivated profiles when compared to the grassland profiles. However, only the Stady soil showed a statistically significant difference between the IC storage in the cultivated and grassland profiles ($P \leq 0.10$). The coarser-textured soil series (Tally, Vebar) showed either no difference or a trend toward higher IC storage in the grassland soils.

These differences may be due to deeper infiltration of stored water in these soils resulting in the deeper accumulation of IC in the Vebar soil compared to the other soils (See Figure 6.1). Combined data of all soils show that IC storage in cultivated soils is significantly higher than in the grassland soils ($P \leq 0.05$).

SUMMARY

Significant differences can occur in IC storage between cultivated and grassland soils, with cultivated soils storing an average 3.2 kg/m^2 more C than grassland soils in this study. However, this phenomenon may vary with location within the Great Plains of North America. It appears that two essential components of increased IC accumulation in cultivated soils are increased infiltration and storage of precipitation, and freezing of the soil profile (Cerling, 1984).

Most cultivated soils in the northern Great Plains freeze during the winter season to a depth of up to 1 m or more, especially when moist. Under fallow conditions, the soil surface is bare of vegetation and traps only a small amount of snow. In contrast, grasslands are capable of readily capturing a 20- to 50-cm layer of snow, which acts as insulation and protects the soil from deep freezing. Grassland soils also will be drier because the root mass of the native vegetation intercepts infiltration precipitation and utilizes it shortly after it falls.

Freezing will cause calcium salts, including the most common of salts in semiarid soils, $CaCO_3$, to rapidly precipitate, thus aiding the accumulation of IC. Due to the limited solubility of $CaCO_3$

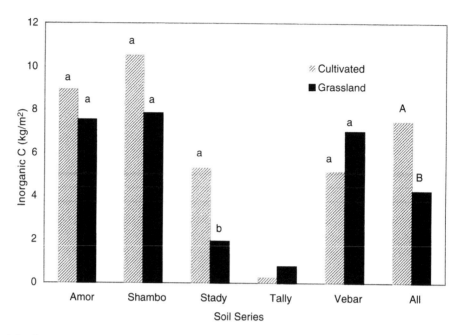

Figure 6.3   Summary of stored inorganic carbon in a 1-m profile of selected soils under cultivated or grassland soils in MLRA 54. Bars annotated with the same lowercase letter are not significantly different within each soil series at $P \le 0.05$ as determined by LSD. Only one site with the Tally soil was sampled and therefore not subjected to statistical analysis. Uppercase letters annotating the combined date (All) show a significant difference at $P \le 0.05$ as determined by LSD.

and seasonally limited moisture in these soils, very little of the $CaCO_3$ resolubilizes; thus it becomes stored as IC in the soil profile.

The significance of soils in sequestering greenhouse gases and reducing global warming may be greater due to C sequestration as inorganic C. Soil IC is a sink for atmospheric CO_2, which may be more resistant to cropping and tillage effects on sequestered soil C and is likely to persist for decades and perhaps centuries after sequestration.

Further research is needed to define soils, soil parameters, and climate parameters conducive to C sequestration as IC, quantify the amount of C sequestered by cropland as IC, and determine residence times of sequestered IC before it is rereleased to the atmosphere.

REFERENCES

Alexander, M., *Introduction to Soil Microbiology*, John Wiley & Sons, New York, 1961.

Blake, G.R. and K.H. Hartge, Bulk density, in Klute, A. (ed.), *Methods of Soil Analysis, Part I*, 2nd ed., Monogr. ASA and SSSA, Madison, WI, 1986, 363–376.

Bundy, L.G. and J.M. Bremner, A simple titrimetric method for determination of inorganic carbon in soils, *Soil Sci. Soc. Am. Proc.*, 36:273–275, 1972.

Cerling, T.E., The stable isotopic composition of modern soil carbonate and its relationship to climate, *Earth Plan. Sci. Let.*, 71:229–240, 1984.

Chadwick, O.A. et al., Carbon dioxide consumption during soil development, *Biogeochemistry*, 24:115–127, 1994.

Cihacek, L.J. and M.G. Ulmer, Estimated soil organic carbon losses from long-term crop-fallow in the northern Great Plains of the USA, in Lal, R. et al. (eds.), *Soil Management and Greenhouse Effect*, CRC Press, Boca Raton, FL, 1995, 85–92.

Cihacek, L.J. and M.G. Ulmer, Effects of tillage on profile soil carbon distribution in the northern Great Plains of the U.S., in Lal, R. et al. (eds.), *Management of Carbon Sequestration in Soil*, CRC Press, Boca Raton, FL, 1997, 83–91.

Gile, L.H., F.F. Peterson, and R.B. Grossman, Morphological and genetic sequences of carbonate accumulation in desert soils, *Soil Soc.*, 101:347–360, 1966.

McFadden, L.D., R.G. Amundson, and O.A. Chadwick, Numerical modeling, chemical and isotopic studies of carbonate accumulations in soils of aerial regions, in W.D. Nettleton, (ed.), *Occurrence, Characteristics, and Genesis of Carbonate, Gypsum, and Silica Accumulations in Soils*, SSSA Spec. Pub. No. 26, Soil Science Society of America, Madison, WI, 1991, 17–35.

Muir, J. et al., Influence of crop management practices on nutrient movement below the root zone in Nebraska soils, *J. Environ. Qual.*, 5:255–259, 1976.

Norum, E.B., B.A. Kranz, and H.J. Haas, The northern Great Plains, in Stefferud, A. (ed.), *Soil*, 1957 Yearbook of Agriculture, USDA, Washington, D.C., 1957, 494–505.

Patterson, D.D. and C.J. Heidt, A taxonomic guide to the soils of North Dakota, North Dakota State University, Dept. Soil Sci. Res. Rep. 20, Fargo, ND, 1987.

SAS Institute, Inc., SAS User's Guide, Statistics, Version 5 edition, Cary, NC, 1985.

USDA Soil Conservation Service, Land resource regions and major land resource areas of the United States, USDA Agric. Hndbk 296, Washington, D.C., 1981.

CHAPTER 7

Climatic Influences on Soil Organic Carbon Storage with No Tillage

A. J. Franzluebbers and J. L. Steiner

CONTENTS

ABSTRACT

No-tillage crop production has become an accepted practice throughout the U.S. The Kyoto Protocol on climate change has prompted great interest in conservation tillage as a management strategy to help sequester CO_2 from the atmosphere into soil organic matter. Numerous reports published in recent years indicate a large variation in the amount of potential soil organic carbon (SOC) storage with no tillage (NT) compared with conventional tillage (CT). Environmental controls (i.e., macroclimatic variables of temperature and precipitation) may limit the potential of NT to store SOC. We synthesized available data on SOC storage with NT compared with CT from published reports representing 111 comparisons from 39 locations in 19 states and provinces across the U.S. and Canada. These sites provided a climatic continuum of mean annual temperature and precipitation, which was used to identify potential SOC storage limitations with NT. Soil organic C storage potential under NT was greatest (~ 0.050 kg $\cdot$ m^{-2} $\cdot$ yr^{-1}) in subhumid regions of North America with mean annual precipitation-to-potential evapotranspiration ratios of 1.1 to 1.4 mm $\cdot$ mm^{-1}. Although NT is important for water conservation, aggregation, and protection of the soil surface from wind and water erosion in all climates, potential SOC storage with NT compared with CT was lowest in cold and dry climates, perhaps due to prevailing cropping systems that relied on low-intensity cropping, which limited C fixation. Published data indicate that increasing cropping intensity to utilize a greater fraction of available water in cold and dry climates can increase potential SOC storage with NT. These analyses indicate greatest potential SOC storage with NT would be most likely in the relatively mild climatic regions rather than extreme environments.

INTRODUCTION

Conservation-tillage crop production has become an accepted practice throughout the U.S. and Canada. Thirty-seven percent of the cropland in the U.S. is now managed with some form of conservation tillage (i.e., no tillage, ridge tillage, and mulch tillage) (CTIC, 1998). The Kyoto Protocol on climate change has prompted the agricultural sector to promote more seriously various forms of conservation tillage as practices to help sequester CO_2 from the atmosphere into soil organic matter.

Numerous reports have been published in recent years concerning the effect of no-tillage crop production (NT) compared with conventional tillage (CT) on potential soil organic carbon (SOC) storage. However, these reports indicate a large variation in the amount of potential SOC storage with NT. For example, SOC in the Ap horizon (0 to 20-cm depth) of a Dark Brown Chernozemic clay loam in Alberta increased at only 0.17 to 0.20 mg $\cdot$ g^{-1} soil $\cdot$ yr^{-1} compared with shallow CT in two studies conducted 9 and 19 years under NT (Dormaar and Lindwall, 1989). In contrast, SOC at a depth of 0 to 7.5 cm during 4 years under NT compared with plowed CT increased at 0.69 mg $\cdot$ g^{-1} soil $\cdot$ yr^{-1} on a Waukegon silt loam in Minnesota (Hansmeyer et al., 1997) and at ~ 1.15 mg $\cdot$ g^{-1} soil $\cdot$ yr^{-1} on a Kamouraska clay in Quebec (Angers et al., 1993). Incorporation of residues below 7.5 cm with plowing would likely reduce this effect when considering the entire plow depth.

Soil organic C accumulation rates between these extremes have also been observed. At a depth of 0 to 5 cm, SOC increased at 0.42 mg $\cdot$ g^{-1} soil $\cdot$ yr^{-1} during 14 years under NT compared with multiple-disk CT on a Norfolk loamy sand in the South Carolina coastal plain (Hunt et al., 1996) and at 0.28 to 0.42 mg $\cdot$ g^{-1} soil $\cdot$ yr^{-1} during more than 20 years under NT compared with plowed CT on a Bertie silt loam in the Maryland coastal plain (McCarty and Meisinger, 1997). On a Hoytville silty clay loam in Ohio, SOC of the 0- to 10-cm depth increased at 0.66 mg $\cdot$ g^{-1} soil $\cdot$ yr^{-1} during 12 years under NT compared with plowed CT (Lal et al., 1990). The large range of changes in SOC with NT compared with CT among the aforementioned studies may be related to differences in cropping system, fertilization, depth of tillage tool, numerous soil characteristics, climatic conditions, and depth of sampling.

When comparing management effects on SOC storage, soil sampling depth is an important consideration. Depth distribution of SOC is altered with NT compared with CT. For example, SOC under NT was $40 \pm 22\%$ greater than under CT at a depth of 0 to 7.5 cm, similar at a depth of 7.5 to 15 cm, and $7 \pm 11\%$ less at a depth of 15 to 30 cm, resulting in a net change of only $9 \pm 7\%$ greater SOC under NT (Doran, 1987). Changes in depth distribution of SOC with tillage systems suggest the need to standardize sampling protocols to collect soil to at least the depth of deepest tillage tool in order to make fair comparisons.

A growing database has accumulated reporting differences in SOC between CT and NT crop management. Recently, efforts have been made to synthesize results from long-term studies on SOC within regions (Paustian et al., 1998). However, efforts to isolate particular regions of North

America with the greatest potential to store SOC with adoption of NT have only begun. Although models have been developed to predict such regional differences in SOC accumulation potential (Smith et al., 1998), long-term observational data could provide verification of such predictions if enough cross-regional data were collected and synthesized. It was hypothesized that macroclimatic conditions would have an influence on the potential of NT to sequester SOC compared with CT. Objectives were to (1) summarize published data from the U.S. and Canada and (2) test whether soil type, cropping intensity, N fertilization, and macroclimatic regime affected the difference in standing stock of soil organic C due to adoption of NT compared with CT.

MATERIALS AND METHODS

Data were obtained from the literature in which SOC was reported for NT crop management compared with some form of CT (Table 7.1; Figure 7.1). Only reports that contained SOC information on an area basis were used in order to avoid misleading interpretations due to management-induced changes in bulk density (Ellert and Bettany, 1995). Differences in standing stock of SOC between NT and CT were standardized to an annual basis. Net annualized change in SOC with NT compared with CT was expected to decline with time, but this did not occur in the available data from three locations (Figure 7.2). Length of time in all comparisons was 10.7 ± 6.1 years, with 85% of comparisons >5 years in length.

"Decomposition potential" of each location was evaluated using several different indices based on climate. Long-term mean monthly precipitation and temperature data from the closest weather station to each of the evaluated locations (i.e., within 30 km) were obtained (Global Historical Climatology Network, 1999). Monthly potential evapotranspiration (PET) was calculated from long-term mean monthly temperatures and latitude using the Thornthwaite equation (Thornthwaite et al., 1957).

Index 1 was calculated as the sum of each monthly precipitation-to-potential evapotranspiration ratio divided by 12. Monthly precipitation exceeding PET was assigned a value of 1, because those locations with subzero mean monthly temperature were calculated to have no PET using the Thornthwaite equation.

Index 2 was calculated as mean annual precipitation divided by mean annual PET (derived from the sum of monthly values). Index 2 was allowed to exceed 1.

Index 3 was calculated as the sum of products from temperature and precipitation coefficients on a monthly basis divided by 12. The temperature coefficient was calculated from a nonlinear function that assumed a doubling of microbial activity for every 10°C change in temperature [$2((°C-30)/10)$] (Kucera and Kirkham, 1971), with 30°C assumed as an optimum (Figure 7.3a). None of the locations had mean monthly temperatures exceeding 30°C. The mean monthly precipitation coefficient was expressed as mean monthly precipitation (mm) divided by 100. It was assumed that 100 mm of precipitation per month would be adequate for maximum decomposition at any temperature. Months with precipitation exceeding 100 mm were assigned coefficients of 1 (Figure 7.3b).

Index 4 was calculated as the product of temperature and precipitation coefficients on an annual basis. The mean annual precipitation coefficient was calculated as mean annual precipitation (mm) divided by 1200. When annual precipitation exceeded 1200 mm, the precipitation coefficient was allowed to exceed 1.

Index 5 was calculated as the sum of the most limiting monthly coefficient (i.e., lowest temperature or precipitation coefficient for each month) divided by 12.

Index 6 was calculated as the most limiting annual temperature or precipitation coefficient.

Because of unequal representation of geographical regions, soil orders, soil textural classes, and fertilization regimes, individual univariate analyses on the net annualized change in SOC with NT compared with CT were conducted for each of these variables separately, using the general linear model procedure (SAS Institute Inc., 1990). Polynomial regressions (i.e., linear plus quadratic

Table 7.1　Characteristics of Locations

Location	Soil Texture	Soil Classification	Years	Crop Intensity	Soil Depth	MAT	MAP	PET	Source
AB Beaverlodge	CL	Cryoboralf	4	0.5	20	2.0	447	489	1
AB Breton	L	Cryoboralf	11	0.5	15	2.0	495	504	2
AB Ellerslie	L	Cryoboroll	11	0.5	15	2.3	483	504	2
AB Lethbridge	SiL	Haploboroll	12 ± 4	0.4 ± 0.1	15 ± 1	5.5	415	556	3, 4
AB Rycroft	C	Natriboralf	6	0.4	20	0.6	369	489	1
BC Dawson Creek	SiL	Cryoboralf	16	0.5	20	1.4	466	489	1
BC Rolla	L	Cryoboralf	7	0.5	20	0.6	466	484	1
CO Akron	SiL	Paleustoll	15	0.25	20	9.4	404	641	5
GA Athens	SL	Kanhapludult	12 ± 3	1.0	20 ± 3	16.5	1230	874	6, 7
GA Griffin	SL	Kanhapludult	10 ± 5	1.0	30	16.5	1230	874	6
GA Watkinsville	SL	Kanhapludult	4	1	15	16.5	1230	874	8
IL DeKalb	SiCL	Haplaquoll	10.5	0.5	30	9.4	874	678	9
IL Elwood	SiL	Ochraqualf	6	0.5	30	9.2	892	660	10
IL Monmouth	SiL	Hapludoll	10.5	0.5	30	10.6	912	716	9
IL Perry	SiL	Argiudoll	10.5	0.5	30	10.6	912	716	9
KY Lexington	SiL	Paleudalf	12 ± 8	0.5	30	12.9	1129	765	10, 11, 12
MD Beltsville	SiL	Hapludult	2 ± 1	0.5	20	12.9	1076	770	13
MI East Lansing	L	Ochraqualf	9 ± 3	0.5	20	8.3	785	617	14
MN Waseca	CL	Haplaquoll	9 ± 3	0.5	23 ± 11	7.2	767	642	10
MT Culbertson	SL	Argiboroll	10	0.5	21	5.6	337	602	15
ND Mandan	SiL	Argiboroll	6	0.4 ± 0.1	30	5.2	402	588	16
NE Lincoln	SiCL	Argiudoll	9 ± 4	0.5	30	10.3	782	712	10, 17
NE Sidney	L	Haplustoll	14 ± 4	0.25	28 ± 5	8.4	468	606	10, 18, 19
OH Wooster	SiL	Fragiudalf	29 ± 1	0.7 ± 0.3	23 ± 8	10.9	1028	695	20, 21

Location	Texture	Classification							Source
ON Delhi	SL	Psamment	4	0.5	60	7.9	950	598	22
ON Harrow	CL	Haplaquoll	11	0.5	60	9.0	831	641	22
ON Ottawa	SL	Eutrochrept	5	0.5	60	5.6	879	587	22
PE Harrington	fSL	Haplorthod	8	0.5	60	5.5	1074	531	22
QC La Pocatiere	C	Humaquept	5 ± 1	0.5	38 ± 32	4.1	944	520	22
QC Normandin	SiC	Humaquept	3	0.5	60	2.2	887	502	22
SK Melfort	CL	Boroll	12	0.25	20	0.8	401	510	3
SK Scott	L	Boroll	2	0.5	10	1.1	356	508	3
SK Swift Current	fSL	Haploboroll	12	0.25	15	3.6	380	545	23
SK Swift Current	SiL	Boroll	12	0.4 ± 0.1	15	3.6	380	545	24
SK Watrous	CL	Boroll	4	0.5	10	0.8	420	505	3
TX Bushland	CL	Paleustoll	9 ± 2	0.4 ± 0.1	17 ± 3	14.0	516	796	25, 26, 27
TX College Station	SiCL	Ustochrept	9 ± 1	0.7 ± 0.2	20	20.2	1027	991	28, 29, 30
TX Corpus Christi	SCL	Ochraqualf	15 ± 1	0.5	20	22.7	713	1033	26, 27, 31
TX Temple	C	Pellustert	10	0.5	20	19.2	871	971	26, 27
WI Lancaster	SiL	Hapludalf	12	0.5	25	8.0	833	634	32

Note: AB = Alberta, BC = British Columbia, CO = Colorado, GA = Georgia, IL = Illinois, KY = Kentucky, MD = Maryland, MI = Michigan, MN = Minnesota, MT = Montana, ND = North Dakota, NE = Nebraska, OH = Ohio, ON = Ontario, PE = Prince Edward Island, QC = Quebec, SK = Saskatchewan, TX = Texas, and WI = Wisconsin. Textures are: C = clay; CL = clay loam; L = loam; SCL = sandy clay loam; SL = sandy loam; SiC = silty clay; SiCL = silty clay loam; SiL = silt loam; fSL = fine sandy loam; MAT = mean annual temperature (°C); MAP = mean annual precipitation (mm); and PET = mean annual potential evapotranspiration (mm). *Sources:* 1 = Franzluebbers and Arshad (1996); 2 = Nyborg et al. (1995); 3 = Carter and Rennie (1982); 4 = Larney et al. (1997); 5 = Halvorson et al. (1997); 6 = Hendrix et al. (1998); 7 = Beare et al. (1994); 8 = Franzluebbers et al. (1999); 9 = Wander et al. (1998); 10 = Mielke et al. (1986); 11 = Blevins et al. (1977); 12 = Ismail et al. (1994); 13 = McCarty et al. (1998); 14 = Pierce et al. (1994); 15 = Pikul and Aase (1995); 16 = Black and Tanaka (1997); 17 = Eghball et al. (1994); 18 = Lamb et al. (1985); 19 = Cambardella and Elliott (1992); 20 = Lal et al. (1994); 21 = Dick et al. (1998); 22 = Angers et al. (1997); 23 = Campbell et al. (1996); 24 = Campbell et al. (1995); 25 = Peterson et al. (1998); 26 = Potter et al. (1997); 27 = Potter et al. (1998); 28 = Franzluebbers et al. (1994); 29 = Franzluebbers et al. (1995); 30 = Franzluebbers et al. (1995); 31 = Salinas-Garcia et al. (1997); 32 = Karlen et al. (1994).

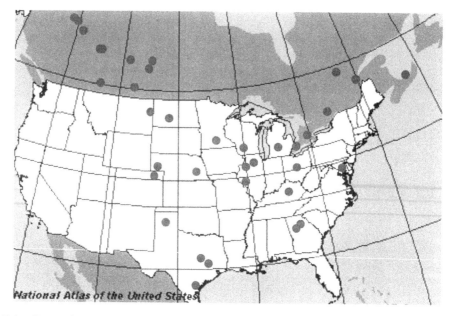

Figure 7.1   Geographical location of studies evaluated.

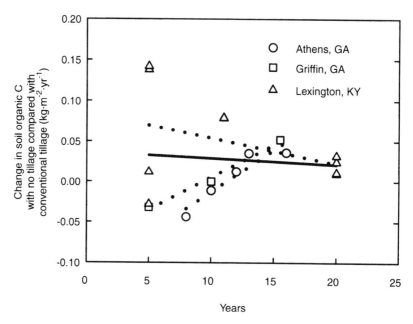

Figure 7.2   Net annualized change in soil organic C with no tillage compared with conventional tillage as affected by number of years under investigation. Data were compiled from Blevins et al. (1977), Mielke et al. (1986), Beare et al. (1994), Ismail et al. (1994), and Hendrix et al. (1998). Dotted lines are for individual locations, while solid line represents the mean of all observations.

functions using the general linear model procedure) were used to test the significance of continuous variables, including cropping intensity, temperature, precipitation, and PET. Cropping intensity was numerically expressed as the fraction of year in cropping, in which the year was divided into two halves of winter and summer cropping. For testing of univariate climatic effects (i.e., indices 1 to 6 composed of temperature, precipitation, and PET variables), mean net annualized change in SOC for each location was computed across cropping systems and N fertilization regimes.

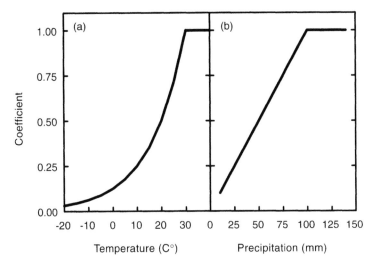

Figure 7.3 Diagrammatic representation of monthly temperature (a) and precipitation (b) coefficients used to characterize decomposition potential of locations in indices 3 and 5.

RESULTS AND DISCUSSION

General

A total of 111 comparisons between NT and CT on SOC were available from 39 locations with a wide range of temperature and precipitation coefficients (Figure 7.4). Temperature coefficients for indices 3 and 4 were highly related to latitude (Figure 7.4c), while precipitation coefficients were closely related to longitude (Figure 7.4b). Unfortunately, locations were not uniformly distributed among these gradients in order to avoid completely the confounding effects of temperature and precipitation. Locations tended to have higher precipitation at lower than at higher latitudes (Figure 7.4a) and higher temperatures at intermediate than at extreme longitudes (Figure 7.4d). To obtain a better distribution of environments in the U.S. and Canada, data from long-term studies are needed, especially from locations west of the Rocky Mountains, in southeastern U.S., and in central Canada.

Without regard to other variables, the net annualized change in SOC with NT compared with CT was normally distributed with a mean of 0.030 kg $\cdot$ m^{-2} $\cdot$ yr^{-1} (P<0.001 of mean = 0) (Figure 7.5). The mode and median of observations were 0.027 and 0.025 kg SOC $\cdot$ m^{-2} $\cdot$ yr^{-1}, respectively. The change in SOC with NT compared with CT was between 0.005 and 0.066 kg $\cdot$ m^{-2} $\cdot$ yr^{-1} for 50% of the observations.

Soil Type

Net annualized change in SOC with NT compared with CT was little affected by soil order (Figure 7.6). Based on a least significant difference comparison, the change in SOC was greater (P = 0.04) only in Inceptisols compared with Mollisols.

Net annualized change in SOC with NT compared with CT was little affected by soil textural class (Figure 7.7). Based on a least significant difference comparison, the change in SOC was greater (P = 0.05) only in silty clay loams compared with loams. Previous observations (Jenkinson, 1988; Amato and Ladd, 1992) and model predictions (Hassink and Whitmore, 1997) have suggested greater potential to store organic C in soils with a greater quantity of fine particles.

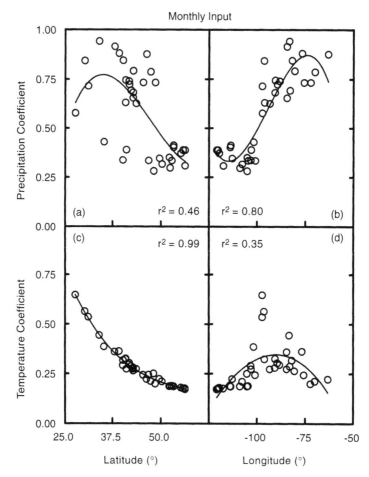

Figure 7.4 Distribution of precipitation (a) and temperature (c) coefficients along a latitudinal gradient and precipitation (b) and temperature (d) coefficients along a longitudinal gradient.

Nitrogen Fertilization

Net annualized change in SOC with NT compared with CT was unaffected by fertilizer application level. However, SOC under CT averaged $0.026 \ kg \cdot m^{-2} \cdot yr^{-1}$ greater ($P = 0.07$) under fertilized ($102 \pm 85 \ kg \ N \cdot ha^{-1} \cdot yr^{-1}$) than unfertilized cropping. Under NT, SOC averaged $0.027 \ kg \cdot m^{-2} \cdot yr^{-1}$ greater ($P = 0.02$) when crops were fertilized rather than unfertilized in 15 comparisons. The additional C stored with fertilization averaged ~$2.5 \ kg \cdot kg^{-1}$ fertilizer-N applied, which is greater than the C cost of manufacturing, distributing, and applying commercial N fertilizer, estimated at $1.23 \ kg \cdot kg^{-1}$ (Izaurralde et al., 1998).

Cropping Intensity

Net annualized change in SOC with NT compared with CT increased ($P<0.001$) with increasing cropping intensity (Figure 7.8). For example, under wheat-fallow (cropping intensity of 0.25), SOC was an average of $0.026 \ kg \cdot m^{-2} \cdot yr^{-1}$ less under NT than under CT. Under continuous sorghum, wheat, or corn (cropping intensity of 0.5), SOC was an average of $0.038 \ kg \cdot m^{-2} \cdot yr^{-1}$ greater under NT than under CT. Under double cropping (cropping intensity of 1.0), SOC was an average of $0.062 \ kg \cdot m^{-2} \cdot yr^{-1}$ greater under NT than under CT. More C input, and less water available to soil microorganisms by crops extracting more water with increasing cropping intensity, would likely

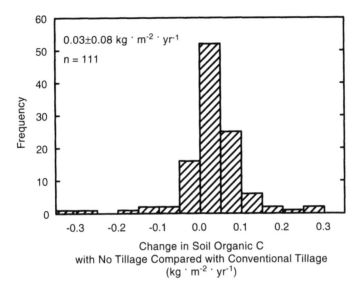

Figure 7.5   Frequency distribution of the net annualized change in soil organic C with no tillage compared with conventional tillage.

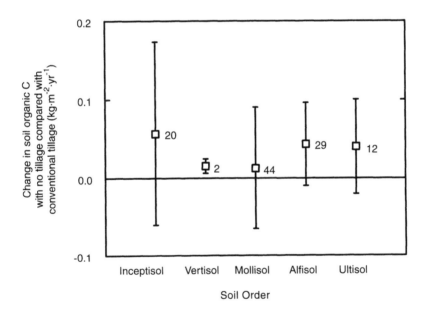

Figure 7.6   Mean and standard deviation of the net annualized change in soil organic C with no tillage compared with conventional tillage as affected by soil order. The number beside the mean is the number of observations.

leave more plant-derived C at the soil surface under conditions less ideal for decomposition than if incorporated. Increasing cropping intensity would utilize available water in winter–spring more effectively to provide more C input.

Climate Indices

Since locations varied in fertilizer rate, cropping intensity, and length of time under investigation of NT compared with CT, a mean difference in SOC storage between tillage regimes across these variables for each location was computed (n = 39). Data were also sorted by rank of each climate index and then a mean computed for each group of three locations to reduce some of the large variation in net annualized change in SOC among locations.

Index 1 (i.e., mean monthly precipitation-to-potential evapotranspiration ratio) was poorly related to net annualized change in SOC with NT compared with CT (Figure 7.9a). Index 1 values

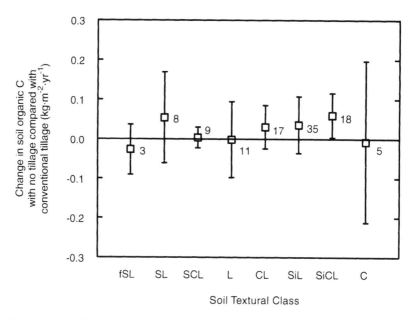

Figure 7.7   Mean and standard deviation of the net annualized change in soil organic C with no tillage compared with conventional tillage as affected by soil textural class. The number beside the mean is the number of observations. (Note: fSL is fine sandy loam, SL is sandy loam, SCL is sandy clay loam, L is loam, CL is clay loam, SiL is silt loam, SiCL is silty clay loam, and C is clay.)

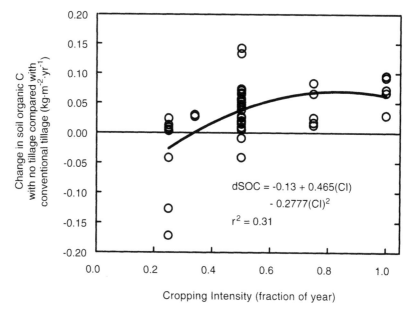

Figure 7.8   Net annualized change in soil organic C with no tillage compared with conventional tillage as affected by cropping intensity.

differed little, partly because precipitation exceeded potential evapotranspiration during winter months at most locations. Across all locations, index 1 values were assigned a value of 1 during $59 \pm 13\%$ of the months. Index 2 (i.e., mean annual precipitation-to-potential evapotranspiration ratio) indicated that maximum potential SOC storage with NT occurred at a ratio of 1.27 mm · mm^{-1} (Figure 7.9b). No benefit of NT on potential SOC storage would be expected at an index 2 level <0.75 mm · mm^{-1}, probably because low precipitation limits the potential of plants to fix C or limits decomposition under both tillage regimes, even when crop residues are mixed with soil using CT. At index 2 levels exceeding 1.75 mm · mm^{-1}, there was also little potential SOC storage with NT. Abundant precipitation would reduce potential SOC storage with NT because surface-placed residues would be moist more frequently or for a longer period of time, leading to rapid decomposition of residues under NT, similar to that under CT.

Indices 3 and 4 (i.e., combined temperature and precipitation coefficients on a monthly and annual basis, respectively) also indicated climatic controls on potential SOC storage with NT (Figure 7.10). Drier and colder locations had poor potential to store additional C with NT compared with CT, whereas mild locations (i.e., neither dry and cold nor wet and hot) had the greatest potential. Interestingly, indices 4 and 2 (i.e., on an annual basis) were better related to the change in SOC with NT compared with CT than were indices 3 and 1 (i.e., on a monthly basis). This indicates that these simple annualized climatic descriptions of locations could more effectively predict potential SOC storage with NT than seasonal descriptions. However, large variation in potential SOC storage occurred among the three locations used to obtain means, suggesting that much more work is needed to elucidate the intricacies of soil organic matter dynamics as affected by management and climate.

Indices 5 and 6 (i.e., most limiting temperature or precipitation coefficient on a monthly and annual basis, respectively) produced climatic responses similar to other indices (Figure 7.11). Most locations had monthly limitations due to temperature, although 28% had at least one month with a precipitation limitation. As an example, Akron, CO, was limited by precipitation rather than by temperature during 7 months. On an annual basis, only Corpus Christi, TX, had a precipitation limitation rather than a temperature limitation.

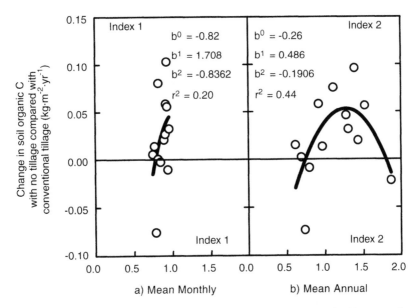

Figure 7.9   Net annualized change in soil organic C with no tillage compared with conventional tillage as affected by precipitation-to-potential evapotranspiration ratio on a mean monthly (a) and a mean annual (b) basis. Points represent the means of 3 consecutively ranked locations. Regression equations are of the form: $\Delta SOC = b_0 + b_1 \cdot (P/PET) + b_2 \cdot (P/PET)^2$.

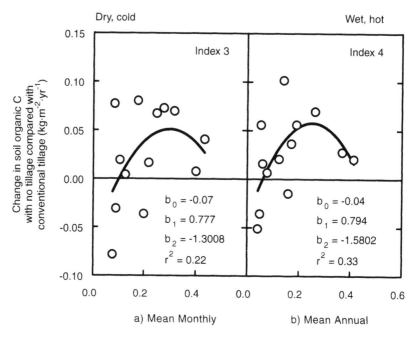

Figure 7.10   Net annualized change in soil organic C with no tillage compared with conventional tillage as affected by the product of temperature and precipitation coefficients on a mean monthly (a) and a mean annual (b) basis. Points represent the means of 3 consecutively ranked locations. Regression equations are of the form: $\Delta SOC = b_0 + b_1 \cdot (TxP) + b_2 \cdot (T \times P)^2$.

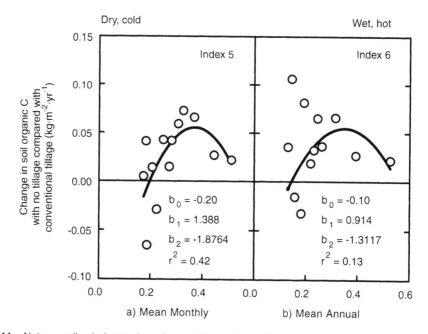

Figure 7.11   Net annualized change in soil organic C with no tillage compared with conventional tillage as affected by the most limiting climatic coefficient (i.e., temperature or precipitation) on a mean monthly (a) and a mean annual (b) basis. Points represent the means of 3 consecutively ranked locations. Regression equations are of the form: $\Delta SOC = b_0 + b_1 \cdot (MLCC) + b_2 \cdot (MLCC)^2$.

Mean annual precipitation-to-potential evapotranspiration ratio (index 2; Figure 7.9b) and mean monthly most limiting climatic coefficient (index 5; Figure 7.11a) were the best predictors of net annualized change in SOC with NT compared with CT. To achieve 90% of maximum net annualized change in SOC, regressions suggested that locations have (1) mean annual precipitation-to-potential evapotranspiration ratios of 1.11 to 1.44 mm · mm^{-1}, (2) mean monthly most limiting climatic coefficients of 0.19 to 0.31, or (3) mean annual temperature × precipitation coefficients of 0.32 to 0.42. Geographical locations in North America that meet these restrictions are in parts of Illinois, Indiana, Ohio, and Kentucky (Figure 7.12). A wider area encompassing one or more of these restrictions extends from the panhandle of Texas in the west to the coastal plain of Maryland in the east and from the piedmont region of Georgia in the south to the prairie region of Minnesota in the north (Figure 7.12). Much lower potential in SOC storage with NT compared with CT was observed in more extreme environments, including the dry Great Plains region and the cold, humid eastern provinces of Canada. However, more data are needed to validate and strengthen the confidence of these relationships.

In a multivariate analysis, none of these climatic variables interacted significantly with cropping intensity. Thus, in all regions the most intensive cropping systems (i.e., greatest C input) would produce the maximum potential SOC storage with NT compared with CT. On a practical level, this might mean shifting from (1) wheat–fallow to a wheat–sorghum–millet opportunity cropping in the central Great Plains, (2) continuous corn to a corn and wheat–soybean and vetch cover cropping system in the Midwest, or (3) continuous cotton to a cotton and clover–sorghum and wheat double cropping system in the southeastern U.S.

The analyses in this review of literature were restricted to the effect of NT compared with CT on SOC storage only and do not imply that NT is an inappropriate technology for semiarid and humid regions. NT offers many other important benefits, including reducing fossil fuel consumption and labor inputs, reducing soil erosion and water runoff, increasing soil aggregation and water infiltration, creating wildlife and soil biotic habitat, etc. These should be considered as incentives for producers to adopt this conservation management system.

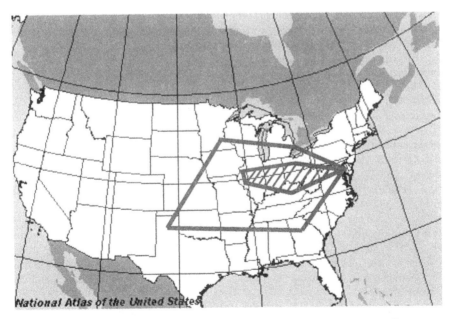

Figure 7.12   Geographical locations of maximum potential soil organic C storage with no tillage compared with conventional tillage meeting each criterion of indices 2, 4, and 5 (inner striated loop) and meeting any criterion of indices 2, 4, and 5 (outer loop).

CONCLUSIONS

It can be concluded from an analysis of available data in the literature that potential SOC storage with NT compared with CT was greatest (~0.050 kg $\cdot$ m^{-2} $\cdot$ yr^{-1}) in mesic, subhumid regions of North America with mean annual precipitation-to-potential evapotranspiration ratios of 1.1 to 1.4 mm $\cdot$ mm^{-1}. Much lower potential in SOC storage with NT compared with CT was observed in more extreme environments, including the dry Great Plains region and the cold, humid eastern provinces of Canada. However, more data are needed to validate and strengthen confidence in these relationships.

Soil order and soil textural class had little effect on potential SOC storage with NT. Interaction of tillage regime with other management variables on potential SOC storage occurred with cropping intensity, but not with level of fertilization. Potential SOC storage with NT compared with CT increased when cropping intensity increased, regardless of climatic conditions. Published data from North America were summarized so that policies to encourage or discourage land use for enhancing SOC storage could be developed on an objective basis.

ACKNOWLEDGMENT

We appreciate the assistance of Deborah Stark with PET calculations.

REFERENCES

Amato, M. and J.N. Ladd, Decomposition of 14C-labelled glucose and legume material in soils: properties influencing the accumulation of organic residue C and microbial biomass C, *Soil Biol. Biochem.*, 24:455–464, 1992.

Angers, D.A. et al., Microbial and biochemical changes induced by rotation and tillage in a soil under barley production, *Can. J. Soil Sci.*, 73:39–50, 1993.

Angers, D.A. et al., Impact of tillage practices on organic carbon and nitrogen storage in cool, humid soils of eastern Canada, *Soil Tillage Res.*, 41:191–201, 1997.

Beare, M.H., P.F. Hendrix, and D.C. Coleman, Water-stable aggregates and organic matter fractions in conventional- and no-tillage soils, *Soil Sci. Soc. Am. J.*, 58:777–786, 1994.

Black, A.L. and D.L. Tanaka, A conservation tillage-cropping systems study in the northern Great Plains of the United States, in Paul, E.A. et al. (eds.), *Soil Organic Matter in Temperate Agroecosystems*, CRC Press, Boca Raton, FL, 1997, 335–342.

Blevins, R.L., G.W. Thomas, and P.L. Cornelius, Influence of no-tillage and nitrogen fertilization on certain soil properties after 5 years of continuous corn, *Agron. J.*, 69:383–386, 1977.

Cambardella, C.A. and E.T. Elliott, Particulate soil organic-matter changes across a grassland cultivation sequence, *Soil Sci. Soc. Am. J.*, 56:777–783, 1992.

Campbell, C.A. et al., Carbon sequestration in a Brown Chernozem as affected by tillage and rotation, *Can. J. Soil Sci.*, 75:449–458, 1995.

Campbell, C.A. et al., Tillage and crop rotation effects on soil organic C and N in a coarse-textured Typic Haploboroll in southwestern Saskatchewan, *Soil Tillage Res.*, 37:3–14, 1996.

Carter, M.R. and D.A. Rennie, Changes in soil quality under zero tillage farming systems: distribution of microbial biomass and mineralizable C and N potentials, *Can. J. Soil Sci.*, 62:587–597, 1982.

CTIC, 1998 crop residue management executive summary, Conservation Technology Information Center, West Lafayette, IN, 1998.

Dick, W.A. et al., Impacts of agricultural management practices on C sequestration in forest-derived soils of the eastern Corn Belt, *Soil Tillage Res.*, 47:235–244, 1998.

Doran, J.W., Microbial biomass and mineralizable nitrogen distributions in no-tillage and plowed soils, *Biol. Fertil. Soils*, 5:68–75, 1987.

Dormaar, J.F. and C.W. Lindwall, Chemical differences in Dark Brown Chernozemic Ap horizons under various conservation tillage systems, *Can. J. Soil Sci.*, 69:481–488, 1989.

Eghball, B. et al., Distribution of organic carbon and inorganic nitrogen in a soil under various tillage and crop sequences, *J. Soil Water Conserv.*, 49:201–205, 1994.

Ellert, B.H. and J.R. Bettany, Calculation of organic matter and nutrients stored in soils under contrasting management regimes, *Can. J. Soil Sci.*, 75:529–538, 1995.

Franzluebbers, A.J. and M.A. Arshad, Water-stable aggregation and organic matter in four soils under conventional and zero tillage, *Can. J. Soil Sci.*, 76:387–393, 1996.

Franzluebbers, A.J., G.W. Langdale, and H.H. Schomberg, Soil carbon, nitrogen, and aggregation in response to type and frequency of tillage, *Soil Sci. Soc. Am. J.*, 63:349–355, 1999.

Franzluebbers, A.J., F.M. Hons, and D.A. Zuberer, Long-term changes in soil carbon and nitrogen pools in wheat management systems, *Soil Sci. Soc. Am. J.*, 58:1639–1645, 1994.

Franzluebbers, A.J., F.M. Hons, and D.A. Zuberer. 1995. Soil organic carbon, microbial biomass, and mineralizable carbon and nitrogen in sorghum, *Soil Sci. Soc. Am. J.*, 59:460–466, 1995.

Franzluebbers, A.J., F.M. Hons, and D.A. Zuberer, *In situ* and potential CO_2 evolution from a Fluventic Ustochrept in south-central Texas as affected by tillage and cropping intensity, *Soil Tillage Res.*, 47:303–308, 1998.

Global Historical Climatology Network, ftp://ftp.ncdc.noaa.gov/pub/data/ghcn, 1999.

Halvorson, A.D. et al., Long-term tillage and crop residue management study at Akron, Colorado, in Paul, E.A. et al. (eds.), *Soil Organic Matter in Temperate Agroecosystems*, CRC Press, Boca Raton, FL, 1997, 361–370.

Hansmeyer, T.L. et al., Determining carbon dynamics under no-till, ridge-till, chisel, and moldboard tillage systems within a corn and soybean cropping sequence, in Lal, R. et al. (eds.), *Management of Carbon Sequestration in Soil*, CRC Press, Boca Raton, FL, 1997, 93–97.

Hassink, J. and A.P. Whitmore, A model of the physical protection of organic matter in soils, *Soil Sci. Soc. Am. J.*, 61:131–139, 1997.

Hendrix, P.F., A.J. Franzluebbers, and D.V. McCracken, Management effects on C accumulation and loss in soils of the southern Appalachian piedmont of Georgia, *Soil Tillage Res.*, 47:245–251, 1998.

Hunt, P.G. et al., Changes in carbon content of a Norfolk loamy sand after 14 years of conservation or conventional tillage, *J. Soil Water Conserv.*, 51:255–258, 1996.

Ismail, I., R.L. Blevins, and W.W. Frye, Long-term no-tillage effects on soil properties and continuous corn yields, *Soil Sci. Soc. Am. J.*, 58:193–198, 1994.

Izaurralde, R.C. et al., Scientific challenges in developing a plan to predict and verify carbon storage in Canadian prairie soils, in Lal, R. et al. (eds.), *Management of Carbon Sequestration in Soil*, CRC Press, Boca Raton, FL, 1998, 433–446.

Jenkinson, D.S., Soil organic matter and its dynamics, in Wild A. (ed.), *Russell's Soil Conditions and Plant Growth*, 11th ed., Longman, New York, 1988, 564–607.

Karlen, D.L. et al., Long-term tillage effects on soil quality, *Soil Tillage Res.*, 32:313–327, 1994.

Kucera, C.L. and D.R. Kirkham, Soil respiration studies in tallgrass prairie in Missouri, *Ecology*, 52:912–915, 1971.

Lal, R., T.J. Logan, and N.R. Fausey, Long-term tillage effects on a Mollic Ochraqualf in northwest Ohio. III. Soil nutrient profile, *Soil Tillage Res.*, 15:371–382, 1990.

Lal, R., A.A. Mahboubi, and N.R. Fausey, Long-term tillage and rotation effects on properties of a central Ohio soil, *Soil Sci. Soc. Am. J.*, 58:517–522, 1994.

Lamb, J.A., G.A. Peterson, and C.R. Fenster, Wheat-fallow tillage systems' effect on a newly cultivated grassland soils' nitrogen budget, *Soil Sci. Soc. Am. J.*, 49:352–356, 1985.

Larney, F.J. et al., Changes in total, mineralizable and light fraction soil organic matter with cropping and tillage intensities in semiarid southern Alberta, Canada, *Soil Tillage Res.*, 42:229–240, 1997.

McCarty, G.W., N.N. Lyssenko, and J.L. Starr, Short-term changes in soil carbon and nitrogen pools during tillage management transition, *Soil Sci. Soc. Am. J.*, 62:1564–1571, 1998.

McCarty, G.W. and J.J. Meisinger, Effects of N fertilizer treatments on biologically active N pools in soils under plow and no tillage, *Biol. Fertil. Soils*, 24:406–412, 1997.

Mielke, L.N., J.W. Doran, and K.A. Richards, Physical environment near the surface of plowed and no-tilled surface soils, *Soil Tillage Res.*, 5:355–366, 1986.

Nyborg, M. et al., Fertilizer N, crop residue, and tillage alter soil C and N content in a decade, in Lal, R. et al. (eds.), *Soil Management and Greenhouse Effect*, Lewis Publ., CRC Press, Boca Raton, FL, 1995, 93–99.

Paustian, K., E.T. Elliott, and M.R. Carter. Tillage and crop management impacts on soil C storage: use of long-term experimental data, *Soil Tillage Res.*, 47:vii–xii, 1998.

Peterson, G.A. et al., Reduced tillage and increasing cropping intensity in the Great Plains conserves soil C, *Soil Tillage Res.*, 47:207–218, 1998.

Pierce, F.J., M.-C. Fortin, and M.J. Staton. Periodic plowing effects on soil properties in a no-till farming system, *Soil Sci. Soc. Am. J.*, 58:1782–1787, 1994.

Pikul, J.L. Jr. and J.K. Aase, Infiltration and soil properties as affected by annual cropping in the Northern Great Plains, *Agron. J.*, 87:656-662, 1995.

Potter, K.N. et al., Crop rotation and tillage effects on organic carbon sequestration in the semiarid southern Great Plains, *Soil Sci.*, 162:140–147, 1997.

Potter, K.N. et al., Distribution and amount of soil organic C in long-term management systems in Texas, *Soil Tillage Res.*, 47:309–321, 1998.

Salinas–Garcia, J.R., J.E. Matocha, and F.M. Hons, Long-term tillage and nitrogen fertilization effects on soil properties of an Alfisol under dryland corn/cotton production, *Soil Tillage Res.*, 42:79–93, 1997.

SAS Institute Inc., SAS user's guide: statistics, Version 6 ed., SAS Institute, Cary, NC, 1990.

Smith, P. et al., A comparison of the performance of nine soil organic matter models using seven long-term experimental data sets, *Geoderma*, 81:153–225, 1998.

Thornthwaite, C.W., J.R. Mather, and D.B. Carter. Instructions and tables for computing potential evapotranspiration and the water balance, *Publ. Climatol.*, 10(3) Drexel Inst. Technol. Lab. Climatol., Centerton, NJ, 1957.

Wander, M.M., M.G. Bidart, and S. Aref, Tillage impacts on depth distribution of total and particulate organic matter in three Illinois soils, *Soil Sci. Soc. Am. J.*, 62:1704–1711, 1998.

Long-Term Effect of Moldboard Plowing on Tillage-Induced CO_2 Loss

D. C. Reicosky

CONTENTS

ABSTRACT

The possibility of global greenhouse warming due to rapid increase in carbon dioxide (CO_2) is receiving attention. Agriculture's role in sequestering carbon (C) is not clearly understood. Intensive tillage reportedly has caused between a 30 to 50% decrease in soil C since many soils were brought into cultivation; recent studies have shown a large short-term pulse of CO_2 released immediately following tillage, which partially explains C loss from soils. The objective of this work was to evaluate the long-term effects (3 months) of moldboard plowing on CO_2 loss from a Barnes loam (Udic Haploborolls, fine-loamy, mixed) in west central Minnesota (45°41′14″ N latitude and 95°47′57″ W longitude).

Treatments were weed-free replicated plots that were moldboard plowed and not tilled. Tillage-induced CO_2 loss was periodically measured using a large portable chamber for 87 days during summer of 1998. The soil CO_2 concentration was measured at 5-, 10-, 20-, 30-, 50-, and 70-cm depths in the not-tilled plots and 30-, 50-, and 70-cm depths in plowed plots. Soil gas samples were drawn once or twice weekly throughout the growing season from stainless-steel mesh sampling tubes; they were drawn with gas-tight syringes and run through an infrared gas analyzer using a computer-controlled data system and numerical integration techniques to determine the concentration. The initial flush of CO_2 immediately after moldboard plowing was nearly 100 g CO_2 m^{-2} h^{-1}, while that from the not-tilled treatment was less than 0.9 g CO_2 m^{-2} h^{-1}. The flux from the plowed plots declined rapidly during the first 4 hours to 7 g CO_2 m^{-2} h^{-1} and yielded a cumulative CO_2 flux from the moldboard plow treatment 14 times that from not-tilled plots. For the 85-day period following tillage, the cumulative CO_2 flux from the plowed treatment was 2.4 times higher that from not tilled.

Both treatments showed a seasonal trend as soil temperature cooled and fluctuations occurred associated with rainfall events.

Soil gas samples showed a similar trend with an increase in the CO_2 concentration with depth on the not-tilled plots until day 200, then gradually declined as soil C resources were used and temperatures decreased. Carbon dioxide concentrations were highest at the 30-, 50-, 70-cm depths. The decline in soil CO_2 concentration on the moldboard plow treatment was more dramatic than on the not-tilled treatment suggesting that higher air permeability in the tilled layer resulted in the higher gas exchange.

Results support previous short-term fluxes and confirm the role of intensive tillage in long-term soil C decline. The large differences in CO_2 loss between moldboard plow and not-tilled treatments reflect the need for improved soil management and conservation policies that favor conservation tillage to enable C sequestration in agricultural production systems.

INTRODUCTION

The possibility of global greenhouse warming due to rapid increase in CO_2 is receiving increased attention (Wood, 1990; Post et al., 1990); concern is warranted because potential climate change could result in increased temperature and drought over present agricultural production areas. Agriculture and intensive tillage have caused a 30 to 50% decrease in soil organic carbon (SOC) since many soils were brought into cultivation more than 100 years ago (Schlesinger, 1985). Examples of how intensive tillage has impacted agricultural production systems and SOC are illustrated by the long-term soil C trends in the Morrow plots in Champaign, IL, (Odel et al., 1984; Peck, 1989) and in Sanborn Field at the experiment station of the University of Missouri, Columbia (Wagner, 1989). Both locations show similar decreases in SOC over the last 100 years; thus, it is necessary to understand better the tillage processes in agricultural production and the mechanisms leading to C loss. Carbon loss can be linked to soil production, soil quality, C sequestration, and, ultimately, crop production (Paustian et al., 1997).

Agriculture's role in sequestering C is not clearly understood; there is a definite need for direct measurements to quantify CO_2 fluxes influenced by agricultural management practices (Houghton et al., 1983). Understanding these processes will lead to enhanced soil management techniques and new technology for increased food production efficiency with a minimum impact on environmental quality and greenhouse gas emissions. Soil disturbance by tillage may alter environmental conditions and soil structure to enhance production of CO_2.

Work using small static chambers showed that tillage influenced soil respiration in a small way compared to other variables such as crop species, growth stage, soil temperature, and soil water (Hendrix et al., 1988; Franzlubbers et al., 1995a,b). Results from simulated tillage led Roberts and Chan (1990) to conclude that a tillage-induced increase in microbial activity was not a major cause of soil C loss. More recently, Reicosky et al. (1997) reported differences as large as tenfold in tillage-induced flushes of CO_2 measured with a large chamber vs. a small soil respiration chamber. Flux differences from the different chambers were not resolved; however, the larger mixing rate in the large canopy chamber may have been partly responsible for the higher tillage-induced fluxes. Both chambers gave essentially the same low flux for the undisturbed soil (not tilled) in three cropping systems. Concern for pressure fluctuations associated with turbulent mixing required for sampling uniformity, and the tillage-induced change in soil air permeability bring uncertainty to the meaning of soil gas fluxes measured with dynamic chambers (Conen and Smith, 1998; Lund et al., 1999).

As the need for soil gas emission data increases, understanding chamber effects and interpreting data from static and dynamic chambers requires careful analysis. While chambers are convenient and relatively inexpensive, their use requires an appreciation of methodological limitations that may permit only qualitative assessment and limited quantitative data. Recent studies involving a dynamic chamber and tillage methods and the associated incorporation of the residue in the field indicate major gaseous loss of C immediately following tillage (Reicosky and Lindstrom, 1993). Immediately following tillage, a pulse of CO_2 was measured using a large portable dynamic chamber technique. Short-term impact of various tillage methods on loss of CO_2 from the soil surface was measured using a portable chamber designed to measure canopy photosynthesis.

Reicosky and Lindstrom (1993) found that the moldboard plow had the roughest soil surface and the highest initial CO_2 flux and maintained the highest flux throughout the 19-day study. High initial CO_2 fluxes correlated to deep soil disturbance and resulted in a rougher surface and larger voids. Lower CO_2 resulted from less soil disturbance and small voids with not tilled showing the least CO_2 loss during the 19-day study. Ellert and Janzen (1999) used a single pass with a heavy-duty cultivator (0.075 m deep) and a smaller dynamic chamber to show CO_2 fluxes 0.6 h after tillage were 2- to 4-fold above pretillage fluxes and rapidly declined to similar values within 24 h of cultivation. They concluded that short-term influence of tillage on soil C loss was small under semiarid conditions. On the other hand, Reicosky and Lindstrom (1993) concluded that tillage methods, especially moldboard plowing to 0.25 m deep in humid climates, affected the initial CO_2 flux differently and suggested that improved soil management techniques can minimize agriculture's impact on global CO_2.

The impact of broad area tillage on soil C and CO_2 loss suggests possible improvements with mulch between the rows and less intensive strip tillage to prepare a narrow seedbed. Reicosky (1998) quantified the short-term, tillage-induced CO_2 loss caused by several strip tillage tools and a moldboard plow. Not tilled had the lowest CO_2 flux during the study and moldboard plow had the highest immediately after tillage, which declined as the soil dried. Forms of strip tillage had initial flushes related to tillage intensity and were intermediate between moldboard plow and not-tilled extremes with both 5-h and 24-h cumulative losses related to disturbed soil volume. These results suggested that a direct interaction of the disturbed volume and amount of shattering of the soil were related to gas exchange, i.e., CO_2 loss and oxygen entry into soil enhances short-term microbial activity while energy resources are available. Reducing the volume of the soil disturbed by tillage should enhance soil and air quality by increasing soil C, therefore enhancing the overall environmental quality.

Historically, intensive tillage of agricultural soils has led to substantial losses of soil C as indicated by long-term and short-term analyses. Even though atmospheric CO_2 enrichment has occurred due to C emissions from soil, a potential to reverse the trend and sequester C in the world's soils exists. Adoption of best management practices that include modifying tillage techniques can lead to increasing soil reserve at the expense of decreasing the atmospheric C pool. Long-term studies show continued decline in SOC with intensive agriculture, but few suggest when the major C loss takes place. The objective of this work was to characterize the long-term impact of intensive tillage (moldboard plowing) on cumulative CO_2 loss. An experiment was designed to characterize the short-term, tillage-induced CO_2 fluxes and determine the impact duration by comparing not tilled (no soil disturbance in the recent past) and moldboard plow for a 3-month period without plant growth. It was hoped that these results would contribute to understanding using appropriate land usage to sequester more C in agricultural production systems.

METHODS AND MATERIALS

This work was conducted on a Barnes loam (Udic Haploboroll, fine-loamy, mixed) at the Barnes–Aastad Swan Lake Research Farm in west central Minnesota (45°41′14″ N latitude and

95°47'57″ W longitude). The surface horizon is generally dark colored with typically 28 to 32 g kg^{-1} C and developed over subsoil high in free calcium carbonate. The cropping history for the last 80 years has been corn (*Zea mays* L.), soybean (*Glycine max* L.), and spring wheat (*Triticum aestivum* L.) with conventional tillage.

The long-term study began July 14 (day 195) and ended October 9, 1998 (day 285) and was conducted on an area planted to soybeans on May 6 (day 126). Two weeks before tillage, soybeans were killed with glyphosate and then mowed to minimize the effect of residue decomposition on surface CO_2 fluxes. On July 14 (day 195), four replicate plots were plowed (0.25 m deep) with a commercially available, four-bottom moldboard plow (each bottom was 46-cm wide) to establish alternating plots of moldboard plow and not tilled. Plots had no prior soil disturbance other than the initial spring tillage and soybean planting. The plot size was 20 m long by 3.7 m wide as a result of two passes with the four-bottom plow. The moldboard plowing resulted in a complete inversion of the surface and nearly 100% incorporation of the residue. Both plowed and adjacent not-tilled plots had negligible surface residue from the previously killed soybean crop. Throughout the remainder of the experiment, the moldboard plow and not-tilled plots were kept weed-free by routine bi-weekly applications of herbicide.

The short-term, tillage-induced CO_2 release was measured with a large portable chamber (area = 2.71 m^2) designed to measure canopy photosynthesis and mounted on a four-wheel drive, rough-terrain forklift for portability. Details of the chamber and operations used to measure CO_2 flux from the soil surface are presented by Reicosky and Lindstrom (1993) with further modifications and improvements described by Wagner et al. (1997).

Immediately (within 1 min) after tillage of the first plot, four measurements were taken to obtain initial CO_2 fluxes followed by measurements at the corresponding not-tilled plot. Measurements were then repeated on the next plowed plot and the adjacent not-tilled plot for all four replicates. The cycle was repeated six times over 7 hours the first day. Subsequent measurements were delayed 48 hours due to intense rainfall (49 mm), which caused some reconsolidation of the plowed soil. Under these conditions, the moldboard plow CO_2 flux was still substantially larger than that from not tilled. Measurements were then made periodically once or twice a week throughout the remainder of the season until termination on October 9 (day 282) prior to a killing frost.

Soil CO_2 concentrations were measured periodically throughout the growing season to quantify the relation of soil CO_2 to CO_2 loss through the surface. Sampling tubes, plumbed with nylon tubing, were installed in the not-tilled and plowed plots on days 161 and 195, respectively. Glass gas-tight syringes were used to draw gas from the stainless steel screen sampling tubes installed at 5-, 10-, 20-, 30-, 50-, and 70-cm depths from the soil surface in the not-tilled plots. Because of the rough, porous nature of the plowed layer, gas-sampling tubes were only installed at 30-, 50-, and 70-cm depths in the plowed plots shortly after tillage. Samples were taken and analyzed with an infrared gas analyzer flow-through system. Generally 1-cc samples were drawn to provide adequate sensitivity for concentrations at all depths. Soil CO_2 concentrations in the not-tilled plots were monitored 3 weeks previous to plowing.

The climate during the experiment was characterized at a weather station (200 m away) that recorded daily measurements of solar radiation, soil and air temperature, and rainfall. The daily minimum and maximum air temperature and rainfall during the study are summarized in Figure 8.1. Air temperature followed a typical seasonal trend. While the air temperatures at the weather station might not have truly reflected soil temperatures at the site, they did represent relative temperature trends. Gas exchange measurement sensitivity to solar radiation and air temperature at the time of measurement and tillage was reflected in the magnitude of the fluxes. This suggests that CO_2 loss as an indicator of biological activity is highly dependent on water content and temperature. Tillage was completed on July 14 (day 195), one of the warmest days of the year with maximum air temperature of 32.3°C and minimum of 17.5°C. Some cooler weather in which daily maximum and minimum air temperature was 23.1°C and 8.6°C, respectively, preceded the end of the study.

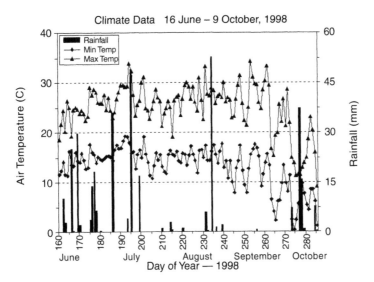

Figure 8.1 Daily minimum and maximum air temperatures and rainfall amounts from June through October, 1998.

Rainfall intensities were expressed only on an hourly basis. Presumably the short-term intensity could have been larger, but these data reflect the relative impact of rainfall events on the magnitude of the CO_2 fluxes and the amount of soil reconsolidation and surface crusting that took place. A heavy rain of 49 mm (42 mm h^{-1} max. intensity) occurred the night following tillage, effectively sealing and consolidating the soil surface of the plowed plots. Another substantial rain of 17 mm (7 mm h^{-1} max. intensity) occurred on day 199, followed by a heavy rain of 53 mm (39 mm h^{-1} max. intensity) on day 234. The major rainfall events partially leveled the plowed surfaces and consolidated the soil. After a few days of drying, both plowed and not-tilled plots showed essentially the same extent of surface soil cracking and drying.

RESULTS AND DISCUSSION

The daily max and min air temperatures and rainfall distribution in Figure 8.1 show typical seasonal trends with a gradual decline in air temperatures towards the end of period. From days 160 to 180 significant rainfall was fairly well distributed prior to tillage. The last significant rainfall (36 mm) prior to tillage occurred on day 186. After tillage, there were five major rainfalls with more than 15 mm. Three of these had 40 mm or more rainfall in 24 h; the rainfall intensity during these times was as high as 39 mm h^{-1} on day 235. This energy reconsolidated the moldboard plowed surface to the extent that both plowed and not-tilled surfaces appeared similar at season's end. The only visible difference was that the moldboard plow had a few larger clumps of soil that were higher compared to the relatively smooth not-till surface. The extent of cracking as a result of drying later in the study was nearly the same in both treatments.

The instantaneous CO_2 flux after the initial tillage, averaged for four measurements on four reps, is shown in Figure 8.2. Error bars represent +/- one standard error of the mean. Prior to tillage, the CO_2 fluxes were low for both treatments. However, in the moldboard plow treatment, the initial flux after tillage was nearly 100 g CO_2 m^{-2} h^{-1}, rapidly declined to about 25 g CO_2 m^{-2} h^{-1} within the first 15 minutes, then continued to decline to about 15 g CO_2 m^{-2} h^{-1} 1 hour after tillage. The gradual decline continued and the CO_2 flux was 7 g CO_2 m^{-2} h^{-1} 4 hours after tillage. During this 4-hour period, the cumulative CO_2 flux for the not-tilled treatment was 3.7 g CO_2 m^{-2}

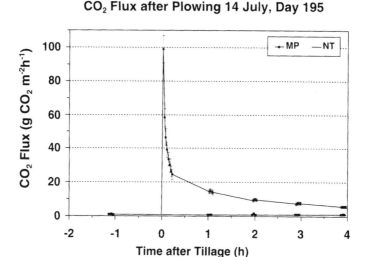

CO₂ Flux after Plowing 14 July, Day 195

Figure 8.2   Short-term CO₂ flux as a function of time after tillage. Each data point is the average of four replicates at the same time in each measurement sequence.

while the corresponding cumulative flux for the moldboard plow treatment was 51.0 g CO_2 m^{-2}, a 14-fold difference.

This short-term high flux of CO_2 immediately following a tillage event agreed with the earlier results of Reicosky and Lindstrom (1993 and 1995). The same pretillage flux on both treatments demonstrates the impact of moldboard plowing on tillage-induced CO_2 loss that showed a rapid decline as a function of time. The change in tillage-induced soil properties allows more gas exchange initially, which is moderated as the soil reconsolidates and C reserves are used up by microbial activity.

The flux measurements on the not tilled and moldboard plow plots were continued throughout the season at weekly to twice-weekly intervals after tillage. The CO_2 fluxes from days 197 to 282 for both moldboard plow and not-tilled treatments are summarized in Figure 8.3. Note the change in the scale of the y-axis with a maximum span of 0 to 3 g CO_2 m^{-2} h^{-1} for the long-term trends as opposed to 0 to 100 g CO_2 m^{-2} h^{-1} for the short-term burst in Figure 8.2. The moldboard plow treatment showed a consistently higher, more erratic trend that appeared to be related to rainfall events and temperature fluctuations. Both treatments tended to show a gradual decline with time as soil and air temperatures decreased.

The similar temporal trends suggest that the CO_2 fluxes measured on moldboard plow and not-tilled treatments were biologically controlled because of the strong dependence of the CO_2 flux on water content and temperature. The seasonal decline could also be related to decreased C resources as a result of microbial activity with no C replenishment. The gradual decline continued through the end of the season; the soil CO_2 concentration differences between the two treatments grew smaller.

The cumulative CO_2 flux for the not-tilled plots from days 197 to 282, not including the first 4 hours after tillage, was 1035 g CO_2 m^{-2}, while the corresponding value for the moldboard plow was 2468 g CO_2 m^{-2}. The difference indicates 2.38 times as much CO_2 loss from the moldboard plow treatment as from the not-tilled treatment. This is only slightly different from the 2.48 times as much CO_2 from the moldboard plow when the first 4 hours after tillage are included. For the long term (87 days), the moldboard plow caused more than 2.4 times as much CO_2 loss as that from not-tilled treatment. To minimize complications in the CO_2 exchange rates, both treatments were maintained weed-free by spraying every 2 weeks so that there was no additional C input during the study.

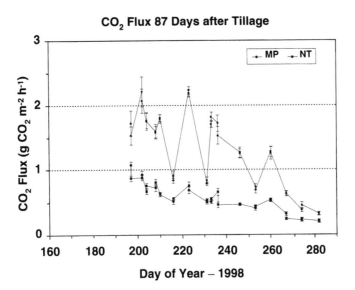

Figure 8.3   The average CO_2 flux from moldboard plow and not-tilled treatments as a function of time during the 1998 season. Note the scale change on the y-axis relative to Figure 8.2.

The mean soil CO_2 concentrations at the six depths sampled in the not-tilled plots are summarized in Figure 8.4a and for three depths in moldboard plow plots in Figure 8.4b. The error bars for each data point represent +/- one standard error of the mean. The data in Figure 8.4a show that, at 5- and 10-cm depths, the CO_2 concentrations never exceeded 15 mL L^{-1} during any time in the summer; but show a gradual decline with time as air temperatures cooled. The CO_2 concentrations at 30-, 50-, and 70-cm of the not-tilled plots showed a peak around day 200, then a gradual decline related to seasonal trends in soil temperature and water content. Some of the variation in CO_2 concentration was apparently related to the rainfall events, but it was very difficult to make a definitive statement because the times between rainfall and actual soil CO_2 measurement were not consistent.

The CO_2 concentration at 50 cm in the moldboard plow plots was nearly 35 mL L^{-1} on Day 200 shortly after tillage, and then showed a rapid decline apparently related to the tillage event. This apparent short-term negative gradient at the 50-cm depth likely reflects the interaction of the plow event 5 days earlier and the combined effect of 49 mm rainfall following tillage and 17 mm one day before the gas measurement on day 200. Intensity and amounts of rainfall acted to "seal" the soil surface and slow loss of CO_2 from the deeper depths. As the plowed and soil surface dried, the soil air permeability increased and CO_2 concentrations at the lower depths decreased below those of not-tilled plots. The decline in soil CO_2 concentration was more dramatic on the moldboard plow plots, decreasing to less than 20 mL L^{-1} at the 70-cm depth between days 220 and 240 and suggesting that the air permeability of the plowed surface layer was higher than the same thickness of the not-tilled treatment. The CO_2 concentrations in the plowed plots were substantially lower than in the not-tilled plots. The gradual decline in both treatments was most likely related to soil temperature and water content decline, as well as to possibly depleted C resources due to microbial activity similar to results of Buyanovsky and Wagner (1983).

Soil CO_2 concentration profiles with depth at selected times in the season are summarized in Figure 8.5. The first (day 197) was shortly after tillage (2 days); then the second and third occurred on days 216 and 274, midway and at the end of the experiment, respectively. The highest CO_2 concentrations during the season were related to the highest soil temperatures and presumably highest microbial activity. The differences between the moldboard plow and the not-tilled treatments from the 30- to 70-cm depth were very small early in the season, but grew larger; on day 216, the not-tilled treatment concentrations were nearly twice as high as moldboard plow. On day 274, both

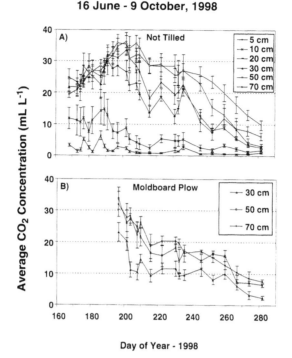

16 June - 9 October, 1998

Day of Year - 1998

Figure 8.4 (a) Soil CO_2 concentration at six depths as a function of time during the growing season in the not-tilled plots; (b) soil CO_2 concentration as a function of time in the moldboard plow plots at 30-, 50-, and 70-cm depths.

magnitudes were generally lower than 10 mL L^{-1} with small differences between the moldboard plow and not-tilled. These results suggest that a change in soil air permeability of the plowed layer above the 30- to 70-cm depth resulted in higher CO_2 fluxes and lower soil CO_2 concentrations compared to the not-tilled treatment. This trend continued throughout the season and the differences between the two treatments grew smaller as the source of CO_2 decreased related to declining temperature.

In general, the soil gas concentrations in plowed plots after tillage were lower than the corresponding depths of not-tilled plots. At the end of the experiment, soil gas concentrations were essentially the same low value as microbial activity declined with temperature. While there was considerable variation in the CO_2 flux measurements in Figure 8.3, there appeared to be a general correspondence in the soil CO_2 flux decline in moldboard plow and not-tilled treatments, with the gradual decline in soil CO_2 concentrations similar to results of Buyanovsky and Wagner (1983) under wheat.

The results suggest that CO_2 loss was a result of a combination of a change in soil properties and current biological activity. If one assumes that the microbial respiration generating CO_2 was similar in both treatments, then the CO_2 concentrations in the not-tilled treatment would build up to a higher concentration than that in the plowed plots. Gas flow by diffusion and convection through the not-tilled surface was slower than that in the moldboard plow treatment; the loose open soil in the plowed plots allowed better gas exchange and resulted in lower CO_2 concentrations. While these observations may seem contradictory, CO_2 concentration with depth and the surface flux were both affected by plow tillage. The lower CO_2 concentration possibly also suggests higher oxygen concentration (not measured) that might result in higher aerobic decomposition of the soil organic matter.

The consistent differences in CO_2 fluxes from the plowed and not-tilled surfaces suggest that tillage-induced soil changes can cause different CO_2 fluxes. These results differ with those of

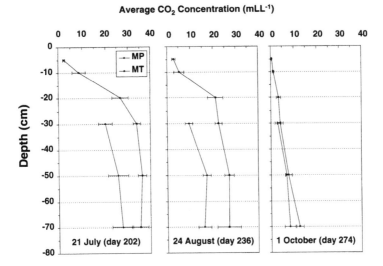

Figure 8.5   Soil CO_2 concentration profiles on the not-tilled and moldboard plow plots early-, mid-, and late season during 1998.

Hendrix et al. (1988) and Franzlubbers et al. (1995a,b) using small soil chambers. Results from simulated tillage led Roberts and Chan (1990) to conclude that a tillage-induced increase in microbial activity was not a major cause of soil C loss. While the absolute magnitudes of the fluxes in this work may be open to question due to the turbulent mixing inside the chamber that creates small negative pressures (Conen and Smith, 1998; Lund et. al., 1999), the relative difference between the plowed and not-tilled treatments should remain the same.

The role of natural wind in the magnitude of gas fluxes is not clear, but presumably the natural turbulence outside the chamber can affect the natural flux similarly. The changes in soil air permeability with tillage would impact the convective component of the total gas flux more than the diffusive component and partially explain the observed treatment differences. The dramatic change in soil properties following the plow are likely related to the large short-term flux differences and to the smaller long-term flux differences for 87 days after tillage. These data differ from Ellert and Janzen (1999), who showed that CO_2 fluxes 0.6 h after shallow cultivation were two- to four-fold above pretillage fluxes and rapidly declined to similar values within 24 h of cultivation.

This longer-term study shows that the moldboard plow treatment with no plants growing on the soil maintains a significantly higher CO_2 flux than plots not tilled throughout the growing season. While this may be an artificial situation, differences between the moldboard plowing and not-tilled plots were maintained for 87 days, even though the appearance of surface sealing from rainfall may have had some affect. The magnitude of the CO_2 fluxes did not reflect any differences in surface sealing or crusting that appeared similar on both treatments in the latter part of the study. The data suggest that soil air permeability and gas exchange from the moldboard plow system were higher supporting the lower CO_2 concentrations with soil depth. Possibly higher oxygen concentrations may be responsible for higher aerobic microbial activity that results in higher respiration and loss of CO_2.

CONCLUSIONS

In summary, results support earlier work on short-term, tillage-induced CO_2 loss that the impact of moldboard plow tillage on CO_2 loss continues to affect soil and air quality for nearly 3 months. To minimize the environmental impact, one must minimize the volume of soil disturbed, till only the soil volume necessary to get an effective seedbed, and leave the remainder of the soil undisturbed to conserve water and C, as well as to minimize erosion and CO_2 loss.

Conservation tillage can be beneficial by improving soil and air quality, minimizing runoff, enhancing water quality, and minimizing contributions to the greenhouse effect. Decreased CO_2 released from the soil to the atmosphere will require less intensive tillage than the moldboard plow. The associated fossil fuel energy savings represent an additional economic benefit connected with less disturbance of the soil. The combination of short- and long-term environmental benefits of conservation tillage and the economic benefits relative to broad area tillage need to be considered when making soil management decisions. These technical considerations need to be identified and placed into policy decisions used for developing and evaluating methods for C sequestration in agricultural systems.

ACKNOWLEDGMENTS

The author would like to acknowledge the technical support of Chris Wente, Steve Wagner, Ryan Bright, and Alan Wilts in data collection and analysis and Chuck Hennen and Laurel Maanum for their help in field operations.

REFERENCES

Buyanovsky, G.A. and G.H. Wagner, Annual cycles of carbon dioxide level in soil air, *Soil Sci. Soc. Am. Proc.*, 27:1139–1145, 1983.

Conen, F. and K.A. Smith, A re-examination of closed flux chamber methods for the measurement of trace gas emissions from soils to the atmosphere, *Eur. J. Soil Sci.*, 49:701–707, 1998.

Ellert, B.H. and H.H. Janzen, Short-term influence of tillage on CO_2 fluxes from a semi-arid soil on the Canadian Prairies, *Soil Till. Res.*, 50:21–32, 1999.

Franzluebbers, A.J., F.M. Hons, and D.A. Zuberer, Tillage and crop effects on seasonal dynamics of soil CO_2 evolution, water content, temperature and bulk density, *Appl. Soil Ecol.*, 2:95–109, 1995a.

Franzluebbers, A.J., F.M. Hons, and D.A. Zuberer, Tillage-induced seasonal changes in soil physical properties affecting soil CO_2 evolution under intensive cropping, *Soil Till. Res.*, 34:41–60, 1995b.

Hendrix, P.F., H. Chun-Ru, and P.M. Groffman, Soil respiration in conventional and no-tillage agroecosystems under different cover crop rotations, *Soil Till. Res.*, 12:135–148, 1988.

Houghton, R.A. et al., Changes in the carbon content of terrestrial biota in soils between 1860 and 1980: a net release of CO_2 to the atmosphere, *Ecol. Monogr.*, 53:235–262, 1983.

Lund, C.P. et al., The effects of chamber pressurization on soil-surface CO_2 flux and the implications for NEE measurements under elevated CO_2, *Global Change Biol.*, 5:269–281, 1999.

Odel, R.T., S.W. Melsted, and W.M. Walker, Changes in organic carbon and nitrogen of Morrow Plot soils under different treatments 1904–1973, *Soil Sci.*, 137:160–161, 1984.

Paustian, K., H.P. Collins, and E.A. Paul, Management controls on soil carbon, in Paul, E.A. et al. (ed.), *Soil Organic Matter in Temperate Agroecosystems; Long-Term Experiments in North America*, CRC Press, Boca Raton, FL, 1997, 15–49.

Peck, T.R., Morrow Plots: long-term University of Illinois field research plots, 1876 to present, in Brown, J.R. (ed.), *Proc. Sanborn Centennial: A Celebration of 100 Years of Agricultural Research*, University of Missouri, Columbia, Publ. No. SR-415, 49–52, 1989.

Post, W.M. et al., The global carbon cycle, *Am. Sci.*, 78:310–326, 1990.

Reicosky, D.C., Strip tillage methods: impact on soil and air quality, in Mulvey, P. (ed.), *Environmental Benefits of Soil Management*, Proc. ASSSI National Soils Conf., Brisbane, Australia, 1998, 56–60.

Reicosky, D.C., W.A. Dugas, and H.A. Torbert, Tillage-induced soil carbon dioxide loss from different cropping systems, *Soil Till. Res.*, 41:105–118, 1997.

Reicosky, D.C. and M.J. Lindstrom, Fall tillage method: effect on short-term carbon dioxide flux from soil. *Agron. J.*, 85:1237–1243, 1993.

Reicosky, D.C. and M.J. Lindstrom, Impact of fall tillage on short-term carbon dioxide flux, in Lal, R. et al. (eds.), *Soils and Global Change*, Lewis Publishers, Boca Raton, FL, 1995, 177–187.

Roberts, W.P. and K.Y. Chan, Tillage-induced increases in carbon dioxide loss from soil, *Soil Till. Res.*, 17:143–151, 1990.

Schlesinger, W.H., Changes in soil carbon storage and associated properties with disturbance and recovery, in Trabalha, J.R. et al. (ed.), *The Changing Carbon Cycle: A Global Analysis*, Springer-Verlag, New York, 1985, 194–220.

Wagner, G.H., Lessons in soil organic matter from Sanborn Field, in Brown, J.R. (ed.), *Proc. Sanborn Centennial: A Celebration of 100 Years of Agricultural Research*, University of Missouri, Columbia, Publ. No. SR-415, 1989, 64–70.

Wagner, S.W., D.C. Reicosky, and R.S. Alessi, Regression models for calculating gas fluxes measured with a closed chamber, *Agron. J.*, 89, 279–284, 1997.

Wood, F.P., Monitoring global climate change: the case of greenhouse warming, *Am. Meteorol. Soc.*, 1(1):42–52, 1990.

Tillage — Soil Organic Matter Relationships in Long-Term Experiments in Hungary and Indiana

Erika Michéli, Beata Madari, Etelka Tombácz, and Cliff T. Johnston

CONTENTS

INTRODUCTION

The importance of soil organic matter (SOM) and its effect on soil properties is well known. In recent years significant amounts of research have been carried out on the composition and structure of humic substances (Stevenson, 1994; Schulten, 1995). It is believed that the molecular composition of humic substances determines their function in the soil (Hayes et al., 1989; Stevenson, 1994; Machado et al., 1993; Madari et al., 1997).

Much information exists in the literature on the influence of tillage on SOM level; the general conclusion is conservation tillage, especially no-till (NT), leads to the accumulation of SOM in the top layer of the soil and that distribution of soil properties is vertically stratified throughout the soil profile (Dick, 1983; Mann, 1986; Havlin et al., 1990; Alvarez et al., 1995; Campbell et al., 1995). Several authors have mentioned no-till as an efficient strategy to enhance C sequestration (Monreal et al., 1997; Lal et al., 1998). According to other publications, NT does not increase C and N sequestration throughout the whole soil profile, assessed on an equivalent mass basis. These publications have emphasized that the effect of NT may depend on soil properties such as texture, bulk density and the original SOM content of the soil (Wander et al., 1998; Franzluebbers et al., 1996, 1999). Less information is available on the influence of tillage on the composition of SOM. Arshad et al. (1990) reported more carbohydrates, amino acids, and alkyls, and less aromatic C, in NT, than the conventionally plowed soils. Schulten et al. (1990) also found, that intensive soil management caused the decrease of carbohydrates and phenols in SOM.

Conservation tillage, especially no-till, has been rapidly adopted on the American continents and in Australia, but not in Europe. In Hungary, no-till is still applied only on an experimental basis and is not yet an accepted practice. The influence of different tillage systems on SOM content, composition, and related soil parameters were investigated in three long-term tillage experiments in Indiana and Hungary.

MATERIALS AND METHODS

Experimental Sites and Sampling

Hungary and Indiana have the same area and similar environmental circumstances, except that precipitation is almost double in Indiana. The site experiments studied have been conducted for 21 years (Griffith plots, Agronomy Research Center, Purdue University), 15 years (Throckmorton Agronomy Center, Purdue University), and 10 years (Bajna Farmers' Coop., Hungary) on conventional tillage (autumn moldboard plow) (PL) and no-tillage (NT) plots under continuous corn. Parameters of the experimental sites are given in Table 9.1. Soil samples were obtained in summer 1997 from depths of 0 to 15 cm and 15 to 30 cm. Twenty-one single samples per plot were collected to study the spatial variability of SOM content, while four-subcomposite samples (each containing four single samples) for the investigations of SOM composition were collected. Based on the very low spatial variability, two composite samples per plot (obtained by combining single samples and subcomposite samples) were used for the extraction and characterization of humic substances.

Table 9.1   Location and Parameters of Experimental Sites

Parameters	Experimental Sites		
Location	Agronomy R. Center Purdue Univ., IN	Throckmorton R. C. Purdue Univ., IN	Bajna Farmers' Coop. Bajna, Hungary
Duration	21 years	10 years	10 years/3 years
Abbreviation	ARC	TAC	BAJNA
Tillage system	PL (autumn mold-board plow to 20 cm) NT (no-till, coulter)	PL (autumn mold-board plow to 20 cm) NT (no-till, coulter)	PL (autumn mold-board plow to 30 cm) NT (no-till, coulter)
Cropping system	Corn monoculture	Corn monoculture	Corn monoculture
Soil — FAO/ U.S. soil taxonomy	Gleyic Pheozem Typic Haplaqoll	Mollic Luvisol Udollic Endoaqualf	Mollic Luvisol Mollic Hapludalf

Note: Autumn mold-board plow of NT plots every 3 years to 30 cm.

Characterization of Bulk Soil Samples

Air-dried samples were ground to pass through the 2-mm sieve before chemical analyses and extraction of humic substances. The organic carbon (OC) content of the soil samples was measured after removing the organic light fraction by the Walkley and Black procedure (Nelson and Sommers, 1982). The cation exchange capacities of the bulk soil samples were measured by the modified Melich method ($BaCl_2$) (Buzás, 1988). Microaggregate stability of concentrated suspensions of collected samples was determined by the rheological method, using a Haake Rotovisco RV-20, CV-100 apparatus. The measurements were performed over the shear rate range of 0 to 100 1/s at 25 + 0.1°C. The evaluation of pseudoplastic flow curves according to the classical Bingham model ($\tau = \tau_B + \eta_{pl}\,\gamma$, where τ is the shear stress, τ_B is the yield value, η_{pl} is the plastic viscosity, and γ is the shear rate, results in yield value (τ_B) by extrapolation of the linear part of the flow curve back to zero shear rate (Barnes et al., 1989; Tombacz et al., 1998). This rheological parameter is related to the strength of the physical network built up from particles. The mineral composition of the soil samples was compared with x-ray powder diffraction (XRD) on air-dry soil samples using a PHILIPS XRG 3100 x-ray generator.

Extraction and Characterization of SOM

Humic substances were extracted from composite samples in six replications using the IHSS protocol: the soil samples were pretreated by 2% HCl prior to extraction by 0.5 M NaOH (soil to liquid ratio of 1:10) under N_2, shaking for 24 hours. The extraction was repeated under the same conditions, but with a 1:5 soil to liquid ratio and overnight shaking.

The humic acid (HA) fraction was separated from the fulvic acid (FA) fraction, purified with the mixture of 0.1 M HCl and 0.3 M HF, and subsequently dialyzed against 0.1 M HCl and DI water. Six replicates of extracts were then lyophilized, combined. Because of the heterogeneous composition and nature of humic substances, several analytical techniques were used to characterize SOM.

Carbon-13 (^{13}C) cross polarization magic angle spinning (CP/MAS) nuclear magnetic resonance (NMR) spectroscopy: NMR spectra were taken by a Bruker 250 MHz spectrometer at 62.9 MHz. The parameters of the measurements were the following:

Pulse delay: 5.5 µs
Contact time: 3 ms
Acquisition time: 0.051 ms
Line broadening function: 50 Hz
Number of scans: 1200 to 6048
Amount of sample material: 200 to 300 mg

The chemical shifts were measured relative to tetramethylsilane ($Si(CH_3)_4$). The peak assignments of ^{13}C CP/MAS NMR spectra of the humic substances were determined after Schnitzer (1990), Piccolo et al. (1990), Hatcher et al. (1981), and Niemeyer et al. (1992). The peak areas were then integrated and expressed in percentage of total spectrum area for each sample.

Potentiometric acid-base titration of HAs: Equilibrium acid-base potentiometric titration was performed on the HA samples. An amount of 0.1 g humic acid was redissolved in 5 mL 0.1 M NaOH solution. The obtained suspension was passed through a column of hydrogen cation-exchange resin (Varion KSM) prior to titration. The HA suspension was titrated with 0.1 M NaOH. The titration range was set for between pH 3 and 11. The titration was performed by the GIMET-1 titration system, in a closed, inert environment (N_2). The total acidity was expressed as the net proton surface excess (Δq, mmol/g) that is defined as the difference of H^+ (Γ_{H+}) and OH^- (Γ_{OH-}) surface excess amounts related to unit mass of solid: $\Delta q = \Gamma_{H+} - \Gamma_{OH-}$.

The values Γ_{H+} and Γ_{OH-} were calculated at each point of titration from the electrode output using the actual activity coefficient from the slope of H^+/OH^- activity vs. concentration straight lines for background electrolyte titration.

RESULTS AND DISCUSSION

The organic carbon content is given in Table 9.2. The OC contents of the NT treatments in the Indiana experiments are significantly higher compared to the PL ones at 0 to 15 cm depths. There is no significant difference in the soil OC content at 15 to 30 cm depths of the NT and PL treatments. These data are in good agreement with those found in the literature concerning tillage effects and indicate that NT induces the accumulation of OC in the upper soil layer, but this effect is not found

Table 9.2   The Organic Carbon Content of the Examined Soils

| Soil | Sampling Layer (cm) | Soil Organic Carbon (%) Content | |
		Conventional Tillage (PL)	No-Till (NT)
ARC	0–15	2.07 ± 0.09	2.89 ± 0.04
ARC	15–30	1.67 ± 0.05	1.81 ± 0.03
TAC	0–15	1.17 ± 0.07	1.63 ± 0.05
TAC	15–30	0.64 ± 0.02	0.87 ± 0.05
BAJNA	0–25	1.43 ± 0.06	1.47 ± 0.04
BAJNA	25–50	0.77 ± 0.03	0.77 ± 0.03

Note: ± SD.

Table 9.3   Cation Exchange Capacity of the Soil Samples

| Soil | Sampling Layer (cm) | Cation Exchange Capacity (CEC), cmol(+)/kg | |
		Conventional Tillage (PL)	No-Till (NT)
ARC	0–15	39.2 ± 0.00	41.8 ± 0.29
ARC	15–30	37.7 ± 0.29	37.3 ± 0.29
TAC	0–15	28.1 ± 0.29	27.7 ± 0.29
TAC	15–30	27.1 ± 0.00	27.1 ± 0.00
BAJNA	0–25	29.8 ± 0.29	30.0 ± 0.00
BAJNA	25–50	28.0 ± 0.88	28.0 ± 0.29

Note: ± SD.

in the deeper soil layer. The spatial variability of OC content within one experimental plot and also among four replications of the treatments was not significant in the PL treatment. It was also not significant in the NT treatments. Therefore, composite soil samples were used for the extraction and examination of the HAs.

In the Bajna experiment slight changes occurred, most probably because of the short period of the experiment and the mold-board plowing of NT plots applied every 3 years. Therefore the Bajna samples were not used for the characterization of composition of SOM.

The cation exchange capacity (CEC) values (Table 9.3) in the 0 to 15 cm sampling layers of the NT soil of the ARC experiment are significantly higher than those of the PL treatment. The differences are not significant in the TAC and Bajna experiments. The CEC values for the deeper sampling layers show almost no difference between NT and PL at all three sites.

Microaggregate stability: Concentrated suspensions of samples from NT and PL were investigated. Figures 9.1 and 9.2 show the flow curves of 0- to 15-cm samples from the Purdue ARC and the Bajna experiments. For the ARC plots coprolite-rich samples were selected and measured in addition to "normal" average samples. The higher yield values for the NT samples suggest greater strength of physical network between the particles in the no-till samples. This might be related to the higher amount of organic matter and the influence of the higher biological activity of the no-till plots. The mineral composition of the soil samples was compared with x-ray powder diffraction (XRD) on air-dry soil samples. The spectra of the mineral matrix of NT and PL samples from the same layers of each experiment were practically identical.

^{13}C *CPMAS NMR characterization of the humic acids:* The percentages of the compositional parts of the humic acid fractions based on the characteristic chemical shift regions of ^{13}C CPMAS NMR spectra are presented in Tables 9.4a and b. The distribution of different structural compounds, as well as the aromatic and carboxyl ratios of HAs of samples provide indications of the tillage effects on the composition SOM. The samples from the PL treatments of the 0- to 15-cm layers

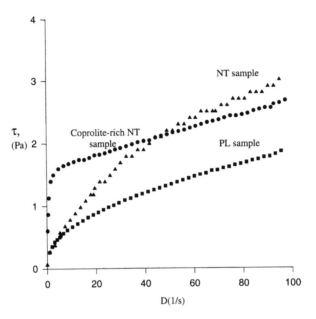

Figure 9.1   Flow curves of PL, NT, and coprolite-rich NT samples from 0- to 15-cm layer of the ARC experiment.

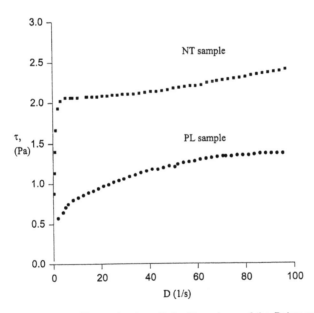

Figure 9.2   Flow curves of PL and NT samples from 0- to 15-cm layer of the Bajna experiment.

contain more carboxyl C in both experimental sites than the samples from the NT treatments. These values are similar at 0- to 15- and 15- to 30-cm depths in the PL treatments; however, in the NT treatments the soils from 0- to 15-cm layers contain significantly less carboxyl C than those from the deeper layers (15 to 30 cm).

The total aliphatic C contents of the PL HAs in the 0- to 15-cm layers are significantly lower than those of the NT HAs in the corresponding sampling layers. The corresponding values in the deeper layers show no significant differences between the treatments. The contribution of the total

Table 9.4   Amounts of Characteristic Structural Compounds of HAs from 13C CPMAS NMR Spectra of the Extracted Humic Acids, in Percentage of Total Area

a. Agronomy Research Center (ARC: 21 years)

Compound	0–15PL ± 0.94%	0–15NT ± 0.87%	15–30PL ± 1.24%	15–30NT ± 1.20%
Carbonyl (190–220 ppm)	8.44	3.87	7.97	9.5
Carboxyl (165–190 ppm)	16.66	11	18.52	17.2
Phenolic–OH (145–165 ppm)	1.63	2.62	1.05	5.07
Aromatic (105–145 ppm)	28.66	25.87	35.05	32.68
Total aliphatic (105–0 ppm)	44.6	55.64	37.42	35.59
Aromatic/Carboxylic	1.72 ± 0.03	2.35 ± 0.04	1.89 ± 0.05	1.90 ± 0.05

b. Throckmorton Agronomy Center (TAC: 10 years)

Compound	0–15PL ± 1.31%	0–15NT ± 1.51%	15–30PL ± 0.90%	15–30NT ± 0.96%
Carbonyl (190–220 ppm)	8.7	2.22	9	8.98
Carboxyl (165–190 ppm)	16.85	10.22	17.72	19.95
Phenolic–OH (145–165 ppm)	0.7	9.34	4.66	3.43
Aromatic (105–145 ppm)	28.91	20.32	26.55	29.44
Total aliphatic (105–0 ppm)	44.85	57.9	42.08	38.19
Aromatic/Carboxylic	1.72 ± 0.05	1.99 ± 0.06	1.50 ± 0.03	1.48 ± 0.03

aliphatic C to the whole HA fractions is higher in the 0- to 15-cm layers in the ARC and TAC experiments than in the deeper sampling layers. Arshad et al. (1990) made similar observations on tillage effects on SOM quality. Stearman et al. (1989) also mentioned that the PL system always contained more aromatic C than the NT system.

The aromatic and carboxyl ratios are lower in the PL treatments in both experimental sites in the 0- to 15-cm layers; there is no difference in this ratio between the treatments in the 15- to 30-cm layers. This result suggests that HAs under PL are more reactive in nature in the tilled layer than under NT. The reason for this effect is probably the higher oxidation state of the soil under PL due to aeration by physical disturbance of the soil by plowing.

Potentiometric titration of the humic acids: Potentiometric acid-base titration was carried out on the humic acid samples of the ARC and TAC sites. The adsorption curves of the humic acids of samples from 0- to 15-cm layers are in Figure 9.3. The adsorption curves show that the NT samples in the tillage-affected layers contain smaller numbers of acidic functional groups than the PL samples along the whole pH range of the titration. The outfit of the adsorption curves of the NT and PL humic acids in the 15- to 30-cm layer is very similar and almost overlapping, suggesting that the numbers of the acidic functional groups in deeper layer are very similar in the two treatments.

Total acidity values — the amount of negatively charged functional groups per gram of substance — of the humic acids were read from the adsorption curves at pH 10.5, and are given in Table 9.5. The values correspond with the average total acidity values cited in the literature for humic acids: between 400 to 500 cmol/kg or 4 to 5 mmol/g of substance (see Stevenson, 1994; Tan, 1993). The humic acids in the 0- to 15-cm layer of the PL samples contain more acidic functional groups than the humic acids under NT at both experiments; however, the difference is greater for the ARC samples. This difference can also be found in the 15- to 30-cm layers, but to a lesser extent.

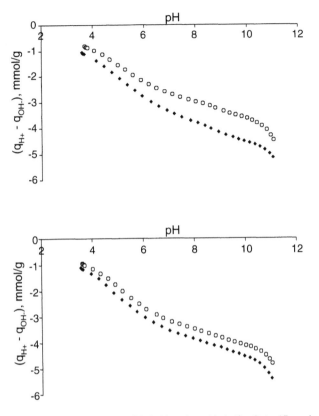

Figure 9.3   Adsorption curves of the NT (o) and PL (♦) humic acids in the 0- to 15-cm layer of the ARC (upper diagram) and TAC (lower diagram) experiments.

Table 9.5   Total Acidity of Humic Acids

		Total Acidity of Humic ACIDS, mmol/g	
Soil	Sampling Layer (cm)	Conventional Tillage (PL)	No-Till (NT)
ARC	0–15	4.726	3.852
ARC	15–30	4.734	4.487
TAC	0–15	4.716	4.320
TAC	15–30	4.540	4.176

CONCLUSION

From the results of investigations of soils from long-term tillage experiments it was concluded that tillage practice influences the level and molecular composition of the SOM. Measurements of SOM content and CEC of the bulk soil samples indicate higher values for both parameters under NT management. The differences in values between NT and PL are greater in the longer experiments. Rheology studies of samples from the experiments indicate greater strength of physical network between the particles in the no-till samples, suggesting higher microaggregate stability. These definitely suggest that NT positively influences the examined soil properties and should be more broadly introduced in Hungary. The results on the differences in the structural compounds are more difficult to interpret and need further investigations on samples from more experiments.

REFERENCES

Alvarez, R. et al., Soil carbon, microbial biomass and CO_2-C production from the three tillage systems, *Soil Tillage Res.*, 33:17–28, 1995.

Arshad, M.A. et al., Effects of till vs. no-till on the quality of soil organic matter, *Soil Biol. Biochem.*, 22(5):595–599, 1990.

Barnes, H.A., J.F. Hutton, and K.Walters, *An Introduction to Rheology*, Elsevier, Amsterdam, 1989, 115.

Buzás, I., Talaj- es agrokemiai vizsgálati módszerkönyv 2. A talajok fizikai-kemiai es kemiai vizsgálati módszerei. (Methods of soil and agricultural chemistry analysis 2. Physical-chemical and chemical methods of soil analysis.) Mezögazdasági Kiadó, Budapest, Hungary, 1988.

Campbell, C.A. et al., Carbon sequestration in a Brown Chernozem as affected by tillage and rotation, *Can. J. Soil Sci.*, 75:449–458, 1995.

Dick, W.A., Organic carbon, nitrogen and phosphorus concentrations and pH in soil profiles as affected by tillage intensity, *Soil Sci. Soc. Am. J.*, 47:102–107, 1983.

Franzluebbers, A.J. and M.A. Arshad, Soil organic matter pools during early adaptation of conservation tillage in northwestern Canada, *Soil Sci Soc. Am. J.*, 60(5):1423–1427, 1996.

Franzluebbers, A.J., G.W. Langdale, and H.H. Schomberg, Soil carbon, nitrogen, and aggregation in response to type and frequency of tillage, *Soil Sci. Soc. Am. J.*, 63:349–355, 1999.

Hatcher, P.G. et al., Aromaticity of humic substances in soils, *Soil Sci. Soc. Am. J.*, 54:448–452, 1981.

Havlin, J.L. et al., Crop rotation and tillage effects on soil organic carbon and nitrogen, *Soil Sci. Soc. Am. J.*, 54:448–452, 1990.

Hayes, M.H.B. et al., *Humic Substances II. In Search of Structure*, John Wiley & Sons, Chichester, U.K., 1989.

Lal, R. et al., *The Potential of U.S. Cropland to Sequester Carbon and Mitigate the Greenhouse Effect*, CRC Press LLC, Boca Raton, FL, 1998.

Machado, P.L.O. de A. and M.H. Gerzabek, Tillage and crop rotation interactions on humic substances of Typic Haplorthox from southern Brazil, *Soil Till. Res.*, 26:227–236, 1993.

Madari, B. et al., Long-term effects of tillage on the composition of soil organic matter: spectroscopic characterization, *Agrokemia es Talajtan (Agric. Chem. Soil Sci.)*, 46(1–4):127–134, 1997.

Mann, L.K., Changes in soil carbon storage after cultivation, *Soil Sci.*, 142(5):279–288, 1986.

Monreal C.M. et al., Impact of Carbon sequestration on functional indicators of soil quality as influenced by management in sustainable agriculture, in Lal, R. (eds.), *Management of Carbon Sequestration in Soil*, Advances in Soil Science, CRC Press, Boca Raton, FL, 1997.

Nelson, D.W. and L.E. Sommers, Total carbon, organic carbon, and organic matter, in Page, A.L., R.H. Miller, and D.R. Keeney (eds.), *Methods of Soil Analysis*. Part 2. *Chemical and Microbiological Properties*, 2nd ed., ASA, SSSA. Madison, WI, 1982, 539–579.

Niemeyer, J., Y. Chen, and J.M. Bollag, Characterization of humic acids, composts, and peat by diffuse relfectance Fourier-transform infrared spectroscopy, *Soil Sci. Am. J.*, 56:135–140, 1992.

Piccolo, A., L. Campanella, and B.M. Petronio, Carbon-13 nuclear magnetic resonance spectra of soil humic substances extracted by different mechanisms, *Soil Sci. Soc. Am. J.*, 54:750–756, 1990.

Schnitzer, M., Selected methods for the characterization of soil humic substances, in MacCarthy, P. et al. (ed.), *Humic Substances in Soil and Crop Sciences. Selected Readings*, ASA, SSSA, Madison, WI, 1990, 65–89.

Schulten, H.R., The three dimensional structure of humic substances and soil organic matter studied by computational analytical chemistry, *Fresenius J. Anal. Chem.*, 351:62–73, 1995.

Schulten, H.R. et al., *Characterization of Cultivation Effects on Soil Organic Matter*, 2. Pflanzenernaehrung und Bodenkde, 153, 97–105, 1990.

Stearman, G.K. et al., Characterization of humic acid from no-tilled and tilled soils using Carbon-13 nuclear magnetic resonance, *Soil Sci. Soc. Am. J.*, 53:1690–1694, 1989.

Stevenson, F.J., Humus Chemistry: Genesis, Composition, Reactions, 2nd ed., John Wiley & Sons, Toronto, Canada, 1994.

Tan, K.H., *Principles of Soil Chemistry, 2nd Ed.*, Marcel Dekker, New York, 1993.

Tombácz, E., Szekeres, and E. Michéli, Surface modification of clay minerals by organic. Colloids and surfaces, *A. Physicichem. Eng. Asp.*, 141:379–384, 1998.

Wander, M.M., M.G. Bidart, and S. Aref, Tillage impacts on depth distribution of total and particulate organic matter in three Illinois soils, *Soil Sci. Soc. Am. J.*, 62:1704–1711, 1998.

Effect of Soil Management Practices on the Sequestration of Carbon in Duplex Soils of Southeastern Australia

W. J. Slattery and A. Surapaneni

CONTENTS

ABSTRACT

In this chapter, we examine the effect of soil management practices on the sequestration of carbon (C) in duplex soils of southeastern Australia, not only in the cropping soils, but also in other farming soils (pasture and horticultural soils). Results from this study show that no-till practices that retain plant residues have the capacity to reduce the rate of organic C depletion by up to four times that of burning or mulching. Organic C levels decreased by only 0.01% per year for no-till practices, indicating that, on these strongly acidic soils (pH_w <5.8) with low levels of organic C (1.02%), the sequestration of C through conservation farming is still unable to keep pace with C losses. The application of green mulches onto vineyard soils was shown to be an effective way of increasing organic C. However, where soil disturbance was used in an intensive industry, such as growing tomatoes, the losses of organic C were high (0.3 gC kg^{-1} soil yr^{-1}). The use of deep-rooted perennial pastures showed an increase in C sequestration by up to 0.8 gC kg^{-1} soil yr^{-1}, and will also help in mitigating the effects of acidity and salinity in environments where annual rainfall exceeds 500 mm. Based on current cropping, grazing, and horticultural practices, estimates of potential C sequestration in Australian agricultural soils identified that up to 12.2×10^6 t C yr^{-1} would be sequestered.

INTRODUCTION

The capacity of a soil to sequester organic C is dependent mainly on climate, soil type and landscape (edaphic conditions), type of vegetation, and soil management imposed by agricultural practice (Carter, 1996). Reductions in soil organic carbon (SOC) have commonly been attributed to traditional agricultural practices (e.g., cultivation and stubble burning) in cropping soils that have left the soil vulnerable to the effects of wind and water erosion (Roberts and Chan, 1990). No-till operations (conservation farming) in arable farming systems, either alone or in conjunction with cover crops, offer hope for the build-up of organic C levels (Chan et al., 1988; Arshad et al., 1990; Grace et al., 1995; Dalal et al., 1995; Lal et al., 1999).

Burning straw residues has been shown to reduce SOC levels below those achieved with stubble retention and direct drilling of seed (Chan et al., 1992). It should be noted, however, that changes in SOC will be influenced by the quantity of stubble, depth and intensity of stubble incorporation, and the efficiency of the burning procedure (Carter and Mele, 1992). The population diversity of the soil microflora may also be important in determining the rate of decomposition of organic matter and therefore total respiration of CO_2. For example, in a stubble-standing treatment, the loss of C as respired CO_2 has been shown to be up to 45% higher than where stubble was removed (Saffigna et al., 1989), due mainly to the higher microbial biomass (Haines and Uren, 1990; Carter and Mele, 1992). In stubble-burnt treatments, the loss of C is in the form of CO_2 emissions during combustion; a smaller amount of respired CO_2 is lost due to microbial respiration and water and wind-blown erosion (Roberts and Chan, 1990).

Other studies have shown that long-term cultivation decreases SOC and that direct drilling and stubble retention increase the total amount of organic C (Dalal and Mayer, 1986; Chan et al., 1988; Saffigna et al., 1989; Arshad et al., 1990; Bowman et al., 1990; Carter and Steed, 1992), increase the particulate organic matter pools (Cambardella and Elliot, 1993; Franzluebbers and Arshad, 1997), and improve soil structure by increasing aggregate stability in the surface 2.5 cm (Chan and Mead, 1988). Although cultivation enhances the decomposition of organic matter occluded within the soil aggregates, the fraction associated with clay particles is only slightly altered. In fact, much of the clay-bound organic matter is protected from chemical or microbial decomposition reactions and represents the pool of organic matter stabilized through decades of cultivation (Golchin et al., 1995).

Conservation farming methods have been promoted for many soil types used for the cropping industries in southeastern Australia to improve soil stability, increase rainfall infiltration, and minimize soil compaction (Carter and Steed, 1992). On many of the soils where no-till farming is promoted in southeastern Australia, soil acidity and sodicity are also a major concern. A recent study (Surapaneni et al., 2000), showed that high SOC levels reduced the effects of sodicity by increasing soil aggregation in pasture soils. Although overwhelming evidence of the importance for SOC in the improvement of soil conditions in cropping and pasture soils exists, very little work has been done in other agricultural enterprises to identify the potential build-up or loss of SOC.

In addition to understanding the impacts of increasing soil organic matter on soil condition and plant production, there is a need to quantify the ability of different farming systems to sequester SOC and thus provide a sink for greenhouse gases. In this chapter, a range of soil data has been collated from long-term (>10 years) and some short-term (2 to 3 years) studies in an attempt to identify the impact of farming practices on the ability of crop, pasture, and some horticultural industries to sequester C in duplex soils of southeastern Australia. These sequestration rates are discussed in terms of their potential to increase soil stores of C.

MATERIALS AND METHODS

Site and Soil Description

Soil management practices were examined in vineyard, broad-acre crop, permanent pasture, and tomato crop soils. Soil types included those of (1) duplex structure consisting of a sandy loam A-horizon, overlaying a heavy clay to clay loam throughout the B-horizon, classified as Chromosol (*Rhodoxeralf*), Sodosol (*Natrixeralf*), and Kurosol (*Alfisol*) soils, and (2) uniform or gradational soil structure consisting of a sandy loam in the A-horizon, overlaying a sandy clay B-horizon, classified as Vertisol (*Vertisol*) and Dermosol (*Alfisol*). These details are summarized in Table 10.1.

All cropping sites contained treatment plots that were replicated at least four times in a random block design. Management practices of plant residues, including burning, mulching, and leaving standing stubble, were imposed as treatments on the cropping soils immediately prior to the autumn break. (Details are given in Table 10.3.) These experimental sites were located in the Rutherglen and Tatura regions of northern Victoria (Figure 10.1).

Table 10.1   Soil Texture and Classification for Long-Term Sites Located in the Rutherglen and Tatura Regions of Northern Victoria

Soil Types	Soil Texture A-Horizon	B-Horizon
Eutrophic yellow Dermosol[a] (*Alfisol*)[b]	Fine sandy loam	Light clay
Brown Kurosol (*Alfisol*)	Sandy loam	Sandy clay loam
Red Kurosol (*Alfisol*)	Fine sandy loam	Medium clay
Yellow Dermosol (*Alfisol*)	Fine sandy loam	Light/medium clay
Red Chromosol (*Rhodoxeralf*)	Fine sandy clay loam	Fine sandy clay
Subnatric Sodosol (*Natrixeralf*)	Clay loam	Medium heavy clay
Vertisol (*Vertisol*)	Sandy loam	Sandy clay loam
Red Chrornosol (*Rhodoxeralf*)	Sandy loam	Clay loam
Red Sodosol (*Natrixeralf*)	Clay loam	Medium heavy clay

[a] Isbell (1996).
[b] Soil Survey Staff (1975).

Table 10.2   General SOC Values (gC kg^{-1} soil) Considered to be Low, Normal, or High for Soils Used for Crop and Pasture Production in Areas of High and Low Rainfall in Victoria

Soil Organic C Status	Low Rainfall Crops	Pastures	High Rainfall Crops	Pastures
Low	<9	<17.4	<14.5	<29.0
Normal	9–14.5	17.4–26.2	14.5–29.0	29.0–58.1
High	>14.5	>26.2	>29.0	>58.1

Source: Peverill et al., 1991.

Figure 10.1   Location of the experimental sites in the Rutherglen and Tatura regions and the locations of the
117 survey sites (λ) across the northern cropping region of Victoria.

In addition to the above experimental sites, another 117 survey sites were sampled in 1997 across the northern cropping region of Victoria (Figure 10.1). The soil types for these survey sites included Calcarosols, Dermosols, Chromosols, Vertosols, Ferrosols, Sodosols, and a Tenosol. Characterization of these 117 survey sites involved extensive sampling (50 soil cores [25 × 10 cm] for the 0 to 2.5 , 2.5 to 5, and 5 to 10 cm soil layers and 25 cores for all soil layers 10 cm below the surface) within a 5-m radius located by GPS and buried transponders.

Soil Sampling

Soil was sampled from each site during the growing season. Ten soil cores (2 by 10 cm) were taken randomly over each plot and bulked for each replicated treatment at each site. All soils were dried immediately after sampling from the field in a fan-forced oven at 40° C for 24 h, rumbled to pass through a 2-mm sieve, and stored in sealed plastic bags at room temperature. A subsample of soil, milled to pass through a 0.25-mm screen, was used for SOC measurements.

Soil Carbon Analysis

Organic carbon was measured by the method of Heanes (1984), which requires a chromic acid digestion of soil heated to 135°C for 30 min, followed by colorimetric determination of the resulting solution at an absorbance at 600 nm. Values for SOC determined in 1998 were compared with those determined at the start of the 3- to 15-year continuous cropping or continuous pasture program, using the same method.

RESULTS AND DISCUSSION

Factors Related to Changes in SOC in Southeastern Australia

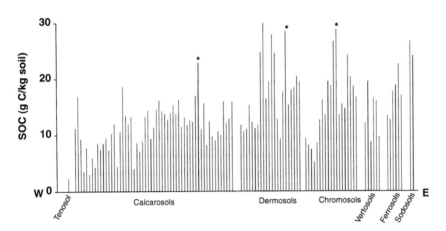

Figure 10.2   Soil organic carbon values in the surface 10 cm of soil for a range of soil types sampled in 1997 across the northern cropping region of Victoria, Australia, west (W) to east (E). * = native bush sites.

Soil organic carbon values for the top 10 cm of surface soil in temperate Australian environments are generally low and rarely exceed 40 gC kg^{-1} soil (Baldock and Skjemstad, 1999), with a typical mean value of around 13.7 gC kg^{-1} soil (Geeves et al., 1995). This variation in SOC can be related to natural phenomena such as climate, soil mineral composition, natural vegetation cover, and soil microbial flora, as well as human-induced changes such as soil disturbance through different farm management practices, flooding, and bushfires (Baldock and Skjemstad, 1999).

Typical SOC values for both low and high rainfall environments in Victoria are given in Table 10.2. These values indicate that high values of SOC in farming enterprises exist that have a long-term history of permanent pasture. They are located in regions that receive a high annual rainfall, compared with the lower values from soil of long-term crop rotations and in low rainfall environments. This observation is further exemplified by data collected in cropping soils (117 survey sites) in the northern region of Victoria (< 600 mm rainfall), which showed an average SOC value of 12.8 gC kg^{-1} soil for soils with a long history (>5 years) of continuous cropping. Values for SOC were generally higher for native bush samples, which were dominated by native annual and perennial grasses (Figure 10.2). In addition, the average value for SOC declined when soil data from the east (E) through to the west (W) in cropping regions of southeastern Australia were compared.

For example, the average SOC value for Sodosols was 26.6 gC kg^{-1} soil compared with Calcarosols, which had an average SOC value of 11.5 gC kg^{-1} soil. The most obvious difference between these two soil types was the amount of annual rainfall that would be received in each region. Calcarosols rarely receive more than 300 mm of annual rainfall whereas Sodosols, Chromosols, and Dermosols receive a minimum annual rainfall of 500 mm. These data add further support to that provided by Peverill et al. (1991) in Table 10.2, even though Peverill's SOC status was based on the whole of Victoria, including high rainfall (> 600 mm) areas in the southern regions (e.g., Gippsland).

Elevated values of SOC in the higher rainfall environments can be associated with more vegetative cover through longer periods of the year, leading to an increased amount of biomass above and below ground, and thus leading to a higher amount of organic matter residue sequestered into the soil (see SOC status of high rainfall rates, Peverill et al., 1991). In general, soils in these higher rainfall environments will retain more soil moisture for a longer period than soils with an annual average rainfall of less than 600 mm, thereby providing an energy-rich matrix for microbial decomposition processes to continue at a more rapid rate. It has been shown previously that SOC contents are positively correlated with rainfall (Post et al., 1982; Parton et al., 1987).

Tillage Effects on the Sequestration of C in Cropping Soils

Soil organic C levels decreased by only 0.07 gC kg^{-1} soil yr^{-1} for treatments that retained stubbles on the surface (DD-standing) and by only 0.04 gC kg^{-1} soil yr^{-1} for treatments that shredded the standing stubble (DD-shredded; Table 10.3). This indicated that, on these strongly acidic soils (pH <5.5) with low levels of SOC (10.2 gC kg^{-1} soil), the sequestration of C through conservation tillage (no-till) almost kept pace with C losses due to respiration after microbial decomposition reactions. Losses in organic C resulting from the incorporation of stubble with some form of soil disturbance (one or two passes with a tined implement), and from the burning of stubble, were substantially higher (0.43 to 0.52 gC kg^{-1} soil yr^{-1}) than no-till systems (Table 10.3). These results are comparable to C losses found in other long-term cereal-fallow rotations (Chan et al. 1992; Dalal et al., 1991). It is worth noting that, when a cover crop of subterranean clover was grown with a cereal crop, the soil carbon levels did not decrease, but showed a small increase.

No-till systems receiving high rates of applied lime (1.5 to 6.0 t ha^{-1}) had significantly decreased SOC values of up to 0.4 g C kg^{-1} soil yr^{-1} (Table 10.3), which is probably related to the higher rates of mineralization. It is important to note that the application of CaCO$_3$ on these strongly acidic soils is essential for continued production, and the cost of C losses must be countered by adequate crop and pasture rotations.

Table 10.3 Changes in SOC (gC kg^{-1} soil yr^{-1}) Values for Different Agricultural Enterprises (0–10 cm) from Different Soil Types with Different Soil and Residue Management Practices

Crop/Soil Types	Management Practice	ΔC g kg^{-1} Soil yr^{-1}
Continuous crop (acid soils pH$_{Ca}$ <5.0)	No-till or minimum-till	
Brown Kurosol (*Alfisol*)	Standing stubble	−0.16
	Burnt stubble	−0.08
	Mulched stubble	−0.20
	Standing stubble + lime	−0.30
	Standing stubble + clover uncover	+0.001
Yellow Dermosol (*Alfisol*)	Standing stubble	−0.15
	Burnt stubble	−0.13
Red Kurosol (*Alfisol*)	Standing stubble	−0.05
	Burnt stubble	−0.36
	Mulched stubble	−0.56
	Shredded stubble	−0.05
Eutrophic yellow Dermosol (*Alfisol*)	Standing stubble	−0.07
	Burnt stubble	−0.43
	Mulched stubble	−0.52
	Cultivation	−1.0
Vineyard soil		
Red Chromosol (*Rhodoxeralf*)	Mulched green waste	+1.7
	Ryegrass cover	+1.3
Subnatric Sodosol (*Natrixeralf*)	Mulched green waste	+3.5
	Ryegrass cover	+1.0
Tomato soil		
Vertisol (*Vertisol*)	Plastic cover over raised beds	0 to − 0.3
Pasture soils		
Red Chromosol (*Rhodoxeralf*)	Annual pasture	+0.4 to +0.8
Red Sodosol (*Natrixeralf*)	Perennial pasture with irrigation	−0.3
Yellow Dermosol (*Alfisol*)	Perennial pasture with irrigation	−1.3
	Annual pasture	+0.2

Other studies (Chan et al., 1992, Dalal et al., 1995; Heenan et al., 1995; Poulton, 1995) have also shown a decline in SOC between 0.04 and 0.3 gC kg^{-1} soil yr^{-1} for continuous cereal or cereal-fallow rotations. Results from this study demonstrate that no-till practices that retain standing stubble have the capacity to reduce the rate of soil C depletion by up to four times that of stubble burning or stubble mulching and by a factor of 10 compared with cultivation practices (Table 10.3). It has been shown that the introduction of fertilizer nitrogen on a regular basis will lead to an increase in SOC content (Paustian et al., 1992). In the experiments under Australian temperate environments described in this chapter, applications of fertilizer nitrogen are low, and it is therefore unlikely that sequestration of SOC will be high.

Sequestration of C in Tomato Soils

Tomato cropping practices led to a decrease in total SOC content by up to 0.3 gC kg^{-1} soil yr^{-1} (Table 10.3). This loss in SOC could be related to a low biomass production for tomatoes, but is also consistent with the view that soil disturbance leads to enhanced oxidation of SOC by increasing soil aeration and soil residue contact, and accelerated soil loss through wind and water erosion (Lal et al., 1999).

Sequestration of C in Vineyard Soils

An increase in SOC content ranged from 1.0 to 3.5 gC kg^{-1} soil yr^{-1} in vineyard soil, which could be attributed to the addition of mulched green organic material to the surface of these soils (Table 10.3). In another study involving the use of straw mulch on vineyards (Whiting and Orr, 1994), there was no observable difference in SOC after 3 years and yield increased in only the first year of mulch application. The application of dried straw in the study by Whiting and Orr (1994) is analogous to that of stubble-standing treatments in continuous cropping studies where SOC did not increase even after 15 years. The rate of C sequestration will be variable for vineyards in different climatic regions and will be a function of the quantity and type of organic material returned to the soil, as well as their decomposition rates for different soil environments (microbial populations; Lal et al., 1999).

Sequestration of C in Pasture Soils

For different agro-ecosystems in Australia, the quantity and quality of pasture in any farming enterprise will vary; hence, the type of organic residue entering the soil will also vary. In addition, the amount of nitrogen in pastures will be largely dependent upon the composition of legume, thus influencing the total C:N ratio in soil and the rate of incorporation of mineral and organic C into the soil matrix. It has also been suggested that the total amount of clay in the surface soil depths will influence the degree of decomposition of organic matter, such that a soil low in clay content will have a higher rate of C decomposition (Grace et al., 1995). Data presented in this chapter were predominantly obtained from dry temperate environments on a range of soil types, and showed that SOC increases were highest in soil of annual pastures with increases ranging from +0.2 to +0.8 gC kg^{-1} soil yr^{-1}. These results are consistent with other Australian experiments in which long pasture phases were included in crop rotations and increases in SOC ranging from +0.4 to 1.3 gC/kg soil yr^{-1} were detected (Dalal et al., 1994; Dalal et al., 1995; Grace et al., 1995).

Irrigated pasture data showed that there was a loss in SOC of between –0.3 to –1.3 gC kg^{-1} soil yr^{-1}. It has been noted in other work (Amato et al., 1984; Crawford et al., 1997), that both perennial and annual pastures are likely to sequester more below-ground C if they are not heavily grazed or subjected to extremes of wetting and drying. The data presented here for perennial irrigated pastures were derived from pastures subjected to heavy grazing and exposed to extremes in soil moisture content, thus resulting in high decomposition rates. This soil would not be expected to accumulate SOC to the same extent that a lightly grazed and nonirrigated pasture would.

Carbon Budget for Agricultural Soils in Australia

Total land areas used for agricultural production in Australia (excluding the 426 M ha of land used for unimproved grazing) are given in Table 10.4 and represent the range of major farming enterprises across many agro-ecological regions. These land areas were used to calculate the total C currently lost or sequestered into soil on an annual basis, using data in Table 10.3. There is a large amount of variability within the data sets for different agricultural enterprises, and consequently the overall budget for C sequestered into soil varied from 2.7 to 12.2 M t C yr^{-1} for those industries in Australia (Table 10.5). Significant opportunities exist in agricultural-based industries to increase this amount of sequestered C, including forestry, mining rehabilitation, and waste reuse, all of which have not been taken into account in this soil C budget.

What is the optimum level of soil C for the wide range of agricultural soils in Australia? Estimates of equilibrium rates for SOC will vary according to soil type, climatic region, fertility management practice, and type of vegetation (Grace et al., 1995; Baldock and Skjemstadt, 1999; Lal et al., 1999). Values of SOC for a range of soil types and fertility management practices for cropping systems in Victoria (Figure 10.2) indicate that 20 gC kg^{-1} soil could be achievable for most soils. It has been noted previously that, on a red Chromosol with wheat-pasture rotations, a value of 13.2 gC kg^{-1} soil would represent a steady-state system (Grace et al., 1995). Similarly on a Vertisol with a grass–wheat–legume rotation, 13.0 gC kg^{-1} soil represented a steady state system

Table 10.4   Agricultural Farming Enterprises and Approximate Land Areas for Those Activities in Australia in 1998

Agricultural Enterprise	Land Area in Australia (1997)
Crops (cereals, oilseeds, legumes)	21.1 M ha
Pastures (sown)	26 M ha
Fruits (vines, apples, pears, etc.)	227,000 ha
Vegetables (carrots, tomatoes, potatoes)	57,000 ha

Table 10.5   Annual C Sequestration Budgets for Some Major Australian Agricultural Enterprises, Based on Rates of Change for SOC Given in Table 10.3

Current Farming Practice	Current Rate of C Sequestration (t C yr^{-1} × 10^6)
Crops burning (22%)[a]	−0.37 to −1.99
Mulching (25%)[b]	−1.06 to −2.95
Cultivation (20%)[c]	−4.22
Standing (28%)[d]	−0.41 to −0.94
Standing + pasture (5%)[e]	+0.009
Vegetables	0 to −0.02
Pastures (sown)	+9 to +18
Fruits (vines, apples, etc.)	+0.2 to +0.7
Annual budget	+3.2 to +8.6

[a] 22% of 21.1 M ha and using values of −0.08 to −0.43 ΔC g kg^{-1} soil yr^{-1}.
[b] 25% of 21.1 M ha and using values of −0.20 to −0.56 ΔC g kg^{-1} soil yr^{-1}.
[c] 20% of 21.1 M ha and using a value of −1.00 ΔC g kg^{-1} soil yr^{-1}.
[d] 28% of 21.1 M ha and using values of −0.07 to −0.16 ΔC g kg^{-1} soil yr^{-1}.
[e] 5% of 21.1 M ha and using a value of +0.001 ΔC g kg^{-1} soil yr^{-1}.

(Dalal et al., 1995). It would be unrealistic to expect that values exceeding 20 gC kg^{-1} soil could be achieved for all soils. However, given the low priority that C sequestration has attracted in current agricultural farming practices, there is significant scope for improvement and these values could be obtained, particularly if a C credit process were adopted.

What then are the strategies to increase soil C in Australian soils? Unfortunately, current conservation tillage practices, which include a small but minimal amount of tillage, under continuous cropping are unable to make a large impact upon the sequestration of C into soil; at best the annual increase might be 0.001 gC kg^{-1} soil. As a consequence, many continuously cropped soils have declined to very low soil C levels and are still declining. However, large gains in total soil C budgets can be achieved by including pastures or organic composted material into the soil. For example, if the organic matter content of the top 10 cm of surface soil could be increased by 10 gC kg^{-1} soil over a period of 30 years, it is estimated that, for some 40 million hectares of highly productive agricultural land in Australia, this would equal savings of 11 M t of C per year. This would effectively double the amount of C sequestered into soil from current agricultural practices.

CONCLUSIONS

It is quite clear that agricultural practices that involve burning biomass or that persist with cultivation will not be sustainable in the long term and will lead to C losses as well as soil degradation. The real challenges for Australian agriculture will be to eliminate stubble burning and cultivation activities that exist in 67% of the broad acre cropping soils. The use of no-till farming practices will need to be supplemented with undersown pastures or long pasture phases (3 to 5 years) in the crop rotation in order to accumulate C in the soil. In particular, the use of deep rooted perennial pastures will not only increase soil C but also will help mitigate the effects of acidity and salinity in environments where annual rainfall exceeds 600 mm.

Opportunities exist for the use of organic waste products as a source of nutrients and C in agricultural soils. Horticultural enterprises are already using these materials but because of their high cost in processing and transportation, these materials are of limited use in broad acre farming (cropping and dairying). If a C credit system can operate in an equitable manner for all production systems, then it is likely that the agricultural use of waste organic materials may well become cost neutral.

Best management practices are needed to promote the conservation of soil and reduce soil erosion and other land degradation processes such as acidification and sodification, which are prevalent in southeastern Australia. The retention of soil C is one way to increase the soil's resilience to structural decline and must be seen as a key spillover benefit from the mitigation of greenhouse gases. Soil C provides an important link between sustainability and productivity within agricultural systems; strategic efforts must be made to identify where the most benefit can be obtained from increases in soil C, rather than taking a blanket approach to appease the greenhouse gases target.

ACKNOWLEDGMENTS

The authors wish to thank Bernadette Carmody for collection and analysis of soils and the Grains Research and Development Corporation (GRDC) and the Grains Program and Catchment and Water of Victorian Department of Natural Resources and Environment (DNRE) for financially supporting this research.

REFERENCES

Amato, M. et al., Decomposition of plant material in Australian soils. II. Residual organic ^{14}C and ^{15}N from legume plant parts decomposing under field and laboratory conditions, *Aust. J. Soil Res.*, 22, 331–341, 1984.

Arshad, M.A. et al., Effects of till vs. no-till on the quality of soil organic matter, *Soil Biol. Biochem.*, 22, 595–599, 1990.

Baldock, J.A. and Skjemstad, J.O., Soil organic carbon/ soil organic matter, in *Soil Analysis: An Interpretation Manual*, K.I. Peverill, L.A. Sparrow, and D.J. Reuter, Eds., CSIRO Publishing, Collingwood, Vic. Australia, 1999, 159–170.

Bowman, R.A., Reeder, J.D., and Lober, R.W., Changes in soil properties in a central plains rangeland soil after 3, 20, and 60 years of cultivation, *Soil Science*, 150, 851–857, 1990.

Cambardella, C.A. and Elliot, E.T., Carbon and nitrogen distribution in aggregates from cultivated and native grassland soils, *Soil Sci. Soc. Am. J.*, 57, 1071–1076, 1993.

Carter, M.R., Analysis of soil organic matter storages in agroecosystems, in *Advances in Soil Science: Structure and Organic Matter Storage in Agricultural Soils*, M.R. Carter and B.A. Stewart, Eds., Lewis Publishers, New York, 1996, 3–11.

Carter, M.R. and Mele, P.M., Changes in microbial biomass and structural stability at the surface of a duplex soil under direct drilling and stubble retention in north-eastern Victoria, *Aust. J. Soil Res.*, 30, 493–503, 1992.

Carter, M.R. and Steed, G.R., The effects of direct-drilling and stubble retention on hydraulic properties at the surface of duplex soils in north-eastern Victoria, *Aust. J. Soil Res.*, 30, 505–516, 1992.

Chan, K.Y., Bellotti, W.D., and Roberts, W.P., Changes in surface soil properties of vertisols under dryland cropping in a semi-arid environment, *Aust. J. Soil Res.*, 26, 509–518, 1988.

Chan, K.Y. and Mead, J.A., Surface physical properties of a sandy loam soil under different tillage practices, *Aust. J. Soil Res.*, 26, 549–559, 1988.

Chan, K.Y., Roberts, W.P., and Heenan, D.P., Organic carbon and associated soil properties of a red earth after 10 years of rotation under different stubble and tillage practices, *Aust. J. Soil Res.*, 30, 71–83, 1992.

Crawford, M.C., Grace, P.R., and Oades, J.M., Effect of defoliation of medic pastures on below-ground carbon allocation and root production, in *Management of Carbon Sequestration in Soil*, R. Lal, J.M. Kimble, R.F. Follett, and B.A. Stewart, Eds., CRC Press, New York, 1997, 381–389.

Dalal, R.C., Henderson, P.A., and Glasby, J.M., Organic matter and microbial biomass in a vertisol after 20 years of zero-tillage, *Soil Biol. Biochem.*, 23, 435–441, 1991.

Dalal, R.C. and Mayer, R.J., Long-term trends in fertility of soils under continuous cultivation and cereal cropping in southern Queensland. I. Overall changes in soil properties and trends in winter cereal yields, *Aust. J. Soil Res.*, 24, 265–279, 1986.

Dalal, R.C. et al., Evaluation of forage and grain legumes, no-till and fertilisers to restore fertility degraded soils, *Trans. 15th Int. Congr. Soil Sci.*, 5a, 62–74, 1994.

Dalal, R.C. et al., Sustaining productivity of a Vertisol at Warr, Queensland, with fertilisers, no-tillage, or legumes. 1. Organic matter status, *Aust. J. Exp. Agric.*, 35, 903–913, 1995.

Franzluebbers, A.J. and Arshad, M.A., Particulate organic carbon content and potential mineralization as affected by tillage and texture, *Soil Sci. Soc. Am. J.*, 61, 1382–1386, 1997.

Geeves, G.W. et al., The physical, chemical and morphological properties of soils in the wheat-belt of southern NSW and northern Victoria, NSW Department of Conservation and Land Management/CSIRO Division of Soils Occasional report, 1995.

Golchin, A. et al., The effects of cultivation on the composition of organic matter and structural stability of soils, *Aust. J. Soil Res.*, 33, 975–993, 1995.

Grace, P.R. et al., Trends in wheat yields and soil organic carbon in the permanent rotation trial at the Waite Agricultural Research Institute, South Australia, *Aust. J. Exp. Agric.*, 35, 857–864, 1995.

Haines, P.J. and Uren, N.C., Effects of conservation tillage farming on soil microbial biomass, organic matter and earthworm populations in north-eastern Victoria, *Aust. J. Exp. Agric.*, 30, 365–371, 1990.

Heanes, D.L., Determination of total organic-C in soils by an improved chromic acid digestion and spectrophotometric procedure, *Commun. Soil Sci. Plant Anal.*, 15, 1191–1213, 1984.

Heenan, D.P. et al., *Aust. Exp. Agric.*, 35, 877–884, 1995.

Isbell, R.F., *The Australian Soil Classification*, CSIRO Publishing, Collingwood, Vic., 1996.

Lal, R. et al., Managing U.S. cropland to sequester carbon in soil, *J. Soil Water Conserv.*, 54, 374–381, 1999.

Parton, W.J., Analysis of factors controlling soil organic matter levels in Great plains grasslands, *Soil Sci. Soc. Am. J.*, 51, 1173–1179, 1987.

Paustian, K., Parton, W.J., and Persson, J., Modeling soil organic matter in organic-amended and nitrogen-fertilized long-term plots, *Soil Sci. Soc. Am. J.*, 56, 476–488, 1992.

Peverill, K.I. et al., The analytical and interpretative service, State Chemistry Laboratory, Vic, Australia, 1991.

Post, W.M. et al., Soil carbon pools and world life zones, *Nature*, 298, 156–159, 1982.

Poulton, P.R., The importance of long-term trials in understanding sustainable farming systems: the Rothamsted experience, *Aust. J. Exp. Agric.*, 35, 825–834, 1995.

Roberts, W.P. and Chan, K.Y., Tillage-induced increases in carbon dioxide loss from soil, *Soil Tillage Res.*, 17, 143–151, 1990.

Saffigna, P.G. et al., Influence of sorghum residues and tillage on soil organic matter and soil microbial biomass in an Australian vertisol, *Soil Biol. Biochem.*, 21, 759–765, 1989.

Soil Survey Staff, *Soil Taxonomy — a Basic System of Soil Classification for Making and Interpreting Soil Surveys*, USDA, Washignton, D.C., 1975.

Surapaneni, A. et al., Water stability behaviour of sodic soils in relation to organic matter, *Proc. 9th Int. Meeting Int. Humic Substances Soc.*, Adelaide, Australia, September 21–25, 2000.

Whiting, J. and Orr, K., Evaluating the use of organic mulch in a vineyard, *Proc. Workshop Nutrient Fertiliser Manage. Perennial Horticulture*, Tatura, Victoria, Australia, June 15–16, 1994.

CHAPTER 11

Exchangeable Aluminum in Composts Rich in Organic Matter during the Maturation Process

M. Jokova and O. Kostov

CONTENTS

INTRODUCTION

Managing soil fertility with organic materials requires much study of soil characteristics, changes with time, and use. Lignin and cellulose wastes are available in large amounts in Bulgaria; aerobically composting these wastes reduces some of their disadvantages (phytotoxicity, long-term nitrogen immobilization, and structural incompatibility). Such composts are sources of nutrients for plants and are used for cucumber and tomato production (Atanasova et al., 1990; Kostov et al., 1995; Tzvetkov, 1993).

Studying the changes in compost constituents during maturation (i.e., the development of negative charge, appearance of reductive substances, and their reductive power) is important. Cation exchange and reducing capacities have been found to be reliable criteria for compost quality (Jokova et al., 1997). Further studies are needed on the presence of toxic compounds, pH changes, and exchangeable aluminum, which are important for plant growth. This chapter discusses the dynamics of exchangeable aluminum in composts rich in organic matter during maturation processes.

MATERIALS AND METHODS

Studied composts included vine branches, grape pruning, husks, and seeds, and sawdust. All composts were treated before composting with N and P. Treatments were carried out with either

1-56670-581-9/02/$0.00+$1.50
© 2002 by CRC Press LLC

powdered marble (for pH adjustment) or *Cephalosporium* sp (for inoculation) or both. The composts (1 m^3 in size) were wet to 60% (w/w) and kept in mesophilic conditions.

Cation exchange capacity (CEC) was determined by the method of Harada and Inoko (1980). After a sequential washing of milled compost samples (mesh size <1 mm) with HCl and distilled water and saturation with Ba(CH$_3$COOH)$_2$ at pH 7, the solutions were titrated with 0.05 N NaOH and phenolphthalein as an indicator. The amounts of exchangeable aluminum (Al$_e$) extractable in the acetate solutions were determined by AAS method, as well as amounts of Ca^{2+} and Mg^{2+}.

Reducing capacity (RC) of composts was determined by the method of Mirchev and Jokova (1986). After adding 1 N H$_2$SO$_4$ to the milled compost sample (mesh size < 1 mm) and shaking for 1 hour, the solutions were titrated with K$_2$Cr$_2$O$_7$ with standard potential E^0 = 1.33 V; phenyl-antranylic acid (E^0 = 1.08 V) was used as an indicator. Under these conditions, the dissolved reductive organic compounds (and Fe^{2+}) react mainly with the oxidizing agent. Solutions were also titrated with 0.01 N KMnO$_4$ (E^0 1.51 V) for the determination of reductive compounds with higher electrode potentials, which can react with stronger oxidizers.

Some chemical characteristics of the substrates, determined according to Page et al. (1982), appear in Table 11.1. The analyses of composts described above, as well as pH measurements, were carried out monthly for a year in three replicates. At the beginning, the pH of the substrates was 5.5 for vine branches, 6.6 for grape pruning, husks, and seeds, and 4.8 for sawdust. Parts of the results appear in Tables 11.2, 11.3, and 11.4.

Table 11.1 Content of Organic Matter Constituents of Composted Substrates (as % of dry matter)

Substrate	pH	Total C	Cellulose	Lignin	Sugars
Vine branches	5.5	45.0	51.0	28.0	17.7
Grape prunings, husks, seeds	6.6	31.0	32.0	21.0	16.4
Sawdust	4.8	42.6	48.0	26.0	0.0

RESULTS AND DISCUSSION

Organic matter in soils is thermodynamically unstable. Changes in organic constituents of composts during the maturation process cause development of a negative charge and CEC dynamics. During the decomposition of lignin, cellulose, organic acids, etc., electrons and protons are released and react with surrounding substances. Therefore, reductive organic compounds appear and are oxidized during the maturation process.

The results show that the content of total carbon, cellulose, and lignin in vine branches and sawdust substrates is higher than that in grape prunings, husks, and seed substrates. The content of sugars in vine branches and grape prunings, husks, and seed substrates is close, and in sawdust substrates is absent (Table 11.1). These differences in organic matter constituents of the substrates lead to differences in the rate of decomposition and values of cation exchange and reducing capacities of the composts.

Jokova et al. (1997) established the maturity of these composts in their studies of CEC dynamics. The points where the slopes of curves CEC vs. time changed, i.e., the rates of the development of the negative charge decreased, showed when the studied compost was nearly matured. The vine branches, grape prunings, husks, and seed composts were nearly stabilized after 120 days of composting, and their quality was very good. The sawdust composts matured for about 360 days of composting, since compounds with higher electrode potentials were found by KMnO$_4$ titration (RC^b), because some components in sawdust were not readily oxidizable. These composts needed to be used with caution during that period. Obviously, the high amounts of reductive compounds delayed the maturation process.

Table 11.2   Chemical Characteristics of Vine Branch, Grape Prunings, Husk, and Seed Composts

Treatments	Composting Period (days)	CEC cmol. kg^{-1}	RC[a] cmol. kg^{-1}	Al$_e$[c] cmol. kg^{-1}	pH (H$_2$O)
Vine branches					
NP	30	28.0	1.00	1.1	6.0
	90	54.9	1.24	3.3	5.1
	120	63.6	1.00	0.0	6.3
NP + marble	30	28.9	1.00	2.0	5.8
	90	52.0	1.22	5.8	6.1
	120	60.7	0.90	10.0	6.2
NP + Cephalosporium sp.	30	23.1	0.80	0.0	5.4
	90	54.9	0.90	0.0	6.2
	120	66.5	1.04	5.0	6.1
NP + Cephalosporium sp. + marble	30	23.1	0.80	1.4	5.4
	90	49.1	0.90	4.2	6.2
	120	49.1	0.92	10.0	6.5
Grape prunings, husk, and seed composts					
NP	30	138.7	0.40	0.05	5.4
	90	115.6	0.94	5.78	5.4
	120	127.2	1.90	2.52	5.6
NP + marble	30	98.3	0.40	1.84	6.1
	90	115.6	0.96	5.05	5.5
	120	112.7	1.86	3.34	6.1
NP + Cephalosporium sp.	30	109.8	0.44	1.74	5.8
	90	118.5	1.16	8.36	5.3
	120	112.7	1.00	1.74	5.2
NP + Cephalosporium sp. + marble	30	83.8	0.46	2.75	5.6
	90	112.7	1.22	7.53	5.7
	120	115.6	1.04	2.72	5.7
LSD (p = 0.05)		3.1	0.04	0.03	

[a] Reducing capacity determined with $K_2Cr_2O_7$.
[c] Exchangeable aluminum extractable in $Ba(CH_3COOH)_2$ solution.

Table 11.3   Chemical Characteristics of Sawdust Composts

Treatments	Composting Period (days)	CEC cmol. kg^{-1}	RC[a] cmol. kg^{-1}	RC[b] cmol. kg^{-1}	Al$_e$[c] cmol. kg^{-1}	pH (H$_2$O)
NP	60	14.4	0.00	1.43	0.00	—
	120	14.4	0.00	1.40	1.72	5.8
	180	10.1	1.84	0.00	0.00	5.9
	360	83.1	0.00	0.08	18.30	5.9
NP + Cephalosporium sp.	60	8.7	0.00	0.53	0.00	—
	120	8.7	0.00	0.50	0.00	6.0
	180	11.9	2.22	0.00	0.00	6.2
	360	83.8	—	0.33	15.82	6.0
NP + marble	60	14.4	0.00	1.40	11.72	—
	120	14.4	0.00	1.40	0.00	6.0
	180	5.8	2.20	0.00	0.00	6.1
	360	80.9	0.00	0.29	16.62	6.2
NP + marble + Cephalosporium sp.	60	2.9	0.00	1.00	—	—
	120	14.4	0.00	1.00	0.00	6.0
	180	11.6	4.53	0.00	0.00	6.4
	360	83.9	0.00	0.48	19.00	6.2
LSD (p = 0.05)		5.1	0.04	0.04	0.03	

[a] Reducing capacity determined with $K_2Cr_2O_7$.
[b] Reducing capacity determined with $KMnO_4$.
[c] Exchangeable aluminum extractable in 1M $Ba(CH_3COOH)_2$ solution.

Table 11.4 Relationships Between Al_e and CEC

Composts	Treatments	Relationships	R^2
Vine branches	NP	$y = -2.75x^2 + 10.45x - 6.6$	1
	NP + marble	$y = 0.2x^2 + 3.2x - 1.4$	1
	NP + *Cephalosporium* sp.	$y = 2.5x^2 - 7.5x + 5$	1
	NP + marble + *Cephalosporium* sp.	$y = 1.5x^2 - 1.7x + 1.6$	1
Grape prunings, husks, and seeds	NP	$y = -3.495x^2 + 14.215x - 8.67$	1
	NP + marble	$y = -2.46x^2 + 10.59 x - 6.29$	1
	NP + *Cephalosporium* sp.	$y = -6.62x^2 + 26.48x - 18.12$	1
	NP + marble + *Cephalosporium* sp.	$y = -5.095x^2 + 20.065x - 12.22$	1
Sawdust	NP	$y = 3.991x^3 - 25.18x^2 + 499.89x - 28.62$	1
	NP + marble	$y = 2.6333x^3 - 15.8x^2 + 28.967x - 15.8$	1
	NP + *Cephalosporium* sp.	$y = 0.8133x^3 + 0.98x^2 - 20.353x + 30.28$	1
	NP + marble + *Cephalosporium* sp.	$y = 2.9945x^3 - 17.005x^2 + 29.07x - 13.68$	1

The results show that the amounts of exchangeable aluminum (Al_e) in most composts increase with the increase of CEC. In the grape prunings, husks, and seed composts, CEC decreases after maturity, as well Al_e (Table 11.2). Al_e appeared to balance the negative charges together with Ca^{2+} and Mg^{2+}, other cations, positively charged radicals, and molecular species.

Differences between the vine branches and the grape prunings, husks, and seed composts were observed at the end of the maturation period. In all variants of the first composting period, the Al_e percentage of CEC was higher (5.4 to 20.4%) than in the second (1.5 to 6.4%). The sum of exchangeable magnesium and calcium (Ca^{2+} + Mg^{2+}) as a percentage of the CEC remained fairly constant. Therefore, the amount of Al_e and the moment of its appearance depended mainly on the ratio between the values of CEC and the amounts of positively charged radicals of organic constituents and molecular species obtained through the changes in organic matter composition. During the maturation period, pH increases from 5.5 at the very beginning to 6.2 to 6.5 in the vine branch composts, and decreases from 6.6 to 5.2 to 6.1 in the grape prunings, husks, and seeds.

The presence of exchangeable aluminum relates to the compost acidity. The dynamics of both Al_e and pH were linked. At pH 6 to 7, the predominating aluminum species is $Al(H_2O)_4(OH)_2^+$ (Lindsay, 1979). The changes in pH may be expressed by the equation:

$$Al(H_2O)_4(OH)_2^+ = Al(H_2O)_3(OH)_3^0 + H^+.$$

The higher rate of the increase in CEC and the higher part of Al_e of CEC during the maturation process of the vine branch composts did not influence pH (Table 11.2). But in the grape prunings, husk, and seed composts, such influence was observed, even in the marble-treated composts. The vine branch composts possess a better ability to balance H^+ and a higher reducing capacity than grape prunings, husk, and seed composts. During the oxidation-reduction reactions, H^+ was bonded. The conclusion may be that buffering ability depends on the content of the reductive compounds. The rate of proton increases with the increase of reductive compounds.

The dynamics of Al_e in sawdust composts with the presence of the reductive compounds RC^b (determined by $KMnO_4$ titration) is the following. In the compost without marble and *Cephalosporium* sp., Al_e appeared close to day 120 of composting; after that it disappeared, as did RC^b. But in the compost with *Cephalosporium* sp., Al_e was absent during the first 180 days of composting. Al_e was observed at day 60 of composting in the marble-treated composts and then it, as well as RC^b was not found. The dynamics of Al_e during the first 2 to 4 months of the maturation period of the sawdust composts is related not only to the CEC values, but also to the activity of the oxidizing reactions of more slowly oxidizable reductive compounds RC^b (than RC^a). Intermediate products were probably present, and the negatively charged radicals were compensated by the released protons and Al_e. During that period, at the moment of the analysis an equilibrium between the positively charged exchangeable cations (mainly Al_e) and the negative charges (CEC) was still not completely approached.

At the end of 1 year (maturation period), CEC and the amounts of Al_e for all treatments sharply increased, as well as the amounts of the reductive compounds RC^b. During their oxidizing, H^+ was consumed. This increase in exchangeable aluminum did not reflect the composts' acidity. The pH increased from 4.8 in substrate to 6.2 in the composts at the end of the year; this material was then more favorable for plant growth.

Relationships between Al_e and CEC of all treatments of the studied composts (Table 11.4) are polynominal type equations with high correlation coefficient ($R^2 = 1$). Therefore, the buffering ability of the composts depends more on the content of organic matter constituents, and the ratios between them, than on their treatments with powdered marble or *Cephalosporium* sp., or both powdered marble and *Cephalosporium* sp.

Jokova et al. (1997) found relationships among CEC, RC, and CO_2 production of these composts with significant correlations ($R^2 = 0.92$). They are associated with the relationship Al_e-CEC found by the present studies. The results illustrate that the dependence between Al_e (y) and CEC (x) in the sawdust composts is the best expressed. Therefore, along with CEC, RC, and CO_2 production, Al_e and pH are main characteristics of the composts during the processes of organic matter decomposition and maturation.

CONCLUSIONS

Exchangeable aluminum Al_e (extractable in $Ba(CH_3COO)_2$ solution) appears during the maturation period of composts so that the negative charges can be balanced. Al_e causes a decrease in the pH of grape prunings, husks, and seed composts. Vine branch and sawdust composts possess a better buffering ability and higher reducing capacity. Since H^+ is bonded during the oxidation-reduction reactions, the higher reductive compound content causes the higher rate of neutralized protons. CEC, RC, CO_2 production, Al_e, and pH are main characteristics of the composts during the processes of organic matter decomposition and maturation.

ACKNOWLEDGMENTS

The European Union Environmental Program financially supported this study (contract No. EV5VCT920240, supplementary agreement [Bulgaria] No. C1PDCT923039).

REFERENCES

Atanasova, G., O. Kostov, and V. Rankov, Microbiological processing on composting wastes from the wood processing industry, *Soil Science Agrochemistry*, 25:64–71, 1990.

Harada, Y. and A. Inoko, The measurement of the cation exchange capacity of composts for the estimation of the degree of maturity, *Soil Sci. Plant Nutr.*, 26:127–134, 1980.

Jokova, M., O. Kostov, and O.V. Cleemput, Cation exchange and reducing capacities as criteria for compost quality, *Biol. Agric. Hortic.*, 14:187–1997, (UK), 1997.

Kostov, O. et al., Cucumber cultivation on some wastes during their aerobic composting, *Bioresource Techn.*, 53:237–242, 1995.

Lindsay, W., *Chemical Equilibrium in Soils*, John Wiley & Sons, New York, 1979.

Mirchev, S. and M. Jokova, Studies on reduction state of soils, *Proc. 4th Bulgarian Natl. Conf. Soil Sci.*, Sofia, Bulgaria, 118–125, 1986.

Page, A.L., R.H. Miller, and D.R. Keeney (eds.), *Methods of Soil Analyses. Chemical and Microbiological Properties*, Part 2. 2nd ed., American Society of Agronomy, Inc., Soil Science Society of America, Inc., Madison, WI, 1982.

Tzvetkov, Y., Grape pruning, husks and seeds as organic fertilizers, *Agrocompass*, 8:12–13, 1993.

PART III

Monitoring and Assessment

CHAPTER **12**

Analysis and Reporting of Carbon Sequestration and Greenhouse Gases for Conservation Districts in Iowa

John Brenner, Keith Paustian, George Bluhm, Kendrick Killian, Jan Cipra, Brian Dudek, Steve Williams, and Timothy Kautza

CONTENTS

INTRODUCTION

Land managers recognize the importance of soil organic carbon (SOC) in maintaining the productivity of the land; soil organic matter maintenance is a major basis for management recommendations for soil conservation. In recent years, interest in maintaining and enhancing soil organic matter stocks has expanded to include mitigating greenhouse gas emissions via carbon sequestration. Several estimates have been made of the potential for carbon sequestration in cropland soils for the U.S. (Lal et al., 1998, 1999; Bruce et al., 1999; Follett, 2001; Sperow et al., 2001) as well as globally (Paustian et al., 1998; IPCC, 2000). However, most such studies have been based on highly aggregated data and simplified assumptions more suited for large-scale, first-order assessments.

Analysis of current and potential carbon sequestration at state and county levels can provide more detailed information to address needs for planning and implementation of greenhouse gas mitigation strategies tailored to local conditions of climate, soils, and management practices. Current research is taking place in Iowa and other states to provide information to state planners and local land managers using a variety of spatial databases and models (Paustian et al., Chapter 18 this volume). In order to obtain local land use information for the assessment of carbon sequestration and to develop a grass-roots involvement of local land managers, a survey methodology called the carbon sequestration rural appraisal (CSRA) was developed.

The CSRA provides a mechanism to acquire data on a variety of management factors that influence soil carbon dynamics, including cropping and tillage histories, drainage history, irrigation installation and conservation practices that have been applied to the land. Much of this information is not available in any existing databases held by federal or state agencies. In addition, the survey was designed to facilitate reporting of greenhouse gas emissions and mitigation to the U.S. Department of Energy (DOE) greenhouse gases database (DOE, 1992). DOE provides a method for local forms of government (i.e., conservation districts) to report sources and sinks of greenhouse gases and the conservation districts in Iowa have agreed to report to DOE the amounts of carbon (C) sequestered by installation of agricultural conservation practices. All 99 counties in Iowa, represented by their conservation districts, are involved in the project. Preliminary results on land management changes and C sequestration estimates for various conservation practices (e.g., CRP and grass conversions and no till) for three Iowa counties (Adair, Fayette, and Hardin) are presented here.

MATERIALS AND METHODS

Initial Contacts and Expectations

Representatives from the Iowa Department of Natural Resources, Iowa Conservation Districts, Colorado State University Natural Resource Ecology Laboratory (NREL), and USDA Natural Resources Conservation Service (NRCS) met in 1998 to discuss agriculture and greenhouse gases and the potential for Iowa to sequester C in agricultural soils. The Iowa C Storage Project was developed to provide education on C sequestration and collect land use and management data not available in the literature or in existing databases. All 99 counties are involved in the project and local conservation districts are providing data for the CSRA. Estimates of C sequestration for current management systems in each county and for potential carbon sequestration following changes in management practices were to be provided to the conservation districts. The conservation districts would in turn be responsible for reporting to DOE on the CO_2 removed from the atmosphere due to implementation of conservation practices.

Carbon Sequestration Rural Appraisal (CSRA)

After discussion with the Iowa conservation partners, it was determined that some method to educate local land managers and planners and to collect county-specific data on land use and management was needed to complete the state- and county-level soil C appraisals. A five-county pilot study was conducted to determine availability of local data and willingness of conservation districts to provide data and to refine the process of collecting local data for the entire state, i.e., all 99 counties. Five counties, Adair, Clay, Fayette, Hardin, and Wapello, participated in the pilot testing during 1998. These counties are diverse in their agricultural production systems, ranging from predominantly row crops grown on drained prairie soils in Hardin County to the row crop-grass rotations on sloping lands in Wapello County. Their respective locations in Iowa provide a good mix of north-to-south and east-to-west regional differences.

Information and suggestions by the conservation districts and NRCS personnel in the pilot counties provided the basis for refining the initial design of the CSRA. Table 12.1 details the general types of data provided by the conservation districts through the use of the CSRA. The CSRA sheets

Table 12.1   Types of Data Provided by the CSRA

Title	Description
Current land use information	Land use by soil map unit
General land use information (drainage)	Installation by soil map unit
General land use information (irrigation)	Installation by soil map unit
County level farming histories	Cropping, fertilizer and tillage practices
Annual conservation practices	Conservation practices installed

(Appendix 12.1) were supported with background information from published databases including current land use and land cover obtained from the Geographic Analysis Program for Biodiversity, 1991–1992 (GAP) records (Scott and Jennings, 1997), the General Land Office (GLO) surveys of presettlement vegetation (Iowa Department of Natural Resources, 1996), the Soil Geographic (STATSGO) Database (SCS, 1994), Conservation Technology Information Center tillage database (CTIC, 1998) and the National Agricultural Statistics Service (NASS, 1999). For each conservation district, maps of current and historical (presettlement) land cover, overlain with STATSGO soil map units, were provided. These maps were designed to help in estimating the joint distribution of management systems by major soil types within a county.

Interactive training to all conservation districts was provided in early 1999. A 2-hour training period covered material on greenhouse gases, agricultural C sequestration, and the CSRA. All the forms were discussed and information was provided on how to complete and submit the appraisal. The overall process in designing and implementing the CSRA is shown in Figure 12.2.

The information on land use and management practices collected with the CSRA was used as input to the Century Model (Parton et al., 1987; Metherall et al., 1993) to estimate soil C dynamics in each county. The C modeling methodology is described in detail in Paustian et al. (see Chapter 18, this volume). Briefly, simulations were run for baseline histories up to 1974 and then, for each county, a suite of dominant cropping and tillage systems (including conversions to conservation reserve program (CRP) grass planting, starting in 1984) were run for a 20-year period (Table 12.2). Each management system was simulated for all major soil types in the county, defined by soil texture and hydric or nonhydric characteristics. To evaluate responses to other changes in management, the model was run for an additional 20-year period (1994 to 2014), as two 10-year blocks, for all possible transitions between the management systems (Table 12.2), including no changes in management. Rates of soil C change are reported as mean annual rates for 10-year periods.

DOE Reporting

Voluntary reporting of greenhouse gases through the DOE Energy Information Administration 1605-b process was used to allow the conservation districts to report the benefits of applying conservation practices in agriculture. The conservation practices installed were provided by the CSRA, CTIC, and NRCS state specialists. Each conservation district would report CO_2 removed from the atmosphere for its respective county. This information was sent to the National Association of Conservation Districts (NACD) to compile all the counties and submit a statewide summary for Iowa.

RESULTS AND DISCUSSION

The five county pilot studies were completed in early 1999. The spreadsheet version of the CSRA was developed and distributed to each county in Iowa by spring of 1999. All the conservation districts received educational and training material necessary to complete the CSRA through the use of interactive training in early 1999. This allowed conservation districts and field NRCS people to receive training in a uniform and time-efficient manner. It also allowed participants to ask

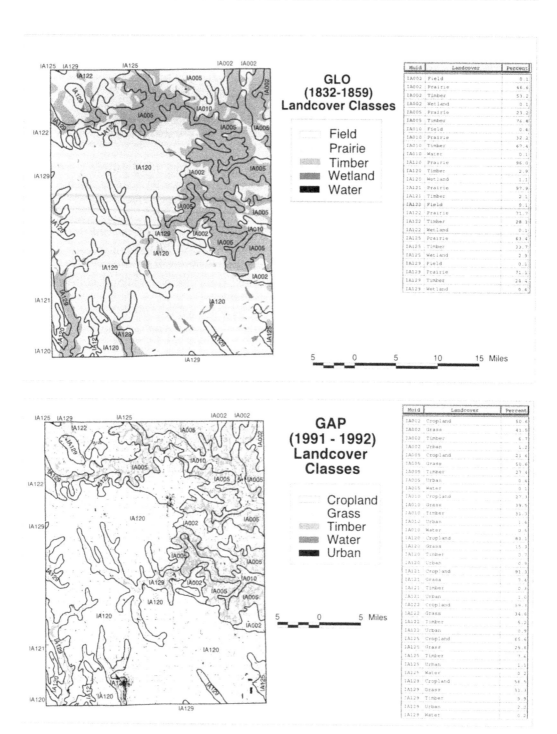

Figure 12.1 Fayette County, Iowa, land cover overlain with STATSGO soil map units.

questions about C sequestration issues and how the CSRA forms were to be completed. Conservation districts and local NRCS offices completed the CSRA over the summer of 1999 with 100% of the counties participating and returning completed CSRAs to NREL.

Since this was the first attempt to collect local land use and other related data for estimating soil C sequestration on agricultural lands, the conservation partners in Iowa felt that some form of

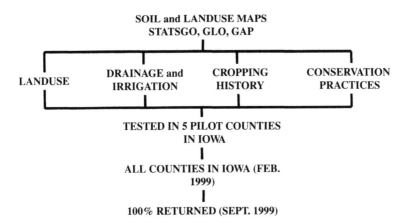

Figure 12.2 Flow chart of CSRA process.

Table 12.2   Crop Rotations and Tillage Interactions Modeled

Scenario	Description
1	Continuous corn, intensive tillage
2	Continuous corn, moderate tillage
3	Continuous corn, no tillage
4	Continuous corn, intensive tillage to CRP (100% grass), no tillage
5	Continuous corn, intensive tillage to CRP (50% legume, 50% grass), no tillage
6	Corn-bean, intensive tillage
7	Corn-bean, moderate tillage
8	Corn-bean, no tillage
9	Corn-bean, intensive tillage to CRP (100% grass), no tillage
10	Corn-bean, intensive tillage to CRP (50% legume, 50% grass), no tillage
11	Corn–bean–oat–alfalfa–alfalfa, intensive tillage
12	Corn–bean–oat–alfalfa–alfalfa, moderate tillage
13	Corn–bean–oat–alfalfa–alfalfa, no tillage
14	Corn–bean–oat–alfalfa–alfalfa, intensive tillage to CRP (100% grass), no tillage
15	Corn–bean–oat–alfalfa–alfalfa, intensive tillage to CRP (50% legume, 50% grass), no tillage
16	Oat–alfalfa, intensive tillage
17	Oat–alfalfa, moderate tillage
18	Oat–alfalfa, no tillage
19	Oat–alfalfa, intensive tillage to CRP (100% grass), no tillage
20	Oat–alfalfa, intensive tillage to CRP (50% legume, 50% grass), no tillage

feedback was necessary from the local participants. Exit questions were developed and sent to all the conservation districts in Iowa asking for feedback about the three parts of the project, including the educational process, the importance of the information to local land managers, and specific questions about the CSRA. Response from the conservation districts was very positive for the education component and the importance to local land managers. The specific questions about the CSRA were also positive, with 73% of the districts responding. The remarks and evaluations show that the CSRA is capable of capturing data at the local level and also of increasing awareness of the C sequestration issue in agriculture (DOE, 1999).

The management data collected using the CSRA, together with other spatial data sets defining soil and climate variables, provided the input to century model simulations of past and current soil C changes (see Paustian et al., Chapter 18, this volume) for management histories and current

Table 12.3   Annual Rates of Soil C Change for a Nonhydric SiCL Soil Texture (0 to 20 cm)

	Year 1–10 after Management Change			Year 11–20 after Management Change		
County	Intensive Tillage (kg ha⁻¹ yr⁻¹)	No Tillage (kg ha⁻¹ yr⁻¹)	CRP (kg ha⁻¹ yr⁻¹)	Intensive Tillage (kg ha⁻¹ yr⁻¹)	No Tillage (kg ha⁻¹ yr⁻¹)	CRP (kg ha⁻¹ yr⁻¹)
Adair	149	625	945	72	350	438
Fayette	132	648	940	63	361	460
Hardin	130	630	966	55	346	454

management systems identified in the county and conservation districts (Table 12.2). The results for selected management systems for three of the five pilot counties are described below.

Soil carbon levels in corn–soybean intensive tillage systems in Adair County (scenario 6 in Table 12.2) on a silty clay loam (SiCL) nonhydric soil were predicted to be increasing slowly (149 kg ha⁻¹ yr⁻¹) for the present period, 1994 to 2004 (Table 12.3). With no changes in management, C sequestration rates decline to 72 kg ha⁻¹ yr⁻¹ for the time period of 2004 to 2014, suggesting that this system is near a steady-state condition. The small increase in soil C, even under the intensive tillage scenario, is due to the trend of increasing crop productivity and crop residue inputs over the past several decades, leading to a recovery of soils depleted of C through past management. It should be pointed out that potential effects of erosion on soil C levels are not included in the analysis.

A comparison of the same system, but changing from intensive tillage to no-tillage system in 1994, shows C increases of 625 kg ha⁻¹ yr⁻¹ from 1994 to 2004, with further increases of 350 kg ha⁻¹ yr⁻¹ during a subsequent 10-year period (2004 to 2014), when maintained under no-till management (Table 12.3). Conversion from intensive tillage to a perennial grass system in 1985 (scenario 9 in Table 12.2) shows an annual C increase of 945 kg ha⁻¹ yr⁻¹ from 1984 to 1994 (Table 12.3). This system also shows increases in C of 438 kg ha⁻¹ yr⁻¹ from 1994 to 2004, which is less than the rate during the first 10 years after making the management change, but suggests that land enrolled in CRP and left in perennial grass continues to sequester C even after 20 years.

The same scenario of a corn–soybean intensive tillage system in Adair County (scenario 6 in Table 12.2), but on a hydric soil, shows the system losing C (–44 kg ha⁻¹ yr⁻¹) annually from 1994 to 2004, reflecting the continuing effects of past soil drainage (Table 12.4). Conversion to a no-till system in 1994 is predicted to reverse the soil C trend, yielding annual increases of 517 kg C ha⁻¹ yr⁻¹ from 1994 to 2004 and 255 kg C ha⁻¹ yr⁻¹ from 2004 to 2014. Conversion from intensive tillage to a perennial grass system in 1985 (scenario 9 in Table 12.2) shows an annual C increase of 1001 kg ha⁻¹ yr⁻¹ from 1984 to 1994 and increases in C of 403 kg ha⁻¹ yr⁻¹ from 1994 to 2004 (Table 12.4). Results for these scenarios involving corn–soybean rotation for nonhydric and hydric SiCL soils are shown for each of the three counties (Tables 12.3 and 12.4).

Estimates of 1998 rates of C sequestration for each of the three counties are given in Table 12.5. The county totals combine the areas under each of the management systems defined by the CSRA (Table 12.2), including CRP, as well as area in other grass conservation practices (grass waterways, terraces, grass seeding, etc.) installed between 1990 and 1994. In addition, small areas involving

Table 12.4   Annual Rates of Soil C Change for a Hydric SiCL Soil Texture (0 to 20 cm)

	Year 1–10 after Management Change			Year 11–20 after Management Change		
County	Intensive Tillage (kg ha⁻¹ yr⁻¹)	No Tillage (kg ha⁻¹ yr⁻¹)	CRP (kg ha⁻¹ yr⁻¹)	Intensive Tillage (kg ha⁻¹ yr⁻¹)	No Tillage (kg ha⁻¹ yr⁻¹)	CRP (kg ha⁻¹ yr⁻¹)
Adair	–44	517	1001	–54	255	403
Fayette	–68	535	1013	–72	263	435
Hardin	–36	491	1032	–46	254	447

Table 12.5 Estimated C Sequestered and CO$_2$ Removed by Conservation Practices for 1998 in Million Metric Tonnes (MMT)

County	C Sequestered (MMT)	CO$_2$ Removed (MMT)
Adair	36,009	132,154
Fayette	55,522	203,764
Hardin	20,402	74,874

biomass C sequestered through the installation of trees (500 kg ha^{-1} yr^{-1}) and C sequestered through the installation of wetlands (250 kg ha^{-1} yr^{-1}) were not simulated by the model, but added to the total (based on per-area rates reported by Lal et. al., 1998). For reporting to DOE, all values were converted to units of CO$_2$ removed from the atmosphere by multiplying by 3.67. The wide range of totals between counties reflects differences in rates of C sequestration due to climate, soil, and management factors as well as the areas of land under different management systems and conservation practices. The total CO$_2$ estimated to have been removed from the atmosphere in 1998 by agricultural soils for these three counties was about 400,000 t.

CONCLUSIONS

Data provided by the conservation districts in Iowa and their use in C sequestration estimates support the following four conclusions:

1. Local participants were interested and willing to provide data for the assessment process, with 100% of the Iowa conservation districts compiling and reporting their data.
2. The CSRA was useful in compiling data on management practices of importance to soil C dynamics not readily obtainable from other sources. These data included:
   - Past and present land use by MUID
   - Drainage and irrigation histories by MUID
   - Past and present cropping and tillage histories
   - Installation of conservation practices by MUID
3. Iowa conservation districts are willing to report the C sequestered by installation of conservation practices to the U.S. DOE through the EIA-1605b reporting procedures.
4. Current agricultural practices are providing a significant sink for C, on the order of 400,000 t of CO$_2$ per year in these three counties in Iowa.

ACKNOWLEDGMENTS

This project is funded by grants from U.S. Department of Energy and USDA/Natural Resource Conservation Service. The authors also recognize the Iowa NRCS, Iowa Department of Natural Resources, Iowa conservation districts, and the National Association of Conservation Districts for their important role in the project.

REFERENCES

Bruce, J.P. et al., Carbon sequestration in soils, *J. Soil Water Conserv.*, 54:382–389, 1999.
CTIC, Crop residue management executive summary. Conservation Technology Information Center. Data available on-line at http://www.ctic.purdue.edu/Core4/CT/CT.html, 1998.
DOE, Title XVI of the Energy Policy Act of 1992, Section 1605(b), U.S. Department of Energy, 1992.
DOE, Iowa Carbon Storage Project, Interim report, U.S. Department of Energy, 1999.

Follett, R.F., Soil management concepts and carbon sequestration in cropland soils (ISTRO Keynote paper — Fort Worth, TX, July 3–5, 2000), *Soil and Tillage Research* (In Press), 2001.

Iowa Department of Natural Resources, Iowa Government Land Office Vegetation 1832–1859, Iowa Department of Natural Resources, Parks and Preserves Division, Des Moines, IA, 1996.

IPCC, *Land Use, Land-Use Change and Forestry*, Special Report of the Intergovernmental Panel on Climate Change (IPCC), Cambridge University Press, Cambridge, U.K., 2000.

Lal, R. et al., (eds.), *The Potential of U.S. Cropland to Sequester Carbon and Mitigate the Greenhouse Effect*, Sleeping Bear Press, Chelsea, MI, 1998.

Lal, R. et al., Managing U.S. cropland to sequester carbon in soil, *J. Soil Water Conserv.*, 54:374–381, 1999.

Metherall, A.K. et al., CENTURY SOM Model Environment, Technical Documentation, Agroecosystem Version 4.0. GPSR Unit, Technical Report No. 4. USDA-ARS. Fort Collins, CO, 1993.

NASS, Published Estimates Database. USDA-National Agricultural Statistics Service. Data available online at http://www.nass.usda.gov:81/ipedb/, 1999.

Parton, W.J. et al., Analysis of factors controlling soil organic matter levels in Great Plains grasslands, *Soil Sci. Soc. Am. J.*, 51:1173–1179, 1987.

Paustian, K. et al., CO_2 mitigation by agriculture: an overview, *Climatic Change*, 40:135–162, 1998.

Paustian, K., J. Brenner, K. Killian, J. Cipra, S. Williams, T. Kautza, and G. Bluhm, State-level analyses of C sequestration in agricultural soils, chap. 18, this volume.

Scott, J.M. and M.D. Jennings, A description of the national gap analysis program, available on-line at http://www.gap.uidaho.edu/gap/AboutGAP/GapDescription/Index.htm, 1997.

SCS, State Soil Geographic Data Base (STATSGO) Data User Guide. U.S. Department of Agriculture Soil Conservation Service, National Soil Survey Center, Lincoln, NE, 1994.

Sperow, M., M. Eve, and K. Paustian, Potential soil C sequestration on U.S. agricultural soils (submitted, 2001).

Appendix 12.1a. CSRA Current Land Use Information

CARBON SEQUESTRATION RURAL APPRAISAL

CURRENT LAND USE INFORMATION FROM LOCAL KNOWLEDGE (SHEET A)

STATE ___IOWA___   COUNTY _____

FOR INDICATED SOILS ON MAP DETERMINE:
MUID (STATSGO ASSOCIATION

LAND USE INFORMATION

TOTAL CROPLAND
  CLASS I & II
  CLASS II & IV
  CLASS V & VI
FOREST OR TREES
GRASS LANDS
WATER/WETLANDS
URBAN/OTHER
TOTAL   0.0%   0.0%   0.0%   0.0%   0.0%   0.0%   0.0%   0.0%

LANDSCAPE DESCRIPTION
FLAT
ROLLING HILLS
STEEP HILLS
FLOOD PLAIN
OTHER
TOTAL   0.0%   0.0%   0.0%   0.0%   0.0%   0.0%   0.0%   0.0%

TOTAL CROPLAND: % OF THIS SOIL IDENTIFIED AS CROPLAND. CLASS I & II, III & IV, AND V & VI MUST ADD TO THIS %.
  CLASS I & II: % OF THIS SOIL THAT IS CLASS I & II CROPLAND.
  CLASS III & IV: % OF THIS SOIL THAT IS CLASS III & IV CROPLAND.
  CLASS V & VI: % OF THIS SOIL THAT IS CLASS V & VI CROPLAND.
FOREST OR TREES: % OF THIS SOILD IDENTIFIED AS FOREST OR TREES.
GRASS LANDS: % OF THIS SOIL IDENTIFIED AS GRASS LANDS.
WATER/WETLANDS: % OF THIS SOIL IDENTIFIED AS WETLANDS.
URBAN/OTHER LANDS: % OF THIS SOIL IDENTIFIED AS OTHER LANDS INCLUDING URBAN, DEVELOPED AND ABANDONED LANDS.
LANDSCAPE DESCRIPTION: % OF THIS SOIL IN EACH LANDSCAPE DESCRIPTION.

Appendix 12.1b. CSRA Drainage Information

CARBON SEQUESTRATION RURAL APPRAISAL

GENERAL LAND USE INFORMATION FROM LOCAL KNOWLEDGE (SHEET B)

STATE IOWA   COUNTY _____

HAS ANY PART OF THE COUNTY BEEN DRAINED (YES/NO):   _____

IF YES, ANSWER THE FOLLOWING.

MUID	OPEN DITCH DRAINAGE			TILE DRAINAGE		
	TIME PERIOD OF INSTALLATION	% OF SOIL DRAINED		TIME PERIOD OF INSTALLATION	% OF SOIL DRAINED	
____	_____	_____		_____	_____	
____	_____	_____		_____	_____	
____	_____	_____		_____	_____	
____	_____	_____		_____	_____	
____	_____	_____		_____	_____	
____	_____	_____		_____	_____	
____	_____	_____		_____	_____	
____	_____	_____		_____	_____	
____	_____	_____		_____	_____	
____	_____	_____		_____	_____	
____	_____	_____		_____	_____	

MUID: SOIL MAP UNIT ID FROM STATSGO. (FROM MAP)
TIME PERIOD OF INSTALLATION: GIVE THE TIME PERIOD WHEN DRAINAGE PRACTICES
WERE INSTALLED. (i.e., 1930–1950, 1940–1960, 1970–1990, ETC.)
% OF SOIL DRAINED: GIVE AN ESTIMATE FOR THESE SOILS OF THE AMOUNT OF DRAINAGE
INSTALLATION.

Appendix 12.1c. CSRA Irrigation Information

CARBON SEQUESTRATION RURAL APPRAISAL

GENERAL LAND USE INFORMATION FROM LOCAL KNOWLEDGE (SHEET C)

STATE <u>IOWA</u>   COUNTY _____

IS 10% OR MORE OF ANY MUID IRRIGATED (YES/NO):   _____

IF YES, ANSWER THE FOLLOWING.

MUID	TIME PERIOD OF INSTALLATION	% OF SOIL IRRIGATED	ANNUAL AMOUNT APPLIED (INCHES)	TYPES OF SYSTEMS
___	_____	_____	_____	_____
___	_____	_____	_____	_____
___	_____	_____	_____	_____
___	_____	_____	_____	_____
___	_____	_____	_____	_____
___	_____	_____	_____	_____
___	_____	_____	_____	_____
___	_____	_____	_____	_____
___	_____	_____	_____	_____
___	_____	_____	_____	_____
___	_____	_____	_____	_____

MUID: SOIL MAP UNIT ID FROM STATSGO. (FROM MAP)
TIME PERIOD OF INSTALLATION: GIVE THE TIME PERIOD WHEN IRRIGATION PRACTICES WERE INSTALLED.
% OF SOIL IRRIGATED: GIVE AN ESTIMATE FOR THESE SOILSOF THE AMOUNT OF IRRIGATION INSTALLED.
ANNUAL AMOUNT APPLIED (INCHES): GIVE AN ESTIMATE OF THE ANNUAL AMOUNT OF IRRIGATION WATER APPLIED IN INCHES. (6 INCHES, 12 INCHES, 15 INCHES, ETC.)
TYPES OF SYSTEMS: TYPICAL TYPE OF IRRIGATION SYSTEM INSTALLED. (CENTER PIVOT, GATED PIPE, ETC.)

Appendix 12.1d. CSRA Cropping and Management Information

CARBON SEQUESTRATION RURAL APPRAISAL

COUNTY LEVEL FARMING AND CROPPING SYSTEM HISTORY FROM PRE 1900 TO PRESENT (SHEET D)

STATE IOWA _____   COUNTY   _____

TIME FRAME   _____

% ESTIMATE OF COUNTY BEING FARMED DURING THIS TIME FRAME: _____

CROP ROTATIONS (SPECIFY 1 TO 3)

1) _____

2) _____

3) _____

FOR INDICATED CROPS

CROP NAME

YIELD (BU OR TONS/AC)

N FERT APPLIED (LBS/AC)

MANURE APPLIED (TONS/AC)

TYPICAL TILLAGE OPERATIONS

Comments:

TIME FRAME: PERIOD OF TIME AS SPECIFIED.
% ESTIMATE OF COUNTY BEING FARMED DURING THIS TIME FRAME: GIVE AN ESTIMATE OF THE COUNTY AREA BEING FARMED DURING THIS TIME FRAME.
TYPICAL CROP ROTATION: CROP ROTATIONS INCLUDE (CORN-CORN; CORN-SOYBEAN; CORN-CORN-OATS-MEADOW-MEADOW; CORN-SOYBEAN-CORN-OATS-MEADOW-MEADOW; ETC.)
FOR INDICATED CROPS: ACTUAL CROP INFORMATION FOR THE INDICATED CROPS IN THE ROATIONS.
CROP: CROP NAME AS SHOWN IN CROP ROTATION.
YIELD: CROP YIELD IN BU/AC FOR GRAINS OR TONS/AC FOR HAY.
N FERT APPLIED: ESTIMATE OF COMMERCIAL NITROGEN FERTILIZER APPLIED ANNUALLY (LBS/AC).
MANURE APPLIED: ESTIMATE OF MANURE APPLIED ANNUALLY (TONS/AC), BY CROP.
TYPICAL TILLAGE OPERATIONS: TYPICAL TILLAGE OPERATIONS USED TO GROW THIS CROP.
(EXAMPLES ARE FALL PLOW; SPRING PLOW; CHIESEL PLOW; DISK; HARROW; CULTIVATOR; DRILL; PLANT; ETC.)

Appendix 12.1e. CSRA Annual Conservation Practice Installed

ANNUAL CONSERVATION PRACTICES INSTALLED

PRACTICES INSTALLED BY COUNTY AND SOIL TYPE

USE IN REPORTING TO DOE FOR CARBON SEQUESTRATION
(USE SEPARATE SHEET FOR EACH SOIL MUID)

STATE IOWA _____   COUNTY _____   MUID _____

| | ACRES OF CONSERVATION PRACTICES INSTALLED (ACRES0 | | | | | | |
| | COMMON CROP ROTATION ($) | | | | GRASS CONVERSIONS | TREE PLANTING | WETLANDS CREATED AND/OR RESTORED |
	REDUCED TILLAGE	NO-TILL	REDUCED TILLAGE	NO-TILL			
1985							
1986							
1987							
1988							
1989							
1990							
1991							
1992							
1993							
1994							
1995							
1996							
1997							
1998							
1999							
2000							

MUID: SOIL MAP UNIT ID FROM STATSGO. (FROM MAP)
CROP ROTATION: PICK THE TWO MOST COMMON CROP ROTATIONS. IF ONE ROTATION IS >90%
OF CROPPED ACRES, REPORT ONLY THAT ROTATION. TOTAL FOR THE COUNTY SHOULD EQUAL
THE CTIC REPORTED VALUES FROM 1989 TO PRESENT. SEE SUPPLEMENTAL INFORMATION.
REDUCED TILLAGE: REDUCED TILLAGE FARMING WHICH LEAVE GREATER THAN 15% RESIDUE
AFTER PLANTING. (INCLUDES MULCH TILL, RIDGE TILL BUT NOT NO-TILL).
NO-TILL: NO-TILL FARMING SYSTEM.
GRASS CONVERSIONS: ALL GRASS PLANTING CONSERVATION PRACTICES.
(WATERWAYS, BUFFERS INCLUDING RIPIARIAN BUFFERS, FILTER STRIPS, TERRACES, CRP).
USE 12' WIDTH FOR TERRACES (LF* 12/43560=ACRE).
USE 40' WIDTH FOR ALL OTHER PRACTICES REPORTED IN LINEAR FEET (LF* 40/43560=ACRE).
TREE PLANTING: ALL CONSERVATION PRACTICES THAT INCLUDE TREE PLANTINGS.
(WINDBREAKS, SHELTERBELTS, AGRO-FORESTRY).
WETLANDS CREATED AND/OR RESTORED: ALL CONSERVATION PRACTICES THAT INCLUDE THE
CREATION OR RESTORATION OF WETLANDS.

CHAPTER **13**

Comparing Estimates of Regional Carbon Sequestration Potential Using Geographical Information Systems, Dynamic Soil Organic Matter Models, and Simple Relationships

P. D. Falloon, P. Smith, J. Szabó, László Pásztor, J. U. Smith, K. Coleman, and S. J. Marshall

CONTENTS

ABSTRACT

In Europe, carbon sequestration potential has previously been estimated using data from the Global Change and Terrestrial Ecosystems Soil Organic Matter Network (GCTE-SOMNET). Linear relationships between management practices and yearly changes in soil organic carbon were developed and used to estimate changes in the total carbon stock of European soils. To refine these semiquantitative estimates, local soil type, meteorological conditions, and land use must also be taken into account. We have previously used the Rothamsted carbon model (RothC) linked to geographical information systems (GIS) to estimate the potential effects of afforestation on soil carbon stocks in central Hungary. Further developments have involved a combined modeling approach. The approach described here is based on the CENTURY model frame and allows either the RothC or CENTURY SOM decomposition model to be used, thus allowing an equal comparison of the models. The GIS-linked system integrates land-use, soil, and weather data with knowledge of land-use history, net primary production, local agricultural practices, and best estimates of current SOC stock.

This chapter describes how these developments have been used to estimate carbon sequestration at the regional level using a dynamic simulation model linked to spatially explicit data. Results of carbon sequestration potential estimated with this system are compared to those obtained using a simple regression-based approach. The system is demonstrated in use for an area of central Hungary.

INTRODUCTION

Soil organic matter (SOM) represents a major pool of carbon within the biosphere, estimated at about 1500 to 1550 Pg globally (Batjes, 1996), roughly twice the size of the atmospheric CO_2 pool. The soil can act as a source and a sink for carbon and nutrients. Changes in agricultural land use and climate can lead to changes in the amount of carbon held in soils, thus influencing the fluxes of CO_2 to and from the atmosphere. Some agricultural management practices will give rise to a net sequestration of carbon in the soil, others may release soil carbon to the atmosphere. Regional estimates of the carbon sequestration potential of these practices are crucial if policymakers are to plan future land uses with the aim of reducing national CO_2 emissions.

The three methods previously used to estimate changes in regional soil organic carbon (SOC) stocks are outlined below. First, Smith et al. (1997a, b, 1998b, 2000a, b), and Gupta and Rao (1994) used regressions of changes in SOC stocks under changes in land use or management based on long-term experimental data to calculate annual percent increases in SOC, and extrapolate future changes in SOC stocks under different scenarios. Second, Kern and Johnson's (1993) approach used regressions based on SOC changes in long-term tillage experiments, but applied their relationships to spatial soil databases, allowing them to take the local soil characteristics into account. The main limitation of approaches based upon linear regressions is that they assume a constant rate of SOC change throughout the time period of the scenario under study. There is almost certainly a period following a change in land use or management when the annual change in SOC is more rapid, followed by a period of much more gradual change (see, for example, the data sets modeled by Coleman et al., 1996).

Third, former studies have used dynamic models linked to spatial databases or geographic information systems (GIS). Examples are given by Williams and Renard (1985), Prentice and Fung (1990), Jenkinson et al. (1991), Lee et al. (1993), Donigian et al. (1994), Parshotam et al. (1995), Post et al. (1996), King et al. (1997), Paustian et al. (1997), Falloon et al. (1998a, 1999), and Smith et al. (1999). This approach allows dynamic estimates, and for local soil characteristics, meteorological conditions, and land use to be taken into account, is amenable to a range of scenarios, and permits geographic areas of particular C sequestration potential to be identified.

However, while SOM models have been compared and evaluated at the site scale (Smith et al., 1998a), there have been few previous SOM model comparisons at the regional scale (Falloon et al., 1999). The objective of this chapter is to compare and improve upon methods used for estimating changes in regional SOC stocks.

The Rothamsted carbon model (RothC; Coleman and Jenkinson, 1996), used previously, has been linked to a geographical information system (GIS) to estimate the potential effects of afforestation on soil carbon stocks in central Hungary (Falloon et al., 1998a). Further developments

(Falloon et al., 1999) have involved using RothC and the CENTURY model (Parton et al., 1987, 1988). The GIS-linked system integrates land-use, soil, and weather data with land-use history, local agricultural practices and best estimates of the current SOC stock. In this study, an area of central Hungary is used to compare model estimates of current C stocks obtained using RothC and CENTURY, and to compare estimates of carbon sequestration potential obtained using this system to those provided by a simple regression-based approach, for a simple afforestation scenario. RothC and CENTURY parameters were set using data from SOMNET (Smith et al., 1996a, b, c; http://www.iacr.bbsrc.ac.uk/res/depts/soils/somnet/tsomnet.html), a global network of long-term experiments and SOM models. Recent changes in land use and management, and the carbon balance for Hungary as a whole, are described by Németh et al. (Chapter 43, this volume).

METHODS AND MATERIALS

Study Area and GIS

The chosen study area (24,804 km^2: 26.7% of the Hungarian land area) was the same as that used by Falloon et al. (1998a). Coordinate details are given in Table 13.1. The GIS platforms used were ArcView and ArcINFO. Details on average soil characteristics, weather variations, and land use are given in the next sections.

Table 13.1 Study Area

	Lower Left of Window		Upper Right of Window	
	X	Y	X	Y
UTM[a] (34)	5183816.64	325048.59	5340133.61	472979.10
Spherical system	46.784	18.708	48.213	20.636
Hungarian EOV[b]	160000	624000	320000	768000

[a] Universal terrain model.
[b] Uniform national projection of Hungary.

Models

The two dynamic SOM models used in this study (RothC and CENTURY) have both been described in detail elsewhere (Coleman et al., 1997; Parton et al., 1987, 1988). Major input variables for the models are shown in Table 13.2. The RothC model is a SOM decomposition model that splits incoming plant residues into two compartments with different decomposition rates, decomposable plant material (DPM) and resistant plant material (RPM). These both decompose to form biomass (BIO), humified organic matter (HUM), and evolved CO_2. Additionally, the model structure includes an inert pool of organic matter (IOM), which allows accurate simulation of the radiocarbon age of soil and SOC dynamics under changes in land use and management. Without the IOM pool, it is impossible to balance the great radiocarbon age of soils or to model certain changes in land use or management.

With the exception of IOM, each compartment decomposes by first-order kinetics, and each has an intrinsic maximum decomposition rate. The actual rate of decomposition is determined using modifiers for soil moisture, temperature, and plant cover operating on the maximum rate. The clay content of the soil affects the apportioning of SOM between the evolved CO_2, the BIO, and the HUM pools (Coleman et al., 1996). The CENTURY model is a general ecosystem model with a multicompartmental SOM decomposition model. The CENTURY SOM model splits incoming residues between structural and resistant plant material, which then decompose to form three pools termed "active," "slow," and "passive" organic matter, along with evolved CO_2. As in RothC, each

Table 13.2   Major Input Variables for the RothC and CENTURY Models

	RothC	CENTURY
Soil variables	Clay content	Sand content
	Inert organic matter content	Silt content
	SOC content	Clay content
	Bulk density	Bulk density
		SOC content
Weather variables	Total monthly precipitation	Total monthly precipitation
	Mean monthly temperature	Mean monthly maximum temperature
	Total monthly evaporation	Mean monthly minimum temperature
Management variables	Residue quality factor (DPM/RPM ratio)	Residue lignin/N ratio
	Soil cover	Plant C and N content
	Residue C input	Atmospheric N deposition
	Manure C inputs	

compartment has an intrinsic maximum decomposition rate, modified by soil water and temperature, with soil clay content affecting SOM transfer to the passive SOM pool.

A regression approach (Smith et al., 1997a, b, 1998b) was also applied, based on the only two woodland regeneration experiments in Europe to estimate SOC changes under afforestation of arable land.

Soils

Soil data were taken from the 1:500,000 Hungarian HunSOTER database, a country-scale database with an integrated hierarchy of point and polygon layers (Pásztor et al., 1996; Szabó et al., 1996). The pilot study window contained 275 representative soil profiles from a data set of 1361 for the whole of Hungary. The FAO soil types in this region are shown in Figure 13.1. The profiles were originally sampled in 1992; only data for the uppermost horizons were used in this study. The application of the data set described here differs from that of Falloon et al. (1998a) in the calculation of SOC stocks and in the treatment of missing data. In this study, SOC stocks were calculated to a standard depth of 20 cm (rather than the 30 cm used by Falloon et al., 1998a). This depth was used because the CENTURY model simulates the top 20 cm of the soil profile. The soil variables used (and their ranges) from this data set were organic carbon content (SOC, 0.3 to 2.3%), clay content (2 to 48%), and bulk density (1.21 to 1.70 Mg m^{-3}).

The profile data were linked to the 351 SOTER unit polygons (representing areas with unique soil, land-form, and lithology characteristics) for the window. These data were used to calculate the soil IOM content for the RothC model in the absence of radiocarbon data, using the method of Falloon et al. (1998b), and SOC to 20 cm depth. Where bulk density data were missing, bulk density was calculated from the SOC value (Howard et al. 1995) and SOC stocks to 20 cm depth, then calculated from depth, bulk density, and SOC percentage. Soil units with no SOC data were excluded. This treatment of missing data differs from former studies, where missing data were replaced with a mean over the whole data set (Falloon et al., 1998a).

Land Use

Land-use data were taken from the CORINE database for Hungary (scale 1:100,000; Büttner et al., 1995a, b; Büttner, 1997), for the representative window, with 6470 polygons. Because the original data set contained multipart polygons, the land-use polygon data were exploded in ArcView using a custom script, giving a total of 7529 polygons. The 44 original land-use codes were rationalized into four codes: (1) arable (six original codes, 21% of total land use); (2) grassland (eight original codes, 8.7% of total land use); (3) forest (six original codes, 20.6% of total land use); and (4) other uses (including marsh, water bodies, urban areas, and so on; 24 original codes, 9.7% of total land use).

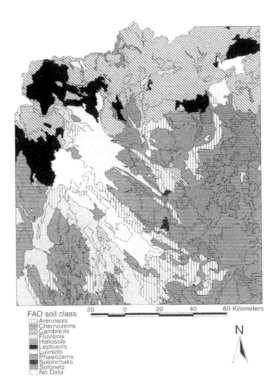

FAO soil class
Arenosols
Chernozems
Cambisols
Fluvisols
Histosols
Leptosols
Luvisols
Phaeozems
Solonchaks
Solonetz
No Data

Figure 13.1 FAO soil types in the study area of central Hungary, from the HunSOTER database.

Meteorology

In an earlier study (Falloon et al., 1998a), long-term averaged mean monthly temperature (14 stations), rainfall (17 stations), and evaporation (6 stations) data from 1931 to 1960 were used (Varga–Haszonits, 1977). For this study, 20 points from the preinterpolated 1961–1990 gridded surface climatology for Europe (Hulme et al., 1995) were used. This data set provided mean monthly temperature and rainfall data. The six stations from Varga–Haszonits (1977) were used to provide evaporation data. Mean annual temperature ranged from 9 to 10°C over the area, total annual precipitation from 461 to 590 mm, and total annual evaporation from 233 to 325 mm. The meteorological data were contained within a point layer.

Layer and Model Linkage

A custom script was written to find the centers of all polygons in the land-use coverage. The polygon centers were then linked to each of the interpolated weather points, using the nearest point to each polygon. This provided a linked land-use and meteorological layer.

The soil layer, based on SOTER units, was overlaid upon the linked land-use and meteorological layer using a custom script and giving 16,360 polygons representing a unique combination of land use, soil, and meteorology. From these, 13,185 polygons were actually used in the modeling exercise; those with land-use code 4 (other uses) and with no underlying SOC data were excluded.

An interface for the CENTURY model (the IGATE system: Figure 13.2) was written to read input data from an ASCII file produced by the GIS in order to create schedule files, run the model, and send results to an ASCII file. The original source code of the RothC model was altered to take input from a fixed-width ASCII file produced by the GIS, to read C inputs derived from the CENTURY model, and to write results to a new ASCII file. The files containing results

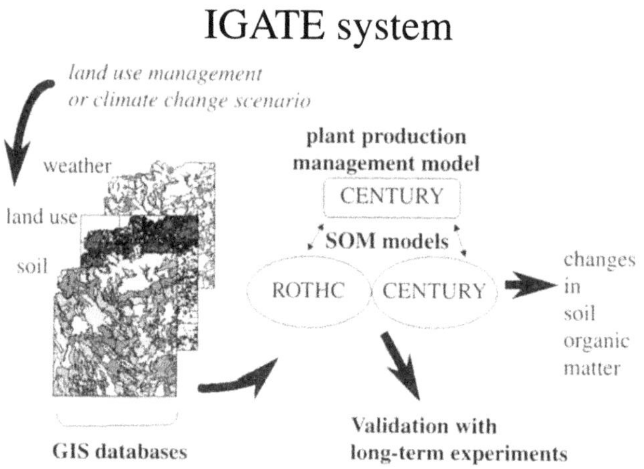

IGATE system

Figure 13.2   Schematic representation of the IGATE system (modified from Parshotam et al., 1995).

from the models were loaded into Excel, Access, and GENSTAT for analysis, and into ArcView for visualization.

Model Initialization and Scenarios

The crop, grass, and forest production parameters for CENTURY were set using data from the GCTE SOMNET database. Soil, management, and weather data were collected, and steady-state SOC levels were simulated at 11 arable sites from Hungary, 4 U.K. grasslands, a Polish grassland, a U.K. forest, and a Hungarian forest (from the IBP woodlands data set; DeAngelis et al., 1981). Sites were chosen to cover a range of management, weather and soil characteristics (Table 13.3).

The regional-scale application of CENTURY was initialized by running for 3000 y under forest, followed by 500 y of current land use (from the CORINE database; Büttner et al., 1995a, b; Büttner, 1997). Available historic information suggests that the natural vegetation of much of the region was likely to have been forest (Marton et al., 1989) and that much of the current land use may have been in existence for less than 500 y (Fusell, 1966; Loczy, 1988; Spoor, 1987; Molnar, 1996; Molnar and Biro, 1996).

As a simple demonstration scenario, changes in SOC stocks over 100 y following afforestation of all current arable land were examined, assuming sequential forestry establishment. Sequential forest growth was accounted for by the CENTURY tree model, which estimated tree growth parameters and C inputs. It should be stressed that this scenario was chosen to demonstrate methodology and does not reflect a realistic option for land-use change. Next, the C input data for each polygon from the CENTURY model outputs were taken and used to run RothC for the same scenario. Finally, changes in the SOC stock of the arable land area were estimated using the regression of Smith et al. (1997a, b, 1998b) and applied to the SOC data of each polygon from the HunSOTER database. Their lower figure of SOC increase (1.17% y^{-1}) was used since earlier studies have shown that the higher figure of 1.66% y^{-1} is too high (Falloon et al., 1999; Smith et al., 2000a, b).

CO_2 emissions for the study area were calculated using 1990 national data (17.5 Tg C) for Hungary from Marland et al. (1999) because 1990 is the baseline level for Kyoto Protocol reporting. The proportion of the Hungarian land area occupied by the study region (26.7%) was then calculated using land-use data from FAOSTAT (1999) and the national 1990 emission level scaled to this study region (4.7 Tg C).

Table 13.3 Sites Used in Calibrating CENTURY

Site	Country	Latitude	Longitude	Clay %	Silt %	Sand %	Mean Annual Temperature (°C)	Mean Annual Rainfall (mm)	Fertilizer Treatment	Crop
Arable Sites										
Nagyhorksok	Hungary	46.90 N	18.52 E	23	27	50	10.54	491.90	Nil	Wheat–maize
Putnok	Hungary	48.30 N	20.43 E	28	37	35	9.43	594.70	N2PK	Wheat–maize
Kompolt	Hungary	47.68 N	20.23 E	41	32	27	10.03	518.00	N3P2K2	Wheat–maize
Mosomnagyarovar	Hungary	47.88 N	17.27 E	12	20	68	12.05	1025.20	Nil	Wheat–maize
Kezthely	Hungary	46.77 N	17.23 E	26	19	59	10.34	632.10	Nil	Wheat–maize
Karcag	Hungary	47.28 N	20.90 E	37	36	27	10.35	502.60	N2P2K2	Wheat–maize
Iregszemcse	Hungary	46.70 N	18.18 E	22	28	50	10.32	557.00	N2Pk	Wheat–maize
Hajdubszormeny	Hungary	47.75 N	21.38 E	35	33	32	10.03	535.20	Nil	Wheat–maize
Bicserd	Hungary	46.03 N	18.08 E	33	32	35	10.28	557.00	Nil	Wheat–maize
Debrecen	Hungary	47.53 N	21.65 E	27.1	68.4	4.5	10.05	562.00	Nil	Maize
Martonvasar	Hungary	47.32 N	18.78 E	31	35	34	10.25	533.10	NPK	Maize
Grassland Sites										
Park Grass	UK	51.82 N	00.35 W	21	23	49	9.30	763.00	Nil	Grass
Long Term Slurry	UK	54.43 N	06.42 W	25.4	42.6	32	9.32	1123.90	Control	Grass
Palace Leas	UK	55.22 N	01.68 W	25.4	24.1	20.3	8.73	671.30	Nil	Grass
Palace Leas	UK	55.22 N	01.68 W	25.4	24.1	20.3	8.73	671.30	FYM	Grass
Czarny Potok	Poland	49.06 N	20.13 E	23	37	40	6.67	646.70	Nil	Grass
Forest Sites										
Geescroft	UK	51.82 N	00.35 W	23	58	19	9.30	763.00	Nil	Forest
Sifokut	Hungary	47.90 N	20.47 E	35	8	53	10.03	518.00	Nil	Forest

RESULTS AND DISCUSSION

Simulation of SOC levels at the calibration sites (Table 13.3) gave good agreement between modeled and measured values (r^2 = 0.83, Figure 13.3). Figure 13.4 shows SOC dynamics in a typical arable polygon, from initialization under native forest through current land use to the afforestation scenario. Under the assumed native vegetation type (forest), the SOC content reached steady state after approximately 3000 y under Hungarian conditions. During the initialization period under forest (until 1500), there was a small decline in SOC every 500 y, since forest burning was simulated every 500 y to roughly mimic natural forest processes. Conversion to arable land use, in a wheat–maize rotation, resulted in a rapid loss of SOC, with an interannual fluctuation due to the difference in C input to the soil between wheat and maize crops. The natural reversion of arable land to forest resulted in an initial loss of SOC because, during the early stages of forest growth, the C inputs are small. SOC then rapidly increased under continued afforestation.

The estimate of the 1999 SOC stock obtained using CENTURY was close to the value obtained from the HunSOTER database (Table 13.4), but the estimate from RothC using the same input data was around 72% of the database value. Table 13.4 also gives the predicted changes in SOC 100 y after afforestation of all arable land, and the percent of annual CO_2 emissions offset. The change in SOC stocks over the study area can be seen in Figures 13.5 and 13.6; darker shaded areas represent a greater change in SOC and lighter shaded areas a smaller change. Predictions of C sequestration under this scenario from RothC and CENTURY predictions were smaller (37 and 55 Tg C) than those predicted by the regression (101 Tg C). This may be attributable to local soil, phenological, and climatic differences influencing tree growth rates, and hence C inputs to soil, and also the decomposition rate. However, more confidence is held in the RothC and CENTURY estimates, since they have explicitly accounted for differences in soil characteristics and climate. The regression was based solely on data (from two sites) collected under U.K. conditions.

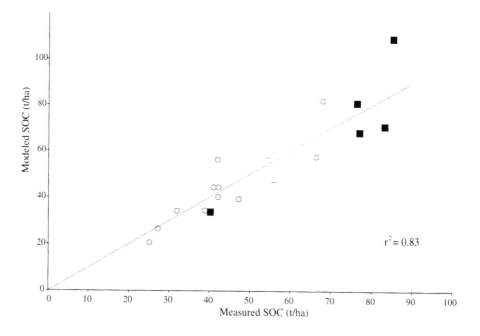

Figure 13.3 Measured SOC values at calibration sites against those calculated using the CENTURY model (■ = grassland site, ○ = arable site, Δ = forest site, — = 1:1 line; r^2 figure is the correlation against 1:1 line).

Assuming 1990 levels of CO_2 emissions (the Kyoto Protocol baseline level), the annual offset of CO_2 emissions is large, so this scenario, although unrealistic, could bring about a significant reduction when considering soils alone. If the additional carbon sequestered in the standing woody biomass is considered, this would further increase the carbon stored in the terrestrial system. In regenerating temperate woodland, Jenkinson (1971) found the standing woody biomass to have accumulated approximately 1.52 times as much carbon as was found in the soil after 86 years. Using this ratio, the above-ground biomass could sequester 111, 165, or 305 Tg C when calculated from the RothC, CENTURY, and regression estimates, respectively. Thus the total C sequestration in the terrestrial system would be 148, 220 or 407 Tg, with an annual offset of CO_2 emissions (1990 levels) of 31.49, 46.81, or 86.64%, when calculated from the RothC, CENTURY, and regression estimates, respectively. If a proportion of land were to be afforested, the biomass accumulated in the trees could be used as a substitute for fossil fuels (Sampson et al., 1993), thus further mitigating fossil fuel-derived CO_2-carbon.

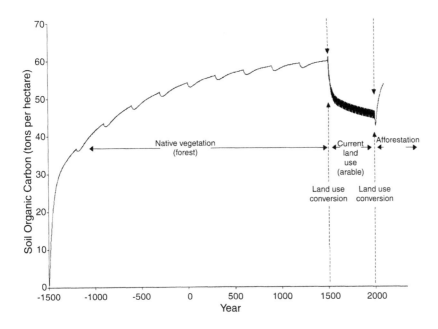

Figure 13.4 SOC dynamics in a typical arable polygon (mean annual temperature 9.8°C, total annual rainfall 562 mm, soil sand content 39%, soil clay content 49%, soil silt content 10%).

Table 13.4   Comparison of SOC Stock Estimates and Changes in SOC Stock after Land Use Change Scenario

	HunSOTER Database	RothC	CENTURY	Regression
Current SOC stock estimate (Tg C)	93.00	67.02	92.00	93.00
SOC stock after land use change (Tg C)	—	104.00	147.03	194.80
Total change in SOC stock (Tg C)	—	37.01	55.03	101.80
Yearly change in SOC stock (Tg C y^{-1})	—	0.37	0.55	1.02
% annual emissions offset[a]	—	7.87	11.8	21.66

[a] Assuming 1990 levels of CO_2 emissions.

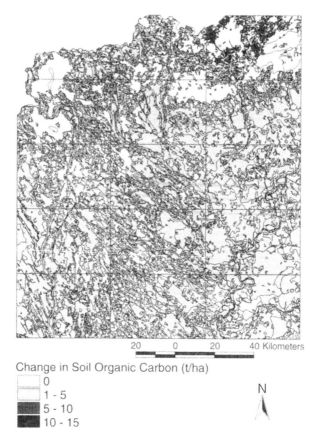

Change in Soil Organic Carbon (t/ha)
- 0
- 1 - 5
- 5 - 10
- 10 - 15

Figure 13.5 Change in soil organic carbon predicted by RothC after 100 y afforestation of all arable land.

There are several limitations in the estimates presented here. First, errors in the RothC estimate may be introduced by estimating the IOM content of the soil, particularly in highly organic and waterlogged soils (Falloon et al., 1998a, b, 2000; Falloon and Smith, 2000). Earlier work has shown that errors in the size of the IOM pool for the RothC model could lead to significant errors in estimates of C sequestration (Falloon et al., 2000). Errors in RothC, CENTURY, and regression estimates may be derived from estimating SOC to 20 cm depth, since the HunSOTER sample depth was variable. The linked GIS–SOM model approach may also be limited by the availability of relevant long-term soil C data sets for validation, the resolution of spatial datasets, and information on historical and current land use, management practices, and production.

This study improved on previous estimates (Falloon et al., 1998a, 1999) by using more recent and interpolated weather data (Hulme et al., 1995), C inputs derived from a dynamic ecosystem model (CENTURY), comparing two SOM models, and accounting for sequential growth during afforestation.

CONCLUSIONS

This preliminary study has shown how coupling a detailed GIS database with dynamic simulation models can allow comparison and refinement of estimates of regional SOC stocks and the potential for C sequestration. The system provides a flexible and powerful method for assessing

Figure 13.6 Change in soil organic carbon predicted by CENTURY after 100 y afforestation of all arable land.

the impact of different scenarios for land-use, management, and climatic change on carbon dynamics at the regional scale.

This study has also shown that, for the demonstration scenario in which afforestation of all arable land took place over 100 y, increases in SOC alone make a significant contribution to reduction of CO_2 emissions. If carbon accumulation in the above-ground biomass is considered, the CO_2-C offset is even larger but, given the extreme measures taken in this demonstration scenario, afforestation is only likely to be a practical or useful option when applied to a portion of the area, for example, set-aside land.

The conclusions drawn here agree with the results from regression-based scenario estimates for the U.K. (Smith et al., 2000a) and Europe (Smith et al. 2000b), which indicated that afforestation could make a significant contribution to carbon sequestration. The regression-based studies did not, of course, take account of local soil or climatic conditions. One possible reason for the difference between model and regression estimates is that the Hungarian study area contains a predominance of sandy soils. These soils have a low potential for accumulating SOC, so generalized treatments, such as the regression method, which use an average SOC accumulation rate covering all soil types, will overestimate the potential in sandy soils. This may account for the greater increase in SOC stock under the afforestation scenario for the Hungarian area calculated from the regression method than by modeling using the appropriate database (Table 13.4). Both RothC and CENTURY account for the effects of soil texture when calculating SOC changes.

ACKNOWLEDGMENTS

We would like to thank György Várallyay, Péter Csathó, and Tamás Németh (RISSAC, Hungary) for assisting with input data and site information, Keith Paustian and Robin Kelly (Colorado State University, U.S.) for advice on the CENTURY model, David Jenkinson (IACR-Rothamsted, U.K.) for assistance with the RothC model and general advice, and David Powlson for comments on an earlier version of this chapter. This work contributes to the following projects: the UK-BBSRC-ASD Project "Modelling SOM Dynamics using SOMNET" (Grant 206/A 06371), and the EU project Low Input Agriculture and Soil Sustainability in Eastern Europe (Grant: EU-COPERNICUS PL97 1006). IACR receives grant-aided support from the Biotechnology and Biological Sciences Research Council of the U.K.

REFERENCES

Batjes, N.H., Total carbon and nitrogen in the soils of the world, *Eur. J. Soil Sci.*, 47:151–163, 1996.

Büttner, G., E. Csató, and G. Maucha, The CORINE land cover — Hungary project, in *Proc.17th Int. Cartography Conf.*, Barcelona, 1813, 1995a.

Büttner G., E. Csató, and G. Maucha, The CORINE land cover — Hungary project, *Proc. Int. Conf. Environ. Informatics*, Budapest, 54–61, 1995b.

Büttner, G., Land cover — Hungary; Final Technical Report to Phare, FOMI, Budapest, (manuscript), 1997.

Coleman, K. and D.S. Jenkinson, RothC-26.3 — a model for the turnover of carbon in soil, in D.S. Powlson, P. Smith, and J.U. Smith (eds.), *Evaluation of Soil Organic Matter Models Using Existing, Long-Term Datasets*, NATO ASI Series I, Vol. 38, Springer-Verlag, Heidelberg, 237–246, 1996.

Coleman, K. et al., Simulating trends in soil organic carbon in long-term experiments using RothC-26.3, *Geoderma*, 81:29–44, 1997.

DeAngelis, D.L., R.H. Gardner, and H.H. Shugart, Productivity of forest ecosystems studied during the IBP: the woodlands data set, in Reichle, D.E. (ed.), *Dynamics of Forest Ecosystems*, Cambridge University Press, Cambridge, 567–672, 1981.

Donigian, A.S. et al., *Assessment of Alternative Management Practices and Policies Affecting Soil Carbon in Agroecosystems of the Central United States*, EPA/600/R-94/067, Environmental Research Laboratory, Athens, GA, 1994.

Falloon, P.D. et al., Regional estimates of carbon sequestration potential: linking the Rothamsted carbon turnover model to GIS databases, *Biol. Fertil. Soils*, 27:236–241, 1998a.

Falloon, P. et al., Estimating the size of the inert organic matter pool from total soil organic carbon content for use in the Rothamsted carbon model, *Soil Biol. Biochem.*, 30:1207–1211, 1998b.

Falloon, P. and P. Smith, Modelling refractory soil organic matter: a review, *Biol. Fertil. Soils*, 30:388–398, 2000.

Falloon, P.D. et al., Linking GIS and dynamic SOM models: estimating the regional C sequestration potential of agricultural management options, *J. Agric. Sci.*, Cambridge, 133:341–342, 1999.

Falloon, P. et al., How important is inert organic matter for predictive soil carbon modelling using the Rothamsted carbon model? *Soil Biol. Biochem.*, 32:433–436, 2000.

FAOSTAT, FAOSTAT Statisitics Database. Food and Agriculture Organization of the United Nations, Rome, <http://apps.fao.org/default.htm>, 1999.

Fussell, G.E., *Farming Technique from Prehistoric to Modern Times*, Pergamon Press, Oxford, 1966.

Gupta, R.K. and L.L.N. Rao, Potential of wastelands for sequestering carbon by afforestation, *Curr. Sci.*, 66:378–380, 1994.

Howard, P.J.A. et al., The carbon content of soil and its geographical distribution in Great Britain, *Soil Use Manage.*, 11:9–15, 1995.

Hulme, M. et al., Construction of a 1961–1990 European climatology for climate change impacts and modelling applications, *Int. J. Climatol.*, 15:1333–1364, 1995.

Jenkinson, D.S., The accumulation of organic matter in soil left uncultivated, Rothamsted Report for 1970, Part 2: 113–137, 1971.

Jenkinson, D.S., D.E. Adams, and A. Wild, Model estimates of CO_2 emissions from soil in response to global warming, *Nature*, 351:304–306, 1991.

Kern, J.S. and M.G. Johnson, Conservation tillage impacts on national soil and atmospheric carbon levels, *Soil Sci. Soc. Am. J.*, 57:200–210, 1993.

King, A.W., W.M. Post, and S.D. Wullschleger, The potential response of terrestrial carbon storage to changes in climate and atmospheric CO_2, *Climatic Change*, 35:199–227, 1997.

Lee, J.J., D.L. Phillips, and R. Liu, The effect of trends in tillage practices on erosion and carbon content of soils in the U.S. Corn Belt, *Water Air Soil Poll.*, 70:389–401, 1993.

Loczy, B., Cultural landscapes in Hungary — two case studies, in Birks, H.H. et al. (eds.), *The Cultural Landscape — Past, Present and Future*, Cambridge University Press, Cambridge, 165–176, 1988.

Marland, G. et al., NDP-030 Global, Regional, and National CO_2 Emission Estimates from Fossil Fuel Burning, Cement Production, and Gas Flaring: 1751–1996 (revised March 1999) in, Trends Online: A Compendium of Data on Global Change. Carbon Dioxide Information Analysis Center, Oak Ridge National Laboratory, Oak Ridge, Tennessee. <http://cdiac.esd.ornl.gov/ndps/ndp030.html>, 1999.

Marton, P. et al. (eds.), *National Atlas of Hungary*, Cartographia, Budapest, 1989.

Molnar, Z., The land-use historical approach to study vegetation history at the century scale, in *Proc. Res., Conserv. Manage. Conf.*, Aggtelek, Hungary, 1–5 May, 1996. VII Session: Vegetation mapping and botanical research on nature conservation purpose, 345–353, 1996.

Molnar, Z. and M. Biro, Vegetation history of the Kardoskut area (S.E. Hungary) I.: Regional versus local history, ancient versus recent habitats, *Tiscia*, 30:15–25, 1996.

Németh, T., L. Pásztor, and E. Michéli, C balance in Hungarian soils, chap. 43, this volume.

Parshotam, A., K.R. Tate, and D.J. Giltrap, Potential effects of climate and land-use change on soil carbon and CO_2 emissions from New Zealand's indigenous forests and unimproved grasslands, *Weather Climate*, 15:3–12, 1995.

Parton, W.J. et al., Analysis of factors controlling soil organic levels of grasslands in the Great Plains, *Soil Sci. Soc. Am. J.*, 58:530–536, 1987.

Parton, W.J., J.W.B. Stewart, and C.V. Cole, Dynamics of C, N, P, and S in grassland soils: a model, *Biogeochemistry*, 5:109–131, 1988.

Pásztor, L., J. Szabó, and G. Várallyay, Digging deep for global soil and terrain data, *GIS Eur.*, 5:32–34, 1996.

Paustian, K. et al., The use of models to integrate information and understanding of soil C at the regional scale, *Geoderma*, 79:227–260, 1997.

Post, W.M. et al., Soil carbon pools and world life zones, *Nature*, 298:156–159, 1982.

Post, W.M., A.W. King, and S.D. Wullschleger, Soil organic matter models and global estimates of soil organic carbon, in Powlson, D.S., P. Smith, and J.U. Smith (eds.), *Evaluation of Soil Organic Matter Models Using Existing, Long-Term Datasets*, NATO ASI Series I, Vol. 38, Springer-Verlag, Heidelberg, 201–222, 1996.

Prentice, K.C. and I.Y. Fung, The sensitivity of terrestrial carbon storage to climate change, *Nature*, 346:48–51, 1990

Sampson, R.N. et al., Biomass management and energy, *Water Air Soil Poll.*, 70:139–159, 1993.

Smith, P. et al., The GCTE SOMNET. A global network and database of soil organic matter models and long-term datasets, *Soil Use Manage.*, 12:104, 1996a.

Smith, P., D.S. Powlson, and M.J. Glendining, Establishing a European soil organic matter network (SOM-NET), in Powlson, D.S., P. Smith, and J.U. Smith (eds.), *Evaluation of Soil Organic Matter Models Using Existing, Long-Term Datasets*, NATO ASI Series I, Vol. 38. Springer-Verlag, Berlin, 81–98, 1996b.

Smith, P., J.U. Smith, and D.S. Powlson, *Soil Organic Matter Network (SOMNET): 1996 Model and Experimental Metadata*, GCTE Report 7, GCTE Focus 3 Office, Wallingford, Oxon, 1996c.

Smith, P. et al., Opportunities and limitations for C sequestration in European agricultural soils through changes in management, in Lal, R. et al. (eds.), *Management of Carbon Sequestration in Soil*, Advances in Soil Science, 143–152, 1997a.

Smith, P. et al., Potential for carbon sequestration in European soils: preliminary estimates for five scenarios using results from long-term experiments, *Global Change Biol.*, 3:67–80, 1997b.

Smith, P. et al., A comparison of the performance of nine soil organic matter models using seven long-term experimental datasets, in Smith, P. et al. (eds.), Evaluation and Comparison of Soil organic matter models using datasets from seven long-term experiments, *Geoderma*, 81:138–153, 1998a.

Smith, P. et al., Preliminary estimates of the potential for carbon mitigation in European soils through no-till farming, *Global Change Biol.*, 4: 679–685, 1998b.

Smith, P. et al., Meeting the U.K.'s climate change commitments: options for carbon mitigation on agricultural land, *Soil Use Manage.*, 16:1–11, 2000a.

Smith, P. et al., Meeting Europe's climate change commitments: quantitative estimates of the potential for carbon mitigation by agriculture, *Global Change Biol.*, 6:525–539, 2000b.

Smith, P. et al., Modelling soil carbon dynamics in tropical ecosystems, in *Global Climate Change and Tropical Soils*, Lal, R. et al. (eds.), Advances in Soil Science, CRC Press, Boca Raton, FL, 341–364, 1999.

Spoor, G. et al., The impact of mechanization, in Wolman, M.G. and F.G.A Fournier (eds.), *Land Transformation in Agriculture*, Scope 32. John Wiley & Sons, Chichester, 133–152, 1987.

Szabó, J., G. Várallyay, and L. Pásztor, HunSOTER, a digitised database for monitoring changes in soil properties in Hungary, in *Proc. Soil Monitoring Czech Republic*, Brno, 150–156, 1996.

Varga–Haszonits, Z., *Agrometeorologia*, Mezogazdasagi Kiado, Budapest, 1977.

Williams, J.R. and K.G. Renard, Assessment of soil erosion and crop productivity with process models (EPIC), in Follett, R.F. and B.A. Stewart (eds.), *Soil Erosion and Crop Productivity*, ASA/CSSA/SSSA, Madison, WI, 67–103, 1985.

CHAPTER 14

Soil C Sequestration Management Effects on N Cycling and Availability

W. R. Horwath, O. C. Devevre, T. A. Doane, A. W. Kramer, and C. van Kessel

CONTENTS

INTRODUCTION

The impact of soil C sequestration on nitrogen (N) cycling and fertilizer use efficiency by crops has received little attention in the context of long-term fertility. It is generally taken for granted that more soil organic matter (SOM) is a key indicator of enhanced soil quality and thus leads to sustainable soil fertility. The importance of SOM in cropping system sustainability is in its ability to store nutrients and improve soil structure. The challenge in determining nutrient availability in cropping systems designed to accumulate SOM is in assessing the (1) temporal availability of nutrients, (2) interaction of fertilizer N with organic N pools, and (3) relationship of SOM turnover dynamics to nutrient availability. For these reasons, soil C sequestration benefits for mitigation of carbon dioxide-related climate change and soil quality enhancement must carefully be weighed against potential N and other nutrient availability as a result of increased SOM.

The increase in stable SOM under soil C sequestration management may lead to the immobilization of N and other essential nutrients such as phosphorus, sulfur, and essential micronutrient metals. However, as SOM increases, components of the soil controlling nutrient cycling, such as the active fraction, also increase (Paul and Clark, 1996). The active fraction is composed of plant residues, root exudation, light fraction, labile SOM, and the microbial biomass (Tisdall and Oades, 1982). As SOM begins to accumulate initially in soil, the ability of the active fraction to supply nutrients for plant uptake is limited, due to mineralization-immobilization reactions associated with changes in microbial substrate use efficiency caused by changes in the quality and increase in C inputs.

As soil C sequestration proceeds, an equilibrium between C inputs and nutrient availability occurs, leading to sustained plant nutrient availability (Doran et al., 1988). This scenario is ideal in natural ecosystems or with perennial plants where nutrient uptake can occur over the growing seasons. However, in agronomic systems, nutrient uptake occurs in a narrow time frame not fully taking advantage of the potential N mineralization from increased SOM. In addition, interaction among nutrient sources and C inputs may affect their availability.

Interaction among fertilizer sources and C inputs has been demonstrated under a variety of field and laboratory conditions (Jenkinson et al., 1985; Hart et al., 1986; Powlson and Barraclough, 1993). The priming of soil N or the added nitrogen interaction (ANI) is often caused by changes in fertilizer sources or quality of plant residue inputs to soil. The ANI often causes changes in the N-use efficiency of crops (Azam et al., 1985; Ehaliotis et al., 1998). The implication of ANI is that the contribution of N from residue management practices or manure additions is underestimated when compared with fertilizer N additions. An increased level of soil C directly influences ANI, increasing the size of the soil microbial biomass and exacerbating its role as a source and sink for essential plant nutrients.

The low recovery of N from organic amendments is in part attributable to not determining belowground N allocation, which often leads to underestimates of N-use efficiency (Juma, 1993). The role of the rotational effect must also be considered (Elliott et al., 1987). Another reason may be that components of organic residues will be mineralized at different rates as the system stabilizes over time (Parr and Papendick, 1978). The size and activity of the microbial biomass are critical factors that regulate the turnover and stabilization of N in SOM. For these reasons, the influence of residue management practices designed to sequester soil C must be understood to determine plant N availability.

The consequences of soil C sequestration management may impact nutrient availability through the immobilization of essential nutrients or interaction with the quality and quantity of soil C inputs. Results are presented that examine the effect of managing for soil C sequestration on available N in soil. In addition, data are presented on the efficiency of plant residue conversion into stable SOM. These results are critical to understanding agronomic performance under soil C sequestration management.

MATERIALS AND METHODS

Site Description

The Sustainable Agriculture Farming Systems (SAFS) Project was established in 1988 at the agronomy farm of the University of California at Davis. The 11.3-ha site is dedicated to the study of agronomic, economic, and biological aspects of conventional and alternative farming systems in California's Sacramento Valley. The soils are classified as Reiff loam (coarse-loamy, mixed, nonacid, thermic Mollic Xerofluvents) and Yolo silt loam (fine-silty, mixed, nonacid, thermic Mollic Xerofluvents).

Description of the Farming System

The study consists of two conventional and two alternative systems that differ primarily in crop rotation and use of external inputs. These include 4-year rotations under conventional (Conv-4), low-input (LI), and organic (Org) management and a conventionally managed 2-year (Conv-2) rotation. The three systems in the 4-year rotations include processing tomatoes, safflower, bean, and corn. In the conventional 4-year treatment, beans are double-cropped with winter wheat. In the low-input and organic treatments, beans typically follow a biculture of oats and vetch that serves as either a cover crop or cash crop. The conventional 2-year treatment is a tomato and wheat rotation typical of farming systems of the region. Table 14.1 shows details of the farming system treatments.

The organic system is managed according to practices recommended by California certified organic farmers (Anonymous, 1990) that do not allow synthetic chemicals. Fertilizer N sources include legume and grass cover crops, composted animal manures, and occasional organic supplements. The low-input system has legume cover crops to reduce the amount of synthetic fertilizers. The conventional systems are managed with standard chemical inputs of pesticides and various N fertilizers. Each cropping system has four replications for each of the possible crop rotation entry points, resulting in a total of 56 plots, each measuring 68 m × 18 m. Treatments are arranged in a split-plot design, with cropping systems as the main plot treatments, and crop point of entry as subplot treatments. Total C and N inputs to the various farming systems over a 10-year period from 1988 to 1998 are summarized in Table 14.2.

Fertilizer and Vetch N Uptake

During the winter of 1997 to 1998, a 9 m^2 area of vetch (*Vicia* spp.) cover crop was labeled with ^{15}N labeled $(NH_4)_2SO_4$ (49 atom% ^{15}N) in the low-input and organic system entry points to be planted with maize in the spring of 1998. In order to ensure uniform labeling of plant components, vetch received ^{15}N-labeled $(NH_4)_2SO_4$ on October 23, 1997, November 22, 1997, and February 27, 1998, totaling a rate of 9 kg N ha^{-1}. In April of 1998, the ^{15}N labeled vetch shoots were harvested

Table 14.1 Description of Treatments, Crop Rotations, and Agronomic Management

Treatment	Crop Rotation	Agronomic Management
Organic (Org)	Tomato, safflower, corn, oats/vetch, bean	Four-year, five-crop rotation using composted manure, legume and grass cover crops, and organic supplements; no synthetic pesticides or fertilizers
Low-input (LI)	Tomato, safflower, corn, oats/vetch, bean	Four-year, five-crop rotation relying on legume and grass cover crops and one half synthetic fertilizer applied
Conventional 4-year (Conv-4)	Tomato, safflower, corn, wheat, bean	Four-year, five-crop rotation using synthetic fertilizer and pesticides
Conventional 2-year (Conv-2)	Tomato, wheat	Two-year, two-crop rotation relying on synthetic fertilizer and pesticides

Table 14.2 Amount of C and N Inputs over a 10-Year Period to the Various Farming Systems

Farming System	Input (Mg ha^{-1})	
	C	N
Organic	42.9 (d)	1.96 (c)
Low-Input	39.4 (c)	1.65 (a)
Conventional-4	36.1 (b)	1.92 (c)
Conventional-2	29.3 (a)	1.74 (b)

and shredded to simulate mowing. Similar sized areas of unlabeled vetch were also harvested and shredded at this time.

Subplots measuring 4 m^2 were established in the maize entry point of the conventional, low-input, and organic cropping systems in April of 1998. To ensure integrity, subplots were established a safe distance from the areas of enriched vetch in the low-input and organic systems. The vetch cover crop was cleared from these subplots in the low-input and organic systems before application of ^{15}N labeled and unlabeled additions. Characteristics and quantities of inorganic and organic N additions to each subplot are recorded in Table 14.3. Incorporation of the vetch was performed 6 days prior to seeding the maize. Side dressing of ^{15}N urea was applied 36 days after seeding.

Table 14.3   Characteristics of N Amendments to Three Cropping Systems at SAFS

Cropping System	Subplot	Amendment	Quantity (kg ha^{-1})
Conventional	1	^{15}N urea	220[a]
	2	—	—
Low-input	1	^{14}N vetch	100
	2	^{15}N urea	90[a]
	3	^{15}N vetch	120
		^{14}N urea	90[a]
		—	—
Organic	1	^{15}N vetch	105
	2	^{14}N manure	330
		—	—

[a] An additional 7 kg unlabeled N ha^{-1} was applied as starter fertilizer at seeding.

Soil and Plant Sampling

The soil collection for C dynamics consisted of taking 30 separate soil cores at depth of 0 to 15 cm from each experimental plot and then compositing them for the analysis of SOM humic fractions. The soils were dried at 35°C, sieved to pass through a 4-mm screen, and stored at 4°C until processed. Plant sampling from the ^{15}N microplot studies consisted of taking punches of leaf material from the centers of all microplots.

Sample Preparation and Analysis

Subsamples of air-dried soil were ball milled. Separated humic fractions were freeze-dried and ground to a fine powder prior analysis for total N and C content and determination of ^{15}N/^{14}N and ^{13}C/^{12}C ratios by dry combustion-continuous flow isotope ratio mass spectrometer GC-IRMS (Europa Scientific, Crewe, England). Ash content of isolated humic fractions was determined after heating at 550°C for 2 hours. Plants were ground and analyzed for N and ^{15}N and analyzed as above.

Chemical Separation of Soil Humic Fractions

The chemical fractionation procedure for the isolation of humic fractions was adapted from Stevenson (1994). Prior to extraction with a 0.4N NaOH solution under N$_2$, air-dried soil samples were washed with a 0.1N HCl solution to remove carbonates and plant debris. The NaOH extraction was repeated until no humic substances could be extracted. The humic acids were separated from fulvic acids by precipitation after acidification to pH 2. The humic acid, fulvic, and humin fractions were freeze-dried and analyzed on a GC-IRMS (Europa Scientific, Crewe, England) for ^{15}N, ^{13}C, and total N and C.

Statistics and Calculations

Statistical analyses were performed using StatView Software (StatView 4.5, Abacus Concepts Inc., Berkeley, CA); significant differences between treatments were measured after Fisher's PLSD at a significance level of 5% and analysis of variance (SNK). Determination of plant C to SOM was done using the isotope mixing model described by Balesdent and Mariotti (1996).

RESULTS AND DISCUSSION

Changes in soil C content will affect N cycling through changes in the amount and quality of soil C inputs. In addition, the conversion of plant residues into stable organic matter is also related to the amount, quality, and timing of soil C inputs. Together these processes affect the sustainability of cropping systems through changes in nutrient availability. A major question to ask is whether the influence on nutrient availability is short-term (occurring during the transition) or long-term. If the change in nutrient availability is long-term, then fertilizer practices would need to be reevaluated to compensate for changes in long-term soil fertility. Substantial increases in fertilizer input could offset the gains in soil C sequestration management efforts. These questions can only be answered by examining long-term experiments to assess changes in soil management practices designed to examine N interactions and SOM dynamics.

Soil Carbon Dynamics

Analysis of the soils from the cropping system treatments showed that the Org and LI treatments experienced significant increases in soil C and N compared to the conventional treatments (Figure 14.1). Soil C levels increased by 36%, 18%, and 13% in the ORG, LI, and CONV-4 treatments, respectively, compared to the CONV-2 treatments in the 0 to 15 cm soil depth. Similar increases were noted for soil N. The increase in soil C and N in the ORG and LI treatments can be attributed to the use of manure and cover crops. The increase in total soil C in the CONV-4 compared to the CONV-2 treatment is most likely attributable to the crop rotation effect. The increase in soil C and N was a result of input rather than changes in tillage. The organic and low input actually received more tillage because of the extra operations associated with incorporating the cover crops and manure. These results show the value of C inputs to soil to sequester C.

Changes in soil C occurred only in the surface 0 to 15 cm depth of soil. The organic system showed the highest increase in total C to a depth of 60 cm. (Figure 14.1). There is no significant difference between the LI (legume-based system) and the CONV-4 (fertilizer-driven system) treatments. These results suggest that the composted manure applied in the organic system is mostly responsible for the increase in SOM-N and SOM-C. The shift from 2-year rotation to 4-year rotation in the conventional system was enough to promote the build-up of total SOM and shows the value of crop rotation in sequestering soil C.

Conversion of C Inputs into SOM Fractions

The proportion of the total C input converted into SOM is equivalent in both Org and CONV-4 systems (Figure 14.2). Significantly less C was converted into SOM in the LI treatment. The LI treatment also had the lowest amount of fulvic acids compared to the other treatments. However, considering the high input of C in the Org system compared with the others, significantly more C (T ha^{-1}) was lost in this system while the Conv-4 treatment lost the least (data not shown). These results indicate that, as C input increases, the efficiency of conversion to stable SOM declines as seen in other studies (Collins et al., 1997). However, the quality of C inputs also can have a great

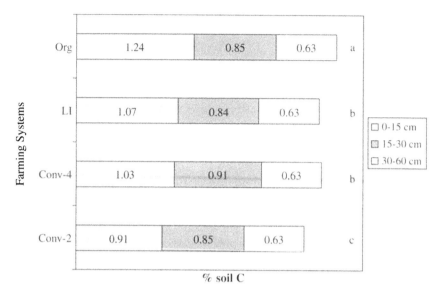

Figure 14.1 Accumulation of soil C (%) by depth in the different farming systems to a soil depth of 60 cm. Significant differences (ANOVA, SNK, P, <0.05) are indicated to the right of the bars and relate only to the 0 to 15 cm soil depth.

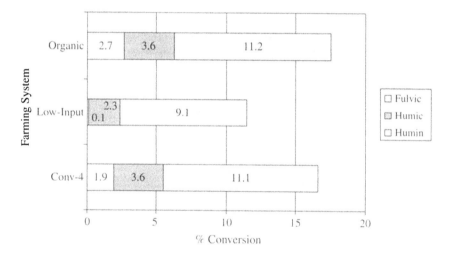

Figure 14.2 The percent conversion of C input into stable SOM fractions.

influence on soil C sequestration. The results show that farming system management can greatly influence the amount of C sequestered into soil. For these reasons, C sequestration strategies must consider fertilizer practices (conventional and alternative) and crops within rotations.

Added N Interactions

Alternative agricultural management, such as organic and low-input, is often applied in agriculture to change or reduce dependence on synthetic fertilizers. These systems are also suited to sequestering soil C. Such systems commonly substitute nitrogen-fixing cover crops, green waste, or manure for all or a large part of a conventional application of fertilizer. Initially, alternative systems receive nitrogen additions in excess of conventional systems in order to produce comparable

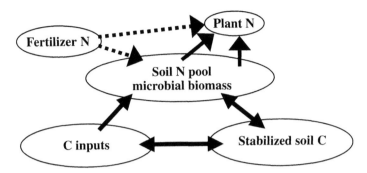

Figure 14.3 Fate of fertilizer N in soil as influenced by the microbial biomass and stable SOM.

yields, possibly leading to adverse environmental impact. However, as soil nitrogen pools increase, these systems retain nitrogen more effectively than conventionally managed systems. It remains a challenge to synchronize soil nitrogen availability with plant uptake in alternative systems. In addition, the management of soil nitrogen availability in alternative systems is complicated because of interactions among different fertilizer sources and soil nitrogen that influence the amount of nitrogen that becomes available to the crop.

The possible fates of fertilizer N in soil are shown in Figure 14.3. The competition for the N between the soil microbial biomass and the crop plant determines the fate of the fertilizer N. The influence of C inputs and stable soil SOM influence microbial activity, thereby influencing the competition for fertilizer N between the microbial biomass and crop plant. In this study, two inputs (manure or mineral fertilizer) were found to increase net recovery of cover crop vetch-derived nitrogen in a corn crop. Manure N significantly increased the uptake vetch N by corn from 13 to 19.6 kg N ha^{-1} (Figure 14.4). Mineral N also induced significantly greater plant uptake of vetch N from 8.9 to 22.7 kg N ha^{-1}. These results show an added N interaction whereby the uptake of vetch N by corn was increased through the addition of another source of N. This N interaction has been demonstrated in numerous other studies (Jenkinson et al., 1985; Azam et al., 1985; Ehaliotis et al., 1998).

As an organic input, greater plant uptake of vetch N could have been the result of increased microbial activity, by which more vetch N would have been mineralized, or increased microbial growth, in which case more vetch-derived N would have potentially been released through turnover of the microbial biomass (Ciardi et al., 1988). The microbial community, stimulated into growth by addition of vetch C, would have been able to capture a considerable part of the fertilizer N applied 1 month after decomposition of vetch residues began, if its N demand was still high. It is more probable, however, that this demand would have been satisfied by that time, although activity resulting from the build-up of microbial biomass could still have been high. More of the (unneeded) N still present in vetch was therefore made available for plant uptake.

The addition of vetch residue to a system receiving fertilizer as urea did not alter either the total N uptake or the amount of fertilizer N taken up by the corn. Since no difference in corn yield was noticed (data not shown), the presence of vetch residue simply reduced the amount of N acquired from SOM by an amount proportional to its own contribution to plant uptake (~14%). In other words, part of plant uptake of soil N was substituted by vetch N. The net availability of fertilizer N, however, remained unchanged. Either the great majority of fertilizer uptake was direct in both treatments, or the indirect pathway through the microbial biomass was unaffected by the previous stimulation of microbial activity following incorporation of vetch residue. Considering that fertilizer was introduced 1 month after vetch residue, the initial surge of activity following addition of high-quality residue had likely subsided by this time, so that, as suggested previously, uptake of fertilizer N by microorganisms was presumably minimal. Organic sources of N such as vetch, applied prior to a crop, provide useful benefits associated with rapid stimulation of microbial

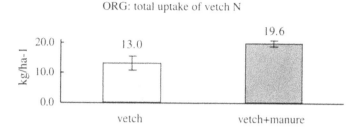

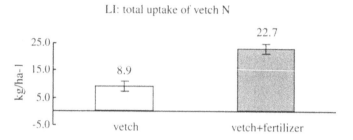

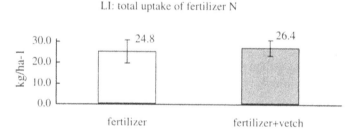

Figure 14.4 Uptake of vetch N as influenced by the interaction with manure and fertilizer urea N in the different farming systems. Stand errors of the mean (n = 4) are shown as line bars.

growth and activity, including the immediate release of N from both this source and enhanced release of nutrients from soil pools otherwise less accessible (Harris et al., 1994). In addition, these residues compliment soil C sequestration management. However, the results show that fertilizer practices need reevaluation when utilizing organic N sources to both build SOM and provide fertility.

CONCLUSIONS

This study addressed the interaction of various fertilizer inputs with relation to plant uptake and the impact of alternative management on soil C sequestration. Under the conditions of this study, two inputs complementary to a vetch cover crop (mineral fertilizer or aged manure) were found to increase net recovery of vetch-derived N in a corn crop. The effect of vetch on uptake of fertilizer N, however, was negligible. Both observations help describe a unique scenario of microbial dynamics; nonetheless, these observations are net results, and the exact progress of microbial dynamics in each system is not certain. The overall effect of one input on another is clearly dependent upon a wide range of circumstances, including site conditions, nature of C inputs, and time of application. Such potential interactions, though complex, provide numerous unique possi-bilities for studying the effects of C sequestration on sources of fertilizer N, which in turn can aid in understanding the processes at work in multiple-input systems.

The most profound changes in SOM occurring during this study were observable by simple analysis of the whole soil (the sum of all pools). Other effects were identified only upon fractionation — in this case, chemical fractionation into humic substances. Similarly, Balesdent and Balabane (1992) used separation by particle size (physical fractionation) to more precisely identify, using ^{13}C measurements, the effect of different C inputs with respect to these fractions. As summarized by Collins et al. (1997), the current theory of humus formation is based on a step-by-step process involving decomposition of plant material to simple C compounds, assimilation and repeated cycling of C through the microbial biomass, and simultaneous joining of microbially synthesized and altered plant-derived compounds (such as lignin) to form large polymers. Results suggest that SOM and its fractions may respond more dynamically than commonly thought to changes in external inputs and conditions.

It is important to understand the impact of different C inputs intended to sequester soil C on influencing sources of fertilizer N uptake by crop plants. Some complications with these sorts of studies lie in the effect of time on both C and N dynamics. Major conclusions drawn from this study were only possible after the management for soil C sequestration was observed for 10 years. For this reason, long-term studies on soil C and N processes are required to determine adequately the effects of soil C sequestration management on the long-term fertility of agricultural systems.

ACKNOWLEDGMENTS

The U.S. Department of Agriculture National Research Initiative grant (9604453) and the Kearney Foundation of Soil Science grants (96-26-D and 98-D-9) provided financial support for this study.

REFERENCES

Anon., California Organic Foods Act of 1990, California Department of Food and Agriculture, Sacramento, 1990.

Azam, F., K.A. Malik, and M.I. Sajjad, Transformations in soil and availability to plants of 15_N applied as inorganic fertilizer and legume residues, *Plant Soil*, 86:3–13, 1985.

Balesdent, J. and M. Balabane, Maize root-derived soil organic carbon estimated by natural ^{13}C abundance, *Soil Biol. Biochem.*, 24:97–101, 1992.

Balesdent, J. and A. Mariotti, Measurement of soil organic matter turnover using ^{13}C natural abundance, in Boutton, T.W. and S. Yamasaki (eds.), *Mass Spectrometry of Soils*, Marcel Dekker, Inc., New York 83–111, 1996.

Ciardi, C. et al., The effect of past fertilization history on plant uptake and microbial immobilization of urea-N applied to a two-course rotation, in Jenkinson, D.S. and K. A. Smith (eds.), *Nitrogen Efficiency in Agricultural Soils*, Elsevier Applied Science, London, 95–109, 1988.

Collins, H.P. et al., Characterization of soil organic carbon relative to its stability and turnover, in Paul, E.A. et al. (eds.), *Soil Organic Matter in Temperate Agroecosystems: Long-Term Experiments in North America*, CRC Press, Boca Raton, FL, 1997.

Doran, J.W. et al., Alternative and conventional agricultural management: influence on the soil microbial processes and nitrogen availability, *Am. J. Alter. Agric.*, 2:99–106, 1988.

Ehaliotis, C., G. Cadisch, and K.E. Giller, Substrate amendments can alter microbial dynamics and N availability from maize to subsequent crops, *Soil Biol. Biochem.*, 30:1281–1292, 1998.

Elliott, L.F., R.I. Papendick, and D.F. Bezdicek, Cropping practices using legumes with conservation tillage and soil benefits, in Power, J.F. (ed.), *The Role of Legumes in Conservation Tillage Systems*, University of Georgia, Athens, Soil Conservation Society of America, 81–89, 1987.

Harris, G.H. et al., Fate of legume and fertilizer nitrogen-15 in a long-term cropping systems experiment, *Agron. J.*, 86:910–915, 1994.

Hart, P.S.B., J.H. Rayner, and D.S. Jenkinson, Influence of pool substitution on the interpretation of fertilizer experiments with ^{15}N, *J. Soil Sci.*, 37:389-403, 1986.

Jenkinson, D.S., R.H. Fox, and J.H. Rayner, Interactions between fertilizer nitrogen and soil nitrogen — the so-called "priming" effect, *J. Soil Sci.*, 36:425–444, 1985.

Juma, N.G., Interrelationships between soil structure/texture, soil biota/soil organic matter and crop production, *Geoderma*, 57:3–30, 1993.

Parr, J.F. and R.I. Papendick, Factors affecting the decomposition of crop residues by microorganisms, in Oschwald, W.R. (ed.), *Crop Residue Management Systems*, Am. Soc. Agron., ASA Spec. Publ. 31, Madison, WI, 101–129, 1978.

Paul, E.A. and F.E. Clark, *Soil Microbiology and Biochemistry*, 2nd ed., Academic Press, Inc., New York, 1996.

Powlson, D.S. and D. Barroclough, Mineralization and assimilation in soil–plant systems, in Knowles, R. and T.H. Blackburn (eds), *Nitrogen Isotope Techniques*, Academic Press, Inc., New York, 209–242, 1993.

Powlson, D.S., J.D.S., G. Pruden, and A.E. Johnson, The effect of straw incorporation on the uptake of nitrogen by winter wheat, *J. Food Agric. Sci.*, 36:26–30, 1985.

Stevenson, F.J., *Humus Chemistry: Genesis, Composition, Reactions*, John Wiley & Sons, New York, 1994.

Tisdall, J.M. and J.M. Oades, Organic matter and water-stable aggregates in soils, *J. Soil Sci.*, 33:141–163, 1982.

CHAPTER 15

Land-Use Effects on Profile Soil Carbon Pools in Three Major Land Resource Areas of Ohio

A. Lantz, Rattan Lal, and John M. Kimble

CONTENTS

ABSTRACT

Conversion from natural to agricultural ecosystems changes soil organic carbon (SOC) pools; the magnitude of the changes depends on land use, management, and ecological factors such as temperature, precipitation, soil type, and native vegetation. Quantification of changes in the SOC pool as a result of agricultural land use provides a reference point regarding sequestration potential of SOC through improved management. A study was initiated in March 1998 to evaluate the differences in SOC pools in cropped, pastured, and forested (native) sites in Ohio. Three major land resource areas (MLRAs) were chosen to analyze land-use effects on SOC pools. These MLRAs are representative of climate, water, soil types, elevation, topography, natural vegetation, and land use for large areas in Ohio (60%) and adjacent states. The SOC pool was quantified by evaluating one soil series in each MLRA to a depth of 170 cm. Established pasture resulted in comparable or higher SOC pool levels than the forest ecosystems. Cultivation reduced the SOC pool in the top

0- to 10-cm layer and increased it in the 10- to 25-cm layer, thus not always decreasing the total SOC pool. The data showed that reduction in SOC in the surface layer does not necessarily imply a decrease in the total SOC pool in the soil profile.

INTRODUCTION

Globally, land-use changes and agricultural activities have released a historic 66 to 90 Pg of soil organic carbon (SOC) and may be currently releasing as much as 1.6 to 2.6 Pg of SOC per year (Lal, 1999). At national and regional levels, land use and its impact on the SOC pool and its dynamics are important knowledge gaps that need to be addressed for making appropriate policy decisions. These knowledge gaps are particularly important now, as nations try to meet commitments under the Kyoto Protocol.

Most studies that have assessed land use effects on SOC have only done so for surface layers of soil profiles. Yet, pedogenesis of lower horizons can be affected by disturbances to the surface horizons. For instance, soil pores, fractures between ped faces and earthworm holes, transport water and SOC to subsoil. Therefore, studies that neglect quantifying SOC content in subsoil may only be getting partial information on how land use affects the SOC pool, and such data may lead to erroneous estimates of the potential of different land uses as source or sink for carbon.

Numerous studies have documented SOC translocation in soil profiles, primarily as dissolved organic carbon (DOC). In Germany, Guggenberger and Zech (1992) found that DOC released from A horizons was translocated and retained in B horizons. The DOC translocation is influenced by different plant species, temperature, sulfate concentrations, and N inputs — all of which are affected by land use (Nambu and Yonevayashi, 1999; Chantingy et al., 1999; Liljeroth et al., 1994; Davidson and Ackerman, 1993; Guggenberger and Zech, 1992). In addition, the disturbance of forest soils and cultivation of native prairies increases DOC concentration in the surface layer, probably due to decomposition processes (Cook and Allen, 1992; Zech et al., 1994; Quideau and Bockheim, 1996). Studies show that assessing SOC in whole soil profiles is important to evaluating effects of land use on the SOC pool and dynamics.

Therefore the objectives of this study were to evaluate the effects of forest, pasture, and cropland land uses on the SOC pool and to determine if land-use change impacts the SOC pool in the subsoil.

MATERIALS AND METHODS

Site Descriptions

Major land resource areas (MLRAs) share common land uses, topography, climate, hydrology, soils, and vegetation. There are 208 MLRAs in the U.S. and the Caribbean, ranging in size from thousands to millions of hectares. The three MLRAs chosen for this study include the Eastern Ohio Till Plain MLRA 139 (1.5 million ha), Erie–Huron Lake Plain MLRA 99 (3.5 million ha), and Indiana and Ohio Till Plain MLRA 111 (8.5 million ha) (Figure 15.1).

Soil samples were obtained from three sites within one soil series in each MLRA, which were chosen to represent forest, pasture, and arable land uses. The sites were also chosen on the basis of pedogenic criteria such as depth of horizons and textural classes typical of each soil series. The three soil series sampled were Mahoning in MLRA 139, Hoytville in MLRA 99, and Miamian in MLRA 111. Site locations and taxonomic classifications of soil series are listed in Table 15.1. The forest sites were sampled to obtain the baseline SOC content of that soil series and thus had never been totally cleared, although all showed some signs of partial logging in the past. Pastoral sites had Ap horizons indicating that they had been affected by previous tillage. The pastoral and arable

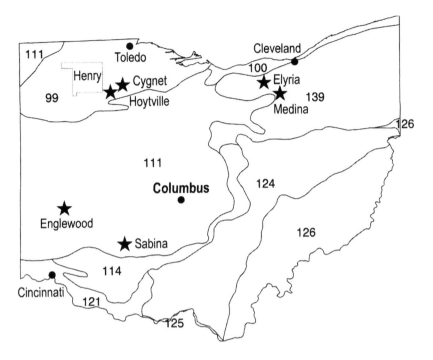

Figure 15.1 Locations of sampling sites and Henry County in Ohio. Starred locations indicate sampling sites. MLRAs: 99 = Erie-Huron Lake Plain, 111 = Indian-Ohio Till Plain, 139 = Eastern Ohio Till Plain.

Table 15.1 Site Locations and Classifications

Site	Series	Classification	Coordinates
Englewood	Miamian	Fine mixed mesic oxyaquic hapludalf	Lat. 39° 52′ Long. 83° 46′
Medina	Mahoning	Fine illitic mesic aeric epiaqualf	Lat. 41° 08′ Long. 82° 04′
Hoytville	Hoytville	Fine illitic mesic mollic epiaqualf	Lat. 41° 31′ Long. 83° 46′

Table 15.2 Site Characteristics

Soil Series	Land Use	Drainage	% Slope	Duration of Land Use (Years)
Miamian	Forest	Well drained	2	Native
	Pasture	Well drained	2	18
	Cropland	Well drained	2	100
Mahoning	Forest	Somewhat poorly drained	1	Native
	Pasture	Somewhat poorly drained	1	37
	Cropland	Somewhat poorly drained	1	35
Hoytville	Forest	Very poorly drained	1	Native
	Pasture	Very poorly drained	1	45
	Cropland	Very poorly drained	1	40

sites ranged in approximate age from 18 to 100 years since conversion to these land uses (Table 15.2). Hoytville and the Mahoning soils were poorly drained and had subsurface drainage; the Miamian soil was well drained. A drainage ditch was in close proximity to the pastoral and arable sites of the Hoytville series.

Sampling and Analysis

Soil profiles were dug on all sites using a backhoe. Three separate soil samples were taken from 0 to 5 cm, 5 to 10 cm, and 10 to 25 cm depths and then by horizon thereafter to a depth of 170 cm. Soil samples were air dried and ground to pass through a 0.5 mm sieve, and were analyzed for total carbon by the dry combustion method and for carbonates by the gas volumetric method (Soil Survey Laboratory Methods Manual, 1996). Soil bulk density measurements were made by the core method for every 10 cm layer to a depth of 60 cm and then for each pedogenic horizon thereafter (Blake and Hartge, 1986).

Data were statistically analyzed separately for each soil series. A one-way analysis of variance (ANOVA) table, according to a randomized design, was used to analyze the data statistically (Minitab, 2000). Statistical significance was calculated using the Turkey's test at the $p < 0.05$ level (Montgomery, 1991).

RESULTS AND DISCUSSION

Soil Bulk Density

Conversion to cropland significantly increased soil bulk density of the 0- to 10-cm layer for all but the Mahoning site (Figures 15.2 to 15.4). The Mahoning series had the highest bulk density in the pasture site for the 0- to 10-cm depth. Forest soil had lower bulk densities than all other land uses for each soil series; cropland soil may be compacted by machinery and pasture soil by animal traffic. Soil bulk densities were similar among land uses below 10-cm depth. Differences in soil bulk densities were partly due to differences in SOC contents (due to erosion and decomposition) and soil compaction. Accelerated erosion can deplete SOC content and alter soil bulk density (Frye et al., 1982; Fahnestock et al., 1996). In this study, however, accelerated soil erosion had the least impact on SOC content because all sites were on slopes of $\leq 2\%$ and soil erosion had not been a serious problem.

Soil Organic Carbon Content

The impact of land use on SOC content was most pronounced in the top 25-cm layer, below which SOC contents were similar among the three land uses. The SOC content in the 0- to 5-cm depths was generally lower in the arable than the forest or pastoral land uses (Figures 15.5 to 15.7). The lower SOC contents in cropland sites may be due to the breakdown of aggregates and the attendant high rate of mineralization (Reicosky, 1997; Lal, 1999). The Miamian and Hoytville soils had considerably higher SOC contents in the forest sites compared to the pasture sites. In the Mahoning soil, the pasture site had a comparable SOC content to that of the forest site. The same trend was observed for the 5- to 10-cm depth except for the Mahoning soil series, in which the SOC content of soil under pasture was equal to that of the cropland (Figure 15.7). In the 5- to 10-cm depth, all forest sites had higher SOC contents than the cropland sites. In the Miamian and Mahoning soil series, the SOC content in pasture sites were comparable to that of the cropland sites for the 5- to 10-cm depth.

For the 10- to 25-cm depth, differences in SOC content between land uses become less clear, indicating that there may be some redistribution of SOC as a result of tillage (Reicosky et al.,

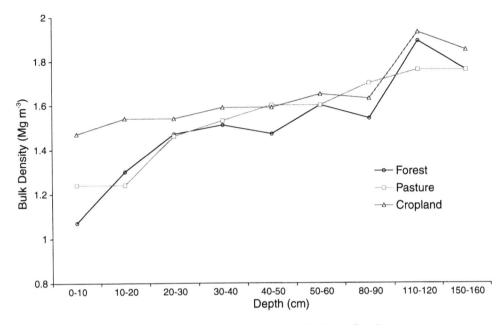

Figure 15.2   Soil bulk density profile for three land uses in the Miamian soil series.

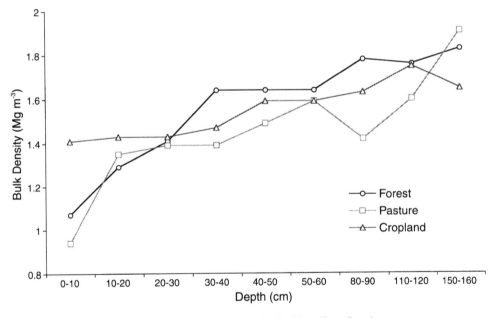

Figure 15.3   Soil bulk density profile for three land uses in the Hoytville soil series.

1995). Redistribution of SOC was apparent for the 10- to 25-cm depth because differences among land uses were less prominent across all soil series in comparison to those in the 5- to 10-cm depth. In addition, deep root penetration in the pasture sites may have a positive effect on SOC content compared to that of the forest sites at the 10- to 25-cm depth (Balesdent and Balbane, 1996).

No clear trends were observed with regard to land use below 25-cm depth. Increases in SOC content of the deeper layer, such as in the Hoytville soil series, may be attributed to interference by carbonates.

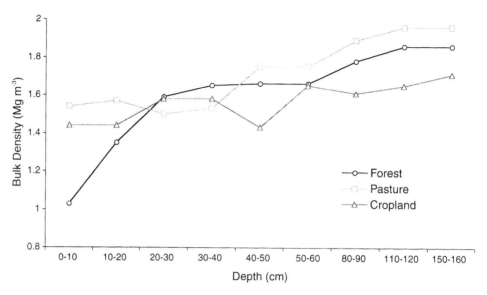

Figure 15.4 Soil bulk density profile for three land uses in the Mahoning soil series.

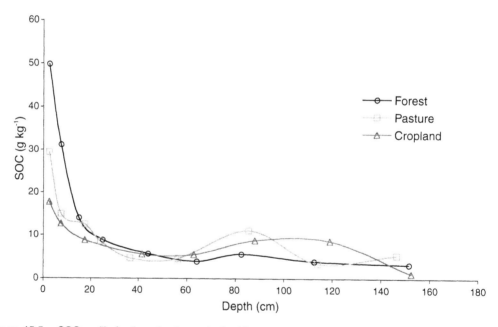

Figure 15.5 SOC profile for three land uses in the Miamian soil series.

Soil Organic Carbon Pool

The cumulative SOC (CSOC) pools for different depths in the profiles were affected by land use. For 5- and 10-cm layers, the least CSOC pool was measured in the cropland. For layers beneath, trends in CSOC pools were not well defined with regard to land use, and there were noticeable differences among soil series.

The data in Tables 15.3 and 15.4 show CSOC and SOC pools for different layers of three soil series. The forest and pasture sites for all soil series contained the highest CSOC pools for 5- and 10-cm layers, although not significantly higher for the Miamian soil for the 5- to 10-cm layer.

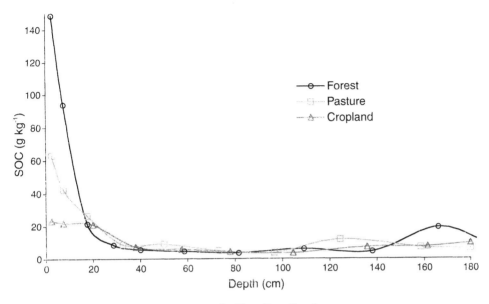

Figure 15.6 SOC profile for three land uses in the Hoytville soil series.

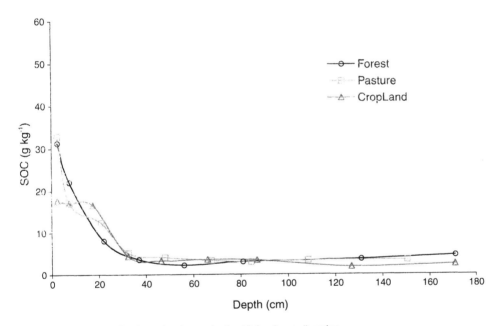

Figure 15.7 SOC profile for three land uses in the Mahoning soil series.

Comparing SOC pools for three soils for the 10- to 25-cm layer shows a significant difference among land uses. Cropland sites had higher or comparable SOC pools to those of forest sites. This trend is largely due to the effects of cultivation on the redistribution of SOC within the plow layer (Reicosky, 1995; Dick et al., 1998). In the Mahoning soil series pasture, land use had higher CSOC for all depths, probably due to dense and prolific root systems (Balesdent and Balbane, 1996). In addition, heavy grazing can also have a positive effect on SOC through dung additions and the stimulation of root growth (Frank et al., 1995; Franzluebbers and Stuedemann, 1999). Lastly, the presence of Ap horizons in pasture sites of all soils is an indication that there may have been a redistribution of SOC to deeper depths as a result of cultivation.

Below 25-cm depth, CSOC pools were affected differently in different soils by land uses. It is thus appropriate to examine the SOC pools for each layer to determine the depth at which changes occur under different land uses. Unlike the Miamian and Hoytville soil series, the Mahoning forest site had lower CSOC pools compared to its cropland site to a depth of 80 cm. The data in Tables 15.3 and 15.4 show that changes in SOC with regards to land use primarily occurred within the 10- to 25-cm layer. The cropland site for the Mahoning soil had an SOC pool larger than and the Hoytville soil one comparable to those of their forest sites for the 10- to 25-cm layer. This was in contrast to the SOC pool in the 0- to 5-cm and 5- to 10-cm layers. The Miamian soil had a lower SOC pool under cropland at this depth; however, it was under alfalfa for 1 year, which may have caused differences in SOC content.

The SOC pools tended to be similar among land uses below 25-cm depth. Hence CSOC also became increasingly similar with increase in depth, although the Mahoning soil series did not follow this trend for the deepest two layers. For the 60- to 140-cm layer the cropland SOC pools for the Miamian soil were considerably (although not always significantly) higher than the SOC pools of its forest site. The Miamian soil is well drained, which may have contributed to an increase in translocation of SOC within the soil profile. Starting at the 60-cm depth clay content largely increases in the Miamian soil, which may impede water infiltration and provide exchange sites for DOC to form organomineral complexes (Guggenberger and Zech, 1992). An increase in clay content in the Mahoning soil at 30-cm depth may have also contributed to some SOC accumulation at this depth. The Mahoning soil series had subsurface drainage and the Hoytville soil series had a drainage ditch in close proximity; this may have increased the translocation of DOC to subsoil.

Comparison of CSOC pools of the 0- to 25-cm and 0- to 170-cm layers among soils shows that depth affected the SOC pools. For the 0- to 25-cm layer, the Miamian cropland site had the lowest CSOC pool, but for the 0- to 170-cm layer, the CSOC pool was similar among land uses. Similar trends occurred in the Mahoning soil, although the CSOC pool in the pasture site was higher than those of the forest and cropland sites. In the Hoytville soil the cropland site had the lowest CSOC pool for the 0- to 25 and 0- to 170-cm layers, but for the 0- to 170-cm layer, forest and pasture land uses had comparable CSOC pools — unlike the CSOC pool of the 0- to 25-cm layer.

CONCLUSIONS

Conversion to cropland reduced the SOC pool in the 0- to 10-cm depth and increased it in the 10- to 25-cm depth. Conversion to pastoral land use has the potential to regain, maintain, and increase SOC pools above levels of SOC pools under native vegetation. The SOC pools below 40-cm depth are generally high in pasture, probably due to deep root systems of grasses.

Conversion to cropland had little impact on the SOC pool of the whole profile; cultivation increased SOC below the plow layer as a result of increased translocation of DOC. Cultivation practices that increase SOC in the surface layer have the potential to attain antecedent SOC levels. Whole soil pedons need to be analyzed for SOC content in order to get an accurate estimation of the current SOC pool.

ACKNOWLEDGMENTS

Help received in site selection and soil sampling and classification from the NRCS (Natural Resources Conservation Service) and ODNR (Ohio Department of Natural Resources) Soil and Water Division is gratefully appreciated. Specific thanks are due to the following soil scientists: John Allen, Doug Dotson, Mark Feusner, Jeff Glanville, Terry Lucht, Don McClure, Larry Milliron, and Rick Robbins. Soil analyses for soil organic carbon and $CaCO_3$ contents were completed by the National Soil Survey Center of the NRCS, Lincoln, NE.

Table 15.3 Cumulative SOC Pools and SOC Pools by Depth Under Three Land Uses for the Miamian and Mahoning Soil Series

Depth (cm)	Cumulative SOC Pools (Mg ha⁻¹) Miamian			Depth (cm)	SOC Pools for Each Depth (Mg ha⁻¹) Miamian		
	F	P	C		F	P	C
5	26.6 a	18.2 b	13.0 c	0–5	26.6 a	18.2 b	13.0 c
10	43.3 a	27.5 b	22.2 b	5–10	16.7 a	9.2 b	9.2 b
25	67.9 a	52.6 b	42.5 c	10–25	24.6 a	25.1 a	20.2 a
40	92.2 a	64.6 b	55.6 b	25–40	24.2 a	12.1 b	13.1 b
60	107.2 a	79.0 b	73.2 b	40–60	15.1 a	14.4 a	17.6 a
80	119.0 a	107.4 ab	94.81 b	60–80	11.8 b	28.4 a	21.6 ab
110	147.3 a	154.9 a	139.4 a	80–110	28.3 a	47.4 a	44.5 a
140	180.6 a	177.9 a	181.8 a	110–140	33.3 a	23.1 a	42.4 a
170	204.9 a	206.1 a	187.1 a	140–170	24.3 ab	28.2 a	5.4 b

Depth (cm)	Cumulative SOC Pools (Mg ha⁻¹) Miamian			Depth (cm)	SOC Pools for Each Depth (Mg ha⁻¹) Miamian		
	F	P	C		F	P	C
5	16.1 b	25.3 a	12.7 c	0–5	16.1 b	25.3 a	12.6 c
10	27.4 b	38.4 a	24.9 c	5–10	11.3 b	13.1 a	12.2 ab
25	45.4 c	66.7 a	60.7 b	10–25	18.0 c	28.3 a	35.9 a
40	56.2 c	80.3 a	70.9 b	25–40	10.8 b	13.6 b	10.2 b
60	67.7 c	94.0 a	80.9 b	40–60	11.4 ab	13.7 a	10.0 b
80	78.6 c	105.7 a	92.5 b	60–80	10.9 a	11.7 a	11.6 a
110	95.1 b	123.9 a	104.9 b	80–110	16.5 a	18.1 a	12.5 a
140	114.0 b	143.6 a	116.9 b	110–140	18.9 ab	19.7 a	11.9 b
170	138.9 b	163.6 a	129.1 b	140–170	24.9 a	20.0 b	12.2 c

Note: Significance at a ($p < 0.05$) level; values with same letters in rows are statistically similar. F = forest, P = pasture, C = cropland.

Table 15.4 Cumulative SOC Pools and SOC Pools by Depth Under Three Land Uses for the Hoytville Soil Series

Cumulative SOC Pools (Mg ha⁻¹)				SOC Pools for Each Depth (Mg ha⁻¹)			
Depth (cm)	Hoytville			Depth (cm)	Hoytville		
	F	P	C		F	P	C
5	79.5 a	29.4 b	15.2 c	0–5	79.5 a	29.4 b	15.2 c
10	129.6 a	48.7 b	28.7 c	5–10	50.1 a	19.3 b	13.5 c
25	171.1 a	102.3 b	60.0 c	10–25	41.5 a	53.6 a	31.2 a
40	186.9 a	121.0 b	83.2 c	25–40	15.8 c	18.7 b	23.2 a
60	203.0 a	135.5 b	97.3 c	40–60	16.1 a	14.5 a	14.1 a
80	216.4 a	146.7 b	109.2 c	60–80	13.4 a	11.2 b	11.9 ab
110	242.1 a	170.8 b	123.0 c	80–110	25.7 a	24.0 b	13.8 b
140	268.8 a	226.9 b	149.4 c	110–140	26.7 b	56.2 a	26.4 b
170	288.8 a	266.9 a	185.1 b	140–170	20.0 a	29.9 a	35.8 a
200	323.9 a	302.3 a	231.3 b	170–200	35.1 a	35.4 a	46.2 a

Note: Significance at a (p<0.05) level; values with same letters in rows are statistically similar. F = forest, P = pasture, C = cropland.

REFERENCES

Balesdent, J. and M. Balbane, Major contribution of roots to soil carbon storage inferred from maize cultivated soils, *Soil Biol. Biochem.*, 28:1261–1263, 1996.

Blake, G.R. and K.H. Hartge, Bulk density, in Klute, A. (ed.), *Methods of Soil Analysis*, Part 1, 2nd ed., Agron. Monogr. 9, ASA and SSSA, Madison, WI, 1986, 363–375.

Chantingy, M.H. et al., Dynamics of soluble organic C and C mineralization in cultivated soils with varying N fertilization, *Soil Biol. Biochem.*, 31:543–550, 1999.

Cook, B.D. and D.L. Allan, Dissolved organic-carbon in old field soils — compositional changes during the biodegradation of soil organic-matter, *Soil Biol. Biochem.*, 24:595–600, 1992.

Davidson, E.A. and I.L. Ackerman, Changes in soil carbon inventories following cultivation of previously untilled soils, *Biogeochemistry*, 20:161–193, 1993.

Dick, W.A. et al., Impacts of agricultural management practices on C sequestration in forest-derived soils of the eastern Corn Belt, *Soil Tillage Res.*, 47:235–244, 1998.

Fahnestock, P.B., R. Lal, and G.F. Hall, Land use and erosional effects on two Ohio Alfisols. I. Soil Properties, *J. Sust. Agric.*, 7:63–84, 1996.

Frank, A.B. et al., Soil carbon and nitrogen of northern Great Plains grasslands as influenced by long-term grazing, *J. Range Manage.*, 48:470–474, 1995.

Franzluebbers, A.J. and J.A. Stuedemann, Soil organic carbon storage under pastures of the southern piedmont USA, USDA-Agricultural Research Service, Watkinsville, GA, 1999.

Frye, W.W. et al., Soil erosion effects on properties of two Kentucky soils, *Soil Sci. Soc. Am. J.*, 46:1051–1055, 1982.

Guggenberger, G. and W. Zech, Retention of dissolved organic-carbon and sulfate in aggregated acid forest soils, *J. Environ. Qual.*, 21:643–653, 1992.

Lal, R., Soil management and restoration for C sequestration to mitigate the accelerated greenhouse effect, *Prog. Environ. Sci.*, 1(4):307–326, 1999.

Liljeroth, E., P. Kuikman, and J.A. Vanveen, Carbon translocation to the rhizosphere of maize and wheat and influence on the turnover of native soil organic-matter at different soil-nitrogen levels, *Plant Soil*, 161:233–240, 1994.

Minitab, MINITAB 13.1, Minitab Inc., State College, PA, 2000.

Montgomery, D.C., *Design and Analysis of Experiments*, 3rd ed., John Wiley & Sons, Inc., New York, 1991.

Nambu, K. and K. Yonevayashi, Role of dissolved organic matter in translocation of nutrient cations from organic layer materials in coniferous and broad leaf forests, *Soil Sci. Plant Nutr.*, 45:307–319, 1999.

Quideau, S.A. and J.G. Bockheim, Vegetation and cropping effects on pedogenic processes in a sandy prairie soil, *Soil Sci. Soc. Am. J.*, 60:536–545, 1996.

Reicosky, D.C., Tillage methods and carbon dioxide loss: fall versus spring tillage, in Lal, R. et al. (ed.), *Management of Carbon Sequestration in Soil*, CRC Press, Boca Raton, FL, 1997, 99–111.

Reicosky, D.C. et al., Soil organic matter changes resulting from tillage and biomass production, *J. Soil Water Conserv.*, 50:253–261, 1995.

Soil Survey Laboratory Methods Manual, USDA-NRCS. U.S. Gov. Print. Office, Washington, D.C., 1996.

Zech, W., G. Guggenberger, and H.R. Schulten, Budgets and chemistry of dissolved organic-carbon in forest soils — effects of anthropogenic soil acidification, *Sci. Total Environ.*, 152:49–62, 1994.

CHAPTER **16**

CQESTR — Predicting Carbon Sequestration in Agricultural Cropland and Grassland Soils

Ron W. Rickman, Clyde L. Douglas, Jr., Stephan L. Albrecht, and Jeri L. Berc

CONTENTS

ABSTRACT

Implementation of a national program for increasing organic carbon levels in agricultural soils will ultimately require specific management plans for individual fields throughout the nation. A tool to help create those plans must provide estimates of change in soil carbon as controlled by crop rotation, tillage practices, local climate, and soil. A program named CQESTR (sequester) has been developed by USDA ARS at Pendleton, OR, to fulfill this need. The program uses a budget of carbon-based residue additions and microbial decomposition losses to estimate rate of gain or loss of soil organic matter (OM). Rotation production and tillage practices are obtained from existing data files created and utilized by the revised universal soil loss equation (RUSLE). Residue nitrogen content and soil layer information including depth, bulk density, and starting organic matter content must be obtained independently from the RUSLE data files. The rate of gain or loss of soil carbon was calibrated to long-term management plots from Pendleton. Comparisons with reported soil carbon changes in long-term plots throughout North America have begun.

INTRODUCTION

Atmospheric carbon dioxide (CO_2) concentration has been steadily increasing for the past several decades. Average global air temperature has also risen during the same period. Physical laws indicate that certain greenhouse gases and global air temperature have a cause-and-effect relationship. Carbon dioxide is one of the greenhouse gasses, which permit the transmission of short wavelength radiation (visible and ultraviolet light received from the sun), and absorb long wavelength radiation (infrared or heat radiation emitted by the earth), providing the same effect as glass in a greenhouse.

It has been predicted that increasing global air temperatures may accelerate melting of ice fields with concomitant sea level rise (DOE/ER/60235-1), aggravate weather pattern shifts (such as those investigated by Nicks et al., 1993) contributing to flooding and drought, and increase the number and intensity of extreme weather events. Human activity substantially contributes to increased atmospheric CO_2 and the production of other greenhouse gasses (Lal et al., 1998). It should therefore be possible to reduce the human contribution to mitigate CO_2 build-up and atmospheric warming.

The increase in atmospheric CO_2 can be slowed by retaining the carbon (C) captured by plant photosynthesis. Soil organic matter (OM) is a natural reservoir of organic C. The amount of C that could potentially be stored in soils in the U.S. has been estimated at between 5 and 10% of the current annual emission (Lal et al., 1998). Historical land-use changes that converted grass and forest land to cropland, as well as many conventional agricultural practices, have caused soil to lose much of its natural reservoir of organic C. Fortunately, improved agricultural conservation management practices, such as conservation no-till systems that keep residues on the soil surface, and utilization of cover crops, crop rotations, and organic amendments, can maintain or increase the soil OM reservoir.

By adopting conservation management systems, production agriculture can sequester C and contribute to the reduction of concentrations of atmospheric CO_2. This contribution will be facilitated if agricultural decision makers have appropriate incentives and specific guidance on the soil C impacts of their management decisions. There is a need for a tool that will estimate how management systems will effect OM storage in soils at the site-specific field level. These estimates could be provided by a field-level C sequestration model sensitive to local soils, climate, crop and tillage management systems, yields, cover crops, and organic amendments (Berc, 1999). It is highly desirable that this model operate in the field, utilizing readily accessible data; it can be applied to assist farm planning efforts to enhance C sequestration, as erosion prediction models have been used to assist farmers to plan management systems to control erosion. In may also be added to national resource inventory protocols to track regional and national scale soil C stocks. Such a tool can help develop policies and programs for C sequestration, as soil loss prediction models have been used in statistical national and regional natural resource inventories to develop and evaluate erosion control policy objectives.

The Agricultural Research Service (ARS) staff at the Pendleton Research Center is currently developing a C sequestration model "CQESTR" that will compute the decomposition rate and residence time in the soil of C from antecedent OM, crop residues, crop roots, and organic C containing amendments such as compost, manure, sewage sludge, or other biosolids.

METHODS

The core of the model is the residue decomposition model "D3R" (Douglas and Rickman, 1992), which uses air temperature and residue nitrogen content as the principal controllers of decomposition rate. Residue location above or below the soil surface, as determined by tillage practices, provides an index for the effect of water on the rate of decomposition. Decomposition computations by D3R have been compared to, and found to accurately predict, decomposition of residues from data sets for a variety of crops from Alaska, Washington, Oregon, Idaho, Missouri, Indiana, North Carolina, Georgia, Texas, Colorado, Saskatchewan, Canada, and Uppsula, Sweden (laboratory study) (Curtin et al., 1998; Douglas and Rickman, 1992; Moulin and Beckie, 1993, 1994; Ma et al., 1999).

The majority of organic C-based residues added to a soil are converted to CO_2 by microbes. A small fraction of residues is consumed by worms, insects, and small mammals. This nonmicrobial consumption depends strongly on climate, the kind and mass of surface residue present, the length of time the residue has been on the soil surface, and the local population of soil fauna. Predicting this consumption would be an independent modeling project not attempted in CQESTR. If this fraction is known to be locally significant relative to microbial oxidation, it should be subtracted from the residue mass input to CQESTR. Physical removal of OM and residue from a field by wind or water erosion is not computed by CQESTR; grazing or mechanical removal of a fraction of harvested residue is also assumed to be accounted for in the values of residue mass provided as input to CQESTR.

Much of the information required by CQESTR can be obtained nationally from existing data files created for use in the universal soil loss equation (USLE) and the revised universal soil loss equation (RUSLE). None of the computations in CQESTR are derived from or utilize routines found in USLE or RUSLE. The nitrogen content of residues may be obtained from local information, literature values, or existing compilations of plant nutrient content, such as the database CPIDS developed for the Water Erosion Prediction Project (WEPP) by Deer–Ascough (1993) or the FAO Tropical Feeds database (Speedy and Waltham, 1998). The FAO database lists crude protein content of hundreds of plant species produced under a wide variety of growing conditions. Nitrogen content for crop residues can be calculated using the 16% nitrogen content for protein. Soil OM content by layer and layer depths can be obtained from the national soil surveys (available from the MUIR database; USDA-NRCS 1997) and local Natural Resources Conservation Service (NRCS) offices. Root distributions are determined from an exponential decay with depth relationship that depends on crop type and local climate (Belford et al., 1987; Gerwitz and Page, 1974).

The decomposition rate of antecedent soil OM and the rate of conversion of composting residues to OM will be calibrated using soil OM data from management treatments with differing amounts and types of added residues. These data are from the long-term management plots on the Pendleton Research Center and other U.S. and international long-term management experiments. Many of these data sets are available from the soil organic matter network (SOMNET) of the Global Change and Terrestrial Ecosystems (GCTE) Project of the International Geosphere–Biosphere Program. Other models that compute soil C dynamics have been collected and compared by the GCTE (Smith et al., 1996). Although several of them can be used to estimate soil OM change, data requirements prohibit national use on a field-by-field scale.

CQESTR will compute C sequestration in a field soil as controlled by climate, crop production, and management practices used in that field. Rate of change of soil C is provided using a windows format computer program that allows point-and-click selection of most input data. Program output includes, for all rotations and management options requested, short- and long-term trends of surface and buried residues and changes in the soil OM content.

RESULTS

Predictions of soil OM content are being compared with observations at several sites across North America. At one of the sites where the comparison has been completed, Watkinsville, GA, better agreement was found with conventionally tilled annual sorghum or sorghum and soybean rotations (slope of the regression line for computed vs. observed soil OM was 1.05 with an R^2 of 0.97) than for no-tilled versions of those treatments (slope of regression 0.97 but R^2 of 0.78). Data used in the comparisons came from Paul et al. (1997). The greatest deviation from measured soil OM was in plots with a history of severe erosion by water, although the erosion occurred before the described management treatments had been applied.

One feature of the model, the computation of the time for conversion of residue into OM, provides an estimate of the lag between residue addition and the resulting change in soil OM. That lag time is difficult to track with conventional measures of soil OM. For CQESTR, this lag time of about 4 years was estimated by calibrating the model to field data from plots with over 60 years of known management and production history at Pendleton. Comparison of the model with both increase and decrease of soil OM following addition of residues of different types will continue at sites with different climatic patterns throughout North America.

Another consistent prediction from CQESTR is a continuing decline in C in soil beneath the tillage layer where roots are the primary source of new carbon. Information about the distribution of root mass with depth for the major agricultural crops, as well as observations of root mass beneath cover crops and range or pasture grasses, will be needed to adjust the root distribution function in the model.

CONCLUSION

A model has been developed that operates on readily accessible data and provides a favorable comparison to observed soil OM. Model predictions of change in soil OM compare more favorably with data from conventionally tilled fields than from those where reduced or no-till was practiced. Where deviations from observed trends on soil OM are greater than the variation expected from the model, parameters not currently considered by the model will be sought to reduce the deviation. Additional data on rate of conversion of residue to soil OM and root-mass distribution with soil depth for all crops will be helpful for this and all other soil C models. If water or wind erosion was a significant component of the carbon budget of a field site, observations of soil OM may need to be adjusted by estimates of those contributions before comparison with CQESTR.

REFERENCES

Belford, B., D. Klepper, and R.W. Rickman, Studies of intact shoot-root systems of field-grown winter wheat. II. Root and shoot developmental patterns as related to nitrogen fertilizer, *Agron. J.*, 79:310–319, 1987.

Berc, J., NRCS roles and needs, presented at USDA-ARS workshop on soil carbon, Baltimore, MD, Jan. 12, 1999.

Curtin, D. et al., Carbon dioxide emissions and transformations of soil carbon and nitrogen during wheat straw decomposition, *Soil Sci. Soc. Am. J.*, 62:1035–1041, 1998.

Deer–Ascough, L.A., Crop Parameter Intelligent Database System, computer program available in WEPP95 USDA-Water Erosion Prediction Project, USDA National Soil Erosion Laboratory, West Lafayette, IN, 1993.

DOE/ER/60235-1, Glaciers, Ice Sheets, and Sea Level: Effect of a CO_2-induced Climatic Change, Workshop Report, Sept. 13–15, 1984, Seattle, WA, 1985.

Douglas, C.L. Jr. and R.W. Rickman, Estimating crop residue decomposition from air temperature, initial nitrogen content, and residue placement, *Soil Sci. Soc. Am. J.*, 56:272–278, 1992.

Gerwitz, A. and E.R. Page, An empirical mathematical model to describe plant root systems, *J. Appl. Ecol.*, 11:773–782, 1974.

Lal, R. et al. (eds.), *The Potential of U.S. Cropland to Sequester Carbon and Mitigate the Greenhouse Effect*, Ann Arbor Press, Chelsea, MI, 1998.

Ma, L. et al., Decomposition of surface crop residues in long-term studies of dryland agroecosystems, *Agron. J.*, 91:401–409, 1999.

Moulin, A.P. and H.J. Beckie, Predicting crop residue decomposition, *Proc. Sask. Soils Crops Workshop*, 104–109, 1993.

Moulin, A.P. and H.J. Beckie, Predicting crop residue in cropping systems, *Proc. Sask. Soils Crops Workshop*, 7–14, 1994.

Nicks, A.D. et al., Regional analysis of precipitation and temperature trends using gridded climate station data, in Wang, S.S.Y. (ed.), *Advances of Hydro-Science & Engineering*, The University of Mississippi, Vol. 1, Part A, 497–502, 1993.

Paul, E.A. et al., *Soil Organic Matter in Temperate Agroecosystems: Long-Term Experiments in North America*, CRC Press, New York, 1997.

Smith, P., J.U. Smith, and D.S. Powlson (eds.), GCTE Report No. 7. GCTE Task 3.3.1 soil organic matter network (SOMNET): 1996 model and experimental metadata. GCTE Focus 3 Office, Wallingford, U.K., 1996.

Speedy, A. and N. Waltham, *Database from Tropical Feeds*, B. Gohl (ed.), FAO of the UN, 1998, [Online] Available at http://www.fao.org/waicent/FaoInfo/AGA/AGAP/FAP/ (verified Nov. 19, 2001).

USDA-NRCS Soil Survey Division, National MUIR database [Online], available at http://www.stat-lab.iastate.edu/soils/index.html/. (verified Nov. 19, 2001), 1997.

CHAPTER **17**

Case Study of Cost vs. Accuracy When Measuring Carbon Stock in a Terrestrial Ecosystem

Gordon R. Smith

CONTENTS

NEED FOR COST-EFFECTIVE MEASUREMENTS

Concern about negative effects of climate warming resulting from increased levels of greenhouse gases in the atmosphere has led nations to establish goals for reduction of these emissions. Initial targets for reductions are stated in the Kyoto Protocol to the United Nations Framework Convention on Climate Change, which allows trading credits that represent verified emissions reductions and removals of greenhouse gases from the atmosphere (United Nations Framework Convention on Climate Change Secretariat, 1997). Emissions trading may make it possible to achieve reductions in net greenhouse gas emissions for far less cost than without trading (Dudek et al., 1997). Storing carbon in soils can help offset greenhouse gas emissions while providing environmental benefits such as increasing site productivity, speeding water infiltration, and maintaining soil flora and fauna diversity (Lal et al., 1998). Storing carbon in forests may provide environmental benefits resulting from increased structural diversity of forests and increased numbers of mature or old trees (Row et al., 1996).

Including terrestrial carbon sinks in greenhouse gas emission trading markets requires reliable and cost effective measurement of amounts stored and released by projects. Without quantified knowledge of the precision of measurements, one runs the risk that sampling designs installed now will fail to detect much of the change in soil carbon occurring between now and remeasurement in 5 to 10 years. At the same time, measurement cost must be less than the market price of the amount of sequestration detected by that measurement. This chapter outlines factors affecting cost effectiveness of measuring terrestrial carbon sequestration, offers a method for predicting cost effectiveness of measurements, and provides an example. The goal is to enable those planning sequestration projects to maximize financial return by optimizing between increasing the amount of sequestration detected by increased precision of measurements, and controlling the cost of measurement.

FACTORS AFFECTING COST EFFECTIVENESS

On-site measurement is generally regarded as more reliable than estimating sinks or emissions from general factors. At the project scale, measurement of change in carbon stocks is more tractable than precisely measuring flows. (*Project scale* is defined as the size of unit at which a carbon sequestration project might be implemented, which is larger than a field or soil mapping unit but smaller than a major land resource area or region, and typically several hundred to several tens of thousands of hectares in size.) Measurement of change in stock is accomplished by measuring the amount of carbon present within the project area at a baseline time, measuring again at a later time, and then finding the difference between the amounts present at the different times. For the purposes of this analysis, measurement errors are assumed to be unbiased and merely to increase the variance of measured changes.

For a given number of samples, tracking individual points through time gives a more precise measurement of change than two independent measurements made at different times. Tracking individual sampling units through time is sometimes called paired sampling (Wonnacott and Wonnacott, 1984; Smith, 2000). A sampling unit is the entity measured during sampling. In opinion surveys the sampling unit is typically an individual person; when sampling forests, the sampling unit is often a plot or quadrat, and when sampling soil, it is sometimes called a microsite. In soil sampling, one or more soil cores may be measured at a single microsite. In paired sampling, a measurement is recorded for each sampling unit at each time measurements are made; then, for each sampling unit, the change over time is calculated. This set of changes over time is then used to make statistical calculations of variance, standard error, and confidence intervals.

Several factors interact to determine the cost of field measurement of change in carbon stocks. For heuristic purposes these factors can be categorized as:

- Variability of physical sequestration
- Variability introduced by plot design
- Proportional change in stock to be detected, and
- Cost per sample

Variability of physical sequestration is the variability that occurs in the world when comparing one microsite to another. Some sites sequester more carbon than others because of soil, weather, climate, vegetation, management practices, and prior history of use. Physical variability exists, and proper measurement quantifies the amount of variability while not allowing it to confound measurement of the variable of interest. Stratification of sampling is a method of enhancing the precision of estimates made from sampling, when values of the variable measured can be estimated in a way that allows grouping sites with similar values for analysis (Avery and Burkhart, 1994). Applying stratification to measurement of carbon sequestration means dividing the sampled area into subareas, or strata, that are believed to have larger and smaller amounts of sequestration, thus leaving less variability within each stratum. Variance is calculated for each stratum; then numbers for the different strata are combined.

Plot design can mitigate or worsen the effects of physical variability occurring in the world. Ecological systems tend to have variability at multiple scales and it is important to choose a plot design to capture the variability of the scale of interest while averaging out variability that occurs at finer scales. Soil carbon measurements typically involve laboratory analysis of a few g of soil. Because of the fine scale variability of soil carbon, very different amounts of carbon may be measured in cores taken one m apart, or adjacent segments within an individual core (Smith et al., 2000). In soil carbon sequestration projects, one generally measures change in carbon stock across ownerships or landscapes and is usually not particularly interested in variability of soil carbon from one soil ped to another, or from one aggregate to the next. It is more useful to average out the fine-scale variability occurring within individual pedons while retaining an understanding of the variability across soil types and treatments. Fine-scale effects could be averaged out by gathering and mixing all the soil of a pedon. However, the goal of averaging can be accomplished more cheaply and easily, and with less disturbance to the site, by taking several cores and compositing them.

When measuring change in carbon stock, the proportion of change in stock that is to be detected interacts with several factors to determine the number of samples required. These additional factors are the variability of the change across sampling units, the proportion of the change to be detected, and the confidence required of measurements. As noted earlier, variability across sampling units is a function of the physical variability of the world and plot design. Equations are available for calculating the number of samples needed to detect a given change in stock with a given confidence, for any given variance.

Cost per sample includes both fixed and variable costs of measurement. Costs of designing a sampling system and calculating results from data are relatively fixed and do not vary linearly with the number of plots measured. The costs of establishing and measuring each plot (including laboratory measurements) are variable, and are essentially the cost of a single plot times the number of plots. Average cost per plot can be calculated by summing all variable and fixed costs and dividing by the number of plots.

In general, as physical variability increases, the number of samples needed to achieve a given level of precision also increases. Plot design can mitigate or exacerbate this difficulty. For larger proportional changes in stock to be detected, a smaller number of samples is needed to detect change within a required level of confidence for a given variability across samples. While total cost of measurement can be calculated by summing component costs, the viability of a carbon sequestration project hinges on the measurement cost per ton of sequestration documented. Cost per ton detected decreases if a given sampling effort can be applied to a larger total change in stock. Ways to increase the total change in stock available for detection include measuring sequestration occurring over a longer time period, or designing a project such that a given number of samples represents a larger area.

COST EFFECTIVENESS OF MEASUREMENT: AN EXAMPLE

A pilot study was conducted to assess variability resulting from a plot design intended to optimize between limiting sampling cost and limiting variability of measurements. This study included calculation of sampling costs. This sampling was an initial measurement, without subsequent remeasurement; consequently, this example does not give information about variability of change over time by plot. Despite lack of data about variability of change over time, this study indicates attributes that a project needs to be financially feasible. Because of substantial interest in forest carbon sequestration projects, the study included measurements of both forest and soil carbon.

The pilot study area is located in southwestern Oregon (43° 35'N, 123° 30'W), with an average elevation of about 60 m. Field sampling was conducted in April 1999, near the beginning of the growing season. Twenty-five sampling plots were distributed across a 6-ha unit within a large ownership. Sampling was not stratified. The sampled area is relatively flat valley bottom bisected by a stream. Although the land use was pasture, the unit was substantially invaded by brush and was reverting to forest. In the portions of the unit with tree cover, the dominant species were *Alnus rubra* (red alder) and *Umbellularia californica* (California laurel). The dominant brush species were the invasive exotics *Cytisus scoparius* (L.) (Scotch broom) and *Rubus procerus* (Himalayan blackberry, also known as *R. discolor*).

For purposes of field measurement and calculation, the carbon on the site was categorized into six pools: soil, vegetative litter on the ground surface, live understory plants, coarse woody debris, standing dead trees, and live trees. Definitions of these pools included all carbon on the site except that in soil more than 50 cm below the top surface of mineral soil. Each pool was measured separately using a nested plot design. Plot design was chosen to limit variability of measurements and sampling effort by using larger plot sizes for less frequently occurring items (such as standing dead trees) and smaller plots for more frequently occurring items (such as understory vegetation). Plot sizes were 0.10 ha for snags, 0.01 ha for larger live trees and coarse woody debris, 0.001 ha for small live trees, and 0.0001 ha for nontree understory vegetation and litter. Soil was sampled by compositing three cores taken 1 m apart and assumed to represent an area of 3 m², or 0.0003 ha. Cores were 2.5 cm in diameter. Allometric equations were used to calculate tree biomass from tree species, diameter at breast height, and total height data.

Size, decay class, and species (when it could be determined) were recorded for snags and woody debris. Density factors for woody material were taken from the literature. Understory vegetation and litter were collected and weighed, with subsamples from each plot dried and weighed to get a factor for dry weight. Carbon content of biomass was calculated using factors from the literature and soil samples were tested for organic carbon content by a commercial laboratory using the Walkley–Black method. Plots were distributed around the study area, with the rule that plot centers were to be not less than 36 m apart or 16 m from the boundary of the study area. The number of tons of carbon per hectare was calculated for each pool. The coefficient of variation (standard deviation divided by the mean) of each pool was calculated (Table 17.1). Using the coefficient of variation allows comparison of variability of pools that have very different means by normalizing the numbers.

Woody debris was distributed very unevenly across the project area, with many plots containing no woody debris; consequently, the coefficient of variation of the woody debris measurement is high. Similarly, live trees, standing dead trees, and understory are unevenly distributed and thus have high coefficients of variation, while litter and soil carbon are less unevenly distributed. To decrease variability, larger plots could be used. However, in this project area, increasing plot size would have limited effect because part of the area was herbaceous vegetation, part brush, and part tree, and larger plots would not have averaged out this heterogeneously. Stratifying the project area into separate blocks of pasture, brush, and trees would substantially reduce the variability observed within each stratum; however, this strategy would also require delineating boundaries and areas of strata, as well as increasing the total number of plots to get a statistically viable number of measurements within each stratum.

The coefficient of variation of a single estimate of the mean carbon stock of a pool does not tell whether change could be detected upon remeasurement, using samples paired in time. Coefficient of variation does tell which change must occur to be detectable with a given number of

Table 17.1   Observed Carbon Stock and Coefficient of Variation of Carbon Stock by Pool

Pool	Coefficient of Variation	Carbon Stock (tons/ha)
Soil (to 50-cm depth)	0.22	197
Litter	0.86	10
Understory	2.38	6.7
Woody debris	4.17	2.9
Standing dead	1.83	0.05
Trees	2.33	21

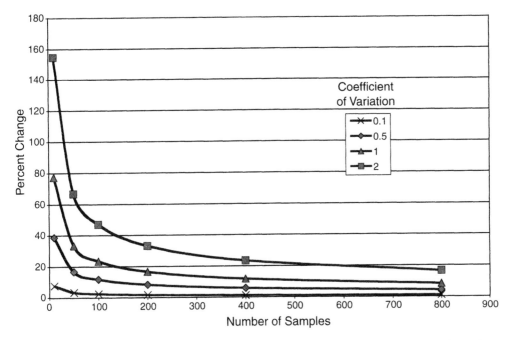

Figure 17.1   Minimum detectable percentage change in stock as a function of the number of samples and coefficient of variation.

samples and a given confidence, using two independent random samples (Figure 17.1). In Figure 17.1, a confidence level of 0.9 is given, and shows the percent change in stock that is likely to be detected by two independent, simple, random samplings using different numbers of samples. This figure shows that proportionally large changes, such as those greater than 40%, are relatively easy to detect with several dozen samples, as long as the coefficient of variation is less than 1 or 2. If the coefficient of variation is small, such as 0.1, changes of a few percent of stock can be detected with a few dozen samples. However, changes of 10% or less are unlikely to be detected if the coefficient of variation is 1 or greater, even if hundreds of samples are taken. The equation used to generate Figure 17.1 is:

$$D = t_{.05} CV \sqrt{\frac{1}{n_1} + \frac{1}{n_2}}$$

where:

D   = Difference likely to be detected between mean estimate of stock at time 1 and mean estimate of stock at time 2

$t_{.05}$ = Critical t value for p = .90, two tailed test, with degrees of freedom = $(n_1 - 1) + (n_2 - 1)$

CV   = Coefficient of variation of sample, which is the sample deviation divided by sample mean

n_1 = Number of sample points at time 1

n_2 = Number of sample points at time 2

The same number of samples is assumed to have been taken before and after the sequestration period, or $n_1 = n_2$. Strictly speaking, the variations observed at times 1 and 2 would be pooled and this pooled variation would be used in place of the coefficient of variation in this equation. However, the variation observed at time 2 is not known. On average, the pooled variation would be slightly

Table 17.2   Expected Change in Carbon Stock, over 10 Years, by Pool

Pool	Expected 10-Year Change (% baseline stock)	Expected 10-Year Change (tons/ha)
Soil	1 to 5	1 to 6
Litter	0 to 50	0 to 5
Understory	−85 to −70	−6 to −5
Woody debris	−25 to 0	1 to 0
Standing dead	−50 to 0	0
Trees	60 to 75	12 to 16

lower than the coefficient of variation observed at time 1 because it is based on a greater number of observations. Consequently, this simplifying assumption is conservative.

Clearly, it is important to estimate the likely proportional change in carbon stock in order to predict how many samples are likely to be needed to detect the change. To determine proportional change in carbon stock, likely change was estimated for each carbon pool within the project area. Change in carbon stock was calculated in tons per hectare, starting with conditions present at the time of the initial measurement, assuming site preparation and planting as planned for the project, and assuming normal plant growth and successional dynamics (Table 17.2). Growth and decay rates were taken from published studies or locally applicable models or tables. Ranges are given because published values vary widely and because the effects of management activities are not strictly determinable.

Changes were predicted for a 10-year period, starting at the time of the baseline measurement. To calculate the expected percent change in stock, the expected change was divided by the baseline stock given in Table 17.1. Soil carbon is expected to increase as forest cover increases. Litter may increase because litter from planted conifers decays more slowly than the present litter and because the ground surface temperature may decrease as tree cover increases. Understory biomass is expected to decrease substantially because of brush-cutting during preparation of the site for tree planting. Woody debris and standing dead biomass are expected to decrease from decay and lack of replacement because the trees on site that could become dead wood are generally young and vigorous and cover only a modest portion of the project area. Existing trees are not yet mature and are expected to grow normally.

Table 17.2 illustrates the effect of the size of the initial stock on the proportion of change likely to occur. Because the initial stock of standing dead biomass is very small, a very small change is proportionately large; a much larger change (in tons per hectare) of litter or soil carbon would yield a much smaller proportional change in those pools.

The number of plots needed to detect change with statistical reliability can be estimated by applying the coefficient of variation and expected percentage change in a particular pool to Figure 17.1 and reading the number of samples likely to be required to detect the given change. Remember that this is the power of independent random samples, not differences between measurements paired on permanent plots. Even assuming the greatest likely changes in stock, the results are not encouraging. No number of plots is likely to detect the likely change in woody debris, and about 400 plots are likely to be needed to detect change in soil carbon. Approximately 100 plots would be required to detect change in understory, standing dead, and tree carbon. Several dozen plots would be needed to detect expected change in litter. These numbers of plots should detect that change has occurred with 90% confidence; to ascertain the specific amount of change, additional plots would be needed.

Fortunately, measuring change on paired measurements taken from permanent plots gives much more power than the difference of two independent random samples described previously. In carbon measurement, a pair of samples would be the measurements taken at time n and time $n + 1$ on a specific plot. One then determines statistics of the set of differences between pairs, rather than comparing an average measured at one time to an average measured at another time. Because this pilot study has not been installed long enough for remeasurement to be due, it does not yet have data on change over time.

Cost per sample varies as a function of a number of factors, which include costs of needed supplies and use of equipment, labor to travel to and locate plots, labor to take measurements at plots, preparation and laboratory analysis of samples, and analysis of data. When a fixed area will be sampled (such as plots totaling a given percentage of the project area or a given number of hectares) and there is a choice between measuring fewer larger plots or more smaller plots, statisticians typically recommend measuring the latter because statistical power is gained by increasing the sample size. However, statistical power is also gained when the samples are less variable, and decreased variability may be obtained by increasing the plot size if the increase is sufficient to average out fine-scale variability. To optimize this choice, one must know the approximate spatial distribution of what is measured. Additionally, labor costs affect this optimization. Often, the major cost in measuring plots is labor, with much of the labor involved in traveling to and locating the plot reference point. When much of the effort is simply to get to plots, one may increase the statistical power obtained for a given cost by shifting labor time from travelling to plots to measuring plots and using fewer, larger plots to sample a larger total area. In soil sampling, this strategy may mean aggregating more cores per microsite, but sampling fewer microsites.

During this pilot project, the amount of time used to perform the different tasks was recorded, as were amounts and costs of supplies. Cost per plot was calculated by the amount of time required to perform each task multiplied by the typical hourly cost of a person doing the task. The amount of time needed to perform a given task varied tremendously from plot to plot. For example, some plot locations were on open level ground near a road and took only a few minutes to reach and establish. Others were in extremely dense brush and took about 2 hours to locate and establish. Also, a skilled field team sometimes performs measurements of different pools at the same time, making it difficult to assign time to the separate pools.

Costs of sample preparation and data entry vary fairly linearly with the number of plots sampled. In contrast, data analysis is relatively insensitive to the number of plots. Once the data are in clean electronic form, it takes essentially the same amount of time to do statistics on data from a dozen plots as for a thousand plots. Exceptions to this low marginal cost of analyzing data from additional plots are cases where adding more plots introduces different things to be calculated or analyzed. For example, adding more plots may increase the number of tree species or range of tree sizes sampled, thus requiring additional work to extend allometric equations used to convert tree size data to biomass.

The sampling data presented here could serve as a pilot project for designing a carbon sequestration project that encompasses the pilot area. Presumably, such a project would use paired measurements of permanent plots to gain the statistical power that the method may provide. While no information is available about plot-to-plot variability of change in carbon stock over time, if one is to implement a measurement program, one must proceed despite imperfect knowledge. Although variability of carbon stock across plots at one time is not necessarily like variability of change in carbon stock across plots, until information about variability of change over time is available, data about variability of stock may be used as a surrogate. When expecting to gain power by switching to analyzing paired data, it is plausible to hope that, with paired plots, the same number of plots needed merely to detect change in a mean will be sufficient to detect much of the expected change in stock. Using the power of test information generated from the pilot measurement data and costs for various tasks, the cost of a full-scale measurement of change in carbon stock can be estimated (Table 17.3).

Costs of remeasurement are likely to be similar to costs of a base line measurement. Although plots will not need to be established during remeasurement, the labor of reaching and locating plots is typically nearly as great as the labor of initial plot establishment. Any savings in plot establishment costs are likely to be more than offset by increased analytical costs of finding the differences between initial measurements and remeasurements.

Table 17.3   Projected Cost of Initial Carbon Stock Measurement by Pool

Pool/Activity	Number of Plots	Field Work ($)	Lab and Data Analysis ($)	Total Cost ($)
Establish plots	400	$10,000	$0	$10,000
Soil	400	$4,000	$6,300	$10,300
Litter	50	$250	$800	$1,050
Understory	100	$500	$700	$1,200
Woody debris	400	$2,000	$400	$2,400
Standing dead	100	$500	$400	$900
Trees	100	$1,000	$1,200	$2,200
Total	400	$18,250	$9,800	$28,050

These costs of measuring change in carbon stocks are substantial. Cost estimates presented here are substantially lower than a commercial service provider is likely to charge to measure change in carbon stock because these costs do not include data cleaning and data entry, supplies, lodging costs of field workers, design costs, mapping or imagery costs, documentation costs, overhead, contingency, or profit.

Before undertaking a project, project directors need to decide that the amount of carbon sequestration likely to be detected is worth the cost and risk of doing the project. Per-ton costs of measurement can be estimated by dividing total tons detected by total costs. For this example, assuming that cost of remeasurement is the same as the cost of the baseline measurement, the total cost of measurement would be $56,000. Using the predicted amounts of sequestration given in Table 17.2, the midrange estimate of total sequestration over 10 years is 13 tons per hectare. If the cost of measurement must be less than $10 per ton of change in carbon stock to be feasible, this project would have to detect 5600 tons. If all change is detected, at the average projected rate of carbon sequestration, a project would have to encompass 430 hectares to sequester 5600 tons of carbon. If the measurement cost must be below $1 per ton, the area would be 4300 hectares; to reduce the measurement cost to $0.10 per ton the project area would need to be 43,000 hectares.

A major challenge of carbon measurement is determining what parts of the total project area are represented by a group of samples. Project areas encompassing several thousand hectares may cover a wide variety of soils, vegetation types, hydrologic regimes, and management regimes that combine to sequester highly variable quantities of carbon. Consequently, measuring areas of several thousand hectares or larger might require substantially more than the 400 plots suggested by this pilot study. As plots are remeasured over time, data will become available about variability of carbon sequestration. If spatial variability of sequestration is sufficiently low, paired sampling might allow measurement of fewer plots to obtain a needed level of precision.

As presented in Table 17.2, woody debris, standing dead, and, possibly, litter may not accumulate much carbon. A project manager could claim total emission of the woody debris and standing dead pools, and no accumulation in the litter pool, and not measure these pools. However, not measuring these three pools would save only about 12% of predicted measurement costs, could make the overall measurement less persuasive, and would fail to detect possible accumulation in the litter pool. Also, if the project is expected to run for several decades, the woody debris and standing dead pools might become large over time, and then baseline measurements of those pools would be desirable.

An alternative strategy for minimizing measurement costs is to test alternative plot designs. If a plot design yields lower variance of measured values, fewer plots would be needed for a given level of measurement precision. Larger plots may average across fine-scale variability. They require more effort to sample, but this increase in effort may be more than offset by a reduction in the number of plots that must be measured. In soil sampling, compositing more cores from a single plot (sometimes called a microsite) is equivalent to increasing the plot size.

SUMMARY

In summary, total measurement cost varies as a function of the physical variability of the world, plot design, proportional change in stock to be detected, and the cost per sample. In a pilot study of afforestation of productive land in southwestern Oregon, observed variability in terrestrial carbon indicates that it will take 100 to 400 samples to detect, with 90% statistical confidence, a substantial portion of carbon sequestration likely to occur on the property within 10 years. Ability to detect change is highly dependent on the variability of the change and the proportional change occurring in a stock. Changes equal to a larger percentage of the baseline stock are easier to detect, even if the absolute number of tons of change is small. If a change is highly variable, from plot to plot, it will be much harder to detect.

It may be possible to choose a plot design that averages out some fine-scale spatial variability in a carbon pool, yielding measurements that are less variable from plot to plot. Cost per sample may not change linearly with the area measured during sampling. The number of samples needed to measure a given change does not vary linearly with the area represented by sampling. With the variability and costs observed in the example presented here, getting measurement costs down to $10 per ton of carbon sequestered would require a project area of 430 hectares. To get measurement costs below $1 per ton would require a project area of 4300 hectares; to reduce the measurement cost to $0.1 per ton the project area would need to encompass 43,000 hectares.

CONCLUSIONS: IMPLICATIONS FOR MEASUREMENT COSTS

There are many conditions that can make the cost per ton of detected carbon sequestration very high. Some of these conditions can be controlled by designing a carbon sequestration project and some cannot. The variability physically occurring within a project area is generally not controlled by someone measuring change in carbon stock. Also, rigor in carbon accounting is enhanced by comprehensiveness, and results may be biased if substantial components of the landscape are not measured because this requires a lot of work. Plot design is within the control of someone doing measurements. While optimal plot design cannot yield absolutely precise answers for no cost, inappropriate plot design can cause failure to get any useful result even after spending a lot of money on sampling. Within limits, someone designing a measurement system may be able to increase the proportion of change in stock to be measured by increasing the length of the time period measured, thus giving more time for slow processes to produce changes. Total measurement costs to detect change in terrestrial carbon stocks are likely to cost several tens of thousands of dollars. If a set of measurements can be applied to a very large area, for sequestration over a number of years, it may be possible to get measurement costs below $1 per ton of carbon sequestration detected.

REFERENCES

Avery, T.E. and H.E. Burkhart, *Forest Measurements*, 4th ed., McGraw-Hill, Inc., New York, 1994.

Dudek, D.J., J. Goffman, and S.M. Wade, Emissions trading in nonattainment areas: potential, requirements, and existing programs, in Kosobud, R.F. and J.M. Zimmerman (eds.), *Market-Based Approaches to Environmental Policy: Regulatory Innovations to the Fore*, Van Nostrand Reinhold, New York, 1997, 151–185.

Lal, R. et al. (eds.), *The Potential of U.S. Cropland to Sequester Carbon and Mitigate the Greenhouse Effect*, Sleeping Bear Press, Inc., Chelsea, MI, 1998.

Row, C., R.N. Sampson, and D. Hair, Environmental and land-use changes from increasing forest area and timber growth, in Sampson, R.N. and D. Hair (eds.), *Forests and Global Change*, Volume 2: *Forest Management Opportunities for Mitigating Carbon Emissions*, American Forests, Washington, D.C., 1996.

Smith, G.R., Towards an efficient method for measuring total organic carbon stocks in a forest, in Lal, R. et al. (eds.), *Assessment Methods for Soil Carbon*, CRC Press, Boca Raton, FL, 2000.

Smith, G., R. Conant, and K. Paustian, Quantifying Precision of Soil Sampling: Limits of Detectable Change in Carbon Stock in a Tilled Field and Forest, poster presented at Carbon: Exploring the Benefits to Farmers and Society, August 29–31, 2000, Des Moines, IA, 2000.

United Nations Framework Convention on Climate Change Secretariat, Kyoto Protocol to the United Nations Framework Convention on Climate Change, United Nations Climate Change Secretariat, Bonn, Germany, 1997.

Wonnacott, T.H. and R.J. Wonnacott, *Introductory Statistics for Business and Economics*, 3rd ed., John Wiley & Sons, Inc., New York, 1984.

CHAPTER 18

State-Level Analyses of C Sequestration in Agricultural Soils

Keith Paustian, John Brenner, Kendrick Killian, Jan Cipra, Steve Williams, Edward T. Elliott, Marlen D. Eve, Timothy Kautza, and George Bluhm

CONTENTS

INTRODUCTION

With concerns about potential adverse effects of changes in atmospheric composition on global climate and ecosystems, interest is growing in various strategies to mitigate greenhouse gas (GHG) emissions, including sequestration of carbon (C) in agricultural soils. To evaluate the current contributions of agricultural activities as a source or sink for carbon dioxide (CO_2) and assess the potential of alternative management practices for GHG mitigation, quantitative assessment methods are needed. An important function of such assessments is to provide guidelines for the design and implementation of mitigation programs. Accordingly, assessments need to consider the temporal and spatial variability of factors — including climate, soils, historical land use as well as current and potential management practices — that control soil C dynamics within a regional or national context. In addition, quantification systems need to be able to project alternative scenarios to evaluate the efficacy of mitigation strategies into the future.

Various assessments of carbon sequestration have been made using simulation models and resource databases, usually involving estimates of sequestration potentials. Donigian et al. (1998) estimated C sequestration potentials for major agricultural regions in the U.S. using the Century model. Paustian et al. (2001) analyzed soil C changes due to the Conservation Reserve Program (CRP) for a 16-state region in the central and western U.S. Similar studies of cropland soil C dynamics have been done in Hungary using the Rothamsted model (Falloon et al., 1998). Mitchell et al. (1998) analyzed C sequestration in a 12-state region in the central U.S. They used representative management practices simulated by the EPIC model to develop a metamodel, which was then combined with an economic assessment to evaluate policy alternatives for promoting C sequestration.

A more detailed approach, applicable for individual states, has been developed to assess current and future potential for sequestration of carbon in agricultural soils. The methodology incorporates a variety of spatial databases of climate, soils and management, data from long-term experiments, simulation models, and county-level surveys of management and conservation practices. The state and county scale of analysis provides a spatial resolution commensurate with the needs of many resource planners and decision makers, while still small enough to capture more localized heterogeneity and engage local participants in the assessment process.

The use of this approach to estimate the current contributions of agricultural soils in Iowa to sequester carbon is described. In Chapter 12, Brenner et al. describe the design and implementation of a survey approach to collect county-level management and resource data to feed into the state-level analysis.

MATERIAL AND METHODS

Data Sources

Data on climate, soils, land use, and management practices used in the analysis were assembled from a variety of sources. For the state-level analysis individual counties were chosen as the spatial unit for representing climate factors, i.e., counties were assumed to be homogeneous with respect to the temperatures and precipitation-driving variables.

Temperature (mean monthly maximum and minimum) and precipitation (monthly total) were obtained from the PRISM monthly climate data set (Daly et al., 1994). PRISM (parameter-elevation regressions on independent slopes model) uses point data from the U.S. network of weather stations and a digital elevation model (DEM) to orographically adjust climate variables for 4-km grid cells across the coterminous U.S. The data used in this analysis consisted of long-term (1961 to 1990) monthly averages (Figure 18.1). For each county, area-weighted mean values of monthly temperature and precipitation variables were calculated.

Soils data were derived from analysis at the component level (i.e., soil series) within soil associations of STATSGO (SCS, 1994). For each county, area-weighted frequency distributions of soil types were determined based on the relative proportion of component soils within each soil association. Soil types for application in the model were grouped according to surface texture (0 to 20 cm) and whether soils were classified as hydric or nonhydric (Figure 18.2). If a particular soil component accounted for less than 1% of the area of the county it was not included in the analysis. The procedure to identify major soil types yielded six to ten distinct soil types per county.

Land-use and management data were compiled from a variety of other sources including data on CRP contract acreage (USDA/FSA — A. Barbarika, personal communication), state and county areas for crops grown by year compiled by the National Agricultural Statistics Service (NASS, 1999), area by tillage practice and crop compiled by the Conservation Technology Information Center (CTIC) (CTIC, 1998), and field operation scheduling and fertilizer use provided by USDA's Natural Resource Conservation Service (NRCS) state office in Iowa. In addition, information was gathered for each county in the state, using a survey instrument — the carbon sequestration rural appraisal (CSRA) — described by Brenner et al. in Chapter 12. Briefly, the CSRA consists of a series of data sheets detailing historical land use, dominant management practices (drainage, irrigation, crop rotations, tillage, fertilization) over time, and installation of conservation practices (e.g., CRP, grassed waterways, buffers) compiled by local experts in each county.

Mean Annual Temperature

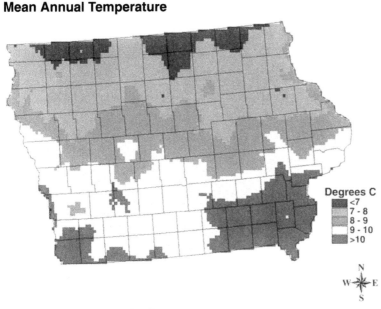

Mean Annual Precipitation

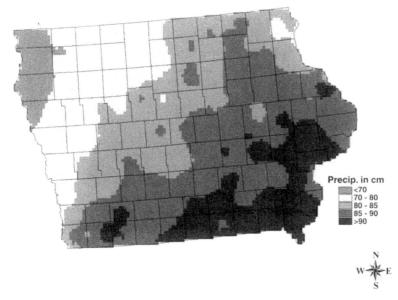

Figure 18.1   Maps of long-term mean annual temperature (°C) and precipitation (cm) for Iowa from PRISM (Daly et al., 1994). For the simulations, the climate data used were mean monthly minimum and maximum temperatures and monthly precipitation from PRISM aggregated by county.

Model Initialization

Soil carbon dynamics were simulated using the Century ecosystem model, with the model inputs derived from data sources listed above. Century is a generalized soil and ecosystem model that simulates soil organic matter dynamics, soil N, P, and S cycling, plant production, and soil water dynamics (Parton et al., 1987, 1994; Metherell et al., 1993). Monthly temperature and precipitation, soil physical characteristics, and management practices are the main data inputs. Originally developed for grassland systems, the model has been used extensively to simulate organic

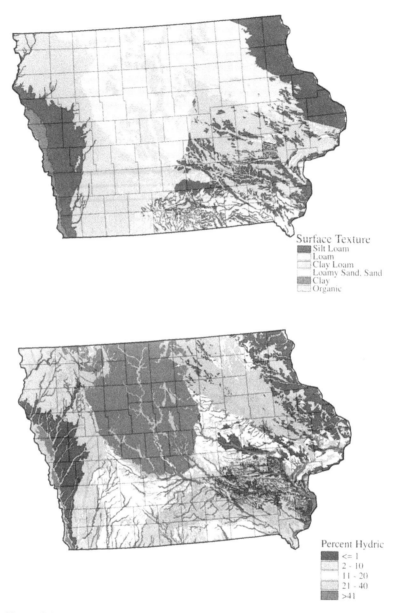

Figure 18.2 Maps of dominant surface textures and percent occurrence of hydric soils within STATSGO soil associations for Iowa.

matter and nutrient dynamics in agricultural cropping systems (e.g., Paustian et al., 1992, 1996, 2001; Carter et al., 1993; Parton and Rasmussen, 1994). A variety of management options can be specified on a monthly basis. These include crop type, tillage, fertilization, organic matter addition (e.g. manuring), harvest (with variable residue removal), drainage, irrigation, fire and grazing intensity. The desired cropping sequence is simulated by specifying crop type and management options in the management schedule file.

The model was initialized by simulating past land use and management histories derived from the CSRA. Briefly, the model was run to equilibrium soil C levels under the dominant native vegetation (tall-grass prairie) for 6000 years for each soil type in each county. The model then simulated changes in soil C as a function of past agricultural practices based on dominant crop rotations and management practices reported in the CSRA. Average date for onset of cultivation was reported as occurring in the 1860s to 1880s for most counties; cropping histories were divided into periods between 1860 to 1920, 1920 to 1950, and 1950 to 1974. Crop production potentials were also varied over time to mimic long-term changes in crop yields as reported by NASS, with yields increasing by 1 to 2% per year since the 1950s. For each time period, the local experts completing the CSRA specified crop rotations and accompanying management practices (i.e., tillage, fertilization, manuring) that were representative for their area. Each county reported a single representative history prior to 1974, which varied between counties (Brenner et al., 2001). However, most counties reported similar trends in dominant cropping practices, with corn, oats, wheat, and hay as dominant crops prior to 1950, followed by a rapid shift towards feed-grain dominated rotations, i.e., corn and soybean, and a substantial reduction of hay in rotation.

Starting in 1974, four crop rotations (continuous corn, corn–soybean, corn–soybean–oat–alfalfa–alfalfa, and oat–alfalfa), for each of three tillage regimes (intensive tillage, moderate tillage, and no tillage), were simulated for a 20-year period (1974 to 1994) in each county. Intensive tillage was defined as multiple tillage operations every year, entailing significant soil inversion (i.e., plowing, deep disking) and low surface residue coverage, corresponding to the intensive tillage and reduced tillage systems defined by CTIC (CTIC, 1998). Moderate tillage was defined as few tillage operations (or the absence of tillage for 1 year when growing soybeans in a corn–soybean rotation) with less soil disturbance, corresponding to mulch tillage and ridge tillage as defined by CTIC. No tillage was defined as not disturbing the soil except through the use of fertilizer and seed drills.

Cropland area by tillage system for each county was derived from the CTIC 1998 database. Estimates of the percent of total cropland in each of the four crop rotations were provided as part of the information collected in the CSRA. CTIC reports the area in various tillage systems by individual crops on an annual basis; however, it does not differentiate between long-term, no-till practices and intermittent or rotational no-till (i.e., tilled corn–no-tilled soybean rotations). For agronomic reasons, i.e., low residue amounts under soybean and use of herbicide-resistant soybeans, the percent of area of soybeans managed under no-till was generally higher than for corn. Thus, to estimate the area of continuous no-till as opposed to rotational no-till, the % area of continuous no-till was based on the acreage of corn under no-till, assuming that, if corn were no-tilled, it was likely that other crops in the rotation (e.g., soybean or oats) would also be no-tilled. The remaining area reported as no-till was assumed to represent rotational no-till and was included as part of the moderate tillage category, which also included area reported as mulch-till and ridge-till by CTIC. The area under intensive tillage was then calculated by difference.

Area under the different tillage systems was estimated from the CTIC database that has reported area by tillage system and county on an annual basis since 1989. To represent changes over time in tillage practices within time blocks simulated in the model it was assumed that (1) all annual row crop area had been managed with intensive tillage prior to 1974, and (2) 75% of the moderate till area reported in 1998 represented area converted to moderate till during the period from 1974 to 1994, with the remaining area converted from intensively tilled systems since 1994. Since the area under continuous no-till prior to 1989 was minimal, the soil C change rates calculated for adoption of no-till beginning in 1994 were utilized for the estimates of current C sequestration rates under no-till. Since no information was available on the relative distribution of tillage systems by crop rotations, it was necessary to apply the relative distribution estimated for total cropland to each of the four crop rotations simulated.

To simulate changes due to CRP, all four crop rotations under intensive tillage were modeled with a change to CRP (grass or grass and legume mixtures) for a 10-year period, starting in 1985. In 1994, all management options were simulated for an additional 20 years, including all combinations of changes among crop rotations, CRP, and tillage regimes (Table 18.1). This provided 280 simulations for each soil texture and hydric combination in each county, or approximately 203,000 simulations for the entire state.

**Table 18.1   Cropping Practices Simulated for Each Soil Type and
County Combination for the Period 1974 to 1994**

Crop Rotation	Tillage Practice	CRP Conversion
Continuous corn	Intensive	• Grass/legume (50/50) • Grass
	Moderate No-till	NA
Corn–soybean	Intensive	• Grass/legume (50/50) • Grass
	Moderate No-till	NA
Corn–soybean–oat–hay	Intensive	• Grass/legume (50/50) • Grass
	Moderate No-till	NA
Oat–hay	Intensive	• Grass/legume (50/50) • Grass
	Moderate No-till	NA

Note: For intensively tilled rotations, conversion to CRP in 1985 comprised
additional scenarios. For the period 1994 to 2014, all of the potential
cropping scenarios (including continuation of CRP) were run as an
extension of each of the cropping and tillage combinations for 1974
to 1994. CRP conversions were not applied (NA) to cropland already
under moderate or no tillage.

Additional information was compiled from the literature to estimate net soil carbon changes
for minor land-use practices not modeled by Century, including changes associated with wetland
restoration of former cropped land and cultivation of organic soils. Mean rates of carbon change
on a per-hectare basis for cropland conversion to wetland and for cultivation of organic soils were
taken from Armentano and Menges (1986).

RESULTS

The principal management trends affecting simulated soil C stock changes for the state of
Iowa were the increase in adoption of moderate-till and no-till systems for row crop production
and the introduction of CRP. No-till and moderate-till systems have increased in Iowa over the
past several years, from <3 and 20% of annual cropland, respectively, in 1989 to 11 and 28%,
respectively, in 1998. Changes in C stocks of soils under continuous no-till averaged about 0.6
Mg ha^{-1} yr^{-1} for the state as a whole. The gain of soil C on reduced (i.e., moderate) tilled soils
averaged about 0.25 Mg ha^{-1} yr^{-1} across the state. Examples of mean rates of soil C change for
different tillage systems and selected soil types under corn–soybean rotations are shown in
Figure 18.3. Rates for moderate-till and no-till are averages for a 10-year period (1994 to 2004)
following conversion from intensive tillage, together with projected rates of change with continuation of intensive tillage practices.

For nonhydric (well-drained) soils, even intensively tilled show a low rate of increase, which is
driven by the trend of increasing residue inputs associated with productivity gains (on the order of
1 to 1.5% per year) since the 1950s (Reilly and Fuglie, 1998). In contrast, hydric soils under
conventional, intensive cultivation are predicted to be losing C in most locations due to the stimulus
of soil drainage on decomposition rates that overrides the positive effects of increasing residue inputs.
In summary, the estimates of current rates of C change under the predominant crop rotation in Iowa,
corn–soybean, are largely due to changes in tillage practices, with reduced tillage intensity acting
to increase C storage, but with underlying effects (operating across tillage practices) of long-term

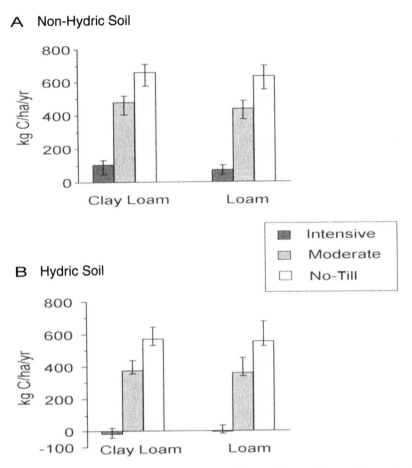

Figure 18.3 Mean annual rates of soil C change for corn–soybean rotations, for two soil types, by tillage. Rates are state averages over the period 1994 to 2004. Bars within columns show the range of values across counties within the state. Rates for no-till and moderate-till represent conversion from intensive-till corn–soybean in 1994.

trends of increasing productivity and residue inputs, along with soil-specific controls such as drainage practices and soil texture, which modify decomposition and soil C stabilization.

Carbon sequestration rates predicted for Iowa soils with adoption of no-till are in line with results from several long-term studies in the Corn Belt region. For example, Ismail et al. (1994) reported increases of about 1 Mg ha^{-1} yr^{-1} (0 to 30 cm), averaged across four nitrogen fertilization treatments, between years 10 and 20 in no-till continuous corn rotations in Kentucky. Similarly, Dick et al. (1998) found that C levels under no-till averaged 17 Mg ha^{-1} higher than under plow tillage, averaged for corn–soybean, continuous corn, and corn–oat–hay rotations: a difference equivalent to about 0.6 Mg ha^{-1} yr^{-1} over a 30-year period. Numerous other studies (see reviews by Paustian et al., 1997; West and Marland, 2001) of tillage impacts illustrate the general trend of increased C sequestration resulting from reducing or eliminating tillage, although rates vary considerably according to soil, climate, and management variables. In some cases negligible effects of tillage reduction on soil C have also been reported (e.g., Wander et al., 1998). The variability of the modeled response of soil C to adoption of no-till is less than might be expected based on comparison across different field studies, with simulated rates of increase with no-till adoption varying between about 0.5 and 0.7 Mg ha^{-1} yr^{-1} across all soils. Additional sources of variability in response to tillage changes important at a site-specific level, such as tillage and productivity interactions and effects of erosion, are not captured in the model application at county and state scales.

Conversion of annual cropland to CRP grasslands was estimated to yield C sequestration rates of about 1.3 Mg ha^{-1} yr^{-1}, averaged across the state. Simulated rates varied across counties and soil types, ranging from about 1.2 to 1.7 Mg ha^{-1} yr^{-1} (Figure 18.4). In comparison, Follett et al. (2001) estimated rates of C sequestration for CRP averaging 0.9 Mg ha^{-1} yr^{-1} for 14 sites in the central U.S., based on field sampling of paired sites. Sites located in Iowa had the highest rates of C increase, ranging up to 4 Mg ha^{-1} yr^{-1} (Follett et al., 2001). Jastrow (1996) measured rates of about 1 Mg ha^{-1} yr^{-1} increase in soil carbon with prairie restoration of former annual cropland in Illinois. A number of other field studies (e.g., Gebhart et al., 1994; Huggins, et al., 1998; Reeder et al., 1998) have documented increases in soil C attributable to CRP, but with varying rates across soil and climate gradients and varying uncertainty depending on the methods used to assess C seques-tration under CRP (Paustian et al., 2001). As was the case for reduced tillage effects, the model probably does not adequately reflect the range of likely values of C change under CRP that would be expected through site-specific effects (e.g., poor stand establishment, high residual nutrient levels, pest effects) that cannot be captured in a regional assessment. It should also be noted that the assumptions used in simulating CRP conversions have a significant impact on the predicted response of CRP. For the present simulations, we assumed CRP planting included a legume component to help meet demands for N by the perennial vegetation. Simulations with pure grass and minimal pre-CRP residual nitrogen were predicted to yield only about half of the rates reported here (unpublished data).

The simulation results were also mapped back to soil polygons to show how predicted changes in soil carbon storage vary geographically across the state (Figure 18.5). For the example of con-version of corn–soybean under intensive tillage to no-till, average rates of increase of carbon stocks during the first 10 years varied from less than 0.5 Mg ha^{-1} yr^{-1} to greater than 0.65 Mg ha^{-1} yr^{-1}. The spatial patterns that emerge are a function of a number of interacting factors, i.e., soil texture and hydric characteristics, climate influences on productivity and decomposition, and regional differences in management practices such as the relative dominance of grain vs. hay crops and the extent and timing of soil drainage. Thus interpretation of the regional patterns is complex.

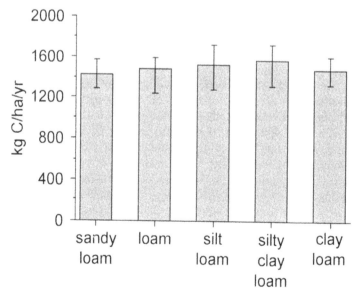

Figure 18.4 Mean annual rates of soil C change for CRP lands for the period 1985 to 1994, converted from intensive-till corn–soybean in 1984, for the state. Bars within columns show the range of values across counties within the state.

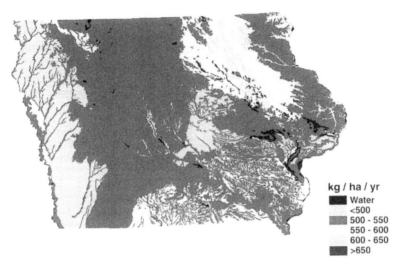

kg / ha / yr
- ■ Water
- <500
- ▨ 500 - 550
- 550 - 600
- 600 - 650
- ■ >650

Figure 18.5 Spatial distribution of mean annual rates (1994 to 2004) of soil C increase following conversion of intensive-till corn–soybean rotations to no tillage.

Rates of increase under no-till are somewhat lower along the western side of the state where productivity tends to be lower due to less precipitation. The greatest response to no-till locally is predicted to occur on the lighter textured soils, predominantly in the eastern half of the state (shown in red). Although carbon stocks on these soils are considerably lower than on surrounding silt- and clay-loam soils, the predicted increase in soil C following adoption of no-till is predicted to be higher. The next highest response is shown on scattered soil polygons with a low incidence of hydric soils (<10%), dominated by loam textures, in the northeastern part of the state. As shown in Figure 18.3, C increases under no-till are predicted to be higher on nonhydric compared to hydric soils, all else being equal. In the south-central part of the state greater increases due to no-till are shown occurring on loam compared to silt-loam soils. Most of the rest of the state, is predicted to show an intermediate response to no-till adoption on the order of 0.5 to 0.6 Mg ha^{-1} yr^{-1} increase.

The model results for each combination of climate conditions (i.e., county average), soil type, and management sequence were compiled and entered into a distributed database that can be used to estimate current soil carbon changes as well as potential carbon sequestration rates for the state as a whole. To provide a planning and assessment tool for land managers, model simulation results were organized into an Access (Microsoft Corp.) database, named COMET (CarbOn Management Evaluation Tool), with facilities to query and graph the results (Figure 18.6). The user selects the desired county and major soil types within the county and then selects from the menu crop rotations and tillage management sequences for each of two time periods (1974 to 1994 and 1994 to 2014). Two contrasting scenarios can be specified and are displayed at the same time, allowing comparison of management alternatives. A table is produced showing the difference in C stock change (for soil organic matter and crop residues) between scenarios. The data are configured to display the relative changes since the base year of 1974, but actual simulated carbon stocks are given in the accompanying data sheets.

To estimate current changes in soil C storage under present management systems, we used the mean annual rates of carbon change for the period 1994 to 2004 for each management sequence × soil type × county combination, multiplied by the area represented by that combination. Compiling all of the model-based estimates for managed cropland and grass, with separate calculations for wetland restoration and cultivated organic soils, it was estimated that Iowa soils are currently (i.e., based on 1998 data) a net sink for CO_2, accumulating soil C at a rate of about 3.1 MMTC per year (Table 18.2). The largest contributions to this carbon sequestration are attributed to the reduction in area under intensive tillage over the past 10 to 20 years and the conversion of former annually cropped area to perennial grasses through the CRP, as well as the increased installation of grass waterways, field buffers, filter strips, terrace walls, and other grassed conservation practices.

More than one half of Iowa's 10.5 Mha of cropland is still managed using conventional tillage practices, predominately under corn–soybean rotations. While many conventionally managed soils may be net sources of CO_2 (particularly artificially drained hydric soils), analysis predicts an overall slow rate of increase of soil carbon for the conventionally managed cropland in the state due to increasing amounts of crop residues added to soil over the past 3-4 decades. Others (Cole et al., 1993; Allmaras et al., 2000) have similarly suggested that the general trends in crop productivity since WWII have moved agricultural soils from net C sink to net C source in the U.S. Wetland restoration is projected to represent a net carbon sink, but the overall effects on the C balance for

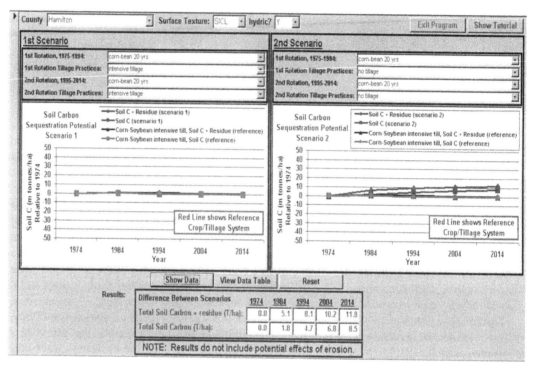

Figure 18.6 Example of location, soil, and scenario options for displaying simulated soil C changes from the database of modeled results.

Table 18.2 Annual C Stock Changes, Per-Hectare Rates and Totals (million metric tonnes — MMT)

System	Area (10^6 ha)	C Change Rate (Mg ha^{-1} yr^{-1})	C Stock Increase (MMT yr^{-1})
Intensively tilled cropland	5.2	0.08	0.43
Moderate tilled cropland	3.0	0.25	0.82
No-till cropland	1.1	0.6	0.66
CRP and grass buffers	1.1	1.3	1.3
Wetland restoration	0.04	0.5	0.02
Cultivated organic soils	0.02	−8.5	−0.17
Total	10.5		3.1

Note: Summarized by major land-use categories for Iowa based on data for 1998.

the state are minor due to the relatively small (40,000 ha) area. Intensive cultivation of organic soils represents a significant source of CO_2 emissions, particularly considering the limited extent of this practice (~20,000 ha) in the state.

CONCLUSION

Analysis suggests that Iowa's agricultural soils are presently a sink for CO_2 on the order of 3.1 MMTC yr^{-1}. For comparison, this amount is equivalent to approximately 15% of the 1997 CO_2 emissions from fossil fuel combustion in the state (EPA, 2001). The analysis suggests that C sequestration could be increased two- to threefold, since more than 50% of the state's annual cropland remains under conventional, intensive tillage practices and only 10% is managed with no-till practices. Analysis also suggests that the installation of conservation practices using permanent grasses and legumes is among the most effective measures on a per-hectare basis; for Iowa, it contributes about one third of the current estimated carbon sequestration.

The analysis framework provides a means for local involvement to provide locally relevant data on past and present management practices that can be incorporated into a user-friendly database of carbon sequestration rates specific to management and soil conditions within a particular county. This provides a means of assessing alternative mitigation strategies, including making future projections, which can be useful for local as well as state-level managers and policymakers. Further development of the analysis framework could entail a more explicit inclusion of topographic effects, particularly with respect to erosion and depositional processes and their effects on soil carbon balance on the regional scale.

REFERENCES

Allmaras, R.R. et al., Soil organic carbon sequestration potential of adopting conservation tillage in U.S. croplands, *J. Soil Water Conserv.*, 55:365–373, 2000.

Armentano, T.V. and E.S. Menges, Patterns of change in the carbon balance of organic soil-wetlands of the temperate zone, *Ecology*, 74:755–774, 1986.

Brenner, J., K. Paustian, G. Bluhm, K. Killian, J. Cipra, B. Dudek, S. Williams, and T. Kautza, Chap. 12, this volume.

Brenner, J. et al., Quantifying the change in greenhouse gas emissions due to natural resource conservation practice application in Iowa, Final Report to USDA/NRCS state office, Des Moines, IA, 2001.

Carter, M.R. et al., Simulation of soil organic carbon and nitrogen changes in cereal and pasture systems of Southern Australia, *Aust. J. Soil Res.*, 31:481–491, 1993.

Cole, C.V. et al., Agricultural sources and sinks of carbon, *Water Air Soil Pollut.*, 70:111–122, 1993.

CTIC, Crop residue management executive summary, conservation technology information center, data available online at http://www.ctic.purdue.edu/Core4/CT/CT.html, 1998.

Daly, C., R.P. Neilson, and D.L. Phillips, A statistical-topographic model for mapping climatological precipitation over mountainous terrain, *J. Appl. Meteorol.*, 33:140–158, 1994.

Dick, W.A. et al., Impacts of agricultural management practices on C sequestration in forest-derived soils of the eastern Corn Belt, *Soil Tillage Res.*, 47:235–244, 1998.

Donigian, A.S. et al., Modeling soil carbon and agricultural practices in the central U.S.: an update of preliminary study results, in Lal, R. et al. (eds.), *Soil Processes and the Carbon Cycle*, CRC Press, Boca Raton, FL, 1998, 499–518.

EPA, Energy CO_2 inventories, U.S. Environmental Protection Agency, data available online at http://yosemite.epa.gov/globalwarming/ghg.nsf/emissions/CO2EmissionsBasedOnStateEnergyData, 2001.

Falloon, P.D. et al., Regional estimates of carbon sequestration potential: linking the Rothamsted carbon model to GIS databases, *Biol. Fert. Soil*, 27:236–241, 1998.

Follett, R. et al., Carbon sequestration under the CRP in the historic grassland soils in the USA, in Lal, R. and K. McSweeney (eds.), *Soil Management for Enhancing Carbon Sequestration*, SSSA Special Publ., Madison, WI, 27–40, 2001.

Gebhart, D.L. et al., The CRP increases soil organic carbon, *J. Soil Water Conserv.*, 49:488–492, 1994.

Huggins, D.R. et al., Enhancing carbon sequestration in CRP-managed land, in Lal, R. et al. (eds.), *Management of Carbon Sequestration in Soil*, CRC Press, Boca Raton, FL, 1998, 323–350.

Ismail, I., R.L. Blevins, and W.W. Frye, Long-term no-tillage effects on soil properties and continuous corn yields, *Soil Sci. Soc. Am. J.*, 58:193–198, 1994.

Jastrow, J.D., Soil aggregate formation and the accrual of particulate and mineral-associated organic matter, *Soil Biol. Biochem.*, 28:665–676, 1996.

Metherell, A. K. et al., CENTURY soil organic matter model environment, Agroecosystem version 4.0, technical documentation, GPSR Tech. Report No. 4, USDA/ARS, Ft. Collins, CO, 1993.

Mitchell, P.D. et al., The impact of soil conservation policies on carbon sequestration in agricultural soils of the central United States, in Lal, R. et al. (eds.), *Management of Carbon Sequestration in Soil*, CRC Press, Boca Raton, FL, 1998, 125–142.

NASS, Published estimates database, USDA-National Agricultural Statistics Service, data available online at http://www.nass.usda.gov:81/ipedb/, 1999.

Parton, W.J. et al., Analysis of factors controlling soil organic matter levels in Great Plains grasslands, *Soil Sci. Soc. Am. J.*, 51:1173–1179, 1987.

Parton, W.J. et al., A general model for soil organic matter dynamics: Sensitivity to litter chemistry, texture and management, in *Quantitative Modeling of Soil Forming Processes*, Special Publication 39, Soil Science Society of America, Madison, WI, 147–167, 1994.

Parton, W.J. and P.E. Rasmussen, Long-term effects of crop management in wheat-fallow: II. CENTURY model simulations, *Soil Sci. Soc. Am. J.*, 58:530–536, 1994.

Paustian, K., W.J. Parton, and J. Persson, Modeling soil organic matter in organic-amended and N-fertilized long-term plots, *Soil Sci. Soc. Am. J.*, 56:476–488, 1992.

Paustian, K. et al., Modelling climate, CO_2 and management impacts on soil carbon in semi-arid agroecosystems, *Plant Soil*, 187:351–365, 1996.

Paustian, K., H.P. Collins, and E.A. Paul, Management controls on soil carbon, in Paul, E.A. et al. (eds.), *Soil Organic Matter in Temperate Agroecosystems: Long-Term Experiments in North America*, CRC Press, Boca Raton, FL, 1997, 15–49.

Paustian, K. et al., Modeling and regional assessment of soil carbon: a case study of the Conservation Reserve Program, in Lal, R. and K. McSweeney (eds.), *Soil Management for Enhancing Carbon Sequestration*, SSSA Special Publ., Madison, WI, 207–225, 2001.

Reilly, J.M. and K.O. Fuglie, Future yield growth in field crops: what evidence exists? *Soil Tillage Res.*, 47:275–290, 1998.

Reeder, J., G.E. Schuman, and R.A. Bowman, Soil C and N changes on conservation reserve program lands in the central Great Plains, *Soil Tillage Res.*, 47:339–350, 1998.

SCS, State soil geographic data base (STATSGO) data user guide, United States Department of Agriculture Soil Conservation Service, National Soil Survey Center, Lincoln, NE, 1994.

Wander, M.M., M.G. Bidart, and S. Aref, Tillage impacts on depth distribution of total and particulate organic matter in three Illinois soils, *Soil Sci. Soc. Am. J.*, 62:1704–1711, 1998.

West, T.O. and G. Marland, A synthesis of carbon sequestration, carbon emissions, and net carbon flux in agriculture: comparing tillage practices in the United States, *Agric. Ecosys. Environ.* (in press).

PART **IV**

Soil Management

CHAPTER 19

Changes in the Thermal and Moisture Regimes in the Plow Zone under Corn Following No-Till Implementation

Kenneth M. Hinkel

CONTENT

ABSTRACT

Comparative soil temperature and soil tension measurements were obtained within the plow zone of two adjacent fields on Patton silty clay loam soil (fine-silty, mixed, mesic Typic Hapludalf). Both fields were planted to corn (*Zea mays* L.), one using conventional tillage techniques (CT) and the other implementing no-tillage (NT) for the first time. Temperature time series were collected at depths of 2, 9, 16, and 23 cm at hourly intervals for the period of July 9 to October 28, 1989. Soil moisture tension at depths of 10, 18, and 26 cm, and relevant meteorological data, were recorded daily. The average daily temperature in the NT plow zone was consistently 0.50 to 0.75°C warmer throughout the growing season. When the soil was relatively dry, the daily temperature range was greater in the NT field near the base of the plow zone, but damped near the surface. This is attributed to the high thermal conductivity of a 2- to 7-cm-thick compacted surface layer in the nontrafficked interrows of the NT field. The compacted layer also inhibited precipitation infiltration, delayed soil water recharge, and caused surface ponding, which further dampened diurnal temperature amplitude at the surface of the NT field. Conductive heat transfer was enhanced by nonconductive processes near the surface of the CT plow zone. In the NT field, it appears that soil compaction blocked water vapor pathways to the atmosphere and restricted coupled-flow of heat and water vapor; heat transfer across the compacted layer occured primarily by conduction.

INTRODUCTION

In the past several decades, conservation tillage (CN) practices have been promoted as the primary strategy to reduce nonpoint sediment source pollution of U.S. waterways by eroded soils. Minimizing the surface area disturbed by tilling reduces soil losses by restricting the amount of soil exposed to erosion processes. Returning crop residue to the soil effectively roughens the surface, which enhances ponding, reduces sheet flow, and limits the entrainment of soil particles.

Other benefits associated with using CN practices have been noted by researchers and farmers alike. Residual crop stubble effectively captures snow, thus inhibiting the loss of heat from the ground in winter. At planting, more soil water is available due to the enhanced snowmelt and by restricted evaporative losses from wind-induced desiccation (Aase and Siddoway, 1980). During the growing season, the surface mulch is an effective thermal barrier; low thermal conductivity of this material inhibits conductive heat transfer into the ground and reduces evaporative losses by soil water vapor diffusion into the atmosphere (Thomas, 1985). Finally, minimizing the number of tillage operations lowers operational expenses and increases net profit (Larson and Osborne, 1982; Phillips and Phillips, 1984).

Conservation tillage practices are not, however, a panacea. Although results are varied and depend on a number of factors, several disadvantages have been noted with the implementation of CN. First, increased crop losses due to specific insects and many weeds may require the additional application of pesticides. Second, reduced soil temperatures in CN plots have been consistently observed (Burrows and Larson, 1962; Allmaras et al., 1964; Blevins and Cook, 1970; Griffith et al., 1973; Willis and Amemiya, 1973; Mock and Erbach, 1977; Johnson and Lowery, 1985). This is often attributed to the increased surface albedo with the addition of crop residue (McCalla, 1947; van Wijk et al., 1959). However, the low thermal conductivity of the mulch layer requires that, given the same energy input at the surface, less heat will be conducted to depth and greater temperature ranges will be experienced at and near the surface (Thomas, 1985). In addition, the greater soil water content associated with CN techniques dictates that a greater proportion of the conductive heat flux be expended on heating and evaporating the soil water, thus buffering temperature changes.

These considerations are critical. Willis and Amemiya (1973) report that corn germination takes place when ambient temperature at the seed level is about 24°C, with an optimum range of 24 to 29°C (Willis et al., 1957). Depressed temperatures in CN fields relative to those in conventionally tilled (CT) fields can result in reduced germination rates, greater seedling losses, and shorter and thinner stands, factors which contribute to reduced yields (Willis et al., 1957; van Wijk et al., 1959; Larson et al., 1961; Burrows and Larson, 1962; Griffith et al., 1973; Mock and Erbach, 1977; Al-Darby and Lowery, 1987). In the northern Corn Belt, research has shown that, by delaying seeding in no-till (NT) fields by approximately 2 weeks, emergence rates and yields are comparable to those for CT fields (Eckert, 1984; Herbek et al., 1986; Imholte and Carter, 1987).

A third drawback commonly noted with the implementation of CN practices is the formation of a compacted layer at or near the surface (Tollner et al., 1984; Spilde and Deibert, 1986; Kaspar et al., 1995). Mechanical impedance, as measured by penetration resistance, can occur in trafficked and nontrafficked interrows. Compaction enhances surface detention of precipitation and decreases infiltration rates, thus delaying soil water recharge (Kayambo et al., 1986). Conversely, moderate compaction may actually be beneficial during dry years since enhanced soil continuity promotes capillary rise (Lipiec and Simota, 1994). In the near-surface region, compaction restricts root development and has an adverse effect on plant growth and field productivity (e.g., Bauder et al., 1985). Studies conducted over long durations (5 to 10 years) suggest that the compacted layer persists over time (Bauder et al., 1981; Duval et al., 1989). The effects of axial load and tillage method on crop yield appear to depend on soil texture. Soil bulk density and penetration resistance in a silt loam soil show no significant impact after an 11-year study (Lal and Ahmadi, 2000), but axial load can have a significant adverse effect on grain yields in clayey soils (Lal, 1996). In Mollisols, Logsdon and Cambardella (2000) report no significant changes in depth-distributed bulk density measurements following conversion to no-tillage.

In an effort to reduce soil erosion rates, enhance soil water retention in fields, and limit production expenses, many small-scale farmers are experimenting with different crop rotation schemes and tillage systems. Of primary concern is the impact of NT on soil temperatures, particularly during the critical several years immediately following conversion. Different crops provide varying amounts of mulch that take time to accumulate and influence surface albedo and bulk thermal properties (Burrows and Larson, 1962). Often, the advantage of delayed planting must be weighed against the risk entailed by entering water stress periods during crucial stages in the plant life cycle.

This study (1) examines the immediate impact on soil temperature and soil moisture which results from converting from CT (moldboard plow) techniques to NT corn, and (2) reports the effect of soil compaction on the thermal properties of the plow zone. It should be noted that, strictly speaking, the NT field has no plow zone; the term is retained here solely for clarity. Finally, although this study was originally planned to last several years, economic factors forced the participating farmer to abandon NT methods after the first year.

METHODS

A 28-ha field in southwestern Ohio of relatively uniform soil type and slope was selected for the study. The field was divided into two test plots of roughly equal size. The test plots were established on a Patton silty clay loam soil (fine-silty, mixed, mesic Typic Hapludalf) with a 1% slope. A corn–soybean rotation had been practiced for the previous 4 years. In autumn 1988, a rye cover crop was aerially broadcast into the NT plot and cut in spring 1989. Only about 5% of the surface of the NT field was covered with residue when the study commenced.

Bed preparation and planting were delayed in 1989 due to unusually wet conditions. The CT field was cultivated with a moldboard plow, followed by disking. A John Deere 7000 conservation planter was used on both fields to deliver 65,450 seeds per hectare (26,500 ac^{-1}) of Pioneer 3343 variety. Planting of both fields commenced on June 4. At that time, 225 kg ha^{-1} of (10-34-0) fertilizer and 150 kg ha^{-1} of potash were applied to each field. In addition, pre-emergence anhydrous nitrogen (225 kg ha^{-1}) and 28% post-emergence nitrogen (170 kg ha^{-1}) were applied to the CT and NT fields, respectively. Finally, BICEP™ herbicide was distributed to both plots and ROUNDUP™ was applied to the NT field. Fields were harvested on December 5. Both the CT yield (9.43 t ha^{-1}) and NT yield (8.46 t ha^{-1}) exceeded the combined previous annual field average of 7.86 t ha^{-1}.

The automatic data acquisition system (ADAS) was built around an Elexor 1000H with 16 A/D channels. Yellow Springs Instruments #44008 (30 K @ 25°C) thermistors were employed. The system has a resolution of 1 mV across the range of ± 4.096 V, and yields temperatures that are accurate to within ± 0.01°C over the environmental temperature range (–40 to 50°C). Each thermistor is embedded within an 8-cm length of copper tubing (7 mm OD), which is filled with a high thermal conduction gel and sealed. The thermistors, with attached cable leads, were calibrated in an ice slush bath using vendor-supplied data.

The ADAS was housed in an instrument shelter and placed at the CT-NT field boundary. Each probe stack, consisting of four thermistors, was situated nine rows (6.8 m) into each field to alleviate boundary effects. The distance between the probe stacks (13.6 m) was minimized to ensure soil homogeneity. The probe stack was located in the interrow, and the probes buried at depths of 2, 9, 16, and 23 cm, i.e., within the plow zone. Channels were scanned on the hour, the voltage measurements converted to temperature using calibration-derived parameters, and the temperature readings written to the Elexor RAM. The limited buffer capacity required daily downloading of the data.

In addition to the ADAS, tensiometers were installed to record the soil water tension in the fields. Three units, located 0.5 m apart in row 11, were inserted so that the center of the porous ceramic tip reached depths of 10, 18, and 26 cm; these were read and serviced daily. Finally, two precipitation gauges, situated atop 2.5 m poles, were read daily. The mean value is reported here.

The instruments were installed and the data acquisition system initiated on June 26, 1989. At this time, the corn in both fields was 15 cm tall. Because a soil-water slurry was used to backfill the pits excavated to install the thermistors, a 2-week period was allotted for the resulting thermal disturbance to dissipate.

The data reported here commences on July 9, following a significant precipitation event. On this date, corn in the CT and NT fields was about 1.0 and 0.8 m tall, respectively. Data collection ended on October 28, just prior to the expected harvest date. A gap exists in the thermal record for the period of August 14 to September 2, when the author was absent. During this period, the participating farmer collected tensiometer and precipitation data and serviced the instruments.

RESULTS

Throughout the duration of the growing season, two patterns were observed that had an impact on the evolution of the thermal and soil moisture regimes. First, although germination and emergence in both fields were coincident, the height of the NT corn was consistently about 80% the height of corn in the CT field. In addition, silking and pollination was delayed 7 to 10 days in the NT field. These observations are in agreement with those of other researchers (Willis et al., 1957; Mock and Erbach, 1977; Imholte and Carter, 1987).

Second, surface compaction was clearly a factor in the upper 2 to 7 cm of the NT field. The visual evidence included enhanced surface detention following precipitation events and the presence of a discontinuous algal mat on the bare soil throughout the growing season. These conditions were observed in both trafficked and nontrafficked interrows. Following harvest, numerous plants were uprooted in both fields. The root balls of the CT plants were generally circular in plan view, while those from the NT field were elongated along the seed insertion cut. Preferential root growth had clearly occurred where the soil had been fragmented; rootlets were observed only between peds and were concentrated near the surface.

A soil penetrometer was used to measure the unconfined compressive strength at the surface of each field. Fifty measurements were made in nontrafficked interrows in each field in late October. The average value for the CT field was 68 kPa, with little deviation around the mean. In the NT field, four to seven times the pressure was required; in one third of the samples, the necessary penetrating pressure exceeded the maximum value for the instrument (440 kPa). Surface compaction associated with conservation tillage techniques has been reported by other researchers (Tollner et al., 1984; Bauder et al.,1985; Flowers and Lal, 1998) in clay loams, but the impact may be temporary; results from long-term studies show no significant long-term effect on bulk density and hydraulic conductivity (Lal, 1999).

Summary temperature data for the CT field (July 9 to October 28) is presented in Figure 19.1. Average daily temperatures, based on 24 hourly readings, are plotted for each depth as are the minimum and maximum temperatures recorded for that day. The temperature magnitude and temporal pattern for each depth are very similar; only the temperature range, as expected, shows significant attenuation with depth. Maximum temperatures were recorded on July 10 to 11.

The soil water tension (kPa), as measured in both fields at depths of 10, 18, and 26 cm, is presented as Figure 19.2a to c. Tensiometers were chosen primarily for their ease of use in obtaining daily measurements and, given identical soils, were useful in determining the relative differences

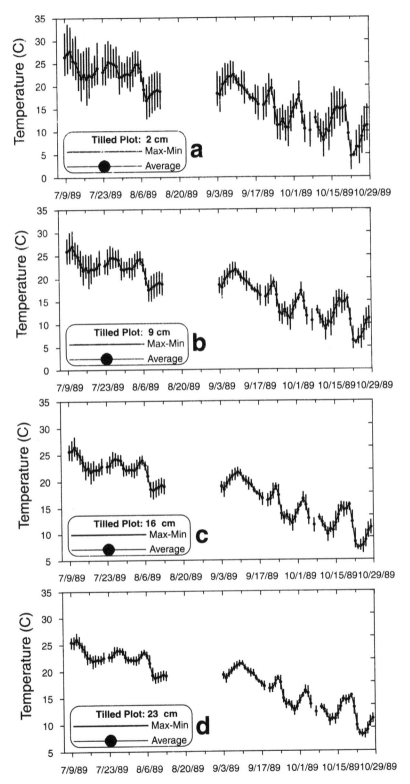

Figure 19.1 Average, maximum and minimum daily temperatures in the plow zone of the CT field. Data series extends from July 9 to October 28, 1989; the period August 14 to September 2 is not represented.

in soil water content within the plow zone. Precipitation (mm) measured above the crop canopy is illustrated in Figure 19.2d. Note that, during the period of August 26 to September 2, 149 mm (5.9 in) of rain fell. This event forms the basis for dividing the growing season into two distinct soil moisture regime periods, hereafter referred to as the *precharge* and *postrecharge* periods.

The precharge period is characterized by high soil water tensions in both fields, which tends to increase with depth. This suggests the predominance of biological (transpiration) over physical (evaporation) processes in the lower region of the plow zone. Response to precipitation infiltration is rapid, though attenuated with depth. In general, tensions in the NT field (dashed line) appear to be slightly less than or equal to those in the CT field (solid line) in July and August. This may relate to reduced soil water utilization associated with the less developed NT stand. However, compaction near the surface could also produce a similar result by restricting evaporative losses to the atmosphere.

During the heavy rains of late August, infiltration is clearly enhanced in the CT field. Surface compaction in the NT field resulted in significant precipitation detention and reduced infiltration rates, thus delaying soil water recharge by 7 to 10 days.

The postrecharge period is characterized by low soil water tensions and rapid fluctuations associated with precipitation events. However, tension values in the NT field are significantly elevated relative to the CT field, and this difference increases with depth. Field capacity in both fields was not achieved until mid-October, when subfreezing temperatures and a mixed rain-snow event occurred.

Using identical graphic parameters, a plot of the thermal regime in the NT field would be visibly indistinguishable from Figure 19.1; however, statistical comparisons between the temperature observations in each field, and for each level, were conducted. The mean and variance were computed for the 2149 temperature observations gathered for each field at depths of 2, 9, 16, and 23 cm. An F-test was performed between identical levels in different fields. As indicated in Table 19.1, in all cases the variance is not significantly different at the 0.05 level of significance (one-tailed test). A t-test ($\alpha = .05$, two-tailed) was conducted to compare the mean temperature between identical levels in the fields. The results are presented in Table 19.1 and suggest that the two samples were drawn from populations having different means, i.e., they are significantly different.

Interfield variations in the thermal regimes can be illustrated by comparing differences between daily temperature values. In Figure 19.3, the solid line represents the difference between average daily temperatures for each depth ($T_{NT} - T_{CT}$); positive values indicate warmer mean temperatures in the NT field. Triangles denote differences in the average daily temperature amplitude or range ($R_{NT} - R_{CT}$); positive values indicate larger temperature fluctuations in the NT field.

DATA ANALYSIS AND DISCUSSION

In general, average daily temperatures within the plow zone of the NT field are consistently 0.50 to 0.75°C warmer than the CT field over the entire period. This suggests enhanced solar radiation absorption at the ground surface of the NT field due to (1) the shorter and thinner corn stand or (2) relatively lower albedo of the NT ground surface; additionally, (3) an alteration of the tillage system could affect the hydraulic and thermal properties of the NT soil so as to increase heat retention. Since preplanting temperature data were not collected, it is not possible to determine which factors promote higher average temperatures. However, given the negligible mulch coverage (5%) and findings of other researchers (e.g., Amemiya, 1970; Van Bavel, 1972), the first factor is believed to be predominant.

The pattern of temperature range differences is more complex. At depth of 16 and 23 cm (Figures 19.3c and d), consistently larger temperature ranges are experienced in the NT field. This would be expected if the absorbed radiation load and resulting conductive flux in the NT field

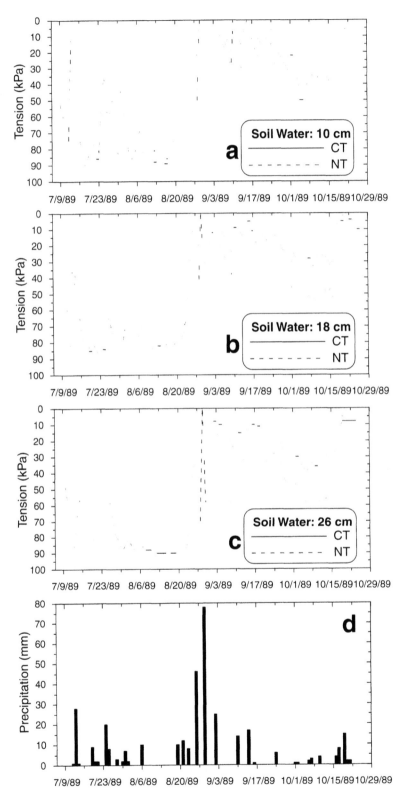

Figure 19.2   (a–c) Soil moisture tension (kPa) measured daily with tensiometers in the plow zone of CT and NT field; (d) Daily precipitation (mm) above crop canopy.

Table 19.1   Statistical Comparison between No-Till and Tilled Temperature Data Sets (n = 2149)

Depth (cm)	Tilled Field (C)	No-Till Field (C)	F Test (α =.05) 1-Tailed CV = 1.46	t-Test (α =.05) 2-Tailed CV = 1.96
2	X = 17.31	X = 17.88	1.01	2.81
	SD = 6.62	SD = 6.66	ns	s
9	X = 17.34	X = 18.07	1.03	3.93
	SD = 6.05	SD = 6.13	ns	s
16	X = 17.57	X = 18.08	1.04	2.93
	SD = 5.65	SD = 5.76	ns	s
23	X = 17.67	X = 18.21	1.02	3.25
	SD = 5.42	SD = 5.48	ns	s

Note: Based on hourly readings at depths of 2, 9, 16, and 23 cm. Over the growing season, the no-till
         field tended to be warmer (0.51° to 0.73°C) than the tilled field, with slightly higher temperature
         variance. X = mean; SD = standard deviation; CV = critical value; ns = not significantly different;
         s = significantly different.

are greater, since the amplitude of the diurnal temperature wave would also be larger at the same depth (Sellers, 1965).

Near the surface (Figures 19.3a and b), the average temperature range differences were predominantly negative in the precharge period. At a depth of 10 cm (Figure 19.2a), both fields tended to be equally dry during this time. Thus, despite the greater temperatures and enhanced conductive flux to the upper region of the NT plow zone, the diurnal temperature amplitude was slightly damped in July and August. This suggests that the thermal conductivity and thermal diffusivity of the soil near the surface of the NT field are larger, allowing the thermal disturbance to propagate more rapidly and with less attenuation then in the CT field (van Wijk and de Vries, 1963). This probably results from soil compaction in the NT field, which would, by decreasing the void space, effectively increase the bulk thermal conductivity in the near-surface layer.

Throughout the postrecharge period, mean daily temperatures were consistently warmer near the surface of the NT field, and temperature range differences were generally positive (Figures 19.3a and b). During this period, the NT field was significantly drier (Figure 19.2a); therefore, warmer temperatures and greater temperature ranges near the surface occurred when the NT soil was dryer than the CT soil. Higher temperatures and amplitude are associated with materials of lower thermal conductance and probably reflect the insulating influence of gases in the void spaces of the NT soil. This relationship is summarized in Figure 19.4. Under drier NT surface conditions (tension (NT–CT) > 0), daily temperature range differences at the surface tend to be positive. There is no pattern in temperature amplitude differences when the CT field had a higher near-surface moisture content.

It is expected that, in dry soils, reduced temperature amplitudes would be associated with compacted soils of high thermal conductance. However, the ponding that occurs at the surface of the NT field would also dampen the temperature range as more energy is dissipated on heating and evaporating the ponded water. Figure 19.5 illustrates these factors. Because average temperatures in the NT field were consistently warmer, it is necessary to make comparisons relative to the average seasonal temperature range difference (0.48°C) at the 2-cm depth. In general, higher average temperatures (23 to 30°C) tend to be associated with reduced amplitudes in the NT field, presumably due to the greater conductivity of compacted NT soils.

Precipitation events are also highlighted in Figure 19.5. All days with measured rainfall are symbolized with a filled circle. To account for ponding effects, the circle for the following day is also filled if more than 5 mm of rainfall was recorded; if precipitation exceeded 10 mm then the subsequent two days have filled circles. Precipitation and ponding are clearly correlated to preferential amplitude damping in the NT field at the 2-cm level; about 60% of the lower-than-normal temperature range differences are associated with these events.

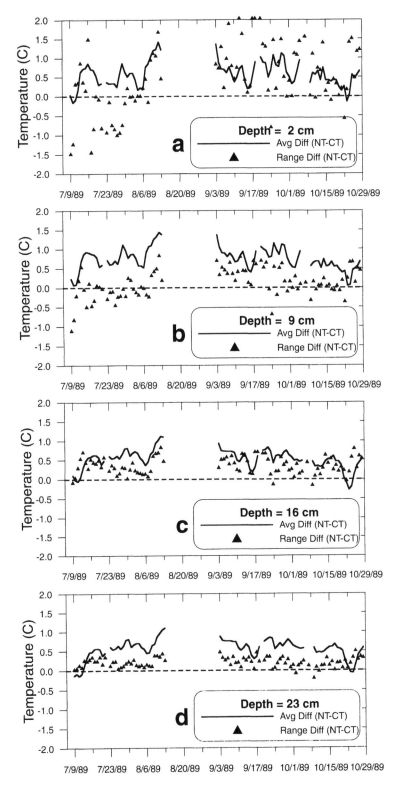

Figure 19.3 Comparison of thermal regimes in plow zone of CT and NT fields. Solid line denotes difference between mean daily temperatures ($T_{NT} - T_{CT}$); triangles represent differences in the daily temperature range ($R_{NT} - R_{CT}$).

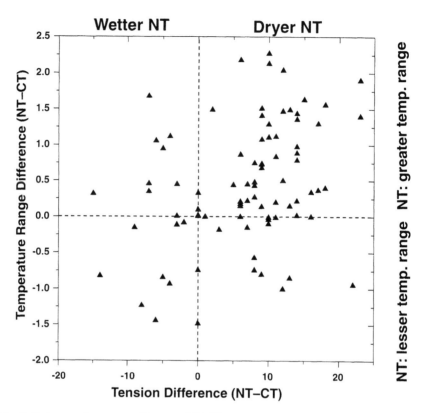

Figure 19.4 Daily temperature range difference (NT–CT) at 2-cm level plotted against daily soil water tension difference (NT–CT) at 10-cm level.

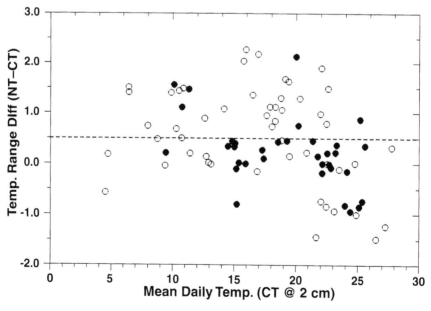

Figure 19.5 Temperature range difference (NT–CT) plotted against mean daily temperature at 2-cm level in CT field. Dashed line at 0.48°C is seasonal average range difference. Observations with filled circles indicate days with precipitation or ponding.

The results suggest that soil compaction at the surface of the NT field has altered the hydraulic and thermal properties of the affected material, with consequent impact on the thermal and moisture regimes in the plow zone. The analysis of thermal property variation within the plow zone is facilitated by the finite-difference calculation of the thermal diffusivity from the time-space temperature matrix. The Biot–Fourier one-dimensional transient heat diffusion equation is:

$$\frac{\delta T}{\delta t} = \alpha \frac{\delta^2 T}{\delta Z^2}$$

(19.1)

where T is temperature (°C), Z is depth (m) in the soil, and α is the conductive thermal diffusivity (m^2 s^{-1}), a constant of proportionality in a conductive system numerically equal to the thermal conductivity divided by the volumetric heat capacity. This physical property describes the facility with which a material changes temperature. In a purely conductive system, α must be positive and constant. In soils, the thermal conductivity and heat capacity (and thus α) will vary with the relative proportion of gas, water, and ice occupying the void space (de Vries, 1963), and are thus affected by compaction, precipitation, infiltration, and desiccation. For example, increasing soil density by compaction reduces porosity, which increases the thermal conductivity and diffusivity. Initially, when water replaces gases in the void spaces, the bulk conductivity of the material will increase. However, beyond a certain moisture content (usually 8 to 20% by volume), α will be reduced due to the influence of the large volumetric heat capacity of water relative to soil gases (Farouki, 1981).

Given nonsteady-state conditions in a heterogeneous soil, it is unrealistic to define a diffusivity value typical of the system. For illustrative purposes, however, data presented in Farouki (1981) were used to define an expected α value in the realm of 5×10^{-7} m^2 s^{-1}.

Using the amplitude ratio method, Johnson and Lowery (1985) found that α in the 5 to 15 cm zone of an NT field was 20 to 25% greater compared to the CT condition. The study was conducted under sunny, precipitation-free conditions for periods of about 3 days. Similarly, Hay et al. (1978) reported α values for CT fields 37% lower between depths of 5 to 25 cm. However, given (1) the variable nature of soil constituents with time and depth and (2) the influence of heat transfer by nonconductive processes (particularly water and vapor migration), the assumption of homogeneity required in these analytical solutions can rarely be met (Sellars, 1965). This study utilized a finite-difference scheme developed by McGaw et al. (1978). Because the diffusivity is calculated from the observed temperatures and includes the thermal impact of nonconductive heat transfer, it is more properly referred to as *apparent thermal diffusivity* (α_a). By rearranging Equation 19.1, one obtains:

$$\alpha_a = Tt / Tzz$$

(19.2)

where Tt and Tzz are the finite-difference approximations of $\delta T/\delta t$ and $\delta^2 T/\delta Z^2$, respectively. The time and space derivatives can be approximated by:

$$Tt = [T_i^{j+1} - T_i^{j-1}] / 2\Delta t$$

(19.3)

and

$$Tzz = [T_j^{i-1} - 2T_j^i + T_j^{i+1}] / \Delta Z^2$$

(19.4)

where Δt and ΔZ are time and space increments, and j and i refer, respectively, to the temporal and depth positions of nodes in the time-space mesh. Substitution of Equations 19.3 and 19.4 into Equation 19.2 yields

$$\alpha_a = [\Delta Z^2 / 2\Delta t] \times [(T_i^{j+1} - T_i^{j-1}) / (T_j^{i-1} - 2T_j^i + T_j^{i+1})],$$

(19.5)

which was used in this study to calculate diffusivity for internal nodes at the 9- and 16-cm levels ($\Delta Z = 0.07$ m). In all cases, the time step is 1 hour.

Underlying these procedures is the assumption that the system is driven by conductive heat transfer processes only. If no phase change occurs and the radiative and convective heat-transfer mechanisms are absent or negligible, diffusivity will not vary greatly over time. However, large temporal fluctuations in α_a reflect the impact of nonconductive heat-transfer processes operating within the plow zone. Suspect mechanisms include phase transformations due to evaporation and condensation (which entails the consumption or liberation of 2.5 MJ kg^{-1}, respectively), vapor transport and water advection to the evaporating front. Several of these mechanisms can operate simultaneously in different regions of the soil column (Farouki, 1981). Further details on the scheme and interpretation are available in Nelson et al. (1985), Outcalt and Hinkel (1989), Hinkel et al. (1990), Hinkel (1997), and Hinkel et al. (in press).

Figure 19.6 presents the average daily α_a for the CT (Figure 19.6a) and NT (Figure 19.6b) fields calculated for the intermediate levels of 9 and 16 cm. In the CT field, at a depth of 16 cm, the system appears to behave conductively, with little variation around the anticipated value of 5×10^{-7} m^2 s^{-1}. Depressions in the trace are clearly associated with precipitation events. An increase in the bulk heat capacity of the soil, in combination with the infiltration of relatively cooler rain water, which opposes the conductive tendency to warm, results in temporarily reduced α_a values.

Near the surface of the CT field, at a depth of 9 cm, the influence of nonconductive heat transfer processes is more pronounced. The combined effects of conductive and nonconductive heat transfer results in the higher average α_a values. Depressions in the trace are also correlated to precipitation events, but are highly amplified since the porosity is larger and the impact on the bulk thermal properties is greater. Ascensions along the trace imply enhanced heat transfer and temperature change. This is most likely attributable to internal condensation of water vapor, which increases the local enthalpy due to the release of latent energy. Thus, in the CT field, precipitation infiltration and vapor diffusion strongly influence the evolution of the thermal profile near the surface, and can be related to the presence of water pathways in the porous and permeable tilled soil. At a depth of 16 cm, the system appears to be primarily conductive or in a steady state with respect to vapor diffusion across this level.

The pattern for the NT field (Figure 19.6b) appears to be similar. Note, however, that the 9-cm level behaves more like a conductive material while nonconductive processes appear to predominate at depth, a situation opposite to that found in the CT field. This indicates that internal evaporation and condensation are restricted to deeper regions of the plow zone, and implies that surface compaction has reduced porosity and permeability to the point where vapor diffusion to the atmosphere is restricted. Conduction appears to be the primary mode of heat transfer near the surface.

Calculation of the α_a at the 9-cm level relies on temperature measurements at the 2-, 9-, and 16-cm level. In this regard, the depth spacing is not optimal for calculating the α_a of a compacted zone that extends from the surface to a depth of 2 to 7 cm. Refinement of estimates of α_a can be achieved by reducing the probe spacing to about one half of the thickness of the compacted layer.

Despite this drawback, Figure 19.6 graphically illustrates one mechanism whereby soil water retention is enhanced under NT systems. Physical changes in the soil impede the flow of water vapor to the atmosphere despite the higher enthalpy of the NT system. The influence of mulch, often cited for this behavior, is negligible in this case.

CONCLUSIONS

In the initial year following conversion to NT corn, temperatures in the plow zone averaged 0.50 to 0.75°C warmer than those in the CT field during the growing season. It is thought that this is due to enhanced solar radiation absorption at the ground surface resulting from shorter and thinner NT corn stands.

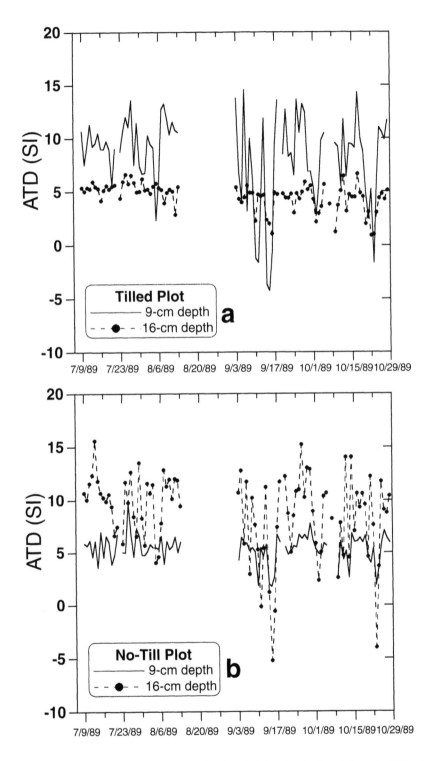

Figure 19.6 Apparent thermal diffusivity ($\times 10^{-7}$ m^2 s^{-1}) calculated for depths of 9 cm (solid line) and 16 cm (dashed line) for (a) CT and (b) NT fields.

The daily range of temperature was influenced by the amount of soil water. When the soil in both fields was equally dry, as is often the case in the early summer months, the enhanced conductive flux to depth in the NT field resulted in warmer temperatures and slightly larger temperature ranges. Near the surface, however, lower daily temperature amplitudes suggest higher thermal conductivity resulting from surface compaction.

Near-surface soil compaction also retarded infiltration and delayed soil water recharge in the NT field. Surface detention and ponding following precipitation events reduced the flow of heat to depth and damped the diurnal temperature amplitude in the NT plow zone.

Numerical calculation of the average daily apparent thermal diffusivity at depths of 9 and 16 cm for each field illustrates the effect of compaction on the conductive and nonconductive heat transfer processes. The compacted surface layer in the NT field behaves more like a conductive material and acts as a semipermeable seal by limiting vapor and water movement through this layer; the effects of water advection and vapor diffusion appear to be restricted to the lower region of the plow zone. The large variation of the α_a within the plow zone suggests that the system cannot realistically be modeled as a conductive solid.

The results from this study are based on data collected during only one growing season, albeit during the critical initial transition period from CT to NT. It is clear that long-term system monitoring is necessary to access the impact on soil physical, chemical, and biological components. This study demonstrates the need to further evaluate the impact of soil structural changes on heat transport and water movement near the surface, and to determine the role these factors play on soil organic carbon sequestration.

ACKNOWLEDGMENTS

Thanks to the participating landowners, Mr. and Mrs. A. Totteroff, the participating farmer, J. Steiner, SCS Agents R. Gehring, D. McElroy, and J. Tkatschenko, CES Agent E. Winkle, and Sheri Farquer. Insightful remarks by three anonymous reviewers are appreciated.

REFERENCES

Aase, J.K. and F.H. Siddoway, Microclimate of winter wheat grown in three standing stubble heights, in Fanning, C. (ed.), *Proc. Tillage Symp.*, 9–11 Sept, 1980, North Dakota State University, Fargo, ND, 1980.

Al-Darby, A.M. and B. Lowery, Seed zone soil temperature and early corn growth with three conservation tillage systems, *Soil Sci. Soc. Am. J.*, 51:768–774, 1987.

Allmaras, R.R., W.C. Burrows, and W.E. Larson, Early growth of corn as affected by soil temperature, Soil *Sci. Soc. Am. Proc.*, 28:271–275, 1964.

Amemiya, M., Tillage alternatives for Iowa, Iowa State Univ. Coop. Ext. Serv. Pamph. pm-488, 1970.

Bauder, J.W., G.W. Randall, and R.T. Schuler, Effects of tillage with controlled wheel traffic on soil properties and root growth of corn, *J. Soil Water Conserv.*, 40:382–385, 1985.

Bauder J.W., G.W. Randall, and J.B. Swann, Effect of four continuous tillage systems on mechanical impedance of a clay loam soil, *Soil Sci. Soc. Am. J.*, 45:802–806, 1981.

Blevins, R.L. and D. Cook, No-tillage — its influence on soil moisture and soil temperature, Progress Rep. 187, University of Kentucky, Lexington, KY, 1970.

Burrows, W.C. and W.E. Larson, Effect of amount of mulch on soil temperature and early growth of corn, *Agron. J.*, 54:19–23, 1962.

de Vries, D.A., Thermal properties of soils, in *Physics of Plant Environment*, North-Holland Publ. Co., Amsterdam, Netherlands, 1963, 210–235.

Duval, J. et al., Residual effects of compaction and tillage on the soil profile characteristics of a clay-textured soil, *Can. J. Soil Sci.*, 69:417–423, 1989.

Eckert, D.J., Tillage system × planting date interactions in corn production, *Agron. J.*, 76:580–582, 1984.

Farouki, O.T., Thermal Properties of Soils, CRREL Monograph 81-1, U.S. Army Corps of Engineers, Cold Regions Research and Engineering Laboratory, Hanover, NH, 1981.

Flowers, M.D. and R. Lal, Axle load and tillage effects on soil physical properties and soybean grain yield on a mollic ochraqualf in northwest Ohio, *Soil Tillage Res.*, 48:21–35, 1998.

Griffith, D.R. et al., Effect of eight tillage-planting systems on soil temperature, percent stand, plant growth and yield of corn in five Indiana soils, *Agron. J.*, 65:321–326, 1973.

Hay, R.K.M., J.C. Holmes, and E.A. Hunter, The effects of tillage, direct drilling and nitrogen fertilizer on soil temperature under a barley crop, *J. Soil Sci.*, 29:174–183, 1978.

Herbek, J.H., L.W. Murdock, and R.L. Blevins, Tillage system and date of planting effects on yield of corn on soils with restricted drainage, *Agron. J.*, 78:824–826, 1986.

Hinkel, K.M., Estimating seasonal values of thermal diffusivity in thawed and frozen soils using temperature time series, *Cold Regions Sci. Techn.*, 26:1–15, 1997.

Hinkel, K.M., S.I. Outcalt, and F.E. Nelson, Temperature variation and apparent thermal diffusivity in the refreezing active layer, *Permafrost Periglacial Proc.*, 1(4):265–274, 1990.

Hinkel, K.M. et al., Patterns of soil temperature and moisture in the active layer and upper permafrost at Barrow, Alaska: 1993–1999, *Global Planetary Change*, (in press).

Imholte, A.A. and P.R. Carter, Planting date and tillage effects on corn following corn, *Agron. J.*, 79:746–751, 1987.

Johnson, M.D. and B. Lowery, Effect of three conservation tillage practices on soil temperature and thermal properties, *Soil Sci. Soc. Am. J.*, 49:1547–1552, 1985.

Kaspar, T.C., S.D. Logsdon, and M.A. Prieksat, Traffic pattern and tillage system effects on corn root and shoot growth, *Agron. J.*, 87:1046–1051, 1995.

Kayambo, B., R. Lal, and G.C. Mrema, Traffic-induced compaction in maize, cowpea and soya bean production on tropical Alfisol after ploughing and no-tillage: soil physical properties, *J. Sci. Food Agric.*, 37:969–978, 1986.

Lal, R., Axle load and tillage effects on crop yields on a Mollic Ochraqualf in Northwest Ohio, *Soil Tillage Res.*, 37:143–160, 1996.

Lal, R., Soil compaction and tillage effects on soil physical properties of a Mollic Ochraqualf in Northwest Ohio, *J. Sustainable Agric.*, 14(4):53–65, 1999.

Lal, R. and M. Ahmadi, Axle load and tillage effects on crop yield for two soils in central Ohio, *Soil Tillage Res.*, 54:111–119, 2000.

Larson, W.E., W.C. Burrows, and W.O. Willis, Soil temperature, soil moisture, and corn growth influenced by mulches of crop residues, in F.A. van Baren (ed.), *Trans. Int. Cong. Soil Sci. 7th*, Madison, WI, 14–24 August 1960, Elsevier, Amsterdam, 1961, 629.

Larson, W.E. and G.J. Osborne, Tillage accomplishments and potential, in P.W. Unger and D.M. Van Doren, Jr. (ed.), *Predicting Tillage Effects on Soil Physical Properties and Processes*, American Society of Agronomy, Madison, WI, Pub. 44, 1982, 1–11.

Lipiec, J. and C. Simota, Role of soil and climate factors influencing crop responses to soil compaction in Central and Eastern Europe, in B.D. Soane and C. van Ouwerkerk (ed.), *Soil Compaction in Crop Production*, Elsevier Science, Amsterdam, 1994, 365–390.

Logsdon, S.D. and C.A. Cambardella, Temporal changes in small depth-incremental soil bulk density, *Soil Sci. Soc. Am. J.*, 64:710–714, 2000.

McCalla, T.M., Light reflection from stubble mulch, *J. Am. Soc. Agron.*, 39:690–696, 1947.

McGaw, R.W., S.I. Outcalt, and E. Ng, Thermal properties of wet tundra soils at Barrow, Alaska, in *Proc. 3rd Conf. Permafrost*, National Research Council of Canada, Ottawa, 1978, 47–53.

Mock, J.J. and D.C. Erbach, Influence of conservation-tillage environments on growth and productivity of corn, *Agron. J.*, 69:337–340, 1977.

Nelson, F.E. et al., Diurnal thermal regime in a peat-covered palsa, Toolik Lake, Alaska, *Arctic*, 38(4):310–315, 1985.

Outcalt, S.I. and K.M. Hinkel, Night-frost modulation of near-surface soil-water ion concentration and thermal fields, *Phys. Geogr.*, 10:336–348, 1989.

Phillips, R.E. and S.H. Phillips (eds.), *No-Tillage Agriculture, Principles and Practices*, Van Nostrand Reinhold, New York, 1984.

Sellers, W.D., *Physical Climatology*, University of Chicago Press, Chicago, 1965.

Spilde, L.A. and E.J. Deibert, Crop yield, water use and soil property changes with conventional, minimum, and no-till systems in the Red River Valley, *ND Farm Res.*, 43:22–25, 1986.

Thomas, G.W., Managing minimum-tillage fields, fertility, and soil type, in *Weed Control in Limited-Tillage Systems*, Weed Science Society of America, Champaign, IL, 1985, 211–226.

Tollner, E.W., W.L. Hargrove, and G.W. Langdale, Influence of conventional and no-till practices on soil physical properties in the southern Piedmont, *J. Soil Water Conserv.*, 39:73–76, 1984.

Van Bavel, C.H.M., Soil temperature and crop growth, in D. Hillel (ed.), *Optimizing the Soil Physical Environment toward Greater Crop Yields*, Academic Press, New York, 1972.

van Wijk, W.R. and D.A. de Vries, Periodic temperature variations in a homogeneous soil, in *Physics of Plant Environment*, North-Holland Publ. Co., Amsterdam, Netherlands, 1963, 102–143.

van Wijk, W.R., W.E. Larson, and W.C. Burrows, Soil temperature and the early growth of corn from mulched and unmulched soil, *Soil Sci. Soc. Am. Proc.*, 23:428–434, 1959.

Willis, W.O. and M. Amemiya, Tillage management principles: soil temperature effects, in *Conservation Tillage*, Proc. Natl. Conserv. Tillage Conf. Des Moines, IA, 28–30 March 1973, Soil Conservation Society of America. Ankeny, IA, 1973, 22–41.

Willis, W.O., W.E. Larson, and D. Kirkham, Corn growth as affected by soil temperature and mulch, *Agron. J.*, 49:323–328, 1957.

CHAPTER 20

Soil Fertility Management with Zeolite Amendments. I. Effect of Zeolite on Carbon Sequestration: A Review

Ekaterina Filcheva and K. Chakalov

CONTENTS

INTRODUCTION

The problem of soil carbon sequestration cannot be discussed in isolation from soil fertility. In this aspect, complex impacts within agroecosystems are established. The main natural emitters of greenhouse gasses (CO_2, CH_4, N_2O, etc.) are organic matter and soils, after their destruction by anthropogenic processes (land use, land-use change, deforestation, and intensive agricultural activities). Denitrification, CO_2 and CH_4 emissions, etc. from undisturbed soils are in equilibrium with organic matter syntheses and nitrogen fixation. In some cases, the accumulation process is more active. All soil types have oxidation-reduction ion units: $Fe^{2+} - Fe^{3+}$; $Mn^{2+} - Mn^{4+}$; $NO_3^- - N_2O$ or NH_3; $SO_4^{2-} - H_2S$, etc. However, aeration, compaction, percolation, humus content, and heavy metals pollution have an important effect on soil oxidation-reduction charges (Kudeyarov et al., 1989; Menyailo, 1996).

This chapter describes ways zeolite meliorants may help sequester soil carbon by limiting factors, which lead to destruction of organic compounds in areas polluted by heavy metals.

ZEOLITES' EFFECT ON SOIL FERTILITY AND CARBON SEQUESTRATION

Direct Influence

Moshev and Mirchev (1989) considered organic matter composition, Fe^{2+} – Fe^{3+}, and Mn^{2+} – Mn^{4+}, the base elements which control soil Eh. Nonsilicate compounds and mobile forms of Fe, Mn, Cu, and Co take part in the process of reduction. The seasonal ozone destruction obviously correlates with the seasonal character of the soil reduction charge (rH). The charge and ozone destruction increase, and the highest levels are measured for both in the spring and autumn, i.e., during waterlogging (Klemenko et al., 1988; Jokova and Andonov, 1989; Moshev and Mirchev, 1989). This is in consequence of the so-called "mirror effect" for moisture and Eh curves (Vosbutskaia, 1969). These abiotic factors are in connection with soil oxygen, available ions with an oxidation-reduction function, temperature regime, organic matter composition, etc. Naturally, in this case, one of the limiting factors is soil microbiological activity. Incorporation of plant residues decreases Eh, which depresses root system development and crop yield production (Kudeyarov et al., 1967).

The transition of ion units may relate to the transformation of organic matter to humic substances, which is an oxidation process. Orlov (1988) reported that transition of Fe_2O_3 to FeO is determined by organic compounds that influence a complex forming reaction. Ding Chang-pu (1986) proved this thesis, discussing reduction possibilities in several soil types. In his opinion, water soluble iron ions play an important role, but Fe^{2+} is not more than 1% in the main soil types, except for organic soils (Histosols, FAO-UNESCO, 1990), where it is about 25%. The influence on oxidation processes is determined by biological effects in peat soils, but reduction charges are controlled by physico-chemical properties (Staricova, 1984).

The changes of O_2 quantities in the soil's gasous phase influence substantially the level of Eh. For example, in the case of O_2 – 0% , the Eh = 170 to 180 mV and O_2 – 20% is more than 600 mV. Nitrification and humus formation predominate when Eh charges are over 600 mV, and nitrification should be stopped under 480 mV (Vosbutskaia, 1968). Evidently, the base methods for controlling organic matter destruction are increasing Eh charges in soils. These methods include maintaining drainage, aeration, fertilization balance, acidity, and toxicant mobility. The known meliorants that depress mobility of heavy metals are lime materials, organic composts and farmyard manure, phosphoric rocks, etc. They cannot influence positively all factors that control the level of Eh charges in soils, and cannot reduce greenhouse gas emission from soils. Liming, as a main method for acidity neutralization, increases the number of bridges between toxicants and colloid systems.

Applying organic mater such as manure, plant residues, composts, peat, etc. without another soil amendment increases rH, and creates a risk for complexation of water-soluble toxicants. These affect the environment (soil–plants–underground water–air). Applying natural adsorbents (bentonites, vermiculite or other hydromicaceus, natural or modified zeolites, etc.) also is discredited, because they are composed of unsuitable materials (with noneligible chemical and physicochemical composition).

Classic soil science considers exchangeable acidity to be formed by the presence of all amphoteric elements, which constitute hydrolytic-acid compounds with soil acid-exchangers, i.e., they organize phytotoxic exchangeable acidity (which may be copper acidity, leads acidity, etc., and may predominate over the H^+ toxicity). When the soils' oxidation charges are deficient for oxidation of certain amphoteric elements to its extreme valence and reach some mediate valence, then the ion-hydroxide equilibrium organizes modifications because of their slight solubility. A typical example is manganese (Ganev, 1990).

Cheteropolaric and complex metal-humus compounds play an important role for soil fertility. Metal cations constitute donor-accepting positions with several functional groups as humic acids and should be able to localize these positions as a heteropolar salt or complex compounds. There are possibilities for organizing intermolecular bridges. Confirmations of this are constituted poly-nuclear complexes, capable of forming ferro- or ferrichumic complexes. (Orlov, 1990).

In previous work (Chakalov et al., 1993, 1995a, b, 1996, 1999; Filcheva et al., 1997, 1998a, b, c, 1999, 2000a, b; Popova et al., 1996a, b), the influence of natural zeolite, clinoptilolite type, on the physicochemical properties and transformation of organic matter to humic substances (organo-zeolite composts and organozeolite amended soils) was proved. The natural zeolites change media reactivity by controlling exchangeable ions with acid function: Al, Fe, H, and Mn Cu. The potassium form of zeolite particularly influences the limitation of exchangeable acidity. The ammonium and calcium form of clinoptilolite accumulate relatively high levels of exchangeable acidity on a mineral adsorbent.

A composted mixture — organic materials and clinoptilolite — acts as a matrix for humus formation with a more complicated structure, which accelerates maturity. This condensation and aromatization is more effective in the presence of zeolite than are the composts without mineral amendments. The stable zeolite structure and relatively uniform desorbtion of base cations maintain humus-forming processes. The differentiated natural ferritization of the zeolite component deter-mines clinoptilolite's ability for various grades of amphoteric hydroxide ionization. Ion-hydroxide equilibrium is determined by the degree of clinoptilolite influence on the exchangeable acidity translocation of exchangeable H^+, Al^{3+}, and Fe^{3+} ions between organic and mineral adsorbent (by maturity of humic substances).

Exchangeable Fe^{2+}– Fe^{3+} ions play the most important role in this process. This correlates with values of variable charges (T_A) for composts, the level of which increases with the accumulation of exchangeable iron ions. The variable charges of substrate adsorbent are responsible for the organization of covalent connections — bridges with several elements. Composts with a low level of T_A accelerate T_A in meliorated soils or the opposite, i.e., the observed "mirror effect," which is well expressed in the case of measured phosphate buffer capacity (PBC). Practically, the most important point to be explained is the influence of natural and ferritizied clinoptilolite on physico-chemical properties of heavy metal and as-polluted soils.

Indirect Influence

Authors who discuss zeolite melioration of polluted or acidic soils argue about the ion-exchange nature of zeolite's influence on soil properties (Mumpton, 1984; Weber et al., 1984; Tsitsiashvili et al., 1988; Iacovlev et al., 1990; Tsadilas et al., 1997). However, Mitchell and Atkinson (1993) promoted the concept that principal organic compounds influence the processes. For example, copper and zinc, adsorbed by clinoptilolite in zeolite-amended media, are more available for a plant's uptake (Ebrle et al., 1995). A discussion of the toxicant's demobilization in soil in connection with the media rH leads to the considerable role of natural zeolites on the availability of biogenic elements for plants' nutrition in amended soils.

Tukvadze (1984), Tzhakaia and Kvashali (1985), and Grigorra (1984) reported results on nitro-gen accumulation in zeolite–amended soils (two to three times more). Weber et al. (1984) showed the negative effect of zeolite application in sewage sludge-amended soils because of the high sodium content (4%) in the zeolites used, and possible nitrogen fixation. However, other authors consider that zeolite application improves plants nitrogen nutrition (Daskalov and Stoilov, 1986; Stoilov et al., 1982; Pirella et al., 1984; Allen and Ming, 1995; Ebrle et al., 1995; Tsadilas, 1997).

Boettinger et al. (1997) proposed slow-release fertilizers, prepared by zeolite saturation with liquid manure or sewage sludge. At the same time, Lego (1997) established a significant rate of ammonium nitrogen decrease in organozeolite-amended substrates, after a 2-week vegetation period. This experiment with a poultry manure-zeolite mixture established the mixture's negative influence on plant development because of NH_3 toxicity, but low doses of fertilization assist plant growth and plants to develop from 2.6 to 2.7 times more biomass. Perhaps, organo-zeolite fertilizers improve soil microbiological activity (Andronicashvili et al., 1995; Lego, 1997).

Formation of stronger plant root systems on zeolite-meliorated soils, especially in reclaimed territories, assumes accumulation of two to three times more plant residues in the formation of more "young" humic substances. For example, zeolite reclamation of yellow-green clay landscapes in the Maritza Iztok JSComines region, Bulgaria, accelerates humus-forming processes and ground content of two times more "young" humic substances (Dimitrov et al., 1990). Possibilities for anthropogenic formation of humus horizon in this region are well known (Krastanov et al., 2000).

In the past 10 years, a method for zeolite ferritization and manganesenation (Kallo et al., 1989; Papp et al., 1995; Nikashina et al., 1997) became more popular. These modifications improve zeolite properties for use in water purification or detoxification of soils polluted by heavy metal and radionucleides, with a positive effect of about 90 to 95%. Combining this method with natural zeolite influence on humus-forming processes should guarantee soil carbon sequestration by:

1. Increasing biological accumulation of atmospheric CO_2 by improving plant development and photosynthesis
2. Accumulating more organic matter in soil from plant residues
3. Transforming organic matter to humic substances with a high level of maturity

Discussing the microbiological activity of zeolite-amended peat substrates used as a media for rooting lemon cuttings, Gamisonija et al. (1987) prove a positive clinoptilolite influence on Eh of the media. Uniquely established properties of organozeolite meliorants to induce a mirror effect for formation of Eh, T_A, and PBC^P in treated soils (i.e., to influence a range of zones with a heteropolar metallic connection) proved their potential to control oxidation-reduction processes in clean or polluted soils and to limit destruction of organic mater — the first step for soil carbon sequestration.

The different types of induced natural ferritization of clinoptilolite determine the selection of zeolite for sorption of heavy metals or their translation to organic or mineral adsorbents in the media. Accumulation of fixed cations (potassium, iron, manganese, etc.) increases the effects of zeolite on managing soil-forming processes. The oxidation character of humification increases the Eh in treated media, limiting the reduction of nitrate nitrogen. At the same time, ammonium fixation by zeolite s components should be maintained at a slow rate of nitrification, without seasonal peaks. This zeolite influence on soil properties correlates with the ion-hydroxide equilibrium of some amphoteric elements, by establishing the mirror effect of meliorants.

CONCLUSIONS

Melioration of soils by zeolite products (natural-, modified-, or organozeolite forms) improves their natural capacity for carbon sequestration. Zeolite influence on Eh of media is a consequence of the variability of releasing exchangeable ion unites ($Fe^{2+} - Fe^{3+}$, $Mn^{2+} - Mn^{4+}$, $Cu^{2+} - Cu^+$, etc.), and may help to manage soil-forming processes and levels of humus maturity. Improved soil fertility induces production of more quantities of plant residues. Zeolite's use as a soil amendment is one of the environmentally clean methods for soil carbon sequestration.

REFERENCES

Allen, E.R. and D.W. Ming, Recent progress in the use of natural zeolite in agronomy and horticulture, in D. Ming and F. Mumpton, eds., *Natural Zeolites '93*, Int. Comm. Natural Zeolites, Brockport, New York, 1995, 477–490.

Andronicashvili, T., M. Kardava, and M. Gamisonija, Effects of natural zeolite on microbe landscape of some soils in Georgia, *Proc. Int. Simp.*, Sofia Zeolite Meeting '95, June 18–25, 111–112, 1995, Sofia, Bulgaria, 1995.

Boettinger, J.L. et al., Developing sustainable agriculture systems based on nitrogen cycling with clinoptilolite, *Proc. 5th Int. Conf. Occurrence, Properties Utilization Natural Zeolites*, ZEO '97, September 21–29, 1997, Ischia, Naples, Italy, 74–75, 1997.

Chakalov, K., J. Yoneva, and T. Popova, Influence of cation exchange composition of clinoptilolite on organozeolite composts quality, *Proc. 4th Int. Conf. Occurrence, Properties and Utilization of Natural Zeolites '93*, Boise, ID, 62–63, 1993.

Chakalov, K., K. Dimitrov, and T. Popova, Organozeolite composts as a substrate and soil conditioners for growing ecologically clean vegetables, *Proc. Int. Simp.* Sofia Zeolite Meeting '95, June 18–25, 1995, Sofia, Bulgaria, 101–108, 1995a.

Chakalov. K. et al., Influence of natural clinoptilolite with variable cation exchange composition on ion-hydroxide equilibrium of amphoteric elements in organozeolite media for zeoponic, Ministry of Education and Science of Bulgaria, National Science Fund, Research Project — CC418, 1995, 1996, and 1998, Research reports. (Bulgarian, English abstracts), 1995b.

Chakalov K., T. Popova, and E. Filcheva, Influence of cation exchange composition of organozeolite composts on the ion-hydroxide equilibrium of copper, *Proc. 2nd Int. Congr., ESSC*, Development and Implementation of Soil Conservation, Strategies for Sustainable Land Use, 1–7 September, 1996, Munich, Germany, 50, 1996.

Chakalov, K., T. Popova, and E. Filcheva, Soil fertility management with zeolite amendments. II. Zeolite effect on the maintaining the soil Eh. Int. Symp. Agricultural Practices and Policies for Carbon Sequestration in Soils, July 19–23, 1999, Columbus, OH, 229–235, 1999.

Chang-pu, Ding, Reducing substances in soil, *Soil Res. Revert*, 14, 1–9, 1986.

Daskalov, V.G. and G.P. Stoilov, Natural zeolites, *Bulgarian Acad. Sci.*, Sofia, 373–378 (Bulgarian), 1986.

Dimitrov, K., K. Chakalov, and T. Popova, Final research report: zeolite reclamation of landscape in mines Maritza Istok JSCo cooling region, Energy Comm. Bulgarian Government, 1989–1990, (Bulgarian, English abstracts), 1989.

Ebrle, D., K. Barbarick, and I. Lai, in *Natural Zeolite '93*, D. Ming, and F. Mumpton, eds., Int. Comm. Natural Zeolites, Brokport, NY, 1995, 491–504.

FAO-UNESCO, Revised legend, Rome, Italy, 1990.

Filcheva, E., K. Chakalov, and T. Popova, Modeling soilforming processes in artificial zeolite soils, in *Soil Properties and Their Management for Carbon Sequestration*, R. Lal, J. Kimble, and R. Follet, eds., USDA-Natural Resources Conservation Service, National Soil Survey Center, Lincoln, NE, 113–117, 1997.

Filcheva, E., K. Chakalov, and T. Popova, Model approach to improve soil quality using organo-zeolite compost, *J. Balkan Ecol.*, 1, 72–78, 1998a.

Filcheva, E., K. Chakalov, and T. Popova, Possibility to manage soil-forming processes, *16th World Cong. Soil Sci.*, 20–26 August, 1998, Montpellier, France, 1998b.

Filcheva, E., K. Chakalov, and T. Popova, Possibility to manage soil-forming processes in ecosystems. Concept for humus conservation in agriculture, *J. Balkan Ecol.*, 1, 72–78, 1998c.

Filcheva, E. et al., Soil amendment with sewage sludge and zeolite. I. Influence on soil organic matter, 6th Int. Meeting Soils Mediterranean Type Climate, July 4–9, 1999, Barcelona, Spain, 870–872.

Filcheva, E. et al., Influence of humus substances on plant growth in organozeolite media, in *Proc. 1st Natl. Conf. Humic Substances Soil Tillage* (BHSS-BSTRS Conference '2000), S. Rousseva, E. Filcheva, and I. Stefanova (Eds.), 11–12 May, Borovec, Bulgaria, 51–54, 2000a.

Filcheva E. et al., Influence of zeolite and barks on the organic matter composition and physico-chemical properties of composts produced by sewage sludge, in *Proc. 9th Natl. Conf.*, Kavala, Greece, 349–358, 2000b.

Gamisonija, M.K., T.G. Andronikashvili, and A.B. Russadtze, Investigation the influence of clinoptilolite tuffs on the physico-chemical properties and biological activity of acidic soils in moist subtropics. Application of clinoptilolite tuffs in agronomy, Tbilisi, Metsniereba, (Russian), 85–116, 1988.

Ganev, St., Modern soil chemistry, *Nauca i Iskustvo*, Sofia, 371 (Bulgarian), 1990.

Grigorra, G.I., Agriculture, Urogai, Kiev, 60 (Russian), 1985.

Iacovlev, E.N. et al., Zeolite containing rocks–soil meliorants of sorption type, *Agrochemistry*, Moscow, 6, 68–79 (Russian), 1990.

Jokova, M. and T. Andonov, Some forms of the mobile Fe, Al, and Mn compounds and oxidation-reduction processes in pseudo-podzolic soil, *Soil Sci. Agrochem.*, 24, 24–33 (Bulgarian), 1989.

Kallo, D., Waste water purification in Hungary using natural zeolite, Natural zeolite '93, D.W. Ming and F.A. Mumpton, eds., *Int. Com. Nat. Zeolite*, Brockport, NY, 341–350, 1995.

Klimenco, N.A. and S.I. Veremenco, *Soil Sci.*, Moscow, 4, 31–37 (Russian), 1988.

Krastanov, S. et al., Humification as basis process for biological reclamation of humus free materials, *J. Balkan Ecol.*, 3, 21–24, 2000.

Kudeyarov, V.N., T. S. Demkina, and E.F. Egorova, Activity of denitrification process in soils under nitrogen fertilization, *Agrochemistry*, Moscow, 4, 3–10, (Russian), 1989.

Kuraev, V.N., Oxidation-reduction conditions in gleyed sod-podzolic soils, *Soil Sci.*, Moscow, 3 (Russian), 50–61, 1969.

Leggo, P. 1997. Experiments with wheat: an investigation of plant growth in an organozeolite substrate, *Proc. 5th Int. Conf. Zeo '97*, 21–27 September, 1997, Ischia, Naples, Italy, 206–207, 1997.

Menyailo, O.V., Influence of heavy metals and soil acidification on the N_2O production from arctic soils, in *Proc. 2nd Int. Cong. ESSC*, 1–7 September, 1996, Munich, Germany, 41, 1996.

Mirchev, S. and M. Jokova, Studies on reduction state of soils, *Proc. 4th Bulgarian Natl. Conf. Soil Sci.*, Sofia, Bulgaria, 118–125, 1989.

Mitchell, P. and K. Atkinson, *In situ* control of heavy metal dispersion, possible use for natural zeolites, *Proc. 4th Int. Conf. Occurrence, Properties and Utilization of Natural Zeolites, Zeo '93*, June 20–28, 1993, Boise, ID, 143–144, 1993.

Moshev, D. and S. Mirchev, *Soil Sci. Agrochem.*, 4, 20–25 (Bulgarian), 1989.

Mumpton, F., ed., Natural zeolites, *Zeoagriculture*, Westview Press, Boulder, CO, 33–43, 1984.

Nikashina, V., I. Serova, and A. Rudenco, Recovery of metal pollutants from the soil and silts by ferro-magnetic natural and synthetic zeolites, *Proc. 5th Int. Conf. Zeo '97*, 21–27 September, 1997, Ischia, Naples, Italy, 232–233, 1997.

Orlov, D.S., *Soil Sci.*, Moscow, 9, 43–52 (Russian), 1988.

Papp, J., K. Heinzel, and S. Adams, Application of manganese-modified clinoptilolite in the treatment of land-shop effluents, Ming, D. and F. Mumpton, eds., *Natural Zeolites '93. Int. Comm. Natural Zeolites*, Brokport, New York, 1995.

Pirella, H.J. et al., Use of clinoptilolite in combination with nitrogen fertilization to increase plant growth, *Zeo-Agriculture*, W.G. Pound and F.A. Mumpton, eds., Westview Press, Boulder, CO, 112–122, 1984.

Popova, T. and K. Chakalov, Method for determining of permanent charges on substrate adsorbent in organozeolite media, in *Proc. 2nd Int. Cong., ESSC*, 1–7 September, 1996, Munich, Germany, 1996a.

Popova, T., K. Chakalov, and K. Dimitrov, Zeolite influence on PBC^p (phosphate buffer capacity) in soils, Ministry of Education and Science of Bulgaria, National Science Fund, Research Project — CC 589, 1996, 1997, and 1998. Research reports (Bulgarian, English abstracts), 1996b.

Staricova, V.G., Oxi-reduction conditions in drained alluvial peat soils, *Agric. Sci., News of Sibirian*, 2, 49–55 (Russian), 1984.

Stoilov, G. et al., Substrate for plant cultivation, BG Pat, No 27106 (Bulgarian), 1982.

Tsadilas, C., D. Dimoyiannos, and V. Samaras, Effect of zeolite application and soil pH on Cd sorbtion in soil, *Soils Commn. Soil Sci. Plant Anal.*, 28, 1591–1602, 1997.

Tzhakaija, N.Sh. and N.F. Kvashali, *Japan's Experience in Zeolite Utilization*, Metsniereba, Tbilisi, Georgia, 126, 1985.

Tsitsishvili, G.V. and T.G. Anjdronicashvili, Natural clinoptilolite rocks in plant cultivation, Metzniereba, Tbilisi, Georgia, 5–33 (Russian), 1988.

Tukvadze, E., in *Proc. Natural Zeolites Utilization in Stock and Plant Breeding*, Tbilisi, Metsniereba, 223–225, 1984.

Vosbutsaija, A.E., Soil chemistry, Graduate School, Moscow, 127 (Russian), 1968.

Weber, N.A., K.A. Barbarick, and P.G. Westfall, *Zeoagriculture*, Westview Press, Boulder, CO, 263–371, 1984.

CHAPTER 21

Soil Fertility Management with Zeolite Amendments: II. Zeolite Effect on Maintaining Soil Eh

K. Chakalov, T. Popova, Ekaterina Filcheva

CONTENTS

INTRODUCTION

All soil types have both oxidation and reduction charges (Eh). Many united components have an oxide-reduction function: $Fe^{3+}-Fe^{2+}$; $Mn^{4+}-Mn^{2+}$; $Cu^{2+}-Cu^{+}$; $SO_4^{2+}-H_2S$; $NO_3^{-}-NO_2-N_2O$ or NH_3; CO_2-CH_4. In general, because of heavy metal pollution, these components T^0 regime, soil compaction, soil percolation etc. (Menyailo, 1996). However, soil water capacity, percolation, and compaction have a more important role. Ferric and ferrous ions previously maintained soil Eh (Moshev and Mirchev, 1989).

In previous work, possibilities for maintaining soil forming processes in zeolite-amended soil were described in connection with their physical and physicochemical properties, which contribute to carbon sequestration in soil (Chakalov et al., 1993, 1995a, b; Filcheva et al., 1997, 1998a, b, c). This work demonstrates the possibilities for managing soil Eh charges, and soil carbon sequestration, especially in media rich in heavy metals influenced differently by Eh.

MATERIALS AND METHODS

The experiments were carried out with organozeolite composts (lignin, barks, and peat) with modified clinoptilolite (Na – Cp, K – Cp, Ca – Cp, NH_4 – Cp form). Methods for preparation, humus composition, and some of the physicochemical properties have been described previously (Filcheva et al., 1997, 1998a, b, c). The influence of meliorants on soil Eh was investigated in the Chromic Luvisol region near the Kremikovtzy metallurgy plant smelter in Bulgaria (Filcheva et al., 1998c). Eh and pH were determined for each substrate and soil suspension potenciometrically, after an inoculation period of 24 days (20°C). The available forms of iron and manganese were determined after extraction with 0.1 N H_2SO_4 (1:10), and measured photocolorimetrically. The theoretical influence of ferrous or ferric ions was evaluated using the Bohn (1968, 1969) formula: Eh = 0.058 – 0.59 × lg (Fe^{2+}) –0.177 pH. Copper ion–hydroxide equilibrium was determined in a model experiment by the method of Gateva (1986), with modifications (Chakalov et al., 1996, Popova et al., 1996a, b).

RESULTS AND DISCUSSION

The clinoptilolite influence on humus forming processes and the consequent changes in the physicochemical properties of organozeolite media (Chakalov et al., 1993, 1995a, b; Filcheva et al., 1997, 1998a, b, c) were established earlier. The data in Table 21.1 show the ion-hydroxide equilibrium of the copper consequence, determined by the physicochemical properties of the model media.

Organic mass composts accumulate a toxic exchangeable acidity in this order:

$$K - C_p > Na - C_p > Ca - C_p >>> NH_4 - C_p$$

This determines the role of zeolite in the translocation of exchangeable acidity on the organic absorbent:

$$NH_4 - C_p > K - C_p > Na - C_p > Ca - C_p$$

Rearrangement of permanent charges (T_{CA}) determined the Cu^{2+} adsorption of the mineral component in the media: $NH_4 - C_p$ — 60.99%; Ca – C_p — 63.14%; K – C_p — 51.88%; Na – C_p — 48.72% (lignin zeolite media). The clinoptilolite influence in the bark substrate follows the same order. The ionic state of copper in this compost is affected by the presence of Al^{3+} and H^+.

Peat composts have more condensed humic substances (Filcheva et al., 1997). The exchangeable acidity is relatively low and the organic mass influence is strong. Copper is present in the ionic form in the case of $NH_4 - C_p$ in the media. The Fe and Mn ions from peat were sorbed by mineral adsorbent. The influence of Al^{3+} and Fe^{3+} ions on copper mobility demonstrates the zeolite role in this process. Manganese ions affect the accumulation of Cu^{2+} on the mineral adsorbent (r = 0,713) and weakly influence Cu^{2+} translocation on the organic adsorbent (r = 0,415). The exchangeable Al^{3+}, Fe^{3+}, and Mn^{2+} ions decreased permanent charges of the model media, but they increased the cation exchange capacity (CEC) of zeolite for Cu^{2+} adsorption.

These processes are connected with the oxidation-reduction charges of the media (Eh) and consequence with united elements (Fe^{2+}/Fe^{3+}; Mn^{2+}/Mn^{4+}; Cu^{2+}/Cu^+, etc.). A high level of Eh charges was measured in studied peat substrates. One of the reasons for rapid decrease of the Eh curve may be due to the formation of H_2 on the Pt electrode and O_2 deficiency. Since antiseptic was used, the influence of microbiological activity of the media cannot be discussed, but the measured Eh values show more correctly the influence of all united ions. The low levels of Eh in peat media were established, due to fixation of more united ions, especially $Cu^{2+} - Cu^+$ on the zeolite surface. This is the paradoxical effect: the Eh decreased proportionally with available Fe^{2+}/Fe^{3+} content in media.

Table 21.1 Some Physicochemical Properties and Copper Chemical Status in Organozeolite Composts

Samples	$T_{8.2}$	T_{SA} org.	T_{SA} min.	$H_{8.2}$	Al^{3+} meq/100 g	Fe^{3+} org.	Fe^{3+} min.	Mn^{2+} org.	Mn^{2+} min.	Cu^{2+} org. ads. meq/100 g	Cu^{2+} org. ads. %-$Cu(OH)_2$	Cu^{2+} min. ads. meq/100 g	Cu^{2+} min. ads. %-$Cu(OH)_2$	Cu^{2+} Compost meq/100 g	pH
Peat(P)	34.17	32.14-in compost		5.25	0.25	1.81	0.22	0.034	—	0.255	2.55	—	—	—	6.35
P+NH_4 zeolite	103.61	94.71-in compost		13.05	3.58+0.57H^+	1.66	3.90	0.091	0.59	2.40	24.00	5.85	58.51	8.25	3.88
P+Ca^{2+} zeolite	98.43	92.75-in compost		7.20	1.52	0.27	1.21	0.040	0.27	1.003	10.03	1.14	11.40	2.14	4.85
P+K^+ zeolite	107.43	102.94-in compost		5.42	0.93	0.43	1.34	0.10	0.30	1.357	13.57	0.94	9.40	2.29	5.25
P+Na^+ zeolite	97.79	92.66-in compost		7.39	1.13	0.48	1.65	0.05	0.61	0.94	9.40	1.07	10.70	2.01	5.10
Lignin(L)	32.36	9.72	0.15	25	2.5					2.18	21.80	1.46-water soluble form		3.64	3.4
L+NH_4 zeolite	41.50	9.08	18.41	15.75	1.75					1.49	14.9	2.33	23.30	3.82	3.8
L+Ca^{2+} zeolite	57.51	12.20	34.81	11.25	0.75					1.01	10.1	1.73	17.30	2.74	4.65
L+K^+ zeolite	66.58	9.33	46.00	11.00	0.75					1.54	15.4	1.66	16.60	3.20	4.75
L+Na^+ zeolite	74.00	12.68	49.32	13.00	1.50					1.54	15.4	1.46	14.60	3.00	4.75
Bark(B)	28.75	20.65	—	8.75	0.32									0.855	5.75
B+NH_4 zeolite	49.50	15.75	23.35	11.50	1.13									2.81	4.32
B+Ca^{2+} zeolite	61.53	18.56	37.10	6.25	0.38									1.33	5.65
B+K^+ zeolite	67.26	15.74	46.77	5.00	0.25									1.11	6.05
B+Na^+ zeolite	63.95	13.51	45.69	5.00	0.25									0.90	6.00

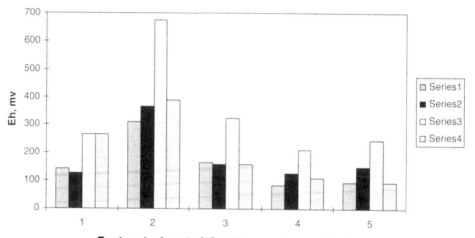

Treatments: 1-control; 2-peat+ammonium zeolite; 3-peat+Cazeolite; 4-peat+Kzeolite; 5-peat+Nazeolite; columns: a-Eh Fe(II); b-Eh Mn(II); a+b; d-Eh potenciometrically

Figure 21.1 Eh values for the peat composts studied.

The maturity of humic substances may play the most important role in this case. Humification, as an oxidation process, obviously correlates with the balance of united ions. However, the relatively slow release of ferric ions from natural zeolites showed a relatively constant Eh, i.e., the media accumulated less exchangeable acidity and had low Eh. Simultaneously, this paradoxical effect corresponds with available ferrous ions. In peat compost, without a zeolite component (control), a mutual influence of manganese and iron was established but, in the organozeolite media, the data had an average value, as a consequence of humus maturity. Obviously, the microbiological acidity stopped because of the addition of the antiseptic. Natural zeolite conserves reduction processes, which are mainly physicochemical, not microbiological, processes.

The theoretical evaluation of Eh proves this thesis (Figure 21.1). The established influence on the copper ion state by zeolites additionally supports the idea that $r = 0.890$ (Cu^{2+} on mineral adsorbent) and $r = 0.709$ (Cu^{2+} on organic adsorbent). Ferrous ions decreased copper mobility in peat-zeolite mixtures, but ferric ions increase this as a part of the explained paradoxical effect: $r = -0.22$ (on mineral adsorbent) and $r = -0.625$ (on organic adsorbent). The iron helatization affects the copper mobility. Bark composts, with more "young" humic acids and a high level of pH, have the usual influence of Fe^{2+} on a copper ion state: ferrous ions increase copper mobility. This proves again the thesis for zeolites' influence on soil-forming processes and the possibilities of managing physicochemical properties of the media, stimulating the maturity of humic substances.

Evidently, the influence of ferrous and ferric ions on zeolite sorption of copper is determined by its natural fertilization. This may modulate the state of Eh in soil by the so-called "mirror effect" (Figures 21.2a, b, c, and d).

The zeolite component of the media influences Eh in the following order:

$$NH_4 - C_p > peat > Ca - C_p > K - C_p > Na - C_p$$

Zeolite melioration of soil polluted by heavy metal (total content of metals, mg/kg: Cu — 45.00; Zn — 119; Mn — 1395; Pb — 157.5; Cd — 0.62; N — 27; Co — 8.25; Fe — 24000; pH — 5.7) influences previous Zn mobility. Ammonium type of clinoptilolite increases exchangeable Zn^{2+} ($NH_4 - Ac$ extraction, pH 7) from 1.47 to 2.05 mg/kg (41%); the potassium form decreases it to 1.35 mg/kg, but natural ferritization of zeolite decreases Zn^{2+} mobility to 0.90 mg/kg (more than 40%), a direct consequence of the mirror effect.

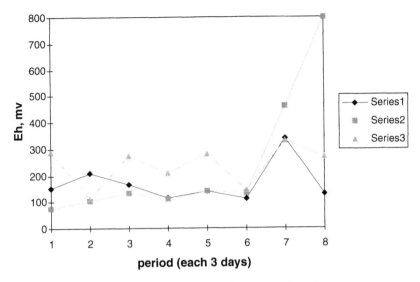

Figure 21.2a Eh changes in the samples studied: soil, soil + peat, and peat.

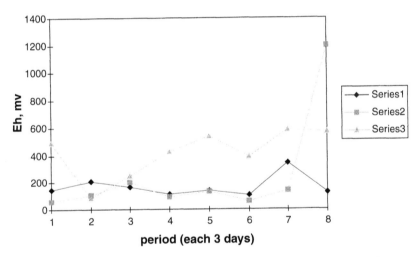

Figure 21.2b Eh changes in samples studied: soil, soil + peat + ammonium zeolite, and peat + ammonium zeolite.

The cation exchange composition of natural clinoptilolite selectively determines zeolites' cation exchange with ferric ions and appropriate ability for natural ferritization. This demonstrates that clinoptilolite has the capacity for stabilizing exchangeable Al^{3+}, a consequence of the available Fe^{2+}/Fe^{3+} and Mn^{2+}/Mn^{4+} concentration in the media. The maturity stage of humic substances influences the state of their equilibrium with organic component, thus creating conditions for managing oxidation-reduction processes in the substrates.

CONCLUSIONS

The differentially natural modification of the clinoptilolite (ferritization and dealuminization) consequence of organic acid influence on a mineral adsorbent changes the usually known selectivity of clinoptilolite. This is the base circumstance for arranging soil-forming processes in organozeolite

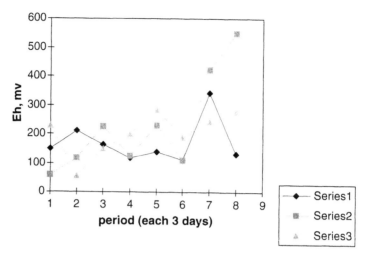

Figure 21.2c Eh changes in samples studied: soil, soil + peat + Cazeolite, and peat + Cazeolite.

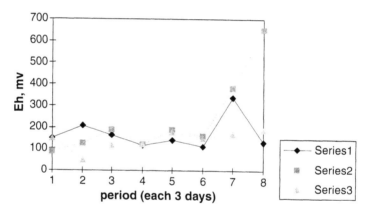

Figure 21.2d Eh changes in samples studied: soil, soil + peat + Kzeolite, and peat + Kzeolite.

media in correlation with its reactivity, especially with amphoteric elements. The first step for soil carbon sequestration is the managing of Eh to maintain ion mobility, which is possible by using zeolite materials as soil conditioners.

ACKNOWLEDGMENT

This work is supported by the National Science Fund, Ministry of Education, Bulgaria, CC-589.

REFERENCES

Bohn, H.L., Electromotive force of inert electrodes in soil suspension, *Soil Sci. Soc. Am. Proc.*, 32, 2, 211–215, 1968.

Bohn, H.L., The EMT of platinum electrodes in dilute solution and its relation to soil pH, *Soil Sci. Soc. Am. Proc.*, 33, 4, 639–640, 1969.

Chakalov, K., J. Yoneva, and T. Popova, Influence of cation exchange composition of clinoptilolite on organozeolite composts quality, *Proc. 4th Int. Conf. "Occurrence, Properties and Utilization of Natural Zeolites Zeo'93,"* Boise, ID, 1993.

Chakalov, K., K. Dimitrov, and T. Popova, Organozeolite composts as a substrate and soil conditioners for growing ecologically clean vegetables, *Proc. Int. Symp. Sofia Zeolite Meetings 95,* June 18–25, 1995, Sofia, Bulgaria, 101–108, 1995a.

Chakalov, K. et al., Influence of natural clinoptilolite with variable cation exchange composition on ion-hydroxide equilibrium of amphoteric elements in organozeolite media for zeoponic, Ministry of Education and Science of Bulgaria, National Science Fund, Research Project — CC418, 1995, 1996, and 1998. Research reports (Bulgarian, English abstracts), 1995b.

Chakalov, K., T. Popova, and E. Filcheva, Influence of cation exchange composition of organozeolite composts on the ion-hydroxide equilibrium of copper, in *Proc. 2nd Int. Congr., ESSC. Development and Implementation of Soil Conservation, Strategies for Sustainable Land Use,* 1–7 September, 1996, Munich, Germany, 1996, 50.

Filcheva, E., K. Chakalov, and T. Popova, Modeling soil forming processes in artificial zeolite soils, in Lal, R., J. Kimble, and R. Follett, eds., *Soil Properties and Their Management for Carbon Sequestration,* USDA-Natural Resources Conservation Service, National Soil Survey Center, Lincoln, NE, 113–117, 1997.

Filcheva, E., K. Chakalov, and T. Popova, Model approach to improve soil quality using organo-zeolite compost, *J. Balkan Ecol.,* 1, 72–78, 1998a.

Filcheva, E., K. Chakalov, and T. Popova, Possibility to manage soil forming processes, *16th World Cong. Soil Sci.,* 20–26 August, 1998, Montpellier, France, 1998b.

Filcheva, E., K. Chakalov, and T. Popova, Possibility to manage soil-forming processes in ecosystems. Concept for humus conservation in agriculture, *J. Balkan Ecol.,* 1, 72–78, 1998c.

Gateva, A., Classification aspects of ion-hydroxide equilibrium of copper in different soils, *Soil Sci. Agrochem. Plant Protection,* Sofia, Bulgaria, XXI, 46–57, 1986.

Menyailo, O.V., Influence of heavy metals and soil acidification on the N_2O production from arctic soils, in *Proc. 2nd Int. Cong. ESSC,* 1–7 September, 1996, Munich, Germany, 1996.

Moshev, D. and S. Mirchev, Oxidation and reduction state and changes in the iron and manganese in grey forest soils, *Soil Sci. Agrochem.,* No. 4, 20–25, (Bulgarian), 1989.

Popova, T. and K. Chakalov, Method for determining of permanent charges on substrate adsorbent in organozeolite media, in *Proc. 2nd Int. Cong. ESSC,* 1–7 September, 1996, Munich, Germany, 1996a.

Popova, T., K. Chakalov, and K. Dimitrov, Zeolite influence on PBC[p] (phosphate buffer capacity) in soils, Ministry of Education and Science of Bulgaria, National Science Fund, Research Project — CC589 (1996, 1997, and 1998). Research reports (Bulgarian, English abstracts), 1996b.

CHAPTER 22

Effect of Crop Rotation on the Composition of Soil Organic Matter

Ekaterina Filcheva and T. Mitova

CONTENTS

INTRODUCTION

Humus content is one of the most important factors determining the productivity of soil because of the multiple effects of humic substances, such as the release of plant nutrients, improvement of soil physical and biological conditions, prevention of soil erosion, increase in the efficiency of applied fertilizers, and environmental protection. Soil organic matter has been identified as an indicator of soil quality and also as a key component of sustainable crop production.

Several researchers have reported data on soil organic matter's improving with the adoption of practices such as elimination of summer fallow, regular inclusion of forages and legumes in crop rotation, improved fertility management, and reduced tillage intensity (Jansen et al., 1998; Rasmussen et al., 1998). The decline in soil organic matter generally is greater with intensive than with reduced or no-tillage methods, with cropping systems involving fallow compared to those without fallow, and with row crops than with small grain crops (Grace et al., 1995; Campbell et al., 1996; Campbell et al., 1998; Cihacek and Ulter, 1997; Hansmeyer et al., 1997; Grant et al., 1997). No-tillage is the most efficient management practice for sequestering carbon (C) in soil when it is combined with cover crops, crop rotation, fertilizer strategies, and supplemental C input (Dick et al., 1998). After 100 years, results from "old rotation" indicate that winter legumes and crop rotations result in a larger amount of both C and N (Mitchell and Entry, 1998). Ionita et al. (1995) have reported data for increasing total soil carbon content when crop rotation with cereals and leguminous plants is used. The more intensive cropping systems not only produce more grain per unit, but also produce and leave more crop residues on the soil surface (Peterson et al., 1998). Soil organic matter storage capacity in agroecosystems varies with soil type, climate, and agricultural management practices, which determined the importance of this study. This chapter reports the effect of different 2-year crop rotations and fertilization on soil organic matter content and composition.

MATERIALS AND METHODS

The humus content and soil organic matter composition were determined in the long-term experiment started in 1962 at the town of Kostinbrod, Sofia district (Figure 22.1). The field experiments were carried out at the experimental station on a Leached Smolnitza (Vertisol, FAO; Hapluderts, Soil Taxonomy, Boyadgiev, 1994). The soil samples were taken in September 1988 and 1989 from the experimental plots in the 0- to 20-cm and 20- to 40-cm depths after harvesting the plants.

The present study includes monoculture of soybean and 2-year crop rotation of barley–wheat, soybean–wheat, and soybean–maize. The plants included in crop rotations were grown with no fertilization (T_0) and with mineral fertilization (T_1) as $N_{120}P_{80}K_{60}$ kg ha^{-1} for maize and wheat and $N_{50}P_{80}K_{60}$ for soybean. The rotations were set out on plots 7.5 m by 7.75 m in randomized block designs with four replicates, without irrigation.

The investigated soil is related to the heaviest variety of the Leached Smolniza (Vertisol, FAO; Hapluderts, Soil Taxonomy, Boyadgiev, 1994), characterized with a deep humus horizon, heavy texture, $pH_{(KCl)}$ 5.9, total nitrogen 0.234%, and total phosphorous 0.078%. The soil structure is water-stable and compact. Soil particle density is 2.68 and the bulk density does not exceed 1.24-g cm^{-3} (Soils in Bulgaria, 1960).

Organic carbon was determined according to Turin's (Kononova, 1966) modified method (acid-dichromate digestion, 120°C, 45 min. in a thermostat, catalyst Ag_2SO_4). Organic matter composition was specified by Kononova and Belchikova's method (Kononova, 1966) as follows: total humic

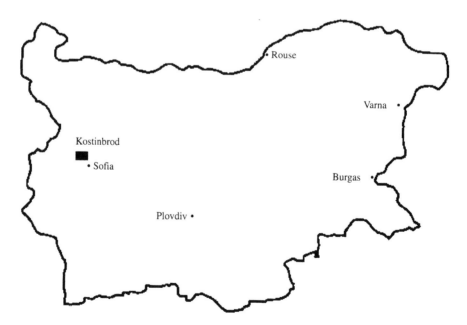

Figure 22.1   Location of the experimental field, Kostinbrod, Sofia region.

and fulvic acids (C_{ext}) after extraction with mixed solution of 0.1M $Na_4P_2O_7$ and 0.1N NaOH, "free," and R_2O_3 bonded humic and fulvic acids (C_{NaOH}) after extraction with 0.1N NaOH in a ratio of soil:solution 1:20 for both extractions. Carbon in both cases was measured by titration using ferrus–ammonium–sulfate, with phenyl-antranilic acid as an indicator. In subsamples of both C_{ext} and C_{NaOH} humic and fulvic acids were separated by acidifying the solution with sulfuric acid. The ratio of humic and fulvic acids in the C_{ext} characterizes the type of humus. Humic acids in C_{NaOH} are usually called "free" and R_2O_3 bonded, and they are considered the more mobile carbon fraction. The amount of humic acids calculated by subtraction of humic acids in C_{ext} and C_{NaOH} comprises the portion of humic acids complexed with alkali earth metals while the ratio of humic acids in C_{ext} and the total organic carbon are an indication of the degree of humification.

RESULTS AND DISCUSSION

Table 22.1 presents results obtained on the soil organic matter composition of the studied crop rotations. The content of humus declined under cultivation and crop rotation systems from 4.5% (Soils in Bulgaria, 1960) to 2.53% for the studied treatments (1988 and 1989). The greatest decrease was found under soybean and maize in 1988. A similar trend also was found in the other observations, due to intensive tillage in this type of crop rotation, including row crops (Ionita et al.,1995). The humus content values ranges without and with fertilization were 3.3 and 3.6, respectively. Results obtained for the studied crop rotations show an increase of humus content in 1989 compared to 1988.

Generally, the humus content was lower in the depth of 20 to 40 cm compared to the surface 0 to 20 cm. The favorable effect on the organic matter content was found in the crop rotation with soybean as a monoculture for 1988 and 1989. This is probably due to the positive effect of leguminous plants on the transformation of organic residues. Humic acids predominated over the fulvic acids, and only humic acids presented in the soil organic matter composition in most of the studied crop rotation, which determined the type of humus as humic, according to the classification of Grishina and Orlov (Orlov, 1985). This is a base for favorable soil physical properties and plant growth conditions. An exception was found under soybean and maize and soybean and wheat (1988), where the fulvic acids predominated over the humic acids and the ratio Ch:Cf characterized a fulvic type of humus (Orlov, 1985). This observation could be explained with the transformation of different plant residues. It is important to note that the maximum amount of alkali extractable organic carbon (mixed solution of $Na_4P_2O_7$ and NaOH) has been determined in the soybean and maize crop rotation. This could be explained by unfavorable conditions (no irrigation) in the heavy soil (Vertisol, FAO, Hapluderts, Soil Taxonomy, Boyadgiev, 1994), where newly formed humic substances were not able to bind with soil minerals.

Most of the data on the degree of humification were lower than others reported in previous studies (Mitova and Filcheva, 1996; Filcheva and Mitova, 1997). Crop rotations that included cereals only (barley and wheat) were characterized with the highest values of humus content and degree of humification in 1988. This is probably due to less intensive tillage operations during the vegetation period, the nature of the root system, and the quantity and quality of organic matter remaining after harvesting of winter wheat early in the summer when the level of organic matter synthesis was higher. The latest study is in agreement with earlier observations (Ionita et al., 1995). Similar results were obtained for soybean as a monoculture.

The aggressive fulvic acids fraction (1[a]) — the most dynamic, low molecular fraction — does not take part in the humus-forming processes in the studied soil under the studied crop rotation systems. In the studied crop rotations, humic acids are heavy bonded with calcium (100%), which causes stabile structure and favorable physical properties, and protects soluble organic matter from leaching.

Table 22.1 Content and Composition of Soil Organic Matter

1988

Depth in cm Fertilization 1	Humus, % 2	Total Carbon, % 3	Organic Carbon (%), C_{ext} Extracted with 0.1M $Na_4P_2O_7$+0.1M NaOH			C_{ha}/C_{fa} 7	Organic Carbon (%) Humic Acid Fractions		Unextracted Organic Carbon (%) 10	Extracted with 0.1N H_2SO_4 (%) 11
			Total 4	Humic Acids 5	Fulvic Acids 6		Free and R_2O_3 Complexed 8	Ca-compl. 9		
Barley and Wheat										
T_0 0–20	3.34	1.94	1.09[a] / 56.19[b]	0.71 / 36.60	0.38 / 19.59	1.87	0.00	100.00	0.85 / 43.81	0.05 / 2.57
T_0 20–40	3.34	1.94	0.96 / 49.48	0.62 / 31.96	0.34 / 17.52	1.82	0.00	100.00	0.98 / 50.52	0.05 / 2.50
T_1 0–20	3.03	1.76	0.96 / 54.55	0.70 / 39.77	0.26 / 14.77	2.69	0.00	100.00	0.80 / 45.45	0.03 / 1.70
T_1 20–40	4.55	2.64	0.88 / 33.33	0.50 / 18.94	0.38 / 14.39	1.32	0.00	100.00	1.76 / 66.67	0.05 / 1.89
Soybean and Wheat										
T_0 0–20	2.81	1.63	0.71 / 43.56	0.71 / 43.56	0.00 / 0.00	—	0.00	100.00	0.92 / 56.44	0.06 / 3.68
T_0 20–40	2.88	1.67	0.68 / 40.72	0.68 / 40.72	0.00	—	0.00	100.00	0.99 / 59.28	0.05 / 2.99
T_1 0–20	3.55	2.06	0.96 / 46.60	0.35 / 16.99	0.61 / 29.61	0.57	0.00	100.00	1.10 / 53.40	0.05 / 2.43
T_1 20–40	3.10	1.80	0.92 / 51.11	0.24 / 13.33	0.68 / 37.78	0.35	0.00	100.00	0.88 / 48.89	0.05 / 2.78
Soybean and Maize										
T_0 0–20	2.81	1.63	0.96 / 58.90	0.32 / 19.63	0.64 / 39.26	0.50	0.00	100.00	0.67 / 41.10	0.05 / 3.07
T_0 20–40	2.65	1.54	0.92 / 59.74	0.44 / 28.57	0.48 / 31.17	0.92	0.00	100.00	0.62 / 40.26	0.02 / 1.30
T_1 0–20	2.84	1.65	0.96 / 58.18	0.20 / 12.12	0.76 / 46.06	0.26	0.00	100.00	0.69 / 41.82	0.04 / 2.42
T_1 20–40	2.53	1.47	0.92 / 62.59	0.32 / 21.77	0.60 / 40.82	0.53	0.00	100.00	0.55 / 37.41	0.04 / 2.72

1989

Soybean and Soybean

T_0 0-20	3.09	1.79	0.76 / 42.46	0.60 / 33.52	0.16 / 8.94	3.75	0.00	100.00	1.03 / 57.54	0.06 / 3.35
T_0 20-40	3.57	2.07	0.73 / 35.27	0.56 / 27.05	0.17 / 8.21	3.29	0.00	100.00	1.34 / 64.73	0.07 / 3.38
T_1 0-20	3.76	2.18	0.81 / 37.16	0.63 / 28.90	0.18 / 8.26	3.50	0.00	100.00	1.37 / 62.84	0.07 / 3.21
T_1 20-40	3.57	2.07	0.81 / 39.13	0.57 / 27.54	0.24 / 11.59	2.38	0.00	100.00	1.26 / 60.87	0.07 / 3.38

Barley and Wheat

T_0 0-20	4.01	2.33	0.71 / 30.47	0.39 / 16.74	0.32 / 13.74	1.22	0.00	100.00	1.62 / 69.53	0.05 / 2.15
T_0 20-40	3.89	2.26	0.71 / 31.42	0.39 / 17.26	0.32 / 14.16	1.22	0.00	100.00	1.55 / 68.58	0.06 / 2.65
T_1 0-20	3.87	2.25	0.65 / 28.89	0.30 / 13.33	0.35 / 15.56	1.00	0.00	100.00	1.49 / 66.21	0.05 / 2.22
T_1 20-40	3.83	2.22	0.76 / 34.23	0.32 / 14.41	0.44 / 19.82	1.00	0.00	100.00	1.46 / 65.77	0.05 / 2.25

Soybean and Wheat

T_0 0-20	3.45	2.00	0.79 / 35.50	0.56 / 28.00	0.22 / 11.00	3.59	0.00	100.00	1.21 / 60.50	0.05 / 2.50
T_0 20-40	3.31	1.92	0.72 / 37.50	0.57 / 29.69	0.15 / 7.81	4.80	0.00	100.00	1.20 / 62.50	0.04 / 2.08
T_1 0-20	3.93	2.28	0.72 / 31.58	0.43 / 18.86	0.29 / 12.72	1.47	0.00	100.00	1.56 / 68.42	0.04 / 1.75
T_1 20-40	3.38	1.96	0.55 / 28.06	0.38 / 19.39	0.17 / 8.67	3.23	0.00	100.00	1.41 / 71.94	0.05 / 2.55

Table 22.1 (Continued)   Content and Composition of Soil Organic Matter

1989

Depth in cm Fertilization	Humus, %	Total Carbon, %	Organic Carbon (%), C_{ext} Extracted with 0.1M $Na_4P_2O_7$+0.1M NaOH				Organic Carbon (%) Humic Acid Fractions		Unextracted Organic Carbon (%)	Extracted with 0.1N H_2SO_4 (%)
			Total	Humic Acids	Fulvic Acids	C_{ha}/C_{fa}	Free and R_2O_3 Complexed	Ca-compl.		
1	2	3	4	5	6	7	8	9	10	11

Soybean and Maize

T_0 0-20	3.58	2.08	0.96 / 46.15	0.72 / 34.61	0.24 / 11.54	3.00	0.00	100.00	1.12 / 53.84	0.06 / 2.88
T_0 20-40	3.38	1.96	1.05 / 53.57	0.72 / 36.67	0.33 / 16.84	2.18	0.00	100.00	0.91 / 46.43	0.05 / 3.06
T_1 0-20	3.24	1.88	1.03 / 54.79	0.77 / 40.96	0.26 / 13.83	2.96	0.00	100.00	0.85 / 45.21	0.05 / 2.66
T_1 20-40	3.45	2.00	1.04 / 52.00	0.82 / 41.00	0.22 / 11.00	3.72	0.00	100.00	0.96 / 48.00	0.05 / 2.50

Soybean and Soybean

T_0 0-20	3.38	1.96	0.88 / 44.89	0.88 / 44.89	0.00	—	0.00	100.00	1.08 / 55.11	0.06 / 3.06
T_0 20-40	3.38	1.96	0.79 / 40.31	0.58 / 29.59	0.21 / 10.71	2.76	0.00	100.00	1.17 / 59.69	0.05 / 2.55
T_1 0-20	3.24	1.88	0.91 / 48.40	0.91 / 48.40	0.00	—	0.00	100.00	0.97 / 51.59	0.06 / 3.19
T_1 20-40	3.45	2.00	1.02 / 50.50	0.75 / 37.50	0.27 / 13.50	2.78	0.00	100.00	0.98 / 49.50	0.05 / 2.50

[a] % of the soil sample.
[b] % of the total carbon.

CONCLUSIONS

Types of crops (row crops, cereals, and legumes) and agrotechnical practices affect the soil organic matter content and composition. The highest decrease occurred under soybean and maize crop rotation in 1988. Humic acids predominated over the fulvic acids, or only humic acids presented in the soil organic matter composition, in most of the studied crop rotation, which determined the type of humus as humic, according to the classification of Grishina and Orlov (Orlov, 1985). The favorable effect on the organic matter content was found in the crop rotation with soybean as a monoculture for 1988 and 1989, probably due to the positive effect of legumes on the transformation of organic residues. In the studied crop rotations, humic acids were heavily bonded with calcium, which caused a stabile structure and favorable physical properties, and protected soluble organic matter from leaching.

REFERENCES

Boyadgiev, T., Soil map of Bulgaria according to the soil taxonomy — explanatory notes, *Soil Sci. Agrochem. Ecol.*, 29:43–51, 1994.

Campbell, C.A. et al., Long-term effects of tillage and crop rotations on soil organic C and total N in a clay soil in southwestern Saskatchewan, Canada, *J. Soil Sci.*, 76 (3):395–401, 1996.

Campbell, C.A. et al., Long-term effects of tillage and fallow-frequency on soil quality attributes in a clay soil in semiarid southwestern Saskatchewan, *Soil Tillage Res.*, 47:135–144, 1998.

Cihacek, L.J. and M.J. Ulter, Effects of tillage on profile soil carbon distribution in the northern great plains of the U.S., in Lal, R. et al., eds., *Management of Carbon Sequestration in Soil*, Advances in Soil Science, CRC Press, Boca Raton, FL, 1997, 83–92.

Dick, W.A. et al., Impacts of agricultural management practices on C sequestration in forest-derived soils of the eastern Corn Belt, *Soil Tillage Res.*, 47:235–244, 1998.

FAO-UNESKO. 1990. Revised Legend. Rome, Italy.

Filcheva, E. and T. Mitova, Organic matter characteristics in crop rotation system, *Nat. Conf. Problems Concerning Genesis, Evolution, Use and Protection of Soils from the South Eastern Region of Romania*, 26–30 August, Bucuresti, 1997 Publicatiile, S.N.R.S.S., 29A, 217–222, 1997.

Grace, P.R. et al., Trends in wheat yields and soil organic carbon in the permanent rotation, *Austr. J. Exp. Agric.*, 35 (7):857–864, 1995.

Grant, R.F. et al., Modeling tillage and surface residue effects on soil C storage under ambient vs. elevated CO_2 and temperature, in Lal, R. et al., eds., *Ecosystems, Soil Processes and the Carbon Cycle*, Advances in Soil Science, 527–545, CRC Press, Boca Raton, FL, 1997.

Hansmeyer, T.L. et al., Determining carbon dynamics under no-till, ridge-till, chisel, and moldboard tillage systems within a corn and soybean cropping sequence, in Lal, R. et al., eds., *Management of Carbon Sequestration in Soil*, Advances in Soil Science, 93–98, CRC Press, Boca Raton, FL, 1997.

Ionita, S, Gh. Sin, and I. Chirnogeanu, The modification of some agrochemical factors depending on crop rotation and fertilization, *Proc. 2nd Int. Sci. Conf.*, Bucharest, 9–14 October, 1995, 44–48, University of Agricultural Sciences and Veterinary Medicine, Bucharest, Romania, 1995.

Jansen, H.H. et al., Management effects on soil C storage on the Canadian prairies, *Soil Tillage Res.*, 47:181–195, 1998.

Kononova, M.M., *Soil Organic Matter: Its Nature, Its Role in Soil Formation and Soil Fertility*, 2nd English Ed., Pergamon Press, New York, 1966, 544.

Mitchell, C.C. and J.A. Entry, Soil C, N and crop yields in Alabama's long-term "old rotation" cotton experiment, *Soil Tillage Res.*, 47:331–338, 1998.

Mitova T. and E. Filcheva, Characteristics of soil organic matter after long-term cropping, Book of Abstracts, 4th Cong. Eur. Soc. Agron., I, 266–267, Veldhoven-Wageningen, The Netherlands, 1996.

Orlov, D.S., *Soil Chemistry*, Moscow University, Moscow, 1985.

Peterson, G.A. et al., Reduced tillage and increasing cropping intensity in the Great Plains conserves soil C, *Soil Tillage Res.*, 47:207–218, 1998.

Rasmussen, P.E., S.L. Albrecht, and R.W. Smiley, Soil C and N changes under tillage and cropping systems in semi-arid Pacific Northwest agriculture, *Soil Tillage Res.*, 47:197–205, 1998.
Soils in Bulgaria, Monograph, Zemizdat Sofia, 532, 1960.

CHAPTER **23**

Application of a Management Decision Aid for Sequestration of Carbon and Nitrogen in Soil

Alan Olness, Dian Lopez, Jason Cordes, Colin Sweeney, Neil Mattson, and W. B. Voorhees

CONTENTS

INTRODUCTION

Carbon (C) sequestration in agricultural lands will require complex management of soil. The U.S. Cornbelt is one of the major areas of potential C sequestration in agricultural soils. This region has about 53 million hectares of land under cultivation annually (Table 23.1; USDA-NASS, 1999). This represents about a third of all cropland in the U.S. In the early years of management, most farms had cattle, pasture, and forage production and this allowed farmers to rotate their crops. Today, much of the land is under annual crop production with some type of tillage to aid seedbed preparation and weed control. Tillage accelerates oxidation of soil organic matter and releases organically bound carbon, nitrogen (N), phosphorus, and sulfur (Schlesinger, 1986; Houghton, 1995; Lal, 1997; Reicosky, et al., 2000). Thus, soils are continually being mined of mineral nutrition for grain and oilseed crop production. In order to successfully sequester C in soil, careful management and co-sequestration of several elements are also essential. Nitrogen is the nutrient most often limiting crop production; it is also the element, aside from C, that is needed in the greatest amounts for C sequestration.

In the 1800s, much of the C and N in crops was recycled within the farm unit. However, over time, export of grain has been removing an increasing percentage of fixed C from the cycle. Today nearly 50% of the C fixed annually is removed and proposals are continually being made to remove an even larger fraction. This exportation removes C from the pool of material exploited by resident biology and gradually removes organic C from use by soil organisms. Today's crop production also exports even larger proportions of N from the agricultural landscape.

Table 23.1   Average Crop Acreage Planted to Corn, Oat, Soybean, and Small Grains (per 1,000 ha) in the U.S. Corn Belt during 1996 to 1998

State	Maize	Soybean	Wheat	Other Grain	Total
Illinois	4,425	4,128	546	58	9,157
Indiana	2,334	2,219	304	27	4,883
Iowa	5,045	4,115	16	121	9,296
Kansas	1,113	944	4,573	80	6,710
Michigan	998	731	244	79	2,053
Minnesota	2,941	2,631	21	356	5,949
Missouri	221	1,902	553	15	2,692
North Dakota	337	479	30	1,230	2,076
Ohio	1,396	1,788	510	62	3,756
South Dakota	1,578	1,275	695	238	3,786
Wisconsin	1,545	420	60	250	2,274
Total	21,933	20,630	7,553	2,516	52,632

Note: Recompiled from USDA-NASS Agricultural Statistics, 1999.

Figure 23.1   A native prairie site on the Great Plains.

Most soil C comes from root mass. However, under natural conditions (Figure 23.1), above-ground biomass is continuously added to the surface mat of dead vegetation, which decays gradually. Rodents and other burrowing animals continuously till the soil and bury a portion of the above-ground biomass. This tillage is temporally and spatially diffuse, but effective in incorporating some aerial plant material into the soil. Under natural conditions, plants are usually N-limited and the C:N ratios of root and above-ground portions are quite large. Seeds and leaves, which generally have lesser C:N ratios, are quickly consumed, the carbohydrate and protein are extracted, and the more resistant material is returned as feces and gradually incorporated into the soil.

Agriculture tends to use plants with a smaller fraction of roots than native plants (Table 23.2). As root systems increase in protein content or fineness, they decay more rapidly and, as a consequence of smaller root fractions and greater ease of decay, the content of soil organic C decreases. Fertilization with N further decreases the root:shoot and root C:N ratios (Geisler and Krutzfeldt, 1984, cited by Klepper, 1991) and drives the system to greater and greater removal of fixed C as CO_2.

Natural systems are often both water- and N-limited. Soil texture (clay content) plus organic matter determines the range of soil water contents between –33 and –1500 kPa suction; this, plus

Table 23.2   Selected Root:Shoot Ratios and Total Plant Production for Selected Species

Environment	Root:Shoot Ratio	Total Aerial Biomass (g m^{-2})	Total Root Carbon (g m^{-2})	Reference	
Agriculture					
Cotton, annual	0.18	972[a]	70[a]	De Souza and Vieira Da Silva	1987
Cotton, perennial	0.49	972[a]	190[a]	with data from Mullins et al.	1990
Barley	0.1			Anderson–Taylor and Marshall	1983
Barley	0.28 to 0.60			Geisler and Krutzfeldt 1984 as recalculated by Klepper	1991
Wheat	0.09 to 0.12			Barraclough	1984
Bean	0.23 to 0.56			Geisler and Krutzfeldt 1984 as recalculated by Klepper	1991
Maize	0.18 to 0.92			Geisler and Krutzfeldt 1984 as recalculated by Klepper	1991
Maize	0.09	1370	130	Foth	1962
Maize	0.21[c]	1740[c]	<152	Balesdent and Balabane	1996
Native Plant					
Sideoats grama	0.35 to 0.60[b]	1080	80 to 130[b]	Kiniry et al.	1999
Switchgrass	0.30 to 0.73[b]	6210	380 to 930[b]	Kiniry et al.	
Eastern gamagrass	0.62 to 0.81[b]	2610	470 to 710[b]	Kiniry et al.	
Big bluestem	0.53 to 0.83[b]	2040	300 to 460[b]	Kiniry et al.	

[a] Assumed 16.2 plants m^{-2} with total mature plant dry weight of 60 g plant^{-1}.
[b] Determined at the end of the second season of growth.
[c] Constructed from the data.

the net difference between rainfall and evaporation, is the water available for crop production. Under natural conditions a small amount of N is received in rainfall, which is about equal to the amounts lost as ammonia or through denitrification. Thus, because C fixation depends largely on N supply, the largest amounts of N in natural soil organic matter are contributed through fixation by symbiotic microorganisms. With continuous plant cover, soils remain cooler, microbial decay of soil organic matter proceeds gradually, and organic N is continuously taken up by the new vegetation. Little N is lost through leaching events.

THE DELICATE BALANCE

The accumulation of soil organic C is the result of a delicate balance between C fixation and microbial decay of senescent vegetation (mainly root mass). Productivity, or carbon fixation, of any site depends on availability of water and nitrogen. The decay of organic matter is described by a negative general energy model for limited systems, GEMLS (Olness et al, 1998):

$$Y_t = Y_o(b - ((e^{k(t-to)} - e^{-k(t-to)})/(e^{k(t-to)} + e^{-k(t-to)})))$$

in which

Y_t = the amount of organic carbon remaining at any time, t
t = time (a substitute expression for internal energy relationships of microbial decay)
Y_o = the amount of organic carbon originally added
k = the time coefficient (a composite of all other energy forms, affecting the system; for example, oxygen, water, other nutrition, etc.)
t_o = that time at which the decay begins (in this case $t_o = 0$), and b = a resistant base level coefficient (here assumed to be zero)

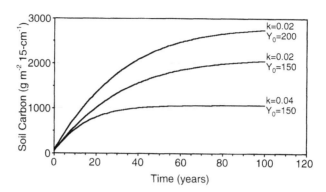

Figure 23.2   Accumulation of C in soil as a function of the time coefficient or decay rate, k, the amount of C in root biomass, and time.

Typical time coefficients range between 0.01 to 0.1 (ignore, for the moment, multicompartment models; see Molina et al., 1980). Using representative data from Table 23.2 and integrating the decay model over time, a relative accumulation of soil organic matter (Figure 23.2) can be described. One of the consequences of decay models is the achievement of some maximal equilibrium concentration of C in soil. As the decay constant increases, this limit is reached more rapidly; in the case of $k = 0.04$ and $Y_0 = 150$ g biomass (60 g of C), the limit is nearly reached in 40 years. As the decay coefficient decreases, the time required to approach equilibrium increases and the amount of C accumulated increases. Decay models generally predict that the largest increases in soil C occur early (20 to 40 years) within a restoration period. Thus a key to C sequestration in soil is managing the rate of decay of soil organic matter.

An accumulation of 2 kg of C m^{-2} 15-cm^{-1} depth increment with a soil bulk density of 1.0 g cm^{-3} will effect a C concentration of about 1.33%. Soil organic C contents of three to four times this amount are common in native prairies. This suggests that root masses of native prairie were much larger than those reported in Table 23.2, the decay constants are smaller than 0.02, or above-ground biomass was incorporated into the soil, or some combination of the above. Perennial grasses, for which root mass lives for several years, could effect an apparent decay constant of less than 0.02.

Once established in the soil, the equilibrium organic C levels are lost through increasing the decay constant. In this regard, tillage is a major factor in soil C loss. With tillage, C fixation in soil is interrupted, often for 2 to 3 months at a time (Figure 23.3). Soil microbial activity, however, continues and is often accelerated because the insulating effect of plant residues has been removed and soil temperatures are increased. With continued mineralization comes production of nitrates easily leached from the soil profile or, in the case of saturated conditions, eliminated through denitrification.

FACTORS OF SOIL CARBON SEQUESTRATION AND DEVELOPMENT OF A MODEL

Five major abiotic factors affect mineralization of soil organic matter (Olness et al., 1998): clay content, soil pH, soil bulk density, rainfall (water balance), and temperature (Figure 23.4). The first four factors affect soil aeration or oxygen supply; temperature is perhaps the most important factor in that it can be manipulated by cultural practice. The general effect of temperature on biological activity is shown in Figure 23.4. While any number of biological data sets show the same general nature of the relationship, the data were extracted from Blacklow (1972) for maize roots and fit with a GEMLS model. Manipulation of temperatures greater than about 30°C or less than about 9°C have little effect on biological activity. However, in the range of 9 to 30°C, a change in soil temperature has an important effect on relative biological respiration.

Figure 23.3   A site similar to that in Figure 23.1 after fall cultivation. Photograph taken in the spring.

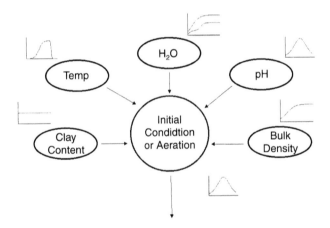

Figure 23.4   The five main factors affecting soil microbial production of nitrate–nitrogen in soil. Mathematical descriptions of these factors are integrated into the USDA-ARS nitrogen fertilizer decision aid.

The effect of soil aeration on microbial activity was well described by Skopp et al. (1990) as a delicate balance between having sufficient water for substrate diffusion and microbial movement and adequate oxygen for respiration. This balance is illustrated as a combination of two opposing GEMLS (Figure 23.5), using data obtained by Doran et al. (1990). Microbial respiration is maximized when water-filled pore space ranges from about 50 to 75%. The sensitive ranges for respiration are water-filled pore spaces < 50 and > 75%, but these ranges are of little value to cultivated agriculture (paddy culture excluded). Thus, manipulation of water would seem to offer less opportunity for control of respiration than that of temperature.

Soil water-filled pore space is a function of soil water content and is largely controlled by soil clay content (texture) and soil organic matter content (see also Hudson, 1994). In this situation, two supplemental GEMLS functions provide a reasonable description of the effects of soil clay content and soil organic matter content on the total water-holding capacity of the soil. (The models were developed using data [not shown] from Olson, 1970; Figure 23.6.)

The other factor needed to determine soil water-filled pore space is total soil porosity or soil bulk density. Along with the combination of texture and clay content, it determines relative soil

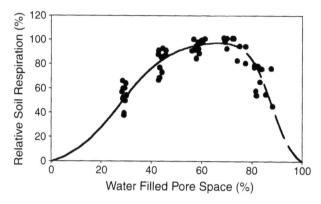

Figure 23.5   Relative microbial respiration as a function of water-filled pore space. Data from Doran et al., 1990, and modeled using two opposing GEMLS.

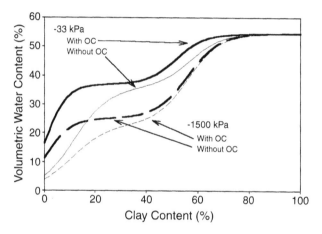

Figure 23.6   Soil volumetric water content at field capacity (–33 kPa) and permanent wilting point (–1500 kPa) as a function of clay content and organic matter. Data from Olson, 1970, and modeled using two supplementary GEMLS.

aeration, which controls soil microbial respiration. When these factors are combined, we can see how complex interactions affect respiration (microbial oxidation) and mineralization of soil organic matter (Figure 23.7). Many native prairie soils have bulk densities of about 1.0, which tends to be too well aerated (too dry) for optimal microbial respiration; this aids accumulation of soil organic C. For many soils, a bulk density of about 1.2 is achieved at planting; this tends to be ideal for microbial mineralization of organic matter. It is no coincidence that crop producers are taking advantage of the mineralization of soil organic matter as a source of N for crop production. Interestingly, in the initial years of conversion to no-tillage, a bulk density of 1.4 is common for medium textured soils, but this tends to be too wet for maximal microbial activity. The common observance is that no-tilled soils tend to be N deficient relative to tilled soils. This slowed N production rate is a measure of the potential conserving ability of no-tillage to sequester C and N.

Soil pH (hydrogen ion activity) controls microbial enzymatic efficiency (Olness, 1999). Organic N is mineralized and nitrified most rapidly at a pH optimum of about 6.7 (Figure 23.8). This observation suggests that both hydroxyl and hydrogen ions are inhibitors of microbial respiration. Liming acid soils or additions of ammoniacal fertilizers effects an acceleration of microbial decay of soil organic matter due simply to the change in hydroxyl or hydrogen ion concentrations.

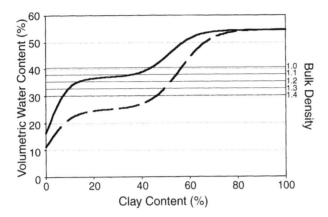

Figure 23.7   The effect of soil bulk density and volumetric water content on water-filled pore space. The intersection between the horizontal bulk density line with the field capacity curve gives water-filled pore space optimal for maximal microbial respiration. Modified from Olness et al., 1997.

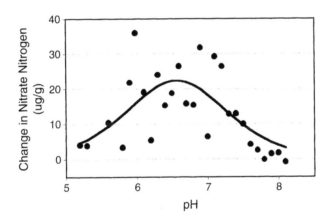

Figure 23.8   The general effect of pH on change in nitrate–N concentration in the upper 30 cm of soil. Reprinted with permission; Olness et al., 1997 (see further, Olness, 1999).

NITROGEN: THE SUBTLE COST OF SOIL CARBON SEQUESTRATION

Nitrate formation in soil is a very sensitive measure of soil microbial respiration as long as leaching is taken into account or prohibited. By combining the abiotic factors, a model of activity can be constructed. When this is done, rather close agreement between observed and predicted nitrate formation is obtained (Table 23.3). The model seems to accommodate a range of soil pH values, soil textures and organic matter contents. A production of 10 µg N g^{-1} of soil in the upper 60 cm of the profile in a 60-day period equals about 80 kg of N converted from organic amine to mineral nitrate-N. Because soil C:N ratios vary in a narrow range around 10.0, a reliable estimate is that about 800 kg of C ha^{-1} were digested and most likely lost as CO_2.

This brings us to a critical aspect of C sequestration in soil. Soil organic matter contains about 9% N, with some additional phosphorus and sulfur. In order to aid C sequestration, a source of N must be sacrificed or stored with the C. The opportunity cost of storing this C in terms of N is shown in Table 23.4. A conservative estimate yields a cost of 136 to 164 kg of N ha^{-1} needed to increase soil organic C by 0.1% if the C:N ratio is 11.0 and the system is 100% efficient with the

Table 23.3 Prediction of Changes in Soil Nitrate-N Concentration

	pH (g kg⁻¹)	Clay (g kg⁻¹)	Organic C (g kg⁻¹)	Change in Nitrate-N (µg kg⁻¹) Observed	Predicted
Minimum	5.81	242	14.1	−0.58	0.81
Mean	6.53	311	20.2	5.68	2.37
Maximum	7.91	373	44.7	11.70	10.7

Table 23.4 Opportunity Cost of Sequestered N Required to Increase Soil Organic C by 0.1%

Bulk Density (Mg m⁻³)	Total OC Increase (kg ha⁻¹ 15-cm⁻¹)	Nitrogen Required[a] (kg ha⁻¹) 100% Efficiency	50% Efficiency	Cost[b] ($ ha⁻¹)
1.0	1500	136	272	52 → 104
1.2	1800	164	328	63 → 126

[a] Assumed C:N ratio of 11.0.
[b] Assumed cost of N = $ .386 kg⁻¹.

N. The value of this N can be reasonably estimated from current market prices at about $52 to $63 ha⁻¹. Nitrogen use efficiency is rarely 100%; literature citations usually quote a range of efficiencies from < 40 to about 75%. Assuming an N use efficiency of 50%, the opportunity cost of N required to increase soil C by 0.1% rises to about $104 to $126 ha⁻¹. Within the U.S. Corn Belt, this would amount to about $27 to $66 billion for N to increase the soil organic C by 1%.

Increasing soil organic C will increase the plant available water content of the soil (Hudson, 1994), which will increase crop yield in areas where water is a limiting factor in crop production. However, the increased available water-holding capacity varies with soil texture. Also, the value of the increased available water-holding capacity, in terms of yield, varies with the climatic zone and likelihood of realizing a loss of yield potential due to drought stress. Thus the opportunity cost of N in sequestered C is affected by the relative recovery of cost with increased yield of crops.

These costs for N virtually guarantee that increasing soil organic C will have to be effected initially through symbiotic N fixation with legumes. As the soil N status increases, legumes tend to become less effective in N₂ fixation and grasses tend to invade the landscape. At present, the most likely candidates for increasing soil organic C appear to be alfalfa (*Medicago sativa* L.), hairy vetch (*Villa vilossa* L.), or, perhaps, woody legumes because of their ability to fix prodigious amounts of N. The latter would seem to be less compatible with current crop production.

CONCLUSIONS

Extensive agricultural production of U.S. soils with tillage encourages continuous mining of soil organic C and release of N. Additionally, grain crops often have lesser root production than native perennial grasses; this further encourages depletion of soil organic C. Decay models predict that the amount of C that can be sequestered in the soil has some natural limit that depends on the type of plant grown as well as five abiotic factors.

The abiotic factors are soil clay content, soil bulk density, soil pH, soil water content, and temperature. Of these factors, manipulation of temperature would seem to offer the most effective means of increasing soil organic C, which will have the beneficial effect of increasing water available for crop production; this will partially offset the cost of increasing C. Manipulations of soil bulk density, aeration, or soil pH are less reasonable alternatives for aiding sequestration of C.

Because soil organic matter has a C:N ratio of about 11, sequestration of soil organic C will require an opportunity cost of N. This cost is estimated to range from $27 to $66 billion for each 1% increase in soil C in the U.S. Corn Belt. Symbiotic fixation of N₂ through use of legumes seems the most likely cost-effective means of achieving this N input.

REFERENCES

Anderson–Taylor, G. and C. Marshall, Root-tiller interrelationships in spring barley (*Hordeum distichum* (L.) Lam.), *Ann. Bot.*, 51:47–58, 1983.

Balesdent, J. and M. Balabane, Major contribution of roots to soil carbon storage inferred from maize cultivated soils, *Soil Biol. Biochem.*, 28:1261–1263, 1996.

Blacklow, W.M., Influence of temperature on germination and elongation of the radicle and shoot of corn (*Zea mays* L.), *Crop Sci.*, 12:647–650, 1972.

Barraclough, P.B., The growth and activity of winter wheat roots in the field: root growth of high-yielding crops in relation to shoot growth, *J. Agric. Sci.*, 103:439–442, 1984.

DeSouza, J.G. and J. Vieira Da Silva, Partitioning of carbohydrates in annual and perennial cotton (*Gossypium hirsutum* L.), *J. Exp. Bot.*, 38:1211–1218, 1987.

Doran, J.W., L.N. Mielke, and J.F. Power, Microbial activity as regulated by soil water filled pore space, *Trans. 14th Int. Congr. Soil Sci.*, Kyoto, Japan, Aug. 12–18, 1990, 3, 94–99, 1990.

Foth, H.D., Root and top growth of corn, *Agron. J.*, 54:49–52, 1962.

Geisler, G. and B. Krutzfeldt, Wirkungen von Stickstoff auf die Morphologie und die Trockenmassebildung der Wurzelsysteme von Mais-, Sommeregersten- und Ackerbohonen-Sorten unter Berücksichtigung der Temperatur: 2. Trockenmassebildung, *J. Agron. Crop Sci.*, 153:90–104, 1984.

Houghton, R.A., Changes in the storage of terrestrial carbon since 1850, in *Soils and Global Change*, Lal, R. et al., eds., CRC Press, Boca Raton, FL, 1995, 45–66.

Hudson, B.D., Soil organic matter and available water capacity, *J. Soil Water Conserv.*, 49:189–194, 1994.

Kiniry, J.R., C.R. Tischler, and G.A. Van Esbroeck, Radiation use efficiency and leaf CO_2 exchange for diverse C_4 grasses, *Biomass Bioenergy*, 17:95–112, 1999.

Klepper, B., Root-Shoot relationships, in *Plant Roots: the Hidden Half*, Waisel, Y., A. Eshel, and U. Kafkafi, eds., Marcel Dekker, Inc., New York, 1991, 265–286.

Lal, R., Residue management conservation tillage and soil restoration for mitigating greenhouse effect by CO_2 enrichment, *Soil Tillage Research*, 43:81–107, 1997.

Molina, J.A.E., C.E. Clapp, and W.E. Larson, Potentially mineralizable nitrogen in soil: the simple exponential model does not apply for the first 12 weeks of incubation, *Soil Sci. Soc. Am. J.*, 44:442-443, 1980.

Mullins, G.L. and C.H. Burmester, Dry matter, nitrogen, phosphorus, and potassium accumulation by four cotton varieties, *Agron. J.*, 82:729–736, 1990.

Olness, A. et al., Biosolids and their effects on soil properties, in *Handbook of Soil Conditioners*, Terry, R.E., ed., Marcel Dekker, New York, 1997, 141–165.

Olness, A. et al., Predicting nitrogen fertilizer needs using soil and climatic data, in *Proc. 11th World Fertilizer Cong.*, Vermoesen, A., ed., Gent, Belgium, Sept. 7–13, 1997, International Centre of Fertilizers, Ghent, Belgium, 1998, 356–364.

Olness, A. et al., A description of the general effect of pH on formation of nitrate in soils, *Plant Nutrition Soil Sci.*, 162. (Accepted for publication July 18, 1999).

Olness, A. et al., A Nitrogen Fertilizer Decision Aid, (www.morris.ars.usda.gov/.), October 1999, 1999b.

Olson, T.C., Water storage characteristics of 21 soils in eastern South Dakota, USDA-ARS Bull. #ARS-41-166, 1970.

Reicosky, D.C., J.L. Hatfield, and R.L. Sass, Chapt. 3. Agricultural contributions to greenhouse gas emissions, in *Climate Change and Global Crop Productivity*, Reddy, K.R. and Hodges, H.F., eds., CAB International Pub., New York, 2000, 37–55.

Schlesinger, W.H., Changes in soil carbon storage and associated properties with disturbance and recovery, in *The Changing Carbon Cycle: A Global Analysis*, Trabalka, J.R. and Reichle, D.E., eds., Springer-Verlag, New York, 1986, 194–220.

Skopp, J., M.D. Jawson, and J.W. Doran, Steady-state aerobic microbial activity as a function of soil water content, *Soil Sci. Soc. Am. J.*, Madison, WI, 54:1619–1625, 1990.

USDA-NASS, *Agricultural Statistics 1999*, U.S. Government Printing Office, Washington, D.C., 1999.

CHAPTER 24

Soil Carbon Turnover in Residue Managed Wheat and Grain Sorghum

Sven U. Bóhm, Charles W. Rice, and Alan J. Schlegel

CONTENTS

ABSTRACT

Management of crop residues influences the ability of soils to sequester carbon (C). The site of a 28-year study of residue management was used to estimate rates of carbon turnover. The study was located on a Richfield silt loam (fine montmorillonitic, mesic, Aridic Argiustoll). Two experiments, both irrigated, included continuous winter wheat (*Triticum aestivum* L.) and continuous grain sorghum (*Sorghum bicolor* (L.) Moench). After harvest the aboveground residue was: (1) physically removed, (2) left, (3) twice the amount, or (4) burned. The treatments were then disked to incorporate residue, plowed, disked, and finally harrowed before planting. Soil samples were taken to 120 cm, and the $\delta^{13}C$ value and content of carbon were measured using mass spectrometry. Burning and physical removal of wheat and grain sorghum residue did not decrease soil C levels when compared with normal residue levels. Carbon from roots appears to contribute more to soil organic matter formation than straw. Organic C >30 cm was affected little by cultivation and residue management. Comparison to a study conducted at 19 years showed that the soil had lost about 20% of the carbon under continuous wheat, but only 3% under continuous sorghum. Total nitrogen losses for both crops averaged 17% during the ensuing 19 years. Based on ^{13}C values, the amount of old soil C was slightly higher with sorghum than with wheat. The burning, removal, and normal treatments had similar levels of old C. The rates of crop C loss were higher for the residue treatments than for the removal treatment, suggesting that root C turns over more slowly than C derived from straw.

INTRODUCTION

Soil organic matter participates as a source or sink in the global CO_2 budget (Tans et al., 1990; Lugo, 1992). Increasing atmospheric CO_2 can lead to increased production of plant biomass, a portion of which can potentially be stored in the soil. However, traditional cultivation of grassland soils in the U.S. has resulted in 30 to 50% loss of soil C (Haas et al., 1957; Tiessen et al., 1982). Recent studies have shown that the decline in soil organic C associated with cropping is not universal and irreversible (Coleman et al., 1984). Cropping intensity (Havlin et al., 1990; Campbell et al., 1991), tillage intensity (Havlin et al., 1990; Reicosky et al., 1995; Potter et al., 1998), and fertilization (Campbell et al., 1991; Raun et al., 1998) can increase soil organic C levels. There is a strong positive relationship between the amount of C returned to the soil and soil organic C content (Havlin et al., 1990; Rasmussen and Collins, 1991; Paustian et al., 1992). Therefore, retention and management of crop residues is an important consideration in the soil's ability to sequester atmospheric CO_2 (Rasmussen and Parton, 1994). Crop type appears to have little impact on soil C (Larson et al., 1972; Havlin et al., 1990).

Long-term studies of residue management provide an opportunity to measure the response of the soil to additional or lack of crop-residue inputs. Straw return has been shown to be beneficial to soil quality (Black, 1973); however, incorporating residue requires additional management inputs, i.e., tillage. Burning is a common method of residue control where a clean seedbed is needed and where furrow irrigation is practiced; however, it can deplete the soil organic C (Smith, 1970; Biederbeck et al., 1980). Straw removal by baling has been shown to have the same effect as burning in the northwestern U.S. (Rasmussen et al., 1980) and in Denmark (Christensen, 1986). On the other hand, Campbell et al. (1991) reported little effect of straw removal on soil organic C content in the northern Great Plains. They hypothesized that C from the roots contributes more to soil organic C than carbon from the straw. They postulated several reasons for this

1. The roots are "well placed within the soil and are continually sloughing, dying and exuding materials," and the C would more likely be incorporated into the soil organic matter.
2. The straw C needs to be incorporated, and the resultant disruption by tillage would accelerate the breakdown of the soil organic matter.
3. The additional energy supplied by the straw (wide C:N ratio) results in more organic matter breakdown.

Buyanovsky and Wagner (1987) estimated that 16% of straw C and 24% of root C from wheat was transferred to soil organic matter. If roots do indeed act differently than straw in the dynamics of soil organic matter, models on soil C will need to account for this difference.

The isotopic signature of soil organic C can be a useful tool for estimating soil C dynamics in long-term studies (Gregorich et al., 1995; Cadish et al., 1996; Follett et al., 1997). The differential discrimination of the ^{13}C isotope by C_3 and C_4 plants provides a natural signature to trace plant C into soil organic C. The introduction of a crop with a different photosynthetic pathway can determine the contribution of the crop to soil organic C and the turnover of the soil organic C. Using ^{13}C techniques, Gregorich et al. (1995) were able to determine the amount and turnover of soil organic C derived from corn compared to the native organic C derived from a forest in eastern Canada.

The objectives of this study were to determine the (1) soil organic C levels after 28 years of residue management, and (2) the turnover of C from roots and straw in continuous wheat and sorghum using isotopic techniques.

METHODS

A residue-management study initiated in 1969 was located near Holcomb, Kansas, details of which have been described by Hooker et al. (1982). Briefly, grain sorghum (*Sorghum bicolor* (L.) Moench) and winter wheat (*Triticum aestivum* L.) were grown on separate field experiments. The treatments consisted of

1. Complete removal of aboveground residue (removal)
2. Aboveground residue retained (normal)
3. Twice the amount of aboveground residue (double)
4. Burning of aboveground residue (burn)

Each treatment was replicated four times for each crop. The residue treatments were imposed after each harvest, i.e., in June and July for wheat and in September and October for sorghum. The plots then were disked to incorporate the residue, plowed, disked, and, finally, harrowed to provide a better seedbed until 1986. After 1986, the plots were only disked for the primary tillage. Furrow irrigation was used in times of water stress. Nitrogen fertilizer was applied to half of each treatment plot. This produced a high- and low- (or no-) N split plot. The N rates for wheat were high N (prior to 1980, 112 kg N ha^{-1}; 1980 to 1992, 90 kg N ha^{-1}, and after 1992, 134 kg N ha^{-1}) and low N (prior to 1980, 68 kg N ha^{-1} and 0 kg N ha^{-1} thereafter). The N rates for grain sorghum were high N (prior to 1980, 179 kg N ha^{-1}; 1980 to 1992, 90 kg N ha^{-1}, and after 1992, 134 kg N ha^{-1}) and low N (prior to 1980, 90 kg N ha^{-1} and 0 kg N ha^{-1} thereafter). The soil was mapped as a Richfield silty clay loam (fine, montmorillonitic, mesic, Aridic Argiustoll).

Soil samples of the top 120 cm were taken with a hydraulic soil probe (Giddings) on March 7, 1997. One core (25-cm dia.) was taken from each plot to 120 cm, and a second core to 30 cm. The soil was separated into increments of 0 to 5, 5 to 15, 15 to 30, 30 to 60, 60 to 90, and 90 to 120 cm. Soil samples were passed through a 6-mm sieve and stored field moist at 4°C until analysis. Bulk density averaged 1.2 Mg m^{-3} with no treatment or depth effects. Soil moisture was determined by weight loss on drying at 105°C for 24 h.

Microbial biomass was determined by the fumigation and incubation technique (Jenkinson and Ladd, 1981). Duplicate 5-g samples of soil were placed in a 160 mL serum bottle and preincubated for 5 days at 25°C. One of the duplicate samples then was fumigated for 24 h. Both samples were capped and incubated for 5 days at 25°C. The CO_2 accumulated in the headspace was measured by taking a 0.5 mL sample and analyzing it on a Shimadzu GC-8A gas chromatograph equipped with a thermal conductivity detector (Shimadzu, Scientific Instruments Inc., Columbia, MO). The column was a 1-m Porapak 80 column operated at 70°C, with He as a carrier gas at 14 mL min^{-1}. Microbial biomass C was calculated as the difference between fumigated and control samples with a Kc = 0.41. (Voroney and Paul, 1984).

A ≈40-g subsample of the soil was air dried and ground with a mortar and pestle. About 2 g of the dried and ground soil was allowed to react with a 6-mL aliquot of a 1 M H_2SO_4/FeO solution to remove inorganic carbonates from the soil. Carbonate removal was tested by adding a second aliquot to several samples from the subsoil, with no additional effervescence serving as an indication of complete removal of carbonate. The soil was analyzed for ^{13}C on a single inlet isotope mass-spectrometer (Europa Scientific, UK). Soil samples from the 90- to 120-cm depth were not included in this analysis because the low levels of organic C made determinations of δ^{13}C and C uncertain.

Whole aboveground plant samples were ground in a Wiley mill and pulverized using a centrifugal mill. Then 15-mg subsamples were analyzed to determine the δ^{13}C values. Because the values were similar for all treatments, a single value for δ^{13}C (–24.28‰) was used for wheat in the calculations.

MODEL

In a binary mixture of C from two sources with differing ^{13}C values (Balesdent et al., 1987; Vitorello et al., 1989):

$$\delta^{13}C_f \cong x\delta^{13}C_i + (1 - x)\, \delta^{13}C_p \tag{24.1}$$

where the subscripts $_{f,i,p}$ represent the final, initial, and plant δ^{13}C, respectively. The amount of C from the two sources can be represented by $C_i = xC$ and $C_p = (1 - x)C$. Substituting this into Equation 24.1 and rearranging gives (Balesdent et al., 1990):

$$\delta^{13}C_f \cong \frac{1}{C}\, (\delta^{13}C_i - \delta^{13}C_p)\, C_p + \delta^{13}C_p \tag{24.2}$$

So a plot of the soil δ^{13}C vs. the reciprocal of the C content should be a straight line, with an intercept at the δ^{13}C value of the plant material. Furthermore, when mixed C$_3$ and C$_4$ vegetation develops on the same soil, i.e., the initial δ^{13}C of the soil profile is uniform, then the δ^{13}C value at the intersection of the two lines of the types of vegetation represents the initial ^{13}C value of the soil (Balesdent et al., 1990).

Estimates of decomposition rates were obtained by assuming the addition of the average residue amount (r) for each of the 28 years. After the addition, each residue cohort is assumed to decompose independently according to first order kinetics:

$$C = \sum_{t=1}^{28} re^{-kt} \tag{24.3}$$

where t is the time in years, and k is the decomposition rate constant. This constant was computed by setting C equal to the current new C store and r equal to the annual residue input.

Residue amounts were estimated by taking the average grain yields and applying a 6-year average of the straw:grain ratio (measured from 1980 to 1986). The 6-year average straw:grain ratio was 1.53 for wheat and 0.66 for grain sorghum. Root inputs were estimated by using a root:straw ratio of 0.59 (van Veen and Paul, 1981). The estimated annual residue inputs are summarized in Table 24.1. The estimated residue input for wheat was similar to that reported by Buyanovsky and Wagner (1987).

The data from the experiment were analyzed using a split-plot analysis. Data presented in tables are the average of the two N rates since, in most instances, there were no significant differences due to the rate. Where significant differences did occur, the differences were magnified by the high N rate.

Table 24.1   Estimated Annual Residue Input Based on Measured Values from 1980 to 1986

Crop	Treatment (kg/ha/yr)			
	Normal	Double	Removal	Burn
Wheat	10,758	17,494	3,925	3,997
Sorghum	7,216	11,745	2,578	2,675

Table 24.2   Treatment Effects (g C kg⁻¹) on the Organic C Contents of the Soil Profile after 28 Years

| Crop | | \multicolumn{5}{c}{Depth (cm)} | | | | |
		0–5	5–15	15–30	30–60	60–90
Wheat	Burn	0.74a[a]	0.70ab	0.58a	0.34a	0.24a
	Removal	0.72a	0.62a	0.50b	0.32a	0.24a
	Normal	0.76a	0.69ab	0.49b	0.33a	0.24a
	Double	0.86b	0.71b	0.52b	0.32a	0.22a
Sorghum	Burn	1.22ab	0.78a	0.62a	0.36a	0.25a
	Removal	1.09a	0.75a	0.61a	0.37a	0.24a
	Normal	1.25ab	0.83ab	0.64a	0.39a	0.27a
	Double	1.36b	0.87b	0.57a	0.39a	0.25a

[a] Means with the same letter are not significantly (P<0.05) different from the other treatments.

RESULTS AND DISCUSSION

Soil organic C decreased with depth (Table 24.2) and no treatment differences were found below 30 cm. With sorghum, the double-residue treatment resulted in higher organic C levels in the surface soil. Burning and residue removal did not significantly depress organic C levels below those of the normal-residue treatment. The same pattern held for the wheat in the 0- to 15-cm depth. The results of this study disagree with those of studies in which soil organic matter levels were positively related to the amount of residue returned to the soil (Larson et al., 1972; Rasmussen et al., 1980; Havlin et al., 1990). Rasmussen and Parton (1994) suggested that the C remaining after burning is less bioavailable and thus has a longer turnover time. No measurements for charcoal were made; however, in this study, there was no difference in soil organic C levels whether the straw residue was burned or physically removed.

The lack of significant reductions in the organic C contents in the burn, removal, and normal-residue treatments compared to the nonremoval treatments, even after 28 years, supports the conclusion of Campbell et al. (1991). However, their hypothesis of differential breakdown of soil organic matter caused by differences in disturbance is not applicable because all of the plots received the same tillage treatment. Thus, it appears that root turnover (sloughing, exudation) during the growing season and differential decomposition of root biomass contributes more to soil organic matter formation than straw (Buyanovsky and Wagner, 1987). Further research should be directed at the role of roots in soil organic matter formation.

Microbial biomass C decreased with depth (Table 24.3), but no treatment effects could be detected, in contrast to Biederbeck et al. (1980), who found that long-term burning decreased the microbial biomass. The sharp decrease in microbial C below 30 cm suggests lower microbial activity and C turnover in the subsoil, which fits with the lack of cultivation effect observed in the $\delta^{13}C$ data below 30 cm.

Table 24.3   Microbial Biomass C of the Soil Profile in the Wheat Experiment

Depth (cm)	Wheat (mg C kg⁻¹)
0–5	235
5–15	254
15–30	193
30–60	17–6
60–90	16–4
90–120	10–9
LSD$_{0.05}$	40

Table 24.4   Changes in N and Organic C Contents in the Top 15 cm of the Soil after 9[a] and 28 Years

Study	Treatment	Organic C (g C kg⁻¹)		Total N (mg N g⁻¹)	
		9 Years	28 Years	9 Years	28 Years
Wheat	Burn	0.85 a[b]	0.708 a	1170 a	962 a
	Removal	0.85 a	0.654 a	1200 a	990 a
	Normal	0.90 ab	0.716 a	1270 b	1001 ab
	Double	0.95 b	0.761 b	1380 b	1053 b
Sorghum	Burn	0.90 a	0.925 a	1310 a	1053 a
	Removal	0.90 a	0.864 a	1270 a	1059 a
	Normal	1.00 ab	0.967 a	1330 b	1157 b
	Double	1.10 b	1.035 b	1420 c	1175 b

[a] Data from Hooker et al. (1982).
[b] Means with the same letter are not significantly (P<0.05) different from the other treatments.

Organic C and total N levels in the top 15 cm of soil declined between 1979 and 1997, even where twice the normal residue was applied (Table 24.4). Both sorghum and wheat crops resulted in a reduction in total N, whereas the organic C reduction was less under sorghum than under wheat. Total N loss under the wheat was slightly more in the removal treatments (17%) than in the incorporated treatments (22%). Under the sorghum, total N losses averaged 16% with no clear trends in response to the treatments. Under continuous wheat, the organic C declined 23% in the removal treatment, 20% in the double-residue treatment, and 17% in the burn treatment. Under the sorghum, the average decline in organic C levels was only 3%, consistent with the lower rates of C loss estimated for the sorghum study. These results suggest that, even though the amounts of organic C present did not differ much among the treatments, a significant amount of C had been lost from the continuous wheat during the previous 19 years. This continued decline of soil C for wheat is somewhat surprising because the area has been in cultivation since the 1890s, and C content normally reaches a new equilibrium level after about 60 years (Haas et al., 1957; Marel and Paul, 1974).

A plot of Equation 24.2 shows the average initial $\delta^{13}C$ value of the surface soil (0 to 15 cm) to be near -15.7‰ (Figure 24.1). Soil below 30 cm apparently was unaffected by the cultivation because the $\delta^{13}C$ value was similar to that of the native C_4-dominated vegetation. The 15- to 30-cm depth was omitted from the graph for clarity, but it showed a transition between the near-surface soil affected by cultivation and cropping and the deeper soil not affected by the cultivation. This transition zone probably was caused by plowing incorporating various amounts of residues into this soil depth and movement of soluble C. This lack of effect on the soil below 30 cm was somewhat surprising, given that the area had been in cultivation since the 1890s. However, a similar lack of effects on organic C in the deeper soil layers with over 100 years of cultivation also has been shown at Sanborn Field in Missouri by Balesdent et al. (1988).

Turnover of soil organic matter has been shown to result in a slight increase in the $\delta^{13}C$ values with depth because of a preferential loss of the lighter isotope (O'Brien and Stout, 1978; Balesdent and Mariotti, 1996). The decrease in the $\delta^{13}C$ values with depth in this study might have been due to a trend to a more arid and warmer environment favoring C_4 species over C_3 species.

The amount of old soil C was slightly higher with sorghum than with wheat (Table 24.5), suggesting lower rates of C turnover in sorghum than in wheat. The double-residue treatments appear to have lost more old C than the other treatments, which would give credence to the hypothesis that the additional energy supplied by the straw accelerates organic C loss of the native soil organic matter (Campbell et al., 1991). The burning, removal, and normal treatments had similar levels of old C, indicating that the breakdown of the old soil organic matter was not affected by the treatments. Because the amount of tillage disturbance was the same for all treatments, this seems likely.

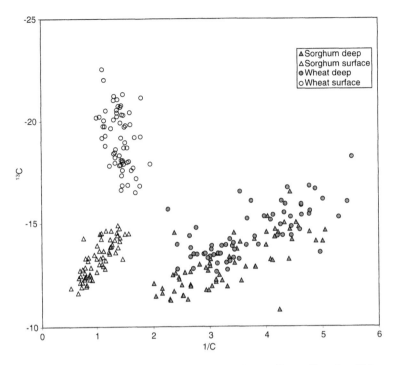

Figure 24.1 $\delta^{13}C$ values vs. the inverse of the C content plotted according to Equation 24.2.

Table 24.5   Carbon Turnover (in kg/ha) in the 0- to 5-cm Depth

Crop	Treatment	C added	Old C	New C	k (year^{-1})
Wheat	Burn	50,357	7,733	8,972	0.182
	Removal	49,455	8,465	7,810	0.204
	Normal	135,550	8,018	8,736	0.441
	Double	220,428	7,417	12,015	0.504
Sorghum	Burn	33,705	9,217	18,554	0.046
	Removal	32,488	10,123	14,626	0.064
	Normal	90,920	11,418	16,944	0.174
	Double	147,993	7,725	23,115	0.205

The calculated rates of C loss were higher for the residue incorporation treatments than for the residue removal treatments. This suggests that less root C than straw C was lost and agrees with the hypothesis that C from roots contributes more to soil organic C (Buyanovsky and Wagner, 1987; Campbell et al., 1991). It appears that roots contribute more to the maintenance of soil organic carbon stores than straw. The differential contribution is likely due to the "well placed" nature of the root carbon or the accelerated soil organic matter turnover due to increased energy inputs from the wide C:N ratio straw.

CONCLUSIONS

Burning and physical removal of wheat and grain sorghum residue after 28 years did not decrease soil C levels when compared with normal residue levels. This may be due to soil texture and the semiarid environment. Carbon from roots appeared to contribute more to soil organic matter formation than straw. Organic C >30 cm was affected little by cultivation and residue management. Comparison to a study conducted 19 years ago showed that the soil had lost about 20% of the carbon under continuous wheat, but only 3% under continuous sorghum. Total nitrogen losses for both crops averaged 17% during the ensuing 19 years. Based on ^{13}C values, the amount of old soil C was slightly higher with sorghum than with wheat; burning, removal, and normal treatments had similar levels of old C. The rates of crop C loss were higher for the residue treatments than for the removal treatment, suggesting root C turns over more slowly than C derived from straw, which also supports the notion that roots contributed more to soil C than the aboveground residue did.

REFERENCES

Balesdent, J. and A. Mariotti, Measurement of soil organic matter turnover using ^{13}C natural abundance, in Boutton, T.W. and S. Yamasaki (ed.), *Mass Spectrometry of Soils*, Marcel Dekker, New York, 1996, 83–111.

Balesdent, J., A. Mariotti, and D. Boisgontier, Effect of tillage on soil organic carbon mineralization estimated from ^{13}C abundance in maize fields, *J. Soil Sci.*, 41:587–596, 1990.

Balesdent, J., A. Mariotti, and B. Guillet, Natural ^{13}C abundance as a tracer for studies of soil organic matter dynamics, *Soil Biol. Biochem.*, 19:25–30, 1987.

Balesdent, J., G.H. Wagner, and A. Mariotti, Soil organic matter turnover in long-term field experiments as revealed by carbon-13 natural abundance, *Soil Sci. Soc. Am. J.*, 52:118–124, 1988.

Biederbeck, V.O. et al., Effect of burning cereal straw on soil properties and grain yields in Saskatchewan, *Soil Sci. Soc. Am. J.*, 44:103–111, 1980.

Black, A.L., Soil property changes associated with crop residue management in a wheat-fallow rotation, *Soil Sci. Soc. Am. Proc.*, 37:943–946, 1973.

Buyanovsky, G.A. and G.H. Wagner, Carbon transfer in a winter wheat (*Triticum aestivum*) ecosystem, *Biol. Fertil. Soils*, 5:76–82, 1987.

Cadish, G. et al., Carbon turnover (δ^{13}C) and nitrogen mineralization potential of particulate light soil organic matter after rainforest clearing, *Soil Biol. Biochem.*, 28:1555–1567, 1996.

Campbell, C.A. et al., Influence of fertilizer and straw baling on soil organic matter in a thin black chernozem in western Canada, *Soil Biol. Biochem.*, 232:442–446, 1991.

Christensen, B.T., Straw incorporation and soil organic matter in macro-aggregates and particle size separates, *J. Soil Sci.*, 37:125–135, 1986.

Coleman, D.C., C.V. Cole, and E.T. Elliott, Decomposition, organic matter turnover, and nutrient dynamics in agroecosystems, in Lowrance, R., B.R. Stinner, and G.J. House (eds.), *Agricultural Ecosystems — Unifying Concepts*, John Wiley & Sons, New York, 1984, 83–104.

Follett, R.F. et al., Carbon isotope ratios of Great Plains soils and in wheat-fallow systems, *Soil Sci. Soc. Am. J.*, 61:1068–1077, 1997.

Gregorich, E.G., B.H. Ellert, and C.M. Monreal, Turnover of soil organic matter and storage of corn residue carbon estimated from natural ^{13}C abundance, *Can. J. Soil Sci.*, 75:161–167, 1995.

Haas, H.J., C.E. Evans, and E.F. Miles, Nitrogen and carbon changes in Great Plains soils as influenced by cropping and soil treatments, USDA Technical Bulletin 1164, Washington, D.C., 1957.

Havlin, J.L. et al., Crop rotation and tillage effects on soil organic carbon and nitrogen, *Soil Sci. Soc. Am. J.*, 54:448–452, 1990.

Hooker, M.L., G.M. Herron, and P. Penas, Effects of residue burning, removal, and incorporation on irrigated cereal crop yields and soil chemical properties, *Soil Sci. Soc. Am. J.*, 46:122–126, 1982.

Jenkinson, D.S. and J.N. Ladd, Microbial biomass in soil — measurement and turnover, in Paul, E.A. and J.N. Ladd (ed.), *Soil Biochemistry*, Vol. 5, Marcel Dekker, New York, 1981.

Larson, W.E. et al., Effects of increasing amounts of organic residues on continuous corn. II. Organic carbon, nitrogen, phosphorus, and sulfur, *Agron. J.*, 64:204–208, 1972.

Lugo, A.F., The search for carbon sinks in the tropics, *Water Air Soil Pollut.*, 64:3–9, 1992.

Marel, Y.A. and E.A. Paul, Effects of cultivation on the organic matter of grassland soils as determined by fractionation and radiocarbon dating, *Can. J. Soil Sci.*, 54:419–422, 1974.

O'Brien, J. and J.D. Stout, Movement and turnover of soil organic matter as indicated by carbon isotope measurements, *Soil Biol. Biochem.*, 10:309, 1978.

Paustian, K., W.J. Parton, and J. Persson, Modeling soil organic matter in organic-amended and nitrogen-fertilized long-term plots, *Soil Sci. Soc. Am. J.*, 56:476–488, 1992.

Potter, K.N. et al., Distribution and amount of soil organic C in long-term management systems in Texas, *Soil Tillage Res.*, 47:309–321, 1998.

Rasmussen, P.E. et al., Crop residue influences on soil carbon and nitrogen in a wheat-fallow system, *Soil Sci. Soc. Am. J.*, 44:596–600, 1980.

Rasmussen, P.E. and H.P. Collins, Long-term impacts of tillage, fertilizer, and crop residue on soil organic matter in temperate semi-arid regions, *Adv. Agron.*, 45:93–134, 1991.

Rasmussen, P.E. and W.J. Parton, Long-term effects of residue management in wheat-fallow. I. Inputs, yield, and soil organic matter, *Soil Sci. Soc. Am. J.*, 58:523–530, 1994.

Raun, W.R. et al., Effect of long-term N fertilization on soil organic C and total N in continuous wheat under conventional tillage in Oklahoma, *Soil Tillage Res.*, 47:323–330, 1998.

Reicosky, D.C. et al., Soil organic matter changes resulting from tillage and biomass production, *J. Soil Water Conserv.*, 50:253–261, 1995.

Smith, D.W., Concentrations on soil nutrients before and after fire, *Can. J. Soil Sci.*, 50:17–29, 1970.

Tans, P.O., I.Y. Fung, and T. Takahashi, Observational constraints on the global atmosphere CO_2 budget, *Science*, 247:1431–1438, 1990.

Tiessen, H., J.W.B. Stewart, and J.R. Betany, Cultivation effects on the amounts and concentration of carbon, nitrogen, and phosphorus in grassland soils, *Agron. J.*, 74:831–835, 1982.

van Veen, J.A. and E.A. Paul, Organic carbon dynamics in grassland soils I. Background information and computer simulation, *Can. J. Soil Sci.*, 61:185–201, 1981.

Vitorello, V.A. et al., Organic matter and natural carbon ^{13}C distribution in forested and cultived Oxisols, *Soil Sci. Soc. Am. J.*, 53:773–778, 1989.

Voroney, R.P. and E.A. Paul, Determination of k_C and k_N *in situ* for calibration of the chloroform fumigation-incubation method, *Soil Biol. Biochem.*, 16:4–14, 1984.

PART V

Soil Structure and Carbon Sequestration

CHAPTER **25**

Organic Carbon Sequestration by Restoration of Severely Degraded Areas in Iceland

Olafur Arnalds, Asa L. Aradottir, and Gretar Gudbergsson

CONTENTS

INTRODUCTION

Special Governmental Effort to Sequester Carbon in Iceland

During the past 1100 years, Icelandic ecosystems have been severely degraded as a result of environmental change and human impact on fragile environment. Fertile and organic carbon-rich systems have been destroyed, leaving vast barren deserts with very low carbon levels. By restoration of some of these severely degraded areas, sequestration of carbon can be combined with the recovery of the ecological value of the land.

Anthropogenic CO_2 emissions from Iceland were estimated at 2.147 million tons in 1990 and 2.281 million tons in 1995 (Ministry for the Environment, 1997). The government aimed to limit emissions of carbon dioxide and other greenhouse gases to 1990 levels in 2000; however, in 1990, greenhouse gas emissions had already been reduced substantially by replacing fossil fuels for house heating with geothermal energy. Reducing the emissions still further will be difficult without considerable impact on Iceland's economy (Ministry for the Environment, 1997). The Icelandic government has therefore set a goal to increase 1990 carbon sequestration levels in biomass by 100,000 tons CO_2. To reach this goal, designated additional funds were made available for reclamation and reforestation projects over the period from 1997 to 2000.

In 1998, a research effort was initiated to document carbon sequestration in biomass and soils associated with this special governmental sequestration effort. The results presented here are preliminary, based on data from the first year of a research project expected to last several years.

The Importance of Ecosystem Approach for Carbon Balance

Soils are carbon reservoirs exceeded only in importance by fossil fuels and the oceans. Soil carbon (C) plays an important role in cycling of nutrients in ecosystems and is a major factor determining the fertility of ecosystems. Several countries have constructed detailed databases for soil organic C and carbon stored in forests and other vegetation, which can be used for management of the soil C to reduce CO_2 levels in the atmosphere. In many countries, the carbon pool has been depleted in vast areas because of environmental change and nonsustainable management. Many severely degraded areas are important future reservoirs for carbon to help reduce increased levels of CO_2 in the atmosphere. In the case of severely degraded lands, restoring ecosystem fertility and reducing harmful effects of greenhouse gasses can be combined for double benefit.

Iceland has vast areas of severely degraded land and deserts (40,000 to 50,000 km²; Arnalds et al., 1997; Arnalds, 1998; see also www.rala.is/desert). An example of Icelandic desert is presented in Figure 25.1. During efforts to restore the ecological value of degraded areas, considerable amounts of carbon are sequestered. This is enhanced by the nature of Icelandic volcanic soils, which have a natural tendency to immobilize and bury organic C.

CARBON STORAGE OF ICELANDIC SOILS

Extensive databases have been constructed for organic carbon in several countries such as the U.S. (Lal et al., 1998), Canada (Lacelle, 1997; Tarnocai, 1997), Indonesia (Reich et al., 1997), and Albania (Zdruli et al., 1997). Soil maps, soil classification, and geographical information systems (GIS) are used to combine the geographical data with laboratory analysis of the soils.

Organic carbon content varies considerably with soil type and environmental conditions as exemplified by a review published by Eswaran et al. (1993) for world soils (Table 25.1). This table shows considerable difference in average carbon storage of the soil orders. Organic soils (Histosols) are the greatest sinks for soil carbon, but these areas include the vast circumpolar areas of the arctic regions of Alaska, Canada, and Russia.

Figure 25.1   An Icelandic desert. The surface is dark and with very sparse vegetation. The soil is sandy with high volcanic glass content.

Table 25.1   Carbon Storage of the Soil Orders

Soil Order	kg C m^{-2}	1000 km^{-2}	PgC
Histosols	205	1,745	357
Andisols	31	2,552	78
Spodosols	14	4,878	71
Mollisols	13	5,480	72
Vertisols	6	3,287	19
Ultisols	9	11,330	105
Alfisols	7	18,283	127
Oxisols	10	11,772	119
Aridisols	3	31,743	110
Inceptisols	16	21,580	352
Entisols	10	14,921	148
Other	2	7,644	18

Source: From Eswaran et al., 1993.

Andisols (U.S. Soil Taxonomy) or Andosols (FAO system) are soils that form in volcanic ejecta. Andisols store noticeably more organic carbon than other dryland soils due to their chemical properties: 31 kg m^{-2} according to the world database (Table 25.1; Eswaran et al., 1993). The colloidal properties of these soils give rise to chemical bonding between organic molecules and inorganic materials, which leads to immobilization of carbon in the soils (Shoji et al., 1993; Wada, 1985).

General Soil Types in Iceland

The Agricultural Research Institute of Iceland is currently preparing a soil map of Iceland in the scale of 1:500,000. This map will be made available on the Internet and through international databases later this year; Figure 25.2 shows a preliminary soil map based on this work.

A steady input of eolian materials and occational tephra additions causes the organic carbon content to be lower than 25% in many of the wetland soils, resulting in Andisol rather than Histosol classification. The desert soils are also often classified as Andisol because of their high volcanic glass content. The current edition of the U.S. Soil Taxonomy (Soil Survey Staff, 1998) fails to separate the major soil types of Iceland at the highest taxonomic level. This, however, demonstrates the dominating influence of andic soil properties of Icelandic soils. Applying the FAO soil classification (FAO, 1998) system involves similar problems as those of the U.S. system in that many contrasting soils are classified the same as Andosols at the highest level.

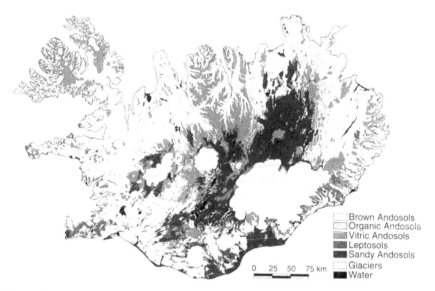

Figure 25.2   Preliminary soil map of Iceland. Based on FAO/Iceland classification system. Gleysols are not separated as a special soil type.

Table 25.2   Selected Properties of Major Soil Types in Iceland, Common Range of Values

Type[a]	Area (km²)	Depth (m)	OC (%)	pH	Clay (%)	CEC (meq 100 g⁻¹)	15 bar H₂O
Brown Andosol	43,770	0.5–1.5	2–10	5.5–6.5	15–40	10–40	30–60
Organic Andosol	8,660	0.5–5	5–30	4.5–5.5	—	30–60	60–120
Vitric Andosol	18,600	0.2–0.5	0.5–1.5	6.5–7.0	5–15	5–15	5–15
Sandy Andosol[b]	18,090	0.1–2	0.2–1.0	6.5–7.0	1–5	2–10	1–10

[a] Icelandic adoption of the FAO classification (Gudmundsson, 1994a and Agricultural Research Institute work). Gleysols are not separated and are included with Andosols.
[b] Includes also the Sandy Andosol/Leoptosol complex.

Soil categories shown in Figure 25.2 represent an Icelandic version of the FAO soil classification system (FAO, 1998), which has been adapted to Icelandic soils to generate a "FAO/Iceland" soil classification system (Gudmundsson, 1994a). The names of the soil types result from present work at the Agricultural Research Institute, but they are only suggestive at this stage and are subject to further development (see also Arnalds, 1999a). General properties of the soils are presented in Table 25.2.

Brown Andosols

Areas dominated by Brown Andosols (FAO/Iceland system) are estimated to be 43,770 km². Other soil types occur within these areas, but they cannot be separated from the Andosols in such a coarse mapping scale. This aerial extent represents, therefore, a maximum cover of Andosols in Iceland.

Arnalds et al. (1995) presented detailed soil analysis of four representative pedons for Icelandic Brown Andosols. Integrating for the profiles resulted in 30 to 50 kg m⁻² organic carbon. Other data, such as those presented by Johannesson (1960), Helgason (1968), Olafsson (1974) and Haraldsson (1996), support this range of carbon levels, but data from East Iceland show considerably lower numbers (Arnalds and Gudbergsson, unpublished data). The 30 to 50 kg m⁻² carbon levels are in good agreement with numbers reported for Andosols in the world (Eswaran et al., 1993). Assuming

Table 25.3   Soil Types (FAO/Icelandic Classification) and Carbon Stored in Icelandic Soils

Soil Type[a]	Storage (kgC m^{-2})	Area (km^2)	Total Storage Million t C	Desert Surfaces
Brown Andosol	40	43,770	1,751	
Organic Andosol	90	8,600	780	
Desert soils				
Sandy Andosol	0.5	15,090	7	Sandy deserts
Leptosols	0.3	4,540	1	Scree and lava fields
Vitric Andosol	2	18,600	37	Lag gravel/till
Total			2,546	

[a] Icelandic adoption of the FAO classification (Gudmundsson, 1994a and Agricultural Research Institute work). Gleysols are not separated and are included with Andosols.

40 kg m^{-2} on average, about 1750 million tons C could be stored in Brown Andosols in Iceland (Table 25.3). This represents a maximum figure, as the actual extent of Brown Andosols is somewhat lower than assumed in this calculation and carbon levels of some areas may be lower than 40 kg m^{-2}.

Organic Andosols

Less detailed data are available for Organic Andosols than for Brown Andosols. Due to eolian flux and tephra additions, organic levels of these soils are lower than usually found for organic soils at northern latitudes (Johannesson, 1960; Helgason, 1968; Gudmundsson, 1978; Arnalds et al., 1995). Organic Andosols are mostly classified as Andisols according to the U.S. Soil Taxonomy because the average organic content is usually lower than 25%. The eolian and organic additions cause burial of organic-rich horizons as the surface is gradually rising, sometimes resulting in soils several meters thick. The available data indicate values from 5 to 30% C, but it is difficult to assume a mean value for Iceland at this point. An average of 15% C, 2-m depth, and a bulk density of 0.3 g cm^{-3} seems reasonable first approximation. This bulk density is based on Gudmundsson (1994b); Bjarnason (1966) estimated that organic soils have an average depth of 2.5 m. The shallower depth was selected as it is likely that the new aerial estimate (Tables 25.2 and 25.3) accounts for some shallower soils not included in Bjarnason's survey.

This results in average organic content of the Icelandic Organic Andosols of 90 kg m^{-2}. This value is lower than value reported for Canadian Histosols (Tarnocai, 1997), tropical Histosols in Indonesia (Reich et al., 1997) and in the Mediterranean (Zdruli et al., 1997), but higher than for organic permafrost soils in Alaska (Eswaran et al., 1993). Combining an estimate for C per m^2 and the area extent of Organic Andosols in Iceland results in about 780 million tons of C stored.

Desert Soils

The desert soils (Vitric Andosols, Sandy Andosols, and Leptosols) dominate about 40,000 km^2 of Icelandic landscapes, but additional desert patches also occur within the Brown and Organic Andosol areas. Desert soils are characterized by low organic levels (Table 25.2), and high proportions of poorly weathered volcanic glass and basaltic crystalline materials. Allophane can be up to 15% in some of the old glacial till surfaces (Vitric Andosols in Tables 25.2 and 25.3). The desert soils were studied by Arnalds et al. (1987), Arnalds (1988, 1990) and Gudmundsson (1991). Additional data can be drawn from Johannesson's monograph on Icelandic soils (1960). The most detailed study of Icelandic desert soils was done by the Icelandic Agricultural Research Institute and the USDA-NRCS (Arnalds and Kimble, unpublished data); it includes chemical and mineralogical properties and will be published in the near future.

In spite of a large aerial extent, the desert soils contain a fraction of the carbon stored in Icelandic soils (45 million tons of a total of 2550 million tons stored). This difference indicates the potential of carbon sequestration in Icelandic soils as a result of revegetation and restoration. Restoration efforts can gradually replace organic-poor desert soils with organic-rich Organic and Brown Andosols.

METHODS

Revegetation and Restoration Methods

Revegetation of desert sites in Iceland usually involves seeding grasses and fertilizing or seeding a fast-growing legume, the Alaskan lupine (*Lupinus nootkatensis*). Most revegetation sites are protected from grazing. Areas seeded with lupine often form dense stands that expand with time; the lupine can dominate the vegetation for decades (Magnusson, 1995). Productivity in lupine stands can be very high (Magnusson, 1999).

Areas seeded with grasses are generally fertilized for 2 years and then left alone after the fertilization is discontinued. The cover of seeded grasses usually decreases with time (Gunnlaugsdottir, 1985; Thorsteinsson, 1991) and cryptogams cover increases, especially mosses and lichens. The vegetation cover prevents soil movement by wind and water and the cryptogamic cover is especially important for colonization of native plant species (Aradottir, 1991). The final goal of these efforts is restoration of the native vegetation and ecosystem potential of these sites.

Afforestation is done by planting tree seedlings at vegetated sites (Brown Andosols) and severely degraded sites or deserts. Direct seeding of trees, especially the native birch (*Betula pubescens*), has also been used on a small scale. Organic amendments or seeding of annual grasses around the seedlings in order to impede mortality due to frost heaving are often used on the desert sites (Oddsdottir et al., 1998). Common species used for afforestation are Sitka spruce (*Picea sitchensis*), larch (*Larix sibirica* and *L. sukaczewii*), the native birch, and lodgepole pine (*Pinus contorta*). For plantations on desert sites, it is important to look at total sequestration in the trees, understory vegetation, and soil. The afforestation part of this research project is presented elsewhere (Snorrason et al., 2000).

Sampling Sites and Methods

Desert Soils

Characteristic desert soil types were sampled to determine site variability. A total of seven sites, two to three depths, and up to 20 cores per site were sampled, resulting in a total of 186 samples that were analyzed for carbon. Subsequent sampling was based on results from this sampling. At other sites, five cores were combined for each sample at each site, and three such samples were gathered to represent the site (3 × 5 cores, resulting in three carbon values obtained and averaged to reach one average carbon value for each depth interval at each site).

Samples were taken from 40 sampling sites in 31 areas with restoration activities (see Figure 25.3). For comparison, an attempt was made to sample desert sites where no revegetation had been undertaken that were adjacent to restoration sites. The 40 sampling sites were divided into 24 revegetation and restoration sites, six of which had been protected from sheep grazing but without seeding and fertilizing, and ten control sites where no activity had been undertaken.

The revegetation and restoration sites sampled included areas involved in the special governmental carbon sequestration program, but emphasis was also made to sample a variety of land conditions. Land type and general characteristics of each site were noted during the sampling. Information on the age of revegetation efforts and the methods used were gathered from the Icelandic Soil Conservation Service (SCS), farmers, and local SCS representatives.

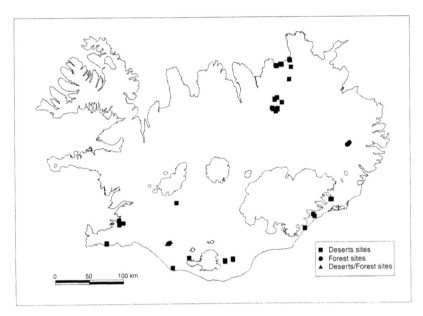

Figure 25.3 Sampling sites for the carbon sequestration project.

Effort was made to estimate the volume of coarse fragments at all sampling sites. Bulk density samples were taken at each sampling depth at each site by driving a 300-cm³ tin can into the profile; these samples were dried and weighted. This method works well for undisturbed Brown and Organic Andosols, but did not work well enough for the desert sites due to coarse fragments. Other methods will be tested in the future.

The amount of carbon in the soils was found by adding up the carbon for the two to three depth increments sampled, using 0.9 g cm⁻³ for bulk density. This density is based on measurements of fully developed Brown Andosols (often 0.75 g cm⁻³) by Arnalds and Kimble (unpublished data).

All samples were sieved through a 2-mm sieve and both fractions weighted. The volume of the coarse fragments was calculated using average density of 2.85 g cm⁻³ for the basaltic rock and volcanic glass fragments. This volume was subtracted from the total volume available for carbon sequestration.

Above-Ground Biomass of Vegetated Sites

In this first year of the research program, there were only preliminary measurements of carbon sequestration of the biomass of revegetated areas. Two representative desert sites were sampled. At site 1, plots were laid out in locations fertilized and seeded with grasses 45 and 17 years ago, and at an adjacent control. At site 2, plots were laid out in an area colonized by lupine about 20 years ago and at an adjacent control. Both sites were protected from grazing and the controls had very low vegetation cover. Two replicate plots (10 × 10 m) were established at each revegetated location and one plot at each of the controls. Five 0.25 m² subplots were randomly laid out within each larger plot. The sample from each subplot was sorted into different components (grasses, herbs, dwarf shrubs, mosses, fine litter, and coarse litter), and the components dried, weighted, and analyzed for carbon. Soil cores were taken within each replication for determination of fine roots and soil carbon. The roots were washed carefully, then dried and their carbon content measured. Soils were treated in the same manner as those in the desert sites. This sampling process will be described in more detail during later stages of the project.

Analysis

Soil samples were oven-dried and sieved through a 2-mm sieve. Both fractions (larger and finer than 2 mm) were weighted to be able to exclude the coarse fraction from the carbon budget for each site. All carbon analyses were done using a Leco CR-12 instrument. Each sample was measured in duplicate; if the difference between two measurements was >10%, additional determinations were made.

RESULTS

Sequestration in Desert Soils

The age of the revegetation areas and current carbon levels were used to calculate carbon sequestration in the soils. The revegetation efforts at these sites are from 1-year-up to about 50-years-old at the oldest site. The overall results are presented in Table 25.4. The values ranged from 0.03 kg C m^{-2} in newly formed alluvial sand to 7.1 in old desertified Brown Andosols. The carbon levels in desert soils showed more variability between areas than previously expected (Figure 25.4). The sources of variability are many, including the type of land or soils revegetated, the type of revegetation activity (plant species, fertilizers, grazing protection, machinery, etc.), and external factors such as climate (temperature and rainfall) and elevation.

There are not enough data from the first year of study to model the influence of each of these factors. However, separating one land type in the data (sandy lag gravel, Sandy Andosol) gives relatively straight line with a slope number close to average calculated sequestration for all land types (Figure 25.5).

Table 25.4 Overall Results for Carbon Sequestration in Desert Soils by Revegetation Efforts

Activity	n[a]	kg Cm^{-2}	tC ha^{-1}	tC ha^{-1}yr^{-1}	tCO$_2$ ha^{-1}yr^{-1}
Revegetation	24	1.81	18.1	0.9	3.2
Protection from grazing	6	0.82	8.2	0.2	0.8
No activity (control)	10	0.36	3.6		

[a] Each number is a result of three samples, each consisting of five sampled cores.

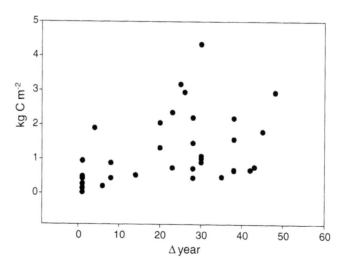

Figure 25.4 Carbon levels as a function of duration of revegetation efforts, all sites.

The comparison of 15 data pairs (revegetation vs. control) revealed yearly accumulation of 0.6 t C yr^{-1} (2.2 t CO$_2$ yr^{-1}). This is a considerably lower number than the average 0.9 tC ha^{-1} yr^{-1} calculated for all sites (Table 25.4). However, some of these controls were not true controls, resulting in too high C values. The data indicate lower sequestration rates in North Iceland than in South Iceland, which may be explained by lower annual temperature and precipitation rates. More research is needed to confirm this tendency. It is important to emphasize that carbon accumulation in vegetation (trees and ground cover) is not included in these numbers. Thus, total sequestration is considerably higher, as discussed in the next section.

The within-site variability is presented in Table 25.5, which indicates considerable variability and similar levels of SD for %C values and total carbon accumulated for both depths. However, this variability can be minimized by combining several cores at each site. The determination of the volume and weight of coarse fragments by sieving seems to be a relatively good method for accounting for this factor when the fragments are relatively small. If larger boulders occur within the soils, additional subtraction is needed to account for decreased volume of active soil materials. Better methods are needed for these conditions; this will be studied more closely during the next phase of this research project.

Bulk density is another problem that needs to be addressed more carefully because the results are sensitive to it. It is unlikely that bulk density of the soil materials is lower than 0.9, but higher values are possible, which would result in higher sequestration rates than presented here.

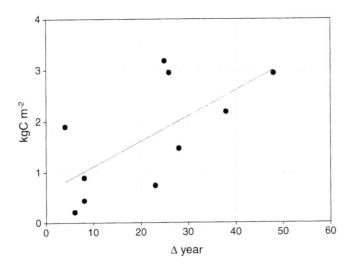

Figure 25.5 Carbon levels as a function of duration of revegetation efforts, one land type.

Table 25.5 Carbon Variability within Sites

	Depth (cm)	N	%C	SD	C^a (kg m^{-2})	SD
Bolholt	0–10	20	2.54	0.86	7.10	1.31
Bolholt	10–30	20	2.68	0.52		
Langibakki	0–10	20	1.84	0.50	2.99	1.29
Langibakki	10–30	18	1.29	0.76		
Selsund	0–10	20	0.17	0.05	0.49	0.12
Selsund	10–30	20	0.24	0.11		
Svínhagi	0–5	12	0.15	0.03	0.15	0.03
Svínhagi	5–10	12	0.24	0.06		
Vesturhraun	0–10	20	1.24	0.89	2.36	1.61
Vesturhraun	10–30	20	0.86	0.71		

^{a} Total for both depths. Coarse fragments subtracted, BD of 0.9 g/cm^3 estimated.

An alternative method to be investigated is to assess the weight of C per unit area. This involves sampling all soil under a given area, e.g., 0.25 m², down to bedrock, sieved *in situ* to exclude fragments > 2 mm, and C determined in the <2 mm fraction. This method bypasses the need for bulk density measurements and accurate determination of % coarse fragments. The desert soils are noncohesive and shallow, making this method potentially easy to employ and worth exploring.

Sequestration in Biomass Other Than Trees

The carbon content of different vegetation and litter components (Figure 25.6) was used to calculate the total C in biomass for each treatment. The lowest carbon concentrations were found in fine litter, which was in some cases difficult to separate from the uppermost soil layer and in mosses on revegetated sites, which were probably affected by eolian depositions.

The results for carbon sequestration in standing biomass (above and below ground) and litter at revegetated sites are shown in Table 25.6. The data shown do not include the soil component. Caution should be taken in interpretation of these data because of small sample size; however, results suggest that the biomass and litter component may have a significant role in carbon sequestration at revegetated sites. More emphasis will be placed on this factor during the next phase of the research.

The potential of carbon sequestration by biomass and litter was not taken into account in planning the governmental carbon sequestration effort. Combining the carbon accumulation in soils and the accumulation in biomass (above and below ground; no trees), gives considerably higher sequestration rates than what was assumed earlier (3.3 to 5.5 t CO_2 ha^{-1} yr^{-1} compared to 2.6 t CO_2 ha^{-1}yr^{-1}).

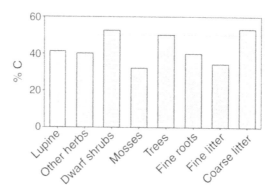

Figure 25.6 Percent carbon content in different components of vegetation and litter.

Table 25.6 Results for Carbon Sequestration in Biomass at Revegetated Sites

Area	Biomass (t C ha^{-1})	Age (yr)	Sequestration t C ha^{-1}yr^{-1}	Sequestration CO_2ha^{-1}yr^{-1}
Site 1, revegetated desert (seed + fertilizer)	13	45	0.3	1.0
Site 1, revegetated desert (seed + fertilizer)	11	17	0.6	2.1
Site 1, control (nearly barren desert)	1.5			
Site 2, lupine stand on desert	10	20	0.5	1.8
Site 2, control, desert	0.2			

The boundary between the soil and vegetation is not well defined, so it is important to sample both these elements at the same time. This problem and methods to prevent double sampling of some materials or omission of some portion of the biomass will be studied in the second year of the project.

Accounting and GIS Databases

One of the most important issues related to the recognition of additional sequestration activities is the reliability of documentation of carbon sequestration in soils. All of the designated carbon sequestration sites in Iceland are being mapped in detail and the data entered into a GIS system. By combining the geographical data with measurements of the carbon sequestration rates, overall carbon sequestration can be assessed.

Table 25.7 shows results for all the designated revegetation activity areas to give an example of the background data for the government project. It shows that after 3 years of activity about 14,000 t CO_2 are sequestered per year by reclamation of deserts as a direct result of governmental effort to reduce greenhouse gas levels. Similar methods have been employed to estimate sequestration at afforestation sites.

Table 25.7   Overview of Sites of the Special Governmental Sequestration Effort by Revegetation and Estimated Total C Sequestration

Site	Size (ha)	Soil (tC/ha)	Vegetation (t C ha⁻¹)	Total (t C ha⁻¹)	Seq. Soil (t C)	Seq. Veg. (t C)	Seq. Total (t C)	Seq. Total (t CO₂)
Höfði	57	0.7	0.4	1.1	40	22.8	63	230
Skagafjörður	47	0.6	0.4	1	28	18.8	47	172
Jarlstaðir	44	0.6	0.4	1	26	17.6	44	161
Grænavatn	111	0.6	0.4	1	67	44.4	111	407
Villingafell	133	0.6	0.4	1	80	53.2	133	488
Ássandur	130	0.5	0.4	0.9	65	52.0	117	429
Ærlækjarsel	130	0.5	0.4	0.9	65	52.0	117	429
Hólsfjöl	734	0.5	0.4	0.9	367	293.6	661	2422
Héraðssandur	78	0.5	0.4	0.9	39	31.2	70	257
Eldhraun	47	0.85	0.4	1.25	40	18.8	59	215
Atley	183	0.85	0.4	1.25	156	73.2	229	839
Geitasandur	105	0.85	0.4	1.25	89	42.0	131	481
Keldnahraun	171	0.85	0.4	1.25	145	68.4	214	784
Stóri-Klofi	139	0.85	0.4	1.25	118	55.6	174	637
Árskógar	397	0.85	0.4	1.25	337	158.8	496	1820
Landskógar	215	0.85	0.4	1.25	183	86.0	269	985
Hafið	64	0.85	0.4	1.25	54	25.6	80	293
Tunguheiði	358	0.7	0.4	1.1	251	143.2	394	1444
þorlákshöfn	159	0.7	0.4	1.1	111	63.6	175	641
Other	185	0.8	0.4	1.2	148	74.0	222	814
Total	3487				2410	1395	3804	13950

Source: SCS planning department, 1999.

DISCUSSION

Overall Rates

The special governmental carbon sequestration effort assumed that 2.6 t CO_2 ha⁻¹yr⁻¹ could be sequestered by revegetation of deserts (depending on type of revegetation activity); these estimates ignored possible additions in above-ground biomass. The results presented in Table 25.4 and the previously assumed sequestration rates are of the same order but present results are slightly higher. It should be emphasized that these are preliminary results, but they strongly support the potential for carbon sequestration in these soils suggested earlier (e.g., Jonsson and Oskarsson, 1996).

The rates of carbon sequestration in soils and above-ground biomass are summarized in Table 25.8. The time frame for carbon additions varies among the different ecosystem components, thus calling for a multicompartment model for carbon sequestration as suggested by Izaurralde et al. (1998) or Liski et al. (1998). It is difficult at this point to hypothesize such a conceptual model for the study sites, but several known points should be noted.

Table 25.8   Overall Rates of Carbon Sequestration in Soils and Above-Ground Biomass

Type of Activity	Soil[a]	t C ha^{-1}yr^{-1}		Total
		Vegetation	Trees	
Revegetation of desert sites	0.8	0.4	0	1.2
Afforestation of desert sites	0.8	0.4	2.0[b]	3.2
Afforestation of vegetated sites	0	0	2.0	2.0
Restoration of birch on deserts	0.8	0.4	1.1	2.2

Note: Values are reported in t ha^{-1}yr^{-1}. Preliminary results. Forested sites after Arnorsson et al., 1999.

[a] Values conservative.
[b] Average for 80-year period, maximum sequestration is higher.

Most of the desert soils will evolve to classical Brown or Organic Andosols with time if vegetation succession continues. Icelandic Brown Andosols commonly contain 30 to 50 kgC m^{-2} after about 10,000 years' buildup. It is likely that, in 100 years, total C accumulated could be of the order of 3 to 8 kgC m^{-2}, based on data now available.

Undisturbed Brown and Organic Andosols continue to sequester carbon for an infinite time due to continuous eolian and tephra influx (Arnalds, 1990, 1999b). The surface is gradually rising, steadily burying soil and carbon at lower depths. Common accumulation rate is 0.5 mm yr^{-1}, which would result in a sequestration rate of 0.1 to 0.2 tC ha^{-1} yr^{-1} (assuming 4% carbon content of the buried soil). This may explain relatively high carbon sequestration rates (0.2 tC ha^{-1} yr^{-1}) for areas with no activity other than protection from grazing.

Carbon accumulation in vegetation other than trees will level off with time. Accumulation time varies greatly between sites, depending on site conditions and successional processes.

Sequestration rates of 0.6 to 0.9 tC ha^{-1} yr^{-1} in soils and 0.4 tC ha^{-1} yr^{-1} in vegetation other than trees for 0 to 50 years were measured. These values might be representative for the first decades until the biomass reaches a steady state, but accumulation in soils is expected to continue much longer (>100 years). Work is currently underway to establish better relationships between carbon sequestration and environmental and management factors associated with restoration of Icelandic deserts.

The sequestration rates measured are high compared to values reported in the literature. As an example, Izaurralde et al. (1998) reported maximum theoretical rates for Canadian prairie soils of about 0.15 t ha^{-1} yr^{-1}, almost one order of magnitude lower than reported for Icelandic soils. In Iceland, carbon levels of the deserts before revegetation activities are started are very low, but the undisturbed Brown and Organic Andosols in Iceland contain large quantities of carbon due to andic soil properties and cold climate; hence there is room for a large increase.

Added Benefit and a Link to CCD

Carbon sequestration by restoration of severely degraded lands can have a significant impact on carbon levels of the atmosphere. In addition, restoration of ecosystems in degraded areas is important for reversing desertification worldwide.

In the Kyoto Protocol, credits for sink enhancement are limited to afforestation, reforestation, and deforestation, thus limiting countries' options to exploit their potential for carbon sequestration. It is particularly unfortunate that no nonforest categories are included. It can be argued that the prevalent focus on carbon sequestration by reforestation and afforestation is biased towards areas where environmental conditions are favorable for tree growth. If sink activities to reduce carbon levels are limited to sequestration by tree growth only, areas outside the humid regions of the world do not have the same opportunities to compensate for industrial emissions.

The Protocol sets up a process for activities to be added to the limited list. The modalities, rules, and guidelines for adding activities should be based on the principle that activities to sequester carbon can be documented in a transparent manner and verified. It is also important that additional activities can be used to foster research for improved understanding of the carbon budget on national and regional scales, with better understanding of the global carbon budget as an overall goal.

Iceland provides an example of a country with severely degraded landscapes where large quantities of carbon can be sequestered in soils and vegetation by restoration efforts. Actions can be monitored closely and success recorded. Severe degradation of land ecosystems is the subject of the recent UN Convention to Combat Desertification (CCD). It is vital that the FCCC and the CCD conventions are linked together. The FCCC could possibly provide financial incentive for restoring the value of many of the world's degraded lands and reducing the harmful effects of greenhouse gases at the same time.

ACKNOWLEDGMENTS

Several people made important contributions to the research reported in this chapter. We especially thank Arnor Snorrason of the Forest Research Station, Kristin Svavarsdottir and Anna Maria Agustsdottir of the Soil Conservation Service, Thorbergur Hjalti Jonsson of the Institute of Natural History, and Jon Gudmundsson of the Agricultural Research Institute for their contributions and cooperation.

REFERENCES

Aradottir, A.L., Population biology and stand development of Birch (*Betula pubescens* Ehrh.) on disturbed sites in Iceland, Ph.D. dissertation, Texas A&M University, College Station, 1991.

Arnalds, O., Soils of denuded areas in Iceland. (in Icelandic, English summary). *Náttúrufræðingurinn*, 58:101–116, 1988.

Arnalds, O., Characterization and erosion of Andisols in Iceland, Ph.D. dissertation, Texas A&M University, College Station, 1990.

Arnalds, O., Desertification in Iceland, *Desertification Control Bull.*, 32:22–24, 1998.

Arnalds, O., Soil survey and databases in Iceland, in Bullock, P. et al. (ed.), *Soil Resources in Europe*, European Soil Bureau, Research Report No. 6, Ispra, Italy, 1999a, 91–96.

Arnalds, O., The Icelandic "rofabard" erosion features, *Earth Surface Process Landforms*, 25:17–28, 2000.

Arnalds, O., A.L. Aradottir, and I. Thorsteinsson, The nature and restoration of denuded areas in Iceland, *Arct. Alp. Res.*, 19:518–525, 1987.

Arnalds, O., C.T. Hallmark, and L.P. Wilding, Andisols from four different regions of Iceland, *Soil Sci. Soc. Am. J.*, 59:161–169, 1995.

Arnalds, O. et al., Soil erosion in Iceland, (in Icelandic; available in English in late 1999). Soil Conservation Service and Agricultural Research Institute, Reykjavik, 1997.

Bjarnason, O.B., *Icelandic Peat*, (in Icelandic), University Research Institute, University of Iceland, Reykjavik, 1966.

Eswaran, H., E. Van den Berg, and P. Reich, Organic carbon in soils of the world, *Soil Sci. Soc. Am. J.*, 57:192–194, 1993.

FAO, World reference base for soil resources, FAO, ISRIC, ISSS, Rome, 1998.

Gudmundsson, T., Soil research on reclamation sites in central Iceland, (in Icelandic), in Thorsteinsson, I. (ed.), *Reclamation on the Audkuluheidi and Eyvindarstadaheidi Commons*, Agricultural Research Institute, Reykjavik, 1991, 51–70.

Gudmundsson, T., Pedological studies of Icelandic peat soils, Ph.D. thesis, University of Aberdeen, 1978.

Gudmundsson, T., The FAO soil classification in relation to Icelandic soils, (in Icelandic), *RALA Report* 167, Agricultural Research Institute, Reykjavik, Iceland, 1994a.

Gudmundsson, T., Soil science, The Agricultural Society of Iceland, Reykjavik, 1994b.

Gunnlaugsdottir, E., Composition and dynamical status of heathland communities in Iceland in relation to recovery measures, *Acta Phytogeogr. Suec.*, 75, 1985.

Haraldsson, H.V., Carbon in Icelandic soils, (in Icelandic), B.Sc. thesis, University of Iceland, Reykjavik, 1996.

Helgason, B., Basaltic soils of southwest Iceland, II. *J. Soil Sci.*, 19:127–134, 1968.

Izaurralde, R.C. et al., Scientific challenges in developing a plan to predict and verify carbon storage in Canadian prairie soils, in Lal, R. et al. (eds.), *Management of Carbon Sequestration in Soil*, CRC Press, New York, 1998, 433–446.

Johannesson, B., The soils of Iceland, *Dept. of Agric. Rep. Series B*, 13, University Research Institute, Reykjavik, 1960.

Jonsson, T.H. and U. Oskarsson, Forestry and reclamation for carbon sequestration, (in Icelandic; English summary), *Forestry Yearbook*, 1996:65–87, Icelandic Forest Association, Reykjavík, 1996.

Lacelle, B., Canada's soil organic carbon database, in Lal, R. et al. (eds.), *Soil Processes and the Carbon Cycle*, CRC Press, Boca Raton, FL, 1997, 93–101.

Lal, R. et al., *The Potential of U.S. Cropland to Sequester Carbon and Mitigate the Greenhouse Effect*, Sleeping Bear Press, Ann Arbor Press, Chelsea, MI, 1998.

Liski, J. et al., Model analysis of the effects of soil age, fires and harvesting on the carbon storage of boreal forest soils, *Eur. J. Soil Sci.*, 49:407–416, 1998.

Magnusson, B., Biological studies of Nootka lupine (*Lupinus nootkatensis*) in Iceland. Growth, effect of cutting, see set and chemical composition, (in Icelandic; extended English summary), *RALA Report* 178, Agricultural Research Institute, Reykjavik, 1995.

Magnusson, B., Biology and utilization of Nootka lupin (*Lupinus nootkatensis*) in Iceland, *Proc. 8th Int. Lupin Conf.*, Asilomar, CA, 11–16 May, 1996. G.D. Hill, Cantebury, NZ, 42–48, 1358, 1996.

Ministry for the Environment, Status report for Iceland. Pursuant to the United Nations framework convention on climate change, Ministry for the Environment, Reykjavik, 1997.

Oddsdottir, E.S. et al., Methods to prevent frost-heaving of plants, (in Icelandic; English summary), *Forestry Yearbook* 1998:72–83. Icelandic Forest Association, Reykjavík, 1998.

Olafsson, S., Physical and physio-chemical studies of Icelandic soil types, Licentiat thesis, Agricultural University, Copenhagen, 1974.

Reich, P., M. Soekardi, and H. Eswaran, Carbon stocks in soils of Indonesia, in Lal, R., J. Kimble, and R. Follet (eds.), *Soil Properties and Their Management for Carbon Sequestration*, USDA-NRCS, Lincoln, NE, 1997, 121–127.

Snorrason, A. et al., Biomass and carbon sink in forests and open lands in Iceland — a preliminary study, (in Icelandic), Forest Yearbook, 2000, 71–89, Forestry Research Station, Mogilsa, Iceland.

Shoji, S., M. Nanzyo, and R.A. Dahlgren, *Volcanic Ash Soils, Genesis, Properties and Utilization*, Elsevier Science Publishers, Amsterdam, 1993.

Soil Survey Staff, *Keys to Soil Taxonomy*, 8th ed., USDA-NRCS, Washington, D.C., 1998.

Tarnocai, C., The amount of organic carbon in various soil orders and ecological provinces in Canada, in Lal, R. et al. (eds.), *Soil Processes and the Carbon Cycle*, CRC Press, Boca Raton, FL, Advances in Soil Science, 1997, 81–92.

Thorsteinsson, I. (ed.), Land reclamation at Audkuluheidi and Eyvindarstadaheidi commons 1981–1989, *RALA Report* 151, Agricultural Research Institute, Reykjavik, in Icelandic, 1991.

Wada, K., The distinctive properties of Andosols, *Adv. Soil Sci.*, 2:173–229, 1985.

Zdruli, P. et al., Organic carbon stocks in soils of Albania, in Lal, R., J. Kimble, and R. Follet (eds.), *Soil Properties and Their Management for Carbon Sequestration*, USDA-NRCS, Lincoln, NE, 1997, 129–141.

CHAPTER 26

Influence of Nitrogen Fertility Management on Profile Soil Carbon Storage under Long-Term Bromegrass Production

L. J. Cihacek and D. W. Meyer

CONTENTS

INTRODUCTION

Long-term management of crops and cropping systems can influence profile carbon (C) storage in soils. Cihacek and Ulmer (1997) reported that profile carbon storage was increased under long-term crop-fallow management as compared with adjacent continuous grasslands. Paustian et al. (1997) have reviewed the controls that management places on soil C sequestration and have summarized the impacts that nitrogen fertilization has on increasing C sequestration by soils. However, most studies have examined C changes in the surface 15- to 30-cm depth so little is known about storage of sequestered C deeper in the soil profile.

Several long-term historical plots have been maintained by the North Dakota State University Agricultural Experiment Station under monoculture management near Fargo, ND. These plots were studied in order to (1) examine the effects of long-term nitrogen treatments on profile C storage under long-term bromegrass (*Bromis inermis* Leyss.) culture, and (2) compare profile C storage under bromegrass culture with C stored in the profile of soils under long-term continuous spring wheat (*Triticum aestivum* L.) and contiunuous flax (*Linum usitatissimum* L.) culture.

MATERIALS AND METHODS

Soils: The soils were a Fargo silty clay (fine, smectitic, frigid Typic Epiaquerts) with a nearly level slope. These soils are formed in very deep, poorly drained and slowly permeable, calcareous, clayey lacustrine sediments deposited in ancient glacial Lake Agassiz.

Grass Site: The bromegrass site was seeded prior to 1929. Present fertility treatments were initiated in 1954 and the first forage crop harvested in 1955. Nitrogen treatments of 0, 37, 74, 149, 224, and 298 kg N/ha were applied each fall to the 4.56 by 3-m plots. A single application of 45 kg/ha P_2O_5 was applied in 1971. Each treatment is replicated two times. Forage growth has been harvested annually for yield. Previous research on grass productivity at this site has been reported by Carter (1961), Meyer et al. (1977), and Meyer and Norby (1997). Changes in soil nutrient status have been reported by Larson et al. (1971) and Meyer and Dahnke (1981).

Wheat Sites: The wheat site has been continuously seeded to wheat since 1882. The plot was formally established in 1892 to study the effect of root diseases on the crop. Fertilizer history of this site is unknown, but very little fertilizer is believed to have been applied over the life of the plot. After grain harvest, residue has been returned to the soil with conventional moldboard plow (black) tillage. This site has been registered in the National Registry of Historical Sites since 1992.

Flax Site: This site has been continuously seeded to flax since 1894; it was initially established to study the effect of bacterial wilt on the crop. Fertilizer history on this site is unknown, but very little fertilizer is believed to have been applied over the life of the plot. After seed harvest, all residue has been returned to the soil with conventional moldboard plow (black) tillage. This plot also has been registered in the National Registry of Historical Sites since 1992.

Sampling: Three soil cores were collected from each treated grass plot using a hydraulic soil coring machine with a 6-cm-diameter steel tube 120 cm long. The steel tube is designed to hold an acrylic contamination liner in place during sampling. Once the core sample was taken, the acrylic liner with soil core was removed and capped with plastic caps after the acrylic tube was trimmed to the length of the core and sealed with duct tape. The encased cores were placed in cold storage until they were processed.

The samples from the grass site were removed from the acrylic tubes and the cores were divided into 15-cm segments. Segments from one core were weighed for bulk density determination and subsampled for moisture. Then they were composited with segments of the other two cores from each treated plot for each treatment. The soils were hand crushed, air dried, and ground to pass through a 2-mm screen for analysis.

Because of the larger area of the wheat or flax plot, six individual cores were collected from each plot to a depth of 1 m using the same procedure as for the grass plots. The cores were then divided into 15-cm segments. The individual segments from each core were weighed for bulk density determination, subsampled for moisture, hand crushed, and air dried. Dry soils were then crushed to pass through a 2-mm screen for analysis.

Fifteen-gram subsamples were taken from all soil samples and ball milled to pass through a 150 μm screen for C analysis. Carbon analysis was done with a solid C analyzer that was capable of determining total C (TC) and inorganic C (IC). Total C was determined by ignition at 1000°C and IC was determined by CO_2 evolution after addition of 20% phosphoric acid at room temperature. Organic C (OC) was determined by subtracting IC from TC. Carbon mass for each 15-cm-depth increment was divided by three and C mass was plotted as kg C/m^2/5-cm depth.

DISCUSSION

Nitrogen Effects of Profile C Distribution

A comparison of each N rate treatment with the unfertilized control is shown in Figure 26.1. At the 37 kg N/ha rate (Figure 26.1a), profile C distribution nearly parallels that of the O kg N/ha rate. Where 74 kg N/ha has been applied (Figure 26.1b), profile C parallels that of the O kg N/ha rate to a depth of 50 cm, where increased C is found under the fertilized treatment with a notable C bulge at a depth between 60 and 95 cm. Long-term application of 149 kg N/ha (Figure 26.1c), shows increased profile C under the fertilized treatment between 35 to 45 cm and a smaller increase between 65 and 75 cm. With applications of 224 kg N/ha (Figure 26.1d), C in the upper 30 cm of the profile is somewhat less than for the unfertilized treatment, but a small increase is noted between 35 and 45 cm and a larger increase between 75 and 95 cm in depth. Figure 26.1e (298 kg N/ha) shows a large increase in stored profile C between 45 and 95 cm in depth.

Although increase in subsurface C is not consistent in magnitude between treatments, 39 years of N fertilizer application appears to increase the C stored deep in the soil profile. The variability in magnitude is to be expected in these soils (vertisols). They are subject to deep cracking (> 1 m) during periods of low precipitation, which allows C-rich surface materials to slough into the cracks and become entrapped at some depth below the soil surface. This enhances storage of C in the profile as well as downward movement of soluble C compounds.

Another feature of these soils is their high capacity for shrinking and swelling, freezing and thawing, and wetting and drying — processes that result in formation of sand-grain-sized aggregates over the winter months. During dry spring periods, these soils act as sandy soils and are subject to wind erosion, thereby redistributing surface C (Cihacek et al., 1993). An examination of the soil profiles in these plots showed that up to 30 cm of soil or more may have been trapped by the grass during wind erosion events over the period of time that these plots have been in existence. This may explain the apparent uniformity of C distribution between treatments in the surface 30 cm of the profiles.

Soil Organic C (kg/m^2/depth)

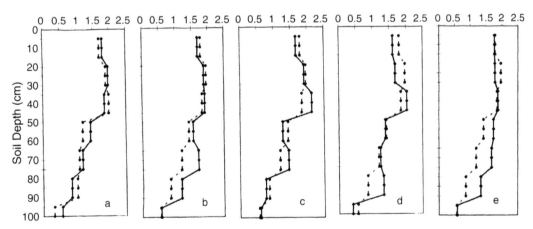

Figure 26.1 Effects of N fertilization on soil profile organic C storage under unfertilized (dashed line) and fertilized (solid line) bromegrass for rates of (a) 37 kg N/ha, (b) 74 kg N/ha, (c) 149 kg N/ha, (d) 224 kg N/ha, and (e) 298 kg N/ha.

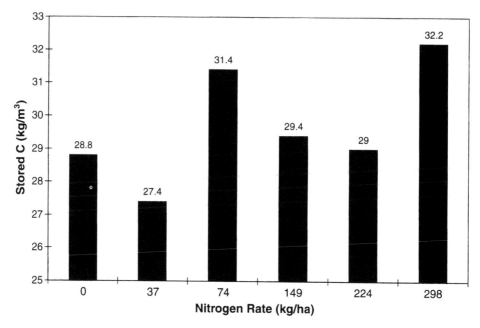

Figure 26.2 Total organic C storage as influenced by long-term N applications in long-term bromegrass plots. Stored C is based on a 1-m profile depth.

Long-Term Grass Plot Soil Carbon Storage

Profile C storage for the differing N treatments is shown in Figure 26.2. Carbon storage for the 0-, 37-, 149-, and 224-kg N/ha rates are similar and not significantly different from each other. The 74- and 298-kg N/ha rates are significantly higher than the other rates with the 298-kg N/ha rate showing the highest soil profile C storage. The high C storage rate for the 74-kg N/ha treatment may be the result of sloughing of surface material into cracks extending below the surface of the soil as previously discussed and may be unrelated to the N rate.

Comparison of C Storage with Wheat and Flax Cultures

Profile carbon distribution for the long-term flax (104 years), wheat (116 years), and unfertilized bromegrass (70 years) plots is shown in Figure 26.3. Bromegrass showed the highest amount of C distributed throughout the soil profile; flax showed the lowest. Wheat was intermediate to the flax and bromegrass.

The actual quantities of C stored in the soil profile are shown in Figure 26.4. Stored C reflects the quantity of biomass produced and retained on and in the soil under each culture. Bromegrass produced the highest quantity above and below the soil surface, while the wheat and flax produced progressively less quantities.

SUMMARY

Although Vertisols are difficult to study due to their natural inversion upon wetting and drying, addition of nitrogen fertilizer to long-term grass plots appears to increase the C stored in a 1-m soil profile, especially at high rates of N.

Deep stored sequestered C (>30 cm deep) is much less likely to be affected by surface perturbation and rereleased into the atmosphere as CO_2. Further research using isotope techniques

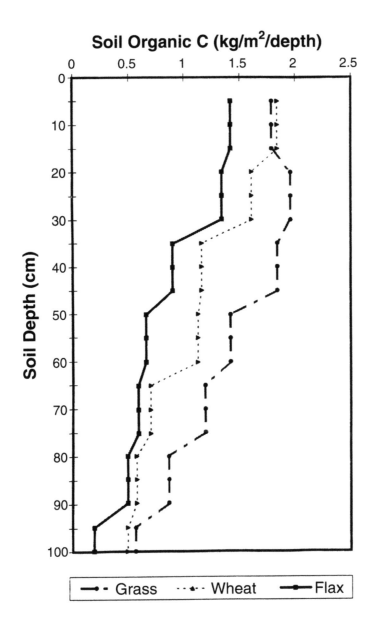

Figure 26.3 Profile organic C storage under long-term cultures of continuous bromegrass, spring wheat, and flax.

needs to examine the longevity of the sequestered C in the lower profile. Heavy clay soils such as these tend to remain saturated below 30 cm throughout most of the year with the exception of dry periods when they can dry out and crack. A saturated environment will prohibit the sequestered C from oxidizing and evolving as CO_2. Deep storage of sequestered C in soils may be an important mechanism in offsetting global CO_2 emissions and reducing the potential effects of global warming.

Although specific cases of deep-stored sequestered C have been found (Cihacek and Ulmer, 1997; Chapter 6 in this book), this phenomenon may occur under a wide range of conditions. However, further research is necessary to identify the conditions promoting deep storage of C and to identify soil management practices that may enhance this phenomenon.

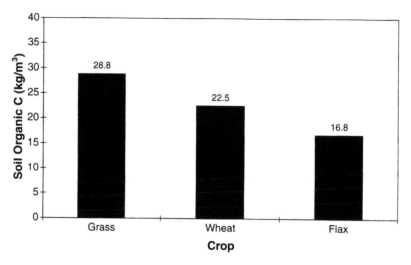

Figure 26.4   Total organic C storage under long-term cultures of bromegrass, spring wheat, or flax in a 1-m profile depth.

ACKNOWLEDGMENTS

We thank Drs. Burton Johnson and James Hammond for allowing us to sample their historical plot areas and Lynn Foss for assistance in sample collection. We also thank Keith Jacobson, Elizabeth Hafner, and Brian Cihacek for their assistance in the processing and analysis of the soils.

REFERENCES

Carter, J.F., Nitrogen fertilizer increases yields of pure grass pastures and meadows, *ND Farm Res.*, 21(2):4–8, 1961.

Cihacek, L.J. and M.G. Ulmer, Effects of tillage on profile soil carbon distribution in the northern Great Plains of the U.S., in Lal, R. et al. (eds.), *Soil Management and Greenhouse Effect*, CRC Press, Boca Raton, FL, 1997, 1995, 85–92.

Cihacek, L.J., M.D. Sweeney, and E.J. Deibert, Characterization of wind erosion sediments in the Red River Valley of North Dakota, *J. Environ. Qual.*, 22:305–310, 1993.

Larson, K.L., J.F. Carter, and E.H. Vasey, Nitrate–nitrogen accumulation under bromegrass sod fertilized annually at six levels of nitrogen for 15 years, *Agron. J.*, 63:527–528, 1971.

Meyer, D.W., J.F. Carter, and F.R. Vigil, Bromegrass fertilization at six nitrogen rates: long and short term effects, *ND Farm Res.*, 34(6):13–17, 1977.

Meyer, D.W. and W.C. Dahnke, Selected chemical properties of Fargo clay soil after 25 years of nitrogen fertilization of bromegrass at six rates, *Agron. Abstr.*, ASA, Madison, WI, 184, 1981.

Meyer, D.W. and W.E. Norby, Forty years of nitrogen fertilization of bromegrass, in Bergland, D.R. (ed.), 1997 Ag. Production Guide, North Dakota State Univ. Ext. Serv., Fargo, ND, 1997, 414–416.

Paustian, K., H.P. Collins, and E.A. Paul, Management controls on soil carbon, in Paul, E.A. et al. (eds.), Soil Organic Matter in Temperate Agroecosystems, CRC Press, Boca Raton, FL, 1997, 15–49.

CHAPTER 27

Short-Term Crop Rotation and Tillage Effects on Soil Organic Carbon on the Canadian Prairies

B. C. Liang, B. G. McConkey, C. A. Campbell, A. M. Johnston, and A. P. Moulin

CONTENTS

INTRODUCTION

On the Canadian prairies, approximately 46 million hectares of land are currently under agricultural production (excluding permanent pasture and forage). This accounts for more than 80% of the total arable agricultural land base in Canada. Historically, agricultural land use on the Canadian prairies employed intensive tillage for seed-bed preparation and weed control. Frequent summer-fallowing was also practiced to restore soil moisture and nutrients, and to reduce agronomic and economic uncertainties in response to severe drought (McConkey et al., 1996). These practices have contributed to soil degradation resulting in loss of soil organic carbon (SOC). Reducing conventional tillage and the frequency of summer-fallowing on the Canadian prairies has the potential to restore some of the lost SOC while also reducing CO_2 emissions from soil (Bruce et al., 1999).

A number of tillage and crop rotation studies have been conducted in Saskatchewan, ranging in length from 15 to 25 years (Campbell et al., 1995; Campbell et al., 1996a, b; Liang et al., 1999a). From these studies it was concluded that, with elimination of tillage and summer-fallow, SOC gains varied from 0.2 to 0.6 Mg C ha^{-1} yr^{-1}, depending on soil zone and texture. Continuous cropping, compared with crop rotations containing frequent fallow, sequestered SOC ranging from 0.04 to 0.45 Mg C ha^{-1}yr^{-1}, depending on frequency of fallow and soil texture. The potential for sequestering SOC with continuous cropping was greater in the Dark Brown and Black soil zones than in the

Brown soil zone, perhaps because the frequency of fallow was greater in the brown soil zone. No-tillage management compared with conventional tillage also resulted in sequestration of SOC ranging from 0.05 to 0.5 Mg C ha^{-1}yr^{-1} (Liang et al., 1999a). Relative annual increase in SOC under NT compared to tilled systems (RAISOC) has been shown to be a positive linear function of soil clay content (Liang et al., 1999a).

Despite numerous field studies of C sequestration on the Canadian prairies, there is a lack of information on changes in SOC as influenced by tillage and crop rotations in the short-term (i.e., typically less than 5 years). This information will be important in the short term quantification and verification of C sequestration as countries strive to implement the Kyoto Protocol on climate change, especially if agricultural soils are included as a sink and ratified under the Protocol. It is also not clear how inclusion of pulse and oilseed crops in a cereal-dominant rotation would affect C sequestration.

The objectives of this study were to (1) evaluate the feasibility of quantifying SOC changes as influenced by tillage and crop rotations in three short-term studies in Saskatchewan, and (2) compare the influence of inclusion of pulse and oilseed crops vs. monocultural cereals in crop rotation on SOC.

MATERIALS AND METHODS

Three field experiments were conducted at Melfort, Tisdale, and Swift Current, Saskatchewan. A field study with a four-replicate, split-plot design was established in 1994 at Melfort (silt clay, Orthic, Thick Black Chernozem; U.S. Taxonomy: Udic Haploustoll) and Tisdale (heavy clay, Orthic, Dark Gray Chernozem; U.S. Taxonomy: Typic Haplocryoll). Tillage was the main plot treatment and crop rotations were the subplot treatment. Tillage systems for these studies consisted of conventional tillage (CT), minimum tillage (MT), and no tillage (NT). Conventional tillage included fall plus spring tillage with field cultivator and harrow packed prior to seeding. Spring tillage was used for MT and no preseeding tillage for NT. Crop rotations at the sites included three 4-year rotations; namely, canola (*Brassica napus* L.)–wheat (*Triticum aestivum* L.)–barley (*Hordeum vulgare* L.)–barley, canola–barley–pea (*Pisum sativum* L.)–wheat, and canola–pea–flax (*Linum usitatissum* L.)–barley. A two-factor factorial field experiment with three replicates was established on a Swinton loam at Swift Current, Saskatchewan, in 1994 (loam, Orthic Brown Chernozem; U.S. Taxonomy: Typic Haploustoll). Crop rotations varied in fallow frequency and frequency of inclusion of pulse or oilseed crops in cereal dominant rotations. Detailed crop rotations used in this study are listed in Table 27.1. Two of the tillage systems included MT with preseeding tillage involving one operation of heavy-duty sweep cultivator and attached rodweeder, or mounted harrow; NT had no preseeding tillage and used herbicides for weed control.

Soil samples were taken from the 0- to 7.5- and 7.5- to 15-cm depths for the three soils during the fall following four years of experiments. In addition, soil samples were also obtained from the same depths in the Swinton loam at the beginning of the experiment and archived. Four soil cores per plot were taken using a hydraulically operated sampling tube and composited by depth. The resulting samples were air dried and sieved through a 2-mm sieve. Crop residues remaining on the sieve were discarded. Representative subsamples were ground with a rollmill (<153 μm) and analyzed for SOC using an automated combustion technique (Carlo Erba™, Milan, Italy).

Analysis of SOC involved pretreating the soil with phosphoric acid in a tin capsule to remove inorganic C, then drying the sample for 16 hours at 75°C prior to analysis for C. The amount of SOC in the 0- to 7.5- and 0- to 15-cm depths was calculated using the method of Ellert and Bettany (1995) based on an equivalent lightest-mass basis. Light fraction organic C was determined for the Swinton loam by density separation on soil samples sieved through a 2-mm sieve. For this analysis, 10 g of air-dried soil was suspended in 40 ml of NaI solution with a specific gravity of 1.7 g cm^{-3}

Table 27.1   Amounts of Soil Organic C in 0- to 7.5-cm Depth at the Beginning and End of a 4-Year Rotation and Tillage Experiment in a Swinton Loam at Swift Current, Saskatchewan

Rotation[a]	Tillage[b]	Soil Organic C (Mg C ha⁻¹)		Difference (Mg C ha⁻¹)
		1995	1998	
C–W–F–(C)	MT	10.95	11.66 ab	0.72
W–W–F–(W)	MT	11.09	10.18 e	−0.91
F–W–W–(F)	MT	11.68	11.25 bc	−0.43
F–X–F–(C)	MT	10.68	10.57 cde	−0.12
P–W–X–(P)	MT	11.87	11.08 bc	−1.04
C–W–F–(C)	NT	11.61	10.53 de	−1.08
W–W–F–(W)	NT	12.08	11.40 abc	−0.68
W–F–W–(W)	NT	12.06	10.53 de	−1.53
F–W–W–(F)	NT	10.62	11.42 abc	0.81
F–X–F–(C)	NT	11.54	12.01 a	0.47
P–W–X–(P)	NT	12.05	11.01 bcde	−1.04
P–F–W–(P)	NT	10.97	11.37 abcd	0.40
Tukey HSD		ns	0.85	ns
Contrast				
NT vs. MT within rotation		ns	$P = 0.03$	ns

Note: Means followed by the same letters within the same column are not significantly different at $P = 0.05$ probability level.

[a] Letters in parentheses indicate the rotation phase sampled: W = spring wheat, F = fallow, X = flax, P = pea, C = canola.

[b] MT = minimum tillage (preseeding tillage, herbicides, and tillage used for weed control during fallow); NT = no tillage (low disturbance direct seeding, all weed control with herbicides).

and dispersed for 30 sec using a Virtis homogenizer (Virtis Co., Gardiner, New York). After a settling period of 48 h, the suspended material was transferred by suction to a filtration unit. The soil was resuspended in NaI solution and the procedure was repeated to ensure complete recovery of light fraction materials. The composited light fraction materials were washed in $CaCl_2$ solution followed by distilled water, dried at 70°C, and weighed. The light fraction materials were analyzed for C using the automated combustion technique described earlier.

The relative annual increase in SOC for NT compared to tilled systems was calculated as

$$RAISOC = \frac{SOC_{NT} - SOC_{CT}}{SOC_{CT} \times Year} \times 100$$

where $RAISOC$ (% yr⁻¹) was relative annual increase in SOC due to adoption of NT, SOC_{NT} was the amount of SOC under NT (Mg C ha⁻¹ in the 0- to 15-cm soil), SOC_{CT} was the amount of SOC under tilled systems (Mg C ha⁻¹ in the 0- to 15-cm soil), and *Year* was number of years since the establishment of tillage experiments.

RESULTS AND DISCUSSION

In order to reduce the spatial variability of SOC within the field at Swift Current and increase the precision of experimental results, soil samples from each plot were taken at the beginning of the experiment and archived. The mean value of SOC in the 0- to 7.5-cm soil across all plots was 11.4 Mg C ha⁻¹ (C.V. = 10%). Because this variation is small, statistical analysis of results using values of SOC at the start of experiment as a covariable generally did not improve the precision of experimental results (Table 27.1). However, when the spatial variability of SOC within the experimental site is high, archived soil samples at the beginning of experiments may be required to reduce experimental errors and increase precision for variables such as SOC, in which changes are usually small within a short term compared to the quantity in the soil. For all sites, neither tillage nor crop rotation affected SOC below the surface 7.5-cm soil depth; thus the remaining discussion is focused on the 0- to 7.5-cm soil. Relative annual increase in SOC under NT, however, was calculated for the 0- to 15-cm soil.

In the Swinton loam, tillage did not affect the amount of light fraction organic C or the ratio of light fraction organic C to SOC (Table 27.2), but NT had a higher amount of SOC compared with MT (Table 27.1). At the Melfort and Tisdale sites, there was a net gain of about $2 \sim 3$ Mg C ha^{-1} under NT compared with CT, but these differences were not significant (P>0.05), reflecting relatively large amounts of SOC in these soils. Thus, in this study, light fraction organic C was no more sensitive as an index of early change in SOC from tillage effects than SOC itself. This result supports the findings of Liang et al. (1999b) in Saskatchewan. Total SOC had been expected to increase with decreasing frequency of fallow, but results did show such a trend (Table 27.1).

Table 27.2   Amounts of Light Fraction Organic C in 0- to 7.5-cm Depth and Ratios of Light Fraction Organic C (LFOC) to Soil Organic C (SOC) after a 4-Year Rotation and Tillage Experiment in a Swinton Loam at Swift Current, Saskatchewan

Crop Rotation[a]	Tillage[b]	Light Fraction Organic C (Mg C ha^{-1})	LFOC/SOC (%)
C–W–F–(C)	MT	1.69 abc	14.3 bcd
W–W–F–(W)	MT	1.68 abc	16.5 abc
F–W–W–(F)	MT	1.21 c	10.8 d
F–X–F–(C)	MT	1.23 c	11.4 cd
P–W–X–(P)	MT	2.09 ab	19.3 ab
C–W–F–(C)	NT	1.46 bc	13.7 bcd
W–W–F–(W)	NT	1.52 bc	13.4 cd
W–F–W–(W)	NT	1.38 c	13.1 cd
F–W–W–(F)	NT	1.18 c	10.4 d
F–X–F–(C)	NT	1.41 c	11.7 cd
P–W–X–(P)	NT	2.23 a	20.4 a
P–F–W–(P)	NT	1.44 bc	12.7 cd
Tukey HSD		0.65	5.64
Contrast			
NT vs. MT within rotation		ns	ns

Note: Means followed by the same letters within the same column are not significantly different at *P* = 0.05 probability level.

[a] Letters in parentheses indicate the rotation phase sampled: W = spring wheat, F = fallow, X = flax, P = pea, C = canola.
[b] MT = minimum tillage (preseeding tillage, herbicides, and tillage used for weed control during fallow); NT = no tillage (low disturbance direct seeding, all weed control with herbicides).

Relative annual increase in SOC under NT compared to tilled systems (RAISOC) was calculated for each site. The RAISOC was 0.34% yr^{-1} for the Swinton loam, 0.70% yr^{-1} for the Melfort soil, and 0.67% yr^{-1} for the Tisdale soil. These results generally support early findings that RAISOC was a linear function of soil clay content (Figure 27.1), thus suggesting that the potential for sequestering SOC under NT is greater in heavy-textured soils on the Canadian prairies (Campbell et al., 1996a; Liang et al., 1999a). Janzen et al. (1998) suggest that the ability to increase SOC depends not only on the amount of residue C inputs, but also on the C content of the soil at the time of initiation of a study. Results also suggest heavy-textured soils sequestering more organic C may have a greater storage capacity under improved management practices such as conservation tillage or reduced summer-fallow (Liang et al., 1998).

The amount of light fraction organic C and the ratio of light fraction organic C to SOC were inversely related to fallow frequency (Table 27.2; Figure 27.2), indicating that fallow is reducing

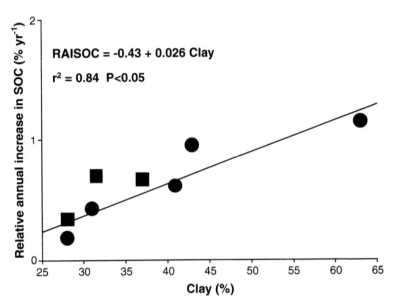

Figure 27.1   Relative annual increase in soil organic C (RAISOC) in the 0- to 15-cm soil under no-till compared with tilled systems as influenced by clay content of soil. Regression equation was derived from observations of mid- to long-term tillage studies conducted on various soil zones of Saskatchewan (●) (Lang et al., 1999a). Observations (■) were made from the current short-term tillage and crop rotation studies.

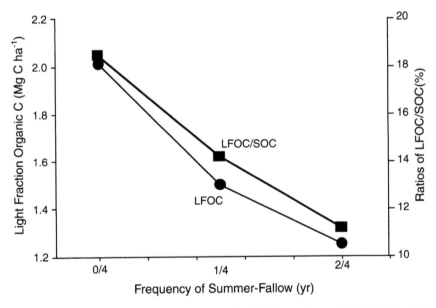

Figure 27.2   Amounts and ratios of light fraction organic C (LFOC) to soil organic C (SOC) as influenced by the frequency of summer-fallow in the Swinton loam at Swift Current, Saskatchewan.

this more labile fraction of SOC. Summer-fallow is used by farmers to reduce economic risk by increasing soil water and nutrient storage on the Canadian prairies; however, increased farm income stability is at the risk of increased soil degradation. Thus, given sufficient time, the reduction of both tillage and fallow frequency would reverse some of the effects of such past degradation on the quantity and quality of SOC.

Crop diversification by inclusion of pulse or oilseed crops has so far not greatly influenced the amount of SOC at Tisdale and Melfort (Table 27.3). At Swift Current, although some significant differences among systems existed, there were no consistent trends for crop types or rotations (Table 27.1). The general lack of differences between crop types across sites may be a reflection of similar crop residue inputs for the different crops. In a long-term study, Collins et al. (1992) found that a wheat–pea rotation had the same residue inputs and SOC as continuous wheat. In a 15-year study of three distinct corn (Zea mays L.) and soybean (Glycine max (L.) Merr.) agroecosystems, Drinkwater et al. (1998) reported that quantitative differences in net primary productivity did not account for the observed changes in soil carbon in a Pennsylvania study; they speculated that the use of low carbon-to-nitrogen organic residues to maintain soil fertility, combined with greater diversity in cropping sequences, significantly increases the retention of SOC.

Table 27.3   Effects of Tillage and Crop Rotations on Soil Organic C in 0- to 7.5-cm Depth after a 4-Year Experiment at Melfort and Tisdale, Saskatchewan

Soil	Tillage[a]	Soil Organic C (Mg C ha^{-1})	Crop Rotations	Soil Organic C (Mg C ha^{-1})
Melfort	NT	45.9a	Canola–Wheat–Barley–Barley	44.5a
	MT	44.2a	Canola–Barley–Pea–Wheat	44.3a
	CT	43.1a	Canola–Pea–Flax–Barley	44.3a
Tisdale	NT	40.1a	Canola–Barley–Pea–Wheat	40.8a
	MT	40.7a	Canola–Pea–Flax–Barley	39.1a
	CT	37.6a	Canola–Wheat–Barley–Barley	38.6a

Note: Means followed by the same letters within the same column for the same soil are not significantly different at $P = 0.05$ probability level.

[a] CT = conventional tillage (fall tillage after crop, preseeding tillage, tillage as required for weed control during fallow); MT = minimum tillage (preseeding tillage, herbicides, and tillage used for weed control during fallow); NT = no tillage (low disturbance direct seeding, all weed control with herbicides).

CONCLUSIONS

Although numerous long-term crop rotation and tillage experiments have been conducted on the Canadian prairies, and the impacts of conservation tillage and intensification of cropping systems on agronomic, economic and environmental issues have been substantial, information on the short-term impact of these management strategies on C sequestration is generally limited. This information is required for quantifying and verifying C sequestration if the Kyoto Protocol is to be successfully implemented, and especially if agricultural soil C sink is included in the Protocol. The results show that tillage influenced SOC after four years in soils at Swift Current, but not at Melfort and Tisdale, Saskatchewan, even though no tillage tended to increase SOC from 0.1 to 0.7 Mg C ha^{-1} yr^{-1} with higher values of sequestration in more moist regions of the prairies and fine-textured soils.

ACKNOWLEDGMENTS

Funding for the initial phase of study was provided by the Canada–Saskatchewan Agricultural Green Plan. GEMCo and MII of Agriculture and Agri-Food Canada provided funding for the present study.

REFERENCES

Bruce, J.P. et al., Carbon sequestration in soils, *J. Soil Water Conserv.*, 54:382–389, 1999.

Campbell, C.A. et al., Long-term effects of tillage and crop rotations on soil organic C and total N in a clay soil in southwestern Saskatchewan, *Can. J. Soil Sci.*, 76:395–401, 1996a.

Campbell, C.A. et al., Tillage and crop rotation effects on soil organic C and N in a coarse-textured Typic Haploboroll in southwestern Saskatchewan, *Soil Tillage Res.*, 37:3–14, 1996b.

Campbell, C.A. et al., Carbon sequestration in a Brown Chernozem as affected by tillage and rotation, *Can. J. Soil Sci.*, 75:449–458, 1995.

Collins, H.P., P.E. Rasmussen, and C.L. Douglas, Jr., Crop rotation and residue management effects on soil carbon and microbial dynamics, *Soil Sci. Am. J.*, 56:785–788, 1992.

Drinkwater, L.E., P. Wagoner, and M. Sarrantonio, Legume-based cropping systems have reduced carbon and nitrogen losses, *Nature*, 396:262–265, 1998.

Ellert, B.H. and J.R. Bettany, Calculation of organic matter and nutrients stored in soils under contrasting management regimes, *Can. J. Soil Sci.*, 75:529–538, 1995.

Janzen, H.H. et al., Management effects on soil C storage in the Canadian prairies, *Soil Tillage Res.*, 47:181–195, 1998.

Liang, B.C. et al., Crop rotation and tillage impact on carbon sequestration in Saskatchewan soils, in *Proc. Saskatoon Soils Crops '99*, February 25–26, 1999, Saskatoon, Saskatchewan, 1999a, 142–154.

Liang, B.C. et al., Effect of tillage and crop rotations on the light fraction organic carbon and carbon mineralization in Chernozemic soils of Saskatchewan, in *Proc. Saskatoon Soils Crops '99*, February 25–26, 1999, Saskatoon, Saskatchewan, 1999b, 552–569.

Liang, B.C. et al., Retention and turnover of corn residue carbon, *Soil Sci. Soc. Am. J.*, 62:1361–1366, 1998.

McConkey, B.G. et al., Long-term tillage on spring wheat production on three soil textures in the Brown soil zone, *Can. J. Plant Sci.*, 76:747–756, 1996.

PART VI

Economics of Carbon Sequestration

CHAPTER 28

Soil Organic Carbon Sequestration Rates in Reclaimed Minesoils

V. Akala and Rattan Lal

CONTENTS

INTRODUCTION

Soil is a dynamic entity easily disturbed by natural and anthropogenic perturbations. Mining is one such anthropogenic activity that causes drastic perturbations in the antecedent soil profile, properties, and processes. The sudden perturbation gives little time for soil's inherent resilience to respond (Lal, 1997a). Soil degradation leads to loss of soil quality, with a net decrease in soil capability and functions (Lal, 1997b), with consequent net loss of soil organic carbon (SOC) due to enhanced mineralization, erosion, and leaching (Lal et al., 1998).

Anthropogenically disturbed soils which retain some potential to be productive when reclaimed are referred to as "anthropic" soils (Eswaran, 1997). In the context of mining, anthropic soils are referred to as "minesoils" (Sencindiver and Ammons, 2000). The characteristics, processes, and profiles of minesoils differ from those present prior to mining and disturbance. Reclamation of disturbed soil is the process of making it fit for cultivation (Barnhisel and Hower, 1997; Bradshaw, 1997); it implies returning soil to a useful, but not necessarily original, state.

The heterogeneous mass, comprising a mixture of topsoil with subsoil and bedrock produced by mining activity, is called mine spoil. Freshly exposed mine spoil is a minesoil at time zero, in respect to formation, because it consists only of rock and pulverized rock material (Kohnke, 1950).

Mine spoil reflects properties of the parent material more closely than natural soils because of its initial stage of pedogenic development (Smith and Sobek, 1978). The rate of minesoil development over time may be an important edaphic factor determining the success of the reclamation plan.

Although there are different theories about the rate of minesoil development, consensus is that soil weathering is rapid in the early stages, but the rate of change decreases over time (Struthers, 1964; Schafer et al., 1979). Very rapid chemical reactions occur in spoil as the exposed material attains equilibrium with its new environment (Struthers, 1964). Also, minesoils are exposed to processes of physical weathering such as cycles of wetting and drying, freezing and thawing, and mechanical disruption by roots (Schafer et al., 1979). A rapid change in particle size distribution occurs as the spoil material weathers (Van Lear, 1971). Schafer et al. (1979) compared development processes in 1- to 50-year-old minesoils in southeastern Montana and found an increase in SOC content in the upper 5-cm layer in young minesoil over an 8- to 10-year period. Older minesoils (50 years) contained more SOC at all depths than new minesoils. The time required to attain the equilibrium SOC pool in the top 1-cm depth was considerably shorter than that required for depths below 10 cm. It was surmised that 200 to 400 years would be required for SOC to reach a steady-state level in the top 10-cm depth.

The potential of SOC enhancement and sequestration in minesoil depends on biomass productivity, root development in subsoil, and changes in minesoil properties resulting from overburden (the earth material above the seam) weathering (Haering et al., 1993). Temporal changes in SOC pools indicate improvements in soil quality and soil's potential to sequester C. A chronosequence study was conducted on reclaimed minesoils in southeastern Ohio to assess the rate of SOC sequestration. Specific objectives of this study were to assess the sink capacity of reclaimed minesoils to sequester SOC and to determine the rate of SOC sequestration.

MATERIALS AND METHODS

The study sites are located at southeastern Ohio under the jurisdiction of American Electric Power Company (AEP). Rich in mineral resources, the area has been subjected to mining during most of the 20th century. Mining for coal was done on the upper seams during the early part of the century, with no planned reclamation. The soil was restored due to natural succession and revegetation (Daniels and Zipper, 1995; Hopps, 1994). Improvement in mining technology resulted in deep mining for coal, which led to further soil disturbance. In the mining process, secondary forests were cleared before removal and storage of topsoil. The overburden was then excavated and back filled into mined pits. Minesoil reclamation in some of these areas started much earlier than the State Law of 1972 and Federal Surface Mining Control and Reclamation Act of 1977, wherein there was no requirement for topsoil replacement. After 1972, it became mandatory to replace topsoil above the overburden and spoil material prior to reclamation.

Reclaimed minesoil chronosequences of 0 to 50 years were identified for similar ecological characteristics. Variability due to topography, climate, vegetation, and soil type was minimized through careful site selection. The reclaimed sites were managed by four different treatments comprising pasture and forest, with (post-1972) and without (pre-1972) topsoil application. The average depth of the topsoil above the overburden was 30 cm in pasture and forest treatments that received topsoil. In contrast, no specific topsoil was used, and it was completely mixed with the overburden in treatments without topsoil application. The maximum duration was 25 years for sites with topsoil application and between 30 and 50 years for sites without topsoil application. For the study, a 70-year-old marginal agricultural land under pasture and a 65-year-old reclaimed forest were considered as pasture and forest control sites respectively. Minesoil samples were obtained from 0- to 15- and 15- to 30-cm depths and analyzed for soil bulk density (ρ_b) by the core method (Blake and Hartge, 1986) and SOC content by the dry combustion technique (Nelson and Sommers, 1986).

A sigmoid fit to the data on SOC sequestration for treatments with topsoil application was determined on the basis of regression values. For the treatments without topsoil application, (a minimum of 30 years old), the SOC pool was seemingly close to achieving equilibrium. Thus, a straight-line fit was deemed appropriate. Table Curve, Version 2, was used to fit the regression curve. The change in SOC pool during the first 5 years of the treatment with topsoil application was neglected in order to arrive at a standard curve and on the assumption that significant SOC enhancement takes place only after 4 to 10 years of reclamation (Ross et al., 1992). The rate of change of the SOC pool during the first 5 years was calculated independently using a straight line between the time periods. The rate of SOC sequestration for the other time periods was determined using the regression models. Variation in the data was reduced using Lowess interpolation with 40% smoothing level.

RESULTS AND DISCUSSION

There was a decrease in ρ_b with increase in reclamation duration in the treatments for 0- to 15- and 15- to 30-cm depths (Akala and Lal, 1999, 2000). The SOC pool of the control pasture site was 42.6 Mg ha^{-1} for 0- to 15-cm depth and 23.7 Mg ha^{-1} for 15- to 30-cm depth. The SOC pool of the control forest site was 43.7 Mg ha^{-1} for 0- to 15-cm depth and 12.9 Mg ha^{-1} for 15- to 30-cm depth. The wide variation in the SOC pool may be attributed to differences in the origin of the topsoil, antecedent SOC content, and minor variations in management methods.

Pasture Treatment with Topsoil Application

The SOC pool of the pasture with topsoil application increased from 15.3 Mg ha^{-1} in the beginning of reclamation to 44.4 Mg ha^{-1} after 25 years for 0- to 15-cm depth and from 10.6 Mg ha^{-1} to 18.3 Mg ha^{-1} for 15- to 30-cm depth during the same duration (Figures 28.1 and 28.2, respectively; Akala and Lal, 2000). The equilibrium SOC pool of 44.5 Mg ha^{-1} was achieved in 27 years for the 0- to 15-cm depth. The maximum rate of SOC sequestration for the 0- to 15-cm depth, measured during the 11th year was 3.1 Mg ha^{-1}yr^{-1} (Figure 28.1). The equilibrium level of the SOC pool for the 15- to 30-cm depth has yet to be achieved. The maximum rate of SOC sequestration for the 15- to 30-cm depth observed during the 21st year was 0.7 Mg ha^{-1}yr^{-1} (Figure 28.2). The present rate of SOC sequestration for the 15- to 30-cm depth is 0.5 Mg ha^{-1}yr^{-1}, which may tend towards an equilibrium pool of 27.5 Mg ha^{-1} between 45 and 50 years on the basis of the regression model.

Forest Treatment with Topsoil Application

The SOC pool of the forest treatment with topsoil application increased from 12.7 Mg ha^{-1} in the beginning of reclamation to 45.3 Mg ha^{-1} after 21 years for 0- to 15-cm depth and from 9.1 Mg ha^{-1} to 13.6 Mg ha^{-1} for 15- to 30-cm depth during the same duration (Figures 28.3 and 28.4, respectively; Akala and Lal, 2000). The equilibrium levels of the SOC pool for 0- to 15- and 15- to 30-cm depths were not achieved during the 25 years of reclamation; however, the highest rate of SOC sequestration for the 0- to 15-cm depth measured during the 13th year was 2.6 Mg ha^{-1}yr^{-1} (Figure 28.3). The highest rate of SOC sequestration for the 15- to 30-cm depth, measured during the 15th year, was 0.8 Mg ha^{-1}yr^{-1} (Figure 28.4). The current rate of SOC sequestration is 0.2 Mg ha^{-1}yr^{-1} for the 0- to 15-cm depth and 0.03 Mg ha^{-1}yr^{-1} for the 15- to 30-cm depth. Based on the regression model, an equilibrium SOC pool of 46.1 Mg ha^{-1} may be achieved in 35 to 40 years for the 0- to 15-cm depth and an SOC pool of 15 Mg ha^{-1} may be achieved in 30 to 35 years for the 15- to 30-cm depth.

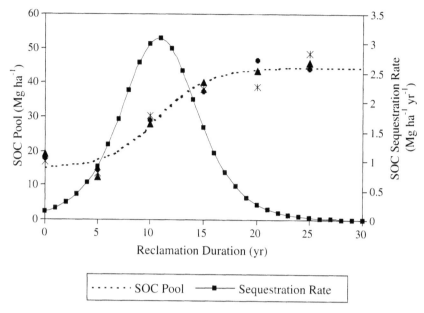

Figure 28.1   Temporal changes in the SOC pool and sequestration rate of pasture with topsoil application for
0- to 15-cm depth.

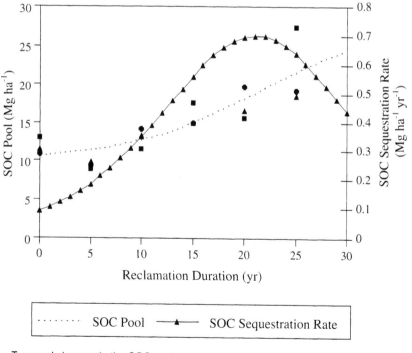

Figure 28.2   Temporal changes in the SOC pool and sequestration rate of pasture with topsoil application for
15- to 30-cm depth.

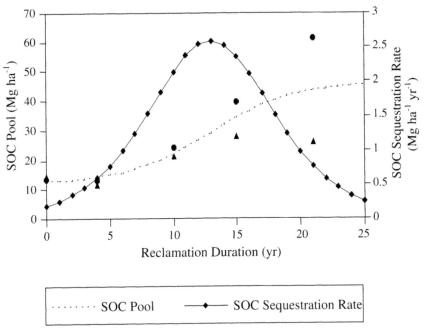

Figure 28.3   Temporal changes in the SOC pool and sequestration rate of forest with topsoil application for 0- to 15-cm depth.

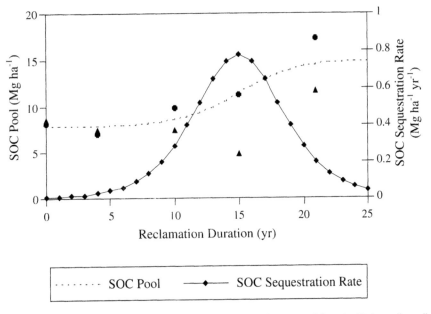

Figure 28.4   Temporal changes in the SOC pool and sequestration rate of forest with topsoil application for 15- to 30-cm depth.

Table 28.1   Regression Equations for SOC Pool (Mg ha^{-1}) with Time in Years (Y)

Treatment System/Depth	Equation	R^2
Pasture with topsoil		
0–15 cm	Mg ha^{-1} = 15+(29.5/(1+exp(−(Y−10.8)/2.39)))	0.95[a]
15– 30 cm	Mg ha^{-1} = 10+(17.6/(1+exp(−(Y−21.1)/6.26)))	0.75[a]
Forest with topsoil		
0–15 cm	Mg ha^{-1} = 12+(34.1/(1+exp(−(Y−13)/3.28)))	0.67[a]
15– 30 cm	Mg ha^{-1} = 7.79+(7.19/(1+exp(−(Y−15)/2.31)))	0.76[a]
Pasture without topsoil		
0–15 cm	Mg ha^{-1} = 0.15Y + 34.44	ns
15– 30 cm	Mg ha^{-1} = 0.0009Y + 19.79	ns
Forest without topsoil		
0–15 cm	Mg ha^{-1} = 0.12Y + 29.64	0.98[a]
15– 30 cm	Mg ha^{-1} = 0.04Y + 16.72	ns

Note: ns = not significant.

[a] Significant at α = 0.05.

Statistical functions for the best fits to the temporal changes in the SOC pool of the different treatments are shown in Table 28.1, where the SOC pool (Mg ha^{-1}) is the independent variable and time (Y in years) is the dependent variable. On interpolating different values of time, the functions show that the SOC pool remains fairly constant during the initial phase of the reclamation and starts to increase between periods of 5 to 10 years, depending on the type of treatment. Thereafter, the SOC pool follows an exponential phase during which most of the C is enhanced and sequestered; the highest rates of SOC sequestration were measured during this phase.

The function indicates equilibrium level of SOC sequestration in the time period after 20 to 25 years, depending on the treatment. The correlation coefficients for the forest treatment were less than those of the pasture treatment, suggesting that soil reclamation under forest may not have attained equilibrium SOC level. The regression models for the treatment with topsoil application did not truly represent the changes in the SOC pool during the first 5 years of reclamation. The rate of decrease (or loss) of the SOC pool during this time period was 0.9 Mg ha^{-1}yr^{-1} for 0- to 15-cm depth and 0.5 Mg ha^{-1}yr^{-1} for 15- to 30-cm depth for pasture treatment with topsoil application. The rate of decrease (or loss) was 0.4 Mg ha^{-1}yr^{-1} and 0.3 Mg ha^{-1}yr^{-1} for 0- to 15- and 15- to 30-cm depths for forest treatment with topsoil application.

Pasture and Forest Treatments without Topsoil Application

Because no topsoil was applied prior to 1972, the reclamation duration sampled for the pasture and forest treatment without topsoil application was 30 to 50 years. The SOC pool remained relatively constant during this time and the rates of SOC sequestration were close to zero. The SOC sequestration rate for the pasture treatment without topsoil application was 0.15 Mg ha^{-1}yr^{-1} for 0- to 15-cm depth. The SOC pool of the pasture treatment without topsoil application did not show an increase for 15- to 30-cm depth; therefore, the sequestration rate was close to zero. The rates of SOC sequestration for the forest treatment without topsoil application were 0.12 Mg ha^{-1}yr^{-1} for 0- to 15-cm depth and 0.04 Mg ha^{-1}yr^{-1} for 15- to 30-cm depth. As an example, the temporal changes in the SOC pool of forest treatment without topsoil application for 0- to 15-cm depths is shown in Figure 28.5; the regression equations are shown in Table 28.1.

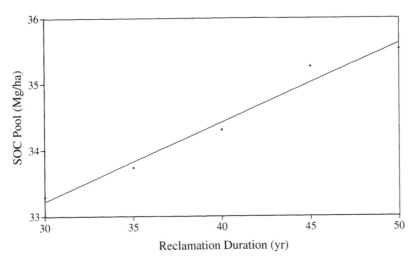

Figure 28.5   Temporal changes in the SOC pool and sequestration rate of forest without topsoil application for 0- to 15-cm depth.

CONCLUSIONS

The data presented support the following conclusions:

1. The SOC sequestration potential of reclaimed minesoils range between 50 to 60 Mg ha^{-1} for 0- to 15-cm depth over a 25-year period.
2. The maximum rate of SOC sequestration ranged between 2 to 3 Mg ha^{-1}yr^{-1} for 0- to 15-cm depth.
3. The SOC pool of an aggrading system does not increase infinitely and reaches equilibrium within 25 to 35 years.
4. Systems with topsoil application reached equilibrium SOC level 15 to 20 years earlier than systems without topsoil application.

ACKNOWLEDGMENTS

Research was sponsored by the Office of Solar, Thermal, Biomass Power and Hydrogen Technologies within the Office of Utility Technologies, U.S. Department of Energy, under contract DE-AC05-96OR22464 with Lockheed Martin Energy Research Corporation. Special thanks to Mark Downing, Paul Loeffelman, Tom Archer, Gary Kaster, and Art Boyer of American Electric Power (AEP).

REFERENCES

Akala, V. and R. Lal, Mineland reclamation and soil organic carbon sequestration in Ohio, in Bengson, S. and D. Bland (eds.), *Proc. 16th Ann. Nat. Meet. Am. Soc. Surface Mining Reclamation*, Aug 13–19, Scottsdale, AZ, 322–331, 1999.

Akala, V.A. and R. Lal, Potential of mineland reclamation for soil organic carbon equestration in Ohio, *Land Degrad. Develop.*, 11:289–297, 2000.

Barnhisel, R.I. and J.M. Hower, Coal surface mine reclamation in the eastern United States: the revegetation of disturbed lands to hayland/pasture or cropland, in Sparks, D.L. (ed.), *Advances in Agronomy*, 233–275, Academic Press, New York, 1997.

Blake, G.R. and Hartge, K.H., Bulk density, in Klute, A. (ed.), *Methods of Soil Analysis*, Part 3, ASA, Madison, WI, 363–375, 1986.

Bradshaw, A., Restoration of mined lands — using natural processes, *Ecol. Eng.*, 8: 255–269, 1997.

Daniels, W.L. and C.E. Zipper, Improving coal surface mine reclamation in the central Appalachian region, in Cairns, J., Jr. (ed.), *Rehabilitating Damaged Ecosystems*, 187–217, Lewis, Boca Raton, FL, 1995.

Eswaran, H., Preliminary thoughts about anthropic soil materials and anthropic processes, unpublished paper, USDA-NRCS World Soil Resources, Washington, D.C., 1997.

Haering, K.C., W.L. Daniels, and J.A. Roberts, Changes in minesoil properties resulting from overburden weathering, *J. Environ. Qual.*, 22(1):194–200, 1993.

Hopps, M., Reforesting Appalachia's coal lands, *Am. For.*, 100:40–45, 1994.

Kohnke, H., The reclamation of coal minesoils, *Adv. Agron.*, 2:317–349, 1950.

Lal, R., Degradation and resilience of soils, *Phil. Trans. R. Soc. Lond.*, B, 327:997–1010, 1997a.

Lal, R., Soil quality and sustainability, in Lal, R. et al. (eds.), *Methods of Assessment of Soil Degradation*, CRC Press, Boca Raton, FL, 17–30, 1997b.

Lal, R. et al., *The Potential of U.S. Cropland to Sequester Carbon and Mitigate the Greenhouse Effect*, Ann Arbor Press, Chelsea, MI, 1998.

Nelson, D.W. and Sommers, L.E., Total carbon, organic carbon, and organic matter, in Klute, A. (ed.), *Methods of Soil Analysis*, Part 1. ASA, Madison, WI, 961–1010, 1986.

Ross, D.J., J.C. Cowling, and T.W. Speir, Soil restoration under pasture after lignite mining: management effects on soil biochemical properties and their relationships with herbage yields, *Plant Soil*, 14(1):85–97, 1992.

Schafer, W.M. et al., Soil genesis, hydrological properties, root characteristics and microbial activity of 1 to 50 year old strip mine spoils, EPA-600/7-79-100. USEPA, Cincinnati, OH, 1979.

Sencindiver, J.C. and J.T. Ammons, Minesoil genesis and classification, in Barnhisel, R.I., R.G. Darmody, and W.L. Daniels (eds.), *Reclamation of Drastically Disturbed Lands*, Agronomu Monograph 41, American Society of Agronomy, Madison, WI, 2000.

Smith, R.M. and A.A. Sobek, Physical and chemical properties of overburdens, spoils, wastes, and new soils, in Schaller, F.W. and P. Sutton (eds.), *Reclamation of Drastically Disturbed Lands*, American Society of Agronomy, Madison, WI, 149–172, 1978.

Struthers, P.H., Chemical weathering of strip minesoils, *Ohio J. Sci.*, 64:125–131, 1964.

Van Lear, D.H., Effects of spoil texture on growth of K-31 tall fescue, Research Note NE-141, USDA Forest Service, 1971.

CHAPTER **29**

Efficiencies of Conversion of Residue C to Soil C

C. A. Campbell, B. G. McConkey, S. Gameda, R. C. Izaurralde, B. C. Liang,
R. P. Zentner, and D. Sabourin

CONTENTS

ABSTRACT

Accurate estimates of the impact of crop management practices on soil organic carbon (SOC) changes are most likely made using process-based models, such as CENTURY. However, quick, first approximations of SOC changes can be made if measurements of grain yield, grain-to-straw, and straw-to-root ratios are available, and if efficiencies of conversion of residue C to soil C for the cropping system are known or can be estimated. This chapter presents C conversion efficiency values derived from various long-term crop-rotation, tillage-management, and fertilizer studies conducted under differing weather conditions in the major soil zones of the Canadian prairies. Use of coefficient of conversion (%) (i.e., ratio of soil organic C change and input residue C) of 0, 10, 15, 20, 0, and 25% for the brown, dark brown, dark gray, thin black and thick black Chernozems, and gray Luvisolic soils, respectively, are proposed; it is suggested that these efficiencies can be used to make first approximations of the impact of crop management on C sequestration in soils.

INTRODUCTION

The contribution of CO_2 emission to the greenhouse gas effect and to global warming is well known. Difficulties in quantitatively assessing CO_2 dynamics compared to the relative ease of measuring, monitoring, and modeling soil C changes, which is inversely related to atmospheric

1-56670-581-9/02/$0.00+$1.50
© 2002 by CRC Press LLC

305

CO_2 dynamics, makes estimation of soil C changes an important current pursuit for soil scientists. Various process-based models that will allow quantitative estimation of SOC changes currently exist. However, most of these require complex calculations and many of the model inputs are not commonly measured. The CENTURY and EPIC models fall into this category. Further, as shown by Campbell et al. (2000a) and McConkey (1998), CENTURY, which is the best of these models for this purpose, does not always provide accurate estimates of SOC changes.

Because of a generally direct relationship between SOC change and residue C inputs to soil, it might be possible to estimate the former from knowledge of the latter. Lal (1997) calculated C sequestration in soils throughout the world by estimating residue C inputs based on crop yield and assuming a 15% conversion of residue C to SOC. However, research to support this approach has been scant. Follett et al. (1997) reported coefficients of 5.4% for an 84-year small grain study at Akron, CO, and 10.5% for a 20-year study at Sidney, NE. Although crop residue values are not always available, it is possible to estimate them from grain yields, straw:grain ratios, and straw:root ratios. The C concentration of crop residues is usually about 45% (Millar et al., 1936); these ratios and carbon conversion coefficients will vary with crop management practices and weather conditions.

The objective of this chapter is to summarize C conversion efficiency values derived from various long- and medium-term agronomic research studies conducted on the Canadian prairies, and to demonstrate how such values can be used to make first estimates of soil carbon changes with time.

MATERIALS AND METHODS

Data on C inputs and changes in SOC taken from the scientific literature were summarized and then used to calculate the efficiency of conversion (%), defined as the ratio of change in SOC to total C inputs in crop residues (Campbell et al., 2000b,c; Table 29.1). Crop residues included straw (i.e., above-ground material minus grain) measured or estimated from straw:grain ratios and roots estimated as 59% of straw yield. All data from the Breton plots and Ellerslie, Alberta, were recalculated by R. C. Izaurralde using the ratio for roots and straw mentioned earlier. Carbon concentration in crop residues was assumed to be 45% (Millar et al., 1936). At the Breton plots, plant C inputs were primarily root-C because straw was removed from plots. C in manure was estimated (Izaurralde et al., 2001) as well.

In Table 29.1, where the fertilizer N and P rates were not specified but treatments were shown to receive N+P, treatments received N based on soil NO_3-N test and P based on general recommendations for the crop and soil provided by soil testing laboratories. For further details of management of the various experiments from which these data were collected, consult the original scientific publication.

This chapter will briefly demonstrate how coefficients such as those presented can be used for practical solutions. For example, Campbell and Coxworth (1999) used such coefficients to assess the feasibility of sequestering C through use of crop residue for industrial products. They summarized grain yield data (from Statistics Canada) for crops grown in the various crop districts of the Prairie Provinces for the period of 1976 to 1994, expressing them as a function of crop, soil zone, and year. They used the grain yields to estimate straw and root C inputs to the soil and the likely changes in SOC that would occur over time if straw was harvested and used for making strawboard.

RESULTS AND DISCUSSION

Due to the large variability often associated with SOC measurements in field experiments (Campbell et al., 2000b, c), and because change in SOC is calculated as the difference between two such measurements over time, large apparent discrepancies in some estimates of efficiency of conversion are often observed. For example, sometimes negative efficiencies are obtained, not because SOC decreased due to unfavorable weather and low C inputs, but because of spatial or sampling variability. Similarly, conversion efficiencies may be unreasonably high or low for the latter reason. Some of the changes in SOC, though sizeable, were not statistically significant (e.g.. at Ellerslie); thus, although large efficiencies of conversion are shown, it could be argued that these values should be near zero. Consequently, caution should be exercised in how the values in Table 29.1 are used in estimating SOC changes from residue C inputs. In other publications another. more rigorous but simple empirical equation has been provided that could be used to achieve this goal with greater confidence (Campbell et al., 2000a, b, c).

The conversion efficiencies showed that, in general, values increased with cropping frequency. fertilization, manure application, and conversion from conventional to no-tillage and were much higher under more favorable weather conditions. The efficiencies were greater when a degraded soil was subjected to improved management conditions than if the soil was already inherently fertile.

Although the values in Table 29.1 may allow more flexibility in choosing an appropriate efficiency of conversion factor to use in estimating SOC changes, coefficients for average weather and management conditions are often required. Consequently, values that can generally be used to make such conversions for each soil zone of the Canadian prairies (Table 29.2) have been prepared. The brown soils, with very frequent fallowing, are unlikely to result in SOC gains because of low C inputs. Additionally, the very fertile "thick" black Chernozems are unlikely to provide an increase in SOC even with good management because their SOC is already so very high (Figure 29.1. Melfort). In contrast, the "thin" black Chernozems and gray Luvisols (low to medium organic matter content but high crop productivity and input C) show potentially the largest increases in SOC (Table 29.2 and Figure 29.1). One difficulty is to try to decide which black soils will respond like the "thin" black Chernozems at Indian Head, and which will respond like the "thick" black Chernozems at Melfort or Ellerslie. Experimental data are lacking in this regard; however, a criteron of 4% C (7% organic matter) has arbitrarily been chosen as the critical level. On this basis it was hypothesized that 74% of the black soils in Alberta, 27% in Saskatchewan, and 11% in Manitoba were likely to be unresponsive like "thick" black Chernozems, while the remainder would respond like "thin" black Chernozems (Table 29.3).

Estimating Relative C Sequestration from Straw Used for Strawboard Compared to Its Return to Soil

Isobord Enterprises has located a strawboard plant at Elie, Manitoba, in the black Chernozemic soil zone, with a capacity of about 200,000 tons straw/year. At full capacity this plant has the ability to sequester 90,000 tonnes C/year (no allowance was made for C used in transporting straw to plant, or in manufacturing strawboard). Campbell and Coxworth (1999) conducted analysis which showed little possible impact on soil carbon sequestration if the straw required for manufacturing strawboards were to be taken from "thick" black Chernozems, since these soils respond only slightly to carbon-sequestering practices (Table 29.2 and Figure 29.1). Conversely, if the straw was removed from "thin" black Chernozems, the analysis reveals a loss in opportunity to sequester about 18,000 tonnes C/year (20% of 90,000 tonnes C/year) in soil. A more complete systems analysis would be required to make decisions of regional straw management by considering (1) advantages of using straw for industrial purposes, (2) impacts of straw removal or retention on soil quality, and (3) environmental impacts of straw management including influence on erosion and, if burnt, as is often the case in Manitoba, air pollution.

Table 29.1 Efficiencies[z] of Conversion of Residue C[y] to Soil Organic C in Various Rotations[x]

Study and Soil	Crop Rotation	Tillage Method	Fertilizer Treatment	No. of Years	Input C (Mg ha^{-1})	Change in SOC (Mg ha^{-1})	Efficiency of Conversion[w] (%)	Typical Weather Conditions	Source of Data
**Brown Chernozem (Aridic Haploboroll)**									
Swift Current (SC) Tillage Study									
Hatton fsl (degraded)[v]	Cont W	MT	N & P	11 (1983–94)	12.46	0.10	1	Average	Campbell et al. (1996a)
	Cont W	NT	N & P	11	12.81	1.57	12	Average	
	F–W	MT	N & P	11	10.60	–1.40	–13	Average	
	F–W	NT	N & P	11	11.33	0.75	7	Average	
Swinton Sil (degraded)[v]	Cont W	MT	N & P	12 (1982–94)	24.69	2.00	8	Average	Campbell et al. (1995)
	Cont W	NT	N & P	12	24.61	3.17	13	Average	
	F–W	CT	N & P	12	14.51	0.40	3	Average	
	F–W	NT	N & P	12	13.64	0.95	7	Average	
Sceptre C (degraded)[v]	Cont W	MT	N & P	11 (1983–94)	16.51	0.58	4	Average	Campbell et al. (1996b)
	Cont W	NT	N & P	11	17.00	2.78	16	Average	
	F–W	MT	N & P	11	12.87	–0.38	–3	Average	
	F–W	NT	N & P	11	12.20	3.85	32	Average	
New Rotation Study at S.C.									
Swinton Sil	Crested wheatgrass (CWG) Hay	NT	N & P	10 (1987–96)	5.08	1.33	26	Moist	Campbell et al. (2000c)
(degraded)[v]	Cont–W	MT	N & P	10	20.85	4.71	23	Moist	
	F–W–W–W	MT	N & P	10	17.09	4.88	29	Moist	
	F–W–W	MT	N & P	10	15.57	3.47	22	Moist	
Old Rotation Study at S.C.[u]									
Swinton (degraded)	Cont W/W–Lent	CT	N & P	9 (1967–75)	12.82	4.90	38	Average	Campbell et al. (2000b)
	Cont W/W–Lent	CT	N & P	14 (1976–89)	20.07	–0.77	–4	< Average	

	Tillage	Fertilizer	Years				Moisture	Reference
ContW/W–Lent	CT	N & P	6 (1990–96)	16.66	4.41	27	Moist	
F–W–W	CT	N & P	9	9.54	1.03	11	Average	
F–W–W	CT	N & P	14	17.28	0.65	4	< Average	
F–W–W	CT	N & P	6	12.44	2.20	18	Moist	
F–W/CONT W	CT	P	9	9.21	0.84	9	Average	
F–W/CONT W	CT	P	14	15.60	−0.30	−2	< Average	
F–W/CONT W	CT	P	6	10.61	3.74	35	Moist	

Dark Brown Chernozem (Typic Boroll)

Lethbridge Rotation Study

	Tillage	Fertilizer	Years				Moisture	Reference
Lethbridge L F–W	CT	None	37 (1955–92)	44.5	−5.3	−12	Average	Campbell et al. (2000a)
(fertile)[y] Cont W	CT	None	37 (1955–92)	53.5	−2.3	−4	Average	

Thin Black Chernozem (Udic Boroll)

Indian Head Rotation Study

	Tillage	Fertilizer	Years				Moisture	Reference
Indian Head C F–W	CT/NT	None	10 yr (yr 30 to yr 40)	9.33	2.60	28	Average	Campbell et al. (2001)
(degraded) F–W	CT/NT	N & P	10 yr (yr 30 to yr 40)	15.23	3.93	26	Average	
(degraded) F–W–W	CT/NT	None	10 yr (yr 30 to yr 40)	9.63	−1.78	−18	Average	
(degraded) F–W–W	CT/NT	N & P	10 yr (yr 30 to yr 40)	20.47	5.21	26	Average	
(degraded) Cont W	CT/NT	None	10 yr (yr 30 to yr 40)	9.59	0.02	0	Average	
(Fertile) Cont W	CT/NT	N & P	10 yr (yr 30 to yr 40)	27.23	1.95	7	Average	

Thick Black Chernozem (Udic Boroll)

Melfort Rotation Study

	Tillage	Fertilizer	Years				Moisture	Reference
Melfort Sicl F–W(N+P)/ F–W–W (none)	CT	(N+P/none)	31 yr (1957–1987)	48	0	0	Average	Campbell et al. (1991) (We assumed that SOC of F–W & F–W–W had not changed in 31yr)
(Fertile) F–W–W	CT	N & P	31 yr	58	0	0	Average	
Cont W	CT	None	31	44	4.0	9	Average	
Cont W	CT	N + P	31	61	4.0	7	Average	

Table 29.1 (Continued)　Efficiencies[z] of Conversion of Residue C[y] to Soil Organic C in Various Rotations[x]

Study and Soil	Crop Rotation	Tillage Method	Fertilizer Treatment	No. of Years	Input C (Mg ha^{-1})	Change in SOC (Mg ha^{-1})	Efficiency of Conversion[w] (%)	Typical Weather Conditions	Source of Data
Ellerslie Fertilizer & Tillage Study (kg N ha^{-1})									
Malmo L (Fertile)	Cont B	CT	0	11 yr	14.39	3.85	27	Average	Dr. C. Izaurralde (Personal Communication)[t] Recalculated from Nyborg et al. (1995)
	Cont B	CT	56	11	25.73	5.83	23	Average	
	Cont B	NT	0	11	12.41	5.55	45	Average	
	Cont B	NT	56	11	21.88	4.24	19	Average	
Ellerslie Fertilizer Rate (kg N ha^{-1})									
Malmo L (Fertile)	Cont B	CT	0	13 yr	20. 67	0	0	Average	Dr. C. Izaurralde (Personal Communication)[t] Recalculated from Solberg et al. (1998)
	Cont B	CT	25	13	26.55	2.40	9	Average	
	Cont B	CT	50	13	32.75	1.60	5	Average	
	Cont B	CT	75	13	33.07	3.60	11	Average	
Gray Luvisolic (Alfisol)									
Breton Rotation Study[s]									
Breton L (Degraded)	F–W	CT	None	50 yr	7.95	9.05	–114	Average	Dr. C. Izaurralde (Personal Communication)[t] See Izaurralde et al. (2001) for description of treatments
	F–W	CT	Fert	50	11.05	8.60	–78	Average	
	F–W	CT	Manure	50	50.55	7.25	14	Average	
	W–O–B–H–H	CT	None	50	17.65	2.65	15	Average	
	W–O–B–H–H	CT	Fert	50	25.75	5.15	20	Average	
	W–O–B–H–H	CT	Manure	50	76.15	17.30	23	Average	

Breton Fertilizer & Tillage Study (kg N ha⁻¹)

Breton L	Cont B	CT	0	11 yr	8.03	-1.41	-17	Average	R. C. Izaurralde (Personal Communication)ᵗ Recalculated from Nyborg et al. (1995)
(Degraded)	Cont B	CT	56	11	19.79	3.88	20	Average	
	Cont B	NT	0	11	7.29	1.65	10	Average	
	Cont B	NT	56	11	16.51	10.70	65	Average	

Breton Fertilizer Rate Study (kg N ha⁻¹)

Breton L	Cont B	CT	0	13 yr	10.02	0	0	Average	R. C. Izaurralde (Personal Communication)t Recalculated from Solberg et al. (1998)
(Degraded)	Cont B	CT	25	13	16.22	1.70	10	Average	
	Cont B	CT	50	13	22.58	5.40	24	Average	
	Cont B	CT	75	13	26.87	6.90	26	Average	

z Increase or decrease in soil C divided by residue C inputs.

y Includes straw and estimated root C except CWG (Agropyron cristatum (L.) Gaeertn.) which is only estimated root C.

x F = fallow; Cont = continuous; W = spring wheat (Triticum aestivum L.); Lent = lentil (Lens culinaris medikus); Barley (Hordeum vulgare L.); O = oats (Avena sativa L.); H = grass-legume hay; CT = conventional till; MT = minimum till; NT = no till.

v Refers to soil organic C status at initiation of experiment. Degraded means land in poorly fertilized, frequently fallowed systems for decades after breaking. Fertile means land in legume-containing rotations with added manure or well fertilized for decades after breaking.

u The efficiencies for the 30-year study were calculated for 3 selected periods representing differing weather conditions.

t R. Izaurralde recalculated input C by assuming root:straw ratio 0.59 as used by Campbell et al. in the other studies.

s Most of the above-ground crop material was harvested and removed each year.

Table 29.2   Coefficient for Estimating Soil C Gains from Crop Residue
              (Straw and Root) Inputs of C[z]

| Soil Zone[y] | Coefficient of Conversion of Input C to Soil C (%) | |
	Business as Usual	Adopt Best Management in Future[x]
Brown Chernozem	0	5
Dark Brown Chernozem	10	15
Dark Gray Chernozem	15	20
Thin Black Chernozem	20	25
Thick Black Chernozem	0	5
Gray Luvisols	25	30

[z] We recommend against using these coefficients for predictions beyond 30 years.
[y] Assume cropping frequency increases from fallow-wheat in brown soils to fallow-wheat-
   wheat in dark brown, to continuous cropping in black and gray.
[x] Assume adoption of reduced tillage and more efficient fertilization in all soil zones, greater
   use of snow management in brown and dark brown, and more varied crop rotations
   including legumes and oilseeds in all zones.

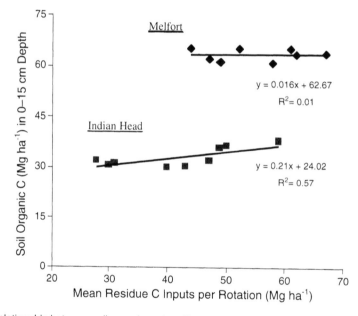

Figure 29.1   Relationship between soil organic carbon (0- to 15-cm depth) and estimated crop residue (includ-
              ing roots) carbon returned to land in 30 years at Indian Head and 31 years at Melfort, Sask.
              Source: Campbell et al., 1991.

Table 29.3   Proportion of High[z] Organic Matter and Lower[z] Organic Matter Soils
              among the Black Chernozems of Prairie Provinces

| Province | Thick Black (>4% C) | | Thin Black (<4% C) | | Total Black |
	Million ha	% of Total	Million ha	% of Total	Million ha
Alberta	3.902	73.9	1.375	26.1	5.278
Saskatchewan	1.847	27.0	4.987	73.0	6.828
Manitoba	0.512	10.5	4.345	89.5	4.858
Total	6.256	36.9	10.709	63.1	16.964

[z] We arbitrarily designated soils with % C > 4% (i.e., 7.0% organic matter) as very fertile "thick black
   Chernozems" in which it might be difficult to increase C more with normal cropping practices.

CONCLUSIONS

Scientists currently desire to make quick approximate estimations of the amount of soil carbon to be sequestered from known or estimated quantities of crop residue inputs. Although several process-based models currently exist that will allow quantitative solution of the latter problem, most of these models require complex calculations and many of the model inputs are not commonly measured.

This chapter used results from several long-term experiments conducted on the Canadian prairies to develop conversion coefficients allowing estimation of soil C changes as a function of residue C inputs. These coefficients were expressed as a function of soil zone, crop management (tillage, rotation, fertilizer treatment), and weather conditions. Because research data exist that allow estimation of straw yields from grain yields, and root yields from straw yields, it is possible to estimate C inputs if statistics of grain production are known. Campbell and Coxworth (1999) used this approach to estimate the feasibility of harvesting some wheat straw for industrial uses in Manitoba; possible consequences for soil organic carbon were projected.

REFERENCES

Campbell, C.A. et al., Effect of crop rotations and fertilization on soil organic matter and some biochemical properties of a thick black Chernozem, *Can. J. Soil Sci.*, 71: 377–387, 1991.

Campbell, C.A. and E. Coxworth, Final report on feasibility of sequestering carbon through use of crop residue for industrial products, Report to Options Table of Agriculture and Agri-Food Table on Climate Change: Analysis of Greenhouse Gas Mitigation Practices, 1999.

Campbell, C.A. et al., Carbon sequestration on the Canadian prairies — quantification of short-term dynamics, in Lal, R. and K. McSweeny (eds.), *Soil Management for Carbon Sequestration to Mitigate the Greenhouse Effect II.* Special Publication (In Press).

Campbell, C.A. et al., Carbon sequestration in a brown Chernozem as affected by tillage and rotation, *Can. J. Soil Sci.*, 75:449–458, 1995.

Campbell, C.A. et al., Tillage and crop rotation effects on soil organic matter in a coarse-textured Typic Haploboroll in southwestern Saskatchewan, *Soil Tillage Res.*, 37:3–14, 1996a.

Campbell, C.A. et al., Long-term effects of tillage and crop rotations on soil organic C and N in a clay soil in southwestern Saskatchewan, *Can. J. Soil Sci.*, 76: 395–401, 1996b.

Campbell, C.A. et al., Adopting zero tillage management; impact on soil C and N under long-term crop rotations in a thin black Chernozem, *Can. J. Soil Sci.*, (In Press).

Campbell, C.A. et al., Organic C accumulation in soil over 30 years in semiarid southwestern Saskatchewan — effect of crop rotations and fertilizers, *Can. J. Soil Sci.*, 80:179–192, 2000b.

Campbell, C.A. et al., Quantifying short-term effects of crop rotations on soil organic carbon in southwestern Saskatchewan, *Can. J. Soil Sci.*, 80:193–202, 2000c.

Follett, R.F. et al., Carbon isotope ratios of Great Plains soils and in wheat-fallow systems, *Soil Sci. Soc. Am. J.*, 61:1068–1077, 1997.

Izaurralde, R.C. et al., Carbon balance of the Breton Classical Plots over half a century, *Soil Sci. Soc. Am. J.*, 65:431–441, 2000.

Lal, R., Residue management, conservation tillage and soil restoration for mitigating greenhouse effect by CO_2-enrichment, *Soil Tillage Res.*, 43:81–107, 1997.

McConkey, B., Report on the Prairie CENTURY research workshop, Aug. 24–26, 1998, Saskatoon, SK, 1998.

Millar, H.C., F.B. Smith, and P.E. Brown, The rate of decomposition of various plant materials in soils, *Am. Soc. Agron. J.*, 28:914–923, 1936.

Nyborg, M. et al., Fertilizer N, crop residue, and tillage alter soil C and N content in a decade, in Lal, R. et al. (eds.), *Soil Management and Greenhouse Effect*, Advances in Soil Science, Lewis Publishers, CRC Press, Boca Raton, FL, 1995, 93–100.

Solberg, E.D. et al., Carbon storage in soils under continuous cereal grain cropping: N fertilizer and straw, in Lal, R. et al. (eds.), *Management of Carbon Sequestration in Soil*, CRC Press, Boca Raton, FL, 1998, 235–254.

CHAPTER **30**

Changes of Organic Matter and Aggregate Stability of the Arable Surface Waterlogged Soils

Raina Dilkova, Ekaterina Filcheva, George Kerchev, and Milena Kercheva

CONTENTS

INTRODUCTION

One major problem in East European countries is the anthropogenic diminishing of the organic matter and deterioration of soil aggregate stability (Contributions of the participants in the Conference on Soils in CEESs-NIS-CACs-M, 2000). The ecological consequences are soil compaction and soil crust formation observed in almost all countries, soil surface waterlogging (Moldavia, Russia, Romania, Bulgaria), soil erosion (Moldavia, Romania, Yugoslavia, Polsha, Czech Republic, Poland, Republic of Macedonia).

Dilkova et al. (1998) studied the significance of organic matter to soil aggregation and water stability for the virgin surface waterlogged soils in Bulgaria. Statistically proved predictive models were obtained for the aggregate stability, with the ratio total carbon to clay for the waterlogged layers of Eutric Planosols, Vertic Chernozem, and Vertisol and with extractable organic carbon with mixed solution (sodium pyrophosphate and sodium hydroxide) for Distric Planosols.

The presumption is that, in the arable soils, the organic matter is the most dynamic component compared to the other two main soil structure factors (clay and oxalate-extractable iron). The question then becomes how these opposite impacts — the anaerobic conditions created during frequent spring soil water logging and the annual aeration of the arable layer by the traditional tillage systems in Bulgaria — change the soil organic matter and soil structure. The answer to this question could help explain the impacts and risk factors on the soil quality changes. The objective of the study reported here was to establish the anthropogenic and hydromorphic changes of organic matter in the main soil waterlogged units in the country and to determine their effect on the soil aggregate stability of the arable layer.

1-56670-581-9/02/$0.00+$1.50
© 2002 by CRC Press LLC

315

MATERIALS AND METHODS

The upper 0- to 20- to 25-cm soil layers of virgin and arable Vertisols (Hapluderts, Soil Taxonomy, Boyadgiev, 1994), Haplic Luvisol (Hapludalfs), Eutric, and Distric Planosols (Albaqualfs) are used in this study (Figure 30.1). The following properties have been determined to explain changes of the soil organic matter: total organic carbon (C_t) (Tjurin, 1965), content of humic (C_h) and fulvic acids (C_f), and their ratio (C_h:C_f), characterizing the type of humus, degree of humification ($C_h/C_t 100\%$), aggressive fulvic acid's low molecular mobile fraction (1^a), and carbon in the residuum (C_{res}) (Kononova, 1966).

Soil physical characteristics analyzed were soil particle size distribution (Katchinski, 1958) and soil aggregate stability, expressed by the ratio (MWDR) of the meanweight diameters of the air-dried soil aggregates smaller than 10 mm (MWD = 3.53 mm) after and before rapid wetting (wet sieving method of Vershinin and Revut, 1952).

RESULTS AND DISCUSSION

The Vertisol and Vertic Chernozem are swelling soils; the prevailing clay mineral is montmorillonite (33%) in the mineral composition. Their clay content is high (45 to 60%) along the whole nondifferentiated profiles (texture coefficient 1.1) and does not change under arable conditions. Their humic characteristics at virgin conditions are typical for soils with vertic properties and morphology. The decreasing coefficients of total carbon are low (1.0 to 1.3); they have a humic type of humus (C_h:C_f = 3.0 ÷ 3.9) and high stability of humic acids (more than 90% are Ca-complexed; Dilkova et al., 1998). The virgin Vertisols have a good aggregate stability (MWDR = 0.44).

The Haplic Luvisol, Eutric Planosols, and Distric Planosols have an elluvial–illuvial type soil profile with a different degree of differentiation. (Texture coefficients are 2.2, 2.2 and 4.0, respectively.) The surface layers under virgin conditions have a low clay content (20 to 30%) and low swelling (<9%), due to the low montmorillonite content (<7%). The soil structure of the upper layers of these soils under virgin conditions is characterized with a good aggregate stability (MWDR = 0.39 to 0.52).

There are significant changes of the physical properties in the upper soil layer under cultivation. Upon mixing, the underlying Bt soil layer usually enriches the clay content (up to 100%), which is one of the reasons for the strong decrease of total carbon. The relative increase of fulvic acids (1^a) with the depth and instability of the humus system in elluvial horizons (more than 50% of humic acids are "free") is a common characteristic for this group (Table 30.1). The mean ratios

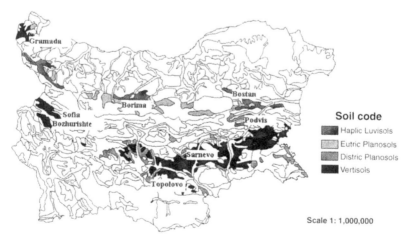

Figure 30.1   Soil map of surface waterlogged Bulgarian soils and studied sites.

Table 30.1 Organic Carbon Content and Composition of Soil Organic Matter at Virgin and Arable Conditions of Representative Soil Profiles

Soils	Land Use	Depth (cm)	C_t % Soil	C_{org} % Soil	C_h % Soil	C_f % Soil	C_h:C_f	1^a % Soil	C_{res} % Soil	% C_t
Vertic Chernozem v. *Gramada*	Virgin	0–5	2.46	0.79	0.59	0.20	3.00	0.07	1.67	68
		10–15	1.78	0.53	0.39	0.14	2.75	0.03	1.25	70
	Arable	0–5	1.37	0.59	0.40	0.19	2.11	0.01	0.78	57
		10–15	1.42	0.54	0.37	0.17	2.18	0.03	0.88	62
	Ar/vir, %	0–5	56	75	68	97	70	15	47	84
		10–15	80	101	94	119	79	99	71	89
Haplic Luvisols v. *Bostan*	Virgin	0–5	4.91	2.06	0.98	1.08	0.91	0.12	2.85	58
		10–15	1.26	0.50	0.21	0.29	0.74	0.10	0.76	60
	Arable	0–5	1.33	0.56	0.24	0.32	0.73	0.07	0.77	58
		10–15	1.10	0.51	0.26	0.25	1.04	0.09	0.59	54
	Ar/vir	0–5	27	27	24	30	83	57	27	100
		10–15	87	101	121	86	141	90	78	89
Eutric Planosols v. *Glavatzi*	Virgin	0–5	1.95	0.81	0.33	0.48	0.69	0.08	1.14	58
		10–15	0.88	0.43	0.16	0.27	0.59	0.07	0.45	51
	Arable	0–5	0.80	0.43	0.17	0.26	0.65	0.06	0.37	46
		10–15	0.69	0.41	0.18	0.23	0.78	0.06	0.28	41
	Ar/vir	0–5	41	53	52	54	95	75	32	79
		10–15	78	95	113	85	132	86	62	79
Distric Planosols v. *Surnevo*	Virgin	0–5	1.61	0.77	0.40	0.37	1.07	0.05	0.38	48
		10–15	0.96	0.45	0.32	0.13	1.39	0.03	0.18	42
	Arable	0–5	0.96	0.45	0.32	0.13	2.46	0.03	0.13	29
		10–15	0.78	0.42	0.28	0.14	2.00	0.04	0.14	30
	Ar/vir	0–5	60	58	80	35	231	60	34	60
		10–15	80	98	112	78	144	133	78	80

$C_h:C_f$ of surface waterlogged layers (A, A_1A_2, A_2l) are 0.7, 1.0, 1.3, and 3.5 for Eutric Planosol, Haplic Luvisols, Distric Planosols, and Vertic Chernozem and Vertisol, respectively; this corresponds to the different degree of surface water logging of these soils (Dilkova et al., 1998).

The common anthropogenic effects due to the annual soil tillage in all studied arable layers are the reduction of the total organic matter content (Table 30.1) and drastic deterioration of water aggregate stability (MWDR = 0.14 to 0.27). The negative organic matter balance is a very representative parameter of the anthropogenic "wasting" of the arable part of the soil profile. It is important to note that the total organic matter content strongly decreases, especially in the upper 0- to 5-cm layer. The reduction of the total organic matter in the arable conditions is due mainly to a decrease in the unextractable part of the organic matter (Table 30.1). In the upper 0- to 5-cm layer, 65 to 75% of the changes of total organic matter content are due to changes of the unextractable organic matter, and in the 10- to 15-cm layer more than 80%. The slower decreasing of the extractable organic matter content in the studied soils could be explained with slower mineralization and, in some cases, syntheses of humic acids under anaerobic conditions, which are frequently observed in the spring (Dilkova et al., 1998). The influence of humus, the total soil carbon, on the aggregate stability (MWDR) of the waterlogged layers of arable soils is confirmed by similar statistical models found (Dilkova et al., 1998) for virgin soils (Figures 30.2 and 30.3).

CONCLUSIONS

Studying the organic matter content and composition and the MWDR as an index of aggregate stability of the genetically different surface waterlogged soils under virgin and arable conditions allows one to determine the anthropogenic and hydromorphic changes of the humus and their influence on the soil structure. The strong reduction of the organic matter content and water stability of the soil aggregate are considered anthropogenic changes due to annual soil aeration by tillage and use of heavy machinery. The relative increase of the extractable organic matter (as a percent of total carbon content) is accepted as a hydromorphic change of the organic matter of the arable layer of the surface waterlogged soils; this is explained with slower mineralization of the humic acids.

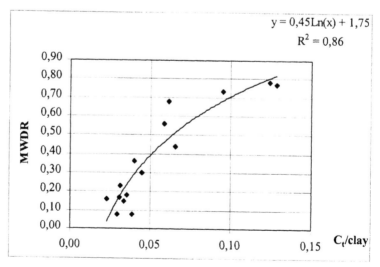

Figure 30.2 Relationship between the aggregate stability (MWDR) and the ratio total Carbon (Ct) and clay (virgin and arable Vertisol, Vertic Chernozem, Haplic Luvisol and Eutric Planosol).

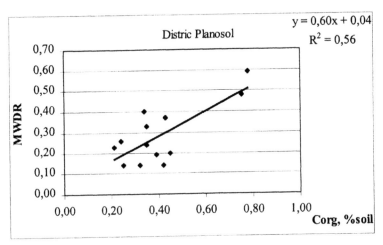

Figure 30.3 Relationship between the aggregate stability (MWDR) and the extractable organic carbon (Corg) for virgin and arable Distric Planosol.

REFERENCES

Boyadgiev, T., Soil map of Bulgaria according to the soil taxonomy, *Soil Sci. Agrochem. Ecol.*, 4-6, 43–51, 1994.

Contributuons of the participants in the conference *Soils in Central European Countries, New Independent States, Central Asian Countries and in Mongolia. Present Situation and Future Prospects*, 26–29 Aug. 2000, Prague, Czech Republic.

Dilkova, R. et al., Humus peculiarities of the virgin surface waterlogged soils, in *Congres Mondial de Sci. du sol*, 16e, Montpellier, France, 20–26 August 1998, *Resumes*, Vol. 1, Symp. N 18, p. 378, 1998.

Katchinski, N.A., Soil texture and soil microaggregate contents, methods for their study, *Acad. Sci. USSR*, Moscow, 1958.

Kononova, M.M., *Soil Organic Matter. Its Nature, Its Role in Soil Formation and in Soil Fertility*, 2nd English ed., Pergamon Press. Inc., Elmsford M.V., 1966.

Tjurin, I.V., Agrochemical methods of soil analysis, *Nauka*, Moscow, 1965.

Tisdall, J.M. and J.M. Oades, Organic matter and water-stable aggregates in soils, *J. Soil Sci.*, 33, 141–163, 1982.

Vershinin, P.V. and I.B. Revut, Methods of soil texture analysis, *Bull. Sci.-Tech. Inform. Agrophys.*, 3, Agrophysical Institute, Leningrad, 1952.

PART VII

Policy Issues and Industrial and Farmer Viewpoints

CHAPTER 31

Designing Efficient Policies for Agricultural Soil Carbon Sequestration

John M. Antle and Siân Mooney

CONTENTS

INTRODUCTION

Background

This paper addresses issues that arise in designing policies to increase carbon (C) sequestration in agricultural soils. There are several motivations for such policies. First, countries may unilaterally desire to undertake policies that have beneficial effects on the productivity and long-term sustainability of agricultural production systems. Second, under the proposed Kyoto Protocol, industrialized countries are required to reduce greenhouse gas (GHG) emissions. The U.S. is requested to reduce net emissions of GHGs such as carbon dioxide, methane, and nitrous oxide by 7% below 1990 emission levels by the period of 2008 to 2012. Both emissions reductions and sequestration can be used to meet this target (UNCTAD, 1998) if the protocol is ratified. Total U.S. emissions of GHGs are estimated at between 1485 and 1709 million metric tons of carbon equivalent per year (MMTCE yr^{-1}), of which emissions from agriculture are thought to be between 109 and 123 MMTCE yr^{-1} (Lal et al., 1998). Estimates indicate that between 75 to 208 MMTC yr^{-1} (up to 8% of U.S. emissions) can be sequestered in the soils of U.S. cropland agriculture (Lal et al., 1998).

Policy Design

Soil scientists have shown that, by changing land use and management practices, agricultural producers can sequester additional soil C. However, as producers change their management practices they also change the overall economic profitability of their business. There are joint roles for economic and soil science research in the design of policies to increase soil C. From an economist's point of view, policies are mechanisms that attempt to induce socially desired behavior. The premise of this chapter is that economic efficiency is the principal objective of policies designed to encourage soil C sequestration, i.e., policies strive to maximize the social benefits generated per unit of resource used to implement the policy.

Agricultural production exhibits certain features that complicate the design of efficient policies to achieve agricultural sustainability objectives or to meet commitments to international environmental agreements such as the Kyoto Protocol. This chapter focuses on the spatial variability of land and farm characteristics and their influence on the measurement of changes in soil C in response to changes in land-use and management practices. The aim is to discuss options for C sequestration policies in agriculture, including existing environmental programs, and to assess the economic efficiency of these policies. This analysis leads to a discussion of additional information needed to evaluate the efficiency of alternative policy designs and some suggestions for further multidisciplinary work.

Agricultural Production Practices and Soil Carbon Sequestration

Historically, C has been released from the soil as a result of converting land to cultivation. Various processes related to agricultural production, including soil erosion and leaching of agricultural chemicals, lead to the release of soil C and other GHGs. Tiessen, Stewart, and Betany (1982), Mann (1986), and Rasmussen and Parton (1994) estimate that 20 to 50% of soil C is lost from the soil during the initial 20 to 50 years of cultivation.

Increases in soil C can be achieved through the adoption of various land-use and management practices (Lal et al., 1998; CAST, 1992), which include removing highly erodible land from production and restoring it to conserving uses such as grassland or wetland. Management practices that facilitate soil C sequestration include conservation tillage, management of crop residue, cover crops, and improved water management (Lal et al., 1998). If the producer subsequently reverts to conventional management practices, however, the stored C is lost. The effectiveness of these

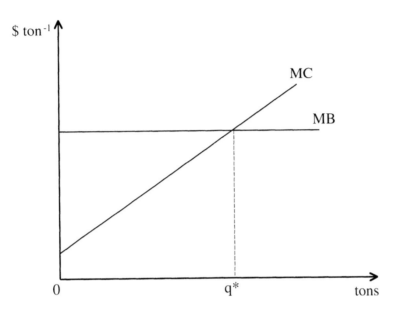

Figure 31.1   Economically efficient private production decision.

production changes in sequestering C depends on both cropping intensity and tillage practice. No single land-use or management practice will be effective at sequestering C in all regions. It is estimated that 49% of agricultural C sequestration can be achieved by adopting conservation tillage and residue management, 25% by changing cropping practices, 13% by land restoration efforts, 7% through land-use change, and 6% by better water management (Lal et al., 1998).

WHAT IS MEANT BY AN EFFICIENT POLICY?

Agricultural producers try to make production decisions that maximize their profits. When producers consider what crops to plant and how to mange them, they take into account factors such as the expected price at harvest, in addition to the cost of inputs such as fertilizer, seed and labor. Based on this information, the producer determines what crops to grow and in what quantities. Consider the case of production decisions for a single crop. The producer will expand production up to the point where the price received for the last unit produced is equal to the additional cost incurred in producing that unit,* i.e., an individual producer will produce up to the point where the marginal private benefit (MB) from production equals the marginal private cost (MC). Figure 31.1 shows the MB and MC functions faced by an individual producer; equating MB with MC, the producer will choose quantity q*. At production levels below q*, MB > MC; therefore, each additional unit contributes more to benefits than to costs and as such production should be expanded to the point, q*, where no addition to profits is possible. At production levels greater than q*, MPB < MPC, indicating that additional units cost more to produce than they provide in benefits. Additional production beyond q* decreases profits.

The production level q* is economically efficient for a private individual. However, if farmers' decisions have off-farm impacts on environmental quality (externalities), decisions to equate marginal private benefits to marginal private costs are not efficient from the point of view of society. Socially efficient decisions equate the marginal *social* benefits (MSB) and marginal *social* costs

---

* The following describes the theoretical ideal of producer decision making assuming that the producer has perfect knowledge. In practice, producers may not attain the theoretical ideal; however, their behavior is consistent with this description.

(MSC) associated with production activities. Efficient policies are designed to provide incentives that encourage farmers to make decisions that equate MSB to MSC.

EXISTING FARM POLICY AND ALTERNATIVE POLICY DESIGNS

At present, many farm policies exist that encourage producers to change production and management decisions to protect or increase environmental quality. It may be possible to alter existing policies to achieve an increase in soil C; however, each policy is designed to elicit specific actions and may not be efficient for C sequestration. This section describes three general policy designs that can be used to affect the environment through agricultural practices and highlights existing policies that use each design.

Command-and-control regulatory policies (also referred to as design standards), could be used to increase soil C by directly mandating that farmers use prescribed land-use or management practices (often referred to as best management practices). Command-and-control regulation is generally an inefficient approach to achieving environmental goals whenever the population of farms regulated is heterogeneous because of the limited ability of regulators to tailor mandated technologies to each producer (Council of Economic Advisers, 1990). Thus, because of the spatial heterogeneity of agricultural land resources and the heterogeneity of farm managers and their human and physical capital, command-and-control regulation cannot be expected to achieve the efficient resource allocation wherein marginal social cost is equated to marginal social benefit on each land unit.

The second broad class of policies alter the economic incentives farmers face for adopting alternative production activities. These policies include subsidies to encourage beneficial practices and taxes to discourage harmful ones. Typically, these policies aim to induce farmers to operate where the marginal social benefit of their actions equals the marginal social cost by altering the economic returns to alternative land-use or management activities. However, this policy design also is likely to be inefficient for reducing an environmental externality because it is based on management decisions rather than on environmental externality. For example, a tax on the use of an input causing water pollution, e.g., nitrogen, will reduce its application but does not necessarily lead farmers to use the input so that the marginal social benefit of reducing water pollution is equated to the marginal social cost. For the tax to produce an efficient outcome, the price of the input inclusive of the tax would have to equal the marginal social value of the pollution created by the input use. This marginal social value varies spatially, depending on where a farm field is located in relation to ground or surface water, and is not likely to equal a fixed tax rate on inputs.

Several existing environmental policies for U.S. agriculture, such as the conservation reserve program (CRP), wetland reserve program (WRP), and the environmental quality incentives program (EQUIP), are incentive policies that offer payments and other support services in exchange for voluntary changes in land-use and management practices (Osborn, 1997; Lynch and Smith, 1994). The economic efficiency of these programs depends on the program design. The original CRP program, for example, had criteria for participation based on soil erosion but did not account for the spatial variation in benefits associated with erosion reduction. As a result, much of the land placed into the CRP in the initial phase of the program was low-productivity land in the Great Plains where the damages associated with erosion are relatively low, whereas relatively little higher-valued land in the Midwest was placed in the program even though the damages from erosion in that region are much higher (Heimlich, 1994).

It was quickly recognized that the CRP produced multiple environmental benefits in addition to reducing soil erosion. In recognition of this fact, eligibility for the program in the 1990s was based on an index of environmental benefits in an effort to increase the environmental benefits received from lands within the CRP. Although this was an attempt to increase the efficiency of the program, it is unlikely that marginal social benefit is equated with marginal social cost due to the heterogeneity of spatial environmental variables and the omission of the full range of benefits from the index. For example, activities under the CRP increase soil C (Gebhart et al., 1994; Barket et al., 1995; Barket et al., 1996); however, this benefit is not accounted for in the payment scheme.

Although arguably more efficient than a strict command-and-control policy, an incentive-based payment program may be inefficient because the policy design fails to provide incentives that lead to the most efficient allocation of land and other resources. Research shows that better targeting of program eligibility through criteria based on site-specific benefits would substantially increase the efficiency of the CRP (Feather, Hellerstein, and Hansen, 1999).

A third class of policy instrument is market-based trading. Market mechanisms for trading environmental goods can be encouraged by establishing institutions that facilitate market transactions. A trading scheme for sulfur dioxide emissions has been in place in the U.S. since 1990 (Petsonk, Dudek, and Goffman, 1998; Joskow, Schmalensee, and Bailey, 1998) and the agricultural sector is already participating in emerging markets for other environmental goods such as water and water rights (Landry, 1996; Colby, Crandall, and Bush, 1993). A market trading mechanism for GHG emissions has been proposed under the Kyoto Protocol as a means to reduce global GHGs (UNCTAD, 1998) and could be one means to encourage soil C sequestration. Market trading schemes have advantages in that they encourage innovation and faster adoption of new technology by not limiting the means by which C is sequestered in the soil, and can result in lower costs than traditional command- and control-systems (Joskow, Schmalensee, and Bailey, 1998).

ANALYSIS OF EFFICIENT POLICY DESIGN FOR SOIL CARBON SEQUESTRATION

Description

The previous section indicates that several policy designs can be used to encourage producers to change their land-use or management practices. Issues that complicate the design of efficient policies are the spatial heterogeneity of land and farm characteristics and their influence on the measurement of changes in soil C associated with alternative land use and management practices. This section uses a simple analytical framework to examine the economic efficiency of alternative policies designed to encourage soil C sequestration.

The first policy provides incentive payments to producers for each hectare of land switched from conventional tillage to no-till. For simplicity, only two management choices are presented; however, this example can be expanded to allow the producer to choose among many management practices and land-use decisions. This per-hectare payment scheme is similar to existing policies, such as the CRP, that provide payments on an area basis to producers that adopt recommended land-use or management practices. The second policy is designed so that producers receive incentive payments for each ton of C sequestered. Under this per-ton payment scheme, farmers are paid a price for each unit of C they sequester. This discussion considers only these two management choices and assumes that producer participation under either policy is voluntary as in existing farm programs.

Per-Hectare and Per-Ton Payments — Spatially Homogeneous Resources

The following stylized framework is used to examine management decisions faced by producers who choose between engaging in crop production using conventional or no-till. The following assumptions are maintained:

1. Each hectare is managed independently and production decisions made on one hectare do not influence production in other areas.
2. Each producer grows a single crop.
3. Producers have a given, fixed amount of capital, z, sufficient to allow production using either conventional or no-till technology.*

For this discussion to focus on the issue of spatial variability, a simple static framework is maintained in which the producers choose inputs, represented by the vector $\mathbf{x}_i$ (e.g., fertilizer, fuel, and labor), to maximize the profit from crop production on each hectare. This decision can be expressed as:

$$\pi_i(p,\mathbf{w}|z) = \underset{\mathbf{x}_i}{\text{Max}} \; p \; y_i(\mathbf{x}_i|z) - \mathbf{w} \; \mathbf{x}_i \tag{31.1}$$

where:

p	=	price of the crop in dollars per-ton
$\mathbf{w}$	=	vector of input prices
$y_i(\mathbf{x}_i|z)$	=	the production function of the crop, for $i = c, n$, where c = conventional till and n = no-till technology
$\mathbf{x}_i$	=	vector of variable inputs (representing fertilizer, pesticides, fuels, labor, and other variable inputs) required to produce the crop given a level of capital z

Profits, $\pi_i = \pi_i(p,\mathbf{w}|z)$, represent the combination of inputs and management decisions that solve the farmer's profit maximizing decision problem. To distinguish between profits under each management practice, let $\pi_c = \pi_c(p,\mathbf{w}|z)$ if the producer chooses to use conventional till and $\pi_n = \pi_n(p,\mathbf{w}|z)$ if the producer uses no-till. Producers will select the tillage practice that maximizes profits from each hectare.

First examine producer decisions in the case where all land units in a region are assumed to be homogeneous, e.g., have the same weather, soil, nutrient, and other conditions, and thus all producers make the same management decisions. Later this case is compared to one of resources that vary spatially. Assume initially that profits from conventional tillage are higher than no-till, so all producers choose to produce using conventional tillage.**

When profits from conventional till are greater than no-till, any government or private entities that wish producers to switch to no-till and sequester soil C will need to provide an incentive payment, denoted by g_n, to induce producers to change their management practices.

First consider a policy that provides payments for each hectare switched from conventional to no-till. If $\pi_c < \pi_n + g_n$, the producer will switch to no-till; however, if $\pi_c > \pi_n + g_n$, the producer will remain in conventional till as the incentive payment, g_n, is not sufficient compensation for losses incurred by changing tillage practices. Essentially, the incentive payment must be at least equal to the difference in profit per-hectare under each management regime before the producer will switch management practices. Thus, $g_n^* = \pi_c - \pi_n$ is referred to as the per-hectare switch price.

Assume that, when producers switch from conventional to no-till, the quantity of soil C increases. Soil C is a function of management practices over a specified time horizon. For this static example, abstract from the dynamics of soil C sequestration and assume that, when a management practice is used for a sufficiently long period of time, the quantity of soil C per-hectare approaches an equilibrium value $C_i = C(\mathbf{x}_i|z)$, $i = c, n$. Due to the assumed homogeneity of soil

---

* The capital requirements for conventional and no-till will differ in reality. Incorporating these differences would not change the main results of the analysis.
** Existing research indicates that yields may decline as producers switch from conventional to no-till, but may then return to higher levels after an initial adjustment period. The effect on profits is a combination of changes in yield and input costs as a result of adopting a new management technique. Such dynamic behavior of yields is not accounted for in this static analysis, but this does not affect the principles discussed.

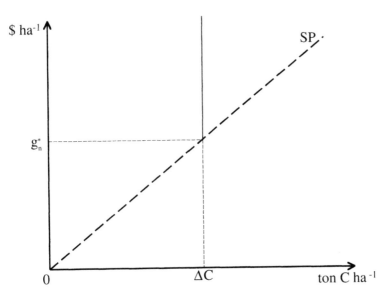

Figure 31.2   Soil carbon change from a no-till payment per hectare with homogeneous land.

and climate conditions in the region, the equilibrium quantity of soil C is determined by the management regime most profitable for farmers.

Figure 31.2 illustrates the case where a per-hectare incentive payment equal to g_n^* is offered to induce producers to switch from conventional to no-till and increase soil C sequestration by $\Delta C = (C_n - C_c)$. Incentives less than g_n^* are not successful in encouraging any producers to switch from conventional to no-till; payments greater than or equal to g_n^* encourage all producers to switch to no-till.

Now consider a policy that offers producers a payment for each ton of soil C sequestered. Similar to the above per-hectare example, producers will switch to no-till and sequester more C if the profits from no-till plus the incentive payment exceed the profits from conventional till. The per-hectare switch price can be used to calculate the minimum payment per-ton of sequestered C required to induce producers to switch from conventional to no-till. The per-ton switch price is found by dividing the per-hectare switch price by the quantity of C sequestered on each hectare, that is, $g_n^*/\Delta C$. The per-ton switch price is represented by line SP (Figure 31.2) that has a slope $g_n^*/\Delta C$. If $\pi_c < \pi_n + (g_n^*/\Delta C)\Delta C$, the producer will switch to no-till and sequester more C; otherwise, the producer will remain in conventional till.*

Spatially Heterogeneous Resources

Now suppose that land in the region is heterogeneous, that is, soil and climate characteristics vary so that productivity and soil C vary spatially. To account for this heterogeneity, a site-specific vector of environmental characteristics, e_j, is introduced for $j = 1,...,s$ hectares (or land units) in the region. The profit function for each hectare can now be represented as $\pi_i(p,w,e_j|z)$, where $i = c, n$, indicating that the profit on each hectare will vary spatially as e_j varies. In addition, the soil C sequestered under each tillage practice will also vary spatially. The C per-hectare can now be expressed as $C_{ij} = C(x_{ij},e_j|z)$.

The solution to the tillage system choice problem is now a function of the environmental characteristics of the site. Figure 31.3 presents the per-hectare switch price and per-ton switch price for four land types with different biophysical characteristics. It is assumed that hectare type

---

* Note that $\pi_n + (g_n^*/\Delta C)\Delta C = \pi_n + g_n^*$.

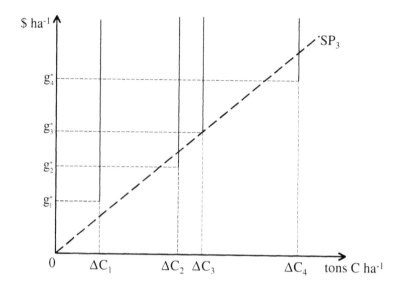

Figure 31.3   Soil carbon change under a no-till payment with spatially varying resource endowments.

1 is the least economically productive and also sequesters the least C, while hectare type 4 is the most economically productive and sequesters the greatest quantity of C. Figure 31.3 demonstrates that a higher per-hectare switch price is required to induce economically productive areas to switch from conventional to no-till in comparison to less productive areas, i.e., $g_{n4}^* > g_{n3}^* > g_{n2}^* > g_{n1}^*$. Unlike the previous case of homogeneous resources, the per-hectare switch price varies by land unit. Although only four different resource types are presented in Figure 31.3, in reality there is likely to be a continuum of types, indicating that, as g_n^* is increased or decreased, there will always be land types ready to switch to no-till or move out of no-till.

The per-ton switch price also varies for each hectare under the assumption of spatially variable resources. The line (SP$_3$), with slope $g_{n3}^*/\Delta C_3$, identifies the switch price per-ton of C for unit 3. The other units will generate a per-ton switch price with different slopes, indicating that the price per-ton of C required to induce producers to switch from conventional to no-till varies in response to resource endowments. Producers with a per-ton switch price with a steeper slope than SP$_3$ (for example, hectare type 1) require a higher price per-ton C to switch from conventional to no-till; those areas with shallower slopes demand a lower per-ton switch price (i.e., hectare types 2 and 4).

Efficiency of a Per-Hectare Payment vs. a Per-Ton Payment

Theoretical Analysis and Results

Figure 31.3 and the related discussion clearly show that spatially varying land characteristics influence the relative size of incentives that must be offered to producers before they will voluntarily adopt management changes that increase soil C. The following discussion builds upon these results to compare the efficiency of a per-hectare payment policy vs. a policy that provides payments per-ton of C sequestered.

The per-hectare and per-ton switch prices for two hectare types are shown in Figures 31.4a and b. Similar to the discussion above, assume that hectare type 2 is more productive than hectare type 1. In both figures, a per-hectare payment, g_n, satisfying $g_{n1}^* > g_n > g_{n2}^*$ is sufficient to induce producers on hectare type 1 to switch to no-till, but is not large enough to induce producers on hectare type 2 to switch; as such, these units remain in conventional till.

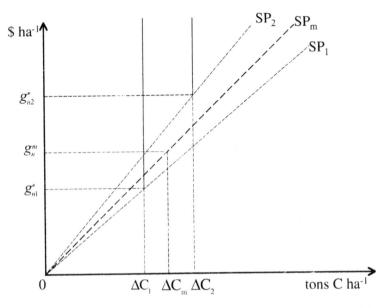

Figure 31.4a   Soil carbon change from no-till payments with heterogeneous land: field 1 with lower switch price.

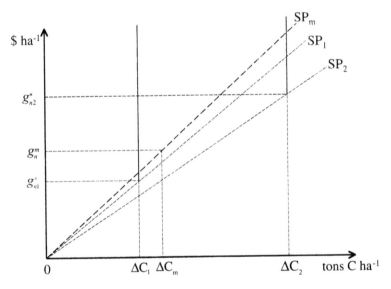

Figure 31.4b   Soil carbon change from no-till payments with heterogeneous land: field 2 with lower switch price.

In Figure 31.4a, the per-ton switch price for hectare 2 is greater than $g_{n1}^*/\Delta C_1$; Figure 31.4b shows the opposite case. Previously, an efficient policy was defined as one that equates the marginal social benefit of actions to sequester soil C to their marginal social cost. Assume that the marginal social benefit of a ton of C is m, represented by the line SP_m with slope $g_n^m/\Delta C_m = m$. If the marginal social benefit, m, of a ton of C satisfies $g_{n1}^*/\Delta C_1 < m < g_{n2}^*/\Delta C_2$, then in Figure 31.4a land type 1 will switch to no-till while land type 2 remains in conventional till (the same outcome as under the per-hectare payment scheme). However, in Figure 31.4b the efficient outcome is for both land types 1 and 2 to switch to no-till. This is different from the outcome suggested in Figure 31.4b when a per-hectare payment is considered. This result demonstrates the potential inefficiency of a per-hectare payment scheme for purchasing environmental services such as soil C.

Empirical Case Studies

Pautsch et al. (2001) use a discrete choice econometric model in conjunction with soil C estimates from EPIC to estimate the quantity of C sequestered by Iowa producers over payments that range between \$12 to \$1,800 ton^{-1} C. Under a per-ton policy* producers are paid for each additional ton of C sequestered as a result of switching tillage practices from conventional to no-till. Under a per-hectare policy, producers are paid for each hectare switched from conventional to no-till. Per-hectare payments are subsequently converted into equivalent payments per-ton C depending on the quantity of C sequestered on each hectare. Their results show that C is sequestered at a lower cost under a per-ton payment policy than a per-hectare payment policy.

For example, 0.5 MMT C can be purchased for an average cost of \$80 ton^{-1} under a per-ton policy, but the same amount would cost approximately \$130 ton^{-1} under a per-hectare design. (Antle et al. (2001a) demonstrate the equivalence between average cost per unit to the purchasing agent and marginal cost of producing that unit to the producer). At all C quantities, the per-ton policy is more efficient than the per hectare policy. Antle et al. (2001b) obtain similar results in a study comparing the relative efficiency of per-ton and per-hectare payments to encourage Montana producers to switch from a crop-fallow system to continuous cropping to sequester more soil C.

These empirical studies support the theoretical results discussed earlier in this chapter. Per-ton policies account for the spatial heterogeneity of land characteristics, and result in an efficient allocation of resources if farmers are paid a price for C equal to its marginal social value.

Design of a Tradable Emissions Program for Carbon

The previous analysis relates directly to C emissions trading schemes. The market for emissions credits sets the price per unit of carbon (m in the above discussion); then, farmers who can efficiently sequester carbon enter into appropriate contracts with firms that want to buy C credits.

Successful domestic trading of C credits between agriculture and industry and also within industry will require contracts for carbon that are fungible and verifiable, and operate within a transparent framework, consistently applied and holding parties accountable for their claims (Petsonk, Dudek, and Goffman, 1998). Difficulty in verifying the quantity of C sequestered will create uncertainty regarding the amount of C bought and sold and may impede market development.

IMPLICATIONS OF SPATIAL HETEROGENEITY FOR C POLICY DESIGN

An efficient policy to sequester C pays producers the marginal social value of each unit of C they sequester. Analysis shows that a per-ton payment scheme is generally more efficient than a per-hectare payment scheme when land characteristics are spatially heterogeneous. These results have several implications for the design of a C sequestration policy.

Spatially heterogeneous land characteristics also mean that the profitability of production systems varies spatially. For example, no-till may be more attractive in Iowa and other midwestern states where land is highly productive, whereas conversion to grass may be attractive in some areas, such as eastern Montana, where land is less productive for crops. An efficient policy would account for these regional differences. In addition, the amount of C sequestered in each region will depend on both the opportunity cost of changing practices and the rates at which soil C can be sequestered. Under an efficient policy, producers in regions with a high capacity to sequester C will be paid more on a per-hectare basis than those in regions with low C sequestration rates, resulting in spatially heterogeneous payments.

---

* Pautsch et al. (2001) call this a discriminating per-acre scheme.

Under a market mechanism or government policy designed to provide payments per-ton C, only producers that can sequester soil C efficiently will participate in the market for C credits or in the government program. Under a per-hectare scheme (or payment for adopting a specified management practice), some inefficient producers of soil C will participate, so the average cost per-ton of C produced will be higher than with a per-ton payment scheme. A policy that provides payments per-ton C or a functioning market for C will require a means to assess the quantity of sequestered C. Models and data are required to estimate a baseline level of C under existing practices and under alternative practices, in addition to a system of verifying compliance with practices specified in contracts. The amount of carbon sequestered may need to be verified to ensure that producers are paid for the amount of soil C in their contract, and also to account for the total amount of C sequestered on a national level.

INFORMATION REQUIREMENTS AND ADDITIONAL ISSUES

If an efficient policy for soil C sequestration requires a payment scheme that equates marginal social benefits and costs, why does the government not immediately set up such a program? The answer lies in the information requirements concerning the relationship among land quality, land use, management practices, and the function C(.) in addition to other benefits.

Scale and Measurement Issues

Policies to increase soil C will require economic and ecological models to estimate the effects of alternative land-use and management decisions on farmer behavior, profitability, and soil C. The preceding analysis showed that the spatial and temporal measurement of changes in soil C are key to designing efficient policies. However, measurements of spatially variable processes are scale dependent (Cao and Lam, 1997; Bian and Walsh, 1993), meaning that the results of similar analyses may vary if they are conducted at different scales (Davies et al., 1991; Wong and Amrhein, 1996). Policy analysis conducted at an aggregate scale may result in unreliable estimates. Analyses conducted using spatially explicit, disaggregate data may provide better estimates, but at a greater cost (Antle, Capalbo, and Mooney, 1999). Reliable estimates of the cost and likely benefits (in terms of tons of C sequestered) are important for successful policy design and implementation.

Dynamic Considerations — Investment and Contract Design

The time path of sequestration is an important factor influencing producer investment decisions and contract duration under efficient C payment and market trading schemes. The long-term decision to produce using conventional till or no-till systems (or convert to any recommended management practice) can be examined as an investment problem where producers choose the physical and human capital requirements for each land use that maximize the net present value of the investment. A producer will choose to invest if the present value of future profits is larger than the cost of the investment. Franzluebbers and Arshad (1996) suggest that soil C will increase slowly over the first 2 to 5 years of improvements in soil management, with larger increases between 5 to 10 years flattening off thereafter and reaching a finite limit after about 50 years (Lal et al., 1998). The time path of C sequestration will be a factor in long-run producer investment decisions as it determines the benefits over time from investment in sequestration efforts.

In addition to producer investment decisions, the time path of soil C sequestration will influence the optimal duration and design of contracts for sequestered C. Soil C rates generally vary according to management and land-use practices, so the duration of contracts needed to sequester a given amount of carbon will vary. Detailed information regarding the response of soil C to alternative management practices in a variety of locations will be needed to support the design of contracts for C sequestration.

Another issue is the permanence of soil C sequestered by changing agricultural practices. If one recognizes that farmers are providing soil C sequestration services for the time specified in a contract, then the permanence issue is resolved by contracting for these services for whatever period society wants to ensure that C remains in the soil. Thus, if a change from conventional to no-till causes soil C to increase for 20 years, but society determines that to reduce climate change risk, soil C should remain sequestered for 50 years, then contracts will have to be written for 50 years to ensure that farmers do not revert to practices that would release the C sequestered during the first 20 years of the contract.

Multiple, Spatially Variable Environmental Benefits (Co-Benefits)

A unique feature of the benefits afforded by each ton of C sequestered is that they are spatially invariant, that is, a ton sequestered in Minnesota provides the same societal benefits as a ton in Florida, California, Iowa, or any other state. This feature led to the assumption that the marginal social value of an additional unit of C removed from the atmosphere is constant at $m ton^{-1}. Spatially invariant benefits cannot be assumed for most other environmental changes that result from altering land use and management. For example, Feather, Hellerstein, and Hansen (1999) indicate that benefits from clean water are spatially heterogeneous, tending to be higher at recreation areas close to large centers of population.

Spatially heterogeneous benefits (or costs) are also likely to accrue from practices that increase soil C, which are expected to increase crop yields (Bauer and Black, 1994), reduce soil erosion, and decrease the need to add nutrients and water (Pimentel, Harvey, and Resosudarmo, 1995). These factors raise producer profits by increasing the returns from crop sales and decreasing input costs on each hectare. The monetary value of sequestered C to U.S. agricultural production is estimated at $15 to $42 billion (Lal et al., 1998).

In addition to production benefits, management changes to increase soil C may modify the habitats of terrestrial and aquatic organisms. For example, decreased soil erosion may decrease sedimentation in rivers, benefiting fish and other aquatic organisms and enhancing recreation opportunities. Thus, one can expect multiple, spatially varying benefits associated with changes in farming practices that increase soil C. In principle, an efficient policy would structure payments so that producers are compensated for all benefits provided by management changes. In practice, this ideal is difficult to achieve, but could be approximated using data, models, and other scientific tools now available to quantify these benefits on a spatially explicit basis.

Base Line Issues

A baseline level of soil C is needed in order to measure actual changes caused by changes in management practices. If incentives are provided to producers only for additional C sequestered when a contract is entered, producers who have already adopted practices that increase soil C would be penalized. They would not receive credit for C already sequestered, whereas producers who have depleted soil C would be rewarded because they could receive payments for putting C back into their soil. A key issue for the design of contracts or policies for soil C sequestration is how to avoid these kinds of perverse incentives (Sandor and Skees, 1999).

CONCLUSIONS

This chapter has showed that the spatial and temporal variability characterizing land resources affect the efficient design, distributional consequences, and information needs of policies to sequester soil carbon. Policies based on payments per-ton C sequestered are more efficient than those based on per-hectare payments for changes in production practices.

Soil C sequestration will vary across land units using similar production practices and with management or land-use decisions upon the same area. Efficient policies will encourage sequestration in areas of the country where production losses from management changes are offset by the sale of sequestered C. It is likely that different production practices will dominate in separate regions of the country, reflecting the spatial heterogeneity of resources.

Although policies based on payments per-ton C or market sales of C are efficient, enforcement and verification could be problematic because they require knowledge about the relationship between carbon sequestration and land quality, use, and management in addition to estimates of base line carbon levels. Soil carbon is spatially variable and its measurement is scale dependent, suggesting that analyses must be carefully designed to provide reliable estimates of C sequestered. There is scope for further research into the implications of combining economic and biophysical models over a range of scales.

The benefits of sequestering soil C, in terms of reducing the risk of climate change, are not spatially variable. However, most other environmental benefits (or costs) associated with changes in management practices are spatially variable, e.g., reduced erosion. This spatial variability in benefits has been recognized as a problem in existing policies such as the CRP. Measuring these benefits poses similar problems to the measurement of soil C.

REFERENCES

Antle, J.M. et al., A comparative examination of the efficiency of sequestering carbon in U.S. agricultural soils, Research discussion paper, Program on Climate Change and Greenhouse Gas Mitigation, Montana State University, www.climate.montana.edu, 2001a.

Antle, J.M. et al., Spatial heterogeneity and the design of efficient carbon sequestration policies for agriculture, Research discussion paper, Program on Climate Change and Greenhouse Gas Mitigation, Montana State University, www.climate.montana.edu, 2001b.

Antle, J.M., S.M. Capalbo, and S. Mooney, Optimal spatial scale for evaluating economic and environmental tradeoffs, selected paper, AAEA Ann. Meetings, Nashville, TN, 1999 (www.climate.montana.edu).

Barket, J.R. et al., Carbon dynamics of the conservation and wetland reserve program, *J. Soil Water Conserv.*, 51(4):340–346, 1996.

Barket, J.R. et al., Potential carbon benefits of the conservation reserve program in the U.S., *J. Biogeogr.*, 22:743–751, 1995.

Bauer, A. and A.L. Black, Quantification of the effect of soil organic matter content on soil productivity, *J. Soil Sci. Soc. Am.*, 58:185–193, 1994.

Bian, L. and S.J. Walsh, Scale dependencies of vegetation and topography in a mountainous environment of Montana, *Prof. Geogr.*, 45, 1–11, 1993.

Cao, C. and N.S. Lam, Understanding the scale and resolution effects in remote sensing, in Quattrochi, D.A. and M.F. Goodchild (Eds.), *Scale in Remote Sensing and GIS*, Lewis Publishers, New York, 1997.

Colby, B.G., K. Crandall, and D.B. Bush, Water right transactions: market values and price dispersion, *Water Resour. Res.*, 29-6:1565–1572, 1993.

Council for Agricultural Science and Technology, *Preparing U.S. Agriculture for Global Climate Change*, CAST, Task Force Report No. 199, Washington, D.C., 1992.

Council of Economic Advisors, Economic Report of the President, Washington, D.C., Government Printing Office, 1990.

Davies, F.W. et al., Environmental analysis using integrated GIS and remotely sensed data: some research needs and priorities, *Photogrametry Eng. Remote Sensing*, 57, 689–697, 1991.

Feather, P., D. Hellerstein, and L. Hansen, *Economic Valuation of Environmental Benefits and the Targeting of Conservation Programs*, Economic Research Service, USDA, Agricultural Economic Report No. 778, 1999.

Franzluebbers, A.J. and M.A. Arshad, Soil organic matter pools during early adoption of conservation tillage in northwest Canada, *J. Soil Sci. Soc. Am.*, 60:1422–1427, 1996.

Gebhart, D.L. et al., The CRP increases soil organic carbon, *J. Soil Water Conserv.*, 49(5):488–492, 1994.

Heimlich, R.E., Targeting green support payments: the geographic interface between agriculture and the environment, in Lynch, S. (Ed.), *Designing Green Support Programs*, Policy Studies Program Report No. 4, Henry A. Wallace Institute for Alternative Agriculture, 1994.

Joskow, P.L., R. Schmalensee, and E.M. Bailey, The market for sulfur dioxide emissions, *Am. Econ. Rev.*, 669–685, 1998.

Lal, R. et al., *The Potential of U.S. Cropland to Sequester C and Mitigate the Greenhouse Effect*, Ann Arbor Press, Chelsea, MI, 1998.

Landry, C.J., Giving color to Oregon's gray water market: an analysis of price determinants for water rights, unpublished M.S. Thesis, Department of Agricultural and Resource Economics, Oregon State University, Corvallis, OR, 1996.

Lynch, S. and K.R. Smith, Lean, mean and green... designing farm support programs in a new era, Policy Studies Program Report No. 3, Henry A. Wallace Institute for Alternative Agriculture, 1994.

Mann, L.K., Changes in soil C storage after cultivation, *Soil Sci.* (142):279–288, 1986.

Osborn, T., New CRP criteria enhance environmental gains. *Agric. Outlook*, October, Economic Research Service, USDA, 15–18, 1997.

Pautsch, G.R. et al., The efficiency of sequestering carbon in agricultural soils, *Contemporary Econ. Policy*, 19(2): 123–134, 2001.

Petsonk, A., D.J. Dudek, and J. Goffman, Market mechanisms and global climate change: an analysis of policy instruments, Environmental Defense Fund in cooperation with the Pew Center on Global Climate Change, 1998.

Pimental, D.C., P. Harvey, K. Resosudarmo, Environmental and economic costs of soil erosion, *Science*, 267:1117–1120, 1995.

Rasmussen, P.E. and W.J. Parton, Long-term effects of residue management in wheat fallow: I. Inputs, yield and soil organic matter, *Am. J. Soil Sci.*, (58):523–530, 1994.

Sandor, R.L. and J.R. Skees, Creating a market for carbon emissions: opportunities for U.S. farmers, *Choices*, First Quarter, 3–17, 1999.

Tiessen, H., J.W.B. Stewart, and J.R. Betany, Cultivation effects on the amount and concentration of C, N and P in grassland soils, *J. Agron.*, 74:831–835, 1982.

UNCTAD, Greenhouse gas emissions trading: defining the principles, modalities, rules and guidelines for verification reporting and accountability, United Nations Conference on Trade and Development, 1998.

Wong, D. and C. Amrhein, Research on the Maup: old wine in a new bottle or real breakthrough? *Geogr. Syst.*, 3, 1996.

CHAPTER 32

Economic Feasibility of Soil C Sequestration in U.S. Agricultural Soils

M. Sperow, R. M. House, Keith Paustian, and M. Peters

CONTENTS

INTRODUCTION

Concern over the influence of greenhouse gases (GHGs) on global warming and climatic variability has focused scientific and political attention on efforts to mitigate net GHG emissions. Agriculture, among other industries, could be involved in activities that reduce GHG emissions through, for example, soil carbon (C) sequestration. Cultivated U.S. agricultural soils have been estimated to contain 20 to 40% less soil C than soils that were never cultivated (Davidson and Ackerman, 1993; Mann, 1985). The loss of soil C resulting from conversion of undisturbed land to annual cropping may be reversed through certain land use and management practices (Lal et al., 1998). The analysis of Allmaras et al., (2000) suggests that U.S. cropland soils shifted from a C source to a C sink during the last 15 years, primarily due to changes in tillage practices away from moldboard plow. Eve et al. (2001) estimated that soil C in U.S. agricultural soils increased by over 8 million metric tons per year (MMT yr^{-1}) between 1982 and 1992 due to changes in land use and agricultural management practices.

Currently, sequestered soil C has no market price and has therefore been a "free" output of the production process. As suggested by economic theory, free production outputs will not influence production decisions. If a price for carbon is now established, economic theory says that it will influence crop production and management decisions, thereby leading to additional C increases in U.S. agricultural soils. U.S. implementation of a GHG emission reduction program would have a broad impact across economic sectors and actors. In determining such programs, the policy process would depend greatly on comparative economic and feasibility analyses.

From a national policy perspective, the viability of a policy to encourage soil C sequestration to mitigate GHG emissions will depend on the economics and environmental or social benefits of carbon-friendly practices. Programs offering incentives to increase carbon sequestration will be financially efficient only if marginal costs of inducing landowners to change land uses and sequester C are competitive with marginal costs of GHG mitigation through direct emission reductions in power generation, transportation, industrial, or other sectors. Additionally, policy to expand soil carbon sequestration may create positive externalities, which add to social well-being.

A key question to be addressed is the cost of GHG emission reduction through soil C sequestration, or how much soil C sequestration and GHG emission reduction will arise from any given price of soil carbon. Preliminary analyses of the economic potential of soil C sequestration have been performed at the farm level in Montana (Antle et al., 2000) and Iowa (Pautsch et al., 2000 and McCarl et al., 2001). Regional analyses have also been completed for the Lake States of Michigan, Wisconsin, and Minnesota, Corn Belt States of Illinois, Indiana, Iowa, Missouri, and Ohio, and Plains States of North Dakota, South Dakota, Nebraska, and Kansas (Pautsch and Babcock, 1999).

Whether the economic analyses are at the farm, regional, or national scale, understanding the relationships between a limited set of key driving variables is critical. Therefore, analysis was conducted on the sensitivity of economic outcomes to the expected rate of soil C sequestration, the price offered for sequestered soil C, the ease with which producers can alter land use and management activities to increase soil C, and the effects of targeting regions with higher soil C sequestration potential.

BACKGROUND

For management of U.S. agricultural soils to be an effective GHG mitigation strategy, their ability to sequester C must be ecologically and economically feasible. The economic feasibility of soil C sequestration must be addressed at the farm and national levels. Farmers and landowners will adopt new practices or change agricultural land uses to increase net soil C sequestration if they perceive sufficient economic and stewardship benefits from the change. Landowners will base their decisions on market revenues for alternative land uses, possible public program revenues and costs related to net carbon emissions, and the direct and indirect effects among commodity and carbon market revenues. Soil C sequestration can also generate additional private benefits to landowners, including improved soil tilth, water-holding capacity, and drainage, reduced soil erosion, improved water and air quality, etc., that could improve crop yields and maintain the soil for future generations. At the same time, society captures the offsite benefits of reduced levels of atmospheric CO_2 concentrations, decreased soil erosion, improved water and air quality, and increased food security.

The amount of soil C that may be sequestered is a function of the physical limits of the soil (soil characteristics), climate, plant residue inputs, tillage practices, and other biophysical characteristics. Soil C may be increased through reduced tillage intensity, more intensive cropping systems, adoption of yield promoting practices, permanent perennial vegetation (Bruce et. al, 1999), soil erosion management, production of biofuels, and land conversion and restoration (Lal et. al, 1999).

In addition, agricultural producers may switch to crops that sequester more C, improve residue management activities, or eliminate fallow operations to increase soil C. These practices increase annual subsurface and surface biomass production and reduce carbon emissions from decomposition.

Lal et al., (1998, 1999) estimated potential soil C sequestration from improved management on U.S. agricultural soils of between 75 and 208 (MMT yr^{-1}), derived primarily from conservation tillage (24 to 40 MMT yr^{-1}) and improved cropping systems (19 to 52 MMT yr^{-1}). Bruce et al. (1999) estimated potential soil C sequestration of 75 MMT yr^{-1}, assuming that the best carbon-conserving practices are adopted on cropland, set-aside land, grassland, and degraded soils. The potential soil C gain was 30.8 MMT yr^{-1} from nondegraded cropland and 14.3 MMT yr^{-1} from eroded cropland based upon an overall average gain of 0.3 Mgha^{-1}yr^{-1} from long term data. These scientific assessments of potential soil C sequestration were derived from long-term studies and modeling activities based on assumed changes in land use and agricultural land management (Bruce et al., 1999; Lal et al., 1998; Paustian et al., 1997a; Paustian et al., 1997b).

MODELS, DATA, AND METHODS

Soil C sequestration rates derived from existing literature (Table 32.1) are combined with the U.S. agriculture sector mathematical programming (USMP) economic model in this analysis (House et al., 1999). USMP is a general purpose, partial-equilibrium economic model that accounts for total domestic supply and demand, import, and export of eight major agricultural crops: barley (*Hordeum vulgare L.*), corn (*Zea mays L.*) (grain and silage), cotton (*Gossypium hirsutum L.*), oats (*Avena sativa L.*), rice (*Oryza sativa L.*), sorghum (*Sorghum bicolor L.*), soybean (*Glycine max L.*), and wheat (*Triticum aestivum L.*) and 44 products comprising the remaining principal U.S. crop and livestock commodities. Hay, primarily alfalfa (*Medicago sativa L.*), is also included as a multiyear crop activity. The spatial resolution of the USMP is the 45 regions formed by the intersection of ten USDA Farm Production Regions and 20 Land Resource Regions (USDA, 1981). Sixty-five multiyear crop rotations (e.g., continuous corn, corn–beans, wheat-fallow, and so on) and five tillage methods (conventional with moldboard, conventional, mulch, ridge, and no-till) are specified in USMP. In each of the 45 production regions, cropping systems with tillage and rotation appropriate for the region (i.e., observed in crop surveys) are specified — summing to about 800 systems across all regions.

Table 32.1   Soil C Sequestration Rates used in USMP Model

Tillage Method[a]	Climatic Region[b]					
	Humid Residue Input[c] (Mg Ha^{-1} Yr^{-1})			Semiarid Residue Input (Mg Ha^{-1} Yr^{-1})		
	High	Medium	Low	High	Medium	Low
NT	0.50	0.30	0.00	0.30	0.10	−0.05
RT	0.20	0.01	−0.10	0.10	0.00	−0.01
CT	0.05	0.00	−0.20	0.05	−0.10	−0.02

[a] NT = No-till, RT = Reduced tillage (>30% residue), CT = Conventional tillage (with and without moldboard plow).
[b] Humid regions include the Lake, Delta, Appalachian, Northeast, Southeast, and Corn Belt Farm Production Regions. Semiarid regions include the Pacific, Northern Plains, Southern Plains, and Mountain Farm Production Regions.
[c] High = High residue input systems (e.g., rotations with alfalfa hay, continuous corn, continuous sorghum, continuous small grains in semiarid regions); Medium = Medium residue input systems (e.g., continuous small grains in humid regions, small grain rotations with fallow in semiarid regions, and rotations with soybean); L = Low residue input systems (e.g., continuous soybean, continuous grain silage, and more frequent fallow periods in small grain rotations).

Source: Adapted from Paustian et. al , 1997.

USMP can be applied to analyze impacts in any desired "base year," that is, a recent historical year or any year in the USDA long-term baseline (1988 to 2012; USDA, 1998). All input–output coefficients, acreage planted, commodity production, and price variables in USMP are calibrated to levels consistent with the base year, using parameters from the USDA Baseline, official/survey price deflators, data such as the National Agricultural Statistics Service (NASS, 1997), National Resources Inventory (NRCS, 1994), and other data. USMP's crop production enterprise economic coefficients are derived from Cropping Practices Surveys (USDA, 1992) and the Agricultural Resource Management Study ([ARMS], USDA, 1996).*

Comparative static analysis is performed by applying a policy, market, or technology shock (e.g., introducing a price for soil carbon) to the fully calibrated USMP model. The model is allowed to adjust to the shock, and then analysis is performed on the changes in performance indicators (economic indicators including prices, acreage, commodity production, and exports; environmental indicators including erosion, nitrogen and other chemical losses to water, and the atmosphere; change in soil carbon sequestered per year by region, etc.).

The USMP model is used to estimate changes in soil C, crop production (tillage practice and rotation), and farm income effects when an economic value for soil C is applied. Conventional tillage (which, with and without moldboard plow, is assumed to have the same effect on soil C), reduced tillage (ridge and mulch tillage) and no-till operations are analyzed.

INDUCING ADOPTION OF AGRICULTURAL CROPPING PRACTICES TO SEQUESTER CARBON

Incentives for Adoption of Crop Management Practices

Revenue and cost effects influence the incentive to change crop management activities to increase soil C sequestration. Direct revenue effects are the direct returns from soil C sequestration that are a function of the market price and quantity (or yield) of soil C. Indirect revenue affects are the price and yield impacts (that affect revenue) on other products that the farmer is already producing. Direct cost effects are the measurable costs incurred from changing cropping practices such as new equipment purchases, changes in applications of herbicides, reduced energy expenses, and lower labor costs. Indirect cost factors may not fit easily into categories. For example, a switch to no-till will likely require an increase in management, which could be a monetary cost if management services are purchased.

Day et al. (1998) found that net returns from conservation tillage were statistically higher, lower, or no different from net returns from conventional tillage in studies throughout the U.S. While costs under conservation tillage were generally lower, net returns were more variable, primarily from increased yield variability (Day et. al., 1998). Thus, new production practices could add implicit risk premia to the ordinary costs of production due to uncertainty associated with expected yields, crop response to severe weather, and adjustments to management skills. These might apply during a transition period in which operators gained familiarity with new practices.

Conservation tillage requires farmers to forego the use of tillage operations to mitigate other risks so yield variability is not the only source of risk. For example, soil temperatures under no-till remain cooler in the spring than those of tilled soils. Exceptionally cool springs may delay planting and seed germination, thus threatening the crop production season. Exceptionally dry springs could result in increased weed competition when moisture may be inadequate for crop seed germination, but adequate for weed seed germination. Chemical control of the weeds may not be possible because the crop may be emerging at the optimal time for weed control.

---

* ARMS is USDA's primary vehicle for data collection of information on a broad range of issues about agricultural resource use and costs, and farm sector financial conditions.

Lack of information or know-how on the part of producers, climate, and physical properties of the soil in the region may act as barriers to switching crop practices. These could include simple lack of experience with no-till planting practices or biases in favor of familiar practices. Policy options that address noneconomic barriers to adopting certain practices include extension, education, training, and cost-sharing types of programs. Sensitivity analysis of factors influencing adoption of carbon sequestration practices may help shape policy development and implementation by defining the most critical barriers to change so that proper incentives can be provided.

Carbon Incentive Price

Antle and Mooney (1999) demonstrated that payments per unit of soil C sequestered or market-based trading systems are more efficient than payments for changes in land management. Pautsch et al. (2000) determined that the lowest payment cost may be achieved with a price discriminating subsidy; however, administrative costs may be prohibitive. Therefore, a single price subsidy scheme for each metric ton (MT) of soil C sequestration rate is applied in this analysis. Five carbon incentive prices were evaluated (in addition to the current zero price): $14, $50, $100, $150, and $200 MT^{-1}. This generic incentive price could result from various explicit policy frameworks, a government price or incentive payment, or from the value of tradable carbon permits.

Sandor and Skees (1999) pointed out that payments must be made for all soil C sequestration activities, including existing ones, to prevent reversion to soil C depleting activities and then back to sequestering activities to gain payment. Therefore, payments are made for all soil C sequestering activities in this analysis, both existing and those that result from land-use or management changes.

Carbon Yield (Carbon Sequestration Rate (CSR) per Ha.)

Kern and Johnson (1993) found that soil C sequestration in dry warm (semiarid) climates is lower than in moist humid areas. Tillage intensity also influences the rate and level of soil C sequestration, with more intensive tillage operations associated with lower soil C sequestration rates (CSRs). Thus, soil CSRs were specified spatially across U.S. land resource regions based upon climate variability, tillage practice, and crop rotation. Holding climate and soil characteristics constant, the rate of C input (primarily plant residue) is the most critical factor in determining the amount of C that can be maintained in the soil (Paustian et al., 1997b). As shown in Table 32.1, high-, medium- and low-residue crops are used to represent crop rotations and assign soil C values across the U.S. by semiarid and humid climatic regions.

High-residue inputs result from crop rotations that include hay, corn, and sorghum in humid and semiarid regions; however, high-residue input systems in humid and semiarid regions are not identical. For example, small grains are considered medium-residue input systems for humid regions and high-residue input systems in semiarid regions. Medium-residue inputs include annual cropping systems such as corn and soybean in both humid and semiarid regions and small grains with short fallow periods in semiarid regions. Low-residue input systems result from more frequent fallow periods with small grains and rotations that include cotton or corn silage.

Ease of Switching to Conservation Tillage Practices

U.S. acreage under no-till cultivation may be relatively constant (CTIC, 1998) because agricultural producers, as profit maximizing entities, are producing crops under systems in which they have a comparative advantage. Research has shown that nonbudget factors may be more influential than economic returns in decisions to adopt conservation tillage practices. Westra and Olson (1997) found that producers adopt conservation tillage practices in Minnesota when they have larger farms, better management skills, believe that it fits into their production goals, obtain information from other producers, are concerned about soil erosion, and when the physical setting of their farm was suitable. Different studies have found the level of education has a positive (Wu and Babcock, 1998) or negative (Westra and Olson, 1997) effect on adoption of conservation tillage. Wu and Babcock (1998) determined that agricultural producers with more experience adopted conservation tillage less frequently.

Whether barriers are economic or noneconomic, there is resistance to converting from conventional to conservation tillage operations. A parameter is applied that serves as a proxy for the flexibility of adopting new crop management practices or barriers to adoption, which may be lowered through education, extension, and other means that make adoption of conservation tillage more likely. In the context of this sensitivity analysis, alternative ease in changing tillage practices denotes the ease with which producers shift from a conventional tillage system to a conservation tillage system, especially no-till. A basic degree of flexibility, moderate flexibility, and high flexibility is specified.

The ability to switch between tillage types in USMP is specified according to constant elasticity of transformation (CET) allocation functions (House et al., 1999: Equation 32.1). A separate CET function controls substitution across tillage types within each multiyear crop rotation production system — one CET function for each of the 341 crop rotation and Land Resource Region strata within USMP. With the CET formulation, alternative degrees of ease in substituting one tillage activity for another are specified by altering the transformation elasticity, ε, controlling the CET functions:

$$Y = A \left[\sum_{i=1..n} \delta_i (X_i)^{-\rho} \right]^{-\frac{1}{\rho}} \tag{32.1}$$

where:

- Y = Total hectares planted
- A = Shift parameter (e.g., technological change), $A > 0$
- i = Crop enterprise or rotation
- δ = Enterprise or crop rotation share of total hectares, $0 < \delta < 1$
- X = Base hectares of enterprise or crop rotation
- ρ = Transformation parameter, $\rho \leq -1$

The tillage CET functions are specified assuming that producers are profit-maximizing operators in a competitive equilibrium environment. Producers choose acreage activity levels of each enterprise X_i (crop production enterprises) so as to minimize the cost P_i (reduced costs of production per hectare) of producing output on acres Y (Equation 32.2).

$$\sum_{i=1..n} P_i X_i \tag{32.2}$$

where:

- P = Reduced costs of production
- i = Crop enterprise or rotation
- X = Crop production enterprises

USMP tillage CET functions are calibrated (specifying their parameters) in a fashion similar to the standard approach of using CES (constant elasticity of substitution) or CET allocation or aggregation functions in computable general equilibrium (CGE) models (House et al., 1999). This approach is widely used in CGE modeling because it makes good use of available information; the functional form is parsimonious in parameters and is empirically reasonable in representing allocation of inputs among competing uses, products among alternative demand sectors, and so on (Dervis et al., 1982). A calibration procedure typically derives the shift parameter, A, and the share parameters, δ_i, from available data while allowing the analyst to specify the transformation (or substitution) elasticity, ε. The parameter ρ (Equation 32.3) determines the shape of the transformation function's contours, where $\rho \geq -1$ is the CES case with contours concave from above, and $\rho \leq -1$ is the CET case with contours concave from below.

$$\rho = \frac{1}{\varepsilon} - 1 \qquad (32.3)$$

where

ρ = Transformation parameter, $\rho \leq -1$
ε = Transformation elasticity, $\varepsilon < 0$

The equilibrium condition (Equation 32.4) has the economic interpretation that the contributory value of each production enterprise i relative to the total value of all n enterprises has the same relationship as the activity level X adjusted by share parameter δ and substitution parameter ρ, relative to the similarly adjusted total sum. The relationships in Equations 32.1 and 32.4 are used in USMP to specify substitution among alternative tillage types within each multiyear crop rotation. The formulation is also used in USMP to specify substitution across alternative multiyear, multicrop rotation systems for production of an individual crop:

$$\frac{P_1 X_1}{\sum_i P_i X_i} = \frac{\delta_1 X_1^{-\rho}}{\sum_i \delta_i X_i^{-\rho}} \qquad (32.4)$$

where:

P = Reduced costs of production
X = Hectares of commodity
i = Index of all commodities grown
δ = Enterprise and crop rotation share of total hectares
ρ = Transformation parameter, $\rho \leq -1$

The USMP model runs embody a complex set of changes across many crop systems in many regions. To demonstrate the effect of alternative tillage elasticity, *ceteris paribus*, the partial effect of changing returns to one tillage method in one cropping system in one region (no-till continuous corn in the central Corn Belt region) was calculated. Continuous corn in the central Corn Belt region is produced using four tillage methods: conventional, conventional with moldboard, mulch till, and no-till. Figure 32.1 presents the base shares of continuous corn, central Corn Belt acreage under these four tillage methods, and the tillage shares that result under three alternative transformation elasticities when the no-till system is subsidized by $26 per acre. Under the basic degree of flexibility ($E_s = -10$) the proportion of conventionally tilled acres decreases from 41 to 36% of total crop acres, while no-till increases from 8 to 19%. When it is relatively easy to switch between tillage methods (high flexibility, $E_s = -50$), conventionally tilled acres are reduced further to 4% and no-till acres increase to 91% of cropped acres.

RESULTS AND DISCUSSION

Conventional tillage operations (with and without moldboard plow) were used on 74%, reduced till on 16%, and no-till on 10% of the U.S cropland planted in the baseline scenario. Most crop

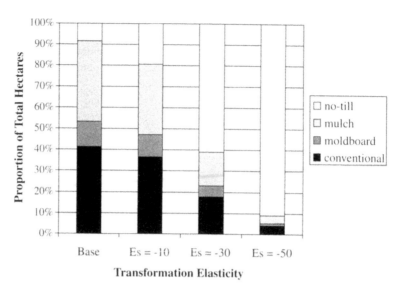

Figure 32.1   Comparison of the proportion of continuous corn produced under various tillage intensities when the ease of switching from one tillage operation to another (transformation elasticity) is varied and the no-till system is subsidized by $26 per acre. The higher the transformation elasticity, the more difficult it is to change from one tillage method to another.

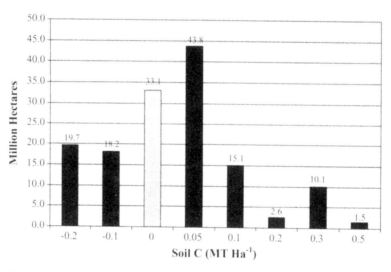

Figure 32.2   Baseline status of U.S. cropland as simulated by the USMP model using soil C sequestration rates derived from literature. (No price is offered for sequestered C.) Soil C rates (x axis) are dependent upon crop residue, tillage, and climate.

production in the baseline scenario occurred under management systems, crop rotations, and climatic regions with soil C sequestration rates of 0.05 and 0.00 MT Ha^{-1}yr^{-1} (Figure 32.2). The soil CSRs were derived from site-specific results and extrapolated to broad geographic regions and crop systems. Because results from this analysis may not be suitable for policy decisions, specific soil C gains from alternative offer prices are not presented. Rather, the purpose of this research is to present the relative influence of various critical factors.

To illustrate the influence of different soil CSRs, soil C supply curves were developed for regions with higher and lower soil CSRs (Figure 32.3). The low CSR response is typical of a drier climatic region predominantly in crop-fallow rotations likely occurring on sandy soils. The higher

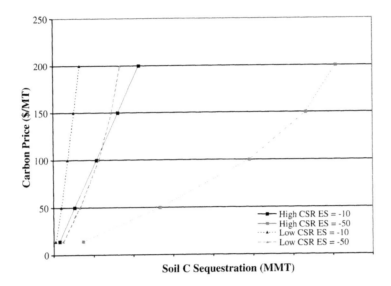

Figure 32.3   Soil C increases in U.S. crop production regions with higher (solid lines) and lower (broken lines) soil C sequestration rates (CSR) under various price scenarios ($14, $50, $100, $150, and $200 MT^{-1}). The figure shows soil C changes when the ability of farmers to change tillage practices is somewhat difficult (darker lines) and easier (lighter lines).

CSR is derived from a moist climatic region with crop rotations more conducive to increased soil C (e.g., continuous corn) on highly productive soils.

More soil C gain results from various offer prices in regions with higher CSR (solid dark line) than areas with lower CSR (broken dark line) when a base transformation elasticity (–10) is used. In economic terms, the supply response to increases in the soil C price is more elastic in regions with higher CSRs than regions with lower CSRs because these areas show more responsiveness to increased prices. When the soil CSR is low and it is somewhat difficult to change tillage operations, higher offer prices result in small increases in soil C sequestration. Conversely, when the soil CSR is higher, higher offer prices lead to greater increases in soil C. However, the soil C supply curve for the region with higher CSR is relatively inelastic when it is somewhat difficult to change tillage operations. Significant increases in the soil C price are required to encourage moderate increases in sequestered soil C.

The ease with which agricultural producers can change tillage practices (represented by the transformation elasticity) is not known with certainty. To determine the sensitivity of results to the ease of changing tillage, the transformation elasticity parameter was varied from somewhat difficult (dark lines in Figure 32.3) to somewhat easy (gray lines in Figure 32.3). Substantial gains in total soil C sequestered can be attained when the transformation elasticity parameter is decreased (from –10 to –50) in the region with the higher CSR (solid lines). At each higher soil C price, and particularly at soil C prices above $150 MT^{-1}, the soil C supply response becomes more inelastic. Crop management practices that increase soil C do not enter the solution until much higher prices are offered in areas where increasing soil C is difficult. A similar decrease in the transformation elasticity parameter in the region with the lower CSR results in very small increases in soil C, despite the ease with which tillage activities may be reduced.

An effective policy mechanism that encourages adoption of crop management practices that increase soil C should consider the total costs and benefits of the program. One approach to determining potential costs is to analyze the incremental increases in soil C resulting from different soil C prices. The effectiveness of a soil C price depends upon the ability of soils to sequester C, the technologies available to enhance soil C sequestration, and the climate, soils, and crop rotations

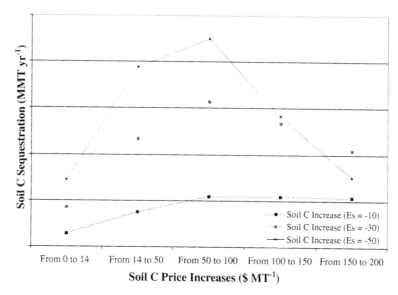

Figure 32.4 Incremental increases in soil C sequestration resulting from various price range increases.

in the region. The incremental increases in soil C that result from different soil C price ranges and different transformation elasticities are shown in Figure 32.4. When it is relatively difficult to change tillage practices (elasticity of –10), increasing the soil C price does not encourage substantial increases in soil C sequestration. The incremental gains in soil C from increased soil C price ranges are larger at lower prices. Above a price range of $50 to $100, the incremental increase in soil C is nearly zero.

When the ease of adopting soil C sequestering activities is increased (e.g., elasticity of –50), the incremental gain in soil C increases. At lower soil C price ranges, substantial incremental increases in soil C are shown, indicating that lower prices induce adoption of soil C sequestering practices early in areas where the potential soil C gain may be higher. Increasing the soil C price beyond a certain level leads to a decline in the incremental gain in sequestered soil C. In this analysis, when the range of soil C offer prices is greater than $50 to $100, the incremental gain in soil C (the additional soil C provided by agricultural soils) is still positive, but declines rapidly. Thus, there may be a soil C offer price where the incremental gains are highest, which would reduce the per unit costs of increasing soil C sequestration.

As illustrated in Figures 32.3 and 32.4, by pricing carbon greater than zero (the current price), policy automatically targets regions where largest gains in sequestered soil C may be obtained at lowest price. In areas where CSR is low, the per hectare payments to producers will be small, even at higher soil C offer prices. In such areas (low CSRs) technological improvements would be necessary before economic incentives could be as effective as areas with high CSRs. In any case, in areas where CSRs are high, per hectare payments will be higher (higher gains for producers), more soil C will be sequestered (increased GHG mitigation), and the overall costs of policy implementation will be lower (more economic efficiency and gains to society), than with a policy that would seek to equalize payments per hectare across the nation.

If there is an important equity reason to equalize carbon payments per hectare across the nation, for example, minimizing changes in income distribution, the government and taxpayers would have to bear the increased costs and reduced efficiency of such an approach. (Maintaining equity is often as important as efficiency in the policy world.) An alternative approach would be funding research to make C sequestration more effective in areas where it is presently less effective.

CONCLUSIONS

The present analysis addresses only crop management practices that enhance soil carbon sequestration and identifies how economic outcomes vary with soil C offer prices, ease of changing tillage practices, and soil C sequestration rates per dollar expended. Ultimately, proper evaluation of carbon sequestration in agriculture as a GHG mitigation strategy requires a comprehensive framework incorporating all types of sequestration options.

Soil C sequestration on agricultural land is influenced by the biophysical limitations of climate, crop rotation and tillage, the C offer price, and the ease with which agricultural producers can switch from one cropping system to another. Analyses revealed that, as expected, increasing the C offer price, relaxing the transformation elasticity, and higher soil CSRs increase the amount of soil C sequestered. The results indicate that making it easier to adopt soil C sequestering crop management practices may be more influential than the CSR in obtaining soil C gains, although both are required. This chapter has identified the framework for more detailed analyses to define soil C sequestration potential on U.S. agricultural soils.

Determining the potential for U.S. soils to sequester C and be a viable GHG mitigation option requires more detailed information than what has been addressed in this chapter. The analyses provide a broad overview of the potential for sequestering soil C in U.S. agricultural soils, but are not sufficient for policy decisions. Using broad-based soil C estimates for various crop management strategies and climatic regions may over- or underestimate the true potential for soil C sequestration. Site-specific data are required to generate soil C potential values and identify changes in cropping patterns and management strategies required to achieve increased soil C values.

In ongoing research, the USMP and Century models are used to estimate changes in crop rotations and management practices given varying prices for sequestered soil C. The Century model is used to estimate soil C under current cropping and management systems and estimate subsequent changes to soil C levels when crop rotations or management practices change. Tillage operations and dates, plant and harvest dates, the quantity and date of fertilizer application, and crop rotations and management practices are derived from the ERS cropping practices surveys (USDA, 1992). The crop rotation and management practice data represent the baseline cropping systems from which subsequent changes will be analyzed.

Century data documenting soil C response to changes in crop production and management decisions will enhance the input data required for the USMP model. Century output data will provide the site or regionally specific data required to improve the confidence level of results provided by this chapter.

Note: Views expressed are those of the authors, and do not necessarily reflect a position of the U.S. Department of Agriculture. This research was performed under a cooperative agreement between Colorado State University and the Resource Economic Division, ERS.

REFERENCES

Allmaras, R.R. et al., Soil organic carbon sequestration potential of adopting conservation tillage in U.S. croplands, *J. Soil Water Conserv.*, 55(3):365–373, 2000.

Antle, J.M. and S. Mooney, Economics and policy design for soil carbon sequestration in agriculture, Research discussion paper No. 36, Trade Research Center, Montana State University, Bozeman, 1999.

Antle, J.M. et al., Economics of agricultural soil carbon sequestration in the Northern Plains, Research discussion paper No. 38, Trade Research Center, Montana State University, Bozeman, 2000.

Bruce, J.P. et al., Carbon sequestration in soils, *J. Soil Water Conserv.*, 54(1):382–389, 1999.

Conservation Technology Information Center (CTIC), Crop residue management executive summary, Conservation Technology Information Center, West Lafayette, IN, 1998.

Davidson, E.A. and I.L. Ackerman, Changes in soil carbon inventories following cultivation of previously untilled soils, *Biogeochemistry*, 20:161–193, 1993.

Day, J.C. et al., Conservation tillage in U.S. corn production: an economic appraisal, select paper for the 1998 Annual Meetings of the American Agricultural Economics Association, Salt Lake City, Utah, August 2–5, 1998.

Dervis, K., J. DeMelo, and S. Robinson, *General Equilibrium Models for Development Policy*, Cambridge: Cambridge University Press, 1982.

Eve, M.D. et al., A national inventory of changes in soil carbon from National Resources Inventory data, in Lal, R. et al. (eds.), *Assessment Methods for Soil Carbon*, CRC Press, Boca Raton, FL, 2001.

House, R.M., M.A. Peters, and F.H. McDowell, USMP regional agricultural model, USDA/ERS/RED working paper 5/26/99, 1999.

Kern, J.S. and M.G. Johnson, Conservation tillage impacts on national soil and atmospheric carbon levels, *Soil Sci. Soc. Am. J.*, 57:200–210, 1993.

Lal, R. et al., *The Potential of U.S. Cropland to Sequester Carbon and Mitigate the Greenhouse Effect*, Sleeping Bear Press, Chelsea, MI, 1998.

Lal, R. et al., Managing U.S. cropland to sequester carbon in soil, *J. Soil Water Conserv.*, 54(1):374–381, 1999.

Mann, L.K., A regional comparison of carbon in cultivated and uncultivated Alfisols and Mollisols in the central United States, *Geoderma*, 36:241–253, 1985.

McCarl, B.A.M. et al., *Economic Potential of Greenhouse Gas Emission Reductions: Comparative Role for Soil Sequestration in Agricultural and Forestry*, paper presented at DOE First National Conference on Sequestration, May 2001.

McDowell, F.H. et al., Reducing greenhouse gas buildup: potential impacts on farm-sector returns, *Agric. Outlook*, Aug. (1999): AGO-263, 19–23, 1999.

NASS (National Agricultural Statistics Service), Crops county data [online] available at http://usda.mannlib.cornell.edu/data-sets/crops/9X100, 1997.

NRCS, 1992 National Resources Inventory digital data, United States Department of Agriculture, Natural Resources Conservation Service (formerly Soil Conservation Service), Washington D.C., 1994.

Paustian, K. et al., Agricultural soils as a sink to mitigate CO_2 emissions, *Soil Use Manage.*, 13:230–244, 1997a.

Paustian, K., H.P. Collins, and E.A. Paul, Management controls on soil carbon, in Paul, E.A. et al. (eds.), *Soil Organic Matter in Temperate Agroecosystems, Long-Term Experiments in North America*, CRC Press, Boca Raton, FL, 1997b, 15–49.

Pautsch, G.R. and B.A. Babcock, Relative efficiency of sequestering carbon in agricultural soils through second best instruments, paper presented at the 3rd Toulouse Conference on Environmental and Resource Economics, Toulouse, June 14–16, 1999.

Pautsch, G.R. et al., The efficiency of sequestering carbon in agricultural soils, working paper 00-WP 246, Center for Agricultural and Rural Development, Iowa State University, Ames, 2000.

Sandor, R.L. and J.R. Skees, Creating a market for carbon emissions, *Choices*, First Quarter: 13–17, 1999.

USDA, Land resource regions and major land resource areas of the United States. Soil Conservation Service, USDA, Washington, D.C. Agricultural Handbook 296, 1981.

USDA, Cropping Practices Survey Data, Economic Research Service, USDA, Washington, D.C., 1992.

USDA, Agricultural Resource Management Study, National Agricultural Statistics Service, USDA, Washington, D.C., 1996.

USDA, USDA Agricultural Baseline Projections to 2007, Interagency Agricultural Projections Committee, Staff Report no. WAOB-98-1, 132 pp., February, 1998.

Westra, J. and K. Olson, Farmers' decision processes and adoption of conservation tillage, Department of Applied Economics, College of Agricultural, Food, and Environmental Sciences, University of Minnesota, staff paper, 97–99, 1997.

Wu, J. and B.A. Babcock, The choice of tillage, rotation and testing practices: economic and environmental implications, *Amer. J. Agr. Econ.*, 80, 494–511, 1998.

CHAPTER 33

Emissions Trading and the Transfer of Risk: Concerns for Farmers

John Bennett and Dave Mitchell

CONTENTS

INTRODUCTION

Agriculture can make two contributions to lowering greenhouse gas (GHG) concentrations in the atmosphere: it can reduce emissions from fossil fuels, fertilizers, and livestock, and it can remove greenhouse gases from the atmosphere with biological sinks. If markets are to provide the incentive to make the soil sequestration of carbon happen, it is necessary to understand the differences between these two contributions and to design markets that address these differences.

Markets for emissions trading have the potential to benefit both greenhouse gas emitters, by lowering the cost of reducing emissions, and farmers, who can increase their farm income by adopting Best Management Practices that reduce emissions and remove GHGs. However, the kind of emissions trading system that may develop may not adequately protect the interests of farmers. This chapter takes a critical look at some possible options for an emissions trading system, paying particular attention to the possible transfer of risk to farmers, and then proposes one possible emissions trading system that could better protect farmers from an unjust and overly heavy burden of risk.

Emissions trading is increasingly gaining attention in the agricultural sector. In October, 1999, GEMCo (a consortium of Canadian utilities and energy companies) struck a deal with Iowa farmers for the purchase of 2.8 million ton of CO_2 equivalent emission offsets.* (McConkey et al., 1999). More and more frequently, farmers are approached with similar offers of money in exchange for adopting farming practices that reduce overall emissions of GHGs as well as practices that sequester or remove atmospheric CO_2. Farmers must only agree that emission reductions or the carbon sequestered under their crops "belongs" to the buyer. Unless farmers take the time to understand the risks that they are being asked to assume, this offer can be just what it seems: too good to be true. GHG removals with soil sinks can be very useful as a bridge to meet short-term GHG targets and to allow emitting industries the time needed to implement permanent emission reductions. However, markets for carbon sinks must not be a mechanism that transfers the liability of emissions from emitters to farmers and land managers.

A WORD ON TERMINOLOGY

When speaking of emission reduction credits (ERCs), it must be noted that there are important differences between credits that represent GHG emission reductions and those that represent GHG removals from the atmosphere. These differences have important implications for determining the price for which the credits are sold:

Emission reductions mean only that a lower level of GHGs is emitted than was previously the case. Significant amounts of GHGs, it must be remembered, are still being emitted.

Removals refer to the actual removal of GHGs from the atmosphere. Biological sinks are currently the only method in place to actually reduce the growing levels of GHGs in our atmosphere.

The distinction between these two methods of generating ERCs reveals some important implications for the price of the credits. In the case of reductions, the price need reflect only the cost of reducing emission levels (through cleaner-burning fuels, increased energy efficiency, fertilizer management, manure management, etc.). For removals, the price must reflect the costs of removing the GHGs from the atmosphere and of maintaining the sink — presumably for as long as global warming is a problem. Farmers should keep these factors in mind when deciding whether to sell their GHG removals, and for how much.

Virtually all farms would have some level of GHG emission and most have some options that would reduce these levels. To illustrate (Figure 33.1), assume emission levels remain reasonably constant, but at time A the farmer implements a fuel consumption reduction and reduces emissions by 10%. Later, at time B the farmer implements a fertilizer management practice that reduces N_2O emissions 10%. The farm has now reached a 20% reduction. If the emission target were 6%, the farmer would have a surplus of 14% to sell in an emission reduction market. The 6% target can change, however, varying the surplus with it. The scientific community is adamant that a Kyoto target of –6% will not address the urgency of global warming. If public opinion follows scientists' lead, so will politicians. Thus, a prudent farm manager should recognize the moving target for emission reduction and not sell off all the surplus credits.

Next assume that this same farmer adopts best management practices (BMP) with reduced or zero tillage and continuous crop production that would remove CO_2 from the atmosphere and sequester it in soil as organic matter, creating a sink. GHG removals would be the greatest in the early years and would level off as the sink becomes saturated (McConkey et al., 2000).

---

* This includes practices that sequester carbon (such as shifting from intensive to minimum or zero tillage farming or planting trees on marginal land) and practices that reduce emissions from, primarily, fossil fuels (CO_2), nitrogen fertilizer (C_2O), and ruminants (CH_4).

The graph in Figure 33.2 shows that GHG removals are the greatest shortly after management changes. Eventually the sink will fill and gains in sequestered carbon will level off. Farmers must also realize that this sink can quickly be eliminated by returning to tillage. Superimposing these two graphs would create an annual farm carbon account as shown in Figure 33.3. The challenges of emissions on the farm will remain long after the benefits of sequestration have leveled off.

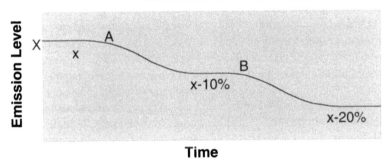

Figure 33.1   Emission level vs. time.

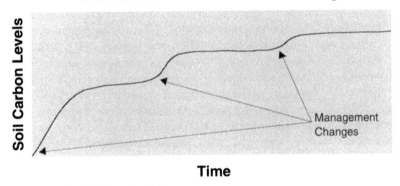

Figure 33.2   Soil carbon level with reduced tillage.

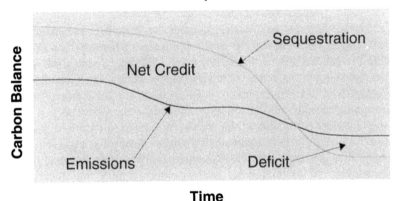

Figure 33.3   Emissions and sequestration levels.

THE GEMCo DEAL

As was mentioned earlier, in October, 1999, GEMCo made a deal with IGF (a U.S. crop insurance provider) to purchase 2.8 million ton of carbon offsets from farmers in Iowa. The exact details of the deal are confidential; however, it is useful to look at what is known, since it may set a precedent for other such agreements. First, the deal is based entirely upon speculation: "The GEMCo deal to buy sequestered carbon in Iowa is based on the assumption by GEMCo that there will be credit for early action in the U.S. that then gives monetary value to any sequestered carbon before 2008" (McConkey et al., 2000). This assumption that governments will attach monetary value to carbon sequestration largely depends on the acceptance of agricultural land as a carbon sink that can be counted against emissions in the Kyoto Protocol. If this protocol is ratified without the inclusion of agricultural sinks, then very little incentive for emitters to pay farmers to sequester carbon exists, since there will be no credit for this sequestration. One hopes that governments would nonetheless encourage use of agricultural land to sequester carbon, since regardless of carbon accounting, this will help to minimize global warming. However, in such a case the price for carbon credits from GHG removals would likely be quite low, and the credits purchased through the GEMCo deal based on GHG removals would be of little or no value to the buyer as an offset against emissions.

The GEMCo-IGF deal, in addition to emission reductions and GHG removals, also identifies biofuels. The GHG removals address (1) shifting from intensive to minimum or no-till farming methods, (2) adopting better crop rotation and soil conserving methods, and (3) planting woodlots or other environmentally appropriate perennial vegetation on marginal land and using the biomass as a biofuel or for storage of carbon for a long period. The emission reduction component could include reduced fuel usage associated with the GHG removals as well as fertilizer and manure management. This manure management includes installing appropriate digesters and other equipment to burn methane generated from animal waste and using animal wastes to reduce nitrogen fertilizer use. The third component includes burning crop residue or sustainable managed wood waste to replace fossil fuel use on the farm as well as utilizing excess biomass to produce ethanol.

Two elements of this deal concerning the question of risk are particularly interesting. The first is the price of carbon, which is highly uncertain, since it is based entirely upon speculation in a commodity of ambiguous value. Farmers who sell their carbon now for the prices GEMCo offers may lose out if the real price of carbon turns out to be substantially higher. On the other hand, if agricultural sinks are not included in the Kyoto Protocol, then the price of carbon would probably be much lower, at least until some means of crediting farmers for sequestering carbon in their soil is developed.

The second element that surrounds sinks is the question of permanence and flexibility. McConkey et al. (2000) suggest that the deal allows "reasonable flexibility for the participating farmers," since farmers are not required to maintain the specified land management practices indefinitely. "After having been paid for the certified emission reduction credit (CERC), the farmer has no obligation to continue any particular practice" (McConkey et al., 2000). If this is true, it would mean that the emission reduction is only temporary, and that the burden of risk is therefore placed on the buyer, who would presumably need to offset any lost carbon by sequestering the same amount (or reducing emissions by the same amount) somewhere else.

However, it is not clear at this time if the burden of risk has in fact been placed on the buyer, or if the question simply has not been addressed. If a buyer meets his emission quotas in any given year with the help of sequestered carbon ERCs purchased from farmers, then that year balances out: the farmers sequester X ton of carbon, and the polluters emit X ton of carbon. However, if a few years later, a farmer is forced to till his or her soil, it is unclear exactly who would be responsible for the carbon emitted. Polluters are likely to argue that it is no longer their obligation, that the farmer's sequestration in the past balanced their emission, and that any obligation was cancelled out at that point. If the emitter succeeds in shirking responsibility for the maintenance of sinks, that is, if no clear mechanisms are in place to assign responsibility, then that responsibility may fall to farmers simply by default.

A close look at the nature of the risks involved in an emission reduction credit market which treats GHG removals interchangeably with emission reductions is now necessary.

TWO SOURCES OF RISK

Price Risk

If farmers sell their carbon now for $3 per ton, they are in effect gambling that the price of carbon will not drastically increase in the future. If this does happen and they find that, for one reason or another, they need to buy back some of the carbon which they had earlier sold for a much lower price, they may be forced to buy it back at the new higher price. For their own security, farmers must ensure that mechanisms are in place to prevent this, or they can become locked into land management practices that may cease to make sense at some point in the future.

The problem here is that any current attempt to set the price of carbon will be based purely on speculation. It will be determined, first, by the level of public concern over global warming and, second, by the cost of reducing emissions. These costs, which in turn depend upon such factors as the inclusion or lack of inclusion of sinks in the Kyoto Protocol, influence the level at which governments set emission caps (the higher the emission cap, the less industry will have to curtail its emissions to meet it, and the less those measures and, consequently, carbon credits, will cost), and the severity of any emission taxes or permits that are imposed.* Only once the Canadian government commits to its Kyoto target and implements a plan to achieve it will a consistent, agreed-upon price for carbon credits develop.

That does not mean, however, that carbon prices are likely to remain stable over time. McConkey et al. (2000) point out that, while agricultural land offers almost immediate offsets of emissions, new technology to accomplish significant emission reductions will eventually be developed. "This could mean," they speculate, "that the value of offsets could be greatest in the near future and decline with time as new technology becomes available that reduces emissions." However, this scenario fails to take into account the fact that 6% reductions is only the first hurdle of efforts to curb global warming. Emission reduction will likely become more strict, requiring further emission reductions. In either case, farmers would be advised to put off selling their carbon until a market-based price of carbon is established, or at least ensure that, if they sell their carbon now, measures are in place to guarantee that they will not have to buy the credits back later at greatly inflated prices. By taking precautions, farmers can ensure that the burden of risk does not fall too heavily on them, the sellers, and that it is shared to some extent with polluters who buy their credits.

Permanent vs. Temporary Agreements

In order for a market for emission reduction credits to develop, buyers must be assured that the credits purchased are reasonably secure. That is, the credits must represent a fixed amount of carbon to be sequestered for an agreed-upon period of time. If the carbon is released during that time, then the credit generated becomes worthless, and must be purchased back by the farmer or simply written off by the buyer as a bad investment. If buyers doubt the security of the credits for sale in a market, the price that they are willing to pay will be significantly lower.

---

* A fuel tax of $0.01 per liter works out to $4.15 per ton of carbon, $0.02 per liter to $8.30 per ton of carbon, etc. Some estimate that an emission tax could be set as high as $0.05 per liter, which would set the price of a ton of carbon at $20.15.

If the credits represent a permanent sequestration of carbon (as opposed to credits that expire after 5 or 10 years, causing the buyer to repurchase them or to buy others), then they will likely require a legal mechanism such as a conservation easement to guarantee this permanence. Conservation easements are permanent and legally binding agreements to maintain certain land management practices (in this case, zero or minimum tillage and continuously cropped). Because they are permanent, they are tied to the land itself, not to the landowner, which means that a transfer in land ownership also involves a transfer in easement obligations to the new owner. If the easement is seen as a liability, then this could depress land values for any land held under easement.

For farmers, there are important implications of permanent agreements to sequester carbon, which are perhaps best demonstrated by looking at two basic facts about soil carbon sinks:

1. Sinks become saturated. When a sink is created after a carbon-sequestering land management practice is adopted, soil carbon begins to accumulate. However, under any given land management practice, soil can only hold a certain amount of carbon. Thus, as carbon accumulates in the soil over 10 or 20 years, the total soil carbon level approaches a new equilibrium and, eventually, no new carbon credits are generated. At this point, unless some additional carbon-sequestering land management practices are adopted, the sink is effectively "full," and therefore ceases to be a revenue generator (unless long-term incentives for the farmer to maintain the sink are included in the agreement) and becomes simply a liability. A cost is associated with maintaining the sink, as well as with establishing it.
2. Carbon sequestration is reversible. If a farmer sequesters 20 ton of carbon in his soil over 15 years after switching to zero-tillage, the bulk of this carbon can be released in only a few years if he reverts to conventional tillage and summer fallow, or in the face of drastic weather conditions. This would effectively render any credits the farmer has sold for sequestered carbon worthless; he would probably then need to buy the credits back from the original purchaser (or buy an equivalent amount from another source) if the soil needs to be tilled to remain productive.

When these two basic facts are taken into consideration, it becomes very clear that farmers need to think long term when deciding whether to sell the carbon in their soils permanently. Some of the possible long term implications of permanent agreements are:

1. Permanent agreements are likely to bring short-term revenue and long-term obligation. Once a sink is full and the farmer has been paid for the carbon sequestered, he or she is left with only the obligation to continue the carbon-sequestering land management practice, perhaps indefinitely, unless the contract includes some mechanism for compensation for maintaining the carbon sink.
2. Permanent agreements reduce future land management flexibility. If the agreement includes signing a conservation easement, then the current and subsequent land owners are obligated to maintain the sink in perpetuity. If the agreement is enforced by some other mechanism, then any subsequent change in land management practices requires the farmer to buy back the credits, perhaps at a higher market value.
3. Permanent agreements reduce farmers' options for total accounting. If governments look at total farm emissions (which they must do for total GHG emission level accounting), and if farmers will be expected to reduce their overall emissions to a set quota, then at some point, farming will cease to be a net sink and become a net source, since soil sinks will eventually become saturated, while farm operations will continue to generate GHG emissions. Once the carbon sinks are full, no more carbon is sequestered; farmers become net emitters of GHGs and subject to required reductions. If farmers have already sold the carbon sequestered in their soils, they must meet these quotas in other ways or, perhaps, pay taxes or fines on the GHGs emitted.

For all of these reasons, a system that employs or implies permanent agreements imposes serious limitations on farmers and places an unfair burden of risk upon them. Temporary agreements, on the other hand, represent an alternative to legally binding long-term contracts.

CARBON BANKING: AN ALTERNATIVE SYSTEM

If carefully designed, temporary agreements can more equitably distribute the burden of risk between farmers and emitters. The key is not to see them as a permanent transfer of carbon from farmers to polluters, but as a lease or loan designed to ease the burden of emission reductions, to be paid back once the industries have their more permanent (and more expensive) measures in place. To follow the banking analogy further, consider the creation of a sequestered carbon bank, which could be run nationally or internationally. Land managers such as farmers or foresters could contribute sequestered carbon to this bank — the bank's "capital." This sequestered carbon could then be leased or loaned on an annual basis to an emitter to allow achievement of a short-term emission reduction target. Then, once the emitter has implemented the necessary reductions to his operation and can show that he has exceeded emission reduction targets, the excess credits (principle) can be returned to the sequestered carbon bank. This system would provide an incentive for the farmer or forester to create and maintain the largest sink possible, without transferring the risk of permanence from the emitter to the land manager. The loan or lease would not substitute for emission reductions: it would simply provide a much needed window to allow industry time to adapt to stricter regulations.

For the farmer or forester who is managing an unpredictable biological process, this "carbon bank" lease-loan system would limit exposure to the risks of permanence. Once a base line carbon level was established, any GHG removal and storage in a sink would have some value for as long as the sink was maintained.

If the sink were to be lost, for example, by fire or disease in the forestry sector or by tillage in the agricultural sector, the land manager would forfeit only the right to loan or lease the value of the sink. Still using the bank analogy, the land manager would be obliged to "withdraw" the principle of his or her sequestered carbon account certified in the bank.

This sequestered carbon bank would also effectively address the issue of price risk for land managers. Currently Canada's Kyoto target is 6% reductions below 1990 emission levels (UNFCCC, 1997). Environmentalists will point out that this is woefully inadequate in terms of actually reducing GHG levels in the atmosphere. As public concern over global warming increases, so will the pressure brought to bear on governments, which in turn will lead to higher targets for emission reductions. The more important the issue becomes, and the harder it becomes to meet new emission reduction targets, the greater the value of mitigation tools. A loan or lease arrangement for GHGs sequestered in biological sinks would address price variability, since it would be reviewed and revised at the end of each term.

Biological sinks are a valuable one-time opportunity to address GHG concerns. Any market system must recognize that sinks are the only GHG removal tool at its disposal. A broad-based market system for emission reductions would establish the value of the sequestered carbon "principle," and any loan or lease rate could be negotiated using this value and the agreed-upon terms. The potential returns of sequestered carbon would add value to land for as long as the sink was maintained. This is in stark contrast to the potential liability a farmer who sells his sequestered carbon permanently as part of an emission reduction trading system would face.

CONCLUSION

GHG removals with sinks need different policies and markets than GHG emission reductions. If soil sinks are an important strategy to help Canada manage its GHG problem, then policies must be developed that encourage land managers to adopt best management practices. Any policy that increases a land manager's risk would, at best, not stimulate adoption and may be a disincentive. A market system for GHG removals that transfers risk from emitters to land managers would likely be a failure. Farmers and foresters already assume tremendous risks due to the uncertainty of managing biological systems as well as volatile commodity markets. A market system for GHG removals with sinks would not provide any incentive for land managers to adopt best management practices if it adds yet another risk to an already risky operation.

If GHG levels in the atmosphere pose a serious environmental threat, one needs to remember that GHG removals with sinks are the only current mitigation tool that actually removes GHGs from the atmosphere. This is a limited opportunity for society and should not be squandered in the early stages of addressing dangerous GHG concentrations in the atmosphere. GHG removals can be very useful as a bridge to meeting short-term GHG targets by allowing emitting industries time to implement permanent emission reductions. GHG emission reductions and GHG removals are very different; therefore, they require different markets to stimulate the desired activities.

REFERENCES

Kyoto Protcol to the United Nations Framework Convention on Climate Change (UNFCCC), http://www.unfccc.de/resource/docs/convkp/kpeng.html, 1997.

McConkey, B. et al., Carbon sequestration and direct seeding, in *Proc. 2000 Saskatchewan Soil Conserv. Assoc. Direct Seeding* Workshop, SSCA, Indian Head, Saskatchewan, Canada, (See web site http://ssca.usask.ca/2000proceedings/McConkey.htm), 2000.

CHAPTER 34

Sequestering Carbon: Agriculture's Potential New Role

Jim Kinsella

Most debates over solutions to global warming have concentrated on the combustion phase of the carbon cycle. This is unfortunate because the biological phase (photosynthesis and mineralization) may offer the greatest opportunity for practical solutions, at least for the next two or three decades.

There are approximately 320 million acres (128 million hectares) of land in the U.S. tilled to grow field crops. Current and past tillage practices have mineralized much of the stored organic matter (humus) in these soils, releasing billions of tons of CO_2 into the atmosphere. This oxidation of humus, along with the increased combustion of fossil fuel (old buried humus), has surely contributed to the increasing concentration of CO_2 in the atmosphere in the 20th century. Using humus-depleted cropland as a biological sink for some of the excess CO_2 appears to be a practical and cost-effective method of reducing atmospheric CO_2 levels, at least until soils can be rebuilt to near pretillage levels of soil organic matter (SOM).

On our home farm in central Illinois we have grown corn and soybeans in a complete no-till system since 1975. We have also no-tilled alfalfa in all land enrolled in government crop acreage reduction programs. During that period, our SOM has more than doubled, going from an average of 1.9% in 1974 to 3.9% in 2000 (Figure 34.1). All samples were taken to a depth of 7 in (17.6 cm) and organic matter was estimated using the Walkley–Black weight loss upon ignition method.

These soil tests indicate that, over the last 26 years, we have taken around 11.6 tons (10.5 MT) of C from the atmosphere and added it to the top 7 in (17.6 cm) of soil on every acre, or 4.2 MTC/Ha. The SOM now appears to be leveling off at near its pretillage level of 4%. Although soil tests have not been taken, it appears that the SOM below 7 in (17.6 cm) is also increasing, probably because of a dramatic increase in earthworm activity and increases in root depth and volume.

The adoption of no-till farming has been slow, primarily because of a current disincentive to no-till called *mineralization*. Tillage temporarily increases the O_2 level in the soil, which increases the mineralization rate of the SOM, thus enhancing the release of CO_2 into the atmosphere. There is a corresponding oxidation of nutrients contained in the organic matter, the most abundant of which is nitrogen. SOM is approximately 58% C and 5.8% N with much smaller amounts of other nutrients. When these nutrients are oxidized, they become available for plant uptake and provide

1-56670-581-9/02/$0.00+$1.50

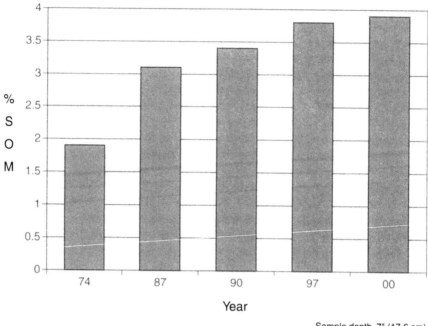

Sample depth. 7" (17.6 cm)

Figure 34.1 Changes in SOC over time under a complete no-till system in central Illinois.

a short-term economic advantage for tillage. Since plant nutrients have a clearly defined value and C does not, soil organic carbon (SOC) is sacrificed to temporarily increase nutrient availability to plants. This is a short-term benefit to the producer but a long-term detriment to the quality of soil, water, and air.

To offset the nutrient availability advantage provided by tillage, a value must be placed on carbon and an economic incentive on storing it. Since there is a wide range of estimates of the value of SOC, the following comparisons may help in determining its true value:

1. Developers around the fringes of the rapidly expanding cities in the Corn Belt will buy a good piece of farmland and immediately scrape off the topsoil and pile it into huge piles. A small portion of this topsoil is used to landscape around the houses or buildings in that development and the remainder is sold to grow lawns and other urban landscape vegetation. The going rate for topsoil around cities is about $10/T ($11/MT) plus hauling. The subsoil without organic matter is usually free for the hauling, so the only real value to the topsoil is the carbon. If this topsoil has 3% SOM, the value of the carbon would be $575/T ($633/MT).
2. There are about 5.5 pounds of carbon per gallon of gasoline (.65 kg/L). If gasoline is selling for $1.50/gal ($0.39/L), then the carbon value in gasoline would equal $545/T ($601/MT). Without carbon, gasoline has no energy and thus has no value.

Averaging these two scenarios would indicate a real value for carbon of $560/T ($617/MT). With all the intrinsic value of carbon in our cropland soils, this value should be at least equal to the value paid to grow a lawn or to propel an auto a few miles down the road with gasoline. Even though the real value of carbon is much greater, I feel many farmers and landowners would be willing to accept much less in order to get an effective carbon sequestration program started. Considering carbon's energy value, the value of humus, and the liability of excess CO_2 in the atmosphere, it would seem a value of $100/T ($110/MT) of C would be quite reasonable.

For once farmers have an opportunity to set the price for something that they produce, in this case SOC. I propose that farmers set $100/ton of C as their "minimum wage" for taking CO_2 out of the atmosphere and storing it in their soils. This value would help offset the cost of equipment and additional management and the risk of yield loss incurred during the transition to less tillage. The most practical method of providing this incentive appears to be a modification of our existing farm programs to pay farmers for the C that they sequester. These incentives could come as cash payments, tax credits, or credits for crop or income loss insurance. A carbon sequestering incentive (CSI) could be similar to other voluntary farm programs such as the conservation reserve program (CRP). The incentive must be high enough to encourage participation and to offset additional expenses and risk.

The amount of carbon sequestered and retained varies considerably between crops and environments. The Agriculture Research Service could use existing information and begin gathering additional data to develop carbon retention models on which payments would be based. Models could also be developed and payments made for SOC added to the soil by various cover crops, as well as from applications of manure, sludge, landscape waste, and other sources.

Carbon payments should stay with the land and be redeemable, at least in part, if and when intensive tillage is resumed. It makes no sense to pay farmers to sequester and store carbon and then allow them, or other farmers, to till the soil, thus sending most of the recently stored carbon back to the atmosphere, without retribution. We should learn from the current exploitation of much of the CRP land, which is going back to crop production. A well managed CRP field sequesters a lot of carbon and vastly improves the quality of the soil while under contract; however, the common practice in the Midwest of plowing this soil prior to putting it back into field crops destroys most of the benefits.

If a CSI program were part of the 2002 Farm Bill, we could buy our nation time to develop new sources of energy and to improve the efficiency of fossil fuel combustion. With this time, a strong economy, and technological expertise, the U.S. could lead the world in developing a more expansive, environmentally friendly energy strategy. Besides improving air quality, an effective CSI would greatly reduce soil erosion, improve soil and water quality, as well as productivity, and make soils sustainable.

For a very minimum cost to taxpayers, farmers now have an opportunity to reward the public with something beyond safe, abundant, and cheap food:

- Reduced risk of global warming
- A better environment
- Long-term food security
- What a bargain!

CHAPTER 35

Public Policy Issues in Soil Carbon Trading

Andrew P. Manale

CONTENTS

INTRODUCTION

Numerous studies have identified the value of soil carbon as a sink in mitigating greenhouse gas emissions (Lal et al., 1998; Follett et al., 2001). Capping the amount of carbon dioxide emitted into the atmosphere by industry, granting property rights or allowances to emit a predetermined amount, and allowing trading in emission reductions or carbon sink credits can create demand for soil carbon, with the market assigning it a positive economic value. However, technical problems relating to the nature of soil carbon can increase market risk and raise transaction costs, hence raising carbon's effective market cost as an offset for the industrial emissions of greenhouse gases (Manale, 2000). Anticipated higher costs associated with managing soils to sequester carbon paid by purchasers can lower the size or number of acres involved in the market in carbon.

Since carbon plays a crucial role in soil quality, with effects on ecological processes at a variety of geographic scales, a cap-and-trade-dependent market in carbon alone may not assign a market value reflective of total ecologic or social value as a component of soils. These multiple environmental benefits include water quality, flood and drought mitigation, habitat, and soil productivity. If the offset value of sequestering carbon for the purpose of mitigating climate change is lower than the social value as soil carbon, a carbon market may result in too few acres under alternative management that increases soil carbon. Managing for maximum social benefit at a given level of social expenditure may require public policies that encompass a broader environmental focus targeted temporally and spatially. Major issues in targeting for maximum environmental benefit are discussed in this chapter.

Unlike a bushel of corn, soil carbon is not a discrete item or set of items that can clearly and unequivocally be measured, given agreement on how the measurement will occur. Any transaction pertaining to soil carbon relates to a promise of performance of a service whereby carbon is sequestered in soils and maintained there for a given period of time. The service involves using agricultural practices scientifically shown to increase the level of carbon in soil and possibly also in surface biomass. Many services, depending upon their set of agricultural or land management practices, can provide the sequestering benefits — and hence value in a potential market — and, perhaps more importantly, the aforementioned additional benefits to society that generally accrue as a consequence of improved soil quality. Conversely, termination of the contract and reversion to agricultural practices that release sequestered carbon can result in environmental damages beyond those associated with the contribution to atmospheric carbon.

The percentage of carbon in soils serves as the key measure of soil quality which, in turn, is ecologically linked not only to atmospheric processes, but also to water quality and landscape functions. Soil quality indicates the capacity of the soil to produce agricultural commodities and to perform a variety of functions simultaneously, including storing nutrients, filtering, purifying, and storing water, storing and degrading waste, and maintaining the stability and resilience of larger landscape ecosystems. Degraded soil quality or soils low in carbon contribute to problems of impairment of water quality, either directly through mineralization and subsequent loss of sediment and nutrients to water bodies, or indirectly by losing the capacity to retard the loss of nutrient-enriched waters (National Research Council, 1993). Reduced water-holding capacity contributes to the severity of downstream flooding and increases the effects of droughts on agricultural productivity. Impaired stability and resilience of landscape ecosystems can lead to a decline in soil and system biodiversity and to increased agricultural demand for pesticides and fertilizers. The latter is often associated with problems of degraded water quality (U.S. Environmental Protection Agency, 1997).

Self-interest alone may not be sufficient to motivate adoption of practices that increase soil carbon and hence improve soil quality (Camboni and Napier, 1994). Changes in soil carbon and soil quality may occur very slowly, depending upon a number of environmental factors such as rainfall; any change in productivity can be scarcely perceptible from year to year. Since the major noticeable economic impact from elevated concentrations of greenhouse gases over time is to users of the surface and ground water resource, wildlife, or future generations, short- and long-term benefits of sequestering or maintaining carbon in agricultural soils and improving soil quality accrue mostly to parties other than those who use land. Though there are examples where improving soil quality has generated immediate rewards to farmers through savings in costs of fuel and time (CTIC, 2001), wider adoption of practices that increase soil carbon levels and, consequently, soil quality generally require changes in public policy that increase the economic incentive to manage the lands so as to provide greater social benefit.

Short-term beneficiaries of maintaining or increasing the carbon in agricultural soils and in so doing, increasing the quality of soils are downstream users of the water resource and communities located in floodplains or subject to flooding. Future generations of farmers benefit from the improved agricultural productivity and resilience of the soil and future generations of the global community benefit from the decreased atmospheric levels of greenhouse gases. The benefits to farmers on an annual basis are small relative to the short-term cost of technology to enable the practices or the demand for additional management expertise, since even long-term loss in productivity and soil resilience can often be compensated with fertilizers and pesticides or improved varieties of crops.

AVOIDING CREATING PERVERSE INCENTIVES

Degraded soils ironically present the greatest economic opportunity for farmers to replenish the carbon in their soils in the event of a market in carbon. Creating a market in carbon based solely on carbon-sequestering potential of soils could potentially reward farmers with degraded soils and penalize farmers who maintain the quality of their soils. This could also create incentives to convert pasture or other marginal cropland to crop production, or to plow good soils to lower the carbon levels in order to then get a credit for putting it back. The consequence could be a decline in environmental quality not only from the loss of soil carbon, but also from the greater use of pesticides, fertilizers, and sediment loadings to surface waters.

PRESERVING THE BENEFITS OF IMPROVED SOIL QUALITY

Sequestered carbon tends to be in a form that is oxidizable and can be rapidly lost with a reversion to tilling. Increasing or even maintaining the level of carbon in the soil requires that a practice be conducted and that it, or a similar practice, continue in future years. How long these practices need to be maintained depends upon the long-term environmental benefit, such as water quality or wildlife habitat, to be achieved through increased soil carbon levels.

If the focus is entirely on mitigating greenhouse gas emissions, the economic value of sequestering carbon in the soil may be temporally limited as new technologies arise which reduce the reliance on fossil fuels. On the other hand, to achieve the full array of environmental benefits associated with improved soil quality, the practice must be maintained in perpetuity (or as long as the downstream user has interest in the resource). Any return to tillage can lead to an immediate, precipitous decline in soil organic matter with concomitant reductions in nutrient and water retention and potentially serious consequences for water quality. Furthermore, continuing to till the soil can lead to constant nutrient loss.

This nondurability of the resource suggests that policies to encourage adoption of practices leading to increased levels of carbon in the soil should incorporate a long-time horizon. Society has traditionally used land retirement to capture public benefits where the resource does not renew within a reasonable time frame. Retiring the land not only allows it to recover many, though probably not all, of its former ecological functions, but also protects those functions for future use. However, a social cost is attached to the loss of economic production associated with the land. Furthermore, retiring all but the most marginally productive land can lead to increases in commodity prices and conversion of other marginal lands to agricultural use, with concomitant environmental degradation.

When the time scale for performance is necessarily long and the primary beneficiary of the policy future generations, legal covenants or easements can serve to conserve the private or societal benefits that accrue from alternative management of the land and discourage use of practices that threaten these benefits in the future. These policies may also have less effect on production and hence the prices of raw agricultural commodities. Examples include easements on farmed wetland purchased by federal and state government agencies that allow farming only in seasonally dry periods. Ownership of the easement or covenant could be assumed by government or by an organization with interests in preserving lands, such as a nonprofit conservation organization.

LEAKAGE

A policy that encourages the adoption of practices to increase soil carbon on specific acreage can inadvertently affect nontargeted lands, an unintended consequence known as *leakage*. For example, a policy that results in a decline in total production of the agricultural commodity can raise agricultural prices, thereby improving the profitability of production on agriculturally marginal lands. Farmers not targeted by the policy may respond to the new economic incentives by converting their land to crop production, resulting in the loss of carbon from the soils and an increase of greenhouse gas emissions. If the objective is to increase net carbon sequestration on agricultural lands, leakage to other lands can potentially reduce, if not offset, any intended benefit. The conservation reserve program (CRP) provides an example of leakage, though its effects on prices have been short term and relatively moderate, and its environmental benefits have likely outweighed its adverse impacts.

WHICH ENVIRONMENTAL BENEFITS TO TARGET

When sequestered in the soil, carbon provides a greater array of environmental benefits than an identical quantity of carbon not emitted, as carbon dioxide, into the atmosphere. The latter serves only to reduce the rate of climate change, whereas the former, as mentioned previously, contributes to soil quality. If the policy objective is to maximize net social benefit in addition to mitigating global warming, the issue of which environmental benefits to target arises. Targeting can affect the preferred array of practices that increase soil organic matter.

For example, policies to encourage soil carbon sequestration solely to mitigate greenhouse gas emissions do not require contiguous plots of land or enrollment of acres of spatial importance within watersheds. A totally market-based or voluntary approach, for example, could lead to "cherry-picking," whereby geographically dispersed farms, or even fields, most suitable for rapid increases in soil carbon are enrolled in contracts. Opportunities for water quality benefits would be constrained.

On the other hand, focusing on water quality benefits in watersheds with impaired water quality resulting from degraded soils could provide ancillary benefits in terms of offsetting greenhouse gas emissions as long as a balance was made for any methane emissions from the wetlands. However, the policies to achieve these benefits require enrolling specific lands within the watersheds that best serve to improve water quality. Any program of spatial targeting for more than one environmental benefit may not be amenable to a total market-based or voluntary approach, such as a market in carbon credits, and may more easily be implemented under a government-managed program restricting negative or positive incentives to lands that meet specific criteria. Alternatively, government can serve to "define" the commodity (or service to be performed) to be traded in a manner that expands the range of benefits to be achieved. In situations in which hydrologic modification, such as tile drains in the midwestern Corn Belt, has led to degradation of soil quality and loss of soil organic matter, collective action, e.g., through a program of government cost-share, may be necessary to restore conditions for maintaining or increasing soil carbon.

EQUITY AND MORAL HAZARD

Lands with good quality soils with high relative organic content provide the least opportunity for sequestering carbon or increasing soil organic matter content. Hence, policies that pay farmers to increase soil carbon, such as a market-based program of carbon trading, may inadvertently penalize producers who have been good stewards. Moreover, the moral hazard posed by a reward for low carbon content, and hence high potential for sequestration, may encourage producers not directly involved in trades to till soils so as to reduce their organic matter levels.

SUMMARY

Agricultural practices can help mitigate global warming by reducing carbon emissions from agricultural land and by sequestering carbon in the soil through regulatory, market incentive, and voluntary or educational means. Public policy can encourage adoption of these practices. However, care should be taken in designing these policies to ensure their success, to avoid unintended adverse economic and environmental consequences, and to provide maximal social benefit.

REFERENCES

Camboni. S.M. and T.L. Napier, Socioeconomic and farm structure factors affecting frequency of use of tillage systems, paper presented at the Sept. 1994 Agrarian Prospects III Symposium in Prague, Czech Republic, 1994.

Conservation Technology Information Center (CTIC), No-till pays bills in Virginia, in *Partners*, May/June, West Lafayette, IN, 2001.

Follett, R.F., J.M. Kimble, and R. Lal, *The Potential of U.S. Grazing Lands to Sequester Carbon and Mitigate the Greenhouse Effect*, Lewis Publishers, Washington, D.C., 2001.

Lal, R. et al., *The Potential of U.S. Cropland to Sequester Carbon and Mitigate the Greenhouse Effect*, Ann Arbor Press, Chelsea, MI, 1998.

Manale, A., Approaches to assessing carbon credits and identifying trading mechanisms, in *Assessment Methods for Soil Carbon*, Lal, R. et al., eds., Lewis Publishers, Washington, D.C., 2000.

National Research Council, *Soil and Water Quality: an Agenda for Agriculture*, National Academy Press, Washington, D.C., 181, 1993.

U.S. Environmental Protection Agency, Fact Sheet: National Water Quality Inventory — 1996 Report to Congress, EPA841-F-97-003, Washington, D.C., 1997.

CHAPTER **36**

The Politics of Conservation

William Richards

CONTENTS

INTRODUCTION

As a farmer and former Chief of the Soil Conservation Service (now NRCS), I would like the politics of conservation to be considered in the debate over global climate change and the potential for carbon sequestration to help in the mitigation of the greenhouse effect. At the beginning of a new millennium, America is awash in prosperity; we are the best fed, best housed, and best entertained people the world has known. Yet billions of people throughout the world suffer from hunger and malnutrition, and we, the food producers, are not sharing these good times because of low commodity prices and rising production costs. There is an increasing concern among the public regarding green space, recreation, wildlife, food safety, and the environment, all luxuries only the well fed and the prosperous can afford. The stage is now set for the farm and environmental debate of our new century: how do we increase profits for farmers and at the same time address the public's concerns for an improved environment?

Freedom to Farm (the last or phase-out farm support program) matures in 2002; only with massive supplemental and disaster assistance has the government supported the agricultural economy through 1999 and 2000, while the rest of the U.S. has enjoyed a booming economy and budget surplus.

TRENDS IN CONVERSION TO NO-TILL

Over the past 8 years federal funds for conservation and environment assistance directed toward agriculture have steadily declined. This is ironic because conservation was used as the excuse to funnel money to an impoverished countryside in the 1930s when farm programs were initiated. This funding and these early programs were the beginning of federal and state conservation movements. Resulting reductions in water and wind erosion and the resulting concept of a conservation ethic have become a model for the world. Conservation cost sharing and free technical assistance became a standard part of federal farm legislation.

This changed in 1985 when erosion control and wetland preservation became requirements for participation in commodity support and other USDA programs called cross compliance. This U-turn from totally "voluntary" to "required" in the politics of conservation sent shock waves through production agriculture; the resulting producer hostility helped to send me to Washington as the SCS chief. With my background in no-till or direct seeding, along with help from chemical and machinery industries, the Farm Press, and other USDA agencies, we were able to sell conservation tillage as the best practice to meet erosion requirements. For the most part we were successful. Erosion dropped to sustainable levels in many regions of the country and conservation tillage peaked at roughly 40% of planted acres.

In the mid 1990s the political climate shifted: the signal from Washington was that conservation tillage was no longer politically correct; the agenda switched from the real measurable problem of soil erosion to the perceived problem of herbicide dependence. Conservation tillage (30% or more surface residue) had leveled out nationally. It was gaining in cotton, wheat, and soybeans but losing in the Corn Belt, especially in highly erodible Iowa, where 1.433 million acres (580,365 hectares) in conservation tillage were lost (10% of the acres in Iowa in 1998 by Conservation Technology Information Center measurements) (CTIC, 1999). In the last year, the area under no-till has decreased even more. Now, with rapidly rising costs for fertilizer, will there be a move to till more land to help liberate the stored nitrogen for a short-term gain resulting in a long-term loss to the idea of sustainable agriculture?

Unlike our major competitors, Canada, Argentina, and Brazil, who have passed us in the percent of cropland direct seeded, the U.S. is reducing the area under no-till (sometimes referred to as *direct seeding*). This competition will be the ultimate factor that dictates adoption. As one of those who started the conservation tillage revolution, I am concerned that we have unleashed a monster. Around the world millions of hectares of new land come into production every year; many of these lands would be too fragile or unprofitable without conservation tillage. Conservation tillage has helped to reduce erosion and long-term no-till has led to increased yields and reduced inputs (chemicals, man-hours, and equipment use). Good crops can be grown under long-term no-till, as shown in Figure 36.1. Soil testing in the no-till vs. conventional-till fields has shown increased infiltration and improved soil structure, with soil under no-till returned to the high quality condition of a native site.

So then why is conservation tillage losing ground in the U.S.? We know of both real and perceived yield problems. We need more encouragement from conservation leaders, especially from those in the NRCS. We need more research and equipment design, especially in the northern Corn Belt. I submit that the major factor is politics. Many believe that the USDA has slacked off on compliance and Congress has reacted to farmer backlash to "required" conservation.

I feel and hope that future conservation programs will be separate, voluntary, and incentive-based. We have learned our lessons on cross compliance. From experience as a farmer and as a past SCS chief, I am convinced that we get conservation on the land and behavioral change with incentives and education, not through requirements and regulations.

Figure 36.1   No-till soybeans on author's (person on right) farm in August, 2000, showed improved soil structure, increased water infiltration, and a thicker A horizon than a conventionally tilled field on the same soil.

POLICY ISSUES AND THE ENVIRONMENT

The environmental political focus is changing. Concern about soil erosion is no longer the only issue. We must now address issues of water quality, air quality, wildlife habitat, etc.; conservation tillage will be in the forefront when we do this. We have always known and understood the immediate fuel, labor, and machine savings of conservation tillage. Because of these savings, my sons found much more time for other activities; this may eventually lead to more young people's willingness to remain in farming. We also captured the management opportunity of spreading our talent over more area with reduced tillage. Then erosion and conservation benefits became political after the 1985 farm bill. Only recently have we understood the long-term soil quality, water quality, and wildlife benefits occurring from continuous direct seeding. No-till or direct seeding will become universal because of the long-term soil quality improvements and raw world competition.

The opportunity to increase organic matter (soil carbon) will, first, enhance productivity or land value and, second, sequester carbon for a world concerned with climate change from rapidly increasing greenhouse gasses. CO_2, for example, has increased from 280 ppm in the 1850s to 365 ppm in 1996 and is continuing to increase at a rate of 0.5%/year (Lal et al., 1998).

Long-term experiments in Ohio (Dick et al., 1997) and elsewhere have shown increases of as much as 1 to 2% in organic matter in fields continuously no-tilled for 10 to 20 years. The bad news is that we have tilled away or eroded approximately 50% of the organic matter from our soils over the last 100 years, which would be as much as 5 to 6 billion metric tons of carbon (MMTC) lost from our nation's farmland (Lal et al., 1998). By converting from plowing to no-till and including cover crops in the rotation cycle, we can return carbon to the soil at a rate of 300 to 500 pounds per acre/year (136 to 227 kg/ha/yr). So the good news is that we have the technology, machinery, and science to put it back. What this organic matter (OM) is worth in terms of actual value is the question facing us now; does it increase the land value (selling price of the land)? In our country, on lighter textured soils the land value increase is probably $500 per acre ($1200/hectare) at current price levels (or 2% OM = 20 Bu/ac at $2.00/bu at 8% = 500/ac): this is something that landowners cannot ignore. The simple answer is that, yes, OM does increase the value of land.

VALUE OF SOIL CARBON

What soil carbon is worth to the public depends on how we address the clean air and greenhouse gas problems. Former President Clinton announced the need for tougher emission controls for new automobiles and SUVs, which may add $200 to the cost of each vehicle. Many feel that agriculture has the potential to sequester the carbon for 50% of auto emissions. Scientists at The Ohio State University and NRCS/ARS have estimated that adoption of recommended agriculture practices on U.S. cropland has a potential to sequester 70 to 200 million tons of carbon per year (Lal et al., 1998). In addition, grazing lands could potentially sequester 30 to 110 millions tons of C per year (Follett et al., 2001). We should encourage our farmers, land managers, and ranchers to do whatever is recommended to achieve these potentials and provide the needed support to carry out the recommended practices.

When the public (and Congress) understands, carbon sequestration will be in the forefront of policy development in the next round of farm and conservation legislation. If given the chance, the public would quickly choose their big autos and support payments to producers for carbon sequestration. A steady increase in CO_2 in the atmosphere over the past 100 to 150 years has been clearly documented (Lal et al., 1998). Whether this is a result of anthropogenic activities does not matter. However, since the world is going to throw money and regulations at global climate change, then agriculture should benefit from some of this money. In the short term, sequestration may be the only game in town to mitigate increases in greenhouse gases and farmers should realize a monetary benefit from their efforts to improve the environment.

The 2001 federal budget contains $4 billion for global climate work, of which agriculture will receive $57 million. In Costa Rica 3 years ago I visited land owners who were buying cheap land, banking on U.S. and European Union-funding to keep it in trees. Ohio power companies are already investing in South America. Worldwide environmentalists are preparing for a harvest of public money. The competition will be keen; will it be trees, grass, and wildlife habitat vs. cropland, grazing land, and wood production? We hope not, but what really is needed is a concentrated joint effort by everyone to improve overall land management so that it is trees, grass, wildlife habitat *and* cropland, grazing land, wood production.

The pressure to take land out of agricultural production will increase because many in the environmental community believe that the solution to overproduction and low commodity prices is simply public control of land or, long term, the conservation reserve program (CRP), which only exports our production and environmental problems to other countries. I frequently hear, "We have bought that land three times in the last 50 years, next time we should keep it." Those of us in agriculture must get our environmental solutions into the public arena and into legislation.

I hope that, in the near future, we will have the opportunity to put in place a comprehensive conservation incentive program to reward farmers and land managers for stewardship. We offer a possible solution to the global climate change greenhouse gas scenario that is a win–win strategy for all concerned. Whether problems are real or perceived, public funding for increased organic matter, improved soil quality, better water quality, and less erosion (all leading to higher productivity) is a good investment for the world's population and the environment and can be sold under the banner of carbon sequestration.

We need a program similar to the concept for payments for conservation practices introduced in the Senate in 2000. It would establish a voluntary program to reward producers for good environmental stewardship; one of the basic practices would be conservation tillage. Grain and livestock producers would contract for 5 years to follow a total resource management plan that calls for a menu of practices pertaining to their operations. Producers could earn up to 40% of their

county land rental rate. Livestock producers could earn up to 10% of sales per year with practices such as conservation tillage, comprehensive soil and nutrient management, precision nutrient application, odor reduction, lagoon management, managed rotational grazing, windbreaks, buffers, or stream improvements. All these practices are profitable for the producer, improve the quality of the land, and benefit the public. We also need to understand better the monetary off-site benefits, some of which needs to be passed back to the farmers.

Such legislation should be voluntary, in addition to but separate from the farm bill due in 2002. A separate conservation bill would bring into focus agriculture's importance to the environment. This bill would move money to the countryside at a time when it is badly needed but, more importantly, it would help production agriculture address concerns of the environmental community and avoid the temptation to increase regulations. USDA and a scaled-back NRCS would have problems delivering such a comprehensive program and may need to look to state and local conservation agencies and private industry for help to service U.S. producers. NRCS does not have the experienced staff and talent that administered the 1985 and 1990 Farm Programs.

CONCLUSIONS

A "freedom to conserve" coupled with a good crop insurance package could help avoid the temptation many have to change the "freedom to farm." Give it time: it has foreign competition worried and producers enjoy the freedom to manage and compete. A freedom to conserve could cost up to $6 billion per year in the out years, which would be expensive, but the taxpayer would directly benefit from soil, water, and air quality and economic stability in the countryside. Conservation stewardship payments are also permitted under the Nafta, Gatt, and WTO rules.

I challenge farmers, ranchers, other land managers, and researchers to get involved in the conservation policy debate. Congress needs to consider a conservation bill that would encourage producers to sequester carbon. Sequestered carbon should be considered as another crop; its monetary value should be based on its direct relation to reductions in greenhouse gases and other environmental benefits that it provides (clean air and water). The time is now. Enforcement of the Clean Air Act is starting and soil carbon will be proven to be a valuable commodity. (I am told that $20 per ton is a reasonable price.) It is possible to sequester 1 ton per hectare per year (1 ton/ac/yr) with well-managed, no-till fields and at the same time reduce the need for herbicides. U.S. agriculture has the potential to sequester 100 million metric tons per year net of the inputs and producer emissions.

The EPA and most environmentalists favor carbon trading so that industry, especially utilities, will finance the carbon reduction. The political issue for agricultural producers is whether we trade our carbon sequestering potential on the market or get our rewards through stewardship payments from a government program. However, we producers should be careful: remember that the market says that those who pollute (emit carbon) will pay those who conserve (sequester); that could include some of us (farm managers, producers). Of course, over time the market will prevail because that is our system. Someday farmers who till may buy carbon credits from farmers who no-till. The public may even decide greenhouse gasses, global warming, etc., are not worth the price of carbon sequestration.

However, a comprehensive, reward-based stewardship conservation program is a good investment for producers, consumers, and the environment and it will sequester carbon. The increase in carbon reserves of the nation's farmland through adoption of conservation tillage provides numerous ancillary benefits to society and the world community. Reduction in water runoff and soil erosion decreases risks of nonpoint source pollution, reduces siltation of waterways and reservoirs, decreases the risks of flooding, and reduces emission of greenhouse gases (Lal et al., 1998; Follett, et al., 2001). Society needs to reward its farmers for taking personal risks while benefiting humanity.

My hope is that we start with a stewardship program and thus give our scientists time to research long-term solutions to the public's real and perceived global concerns.

REFERENCES

CTIC, National survey of conservation tillage practices, *Cons. Tillage Inf. Center*, West Lafayette, IN, 1999.

Dick, W.A., Tillage system impacts on environmental quality and soil biological parameters, *Soil Tillage Res.*, 43, 165–167, 1997.

Follett, R.F., J.M. Kimble, and R. Lal, *The Potential of U.S. Grazing Lands to Sequester Carbon and Mitigate the Greenhouse Effect*, Lewis Publishers, Boca Raton, FL, 442, 2001.

Lal, R. et al., *The Potential of U.S. Cropland to Sequester Carbon and Mitigate the Greenhouse Effect*, Ann Arbor Press, Chelsea, MI, 128, 1998.

PART VIII

Regional Pools

CHAPTER **37**

Growing the Market: Recent Developments in Agricultural Sector Carbon Trading

Michael J. Walsh

CONTENTS

INTRODUCTION

The soil carbon crediting issue is evolving at a steady pace. Many of the important steps necessary for a market to emerge and grow have been advanced domestically and internationally. This chapter reviews selected examples of the most recent technical, policy, and market-building developments. While far more is happening than is cited in this brief summary, the examples nonetheless illustrate the range of activities that will foster the market and highlights steps taken to form an organized carbon market that will include crediting for the agricultural sector. This chaper takes note of international discussions and proposed domestic legislation, identifies important ongoing efforts to quantify soil carbon, and examines proposed parameters of an organized carbon market prepared by the Chicago Climate Exchange[sm].

BUILDING MOMENTUM

A strong international consensus now exists to use market mechanisms to restrict greenhouse gas emissions in an economically optimal manner, as reflected in the emergence of informal and formal carbon credit markets. Those who look to the United Nations process for guidance on the issue risk missing the real action: the private sector has made incredible strides on the issue. Some of the world's largest and most respected corporations, including Ontario Power Generation, Boeing, British Petroleum, Suncor, DuPont and Royal Dutch Shell, and others have taken strong proactive positions in support of greenhouse gas (GHG) reduction efforts. Several of these companies have established voluntary GHG reduction targets, often paired with in-house and external GHG trading initiatives. These leaders have made important contributions to the formation of political momentum and development of the carbon market.

The latest international negotiations to finalize a climate treaty were widely publicized as a failure, but strong signs exist that the U.S. and other countries will act to build a market to address the climate issue. Whether or not the Kyoto Protocol ever enters into force, some form of climate protection program appears likely to emerge, and carbon sequestration and carbon trading will be central features of the package.

TECHNICAL ACTIVITIES THAT CONTRIBUTE TO MARKET FORMATION

This section provides a brief overview of selected technical and modeling work that supports formation of a carbon market incorporating sequestration in agricultural soils.

- *The IPCC special report on land use, land-use change, and forestry (LULUCF)* (IPCC, 2000). This now-completed report, called for by the Parties to the United Nations Framework Convention on Climate Change (UNFCCC), represents an important step forward in improving understanding around the world. Many countries do not have the detailed expertise needed to conduct informed negotiations on these issues. By design, the report is informative but does not advocate any particular approach and, extremely helpful to many involved in the process, does not represent the major source of momentum towards resolution of issues that some seemed to expect. Although the report appears to have limited impact on international negotiations, at a minimum its publication eliminates one excuse for further postponement of the issues addressed.
- *Continued development of soil sequestration inventories and models.* A number of state and regional initiatives have been undertaken (e.g., by Nebraska and South Dakota) (Leonard, 2000) to develop a better understanding of current and potential soil carbon storage in the agricultural sector. A number of other countries with large areas in crops (e.g., Australia, Argentina, and Canada) are expanding their analyses of carbon stocks and sequestration potential. Australia has taken proactive legislative steps to clarify the legal status of forest carbon credits. (New South Wales, 1998). A project focusing on Iowa cropland is assembling detailed soil carbon data for all 99 counties, and is employed to enhance the capability of the CENTURY soil carbon modeling system (Natural Resource Ecology Laboratory, undated). In addition, the U.S. Department of Agriculture has released for beta testing a computer program called CQESTR. The program provides estimates of expected soil carbon increases on an individual farm based on initial carbon levels, location (weather), and changes in soil and crop management practices. These initiatives will help advance the ability of individual farmers to participate in carbon markets.
- *Analysis of whole-farm greenhouse gas impacts and sequestration opportunities.* Robertson et al. (2000) released a study reporting GHG benefits possible through a wide range of changes in

cropping, tillage, and fertilization practices. They examine long-time series of data on soil carbon under various management regimes. Their study found that where typical Midwest soil sequestration potential is present, under a corn–soybean–wheat rotation a switch to no-till would reduce the overall GHG impact of that farm to near zero, due to significant uptake of carbon by the soils. Under the right policy setting, such a major reduction in overall GHG impacts would yield a surplus credit position for the farmer. The study also found new evidence that various cropping practices (e.g., cover crop plantings) can also be highly effective in boosting soil carbon levels.

- *Economic modeling of carbon markets and agriculture.* The World Resources Institute released a report that models several quantitative impacts of carbon constraints, trading, and soil carbon crediting (Faeth and Greenhalgh, 2000). They found that various policy configurations, particularly extending the conservation reserve program (CRP) and implementing a nutrient trading program for nitrogen, could supplement the positive economic impacts from soil carbon crediting. Each of these actions would contribute to realization of a broad set of benefits beyond greenhouse gas reduction, including reduced soil erosion and improved water quality and wildlife habitat. Importantly, the report also identifies fundamental methodological flaws with some earlier economic analyses that concluded that carbon constraints would cause major economic damage to the agricultural sector. Once these flaws are corrected, even the most pessimistic modeling methods conclude that the impacts are rather minor and can be more than offset via crediting for agricultural soil carbon crediting.

INTERNATIONAL AND DOMESTIC POLICY DEVELOPMENTS

Various international developments and domestic U.S. policy initiatives and proposals offer strong signs that agricultural carbon credits will soon move from concept to reality. Contrary to conventional wisdom, important official and unofficial progress was on display at the UN climate negotiations at The Hague (COP-6) in November, 2000. The agenda set for the UN conference was far too broad to achieve major progress on all fronts. Nevertheless, the leading industrialized country negotiating blocks (the European Union and the U.S. and its "umbrella group" partners) nearly agreed on a package deal covering guidelines on carbon trading and rules for carbon sink recognition. Among the items on the table in the final negotiations was the extent to which (not "if") agricultural sequestration would be credited toward the U.S. mitigation commitment. These concrete discussions indicate that the intergovernmental negotiating process is, at last, getting down to the details.

Governmental negotiations take center stage at the UN Climate meetings, but the meetings have also evolved to become the world's biggest event for showcasing initiatives of individual governments, the private sector, and nongovernmental organizations. From these sectors an enormous amount of progress towards widespread carbon markets was on display at COP-6. A notable increase in activity to build a GHG market can be seen among major corporations, and nongovernmental organizations such as the Joyce Foundation, Pew Center on Climate Change, and individual governments. The U.K., Norway, the Netherlands, and others are moving forward with plans for carbon markets, thus helping build a variety of examples that others can learn from and emulate.

Importantly, the agricultural sector has been more fully engaged in the climate policy discussions. While this has been most evident in the U.S. and Canada, preparations are being made in other major crop-producing countries (such as Argentina) and in countries where livestock contributes a significant portion of overall GHG emissions (e.g., Ireland).

Some of these efforts were highlighted in presentations given at COP-6. Experts from Canada as well as others, including Dr. Rattan Lal, provided a concise summary of the massive potential for carbon sequestration in soils. Similarly, presentations made by World Resources Institute and the Union of Concerned Scientists, as well as one made by representatives of the U.S. Departments of Agriculture, Interior, and Energy, highlighted widespread opportunities for the U.S. to realize multiple environmental benefits, in addition to carbon sequestration, through improved land management and increased use of carbon-neutral bio-fuels.

Proposed U.S. Legislation for Rewarding Carbon Sequestration Activities

Three different Senate bills that would provide direct cash support or tax credits for carbon sequestration activity have been prepared. No wording in any of the bills appears to preclude further participation in carbon credit markets by program participants. A bill introduced by Senator Tom Harkin (Democrat, Iowa) in September, 2000, would provide direct annual payments to farmers who undertake a range of eligible conservation activities, including nutrient management, resource-conserving crops and crop rotations, and other resource management systems. The proposed bill (S. 3260) explicitly recognizes that many of these activities will cause "the reduction of greenhouse gas emissions and enhancement of carbon sequestration" (Sec. 1240Q(a)(10)).

Senator Brownback (Republican, Kansas) introduced the Domestic Carbon Storage Incentive Act of 2000 (S. 2540) in May, 2000. He later drafted a complimentary bill (to date unnumbered) called the International Carbon Storage Incentive Act. S. 2540 would allow landowners to enter into contracts with USDA for 10 or more years and receive payments up to 50% of the cost of carrying out sequestration practices. More importantly, an annual "rental payment" possibly as high as $20 per acre (but not to exceed $50,000 to a single person), would be paid to farmers selected on the basis of a bidding and scoring process similar to the CRP program.

This bill would establish transferable tax credits worth $2.50 per ton of carbon sequestered by eligible carbon sequestration projects. Eligible project categories include native forest preservation, reforestation, biodiversity enhancement, and soil erosion management. A 30-year contract is required in order to be qualified to generate the tax credits.

President Bush's Proposed Reductions and Trading in Carbon Dioxide Emissions from Power Plants

The proposed four-pollutant strategy would address sulfur dioxide, nitrogen oxides, mercury, and carbon dioxide (http://www.georgebush.com/issues/energy.html). Citing success with the acid rain trading program established by the 1990 Clean Air Act Amendments, the proposal calls for reductions to be phased in. It also calls for use of emissions trading and carbon credits. The Energy Information Administration (2000) recently released a study of this approach which found that the cost of in-house reductions of CO_2 by power plants significantly raises the cost of the four-pollutant strategy. Clearly the four-pollutant strategy would make far more economic sense if complemented by giving power plants the option to offset CO_2 emissions through carbon sequestration initiatives. The option to offset emissions through sequestration would represent one of the least costly methods for achieving greenhouse gas mitigation.

Broader Industry Appreciation of New Business Opportunities for Agriculture

A November 13, 2000, letter to Secretary of Agriculture Daniel Glickman from various agricultural associations (including the American Farm Bureau Federation and the National Farmers Union) shows that broader segments of the agricultural community are gaining appreciation of opportunities associated with soil carbon sequestration. The two-sided nature their concerns are two-sided:

> If we are to move beyond the deeply held concerns of the agricultural community, it is important that the current negotiations provide the greatest possible flexibility for the U.S. to fully and immediately account for carbon sequestered through agricultural activities.

This message regarding the importance of carbon sequestration opportunities is consistent with the argument that Environmental Financial Products made before the U.S. Senate Environment and Public Works Committee on March 24, 1999 (see Appendix or www.envifi.com).

MARKET FORMATION INITIATIVES FOR SOIL CARBON TRADING

A large number of greenhouse gas trading initiatives are now underway. These include in-house trading programs at major multinational energy companies, individual over-the-counter offset trades, and organized markets formed in the U.S., U.K., and other countries.

Major Energy Companies Have Established In-House GHG Trading Programs to Help Meet Voluntary Emission Reduction Commitments

British Petroleum, which has acquired U.S.-based Amoco and is in the process of acquiring Arco, and Royal Dutch Shell have both taken GHG reduction commitments and established internal emissions trading programs. These programs have provided strong evidence of the viability of the approach, while also highlighting the benefits of initiating GHG trading systems with small, simple programs. The scale of these efforts is significant: each company has GHG emission levels exceeding those of several small western European countries. Another important step has been Shell's strategy of weighing the economic value of GHG emissions when making major capital investment decisions. Other large companies setting internal emission reduction commitments include DuPont, Ontario Power Generation, and United Technologies.

Individual Offset Trades and Trade Support Structures are Proliferating

Individual trades involving various types of sequestration and emission reduction projects have helped build the expertise and institutions needed to broaden the market. A coalition of Canadian companies entered into a contract with groups of Iowa farmers who commited to sequestration and emission reductions. Individual reforestation carbon contracts have been executed in the U.S. and abroad. Environmental Financial Products has executed a multimillion-ton landfill gas reduction GHG transaction between Zahren Alternative Power and Ontario Power Generation.

Numerous mechanisms for facilitating carbon trading are emerging. Insurance companies, agricultural marketing cooperatives, and others are building systems for aggregating offsets and economizing on transaction costs. One example is the Montana Carbon Offset Coalition, which aims to coordinate multiple private landowners wishing to participate in the carbon market. Working with agricultural soil and reforestation efforts, the coalition employs standardized contracts, realistic carbon quantification methods, and a pooled risk management methodology to reduce the costs of transacting and establish a high-quality basis for selling carbon offsets.

These initiatives provide a variety of alternative transaction support structures that must be tested and refined in order to allow widespread market participation. As more trades are executed, methods for aggregating offsets and transaction contracting styles will evolve. In particular, experimentation with various approaches for handling sequestration measurement, ownership, and permanence need to be tested.

Organized Carbon Markets

While government-driven efforts to form organized carbon markets are underway in a number of European countries (U.K., Norway, and Denmark), none feature a major role for carbon sequestration. Responding to the lack of such efforts in North America, the Chicago-based Joyce

Foundation has funded a study, through Northwestern University's Kellogg Graduate School of Management, to examine the feasibility of an organized carbon market in the U.S., starting in the Midwest. That study, conducted by Environmental Financial Products LLC, has yielded a proposed market design that would include a variety of opportunities for participation by agriculture.

The objectives of the pilot program — named the Chicago Climate Exchange[sm] (CCX) — are to:

- Begin reducing net greenhouse gases in a program of manageable size and representative and scalable composition
- Provide initial price signals on the cost of mitigating greenhouse gases
- Provide a basis of experience and learning for participants
- Allow introduction of a phased, economically rational process for achieving additional GHG reductions in the future

Participation in the market is voluntary. The market system is intended to be self-governing and primarily self-funded. Participants in the pilot can gain commercial advantage by building their ability to take advantage of the flexibility in timing, methods and locations for mitigating emissions. Early action will help participants better prepare their long-term environmental management plans, cut costs by slowly phasing in GHG mitigation efforts, and building new domestic and international business relationships.

If participation mirrors the diverse nature of the Midwest economy, the price signals emerging from the market will offer a useful indication of prices possibly emerging under a broader carbon market. In any event, prices in the pilot market will reflect the balance of supply and demand, which will, in turn, reflect the mix of participants, the cost of reducing emissions at various emission sources, and the cost of mitigation through sequestration, renewable energy projects, and other sources of offsets.

As a pilot program, the market is likely to undergo revisions as experience is developed, as the market grows, and as linkages with other GHG markets are developed. The features of successful emissions trading systems and commodity and capital markets, as well as the emerging international standards, are being drawn upon for guidance. A rules-based system is proposed because it provides the necessary transparency and consistency to achieve an easily understood and functional system. Table 37.1 summarizes the initial proposals for design of the market. In order to assure that pilot design conforms to the capabilities of the agricultural sector, a technical committee including farmers, marketing cooperatives, and agricultural insurance companies has been assembled. That committee will help refine the initial proposals for market mechanics described further below.

The mechanics of the process for issuing offsets to participating projects is as follows:

1. A registry account is opened in the name of the project proponent.
2. Proponents of a registered project are eligible to trade standardized forward contracts in the CCX electronic trading system or through privately negotiated contracts.
3. The project must be implemented and maintained as indicated in the project registration form.
4. Annual verification reports indicating total offsets generated during the prior year must be submitted.
5. Offsets are placed into the account of the project proponent.
6. Carbon sink projects may be required to contribute to a long-term performance assurance pool.

The need to assure permanence of carbon storage from carbon sink enhancement projects introduces the need to obtain long-term commitments from project proponents, as well as a method for making the environment whole if commitments are broken. Required carbon storage time frames of 50 to 100 years have been proposed, reflecting both GHG residence times and the time until widespread usage of GHG-free technologies exists. For example, one proposed solution to the

Table 37.1   Indicative Market Architecture for the Chicago Climate Exchange

Geographic coverage	2002: 7 Midwest states (IA, IL, IN, MI, MN, OH, WI) 2003–2005: U.S., Canada, and Mexico Offsets accepted from projects in Brazil in all years, other strategic countries to be determined
Greenhouse gases covered	Carbon dioxide, methane, and all other targeted gases
Emission reduction targets	2002: 2% below 1998 baseline level, falling an additional 1% per year through 2005, which has a target of 5% below 1998
Industries and firms targeted	Combination of "downstream" and "micro" participants: power generation, refineries, manufacturing, vehicle fleets
Tradable Instruments	Emission allowances (original issue) and offsets produced by targeted project types
Eligible Offset Categories (projects implemented by entities with emission levels below participation cut-off of 250,000 tons)	Methane destruction (e.g., agricultural waste, landfills) Carbon sequestration via no-till agricultural soils and grass plantings, afforestation, and reforestation Increased vehicle efficiency in autos, trucks, and buses Conversion to less GHG-intensive fuels Wind, solar, hydro, and geothermal power systems Direct on-site emission reductions from energy efficiency enhancements.
Trading mechanisms	CCX electronic trading system and private contracting
Annual auctions	2% of issued allowances withheld and auction in "spot" and "forward" auctions, proceeds returned *pro rata*

timeframe question has been adopted by the Montana Carbon Offset Coalition, which requires an 80-year commitment from participants who wish to earn carbon credits for reforestation projects.

A compromise approach is to obtain long-term commitments from landowners, require annual attestations that the project is being maintained, and conduct random on-site project inspections. As experience is gained and policy developments continue, mechanisms for assuring that the credited sequestration projects are long-lasting are likely to emerge. In the meantime, successful incorporation of sequestration projects will allow development of practical experience. As experience is gained, and as new policy initiatives emerge, additional provisions that can help assure long-lasting carbon storage will be developed. Eventually a variety of questions related to sale of land committed to sequestration activities, nonperformance, and contracting styles must be addressed.

Accurate measurement of soil carbon can be a costly challenge. Soil tests by farmers do not necessarily provide a characterization of soil carbon content that accurately describes an entire farm. In order to keep the process of quantifying soil carbon sequestration practical, it is imperative to devise a suitable substitute for the costly process of collecting and analyzing a large number of soil samples. One option is to use simplified and transparent proxies, such as conservative, standard carbon sequestration values for each region and each targeted set of sequestration practices. The standard values would be based on established evidence presented in the scientific literature. The option to be credited for higher amounts of carbon sequestration would be exercised by submitting, *ex post*, evidence that individual sequestration rates exceeded the standard level. In any event, the CCX will work with ongoing research efforts to establish more accurate measurements of soil and biomass carbon sequestration rates. Use of carbon accumulation proxies should be based on well-established scientific evidence and be verified through actual experience and field validation.

Another option is to use statistical estimation processes that predict the quantity of carbon sequestration expected to occur at a specific site, based on soil characteristics, crop rotations, and local climate. The CQESTR numerical simulation system developed by the U.S. Department of Agriculture may offer such a system (Rickman et al., 2000). This program, which underwent beta testing during December, 2000, will be examined to assess its suitability for the CCX. Another option being examined is the use of the CENTURY model cited earlier.

CONCLUSIONS

Evidence that the world will continue to move towards adoption of market-based greenhouse management systems is indisputable. Constraints on GHG emissions and a marketplace that rewards carbon sequestration will become an integral part of the economic system.

Many ingredients required to make a market for agricultural soil carbon crediting a reality have advanced significantly during 1999 and 2000. Technical understanding and capabilities have grown, international and domestic policy approaches have been more keenly refined and debated, pilot trades have been conducted, and formation of organized markets involving agricultural carbon credits are being formed. Large amounts of technical information have been disseminated via the IPCC's special report on land-use carbon issues. Focused initiatives to improve the ability to model soil carbon dynamics have progressed, and a number of initiatives to improve carbon inventories are underway.

International climate policy negotiations have, at last, included focused consideration of the rules and modalities for recognizing and trading soil carbon credits. A variety of national and subnational policy proposals designed to encourage increased carbon sequestration in soils have been drafted. The intense pace of government, corporate, NGO, and academic activity to increase the capability to quantify and trade soil carbon has improved the capability to trade soil carbon credits. These activities and other market-building developments make it likely that a carbon market that includes opportunities for agriculture will soon emerge, whether or not an international climate treaty enters into force.

The slow but steady increase in individual carbon trades and the preparation of organized carbon markets signal that broadened participation of agricultural soil carbon credits will present real opportunities in the near term. The Chicago Climate Exchange^{sm} offers one such mechanism for Midwest farmers in 2001, and for farmers throughout the U.S. and Canada starting in 2002.

Movement is now from the concept to the implementation stage of a carbon market. It is critically important that the agriculture community take an active role in developing various carbon trading techniques, improved measurement systems, and risk management practices. All these activities cited represent parts of a broad set of initiatives that will, over time, cause a more sophisticated carbon trading system to emerge.

The agricultural sector faces some of the greatest risks and transition costs associated with a change in the earth's climate. Given the sector's massive potential to help slow and reverse the risks of climate change, it is imperative that climate protection strategy harness the agricultural community's ability in order to provide a wide range of locally and globally important environmental services based on stewardship of the land.

REFERENCES

American Farm Bureau Federation, American Soybean Association, National Cattlemen's Beef Association, National Farmers Union, Letter to U.S. Secretary of Agriculture Dan Glickman, November, 2000.
Faeth, P. and Greenhalgh, S., A climate and environmental strategy for U.S. agriculture, World Resources Institute, November, 2000.
Intergovernmental Panel on Climate Change, *Land Use, Land Use Change and Forestry: Summary for Policymakers*, Cambridge University Press, 377, 2000.
Lal, R. et al., *The Potential of U.S. Cropland to Sequester Carbon and Mitigate the Greenhouse Effect*, Ann Arbor Press, Chelsea, MI, 1998.
Leonard, R., staff member for Nebraska State Senate, personal correspondence, 2000.
Natural Resource Ecology Laboratory, The CENTURY model, http://www.cgd.ucar.edu/vemap/abstracts/CENTURY.html, undated.
New South Wales, Carbon Rights Legislation Amendment Bill 1998, legislation submitted to the Legislature of New South Wales, 1998.

Rickman, R.W. et al., Planning and predicting soil carbon sequestration in farming systems: a new field model, presentation to Carbon: Exploring the Benefits to Farms and Society conference, Des Moines, IA, 2000.

Robertson, G., Paul, E., and Harwood, R., Greenhouse gases in intensive agriculture: contributions of individual gases to the radiative forcing of the atmosphere, *Science,* 289, 1922–1925, 2000.

Sandor, R. and Walsh, M., Kyoto or not: opportunities in carbon trading are here, *Environ. Quality Manage.,* Spring, 53–85, 2001.

APPENDIX

Statement to the U.S. Senate Environment and Public Works Committee
March 24, 1999 Hearing on Credit for Voluntary Early Actions
Dr. Richard L. Sandor
Chairman and Chief Executive Officer, Environmental Financial Products LLC

Mr. Chairman, thank you for inviting our participation in today's hearing. Our company, Environmental Financial Products, is a small business dedicated to designing, launching, and trading new markets. We have had some success with new financial and agricultural markets, and over the past 10 years we have had the privilege of helping the U.S. sulfur dioxide allowance market take root and succeed.

In 1991, while serving as a director of the Chicago Board of Trade, I encouraged the exchange to support the SO_2 emission allowance market. The EPA ultimately selected the CBOT to administer its annual allowance auctions. The results of the 7th auction were announced today in Chicago. The success of that program truly represents a milestone in environmental and financial history. It gave us faster than required pollution cuts at far lower cost than predicted. This success has been realized despite the predictions of the naysayers. We were told the sulfur program would never work. We heard: "It's too complicated, it will cost too much, it's too hard to measure, utilities don't know how to trade, where will the price data come from?" Despite the huge success of the sulfur trading program, those of us who believe we can harness a market to protect against global climate change are now hearing the same protests. People have not learned the lesson: never sell short America's entrepreneurial ingenuity.

The success of emissions trading is further proof that the private sector brings forth enormous creativity in solving social problems if we introduce a profit motive and a price signal. The Credit for Voluntary Early Actions bill offers just the sort of signal that will unleash the creativity and innovation needed to prevent, at low cost, the economic damage that sudden climate changes threaten to bring.

In anticipation of opportunities associated with early action and future export possibilities, we have been working with a wide range of businesses, most of them small ones, that are eager to act for the good of the environment while doing well for themselves. We work with farmers and foresters, with entrepreneurs who collect landfill and coalbed methane to produce electricity, and with large electric power companies. All these businesses are prepared to cut and capture even more greenhouse gas emissions if we give them a signal. These businesses are ready to take action, and they view the emerging international carbon credit market as a new business line. It is not just the producers of carbon credits who will gain. We will see a wide range of new opportunities for small businesses in the fields of new energy efficiency technologies, remote sensing, carbon certification, soil testing, project finance and trading.

Credit for Voluntary Early Actions is the positive signal needed to unleash this new economic sector. We believe that there are three critical components that must be addressed to create a viable greenhouse gas emissions trading system: (1) the development of a homogeneous or fungible greenhouse gas trading unit; (2) a legal instrument or quasi-property right that permits simple conveyance of ownership; and (3) a process for monitoring and verification of emission reductions and sequestration. In addition, the proposed legislation should provide a "no regrets" component that lets those who voluntarily cut and capture emissions know that their early action would not count against them if future regulations are adopted. Such an Early Action bill would provide a common standard that will give American businesses a uniform yardstick if they wish to participate in the various carbon trading programs now emerging at the international, state, and local levels. In conclusion, we would urge that the legislation be implemented using rigorous measurement standards, and that credits be granted only to well documented cases of emission cuts and sink enhancement.

By getting us moving sooner, we can further cut the cost of the global climate insurance policy that the public wants. I should note that some of the model-driven cost estimates for cutting U.S. net emissions that are put forth by consultants such as Wharton Econometrics Forecasting and Charles River Associates run completely counter to all the evidence available in the real world. Their estimates range from $100 to $300 per ton carbon. Our estimate indicates that cutting net U.S. GHG emissions 7% below 1990 would cost $20 per ton carbon and would involve redeployment of resources worth $12 billion per year, which is 2.5% of the annual U.S. energy bill. This is also less than one tenth of 1% of national income in 2010 (i.e., gross domestic product, conservatively projected). Resources shifted to greenhouse gas mitigation do not evaporate into thin air. They will be invested in energy efficiency and process changes that will cut energy bills, and will flow to U.S. businesses that can cut and capture emissions at low cost.

The largest new business opportunity for Americans has been given little attention. The science tells us (and the Kyoto Protocol recognized) that we need to slow the increase in *net* greenhouse gas emissions. This means both cutting emissions and capturing carbon from the atmosphere by enhancing carbon sinks. A recent study by the American Farm Bureau Federation that examined only the potential cost increases to farmers used the American Petroleum Institute/Wharton Econometrics conclusion that energy prices would increase 50%, which they equate to a carbon permit price of $200 per ton. If those studies were consistent, the same price would be applied when analyzing the impacts on revenues from carbon sequestration. The scientific literature reports that adoption by U.S. farmers of best management practices for soils and use of biomass fuels, when combined with relatively modest amounts of reforestation, can capture 300 million tons of carbon per year. If carbon prices really were $200 per ton, annual revenues from sequestering 300 million tons would be $60 billion, which is 142% of U.S. net farm income. It is our firm conviction that the $200 price is completely out of the question. Using a more realistic estimate of $20 per ton means the carbon market could provide $6 billion per year in new income, which is 14% of net farm income. Thus it is critical that soil and forest sinks be recognized in the legislation. Doing so will stimulate the needed advances in monitoring and verification of carbon sinks. However, it should be recognized that there are numerous examples where limitations on the science of monitoring and verification did not prevent the emergence of successful markets.

Parenthetically, it should be noted that the 300 million tons of carbon that can be sequestered by the farm and forestry sector would cover half the projected U.S. emissions gap under Kyoto. Alternatively, American farmers and foresters could export carbon credits to Europe and cover the entire European Community emissions gap. These exports would significantly reduce the U.S. trade deficit.

Attached for the record is an article that will appear in the forthcoming edition of *Choices* magazine, a publication of the American Agricultural Economics Association. The article, which I co-authored with Dr. Jerry Skees of the University of Kentucky, provides further details on the scale of the opportunity for U.S. farmers in the emerging international carbon market.

Of course, the modest cost estimates cited above do not even take into account the many additional environmental benefits that arise from cutting and capturing greenhouse gases. We are likely to also realize improved air quality in our cities, making it easier for children with asthma to enjoy the outdoor activities most of us take for granted. Expanding our forests will give more families a chance to enjoy hunting and hiking. Fishermen win because no-till farming and capturing carbon through grass and tree plantings will improve water quality in our rivers and lakes. The benefits that hunters, hikers, and fishermen bring to small businesses in our rural economies are enormous.

Independent of the Kyoto Protocol, and independent of the climate change science, the Early Action bill is an intelligent step forward. It is clear that many countries around the world will adopt greenhouse gas-cap and trade programs. We cannot pass up the opportunity to give American industry, farmers, and foresters a strong foundation for selling into these new markets. America's capital and agricultural markets are the envy of the world. This bill will allow us to combine the best of these two global powerhouses and keep the U.S. in the lead in setting the standards for the emerging international carbon trading system.

Mr. Chairman, thank you again. We encourage the Committee to move this bill forward. Its passage will lead to major new opportunities for American business while contributing to a safer future for the planet and cleaner future for our country.

CHAPTER 38

Role of Agro-Industries in Realizing Potential of Soil C Sinks

Bruno A. Alesii

CONTENTS

INTRODUCTION

Most of the focus to find technologies to reduce greenhouse gases in the atmosphere over the last few years has been in the areas of energy use, renewable energy sources, manufacturing efficiencies, and motor vehicle efficiency. Recently, however, farmers have learned that the way in which they farm can have a significant impact on climate change. By using a simple but powerful farming technology facilitated by Roundup and Roundup Ready crops, farmers can significantly decrease the rate of enrichment of atmospheric carbon dioxide.

This technology is called *conservation tillage* (Figure 38.1). It is a farming practice that utilizes little if any tillage, leaving the soil relatively undisturbed. This is not a new concept; some farmers have used it for years as a way of reducing erosion, protecting the water quality of streams and lakes, and providing habitat and food for birds and other wildlife. Now, however, farmers are learning that conservation tillage can significantly mitigate global climate change. The concept is simple: through the photosynthesis process, crops pull carbon dioxide from the atmosphere and store it in their stalks and leaves; when their residue is left on the fields to decompose, the carbon pulled from the atmosphere is deposited in the soil.

Unfortunately, traditional intensive-tillage farming practices release virtually all of the carbon back into the atmosphere. Tillage stirs up the soil and exposes the stored carbon to oxygen, which turns it into carbon dioxide and releases it back into the atmosphere. However, if the soil is farmed

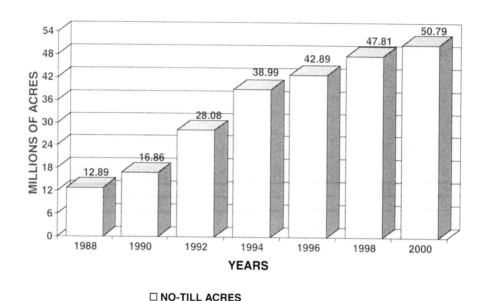

Figure 38.1   Growth of no-till acres in the U.S. (all crops). *Source*: CTIC Tillage Database.

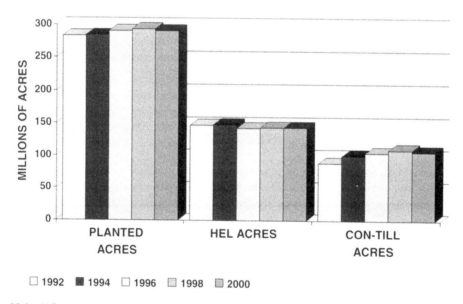

Figure 38.2   U.S. crop acreage, 1992 to 2000. *Source*: CTIC Tillage Database.

with conservation tillage techniques, a significant portion of the carbon sequestered by the crop is returned to the soil profile, thus building up the soil organic matter.

According to the Soil and Water Conservation Society, no-tillage farming methods (a form of conservation tillage) can sequester from 448 to 1120 Kg of carbon per hectare per year (0.2 to 0.5 ton of carbon per acre per year) and that process could continue for the next 30 to 50 years. That yearly amount would offset carbon released by burning 187 to 470 liters of gasoline per hectare (20 to 50 gallons of gasoline per acre) (Bruce et al., 1998; Lal et al., 1999; Paustian et al., 1997; Janzen et al., 1997; Frye, 1995).

In the U.S. alone, widespread adoption of conservation tillage and other management practices, such as buffers and crop intensification, could potentially fulfill 20 to 30% of the U.S. carbon

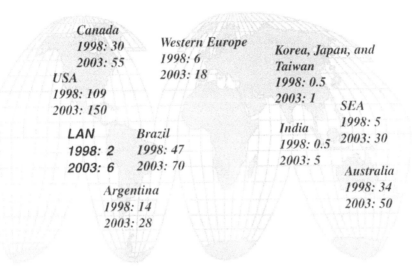

M acres under CT in 1998 and projected by 2003

Figure 38.3   M acres under conservation tillage in 1998 and projected by 2003.

dioxide reductions targeted at the Kyoto meetings (Figure 38.2). Additionally, since heavy tillage with large tractors is not necessary, no-tillage farming saves 32.7 liters of fuel per hectare (3.5 gallons of fuel per acre) (Hill, 1996; Frye, 1995), resulting in less carbon dioxide emitted.

Monsanto's agricultural products are well suited to support and encourage conservation tillage. *Roundup* herbicide is an environmentally friendly tool that farmers can use to control weeds without resorting to the plow or cultivator; Roundup has been instrumental in facilitating conservation tillage around the world (Figure 38.3). The more recent addition of Roundup Ready crops has further encouraged and facilitated the adoption of conservation tillage. These crops are resistant to the Roundup herbicide, which means that, even after crops are up and growing, Roundup herbicide can still be used to control weeds without tillage (Figures 38.4 and 38.5).

Monsanto believes that it can help mitigate climate change through its products, especially by encouraging the adoption of conservation tillage. As a company, Monsanto is also committed to supporting and helping the agricultural community as a whole adopt practices which sequester carbon and develop climate-friendly technologies.

MONSANTO'S POSITION

The sufficient scientific evidence on the issue of climate change justifies prudent action. Monsanto will:

- Measure and publicly report its greenhouse gas emissions and manage its activities to improve material and energy use efficiency continually
- Seek to develop and introduce products, technologies, and services that contribute to stabilizing greenhouse gases
- Work to ensure that agricultural practices, such as conservation tillage, that increase carbon sequestration in soils are recognized as a positive factor in mitigating climate change and that these practices are included in market-driven mechanisms such as CO_2 emission trading programs
- Support worldwide policies for reducing the possibility of global climate change that emphasize flexibility, market-driven mechanisms, and meaningful participation by all

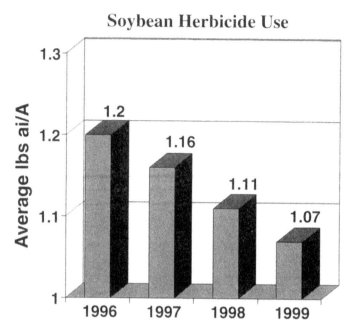

Figure 38.4   Roundup Ready soybeans have grown to over half of the U.S. soybean market since their launch in 1996. During this period, overall soybean herbicide use decreased by 10%. *Source*: Doane Market Research for conventional and Roundup Ready soybeans.

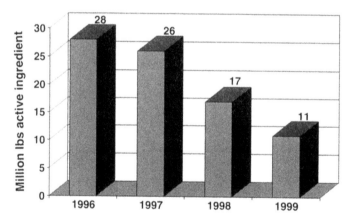

Figure 38.5   Since the introduction of Roundup Ready soybeans in 1996, the use of soybean herbicides with EPA ground water advisory labeling has been reduced by 60% (17 million pounds). *Source*: Doane Market Research.

Soil Carbon Sinks and Con-Till

- Soil carbon sinks can play a significant role in mitigating global climate change by sequestering carbon dioxide out of the atmosphere and storing it in the soil. Soil carbon sinks are natural systems such as farmland or forests.
- Farmers who practice conservation tillage are sequestering carbon. In con-till or no-till farming, crops are planted into the previous year's stubble without plowing. Trapped in the fiber of the crop residue, carbon is returned to the soil in the form of organic matter, enriching the soil in its biological, chemical, and physical properties. The elimination of tillage also prevents the soil organic matter (carbon) from being oxidized and released into the atmosphere where it can become a greenhouse gas that contributes to global warming.

- Monsanto products complement the use of no-till farming techniques.
- No-till farming methods can sequester from 448 to 1120 Kg of carbon per hectare per year (0.2 to 0.5 ton of carbon per acre per year — equivalent to the carbon released from burning 187 to 468 liters of gasoline (20 to 50 gallons). In addition, switching from conventional to no-till farming can reduce fuel use by at least 32.7 L/ha (3.5 gal/ha) (Bruce et al., 1998; Lal et al., 1999; Paustian et al., 1997; Janzen et al., 1997; Frye, 1995; Hill, 1996).
- Soil C sinks can offset 20 to 30% of the emission reduction target of the U.S.
- In addition to sequestering carbon, con-till holds these benefits:
  - Topsoil preservation: 25 billion tons of topsoil are lost each year to runoff. No-till farming decreases soil erosion rates by 90% by holding soil particles on the field (Hill, 1996; Hebblethwaite, 1995).
  - Water quality: nutrient, pesticide, and water runoff are decreased by at least 70% over conventional tillage (Hill, 1996; Fawcett, 1995).
  - Reduced air pollution: crop residues reduce wind erosion and the amount of dust in the air. Lower horsepower requirements and fewer trips also reduce fossil fuel emissions (Hill, 1996).
  - Long-term farm productivity: the less one tills, the more carbon one keeps in the soil to build organic matter and promote future productivity. Intensive tillage speeds the breakdown of organic matter and soil structure (Hill, 1996).
  - Increased wildlife habitat: crop residues provide shelter and food for wildlife such as game birds and small animals (Hill, 1996; Best, 1995; Basore, 1986).

In addition, conservation tillage saves farmers money and time; no-till requires as little as one trip for planting compared to two or more tillage operations plus planting for conventional tillage. This saves an average 450 hours on a 1000-acre (405-hectare) farm. Fuel savings average 32.7 liters a hectare (3.5 gallons an acre) compared to conventional tillage systems. Fewer trips requires less wear on equipment and thus lower maintenance costs, estimated to save $12.35 per hectare ($5 per acre) (Hill, 1996; Frye, 1995).

Need for Unity in the Agriculture Industry on Climate Change

Monsanto does not pretend to speak for the agricultural community on the issue of climate change; however, the company would like to see the leadership of the agricultural community take a more proactive position on the climate change issue. By doing so destiny on this issue is shaped by the agricultural community rather than others. Monsanto believes that agriculture can play a major role in mitigating climate change. The technology and the capacity to begin to impact the climate change issue exist today without having to wait for new technology development. All that is needed to begin to take action is the development of agricultural policies that would be favorable to agriculture, i.e., incentives, research dollars, and tax credits. Monsanto believes that, for the next 20 to 40 years, agriculture can be the bridge to a safer climate; agriculture needs to be at the table on the climate change issue so that it can help to shape the outcome into one good for U.S. agriculture and the environment. This should include:

- A process to build consensus across the agricultural community on a position on climate change
- The establishment of an agricultural community-working group to shape U.S. agriculture position on climate change. This working group would comprise representatives from all key commodity groups, American Farm Bureau, industry representatives (fertilizer, equipment, chemical, food processors, and distributors) as well as representatives from key governmental agencies, i.e., U.S. Department of Agriculture, Department of Energy, etc. The working group would be facilitated by a third-party organization to ensure that every group had an equal voice on the outcome, which would include:
  1. Development of a U.S. agriculture position on climate change agreeable to all participants
  2. Agreement on what kinds of policies would encourage climate-friendly practices in agriculture
  3. Future actions that agriculture can undertake to have an impact on climate change

What Government Can Do to Promote Agricultural Involvement

The government can also play an important role through actions that encourage stewardship and development of new technologies such as incentives for best practices, raising awareness and funding innovative research, and supporting market driven mechanisms such as carbon dioxide emission trading programs. Monsanto also believes that government can do a lot more than it has done to engage agriculture in a more meaningful way, such as:

- More frequent and open dialogue with the agricultural community on climate change other environmental issues
- Tax credits to advance the adoption of new technologies such as biofuels
- Development of farm policies that provide incentives to growers who adopt best management practices (BMPs) that improve the environment
- Funding of research to overcome or address the issue of greenhouse gas methane in the livestock industry and NOx in the rice industry
- Funding to support research and development of new biofuels
- Funding to support research in measurement, monitoring, and verification of soil C as it relates to carbon credit trading
- Early action legislation to reward, not penalize, early adopters for their environmental stewardship
- Assisting in the creation of a working model for carbon credit trading

CONCLUSION

Agriculture can be the bridge to a safer climate and can play a critical role in mitigating atmospheric CO_2 in the next 20 to 40 years. Agriculture has the technology and the know-how to begin to stabilize atmospheric CO_2 immediately with such practices as no-till farming, which has been shown to sequester up to 1123 Kg of carbon per hectare per year (1.87 metric tons of CO_2 per hectare per year). All that would be required to spur farmers into action is:

1. Official recognition regarding the role that agriculture would play in mitigating climate change
2. Monetary incentive to encourage farmers to convert from traditional tillage systems to no-till either through some form of carbon trading mechanism or by the enactment of some national policy that would reward farmers for adopting no-till farming practices
3. Working together in partnership with U.S. government agencies, farm groups, and the private sector to make it happen

REFERENCES

Basore, N.S. et al., Bird nesting in Iowa no-tillage and tilled cropland, *J. Wildlife Manage.*, 50:19–28, 1986.
Best, L.B., Impact of tillage practices on wildlife habitat population, in *Farming for a Better Environment*, Soil and Water Conservation Society. Ankeny, IA, 1995.
Bruce, J.P. et al., Carbon sequestration in soils, Soil and Water Conservation Society's (SWCS) Carbon Sequestration in Soils Workshop in Calgary, Canada, SWCS, Ankeny, IA, 1998.
Fawcett, R.S. et al., Agricultural tillage systems: impacts on nutrients and pesticide runoff and leaching, in *Farming for a Better Environment*, Soil and Water Conservation Society, Ankeny, IA, 1995.
Frye, W.W., Energy use in conservation tillage, in *Farming for a Better Environment*, Soil and Water Conservation Society, Ankeny, IA, 1995.
Hebblethwaite, J.F., The contribution of no-till to sustainable and environmentally beneficial crop production: a global perspective, Conservation Technology Information Center (CTIC), West Lafayette, IN, 1995.
Hill, P.R., Conservation tillage: a checklist for U.S. farmers, Conservation Technology Information Center (CTIC), West Lafayette, IN, 1996.

Janzen, H.H. et al., Soil carbon dynamics in Canadian agroecosystems, in Lal, R. et al. (eds.), *Soil Processes and Carbon Cycles*, CRC Press, Boca Raton, FL, 1997, 57–80.

Lal, R. et al., *The Potential of U.S. Cropland to Sequester Carbon and Mitigate the Greenhouse Effect,* CRC Press, Boca Raton, FL, 1999.

Paustian, K.O. et al., Agricultural soil as a C sink to offset CO_2 emissions, *Soil Use Manage.*, 13:230–244, 1997.

CHAPTER 39

Potential for Carbon Accumulation under *Brachiaria* Pastures in Brazil

R. M. Boddey, S. Urquiaga, B. J. R. Alves, and M. Fisher

CONTENTS

INTRODUCTION

In the Brazilian tropics, African grasses have long been used as pastures for both beef and dairy cattle. The first was *Panicum maximum* (Guinea grass), the seeds of which arrived in straw bedding on slave ships. *P. maximum*, known locally as Colonião, now covers huge areas of degraded land on coffee plantations in the states of Rio de Janeiro and São Paulo, which were created when the Atlantic forests were cleared, but were subsequently abandoned. Other African species such as molasses grass (*Melinis minutiflora*), jaraguá (*Hyparrhenia rufa*), and pangola (*Digitaria decumbens*) were introduced in the early part of this century. In the 1970s, coinciding with the opening up of the Cerrado (central savanna) region of Brazil, *Brachiaria decumbens* was introduced from Australia.

After clearing the Cerrado vegetation, a grain crop, usually rice, was sown with lime, phosphorus (P), and potassium (K) and undersown with *B. decumbens*. The pasture gave animal live weight gains at least 10 times those of the native vegetation and required no further inputs for several years (Thomas et al., 1995). *B. decumbens* was rapidly adopted for pastures in the Cerrados; this and other *Brachiaria* species (principally *B. brizantha* and *B. humidicola*) are now estimated to cover 40 to 50 million ha (M ha) in the region (Macedo, 1995; Zimmer and Euclides–Filho, 1997).

1-56670-581-9/02/$0.00+$1.50
© 2002 by CRC Press LLC

Recently, *B. humidicola* and, to a lesser extent *B. brizantha* have largely replaced *P. maximum* in pastures in Amazonia, now estimated to occupy 20 M ha (Neto and Dias Filho, 1995). There are also large areas of pastures of *Brachiaria* spp. in the coastal Atlantic forest region from São Paulo northwards to Maranhão. The total area occupied by *B. decumbens*, *B. humidicola*, and *B. brizantha* is approximately 80 M ha (Zimmer and Euclides–Filho, 1997) or almost 10% of Brazil (850 M ha).

Most of these pastures received no further fertilizer after establishment and they are also frequently overgrazed. It is estimated that 50 to 80% of *Brachiaria* pastures in Brazil are degraded (Macedo, 1995; Kichel and Miranda, 1999). Degraded pastures produce less than 50 kg live weight gain ha^{-1} year^{-1}, have low ground cover, are generally invaded by unpalatable native plants, and have compacted soil and many termite mounds.

The decline in productivity is mainly caused by deficiencies of nitrogen (N) and P, (de Oliveira et al., 1997; Primavesi and Primavesi, 1997). Recently Boddey et al. (1996; 1999) suggested that the decline is caused by lack of P fertilization, and inefficient N recycling through the animal. Nitrogen losses from urine are particularly high, (Vallis and Gardener, 1984; Ferreira et al., 1995) especially via ammonia volatilization in these hot environments, and increase with stocking rate. At lower stocking rates, a higher proportion of the nutrients are recycled by internal remobilization and through plant litter, so that losses are fewer (Fisher et al., 1999).

Fisher et al. (1994) found that *Brachiaria* pastures on Oxisols of the eastern plains (Llanos) of Colombia accumulate soil organic matter (SOM). They reported that total carbon (C) in the soil to 80 cm in depth under a pasture of *B. humidicola* compared to the native savanna at Carimagua research station increased by 26 Mg ha^{-1} from 197 to 223 Mg ha^{-1} over 9 years. Total soil C to 80 cm under mixed swards of *B. humidicola* with the forage legume *Arachis pintoi* was 267 Mg ha^{-1}. At Matazul Farm, some 200 km west of Carimagua, the soil (0 to 100 cm) below a mixed pasture of *B. dictyoneura* with the legume *Centrosema acutifolium* accumulated 28 Mg ha^{-1} in just 3.5 years.

No similar data are yet available for the Cerrados of Brazil. Data from various sites in the Amazon forest show that, immediately after clearing, soil C stocks decline. However, after a few years, soil C usually exceeds that of the original forest (Eden et al., 1990; Choné et al., 1991; Bonde et al., 1992; Feigl et al., 1995; Neill et al., 1997). This chapter discusses the factors that control C accumulation under *Brachiaria* pastures and the conditions, and hence management practices, that favor it.

ACCUMULATION OF CARBON UNDER *BRACHIARIA* AND NATIVE VEGETATION

For the soil under sown pastures to accumulate more carbon than under the native vegetation two factors must be considered: the rates of deposition of C as organic matter in the soil, and the rates at which this C decomposes or oxidizes. If C is deposited at a higher rate under the *Brachiaria* pasture, and if it oxidizes at the same or lower rate, then the soil under it will accumulate C. Organic matter deposited in soil is derived from senescent aerial tissue (litter), which falls to the ground and becomes incorporated in the surface layers of the soil with the aid of soil fauna, and dead roots.

It is logical to assume that the rate of deposition of organic residues by any vegetation community will be a function of its net primary productivity (NPP). Soil fertility limits NPP of native vegetation in much of the Cerrados, especially in the *campo limpo* and *campo sujo*, which occur on poorer soils and are where most introduced pastures are found (Fisher et al., 1997). Almost nowhere in Brazil are *Brachiaria* pastures regularly treated with chemical fertilizers after establishment. Lime, P, and K are normally applied at planting, either to the pasture or to the pioneer crop, which precedes the introduction of the *Brachiaria*. NPP of the sown pasture should therefore be considerably greater than that of the native vegetation that it replaces.

This hypothesis should be justified with data of NPP of *Brachiaria* pastures compared with NPP of the native vegetation that they replace. However, while there are many estimates for the arial NPP of the Amazon forest (Barbosa and Fearnside, 1996, and the references therein), there are few for *Brachiaria* pastures or for the savanna vegetation that they replace. Moreover, there are serious methodological problems in estimating NPP of savanna and pasture communities (Rezende et al., 1999).

ESTIMATION OF NET AERIAL PRIMARY PRODUCTIVITY OF PASTURES AND NATIVE VEGETATION

Methods to estimate arial NPP have been developed for forests. The standard technique is to sum any increase in total standing biomass and the annual deposition of litter (leaves, twigs, branches, and even whole trees if they die and fall in the year of measurement) (Raich et al., 1991). However, until very recently, techniques based on this principle were not adopted to estimate NPP in pastures, grasslands, or prairies. Historically, peak standing biomass was used as measure of aerial NPP in grassland communities (Boulière and Hadley, 1970). Subsequently, in the 1970s the international biological program (IBP) developed a methodology based on sequential estimates of standing biomass at various times (usually monthly) during the year (Singh et al., 1975). All positive increments in standing biomass were summed and the total was considered to estimate annual NPP.

Although natural grasslands normally contain many plant species, not all of them are grasses. However, because different species have peaks of production at different times of the year, it was suggested that all positive increments of biomass of individual species should be used instead of increases in the bulk yield. A further refinement was to include only statistically significant increases (at $P<0.05$, 0.10, or 0.20) to separate sampling variability from real increases in standing biomass. Finally, Singh and Yadava (1974) estimated annual NNP by summing the positive increments in above-ground live biomass with "positive increments in standing litter, if both increments occurred concomitantly."

Singh et al. (1975) reported the application of 13 variations of these standing biomass techniques at ten diverse grassland sites in the U.S. from latitudes 30 to 50°N, altitudes 390 to 2340 m, and annual rainfall of 158 to 984 mm. The methodologies consisted of different combinations of:

1. Estimating peak biomass of individual species or the whole crop
2. Using positive increments of growth (trough-peak method) instead of peak biomass
3. Including increments of standing dead material or litter, or not

Within methodologies in which positive increments from harvest to harvest were estimated, they applied different statistical criteria (differences significant at $P = 0.05$, 0.10, 0.20, or no constraint) to accept or reject these increments.

From this large array of estimates for each site, they used analysis of variance and cluster analysis to show relationships between the different methods. However, because no absolute standard for comparison existed, they could not recommend any one "best method" and their discussion as to which ones were suitable was extremely subjective. In general, they seem to rank the different techniques based on the precision (standard errors, etc.) and with some preconceived idea of what the result should be.

Long et al. (1989) pointed out the shortcomings in this approach: the IBP methodology is based on the assumption that, if yield of standing biomass increases between any two harvests, death of plant material does not occur. On the other hand, any decrease in standing biomass is assumed to be solely due to tissue death. If species are not separately evaluated, then a species most productive when total biomass of the community is declining will be ignored.

There is a further limitation. In individual grassland species, old leaves and stems die and new ones are produced continuously. Because of this, even if each species that constitutes a significant proportion of the total population is evaluated, or if one species is overwhelmingly predominant (as is often the case with pastures), this error can still lead to serious underestimates of NPP.

In the case of sown pastures, grazing specialists rarely see the need to measure NPP, perhaps because animal weight gain is a good indirect index of plant production. Moreover, grazing management is generally regulated by standing biomass, either forage on offer, (i.e., the total standing yield), or forage allowance (standing yield per animal unit) (Fisher et al., 1999).

The paired plot method (Frame, 1981) is usually used to measure pasture productivity in grazing experiments. Paired areas with identical sward height, plant density, and botanical composition are selected. One of the paired areas is harvested to estimate dry matter yield at the start of the evaluation; the other area is protected from grazing by a cage and harvested some weeks later. The period between harvests is dictated by the growth characteristics of the forage species.

Pastures of *B. humidicola* alone and associated with the forage legume *Desmodium ovalifolium* were grazed at stocking rates of 2, 3, and 4 head ha^{-1} with three replicates at Itabela in the Atlantic forest region in the south of Bahia state (16°39'S, 39°30'W, rainfall 1400 mm year^{-1}). The paired-plot technique was used to estimate pasture growth from January to May, 1994, using a 21-day evaluation period. Ten replicated paired plots per paddock were used for each of the six replicated treatments. Coefficients of variation (CV) of the mean yields of the initial or cage plots within each treatment were acceptable (6 of 8 < 20%). Because the differences between the mean yields were small compared to their magnitude, CVs of the differences between the means were extremely high (Table 39.1).

Approximately half of the estimates of pasture production were negative, suggesting that neither the grass nor the legumes were growing. This conclusion was contradicted by satisfactory mean animal live weight gains during this period of between 1.1 and 1.6 kg ha^{-1} day^{-1}. It was concluded that the method was unsatisfactory for pastures such as these where total standing biomass changed little over the evaluation period because of simultaneous growth and senescence. It was clear that a different approach was needed to estimate NPP.

In a stable ecosystem, that is, one whose total biomass is neither increasing nor decreasing over time, during any one year the rate of disappearance of dead material must equal ecosystem NPP (Wiegert and Evans, 1964). Pastures often reach stable above-ground yields within a year or two. Therefore the sum of total annual litter production, any increase (or decrease) in standing biomass and the forage consumed by the grazing animals, should equal above-ground NPP.

Rezende et al. (1999) used this approach to estimate the aerial NPP of the pastures at Itabela described previously. Existing litter was collected from 1.0×0.5 m quadrats at monthly intervals, and the quantity of litter deposited in quadrats of the same dimensions (previously cleared of litter) in 14-day periods was also evaluated each month (for full details, see Rezende et al., 1999).

In 1995, the mean litter dry matter deposited in each 14-day period was 63 g m^{-2} (630 kg ha^{-1}), while mean existing litter was about 100 g m^{-2} (1000 kg ha^{-1}; Table 39.2). It is evident that the quantity of litter deposited in any one month was much greater than the amount of existing litter, which varied little during the year. This suggests that litter decomposed very rapidly.

The litter decomposition constant, k, calculated for a single exponential decay function (Olson, 1963), gave values of -0.021 to -0.031 g g^{-1} day^{-1} (i.e., between 2.1 and 3.1% of existing litter decomposed each day). These k values estimate litter half life ($t^{1/2}$) between 33 and 22 days, respectively, leading to concern that a significant proportion of the litter deposited during the measured 14-day accumulation period would have decomposed or disappeared before the quadrats were sampled.

A mathematical analysis of the problem showed that a simple equation could be used to correct for the loss, based on the supposition that the rate of decomposition of the litter deposited on the soil surface in the cleared quadrats was equal to the rate of decomposition of the existing litter (see Rezende et al., 1999). The corrected equation gave much higher k values, ranging from -0.071 to

Table 39.1 Evaluation of total dry matter production (g m^{-2}) using the paired-plot (cage) technique, of grass-only *Brachiaria humidicola* (BH) and in mixed *B. humidicola* and *Desmodium ovalifolium* (BH/DO) pastures grazed at three different stocking rates (2, 3, and 4 head ha-1) at Itabela, South Bahia, January to May, 1994. Data are means of 10 paired areas per paddock, 3 replicate paddocks per treatment. Unpublished data from CEPLAC/ Embrapa-Agrobiologia.

Pasture	Stocking Rate (head/ha)	I			II		
		19/01 Initial	09/02 Cage	09/02 Difference	16/02 Initial	09/03 Cage	09/03 Difference
BH	2	465.1	426.6	−38.6	516.3	480.3	−36.0
	3	519.9	557.0	+37.0	505.0	446.8	−58.2
	4	413.3	401.8	−11.4	338.5	380.7	+42.3
	Mean	466.1	469.8	−4.3	453.3	435.9	−17.3
BH/DO	2	472.2	450.3	−21.9	440.3	492.7	+52.4
	3	499.7	438.9	−60.7	512.7	385.5	−127.2
	4	360.1	366.1	+5.9	388.4	342.3	−46.2
	Mean	444.0	418.4	−25.6	447.1	406.8	−40.3
Coefficient of Variation (%)		19	17	−466	17	13	−219

Pasture	Stocking Rate (head/ha)	III			IV		
		16/03 Initial	06/04 Cage	06/04 Difference	16/04 Initial	04/05 Cage	04/05 Difference
BH	2	296.4	401.4	+105.0	439.1	402.3	−36.8
	3	444.4	413.5	−30.9	419.8	434.6	+14.8
	4	316.6	380.1	+63.6	297.3	325.5	+28.2
	Mean	352.5	398.3	+45.9	385.4	387.5	+2.1
BH/DO	2	407.7	350.3	−57.4	388.8	343.9	−45.0
	3	404.3	411.6	+7.3	341.5	339.1	−2.4
	4	338.6	432.1	+93.5	314.5	222.0	−92.5
	Mean	383.5	397.8	+14.5	348.3	301.7	−44.6
Coefficient of Variation (%)		38	16	+437	19	30	−451

−0.097, g g^{-1} day^{-1}, and total annual litter production was calculated to be between 21 and 33 Mg ha^{-1} year^{-1}. Not surprisingly, rates of litter deposition were significantly lower at higher cattle stocking rates.

The other two terms in above-ground NPP, animal consumption and biomass change, were measured directly (Table 39.3). Animal consumption was calculated from total fecal production, estimated using chromic oxide as an external index (Le Du and Penning, 1982) and the *in vitro* digestibility of the consumed ration, estimated from samples taken from animals fitted with esophageal fistulae. As the cattle gained weight during the year, total standing biomass decreased from between 4 to 6 Mg ha^{-1} to 1 to 3.5 Mg ha^{-1}, depending on stocking rate. Total above-ground NPP was estimated to be 31 to 36 Mg ha^{-1} year^{-1} (Table 39.3).

The methodology used here is based on the concept described by Wiegert and Evans (1964) and Bulla et al. (1981) and similar to that used by Long et al. (1989). A similar approach has also been used for estimation of fine root production in forest ecosystems (Santantonio and Grace, 1987).

A further study in the Cerrado region using the method of Rezende et al. (1999) was conducted on pastures of *B. ruziziensis* alone or with the forage legume *Stylosanthes guianensis* at Uberlândia (Minas Gerais state, annual rainfall 1200 mm) (Boddey et al., 1999). In this case the arial NPP of the grass-only pasture was estimated as 13.6 Mg ha^{-1} year^{-1}, and in the mixed pasture 21.8 Mg ha^{-1} year^{-1} (Table 39.4). The much higher above-ground NPP at Itabela can be attributed to uniform

Table 39.2 Annual means of existing litter, litter deposited in 14 days and litter decomposition parameters in pastures of *B. humidicola* in monoculture and mixed *B. humidicola* and *D. ovalifolium* pastures under three stocking rates of crossbred Brahman cattle in the period January to December 1995. The values are means of 12 monthly evaluations, from 10 quadrats per paddock, and 3 replicate paddocks per treatment.

Pasture	Stocking rate (an. ha^{-1})	Means of Existing and Deposited Litter (g m^{-2})		Decomposition Constant k (g g^{-1} day^{-1})	Total Litter Deposited in 12 Months (Mg ha^{-1} year^{-1})	
		Existing	Deposited in 14 days		Estimate[b] 14 days	Corrected[a]
B. hum						
	2	116.6	72.5	0.0706	18.9	29.7
	3	105.8	66.4	0.0734	17.3	27.5
	4	73.7	49.1	0.0797	12.8	21.3
	Mean	98.7	62.7	0.0746	16.3	26.2
Bh/Do						
	2	100.6	69.0	0.0969	18.0	33.1
	3	101.2	63.0	0.0716	16.4	26.0
	4	98.1	58.4	0.0680	15.2	23.6
	Mean	100.0	63.5	0.0788	16.5	27.6
Coef. Variation (%)		20.5	12.7	30.8	12.7	16.6
Analysis of variance[c]						
Factor: Pasture (P)		ns	ns	ns	ns	ns
Stocking rate (S)		ns	**	ns	**	**
Interaction P × S		ns	ns	ns	ns	ns

[a] Calculated using the equation k = −ln [(X$_{eq}$ − X$_N$)/X$_{eq}$]/t$_N$ where X$_{eq}$ is the quantity of existing litter and X$_N$ the quantity of litter deposited in the time t$_N$ (which in this study was 14 days).
[b] Calculated from ((litter deposited in 14 days)/14) × 365.
[c] ns, **, indicate respectively that differences between means were not significant at P = 0.05, or significant at P = 0.01.

Table 39.3 Above-ground net primary production of mixed legume (*B. humidicola* and *D. ovalifolium*) and grass-only pastures of *B. humidicola* grazed at 3 different cattle stocking rates (2, 3, and 4 head/ha) at Itabela, South Bahia. Unpublished data from CEPLAC/Embrapa-Agrobiologia.

Pasture	Stocking Rate (head/ha)	% Legume in Sward	Litter Deposition[b]	[a]Forage Consumption[b]	Change in Standing Biomass[b]	Aerial NPP[b]
B. humidicola	2	0	29.7	8.3	−2.1	35.9
	3	0	27.5	10.0	−3.3	34.2
	4	0	21.3	14.0	−3.0	32.3
	Mean	0	26.2	10.8	−2.8	34.1
B. humidicola/	2	25	33.1	6.8	−3.5	36.4
D. ovalifolium	3	10	26.0	9.8	−3.5	32.3
	4	20	23.6	9.9	−2.2	31.3
	Mean	18	27.6	8.8	−3.1	33.3

[a] Estimate derived from digestibility of forage taken from oesophageal fistulas on two occasions and 3 applications of the Cr$_2$O$_3$ technique.
[b] Mg dry matter ha^{-1}.

distribution of rainfall throughout the year compared to Uberlândia, which has a marked dry season from April to September.

It is not possible to compare these estimates of above-ground NPP of pastures with others in the literature, as to date this type of technique has not yet been used by others. However, Long et al. (1989) used a methodology based on the same principles to estimate the NPP of the native savanna in Mexico, Kenya, and Thailand. They estimated above-ground NPP at, respectively, 8,

Table 39.4   Estimates (in Mg/ha) of above-ground net primary production of mixed grass and legume (*Brachiaria ruziziensis* and *Stylosanthes guianensis*) and grass-only pastures of *B. ruziziensis* under continuous grazing on a ranch (Fazenda Santa Inês) near Uberlândia, Minas Gerais. Cattle stocking rates varied during the year and were adjusted according to forage on offer.

Pastures	Forage on Offer			Litter Production[a]	Animal Consumption[b]	Total
	05/02/97	14/01/98	Difference			
Grass-only	4.12	2.20	−1.92	11.48	4.03	13.59
Mixed	6.12	4.34	−1.78	17.39	6.21	21.82

[a] Sum of the difference between the existing litter at the end of the sampling period and that at the start, plus that deposited during this period.
[b] Dry matter consumed by the animals calculated from = 0.03 × animal live weight × time of grazing (days).

Data from Ayarza et al. (1997) and unpublished data from Embrapa-Cerrados/Embrapa-Agrobiologia.

10, and 15 Mg ha^{-1} year^{-1}. The edapho-climatic conditions at each of the sites studied by Long et al. were very different from the sites in Brazilian Cerrados reviewed here. It is therefore not possible to extrapolate from their data to arrive at a general conclusion of NPP of native savannas nor to determine whether above-ground NNP of the *Brachiaria* pastures is definitely higher than that of the native vegetation at either Itabela or Uberlândia. However, Long et al. (1989) showed that the application of the standard IBP methods underestimated total NPP (aerial and below ground to 15 cm) by between 47 and 74%.

There are four main vegetation communities in the Cerrado (Eiten, 1972): *campo limpo*, where the vegetation is predominantly grasses and trees are totally absent, *campo sujo*, with some scattered trees and shrubs, *campo cerrado*, where small trees and shrubs predominate, and *cerradão*, which is woodland.

The only data for NPP of any of the Cerrado vegetation types are based on the standard IBP method. Meirelles and Henriques (1992) estimated the arial NPP of an area near Brasília of *campo cerrado* immediately after it was burned at 1.7 Mg ha^{-1} year^{-1}, and 1.2 Mg ha^{-1} year^{-1} in an unburned area. These authors cite estimates of aerial NPP published in postgraduate theses from the University of Brasilia that vary from 3.7 Mg ha^{-1} year^{-1} for a burned area of *campo sujo* (Cezar, 1980) and 3.3 and 2.4 Mg ha^{-1} year^{-1} for burned and unburned areas of *campo cerrado*, respectively (Batmanian, 1983). Meirelles (1990) also evaluated the aerial NPP of a sward of *Brachiaria decumbens* growing on the same soil type adjacent to the *Campo cerrado* site (Meirelles and Henriques, 1992) and registered an estimate of 3.9 Mg ha^{-1} year^{-1}. However, this area was not under grazing and there is no information whether this pasture could be regarded as productive, degrading, or already degraded; experience would suggest the latter.

Meirelles and Henriques' (1992) data suggest that the above-ground NPP of *B. decumbens* is at least twice that of the native Cerrado vegetation. However, it must be emphasized that these estimates were made with the standard IBP methodology, which at best underestimate NPP by 25 to 75% (Long et al., 1989) and, for this reason, could be seriously in error in comparing different plant communities. Hence, from the data of Meirelles (1990) and Meirelles and Henriques (1992) alone it cannot be concluded that the NPP (arial or below ground) of the *Brachiaria* pasture was higher than that of the native Cerrado vegetation, although it seems plausible that it should be.

BELOW-GROUND PRODUCTIVITY OF PASTURES AND NATIVE VEGETATION

Below-ground productivity of pastures and native vegetation is much more closely related to the potential of these systems to accumulate C in the soil. The preceding discussion centered on the estimation of arial productivity, as to date there are no data for below-ground production of *Brachiaria* pastures in the neo-tropical savannas. Probably only two techniques are suitable for these studies: the ingrowth mesh bag technique of Steen (1984) or the compartment-flow model technique of Santantonio and Grace (1987), both of which are extremely labor-intensive. No data could be found in the literature for *Brachiaria* pastures nor for either forest or savanna vegetation in Brazil or other tropical regions. However, indirect evidence exists that root turnover under *Brachiaria* pastures is higher than under savanna vegetation in the eastern plains (Llanos) of Colombia. Rao (1998) found that total root mass (0 to 80 cm) under a pasture was 5.7 Mg ha^{-1} in comparison with only 1.4 Mg ha^{-1} for native savanna grassland in this region.

DECOMPOSITION OF ROOTS OF *BRACHIARIA* AND SAVANNA VEGETATION

The evidence of Rao (1998) suggests that below-ground NPP of *Brachiaria* spp. is greater than that of the native vegetation in the Colombian savannas, largely because fertilizer was applied to the *Brachiaria* pastures at sowing and for maintenance. However, higher below-ground NPP will not necessarily lead to higher C accumulation in the soil if the specific rate of decomposition of the roots of *Brachiaria* spp. is proportionately higher than that of the roots of the savanna vegetation. One of the most important factors controlling the decomposition of plant residues is their C:N ratio. Residues with high C:N ratio decompose at lower rates than those of lower C:N ratio (Herman et al., 1977). This has recently been shown to be true for tropical forage species; also, C:N ratios of live *Brachiaria* roots were found to range from 109 to 137 (Gijsman et al., 1997; Urquiaga et al., 1998). The C:N ratios of native vegetation of the Eastern Llanos (Colombia) were found to be lower than in neighboring *B. brizantha* pastures in a recent study by Trujillo (2000; Table 39.5).

Fisher et al. (1995) reported that the C:N ratio of the soil organic matter in the Colombian savannas was approximately 21.5, much higher than the 10 to12 recorded elsewhere. They found that the C:N ratio of soil organic matter under a 9-year-old *Brachiaria* pasture in this region was 33.2, indicating that the C:N ratio of the newly acquired C was very high and suggesting that *Brachiaria* roots decompose at rates lower than those of the native vegetation of the savanna.

Table 39.5   C, N, and Lignin Content in Roots of Native Savanna and from a *Brachiaria dictyoneura* Pasture, with and without the Legume *Arachis pintoi*

Date	Parameter	Native Savanna	*B. dictyoneura* Grass-Only	*B. dictyoneura/A. pintoi* Mixed legume/grass	
				B. dictyon.	*A. pintoi*
June 97	% C	41.2	51.0	39.3	51.0
	% N	0.61	0.44	0.44	2.62
	C:N	68	80	90	20
	% Lignin	15.2	11.9	11.6	n.d.
April 98	% C	45.9	47.8	49.5	50.1
	% N	0.37	0.27	0.22	2.41
	C:N	124	177	225	21
	% Lignin	14.2	10.7	9.5	11.2
April 99	% C	42.1	41.5	38.5	51.5
	% N	0.47	0.39	0.34	1.88
	C:N	89	106	113	27
	% Lignin	n.d.	n.d.	n.d.	n.d.

Data from Wilma Trujillo, CIAT, Cali, Colombia.

CARBON STOCKS UNDER *BRACHIARIA* PASTURES

No specific studies appear to have been performed to assess the potential for C accumulation under *Brachiaria* pastures in the Cerrado region. However, in the study on nutrient cycling in pastures of *B. humidicola* (Boddey et al., 1995; 1999; Rezende et al., 1999) at Itabela mentioned earlier, the C stock in the soil under the pastures was measured, as well as in the adjacent forest and a degraded *B. humidicola* pasture.

The pastures were sown in 1987 on an area cleared from secondary forest. The secondary forest bordering the western boundary of the experiment had been undisturbed for over 20 years, but it is not possible to verify whether the C stock in this area represents that of the original primary forest vegetation. Immediately after the pastures were established in 1988, soil samples were taken from each of the 27 paddocks at 5-cm intervals to a depth of 25 cm. These samples were stored and only rediscovered in 1997, so that two other samplings of this experiment (1994 and 1997) were made at different depth intervals. Only in 1997 were bulk density measurements taken. To calculate total C stock, these values were used for all three samplings. For the 1988 sampling it was assumed that the soil C content in the 25- to 30-cm layer (not sampled) was equal to that in the 20- to 25-cm layer, which is likely to overestimate the C stock in 1988, as C content decreased steadily with depth. The data show a strong trend of increasing C content of the soil to 30-cm depth from 1988 to 1997 (Table 39.6).

In 1997 the sampling was made to 100-cm depth, and the adjacent forest was also sampled for C content and bulk density (three profiles, 50 m into the forest and 100 m apart). The carbon stock was slightly higher under the pastures than under the forest and most of this difference was in the top 20 cm of soil (Table 39.7). This contrasts with the results of Fisher et al. (1994) for the Colombian savanna, where over 75% of the increase in soil C was in the layers below 20 cm (20 to 100 or 20 to 80 cm). Samples taken from a degraded *B. humidicola* pasture on the same soil approximately 300 m from the eastern boundary of the experiment showed much lower C content than either the forest or the productive pastures of the nutrient cycling study.

Data from the Amazon region show that, where forest has been cleared for pastures of *Brachiaria* spp., the immediate effect is to lower the C stocks in the soil. However, after several years, C stocks increase and eventually exceed those found under the forest (Eden et al., 1990; Choné et al., 1991; Bonde et al., 1992; Feigl et al., 1995; Neill et al., 1997). This is the same pattern observed at the

Table 39.6   Changes (in Mg C ha⁻¹ per layer) in Carbon Content in Soils under Pastures of *B. humidicola* in Monoculture and Mixed *B. humidicola* and *D. ovalifolium* Pastures under Three Cattle Stocking Rates

Layer (cm)	Pasture					
	Grass-Only (Year)			Grass/Legume (Year)		
	1988	1994	1997	1988	1994	1997
0–5	9.50	11.66	13.06	8.90	10.70	13.32
5–10	9.24	9.44	10.61	8.12	9.50	10.59
10–15	7.74		15.61	7.08		15.60
15–20	6.65	27.40		6.17	29.08	
20–30	11.85		11.46	10.98		12.19
TOTAL	44.99	48.49	50.74	41.25	49.27	51.69

Unpublished data from CEPLAC/ Embrapa-*Agrobiologia*.

Tabela 39.7 Carbon Concentration and Content of Soils under *B. humidicola* in Monoculture and Mixed *B. humidicola* and *D. ovalifolium* Pastures under Grazing[a] and under Neighboring Secondary Forest and a Degraded Pasture at Itabela, South Bahia. Means of 3 Profiles, 3 Samples per Layer

Depth (cm)	Grass-Only	Grass/Legume	Secondary Forest	Degraded Pasture[b]
		% C		
0–5	1.99	2.10	2.04	1.20
5–10	1.43	1.49	1.25	1.10
10–20	1.02	1.04	0.98	0.91
20–30	0.75	0.81	0.84	0.76
30–40	0.66	0.69	0.78	0.60
40–50	0.57	0.61	0.68	0.53
50–60	0.53	0.54	0.57	0.43
60–80	0.45	0.48	0.47	0.37
80–100	0.38	0.40	0.37	0.35
		Mg C/layer		
0–5	13.06	13.32	11.24	8.19
5–10	10.61	10.59	8.06	7.30
10–20	15.61	15.60	14.06	11.46
20–30	11.46	12.19	11.38	8.87
30–40	9.88	10.21	10.06	6.11
40–50	7.91	8.57	8.80	5.63
50–60	7.40	7.48	7.34	4.59
60–80	11.77	12.63	11.83	7.18
80–100	9.36	10.02	8.95	7.18
Total	97.06	100.60	91.73	66.51

[a] Means of 3 stocking rates (2, 3, and 4 head ha^{-1}).
[b] Samples from 1 profile, 3 samples per layer.

Unpublished data from CEPLAC/ Embrapa-Agrobiologia.

Itabela site in the Atlantic forest region. By definition, degraded pastures have lower productivity, so it is not surprising that C stocks were lower in the degraded pasture than in the original forest or the productive *B. humidicola*-based pastures. Similar results were reported by Trumbore et al. (1995) working in the Amazon forest in the eastern state of Pará, where the carbon stock under a degraded pasture of *B. humidicola* was lower than that below the original forest. However, at the same site where a pasture had been fertilized and replanted with *B. brizantha*, soil C stocks increased above those of the forest.

CONCLUSIONS

The data presented indicate that the NPP of *Brachiaria* pastures is far higher than previously suspected. As yet no comparisons have been made for the NPP of these pastures with that of adjacent native vegetation, but circumstantial evidence suggests that the NPP of productive *Brachiaria* swards is much higher than that of the native vegetation. The evidence also suggests that the C:N ratio of *Brachiaria* roots is higher than that of native savanna vegetation and thus decomposition rates should be lower. If these assumptions are correct and below-ground productivity is proportional to above-ground NPP, then the final equilibrium C content of the soil under *Brachiaria* should exceed that at present in the native vegetation. The data of Fisher et al. (1994) for the Colombian savannas confirm this, and suggest that C accumulation under these pastures is very significant. In the case of the Amazon and Atlantic forest ecosystems, the reason that C stocks increase under *Brachiaria* pastures may be due not so much to any higher NPP of the pastures compared with the native vegetation, but more to the high C:N ratio of the dead *Brachiaria* roots, which leads them to decompose at much lower rates than that of the forest vegetation; hence, soil C accumulates.

These arguments are somewhat fragile because of the lack of data from studies on C stocks under existing pastures in comparison with native vegetation. There are no studies for the Cerrado region, nor are there reliable estimates of both above- and below-ground NPP of the pastures and the native vegetation. Because the above-ground biomass of tropical forest ecosystems in Brazil contains between 100 and 200 Mg C ha^{-1}, there is clearly no possible justification to destroy virgin forest to install pastures with the objective of sequestering C.

In the *campo limpo* and *campo sujo* of the Cerrados, aerial biomass rarely contains more than 10 Mg C ha^{-1}, so a small increase in soil C can readily offset the above-ground C in the native vegetation. Nevertheless, areas in these regions not converted to pasture are important reserves of biodiversity. Moreover, a large proportion of the area sown to pastures is unproductive and degraded. The drive to increase soil C stocks would be much more rational if it recuperated degraded pastures and maintained their productivity, rather than expanding the area of pasture.

Management practices exist to maintain *Brachiaria* pastures in sustainable production, but they have only been adopted by a small number of landowners. It is essential to apply modest quantities of maintenance fertilizer; occasional additions of lime and annual applications of 10 to 20 kg P and K typically are sufficient.

Most critical is grazing management. Stocking rate should be regulated as a function of the forage on offer, in such a way that a large proportion of the nutrients are recycled through the litter pathway and internal remobilization. If pastures are overgrazed, almost all nutrients are recycled via animal excreta, which, in the case of N and K, leads to large losses from the system, especially from urine via volatilization (N) and leaching (N and K) (Fisher et al., 1999). However, the adoption of these practices often represents a great problem for the cattle owner in the dry season, when forage on offer declines and no forage is available for the animals. Where these management practices are adopted, pastures of *Brachiaria* spp. can maintain their productivity for many years without the addition of N fertilizers.

Lascano and Euclides (1996) showed that, on the eastern plains of Colombia, with good management, cattle live weight gains averaged 225 kg ha^{-1} year^{-1} on a *B. decumbens* pasture and production continued at this level even 16 years after pasture was sown. Similar results were reported by Boddey et al. (1996) over 8 years of continuous grazing of *B. humidicola* at Itabela in the south of Bahia.

The introduction of a forage legume into a grass pasture presents various advantages to the cattle owner. Nitrogen supply to the system increases from biological nitrogen fixation by the legume, permitting higher stocking rates and live weight gains (Thomas et al., 1995). In the case of deep rooting legumes, such as *Stylosanthes* spp., protein-rich forage is available in the dry season, which increases live weight gain at that time of year (Boddey et al., 1997; Ayarza et al., 1997). At present this practice has not been adopted on a significant scale in Brazil, partly because the persistence of the legume in the sward requires more careful management and, as yet, few suitable legumes are available for this purpose.

The recovery, or reform, of the vast areas of degraded pastures in Brazil would bring highly significant financial benefits to cattle owners. At the same time it would reduce the susceptibility of these areas to erosion and contribute to the improvement of drainage, replenishment of aquifers, and decreased runoff. Also the transformation of these areas into productive pastures would almost certainly lead to accumulation of C in these soils on a scale which would be highly significant for the Brazilian carbon inventory — perhaps even on a scale significant for the global carbon budget.

ACKNOWLEDGMENT

This study was supported by the Department of International Development (DfID) of the United Kingdom Government.

REFERENCES

Ayarza, M. et al., Introdução de *S. guianensis* cv. Mineirão em pastagens de *Brachiaria ruziziensis*: influência na produção da pastagem e na reciclagem da liteira, Technical Bulletin. EMBRAPA-Agrobiologia, Seropédica, Rio de Janeiro, 1997.

Barbosa, R.A. and P.M. Fearnside, Carbon and nutrient flows in an Amazon forest: fine litter production and composition at Apiau, Roraima, Brazil, *Trop. Ecol.*, 37:116–125, 1996.

Batmanian, G.J., Efeitos do fogo sobre a produção primária e a acumulação de nutrientes do estrato rasteiro de um Cerrado, M.Sc. Thesis, University of Brasilia, 1983.

Boddey, R.M. et al., The nitrogen cycle in pure grass and grass/legume pastures: evaluation of pasture sustainability, in *Nuclear Techniques in Soil-Plant Studies for Sustainable Agriculture and Environmental Preservation*, FAO/IAEA, Vienna, Austria, 1995, 307–319.

Boddey, R.M., I.M. Rao, and R.J. Thomas, Nutrient cycling and environmental impact of *Brachiaria* pastures, in Miles, J.W., B.L. Maass, and C.B. do Valle (eds.), *Brachiaria: The Biology, Agronomy and Improvement*, Publication 259, CIAT, Cali, Colombia, 1996, 72–86.

Boddey, R.M. et al., The contribution of biological nitrogen fixation for sustainable agricultural systems in the tropics, *Soil Biol. Biochem.*, 29:787–799, 1997.

Boddey, R.M. et al., Effect of the introduction of a forage legume (*Desmodium ovalifolium*) on nitrogen cycling in *Brachiaria humidicola* pastures in the extreme South of Bahia, in *Proc. Int. Symp. Pasture Ecophysiology Grazing Ecol.*, 24–26 August, Curitiba, Paraná, Brazil, 1999.

Bonde, T.A., B.T. Christensen, and C.C. Cerri, Dynamics of soil organic matter as reflected by natural ^{13}C abundance in particle size fractions of forested and cultivated oxisols, *Soil Biol. Biochem.*, 24:275–277, 1992.

Boulière, F. and M. Hadley, The ecology of tropical savannas, *Ann. Rev. Ecol. Syst.*, 1:125–152, 1970.

Bulla, L., J. Pacheco, and R. Miranda, A simple model for the measurement of primary productions in grasslands, *Bol. Soc. Venez. Cienc. Nat.*, 136:281–304, 1981.

Cezar, H.L., Efeitos da queima e corte sobre a vegetação de Campo sujo na Fazenda Água Limpa, Distrito Federal, M.Sc. Thesis, University of Brasilia, 1980.

Choné, T. et al., Changes in organic matter in an oxisol from the central Amazonian forest during 8 years as pasture, determined by ^{13}C composition, in Berthelin, J., ed., *Diversity of Environmental Biogeochemistry*, Elsevier, New York, 1991, 307–405.

de Oliveira, O.C. et al., Resposta de pastagem degradada à adubação em condições de campo no cerrado brasileiro, in Dias, L.E. et al. (eds.), *Proc. Simpósio Nacional de Recuperação de Áreas Degradas (III SINRAD)*, Ouro Preto, MG, Universidade Federal de Viçosa, 1997, 110–117.

Eden, M.J. et al., Pasture development on cleared forest land in northern Amazonia, *Geogr. J.*, 156:283–296, 1990.

Eiten, G., Cerrado vegetation of Brazil, *Bot. Rev.*, 38:201–341, 1972.

Feigl, B.J., J. Melillo, and C.C. Cerri, Changes in the origin and quality of soil organic matter after pasture introduction in Rondonia (Brazil), *Plant Soil*, 175:21–29, 1995.

Ferreira, E. et al., Destino do ^{15}N-urina bovina aplicado na superfície de um solo podzólico descoberto, ou sob cultura de *Brachiaria brizantha*, in *Anais XXXII Congresso Anual de Soc. Bras. Zootecnia*, 17–21 Julho, Brasília, D.F., 1995, 109–110.

Fisher, M.J. et al., Carbon storage by introduced deep rooted grasses in the South American savannas, *Nature*, 371:236–237, 1994.

Fisher, M.J. et al., Pasture soils as carbon sink, *Nature*, 376:472–473, 1995.

Fisher, M.J., R.J. Thomas, and I.M. Rao, Management of tropical pastures in acid-soil savannas of South America for carbon sequestration in the soil, in Lal, R., R.F. Follet, and B.A. Stewart (eds.), *Management of Carbon Sequestration in Soil*, CRC Press, Boca Raton, FL, 1997, 405–420.

Fisher, M.J., I.M. Rao, and R.J. Thomas, 1999. Nutrient cycling in tropical pastures, with special reference to the neotropical savannas, in *Proc. 18th Int. Grassland Congr.*, June 8–19, 1997, Winnipeg and Saskatoon, Canada, (in press).

Frame, J., Herbage mass, in Hodgson, J. (ed.), *Sward Measurement Handbook*, British Grassland Society, Maidenhead, UK., 1981, 36–69.

Gijsman, A.J., H.F. Alarcón, and R.J. Thomas, Root decomposition in tropical grasses and legumes, as affected by soil texture and season, *Soil. Biol. Biochem.*, 29:1443–1450, 1997.

Herman, W.A., W.B. McGill, and J.F. Dormaar, Effects of initial chemical composition on decomposition of three grass species, *Can. J. Soil. Sci.*, 57:205–215, 1977.

Kichel, A.N. and C.H.B. Miranda, Recuperação e renovação de pastagens degradadas. Recomendações Técnicas para a Produção de Gado de Corte, Technical Bulletin, Embrapa-Gado de Corte, Campo Grande, MS, Brazil, 1999.

Lascano, C.E. and V.P.B. Euclides, Nutritional quality and animal production of *Brachiaria* pastures, in Miles, J.W., B.L. Maass, and C.B. Valle (eds.), Brachiaria: *Biology, Agronomy and Improvement*, CIAT, Cali, Colombia, 1996, 106–123.

Le Du, Y.L.P. and P.D. Penning, Animal based techniques for estimating herbage intake, in Leaver, J.D. (ed.), *Herbage Intake Handbook*, British Grassland Society, Maidenhead, U.K., 1982, 37–75.

Long, S.P. et al., Primary productivity of natural grass ecosystems, *Plant Soil*, 115:55–166, 1989.

Macedo, M.C.M., Pastagens no ecossistema Cerrados: Pesquisa para o desenvolvimento sustentável, in de Andrade, R.P., A. de O. Barcellos, and C.M.C. da Rocha (eds.), *Proc. Symp. Pastagens nos Ecossistemas Brasileiros: pesquisas para o desenvolvimento sustentável*, Sociedade Brasileira de Zootecnia, Universidade Federal de Viçosa, Viçosa, MG, Brazil, 1995, 28–62.

Meirelles, M.L., Produção primária de pastagem de *Brachiaria decumbens*, *Revista Ceres*, 37:16–24, 1990.

Meirelles, M.L. and R.P. Henriques, Produção primária líquida em área queimada e não queimada de campo sujo de cerrado (Planaltina-DF), *Acta Botânica Brasileira*, 6:3–13, 1992.

Neill, C. et al., Soil carbon and nitrogen stocks following forest clearing for pasture in the southwestern Brazilian amazon, *Ecol. Appl.*, 7:1216–1225, 1997.

Neto, M.S. and M.B. Dias Filho, Pastagens no ecossistema do trópico úmido: pesquisa para o desenvolvimento sustentável, in de Andrade, R.P., A. de O. Barcellos, and C.M.C. da Rocha (eds.), *Proc. Symp. Pastagens nos Ecosistemas Brasileiros: pesquisas para o desenvolvimento sustentável*, Sociedade Brasileira de Zootecnia, Universidade Federal de Viçosa, Viçosa, MG, Brazil, 1995, 76–93.

Olson, J.S., Energy storage and the balance of producers and decomposers in ecological systems, *Ecology*, 44:322–331, 1963.

Primavesi, O. and A.C.P. de A. Primavesi, Recuperação de pastagens degradadas, sob manejo intensiva, sem revolvimento de solo, e seu monitoramento, in Dias, L.E. et al. (eds.), *Proc. Simpósio Nacional de Recuperação de Áreas Degradas (III SINRAD)*, Ouro Preto, MG. Universidade Federal de Viçosa, 1997, 150–155.

Raich, J.W. et al., Potential net primary productivity in South America: application of a global model, *Ecol. Appl.*, 1:399–429, 1991.

Rao, I.M., Root distribution and production in native and introduced pastures in the South American savannas, in Box, J.E., Jr. (ed.), *Root Demographics and Their Efficiencies in Agriculture, Grassland and Forest Ecosystems*, Kluwer Academic Publishers, Dordrecht, Netherlands, 1998, 19–41.

Rezende, C. de P. et al., Litter deposition and disappearance in *Brachiaria* pastures in the Atlantic region of the South of Bahia, Brazil, *Nutr. Cycl. Agroecosyst.*, 54:99–112, 1999.

Santantonio, D. and J.C. Grace, Estimating fine root production and turnover from biomass and decomposition data: a compartment-flow model, *Can. J. For. Res.*, 17:900–908, 1997.

Singh, J.S. and P.S. Yadava, Seasonal variation in composition, plant biomass, and net primary production of a tropical grassland at Kurukshetra, India, *Ecol. Monogr.* 44:351–376, 1974.

Singh, J.S., W.K. Lavenroth, and R.K. Steinhurst, Review and assessment of various techniques for estimating net aerial primary production in grasslands from harvested data, *Bot. Rev.*, 41:181–232, 1975.

Steen, E., Variation in root growth in a grass ley studied with a mesh bag technique, *Swedish J. Agric. Res.*, 14:93–97, 1984.

Thomas R.J. et al., The role of forage grasses and legumes in maintaining the productivity of acid soils in Latin America, in Lal, R. and B.A. Stewart (eds.), *Soil Management — Experimental Basis for Sustainability and Environmental Quality*, CRC Press, Boca Raton, FL, 1995, 61–83.

Trujillo, W., Accretion of organic carbon in the acid soils of the eastern plains of Colombia, Ph.D. thesis, Ohio State University, Colombus, OH, 2000.

Trumbore, S. et al., Below-ground cycling of carbon in forest and pastures of eastern Amazonia, *Global Biogeochem. Cycles*, 9:512–528, 1995.

Urquiaga S. et al., The influence of the decomposition of roots of tropical forage species on the availability of soil nitrogen, *Soil Biol. Biochem.*, 30:2099–2106, 1998.

Vallis, I. and C.J. Gardener, Short-term nitrogen balance in urine-treated areas of pasture on a yellow earth in the subhumid tropics of Queensland, *Aust. J. Exp. Agric. Anim. Husb.*, 24:522–528, 1984.

Wiegert, R.G. and F.C. Evans, Primary production and the disappearance of dead vegetation on an old field in Southeastern Michigan, *Ecology*, 45:49–63, 1964.

Zimmer, A.H. and K. Euclides–Filho, Brazilian pasture and beef production, in Gomide, J.A. (ed.), *Proc. Int. Symp. Animal Production under Grazing*, Viçosa, MG, Brazil, 4–6 Nov. 1997, Universidade Federal de Viçosa, MG, Brazil, 1997, 1–29.

CHAPTER 40

Carbon Content of Desert and Semidesert Soils in Central Asia

E. Lioubimtseva and J. M. Adams

CONTENTS

INTRODUCTION

The amount of organic and inorganic carbon in the soils of arid zones is thought to exceed the total reserve of carbon in terrestrial vegetation. According to some authors (e.g., Schlesinger, 1985, 1995) arid zone soils contain 50% of the amount of organic carbon in soils of the world. According to figures given in the widely used database of Zinke et al. (1986), organic carbon in cooler desert and semidesert areas rivals that in temperate forest soils. In the more recent database of Batjes and Sombroeck (1997), cool desert soils are assigned a value about half that in the Zinke et al. (1984) database, but still a major store of carbon. Since governmental decisions on land use are increasingly concerned with the effect of change on carbon storage and its implications for CO_2 fluxes, it is important to know which, if either, of these figures is correct. If indeed there is as much organic carbon per unit area as either database suggests, cool desert zones may be important in terms of balancing a country's carbon budget. This chapter considers the issue of desert soil carbon storage with detailed reference to the arid zone of central Asia.

Most of the desert soils data discussed and used in the biogeochemical literature comes from the English-speaking world, especially North America. In instances where the large Central Asian desert and semidesert areas are included in these assessments, the range of data sources used is limited, so confusion commonly exists about the nature of the sources and the ways in which the data were gathered. Quite understandably, given the political history of the region and the language barrier presented by Russian or Chinese, the Central Asian deserts remain something of a mystery to Western scientists. The intention here is to begin to lift the veil on the Central Asian deserts by reviewing some of the Russian literature, and then to consider how this fits into the overall context of the carbon cycle and environmental change.

The Central Asian deserts are not a pristine environment; they have been exploited by graziers for thousands of years, and in recent decades large irrigation schemes have been added to these landscapes. It is necessary to consider the extent to which these anthropogenic effects have modified the background level of carbon storage, and whether change in the intensity of either process can potentially take up or release carbon from the desert-zone carbon reservoir.

A BRIEF DESCRIPTION OF THE CENTRAL ASIAN DESERT REGIONS

Arid and semiarid environments cover about 55% of the land in Central Asia and Kazakhstan. This vast desert region of about 3.5 million km^2 comprises the entire Turan Lowland and the southern margin of the Kazakh hills (Figure 40.1). Its southern and southeastern edges are bounded by high mountain chains, such as the Hindu-Kush, Pamiro-Alaï (7450 m), and Tiang-Shan (7440 m). In the southwest, the somewhat lower mountains of the Kopet-Dag (2000 m) allow monsoon precipitation to reach the western slopes of the Tian-Shan and Pamiro-Alaï. In the north, the Turan plain descends progressively northward and westward and opens out towards the Precaspian lowland, which is joined with the West Siberian plain via the Turgaï Valley. The northern boundary of this arid zone is rather poorly defined but lies at approximately 48°N.

From a geopolitical standpoint Central Asia comprises the five states of CIS (Commonwealth of Independent States), formed in December 1991 after the USSR collapsed: Uzbekistan, Kazakhstan, Turkmenistan, Tadjikistan, and Kirgizstan. Desert environments are typical for the first three of these states, especially Turkmenistan, where they cover more that 95% of the total territory. The semidesert of the Precaspian lowland is the continuation of this area on the southeastern European part of Russia.

The CIS deserts and semideserts have a typical continental climate. Summers are hot, cloudless, and dry, and winters are moist and relatively warm in the south and cold with severe frost in the north. There are several types of deserts in this region, varying according to the lithology of parent rocks and the types of soils (Petrov, 1976; Glazovskaya, 1996; Babaev, 1996). They include:

* Sand desert with loose-sandy serozems and greyish-brown soils of ancient alluvial plains
* Pebble-sand desert on gypsiferous, greyish-brown soils of Tertiary plateaus
* Gravelly, gypsiferous deserts on Tertiary plateaus
* Stony deserts on mountain slopes and hills
* Loamy deserts on slightly carbonate, greyish-brown soils of the elevated plains of the Northern Aral Sea
* Loess deserts (semideserts) on serozems of piedmont plains
* Clayey takyrs on plains and ancient river deltas
* Badlands on the Palaeogene (low-mountain deserts and semideserts composed of saliferous marls and clays)
* Solonchak deserts in saline depressions and along the coasts of the inland seas

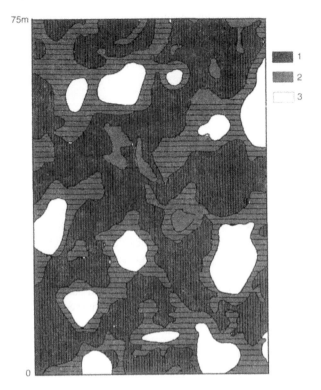

Figure 40.1   Microrelief of the Precaspian Lowland. (Modified from Gordeeva and Larin, 1965, and Olovyan-nikova, 1976.)

ARID ZONE SOIL CHARACTERISTICS AND ORGANIC CARBON CONTENT

Three broad regions can be distinguished within the Central Asian arid zone in terms of soil–vegetation relationships and general environmental conditions: the Precaspian Lowland biome, semideserts and deserts of Central Kazakhstan, and deserts of the Tunranian Lowland, delimited in the south and east by the middle Asian mountains.

The Precaspian Lowland: the Edges of the Desert

Even a relatively flat semiarid environment has significant small-scale variability in soil and vegetation conditions, relating to moister hollows and drier hummocks and the slopes in between them. The vast plains of the Precaspian Lowlands show three basic components: microdepressions with steppe Chernozem-like soils, microuplands with desert solonetz soils, and semidesert microslopes between them with alkalized chestnut earths (Walter and Box, 1983; Babaev et al., 1986). The total range of topography between these areas usually does not exceed 50 cm (see Figure 40.1). About 60% of the plain is occupied by drier, desert-like plant communities on microuplands with the average annual above-ground production about 0.8 t/ha (Bazilevich, 1995). The solonetz soil contains a 5- to 10-cm thick bleached, light-gray layer with humus content of 1.7 to 2.8% (by weight), under which lies a dense alkaline horizon that is tough and chocolate-colored when wet.

The zonal chestnut soils on the microslopes contain up to 3% of humus in the uppermost soil horizon. Conditions for plant growth are much more favorable in the microdepressions, which constitute up to 25% of the area of the plain. The upper 100 to 150 cm of soil is continually wet, the Chernozem-like profile is salt-free, and a humus horizon 35-cm thick with 5 to 6% humus is in the upper layer (Bolshakov and Rode, 1972; Glazovskaja, 1996b). The most extensive depressions today are cultivated land or covered by ruderal vegetation, as is often found on fallow ground, often with *Amygdalus nana*. In past times, these depressions were true fragments of steppe with thick humus horizons and freshwater lenses underneath.

Deserts and Semideserts of Central Kazakhstan

The zonal soil type in the semidesert is a light-chestnut earth, which represents the transition between the true chestnut earths of the dry steppe and the burozems (brown earths) of the desert (Walter and Box, 1983). Most light-chestnut soils are slightly alkaline. The humus horizon is 35 to 40 cm thick; a prismatic or flaky alkaline layer is often found at a depth of 10 to 25 cm. Deeper, at 25 to 35 cm, a calcareous horizon is recognizable and, still deeper at 80 to 100 cm, gypsum horizon (Korovin, 1962). The most widespread soil types are rubbled, poorly developed soils of the slopes of the numerous hills whose characteristics depend entirely on their lithological substrates. Solonetz and solonchak soils also cover large areas. With the sparse vegetation and the rapid decomposition of organic remains, the humus content of the soils does not exceed 2%. Among the humic compounds, fulvic acids predominate. In the western part of the Betpak-Dala, one often finds solonetz-like burozems in complex with solonetz soils. In these, the upper half of the humus horizon is laminated and loose, while the lower half is brown and dense, with a lumpy structure (Korovin, 1962).

In the desert, the subdivision siero-burozem is the most xeromorphic type of Burozem. In the Betpak-Dala region these are shallow, very dry soils in which the dryness is accentuated by the type of substrate material. The organic carbon content of these soils is only 0.8 to 1.6 t C/ha (Maljugin, 1984). The calcareous accumulation occurs on the surface, but sometimes a second carbonate maximum at a depth of about 50 cm can be found.

Turanian Deserts

In conditions of extreme aridity and prevailing high temperatures, combined with a low production rate and rapid decomposition, the soil organic carbon content is much reduced (0.5 to 1.5 t C/ha) and is only to be found in a thin uppermost horizon. These soils are relatively inactive and their carbonate content is high. The high evaporation rate produces a crust on the soil surface; if the groundwater level is high, the surface is kept wet by the capillary fringe. The water in the soil solution evaporates, but the dissolved salts can only accumulate and eventually form a solonchak soil. Only the sandy soils, in which there is almost no capillary movement, avoid becoming saline (Babaev, 1996).

The soil types most often found on the Turanian Lowland are sandy soils, solonchak soils, takyr soils, and dark sierozems (siero-burozems) on the outcropping rock of plateau sites, and light sierozems (Miroshnichenko, 1975; Glazovskaja, 1996b).

- Sandy soils, which are the most widespread in the Turanian deserts, are water permeable and have low salt content. In arid regions these are among the most favorable sites for plants, since they store water well. Psammophyte plants have well-developed root systems, however, because the field capacity of sandy soils is very low.
- Solonchak soils with a heavy accumulation of salts (NaCl, sulphates) in the upper horizons form in depressions with higher groundwater levels. Here one finds only salt-tolerant halophytes.
- Takyr soils form in shallow depressions or on almost level plains flooded temporarily in the spring. These are clayey and, once swollen with water, are water-impermeable. In the dry state these soils crack into polygon-shaped plates. The content of organic carbon is negligible (Kovda, 1977).

- Dark sierozems are characteristic to the rock desert (hammada) and have rocky or gravelly upper horizons, compaction and salinization at a depth of 5 to 15 cm, and a gypsum horizon at 50 to 70 cm. Their humus content is only 0.7 to 1.5%.
- Light sierozems, which arise over loess and carry ephemeral vegetation, are characterized by calcium deposits in the form of fine, white fibers in the upper horizon. The soil is densely permeated by roots and the humus content is relatively high at 1 to 1.5%. These soils are thoroughly wet in the spring but dry out almost completely in the summer.

In the floodplains and river deltas within the arid zone, one can find alluvial meadow and swampy soils. In contrast with surrounding ecosystems, these carry dense floodplain vegetation and contain much more organic carbon (3 to 5%). They are similar to the soils of irrigated, cultivated areas (Rodin, 1972), many of which have rapidly gained carbon on the time scale of decades.

LAND-USE IMPACTS: PAST AND PRESENT

As in many other arid regions of the world (i.e., the Middle East, Northern Africa, and Rajasthan), the influence of humans on the environment of Central Asia goes back a very long way. Numerous artefacts of the Mesolithic and Neolithic settlements are found all over this vast region. Centers of great civilizations repeatedly grew, flourished, and declined during antiquity and medieval history (Baktria, Margiana, Nisa, Merve, Horesm, Bukhara, and Samarkand). Various nomadic tribes (Sarmats, Khosars, Huns, Bulgars, Kazakhs, Kalmyks, and Mongols) successively came through the area into the Middle Ages.

Grazing and Its Effects on Soil Carbon

Except for the large oases alongside the major river valleys, where intensive irrigated agriculture was dominant for millennia, the most important human influences on the desert and semidesert biomes have been those connected with grazing. The domesticated herds of Middle Asia and Kazakhstan are generally dominated by sheep, but camels and goats are also kept.

The effects of grazing by domestic animals on vegetation and soils are thought to be different from those of the wild herds that preceded them. Domesticated herds move slowly and do not stray great distances from watering places (Le Houérou, 1989). They have an intensive impact on pasture around watering places, while areas further away remain almost undisturbed, resulting in concentric patterns around watering places, with increasingly degraded vegetation approaching center. Close to the water hole, where the animals remain longest, is no vegetation at all, and the soil is overfertilized by animal excrement. The negative effect of this overfertilization on plant growth remains noticeable for many years after watering places have been abandoned. There is too little precipitation for leaching to take place. Former watering places are recognizable from a great distance as barchan fields (Walter and Box, 1983).

The destruction of plant cover in intensively grazed areas often leads to the formation of barchan dune fields and chains. In the desert zone, around such a barchan field one can usually find some vegetation cover, composed of such species as *Aristida karelinii, A. pennata, Astragalus chivensis,* etc. (Rustamov, 1994), all of which livestock will not eat. Continuing outward from the center, is a ring with semishrubs, such as *Eremosparton flaccidum, Salsola richteri, Ammodendron coollyi, Haloxylon persicum,* and *Calligonum* species, as well as herbs, although Carex physodes is still poorly represented (Rustamov, 1994; Babaev, 1996). The natural vegetation cover typical for this zone may be found only at a distance of several kilometers from the well. On loamy soil the weed *Peganum harmala (Zygophyllaceae)* is found; can also be found around watering places as far as west as the Ebro Basin in Spain (Walter and Box, 1983).

In semidesert areas under grazing pressure by domesticated herds, the perennial grasses disappear from the vegetation cover, while the number of annuals and biennials, especially *Poa bulbosa* var. *vivipara*, increases.

In order to discover the most rational ways of using desert rangeland, a series of experiments was carried out in the late 1980s to early 1990s near the research station of Repetec on an experimental area in the southeastern part of the Karakumi desert, using a herd of several hundred sheep, goats, and camels. (One of the authors of this chapter (E.U.L.) was involved in the experiments while working on a student project in 1987 and 1988.) The experimental area consisted primarily of a *Haloxyletum caricosum*. The experiments were carried out at varying grazing densities; the average was 0.7 sheep or goat or 0.3 camel per ha. the minimum was 0.1 sheep per ha, and the maximum was twice the average density. Medium occupancy with a correct rotation of grazing seasons proved to give the best results in terms of promoting vegetation cover (in particular, promotion of *Carex physodes*, the best forage plant) (Nechaeva, 1984).

Grazing can thus exert both positive and negative effects on total vegetation cover and, by implication, soil organic carbon (which is derived from plant material). Positive effects involve a slight loosening of the soil surface, facilitation of seed germination by plants and the pressing of seeds into the soil under foot, and fertilization of the soil through excrement. These positive effects are noticeable only at an intermediate grazing intensity. In contrast, with overgrazing, the soil is loosened too much and blown away to start forming barchan fields. With undergrazing, the loosening is too little and unwanted mosses take over the area. Similarly, the pressing of seeds under foot to the optimal depth occurs only with moderate numbers of animals per hectare of pasture.

Grazing the same area twice in the spring proved to be very damaging, especially when done every year, as is often the case; the pastures became degraded after only 2 or 3 years. It was found that arid pasture areas suffer less when divided into many lots, each of which is grazed alternatively for a short period during each season (rotation grazing) (Walter and Box, 1983). This presupposes, however, that all the paddocks have similar plant covers, which is seldom the case. With heterogeneous pastures the rotation must be adapted to the particular local conditions; one must also consider the location of the watering places. One variation, that of alternative grazing of a paddock in the spring and summer one year and in autumn and winter the next year, proved to be useful also (Nechaeva, 1985).

Under rational use with moderate occupancy, optimal utilization could be sustained without damaging the plant and soil cover of the pastures. A short period of grazing during each season most closely resembles the original use by wild herds, which never remained for very long in one place. On the other hand, with the almost total extinction of natural ungulate fauna (as occurred hundreds of years ago in this region), a complete protection of the grazing field would not have a positive effect on overall vegetation cover because annual grasses or moss gradually replace the sparser natural perennial vegetation, and the surface becomes more compacted.

So what, then, are the effects of the artificial grazing regime on soil carbon? The effects are complex, with the direction of influence varying from one place to another. Moderate and rationally managed grazing generally favors sequestration of organic carbon in soils both directly (organic matter from excrement) and indirectly (preserving perennial grasses such as *Carex physodes*, with well-developed root systems and high underground biomass). In contrast, overgrazing is the main reason for losses of vegetation, soil carbon, and nitrogen in arid and semiarid rangelands. In general, soil carbon has been widely lost from Central Asian desert ecosystems since the domestication of ungulates and, especially, the increase in herds over the last century.

Apart from grazing, the impact of humans in the Central Asian desert environments has been relatively insignificant and is restricted to several urban and industrial sites (especially mining and oil extraction regions) and large oases with irrigated agriculture alongside the major river valleys and artificial canals.

OTHER ANTHROPOGENIC EFFECTS ON SOIL CARBON

During the Soviet period, major efforts were focused on irrigation and monoculture development with extensive cotton plantations, especially in the southern desert biome, and cereal cultivation in the semidesert. Irrigation of the Golodnaja Steppe in Uzbekistan and construction of the Karakum Canal, crossing the entire territory of Turkmenistan, are certainly among the most impressive projects in this area. Several types of soil changes related to agricultural land use can be found in these regions.

Irrigation is certainly the major factor of soil transformation in the oasis areas in Central Asia. Although oasis-irrigated agriculture was developed in this region centuries ago, extensive irrigation started only at the beginning of the 20th century. Chestnut and brown soils in the semideserts of the Precaspian Lowland and Central Kazakhstan, as well as relatively fertile light sierozems of Turanian piedmont deserts (previously occupied by extensive pastures with spots of small-size oases) have been converted into vast irrigated plantations, with cotton monoculture on the south and wheat on the north.

Many studies have shown that the carbonate profile of irrigated soils changes noticably under irrigation (Ovechkin, 1984; Schlesinger, 1985; Rozanov, 1984; Kholova et al., 1998); however, the effects of irrigation depend on the natural soil properties. In fertile depressions with chernozem-like soils of the Precaspian Lowland, irrigation causes a general decrease of the diversity of carbonate accumulation forms, and replacement of segregational forms by migrational ones. Zborishchuk (1980) and Kholova et al. (1998) have noted that significant organic carbon loss occurs during the first years of irrigation. After 12 to 15 years the organic carbon content reaches an equilibrium similar to initial values before irrigation. At the same time, the reserves of organic carbon in the irrigated chernozems are less than those in the nonirrigated ones.

In the sierozem soils of the Turanian deserts, which initially had a relatively low content of organic carbon, irrigation and fertilization in the oases have increased the amount of soil carbon. While nonirrigated desert soils have a maximum organic carbon content not exceeding 1.5 tC/ha in the light serozems, irrigated cultivated soils derived from these generally contain at least 3 to 5 tC/ha, although these figures may vary depending on management system.

A number of studies during the last few decades have focused on various aspects of desertification in the FSU Central Asia (see Figure 40.2). The Institute of Deserts of the Turkmen Academy of Science (Babaev et al., 1983) carried out the most detailed and fundamental study on human impacts and desertification processes in Central Asia. Also, the Ecological Map of Northern Caspian Lowland (Mialo, 1993) and the Map of Ecological Conditions for Phytoamelioraton of the Aral Sea Basin (Bahiev et al., 1989; Bahiev, 1993) provide a wide range of information on land degradation processes and ecological conditions of this region.

Analysis of these data provides approximate estimations of typical soil organic carbon content and how it may be affected by anthropogenic processes (Table 40.1). Across the region, one sees a range of soil organic carbon content from chernozem-like soils (meadow steppe) with about 5 to 6 tC/ha, to takyr soils (clayey desert) that have almost no organic content. The predominant soil types in the desert and semidesert regions have organic carbon contents between 0.5 and 2 t/ha. Most of these areas are subject to some sort of land use, either irrigated agriculture or grazing. Grazing may be expected to reduce the soil organic content by reducing litter organic matter flux to the soil, although this is, of course, partly compensated by dunging.

It is likely that the change from wild to domesticated herds has resulted in a change in the "patchiness" of vegetation and organic matter reaching the soil; with wild herds the effects would be more evenly spread across the landscape and mild grazing and soil disturbance might actually promote greater overall vegetation cover and flux to the soils. Domesticated herds are clustered around water holes where grazing and soil disturbance are intense. Soil compaction, probably tending to reduce soil organic content, may also be intense close to water holes. Such sources of spatial heterogeneity must be considered when sampling for representative organic matter content in these desert regions. Taking this broad region as a whole, however, it is unlikely that desert soil organic content has been fundamentally transformed by the switch from wild to domesticated grazing.

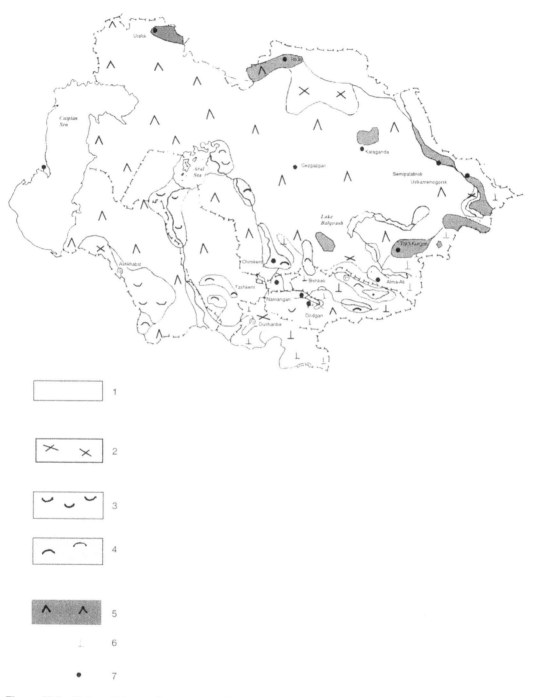

Figure 40.2   Major soil degradation processes in the NIS Central Asia site. (Modified from Glazovskaja, 1996a.)

HOW MUCH CARBON IS CURRENTLY IN THE ARID-ZONE SOILS OF CENTRAL ASIA?

Currently existing global databases (e.g., Zinke et al., 1986; Batjes and Sombroeck, 1997 and Post et al., 1982) tend to overestimate carbon content in the deserts of the former USSR, Mongolia, and China (i.e., 10.0 tC/ha in Zinke et al., 1986, 4.9 tC/ha in Batjes and Sombroeck, 1997). Russian

Table 40.1   A Summary of the Range of Types of Soils Found in the Central Asian Desert Region, together with Typical Organic Carbon Storage and the Range of Land Uses Positively or Negatively Affecting Soil Organic Carbon Content Relative to the "Natural" State

Soil Group	Ecosystem/ Landscape Type	Soil Organic C (t C/ha)	Depth of Humus Horizon (cm)	Dominant Land Use	Human-Induced Processes	Effect on SOC (+ or −)
Chernozem-like meadow soils	Meadow steppe	5–6	30–35	Pastoral, arable	Degradation of vegetation, salinization	− −
Alkanized chestnut earth	Semidesert	2–3	35–40	Pastoral, arable irrigated	Degradation of vegetation, compactness salinization	− − +
Burozem	Rock desert (hammada)	0.8–1.6	1–2	Pastoral	Degradation of vegetation	−
Siero-Burozem	Rock desert (hammada)	0.8	0.5–1.5	Negligible (extensive pastoral)	Degradation of vegetation	−
Light Sierozem	Loess desert	1–1.5	3–5	Pastoral Arable irrigated	Trampling and compactness, fertilization salinization	− + +
Solonchak	Saline desert	0.5–2	1–3	Pastoral	Degradation of vegetation, trampling and compactness, fertilization	− − +
Solonetz	Semidesert	1.7–2.8	5–10	Seasonal pastoral	Salinization, degradation of vegetation, trampling and compactness, fertilization	− +
Sandy desert soils	Sandy desert	0.5–1.5	1–2	Seasonal pastoral	Soil loosening, wind erosion Fertilization	− +
Takyr Alluvial meadow and swampy soils	Clayey desert Floodplains and delta	Negligible 3–4	Negligible 10–20	— Arable or pastoral	— Salinization, water erosion	−

sources indicate much lower values for the same areas (between 1 and 3.5 tC/ha in dark sierozems and sandy soils and relatively high only in the light sierozems — up to 4.5 tC/ha) (Gladishev and Rodin, 1977; Lioubimtseva et al., 1998). This contradiction between global and regional data is discussed by Adams and Lioubimtseva (see Chapter 44).

From their databases, Post et al. (1982) and Zinke et al. (1986) suggest mean global values for cool desert soil carbon of around 10 kg/m²; these figures have been used in most long-term global carbon cycle estimates (Post et al., 1990; Van Campo et al., 1993; Peng, 1994; Peng et al., 1994). Such large amounts of carbon in desert or semidesert soils would seem paradoxical, considering the low rates of input of organic matter from vegetation.

Detailed assessment of the databases suggests that, in fact, the soil carbon values obtained for the FSU Central Asian deserts are inflated due to misassignment of sampling sites. Actually, all sampling sites used by Zinke et al. (1986) as a reference for values of organic carbon in Central Asia are clustered in only two small areas (see Figure 40.3). Unfortunately, both of these areas are nonrepresentative.

The bigger of the two sampling areas cited in Zinke et al. (1986) comprises seven sampling sites; it is situated on the slopes of the Tiang-Shan mountains between altitudes of 1500 and 3000 m. This area receives precipitation of about 400 mm, mainly from the India monsoon, and existence of any desert-like ecosystems here is impossible. This sampling area is dominated by alpine

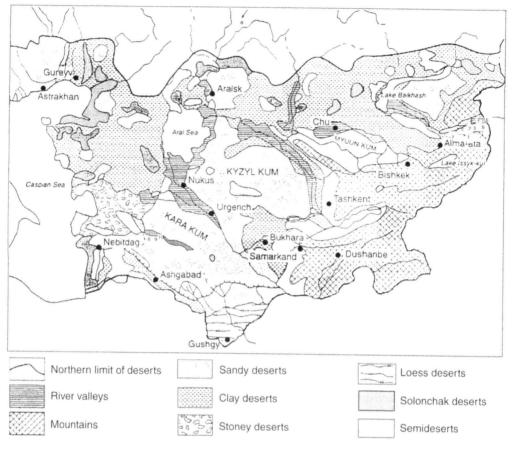

Figure 40.3   Deserts of the CIS Central Asia and sample sites of Zinke et al. dataset (1986). Most of grid
coordinates fall within dry steppe and not within desert zones.

meadows on ruby chernozems and continues in northwestern China, also on the slopes of the
Tiang-Shan Mountains and at even higher altitudes.

There is a second, very small sampling area; actually, it is almost one site with three points
within the same ecosystem. All three sampling points fall within the Uzboj, an ancient river valley,
which once supported a river until Medieval times; it then progressively dried out after an earthquake
changed a flow. The valley is now full of saline swamps, solonchaks, and other soils atypical of
the entire zone intrazonal ecosystems here with, quite naturally, much higher values of vegetation
biomass (succulent halophytes) and higher values of soil carbon. The results of this simple map
analysis are summarized in the Table 40.2.

Spatial distribution of sampling sites from Zinke et al., 1986 global database shows that *none*
of these sites was located in the desert and carried any information about desert soils. From 10
sampling points taken in the FSU Central Asia, five are situated on the mountain slopes and the
Ili river valley eastward from Alma-Ata, Kazakhstan (Figure 40.3). All of them are taken within
mountain steppe or meadow ecosystems, where soil groups vary from light chestnut soils on the
piedmonts to meadow Chernozems in the alpine meadows.

The other three sites taken in western Turkmenistan drop right into the Uzboj ancient river
valley currently occupied by solonchaks with thick halomorphic vegetation. The rest of the area
surrounding the Uzboj valley is covered by sandy dunes, where organic carbon content is close to
zero. Moreover, in addition to these 10 sites in the FSU, the other 14 sites from central Asian

Table 40.2   Results of Identification of "Desert" Sampling Sites from Zinke et al. (1985) Database on Russian Vegetation and Soil Maps

N	Zinke et al., 1985 Database No.	Life Zone Type, According to Zinke et al., 1985 Database	Ecosystem Type According to Map of Vegetation of the USSR, BIN RAS, 1994	Soil Group According to Soil Map of Northern Eurasia, Glazovskaja et al., 1996a	Organic Carbon, According to Zinke et al., 1985
1	2050001	200.13 B 40	Alpine meadow	Mountain meadow dark soils	15.3 1001
2	2050002	200.13 B 40	Halophyte semidesert	Siero-Burozem	6.3 0
3	2050003	200.13 B 40	Piedmont semidesert	Siero-Burozem	8.0 0
4	2050004	200.13 B 40	Alpine steppe	Mountain burozem on light clayey carbonate deposits	5.5 0
5	2051001	200.13 B 40	Alpine meadow	Mountain underdeveloped Chernozems	8.3 0
6	2051002	200.13 B 40	Alpine meadow with halophites	Mountain saline underdeveloped meadow Chernozems	6.6 0
7	2052001	200.13 B 40	Semishrub dry mountain steppe	Light chestnut soils	4.3 0
8	2069001	300 19 B	Halophytic vegetation	Solonchak	8.7 1180
9	2069002	300 19 B	Halophytic vegetation	Solonchak	6.0968
10	2069003	300 19 B	Halophytic vegetation	Solonchak	7.8 1237

"deserts" fall within high mountain ecosystems of northwestern China, dominated by alpine meadow and mountain steppes on mountain chernozems and chestnut soils.

The only reason for such "findings" of high soil organic carbon values is that of wrong interpretation of Russian and Chinese sources.

Another aspect causing inflation of the Zinke et al. figures could be confusion between organic and nonorganic (carbonate) carbon contents in the desert soils. Inorganic carbon (calcrete) is a quite separate soil property. Most of the soils discussed here contain large amounts of inorganic carbon and gypsum in a form of inclusions. Dry-steppe and semidesert chestnut soils, also relatively rich in organic carbon, usually contain a distinct calcerous (nonorganic carbon) horizon between 20 and 40 cm within the humus layer. In more arid soils (burozems and siero-burozems) the calcerous accumulation occurs on the surface, but the second carbonate maximum sometimes occurs at a depth of about 50 cm. Light sierozems, which arise over loess deposits, are characterized by very dense calcium deposits in the form of fine, white fibers in the upper horizons; gypsum crystals accumulate at greater depths.

Most soil descriptions in the Russian literature, except publications in the last few years, usually mean inorganic forms of carbon when they employ the word *carbon*, as opposed to *humus*, the term by which they refer to organic soil carbon (Kovda, 1950; Rodin, 1972). Although a lot of attention was given to estimations of vegetation biomass and production, the question of organic soil carbon is very poorly documented in the Russian literature, particularly as far as arid zones are concerned. Indication of high levels of total (organic and inorganic) carbon in desert and semidesert soils, or inadequate translation from Russian intro English, could be the main reasons why western authors tend to overestimate levels of organic carbon in these soils.

If one accepts a mean present-day figure of around 2.0 to 2.5 tC/ha, between 7.0 and 9.0 Gt of organic carbon would be in the top 50 cm of arid zone soils across Central Asia at present; this is 3 or 4 times less than during the Holocene climatic optimum (Lioubimtseva et al., 1998).

CONCLUSIONS

Considering the data from field studies published in the Russian literature, it seems likely that Central Asian desert soils contain (in total) between 7.0 and 9.0 Gt organic carbon in their top 50 cm. A representative figure on a per-unit-area basis would be in the range of 2 to 2.5 tC/ha, well below the figures given by Zinke et al. (1986) or the Batjes and Sombroek (1997) figure of 4.9 tC/ha for temperate desert (based on the ISRIC-WISE dataset). Separate sampling problems within the North American desert region are already evident (Adams and Lioubimtseva, in press), and evidence from the Russian literature further emphasizes that the figures for desert regions most generally used in global modeling are likely to be in error. Further fieldwork and systematic sampling are needed to check the figures suggested here. However, it is already evident that considerable spatial heterogeneity is present at all scales in the soils of the Central Asian deserts; any nonrandomized sampling process could easily give very inaccurate figures for soil carbon. Alluvial meadow and swampy soils, found in the floodplains and delta areas, as well as the soils of irrigated, cultivated areas, contain much more humus than surrounding them than the predominant zonal arid soils.

In putting forward figures for soil organic carbon storage, the components of time and land use only become evident if one considers the whole history and social geography of the central Asian region. Overall, the shift to domestication and increase in herds over recent decades have probably brought about a slight decrease in mean soil organic carbon density across central Asia, even though there may have been *local* increases in soil carbon, far away from water holes (where the herds do not venture) and very close to them (with manuring of soils). Careless or unconsciously biased sampling could easily result in very misleading figures for desert soil organic content because of these heterogeneities. It is uncertain as to what extent such problems may have occurred in other cool temperate desert zones, but in central Asia at least, database figures have tended to be biased by unrepresentative sampling.

When considering the carbon storage of deserts, it is important not to become confused by the inorganic component in arid-zone soils. Many authors, both Russian and western, point out the great importance of nonorganic components of soils but do not fully consider their behavior as part of the global carbon cycle (Schlesinger, 1990, 1995; Khokhlova et al., 1998). The effect of this soil carbon reservoir on the global carbon cycle is opposite to that of organic carbon; any decrease in calcrete by chemical weathering takes up CO_2 from the atmosphere, at least temporarily (sulphuric acid weathering, which releases CO_2, is virtually always a minor component of the total weathering flux). An oxidative decrease in organic matter in a soil should, by contrast, increase atmospheric CO_2.

The problems pointed out in previously published databases will have affected the various global carbon cycle models and carbon reservoir assessments that have attempted to use them. This is a typical example of the type of problem associated with the "numeric modeling culture" (Adams and Lioubimsteva, Chapter 44). Databases built up in a sweep for huge amounts of data for use in biogeochemical modeling tend to incorporate errors from a range of sources. Yet these databases, because they have the appearance of systematic rigor, are accepted uncritically by modelers, and defended even in the face of better advice from experts in the source disciplines. This has certainly been the case in the field of soil organic carbon content (Adams and Lioubimtseva, Chapter 44), where database values for cool desert soil organic content (e.g., those in the Zinke et al. database) have met with widespread disbelief from soil scientists, but complete acceptance by carbon cycle modelers. This tale serves as a cautionary lesson that accurate modeling of the natural and semi-natural environment is not as easy as it first appears.

REFERENCES

Adams, J.M. and Lioubimtseva, E., Some key uncertainties in the global distribution of soil and peat carbon, *Advances in Soil Science*, CRC Press, Boca Raton, FL, 2002.

Babaev, A.G., *Problemi osvienija aridnij zemel* (*Problems of Arid Land Development*), Moscow University Press, Moscow, in Russian, 1996.

Babaev, A.G. et al., *Pustini* (*Deserts*), Nauka Publishers, Moscow, in Russian, 1986.

Babaev, A.G., N.G. Kharin, and N.S. Orlovsky, Assessment and mapping of decertification processes, Map, 1:1000000, Desert Research Institute, Academy of Science of Turkmenistan, Ashgabat, 1983.

Bahiev, A., Problems of conservation and reconstruction of vegetation in the South Aral region, and prognosis of its development in relation with desertification (Problemi sohranenija i vosstanovlenija rastitelnosti Juznogo Priaralja i prognozirovanije ih razvitija v sviazi s processom opustinivanija), *Bull. Russian Acad. Sci.*, Geography series, 1, 51–60, in Russian, 1993.

Bahiev, A. and E.A. Vostokova, Map of ecological conditions for phytoamelioraton of the Aral Sea Basin, Tashkent, 1989.

Batjes, N.H. and W.G. Sombroeck, Possibilities for carbon sequestration in tropical and subtropical soils, *Global Change Biol.*, 3, 161–173, 1997.

Bazilevich, N.I., *Biomass and Biologic Production of Vegetation Formations of the Former USSR* (Biomassa i bioproductivnost rastitelnih formatsij), Nauka, Moscow, in Russian, 1995.

Bolshakov, A.F. and A.A. Rode, Soils of the solonetz complex in the northern part of the Precaspian lowland and their biological productivity, in Rodin, L.E. (ed.), *Int. Symp. Ecophysiological Found. Ecosyst. Productivity Arid Zones*, Leningrad, Nauka Publishers, 124–126, in Russian, 1972.

Gladishev, A.I. and L.E. Rodin, Structure and distribution of phytomass of riverbed forest communities along the middle stretch of the River Amu-Daria (Turkmenian SSR), *Bot. J.* (*Botanicheskij Zurnal*), 62:3–14, in Russian, 1977.

Glazovskaja, M.A., Soils of Northern Eurasia, Map 1:5 M with explicatory note (in Russian), *Map of Soil Degradation Processes* (1:20 M), Moscow State University Press, Moscow, 1996a.

Glazovskaja, M.A., Rol i functsii pedosferi v geohimicheskom globalnom tsycli (Role and functions of the pedosphere in geochemical carbon cycles). *Pochvovedenie*, Moscow, 2:174–186, in Russian, 1996b.

Gordeeva, T.K. and I.B. Larin, *Natural Vegetation of the Precaspian Semi-Desert as a Grazing Basis for Livestock*, Botanic Institute Acad. Science USSR, Moscow-Leningrad, Nauka Publishers, in Russian, 1965.

Khokhlova, O.S., S.V. Mergel, and I.S. Kovalevskaya, An approach to reconstruction of the CO_2 regime in semiarid soils of Russia on the basis of their carbonate profile data, in Lal, R., J. Kimble, and R. Follet, eds., *Soil Properties and Their Management for Carbon Sequestration*, USDA-NRCS NSSC, Lincoln, NE, 1998.

Korovin, E.P., *Rastitelnost srednej Azii i Yuznogo Kazakhstana* (*Vegetation of Central Asia and Southern Kazakhstan*), Tashkent, 2nd ed., Vol. 1, in Russian, 1962.

Kovda, V.A., *The Soils of the Precaspian Lowland*, Moscow-Leningrad, Nauka Publishers, in Russian, 1950.

Kovda, V.A., ed., *Takyry zapadnogo Turkmenistana i meri ih selskohozajstvennogo osvoenija* (*The Takyrs of Western Turkmenistan and the Means of Their Agricultural Utilization*), Moscow, Nauka Publishers, in Russian, 1977.

Le Houérou, H.N., *The Grazing Land Ecosystems of the African Sahel*, Springer-Verlag, Heidelberg, 1989.

Lioubimtseva, E. et al., Impacts of climatic change on carbon storage in the Sahara–Gobi Desert Belt since the late glacial maximum, *Global Planetary Change*, 16–17:95–105, 1998.

Maljugin, E.A., Agricultural methods for cultivation the sandy deserts of western Kazakhstan, *Pustini SSSR i ih osvoenije* (*Deserts USSR Devel.*), 2:66–134, in Russian, 1984.

Mialo, E.G., ed., *Ecological Map of Northern Caspian Lowland*, Moscow, Moscow State University Press, 1993.

Miroshnichenko, Yu.M., Pecularities of seasonal dynamics of productivity in phytocenoses of the Afghan arid area, *Bot. J.* (*Botanicheski Zurnal*), 60:1164–1178, in Russian, 1975.

Nechaeva, N.T., ed., *Resursy biosphery pustin Srednei Azii i Kazakhstana* (*Biosphere Resources of Deserts in Central Asia and Kazakhstan*), Moscow, Nauka Publishers, in Russian, 1984.

Nechaeva, N.T., Ecologicheskije osnovi reconstructsii sovremennogo rastitelnogo pokrova pustin Tcentralnoj, Azii. (The ecological fundamentals of reconstruction of the present vegetative cover of the central asian deserts), *Problemy Osvoeniya Pustyn* (*Probl. Desert Devel.*), 4:15–21, in Russian, 1985.

Olovyannikova, I.N., The Influence of woodland on the Solonchak-like Solonetz soils, *Forestry Lab. Acad. Sci. USSR*, Moscow, in Russian, 1976.

Ovechkin S.V., Genesis and mineralogical composition of carbonate features in Chernozems of Ukraine and the Middle Vilga Basin, in *Pocvi i pochvennij pokrov lesnoi i stepnoi zon SSSR i ih ratsionalnoje ispolzovanije (Soils and Soil Cover of Forest and Steppe Zones of the USSR and Their Rational Use)*, Moscow, Moscow State University Press, 1984.

Peng, C., Reconstruction du stock de carbone terrestre du passe a partir de donnees polliniques et de modelels biospheriques depuis le dernier maximum glaciaire. Abstract, PhD thesis, Université d'Aix-Marseille II, 1994.

Peng, C.H., J. Guiot, and E. van Campo, Reconstruction of past terrestrial carbon storage in the northern hemisphere from the Osnabroeck biosphere model and palaeodata, *Clim. Res.*, 5, 107–118, 1995.

Petrov, M.P., *Deserts of the World*, Chichester, John Wiley & Sons, 1976.

Post, W.M. et al., Soild carbon pools and world life zones, *Nature*, 317:613–616, 1982.

Post, W.M. et al., The global carbon cycle, *Am. Sci.*, 78:310–326, 1990.

Rodin, L.E., Ecophysiological foundations of ecosystem productivity in arid zones, *Proc. Int. Symp.*, 7–19 June, 1972, Nauka Publishers, Leningrad, 1972.

Rozanov, B.G., Principi i doctrini okruzaychej sredi. (Principles of the doctrine on the environment) *Int. Geogr. Union, Proc.*, Moscow, Moscow State University, 115, in Russian, 1984.

Rustamov, I.G., Vegetation of the deserts of Turkmenistan, in Fet, V. and K.I. Atamuradov, ed., *Biogeography and Ecology of Turkmenistan*, Kluwer Academic Publishers, the Netherlands, 77–104, 1994.

Schlesinger, W.N., The formation of caliche in the soils of the Mojave Desert, California, *Geochim. Cosmochim. Acta*, 49:57–66, 1985.

Schlesinger, W.H., Evidence from chronosequence studies for a low carbon storage potential of soils, *Nature*, 348, 232–234, 1990.

Schlesinger, W.H., An overview of the carbon cycle, in Lal, R. et al. (eds.), *Soils and Global Change. Advances in Soil Science*, CRC/Lewis Publishers, Boca Raton, FL, 1995.

Van Campo, E., J. Guiot, and C. Peng, A data-based re-appraisal of the terrestrial carbon budget at the last glacial maximum, *Global Planetary Change*, 8, 189–201, 1993.

Walter, H. and E.O. Box, Temperate deserts and semi-deserts, in West, I. and E. Neil, (eds.), *Ecosystems of the World*, Vol. 5, Elsevier Publishers, Amsterdam, 1–159, 1983.

Zinke P.J. et al., Worldwide organic soil carbon and nitrogen data (CDIAC NDP-018), Oak Ridge National Laboratory, Oak Ridge, TN, 1986.

CHAPTER **41**

Pastureland Use in the Southeastern U.S.: Implications for Carbon Sequestration

R. T. Conant, Keith Paustian, and Edward T. Elliott

CONTENTS

ABSTRACT

More than 13 Mha of nonfederal land in the southeastern U.S. are devoted to pastureland. Between 1982 and 1992, pastureland increased by 100,000 ha, with nearly 70% converted from cultivated land. We examined the potential for carbon (C) sequestration with improved pasture management and conversion into pastureland from cultivated land. Improved pasture management techniques, such as intensive grazing, fertilization, introduction of improved grass and legume species, and better irrigation systems can lead to sequestration of atmospheric C in soil. Literature values for the influence of changes in pasture management on soil C were summarized for several potential management changes in the Southeast. Soil C sequestration estimates for the Southeast were based on current pasture management practices and evaluated for a range of different adoption rates of improved practices.

Conversion into pasture can also potentially sequester significant amounts of atmospheric C in soils. Land-use data from the National Resources Inventory and literature estimates of soil C changes following conversion to pasture were used to estimate historical (1982 to 1992) soil C sequestration in pastures. Potential future sequestration was estimated based on extrapolation of land-use trends between 1982 and 1992. With continued conversion into pasture and improvement of pasture management, southeastern U.S. pasture soils may be a significant C sink for several years.

INTRODUCTION

Pastureland is an important land resource in the southeastern U.S. In 1992 pastureland covered more than 13 MhA, or 12% of the total land area in the nine states comprising the southeastern region (Alabama, Florida, Georgia, Kentucky, Mississippi, North Carolina, South Carolina, Tennessee, and Virginia) (Ag. Census, 1992). Pastureland in the Southeast supports 6.5 million beef cattle and more than 990,000 dairy cows (Ag. Census, 1992). Sales of dairy ($1.8 billion) and beef ($2.7 billion) products produced in the Southeast are substantial (Ag. Census, 1992).

Between 1982 and 1992 pastureland in the Southeast increased by 100,000 ha (USDA, 1994). Nearly 70% of land converted into pasture over this period came from cultivated land, with net conversion from cultivation to pasture in all nine states (USDA, 1994). Significant amounts of forested, barren, urban, and federal lands were converted into pasture, but more land of each type was created from pastureland. Nearly 100,000 acres of rangeland in Florida and Alabama were converted into pasture (USDA, 1994).

Pastureland is often intensively managed and research has shown that a variety of management improvements including fertilization, sowing improved species of grasses and legumes, and intensive grazing management can lead to increased production (Fink et al., 1933; Watson, 1963; Reeder et al., 2001; Schnabel et al., 2001; Franzleubbers, unpublished data). Pasture improvement intended to increase production can also lead to increased soil organic matter (Conant et al., 2001). In addition to sequestering atmospheric C, increasing soil organic matter can lead to many other benefits (Lal et al., 1998).

The objective of this work was to assess the role of southeastern pastureland in the C budget for the U.S. Three elements of C sequestration in pastureland in the Southeast were evaluated: (1) recent historical conversion into pastureland, (2) the current extent of pastureland and the status of pasture management, and (3) potential future pastureland use and extent based on current trends. Historical conversion rates were developed based on changes to and from pasture between 1982 and 1992. An assessment of the current rate of utilization of improved management for each state in the Southeast was also developed and used to quantify regional C sequestration. Finally, the potential for increased C storage was evaluated by extrapolating current trends in land-use change and evaluating a range of adoption rates of improved pasture management.

MATERIALS AND METHODS

The National Resources Inventory (NRI; USDA, 1994) was used to quantify the areal extent of pastureland and the amount of land area converted to and from pasture. The NRI employed a stratified two-stage area sampling design for collecting land use information collected at hundreds of thousands of points in 1982, 1987, and 1992 (Nusser and Goebel, 1997). Broad land-use data from 1982 and 1992 were primarily used for this analysis. Weighting factors derived as part of the NRI were used to quantify the amount of pastureland in 1982 and 1992 and the area converted into and out of pasture from land uses of various types over this time period.

Substantial information regarding the influence of changes in management on soil C (e.g., Donald and Williams, 1955; Johnston et al., 1980; Gifford et al., 1992) was recently reviewed (Conant et al., 2001). Carbon sequestration rates were summarized by location, native vegetation, and type of management change. These literature review data were the source of all C sequestration rates used in this analysis.

Few of the studies reviewed, however, were based on data collected in the southeastern U.S. In order to relate C sequestration rates from the literature to changes in management, the review data were reevaluated and reclassified. Of the independent variables included in the review, climate and management were identified as primary factors that can influence primary production. Therefore, data were summarized by type of management change and the ratio of precipitation to potential evapotranspiration (PPT/PET) was used to link C sequestration rates from the review with counties in the southeastern U.S. The PPT/PET ratio (Thornthwaite, 1948) was calculated for each of the data points using data compiled during the literature review, and for each county in the Southeast, using mean annual temperature and precipitation data obtained from a $0.5° \times 0.5°$ grid cell climate map developed for use in the POTSDAM project (Schimel et al., 1996). Each county was then linked to a potential C sequestration rate for each type of management change, based on data from the literature review, using the PPT/PET ratio.

Qualitative sensitivity analysis was used to determine the number and size of PPT/PET classes. As many PPT/PET classes as possible were generated while keeping the number of data points within each of the classes equivalent. It was necessary to evaluate the distribution of the number of data points within two, three, four, five, and six PPT/PET classes. Average C sequestration rates were not corrected for differences in sample depth, but average depths for all management changes and PPT/PET classes were reported.

Scant data exist on the current state of pasture management in the Southeast. Since no published source of information for the extent of adoption of improved management practices exists, state coordinators for the grazing lands conservation initiative (GLCI) program within each state were contacted in order to obtain state-by-state assessments. Each coordinator was asked to estimate the portion of pastureland in the state regularly fertilized (to soil test recommendations at least once every 2 years) and the area currently sown with legumes. The assumption was that answers to the questions as phrased would indicate the portion of land area realizing maximum potential production. These data were then weighted by the amount of grazing land within each state and summarized for the southeastern U.S. These data and C sequestration rates from the literature (112 data points for fertilization; 9 for sowing legumes) were used to estimate current regional C sequestration rates due to improved pasture management.

Potential future C sequestration due to land-use change was estimated based on historic trends. The rate of gross land use conversion into pasture from various land uses between 1982 and 1987 and 1987 and 1992 was calculated using NRI data for those years (USDA, 1994). The change in the average annual rate of change between these two periods was then used to estimate trends in net land-use change to and from pasture. This information was then linked to C sequestration rates with land-use change and extrapolated into the future. It was assumed that land converted to pasture would continue to sequester C for 20 years. Potential C sequestration due to increased use of fertilization and sowing of legumes was estimated by setting a range of targets for increased adoption of improved management by 2010. Increases were calculated by directly relating increases in adoption rates to current estimates of C sequestration. Estimates of future C sequestration assume that current land-use conversion trends continue at rates described above, and that C sequestration rates do not change over this period of time.

RESULTS

Counties in the Southeast were split into two classes based on the PPT/PET ratio (Figure 41.1). Qualitative sensitivity analysis indicated that equal and representative size classes could only be generated using two PPT/PET classes for each of the three types of management change evaluated. Counties along the eastern portion of the study area covering most of Virginia, North Carolina, South Carolina, Georgia, and Florida generally had PPT/PET ratios of two or less, while the other counties had PPT/PET ratios of greater than two (Figure 41.1).

Between 1982 and 1992 pastureland area increased by nearly 100,000 ha. Six of the nine states in the Southeast had net increases in pastureland and there was a net shift from cultivated land to pastureland in all nine states (Figure 41.2). Rangeland was converted to pasture in Florida and Alabama, while forested land increased at the expense of pastureland in all states but Florida

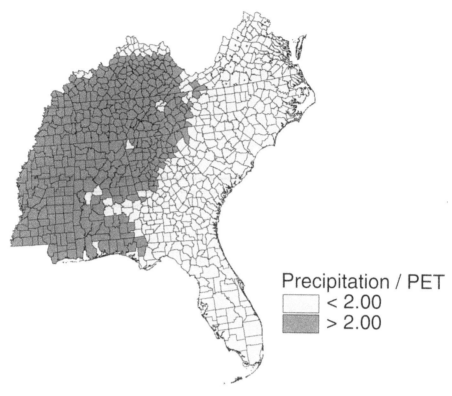

Figure 41.1 Southeastern region indicating the PPT/PET ratio for each county.

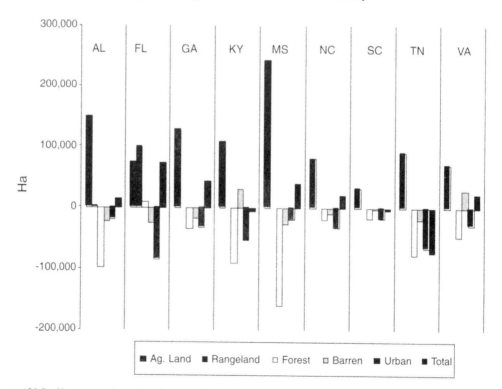

Figure 41.2 Net conversion of various land uses into pastureland for nine states in the southeastern U.S. between 1982 and 1992.

(Figure 41.2). All states showed net conversions from pasture to urban land, and all states but Kentucky and Virginia had net increases in barren land at the expense of pastureland (Figure 41.2).

Since only the net conversion of cultivated land into pasture was positive in all states, only the effects of conversion from cultivation to pasture on soil C were evaluated. The literature review indicated that the average C sequestration rate following conversion from cultivation to pasture was greater for the higher PPT/PET ratio class (Table 41.1). Thirty-five data points from the literature were used and 21 of 35 were in locations with PPT/PET < 2.00. The overall average C sequestration rate was 0.33 MgC ha^{-1} yr^{-1}, and values ranged from 0.00 to 2.37 MgC ha^{-1} yr^{-1} (Table 41.1).

All states had substantial C sequestration between 1982 and 1992 due to conversion from cultivation to pasture (Figure 41.3). South Carolina, North Carolina, and Virginia had the lowest rates of C sequestration due to smaller areas converted from cultivation to pasture and generally low PPT/PET ratios (Figures 41.1 and 41.3); carbon sequestration rates were slightly greater in Florida, Georgia, Tennessee, and Alabama. Greatest rates were estimated in Mississippi and Kentucky, which were wetter and had more land area converted from cultivation to pasture. Interestingly, Alabama had more land area converted from cultivation to pasture than Kentucky and Tennessee, but much of that was in the Blackland Prairie region. The PPT/PET ratio in this region is lower than in the rest of the state, effectively lowering estimates of C sequestration. Annual total C sequestration due to conversion from cultivation to pasture in the Southeast between 1982 and 1992 was approximately 0.22 TgC yr^{-1}.

GLCI coordinators for states in the Southeast reported a range of adoption rates for both fertilization and sowing legumes. Between 25 and 85% of the pastureland is fertilized, with an area weighted average of 45%, but a median value of 33%. Legumes are sown on between 0 and 50% of the pastureland in the southeastern states, with an area-weighted average and mean near 16%.

Table 41.1 Average, Minimum, and Maximum C Sequestration Rates (in MgC ha^{-1} yr^{-1}) for Both Climate Classes

PPT/PET Class	Average	Minimum	Maximum	Depth (cm)	No. of Studies
< 2.00	0.20	0.00	0.78	20.0	21
> 2.00	0.51	0.01	2.37	13.1	14

Note: Average depth and number of studies are also shown. All values are derived from a literature review.

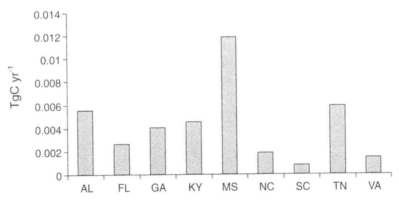

Figure 41.3 Carbon sequestration for the southeastern U.S. due to conversion into pastureland between 1982 and 1992. Positive values changes to pasture from a particular land use.

Table 41.2   Average C Sequestration Rates (and Ranges) for Both Climate Classes for Fertilization and Sowing of Legumes

PPT/PET Class	Fertilization			Sowing Legumes		
	C seq. (MgC ha^{-1} yr^{-1})	Depth (cm)	No. of Studies	C seq. (MgC ha^{-1} yr^{-1})	Depth (cm)	No. of Studies
< 2.00	0.201 (-0.2 – 2.5)	20.1	80	0.298 (0.1 – 0.9)	21.8	4
> 2.00	0.137 (-3.2 – 1.1)	35.5	32	0.382 (0.0 – 0.9)	33.0	5

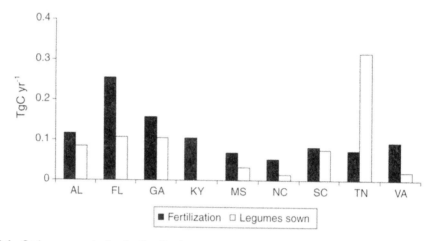

Figure 41.4 Carbon sequestration for the Southeast calculated for current rates of fertilization and sowing of legumes.

Carbon sequestration rates derived from the literature ranged from 0.13 to 0.38 MgC ha^{-1} yr^{-1} for fertilization and sowing legumes (Table 41.2). Soil C increased more when legumes were sown than when fertilizer was added, regardless of the PPT/PET ratio (Table 41.2). For areas sown with legumes, C sequestration was greater in wetter areas, while the opposite was true for fertilized pastures (Table 41.2). Average changes in soil C were not directly related to depth. More data points were used for the estimate of the influence of fertilization on soil C sequestration than for the effect of sowing legumes (Table 41.2).

All states but Tennessee had greater C sequestration due to fertilization than due to sowing legumes (Figure 41.4). Sequestration rates for fertilization ranged from 0.05 to 0.27 TgC yr^{-1}, with the most C sequestered in Florida and Georgia and the least in Mississippi and North Carolina. Sequestration rates due to sowing legumes ranged from zero for Kentucky (due to very low adoption rates) to 0.32 TgC yr^{-1} in Tennessee, which has more area sown with legumes than average and substantial area in pasture. Total C sequestration for the Southeast was 1.00 TgC yr^{-1} due to fertilization and 0.75 TgC yr^{-1} due to sowing legumes.

Estimates of the contribution of land-use change and improved management to current annual C sequestration varied from state to state (Figure 41.5). Conversion from cultivation was responsible for the largest portion of total state C sequestration in Mississippi, accounting for only 40% of the total. Fertilization contributed the majority of C sequestration in all states but Mississippi, where land use change and fertilization accounted for 41% of sequestered C, and Tennessee, where legumes were responsible for the largest portion of C sequestered (76.7%). Across all states fertilization averaged the largest contribution (55%), while sowing legumes (32%) and conversion (13%) contributed less. Total C sequestration for the Southeast due to land-use change and improved management was estimated to equal 1.96 TgC yr^{-1}.

If current land-use change trends continue, significant amounts of C may be sequestered in the soil. Between 1982 and 1987 pastureland area averaged 13.18 Mha, while between 1987 and 1992

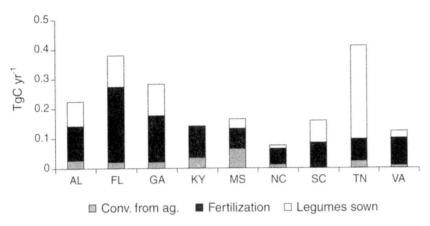

Figure 41.5 Summary of current C sequestration rates due to land-use change and improved management in the southeastern U.S.

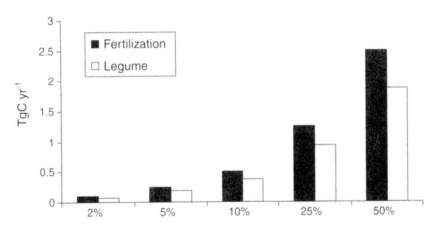

Figure 41.6 Annual C sequestration between 1999 and 2010 due to improved fertilization rates and sowing of legumes for a range of adoption targets.

it was 13.22 Mha. While a net of more than 53,000 ha was converted from cultivation to pasture between 1982 and 1987, there was a net conversion of just more than 45,000 ha between 1987 and 1992. Thus the average annual conversion rate as an average of total pastureland decreased by 4.1% per year. Using this as the average annual decrease in the rate of conversion from cultivation to pasture after 1992, we estimated that total C sequestration between 1999 and 2010 will be nearly 9.0 TgC.

Substantial amounts of C can be sequestered with increased adoption of improved management practices (Figure 41.6). With moderate adoption of between 2 and 5%, improved management practices store a total of less than 0.5 TgC yr^{-1} in the Southeast. However, with wider adoption, improved fertilization and sowing of legumes may result in C sequestration rates of between almost 1 TgC yr^{-1} (with 10% adoption) and more than 4 TgC yr^{-1} (with 50% increase in adoption).

DISCUSSION

Significant amounts of C have been and may be sequestered in southeastern pastures through land-use change (Sobecki et al., 2001). Historic trends in land use change indicate that conversion of cultivated land into pastureland may continue to lead to C sequestration. These results suggest that C sequestration due to land conversion from cultivation to pasture offsets approximately 1.4% of total U.S. CO_2 emissions due to agricultural activities in 1996 (EIA, 1996). Furthermore, extrapolations based on current trends indicate that this type of land-use change may continue to sequester at least 0.6 TgC yr^{-1} until 2020.

Similarly, improved pasture management is currently sequestering substantial amounts of C in soil in the southeastern U.S. Improved management is adopted because it leads to increased production and greater returns. Coordinators of the GLCI program in the southeastern states are enthusiastically optimistic that adoption rates will increase. There are, however, barriers to adoption of improved management such as time, financial resources, and motivation (G. Johnson, personal communication). If only a moderate proportion of producers are motivated to adopt improvement management practices to increase production, significant amounts of C can potentially be sequestered in soils.

The estimated rate of soil C sequestration due to land-use change and improved management was 1.96 TgC yr^{-1}. This estimate is of the same order of magnitude as potential C sequestration rates estimated for the conservation reserve program (8.1 TgC yr^{-1}), the grassland waterways program (1.6 TgC yr^{-1}), and adoption of no tillage (9.4 TgC yr^{-1}; Lal et al., 1998). Due to relatively high potential C sequestration rates and extensive pastureland coverage, improved management is potentially a substantial global sink for atmospheric C.

These estimates for C sequestration rates are limited in three main regards. First, by relying on NRI data for land-use information, this analysis is restricted to a relatively short period of time. National Resources Inventory information before 1982 is not as extensive as that after 1982; thus, historical land use perspective is limited. Also, the 1997 NRI data are not yet available, forcing evaluation of "current" conditions that represent land use 7 years ago. Second, the paucity of information regarding the current state of pasture management requires reliance on informal, uncertain estimates of adoption rates of improved management. Finally, the limited amount of published information regarding the effects of improved management on soil C does not allow adequate assessment of various other types of management improvement, extrapolation of results to a uniform depth for intercomparison, or evaluation of the duration of net C sequestration. Regarding the last point, however, many studies indicate very substantial increases in soil C with improved management (Conant et al., 2001), and some long term studies indicate that soil C increases can continue nearly linearly for as long as 40 to 50 years (Williams and Donald, 1957; Russell, 1960; Barrow, 1969; Johnston, 1986).

Though this study concludes that improved pasture management can sequester considerable amounts of atmospheric C, emission costs are associated with many types of grassland management. For example, Lee and Dodson (1996) found that pasture soils sequestered 0.16 MgC ha^{-1} yr^{-1} with application of 70 kg N ha^{-1} yr^{-1}, which agrees with estimates from the literature. However, approximately 0.82 kg CO_2-C are emitted per kilogram of N manufactured (Cole et al., 1993), reducing net C sequestration to 0.06 MgC ha^{-1} yr^{-1} (Lee and Dodson, 1996). Recent research indicates that CO_2 emissions associated with fertilizer production are lower (0.82 kg C per kg of N; West and Marland, 2001; Lal et al., 1998). Furthermore, it should be noted that fertilizer is used to stimulate pasture production and maximize production, not to sequester atmospheric C. Though sowing legumes and land-use change result in much smaller greenhouse gas production, greenhouse gas emission costs associated with improved management must be considered when estimating C sequestration potential of pastureland.

Due to relatively high potential C sequestration rates and extensive pastureland area, modest increases in the areal extent of improved pasture management can potentially sequester significant amounts of C. Since improvements lead to increased production, they are also beneficial for producers. Therefore, policies encouraging improved management may be an inexpensive method for offsetting a portion of greenhouse gas production by the U.S.

CONCLUSIONS

Historical land use change has been substantial in the southeastern region, with approximately 0.22 MMtC yr^{-1} sequestered due to conversion from cultivation to pasture. Current pasture management is sequestering approximately 1.75 MMtC yr^{-1} due to fertilization (1 MMtC yr^{-1}) and sowing legumes (0.75 MMtC yr^{-1}). Results indicate that land-use change will continue at decreasing rates and will sequester 9.0 MMtC between now and 2010. Modest increases in adoption of improved pasture management could sequester between 0.44 and 4.5 additional MMtC by 2010.

ACKNOWLEDGMENTS

We are grateful to the Southeastern GLCI coordinators, Greg Brann, Pete Deal, Micheal Hall, Walter Jackson, Glenn Johnson, Holly Kuykendall, Kenneth Rogers, and David Stipes, for their assessment of pasture management in the Southeast. John Hall, Gary Peterson, and Jim Green also provided sound opinions on extent of different types of pasture management in Virginia and North Carolina.

REFERENCES

Agricultural Census, Agricultural census data and software, Department of Commerce, Washington, D.C., 1992.

Barrow, N.J., The accumulation of soil organic matter under pasture and its effect on soil properties, *Aust. J. Exp. Agr. Anim. Husb.*, 9:437–445, 1969.

Cole, C.V. et al., Agricultural sources and sinks, *Water Air Soil Poll.*, 70:111–122, 1993.

Conant, R.T., K. Paustian, and E.T. Elliott, Grassland management and conversion into grassland: effects on soil carbon, *Ecol. Appls.*, 11:343–355, 2001.

Donald, C.M. and C.H. Williams, Fertility and productivity of a podzolic soil as influenced by subterranean clover (*Trifolium subterraneaum* L.) and superphosphate, *Aust. J. Agr. Res.*, 5:664–688, 1955.

EIA — Energy Information Agency, Emissions of greenhouse gases in the United States, DOE/EIA-0573(95), U.S. Department of Energy, Washington D.C., 1996.

Fink, D.S., G.B. Mortimer, and E. Truog, Three years results with an intensively managed pasture, *J. Am. Soc. Agron.*, 25:441–453, 1933.

Gifford, R.M. et al., Australian land use, primary production of vegetation and carbon pools in relation to atmospheric carbon dioxide concentration, in *Australia's Renewable Resources: Sustainability and Global Change*, 1992, 151–187.

Johnston, A.E., Soil organic matter, effects on soils and crops, *Soil Use Manage.*, 2:97–105, 1986.

Johnston, A.E., P.R. Poulton, and J. McEwen, The soil of Rothamsted farm: the carbon and nitrogen content of the soils and the effect of changes in crop rotation and manuring on soil pH, P, K, and Mg, *Rothamsted Experimental Station*, Report for 1980, 2:5–20, 1980.

Lal, R. et al., *The Potential of U.S. Cropland to Sequester Carbon and Mitigate the Greenhouse Effect*, Ann Arbor Press, Chelsea, MI, 1998.

Lee, J.J. and R. Dodson, Potential carbon sequestration by afforestation of pasture in the South-Central United States, *Agron. J.*, 88:381–384, 1996.

Nusser, S.M. and J.J. Goebel, The National Resources Inventory: a long-term multi-resource monitoring programme, *Environ. Ecol. Stat.*, 4:181–204, 1997.

Reeder, J.D., C.D. Franks, and D.G. Milchunas, Root biomass and microbial processes, in Follett, R.F., J.M. Kimble, and R. Lal (eds.), *The Potential of U.S. Grazing Lands to Sequester Carbon and Mitigate the Greenhouse Effect*, Lewis Publishers, Boca Raton, FL, 2001, 139–166.

Russell, J.S., Soil fertility changes in the long term experimental plots at Kybybolite, South Australia. I. Changes in pH, total nitrogen, organic carbon and bulk density, *Aust. J. Agr. Res.*, 11:902–926, 1960.

Schimel, D.S. et al., Climate and nitrogen controls on the geography and timescales of terrestrial biogeochemical cycling, *Global Biogeochem. Cy.*, 10:677–692, 1996.

Schnabel, R.R. et al., The effects of pasture management practices, in Follett, R.F., J.M. Kimble, and R. Lal (eds.), *The Potential of US Grazing Lands to Sequester Carbon and Mitigate the Greenhouse Effect*, Lewis Publishers, Boca Raton, FL, 2001, 291–322.

Sobecki, T.M. et al., A broad-scale perspective on the extent, distribution, and characteristics of U.S. grazing lands, in Follett, R.F., J.M. Kimble, and R. Lal (eds.), *The Potential of U.S. Grazing Lands to Sequester Carbon and Mitigate the Greenhouse Effect*, Lewis Publishers, Boca Raton, FL, 2001.

Thornthwaite, C.W., An approach toward a rational classification of climate, *Geogr. Rev.*, 38:55–94, 1948.

USDA, Natural Resources Inventory Digital Data, USDA-Natural Resource Conservation Service, Washington, D.C., 1994.

Watson, E.R., The influence of subterranean clover pastures on soil fertility I. Short term effects, *Aust. J. Agr. Res.*, 14:796–807, 1963.

West, T.O. and G. Marland, Carbon sequestration, carbon emissions, and net carbon flux in agriculture: a comparison of tillage practices, *Agr. Ecosyst. Environ.* (in press).

Williams, C.H. and C.M. Donald, Changes in organic matter and pH in a podzolic soil as influenced by subterranean clover and superphosphate, *Aust. J. Agric. Res.*, 8:179–189, 1957.

CHAPTER 42

On-Farm Carbon Sinks: Production and Sequestration Complementarities

Jeffrey Hopkins

CONTENTS

ABSTRACT

A model of optimal residue management is presented to illustrate the economic tradeoffs associated with residue management adoption. In general, the benefits of residue management are not realized in the present, but in the future, through maintenance of nutrient stocks, soil profile depth, and soil organic carbon (SOC) stocks. However, these future benefits come at a price, realized in the present, of residue management. A dynamic model of optimal residue and fertility management, which values future stocks of nutrients, soil depth, and SOC, is used to arbitrage the intertemporal tradeoff of benefits for costs. SOC is treated as an input in corn production, with minimal substitution possibilities with soil depth and soil nutrients; 0, 5, and 10% yield effects from increased SOC are alternatively modeled. Results indicate that without a yield effect, annual rates of SOC buildup do not exceed 0.2 tons per acre in SOC, although a 10% yield effect would nearly double the rate of SOC buildup.

GLOBAL WARMING POLICY

Soil organic carbon (SOC) sequestration is a biophysical process that has the potential to be manipulated through engineering strategies. The benefit of heightening SOC sequestration is that it could potentially offset anthropogenic emissions of carbon dioxide gas. Carbon dioxide is a "greenhouse" gas because it exacerbates the "greenhouse effect" or the trapping of some of the earth's radiation within the atmosphere. Radiative forcing, or the effect on climate due to changes in greenhouse gas levels, can be precisely predicted (U.S. Environmental Protection Agency, 1999).

As it turns out, predicting radiative forcing is not sufficient to predict future temperature levels. Climate change is an issue with rampant uncertainties. Panelists at a recent climate science forum at Resource for the Future detailed scientific uncertainties surrounding the effects of oceans, aerosols, clouds, glaciers, extreme weather events, and high-risk, low-probability phenomena on climate change (Thatcher, 1999). Because of evidence that these factors play important roles in determining climate, excluding their influence from climate models is not advised. However, including them decreases the precision of predictions, so that reliability of modeling efforts relies on increasing knowledge of the magnitude and direction of these effects. Debate in the past has revolved around the issue of whether to wait until more is known about how to control the problem or act on what is known now. If knowledge accumulates quickly, it might be best to wait; however, there is evidence that knowledge is not accumulating as quickly as earlier believed (Kelly and Kolstad, 1999).

While scientific uncertainty is likely to remain, policy must proceed with the evidence at hand. The Office of the Chief Economist for the U.S. Department of Agriculture recently wrote that a broad scientific consensus exists regarding the threat of climate change. Moreover, agreement exists that climate change could lead to serious adverse consequences, such as intense floods and draughts, disease, and rising sea levels (USDA Office of the Chief Economist, 1999). The realization that scientific uncertainty is not going to go away anytime soon has set the tone for the design of appropriate policy mechanisms. Because policies dealing with global warming are potentially costly, policy efficiency becomes an important concern. Efficient policy must consider how people will react and adapt to policy instruments. In some cases, policy-induced distortions can, in fact, become more socially damaging than the market distortion that they were designed to address (Parry and Oates, 2000).

Flexibilities such as delayed implementation, the creation of emissions markets, and accounting for both sources and sinks were built into the Kyoto Protocol. Flexibilities allow adjustments to occur that limit the overall resource cost of the policy without compromising the economic benefits. While agricultural soils are listed as a *source* of greenhouse gases within the Kyoto Protocol, they are not listed as *sinks*, but may be deemed sinks by the Conference of Parties (Bruce et al., 1999). Within the spirit of minimizing resource costs of a global warming policy, SOC sequestration will become seriously considered as a sink to offset global emissions.

Recently, some scientists and policymakers have depicted on-farm SOC sequestration as a win–win opportunity for farmers (Lal et al.,1998; *The Washington Post*, 1998). Win–win policies would minimize distortions that result from second-best policies, improve efficiency, and raise no equity concerns; therefore, they should always be adopted. A win–win SOC sequestration policy would offer simultaneous gains to society and producers. Social gains result from sequestration itself, as carbon dioxide is removed from the atmosphere, fixed in plant biomass, and later sequestered in soil humus and more stable forms. Farmer gains, on the other hand, are a by-product of adoption. Gains may come from increased yields or decreased costs of production, or some combination of the two leading to increased net returns. Lal et al. (1998) offer hints at where farmer gains may be found, but do not answer the question directly. In practice, the magnitude of any economic benefits and the rate at which benefits accrue will determine the likelihood that practices shown to enhance SOC will be adopted. Despite years of research on soil degradation and productivity (Lowery and Larson, 1995), clear yield impacts have not been demonstrated.

This chapter examines this issue in greater detail and discusses incentives faced by producers intent on maximizing net returns over time. First, it examines the effect of carbon enhancement on decreasing costs, then the case of carbon increasing yields. Finally, it presents a dynamic, forward-looking, firm-level model of optimal SOC sequestration embedded within a more general crop management model. These optimization and simulation results are highly dependent on beliefs regarding the influence of SOC on yield; therefore, several scenarios regarding yield effects and contrast decisions and outcomes are presented.

FIRM-LEVEL MODEL OF SOC SEQUESTRATION USING RESIDUE MANAGEMENT

In examining the case of farmer gains from SOC sequestration, particular attention is focused on the complementarities between increasing the size of carbon pools and increasing on-farm net returns. While most economic analysis to date has treated the social costs and benefits of global warming, analysis of private costs and benefits is important in the case of agriculture because it is a highly competitive industry, and any economic impacts are likely to be passed on to consumers. Also, many of the strategies for increasing rates of SOC sequestration involve management practices not universally used by producers today. Adoption of environmentally friendly technology in agriculture is most rapid when it is compatible with the overall financial management of the firm. Therefore, a firm-level model is a necessary component to a larger social welfare model.

Residue Management: the Cost-Reducing Case

In parts of the eastern Corn Belt, one type of residue management technique, no-till, appears to be tailor-made for SOC sequestration. No-till limits soil disturbances to fertilizer application and planting operations, which are relatively minor. Soil disturbances by mixing and inversion are primary oxidation mechanisms by which SOC is lost to the atmosphere as carbon dioxide.

Residue management systems are often characterized as saving costs, particularly fuel and labor costs (Lal et al. 1999; Lines et al., 1990). In practice, however, overall economic gains have been harder to demonstrate. Recent evidence gathered by Day et al. (1998) show little consistent difference in costs, yields, or returns across farms already using certain tillage practices. Recently, however, no-till adoption rates have declined. The Conservation Technology Information Center has reported declines in conservation technology adoption for corn production since the early 1990s (CTIC, 1998).

Counting quantities and expenses for conventional inputs may be only part of the story. Residue management often requires a different complement of equipment, which could require costly adjustments on the part of the producer. Moreover, there is some evidence that different skills are required for residue management systems over conventional systems. Wu and Babcock (1998) used Agricultural Resource Management survey data for 459 producers in Central Nebraska and found that one of the distinguishing characteristics associated with those who practice no-till production is higher educational attainment than that of those who do not no-till.

One way that more intensive management is needed under no-till is at planting time. Randall Reeder, an agricultural engineer with Ohio State University Extension, theorizes that Ohio producers may be reverting to plow systems because no-till soils warm up later in the spring and can cause planting delays compared to tilled soils (Zolvinski, 1998). Planting delays have been clearly associated with decreasing yields (Ohio State University Extension, 1995). Indeed, one of the only studies using farmer data related to available field time indicated that no-till producers had one less day (5% less time on average) during the first 3 weeks of the planting season than producers practicing plow-till systems (Fletcher and Featherstone, 1978). The smaller window of opportunity places increased importance on coordinating all field operations. Coordination costs may be high; given the high opportunity cost of operator time during planting season, producers may find bearing increased fuel and labor expenditures cheaper than risking delayed planting.

The evidence from Ohio and other parts of the Corn Belt detailing reversals in the rate of adoption of no-till leads to the suspicion that, rather than decreasing costs, residue management increases production costs. However, the link between management costs and other input costs has not been adequately explored in the literature. Financial incentive programs have been somewhat successful in increasing adoption. In Ohio, the Environmental Quality Incentive Program has paid $8 per acre per year over a 3-year period for no-till, mulch-till, and ridge-till adopters. Producers wishing to adopt an intermediate system were eligible for a $5 per acre payment for 3 years (Rausch and Sohngen, 1997).

Residue Management: the Production-Enhancing Case

In theory, the case that increasing levels of SOC can increase crop yields is not as hard to make because, along with hydrogen and oxygen, carbon is essential for plant growth. The case that SOC levels are not high enough already, however, is more difficult. Von Liebig's law of the minimum, which states that crop yield can only be raised by increasing the level of the limiting input, lends some perspective. Under the law of the minimum, increasing an input level only increases crop yield when that input is limiting crop yield.

Historical analysis suggests that SOC stocks may have been seriously depleted after periods of cultivation of newly exposed soils. In the Great Plains region, soil productivity declined by 71% during the 28-year period after Great Plains soils were initially brought into production (Flach et al., 1997). The Great Plains soils story is often cited as an example of how depleted organic matter decreased crop yield. However, this was before the widespread availability of commercial fertilizers; much of the early yield decline was later reversed through new production technology, which substituted fertility derived from SOC stocks to fertility derived from commercial fertilizer. Historical data and simulation analysis make the case that SOC levels in the central U.S. in the 1940s were nearly half those reported in 1907 (Lal et al., 1998). SOC levels, however, have been increasing since the 1970s, as conservation tillage and higher-yielding varieties were adopted (Lal et al., 1999).

Data do not currently permit knowing the extent to which SOC levels are limiting yields. Syers (1998) suggests that chemical degradation from nutrient depletion (including organic matter and, presumably, SOC) severely limits crop yields on only about 5% of agricultural acreage, using FAO data. While FAO may be appropriate for signaling severe degradation, it may not indicate where yields are slightly impacted. For example, wheat biomass has been shown to respond to increasing SOC levels (Bauer and Black, 1994).

In addition to experimental evidence, demonstrations of the effect of SOC on yield under farmer management will determine adoption rates. For corn growers, single-year yield comparisons have not been favorable for no-till. No-till production methods decreased expected yields by 11% for corn farmers who adopted only that practice without also adopting crop rotation with a legume or soil testing practices (Wu and Babcock, 1998). Because SOC stocks were not controlled for in the study, and because these may take years to build up, multiple-year surveys may be necessary in order to attain theorized yield effects.

In practice, there are several aspects of crop residue management to maximize SOC sequestration at odds with maximizing net returns, at least in the short term. For example, tillage is believed to accelerate carbon oxidation by increasing soil aeration and soil-residue contact. Implications for crop yield follow, because more rapid mass and energy transfer between the soil surface and the atmosphere increases soil temperatures, thus improving germination rates, emergence, and vigor (Grant, 1997). Subsurface drainage, used in most parts of the eastern Corn Belt, has also been shown to increase oxidation and decrease SOC stocks. In production areas where cool, moist soil conditions limit crop production, leaving residues on the soil surface may be undesirable to producers.

On the other hand, several aspects of residue management are favorable to crop production. Most of these are realized in the future, however, while the costs are incurred in the present. By increasing SOC stocks, protecting nutrients from erosion, and maintaining soil profile depth, residue management has the potential to increase future returns. Any study of on-farm incentives to adopt conservation tillage needs to be sensitive to future returns, not just react to what is most profitable today. Myopic behavior is suboptimal and will lead to less carbon build-up, lower soil nutrient stocks, and greater losses in soil profile depth than decisions corresponding to forward-looking behavior. Myopic behavior does not prize the future value of these stocks and therefore does not properly assess the financial incentive to conserve soil. Future-regarding behavior is modeled by producers who trade off current returns for expected future returns.

A FORWARD-LOOKING MODEL OF OPTIMAL SOIL CONSERVATION

The present study presents an extension of an optimal soil conservation model developed by Hopkins et al. (2001). The model determines dynamically optimal crop-management strategies, given the biophysical processes determining plant growth. Crop-management decisions in the model include commercial fertilizer use and residue management, where optimal levels of these variables are a function of response functions for three soil states: soil profile depth (sdepth), soil nutrient stocks (snut), and soil organic carbon stocks (SOC). The model maximizes returns net of fertilizer and residue management costs over an infinite horizon. The problem can be restated as:

$$\text{Max}_{\text{fert}_t, \text{res}_t} \sum_{t=0}^{\infty} \rho^t \left(\text{price}^* \text{yield}(\text{sdepth}_t, \text{snut}_t, \text{soc}_t, \text{fert}_t) - \text{cost}(\text{fert}_t, \text{res}_t) \right) \qquad (42.1)$$

s,t,

$$\text{sdepth}_{t+1} = \text{sdepth}_t - \alpha_0 \alpha_1 10^{\alpha_2 \text{res}_t} \qquad (42.2)$$

$$\text{snut}_{t+1} = \text{snut}_t + \beta_0 \text{fert}_t - \beta_1 \text{yield}_t - \beta_2 \text{snut}_t \beta_3 \text{denude}_t \qquad (42.3)$$

$$\text{soc}_{t+1} = \text{soc}_t + \gamma_1 \text{res}_t. \qquad (42.4)$$

where

$$\text{yield}_t = \delta_0 \left((1 - 10^{(-\delta_1 \text{fert} - \delta_2 \text{snut})})(1 - 10^{(\delta_3 + \delta_4 \text{sdepth} + \delta_5 \text{sdepth}^2)})(1 - 10^{(\delta_6 + \delta_7 \text{soc}_t)}) \right) \qquad (42.5)$$

Bellman's equation (42.6) is used to solve Equation 42.1 using dynamic programming. Bellman's principle of optimality states that the maximum attainable net present value starting at any state must be equal to today's maximum current return plus the following year's maximum return, assuming that the optimal control is chosen today.

$$V(\text{sdepth}_t, \text{snut}_t, \text{soc}_t) = \text{Max}_{\text{fert}_t, \text{res}_t} \{ P^* \text{yield}_t - \text{cost}(\text{fert}_t, \text{res}_t) + \rho V(\text{sdepth}_{t+1}, \text{snut}_{t+1}, \text{soc}_{t+1}) \} \qquad (42.6)$$

Maximizing over current and future returns for the producer, then, consists of picking fertilizer rates and residue levels that maximize the stream of returns over an infinite time horizon. Fertilizer rates and residue management are depicted as continuous variables, with residue management taking values between 0 and 100% residue cover at planting. Equation 42.6 is a functional equation showing the value (in dollars per hectare) associated with levels of soil profile depth, soil nutrients, and SOC. The three state variables in the model are assumed to represent soil productivity jointly. Because the state variables can take on an infinite number of values, the problem is simplified by bounding the state space and discretizing a three-dimensional compact grid. Values between the points in the grid are interpolated using the polynomial projection technique (Miranda and Fackler, 2000). The discount factor ρ ($\rho = 1/(1+.05)$) on the right-hand side makes Equation 42.6 convex; therefore it can be solved using function iteration, where iterations on Equation 42.6 proceed until the fixed point is found.

The soil profile depth equation of motion (42.2) showing the effect of increases in residue rates comes from Midwest Plan Service (1992). The soil nutrient equation of motion (42.3) uses parameters on crop uptake and fertilizer transformations reported in Vitosh et al. (1996), as well as enrichment effects for eroded nutrients reported by Menzel (1980). Further discussion of Equations 42.2 and 42.3 are found in Hopkins et al.

Equation 42.4 is the effect of residue management on SOC levels. Lal et al. (1998) report that sequestration rates can vary from 0.2 to 0.5 metric tons per hectare per year. The midpoint of these estimates, 0.35 tons per hectare per year, was assigned to 35% residue cover. Sequestration rate was assumed to respond linearly with residue management, with 100% residue adding 1 ton per hectare of SOC, and 0% residue adding zero SOC. Although carbon would not be expected to increase linearly over a long period of time (at some point an equilibrium is reached), the equation for crop yield accounts for diminishing incentives to increase SOC. Therefore, a constant level of residue management is more or less reached even though a steady state in SOC does not attain.

EMPIRICAL APPLICATION TO MARLETTE SOIL

The model is applied to data available for a Marlette soil in Michigan. Physical and chemical data necessary to parameterize Equation 42.1 through Equation 42.5 were available from several sources. Corn yield is assumed to follow a Mitscherlich–Baule form, this functional form allows limited substitutability between inputs. The yield function is discussed in parts. The coefficient outside the brackets is the maximum attainable yield, derived as 102% of maximum reported yield reported in Schumacher et al. (1995). This yield level is reached only asymptotically. Marginal response to the first set of inputs, soil nutrients and commercial fertilizer, was taken from Johnson and Sheperd (1978). In order to keep the model simple, response for an aggregate nutrient was assumed to follow soil and commercial fertilizer response for phosphorous. Corn yield response to changes in soil profile depth comes from data on irreversible soil degradation and productivity contained in Schumacher et al. (1995).

Corn yield marginal response to the third input, SOC, was simulated under three alternative scenarios. Few data supported a single SOC–corn yield relationship. Scenario 1 was constructed using the assumption that corn yield varied 10% over the range of SOC levels; scenario 2 was constructed assuming that corn yield varied 5% and scenario 3 assumed that corn yield varied 0%. Ranges of SOC were derived from soil data for slightly and severely eroded Marlette soils (Mokma et al., 1996; Cihacek and Swan, 1994; Lowery et al., 1995). A lower bound was taken from SOC stocks reported for a severely eroded Marlette soil. While the upper bound could have been taken from reported carbon stocks for slightly eroded soils, which have most of their topsoil intact, this would have underestimated the SOC holding capacity of the soil because the slightly eroded soil had been cultivated for a number of years. Instead, an upper bound was established at 150% of the SOC of the slightly eroded Marlette soil, assuming that slightly eroded soil had lost 50% of its original stock, the upper bound of a range reported by Lal et al. (1998).

Fertilizer costs were taken from crop enterprise budgets from Ohio State University Extension (Schnitkey and Acker, 1995). The price of corn was assumed to be $2.50 per bushel and the price of fertilizer (nitrogen, phosphorous, and potassium combined) was $0.55 per pound. The cost of crop residue management was zero for zero residue management (i.e., 0% standing and flat crop residue at the time of planting), while 35% residue cost $1.14 per acre. The latter cost comes from the current EQIP contract for Ohio, which pays $8 per acre per year for up to 3 years. An annual payment in perpetuity of $1.14 is equivalent to $8 over 3 years, assuming a 5% discount rate. Because some of the drawbacks associated with residue management (increased moisture and decreased temperature) increase with the level of crop residue maintained on the soil surface, residue management costs were assumed to increase at an increasing rate.

RESULTS

Results for a Marlette soil were obtained by iterating on Equation 42.1, until a fixed point was found. Graphical representation of that value function is difficult because it has four dimensions: value, soil profile depth, soil nutrients, and SOC. Three separate three-dimensional views were examined, holding the fourth dimension fixed at varying levels. The marginal effect of soil profile depth was uniform for most of that variables range, so what follows holds soil profile depth at 90 cm (corresponding to a moderately eroded soil).

Optimization Results: Value Functions and Residue Management

Separate optimizations were completed for each of the three scenarios. Because demonstrated yield effects will likely play a large role in motivating adoption, a range of marginal responses was chosen that would indicate how expected yield effects influence residue management decisions.

Figure 42.1 shows the value function in three dimensions — value (in dollars per hectare), soil nutrients (in parts per million), and SOC (in metric tons per hectare) — for the first scenario, where yield was assumed to vary by 10% over the allowable range of SOC levels. Value is expressed as an annual equivalent measure over an infinite horizon. The annual equivalent sum was determined by multiplying net present value by the discount rate, assumed equal to 0.05. The minimum point of the function is when SOC and soil nutrient levels are at their minimum, and the maximum is when these levels are at their highest. Returns at the high point in the graph are over 13% greater than at the low point, implying that soils at these two poles should differ in value by about 13%. Slopes are steeper in the SOC dimension than in the soil nutrients dimension because commercial fertilizers provide a quick and economical substitute for soil nutrients, compared to SOC, which is built up more slowly. Note the concave shape of the value function in the SOC dimension, resulting from very rapid gains in value for marginal increases at low carbon levels. Also note that the slope of the value function diminishes as SOC increases, implying that the value to the producer of increasing carbon stocks also decreases at the margin.

Figure 42.2 shows the value function over the same state space, but with only a 5% difference in yield over the range of SOC values. Note that the vertical relief corresponding to Figure 42.2 is less than that corresponding to the case with greater yield impacts over the range of SOC values. Figure 42.3 shows the value function under the third scenario, with 0% yield response due to differences in carbon levels. Note that Figure 42.3 is drawn relaxing the assumptions of Figures 42.1 and 42.2 that holds soil profile depth constant. The slope of the value function in the soil profile depth dimension is greater than in the SOC direction because changes in soil profile depth are largely irreversible, as residue management can only slow down, but not reverse, these losses. SOC stocks can be more readily reversed, although it might take some time to do so.

Corresponding to Figures 42.1 and 42.2 are functions showing optimal decisions regarding fertilizer applications and residue management. Fertilizer application optimal decisions are similar across all three scenarios and are not shown (see Hopkins et al., 1999). Discussion of optimal residue management decisions over soil nutrient and SOC states, with soil profile depth held

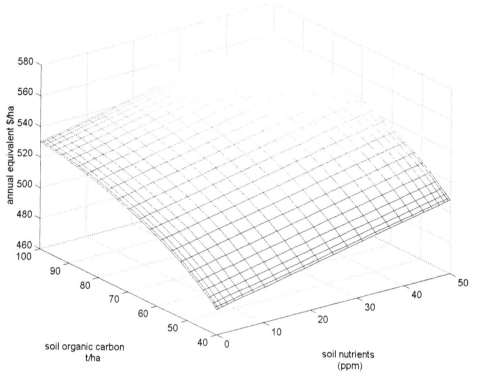

Figure 42.1   Value function: scenario 1.

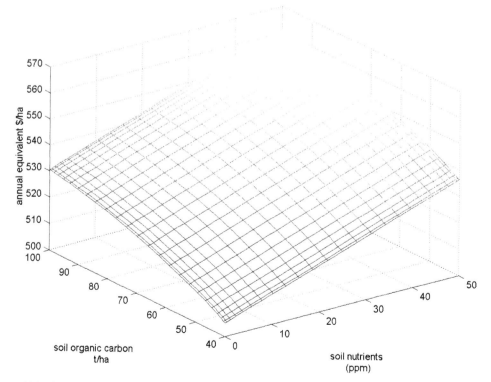

Figure 42.2   Value function: scenario 2.

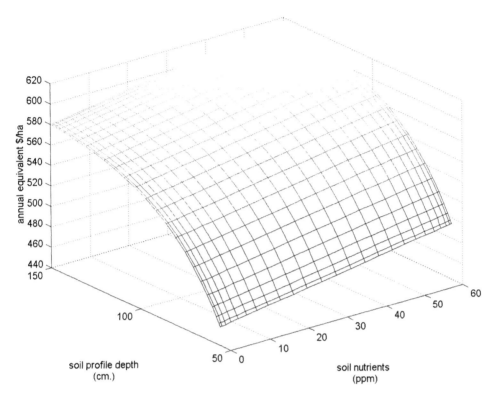

Figure 42.3 Value function: scenario 3.

constant, follows. Again, scenario 1, a 10% yield response over the carbon range, is shown first. Figure 42.4 corresponds to scenario 1 and shows residue management decisions for different stocks of nutrients and carbon; this figure also shows that residue cover should increase as soil nutrients increase. Because high concentrations of soil nutrients are potentially vulnerable to erosion, residue management increases in order to maintain them.

However, residue management decreases in SOC levels, as seen in Figure 42.4 because, as carbon levels increase, yield response decreases as a result of convexity of the yield function. Figure 42.4 shows that residue management more than doubles along the range of the SOC state. Residue management will have the highest payoff when soil carbon levels are very low and therefore have the greatest marginal impact on crop yields. As carbon levels increase, further additions to the stock of SOC through residue management have even smaller marginal yield effects, manifested by decreased willingness to pay for residue management.

Figure 42.5 shows optimal residue management for scenario 2, in which corn yield responds only 5% over the range of SOC values. The optimal decision function is shifted downward and becomes flatter. Figure 42.6 shows the quantitative difference across the state space under scenarios 1 and 2. Note that, because of the greater yield sensitivity to carbon considered under scenario 1, differences in residue management are greatest when carbon levels are lowest.

Simulation Results

The value function and optimal decisions shown in Figures 42.1 through 42.6 are essentially static representations, expressed for a single point in space. In reality, the point in space will move around from year to year. As states evolve over time, decisions will change as well, even though the decision rule is invariant. Simulations of scenarios 1 through 3 will help to illustrate the adjustment path that producers would take if they operated under the assumptions of each scenario.

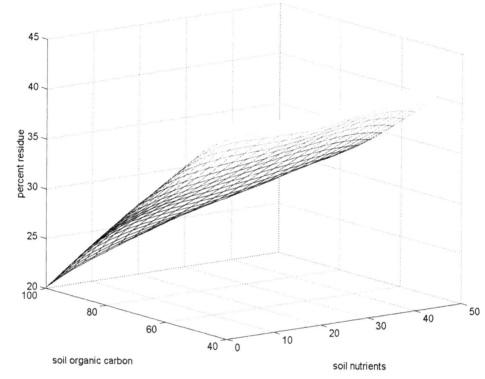

Figure 42.4   Optimal residue management: scenario 1.

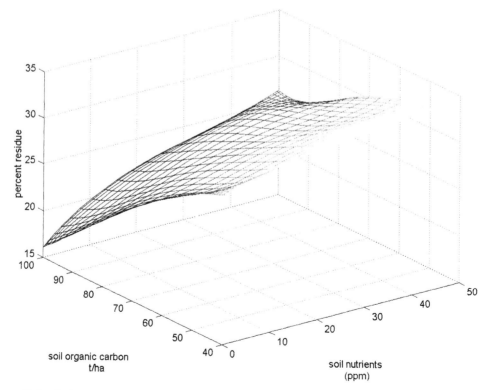

Figure 42.5   Optimal residue management: scenario 2.

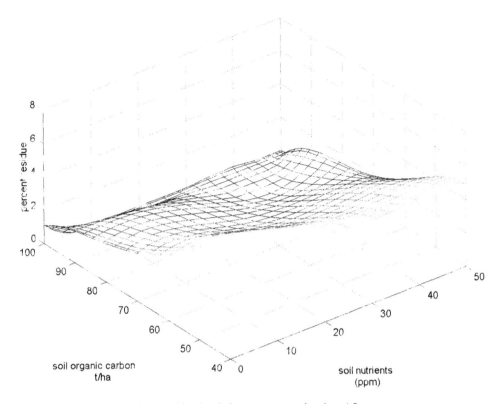

Figure 42.6   Difference in optimal residue levels between scenarios 1 and 2.

Simulations were carried out by moving forward in time and generating decisions off the value function corresponding to each scenario.

A set of initial conditions corresponding to a moderately eroded soil, 40 parts per million of soil nutrients, and with SOC levels at their lower bound (40 tons per hectare) was used for simulation purposes. Figure 42.7 shows how decisions varied over time for all scenarios. Note that the scenario with the high yield response had higher levels of residue management throughout the simulation. Because residue management is costly, there is a greater willingness to build carbon stocks when they have a greater marginal effect on crop yield. Residue decisions are flat for the case where no residue management is chosen because it is only assumed to determine the rate at which soil profile depth is lost and protects soil nutrients. Because the maximum attainable yield does not change among the scenarios, all three eventually converge to the same level of residue management once soil carbon levels are no longer limiting yield.

Figure 42.8 shows how states evolve in reaction to the decisions of Figure 42.7. Over the 100-year simulation period, farmers facing a 10% yield response would increase carbon stocks by 35 tons per hectare. Under a 5% yield response, farmers would increase carbon stocks by 30 tons per hectare. These simulations can be compared to a scenario in which producers would naturally build up carbon levels by over 15 tons per hectare over the same time period as a result of residue management in response to soil profile loss and soil nutrient maintenance. One problem with these scenarios is that carbon dynamics may imply increased oxidation of carbon as residue management declines.

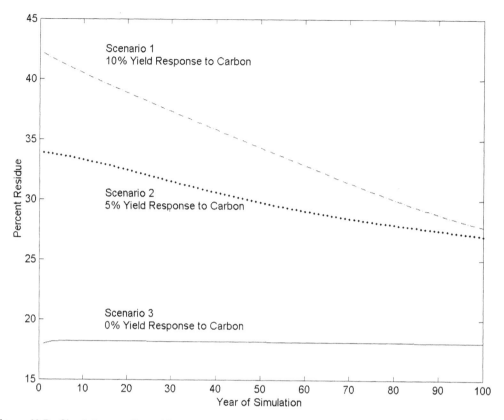

Figure 42.7   Simulation results: residue management over 100 years.

CONCLUSIONS

Policy assumptions can frequently be problematic. The assumption is frequently made that residue management, on-farm carbon sequestration, and increasing profitability are mutually complementary. In fact, adoption rates for residue management decisions are highly dependent on the extent to which technology can be shown to increase net returns. One can expect greater growth in carbon levels when farmers believe that yields will improve as a result of building up carbon stocks, and that a technology such as crop residue management can be used that will result in predictable rates of carbon buildup.

For the Marlette soil modeled, a program to increase soil carbon sequestration, when looked at from the firm's point of view, fits within a scheme of improving soil quality in order to maximize the value of land. From the point of view of society, residue management increases social welfare by fixing atmospheric soil carbon. The model described predicts that on-farm sinks will continue to grow as a result of economical adoption of residue management, even without demonstrated cost and yield effects from carbon increases.

Given the desire for a win–win outcome, the simulation results are favorable. However, rates of carbon gain greater than 0.2 tons per hectare per year impose private costs greater than the private benefits of residue management. Therefore, greater levels of carbon sequestration, even if desired by society, cannot be considered a "win." For more rapid adoption and increases in residue management, in many cases further proof of cost savings or yield effects will be needed. Societal gains may warrant additional subsidies to agriculture, or contractual agreements between carbon emitters and farmers, in order to increase carbon levels.

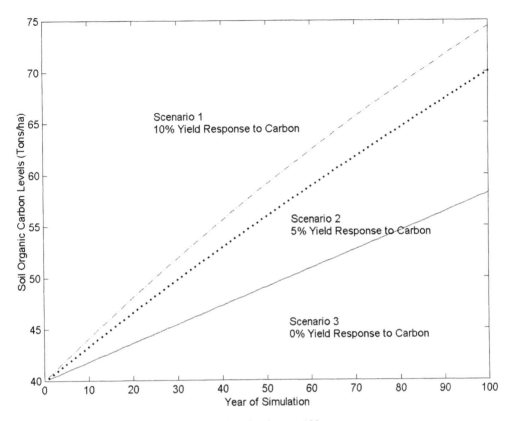

Figure 42.8   Simulation results: soil organic carbon levels over 100 years.

As suggested by Schlesinger (1999), more encompassing research is needed in soil carbon studies in general, a point especially true where practice adoption is presumed to follow solely from firm-level economic incentives. This study further indicates three areas for additional research. First, concentrating on refining technology to address the issues of delayed planting and increased management costs would be one way to reverse the recent declines in adoption in some parts of the Corn Belt. Intermediate systems, e.g., less than no-till, may decrease these costs.

Second, while there is a literature showing marginal carbon response to increased fertilization, the more valuable information for producers is how yield response to fertilizer changes with increased carbon stocks. With this information, economical substitution between carbon and commercial fertilizers could be pursued by producers and production costs may be reduced.

Third, while there is literature showing that increasing crop yields will increase carbon sequestration, it would be more useful to farmers to know the impact of increased carbon stocks on crop yields. With this information, producers could calculate how much they should spend today in order to increase yields in the future. Yield increases depend on the extent to which existing soil carbon stocks are depleted. This chapter compared the effect of a 5 or 10% yield response to carbon, with both resulting in the same maximum attainable yield. If SOC increases the maximum attainable yield, then the incentives might be even greater.

REFERENCES

Bauer, A. and A.L. Black, Quantification of the effect of soil organic matter content on soil productivity, *Soil Sci. Am. J.*, 58:185–193, 1994.

Bruce, J.P. et al., Carbon sequestration in soils, *J. Soil Water Conserv.*, 54:382–389, 1999.

Cihacek, L.J. and J.B. Swan, Effects of erosion on soil chemical properties in the North Central region of the United States, *J. Soil Water Conserv.*, 49:259–265, 1994.

Conservation Technology Information Center (CTIC), Conservation tillage report "troubling" for U.S. agriculture, Press Release, November 4, 1998.

Day, J. et al., Conservation tillage in U.S. corn production: an economic appraisal, 1998 Ann. Meetings Am. Agric. Econ. Assoc., Salt Lake City, Utah, August 2–5, 1998.

Flach, K.W., T.O. Barnwell, Jr., and P. Crosson, Impact of agriculture on atmospheric CO_2, in Paul, E.A. et al. (eds.), *Soil Organic Matter in Temperate Agroecosystems: Long-Term Experiments in North America*, CRC Press, Boca Raton, FL, 3–13, 1997.

Fletcher, J.J. and A.M. Featherstone, An economic analysis of tillage and timeliness interactions in corn–soybean production, *North Cent. J. Agric. Econ.*, 9:207–215, 1987.

Grant, R.F., Changes in soil organic matter under different tillage and rotation: mathematical modeling in ecosys, *Soil Sci. Soc. Am. J.*, 61:1159–1175, 1997.

Hopkins, J.W. et al., Dynamic economic management of soil erosion, nutrient depletion, and productivity in north central USA, *Land Degrad. Devel.*, 12:305–18, 2001.

Johnson, J.W. and N.L. Shepherd. 1978. Columbus, OH: Ohio State University Extension, unpublished manuscript.

Kelly, D.L. and C.D. Kolstad, Bayesian learning, growth, and pollution, *J. Econ. Dyn. Control*, 23:491–518, 1999.

Lal, R. et al., Managing U.S. cropland to sequester carbon in soil, *J. Soil Water Conserv.*, 54:374–381, 1999.

Lal, R. et al., *The Potential of U.S. Cropland to Sequester Carbon and Mitigate the Greenhouse Effect*, Ann Arbor Press, Chelsea, MI, 1998.

Lines, A.R. and D.L. Acker, An economic comparison of tillage systems. Ohio State University Extension AEX 506, 1990.

Lowery, B. et al., Physical properties of selected soils by erosion class, *J. Soil Water Conserv.*, 50:306–311, 1995.

Lowery, B. and W.E. Larson, Symposium preamble: erosion impact on soil productivity, *Soil Sci. Soc. Am. J.*, 59:647–648, 1985.

Menzel, R.G., Enrichment ratios for water quality modeling, in *CREAMS: A Field-Scale Models for Chemicals, Runoff and Erosion from Agricultural Management Systems*, W.G. Knisel, ed., Conservation Research Report No. 26, U.S. Department of Agriculture, 1980, 486–492.

Midwest Plan Service, *Conservation Tillage Systems and Management*, Ames, 1992.

Miranda, M.J. and P. Fackler, Applied computational economics, unpublished manuscript, The Ohio State University, Accessed at http://www-agecon.ag.ohio-state.edu/ae802/lectures.htm, 2000.

Mokma, D.L., T.E. Fenton, and K.R. Olson, Effect of erosion on morphology and classification of soils in the North Central United States, *J. Soil Water Conserv.*, 51:171–175, 1996.

Ohio State University Extension, *Ohio Agronomy Guide*, 13th Ed., Columbus, OH: The Ohio State University, Bulletin 472, 1995.

Parry, I.W.H. and W.E. Oates, Policy analysis in the presence of distorting taxes, *J. Policy Anal. Manage.*, 19:603–613, 2000.

Rausch, J. and B. Sohngen, Environmental quality incentive programs, AE 97-02, Ohio State University Extension Bulletin, Department of Agricultural, Environmental, and Development Economics, Columbus, OH.

Schlesinger, W.H., Carbon sequestration in soils, *Science*, 284:2095, 1999.

Schnitkey, G. and D. Acker, Ohio enterprise budgets, grains 1995, Ohio State University Extension Bulletin, Department of Agricultural, Environmental, and Development Economics, Columbus, OH, 1995.

Schumacher, T.E. et al., Corn yield: erosion relationships of representative loess and till soils in the North Central United States, *J. Soil Water Conserv.*, 49:77–81, 1995.

Syers, J.K., Managing soils for long-term productivity, in *Land Resources on the Edge of the Malthusian Precipice*, D.J. Greenland, P.J. Gregory, and P.H. Nye, eds., CAB International, New York, 1998, 151–160.

Thatcher, J., Uncertainties plague climate forecasting, forum finds, Resources for the Future, accessed at http://www.weathervane.rff.org/features/feature070.html, 1999.

The Washington Post, Cultivating farms to soak up a greenhouse gas, A3, November 23, 1998.

U.S. Department of Agriculture, Office of the Chief Economist, Economic analysis of U.S. agriculture and the Kyoto protocol, May, 1999.

U.S. Environmental Protection Agency, Climate change and impacts — introduction, slide show presentation, http://www.epa.gov/globalwarming/reports/slides/cc&i_toc.html, 1999.

Vitosh, M.L., J.W. Johnson, and D.B. Mengel, Tri-state fertilizer recommendations for corn, soybeans, wheat, and alfalfa, Ohio State University Extension Bulletin E2567, 1996.

Wu, J.J. and B. Babcock, The choice of tillage, rotation, and soil testing practices: economics and environmental implications, *Am. J. Agric. Econ.*, 80:494–511, 1998.

Zolvinski, S., Dampened enthusiasm? While no-till acres in soybeans and wheat are up, a decline in no-till corn concerns experts, *Ohio State Agric.*, 1–2, Spring, 1998.

CHAPTER 43

Carbon Balances in Hungarian Soils

Tamás Németh, Erika Michéli, and László Pásztor

CONTENTS

INTRODUCTION

Maintenance of carbon in soils is a major possibility for land use under long-term sustainable agriculture because declining carbon (C) generally leads to lower yield and reduced crop production, on the one hand, and unfavorable environmental changes on the other. Houghton et al. (1983) stated that, at the end of the 1970s, more carbon went from terrestrial ecosystems into the atmosphere than from fossil fuel combustion. (This phenomenon, i.e., land as a source of C, started when agricultural expansion began in the second half of the last century.) The C originating from the soils had a large impact on the greenhouse effect (Lal et al., 1995); in addition, the loss plays a key negative role in long-term soil fertility and overall sustainable agricultural production.

Increases in the amount of arable land led to soil carbon losses into atmosphere in two principal ways: (1) release of C in the biomass, which is either burnt or decomposed, and (2) release of soil organic carbon (SOC) following cultivation due to enhanced mineralization. The latter is brought about by change in soil moisture and temperature regimes and low rate of return of biomass to the soil (Lal et al., 1998). Accesory effects of land-use change and carbon depletion can be detected in other physical, chemical, and biological processes of arable soils, including soil fertility.

Changes in SOC are primarily effected by human activities (agriculture, forestry, etc.). Many management practices open the way for declines in SOC content, including one of the most important degradation processes — erosion by water and wind. Through different degradation processes, native biologically productive soils became unproductive; this can be characterized by thin humus layer, low SOC, low soil quality, and low biomass productivity.

Table 43.1　Farmyard Manure and Fertilizer Use in Hungary, 1931–1995

Year	Farmyard Manure (million Mg year⁻¹)	Fertilizer Active Ingredients (1000 Mg year⁻¹)				For Arable Lands (kg ha⁻¹ year⁻¹)
		N	P_2O_5	K_2O	Total	
1931–40	22.4	1	7	1	9	2
1951–60	21.2	33	33	17	83	15
1961–65	20.6	143	100	56	299	57
1966–70	22.2	293	170	150	613	109
1971–75	14.8	479	326	400	1205	218
1976–80	14.3	556	401	511	1468	250
1981–85	15.4	604	394	495	1493	282
1986–90	13.2	559	280	374	1213	230
1991	8.0	140	23	33	196	37
1992	7.2	148	21	20	189	36
1993	5.0	160	25	21	206	38
1994	4.9	222	27	31	280	50
1995	4.8	191	29	27	247	47
1996	4.0	203	29	27	270	54

Source: Statistical Yearbooks for Agriculture, KSH.

Soils represent a considerable part of the natural resources in the Central and Eastern European countries as well as in Hungary. Consequently, rational and sustainable land use and proper management practices ensuring normal soil functions have particular significance in the national economy. Soil conservation is an important element of environmental protection (Várallyay, 1994).

As a consequence of improving agricultural practice in Hungary, increased use of fertilizers was characteristic of the early 1960s, reaching a rate as high as 250 kg N+P_2O_5+K_2O /ha arable land units per year from the second half of the 1970s up to the late 1980s (Table 43.1). As a result, the proportion of nutrients from farmyard manure diminished in the Hungarian plant nutrition system. On the other hand, with increased application of mineral fertilizers, average yields doubled or even trebled, resulting in higher amounts of stubble and root remains in the soil, thus increasing the quantity of organic carbon. From the early 1990s, however, fertilizer use dropped dramatically to the level of 30 to 40 kg ha⁻¹ active ingredients (of which 90 to 95% was nitrogen). The same trend was detectable in the farmyard manure application as well because of the dramatic decrease in animal numbers: the animal unit decreased from 3 million to 1.5 million in the past 10 years.

The previous intensive land use practice also had some unfavorable effects on soil carbon sequestration because large fields (100 ha or more) were needed for the efficient use of large machines, rows of trees were cut, causing an increase in erosion, deflation, and soil carbon loss. The large, overweight machines caused a great deal of soil compaction (Németh et al., 1998).

A new land-use zone system is being developed in Hungary in association with discussions for accession to the European Union.

LAND-USE ZONE SYSTEM IN HUNGARY

The basic idea of this development in Hungary is that the decision of the common agricultural practice (CAP) reform in 1992 made it compulsory to form subsidizing systems that would promote integration of environmental and landscape protection and nature conservation into the system of agricultural production for all member states. It was decided during the negotiations that agricultural production would probably be concentrated in those areas where it is most profitable and the comparative ecological advantages are most favorable. The question is how to convert subsidies formerly devoted to producers dependent on production into subsidies supporting rural development and the nonproductive (i.e., environmental, ecological, employment, cultural, etc.) functions of agriculture.

In an analysis of the possibilities of land-use change, Ángyán et al. (1998a) summarized that Hungary can achieve an advantage if the special conditions of the different measures to be taken are precisely determined, i.e., a land-use zone system can be formed. The zonality characterizes both nature conservation and agriculture and can be grouped as follows:

- Basic nature conservation zones — nature reserves, strictly protected areas
- Buffer zone of nature conservation and protection zones for water reservoirs — limited land use, areas with priority for protection
- Mixed zones (ecological and other extensive farming type systems) — land use limitations for protective purposes
- Zones for agricultural production — best agroecological conditions for intensive land use
- Noncultivated land

Along these lines, available nature and land information was collected by the Institute for Environmental and Landscape Management of the Gödöllő Agricultural University (IELM-GAU) and Research Institute for Soil Science and Agricultural Chemistry of the Hungarian Academy of Sciences (RISSAC-HAS). The databases were put into four groups (Ángyán et al., 1998b; Németh et al., 1998) and the variables and databases used as follows:

- Evaluation and qualification of the suitability for agricultural production, i.e., terrain and soil databases, and climatic parameters
- Evaluation of environmental sensitivity, i.e., flora and fauna, soil, and water
- Database of land use and land cover, i.e., CORINE (Coordination of Information on the Environment) land cover, and forest areas
- National Ecological Network (NECONET).

Altogether 28 environmental datasets were classified and weighted according to their role in the determination of agricultural production and environmental sensitivity. (The priority standards were given by the experts and institutes that developed the databases.) The area of the observation unit (cell) was 1 hectare (100 × 100 m grids).

The values of environmental sensitivity (VES) and agricultural suitability (VAS) varied between 0 and 99, respectively. During the calculation, the VES were subtracted from VAS in each cell, then 100 was added to the difference, i.e., (VAS–VES)+100. Using this formula, the values varied between 0 and 198; values under 100 suggested environmental sensitivity and values above 100 agricultural suitability (Table 43.2). The well-determined areas (agricultural and environmental) can be found at the two extremes of this scale, while the mixed areas (areas with extensive production limited by environmental features) are situated in the middle of the scale.

Table 42.2 The Position of Hungary's Areas on a Scale of Environmental Sensitivity and Agricultural Suitability (%)

Standard Categories	Total	Agricultural Land
< 60	0.42	0.04
61–70	1.09	0.10
71–80	2.06	0.56
81–90	5.84	2.53
91–100	11.78	7.96
101–110	18.99	16.76
111–120	18.33	19.44
121–130	15.08	17.91
131–140	12.33	15.62
141–150	10.18	13.65
151–160	3.88	5.42
> 160	0.01	0.01
Total:	100.00	100.00

Table 43.3   Suggestion for Development of a Land-Use Zone System in Three Categories (second scenario)

Land-Use Zone	Total	Agricultural Land
	(%)	
Protection zones	10.38	3.74
Zones for extensive agricultural production	41.15	35.88
Zones for intensive agricultural production	48.47	60.37
Total:	100.00	100.00
	(ha)	
Protection zones	966,095	229,257
Zones for extensive agricultural production	3,827,954	2,196,834
Zones for intensive agricultural production	4,508,952	3,695,909
Total:	9,303,000	6,122,000

The differences were set up between the extensive and intensive agricultural zones according to the extensive rank between 100 and 120, 100 and 125, and 100 and 130. Using the values of this estimation, three scenarios were worked out in order to develop a land-use zone system. The middle scenario was calculated with the following categories (Table 43.3):

- Areas with a value less than 100 were ranked into the protection zone
- Areas with a value between 100 and 125 were ranked into the extensive agricultural zone
- Areas with a value more than 125 were ranked into the intensive agricultural zone

According to this scenario, nearly 4% of Hungary's existing agricultural land (about 230,000 ha) can be turned into a protection zone, more than 35% (~ 2.2 million ha) can be considered for extensive production, and more than 60% (~3.7 million ha) left for intensive agricultural production. Regarding arable land, the same scenario showed that 111,300 ha can be moved from existing arable land (4,714,000 ha) to the protection zone and 1,408,900 ha to extensive agricultural production, while more than 67% (3,193,800 ha) can remain in the intensive agricultural production zone. The following conversions can be suggested:

- 533,000 ha of grassland into forest
- 229,000 ha of arable land into forest
- 788,000 ha of arable land into grassland
- 503,000 ha of intensive arable land into extensive arable land

This change will affect the C balance of the converted soils, resulting in a need for country-scale C balance estimation.

CARBON BALANCE OF THE HUNGARIAN SOILS

The carbon in the soil has a special future through humus formation; humus is considered the most important fraction of soil organic matter (SOM) (as well as SOC). From agricultural production as well as environmental points of view, the quantity and the quality of organic matter play a significant role. Humus plays a significant role in the fertility of the soil and plant nutrition; it can also improve the physical characteristics of soils. The basis of these favorable effects is that molecules of humus belong to the colloid range, so its capacity of binding nutrients and water is remarkable.

Sustainable land use and proper management aiming to ensure normal soil functions can have a significant effect on the Hungarian national economy. Soil conservation is an important element of environmental protection as well. Most carbon leaves agricultural fields in Hungary as a consequence of wind and water erosion. More than 30% of the arable land is sensitive to erosion.

Depth of the Soil

The majority (86%) of Hungarian soils are more than 1 m deep. Soil depth is between 0.7 and 1.0 m, 0.4 and 0.7 m, and 0.2 and 0.4 m in 4 and 5% of Hungarian soils, respectively (Várallyay et al., 1980). Soil depth and soil organic matter content can strongly determine the amount of organic matter resource in a given territorial unit. Figure 43.1 shows the rootable depth of the Hungarian soils (1:100,000).

Soil Organic Matter (OM) Content

The distribution percentage of Hungarian soils according to their organic matter content shows that two thirds of Hungarian soils have between 1 and 3% organic matter. In sandy soils, OM is usually below 1% (15% of the area), while in clay loams it is between 3 and 4% (also 15% of the total area). It is over 4% on about 5% of the territory (Baranyai et al., 1987). This territorial distribution shows that sandy soils with low original organic matter content are situated in the southwestern, central, and eastern parts of Hungary, while those with the highest OM contents are found in the southeastern part of the country.

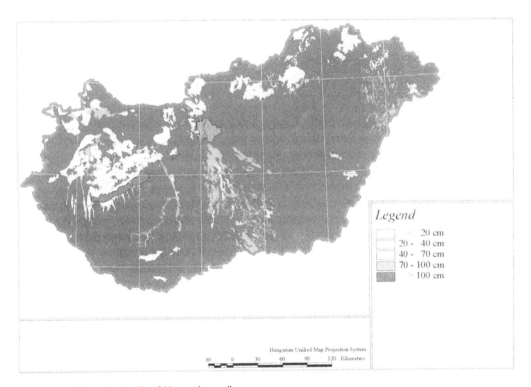

Figure 43.1   Rootable depth of Hungarian soils.

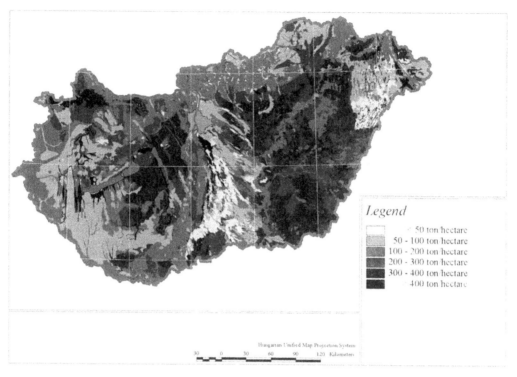

Figure 43.2   Organic matter content of Hungarian soils.

Organic Matter and SOC Resources of Hungary

The distribution of Hungarian soils according to their soil organic matter resource groups is shown in Figure 43.2. In the majority of Hungarian soils, the soil organic matter resource is between 50 and 400 t/ha; it is between 100 and 200 t/ha on about 30% of the total area (Várallyay et al., 1980).

The distribution of organic matter and C in the main Hungarian soil types according to the Hungarian classification (U.S. Soil Taxonomy and FAO classification also) are shown in Table 43.4 for the whole country and Table 43.5 for the arable land. Tables 43.5 through 43.7 contain only the FAO classification from the previously depicted classification systems.

Estimation of OM and SOC contents and pools was based on the calculation on territorial base with the thickness of the OM layer and the average SOC concentration in two layers (upper 20 cm and under) in the given soil. The largest OM, as well as SOC pools, can be found in the chernozem, peat, and meadow soils — 182 t/ha, 180 t/ha, and 104 t/ha OM, respectively, in the above 40 to 60 cm (40 cm for meadow soil and 60 cm for chernozem and peat soils). The same calculation shows an average 105.6 t/ha SOC on chernozem soils, 104.4 t/ha on peat soils, and 60.3 t/ha on meadow soils.

Altogether more than 1102 million t OM and more than 639 million t SOC make up the reserves of the Hungarian soils in the given thickness. Approximately 53% of the OM and SOC can be found in arable land (Tables 43.4 and 43.5).

Predicted changes in the land-use system offers a possibility for calculating the OM and SOC according to the new distribution (Tables 43.6 and 43.7). Table 43.6 shows how this change will affect distribution of the soils in different land-use categories, while Table 43.7 gives a scenario for the SOC balance in the next 20 years.

Table 43.4 Distribution of Organic Carbon in Soils of Hungary

Hungarian Classification	Soil Type U.S. Soil Taxonomy	Soil Type FAO	Area (ha)	Depth of Roots (cm)	OM % in Upper 20 cm	OM % in Below 20 cm	OM (t/ha)	OM in Total Area of Soil Type (t)	OC (t/ha)	OC in Total Area of Soil Type (t)
Skeleton soils	Entisols (Ustipsamments, Ustiorthents)	Regosols/Leptosols	763,750	10	0.5	0	6.5	4,964,375	3.8	2,879,338
Stony soils	Inceptisols (Ohrepts, Umbrepts)	Regosols/Leptosols	262,936	30	2	1	65	194,835,576	37.7	113,004,634
Forest soils	Alfisols (Ustalfs)	Luvisols	3,195,004	40	2	1	78	249,210,312	45.2	144,541,981
Chernozem soils	Mollisols (Ustolls)	Chernozems/Phaeozems	2,064,731	60	3	2	182	375,781,042	105.6	217,953,004
Salt-affected soils	Inceptisols (Halaquepts)/Vertisols (Salaquerts, Natraquerts)	Solonets/Solonchak	562,440	20	2.5	0	65	36,558,600	37.7	21,203,988
Meadow soils	Mollisols/Vertisols	Phaeozems/Vertisols	1,987,554	40	3	1	104	206,705,616	60.3	119,889,257
Peat soils	Histosols (Hemists, Saprists)	Histosols	132,983	60	30	30	180	23,936,940	104.4	13,883,425
Wetland forest soils	Inceptisols (Endoaquerts)	Gleysols	8,087	20	1	0	26	210,262	15.1	121,952
Floodplain soils & sediments	Entisols (Fluvents), Inceptisols	Fluvisols, Regosols	254,511	20	1.5	0	39	9,925,929	22.6	5,757,039
Total			9,231,996					1,102,128,652		639,234,618

Table 43.5   Distribution of Organic Carbon on Arable Land of Hungary

Soil Type FAO	Area (ha)	Depth of Roots (cm)	OM % in Upper 20 cm	OM % in Below 20 cm	OM (t/ha)	OM in Total Area of Soil Type (t)	OC (t/ha)	OC in Total Area of Soil Type (t)
Regosols/ Leptosols	255,392	10	0.5	0	6.5	1,660,048	3.8	962,828
Regosols/ Leptosols	25,961	30	2	1	65	1,687,465	37.7	978,730
Luvisols	1,425,147	40	2	1	78	111,161,466	45.2	64,473,650
Chernozems/ Phaeozems	1,682,508	60	3	2	182	306,216,456	105.6	177,605,544
Solonets/ Solonchak	262,096	20	2.5	0	65	17,036,240	37.7	9,881,019
Phaeozems/ Vertisols	1,280,565	40	3	1	104	133,178,760	60.3	77,243,681
Histosols	50,738	60	30	30	180	9,132,840	104.4	5,297,047
Gleysols	3,908	20	1	0	26	101,608	15.1.	58,933
Fluvisols, Regosols	129,220	20	1.5	0	39	5,039,580	22.6	2,922,956
Total	5,115,535					585,214,463		339,424,389

Table 43.6   Scenario of Land-Use Change of Arable Land for Next 25 Years

Soil Type FAO	Currently Arable Land Area (ha)	Expected Partial Land-Use Change	Change to Area (ha)	Remaining Arable Land Area (ha)
Regosols/Leptosols	255,392	Grassland	74,722	180,670
Regosols/Leptosols	25,961	Grassland	5,324	20,637
Luvisols	1,425,147	Forest	824,501	600,646
Chernozems/Phaeozems	1,682,508	Grassland	1,658,345	24,163
Solonets/Solonchak	262,096	Grassland	171,537	90,559
Phaeozems/Vertisols	1,280,565	Grassland	1,109,887	170,678
Histosols	50,738	Wetland	36,101	14,637
Gleysols	3,908	Wetland forest	1	3,907
Fluvisols, Regosols	129,220	Grassland/forest	64,087	65,132
Total	5,115,535		3,944,505	1,171,029

Table 43.7   Scenario for Organic C Content Due to Land-Use Change and Erosion after 25 Years

Soil Type FAO	Current OC Status (t)	Expected Increase of OC (t) Due to Land-Use Change	Expected Loss of OC Due to Erosion (t) On Remaining Arable	Expected Loss of OC Due to Erosion (t) On "Changed"	Summa Change	Expected OC Status (t) after 25 Years
Regosols/ Leptosols	962,828	3,406	56,340	58,718	−111,652	851,176
Regosols/ Leptosols	978,730	6,224	32,114	53,656	−79,546	899,184
Luvisols	64,473,650	181,155	2,486,695	1,561,680	−3,867,220	60,606,431
Chernozems/ Phaeozems	177,605,544	10,931	3,751,176	37,694	−3,777,939	173,827,606
Solonets/ Solonchak	9,881,019	51,211	646,694	294,317	−889,800	8,991,220
Phaeozems/ Vertisols	77,243,681	77,215	5,021,129	665,644	−5,609,558	71,634,122
Histosols	5,297,047	0	0	0	0	5,297,047
Gleysols	58,933	295	0	0	295	59,227
Fluvisols, Regosols	2,922,956	7,366	289,930	508,030	805,326	3,728,282
Total	339,424,389				−13,530,094	325,894,294

REFERENCES

Ángyán J. et al., *Land Use System for Hungary*, Gödöllő Agricultural University, Gödöllő, Hungary, 1998a, 1–18.

Ángyán, J. et al., *Working Out of the Land Use Zone System for Hungary*, Gödöllő Agricultural University, Gödöllő, Hungary, (in Hungarian), 1998b, 1–38.

Baranyai, F., A. Fekete, and I. Kovács, *Evaluation of the Nutrient Status of Hungarian Soils*, in Hungarian, Mezőgazdasági Kiadó, Budapest, 1987.

Houghton, R.A. et al., Changes in the carbon content of terrestrial biota and soils between 1860 and 1980: a net release of CO_2 to the atmosphere, *Ecol. Monogr.*, 53:235–262, 1983.

Lal, R., J. Kimble, and B.A. Stewart, Towards soil management for mitigating the greenhouse effect, in Lal, R. et al. (eds.), *Soil Management and Greenhouse Effect*, Advances in Soil Science, CRC Press, Boca Raton, FL, 1995, 373–381.

Lal, R., J. Kimble, and R. Follett, Land use and soil C pools in terrestrial ecosystems, in Lal, R. et al. (eds.), *Management of Carbon Sequestration in Soil*, Advances in Soil Science, CRC Press, Boca Raton, FL, 1998, 1–10.

Németh T., P. Csathó, and A. Anton, Soil carbon dynamics in relation to cropping systems in principal ecoregions of Eastern Europe, with particular regard to Hungarian experiences, in Lal, R. et al. (eds.), *Management of Carbon Sequestration in Soil*, Advances in Soil Science, CRC Press, Boca Raton, FL, 1998, 255–283.

Várallyay, G., Soil management and environmental relationships in Central and Eastern Europe, *Agrokémia és Talajtan*, 43:41-66, 1994.

Várallyay, G. et al., Map of soil factors determining the agroecological potential of Hungary, 1:100,000, II, (in Hungarian), *Agrokémia és Talajtan*, 29:35–76, 1980.

CHAPTER 44

Some Key Uncertainties in the Global Distribution of Soil and Peat Carbon

J. M. Adams and E. Lioubimtseva

CONTENTS

INTRODUCTION

Global biogeochemical modeling of soil and peat carbon reservoirs is essential for forecasting future CO_2 levels. A large number of models have attempted this task over the past few years, often integrating a range of global vegetation and soil reservoirs with climatic effects and direct CO_2 fertilization. Examples of recent and current models attempting these sorts of tasks include BIOME-BGC, OBM, TEM, and BIOME 3 (VEMAP members, 1995). However, at present various key uncertainties and basic problems with published data sources are not properly acknowledged in these broad-scale models of the carbon cycle. This chapter examines some of these uncertainties and suggests how they may affect the prospects of realistically modeling the global carbon cycle.

As an example of difficulties which can come about from failing to consider details, standard soil carbon databases tend to concentrate on sites of agricultural potential, biasing the soil carbon content in arid regions in favor of sites with deep and well-watered soils. Global peat carbon estimates vary greatly among authors; therefore, it is important to use the most rigorous and recent studies on northern peatlands that take into account spatial heterogeneity in these bogs. Lack of attention to these difficulties has sometimes led to use of unrealistic figures in certain global reservoir estimates and carbon cycle models.

Other types of serious scaling difficulties also plague the field of carbon cycle modeling. The direct-CO_2 fertilization effect on soil carbon remains an almost intractable problem, possibly varying greatly on very small spatial scales and with little prospect of simple relationships that can be modeled on a global scale.

The greatest uncertainties in existing carbon databases result from scale-dependent errors in estimates of land-use distribution. This problem is related to the coarse resolution of regional, continental, and global datasets, which treat finer-grain features as inherent heterogeneity. Extrapolation for biochemical fluxes to the landscape scale, where different land-cover patches can serve as either sinks or sources, can completely distort the true picture. Produced by localized sources and taken up by nearby sinks, carbon content shows strong dependence on fractal distribution of soil and vegetation patches in the landscape; thus, the source and sink density varies with scale of study and the flux of carbon is scale-dependent. Models need to include greatly improved corrections of scale-dependence while shifting from high- to low-resolution in order to preserve effects of interpatch carbon fluxes.

If biogeochemical modeling is to make the greatest progress, it will be necessary for the community to examine critically all numerical data sources with close reference to the detailed knowledge and experience of field ecologists and pedologists, who understand the spatial graininess of the data.

CARBON CONTENT IN DESERT SOILS

Published databases of the spatial distribution of soil organic carbon are widely used to parameterize global carbon cycle models. To ensure that the assessments of reservoirs and models are well founded, it is necessary to point out some of the problems which may arise from use of existing databases. The major source of data for carbon cycle modeling is still the global database compiled by Zinke et al. in 1986. For the U.S., the updated source is the STATSGO database (1997).

Some paradoxical features emerge if one examines the Zinke et al. (1986) data. For example, their results suggest that as much organic soil carbon is in temperate deserts as in temperate forests and woodlands (9.7 tC/ha in cool desert vs. 10.0 tC/ha in temperate forest); (Zinke et al., 1986).

Most of the data come from desert soils from the southwestern U.S. From the database, Zinke et al. (1986) suggest global values for cool desert soil carbon of around 10 kg/m₂, and these figures have been used in most of the long-term global carbon cycle estimates (Van Campo et al., 1993; Peng, 1994; Peng et al., 1994). The present examination of the more detailed STATSGO database also suggests relatively high organic soil carbon values (around 4.8 tC/ha) for the warm western deserts of North America. Such large amounts of carbon in desert or semidesert soils would seem paradoxical, considering the low rates of input of organic matter. Detailed assessment of the databases suggests that, in fact, the soil carbon values for the southwestern U.S. and especially for former USSR Central Asian desert regions are inflated due to a number of factors.

First, in the face of many thousands of soil data points from around the world, some data points were probably misassigned, particularly for cool deserts. In the Central Asian dataset, many of approximately 30 points labeled as "desert" fall (Table 44.1) within the dry steppe and steppe-marginal zones of the former USSR and Mongolia (Lioubimtseva and Adams, see Chapter 40).

Other problems with existing databases are likely to have arisen due to selective sampling in agricultural prospecting (which was the reason most soil profiles were first taken), and exclusion from the datasets of shallow soil profiles from the 100-cm depth calculation. The separate STATSGO dataset of carbon in soil profiles only to 25- and to 50-cm depth show a greater number of samples, each with much less carbon. Since desert soils are generally shallower than 100-cm depth, a bias exists in the 100-cm database in favor of deep soils in favorable sites such as hollows which collect more water and sediment.

Table 44.1   Sample Sites of Central Asian Dataset

	Database No.	Biome	Organic C	Latitude	Longitude	Life Zone Type
1.	2050001	245.01 SB	15.3 1001	53.0 N	78.6 E	200.13 B 40
2.	2050002	245.03 SB	6.3 0	54.0 N	78.7 E	200. 13 B 40
3.	2050003	245.04 SB	8.0 0	53.3 N	78.3 E	200.13 B 40
4.	2050004	245.05 SB	5.5 0	52.3 N	79.1 E	200. 13 B 40
5.	2051001	248.01 SB	8.3 0	53.5 N	80.2 E	200.13 B 40
6.	2051002	248.02 SB	6.6 0	53.5 N	80.2 E	200. 13 B 40
7.	2052001	251.01 SB	4.3 0	53.4 N	79.0 E	200.13 B 40
8.	2052002	251.02 SB	11.3 0	52.7 N	83.0 E	200. 13 B 40
9.	2053001	254.01 SB	6.4 711	53.8 N	91.4 E	375. 13 B 40
10.	2053002	254.02 SB	3.5 0	53.8 N	91.4 E	375. 13 B 40
11.	2053003	254.03 SB	4.9 394	52.2 N	90.7 E	750. 13 B 40
12.	2053004	254.04 SB	6.7 377	53.2 N	90.5 E	375. 13 B 40
13.	2053005	254.05 SB	5.5 357	53.2 N	90.5 E	375.13 B 40
14.	2053006	254.06 SB	10.6 658	52.1 N	90.7 E	750. 13 B 40
15.	2053007	254.07 SB	6.3 473	53.2 N	90.5 E	375. 13 B 40
16.	2053008	254.08 SB	6.4 938	56.1 N	92.8 E	0200 13 B 40
17.	2054001	257.01 SB	17.4 948	50.3 N	87.6 E	1600 13 B 40
18.	2060001	VV14 SB	5.6 768	50.1 N	88.9 E	2000 42 B 60
19.	2060002	VV87 SB	3.3 225	50.1 N	88.3 E	42 B 60
20.	2060003	VV2 SB	8.7 869	50.1 N	88.9 E	2000 42 B 60
21.	2060004	VV9 SB	4.0271	50.1 N	88.9 E	42 B 60
22.	2069001	SC01 RS	8.7 1180	49.6 N	57.2 E	300 19 B
23.	2069002	SC01A RS	6.0968	49.6 N	57.2 E	300 19 B
24.	2069003	SCO2 RS	7.8 1237	49.6	57.2 E	300 19 B

Source: Zinke et al., 1986. Most of grid coordinates fall within the dry steppe zone.

Another source of confusion in terms of understanding spatial heterogeneity and overall totals of soil carbon is the important and fundamental distinction between organic and inorganic carbon. Calcrete (carbonate precipitated into the soil, often as nodules or a crust near the surface) is a major reservoir of carbon in present-day soils in arid and semiarid regions. Inorganic carbon is excluded from the Zinke et al. (1986) and STATSGO databases, but there is still possibility of confusion from other sources. Many biogeochemists (e.g., Schlesinger, 1982) discuss this soil reservoir in a purely empirical sense, pointing out its importance but not its actual behavior as part of the carbon cycle. Yet its effect on the carbon cycle is opposite to that of organic carbon. Any decrease in calcrete by chemical weathering takes up CO_2 from the atmosphere, at least temporarily. A decrease in organic matter in a soil should increase atmospheric CO_2.

In summary, the extraordinarily high values for soil carbon found in published soil databases from cool desert bush regions of the western U.S. and Central Asia can thus be assigned to a combination of problems:

1. Some samples in the Zinke et al. (1986) database have been misassigned in the biome sense, at least in arid regions; these sampling points fall in steppe regions and not desert.
2. Of samples that do fall within desert regions, some were gathered with an eye to agricultural potential and so favor locally moist pockets within the landscape, no matter how unrepresentative these might be.
3. Most semidesert and desert soils do not reach 100-cm depth and are thus excluded from the 100-cm STATSGO database (allowed for in the Zinke et al. (1986) database, however), which favors locally deeper soil pockets in moister sites, P>4. Even those areas in the western U.S. that fall within "desert," as generally described from that region, are in fact from relatively dense sagebrush semidesert capable of sustaining commercial ranching. On a global scale this would not qualify as desert, yet the high carbon storage values are extrapolated to include all cool desert zones, no matter how dry.

The Zinke et al. (1986) database is an impressive piece of work; when something on this scale is attempted it is inevitable that some errors will enter the system. It is unfortunate that carbon cycle modelers do not always take these problems into consideration, in effect, preferring to treat data from a published source as if it were infallible.

Another lesson in potential problems with spatial sampling of soil carbon is evident from the very extensive STATSGO soil carbon database. Although no one appears to be using the database for the particular purpose of modeling natural ecosystems or broadscale soil carbon coverage, further discussion will point out one potential pitfall within this soil carbon database: the sampling in the STATSGO database is mostly based upon agricultural soil surveys of sites with an existing crop cover. Thus, it is not representative of seminatural ecosystems, nor is it a spatially unbiased picture of soil carbon distribution. This gives, for example, much lower average soil carbon contents for the cool temperate forest climate zone (around 70 tC/ha) than for the Zinke et al. database (about 120 to 140 tC/ha), which selected seminatural forested sites rather than cultivated land, as it did with the desert zones. This is true even in predominantly forested portions of the country; still, most of the STATSGO sites are from cultivated areas within those. It is always necessary to be wary of the land use heterogeneity of each region and the original process by which data points were selected.

PEATLANDS AND THEIR CARBON CONTENT

Various figures are used in the literature to suggest how much carbon is in peat bogs at present. Often in models of the carbon cycle one sees a single figure accepted as "the" figure for the peatland carbon reservoir. Yet there is room for considerable skepticism about closeness to the true figure for peatland carbon storage (Table 44.2). Accepted figures for peatland carbon in the high latitudes have changed over the past 30 years or so; recent attempts dealing more explicitly with spatial heterogeneity in northern peatland regions suggest that there may be much more peat than previously thought. Further careful field sampling and GIS work are needed to back up these assertions. Much of the difficulty at present centers on the varying proportions of hummocks and pools in peat bogs in different regions. If problems with adding up the amount of peat in the high latitudes exist, then the problems are much greater for tropical peats. How much peatland carbon is in the tropics is still not known; it could be hundreds of gigatonnes according to at least one geologist who has worked on tropical peats (H. Faure, personal communication).

The problem of allowing for spatial heterogeneity within peat bogs is difficult enough. A further problem is distinguishing the area covered by peat bogs, as opposed to other forms of carbon-containing ecosystems such as tundra or closed forest, which remains a further uncertainty difficult to allow for in calculating broad scale carbon storage. It is important to subtract the area of land surface covered by peat bog from that covered by better-drained vegetation types, or the same area could be counted twice. Yet one needs a clear definition of each area in order to do this.

In fact, no clear and consistently used definition exists in the literature of exactly how one should distinguish peat bogs from other ecosystems. Hence many studies on global or local carbon storage demonstrate considerable ambiguity as to where the category of peat ends and that of soil begins. One widely accepted definition for peat is that it is a pure organic layer at least 20 cm in thickness; this was used by widely cited studies by Post et al. (1982) and Zinke et al. (1986). However one can take as another example the study of Canadian peatland areas by Tarnocai (1980), which defines peatlands as having peat depths (i.e., an organic matter layer) greater than 40 cm and mineral wetlands as having an organic matter layer of less than 40 cm. The recent wide-ranging study of northern peatlands by Gorham (1991) used a minimum figure of 30 cm of organic matter as its dividing line between peat and nonpeat, so there is no sign of emerging consensus.

For the tropics, the actual area and depth of peatland remains largely unknown (Walter and Box, 1983; Radjagukuk, 1985), thus it has been necessary to exclude tropical peatlands from

Table 44.2   Some Published Estimates for Global Peatland Carbon Storage

Carbon Storage	Description	Source
860 Gt	Peats, present-day	Bohn (1982)
300 Gt	Peats	Sjors (1981)
202 Gt	Peats	Post et al. (1982)
377 Gt	Peats	Bohn (1985)
180–227 Gt	Peats	Gorham (1991)
500 Gt	Global peats	Markov et al. (1988)

estimates (instead ascribing to these areas a normal forest and soil cover). This probably leads to present studies underestimating carbon accumulated in the tropics since the early Holocene, although the underestimate will be compensated to some extent by the thinner forest cover which tropical swamp areas tend to have (Walter and Box, 1983).

From the values in Table 44.2, one can see that peatland carbon mass is still an unknown and it may be much more than has generally been suspected.

Of the estimates for global peatland, the recent one of 461 Gt compiled by Gorham (1991) is generally seen as the most robust. It was based on a wide range of data sources including much recently gathered data and compiled using a GIS-based approach. However, the latest estimates currently emerging from studies on Canadian peatlands seem to suggest figures of around 200 Gt for Canada alone; if one considers Russian and Scandinavian peatlands together as containing about twice this amount of carbon (a conservative estimate), this would give a total of around 600 Gt. This figure also does not include any tropical peatlands, which might well contain 100 Gt or more of peat carbon (H. Faure, unpublished calculations).

What is evident from the range of values in Table 44.2 is that, at present, there is little justification in selecting any one particular value from the literature to represent the size of the global peat carbon reservoir. If anything, all previous published values for global peat are likely to be serious underestimates.

DIRECT CO_2 EFFECTS ON SOIL CARBON STORAGE

A fair amount of discussion has occurred during recent years concerning how rising ambient CO_2 levels might affect soil carbon storage through the plants that root into the soils and supply carbon to the soil reservoir. Various models have attempted to allow for aspects of the CO_2 fertilization effect on ecosystem processes. At present it seems that those models that include direct CO_2 fertilization factors on plant growth do so inadvisably. It seems that soil effects will be even harder to model because of the spatial complexity of direct CO_2 effects.

Some clues to the way in which higher CO_2 levels might affect soil processes can be gleaned from experiments carried out on plant growth. Under higher future CO_2 levels, a higher photosynthetic rate of vegetation could have mean changes in the net flux of primary production reaching the soil as dead leaves, roots, branches, etc. The actual extent to which CO_2 fertilization of plant growth occurs in natural and seminatural ecosystems is highly uncertain (Koerner and Arnone, 1992; Mooney and Koch, 1994; McConnaughy et al., 1993). The rapid rise of 80 to 90 ppm CO_2 (of the same order as the glacial–interglacial increase) over the last 200 years has provided no clear evidence of a direct CO_2 effect on vegetation productivity or biomass (Adams and Woodward, 1992; Shiel and Phillips, 1995) that cannot possibly be explained by climate fluctuation and increased sulfur and nitrogen fertilization by air pollution. Basically, the direct CO_2 effect must be present in natural vegetation to some extent, but its importance in affecting the productivity of vegetation is largely a matter of guesswork.

Quite possibly, even with no significant change in plant growth rate above ground, the result of higher CO_2 in terms of faster turnover time of roots below ground would constitute a substantial increase in the amount of organic carbon to the soil, without any corresponding increase in microbial decomposition rates. For example, some experiments on CO_2 fertilization in artificial tropical microcosms have found that raised CO_2 levels above 600 ppm give faster turnover rates of fine roots than under present-day ambient CO_2 levels (Mooney and Koch, 1994). Many experiments on both tropical and temperate plants (Mooney and Koch, 1994; Rogers et al., 1994) also indicate that, at higher CO_2 levels, the root mass is increased much more than the above-ground material; this might have implications for the supply rate of organic matter directly into the soil from dead root material. Once again, however, the subject is potentially so complex and so poorly understood that any attempt to model it on a global scale seems almost unwarranted.

Bazzaz (1990) has suggested that changes in the carbon to mineral ratios in plant materials reaching the soil surface could also have far reaching effects on levels of long-lived carbon in soils. Under higher CO_2, for instance, a higher ratio of carbon to minerals might be in the soil litter (due to relative carbon starvation of the plants), slowing the rate of fungal and bacterial decomposition. This would tend to result in more carbon accumulating in the soil.

Others have suggested that the opposite could occur, with less carbon accumulating in soils despite the greater input of plant material from CO_2-fertilized plants, due to a "priming" effect, in which microbial activity on the new organic matter is increased. This is because, at higher growth rates, each resource type becomes more steadily available (due to increased total input of materials from each plant type) and better able to maintain a specialized microbial community that will break it down quickly (Schimel et al., 1995).

In their experiment on several artificial tropical ecosystems exposed to high CO_2 levels, Koerner and Arnone (1992) found a decrease in soil carbon at higher-than-present CO_2 concentrations (i.e., lower CO_2 = more soil carbon), which is what the priming hypothesis predicts. However, the result may vary greatly on a spatial basis, with soil type, and with overlying crop and vegetation type, as a recent experiment comparing C_4 sorghum with C_3 soybean has shown (Torbert et al. 1998). Under sorghum, the soil carbon increased under the higher CO_2 regime, whereas under soybean it decreased. This experiment by Torbert et al. contrasts two very different photosynthetic metabolisms, and it is not clear how more subtle differences in metabolism and growth characteristics might affect the soil carbon response to raised CO_2. However, this degree of complexity is a bad omen for any real prospect of global-scale modeling of direct CO_2 effects on soils. In natural plant communities, as well as cropland mosaics, the direction of change in soil carbon might vary on a scale of a few centimeters depending on the individual plant species growing above.

Despite certain attempts to model CO_2 effects on the present-day and future biosphere (Esser, 1984, 1987), considerable awareness exists that, at the present state of knowledge, one can quantify the effects of the present anthropogenic phase of CO_2 increase on broadscale ecosystem processes (Mooney and Koch, 1994; Amthor and Koch, 1996). These problems apply to plants and, even more severely, to the soils underneath them, which are only indirectly affected by the CO_2 rise. Soils are, of course, heterogeneous; responses of the plants to CO_2 changes are very poorly understood and soil carbon may take many decades and centuries to equilibrate (see next section). These complexities of spatial and temporal patterning in soil and vegetation responses may well mean that the soil carbon response to direct CO_2 effects is unquantifiable for the foreseeable future at least.

IMPACT OF SCALE ON CARBON ESTIMATIONS IN ECOSYSTEMS

Serious uncertainties in carbon databases are caused by scale-dependent errors in the estimations of land-cover and land-use proportions. Generally, rather poor understanding of the impact of spatial resolution ecosystem heterogeneity results in unacceptable assumptions in estimations of vegetation, soil, and peat carbon. Until now little attention was given to scaling issues related to carbon estimations and modeling and effects of errors resultant from extrapolation in databases.

No consensus or general rules exist on the optimum size of experimental units, sampling design, and baseline sampling of soil carbon. It is not clear if soil samples should be taken completely at random or randomly within each soil strata (Izauralde et al., 1998).

Like many other spatially distributed variables, soil carbon content has at least three-dimensional distribution in the landscape, although usually its representation is reduced to two dimensional. Moreover, carbon content is a fuzzy variable — no sharp borders and no constant values exist in nature. In databases, fuzzy values are generally represented either by raster (grid) or vector models, which both have considerable shortcomings. In most cases, carbon mapping relies on the raster model, which partly resolves problems of fuzziness but does not satisfy scaling problems.

Extrapolation results in narrowing fuzzy borders and their replacement by sharp borders but, again, there is still no clear understanding of how values' fuzziness behaves with a changing scale.

Scale-dependent errors in the estimation of land-cover characteristics have been fairly well explored in the remote sensing literature (Moody and Woodcock, 1994; Woodcock and Strahler, 1987) and in the landscape ecology (Turner and Gardner, 1991). Remote sensing is a very useful complement to sampling and modeling soil carbon. Levine et al. (1993) proposed using the NDVI to infer plant C inputs to soil from NDVI(Normalized Deviation Vegetation Index)-derived greenness values. Combinations of data from different remote sensors may be used to detect such biophysical characteristics as biomass, soil moisture, and surface temperature and their indices, together with principal component analyses and partial regression analyses, to help predict soil carbon and monitor its dynamics. Woodcock and Strahler (1987) assessed the relationship between resolution and local variance for a variety of cover types. The local variance was found to be a function of the relationship between the size of scene objects and the sampling resolution. Turner et al. (1989) investigated the effect of changing map scales on apparent landscape pattern and attempted to relate some measures of landscape diversity, dominance, and contagion to the observed changes in spatial pattern.

Moody and Woodcock (1994) carried out experiments with degraded Landsat TM imagery for MODIS data simulation. They found that coniferous forest, which dominated the Plumas National Park landscape, had considerably larger proportional errors than other classes and was increasingly overestimated with progressive aggregation. The conifer class rapidly grows in proportion because it is frequently the most common class, against which smaler, more isolated patches are compared through the aggregation procedure. Lioubimtseva (1998) carried out experiments with high-resolution satellite data from MK-4 instrument onboard Russian satellite Resource F, progressively degraded to 25-, 100-, 400-, 600-, and 1200-m resolution. On the boreal forest-dominated test scene the proportion of coniferous forest increased from 65 to 88% at 1-km resolution due to the proportional error and mixed pixel effect. In the forest steppe test site, the proportion of cover classified as grassland considerably increased after a threshold of 500 m; this category included the major part of misclassified agricultural lands. Land-cover classes at 1-km or coarser resolution might possibly be better defined explicitly as mixes of component land covers, and that it may be possible to calibrate these kinds of component mixes for land-cover classes for various ecoregions worldwide.

Such over- and underestimations of biome or land-use categories due to upscaling should always be considered by modelers working with regional or global soil and vegetation databases. In order to reduce scaling errors, it is necessary to develop regression methods of calculating the original proportions estimated from coarse-resolution datasets.

HOW RAPIDLY WILL SOIL CARBON INCREASE WITH SUCCESSION TO NATURAL ECOSYSTEM COVER?

There is a tendency to assume that, as any area of disturbed vegetation is left alone, with time the soil carbon will build up. So if one stops cultivating an area, soil carbon will accumulate as it goes to natural vegetation. Some general understanding of these processes has already been attained but much further consideration will be necessary before realistic understanding can be reached. Nevertheless, various complexities are evident.

It is generally accepted that, with succession from cultivated to natural climax vegetation in humid climates, soil organic carbon will build up. This generally seems to hold true in the temperate zones (Johnson, 1992), but is not always the case in the tropics. For example, a study in Hawaii found that going from sugar cane fields to secondary tropical forest resulted in a net decrease in soil carbon (Bashkin and Binkley, 1998). This seems reasonable if one compares natural prairies with other ecosystems. Such a result shows once again that the direction in which soil carbon storage moves depends on where one is and precise details of the agricultural or natural system. One must consider the details before one can understand generalities.

In order to gain further understanding of how soil carbon may vary in response to future climate and land-use change, one may need to zoom out to look at a very different scale over broad spans of time rather than spatially. This perspective comes from the Quaternary sciences, which study changes over thousands and tens of thousands of years, rather than the years, decades, and centuries in which soil scientists and biogeochemists most commonly think.

Well-dated studies indicate that a natural disequilibrium in soil carbon can last thousands of years following climatic changes or other more localized disturbances in the environment (Adams and Woodward, 1992; Schlesinger, 1990; Lioubimtseva et al., 1998). In addition to examples of increasing carbon storage cited by Schlesinger (1990), another more recent example found by Schwartz (1991) shows how the "imprint of the past" can persist in a soil's carbon reservoir for thousands of years. Schwartz found that central African savanna soils still contain some carbon at depths bearing the isotopic imprint of forest vegetation, and that, the deeper into the soil profile one goes, the more forest carbon one finds. In a large total column depth (many tropical soils are very deep), the overall amount of "old" carbon from the previous ecosystem could be quite large. No one, however, has yet attempted to assess this factor systematically on a global scale.

From dating this carbon it seems that these areas were covered by forest during the early Holocene period and that this relatively small, deep soil reservoir still persists thousands of years after the forest has retreated. Ideally, one should consider the possibility that the soil carbon density observed in natural sites nowadays might differ greatly from levels 8000 or 5000 years ago, when the soil carbon might not have had as much chance to equilibrate with the vegetation conditions (Klinger, 1996; Adams and Faure, 1998). A similar situation affecting the patterns in soil carbon storage could exist in a future scenario of climate change.

CONCLUSIONS

This brief review has attempted to point out some of the difficulties that remain in assessing and forecasting behavior of soil carbon reservoirs. It is often important to zoom in to the small spatial scale for further understanding of soil carbon processes and to check generally accepted figures for reservoirs and responses. Many broad extrapolations published in the literature do not subdivide the environment enough, or in a clearly defined way. Extensive soil databases do not always contain data reliable for the purposes one might want to use them for because of the precise ways in which the samples were taken. Equations, which predict the rate of carbon buildup in a soil over time, or the CO_2 fertlization effect on total soil carbon, are unlikely to be reliable under anything more than the most spatially restricted and controlled conditions. Databases of peat carbon storage may well be unreliable if they make incorrect assumptions about small-scale spatial patterns within the peat bogs. Thus, simply standing back and looking at global soil and peat carbon on a regional and global scale — as is necessary for global modeling — can lead to very unreliable conclusions if one does not also carefully allow for local scale complications.

Paradoxically, standing back still further can sometimes give a better perspective for current understanding of soil carbon storage, including the extra dimension of long time scales, where ecosystems have continually retreated and returned in the face of large global climate changes. It is continually necessary to look across a range of spatial and temporal scales if one is to attain a realistic understanding of present and future patterns and processes.

Sometimes, all one gains from this added perspective of different spatial and temporal scales is a pessimistic outlook. One may have to acknowledge that certain aspects of the soil carbon cycle cannot realistically be modeled in the foreseeable future because small-scale heterogeneity in reservoirs and responses is too great. This includes the direct CO_2 effect on soil carbon, which can be affected by an array of factors, including the features of individual plants rooted into the soil. A position of well-founded pessimism is much better than poorly founded optimism; in science, it is always better to know with good foundation that one may be wrong, than to believe unrealistically that one knows the answer.

Searching for simplicity and numerical rigor in biogeochemistry is desirable and necessary for the full progress of the field. However, it is important not to let the desire to find broad, consistent spatial patterns deteriorate into wishful thinking. For example, a published numeric database of soil is often regarded as infallible with "no questions asked" about exactly how the data were gathered. CO_2 fertilization parameters are confidently extrapolated from tiny amounts of shaky data, without any real attempt to acknowledge the huge bounds of the uncertainties. Published peatland estimates are taken and used without an analysis of the uncertainties and contradictions within this literature. In the current culture of global biogeochemistry, more anecdotal evidence and opinion from the community of ecologists, foresters, and soil scientists is regarded as suspect and rather unworthy; yet, hopefully, this chapter has shown that it is vital to the subject.

It is necessary to accept that global biogeochemistry is not currently in the position to be a hard science like physics and chemistry because ecosystem patterns and processes are too complex and still too poorly understood. This is not a reason to give up; real progress in understanding the world's ecosystems has been and will continue to be made. However, only when the importance of continually adding a touch of natural history is acknowledged will biogeochemical modeling be able to make the greatest progress.

ACKNOWLEDGMENTS

We are grateful to Professor Andrew Goudie (Oxford School of Geography) and two anonymous reviewers for their helpful remarks.

REFERENCES

Adams, J.M. and Woodward, F.I., The past as a key to the future: the use of palaeoenvironmental understanding to predict the effects of man on the biosphere, *Adv. Ecol. Res.*, 22, 257–314, 1992.

Adams, J.M. and Faure, H., A new estimate of changing carbon storage on land since the last glacial maximum, based on global land ecosystem reconstruction, *Global Planetary Change*, 16, 3–24, 1998.

Bashkin, M.A. and D. Binkley, Changes in soil carbon following afforestation in Hawaii, *Ecology*, 79:828–833, 1998.

Bazzaz, F.A., The response of natural ecosystems to the rising global CO_2 levels, *Ann. Rev. Ecol. Syst.*, 21:167–196, 1990.

Bohn, H.L., Organic carbon in world soils, *Soil Science J. America*, 46:1118–1119, 1982.

Bohn, H.L. et al., *Soil Chemistry* (2nd ed.), John Wiley & Sons, New York, 1985, 341.

Esser, G., The significance of biospheric carbon pools and fluxes for atmospheric CO_2: a proposed model structure, *Prog. Biometeorol.*, 3, 253–294, 1984.

Esser, G., Sensitivity of global carbon pools and fluxes to human and potential climatic impacts, *Tellus*, 39B, 245–260, 1987.

Franzen, L., Are wetlands the key to the ice-age cycle enigma? *Ambio*, 23, 300–308, 1994.

Gorham, E., Northern peatlands: role in the carbon cycle and probable responses to climate warming, *Ecological Appl.*, 1:182–195, 1991.

Izauralde, R.C. et al., Carbon storage in eroded soils after 5 years of reclamation techniques, in Lal, R. et al. (eds.), *Soil Processes and Carbon Cycle*, Advances in Soil Science, CRC Press, chap. 26, 369–385, 1998.

Johnson, H.B. and Mayeux, H.S., Viewpoint: a view on species additions and deletions and the balance of nature, *J. Range Manage.*, 45:322–333, 1992.

Klinger, L.F., Potential role of peatland dynamics in Ice-Age initiation, *Quaternary Res.*, 45, 89–92, 1996.

Koerner, C. and Arnone, J.A., Responses to elevated carbon dioxide in artificial tropical ecosystems, *Science*, 257, 1672–1674, 1992.

Lottes, A.L. and Ziegler, A.M., World peat occurrence and the seasonality of climate and vegetation, *Palaeogeogr., Palaeoclim., Palaeoecol.*, 106, 23–37, 1994.

Levine, E. et al., Forest ecosystem dynamics: linking forest succession, soil process, and radiation models, *Ecol. Modeling*, 65:199–219, 1993.

Lioubimtseva, E. and Adams, J., Carbon content of desert and semidesert soils in central asia, in Kimble, J. Lal, R., and Follett, R. (eds.), *Agricultural Practices and Policies for Carbon Sequestration in Soil*, Lewis Publishers, Boca Raton, FL, 409–422, 2002.

Lioubimtseva, E. et al., Impacts of climatic change on carbon storage in the Sahara-Gobi Desert Belt since the late glacial maximum, *Global Planetary Change*, 16–17, 95–105, 1998.

Lioubimtseva, E., Interpretation and mapping landscape pattern in forest and forest-steppe zones of Russia using RESURS-F/MK-4 and RESURS-01/MSU-SK satellite imagery, *Proc. 24th Int. Conf. Remote Sensing Soc.*, Chatham Maritime, Sept. 1998.

Markov, V.D. et al., *World Peat Resources*, Nedra, Moscow, 1988, 383 pp. (in Russian).

McConnaughay, K.D.M., Bernston, K.D.M., and Bazzaz, F.A., Plant responses to carbon dioxide, *Nature*, 361, 24, 1993.

Moody, A. and C.E. Woodcock, Scale-dependent errors in the estimation of land-cover proportions: implications for global land-cover datasets, *Photogrammetric Eng. Remote Sensing*, 60, 585–594, 1994.

Mooney, H.A. and Koch, G.W., Impact of rising CO_2 concentrations on the terrestrial biosphere, *Ambio*, 23 74–76, 1994.

Peng, C., Reconstruction du stock de carbone terrestre du passe a partir de donnees polliniques et de modelels biospheriques depuis le dernier maximum glaciaire, Abstract, PhD thesis, Université d'Aix-Marseille II, 1994.

Peng, C.H., Guiot, J., and van Campo, E., Reconstruction of past terrestrial carbon storage in the Northern Hemisphere from the Osnabroeck biosphere model and palaeodata, in press, *Clim. Res.*

Phillips, O.L. et al., Changes in the carbon balance of tropical forests: evidence from long-term plots, *Science*, 282:439–442, 1994.

Post, W.M. et al., Soil carbon pools and world life zones, *Nature*, 29:156–159, 1992.

Radjagukuk, B., Prospects of peat exploitation in Indonesia, 99–112, in *Tropical Peat Resources — Prospects and Potential*, International Peat Society, *Proc. Symp.*, Kingston, Jamaica, 1985. IPS, Helsinki, 1985.

Rogers, H.H., Runion, G.B., and Krupa, S.V., Plant responses to atmospheric CO_2 enrichment with emphasis on roots and the rhizosphere, *Environ. Pollut.*, 83, 155–189, 1994.

Schimel, D.S., Terrestrial ecosystems and the carbon cycle, *Global Change Biol.*, 1:77–91, 1995.

Schlesinger, W.H., Carbon storage in the caliche of arid soils: a case study from Arizona, *Soil Sci.*, 133, 247–255, 1982.

Schlesinger, W.H., Evidence from chronosequence studies for a low carbon storage potential of soils, *Nature*, 348, 232–234, 1990.

Schwartz, D., Intéret de la mesure du delta-13C des sols en milieu naturel équatorial pour la connaissance des aspects pédologiques et écologiques des relations savaneforet, *Cah. Orstom, sér P'dol.*, 16:327–341, 1991.

Shiel, D. and Phillips, O., Evaluating turnover in tropical forests, *Science*, 268:894, 1995.

Sjors, H., The zonation of northern peatlands and their importance for the carbon balance of the atmosphere, *Int. J. of Ecol. and Environ. Science*, 7:11–14, 1981.

STATeSoilGeOgraphic data base (STATSGO) 1:250.000 scale, National Soil Survey Center, USA (http://www.or.nrcs.usda.gov/soil/statgo-i.html), 1997.

Tarnoaci, C., *Peat Resources of Canada*, National Research Council of Canada of Energy, Peat Program, Halifax, Nova Scotia, 17 pp., 1984.

Torbert, H.A. et al., Crop residue decomposition as affected by growth under elevated atmospheric CO_2, *Soil Science*, 163:412–419, 1998.

Turner, S.J. et al., Pattern and scale statistics for landscape ecology, in M.G. Turner and R.H. Gardner (eds.), *Quantitative Methods in Landscape Ecology*, Springer-Verlag, New York, 19–49, 1991.

Turner, M.G. and R.G. Gardner, Quantitative methods in landscape ecology, an introduction, in Turner, M.G. and Gardner, R.G., eds., *Quantitative Methods in Landscape Ecology*, Springer-Verlag, 3–14, chap. 1, 1991.

Van Campo, E., Guiot, J., and Peng C., A databased reappraisal of the terrestrial carbon budget at the last glacial maximum, *Global Planetary Change*, 8, 189–201, 1993.

VEMAP members, Vegetation/ecosystem modeling and analysis project: comparing biogeography and biogeochemistry models in a continental-scale study of terrestrial ecosystem responses to climate change and CO_2 doubling, *Global Biogeochem. Cycles*, 4, 407–437, 1995.

Walter, H. and Box, E.O., Temperate deserts and semi-deserts, in I. West and E. Neil (eds.), *Ecosystems of the World*, Vol. 5, Elsevier, Amsterdam, 1–159, 1983.

Woodcock, C.E. and A.H. Strahler, The factor of scale in remote sensing, *Remote Sens. Environ.*, 21:311–332, 1987.

Zinke, P.J. et al., *Worldwide Organic Soil Carbon and Nitrogen Data*, (CDIAC NDP-0181), available online at [http://www.daac.ornl.gov?] from ORNL Distributed Active Archive Center, Oak Ridge National Laboratory, Oak Ridge, TN, 1986.

CHAPTER 45

Organic Carbon Stocks in Soils of Bulgaria

Ekatarina Filcheva, S. V. Rousseva, A. Kulikov, S. Nedyalkov, and T. Z. Chernogorova

CONTENTS

INTRODUCTION

Soils play an important role in the earth's carbon cycle because they are the largest carbon pool other than the oceans. This is important not only to the global carbon balance, but also to the present and future potential of the soil to produce sufficient food and fiber to feed and clothe the world and to meet the demand for wood for fuel, building, and other domestic uses. Human conversion of natural ecosystems to agricultural use has a strong effect on the fate of this stored carbon. Agricultural soils show large changes in carbon concentrations, carbon stock, and associated properties as bulk density and soil structure (Schlesinger, 1986).

Kobak (1988) estimated the global reservoir of terrestrial stored carbon at 5075 Pg yr^{-1}, while Greenland (1995) reported estimates of soil carbon from 700 to 3000 10^9 t. Globally, the upper meter of mineral soils contains 1300 to 1600 Gt carbon (Eswaran et al., 1993; Neill et al., 1997) — twice the carbon stored in terrestrial biomass (Deevy, 1979; Post et al., 1982; Schlesinger, 1986). Zdruli et al. (1997) estimated the organic carbon stocks for the Albanian territory of 28.6 km^2 at 253 Mt.

Several papers on the carbon reserve in soils have been published recently (Bohn, 1976, 1982; Kimble et al., 1990; Tarnocai and Ballarr, 1994; Zhong et al., 1997; Reich et al., 1997; Zdruli et al., 1997; Kern et al., 1997; Tarnocai, 1997; Lioubimtseva, 1997; Kolchugina et al., 1995; Eswaran et al., 1993). Some of these papers concluded that additional data were needed to improve the global estimates of carbon reserves.

The organic carbon reserve and energy reserve in Bulgaria also have been estimated (Boyadgiev et al., 1994; Raichev and Filcheva, 1989; Simeonova et al., 2001). Knowledge on carbon stocks is desirable for environmental protection and agricultural production. This chapter reports estimates of the organic carbon stocks of the Bulgarian soils, based on available data of organic carbon and bulk density measurements along the profile depths of various soil types.

MATERIALS AND METHODS

The soil cover of Bulgaria is very complicated and not highly correlated to modern climate conditions. Nine soil orders (Soil Taxonomy, Table 45.1) can be distinguished within Bulgaria, an area of 11.1 10^6 ha (Boyadgiev, 1994), with agricultural land comprising 6.85 10^6 ha, woodland 3.85 10^6 ha, and urban areas 0.4 10^6 ha. Cultivated lands cover 4.6 10^6 ha, 83.1% of which are occupied by cropland, 10.5% by grassland, and 13.7% by pastures.

Table 45.1 Soil Cover of Bulgaria

Soil Order	Area (million ha)	Number of Samples
Histosols	0.011	7
Vertisols	0.739	19
Aridisols	0.051	3
Ultisols	0.026	2
Mollisols	3.407	19
Alfisols	2.459	151
Inceptisols	3.838	6
Entisols	0.468	65
Spodosols	0.044	2
Total	11.044	274

Soil organic carbon stocks were estimated for layers of 0 to 25, 0 to 50, and 0 to 100 cm, considering the area occupied by each soil type and the respective average soil organic carbon density, calculated from measured values of soil organic carbon and soil bulk density. Soil organic carbon was determined by Turin's method (Kononova, 1966) modified in the laboratory (dichromate digestion at 125°C, 45 min, in presence of Ag_2SO_4 and $FeSO_4$ titration). Soil bulk density was determined by soil cores of volume 200 cm^3 at field capacity (Revut and Rode, 1969).

Relevant information was organized in a database including data for both virgin and cultivated soils, where measurements of soil organic carbon and soil bulk density were available along the soil profile depths. Soils were classified according to their organic carbon density (Table 45.2).

RESULTS

Tables 45.3, 45.4, and 45.5 show the average carbon density and total organic carbon calculated for each of the nine soil orders. The greatest carbon density occurs in Histosols and Spodosols for layers 0 to 25 and 0 to 50 cm. Histosols are characterized by greatest carbon density of the whole soil (0 to 100 cm). The average carbon densities of the remaining soils are much lower. Aridisols have the lowest average carbon densities for all depths studied.

Average organic carbon density ranges from 3.1 kg m^{-2} (Aridisols) to 18.6 kg m^{-2} (Spodosols) for the surface soil, from 4.6 kg m^{-2} (Aridisols) to 23.7 kg m^{-2} (Spodosols) for the 0- to 50-cm layer, and from 6.7 kg m^{-2} (Aridisols) to 33.0 kg m^{-2} (Histosols) for the entire soil profile. The average organic carbon density of the most widely distributed soils, Inceptisols, Mollisols, and Alfisols, are 12.1, 9.6, and 6.3 kg m^{-2}, respectively, for the 0- to 50-cm layer. Discussed values are similar to the respective data reported by Zdruli et al. (1997) for the soil orders in Albania. They estimated the organic carbon density for Inceptisols from 5 to 11 kg m^{-2}, for Mollisols from 8 to 27 kg m^{-2}, and for Alfisols from 6 to 8 kg m^{-2}.

Table 45.2   Classes of Organic Carbon Density

Class	Organic Carbon Density (kg m^{-2})
I	<0.5
II	0.5–1
III	1–1.5
IV	1.5–2
V	2–2.5
VI	2.5–3
VII	3–3.5
VIII	3.5–4
IX	4–4.5
X	4.5–5
XI	5–6
XII	6–7
XIII	7–8
XIV	8–9
XV	9–10
XVI	10–15
XVII	15–20
XVIII	20–30
XIX	30–40
XX	40–50
XXI	50–75

Table 45.3   Organic Carbon Stocks in the 0- to 25-cm Layer

Soil Order	Average Organic Carbon Density (kg m^{-2})	Standard Deviation	Classes of Organic Carbon Density	Total Organic Carbon (Mt)
Histosols	13.3	3.6	XI–XVII	1.5
Vertisols	5.8	2.3	V–XVI	43.0
Aridisols	3.1	0.5	V–VII	1.6
Ultisols	3.7	0.0	VIII	1.0
Mollisols	5.6	1.6	III–XII	189.6
Alfisols	3.8	1.7	III–XVI	93.3
Inceptisols	8.3	3.8	VII–XVI	320.3
Entisols	3.5	1.9	II–XVI	16.6
Spodosols	18.6	0.5	XVII	8.2
Total				675.1

Table 45.4   Organic Carbon Stocks in the 0- to 50-cm Layer

Soil Order	Average Organic Carbon Density (kg m^{-2})	Standard Deviation	Classes of Organic Carbon Density	Total Organic Carbon (Mt)
Histosols	22.3	8.8	XI–XIX	2.6
Vertisols	10.3	3.9	VIII–XVIII	76.1
Aridisols	4.6	0.6	VIII–XI	2.3
Ultisols	4.8	0.4	IX–X	1.2
Mollisols	9.6	3.1	VI–XVII	328.3
Alfisols	6.3	2.5	V–XVII	154.3
Inceptisols	12.1	5.4	XII–XVIII	464.4
Entisols	6.0	3.1	III–XVI	28.1
Spodosols	23.7	0.4	XVIII	10.4
Total				1067.7

Table 45.5 Organic Carbon Stocks in the 0- to 100-cm Layer

Soil Order	Average Organic Carbon Density (kg m^{-2})	Standard Deviation	Classes of Organic Carbon Density	Total Organic Carbon (Mt)
Histosols	33.0	21.5	XI–XXI	3.8
Vertisols	18.1	8.0	XII–XX	134.1
Aridisols	6.7	1.5	X–XII	3.4
Ultisols	6.8	0.3	XII	1.8
Mollisols	14.3	4.8	VII–XVIII	487.5
Alfisols	9.7	3.7	VII–XVIII	238.0
Inceptisols	15.3	8.0	XIV–XVIII	586.3
Entisols	9.5	5.3	III–XVIII	44.3
Spodosols	23.7	0.4	XVIII	10.4
Total				1509.6

The standard deviations for the average organic carbon densities show significant variation among the soil profiles. The reason for this variability can be attributed to the numerous combinations of climatic conditions, native vegetation (about 10% of the soil profiles included in this study represent virgin soils), crops grown, soil characteristics, land use, and land management. Comparing virgin and arable lands, Boyadgiev et al. (1994) estimated organic carbon reserves coefficients for the arable land ranging from 1.1 to 2.3, depending on the quality and quantity of soil organic matter.

The total organic carbon stock of the soils in Bulgaria is estimated at 1.5 Gt, which is about 1/1000 of the world reserve estimate of 1576 Gt (Eswaran et al., 1993). The distribution of organic carbon stocks among soil orders for the studied soil layers is shown in Figure 45.1. Inceptisols, which occupy about 34.7% of the territory, contribute about 38.8% of the total carbon stocks. Mollisols, Alfisols, and Vertisols, which cover 30.8, 22.2, and 6.7% of the territory, respectively, contribute 32.3, 15.8, and 8.9%. The other soils do not contribute greatly to the carbon stock. Spodosols occupy about 0.4% of the total area, but contribute 2.9% of the national carbon stock; part of the reason for this relatively high soil organic carbon is related to their geomorphic association with Histosols.

CONCLUSION

Nine soil orders (Histosols, Vertisols, Aridisols, Ultisols, Mollisols, Alfisols, Inceptisols, Entisols, and Spodosols) occur in the territory of Bulgaria. Average organic carbon density of the entire soil profile ranges from 6.7 to 33.0 kg m^{-2}. Inceptisols, Mollisols, Alfisols, and Vertisols contain 95.7% of the total national organic carbon stock, which is estimated at 1.5 Gt. Even though Bulgaria is a small country, the soil organic carbon stock of different soils varies considerably. This high variability in a small part of the world points to the need for more detailed national assessments to obtain a more reliable estimate of soil organic carbon stocks.

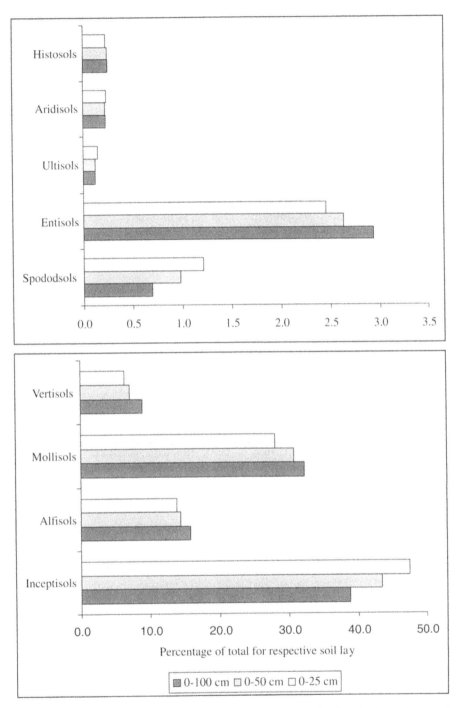

Figure 45.1 Percentage distribution of organic carbon stocks among soil orders for the soil layers studied.

REFERENCES

Bohn, H.L., Estimate of organic carbon in world soils, *Soil Sci. Soc. Am. J.*, 40:468–470, 1976.

Bohn, H.L., Estimate of organic carbon in world soils II. *Soil Sci. Soc. Am. J.*, 46:1118–1119, 1982.

Boyadgiev, T., E. Filcheva, and L. Petrova, The organic carbon reserve of Bulgarian soils, in Lal, R. et al., eds., *Soil Processes and Greenhouse Effect*, 19–23, 1994.

Boyadgiev, T., Soil map of Bulgaria according to the Soil Taxonomy — explanatory notes, *Soil Sci. Agrochem. Ecol.*, 29 (No 4-6):43–51, 1994.

Deevy, E.S., Mineral cycles, *Sci. Am.*, 223, 148–158, 1979.

Eswaran, H., E. Van Denberg, and P. Reich, Organic carbon in soils of the world, *Soil Sci. Soc. Am. J.*, 57, 192–194, 1993.

Greenland, D.G., Land use and soil carbon in different agroecological zones, in Lal, R. et al., eds., *Soil Management and Greenhouse Effect*, Advances in Soil Science, 9–24, 1995.

Kern, J.S., D.P. Turner, and R.F. Dotson, Spatial patterns of soil organic carbon pool size in the north-western United States, in Lal, R. et al., eds., *Soil Processes and the Carbon Cycle*, Advances in Soil Science, 1, 29–43, 1997.

Kimble, J.M, H. Eswaran, and T. Cook, Organic carbon on a volume basis in tropical and temperate soils, *Proc. 14 Int. Congr. Soil Sci.*, 5, 248–253, 1990.

Kobak, K.I., *Biotical Components of Carbon Cycles*, Hydrometeroizdat, Leningrad, 1988.

Kolchugina, T.P. et al., Carbon pools fluxes and sequestration potential in soils of the former Soviet Union, in Lal, R. et al., eds., *Soil Management and Greenhouse Effect*, Advances in Soil Science, 25–40, 1995.

Kononova, M.M., *Soil Organic Matter*, 2nd ed., Pergammon Press, Inc., New York, 1966.

Lioubimtseva, E., Impacts of climatic change on carbon storage variations in African and Asian deserts, in Lal, R. et al., eds., *Soil Processes and the Carbon Cycle*, Advances in Soil Science, 1, 561–576, 1997.

Neill, C. et al., Stocks and dynamics of soil carbon following deforestation for pasture in Rondonia, in Lal, R. et al., eds., *Soil Processes and the Carbon Cycle*, Advances in Soil Science, 1, 9–28, 1997.

Post, W.M. et al., Soil carbon pools and world life zones, *Nature*, 298, 156–159, 1982.

Raichev, T. and E. Filcheva, Energetic stocks in some soils in Bulgaria, *Soil Sci. Agrochem. Plant Protection*, 24 (No 4):9–13, 1989.

Reich, P., M. Soekardi, and H. Eswaran, Carbon stocks in soils of Indonesia, in Lal, R., J. Kimble, and R. Follett, eds., *Soil Properties and Their Management for Carbon Sequestration*, USDA, National Resources Conservation Services, National Soil Survey Center, Lincoln, NE, 121–127, 1997.

Revut, I.V. and A.A. Rode, eds., *Methodological Handbook for Soil Structure Studies*, Kolos Publ., Leningrad, 1969.

Schlesinger, W.H., Changes in soil carbon storage and associated properties with disturbance and recovery, in Trabalka, J.R. and D.E. Reichle, eds., *The Changing Carbon Cycle*, Springer, NY, 1986, 194–220.

Simeonova, N. et al., Soil organic matter of Bulgarian soils, *Soil Res. Bulgaria*, 1. (in press), 2001.

Tarnocai, C., The amount of organic carbon in various soil orders and ecological provinces in Canada, in Lal, R. et al., eds., *Soil Processes and the Carbon Cycle*, Advances in Soil Science, 1, 81–92, 1997.

Tarnocai, C. and M. Ballarr, Organic carbon in Canadian soils, in Lal, R., J. Kimble, and E. Levine, eds., *Soil Processes and the Greenhouse Effect*, USDA, SCS, National Soil Survey Center, Lincoln, NE, 31–45, 1994.

Zdruli, P. et al., Organic carbon stock in soils of Albania, in Lal, R., J. Kimble, and R. Follett, eds., *Soil Properties and Their Management for Carbon Sequestration*, USDA, National Resources Conservation Services, National Soil Survey Center, Lincoln, NE, 129–141, 1997.

Zhong, L., Z. Qiugue, and J. Xiao, Organic carbon contents and storage in soils of northeastern China, in Lal, R., J. Kimble, and R. Follett, eds., *Soil Properties and Their Management for Carbon Sequestration*, USDA, National Resources Conservation Services, National Soil Survey Center, Lincoln, NE, 79–87, 1997.

PART IX

Summary

CHAPTER 46

Assessment of Soil Organic Matter Layers Deposited at Open Pit Coal Mines

L. Petrova, M. Banov, and V. Somlev

CONTENTS

INTRODUCTION

In 1952 the power complex Maritsa–Iztok was established on the Southeast Trace lowland. The complex includes three open-pit lignite mines, a power station, which uses lignite coal and produces about 30% of the country's electric power, and the only briquette factory in Bulgaria. Stripping the surface soil and geological layers precedes the process of open-pit mining.

The removed layers are deposited so that they can be used in land reclamation after the exhaustion of the coalfields. Reclamation of the disturbed lands at Maritsa–Iztok region began in 1965. With the efforts of many Bulgarian soil scientists an experimental station for studying soil reclamation was organized. Data from field and pot trials clarified the problems and helped to suggest the most suitable methods of land reclamation (Banov, 1989; Banov et al., 1998; Christov and Banov, 1996; Christov et al., 1998; Garbuchev et al., 1973; Garbuchev and Treikjashky, 1974; Garbuchev et al., 1975). Deposited soil and geological materials are the most suitable to be used in biological land reclamation (Banov, 1989).

The purpose of the study described in this chapter was to assess the physicochemical status of the organic matter layers stockpiled at the coal mine in the Maritsa-Iztok district, and to determine the possibility of using these materials in land reclamation.

MATERIALS AND METHODS

The study included two coal-mining dumps containing soil and geological materials: the first at the village of Ovchartsy and the second at the village of Kovachevo, both located in the Maritsa–Iztok region of Southern Bulgaria. The dump at Ovchartsy was 7 to 8 years old and 500 cm in depth. The one at Kovachevo was 4 to 5 years old with a depth of 400 cm. Soil samples were collected at 100-cm depth increments at both dumps and analyzed to determine their texture (Guidelines for Soil Profile Description), particle size distribution (Kachinsky, 1958), bulk density (Methods of Soil Analysis, 1965), total and available nitrogen (N), phosphorus (P), and potassium (K) (Methods of Soil Analysis, 1965), amount and quality of soil organic matter with the method of Kononova and Belchicova (Kononova, 1966), and soil organic carbon (SOC) reserve, using a formula suggested by Orlov (1985). The most abundant soils in the area were Vertisols and these were used for comparison to the stockpiled material. Apart from the Vertisols, Chromic Luvisols, Alluvial soils and small areas of other soils are found in the Maritsa–Iztok district; the geological materials are mainly green or yellow-green calcareous clays. At the mine spoils, there are some black acidic clays, but those were not found in the Ovchartsy and Kovachevo dumps materials.

RESULTS AND DISCUSSION

In general, the dumps do not possess proper soil morphology because they are man-made heaps, with mixtures of soil and geological materials. Nevertheless, the Ovchartsy and Kovachevo dumps have well distinguished layers, suggesting a favorable base for soil sampling and assessment of their properties. The texture of the dumps' layers is very much related to that of the deposited soil and geological materials; in fact, almost all properties of those formations are, at least in the first 20 years after dumping. The Vertisols and the Chromic Luvisols are heavy textured; the Alluvial soils sometimes comprise lighter layers and the mixed geological materials are generally clays. Particle size distribution analysis showed that the texture of the layers at Ovchartsy dump was clay loam with the exception of that at 200 to 300 cm, which was loam. The texture of the dump of Kovachevo was loam in all but the 100- to 200-cm layer, where it was clay loam (Table 46.1). The texture of the standard Vertisol is clay loam to 60 cm, clay in the B-horizon, and sandy loam in the deeper horizon (Table 46.1).

The bulk density of the deposited materials varies from 1.79 to 1.96 g/cm^2 at Ovchartsy dump and 1.90 to 2.06 g/cm^2 at Kovachevo, slightly higher than that of the Vertisols. The dumps are too young to explain the higher bulk density of their layers compared with other processes except to note mixing the soil with the clayey geological materials and increased internal pressure due to the increased depth of the dumps as compared to that of the Vertisol. In this aspect the compaction of the material from the heavy equipment at storing must also be considered.

Total SOC at the Ovchartsy dump varied from 0.74 to 1.12% in the different layers. The Kovachevo dump had a higher total SOC content in the upper 300 cm; in the 300- to 400-cm layer, this amount was 0.65%. The Vertisols are one of the most fertile soils in Bulgaria. In the surface horizons (10 to 15 cm) of their virgin variants, the total SOC content is about 2.30 to 3.50%, and in the arable variants it is about 1.20%. The SOC reserve of these soils to the depth of 100 cm usually is 174 to 232 t/ha, but the values vary depending on the soil thickness. As a standard of this study, a Vertisol with average SOC reserve was chosen, i.e., 250.65 t/ha to the depth of 100 cm (Table 46.2). The data showed a good reserve of organic carbon in every layer at both of the dumps, but in the surface layer at Ovchartsy and the two deepest layers of Ovchartsy and Kovachevo.

Extractable SOC, the amount of humic and fulvic acids in the dumps, is significantly lower than those in the Vertisols (Table 46.3). The unextractable SOC amount is higher, about 80%, in the dumped soil and geological materials than in the Vertisols, where it is about 50 to 69% from

Table 46.1   Mechanical Composition of the Vertisols and the Material from the Two Dumps

| Horizon Depth in cm | Size of Particles (mm) | | | | | | | Total <0.01 |
	Total 1>1	1–0.25	0.25–0.05	0.05–0.01	0.01–0.005	0.005–0.001	<0.001	
Vertisols								
A₁ 0–13	4.8	16.7	13.3	13.4	3.8	7.7	38.2	49.7
A₂ 13–36	2.1	18.1	16.4	15.5	4.3	8.3	35.1	47.7
AB 36–56	4.1	15.7	13.3	12.2	3.5	6.6	43.0	43.1
B₁ 56–81	3.1	14.5	14.1	13.1	4.0	6.7	42.4	53.1
C₁ 81–108	0.9	12.4	11.8	8.5	1.5	4.9	25.5	33.3
Ovchartsy Dump								
0–100	0.0	8.6	12.7	15.9	8.4	5.6	43.9	57.9
100–200	0.0	13.4	14.3	18.0	7.0	9.2	33.9	50.1
200–300	0.0	12.0	26.1	17.0	6.1	10.6	24.2	40.9
300–400	0.0	18.9	12.2	17.9	6.5	5.9	33.2	45.6
400–500	0.0	12.7	13.6	18.9	5.5	7.1	38.2	50.8
Kovachevo Dump								
0–100	0.0	23.2	17.0	16.4	5.0	7.0	27.2	39.2
100–200	0.0	18.0	17.9	13.1	6.3	6.1	33.6	46.0
200–300	0.0	19.7	18.5	11.0	7.6	5.3	31.5	44.6
300–400	0.0	20.2	15.9	14.3	6.0	4.7	34.0	44.7

Table 46.2   Soil Organic Carbon Reserves (t/ha) in the Vertisols and the Material from the Two Dumps

Horizon Depth in cm	Bulk Density	C (%)	Org. C Reserve (t/ha)
Vertisols			
A₁ 0–13	1.72	1.66	37.12
A₂ 13–36	1.75	1.87	75.27
AB 36–56	1.68	1.22	40.99
B₁ 56–81	1.92	1.72	82.56
C₁ 81–100	1.80	0.43	14.71
Ovchartsy Dump			
0–100	1.96	0.74	145.04
100–200	1.91	1.05	200.55
200–300	1.95	1.12	218.40
300–400	1.84	1.10	202.40
400–500	1.79	0.86	153.94
Kovachevo Dump			
0–100	2.06	1.53	315.18
100–200	1.98	1.48	293.04
200–300	1.90	1.53	290.70
300–400	2.11	0.65	137.15

Source: Orlov (1985).

Table 46.3   Humus Composition

Horizon, Depth cm	Total Organic C %	Extractable Organic Carbon			Cha/Cfa	Carbon % of Ca-Compl. HA	Unextactable Organic C %
		Total	HA	FA			
Vertisols							
A₁ 0–13	2.18	0.92[a]	0.52	0.40	1.30	78.80	1.26
		42.20[b]	23.85	18.35			57.80
A₂ 13–36	1.10	0.45	0.29	0.16	1.81	100.00	0.65
		40.91	26.36	14.54			59.09
AB 36–56	0.82	0.25	0.19	0.06	3.17	100.00	0.57
		30.49	23.17	7.32			69.50
B₁ 56–81	0.70	0.33	0.19	0.10	2.30	100.00	0.37
		47.14	23.17	14.28			52.86
C₁ 81–108	0.66	0.26	0.22	0.04	5.50	100.00	0.40
		39.39	33.33	6.06			60.61
Ovchartsy Dump							
0–100	0.74	0.04	0.00	0.04	0.00	0.00	0.70
		5.41		5.41			94.59
100–200	1.05	0.17	0.14	0.03	4.66	100.00	0.88
		16.19	13.33	2.86			83.81
200–300	1.12	0.24	0.12	0.12	1.00	100.00	0.88
		21.43	10.71	10.71			78.57
300–400	1.10	0.21	0.14	0.07	2.00	100.00	0.89
		19.09	12.73	6.36			80.91
400–500	0.86	0.17	0.09	0.08	1.13	100.00	0.69
		19.77	10.47	9.30			80.23
Kovachevo Dump							
0–100	1.53	0.31	0.22	0.09	2.44	95.46	1.22
		20.26	14.38	5.88			79.74
100–200	1.48	0.25	0.18	0.07	2.57	83.33	1.23
		16.89	12.16	4.33			83.11
200–300	1.53	0.25	0.13	0.12	1.08	69.23	1.28
		16.34	8.50	7.84			83.66
300–400	0.65	0.04	0.02	0.02	1.00	100.00	0.61
		6.15	3.08	3.08			93.85

[a] Percent of dry matter.
[b] Percent of total organic carbon.

Source: Kononova (1966).

the total organic carbon. The absolute amounts (percent to dry matter) of the unextractable SOC in the surface horizon of the Vertisols and in Kovachevo dumps layers are almost equal, 1.22, 1.23, and 1.28% in the dump and 1.26% in the surface horizon of the Vertisol. The explanation of this high amount of unextractable organic carbon is obvious: mixing the soil and geological materials in the dumps decreased the amount of total SOC in the new substrate. In addition, this amount was decreased due to soil stockpiling; consequently, plant remains and biological activity — the main sources and agents of soil organic compound production — decreased. With the increased duration of the dumped materials' preservation, the amount of total and extractable forms of SOC decreased more and more, thus increasing the relative amount of unextractable carbon compounds.

The SOC type of the dumped materials can be determined as humic to fulvic-humic with Cha/Cfa = 1 to 2, and only in the 0- to 100-cm Ovchartsy layer as fulvic, Cha/Cfa < 0.5, or as in the case, there were only fulvic acids, but this layer is made of almost pure yellow or yellow-green clays. The type of SOC in the Vertisols, is also humic or fulvic-humic. The humic acids of the deposited materials, like those of the Vertisols are predominantly complexed by Ca in the surface horizons and 100% complexed by Ca below the surface horizons (Table 46.3). The strong similarity of the dumped layers with those of the Vertisols predetermines some very well known disadvantages: these soils are very sticky when moist and very hard when dry. The same holds true for materials dumped near Kovachevo and Ovchartsy, the Maritsa–Iztok district, and at open mining of lignite coal.

CONCLUSION

This study showed that stripped and stockpiled materials at open-pit coal mines were similar in texture, bulk density, pH, soil humus characteristics, and those of the SOC reserves to Vertisols, which are some of the most fertile soils in Bulgaria. The materials from the dumps can be used in biological reclamation of the disturbed lands at the Maritsa–Iztok district, preferably mixed with some inert additives, such as those from the power station cinder dumps. This will improve the physical properties of the deposited substrates. Fertilization, tillage, and crops grown on the reclaimed soils should be similar to those used for the Vertisols.

REFERENCES

Banov, M., Study of some soil-genesis changes in reclaimed lands without use of humus layer in the region of Maritza–Iztok, PhD Thesis, Poushkarov Institute, Sofia, 1989.

Banov, M. and V. Somlev, Contemporary problems at land reclamation in the region of Maritsa–Iztok coal basin, *Ecol. Ind.*, 1, 1-3, 9–11, 1998.

Boyadjiev, T., L. Petrova, and M. Jokova, Relationship between some soil diagnostic characteristics, *Soil Sci. Agrochem. Ecol.*, 4-6, 57–63, 1994.

Christov, B. and M. Banov, On some changes of the mineral mass of the reclaimed without humus layer lands, *Soil Sci. Agrochem. Ecol.*, 31, 3, 31–35, 1996.

Christov, B. and M. Banov, Comparative characteristic of reclaimed lands, *Soil Sci. Agrochem. Ecol.*, 33, 3, 8–14, 1998.

Garbuchev, I., P. Treikjashky, and O. Mateeva, Chemical composition of the geological materials in the lignite coal basin Maritsa–Iztok, *Soil Sci. Agrochem.*, 1, 67–76, 1973.

Garbuchev, I. and P. Treikjashky, Protection and reclamation of the disturbed soils, in *Conservation and Increasing Soil Fertility*, 462–471, Zemizdat, Sofia, Bulgaria, 1974.

Garbuchev, I. et al., *Suitability of the Substrates for Land Reclamation in Maritsa-Iztok Region*, BAS, Sofia, Bulgaria, 178, 1975.

Guidelines for Soil Profile Description (2nd ed.), FAO, 66, 1977.

Kachinsky, N.A., Mechanical and microaggregates composition of soil, methods of their determination, Moscow, *Acad. Sci.*, 192, 1958.

Kononova, M.M., *Soil Organic Matter: Its Role in Soil Formation and in Soil Fertility*, Pergamon Press. Inc., New York, 544, 1966.

Orlov, D.S., *Soil Chemistry*, Moscow University Press, Moscow, 376, 1985.

CHAPTER 47

Organic Carbon Stores in Alaska Soils

C. L. Ping, G. J. Michaelson, John M. Kimble, and L. R. Everett

CONTENTS

INTRODUCTION

The global soil organic matter (SOM) pool contains approximately three times as much carbon (C) as terrestrial vegetation (Schlesinger, 1995). The soil organic carbon (SOC) component of SOM is very large and generally has long residence time. In regions where SOM accumulation exceeds decomposition, SOC serves as a sink in the global C cycle. This is especially true in the Arctic and the Subarctic (Oechel et al., 1993). Natural and anthropogenic environmental changes, such as wildfire, forest harvest, cultivation, and hydrological shifts caused by climate change alter the dynamics of SOC. Decreasing the SOC pool leads to an increase in greenhouse gas input to the atmosphere. In the permafrost zone, climatic changes may have a major impact on active layer dynamics that can affect SOC fluxes. Alaska has one third of its landmass in the arctic with continuous permafrost and the rest in the boreal (subarctic) with discontinuous or sporadic permafrost.

Carbon dynamics in the Arctic has received more attention because of the integrated NSF-sponsored Land–Atmosphere–Ice Interaction study (Michaelson et al., 1996), with only limited attention paid to C stores in the rest of the state (Ping et al., 1997; Van Cleve et al., 1993). The vastness of the boreal forest ecosystem and its impact on C stores and C fluxes cannot be overlooked. Assessment of the impact of C flux and ecosystem changes, which may result from climate change, can only be done with a systematic estimation of C stores in all of Alaska. C stores estimated by STATSGO (Soil Survey Staff, 1996) are considerably lower than those estimated by the whole pedon approach (Michaelson et al., 1996; Ping et al., 1997). The objectives of this chapter are to: (1) summarize C-store data on a pedon basis among different ecosystems to provide a basis for a regional view of the present level of C stores in Alaska soils, and (2) summarize the patterns of C stores within soil profiles of the different ecosystems of Alaska.

MATERIALS AND METHODS

SOC Analysis

All soil organic carbon data collected in and after 1992 were described and sampled according the Soil Survey Manual (Soil Survey Division Staff, 1951, 1993) and shipped to the USDA National Soil Survey Laboratory for characterization analysis. SOC was determined by LECO combustion C analyzer for samples taken after 1989 and by the Walkley–Black chemical oxidation method (Nelson and Sommers, 1982) for samples prior to 1989. The samples were air-dried and disaggregated to pass through a 2-mm sieve. Soil organic carbon was determined after the soil sample was treated with dilute acids to remove inorganic C.

Soil C Stores

Soil organic carbon store data were obtained from published sources (Ping et al., 1997a, 1998a; Michaelson et al., 1996) or calculated from the National Soil Survey Center (NSSC) soils database for those pedons sampled prior to 1991. In many of the early samples, organic carbon contents and bulk densities of the organic horizons were not analyzed and only horizon thickness was recorded. Therefore the C stores were estimated based on the correlation between SOC and bulk density determined for soils of each region (Michaelson, G.J. et al., 1999). (Estimated bulk densities are in bold type in Appendix A.) Soil classification of pedons listed in NSSC database prior to 1997 was correlated according to the latest version of Soil Taxonomy (Soil Survey Staff, 1998). The general locations of sample sites are shown in Figure 47.1.

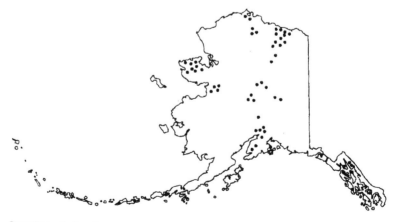

Figure 47.1 Sampling site locations.

RESULTS AND DISCUSSION

C Stores

Arctic Alaska

The organic carbon (OC) content and C stores of tundra soils vary according to genetic horizon and ecosystem; the results are summarized in Tables 47.1 through 47.5. On the coastal plain the vegetation is dominantly sedge and the dominant soils are Historthels and Histoturbels. The OC contents of soils range from 12 to 43% in the O horizons, 14 to 20% in the A horizons, and 6 to 7% in the B and Cf horizons. When partitioned between the organic and mineral horizons, 21 to 84% of the C stores are in the organic horizons and 16 to 79% in the mineral horizons. When partitioned between the active and permafrost layers, 54 to 60% of the C stores are in the active layers and 40 to 46% in the upper permafrost (Table 47.1). The difference in partitioning between the active and permafrost layers is greater than that between the organic and mineral horizons, indicating the important role of permafrost in carbon sequestration. This is the result of a physical equilibrium between the active and permafrost layers maintained by the annual freeze–thaw cycle.

In the arctic foothills, the main soil types are Aquiturbels, with organic and mineral horizon partitioning of C stores of 1 to 3; however, the partitioning between the C stores in active and permafrost layers is similar to that of the coastal plain soils. C stores in soils not affected by

Table 47.1   Carbon Storage to 1 m in Soils of Arctic Alaska Including the Seward Peninsula

Landform Great Group (n)	C Stores (range) (kg C m^{-2})	Organic Horizons (%)	Mineral Horizons (%)	Active Layer (%)	Permafrost (%)
Arctic Coastal Plain					
Histoturbels (3)	80 (59–94)	60	40	60	40
Historthels (7)	62 (36–82)	55	45	56	44
Sapristels (4)	70 (61–78)	84	16	54	46
Aquiturbels (4)	52 (48–60)	21	79	58	42
Arctic Foothills					
Histoturbels (2)	55 (32–78)	69	31	78	22
Sapristel (1)	70	100	0	54	46
Aquiturbels (9)	46 (16–88)	26	74	57	43
Dystrocryept (1)	3	0	100	100	0
Cryorthent (1)	4	25	75	100	0
Aquorthel (1)	30	30	70	57	43
Haplorthel (1)	10	20	80	100	0
Arctic Seward Penninsula					
Histoturbels (2)	76 (61–91)	55	45	82	18
Historthel (1)	102	12	88	18	82
Aquiturbel (1)	94	60	40	69	31
Haplorthel (1)	52	63	37	85	15
Fibrist/Sapristel (3)	96 (81–109)	88	12	66	34

Source: Hoefle, C. et al., 1998; Michaelson, G.J. et al., 1996; Ping, C.L. et al., 1998a.

permafrost are very low, ranging from 3 to 4 kg C m^{-2}, with more than 75% in the mineral horizons. Permafrost-affected organic soil, the Sapristel, forms in deep organic matter deposit due to its landscape position; thus all of its C stores are in the organic horizons as compared with the 84% for the Sapristel on the Arctic Coastal Plain where the subsoils consist of mineral sediments. However, the Sapristels in the arctic foothills have slighter higher C stores in the permafrost layer as compared to that of the arctic coastal plain due to rising permafrost tables following the organic deposits. The partition of SOC between the organic and mineral horizons in the Arctic Seward Peninsula mineral soils is nearly equal, but between the active and permafrost layers it is 2 to 1. The C stores of the Seward Peninsula soils are higher than those of the rest of Arctic Alaska but lower in the permafrost, reflecting the lesser degree of cryoturbation and increased active layer depths due to warmer climate.

The C stores in the coastal plain mineral soils range from 36 to 94 kg C m^{-2}, with an average of 65 kg C m^{-2}; in the Arctic Seward Peninsula they range from 52 to 102 kg C m^{-2} with an average of 80 kg C m^{-2} (Hoefle et al., 1998; Michaelson et al., 1996; Tables 47.1 and 47.5.). The tundra soils on the Arctic Foothills have a wide range of C stores, from 10 to 88 kg C m^{-2} with the exception of these newly formed alluvial soils and barren ridge tops that have C stores less than 4 kg C m^{-2}. These values are 30 to 100% higher than previously reported (Billings, 1987; Chapin and Matthews, 1993; Post et al., 1982). On average, nearly half of the total C is stored in the upper permafrost, a phenomenon that indicates the importance of permafrost in carbon sequestration and is vulnerable to climate change.

Interior Alaska

In estimating C storage of Alaskan soils, Ping et al. (1997) found that values in interior Alaska range from 17 to 79 kg C m^{-2} in mineral soils and more than 130 kg C m^{-2} in organic soils. These estimates generally agree with those of Chapin and Matthews (1993) and Kimble et al. (1993). There is less cryoturbation in most interior soils compared to arctic tundra soils and most OC is concentrated near the soil surface. However, the spatial variability of OC accumulation in this setting has not been adequately defined and deserves more attention. Fire cycles and vegetation succession play a key role in SOC dynamics (Powers and Van Cleve, 1991). Van Cleve et al. (1993) have shown that the hardwood communities generate more C in surface organic horizons than spruce communities, but that the latter accumulate more C in mineral soils.

On the Tanana River floodplain near Fairbanks, Van Cleve et al. (1993) and Ping (unpublished data) measured C stores increasing from less than 0.5 kg C m^{-2} in Entisols on the youngest river terrace (stage I) to 11 kg C m^{-2} in soils under mature spruce (stage VIII). The landscape of interior Alaska is characterized by rolling hills and broad floodplains. Thus slope, aspect, and terrace age play key roles in soil formation and vegetation community development (Furbush and Schoephorster, 1977; Péwé, 1954). On the relatively warm, well drained south facing slopes, Typic Dystrocryepts are the dominant soils with C stores ranging from 6 to 88 kg C m^{-2} and an average of 44 kg C m^{-2} (Table 47.2). In these well-drained soils, OC in the litter layer recycles rapidly, thus the surface organic horizons only account for 25% of the total pedon C stores.

The north-facing slopes tend to be poorly drained due to the presence of permafrost within 50 cm of the surface with vegetation dominated by black spruce and mosses. Here the Historthels are the dominant soils with C stores ranging from 37 to 65 kg C m^{-2}, averaging 49 kg C m^{-2}. These soils have thick organic horizons containing 80% of the pedon C stores. Since the mineral horizons are mostly shallow over fractured bedrock, only 10% of the pedon C stores are in the permafrost. Aquorthels form on floodplains, terraces, and toe slopes in the boreal forest zone with permafrost rising to within 100 cm of the soil surface due to advanced stages of vegetation succession. C stores range from 14 to 48 kg C m^{-2} with an average of 35 kg C m^{-2}. In these soils, surface organic horizons contain 71% of the total pedon C, with 29% of C stores partitioned into the upper permafrost. Such an increase in C-stores in upper permafrost of these soils as compared with the

Table 47.2   Carbon Storage to 1 m in Soils of Interior Alaska

Great Group (n)	C Stores (range) (kg C m⁻²)	Organic Horizons (%)	Mineral Horizons (%)	Active Layer (%)	Permafrost (%)
Historthels (5)	49 (37–65)	80	20	90	10
Cryaquepts (2)	18 (16–20)	33	67	—	—
Aquorthels (4)	35 (14–48)	71	29	86	24
Dystrocryepts, Typic (11)	44 (6–88)	25	75	—	—
Cryofluvent (1)	11	36	64	—	—
Dystrocryepts, Humic (4)	26 (10–49)	42	58	—	—
Haplocryods (1)	8	63	37	—	—
Histoturbel (1)	107	81	19	83	17
Cryorthent (1)	7	57	43	—	—
Hemistel (1)	119	49	51	55	45
Haplocryoll (1)	18	33	67	—	—

Source: Data from USDA-Natural Resources Conservation Service-(NSSC) laboratory database; Ping, C.L. unpublished data.

Historthels on the north-facing slope is due to the alluvial processes, which result in carbon sequestration in the raised permafrost tables.

Cryaquepts occur along drainages and in depressions most frequently in the southern extent of the boreal zone. In these soils, C stores range from 16 to 20 kg C m⁻², with 67% stored in the mineral horizons. Humic Dystrocryepts contain lower C stores because they are well drained and have mostly shrub vegetation. Hemistels and Histoturbels occur in the broad basins and valleys and have C stores of 119 and 107 kg C m⁻², respectively; these organic soils or organic horizons sequester 45% carbon, which is more than the mineral soils. C stores in the Typic Dystrocryepts, Humic Dystrocryepts, Historthels, and Aquorthels vary greatly.

Boreal regions experience little variation in distribution of precipitation and thus C-store variability is largely attributed to landscape position (drainage, aspect, and slope) and vegetation succession as affected by fire. Even though C stores vary widely across the boreal forest zone, the relative amount of C stores partitioned into the mineral horizons is much less than for the organic horizons (Table 47.2). This further illustrates the effect of vegetation succession as a result of fire history and its effect on C stores within interior Alaska. Further study should be made of soil organic carbon storage within component members of soil catena across known vegetation successions.

Southern Coastal Regions

The mild, more humid south-central region is affected by tectonic activity and Andisols (volcanic ash-derived soils) are widespread (Ping et al., 1989). In the Aleutian Islands and the Cook Inlet region, Haplocryands have developed under grass and herbaceous vegetation. For these soils, nearly all of the pedon C stores are in the mineral matrix that contains from 22 to 23 kg C m⁻² (Table 47.3). The Haplocryods are also volcanic ash-derived soils but form under white spruce forest; they have comparable pedon C stores, but 38% of the C stores are in the organic horizons. C stores in these soils vary with microrelief and substratum, but are generally increasing from the north to the south.

Humicryods form in the same parent material as other Spodosols in the region but with increased precipitation and under Sitka spruce. Thus the pedon C stores increase to 51 kg C m⁻² with 22% stored in surface organic horizons (Ping et al., 1989; USDA-NRCS database). The increased partition of pedon C into the mineral horizons in Humicryods is the result of increased leaching. Typic Dystrocryepts and Cryorthents in South-Central Alaska have similar pedon C stores, ranges, and partitioning between organic and mineral horizons due to their similar genetic background: both form in loess material and under mixed forest. The high C stores (49 kg C m⁻²) in Cryaquepts are due to their occurrences on depressional landscape

Table 47.3   Carbon Storage to 1 m in Soils of Southcentral Alaska[a]

Great Group (n)	C Stores (range) (kg C m^{-2})	Organic Horizons (%)	Mineral Horizons (%)
Haplocryands (3)	22 (22–23)	0	100
Cryaquepts (2)	49 (23–75)	18	82
Haplocryods (6)	21 (11–31)	38	62
Humicryods (5)	51 (30–88)	22	78
Borosaprist (1)	219	100	0
Cryorthents (2)	30 (27–33)	30	70
Dystrocryepts (3)	27 (20–41)	22	78

[a] Includes Aleutian Islands.

Source: Ping, et al., 1989; USDA-Natural Resources Conservation Service-NSSC laboratory database.

Table 47.4   Carbon Storage to 1 m in Soils of Southeast Alaska

Great Group (n)	C Stores (range) (kg C m^{-2})	Organic Horizons (%)	Mineral Horizons (%)
Haplaquod (1)	99	70	30
Humicryod (1)	50	26	74
Cryohemist (2)	114 (72–156)	95	5
Haplocryod (2)	42 (34–50)	71	29

Source: Ping, C.L. et al., 1989.

Table 47.5   Cabon Storage in Soils of the Various Regions of Alaska

Geographic Region (n)	C Stores (range) (kg C m^{-2})	Organic Horizons (%)	Mineral Horizons (%)	Active Layer (%)	Permafrost (%)
Arctic (42)	59 (3–110)	56	44	61	39
Coastal Plain (18)	65 (36–94)	57	43	57	43
Foothills (16)	40 (3–88)	43	57	63	37
Seward Penn. (8)	86 (52–110)	64	36	64	36
Interior (32)	40 (6–119)	50	50	—	—
Southcentral (22)	37 (11–219)	22	78	—	—
Southeast (6)	77 (34–156)	79	21	—	—

Source: Hoefle, C. et al., 1998; Michaelson, G.J. et al., 1996; Ping, C.L. et al., 1997, 1998; and USDA-NRCS-NSSC laboratory database.

positions with saturated conditions, which favor the preservation of organic matter in the mineral horizons. The Borosaprist measures 167 kg C m^{-2} with sampling depth reaching 165 cm; the C stores of this soil would be higher if sampled deeper.

C stores increase concordant to precipitation in the southern coastal region from the Cook Inlet to the southeast where annual precipitation is more than 300 cm yr^{-1} in the mountains. Here, large amounts of humus accumulate and are incorporated into the mineral matrix through leaching and bioturbation. Burt and Alexander (1996) found that on a 220-year-old glacial moraine deposit, Haplocryods had accumulated 7 to 9 kg C m^{-2} with 60 to 90% in the O horizons. The mineral horizons contain only 1 to 4 kg C m^{-3}. Most areas in the Cook Inlet region and in southeast Alaska were deglaciated about 9000 B.P., with Haplocryods on older glacial moraines and drift having higher C stores (34 to 50 kg C m^{-2}) with 71% of pedon C in the O horizons. The Humicryods have C stores at the high range with only 26% of pedon C in the O horizons (Table 47.4).

Such a reverse trend again illustrates the effects of increased leaching and illuviation of OC into the mineral horizons. The Haplaquod forms in depressions and accumulated 99 kg C m^{-2}. Yet the highest C store is in the Histosols form in this mild and perudic moisture regime; the two Histosols measured 72 to 402 kg C m^{-2} with an average of 95% in the organic horizons. The difference in pedon C stores is attributed to landscape positions. Drainage restriction increases from Haplocryods and Humicryods to Haplaquod with more OC incorporated into the mineral soils with time and with vegetation changes, a demonstration of the progressive C sequestration process.

Carbon Accumulation in Organic Soils of Alaska

Histosols are soils high in organic matter containing large amounts of C and therefore are historically the largest C sinks. C stores in Alaska Histosols ranged from 60 to 402 kg C m^{-2} (Alexander et al., 1991; Michaelson et al., 1996; Ping et al., 1997). Generally, C-store ranges of Alaskan Histosols are differentiated: Folists = 8 to 15 kg C m^{-2}, Saprists = 50 to 167 kg C m^{-2}, Hemists = 72 to 402 kg C m^{-2}, and Fibrists = 100 to 240 kg C m^{-2}. The C density of Histosols depends on the degree of OM decomposition, expressed as texture. In general, C stores of Histosols follow the climate gradient, increasing with rainfall from the arid arctic north to the humid and wet south, but also varying with topographic position. The broad floodplains along the major rivers in interior Alaska, the Matanuska–Susitna valleys, and the Kenai Peninsula commonly have peat bogs more than 3 m deep; raised bogs formed in sphagnum moss in the rainforest of southeast Alaska have depths of more than 2 m.

In studying the stabilization of eolian sand deposits in northwestern Alaska, Carter (1993) identified a Late Pleistocene paleosol consisting of peat more than 1 m thick dated 12,100 to 11,130 B.P., which corresponds to the Younger Dryas chronozone. Undifferentiated sand deposition intervened between the paleosol and overlying a second maximum peat accumulation (dated 8600 to 5200 B.P.). This suggests a relatively dramatic shift in climatic conditions during that interval. Schell and Ziemann (1983) found the ^{14}C dates of basal peats of eight sites across the Arctic Coastal Plain ranging from 8435 to 12,610 with an average of 9663 B.P. In a SOM characterization study, Ping et al. (1998b) found a similar basal date of 7500 B.P. for a Histosol near the Arctic Coast and calculated a C accumulation rate of 9 and 17 g m^{-2} yr^{-1} for two Histosols. They also found an old surface rich in humus with a ^{14}C age of 12,740 B.P. The results from these three studies indicate that peat accumulation started in Late Pleistocene in arctic Alaska.

At similar latitude, Vardy et al. (1997) studied several peat cores from the Tuktoyatuk Peninsula, Northwest Territory of Canada. They found that peat accumulation followed the deterioration of mild climatic conditions with permafrost aggradation about 6300 to 5000 B.P. when the area transformed from an open water mineral wetland to a graminoid fen peatland. This event corresponded with the southward retreat of the tree line from the peninsula and the end of the early Holocene insolation maximum. By 4000 B.P., the peatland had gradually transformed from sedge to Sphagnum dominated with a slower accumulation rate.

Peat deposition in the boreal wetland regions tends to be deeper than those of the arctic and has higher long-term net accumulation rates (Ovenden, 1990; Ping et al., 1998b). On the average, long-term accumulation rates in the arctic and boreal region peatlands were estimated at 7 to 11 and 23 to 41 g m^{-2} yr^{-1}, respectively (Ovenden, 1990). Schell and Ziemann (1983) estimated the average C accumulation rate of 13.3 g m^{-2} yr^{-1} from peat sampled in the Arctic Coastal Plain. Peat accumulation began early in western Alaska and progressed eastward following deglaciation. Peatlands started about 12,000 B.P. in arctic Alaska, 7000 B.P. in northwestern Canada, 6000 to 5000 B.P. in the Hudson Bay Lowlands, and 5500 to 2500 B.P. in Quebec and northern Minnesota (Carter, 1993; Glaser and Janssen, 1986; Ovenden, 1990). Recent studies estimated total carbon storage in the peatlands in the Arctic and the boreal zones of Alaska ranging from 32 to 80 to over

130 kg C m^{-2}, respectively (Michaelson et al., 1996; Ping et al., 1998b). These results indicate that peat accumulation is a direct function of climate and landscape position and is maximized in the boreal region.

Patterns of Carbon Storage

When partitioned by genetic horizons, the percent of OC ranges from 29 to 42% in the O horizons, 6 to 9% in the A horizons, 2 to 4% in the B horizons, 17% in the cryoturbated upper permafrost layer (Bgf/Oaf), and 6% in the underlying Cf horizon. The upper permafrost layer is the part of permafrost within 1 m of the surface that may include the lower Bgf and the upper Cf horizons. The OC content of coastal plain O horizons is lower than its counterparts on foothill tundra soils, due to high mineral content in coastal plain O horizons resulting from frequent inundation. The OC content of the upper permafrost layers appears low as compared with those in the active layers; however, once the higher bulk density and thickness of the mineral horizons are factored in, the C stores in the upper permafrost increase drastically, and account for 40 to 50% of the whole pedon C storage. Much of the storage is due to incorporation of surface organic horizons to the subsurface mineral horizons by cryoturbation (Ping et al., 1997).

In epigenetic soils, i.e., soils developed after the sedimentation of parent materials, soil horizons are stratigraphically concordant and the SOC accumulation is highest in the surface horizons, decreasing with depth. However, patterns of SOC accumulation are different in soils formed in syngenetic or cryogenic environments. In cryogenic environments such as arctic Alaska, cryoturbation due to frost heave and churning often distorts soil morphology and results in broken or discontinuous horizonation. Thus, SOC accumulation is not necessarily highest in the surface horizons. Instead, SOM is often churned into lower horizons and results in a secondary maximum of organic carbon below the surface, usually on top of or within the upper permafrost (Ping et al., 1998a). In another cryogenic case in the boreal region, Ping et al. (1992) described the organic matter not entrained through cryoturbation but encased by rising permafrost due to climate change or fire. Thus multiple accumulation layers of SOC are also found in soils of the boreal region where wildfire frequently impacts the ecosystem.

South-central Alaskan soils often contain multiple layers of SOC sequestration associated with tephra deposit because of the intermittent nature of the tephra fall (Ping et al., 1989). In these soils, buried surface horizons and multisequa soils are very common. The buried surface O horizons eventually are transformed into buried A or Bhs horizons. In the Matanuska Valley of the Cook Inlet region where loess deposition is active, Cryochrepts and Cryorthents are formed in multiple layers consisting of alternating surface sod layers or A horizons and underlying loess layers. These soils have C stores higher than most Cryochrepts. In the coastal forest zone, especially in southeast Alaska, windthrow is common (Boormann et al., 1995), causing localized horizon disruption and SOC accumulations. The surface O and associated subsurface horizons, such E and upper Bh horizons, frequently intracted into the mineral horizons. Consequently, decomposition of OM is retarded by burial and C stores increase in this ecosystem.

CONCLUSIONS

The tundra soils of Alaska contain measurably higher C stores than those of the temperate regions; nearly half of the total pedon C store are sequestered in the upper permafrost layers. In earlier studies most data only reflected sampling of the active layer. In the NSF-Arctic System Science Project, soil sampling included the active layers and the upper permafrost to 1 m. Thus the C stores measured are 30 to 100% higher than previously reported. The C stores within the arctic tundra landscape generally increase northward, due to low decomposition as a result of lower biological activity. Carbon distribution in the tundra soil profiles does not follow a simple depth function due to cryoturbation that often causes a secondary concentration zone of C in upper permafrost.

Soils of the boreal regions show a wide range of C distribution among different ecosystems. Although varied in climate conditions, C stores and distribution patterns in these soils are mainly controlled by landscape position and drainage. Soils on uplands and recent alluvium have a regular pattern of OC decreasing with depth. C stores in the upland soils estimated by the whole pedon approach are twice as high as indicated in the STATSGO database (Soil Survey Staff, 1996) due to its omission of the organic horizons. But in areas affected by permafrost, SOC distribution is either disrupted by cryoturbation or encasement of old surface horizons by rising permafrost levels. Although the C stores in upland soils of the boreal zone are low, upland soils account for less than 20% of the total area in Alaska. The remaining 80% of the area contains the largest acreage of organic and forest tundra soils, most of which is underlain with permafrost with high C stores. In addition, permafrost in the boreal zone is relatively "warm" compared with that of the arctic zone, rendering the former more sensitive to climate change, with a high potential of increased greenhouse gas flux under warming conditions.

C stores in soils of the coastal zone are controlled largely by drainage and by stages of vegetation succession following glacial retreat. Due to the high annual precipitation, bogs and fens are widespread throughout the area, thus contributing to high C stores in the area. As compared to the south-central area of the coastal zone, considerable OC is stored in the boreal rainforest soils of southeast Alaska. These C stores are incorporated into the mineral soil horizons through illuviation and windthrow. The high rainfall and mild environment of this region favors C sequestration through leaching and bioturbation.

REFERENCES

Alexander, E.B. et al., Soils of southeast Alaska as sinks for organic carbon fixed from atmospheric carbon-dioxide, USDA Forest Service, Juneau, AK, 1991.

Billings, W.D., Carbon balance of Alaska tundra and taiga ecosystems: past, present, and future, *Q. Sci. Rev.*, 6:165–177, 1987.

Boormann, B.T. et al., Rapid soil development after windthrow disturbance in pristine forests, *J. Ecol.*, 83:747–757, 1995.

Burt, R. and E.B. Alexander, Soil development on moraines of Mandenhall Glacier, southeast Alaska. 2. Chemical transformations and soil micromorphology, *Geoderma*, 72:19–36, 1996.

Carter, L.D., Late Pleistocene stabilization and reactivation of eolian sand in northern Alaska: implications for the effects of future climatic warming on an eolian landscape in continuous permafrost, in *Permafrost 6th Int. Conf. Proc.* (Vol. 1), July 5–9, 1993, Beijing, China, South China University of Technology Press, Guangzhou, China, 1993, 78–83.

Chapin III, F.S. and E. Matthews, Boreal carbon pools: approaches and constraints in global extrapolations, in T.S. and T.P. Kolchugina (eds.), *Proc. Int. Workshop Carbon Cycling Boreal Forest Subarctic Ecosyst.: Biospheric Responses Feedbacks Global Climate Change*, September 1991, U.S. environmental Protection Agency, Corvallis, OR, 1993, 9–20.

Furbush, C.E. and D.B. Schoephorster, Soil survey of Goldstream–Nenana area, Alaska, USDA Soil Conservation Service, U.S. Government Printing Office, Washington, D.C., 1977.

Glaser, P.H. and J.A. Janssens, Raised bogs in eastern North America: transition in landforms and gross stratigraphy, *Can. J. Bot.*, 64:395–415, 1986.

Hoefle, C. et al., Properties of permafrost soils on the Northern Seward Peninsula, Northwest Alaska, *Soil Sci. Soc. Am. J.*, 62:1629–1639, 1998.

Kimble, J.M. et al., Determination of the amount of carbon in highly cryoturbated soils, in D. Gilichinsky (ed.), *Post-Seminar Proc., Joint Russ.-Am. Semin. Cryopedology Global Change*, November 15–16, 1992, Puschino, Russia, Russian Academy of Sciences, Moscow, 1993, 277–291.

Michaelson, G.J., C.L. Ping, and J.M. Kimble, Effects of soil morphological and physical properties on C store estimation, in Lal, R., J.M. Kimble, and R.F. Follett (eds.), *Assessment Methods for Soil C Pools*, Ann Arbor Press, 1999.

Michaelson, G.J., C.L. Ping, and J.M. Kimble, Carbon storage and distribution in tundra soils of Arctic Alaska, USA, *Arctic Alp. Res.*, 28:414–424, 1996.

Nelson, D.W. and L.E. Sommers, Total carbon, organic carbon, and organic matter, in A.L. Page (ed.), *Methods of Soils Analysis*, Part 2, Chemical and Microbiological Properties, 2nd ed., Agronomy no. 9. American Society of Agronomy, Inc., Madison, WI, 1982, 539–594.

Oechel, W.C. et al., Recent change of Arctic ecosystems from a net carbon dioxide sink to a source, *Nature*, 361:520–523, 1993.

Ovenden, L., Peat accumulation in northern wetlands, *Quaternary Res.*, 33:377–386, 1990.

Péwé, T., Effect of permafrost on cultivated fields, Fairbanks area, Alaska, in Mineral Resources of Alaska 1951–53, *Geol. Survey Bull.* 989, 315–351, 1954.

Ping, C.L. et al., Characteristics of cryogenic soils along a latitudinal transect in arctic Alaska, *J. Geogr. Res.*, 103:(D22) 28917–28928, 1998a.

Ping, C.L. et al., Characterization of soil organic matter by stable isotopes and radiocarbon age of selected soils in arctic Alaska, in J. Drozd et al. (eds.), *The Role of Humic Substances in the Ecosystems and in Environmental Protection*, Polish Society of Humic Substances, Wroclaw, Poland, 1998b, 475–480.

Ping, C.L., G.J. Michaelson, and J.M. Kimble, Carbon storage along a latitudinal transect in Alaska, *Nutr. Cycling Agroecosys.*, 49:235–242, 1997.

Ping, C.L. et al., Characteristics and classification of volcanic ash-derived soils in Alaska, *Soil Sci.*, 148:8–28, 1989.

Ping, C.L., Y.L. Shur, and G.J. Michaelson, Pedological properties of the Dry Creek archaeological site in the Nenana Valley, Alaska, in *Proc. 43rd Arctic Sci. Conf.*, September 9–12, 1992, Valdez, Alaska, University of Alaska, Fairbanks, 1992, 101.

Post, W.M. et al., Soil carbon pools and world life zones, *Nature*, 298:156–159, 1982.

Powers, R.F. and K. Van Cleve, Long-term ecological research in temperate and boreal forest ecosystems, *Agron. J.*, 83:11–24, 1991.

Schell, D.M. and P.J. Ziemann, Accumulation of peat carbon in the Alaska arctic coastal plain and its role in biological productivity, *Permafrost, 4th Int. Conf.*, Washington, D.C., National Academy Press, 1983, 1105–1110.

Schlesinger, W.H., An overview of the carbon cycle, in Lal, R. et al. (eds.), *Soils and Global Change*, CRC/Lewis Publishers, Boca Raton, FL, 1995.

Soil Survey Division Staff, Soil Survey Manual. USDA Handbook no. 18, U.S. Dept. of Agriculture, U.S. Government Printing Office, Washington D.C., 1951.

Soil Survey Division Staff, Soil Survey Manual. USDA Handbook no. 18, U.S. Dept. of Agriculture, U.S. Government Printing Office, Washington D.C., 1993.

Soil Survey Staff, The Alaska State Soil Geographic Data Base (STATSGO), 1996.

Soil Survey Staff, Keys to Soil Taxonomy, 8th ed., USDA Natural Resources Conservation Service, Washington, D.C., 1998.

Van Cleve, K. et al., Control of soil development on the Tanana River Floodplain, interior Alaska, *Can. J. For. Res.*, 23:941–955, 1993.

Vardy, S.R., B.G. Warner, and R. Aravena, Holocene climate effects on the development of a peatland on the Tuktpyktuk Peninsula, Northwest Territories, *Quaternary Res.*, 47:90–104, 1997.

CHAPTER **48**

Agricultural Practices and Policy Options for Carbon Sequestration: What We Know and Where We Need to Go

John M. Kimble, Rattan Lal, and Ronald R. Follett

CONTENTS

INTRODUCTION

The chapters of this book present an overview from scientists, policymakers, economists, and farmers related to what is known about soil carbon (C) sequestration and the relationships of science and policy. They also address areas of concern and what may need to be done in the future. All of these diverse groups have published extensively, given talks, and taken part in policy development, reaching a wide range of audiences. Many times, however, this was only in the context of scientist to scientist or policymaker to policymaker. Many times they did not consult peers and colleagues in fields other than their own, but hoped that what they wrote or said would be considered in the formulation of policies. Sometimes this may have been the case, but in most it was not. In most cases, policymakers are not made aware of the information developed by scientists or even of what farmers' needs are.

Policy development is a series of events that often tries to please people with extremely divergent views and ideas; therefore, the polices developed result from a series of compromises that in the end may leave everyone unhappy. The scientific views are looked at, but many times scientists present their ideas as simple facts based on the principal of scientific truth and then are disappointed when those ideas are not included in the final product or policy.

Producers need to optimize profit and sometimes are accused of degrading the land for economic gain, which, in most cases, is not true. Environmental groups feel that the overriding concern should be the environment and do not adequately consider the needs of producers, which is an oversimplification because most farmers and producers are also interested in the environment and want to be good stewards of the land. In many cases, they were the first environmentalists. Producers live on the land every day and do not like to see high rates of erosion or other types of degradation because the land is their livelihood; they are the first to feel the impacts of improper management. On top of that, most want to pass the "family farm" to the next generation in a state better than the one that it was in when they received it from their parents.

The problem is that, while one group looks at profit, another looks at water quality, another at wildlife habitat ... the list could go on and on. The fact is that all of these are interrelated and must be balanced if we are to develop the most beneficial policies. All too often the concept of holistic management is not considered. The material presented in this book is an attempt to move beyond one-dimensional approaches and ideas into a multidimensional arena where all parties work together with a common goal that takes into consideration concerns of all the divergent groups who have a stake in the land and, particularly, the soil. If this is done, the impacts of management practices will have effects not only on productivity and the soil but also on the overall environment and thus provide a multitude of societal benefits.

Figure 48.1 shows the pedosphere in the center of all other spheres; the soil that is the pedosphere is the real "capital" of farmers. If farmers destroy the soil, they destroy the base of their livelihood and, at the same time, create many other environmental problems. We need to develop policies that ensure development of sustainable, environmentally friendly farming and land management practices. The heart of a farming operation can be considered the soil organic matter, i.e., soil organic carbon (SOC). Without the SOC, soils are just a sterile media; they may supply nutrients, but without SOC they do not support a diverse flora and fauna within the soil and crop production will be very limited. In time, even the ability of the soil to supply nutrients even when added to the soil will be reduced.

Scientific techniques for sequestering C in soils have been reported in many publications. (References to these are contained in all the chapters in this book and will not be repeated here.) Many advances have been made in understanding the science of SOC through these many research activities. We are now going beyond the science of SOC sequestration and considering C credits and trading, which make the soil itself a commodity with the idea that SOC has a defined value. Many have tried to put a value on C in the past, usually in an abstract way.

Costanza et al. (1997) looked at the value of the world's ecosystems as types of natural capital soils were given a value based on their societal importance. It is now accepted that soils have an intrinsic value to humans, and that SOC contributes much to this value. When land is sold, a higher price is paid for soils with more C. We exploit the C in the soil just as we do oil and gas, which are also C-based compounds. The major difference is that we can replace SOC, but not oil and gas.

Therefore policies need to be developed that will prevent the exploitation of SOC and at the same time replace the lost C and establish a value for this C form. Ways of doing this have been discussed in this book and briefly summarized in the following sections, which encapsulate a summary brochure from the meeting. Future activities and concerns are also discussed.

SCIENTIFIC TECHNIQUES FOR SEQUESTERING C IN SOIL

A wide range of options exists for sequestering C in soil. Appropriate options differ for different soils and ecoregions because of differences in biophysical and socioeconomic conditions; no single

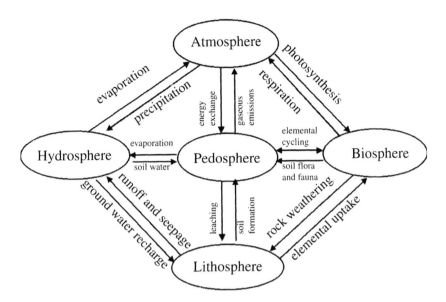

Figure 48.1

option universally applies. Site-specific adaptation is needed to select the most appropriate combination of options. We need to consider the total cost of all inputs and also the benefit of each practice. No-till may be the best way to sequester C, but it may require more use of herbicides. If a minimum tillage operation is done before planting, it may cost less and be more environmentally friendly than a pure no-till system. Cost-benefit analysis needs to be applied to different options and then a decision made as to what is the best solution, based on many different considerations. Some promising options for sequestering SOC include:

1. Adopt conservation tillage, residue management, and mulch farming.
2. Develop and grow crops and plants that have deeper rooting systems.
3. Develop plants that contain high lignin, suberin, and phenol contents, especially in the residue and roots.
4. Replace annual crops with perennials.
5. Manage the water table by changing drainage in noncropping seasons.
6. Reduce cultivation of organic soils as much as possible and better manage the water table under cropping systems to reduce oxidation of these organic soils.
7. Eliminate summer fallow, and incorporate legumes and other appropriate cover crops in rotations.
8. Use nitrogen (N) and other fertilizers more efficiently through integrated nutrient management and site-specific farming practices.
9. Increase the SOC pool in subsoil layers (deep in the profile) through N and phosphorus (P) management and other techniques.
10. Manage rangelands and pastures through improved grazing, vegetation management, fire, and other recommended management practices (RMPs).
11. Increase crop and biomass yields and return them to the soil.
12. Apply organic manure and other biosolids to soils.
13. Conserve soil and water resources.
14. Use water more efficiently through drip irrigation, water harvesting, and other measures that conserve water.
15. Convert marginal and degraded lands to restorative land uses, e.g., adopt CRPs and WRP afforestation, and establish perennial cover.
16. Enhance biological N fixation through the use of legume crops in crop rotation or as cover crops.
17. Consider the nutrient (N, P, sulfur (S), and C) cycles together, not separately (total ecosystem management).

ADVANCES IN SCIENCE OF SOIL C POOL AND DYNAMICS

Research on soil organic matter dynamics has continued throughout the 20th and into the 21st century. There have been many notable advances in basic and applied research. A strong knowledge base exists, especially about:

1. The impact of changes in land use on the SOC pool and fluxes
2. The effect of the quantity and quality of the SOC pool on crop growth and nutrient (N, P, S) dynamics
3. The impact of SOC dynamics on the fluxes of radiatively active gases (e.g., CO_2, CH_4, N_2O)
4. Needed measurement, verification, scaling, and prediction of the SOC pool at different scales (plot, farm, landscape, region, nation, and globe)
5. The rates of SOC sequestration for a range of practices in different soils and ecoregions
6. The humification efficiency of applied crop residue and biosolids
7. The turnover rate and residence time of different fractions of SOC pools
8. An understanding of the mechanism of soil C sequestration
9. The processes that lead to depletion of the SOC pool
10. The interaction between the SOC and soil inorganic carbon (SIC) pools
11. The strong dependence of the SOC pool on temperature and precipitation
12. Relevant and site-specific techniques of SOC sequestration

The most widely accepted rate of SOC sequestration by conversion from plow-till to no-till is 0.3 to 0.5 Mg (metric tons) per hectare per year. The sequestration potential under CRP is 0.5 to 1 Mg/ha/yr. The humification efficiency is 8 to 12% for farmyard manure and compost and 10 to 20% for crop residue and other biomass. The duration of the most active SOC sequestration is 10 to 20 years. The SOC pool in the root zone (30 to 50 cm) can be measured with an accuracy of ± 0.1 Mg/ha at plot scale, ± 1 Mg at farm and landscape scale (50 to 100 ha), and ± 1 Tg/yr at regional scale.

CARBON CREDITS AND TRADING

Making C a commodity necessitates determining its market value, and doing so with rational criteria. Both farmers and society will benefit from sequestering C. Enhanced soil quality benefits farmers, but farmers and society in general benefit from erosion control, reduced siltation of reservoirs and waterways, improved air and water quality, and biodegradation of pollutants and chemicals. Farmers need to be compensated for the societal benefits of C sequestration and mechanisms developed that will allow for C trading and also maintain property rights. Measurement and verification of C options for C sequestration have been developed and policymakers need to be made aware of these procedures.

With a wide range of SOC sequestration rates at 300 to 500 kg/ha/yr (250 to 450 lbs/acre), an appropriate price may be $125/Mg (metric ton), or $0.06/lb, of soil C. At this rate, the farmer may be compensated $15 to 25 per acre for conversion from plow-till to no-till or for adopting other measures discussed. Most farmers participating in the workshop agreed that $10/acre/year was a fair incentive for adopting technologies for sequestering C in soil. But this is a market-driven process and the price will develop if a market exists. Even if there is no trading market, a value is still needed for adoption of RMPs and mechanisms developed to compensate farmers for using these practices based on their societal value. Once a price is established for SOC it can be reflected in the value of the land when it is sold or even rented. Poor management would decrease the value of the land, as it would lead to a lowering of C levels. Maintaining C levels could be considered in rental agreements to prevent the "mining" of this valuable resource.

There are at least four criteria that need to be applied to a C trading or credit system. The criteria must be simple, fair, and easy to implement, and must enhance efficiency and production. A "soil credit kit" is also needed; this could vary among ecoregions and would provide information such as:

1. Rates of SOC sequestration for different management options for specific soils and ecoregions that will allow for the selection of the best options for a given area
2. Defined guidelines and methods for monitoring and verifying C sequestration that are clear, transparent, and accepted by all parties
3. How long the C will remain in the ground, which addresses the residence time or permanence of the C sequestered and is also concerned with property rights
4. Transparent pricing criteria and policies
5. A determination of additional and ancillary effects and how we compensate for them

Soil C sequestration policies can be implemented providing that there is a political will to act that will develop the needed policies and there are supporting institutions and infrastructures for measurement and verification.

RESEARCH ISSUES AND KNOWLEDGE GAPS

In any meeting, you cannot answer all of the questions and new ones are raised. One of the major outcomes of such meetings is that they help to move us forward. Two panel discussions highlighted several knowledge gaps: in agronomic research and for developing policies and providing incentives. The panel discussions were a mix of all the interested parties so they reflected a broader set of ideas more integrative and of interest to all parties.

Research Priorities and Issues Related to Soil Management and Agronomic Techniques

Some major concerns of the farming community about adopting RMPs need to be addressed through on-farm and site-specific research. These include:

1. Techniques to reduce "perceived" risks of no-till farming. Even though in many cases a risk is only perceived, perception is reality in many situations. In the first few years of no-till there may be more risks until the soil quality is improved and the system reaches a sustainable level. This could be addressed through crop insurance or other policies that help to reduce the fiancial risks of conversion to no-till farming.
2. Methods to minimize uncertainties. This also addresses the perceived risk question and problems dealing with wet spring weather delaying planting or limiting seed germination because of cold soil temperatures. One possibility to overcome this may be strip tillage, where only a small percentage of the field is tilled.
3. Methods for controlling herbicide-resistant weed.
4. Farming systems that increase crop yields while reducing emissions.
5. Evaluations of long-term improvement in soil quality due to no-till and other effective conservation measures.
6. Facts about how the SOC pool relates to drought tolerance.
7. Establishment of a network of on-farm applied research in soil C.
8. Facts about how to enhance the efficiency of nutrient (N) use.

Research Topics and Issues Related to Environmental Concerns

The SOC pool and the environment are strongly linked. Some pertinent research issues in addressing environmental concerns include:

1. What is the SOC equivalent of fossil fuel?
2. What is the relationship between the SOC pool and water quality?
3. What are the quantified ancillary benefits of SOC sequestration?
4. What is the relationship between the SOC pool and air quality?

Research Topics and Issues Related to Soil Processes

Despite notable advances in soil science, several important issues need research. Important among these are:

1. Harmonizing the conceptual differences in soil C sequestration among soil and agricultural scientists, foresters, and rangeland managers
2. Identifying all factors that affect sources and sinks of C in soils
3. Delineating soil attributes or indicators that enhance the SOC pool in the subsoil
4. Quantifying the relative importance of dissolved organic carbon (DOC) and SIC dynamics in different soils and ecoregions
5. Improving and developing new tools for simple and routine monitoring of the SOC pool at different scales
6. Quantifying the magnitude of historic and current C emissions from soil at different scales
7. The effect of minimum tillage on killing weeds before planting instead of using chemicals and how much C is lost by this practice

Research Topics and Issues Related to Policy and Economics

The symposium provided an excellent opportunity for interaction among biophysical and socioeconomic and political scientists, farmers, and industry personnel. Some relevant researchable issues that emerged from this interaction include:

1. What is the societal value of soil C?
2. What is the cost of SOC sequestration?
3. Is soil misuse a criminal act? If so, by whom and against whom?
4. Can the consumer solve the problem of soil C sequestration? If so, who is the consumer (farmer or society)?
5. How can an "Endangered Soil Act" be established?
6. How can we develop a just reward system so that we do not reward only the "bad apple"?
7. How should we involve farmers in the decision-making process that affects soil and crop management practices?
8. How should we reach out to the general public and government and create awareness about the importance of soil C?
9. How can we establish a dedicated extension service or even a private extension service to strengthen channels of communication with farmers?
10. How can we involve nongovernment organizations (NGOs) and environmental groups in the dialogue about soil C sequestration?
11. How can we get "soils" on the agenda for national and international programs for sequestering C?
12. How can we enlist legislative support in using soils to sequester C?
13. How can we establish an insurance scheme for agricultural practices to sequester C?
14. How can we reduce the cost of transaction vis-a-vis the cost of C sequestration?

CONCLUSIONS

This volume has highlighted the importance of sequestering C in soil to help in mitigating the potential greenhouse effect, the significance of dialogue among soil scientists, agronomists, economists, policymakers, environmental groups, industry personnel, farmers, and land managers, and the need to reach out to farmers and the public at large about the importance of soil C and agricultural practices in improving the environment and mitigating the greenhouse effect. Yet, some notable gaps and needs remain. We must develop a unified and common opinion among soil scientists, as well as establish a matrix of rates of sequestering C in soil C with different practices for principal soils and ecoregions. We also need to standardize a "soil C monitoring and evaluation" kit and convincingly demonstrate to policymakers and the public that soil is an important, integral part of the environment and that we can measure and verify changes to the largest terrestrial poll of C, i.e., soil C.

Soil scientists must be active in environmental issues, especially the greenhouse effect. A cap on emissions at the 1990 level through reduction in fossil fuel consumption would require a $100/Mg of C tax or $52.5/Mg of coal, $0.02/kwh of electricity, $0.26/gallon of gasoline, $1.49/1000 CF of natural gas. Electric output would decline by $30 billion in 2010, and by $42 billion in 2020. Therefore, soil scientists must be proactive and convey to policymakers the message that soil C sequestration is the most cost-effective option, and a bridge to the future. We must convince the policymakers, environmentalists, and industrialists that soil C sequestration is an additional important benefit of adopting improved and recommended agricultural production systems. This option stands on its own, regardless of the threat of global climate change.

An overworked phrase, but nevertheless worth repeating, is that the sequestration of C in soils is a win–win proposition. This will not change; the sooner the concept is adopted by all parties and policies developed that will encourage this, the better off all of us will be: The science of C sequestration is known and we can measure and verify changes. The next step is the development and implementation of policy options that meet the needs of production agriculture and also allow for a sustainable, environmentally friendly agriculture. Once this is done everyone will benefit from the "critters" that live in the soil and make it one of the most diverse biological systems to the people who eat the food produced on well managed land. We will also have improved water quality and wildlife habitat and even reduced the need for external inputs to the soil. All this is possible: the key is soil organic carbon.

REFERENCES

Costanza, R. et al., The value of the world's ecosystem services and natural capital, *Nature*, 387:253–260, 1997.

Index

A

B

C

IMPORTANT:

W9-DHK-982

HERE IS YOUR REGISTRATION CODE TO ACCESS
YOUR PREMIUM McGRAW-HILL ONLINE RESOURCES.

For key premium online resources you need THIS CODE to gain access. Once the code is entered, you will be able to use the Web resources for the length of your course.

If your course is using **WebCT** or **Blackboard**, you'll be able to use this code to access the McGraw-Hill content within your instructor's online course.

Access is provided if you have purchased a new book. If the registration code is missing from this book, the registration screen on our Website, and within your WebCT or Blackboard course, will tell you how to obtain your new code.

Registering for McGraw-Hill Online Resources

To gain access to your McGraw-Hill web resources simply follow the steps below:

1. USE YOUR WEB BROWSER TO GO TO: **www.mhhe.com/criticalthinking**

2. CLICK ON **FIRST TIME USER**.

3. ENTER THE REGISTRATION CODE* PRINTED ON THE TEAR-OFF BOOKMARK ON THE RIGHT.

4. AFTER YOU HAVE ENTERED YOUR REGISTRATION CODE, CLICK **REGISTER**.

5. FOLLOW THE INSTRUCTIONS TO SET-UP YOUR PERSONAL UserID AND PASSWORD.

6. WRITE YOUR UserID AND PASSWORD DOWN FOR FUTURE REFERENCE.
   KEEP IT IN A SAFE PLACE.

TO GAIN ACCESS to the McGraw-Hill content in your instructor's **WebCT** or **Blackboard** course simply log in to the course with the UserID and Password provided by your instructor. Enter the registration code exactly as it appears in the box to the right when prompted by the system. You will only need to use the code the first time you click on McGraw-Hill content.

Thank you, and welcome to your McGraw-Hill online Resources!

cognitions-65500183

REGISTRATION CODE

* YOUR REGISTRATION CODE CAN BE USED ONLY ONCE TO ESTABLISH ACCESS. IT IS NOT TRANSFERABLE.

0-07-293714-9 T/A MOORE/PARKER: CRITICAL THINKING, 7E

Critical Thinking

Critical Thinking

BROOKE NOEL MOORE ▲ RICHARD PARKER

California State University, Chico

CHAPTER 12
with Nina Rosenstand and Anita Silvers

Mc
Graw
Hill

Boston   Burr Ridge, IL   Dubuque, IA   Madison, WI   New York
San Francisco   St. Louis   Bangkok   Bogotá   Caracas   Kuala Lumpur
Lisbon   London   Madrid   Mexico City   Milan   Montreal   New Delhi
Santiago   Seoul   Singapore   Sydney   Taipei   Toronto

Higher Education

CRITICAL THINKING, SEVENTH EDITION
Published by McGraw-Hill, a business unit of The McGraw-Hill Companies, Inc., 1221 Avenue of the Americas, New York, NY 10020. Copyright © 2004 by The McGraw-Hill Companies, Inc. All rights reserved. Previous editions © 2001, 1998, 1995, 1992, 1989, 1986 by Mayfield Publishing Company. No part of this publication may be reproduced or distributed in any form or by any means, or stored in a database or retrieval system, without the prior written consent of The McGraw-Hill Companies, Inc., including, but not limited to, any network or other electronic storage or transmission, or broadcast for distance learning.

   Some ancillaries, including electronic and print components, may not be available to customers outside the United States.

1 2 3 4 5 6 7 8 9 0 2CUS/2CUS 0 9 8 7 6 5 4 3

Vice president and editor-in-chief: *Thalia Dorwick*
Publisher: *Chris Freitag*
Sponsoring editor: *Jon-David Hague*
Marketing manager: *Lisa Berry*
Production services manager: *Jennifer Mills*
Production service: *Fairplay Publishing Service*
Manuscript editor: *Joan Pendleton*
Art director: *Jeanne M. Schreiber*
Design manager: *Sharon Spurlock*
Cover designer: *Linda Robertson*
Interior designer: *Terri Wright*
Art manager: *Robin Mouat*
Photo researcher: *Connie Gardner*
Illustrators: *Jim Dandy, Lotus Art, Mitch Rigie*
Production supervisor: *Tandra Jorgensen*

The text was set in 10/12 Trump Mediaeval by Thompson Type and printed on acid-free 45# Pub Matte by Von Hoffmann, Owensville.

Cover art by Gary Overacre

The credits for this book begin on page C-1, a continuation of the copyright page.

Library of Congress Cataloging-in-Publication Data

Moore, Brooke Noel.
    Critical thinking / Brooke Noel Moore, Richard Parker. — 7th ed.
      p. cm.
    Includes index.
    ISBN 0-07-281881-6
    1. Critical thinking.   I. Parker, Richard.   II. Title.

  B105.T54M66 2003
  160—dc21   2003051343

INTERNATIONAL EDITION   ISBN 0-07-121491-7
Copyright © 2004. Exclusive rights by The McGraw-Hill Companies, Inc., for manufacture and export. This book cannot be re-exported from the country to which it is consigned by McGraw-Hill. The International Edition is not available in North America.

www.mhhe.com

To Alexander, Bill, and Sherry,
and also to Sydney, Darby, and Peyton Elizabeth

Contents

Preface

J on-David Hague, Ph.D., climbed out of his black, air-conditioned, 1995 Honda into appalling heat. "Nice place," he muttered, kicking aside an empty beer can and looking around. Dr. Hague is the new philosophy editor at McGraw-Hill, in town to hear our ideas for this edition. He was buying us breakfast at the Sin of Cortez, a Chico restaurant. We went inside and ordered Mexican food. Moore got down to business.

"Parts of the book," Moore said, "are starting to read like lectures. We want to get back to making things fun again."

Dr. Hague poured coffee and sat back in his chair. He was still wearing sunglasses.

"Surely," Parker joined in, "there's no reason this book can't be at least as enjoyable to college students as *Survivor.*"

Dr. Hague frowned. "What do you have in mind?" he asked. "Are you going to talk about eating bugs?"

We assured him we'd keep it proper and decent. We planned to add new features, make organizational modifications, rewrite the denser passages, improve the content, add new exercises, and update the political context—and the Editor of Good Taste at McGraw-Hill would not flinch. Much. We spent the next three hours drinking coffee and going over specifics.

Changes in the Seventh Edition

Not every change we discussed at the Sin of Cortez was actually made; in particular, we decided not to have adopters vote on which coauthor to kick off the title page. Here are the changes that we did make.

■ *New Features*

To this edition, we have added two unusual new features:

- ▲ The Top Ten Fallacies of All Time (Appendix 2)
- ▲ The Scrapbook of Unusual Issues (Appendix 3)

"The Top Ten Fallacies of All Time" list is based on our experience during the previous six editions gathering real-life examples of fallacies to use for text illustrations and exercises. The list isn't scientifically derived, but we have a pretty good idea of which fallacies occur most often.

"The Scrapbook of Unusual Issues" is a compendium of topics to generate discussion or to adapt for homework assignments or in-class mini-writing exercises.

■ *Modified Organization*

We've modestly modified the organization of the book and now divide it into Part 1: Claims and Credibility, Part 2: Fallacies and Other Rhetorical Devices, and Part 3: Arguments.

As you can see, we devote an entire part to fallacies and other rhetorical devices. The emphasis in this edition is on traditional fallacies rather than "pseudoreasoning," although much of the change is terminological. We are also using more self-explanatory titles for fallacies. Specific content changes are noted below, where you will observe that we now cover several fallacies that were missing from previous editions.

We no longer have a separate chapter on explanations. If you used the old Chapter 7, don't worry. The essential differences between explanations and arguments are set forth in Chapter 1 (Thinking Critically: It Matters); explanations are covered in their own right in Chapter 11 (Causal Arguments). This change enabled us to devote Part 2 entirely to fallacies and other rhetorical devices. It also eliminated redundancy.

If you've devoted time in the past to the book's appendices, you should be aware that material that was in the third appendix (Some Common Patterns of Deductive Arguments) now appears in boxes in Chapters 8 and 9. Likewise, the material on conflicting claims and on analytic claims that previously was covered in the first and second appendices can be found in Chapters 8 and 9 in the first case and on our Web site in the second. One of us (Moore) doesn't assign the chapters on deductive logic (now Chapters 8 and 9) to his classes (and you needn't feel obliged to, either) and instead just goes over the commonly encountered patterns of deductive reasoning. If you are like Moore and don't try to teach deductive logic from the ground up, you may want to refer to the boxes on pages 292–293 and 329 where these patterns are displayed. (Parker, on the other hand, is a firm believer that learning the logical structure of sentences—both categorical and truth functional—is a pedagogically sound endeavor and so drives his students through Chapters 8 and 9 in a manner reminiscent of a row master with a hold full of galley slaves.)

■ *Content Changes*

In this edition, we have modified several chapters; here are the specifics.

Chapter 1, Thinking Critically: It Matters   We still cover the basics here, but we have improved the discussion of *factual* versus *nonfactual* issues and made it fit better with the section in Chapter 12 on moral reasoning. In Chapter 1, we now explain the difference between *argument* and *persuasion* and between *argument* and *explanation.* Also discussed are *subjectivism* and *relativism.*

Chapter 3, Credibility   We take a new approach to credibility in this chapter. Before, we tried to give sufficient conditions for a claim's being credible; now we discuss ways in which a claim or source can lack credibility. We include a section on advertising (found in Chapter 4 in earlier editions), and we say something about talk radio as well.

Chapter 4, Persuasion Through Rhetoric   We make a clearer distinction now between rhetoric as the venerable study of persuasive writing and rhetoric as a broad category of linguistic techniques people use when their primary objective is to influence beliefs, attitudes, and behavior. We also emphasize the distinction between logical power and psychological force.

Chapter 5, Psychological and Related Fallacies   We have substantially rewritten this chapter, placing more emphasis on fallacies than on pseudo-reasoning, covering several new fallacies, and making minor changes to terminology.

Chapter 6, More Fallacies   In this edition, we offer what we hope is a clearer explanation of the underlying mistake in all versions of ad hominem argument; we have made a few minor changes in terminology as well. The fallacies now covered in Chapters 5 and 6 are these:

the Rush Limbaugh fallacy (the "argument" from outrage)
scapegoating
scare tactics
the "argument" by force
the "argument" from pity
the "argument" from envy
apple polishing
guilt trip
wishful thinking
the peer-pressure argument
the group think fallacy
nationalism
rationalization

the "argument" from popularity
the "argument" from common practice
the "argument" from tradition
relativism
subjectivism
red herring/smokescreen
the personal attack ad hominem fallacy
the circumstantial ad hominem fallacy
the inconsistency ad hominem fallacy
poisoning the well

the genetic fallacy	the line-drawing fallacy
straw man	slippery slope
false dilemma	begging the question
the perfectionist fallacy	misplacing the burden of proof

Chapter 10, Inductive Arguments   We have revised our approach to the relationship between analogical arguments and generalizations. The result, with luck, is a chapter that makes sense intuitively, does not overly complicate things, and is easier to teach.

Chapter 11, Causal Arguments   Several changes were made here. We added a section on how to avoid confusing arguments with causal explanations, a section on the ways in which causal hypotheses can be defective, and a section on explanations and excuses. Much of the latter two parts was included in the former chapter devoted to explanations. We've also simplified the discussion of common mistakes made in causal reasoning. Chapters 10 and 11 cover the most important inductive fallacies:

- ▲ biased sample
- ▲ hasty generalization (including appealing to anecdotal evidence either to establish or to rebut a generalization)
- ▲ post hoc, ergo propter hoc (and related mistakes)

Chapter 12, Moral, Legal, and Aesthetic Reasoning   The main change here was to bring the discussion of the difference between prescriptive and descriptive claims into conformity with what we say about it in Chapter 1. We also try to walk readers through the important conceptual distinctions:

- ▲ factual claims versus nonfactual claims
- ▲ nonfactual claims that are value judgments versus nonfactual claims that are not value judgments
- ▲ moral value judgments versus nonmoral judgments

This finely tuned analysis of these important distinctions enables us to define moral reasoning more precisely.

■ *Other Changes, Including Additional Supplements*

We have added new exercises and removed clunkers. Many new boxes have been added as well and some of the old ones removed. In general, the book has been updated to reflect the current political context.

Also, we have added new real-life essays at the end of the book (Appendix 1) and have even included a reasonably well-argued essay you could use as a model. Further, we offer an exercise that requires using bits and pieces of most of the book and explains how this works. If students asks why it is useful to study critical thinking, you could refer them to this exercise.

In addition, we have improved our interactive student support material. Those who use this text will enjoy a full range of Internet-based services and enhancements.

▲ A new Online Learning Center at www.mhhe.com/criticalthinking provides interactive exercises and resources for students and instructors. It is compatible with McGraw-Hill's free PageOut course management system as well as WebCT and Blackboard, allowing instructors to create customized course Web sites.

▲ Packaged with each new copy of the text is a free subscription to the online resource *PowerWeb: Critical Thinking,* which contains readings from classical sources (Plato, Locke, Bacon, and more), readings on such topics as contemporary issues and perspectives, language, the media, and pseudoscience and the paranormal; and essays for critical analysis. Students can access these readings from the Online Learning Center.

▲ Packaged free with each new copy is also Nickolas Pappas's popular *Study Guide to Accompany Critical Thinking* on CD-ROM.

Finally, as we indicated above, we have tried to make difficult discussions more accessible and fun to read.

Distinguishing Features of This Text

We are gratified by the success of this text. Its basic features may help explain its popularity.

■ *Focus*

Critical thinking includes a variety of deliberative processes aimed at making wise decisions about what to believe and do, processes that center on evaluation of arguments but include much more. We believe the best way to teach critical thinking is to integrate logic, both formal and informal, with a variety of skills and topics useful in making sound decisions about claims, actions, and practices—and to make it all palatable by presenting it in real-life contexts. This book is chatty in tone—the author of another critical thinking text griped about this (his book certainly does not have the problem)—but it doesn't duck important issues.

■ *Illustrations and Examples*

These are taken from or designed to resemble material undergraduates will find familiar. First- and second-year university students generally enjoy the book.

■ *Boxes*

On more than one occasion, we have heard students exclaiming how much they like the boxes. We do too.

■ *Exercises*

The exercises give guided practice in applying important critical skills. There are more than a thousand in the text and many more in *The Logical Accessory* (the instructor's manual).

■ *Answers, Suggestions, and Tips*

Questions marked with a triangle in the book are answered in the answer section in the back (look for the colored page edges), and sometimes discussions that extend material in the text proper are also found there. Instructors may find the answer section useful as a direct teaching aid or as a foil for their own comments.

■ *Collaborative Exercises*

We try to involve students actively in the learning process. Several exercises require students to collaborate with one another; these exercises work pretty well. Sometimes we even use these exercises *before* explaining the material in the chapter in which they occur and before assigning the chapter as homework.

■ *Organization*

Sometimes people support what they assert by providing arguments; other times they don't. In both cases, they may try to add psychological force to what they say by using various rhetorical techniques. Logically, then, you need a critical thinking text that has three parts: one dealing with claims and credibility, another dealing with various rhetorical techniques, and a third dealing with arguments. That's how we've organized things, anyhow.

■ *Discussion of Rhetorical Devices*

Part 2 deals with a large and diversified inventory of persuasive devices, emotional appeals, and irrelevancies that we all use to add psychological power to our assertions. This part of the book helps students distinguish weak reasons from irrelevant considerations, a subtle but important distinction.

■ *Credibility*

In many cases, we must assess the credibility of sources as well as the inherent believability of what they assert. This book devotes an entire chapter (Chapter 3) to credibility, authority, and expertise.

■ *Real-Life Essays*

At the end of the book (Appendix 1) are several essays that can be used in various ways: For analysis, to provide topics for written work or in-class discussion, or as sources of additional exercises of your own design. You can read them just for amusement, too.

■ *Accommodating Alternative Teaching Strategies*

If you want to teach a traditional course in logic, you can do so using nothing more than this book. We advise first covering Chapters 1 and 7, then working through Chapters 8–11 in order. In whatever time remains, take advantage of the material in Part 2, which deals with rhetorical devices of various sorts. These devices are major players in the attempts people in the real world make to persuade each other. Adding this material to a traditional class in logic adds a powerful practical dimension to the course.

On the other hand, if you don't like to teach elementary logic, skip Chapters 8 and 9, which, we might just point out, give pretty complete treatments of categorical and truth-functional logic.

■ *Teaching Writing*

If you teach critical thinking in a basic writing course or teach basic writing in a critical-thinking course, you can adapt this book to your needs. Chapter 2 in particular is devoted to subjects related to argumentative essays.

■ *Critical Thinking Across the Disciplines*

Discussions of moral, legal, and aesthetic reasoning appear in Chapter 12.

■ *Model Essay*

In Appendix 1, we've taken an essay that deals with a controversial topic that we found in a moderately obscure magazine and edited it down to manageable length, and we offer it as an example of a fairly well-reasoned series of arguments designed to support a (probably not too popular) conclusion about the September 11, 2001, terrorist attacks. You can use it as an example of how to adduce arguments to a conclusion, or as a challenge to find rebuttals if one does not want to accept the conclusion.

■ *The Whole Enchilada Exercise*

Also in Appendix 1, this new feature illustrates how critical evaluation of a real-life issue involves skills from several chapters.

■ *A Chart of Rules of Deduction*

The rules of truth-functional deduction are gathered on one page in the form of a handy chart; we're mentioning this again, although we began doing it in the previous edition. We don't know why it took us that long to do it.

■ *Glossary*

The glossary at the end of the book defines important terms, usually correctly.

■ *The Logical Accessory (Instructor's Manual)*

The instructor's manual does not presume to guide instructors. But it does give (our) answers to the hundreds of exercises not answered in the answer section of the text. It also provides we-don't-know-how-many more examples, exercises, and test questions. We continue to include, here and there, various hints, strategies, lecture topics, tangents, and flights of fancy.

■ *Study Guide CD-ROM*

A fine student study guide, by Nickolas Pappas of the City College of New York, comes packaged with each new copy of the book. Professor Pappas has provided stimulating new exercises for each chapter that will help students master the content. We thank him profusely for his wonderful work.

■ *Essay Grading Rubric*

Grading rubrics are widely used in K–12 and are found increasingly on the college scene as well. Students like rubric-based grading. They think it reduces the subjectivity involved in evaluating essays. Our rubric is tucked into *The Logical Accessory.*

Acknowledgments

This isn't talk radio. We don't try to pin every last thing on the left or the right, on some group of poorly identified whackos, or on CNN. And we don't blame our mistakes on Jon-David Hague (our principal editor), on our fine production people, or on our excellent reviewers. We admit it, you'll just have to blame the two of us for the mistakes, as usual.

We want to thank a number of people, especially the various editors referred to above: Jon-Hyphen (as we've come to think of him); April Wells-Hayes, our meticulous production boss and den mother; Jen Mills, who oversaw the entire production process; Marty Granahan, who has the difficult job of talking people into letting us use their material just so we can criticize and poke fun at it; and our sublime copyeditor, Joan "Red Pencil" Pendleton.

Those who have advised us in the preparation of this edition or slipped us material, and to whom we are especially grateful, include

Charles Blatz, University of Toledo

K. D. Borcoman, Coastline Community College

Sandra Dwyer, Georgia State University

The Editor of Good Taste Hard at Work

Ellery Eells, University of Wisconsin–Madison

Geoffrey Frasz, Community College of Southern Nevada

Dabney Gray, Stillman College

Steven Hoeltzel, James Madison University

Eric Parkinson, Syracuse University

James Stump, Bethel College.

We continue to be grateful for the thoughtfulness and insight of those who reviewed the first six editions and who are now too numerous to list here. They helped the book improve with each edition, and we continue to appreciate their efforts.

Special thanks are due to our colleagues at Cal State Chico, Marcel Daguerre and Anne Morrisey; Dan Barnett, Butte College; and Linda Bomstad, Cal Poly–San Luis Obispo.

Finally, we add special thanks to Alicia Alvarez Giles and Marianne Larson for their affection, patience, and support in the ways that really count.

Chapter 1

Thinking Critically: It Matters

One sunny day last spring we saw our friend Ken out mowing his lawn. Ken was in his bare feet.

"Ken," we said, "have you lost your mind?"

Ken had an answer. According to statistics, he said, the overwhelming majority of lawn-mowing accidents involve people wearing shoes. Very few accidents involve people going barefoot.

"So your chances of having an accident are statistically much greater if you wear shoes," he reasoned.

Ken is entitled to his opinion. We are all entitled to our opinions. But that doesn't mean opinions are all equally reasonable. Ken's opinion, for example—and with all due respect—could use a little fine-tuning. We humans are clever enough to send spacecraft beyond the solar system, combine genetic material so as to alter the varieties of life, and build machines that outplay grand masters at chess—yet we frequently make mistakes in logic like Ken's. Despite the impressive accomplishments of the human intellect, one frequently comes face to face with examples of faulty reasoning, error, and misjudgment. In a recent Gallup poll 18 percent of those polled thought the sun revolves around the earth. Our colleague Michael Rich discovered that over half the students in his class think the first person to walk on the moon was Lance Armstrong. As we write these words there is a story in our newspaper about one Robyn Rouse, of Columbus, Ohio, who paid $25 for a pair of green contact lenses at a grocery store to match her tennis shoes. By the next morning she had developed a serious eye infection. It's a safe bet that all of us from time to time make decisions like this—decisions that are uninformed, poorly reasoned, or otherwise defective. Occasionally such decisions are disastrous. Robyn now needs a year of treatment and a cornea transplant to save her vision.

Clear thinking requires an effort and doesn't always come naturally. But one can get better at it if one is willing to work a bit and accept guidance

Vladimir Does Not Like This Book

▲ *The title conveys the idea that Vladimir Putin dislikes this book. In fact, it's safe to say Vladimir Putin has no opinion on this book one way or the other. He does not like it, but he does not dislike it either. The title is ambiguous (discussed in Chapter 2) and could be used as innuendo (discussed in Chapter 4).*

Careful reading is important to thinking critically.

here and there. We hope this textbook, in conjunction with whatever course you may be using it in, will help you develop some of the skills required to form intelligent opinions, make good decisions, and determine the best course of action—as well as recognize when someone else's reasoning is faulty or manipulative. In short, we hope this book helps you become a more careful and critical thinker.

We hasten to caution you that thinking critically is not about attacking people. If you are reading this book for a course, chances are you will be expected to critique others' ideas, and they will be asked to critique yours. Doing this, however, doesn't mean putting people down. Every single one of us makes mistakes, and it can be useful to have others help us see them. We appreciate it when someone points out that we have a low tire or that our lawn looks bad because it lacks iron. Likewise, we can appreciate it if someone suggests that our position, theory, or idea is incomplete, unclear, insufficiently supported, or in some other way not all it could be. And when we are on the other side, we can help others see the holes in their arguments. We don't do Ken a favor by pretending his idea about mowing the lawn is a good one. Critical thinking is more about helping others than attacking them; and to the extent we are able to think critically about our own ideas, it is about helping ourselves. Our goal is knowledge and understanding, not winning or coming out on top.

So you might say this is another one of those self-help books. But unlike self-help books that tell you how to stay young or make a million dollars, this book will help you understand when books like those, and other attempts to sell you on something, have good reasoning going for them—and when they don't.

Good reasoning doesn't happen in a vacuum. When we hear somebody express an opinion, we usually already have information on the topic and can generally figure out where to find more if more is needed. Having both the desire and the ability to bring such information to bear on decisions is part of the critical-thinking process. Critical thinking involves a lot of skills: Among them are the abilities to listen and read carefully and to stay informed. Reading the *New York Times* carefully from start to finish each day for a year might do even more for your ability to think clearly than reading this book. (But it would take a lot more time.)

Basic Critical Thinking Skills

What, concretely and specifically, is critical thinking about? When we take a position on an issue, we assert or claim something. The claim and the thinking on which it is based are subject to rational evaluation. When we do that evaluating, we are thinking critically. To think critically, then, we need to know

- ▲ When someone (including ourselves) is taking a position on an issue, what that issue is, and what the person is claiming relative to that issue—that is, what the person's position is
- ▲ What considerations are relevant to that issue
- ▲ Whether the reasoning underlying the person's claims is good reasoning
- ▲ And whether, everything considered, we should accept, reject, or suspend judgment on what the person claimed

Finally,

- ▲ Doing all this requires us to be levelheaded and objective and not influenced by extraneous factors.

You can fool all of the people all of the time if the advertising budget is big enough.

—Ed Rollins, Republican campaign adviser

Why books like this are important

These points will receive considerable coverage in this book, but we need to consider them briefly here.

Issues

Whenever we have to think critically, the first item of business is to make sure we are focusing on the correct issue. In the broadest sense, an issue is any matter of controversy or uncertainty; an issue is a point in dispute, in doubt, in question, or simply up for discussion or consideration. Should I mow the lawn barefoot? This question raises an issue. Should we get a new car? Is George W. Bush taller than his father? Are pit bulls more apt to bite than golden retrievers? Should we purchase Wal-Mart stock? As soon as we ask one of these questions, we have raised an issue. An **issue** is raised when a claim is in question. In fact, for our purposes, "question" and "issue" can be used pretty much interchangeably.

Many issues can be posed beginning with the word "whether." For example, the issues raised by the questions in the preceding paragraph can be posed as whether we should get a new car, whether George W. Bush is taller than his father, and whether pit bulls are more apt to bite than golden retrievers. But issues can be posed in other ways as well. For example, we might wonder *how many* miles per gallon a car will get, *when* the train arrived, *who* fired the fatal shot, or *what* the root causes of terrorism are. All of these become issues as soon as we consider them.

We don't need a discussion or a dispute between parties to have an issue; an issue can be raised in a single person's thoughts. Did you turn off the air conditioning before you left, you wonder? The issue here is whether you turned off the air conditioning; it's a question in your mind, so it is an issue for you.

It is common these days to call psychological problems "issues." You hear people say things like "She has an issue with this class" or "He has an issue with cats." These statements simply mean that she doesn't like this class and that cats upset him. A couple of years ago there was a popular song named "She's Got Issues," which employed the word in a similar vein. But as we use the term, an issue is never something a person *has*, and it certainly isn't something a person has *got*. There are lots of things you can do with issues—you can raise them, discuss or debate them, try to settle them, and ignore them, but one thing you can't do is simply *have* them. In this book we use "issue" the good, old-fashioned way, as what is raised when you consider whether a claim is true.

Notice also that an issue is different from a topic of conversation. Pet care can be a topic of conversation, but not an issue. How you should take care of your pet is an issue. Whether you are taking proper care of your pet is an issue. Whether dogs or cats are easier to take care of is an issue. But pet care is just a topic.

Knowing Both Sides of an Issue

According to a report in *The New Republic* magazine, written by journalist Karen Alexander, a course titled "The Politics and Poetics of Palestinian Resistance" offered in the spring of 2002 at the University of California, Berkeley, was listed in the course catalogue with this line: "Conservative thinkers are encouraged to seek other sections." The line was eventually removed.

The Roman politician and orator Cicero (106–43 B.C.E.) once advised: The person who knows only his side of an issue knows neither side. Cicero was exactly right, and his advice worth remembering.

We assume, based on the above report, that the UC Berkeley grad student who offered the course never read Cicero. He probably should have, would be our guess.

Arguments

It is all very well to take a position on a given issue, but if the issue matters to us, we want to be able to *support* our position. This is where arguments come in—they contain that support. If the issue is a simple one, like whether this key will open that lock, the argument may just be that we were able to make the relevant observations or that we have reliable information from a reasonable source. If the issue is complicated, like what causes Dutch elm disease, much more is required. Whatever the issue, arguments are what we use to resolve it, along with direct observation and information from reliable sources.

Let us define an **argument** as an attempt to support a claim or assertion by providing a reason or reasons for accepting it. The claim that is supported is called the **conclusion** of the argument, and the claim or claims that provide the support are called the **premises.** The premises, in other words, specify the reason (or reasons) for accepting the conclusion. When we offer an argument, we are reasoning.

To illustrate this familiar and important concept, suppose you see someone mowing his lawn in his bare feet. You might think to yourself,

> What the devil does that fellow think he is doing? He is going to hurt himself seriously if he doesn't watch out! I'd better tell him he should wear shoes.

You are considering the issue of whether you should warn this individual to wear shoes when he is mowing his lawn, and you have, in effect, given yourself the following argument:

> Premise: If this person doesn't wear shoes, he could be badly hurt.
> Conclusion: Therefore, I should tell him to wear shoes.

As you can see from this example, the conclusion of an argument states a position on an issue. The issue in this example is *whether* you should tell the person to wear shoes, and the conclusion of the argument is *that* you should do so.

In a moment we shall offer some advice on identifying issues and recognizing arguments, but first we need to clear up two common misconceptions about arguments.

Misconceptions About Arguments

The first misconception is that arguments are attempts to *persuade* somebody of something. Some writers even define an argument as an attempt to persuade. This definition is not as perverse as mowing the lawn in your bare feet, but it is pretty obviously incorrect. True, when you want to persuade somebody of something, you might *use* an argument. But not all arguments attempt to persuade, and many attempts to persuade do not involve argument. In fact, giving people an argument is probably one of the *least* effective methods of persuading them—which, of course, is why so few advertisers bother with arguments. (If you really want to persuade people, your best bet may simply be to flatter them or make them angry or frightened; but we shall go into this later on.)

So it is false to think all attempts to persuade are arguments. It is equally false to think that all arguments attempt to persuade. Just consider the argument in the example about the person mowing the lawn barefoot. *That* argument is not an attempt to persuade anybody of anything, although having given yourself that argument you might *then* try to persuade the person to go put on shoes. Properly understood, an argument is not an attempt to *persuade,* but an attempt to *prove* or *establish* or *support* some claim. The proper definition of an argument is that it is two or more claims, one of which, the conclusion, is supposed to follow from or be supported by the rest. An attempt to persuade is an attempt to win someone to your point of view.

The second misconception about arguments is that they are attempts to *explain.* Here again, explanation and argument are distinct entities, although at times it is easy to confuse them. An argument attempts to prove *that* some claim is true, while an explanation attempts to specify *how something works* or what *caused* it or *brought it about.* Arguing *that* a dog has fleas is quite different from explaining *how* it came to have fleas. Wondering *whether* violent crimes increased is different from wondering *what caused* violent crimes to increase. Offering an explanation of Dutch elm disease is entirely different from trying to prove that your explanation is correct. Explanations and arguments are different things. However, they are easily confused, and we will include an exercise in this chapter that should help you keep them straight.

Up a Tree

▲   *Big Boy sits atop his favorite branch in Gulfport, Miss. He was blown there three years ago by Hurricane Georges.*

People who don't think critically get tossed around by the winds of fate, like Big Boy here.

Recognizing Arguments

Being able to identify arguments can help you identify issues, so let's discuss arguments first and then turn to identifying issues.

An argument *always* has a conclusion. Always. Without a conclusion, a bunch of words isn't an argument. That, after all, is what an argument does: It tries to prove, establish, or confirm a claim. In other words, it sets forth a conclusion.

Fortunately, specific words and phrases often, though not always, alert us to the fact that a conclusion is about to be set forth. Take the word "therefore":

> The U.S. steel industry believes it needs government protection to be able to be competitive internationally. Therefore, the industry won't support a free trade bill.

The word "therefore" is a **conclusion indicator:** It lets an audience know that what follows (the industry won't support a free trade bill) is a conclusion.

Other fairly reliable conclusion-indicating words and phrases include

- ▲ It follows that . . .
- ▲ We may conclude that . . .
- ▲ This serves to show that . . .
- ▲ Thus . . .
- ▲ Hence, . . .
- ▲ Accordingly, . . .
- ▲ Consequently, . . .
- ▲ So . . .

An argument, then, always has a conclusion. It also always has at least one premise. Premises support the conclusion, and if you don't have support, you don't have an argument. Words and phrases that often indicate a premise is about to be stated **(premise indicators)** include

- ▲ Since . . .
- ▲ For . . .
- ▲ Because . . .
- ▲ In view of . . .
- ▲ This is implied by . . .
- ▲ Given . . .

For example, somebody says:

> It's risky to speed on I-75 in Cincinnati, since the police like to set up speed traps for folks driving up from Kentucky.

The premise of this argument is "the police like to set up speed traps for folks driving up from Kentucky." That claim is given as support for the idea that it is risky to speed on I-75 in Cincinnati. You know what the premise is because it follows the word "since."

Unfortunately, even though an argument always has a conclusion and at least one premise, the conclusion or one or more premises might be implied rather than stated overtly. For example, if François says to Desmonde:

> Des, it's a nice day, we aren't doing anything, and the car needs to be washed.

▲ *Moore and Parker's students discover they passed.*

You can be pretty sure that François is giving Desmonde an argument for the unstated conclusion that they should wash the car.

Many arguments, like François', don't contain indicator words like "therefore" and "because." This means you just have to pay attention to whether someone is trying to establish something and what he or she is trying to establish.

By the time you have finished this book, you will have had much practice in identifying arguments, premises, and conclusions.

Identifying Issues

If you listen to a group of people discussing something, at some point in the conversation you might well hear something like this:

> FIRST PERSON: "Look. The issue is whether . . ."
>
> SECOND PERSON: "No, actually, the issue is . . ."
>
> THIRD SPEAKER: "Hold on . . . what we really need to determine is . . ."

And so on.

Why do people have so much trouble identifying issues? Discussions in the real world (including real-life written essays such as might be found in letters on a newspaper opinion page) are spontaneous, freewheeling, non-sequential, unorganized, haphazard, off the wall, and all over the map. Good writers and speakers certainly try to be clear about what issues they are discussing as well as about their position on them. But in free-flowing discussions and conversations (and in one person's own thinking), several issues can get attention more or less simultaneously; tangents will be followed without hesitation; irrelevancies and asides will be everywhere; and the parties to the discussion will frequently make confusing statements about what *the* issue is. Another problem is that people sometimes purposely confuse issues in order to draw attention away from a claim they don't want to deal with or to make it look as though they have proved a point when they haven't. And still a further problem is that in many conversations different speakers will address entirely different issues without realizing they're doing so. Suppose you hear:

> FIRST PERSON: School vouchers? They may be a good idea.
> They'll give parents an opportunity to get their
> kids out of bad schools. What do you think?
>
> SECOND PERSON: I think the people who want them are just a
> bunch of selfish zealots who want to send their
> kids to religious schools.

What is the issue here? We can't identify *the* issue. The first person's point has to do with vouchers, and the second person's point has to do with the people who want vouchers; they're talking about entirely different issues.

Let's take another example:

> MR. X: They shouldn't go building power plants right on fault zones
> like that! What's the matter with people, anyway?
>
> MR. Y: What's the matter with you? If it weren't for power plants,
> you wouldn't have enough light to read about faults.

These two are not addressing the same issue: Y's being right on his point won't help anyone know whether X is right on his.

So, identifying issues can be confusing. Still, it helps to remember the following two points:

First, since every argument addresses an issue, an excellent way to pin down at least some of the issues in a conversation or passage is to look for the conclusions of any arguments that have been given. Let's say a friend tells you:

> You should donate your old car to United Cerebral Palsy because doing so helps
> out a worthy cause and is tax deductible besides.

Your friend has argued you should donate your old car to United Cerebral Palsy. He or she is therefore taking a position on the issue of whether you

On Knowing What Your Great-Grandparents Knew

Having a reservoir of information in your head helps you avoid being misled. Here are a few things an eighth-grader was expected to know in 1895 in Kansas, along with sample questions from the final examination for that grade:

- Grammar (Sample exam question: Name and define the parts of speech.)
- Arithmetic (Sample question: Find the interest on $512.60 for 8 months and 189 days at 7 percent.)
- U.S. History (Sample question: Explain the causes and results of the Revolutionary War.)
- Orthography (Sample question: Define the following prefixes: bi, dis, mis, pre, semi, post, non, inter, mono, sup.)
- Geography (Sample questions: Name all the republics of Europe and give the capital of each. Give the inclination of the earth.)

This helps explain why our grandparents know so much.

should donate your old car to United Cerebral Palsy. The issue addressed by an argument can *always* be stated by inserting a "whether" in front of the argument's conclusion.

Second, remember that in many conversations there is no such thing as *the* issue. So, instead of trying to find out what *the* issue is, just focus on the issues that have been addressed with an argument. Look for arguments as guides to issues, because in offering a conclusion a speaker or writer is always taking a position on an issue. This strategy will not necessarily disclose what the most important issue is in a conversation or discussion, and it won't reveal every issue that has been raised because sometimes issues are raised just by asking questions. But it will at least tell you which issues are important enough to warrant somebody's providing an argument.

Factual Issues Versus Nonfactual Issues

Before we leave the subject of issues, we must call your attention to a most important distinction, the distinction between **factual** and **nonfactual issues/ questions.** Whether George W. Bush is *taller* than his father is a factual issue, but whether he is *better looking* than his father is not. Whether Dutch elm disease will spread to Oregon is a factual issue, but whether Dutch elms are the most stately trees is not. People sometimes think that only factual issues are worth discussing, but that isn't so. Is it right to run medical experiments on animals? That isn't a factual question, but it's worth discussing.

Should you contribute to the support of your aging parents? That isn't a factual question, but it is important.

People also speak of factual claims. When we take a position on a factual issue we make a **factual claim.** For example, the claim "George W. Bush is really a cleverly disguised space alien" is a factual claim. That isn't a misprint: *It is a factual claim.* It is a *false* claim, but it is indeed a factual claim, since the question of whether George W. Bush is a cleverly disguised space alien is a factual question/issue. *Saying that a claim is factual is not equivalent to saying it is true.* At the risk of being repetitious, we must emphasize: "Factual claim" does not mean "true claim." A factual claim is simply a claim, whether true or false, that states a position on a factual issue.

Obviously, then, you need to be able to tell if an issue/question is factual. *An issue is factual if there are established methods for settling it—* that is, if there are generally accepted criteria or standards on which the issue can be judged or if we can at least describe what kinds of methods and criteria would apply even though it may be impractical or impossible to actually apply them. We can determine whether George W. Bush is taller than his father by placing the two Bushes side by side and observing. We also could come up with criteria for settling whether George W. Bush is a space alien. But what could settle whether Bush Senior or Bush Junior is the better looking of the two?

Notice that, if there are no established methods for settling an issue, then, if two people disagree on that issue, there is no way to determine that either of them is mistaken. This is a mark of a nonfactual issue, since, if an issue is factual, and two people disagree about it, then at least one of them must be mistaken. Thus, if Moore and Parker disagree about whether George W. Bush is taller than his father, one of them must be mistaken. But, if they disagree about whether George W. Bush is better looking than his father, we do not insist that one of them must be mistaken. In fact, it seems odd to say that either one is "mistaken" about something like this.

So, here's how we can boil this down: There are two criteria that sort issues into the categories of factual and nonfactual: A factual issue will meet both of the following criteria:

There are established methods for settling the matter.

If two people disagree about the issue, at least one of them must be mistaken.

Any issue that fails these criteria is nonfactual.[1]

Using these two indicators of when an issue is a factual one, we see that the following questions all raise factual issues:

---

[1]It would be a mistake to think that the line between factual and nonfactual issues can be drawn with absolute clarity and distinctness. One might argue that there are some issues that might meet one of our criteria and fail the other. Difficult-to-determine cases do not affect the usefulness of the distinction, nor of this way of making it, however.

> ## Recognizing Factual Issues
>
> - If two people disagree on the issue, must one of them be incorrect or mistaken?
> - Is there an agreed-upon method or criterion for settling the issue?
>
> A "yes" answer to these questions indicates the issue is factual.

- Is there ice on the moon?
- Does poverty breed terrorism?
- Will it hurt my eyes to stick these cheapo green contact lenses in them?
- What time is it right now in Singapore?
- Hey, how much do you think we can get away with charging for plain water if we put it in a nice bottle and give it a French-sounding name?

And we see that the following questions raise nonfactual issues:

- Is it is okay to break a promise in order to save someone's life?
- Does this sweater make me look funny?
- Is the Italian coast as beautiful as people say?
- Does the death penalty give murderers what they deserve?
- Is it unwise to mow the lawn in your bare feet?

Some people have the idea that "factual" somehow equates to "uncontroversial," that an issue becomes "factual" when it is no longer subject to controversy. This is *not* correct. Does raising taxes cause recession? Did humans evolve from more primitive primates? Does the death penalty deter murder? These questions all raise controversial issues, but they are in each case factual issues.

Relativism and Subjectivism

Two big mistakes in thinking relate to this distinction between factual and nonfactual issues. One mistake comes from those who make the distinction, and the other from those who do not.

The first mistake is to think that, when it comes to nonfactual issues, all opinions are equally reasonable. This view is *not* correct. Whether Bill should turn down a really good job offer in North Carolina to take care of his dying father in California is an important nonfactual issue; it can be intelligently discussed and argued, and the arguments can vary in their merit. A person who argues that Bill should turn down the job offer "because the mosquitos are bad in North Carolina" has not offered as good an argument as the person who says that Bill should turn down the job offer because one's first duty is to one's parents. Chapter 12 of this text is devoted in part to moral judgments like this one, as well as to other types of nonfactual judgments; if you stick it out that far you will get guidance on why some opinions on nonfactual issues are more reasonable and rational than others.

The other mistake is made by those who fail to see a distinction between factual and nonfactual issues. One important version of this mistake is known as subjectivism, and another important version of the mistake is known as relativism. **Subjectivism** is the idea that, just as two people can both be "correct" in their opinion on a nonfactual issue, they can both be correct in their differing opinions on *factual* issues. **Relativism** is the parallel idea that two different *cultures* can be correct in their differing opinions on the same factual issue. We'd be surprised if you didn't encounter versions of subjectivism or relativism in things you read and courses you take, because both ideas are enjoying a measure of popularity these days, especially, maybe, in literature departments. But it would be unwise to take either doctrine literally. Let's say two physicians differ on whether you have cancer—a factual issue. It would be folly to suppose they might both be correct. And you can be sure that if you do have cancer, you won't be cured just by moving to a culture that hasn't heard of the disease.

Relevance, Rhetoric, and Keeping a Clear Head

Mom's opinions are bound to carry extra weight with her kids. They may even carry more weight than the opinions of someone who is better qualified on the subject at hand; after all, she is Mom. It is a fact of life that we are influenced in our thinking by considerations that, logically, are beside the point; a speaker's relationship to us is a good example of such a consideration.

Or let's say an expert is talking about the nutritional value of spinach. Let's also say he or she speaks with a refined British accent. For a lot of people the accent might lend an air of expertise to what the person says, even though a person's accent is unrelated to knowledge of nutrition or spinach.

To take an example that works in the other direction, many people attach *less* weight to what a speaker says if he or she (a) hesitates or stumbles over words, (b) is reluctant to make eye contact, (c) is perspiring or is nervous. Such behaviors may enter into our evaluation of what the person said—which, of course, is why speaking coaches encourage speakers to make eye contact and practice a smooth delivery. It can be really hard to evaluate an argument or claim objectively if the speaker exhibits mannerisms that we associate with evasiveness, insecurity, or dishonesty.

Now, one of the most serious and difficult obstacles to clear thinking is the tendency to confuse extraneous and irrelevant considerations with the

The Talk-Show Syndrome

*Those of us in the lay community don't reason that way; we don't
think in those terms of logic; it doesn't compute with us, that type
of philosophical reasoning; we don't care.*

—Radio talk-show host BRUCE SESSIONS, summarizing his reaction to a
logician giving examples of illogical arguments from Rush Limbaugh

Our suspicions are confirmed.

merits of a claim. The examples we've given deal with such irrelevancies as
a speaker's accent, dress, relationship to us, and mannerisms. Another and
different obstacle is the psychological force of words, which is easy to confuse with their logical force. Consider the different impact of these two
statements:

> The desperate attempts by the hawks in the Bush Administration to link Saddam Hussein to al-Qaeda only highlight the obvious fact that an invasion of Iraq
> has not the slightest relationship to the so-called war on terrorism.

> An invasion of Iraq is unrelated to the war on terrorism.

Logically these two statements say exactly the same thing, but the first
statement has greater psychological impact. The first statement, you might
say, is equivalent to the second statement with a little advertising thrown
in. Being ever alert to the psychological or emotional or rhetorical force of
words can help us respond objectively to assertions and not impulsively.

Earlier in this chapter, we made a distinction between argument and
persuasion. Argument attempts to prove a point; persuasion attempts to
win others to a point of view. It's a subtle distinction, but a very real one.
Often people use argument to persuade others, and there is absolutely nothing wrong with doing that. There also is nothing wrong with being persuaded by an argument if the argument is a good one. But attempts at persuasion often rely not so much on logic as on the psychological or rhetorical
power of words, and one needs to be sensitive to the psychological associations of words to avoid being manipulated. We need to avoid being seduced
by the emotional coloration of the language that surrounds a claim, proposal,
theory, idea, or argument. Newt Gingrich, a former Republican Congressperson and Speaker of the House, is reported to have advised Republicans
to use the words "extreme," "traitor," and "treasonous" when referring to
Democrats or their proposals. Words like these can inflame listeners' passions—and may make it difficult for them to evaluate ideas on their merits.
Demagogues rely on the emotional associations of words to scare us, flatter
us, and amuse us; to arouse jealousy, desire, and disgust; to make good
things sound bad and bad things sound good; and to confuse, mislead, and

Everyone Is Entitled to His or Her Opinion

At first glance, it seems obviously true that everyone is entitled to his or her opinion; and indeed we ourselves stated this casually in the text.

But the statement requires qualification. The *police* don't force us to hold any particular opinion; and in that sense certainly everyone is entitled to his or her opinion. However, that doesn't mean that all opinions are equally intelligent, practical, or humane. Some opinions may be so bad, stupid, or dangerous that it really isn't clear that one *is* entitled to hold them. Is one entitled to the opinion that it's okay to put out a dog's eye for the fun of it? Or that Israel bombed the World Trade Center? It makes as much sense to say that someone who holds such views is entitled to an explanation of why he or she is mistaken.

The remark "Everyone is entitled to his or her opinion" is often just a conversation stopper, a way of saying "Hey, I don't want to argue about this anymore." Now certainly there are times when further discussion is useless, but that does not mean that every opinion is as good—is as "entitled" to be held—as every other. Giving equal worth to people does not require giving equal worth to their opinions.

misinform us. Critical thinking involves recognizing the rhetorical force of language and trying not to be influenced by it. This is so important that we devote three full chapters of this book to the subject.

However, you should understand that psychological and emotional coloration is also present when good and decent people state their honest opinions; persuasion is not limited to demagogues. There is nothing wrong with presenting one's views in the best and most favorable light or in trying to be as persuasive as possible. It's just that as consumers of thoughts and ideas we must refine our ability to distinguish between the thought itself and the psychological packaging in which it is given to us.

None of this should be construed as implying that critical thinking rules out feelings or emotions. Feelings and emotions can supply powerful and often perfectly safe motivations for doing things. But it is important that we also employ our ability to reason—to be swayed by good arguments and to ignore irrelevancies—if we are to live up to our potential as reasonable creatures.

Concepts and Terms

Some of the bedrock concepts in this book use a vocabulary straight from ordinary English. People have *opinions, views, thoughts, beliefs, convictions,* and *ideas,* and for our purposes these things are the same. People may also express these opinions, views, thoughts, and the like, in *statements, judgments, assertions,* or—to use our preferred word—*claims.* "Statement," "judgment," "assertion," and "claim" all mean the same for us. And, as we

Doing Things with Words

You should not get the idea from this chapter that the only important thing you can do with words is make claims or take positions on issues. You can do lots of other important things: You can hypothesize, conjecture, suppose, and propose. You can amuse or entertain. You can try to persuade others (or yourself) of something or attempt to get them (or you) to do something. We use words to pray, promise, praise, and promote; to lie, deceive, insult, and humiliate; to excuse, comfort, and let off steam; and so on indefinitely. (Sometimes we don't know *what* we are up to when we use words.) All these things are subject to critical thinking as to success, efficacy, completeness, legitimacy, authenticity, originality, clarity, and many other qualities. In this book, however, we focus primarily on claim-making and argument-presenting functions of discourse and, to a lesser extent, on hypothesizing and conjecturing functions.

Here are some examples of a few of the many different things people do with words:

> *Red meat is not bad for you. Now blue-green meat, that's bad for you.*
>
> —Tommy Smothers, *amusing us*

> *I want to rush for 1,000 or 1,500 yards, whichever comes first.*
>
> —New Orleans Saints running back George Rogers, *expressing a desire*

> *I enjoyed reading your book and would look forward to reading something else you wrote if required to do so.*
>
> —E-mail from one of our students; *we'd like to think this is praise, but . . .*

> *Do not take this medication within two hours of eating.*
>
> —Caution note on some gunk one of us had to drink. *It's warning us, but notice that you can't tell if you're not supposed to take the medication within two hours before eating or within two hours after eating or both.*

> *Whenever I watch TV and see those poor starving kids all over the world, I can't help but cry. I mean, I'd love to be skinny like that but not with all those flies and death and stuff.*
>
> —Mariah Carey, *expressing compassion*

> *They know so little science that they don't realize how ridiculous they look to others.*
>
> —Marilyn Vos Savant, *offering her explanation of why people claim to be psychic*

> *It's due to the country's mixed ethnicity.*
>
> —National Rifle Association president Charlton Heston, *explaining the country's high murder rate and making it clear he may not know too much about the subject*

> *I did not have sexual relations with that woman.*
>
> —Bill Clinton, *telling a fib*

> *Osama bin Laden is either alive and well or alive and not too well or not alive.*
>
> —Defense Secretary Donald Rumsfeld; *beats us*

have seen, whenever a claim is called into question or its truth or falsity becomes the subject of discussion, an issue has been raised, the issue of whether or not the claim is true. We use the terms "issue" and "question" pretty much interchangeably.

About truth: Philosophers wonder what sort of thing truth is and what sort of things facts are. We must leave these mysteries for your next philosophy course. In a real-life situation, when people say that a claim is true, they are often just agreeing with it. Likewise, when they say that a claim states a fact, they are often just employing an alternative method of agreeing with it. Thus, in real-life conversations, these four claims often serve the same purpose:

> A book is on the table.
> It is true that a book is on the table.
> It is a fact that a book is on the table.
> I agree; a book is on the table.

Finally, let us say something about knowledge: Maybe you wonder, when can a person be said to *know* something? For example, when can I be said to *know* that there is a book on the table? In real-life situations, you have a right to claim to know there is a book on the table if (1) you believe there is, (2) have evidence beyond a reasonable doubt there is, and (3) have no reason to think you are mistaken, such as might be the case if you hadn't slept for thirteen days.

Advice

If you sense there is more than meets the eye to thinking critically, you are on the right track. Spotting issues and arguments and evaluating arguments and claims while striving for clearheadedness isn't easy. People learn these skills through example, illustration, and informed guidance. That's where this book comes in. But if you are to improve these skills, you will have to practice. Like playing baseball or the piano, critical thinking requires constant practice if you are going to be skilled in it. We've supplied lots of exercises throughout the book and you'll find that examples of the material we cover will turn up regularly in everyday life. Put what you learn into practice—critical thinking is not just a classroom activity.

Recap

Critical thinking helps you know when someone (including yourself) is taking a position on an issue, what that issue is, and what the person is claiming relative to that issue—that is, what the person's position is. It helps you

Doubts About Rationality

These days you hear some academics express doubts about rationality. Three fairly widespread ideas are these:

- Relativists maintain one is bound by the perspective of one's own culture. They say there is no meaning in the idea of getting beyond that perspective to see which beliefs are true in some objective sense. What is true, for the relativist, is what one's culture thinks is true.
- Post-Marxist scholars, some of them anyhow, say rationality reduces to power politics. A "strong" or "valid" or "good" argument (they say) is in reality simply an argument advanced by those in society who wield the most power.
- Cognitive scientists, many of them, think reasoning involves mostly unconscious processes. Emotional disengagement, they say, is impossible.

We criticize relativism in the chapter and won't say more about it here.

Cognitive scientists are no doubt right in saying that the conclusions we come to and the opinions we hold are formed in part, or even largely, by influences of which we are not aware. Nevertheless (as we shall explore in detail in Chapter 7 and the following chapters) universally valid distinctions can be made among arguments, and there are indisputably important respects in which one argument may be said to be better than another. It may not be possible to be governed in our opinions and actions entirely by the force of good argument, but it is possible to recognize a good argument; one is rational exactly to the extent one is able to do so.

As for the idea that an argument is good or valid or strong because those in power say it is good or valid or strong, you will find an almost endless supply of examples in this book of poor, invalid, and weak arguments advanced by powerful people, and good and valid and strong arguments not endorsed at all by those in power.

know what considerations are relevant to that issue and whether the reasoning underlying the person's claim is good reasoning. It helps you determine whether, everything considered, you should accept, reject, or suspend judgment on what the person claims. These skills require you to be level-headed and objective and uninfluenced by extraneous factors.

To avoid common misconceptions and confusions connected with critical thinking, we need to remember the following:

- Issues are not psychological problems.
- Not all arguments are attempts to persuade.
- Not all attempts to persuade are arguments.
- If persuasion is an objective, other methods can work as well as or better than arguments.

- ▲ Arguments and explanations are different things.
- ▲ Factual issues and claims may or may not be controversial.
- ▲ Nonfactual issues and claims can be important, intelligently discussed, and reasonably debated.
- ▲ Not all factual claims are true.
- ▲ Subjectivism, the view that two people with conflicting opinions on a factual issue can both be correct, is a defective view.
- ▲ Relativism, the view that two different cultures can be correct in having different opinions on the same factual issue, is a defective view.
- ▲ Two statements can differ in their psychological force yet have the same content.

Exercises

The exercises in this book are designed to provide practice in critical thinking. Some can be answered directly from the pages of the text, but some will require your careful consideration. For many exercises in this book, there is no such thing as *the* correct answer, just as there is no such thing as *the* correct way to serve a tennis ball or to make lasagna. But some answers, like some tennis serves and some lasagnas, are better than others. So in many exercises, answers you give that differ from your instructor's are not necessarily incorrect. Still, your instructor's answers most likely will be rational, sound, and worth your attention. In general, his or her judgment will be very reliable. We recommend you take advantage of your instructor's experience to improve your ability to think critically.

Answers to the exercise items marked with a triangle are found in the answer section (look for the colored edges) at the back of the book. You'll also find the occasional comment, tip, suggestion, joke, or buried treasure map back there.

Exercise 1-1

Answer the questions based on your reading of Chapter 1, including the boxes.

- ▲ 1. What is an argument?
-     2. List five fundamental critical-thinking abilities.
-     3. T or F: Critical thinking provides an opportunity to attack other people.
- ▲ 4. Are there situations when the wisest move is to accept a claim even though we are not entirely certain it is true?
-     5. What is an issue?
-     6. Are topics of conversation the same as issues?

▲ 7. Are all issues posed beginning with the word "whether"?

8. T or F: Whenever a claim has been questioned, an issue has been raised.

9. Do arguments all have conclusions?

▲ 10. Can a conclusion be implied, or must it always be explicitly stated?

11. Explain the connection between an argument and an issue.

12. Are all opinions equally reasonable?

▲ 13. Are all opinions about nonfactual issues equally reasonable?

14. If one of two people disagreeing on an issue must be mistaken, then is the issue factual or nonfactual?

15. Are factual issues ever controversial?

▲ 16. T or F: All arguments are used to try to persuade someone of something.

17. T or F: All attempts to persuade someone of something are arguments.

18. Does every argument address an issue?

▲ 19. T or F: Subjectivism is the idea that two people can both be correct if they disagree on a factual issue.

▲ 20. T or F: Relativism is the idea that two cultures can both be correct if they disagree on a factual issue.

21. T or F: Explanations and arguments serve the same purpose.

22. "Therefore" and "consequently" are conclusion-indicators.

23. T or F: "Rhetorical" or "emotive force" refers to the emotional content or associations of a word or phrase.

24. T or F: The rhetorical force of language can get in the way of clear and critical thinking.

25. T or F: We should not try to put our own position on any issue in the most favorable light.

Exercise 1-2

This exercise is designed to be done as an in-class group assignment. Your instructor will indicate how he or she wants it done. On the basis of a distinction covered in this chapter, divide these items into two groups of five items each such that all the items in one group have a feature that none of the items in the second group have. Describe the feature upon which you based your classifications. Compare your results with those of a neighboring group.

1. You shouldn't buy that car because it is ugly.

2. That car is ugly, and it's expensive too.

3. Rainbows have seven different colors in them, although it's not always easy to see them all.

4. Walking is the best exercise. After all, it is less stressful on the joints than other aerobic exercises.

5. The ocean on the central coast is the most beautiful shade of sky blue. It's more green as you go north.

6. Her favorite color is yellow because it is the color of the sun.

7. Pooh is my favorite cartoon character because he has lots of personality.

8. You must turn off the lights when you leave the room. They cost a lot of money to run and you don't need them on during the day.

9. Television programs have too much violence and immoral behavior. Hundreds of killings are portrayed every month.

10. You'll be able to find a calendar on sale after the first of the year, so it is a good idea to wait until then to buy one.

Exercise 1-3

Some of these items are arguments and some are not. Can you divide them up correctly?

▲ 1. Sampras is unlikely to win the U.S. Open again this year. He has a nagging leg injury, plus he just doesn't seem to have the drive he once had.

2. Hey there, Marco!—don't go giving that cat top sirloin. What's the matter with you, you got no brains at all?

3. If you've ever met a pet bird, then you know that they are very busy creatures.

▲ 4. "'Water resistant to 100 feet,' says the front of this package for an Aqualite watch, but the fine-print warranty on the back doesn't cover 'any failure to function properly due to misuse such as water immersion.'"

Consumer Reports

5. Everybody is saying the president has made us the laughingstock of the world. What a stupid idea! He hasn't made us a laughingstock at all. There's not a bit of truth in that notion.

6. "Is the author really entitled to assert that there is a degree of unity among these essays which makes this a book rather than a congeries? I am inclined to say that he is justified in this claim, but articulating this justification is a somewhat complex task."

From a book review by Stanley Bates

▲ 7. As a long-time customer, you're already taking advantage of our money management expertise and variety of investment choices. That's a good reason for consolidating your other eligible assets into an IRA with us.

8. MOORE: Well, I see where the new chancellor wants to increase class sizes.
PARKER: Yeah, another of his bright ideas.
MOORE: Actually, I don't think it hurts to have one or two extra people in class.
PARKER: What? Of course it hurts. What are you thinking, anyway?
MOORE: Well, I just think there is good reason for increasing the class size a bit.

9. John Montgomery has been the Eastern Baseball League's best closer this season. Unfortunately, when a closer fails, as Montgomery did last night, there's usually not much chance to recover.

▲ 10. Yes, I charge a little more than other dentists. But I feel I give better service. So I think my billing practices are justified.

11. If you want to purchase the house, you must exercise your option before June 30, 2003. Otherwise, you will forfeit the option price.

Exercise 1-4

Determine which of the following passages contain arguments, and, for any that do, identify the argument's conclusion. Remember that an argument occurs when one or more claims (the premises) are offered in support of a further claim (the conclusion). There aren't many hard-and-fast rules for identifying arguments, so you'll have to read closely and think carefully about some of these.

▲ 1. The *Directory of Intentional Communities* lists more than two hundred groups across the country organized around a wide variety of purposes, including environmentally aware living.

2. Carl would like to help out, but he won't be in town. So we'll have to find someone else who owns a truck.

3. In 1976, Washington, D.C., passed an ordinance prohibiting private ownership of firearms. Since then, Washington's murder rate has shot up 121 percent. Bans on firearms are clearly counterproductive.

▲ 4. Computers will never be able to converse intelligently through speech. A simple example proves that this is so. The sentences "How do you recognize speech?" and "How do you wreck a nice beach?" have entirely different meanings, but they sound similar enough that a computer could not distinguish the two.

5. Recent surveys for the National Science Foundation report that two of three adult Americans believe that alien spaceships account for UFO reports. It therefore seems likely that several million Americans may have been predisposed to accept the report on NBC's *Unsolved Mysteries* that the U.S. military recovered a UFO with alien markings.

6. "Like short-term memory, long-term memory retains information that is encoded in terms of sense modality and in terms of links with information that was learned earlier (that is, *meaning*)."

Neil R. Carlson

▲ 7. Fears that chemicals in teething rings and soft plastic toys may cause cancer may be justified. Last week, the Consumer Product Safety Commission issued a report confirming that low amounts of DEHP, known to cause liver cancer in lab animals, may be absorbed from certain infant products.

8. "It may be true that people, not guns, kill people. But people with guns kill more people than people without guns. As long as the number of lethal

weapons in the hands of the American people continues to grow, so will the murder rate."

Susan Mish'alani

9. June 1970: A Miami man gets thirty days in the stockade for wearing a flag patch on the seat of his trousers. March 2004: Miami department stores are selling boxer trunks made up to look like an American flag. Times have changed.

▲ 10. Levi's Dockers are still in style, but pleats are out.

11. There is trouble in the Middle East, there is a recession under way at home, and all the economic indicators have turned downward. It seems likely, then, that the only way the stock market can go is down.

12. Lucy is too short to reach the bottom of the sign.

▲ 13. "Can it be established that genetic humanity is sufficient for moral humanity? I think that there are very good reasons for not defining the moral community in this way."

Mary Anne Warren

14. Pornography often depicts women as servants or slaves, or as otherwise inferior to men. In light of that, it seems reasonable to expect to find more women than men who are upset by pornography.

15. "My folks, who were Russian immigrants, loved the chance to vote. That's probably why I decided that I was going to vote whenever I got the chance. I'm not sure [whom I'll vote for], but I am going to vote. And I don't understand people who don't."

Mike Wallace

▲ 16. "President Clinton's request for $1 billion to create summer jobs for low-income young people was killed in the Senate, forcing him to settle for $166.5 million. Jobs would help make them part of the real community and would represent a beacon of hope—the first step out of poverty and despair."

Christian Science Monitor

17. "Hayek argues that we cannot know enough about each person's situation to distribute to each according to his moral merit (but would justice demand we do so if we did have the knowledge?)."

Robert Nozick

18. The Great Lakes Coastal Commission should prepare regulations that are consistent with the law, obviously. We admit that isn't always easy. But there's no reason for the commission to substitute its judgment for that of the people.

▲ 19. We need to make clear that sexual preference, whether chosen or genetically determined, is a private matter. It has nothing to do with an individual's ability to make a positive contribution to society.

20. "Cinema rarely rises from a craft to an art. Usually it just manufactures sensory blizzards for persons too passive to manage the active engagement of mind that even light reading requires."

George Will

Exercise 1-5 ――――――――――――――――――――――――

For each passage in this exercise, identify which of the items that follow best states the primary issue discussed in the passage. Be prepared to say why you think your choice is the correct one.

▲ 1. Let me tell you why Hank ought not to take that math course. First, it's too hard, and he'll probably flunk it. Second, he's going to spend the whole term in a state of frustration. Third, he'll probably get depressed and do poorly in all the rest of his courses.
   a. Whether Hank ought to take the math course
   b. Whether Hank would flunk the math course
   c. Whether Hank will spend the whole term in a state of frustration
   d. Whether Hank will get depressed and do poorly in all the rest of his courses

2. The county has cut the library budget for salaried library workers, and there will not be enough volunteers to make up for the lack of paid workers. Therefore, the library will have to be open fewer hours next year.
   a. Whether the library will have to be open fewer hours next year
   b. Whether there will be enough volunteers to make up for the lack of paid workers

▲ 3. Pollution of the waters of the Everglades and of Florida Bay is due to multiple causes. These include cattle farming, dairy farming, industry, tourism, and urban development. So it is simply not so that the sugar industry is completely responsible for the pollution of these waters.
   a. Whether pollution of the waters of the Everglades and Florida Bay is due to multiple causes
   b. Whether pollution is caused by cattle farming, dairy farming, industry, tourism, and urban development
   c. Whether the sugar industry is partly responsible for the pollution of these waters
   d. Whether the sugar industry is completely responsible for the pollution of these waters

▲ 4. It's clear that the mainstream media have lost interest in classical music. For example, the NBC network used to have its own classical orchestra conducted by Arturo Toscanini, but no such orchestra exists now. One newspaper, the no-longer-existent *Washington Star,* used to have thirteen classical music reviewers—that's more than twice as many as the *New York Times* has now. H. L. Mencken and other columnists used to devote considerable space to classical music; nowadays, you almost never see it mentioned in a major column.
   a. Whether popular taste has turned away from classical music
   b. Whether newspapers are employing fewer writers on classical music
   c. Whether the mainstream media have lost interest in classical music

5. This year's National Football League draft lists a large number of quarterbacks among its highest-ranking candidates. Furthermore, quite a number

of teams do not have a first-class quarterback. It's therefore likely that there will be an unusually large number of quarterbacks drafted early in this year's draft.

    a. Whether teams without first-class quarterbacks will choose quarterbacks in the draft

    b. Whether there is a large number of quarterbacks in this year's NFL draft

    c. Whether an unusually large number of quarterbacks will be drafted early in this year's draft

6. An animal that will walk out into a rainstorm and stare up at the clouds until water runs into its nostrils and it drowns—well, that's what I call the world's dumbest animal. And that's exactly what young domestic turkeys do.

    a. Whether young domestic turkeys will drown themselves in the rain

    b. Whether any animal is dumb enough to drown itself in the rain

    c. Whether young domestic turkeys are the world's dumbest animal

7. The defeat of the school voucher initiative was a bad thing for the country because now there won't be any incentive for public schools to clean up their act. Furthermore, the status quo perpetuates the private-school-for-the-rich, public-school-for-the-poor syndrome.

    a. Whether there is now any incentive for public schools to clean up their act

    b. Whether the defeat of the school voucher initiative was bad for the country

    c. Two issues are equally stressed in the passage: whether there is now any incentive for public schools to clean up their act and whether the private-school-for-the-rich, public-school-for-the-poor syndrome will be perpetuated

8. From an editorial in a newspaper outside southern California: "The people in southern California who lost a fortune in the wildfires last year could have bought insurance that would have covered their houses and practically everything in them. And anybody with any foresight would have made sure there were no brush and no trees near the houses so that there would be a buffer zone between the house and any fire, as the Forest Service recommends. Finally, anybody living in a fire danger zone ought to know enough to have a fireproof or fire-resistant roof on the house. So, you see, most of the losses those people suffered were simply their own fault."

    a. Whether there were things the fire victims could have done to prevent their losses

    b. Whether insurance, fire buffer zones, and fire-resistant roofs could have prevented much of the loss

    c. Whether the losses suffered by people in the fires were their own fault

9. "Whatever we believe, we think agreeable to reason, and, on that account, yield our assent to it. Whatever we disbelieve, we think contrary to reason, and, on that account, dissent from it. Reason, therefore, is allowed to be the principle by which our belief and opinions ought to be regulated."

                              Thomas Reid, Essays on the Active Powers of Man

    a. Whether reason is the principle by which our beliefs and opinions ought to be regulated

b. Whether what we believe is agreeable to reason

c. Whether what we disbelieve is contrary to reason

d. Both b and c

10. Most people you find on university faculties are people who are interested in ideas. And the most interesting ideas are usually new ideas. So most people you find on university faculties are interested in new ideas. Therefore, you are not going to find many conservatives on university faculties, because conservatives are not usually interested in new ideas.

a. Whether conservatives are interested in new ideas

b. Whether you'll find many conservatives on university faculties

c. Whether people on university faculties are interested more in new ideas than in other ideas

d. Whether most people are correct

11. In pre–civil war Spain, the influence of the Catholic Church must have been much stronger on women than on men. You can determine this by looking at the number of religious communities, such as monasteries, nunneries, and so forth. A total of about 5,000 such communities existed in 1931; 4,000 of them were female, whereas only 1,000 of them were male. Seems to me that proves my point about the Church's influence on the sexes.

a. Whether the Catholic Church's influence was greater on women than on men in pre–civil war Spain

b. Whether the speaker's statistics really prove his point about the Church's influence

c. Whether the figures about religious communities really have anything to do with the overall influence of the Catholic Church in Spain

12. The movie *Pulp Fiction* might have been a pretty good movie without the profanity that occurred all the way through it. But without the profanity, it would not have been a believable movie. The people this movie was about just talk that way, you see. If you have them speaking Shakespearean English or middle-class suburban English, then nobody is going to pay any attention to the message of the movie because nobody will see it as realistic. It's true, of course, that, like many other movies with some offensive feature—whether it's bad language, sex, or whatever—it will never appeal to a mass audience.

a. Whether movies with offensive features can appeal to a mass audience

b. Whether *Pulp Fiction* would have been a good movie without the bad language

c. Whether *Pulp Fiction* would have been a believable movie without the bad language

d. Whether believable movies must always have an offensive feature of one kind or another

▲ 13. "From information gathered in the last three years, it has become clear that the single biggest environmental problem in the former Soviet Union—many times bigger than anything we have to contend with in the United States—is radioactive pollution from nuclear energy plants and nuclear weapons testing and production. Soviet communist leaders seemed to believe they could do anything to hasten the industrialization process and compete with Western countries, and that the land and natural resources they controlled were vast enough to suffer any abuse without serious consequence. The arrogance of the communist leaders produced a burden of misery and death that fell on the people of the region, and the scale of that burden has only recently become clear. Nuclear waste was dumped into rivers from which downstream villages drew their drinking water; the landscape is dotted with nuclear dumps which now threaten to leak into the environment; and the seas around Russia are littered with decaying hulks of nuclear submarines and rusting metal containers with tens of millions of tons of nuclear waste. The result has been radiation poisoning and its awful effects on a grand scale.

"A science advisor to former Russian President Boris Yeltsin said, 'The way we have dealt with the whole issue of nuclear power, and particularly the problem of nuclear waste, was irresponsible and immoral.'"

Adapted from the Washington Post

a. Whether communism failed to protect people from nuclear contamination as well as capitalism did
b. Whether nuclear waste problems in the former Soviet Union are much worse than had been realized until just recently
c. Whether former leaders of the Soviet Union made large-scale sacrifice of the lives and health of their people in their nuclear competition with the West
d. Whether communism, in the long run, is a much worse system than capitalism when it comes to protecting the population from harm

▲ 14. "The United States puts a greater percentage of its population in prison than any other developed country in the world. We persist in locking more and more people up despite the obvious fact that it doesn't work. Even as we build more prisons and stuff them ever more tightly, the crime rate goes up and up. But we respond: 'Since it isn't working, let's do more of it'!

"It's about time we learned that fighting criminals is not the same thing as fighting crime."

Richard Parker, radio commentary on CalNet, California Public Radio

a. Whether we build more prisons than any other country
b. Whether we imprison more people than do other countries
c. Whether reliance on imprisonment is an effective method of reducing crime
d. Whether attacking the sources of crime (poverty, lack of education, and so on) will reduce crime more than just imprisoning people who commit crimes

▲ 15. [As the Clinton administration planned to raise the fees charged for grazing animals on public land, several members of the Senate argued against the proposal.] "Senator Alan K. Simpson of Wyoming said higher prices would 'do those old cowboys in.'

"But a review of the top 500 holders of grazing permits on federal land shows that many of those 'old cowboys' are more likely to wear wingtips than boots. They include the Metropolitan Life Insurance Company, with 800,000 acres under its control, the Mormon Church, a Japanese conglomerate, the Nature Conservancy and some of the wealthiest families in the nation.

"The largest permit holders, the top 10 percent, control about half of the nation's public grazing land, according to Interior Department figures. Only 12 percent of the permit holders are listed by the government as small operators."
a. Whether Senator Simpson understands who holds grazing permits on federal land
b. Whether permits to graze on most public land are held by corporations or by individuals and families
c. Whether permits to graze on most public land are held by small operators or by large, wealthy operators
d. Whether the administration's plan to raise grazing fees is a reasonable idea

▲ 16. Letting your children surf the Net is like dropping them off downtown to spend the day doing whatever they want. They'll get in trouble.
a. Whether letting your children off downtown to spend the day doing whatever they want will lead them into trouble
b. Whether letting your children surf the Net will lead them into trouble
c. Whether restrictions should be placed on children's activities

17. When Rep. Paul McHale became the first Democrat to call for Clinton's resignation, the press pointed out that McHale wasn't running for reelection. But when Barbara Boxer, who was in a pitched battle for her Senate seat, lashed out at the president from the Senate floor, it became clear that attacking the president was politically safe.
a. Whether Barbara Boxer was fighting for her political life in California
b. Whether it was politically necessary for Barbara Boxer to attack the president
c. Whether attacking the president was politically safe
d. Whether Paul McHale's call for Clinton's resignation was sincere

18. Illinois state employees, both uniformed and non-uniformed, have been loyally, faithfully, honorably, and patiently serving the state without a contract or cost-of-living pay increase for years, despite the fact that legislators and governor have accepted hefty pay increases. All public employee unions should launch a signature-gathering initiative to place on the ballot a proposition that the Illinois constitution be amended to provide for compulsory binding arbitration for all uniformed and non-uniformed public employees, under the supervision of the state supreme court.
a. Whether Illinois state employees have been loyally, faithfully, honorably, and patiently serving the state without a contract or cost-of-living pay increase for years
b. Whether public employee unions should launch a signature-gathering initiative to place on the ballot a proposition that the Illinois constitution be amended to provide for compulsory binding arbitration for all

uniformed and non-uniformed public employees, under the supervision of the Illinois Supreme Court

c. Neither of the above

19. That Japan needs reform of its political institutions is hardly in doubt. The country is experiencing the worst recession since the Second World War, with forecasts of up to minus 3 percent growth this year. Japan is not only the world's largest economy, but it also dominates the East Asian economic zone, so recovery through that region depends directly on Japan. Reforms are urgently needed to ensure this recovery will happen.

a. Whether Japan needs reforms of its political institutions

b. Whether Japan is experiencing its worst recession since the Second World War

c. Whether Japan is the world's second largest economy

d. Whether reforms will ensure that recovery will happen

▲ 20. MOORE: So, what do you think of the governor?

PARKER: Not much, actually.

MOORE: What do you mean? Don't you think she's been pretty good?

PARKER: Are you serious?

MOORE: Well, yes. I think she's been doing a fine job.

PARKER: Oh, come on. Weren't you complaining about her just a few days ago?

a. Whether Parker thinks the governor has been a good governor.

b. Whether Moore thinks the governor has been a good governor.

c. Whether the governor has been a good governor

d. Whether Moore has a good argument for thinking the governor has been a good governor

Exercise 1-6

Identify the main issue in each of the following passages.

▲ 1. Police brutality does not happen very often. Otherwise, it would not make headlines when it does happen.

2. We have little choice but to concentrate our crime-fighting efforts on enforcement because we don't have any idea what to do about the underlying causes of crime.

3. A lot of people think that the gender of a Supreme Court justice doesn't make any difference. But with two women on the bench, cases dealing with women's issues are being handled differently.

▲ 4. "The point is that the existence of an independent world explains our experiences better than any known alternative. We thus have good reason to believe that the world—which seems independent of our minds—really is essentially independent of our minds."

Theodore W. Schick, Jr., and Lewis Vaughn, How to Think About Weird Things

5. Sure, some of the hotdoggers get good grades in Professor Bubacz's class. But my guess is that if Algernon takes it, all it'll get him is flunked out!

6. It is dumb to claim that sales taxes hit poor people harder than rich people. After all, the more money you have, the more you spend; and the more you spend, the more sales taxes you pay. So people with more money are always going to be paying more in sales tax than poor people.

▲ 7. If you're going to buy a computer, you might as well also sign up for some lessons on how to use the thing. After all, no computer ever did any work for its owner until its owner found out how to make it work.

8. Intravenous drug use with nonsterile needles has become one of the leading causes of the spread of AIDS. Many states passed legislation allowing officials to distribute clean needles in an effort to combat this method of infection. But in eleven states, including some of the most populous, possession of hypodermic syringes without a prescription is illegal. The laws in these foot-dragging states have to be changed if we ever hope to bring this awful epidemic to an end.

9. The best way to avoid error—that is, belief in something false—is to suspend judgment about everything except that which is absolutely certain. Because error usually leads to trouble, this shows that suspension of judgment is usually the right thing to do.

▲ 10. "[Readers] may learn something about their own relationship to the earth from a people who were true conservationists. The Indians knew that life was equated with the earth and its resources, that America was a paradise, and they could not comprehend why the intruders from the East were determined to destroy all that was Indian as well as America itself."

Dee Brown, Bury My Heart at Wounded Knee

Exercise 1-7

For each item, identify the issue that the first speaker is addressing. Are the two speakers addressing the same issue?

Example

THERESA: I think toilet paper looks better if it unwinds from the back side of the spool.

DANIEL: No way! It looks stupid that way. It should unwind from the front side of the spool.

Analysis

The issue for both Theresa and Daniel is whether the toilet paper looks better if it unwinds from the front of the spool. This issue is not a factual matter.

▲ 1. MR.: Next weekend we go on Standard Time again. We'll have to set the clocks ahead.
      MRS.: It isn't next weekend; it's the weekend after. And you set the clocks back one hour, not ahead.

   2. BELIEVER: Ghosts exist. People everywhere in all cultures have believed in them. All those people couldn't be wrong.
      SKEPTIC: If ghosts exist, it's not for that reason. People once believed Earth was flat, too.

   3. SHE: You don't give me enough help around the house; you hardly ever do anything.
      HE: That's not true. I mowed the lawn on Saturday and I washed both of the cars on Sunday. What's more, I've been cleaning up after dinner almost every night and I've hauled all that stuff from the garden to the dump. So I don't see how you can say I hardly ever do anything.
      SHE: Well, you don't want to hear all that *I* do around here; your efforts are pretty puny compared to mine!

▲ 4. HEEDLESS: When people complain about American intervention in places like Iraq, they tell every tinhorn dictator to go ahead and take over because America will just stand by and watch. I, for one, think people who complain like that ought to just shut up.
      CAUTIOUS: Not me. Complaining like that reminds everyone that it isn't in our best interest to get involved in extended wars abroad.

   5. ONE SPEAKER: Nothing beats summertime. It's sunny and warm, and you can wear shorts, go on picnics, take hikes, and, best of all, take in a ball game.
      ANOTHER SPEAKER: Naw, summer's hot and sticky, and there's no skiing or skating or getting warm 'round a nice cozy fire. Summer's okay—if you're a mosquito.

   6. FITNESS BUFF ONE: Look here, the speedometer cable on this exercise bike is starting to squeak. If we don't fix it, the speedometer is going to stop working.
      FITNESS BUFF TWO: What we need to do is get a new bike. This old thing is more trouble than it's worth.

   7. YOUNG GUY: Baseball players are much better now than they were forty years ago. They eat better, have better coaching, you name it.
      OLD GUY: They aren't any better at all. They just seem better because they get more publicity and play with a livelier ball.

   8. STUDENT ONE: Studying is a waste of time. Half the time, I get better grades if I don't study.
      STUDENT TWO: I'd like to hear you say that in front of your parents!

   9. PHILATELIST: Did you know that U.S. postage stamps are now being printed in Canada?
      PATRIOT: What an outrage! If there is one thing that ought to be made in the United States, it's U.S. postage stamps!

PHILATELIST: Oh, I disagree. If American printing companies can't do the work, let the Canadians have it.

10. GEORGE: I think the United States ought to pull its troops out of every country they are currently stationed in. We simply cannot afford and do not know how to police the entire globe.

    NEVILLE: Boy, there's a laugh, coming from you. You're the one who thought we should be sending more troops to Vietnam in the early seventies; you cheered the invasions of Grenada and Panama in the eighties; and you were for the action in Somalia in the nineties. It's a little too late to be handing out this isolationist stuff now, isn't it?

11. FIRST NEIGHBOR: Look here. You have no right to make so much noise at night. I have to get up early to get to work.

    SECOND NEIGHBOR: Yeah? Well, you have no right to let your idiot dog run around loose all day long.

12. STUDY PARTNER ONE: Let's knock off for a while and go get some pizza. We'll be able to function better if we have something to eat.

    STUDY PARTNER TWO: Not one of those pizzas you like! I can't stand anchovies.

13. FEMALE STUDENT: The Internet is totally overrated. It takes forever to find something you can actually use in an assignment.

    MALE STUDENT: Listen, it takes a lot longer to drive over to the library, find a place to park, and wait in line to use a terminal.

14. CITIZEN ONE: In 2004 it's going to be George W. Bush for the Republicans and John Kerry for the Democrats, what do you want to bet?

    CITIZEN TWO: I doubt it. Kerry's fading. The Democrats will find someone else.

15. CULTURALLY CHALLENGED PERSON: A concert! You think I'm gonna go to a concert when I could be home watching Monday Night Football?

    CULTURALLY CHALLENGED PERSON'S SPOUSE: Yes, if you want dinner this week.

Exercise 1-8

For each of the brief conversations that follow, identify the issue the first speaker is addressing. To what extent does the second speaker address the same issue? Does he or she miss the point? If so, might the misdirection be intentional? Some of these are best suited to class discussion.

Example

> MOORE: I've seen the work of both Thomas Brothers and Vernon Construction, and I tell you Thomas Brothers does a better job.
>
> PARKER: Listen, Thomas Brothers is the highest-priced company in the whole blasted state. If you hire them, you'll pay double for every part of the job.

Analysis

Moore thinks Thomas Brothers does better work than Vernon Construction; Parker thinks Thomas Brothers' work is overpriced. Moore's view is quite compatible with Parker's view: Thomas Brothers may indeed do the best work (i.e., Moore is right) *and* charge wildly excessive prices (i.e., Parker is right, too). However, there is an underlying issue on which Moore and Parker will almost certainly disagree: whether Thomas Brothers should be hired. They have not made this disagreement explicit yet, however.

▲ 1. URBANITE: The new requirements will force people off septic tanks and make them hook up to the city sewer. That's the only way we'll ever get the nitrates and other pollutants out of the ground water.
SUBURBANITE: You call it a requirement, but I call it an outrage! They're going to charge us from five to fifteen thousand dollars each to make the hookups! That's more than anybody in my neighborhood can afford.

2. CRITIC: I don't think it's morally proper to sell junk bonds to anybody without emphasizing the risk involved, but it's especially bad to sell them to older people who are investing their entire savings.
ENTREPRENEUR: Oh, come on. There's nothing the matter with making money.

▲ 3. ONE HAND: What with the number of handguns and armed robberies these days, it's hard to feel safe in your own home.
THE OTHER HAND: The reason you don't feel safe is that you don't have a handgun yourself. It's well known that a criminal would rather hit a house where there's no gun than a house where there is one.

4. ONE GUY: Would you look at the price they want for these recordable DVD machines? They're making a fortune in profit on every one of these things!
ANOTHER: Don't give me that. I know how big a raise you got last year—you can afford *two* of those players if you want!

▲ 5. FED-UP: This city is too cold in the winter, too hot in the summer, and too dangerous all the time. I'll be happier if I exercise my early retirement option and move to my place in Arkansas.
FRIEND: You're nuts. You've worked here so long you'll be miserable if you retire, and if you move, you'll be back in six months.

Exercise 1-9

On the basis of a concept or distinction discussed in this chapter, divide the following issues into two groups, and identify the concept or distinction you used.

▲ 1. Whether George Pataki was as old when he became governor as Mario Cuomo was when he became governor.

2. Whether George Pataki is more liberal than Mario Cuomo was when Mario Cuomo was governor.

3. Whether Willie Mays hit more home runs than Mark McGwire.

▲ 4. Whether Leno tells better jokes than Letterman.

5. Whether your teacher will complain if you wear a baseball cap in class.

6. Whether your teacher should complain if you wear a baseball cap in class.

▲ 7. Whether there has ever been life on Mars.

8. Whether golf is more challenging than tennis.

9. Whether the movie scared me.

▲ 10. Whether I said the movie scared me.

11. Whether the movie is scary to lots of people.

Exercise 1-10 _____

Decide whether each question raises a factual issue or a nonfactual issue. If some items are difficult to decide, explain why you think this is so.

▲ 1. How much does Manuel weigh?

2. Does diet soda taste more bland than regular soda?

3. Does diet soda contain less sugar than regular soda?

▲ 4. Will diet soda help you lose weight better than regular soda?

5. Is it more fun to sail on a cruise ship than to lie on the beach at Hilton Head?

6. Is it expensive to sail on a cruise ship?

▲ 7. Is there life on another planet somewhere in the universe?

8. Would most people find this bath water uncomfortably hot?

9. Is the lake too cold for swimming?

▲ 10. Is Tom Brokaw better looking than Dan Rather?

11. Is Tiger Woods a better golfer than Jack Nicklaus was?

12. Is Al Sharpton smarter than George W. Bush?

▲ 13. Is abortion immoral?

14. Is Elvis still alive?

Exercise 1-11 _____

▲ Four of the following are factual claims, and six aren't. Can you identify them correctly?

1. Rice vinegar tastes a darn sight better than white vinegar.

2. White vinegar removes lipstick stains; rice vinegar doesn't.

3. None of the Supreme Court justices view the Constitution objectively.

4. Nine authors collaborated on that article.

5. Microsoft shares are significantly overpriced.

6. People who go to church regularly live longer than people who don't.

7. The FBI and CIA don't share information as often as they should.

8. The report stated that the FBI and CIA don't share information as often as they should.

9. Okay—enough of this logic stuff. It's Miller Time!

10. That last item is too challenging for this text.

Exercise 1-12

Sometimes factual claims contain nonfactual elements, and sometimes nonfactual claims contain factual elements. For example, here is a factual claim that contains a nonfactual element:

> (A) Tom, a fine lad, is over six feet tall. (The phrase "a fine lad" is a nonfactual element.)

And here is a nonfactual claim that contains a factual element:

> (B) Tom, who is over six feet tall, is a fine lad. (The clause "who is over six feet tall" is a factual element.)

Where you find factual claims in the following list, revise them so that they contain no nonfactual elements.

▲ 1. The house, an imposing two-story colonial-style building, burned to the ground.

2. Assad, who was reluctant to do so, attended the summit that took place in Qatar last March.

3. Hurricane Fran stranded 175 devastated homeowners on barrier islands.

▲ 4. The Coors Brewing Company, long a supporter of a right-wing agenda, began offering gay partners the same benefits as spouses in 1995.

5. Castellanos attended the University of North Carolina but dropped out in disgust during his final semester.

6. On August 22, the leader of a Kurdish faction wrote a desperate plea to Saddam Hussein.

▲ 7. McCovey, who at the time was overwhelmed with grief, was able to finish the assignment on time.

8. In his question, Larry King, the best interviewer in the business, probed Mr. Cheney's business contacts.

9. Senator Byrd announced his proposal to protect the pension funds of struggling workers.

▲ 10. The Reverend Jesse Jackson's provocative visit to Pakistan has been delayed due to lack of support from the Bush administration.

11. Senator Clinton gave a sharp and vigorous response to the Republican challenger.

12. This Sunday, as prescribed by the barely functioning peace accords, Bosnians are to go to the polls to elect a president.

▲ 13. For two years the rapidly expanding global computer matrix had nagged at Gates like a low-level headache.

14. Creative policies have stopped the declining enrollments at Spring Grove Community College.

15. In early 2004 the U.S. Senate tried to reach agreement on the perplexing question of how to rescue Medicare.

Exercise 1-13

Revise each of these press reports to exclude any nonfactual elements. If no such elements appear, leave the original as is.

▲ 1. OROVILLE—A judge Friday ordered an accused gunman held for trial on a felony assault charge stemming from an aborted residential robbery last month during which an Oroville bank executive was wounded.

2. WASHINGTON—As details of a remarkable week of grand jury testimony by President Clinton and a young intern filter into the public domain, Clinton apologists are unwinding a fragile string of explanations to protect him against charges of perjury and obstruction of justice.

3. LHOK SUKON, Indonesia (AP)—Human rights workers on Saturday dug up the skeletons of people whom activists believe were killed by the Indonesian military. Villagers looking on shouted slogans against former President Suharto.

▲ 4. Gunshots were fired at an apartment complex off 10th Street Friday during a "gang bang," which, according to witnesses, may have been contrived to intimidate or retaliate against rival gang members.

5. WASHINGTON—From hideouts in Afghanistan's rugged mountains, Osama bin Laden used his wealth to create cells of Muslim fighters to cleanse the country of its Soviet occupiers. Now bin Laden, who became the most significant sponsor of Islamic extremist activities in the world, is using his wealth in his war on the United States.

6. HARRISBURG, PA. (AP)—With no bathrooms or portable potties in space, NASA has relied on a brand of inexpensive, compressed adult diapers to let astronauts take care of business while they work.

▲ 7. WASHINGTON—The attorney general has launched a 90-day investigation to see if she should seek still another independent counsel, this time to check into the truthfulness of the vice president. On the face of it, suspicion arises that the investigation might be more a means of dealing with political pressure—another stalling tactic—than of enforcing the law, since the probe won't be finished until after the November elections.

8. SACRAMENTO (AP)—State senators on Thursday approved a bill by Assemblywoman Debra Bowen, D–Marina del Rey, that would give

computer owners a way to block unwanted commercial electronic mail, which is commonly known as spam.

9. WASHINGTON—Bill Clinton played kinky games with a girl in a room next to the Oval Office while dignitaries were waiting for him. Independent counsel Kenneth Starr was trying to find out whether Clinton took criminal measures to conceal this weird behavior. Clinton, of course, was merely trying to shield his "family life."

▲ 10. Prior to the 2000 national election, President George W. Bush asked whether character counts. Those who felt character was important went on to elect the feeblest intellect this nation was ever to suffer under as president.

Exercise 1-14

Which of the lettered options serves the same kind of purpose as the original remark?

Example

Be careful! This plate is hot.
a. Watch out. The roads are icy.
b. Say—why don't you get lost?

Answer

The purpose of (a) is most like the purpose of the original remark. Both are warnings.

▲ 1. I'd expect that zipper to last about a week; it's made of cheap plastic.
a. The wrinkles on that dog make me think of an old man.
b. Given Sydney's spending habits, I doubt Adolphus will stick with her for long.

2. If you recharge your battery, sir, it will be almost as good as new.
a. Purchasing one CD at the regular price would entitle you to buy an unlimited number of CDs at only $4.99.
b. I shall now serve dinner, after which you can play if you want.

3. To put out a really creative newsletter, you should get in touch with our technology people.
a. Do unto others as you would have them do unto you.
b. To put an end to this discussion, I'll concede your point.
c. You'd better cut down on your smoking if you want to live longer.

▲ 4. GE's profits during the first quarter were short of GE's projections. Therefore, we can expect GE stock to fall sharply in the next day or so.
a. Senator Torricelli apparently thinks what he does in private is nobody's business but his own.
b. The dog is very hot. Probably he would appreciate a drink of water.
c. The dog's coat is unusually thick. No wonder he is hot.

5. How was my date with your brother? Well . . . he has a great personality.
   a. How do I like my steak? Well, not dripping blood like this thing you just served me.
   b. How do I like the dress? Say, did you know that black is more slimming than white?

6. The wind is coming up. We'd better head for shore.
   a. They finally arrived. I guess they will order soon.
   b. We shouldn't leave yet. We just got here.

▲ 7. Ties are as wide as handkerchiefs, these days. That's why they cost so much.
   a. Belts are like suspenders. They both serve to keep your pants up.
   b. Football is like rugby. Given that, it is easy to understand why so many people get injured.

8. Daphne owns an expensive car. She must be rich.
   a. This dog has fleas. I'll bet it itches a lot.
   b. This dog has fleas. That explains why it scratches a lot.

9. Dennis's salary is going to go up. After all, he just got a promotion.
   a. Dennis's salary went up after he got a promotion.
   b. Dennis's salary won't be going up. After all, he didn't get a promotion.

▲ 10. Outlawing adult Web sites may hamper free speech, but pornography must be curbed.
   a. The grass must be mowed, even though it is hot.
   b. The grass is much too long; that means it must be mowed.

Exercise 1-15 ⎯⎯⎯⎯⎯⎯⎯⎯⎯⎯⎯⎯⎯⎯⎯⎯⎯⎯⎯⎯⎯

Five of the following questions call for an explanation; five call for an argument. Identify which call for which. (It may be possible to imagine strange situations in which all ten questions call for explanations and possible to imagine other strange situations in which all ten questions might call for arguments. Don't try to imagine far-out situations like that. Try to imagine normal, everyday situations in which somebody might ask these questions.)

▲ 1. YOU TO A FRIEND: You really think the dog is overweight? What makes you so sure?

2. YOU TO A FRIEND: Hey! The dang dog got out again! How do you suppose that happened?

3. YOU TO YOUR DENTIST: Yes, yes, I know I have another cavity, but what I don't understand is why. How did I get it—too many Jolly Ranchers?

▲ 4. YOU TO YOUR DENTIST: You're saying I have another cavity? Are you certain?

5. YOU TO YOUR DOCTOR: I haven't been sleeping very well, and I wondered what might account for that.

6. YOU TO YOUR DOCTOR: Doc, I've heard really bad things about that medication. Should I be taking it?

▲ 7. YOU TO A MECHANIC: This Hyundai is always giving me problems. Half the time I can't even get it in gear! What causes something like that?

8. YOU TO A MECHANIC: Well, I certainly don't dispute what you are saying, but can you tell me again why you think I need a new transmission?

9. YOU TO YOUR TEACHER: I don't understand this grade! Are you sure you didn't make a mistake?

▲ 10. YOU TO YOUR TEACHER: I understand this grade; but can you tell me how I can do better next time?

Writing Exercises

1. Turn to the "Essays for Analysis" in Appendix 1. Identify and write in your own words the principal issues in selections 3, 4, or 5. Your instructor may have other instructions in addition to (or in place of) these.

2. Do people choose which sex they are attracted to? Write a one-page answer to this question, defending your answer with at least one supporting reason. Take about ten minutes to write your answer. Do not put your name on your paper. When everyone is finished, your instructor will collect the papers and redistribute them to the class. In groups of four or five, read the papers that have been given to your group. Divide the drafts into two batches, those that contain an argument and those that do not. Your instructor will ask each group to read to the class a paper that contains an argument and a paper that does not contain an argument (assuming that each group has at least one of each). The group should be prepared to explain why they feel each paper contains or fails to contain an argument.

3. Using the issues you identified in Exercise 1 for each of the selections, choose a side on one of the issues and write a short paper supporting it.

Chapter 2

Critical Thinking and Clear Writing

There you are, with keyboard, pen, or pencil. And there it is, the enemy: a blank screen or piece of paper that you must somehow convert into an essay. If you are like many students we know, the kind of essay that causes you more trouble than any other variety is the argumentative essay, in which the purpose is to support a position on an issue. Successful essays achieve their goal by offering good arguments for their author's position. And because arguments consist of claims, a good argumentative essay contains credible claims.

In the previous chapter, we discussed claims at length and introduced the basics of argument. In this chapter, we show how to apply principles of critical thinking to essay writing: how to organize your thoughts, state your claims clearly, and avoid ineffective and counterproductive language. The chapter is not a substitute for a course in composition or writing, but it provides information that will help you write strong essays based on clearly stated arguments and reasonable claims.

As you progress through this book, you will acquire a working understanding of the various principles by which you can evaluate claims and arguments and, accordingly, argumentative essays. You can also apply these principles of critical thinking to your own writing. So working on your critical-thinking skills will help you get better at both appraising and writing argumentative essays.

The benefits work in reverse, too. You can enlist essay writing in service of critical-thinking objectives. For example, suppose you read an editorial you do not agree with. You might try drafting a response. Doing so will probably help you organize your thoughts and clarify your position; it will almost surely disclose to you considerations you had not thought of or aspects of the issue you did not perceive at first. Your thinking on the subject will probably sharpen.

Organization and Focus

A good argumentative essay must first of all be well organized. Every now and then, you encounter pieces of writing in which the words, claims, and arguments are so strangely assembled that the result is unintelligible. Let's just hope that the piece is nothing you yourself have written.

If you come across an argumentative essay that suffers from such serious organizational defects that it cannot be fully understood, then your only option is to suspend judgment on the unintelligible aspects. If, however, your own writing suffers from these defects, then you can benefit from some simple principles of good organization.

■ Principles of Organization

In an argumentative essay, the most natural and common organizational pattern is to state what you are trying to establish and then proceed to establish it by setting forth the considerations that support your position, adding explanations, illustrations, or other elaboration as needed. Here are some guidelines to help you get organized:

1. *Focus.* Make clear at the outset what issue you intend to address and what your position on the issue will be. Of course, nothing is quite so boring as an essay that begins, "In this essay I shall argue that . . ." and then goes on to itemize everything that's going to be said later. As a matter of style, you should let the reader know what to expect without using trite phrases and without going on at length.

2. *Stick to the issue.* All points you make in an essay should be connected to the issue under discussion and should always either (a) support, illustrate, explain, clarify, elaborate on, or emphasize your position on the issue or (b) serve as responses to anticipated objections. Rid the essay of irrelevancies and dangling thoughts.

Who Said What?

Can you tell which of these quotations are from students in community college and which are from the forty-third president of the United States, who happens to be a Yale graduate?

1. "Sometimes the standards of society causes people to wear a mask."
2. "Reading is the basics for all learning."
3. "I played all types of sports that was offered to children then."
4. "More and more, our imports come from overseas."
5. "Children are born as a ball of clay; to mold into what we are able, good or bad. To enhance who we are and to carry on the person we are known to be."
6. "Drug therapies are replacing a lot of medicines as we used to know it."
7. "You teach a child to read and he or her will be able to pass a literacy test."
8. "He uses an abundance of words and phrases both of which I don't believe should have."
9. ". . . if you say you're going to do something and you don't do it, that's trustworthiness."
10. "Male dominance has long been a factor in human existence since our existence."
11. "My pro-life position is I believe there's life. It's not necessarily based in religion. I think there's life there, therefore the notion of life, liberty and pursuit of happiness."

12. ". . . teaching children to read and having an education system that's responsive to the child and to the parents, as opposed to mired in a system that refuses to change, will make America what we want it to be—a literate country and hopefuller country."
13. "Men love women who can cook good."
14. "I have never really thought of myself to be very interesting as a human being or as anything else."

1, 3, 5, 8, 10, 13, and 14—Community college student; 2, 4, 6, 7, 9, 11, 12—Yale graduate

From an article by Jaime O'Neill, a member of the English faculty at Butte Community College, in *San Francisco Chronicle*, September 2, 2001.

3. *Arrange the components of the essay in a logical sequence.* This is just common sense. Make a point *before* you clarify it, for example, not the other way around. Place support for item B next to item B, not next to item F or G.

When supporting your points, bring in examples, clarification, and the like in such a way that the reader knows what you are doing. Your readers should be able to discern the relationship between any given sentence and your ultimate objective, and they should be able to move from sentence to sentence and from paragraph to paragraph without becoming lost or confused. If a reader cannot outline your essay with ease, you have not properly sequenced your material.

4. *Be complete.* You don't have to be exhaustive in your treatment of the issue; many issues are in fact much too large to be treated exhaustively in a single essay. Remember, finally, this basic principle: The more limited your topic, the easier it is to be complete in your coverage. However, do accomplish what you set out to accomplish, support fully and adequately whatever position you take on the issue, and anticipate and respond to possible objections. Also, be sure that there is closure at every level: Sentences should be complete, paragraphs should be unified wholes (and usually each should stick to a single point), and the essay should reach a conclusion. Reaching a conclusion and summarizing are not the same, incidentally. Short essays do not require summaries.

■ *Good Writing Practices*

Understanding these principles is one thing—actually employing them may be more difficult. Fortunately, five practices are almost guaranteed to improve the organization of your essay and to help you avoid other problems.

1. At some stage *after* the first draft, outline what you have written. Then make certain that the outline is logical and that every sentence in the essay fits into the outline as it should. Some writers create an informal outline before they begin, but many do not. Our advice: Just start writing and worry about outlining later.

2. Revise your work. Revising is the secret to good writing. Even major league writers revise what they write, and they revise continuously. Unless you are even more gifted than the very best professional writers, then revise, revise, revise. Don't think in terms of two or three drafts. Think in terms of *innumerable* drafts.

3. Have someone else read your essay and offer criticisms of it. Revise as required.

4. If you have trouble with grammar or punctuation, reading your essay out loud may help you detect problems that your eyes have missed.

5. After you are *completely* satisfied with your essay, put it aside. Then come back to it later for still another set of revisions.

■ *Essay Types to Avoid*

Seasoned instructors know that the first batch of essays they get from a class will most likely include several samples of each of the following types. At all costs, try to avoid these pitfalls.

- ▲ *The Windy Preamble.* Writers of this type of essay avoid getting to the issue and instead go on at length with introductory remarks, often about how important the issue is, how it has troubled thinkers for centuries, how opinions on the issue are many and various, and so on—and on.

- ▲ *The Stream-of-Consciousness Ramble.* This type of essay results when writers make no attempt to organize their thinking on the issue and instead simply list thoughts more or less in the order they come to mind.

- ▲ *The Knee-Jerk Reaction.* In this type of essay, writers record their first reaction to an issue without considering the issue in any depth or detail.

- ▲ *The Glancing Blow.* Writers of this type of essay address the issue obliquely rather than straight on. If they are supposed to evaluate the health benefits of exercise, they will discuss the health benefits of using exercise equipment. If they are supposed to consider the health benefits of using exercise equipment, they will discuss the benefits of bicycling.

- ▲ *Let the Reader Do the Work.* Writers of this type of essay expect the reader to follow them through non sequiturs, abrupt shifts in direction, and huge gaps in logic.

Exercise 2-1 ─────────────────────────────

The sentences in this Associated Press health report have been scrambled. Rearrange them so that the report makes sense.

1. The men, usually strong with no known vices or ailments, die suddenly, uttering an agonizing groan, writhing and gasping before succumbing to the mysterious affliction.

2. Scores of cases have been reported in the United States during the past decade.

3. In the United States, health authorities call it "Sudden Unexplained Death Syndrome," or "SUDS."

4. Hundreds of similar deaths have been noted worldwide.

5. The phenomenon is known as "lai tai," or "nightmare death," in Thailand.

6. In the Philippines, it is called "bangungut," meaning "to rise and moan in sleep."

A "Weakest Link" Quiz for You

Here is a "Weakest Link" quiz that was circulating on the Net a while back. It's a good test of your ability to read carefully and think clearly, two key components of critical thinking.

1. You are competing in a race and overtake the runner in second place. In which position are you now?
   Answer: If you said "first," then you aren't reading carefully (or aren't thinking clearly). The correct answer, of course, is "second."
2. If you overtake the last runner, what position are you in now?
   Answer: Second to the last? Hardly. The question doesn't make sense.
3. Take 1,000. Add 40. Add another 1,000. Add 30. Add 1,000 again. Plus 20. Plus 1,000. And plus 10. What is the total?
   Answer: If you said 5,000 you weren't reading carefully. The correct answer is 4,100.
4. Marie's father has five daughters:
   1. Cha Cha
   2. Che Che
   3. Chi Chi
   4. Cho Cho
   5. ??
   What is the fifth daughter's name?
   Answer: Chu Chu? Wrong, wrong, wrong. It's Marie!

Thanks to Anne Morrissey.

7. Health officials are baffled by a syndrome that typically strikes Asian men in their 30s while they sleep.

8. Researchers cannot say what is killing SUDS victims.

Exercise 2-2

▲ The sentences in the following passage have been scrambled. Rearrange them so that the passage makes sense.

1. Weintraub's findings were based on a computer test of 1,101 doctors twenty-eight to ninety-two years old.

2. She and her colleagues found that the top ten scorers aged seventy-five to ninety-two did as well as the average of men under thirty-five.

3. "The test measures memory, attention, visual perception, calculation, and reasoning," she said.

4. "The studies also provide intriguing clues to how that happens," said Sandra Weintraub, a neuropsychologist at Harvard Medical School in Boston.

5. "The ability of some men to retain mental function might be related to their ability to produce a certain type of brain cell not present at birth," she said.

6. The studies show that some men manage to escape the trend of declining mental ability with age.

7. Many elderly men are at least as mentally able as the average young adult, according to recent studies.

Clarity in Writing

In addition to being well organized, a good argumentative essay must be clearly written. If your objective is to support your position, you must write as clearly as possible. Likewise, before you accept what someone else writes (or says), you should be sure you understand what is being expressed.
Consider these examples:

> When I was in the Marine Corps, I was plainly told that many good men died in the uniform that was issued to me.
>
> —*From a letter to the editor*

> Not every framistan has gussets.

> I am glad to be an American, and I appreciate our system of government. Also, I am for a very strong defense. However, the people protesting the war on all sides are out there because they care about life. Now we are in an awful mess. Why? We need to put ourselves in the other guy's shoes. Going out and killing the other guy may be the way to preserve your own.
>
> —*From a newspaper call-in column*

> Today, morals are breaking down everywhere. There are no longer any absolutes. That is why the Boy Scouts should not permit gays to be scout leaders. The Boy Scouts stand for values that never change or go out of style.
>
> —*From a student paper*

> I am forty-two years old and I'm conservative. I keep reading and hearing in the news that voters are in a bad mood. No, we are not. We are hopeful and up-lifted, but we are mad as hell.
>
> —*Reported by William Endicott*

> Legal laws are fine, but illegal ones should be changed.
>
> —*From a student paper*

The first example does not mean what it first appears to mean; the second can be understood only by those familiar with framistans, whatever those are. The third example is a set of claims that individually mostly make sense—

If openness means to "go with the flow," it is necessarily an accommodation to the present. That present is so closed to doubt about so many things impeding the progress of its principles that unqualified openness to it would mean forgetting the despised alternatives to it, knowledge of which makes us aware of what is doubtful in it.

—From Allan Bloom, *in The Closing of the American Mind*

A Little Gobbledygook from the Defense Secretary

In an Associated Press story in August 2002, Defense Secretary Donald Rumsfeld said that talk of a U.S. strike against Saddam Hussein in Iraq was hypothetical and that the president had not asked the Saudis for use of their territory. Then, in response to a question regarding King Saud's remarks about using Saudi bases, Rumsfeld responded, "The president has not proposed such a thing; therefore, I don't find it really something that has been engaged as such."

We think he just means, "it hasn't been proposed," but he wasn't quite ready to stop talking yet.

The purpose of extending the kindergarten hours is to provide a thinking meaning appropriately centered-based, academic/social program to meet the diverse needs of the kindergarten students entering at Lewis Elementary. The exposure of the students to the world, due to extended travels, has increased their student based experiences.

—From a California public school proposal for extending kindergarten hours

Maybe Bloom went to Lewis Elementary.

For more than almost half a century, the Humane Society has led the fight to protect animals.

—Humane Society solicitation sent to one of us

For how long?

but the assembled package defies comprehension. The fourth and fifth seem illogical and contradictory. (In the fourth, the second sentence asserts there are no longer any absolutes; the last sentence implies there are. In the fifth, it's hard to see how voters can be mad as hell but not in a bad mood). And the last example is nonsensical.

When we can't tell what someone else is claiming or arguing, or when someone else can't tell what we are claiming or arguing, any number of problems may be causing the difficulty. Here we consider ill-defined terms, poorly chosen words, unintentional ambiguity, vagueness, and faulty comparisons.

■ *Defining Terms*

Any serious attempt to support or sustain a position requires a clear statement of what is at issue. Sometimes stating what is at issue involves a careful definition of key terms.

Purposes and Types of Definitions   Definitions can serve different purposes:

- ▲ To introduce unusual or unfamiliar words, to coin new words, or to introduce a new meaning to a familiar word (these are called **stipulative definitions**)
- ▲ To explain, illustrate, or disclose important aspects of difficult concepts (**explanatory definitions**)
- ▲ To reduce vagueness and eliminate ambiguity (**precising definitions**)
- ▲ To influence the attitudes of the reader (**rhetorical definitions**)

Sometimes definitions are intended just to amuse.

Whatever their purpose, most definitions take various common forms. Three of the most common are **definition by example, definition by synonym,** and **analytical definition:**

1. *Definition by example:* pointing to, naming, or describing one or more examples of something to which the defined term

Winning the Lottery

- Each and every combination of six numbers has exactly the same chance of winning the lottery as any other combination of six numbers.
- However, the chance of drawing any sequence of six numbers (such as 1, 2, 3, 4, 5, 6 or 27, 28, 29, 30, 31, 32) is higher than drawing some particular combination of numbers.

Would you be puzzled if we told you both these assertions are correct? If you are, you didn't read the second sentence carefully. Reading carefully is a key to thinking critically.

applies. "By 'scripture' I mean books like the Bible or the Koran." "By 'temperate climate' I mean weather in an area like the mid-Atlantic states."

2. *Definition by synonym:* giving another word or phrase that means the same thing. "'Fastidious' means the same as 'fussy.'" "'Prating' is the same as 'chattering.'" "'Pulsatile' means the same as 'throbbing'"; "to be 'lubricous' is to be 'slippery.'"

3. *Analytical definition:* specifying (a) the type of thing the term applies to and (b) the differences between the things the term applies to and other things of the same type. "A mongoose is a ferret-sized mammal native to India that eats snakes and is related to civets." "A samovar is an urn with a spigot, used especially in Russia to boil water for tea."

Some terms, especially terms for abstractions (e.g., "goodness," "truth," "knowledge," "beauty") cannot be defined in any complete way, so a writer may have to settle for providing mere hints of their subtle meanings. "By 'reality' I mean the things that most of us agree have independent existence apart from our perceptions of them."

However we define a term, we should be aware that most terms convey meaning beyond the literal sense of the written or spoken words. This "meaning" is a term's **emotive, or rhetorical, force**—its tendency to elicit certain feelings or attitudes. The word "dog," for example, has the same literal meaning as "pooch," "mutt," and "cur," but all these terms differ in the attitudes they convey. Or consider the words "child," "dependent minor," "brat," and "little one." What associations do these terms have for you? When people want to manipulate the emotive force of their message, they frequently substitute euphemisms for more pointed terms. This is the origin of such substitutions as "urban camping" for "homelessness" and "food-insecure" for "starving." The emotive, or rhetorical, force of a term, which is subjective and can vary considerably from one person to another, is usually not taken to be a part of the literal meaning.

Autobiography Skewers Kansas' Sen. Bob Dole

—Headline in the *Boulder* (Colo.) *Sunday Camera* (reported by Larry Engelmann)

Say What??...

You don't have to be a national political figure to put your foot in your mouth. Ordinary folks can do it too! These are all actual examples from news programs and other published sources.

The President's energy tax won't even be noticed. Besides, it will discourage consumption.

> *[Hey, if it won't be noticed, it won't discourage consumption.]*

Females score lower on the SAT than males, which right there proves the tests don't show anything. They also demonstrate that teachers do a better job of teaching male students, which is just what you'd expect given the sexual bias that exists in the classroom.

> *[If the SATs don't show anything, then they don't show that teachers do a better job teaching males.]*

We have to liberate discussion on this campus and not be so restrained by the First Amendment.

> *[Right. And we can make people free by sticking them in jail.]*

Once your body gets cold, the rest of you will get cold, too.

> *[On the other hand, if you can keep your body warm, the rest of you will stay warm, too.]*

It's hard to support the President's invasion of Haiti when the American public is so strongly against it. And besides, he's just doing it to raise his standings in the polls.

> *[Hmmm. How's it going to raise his standings if the public is so strongly against it?]*

Has anyone put anything in your baggage without your knowledge?

> *[Asked of a colleague by an airport security employee]*

Keeping Your Word Choices Simple   Good writing is often simple writing: It avoids redundancy, unnecessary complexity, and prolixity (long-windedness and wordiness). These writing characteristics often confuse readers and listeners, and they sometimes make writers (or speakers) look silly. Why write of *armed* gunmen? Gunmen are automatically armed. Why say that something is *completely* full? If it's full, it's completely full. Michael Jordan, it is often said, is a *famous* superstar. The famous part is pointless, if a person is a superstar. In fact, come to think of it, why say *super*star?

Here's another example:

They expressed their belief that at that point in time it would accord with their desire not to delay their departure.

Uh, Would You Mind Repeating That Answer?
(Gobbledygook on the Radio)

The following exchange took place during a call-in program with Dr. Laura Schlessinger (17 January 2002):

Caller: Dr. Laura, I am a Christian, and I solidly believe in Jesus as the Messiah. However, I have recently decided to familiarize myself with Judaism, Islam, etc., in order to strengthen my own commitment to Christianity. I have so often been intrigued by your decision to commit to God through Judaism. Like you, I was not raised in a seriously, genuinely religious home. Your insights would be most appreciated and welcome.

Dr. Laura: I worry that sometimes people study something else to avoid the depth of what they need to be and know right where they are.

You know, we worry a lot about that ourselves.

But all that is necessary is

They said they wanted to leave.

On the other hand, if the briefest way of making a point is to use words that a reader isn't likely to understand, it is probably better to avoid those words in favor of more familiar ones, even if it takes more of the latter to express the information. "His remarks were obfuscatory and dilatory" will be less clear to most readers than "His remarks confused the issue and were unnecessarily time-consuming." Further, because the world is a complicated place, the language we use to describe it often has to be correspondingly complicated. Sometimes it is necessary to be complicated to be clear. But, in general, simplicity is the best policy.

> I'm for abolishing and doing away with redundancy.
> —J. Curtis McKay, of the Wisconsin State Elections Board (reported by Ross and Petras)
>
> We ourselves are also for that too.

Exercise 2-3

In groups (or individually if your instructor prefers) determine what term in each of the following is being defined and whether the definition is a definition by example or by synonym or an analytical definition. If it is difficult to tell which kind of definition is present, describe the difficulty.

▲ 1. A piano is a stringed instrument in which felt hammers are made to strike the strings by an arrangement of keys and levers.

2. "Decaffeinated" means without caffeine.

3. Steve Martin is my idea of a successful philosophy major.

▲ 4. The red planet is Mars.

5. "UV" refers to ultraviolet light.

6. The Cheyenne perfectly illustrate the sort of Native Americans who were plains Indians.

7. Data, in our case, is raw information collected from survey forms which is then put in tabular form and analyzed.

▲ 8. "Chiaroscuro" is just a fancy word for shading.

9. Bifocals are glasses with two different prescriptions ground into each lens, making it possible to focus at two different distances from the wearer.

10. Red is the color that we perceive when our eyes are struck by light waves of approximately seven angstroms.

▲ 11. A significant other can be taken to be a person's spouse, lover, long-term companion, or just girlfriend or boyfriend.

12. "Assessment" means evaluation.

13. A blackout is "a period of total memory loss, as one induced by an accident or prolonged alcoholic drinking." When your buddies tell you they loved your rendition of the Lambada on Madison's pool table the other night and you don't even remember being at Madison's, that is a blackout.

Adapted from the CalPoly, San Luis Obispo, Mustang Daily

14. A pearl, which is the only animal-produced gem, begins as an irritant inside an oyster. The oyster then secretes a coating of nacre around the irritating object. The result is a pearl, the size of which is determined by the number of layers with which the oyster coats the object.

15. According to my cousin, who lives in Tulsa, the phrase "bored person" refers to anybody who is between the ages of sixteen and twenty-five and lives in eastern Oklahoma.

© Grimmy Inc. Reprinted with special permission of King Features Syndicate.

■ *Ambiguous Claims*

A claim is an **ambiguous claim** if it can be assigned more than one meaning and if the particular meaning it should be assigned is not made clear by context. If an accountant rises from her desk on Friday afternoon and says, "My work here is finished," she might mean that she has finished the account she was working on, that her whole week's work is done and she's leaving for the weekend, or that she is fed up with her job and is leaving the company.

Sometimes an ambiguous word or phrase can make the difference in a discussion or a dispute. For example, imagine two people who are disputing whether abortion should be made illegal. Imagine that one person uses the phrase "human being" to refer to any organism that possesses a full complement of human chromosomes and the other person uses the same phrase to refer to a being recognized by statute or Supreme Court decision to have full rights as a citizen under the Constitution. There will be much wasted breath and misunderstanding until they get their two disparate definitions of this crucial phrase sorted out.

Semantic Ambiguity   A claim can be ambiguous for different reasons. Consider these claims:

1. He always lines up on the right side.

2. She is cold.

3. I know a little Greek.

4. She disputed his claim.

5. My brother doesn't use glasses.

The meanings of these claims are unclear because each claim contains an ambiguous word or phrase. For example, "claim" in (4) could mean either a statement or a claim to a gold mine; "glasses" in (5) could mean either eyeglasses or drinking glasses. A claim whose ambiguity is due to the ambiguity of a particular word or phase is called a **semantically ambiguous claim.**

The World Health Organization does not promote a drink a day for health reasons.
—*Chico Enterprise-Record*

What *do* they promote it for?

I am a marvelous housekeeper. Every time I leave a man I keep his house.
—Zsa Zsa Gabor

The Fallacy of Division in Clinton's Impeachment Trial

Assume two-thirds of a jury thinks a defendant is guilty of having lied about W, X, or Y. Which of these options logically follows?

1. Two-thirds of the jury think the defendant lied about W.
2. Two-thirds of the jury think the defendant lied about X.
3. Two-thirds of the jury think the defendant lied about Y.
4. There is at least one matter, W, X, or Y, that two-thirds of the jury think the defendant lied about.
5. None of the above.

The correct answer is 5. Two-thirds of a jury could think the defendant lied about "W, X, or Y" even though less than two-thirds thought he lied about W, and less than two-thirds thought he lied about X, and less than two-thirds thought he lied about Y!

This confusion, which is actually a subtle example of division, explained later in this chapter, was involved in one of the most important trials to ever happen in the United States, Bill Clinton's impeachment trial before the U.S. Senate.

Here's how Article I of the impeachment charges read: "William Jefferson Clinton willfully provided perjurious, false, and misleading testimony to the grand jury concerning one or more of the following: [here, four options were listed]."

Here's how the key part of Article II read: "The means used to implement this course of conduct [obstructing justice] included one or more of the following acts: [here, several options were listed]."

This presented the Clinton legal defense team with a logical dilemma. The Article I and Article II charges were both worded in such a way that, even if two-thirds of the Senate agreed that Clinton was guilty under one of these two articles of impeachment, it would not follow that there was even a single case of lying or a single case of obstructing justice that two-thirds of the Senate agreed the president committed! In other words, Clinton could have been convicted and removed from office despite not being found guilty *on any specific charge of lying or obstructing justice* by the required two-thirds of the Senate!

This, of course, explains why prosecutors normally make each crime alleged to have been committed by a defendant a separate count.

Semantic ambiguity can be eliminated by substituting an unambiguous word or phrase, such as "eyeglasses" for "glasses" in (5). Sometimes the substitution may require several extra words.

Syntactic Ambiguity   Now consider these examples:

1. She saw the farmer with binoculars.

2. Players with beginners' skills only may use court 1.

I wrote a story about a girl making lunch in a skillet.

—GARRISON KEILLOR

A big skillet, perhaps?

"Grammar Does Some Work!"

Notice in a Help Wanted section of a newspaper:

Wanted. Man to take care of cow that does not smoke or drink.

If the nonsmoking and nondrinking creature is the man, the sentence should say "*who* does not smoke or drink," rather than "*that* does not smoke or drink." That the word "that" is used should tell us that it's the cow the speaker means. A little grammatical knowledge would go a long way—if the writer had any.

The Fusco Bros. © Lew Little Ent. Reprinted with permission of Universal Press Syndicate. All rights reserved.

▲ *A lot of jokes depend on ambiguity for humor. Unintentional ambiguities can make what you write sort of funny, too.*

3. People who protest often get arrested.

4. He chased the girl in his car.

5. There's somebody in the bed next to me.

In contrast with those in the preceding list, these claims are ambiguous because of their structures. Even though we understand the meaning of the phrase "with binoculars," for example, we can't tell whether it pertains to the farmer or to the subject of (1). A claim of this sort is called a **syntactically ambiguous claim.** The only way to eliminate syntactic ambiguity is to rewrite the claim.

Sometimes syntactically ambiguous claims result when we do not clearly show what a pronoun refers to. "The boys chased the girls, and they giggled a lot" is an example of this sort of ambiguity; you can't be certain whether "they" refers to the boys or to the girls. "After he removed the trash from the pool, the children played in it" is another example.

Modifying phrases can create syntactic ambiguities if we are careless with them. Consider "He brushed his teeth on the carpet." If you want to

© Grimmy Inc. Reprinted with special permission of King Features Syndicate.

avoid suggesting that a carpet might make a good toothbrush, then you should recast the sentence, perhaps as "He brushed his teeth while he was on the carpet." Or take "She wiped up the water with her younger brother." Yes, of course it is unlikely that she used her brother as a towel. But a clear writer will avoid any possibility of misunderstanding—or ridicule—by recasting the sentence, perhaps as "She and her younger brother wiped up the water." "He was bitten while walking by a dog" should be revised either as "While he was walking, he was bitten by a dog" or as "While he was walking by a dog, he was bitten."

Grouping Ambiguity   A peculiar kind of semantic ambiguity, which we'll call **grouping ambiguity,** is illustrated by this claim:

> Secretaries make more money than physicians.

Is this claim true or false? We can't say. We don't know *what* the claim is because we don't know exactly what "secretaries" and "physicians" refer to. If the claim is that secretaries *as a group* make more money than physicians make *as a group,* then the claim is true because there are many more secretaries than physicians. But if the claim is that secretaries individually make more money than physicians individually, then the claim is of course false. Whenever we refer to a collection of individuals, we must clearly show whether the reference is to the collection as a group or as individuals.

The Fallacies of Composition and Division   Related to grouping ambiguities are mistakes, or **fallacies,** known as composition and division. Suppose something holds true of a group of things individually. To think it must therefore necessarily hold true of the same things as a group is to commit the **fallacy of composition.** Here are two examples of this fallacy:

If I said anything which implies that I think that we didn't do what we should have done given the choices we faced at the time, I shouldn't have said that.
—BILL CLINTON (reported by Larry Engelmann)

The President will keep the promises he meant to keep.
—Bill Clinton's White House senior adviser, GEORGE STEPHANOPOULOS

Slow Recovery Likely in State

This ambiguous headline, from the *Sacramento Bee* of June 9, 2002, is subject to two interpretations, as good news and as bad news:

1. Fortunately, a recovery is likely to happen, although it will be slow.
2. Unfortunately, the recovery we're expecting will be slow.

Sampras and Agassi are the two best tennis players in the United States, so they'd make the best doubles team.

We don't spend *that* much on military salaries. After all, who ever heard of anyone getting rich in the Army? (In other words, we don't spend that much on service personnel *individually;* therefore, we don't spend much on them *as a group.*)

> *The people voted for a Republican Senate.*
>
> —GEORGE BUSH
>
> Actually, no person anywhere voted for a Republican Senate. The question "Do you want a Republican Senate?" appeared on no ballots.

Conversely, to think that what holds true of a group automatically holds true of all the individuals in that group is to commit the error known as the **fallacy of division.** Here are examples:

Congress is incompetent. Therefore, Congressman Benton is incompetent.

The Eastman School of Music has an outstanding international reputation; therefore, Vladimir Peronepky, who is on the faculty of Eastman, must have a good reputation.

In the 1994 congressional elections, the voters (as a group) made it clear that they did not want a Democratic Congress. Therefore, the voters (individually) made it clear that they did not want a Democratic Congress.

Recognizing and Deciphering Ambiguity   Some claims, such as "Women can fish" and "I will put the sauce on myself," can reasonably be diagnosed as either semantically or syntactically ambiguous. So don't spend much time arguing over which category an ambiguous claim belongs to. What is important is recognizing ambiguities when you encounter them and avoiding them in your own writing.

Some ambiguous claims don't fall into any of the categories we have mentioned. For example, a claim that refers to "the fastest woman on the squad" is ambiguous if two or more women on the squad are equally fast and are faster than anyone else. "I cannot recommend him highly enough" might mean that he is even better than my highest recommendation, or it might mean that I cannot recommend him very highly at all. The ambiguity, however, is not clearly semantic or syntactic.

Often, but not nearly often enough, the context of a claim will show which possible meaning a speaker or writer intends. If a mechanic says, "Your trouble is in a cylinder," it might be unclear at first whether she means a wheel cylinder or an engine cylinder. Her meaning will probably become

More Examples of Composition and Division

Fallacy of Division

A balanced daily diet consists of the right proportion of protein, carbohydrates, and fat. Therefore, each meal should consist of the same proportion of protein, carbohydrates, and fat.

—DR. NICHOLAS PERRICONE, author of the best-selling book, The Wrinkle Cure

Fallacy of Composition

The Kings don't have a chance against the Lakers. The Lakers are better at every position except power forward.

—CHARLES BARKLEY

Fallacy of Division

Men are more likely than women to be involved in fatal automobile accidents. So if you are a man, you are more likely than the next woman you see to be involved in a fatal automobile accident.

—Overheard

We will briefly discuss the history of American Indians on this campus.

—From a flyer advertising "Conversations on Diversity" at the authors' university

Somewhere on this globe, every ten seconds, there is a woman giving birth to a child. She must be found and stopped.

—SAM LEVENSON (1911–1980)

clear, though, as you listen to what else she says and consider the entire context in which she is speaking.

Also, common sense often dictates which of two possible meanings a person has in mind. In the example "He brushed his teeth on the carpet," you would probably be safe to assume that "on the carpet" refers to his location and not to an unusual technique of dental hygiene. So, the claim is perhaps not truly ambiguous. But if you wish to be a clear writer (or aren't fond of people laughing at you), you should try to say exactly what you mean and not rely on common sense to make your meaning clear.

Although ambiguous claims can be a fine source of amusement, they can also furnish clever ways of duping people. In advertising, for example, a claim might make wild promises for a product when interpreted one way. But under another interpretation, it may say very little, thus giving the manufacturer a loophole against the charge of false advertising.

Unintentional ambiguity has no place in clear writing. It can defeat the purpose of communication by confusing readers, and it can sometimes make the writer look foolish. You'll find other suggestions for improving your writing elsewhere in this chapter. But we want to note here that, aside from simply learning about the problem, proofreading is probably your best guard against unintentional ambiguity. In fact, failure to proofread is a common problem with student writing. It's a good policy to read your work more than once because there are quite a few things you'll be watching for: coherence, grammar, ambiguity, word choice, vagueness, and so on. You want to make sure you find what you want and *don't* find what you *don't* want.

Waitress Sues Former Employer
Over an Ambiguity!

PANAMA CITY, Fla. (AP) . . . A former Hooters waitress has sued the restaurant where she worked, saying she was promised a new Toyota for winning a beer sales contest.

Instead, she said, she won a new toy Yoda—the little green guy from the *Star Wars* movies.

Jodee Berry, 26, won a contest to see who could sell the most beer in April at the Hooters in Panama City Beach. She said the top-selling waitresses from each Hooters restaurant in the area were entered into a drawing and her name was picked.

She believed she'd won a new car.

She was blindfolded and led to the restaurant parking lot, but when her blindfold was removed she found she was the winner not of a Toyota, but a toy Yoda doll.

. . . She sued Gulf Coast Wings, Inc., owners of the restaurant, alleging breach of contract and fraudulent misrepresentation.

Her lawyer, Stephen West of Pensacola, said he was also looking at false advertising statutes.

Exercise 2-4

Rewrite the following claims to remedy problems of ambiguity. Do *not* assume that common sense by itself solves the problem. If the ambiguity is intentional, note this fact, and do not rewrite.

Example

Former professional football player Jim Brown was accused of assaulting a thirty-three-year-old woman with a female accomplice.

Answer

This claim is syntactically ambiguous because it isn't clear what the phrase "with a female accomplice" modifies—Brown, the woman who was attacked, or, however bizarre it might be, the attack itself (he might have thrown the accomplice at the woman). To make it clear that Brown had the accomplice, the phrase "with a female accomplice" should have come right after the word "Brown" in the original claim.

▲ 1. The Raider tackle threw a block at the Giants linebacker.

2. Please close the door behind you.

3. We heard that he informed you of what he said in his letter.

▲ 4. "How Therapy Can Help Torture Victims"

Headline in newspaper

5. Charles drew his gun.

6. They were both exposed to someone who was ill a week ago.

▲ 7. Susan has Hillary Rodham Clinton's nose.

8. I flush the cooling system regularly and just put in new thermostats.

Watch for Weasel Numbers in Medical Studies

Ever wonder why so many medical studies discover spectacular benefits from new drugs? Maybe it's because researchers report the numbers in a way that makes them sound better than they should.

That's the concern of Dr. Jim Nuovo, professor of family and community medicine at the University of California at Davis. In a study published in the June 5 *Journal of the American Medical Assn.*, Nuovo looked at 359 studies from leading medical journals and found that only 18 of them reported results in terms of absolute risk reduction. The rest reported only relative risk reduction—which can sound far more substantial.

Here's an example from Nuovo to explain the difference: Suppose an existing drug cuts the rate of heart attacks in men to 5%. A new drug cuts that risk to 4%. In terms of relative risk that's an impressive-sounding 20% improvement. In absolute terms, however, there's only a one-percentage-point difference. The risk has gone from low to slightly lower—which doesn't sound very impressive, especially if the new drug costs 10 times as much as the existing drug.

Nuovo's point is that all of these numbers, and more, ought to be reported in every study. "There's no one number that explains everything in an article," he says. "Once you start putting things into context, you start getting a better picture." Proposed guidelines for clinical trials call for researchers to report absolute risk reduction along with relative risks, Nuovo says, but those guidelines are often ignored. Editors should require authors to follow them, he says.

Business Week, June 17, 2002, p. 85.

9. "Tuxedos Cut Ridiculously!"

> *An ad for formal wear, quoted by Herb Caen*

▲ 10. "Police Kill 6 Coyotes After Mauling of Girl"

> *Headline in newspaper*

11. "We promise nothing"

> *Aquafina advertisement*

12. Former governor Pat Brown of California, viewing an area struck by a flood, is said to have remarked, "This is the greatest disaster since I was elected governor."

> *Quoted by Lou Cannon in the* Washington Post

▲ 13. "Besides Lyme disease, two other tick-borne diseases, babesiosis and HGE, are infecting Americans in 30 states, according to recent studies. A single tick can infect people with more than one disease."

> Self *magazine*

14. "Don't freeze your can at the game."

> *Commercial for Miller beer*

15. Volunteer help requested: Come prepared to lift heavy equipment with construction helmet and work overalls.

▲ 16. "GE: We bring good things to life."

Television commercial

17. "Tropicana 100% Pure Florida Squeezed Orange Juice. You can't pick a better juice."

Magazine advertisement

18. "It's biodegradable! So remember, Arm and Hammer laundry detergent gets your wash as clean as can be [pause] without polluting our waters."

Television commercial

▲ 19. If you crave the taste of a real German beer, nothing is better than Dunkelbrau.

20. Independent laboratory tests prove that Houndstooth cleanser gets your bathroom cleaner than any other product.

21. We're going to look at lots this afternoon.

▲ 22. Jordan could write more profound essays.

23. "Two million times a day Americans love to eat, Rice-a-Roni—the San Francisco treat."

Advertisement

24. "New York's first commercial human sperm-bank opened Friday with semen samples from 18 men frozen in a stainless steel tank."

Strunk and White, Elements of Style

▲ 25. She was disturbed when she lay down to nap by a noisy cow.

26. "More than half of expectant mothers suffer heartburn. To minimize symptoms, suggests Donald O. Castell, M.D., of the Graduate Hospital in Philadelphia, avoid big, high-fat meals and don't lie down for three hours after eating."

Self *magazine*

27. "Abraham Lincoln wrote the Gettysburg address while traveling from Washington to Gettysburg on the back of an envelope."

Richard Lederer

▲ 28. "When Queen Elizabeth exposed herself before her troops, they all shouted 'harrah.'"

Richard Lederer

29. "In one of Shakespeare's famous plays, Hamlet relieves himself in a long soliloquy."

Richard Lederer

30. The two suspects fled the area before the officers' arrival in a white Ford Mustang, being driven by a third male.

31. "To travel to Canada, you will need a birth certificate or a driver's license and other photo ID."

American Automobile Association

▲ 32. "AT&T, for the life of your business."

▲ 33. The teacher of this class might have been a member of the opposite sex.

▲ 34. "Woman gets 9 years for killing 11th husband."

Headline in newspaper

35. "Average hospital costs are now an unprecedented $2,063.04 per day in California. Many primary plans don't pay 20% of that amount."

AARP Group Health Insurance Program advertisement

36. "I am a huge Mustang fan."

Ford Mustang advertisement

37. "Visitors are expected to complain at the office between the hours of 9 and 11 A.M. daily."

Sign in an Athens hotel

38. "Order your summers suit. Because is big rush we will execute customers in strict rotation."

Sign in a Rhodes tailor shop

39. "Please do not feed the animals. If you have any suitable food, give it to the guard on duty."

Sign at a Budapest zoo

40. "Our wines leave you with nothing to hope for."

From a Swiss menu

41. Our Promise—Good for life.

Cheerios

42. Thinking clearly involves hard work.

43. Cadillac—Break Through

Exercise 2-5 ⎯⎯⎯⎯⎯⎯⎯⎯⎯⎯⎯⎯⎯⎯⎯⎯⎯⎯⎯⎯⎯⎯⎯⎯⎯

Determine which of the italicized expressions are ambiguous, which are more likely to refer to the members of the class taken as a group, and which are more likely to refer to the members of the class taken individually.

Example

Narcotics are habit-forming.

Answer

In this claim, *narcotics* refers to individual members of the class because it is specific narcotics that are habit-forming. (One does not ordinarily become addicted to the entire class of narcotics.)

▲ 1. *Swedes* eat millions of quarts of yogurt every day.

2. *College professors* make millions of dollars a year.

3. *Our CB radios* can be heard all across the country.

▲ 4. *Students at Pleasant Valley High School* enroll in hundreds of courses each year.

5. *Cowboys* die with their boots on.

6. The *angles of a triangle* add up to 180 degrees.

▲ 7. *The New York Giants* played mediocre football last year.

8. On our airline, *passengers* have their choice of three different meals.

9. On our airline, *passengers* flew fourteen million miles last month without incident.

▲ 10. *Hundreds of people* have ridden in that taxi.

11. *All our cars* are on sale for two hundred dollars over factory invoice.

▲ 12. *Chicagoans* drink more beer than *New Yorkers*.

13. *Power lawn mowers* produce more pollution than *motorcycles*.

14. *The Baltimore Orioles* may make it to the playoffs by the year 2000.

▲ 15. *People* are getting older.

■ Vague Claims

Vagueness and ambiguity are often confused. An ambiguous claim has two or more possible meanings, and the context does not make clear which meaning is intended. A **vague claim,** by contrast, has a meaning that is indistinct or imprecise.

The vagueness of a claim is a matter of degree. "He is old" is more vague than "He is at least 70," and "He is at least 70" is more vague than "He is 73 years and 225 days old."

People tend to think that vagueness is something to be avoided. What's to be avoided, however, is *an undesirable degree* of vagueness. Even though a claim may be less precise than it could be, that does not mean that it is less precise than it should be. For example, if you want to move your car and you ask the usher how long you have until the play begins, the reply "Only a minute or two" is less precise than is possible, but it is not less precise than is desirable for your purposes. You don't have time to move your car. But if you are the lead actor in the play and ask the stage manager exactly how much time you have before the curtain goes up, the same answer is probably not precise enough.

The vagueness of some claims is due to the use of "fuzzy" words, words such as "old" and "bald" and "rich," that apply to lots of borderline cases. Bill Gates, for instance, clearly is rich. But is a person who is worth a half million dollars rich? One worth a quarter million? Where does "rich" end and "well-off" begin? Notice, however, that claims with fuzzy words, like vague claims in general, are not always too vague. "Don't take a course from the bald guy in the philosophy department" may well be sufficiently precise to accomplish its intended effect, even though "bald" is not a very precise word.

Notice, too, that the absence of fuzzy words does not automatically immunize a claim from undesirable vagueness. "Maria plans to bring her sibling to class on Friday" is probably precise enough, depending on the kind

It's wonderful that we have eight-year-old boys filling sandbags with old women.

—LARRY KING, on attempts to stem Mississippi floods (reported by Larry Engelmann)

But, then, old women filled sandbags with eight-year-old boys, too.

of class we're talking about. But "Maria plans to bring her sibling to the slumber party Friday night" may not be precise enough, especially if it's your teenage daughter who is giving the slumber party and Maria's sibling is her twenty-five-year-old brother. It's not that "sibling" is particularly fuzzy; it's just that in the second of these contexts, the claim in which "sibling" appears is too imprecise for the occasion.

As should be clear by now, it makes little sense to insist that a claim be totally free of vagueness. If we had to be absolutely precise whenever we made a claim, we would say and write very little. The appropriate criticism of a claim is not that it is vague, but that it is too vague relative to what you wish to communicate or know.

■ *Claims That Make Comparisons*

Advertisers and politicians seem to be especially fond of claims like these:

If you are won-dering about the boys, they both got gangrene and lost some legs.

—CNN anchor ANDERSON COOPER

Makes sense, sort of, if you view the boys collectively.

Cut by up to half

Now 25 percent larger

Quietest by far

New and improved

Now better than ever

More than 20 percent richer

Such claims cry out for clarification: What does "up to half" mean? Twenty-five percent larger than what? How far is "by far"? New and improved over what? Better in what way? How much more than 20 percent? And you can ask other questions, too. Remember, though, that the amount of vagueness you can tolerate in a comparative claim depends on your interests and purposes. Knowing that attic insulation will reduce your utility bill "by 15 to 45 percent" may be all it takes for you to know that you should insulate.

Of course, some comparisons may be too vague even to be meaningful: "Have more fun in Arizona." "Gets clothes whiter than white." "Delivers more honest flavor." In other words, have fun in Arizona. Gets clothes white. Can be tasted. On the other hand, "Nothing else is a Pepsi" isn't vague; taken literally, it's necessarily true. Nothing else is a turkey, either. What the claim means, of course, is that no other soft drink tastes as good as Pepsi.

Here are some other questions to ask when you are considering claims that make comparisons:

1. *Is important information missing?* Suppose someone says that only 10 percent of child molestation cases are at the hands of alcoholics. The person might then say the statement means that alcoholics are less likely to be child molesters than are nonalcoholics because, after all, 90 percent of child molestation cases are at the hands of nonalcoholics. What's missing is important information about the percentage of the general popula-

Cause for Alarm?

According to the National Household Survey on Drug Abuse, cocaine use among Americans twelve to seventeen years of age increased by a whopping 166 percent between 1992 and 1995. Wow, right?

Except that the increase *in absolute terms* was a little less spectacular: In 1992, 0.3 percent of Americans aged twelve to seventeen had used cocaine; in 1995, the percentage was 0.8 percent of that population.

Be wary of comparisons expressed as percentage changes.

tion that is alcoholic. What if, say, only 3 percent of the general population is alcoholic but that 3 percent commits 10 percent of the child molestations? Then it begins to seem that alcoholics are *more* likely than others to molest children.

2. *Is the same standard of comparison being used? Are the same reporting and recording practices being used?* Check to be sure. In February 1994, the unemployment rate suddenly jumped by half a percentage point (a significant jump). Had more people been laid off? Nope. The government had simply changed the way it figured unemployment.

Here's another example. In 1993 the number of people in the United States with AIDS suddenly increased dramatically. Had a new form of the AIDS virus appeared? No; the federal government had expanded the definition of AIDS to include several new indicator conditions. Overnight, 50,000 people were considered to have AIDS who had not been so considered the day before.

3. *Are the items comparable?* Be wary of comparisons between apples and oranges. Don't place undue faith in a comparison between this April's retail business activity and last April's if Easter fell in March in one of the years or the weather was especially cold.

Or suppose someone cites statistics that show you are safer flying than driving because the accident rate for commercial airplane passengers is lower than that for car drivers. What you should note first is that the comparison is between air *passengers* and car *drivers*. A more telling comparison would be between airplane *pilots* and car drivers. And if you are considering going Greyhound, an even better comparison would be between commercial airplane pilots and commercial bus drivers. Of course, the claims about the safety of flying versus driving sometimes involve comparisons between air *passengers* and car *passengers*. But such comparisons usually still involve apples and oranges. What you really need to know is the *fatality rate* of air passengers versus that of car passengers in cars *driven by drivers who are as cautious, sober, and skilled as commercial airplane pilots.* After all, that is the kind of person who will be driving you, right?

4. *Is the comparison expressed as an average? Again, be sure important information isn't missing.* The average rainfall in Seattle is about the same as that in Kansas City. But you'll spend more time in rain gear in Seattle

There is no excuse for passing children through schools who can't read.
—GEORGE W. BUSH

Apparently, however, there is no problem with schools that don't teach grammar.

Potter's Take May Be Overhyped

Box-Office Figures Need to Be Viewed in Context

If one listens to the hype, it would seem that "Harry Potter and the Sorcerer's Stone" is destined to be the most popular and lucrative movie of all time. That's what the studio, Warner Bros., wants us to believe, and that's why box-office receipts are always touted—if not willfully distorted—for the sale of publicity and promotion.

"Harry Potter" is a smash hit, all right, but the amazing figures that attended its opening—$90 million the first weekend, $220.1 million in 17 days—don't make it the almost box-office champ. What's ignored in all the hoopla over box office are such factors as inflation, population growth and the number of movie screens at which a film opens.

When Hollywood brags about box office, it never mentions, for example, that ticket prices keep rising; that our country's population keeps growing (275 million today versus 226 million in 1980 and 131 million in 1939); or that blockbusters like "Harry Potter" open at hundreds more theaters than movies did even a decade ago.

When one considers those factors, the luster of box-office figures starts rapidly to dim. It could easily be argued that "Gone With the Wind" is the most popular film of all time, based on its huge popularity at the time of its initial release (1939) and subsequent re-releases.

During the long span of its theatrical life, "GWTW" sold 200 million movie admissions, according to Paul Dergarabedian of Exhibitor Relations, a Los Angeles-based box-office tracking firm. But with a cumulative domestic gross of $173 million, "GWTW" fell off the top 10 box-office chart years ago simply because people paid so much less to see it than they do for today's movies.

But one needn't go back 60 years to understand the misleading nature of box-office statistics. "E.T. the Extra-Terrestrial" was a huge hit when released in 1982 and remains the No. 3 box-office champ, even though its opening-weekend gross was $11.9 million.

In 2000, the average movie admission price was $5.39 (including lower rates for seniors and children), according to the National Association of Theater Owners. Dergarabedian estimates that it's climbed to $5.60 this year. Back in 1991, the average admission price was $4.21; in 1976 it was $2.13; in 1960, 69 cents; in 1949, 46 cents; and in 1939, 23 cents.

The math is simple and astonishing. With today's prices, it takes 16.1 million moviegoers to create a $90.3 million opening weekend. In 1939, it would have taken 392.6 million moviegoers to create the same number.

Here's another way of looking at the situation. If the domestic gross on "GWTW" over its entire run were adjusted for inflation, it would jump from $173 million to an estimated $1.1 billion.

Consider that "Harry Potter" opened at a record-breaking 3,672 theaters. Most of those were multiplexes, where the movie played throughout the day on multiple screens. "Harry" occupied a whopping 8,200 screens—roughly 21 percent of the nation's 37,396 total.

because it rains there twice as often as in Kansas City. If Central Valley Components, Inc., reports that average salaries of a majority of its employees have more than doubled over the past ten years, CVC still may not be a great place to work. Perhaps the increases were all due to converting the majority of employees, who worked half time, to full time and firing the rest. Comparisons that involve averages omit details that can be important, simply because they involve averages.

Remember, too, that there are different kinds of averages, or measures of central tendency, as they are also known—the mean, the median, and the mode. Consider, for instance, the average paycheck at Central Valley Components, which happens to be $41,000. That may sound generous enough.

Never try to wade a river just because it has an average depth of four feet.

—MARTIN FRIEDMAN

The wrong average can put you under.

Tom Felton and Daniel Radcliffe play rival wizards Draco Malfoy and Harry Potter in "Harry Potter and the Sorcerer's Stone."

But let's forget about inflation for a minute and ponder this question: Can "Harry Potter" top "Titanic," the current box-office champion with $600.8 million in domestic grosses, $1.2 billion in international and $1.8 billion in total receipts?

"It's too early to tell," says Dergarabedian. "It's doing extraordinarily well, and I think it's going to get a boost at Christmas. But this whole notion of surpassing 'Titanic'—that's no small feat."

Staying power is the magic potion that "Harry Potter" needs to break the record of James Cameron's Oscar-winning epic. "Titanic" took in $28.6 million its first weekend but was so durable, and generated such phenomenal repeat business, that it held the No. 1 position for 15 weeks and was still playing in some theaters a full nine months after its December 1997 release.

"It was in release for 41 weeks," says Dergarabedian. "That's amazing."

By Edward Guttmann
Chronicle Staff Writer

_____

Article from *San Francisco Chronicle,* December 24, 2001.

But that average is the **mean** (total wages divided by the number of wage earners). The **median** wage, an "average" that is the "halfway" figure (half the employees get more than the figure, and half get less), is $27,000. This average is not so impressive, at least from the perspective of a job seeker. And the **mode,** also an average, the most common rate of pay, is only $17,000. So when someone quotes "the average pay" at CVC, which average is it? At CVC a couple of executives draw fat paychecks, so the mean is a lot higher than the other two figures.

In a class of things in which there are likely to be large or dramatic variations in whatever it is that is being measured, be cautious of figures about an unspecified "average."

Misleading Averages

In 2003 George W. Bush proposed a tax cut that, he said, would give the average taxpayer $1,083.

The "average" here is the mean average. However, before you start dreaming about how to spend your $1,083, you might want to check on the modal average. Most taxpayers, according to the Urban Institute–Brookings Institution Tax Policy Center, would receive less than $100 under the Bush proposal.

Exercise 2-6

The lettered words and phrases that follow each of the following fragments vary in their precision. In each instance, determine which is the most vague and which is the most precise; then rank the remainder in order of vagueness between the two extremes. You will discover when you discuss these exercises in class that they leave some room for disagreement. Further class discussion with input from your instructor will help you and your classmates reach closer agreement about items that prove especially difficult to rank.

Example

Over the past ten years, the median income of wage earners in St. Paul
a. nearly doubled
b. increased substantially
c. increased by 85.5 percent
d. increased by more than 85 percent

Answer

Choice (b) is the most vague because it provides the least information; (c) is the most precise because it provides the most detailed figure. In between, (a) is the second most vague, followed by (d).

1. Eli and Sarah
▲   a. decided to sell their house and move
   b. made plans for the future
   c. considered moving
   d. talked
   e. discussed their future
   f. discussed selling their house

2. Manuel
   a. worked in the yard all afternoon
   b. spent the afternoon planting flowers in the yard
   c. was outside all afternoon
   d. spent the afternoon planting salvia alongside his front sidewalk
   e. spent the afternoon in the yard

"The Biggest Tax Increase in History"

Every presidential election seems to see somebody link the opposing candidate to "the biggest tax increase in history." What we have here, however, is an ambiguous comparison. Is the increase the biggest in *absolute dollars* or in *dollars adjusted for inflation*? And notice as well the grouping ambiguity in the claim: One and the same tax increase might not be very large as an increase in the *percentage of an individual's income* that he or she pays to taxes, and yet it might still be the largest increase in terms of the total tax revenues it produces.

3. The hurricane that struck South Carolina
   a. caused more than $20 million in property damage
   b. destroyed dozens of structures
   c. was severe and unfortunate
   d. produced no fatalities but caused $25 million in property damage

4. The recent changes in the tax code
   a. will substantially increase taxes paid by those making more than $200,000 per year
   b. will increase by 4 percent the tax rate for those making more than $200,000 per year; will leave unchanged the tax rate for people making between $40,000 and $200,000; and will decrease by 2 percent the tax rate for those making less than $40,000
   c. will make some important changes in who pays what in taxes
   d. are tougher on the rich than the provisions in the previous tax law
   e. raise rates for the wealthy and reduce them for those in the lowest brackets

5. Smedley is absent because
   a. he's not feeling well
   b. he's under the weather
   c. he has an upset stomach and a fever
   d. he's nauseated and has a fever of more than 103°
   e. he has flulike symptoms

Exercise 2-7

Which of each set of claims is vaguer, if either?

Example

> a. The trees served to make shade for the patio.
> b. He served his country proudly.

Answer

> The use of "served" in (b) is more vague than that in (a). We know exactly what the trees did; we don't know what he did.

▲ 1. a. Rooney served the church his entire life.
    b. Rooney's tennis serve is impossible to return.

2. a. The window served its purpose.
   b. The window served as an escape hatch.

3. a. Throughout their marriage, Alfredo served her dinner.
   b. Throughout their marriage, Alfredo served her well.

▲ 4. a. Minta turned her ankle.
    b. Minta turned to religion.

5. a. These scales will turn on the weight of a hair.
   b. This car will turn on a dime.

6. a. Fenner's boss turned vicious.
   b. Fenner's boss turned out to be forty-seven.

▲ 7. a. Time to turn the garden.
    b. Time to turn off the sprinkler.

8. a. The wine turned to vinegar.
   b. The wine turned out to be vinegar.

9. a. Harper flew around the world.
   b. Harper departed around 3:00 A.M.

▲ 10. a. Clifton turned out the light.
     b. Clifton turned out the vote.

11. a. The glass is full to the brim.
    b. Mrs. Couch has a rather full figure.

12. a. Kathy gave him a full report.
    b. "Oh, no, thank you! I am full."

13. a. Oswald was dealt a full house.
    b. Oswald is not playing with a full deck.

14. a. The pudding sat heavily on Professor Grantley's stomach.
    b. "Set the table, please."

▲ 15. a. Porker set a good example.
     b. Porker set the world record for the 100-meter dash.

Exercise 2-8

Are the italicized words or phrases in each of the following too vague given the implied context? Explain.

▲ 1. Please cook this steak *longer.* It's too rare.

2. If you get ready for bed quickly, Mommy has a *surprise* for you.

3. This program contains language that some viewers may find offensive. It is recommended for *mature* audiences only.

▲ 4. *Turn down the damned noise!* Some people around here want to sleep!

5. Based on our analysis of your eating habits, we recommend that you *lower* your consumption of saturated fat.

6. NOTICE: Hazard Zone. *Small* children not permitted beyond this sign.

▲ 7. SOFAS CLEANED: $48 & *up.* MUST SEE TO GIVE *EXACT* PRICES.

8. And remember, all our mufflers come with a *lifetime guarantee.*

9. CAUTION: *To avoid* unsafe levels of carbon monoxide, do not set the wick on your kerosene stove *too high.*

▲ 10. Uncooked Frosting: Combine 1 unbeaten egg white, ½ cup corn syrup, ½ teaspoon vanilla, and dash salt. Beat with electric mixer until of fluffy spreading consistency. Frost cake. Serve *within a few hours,* or refrigerate.

Exercise 2-9

Read the following passage, paying particular attention to the italicized words and phrases. Determine whether any of these expressions are too vague in the context in which you find them here.

> Term paper assignment: "Your paper *should be* typed, *between eight and twelve pages in length,* and double-spaced. You should *make use of* at least three *sources.* Grading will be based on *organization, use of sources, clarity of expression, quality of reasoning,* and *grammar.*
>
> A *rough draft* is due *before Thanksgiving.* The final version is due *at the end of the semester.*

Exercise 2-10

▲ Read the following passage, paying particular attention to the italicized words and phrases. All of these expressions would be too vague for use in *some* contexts; determine which are and which are not too vague in *this* context.

> In view of what can happen in twelve months to the fertilizer you apply at any one time, you can see why just one annual application may not be adequate. Here is a guide to timing the *feeding* of some of the more common types of garden flowers.
>
> Feed begonias and fuchsias *frequently* with label-recommended amounts or less frequently with *no more than half* the recommended amount. Feed roses with *label-recommended amounts* as a *new year's growth begins* and as *each bloom period ends.* Feed azaleas, camellias, rhododendrons, and *similar* plants *immediately after bloom* and again *when the nights begin cooling off.* Following these simple instructions can help your flower garden to be as attractive as it can be.

Exercise 2-11

Critique these comparisons, using the questions about comparisons discussed in the text as guides.

Example

You get much better service on Air Atlantic.

Answer

Better than on what? (One term of the comparison is not clear.)

In what way better? (The claim is much too vague to be of much use.)

▲ 1. New improved Morning Muffins! Now with 20 percent more real dairy butter!

  2. The average concert musician makes less than a plumber.

  3. Major league ballplayers are much better than they were thirty years ago.

▲ 4. What an arid place to live. Why, they had less rain here than in the desert.

  5. On the whole, the mood of the country is more conservative than it was in the sixties.

  6. Which is better for a person, coffee or tea?

▲ 7. The average GPA of graduating seniors at Ohio State is 3.25, as compared with 2.75 twenty years ago.

  8. Women can tolerate more pain than men.

  9. Try Duraglow with new sunscreening polymers. Reduces the harmful effect of sun on your car's finish by up to 50 percent.

▲ 10. What a brilliant season! Attendance was up 25 percent over last year.

Exercise 2-12

Critique these comparisons, using the questions discussed in the text as guides.

▲ 1. You've got to be kidding. Paltrow is much superior to Blanchett as an actor.

  2. Blondes have more fun.

  3. The average chimp is smarter than the average monkey.

▲ 4. The average grade given by Professor Smith is a C. So is the average grade given by Professor Algers.

  5. Crime is on the increase. It's up by 160 percent over last year.

  6. Classical musicians, on the average, are far more talented than rock musicians.

▲ 7. Long-distance swimming requires much more endurance than long-distance running.

  8. "During the monitoring period, the amount of profanity on the networks increased by 45–47 percent over a comparable period from the preceding year. A clear trend toward hard profanity is evident."

Don Wildmon, founder of the National Federation for Decency

  9. "Organizations such as EMILY's List and the Women's Campaign Fund encourage thousands of small contributors to participate, helping to offset

And While We're on the Subject of Writing . . .

Don't forget these rules of good style:

1. Avoid clichés like the plague.
2. Be more or less specific.
3. NEVER generalize.
4. The passive voice is to be ignored.
5. Never ever be redundant.
6. Exaggeration is a billion times worse than understatement.
7. Make sure verbs agrees with their subjects.
8. Why use rhetorical questions?
9. Parenthetical remarks (however relevant) are (usually) unnecessary.
10. Proofread carefully to see if you any words out.
11. And it's usually a bad idea to start a sentence with a conjunction.

This list has been making the rounds on the Internet.

the economic power of the special interests. The political system works better when individuals are encouraged to give to campaigns."

Adapted from the Los Angeles Times

▲ 10. Which is more popular, the movie *Gone With the Wind* or Bing Crosby's version of the song "White Christmas"?

Exercise 2-13 _____

In groups, or individually if your instructor prefers, critique these comparisons, using the questions discussed in the text as guides.

▲ 1. If you worry about the stock market, you have reason. The average stock now has a price-to-earnings ratio of around 25:1.

2. Students are much less motivated than they were when I first began teaching at this university.

3. Offhand, I would say the country is considerably more religious than it was twenty years ago.

▲ 4. In addition, for the first time since 1960, a majority of Americans now attend church regularly.

5. You really should switch to a high-fiber diet.

6. Hire Ricardo. He's more knowledgeable than Annette.

▲ 7. Why did I give you a lower grade than your roommate? Her paper contained more insights than yours, that's why.

8. Golf is a considerably more demanding sport than tennis.

Making Ambiguity Work for You

Were you ever asked to write a letter of recommendation for a friend who is, well, incompetent? If you don't want to hurt his or her feelings but also don't want to lie, Robert Thornton of Lehigh University has some ambiguous statements you can use. Here are some examples:

I most enthusiastically recommend this candidate with no qualifications whatsoever.

I am pleased to say that this candidate is a former colleague of mine.

I can assure you that no person would be better for the job.

I would urge you to waste no time in making this candidate an offer of employment.

All in all, I cannot say enough good things about this candidate or recommend the candidate too highly.

In my opinion you will be very fortunate to get this person to work for you.

9. Yes, our prices are higher than they were last year, but you get more value for your dollar.

▲ 10. So, tell me, which do you like more, fried chicken or Volkswagens?

Exercise 2-14 _____

Find two examples of faulty comparisons, and read them to your class. Your instructor may ask other members of the class to critique them.

Persuasive Writing

The primary aim of argumentation and an argumentative essay is to establish something, to support a position on an issue. Good writers, however, write for an audience and hope that their audience will find what they write persuasive. If you are writing for an audience of people who think critically, it is helpful to adhere to these principles:

1. Confine your discussion of an opponent's point of view to issues rather than personal considerations.

2. Anticipate and discuss what opponents might say in criticism of your opinion.

3. When rebutting an opposing viewpoint, avoid being strident or insulting. Don't call opposing arguments absurd or ridiculous.

4. If an opponent's argument is good, concede that it is good.

5. If space or time is limited, be sure to concentrate on the most important considerations. Don't become obsessive about refuting every last criticism of your position.

6. Rebut objections to your position before presenting your own positive case for your side.

7. Present your strongest arguments first.

There is absolutely nothing wrong with trying to make a persuasive case for your position. However, in this book, we place more emphasis on making and recognizing good arguments than simply on devising effective techniques of persuasion. Some people can be persuaded by poor arguments and doubtful claims, and an argumentative essay can be effective as a piece of propaganda even when it is a rational and critical failure. One of the most difficult things you are called upon to do as a critical thinker is to construct and evaluate claims and arguments independently of their power to win a following. The remainder of this book—after a section on writing and diversity—is devoted to this task.

Writing in a Diverse Society

In closing a chapter that deals with writing essays, it seems appropriate to mention how important it is to avoid writing in a manner that reinforces questionable assumptions and attitudes about people's gender, ethnic background, religion, sexual orientation, physical ability or disability, or other characteristics. This isn't just a matter of ethics; it is a matter of clarity and good sense. Careless word choices relative to such characteristics not only are imprecise and inaccurate but also may be viewed as biased even if they were not intended to be, and thus they may diminish the writer's credibility. Worse, using sexist or racist language may distort the writer's own perspective and keep him or her from viewing social issues clearly and objectively.

But language isn't entirely *not* a matter of ethics, either. We are a society that aspires to be just, a society that strives not to withhold its benefits from individuals on the basis of their ethnic or racial background, skin color, religion, gender, or disability. As a people we try to end practices and change or remove institutions that are unjustly discriminatory. Some of these unfair practices and institutions are, unfortunately, embedded in our language.

Some common ways of speaking and writing, for example, assume that "normal" people are all white males. It is still common practice, for instance, to mention a person's race, gender, or ethnic background if the person is *not* a white male, and *not* to do so if the person *is.* Thus, if we are talking about a white male from Ohio, we are apt to say simply, "He is from Ohio." But if the male is Latino, we might tend to mention that fact and say, "He is a Latino from Ohio"—even when the person's ethnic background is irrele-

"Always" and "never" are two words you should always remember never to use.

—Wendell Johnson

Another tip on writing.

vant to whatever we are talking about. This practice assumes that the "normal" person is not Latino and by implication insinuates that if you are, then you are "different" and a deviation from the norm, an outsider.

Of course, it may be relevant to whatever you are writing about to state that this particular man is a Latino from Ohio, and, if so, there is absolutely nothing wrong with writing "He is a Latino from Ohio."

Some language practices are particularly unfair to women. Imagine a conversation among three people, you being one of them. Imagine that the other two talk only to each other. When you speak, they listen politely; but when you are finished, they continue as though you had never spoken. Even though what you say is true and relevant to the discussion, the other two proceed as though you are invisible. Because you are not being taken seriously, you are at a considerable disadvantage. You would have reason to be unhappy.

In an analogous way, women have been far less visible in language than men and have thus been at a disadvantage. Another word for the human race is not "woman," but "man" or "mankind." The generic human has often been referred to as "he." How do you run a project? You *man* it. Who supervises the department or runs the meeting? The chair*man*. Who heads the crew? The fore*man.* Women, like men, can be supervisors, scientists, professors, lawyers, poets, and so forth. But a woman in such a profession is apt to be referred to as a "woman scientist" or "woman supervisor" or as a "lady poet" (or whatever). The implication is that in their primary signification, words like "scientist" and "supervisor" refer to men.

Picture a research scientist to yourself. Got the picture? Is it a picture of a *woman*? No? That's because the standard picture, or stereotype, of a research scientist is a picture of a man. Or, read this sentence: "Research scientists often put their work before their personal lives and neglect their husbands." Were you surprised by the last word? Again, the stereotypical picture of a research scientist is a picture of a man.

A careful and precise writer finds little need to converse in the lazy language of stereotypes, especially those that perpetuate prejudice. As long as the idea prevails that the "normal" research scientist is a man, women who are or who wish to become research scientists will tend to be thought of as out of place. So they must carry an *extra* burden, the burden of showing that they are *not* out of place. That's unfair. If you unthinkingly always write "The research scientist . . . he," you are perpetuating an image that places women at a disadvantage. Some research scientists are men, and some are women. If you wish to make a claim about male research scientists, do so. But if you wish to make a claim about research scientists in general, don't write as though they were all males.

Most often the problem of unintentional gender discrimination arises in connection with pronouns. Note how the use of "he" and "his" in the following two sentences excludes females.

As *a student* learns to read more critically, *he* usually writes more clearly, too.

What day is the day after three days before the day after tomorrow?

Complicated, but neither vague nor ambiguous

Do you have blacks, too?

—President George W. Bush, during a discussion with Brazilian President Fernando Henrique Cardoso

In fact, Brazil has more black people than does the United States.

Avoiding Sexist Language

Use the suggestions below to help keep your writing free of sexist language.

Instead of	Use
actress	actor
chairman	chair, chairperson, coordinator, director, head, leader, president, presider
congressman	congressional representative, member of Congress
fathers (of industry, and so on)	founders, innovators, pioneers, trailblazers
housewife	homemaker, woman
man (verb)	operate, serve, staff, tend, work
man (noun)	human beings, individuals, people
man and wife	husband and wife
man-hour	operator-hour, time, work-hour
mankind	human beings, humanity, humankind
manmade	artificial, fabricated, manufactured, synthetic, human-made
manpower	crew, personnel, staff, workers, workforce
newsman	journalist, newscaster, reporter
poetess	poet
policeman	police officer
repairman	repairer
spokesman	representative, spokesperson
statesmanlike	diplomatic, tactful
stewardess	flight attendant
waitress	service person, waiter
weatherman	weather reporter, weathercaster
craftsman	artisan
deliveryman	courier
foreman	supervisor; lead juror
freshman	first-year student
heroine	hero
layman	layperson

You can also refer to any of numerous reference works on bias-free language. See, for example, Marilyn Schwartz, *Guidelines for Bias-Free Writing* (Bloomington: Indiana University Press, 1995).

> As *a student* learns to read more critically, *his* writing, too, usually becomes clearer.

Obviously, changing "his" and "he" to "her" and "she" would then exclude males. If the writer means to include people of both sexes, one solution is to use two pronouns: for instance, "he or she" (or "his or her").

> As *a student* learns to read more critically, *she or he* usually writes more clearly, too.
> As *a student* learns to read more critically, *his or her* writing, too, usually becomes clearer.

However, a usually less awkward remedy is to change from singular to plural (because plural pronouns in English do not show gender).

> As *students* learn to read more critically, *they* usually write more clearly, too.
> As *students* learn to read more critically, *their* writing usually becomes clearer, too.

Often the best solution of all is to rephrase the statement entirely, which is almost always possible.

> Critical reading leads to clear writing.
> Learning to read more critically helps students write more clearly, too.

Frequently a pronoun can simply be deleted or replaced by another word, especially "a" or "the":

> The student who develops the habit of reading critically will get more out of *his* college courses.
> The student who develops the habit of reading critically will get more out of college courses.
> The student who develops the habit of reading critically will get more out of *a* [or *any*] college course.

The rule to follow in all cases is this: Keep writing free of *irrelevant implied evaluation* of gender, race, ethnic background, religion, or any other human attribute.

Recap

An argumentative essay is intended to support a position on some issue. Principles of critical thinking can and should be applied both to essays written by others and to those we ourselves write. Such essays must be soundly organized and clearly written and must truly support the position taken by their author.

Check Your Chart

The following are alleged to be actual notations in patients' records. Each has a flagrant ambiguity or other problem. Sometimes the problem interferes with clarity; other times it only produces amusement. Describe what is going on in each case.

- Patient has chest pain if she lies on her left side for over a year.
- She has had no rigors or shaking chills, but her husband states she was very hot in bed last night.
- The patient has been depressed ever since she began seeing me in 1983.
- I will be happy to go into her GI system; she seems ready and anxious.
- The patient is tearful and crying constantly. She also appears to be depressed.
- Discharge status: Alive but without permission. The patient will need disposition, and therefore we will get Dr. Blank to dispose of him.
- Healthy-appearing decrepit 69-year-old male, mentally alert but forgetful.
- The patient refused an autopsy.
- The patient has no past history of suicides.
- The patient expired on the floor uneventfully.
- Patient has left his white blood cells at another hospital.
- Patient was becoming more demented with urinary frequency.
- The patient's past medical history has been remarkably insignificant with only a 40-pound weight gain in the past three days.
- She slipped on the ice and apparently her legs went in separate directions in early December.
- The patient experienced sudden onset of severe shortness of breath with a picture of acute pulmonary edema at home while having sex which gradually deteriorated in the emergency room.
- The patient left the hospital feeling much better except for her original complaints.
- The patient states there is a burning pain in his penis which goes to his feet.

(These were sent around the Internet after they appeared in a column written by RICHARD LEDERER in the *Journal of Court Reporting*. Lederer is famous for his collections of remarks taken from student papers, especially papers on historical subjects.)

Writers can get into trouble in many ways, but you can avoid the most common pitfalls if you are willing to take care. To keep your essay organized, stay focused on your main points, use an outline, and be prepared to revise. To achieve clarity, be sure you clearly define new terms, terms that are especially important to your argument, or terms you are using in an unusual way; and remember to pay attention to the emotive force of your words. Avoid redundancy and unnecessarily complex language. Watch out for ambiguous claims (claims that are insufficiently precise for the purposes at hand). Be especially careful when writing or analyzing claims that make comparisons.

It is acceptable to use persuasive techniques to support a position in an argumentative essay, but don't use them *instead* of reasonable claims and well-constructed arguments. Be careful to avoid irrelevant (and unwarranted) assumptions about people, including those you may overlook because they are embedded in our language.

Remember: If what you want to say is not clear to you, it certainly will remain obscure to your reader.

Having read this chapter and the chapter recap, you should begin to be able to

- ▲ Create a clear and focused organizational plan for an argumentative essay
- ▲ Define terms in your essay clearly and choose the words that best suit your purposes
- ▲ Recognize different types of ambiguity, know the uses and abuses of vagueness, and analyze comparative claims
- ▲ Make a persuasive case for your positions to an audience of people who can think critically
- ▲ Avoid sexist, racist, and other inappropriate writing practices

Additional Exercises

Exercise 2-15

Rewrite each of the following claims in gender-neutral language.

Example

We have insufficient manpower to complete the task.

Answer

We have insufficient personnel to complete the task.

▲ 1. A student should choose his major with considerable care.

2. When a student chooses his major, he must do so carefully.

3. The true citizen understands his debt to his country.

▲ 4. If a nurse can find nothing wrong with you in her preliminary examination, she will recommend a physician to you. However, in this city the physician will wish to protect himself by having you sign a waiver.

5. You should expect to be interviewed by a personnel director. You should be cautious when talking to him.

6. The entrant must indicate that he has read the rules, that he understands them, and that he is willing to abide by them. If he has questions, then he should bring them to the attention of an official, and he will answer them.

▲ 7. A soldier should be prepared to sacrifice his life for his comrades.

8. If anyone wants a refund, he should apply at the main office and have his identification with him.

9. The person who has tried our tea knows that it will neither keep him awake nor make him jittery.

▲ 10. If any petitioner is over sixty, he (she) should have completed form E-7.

11. Not everyone has the same beliefs. One person may not wish to put himself on the line, whereas another may welcome the chance to make his view known to his friends.

12. God created man in his own image.

▲ 13. Language is nature's greatest gift to mankind.

14. Of all the animals, the most intelligent is man.

15. The common man prefers peace to war.

▲ 16. The proof must be acceptable to the rational man.

▲ 17. The Founding Fathers believed that all men are created equal.

18. Man's pursuit of happiness has led him to prefer leisure to work.

19. When the individual reaches manhood, he is able to make such decisions for himself.

▲ 20. If an athlete wants to play for the National Football League, he should have a good work ethic.

21. The new city bus service has hired several women drivers.

22. The city is also hiring firemen, policemen, and mailmen; and the city council is planning to elect a new chairman.

23. Harold Vasquez worked for City Hospital as a male nurse.

▲ 24. Most U.S. senators are men.

25. Mr. and Mrs. Macleod joined a club for men and their wives.

26. Mr. Macleod lets his wife work for the city.

▲ 27. Macleod doesn't know it, but Mrs. Macleod is a women's libber.

28. Several coeds have signed up for the seminar.

29. A judge must be sensitive to the atmosphere in his courtroom.

▲ 30. To be a good politician, you have to be a good salesman.

Exercise 2-16

▲ A riddle: A man is walking down the street one day when he suddenly recognizes an old friend whom he has not seen in years walking in his direction with a little girl. They greet each other warmly, and the friend says, "I married since I last saw you, to someone you never met, and this is my daughter, Ellen." The man says to Ellen, "You look just like your mother." How did he know that?

This riddle comes from Janice Moulton's article, "The Myth of the Neutral Man." Discuss why so many people don't get the answer to this riddle straight off.

Writing Exercises

Everyone, no matter how well he or she writes, can improve. And the best way to improve is to practice. Since finding a topic to write about is often the hardest part of a writing assignment, we're supplying three subjects for you to write about. For each—or whichever your instructor might assign—write a one- to two-page essay in which you clearly identify the issue (or issues), state your position on the issue (a hypothetical position if you don't have one), and give at least one good reason in support of your position. Try also to give at least one reason why the opposing position is wrong.

1. The exchange of dirty hypodermic needles for clean ones, or the sale of clean ones, is legal in many states. In such states, the transmission of HIV and hepatitis from dirty needles is down dramatically. But bills [in the California legislature] to legalize clean-needle exchanges have been stymied by the last two governors, who earnestly but incorrectly believed that the availability of clean needles would increase drug abuse. Our state, like every other state that has not yet done it, should immediately approve legislation to make clean needles available.

   Adapted from an editorial by Marsha N. Cohen,
   professor of law at Hastings College of Law

2. On February 11, 2003, the Eighth Circuit Court of Appeals ruled that the state of Arkansas could force death-row prisoner Charles Laverne Singleton to take antipsychotic drugs to make him sane enough to execute. Singleton was to be executed for felony capital murder but became insane while in prison. "Medicine is supposed to heal people, not prepare them for execution. A law that asks doctors to make people well so that the government can kill them is an absurd law," said David Kaczynski, the executive director of New Yorkers Against the Death Penalty.

3. Some politicians make a lot of noise about how Canadians and others pay much less for prescription drugs than Americans do. Those who are constantly pointing to the prices and the practices of other nations when it comes to pharmaceutical drugs ignore the fact that those other nations lag far behind the United States when it comes to creating new medicines.

Canada, Germany, and other countries get the benefits of American research but contribute much less than the United States does to the creation of drugs. On the surface, these countries have a good deal, but in reality everyone is worse off, because the development of new medicines is slower than it would be if worldwide prices were high enough to cover research costs.

Adapted from an editorial by Thomas Sowell,
senior fellow at the Hoover Institution

Chapter 3

Credibility

We size up people's believability, or credibility, every day. Sometimes we make mistakes.

On July 13, 2002, Charles Keller of Sacramento, California, withdrew $10,000 from his Wells Fargo savings account and gave it to two men he didn't know from Adam. The men had convinced Keller, merely by talking to him for a while, that by giving them $10,000 to demonstrate "good faith" he would receive a check for $60,000 at the end of the day. They told Mr. Keller they represented an organization that located individuals who had received large inheritances but didn't know it and were authorized to pay "finder's fees" to anyone who could help them locate the people. We don't know for sure what else they told Mr. Keller, but it is easy to imagine possibilities. Perhaps they described a neighbor of Mr. Keller as one of the people they were looking for, prompting Keller to think the finder's fee would be easy money.

It may strike you as odd that anyone at least smart enough to accumulate $10,000 in the bank would fall for such an obvious swindle. Yet people both wealthy and intelligent are conned every day. We'd bet in your own city there are intelligent people who are losing their shirt at this very moment. The problem stems not from a lack of intelligence but from other things, including perhaps wishful thinking (which we will say more about in a bit); it also stems from a failure to judge credibility correctly.

Two different things in varying degrees have or lack credibility: *claims* and *sources* (that is, *people*). Some claims lack credibility no matter who says them; other claims depend for their credibility on who says them. Do ducks quack in Morse Code? The idea would lack credibility no matter who were to express it. But the claim that ducks mate for life would depend for its credibility on who said it. It would be more believable coming from a bird expert than from our editor, for example.

Credibility is not an all-or-nothing thing, whether we are talking about claims or sources. If someone claimed that George W. Bush sends weekly bribes to the U.S. Supreme Court, his claim would be rather far-fetched. But it wouldn't be as far-fetched as the claim that George W. Bush is a space alien. A similar variability in credibility attaches to people. An individual known to lie habitually is less credible than an otherwise honest person caught lying to save someone's life. A man charged with armed robbery becomes less credible in his denial if we learn he owns a silencer and an illegal semi-automatic .40-caliber handgun with the serial numbers removed.

Now, we often judge the credibility of people on the basis of irrelevant considerations. Physical characteristics, for instance, are poor indicators of credibility or its lack. Does a person maintain eye contact? (Mr. Keller's swindlers might well have.) Is the person sweating? Does he or she laugh nervously? Does he or she shake our hand firmly? These characteristics are widely used in sizing up a person's credibility. But they are generally worthless.

Other irrelevant features we sometimes use to judge a person's credibility include gender, height, age, ethnicity, accent, and mannerisms. People also make credibility judgments on the basis of the clothes a person wears. A friend told one of us that one's sunglasses "make a statement"; maybe so, but that statement doesn't say much about credibility. A person's occupation certainly bears a relationship to his or her knowledge or abilities, but as a guide to moral character or truthfulness, it is less reliable.

Which considerations are relevant to judging someone's credibility? We shall get to these in a moment, but appearance isn't one of them. You may have the idea that you can size up a person just by looking into his or her eyes. This is a mistake. Just by looking at someone we cannot ascertain that person's truthfulness, knowledge, or character.

Of course, we sometimes get in trouble even when we accept credible claims from credible sources. Many of us rely, for example, on credible advice from qualified and honest professionals in preparing our tax returns. But qualified and honest professionals can make honest mistakes, and we can suffer the consequences. In general, however, trouble is much more likely if we accept either doubtful claims from credible sources or credible claims from doubtful sources (not to mention doubtful claims from doubtful sources). If a mechanic says we need a new transmission, the claim itself may not be suspicious—maybe the car we drive has many miles on it; maybe we neglected routine maintenance; maybe it isn't shifting smoothly. But if we find out that the mechanic has been charged with selling unneeded repairs to customers, we'd get a second opinion about our transmission.

On the other hand, if we look at Mr. Keller's case closely, it seems clear that his mistake was *not* that he did not pay close enough attention to the *swindlers'* credibility. There may well have been no detectable telltale signs that *they* were dishonest; after all, successful swindlers don't exhibit signs of dishonesty. Mr. Keller's mistake lay in accepting something the

Lovable Trickster Created a Monster with Bigfoot Hoax

Ray L. Wallace, 1918–2002 / Relatives finally fess up: Man faked famous jumbo footprints in 1958.

SEATTLE — Bigfoot is dead. Really.

"Ray L. Wallace was Bigfoot. The reality is, Bigfoot just died," said Michael Wallace about his father, who died of heart failure November 26 in a Centralia [California] nursing facility. He was 84.

The truth can finally be told, according to Mr. Wallace's family members. He orchestrated the prank that created Bigfoot in 1958.

Some experts suspected Mr. Wallace had planted the footprints that launched the term "Bigfoot." But Mr. Wallace and his family had never publicly admitted the 1958 deed until now.

"The fact is, there was no Bigfoot in popular consciousness before 1958. America got its own monster, its own Abominable Snowman, thanks to Ray Wallace," said Mark Chorvinsky, editor of *Strange* magazine and one of the leading proponents of the theory that Mr. Wallace fathered Bigfoot.

Pranks and hoaxes were just part of Mr. Wallace's nature.

"He'd been a kid all his life. He did it just for the joke, and then he was afraid to tell anybody because they'd be so mad at him," said nephew Dale Lee Wallace, who said he has the alderwood carvings of the giant humanoid feet that gave life to a worldwide phenomenon.

It was in August 1958 in Humboldt County, California, that Jerry Crew, a bulldozer operator for Wallace Construction, saw prints of huge naked feet circling and walking away from his rig.

The *Humboldt Times* in Eureka, California, ran a front-page story on the prints and coined the term "Bigfoot."

According to family members, Mr. Wallace smirked. He had asked a friend to carve the 16-inch-long feet. Then he and his brother Wilbur slipped them on and created the footprints as a prank, family members said.

His joke soon swept the country, which was fascinated by rumors of Himalayan Abominable Snowmen in the 1950s, Chorvinsky said.

"The Abominable Snowman was appropriated by Ray Wallace. It got into the press, took on a life of its own, and next thing you know there's a Bigfoot, one of the most popular monsters in the world," he said.

Mr. Wallace continued to milk the prank for years. He offered to sell a Bigfoot to Texas millionaire Tom Slick and then backed out when Slick made a serious bid. Mr. Wallace later put out a press release saying he wanted to buy a baby Bigfoot for $1 million, said Loren Coleman, who has written two books about Bigfoot. Mr. Wallace also cut a record of supposed Bigfoot sounds and printed posters of a Bigfoot sitting peaceably with other ani-

swindlers *said* that lacked credibility, such as, at the very least, "Give us $10,000 and we will give you back $60,000."

In light of these considerations a general principle emerges, to this effect:

> It is reasonable to be suspicious if a claim either lacks credibility inherently or comes from a source that lacks credibility.

Thus, we need to ask two questions: When does a *claim* lack credibility inherently—in other words, when does its content lack credibility? And when does a *source* lack credibility?

To begin with claims, the general answer is

mals, said Chorvinsky, who received several hundred pages of correspondence from Mr. Wallace.

But Mr. Wallace's chief contributions to bigfootery were films and photos he supposedly captured of the creature in the wild.

There were depictions of Bigfeet eating elk and frogs, of a Bigfoot sitting on a log, and of a Bigfoot munching on cereal.

"Ray's contribution was study into the actual behavior of Bigfoot, what it eats, how it acts," said Ray Crowe, director of the International Bigfoot Society in Hillsboro, Oregon.

Chorvinsky believes the Wallace family's admission creates profound doubts about leading evidence of Bigfoot's existence: the so-called Patterson film, the grainy celluloid images of an erect apelike creature striding away from the movie camera of rodeo rider Roger Patterson in 1967. Mr. Wallace said he told Patterson where to go—near Bluff Creek, California—to spot a Bigfoot, Chorvinsky said.

"Ray told me that the Patterson film was a hoax, and he knew who was in the suit," Chorvinsky said.

Michael Wallace said his father called the Patterson film "a fake" and said he had nothing to do with it. But he said his mother admitted she had been photographed in a Bigfoot suit. "He had several people he used in his movies," Michael Wallace said.

Mr. Wallace never received proper credit in the Bigfoot community, Chorvinsky said. "He got it off the ground, and he kept getting glossed over. He's been consistently marginalized or ignored by authors," Chorvinsky said.

Why? "Because it hurts the case for Bigfoot if you talk too much about Ray Wallace," he replied.

The Wallace family's revelation does not faze some Bigfoot experts, and the debate about Bigfoot's existence rages on.

"These rumors have been circulating for some time," said Jeff Meldrum, an associate professor of anatomy and anthropology at Idaho State University.

Meldrum said he has casts of 40 to 50 footprints that he concludes, from their anatomical features, come from authentic unknown primates.

"To suggest all these are explained by simple carved feet strapped to boots just doesn't wash," he said. Even if the Wallace family's claims are true, Meldrum added, there are historical acounts of Bigfootlike creatures going back to the 1800s. "How do you account for that?"

It's easy, replied Chorvinsky; the historical accounts were mistakes, myths, or hoaxes. "I would like to see the evidence beyond the anecdotal. Jeff Meldrum's job is to show us the beef, something beyond old newspaper articles."

As for Meldrum's claim about authentic footprints, Chorvinsky said: "Jeff Meldrum is not an expert in creating hoaxes. I was a professional magician and special-effects film director; anything can be faked."

Michael Wallace said family members knew about his father's hoax but never let on.

—**Bob Young,** *Seattle Times,* **December 5, 2002. © 2002 Seattle Times Company. Used with permission.**

A claim lacks inherent credibility to the extent it conflicts with what we have observed or what we think we know—our background information—or with other credible claims.

Assessing the Content of the Claim

■ Does the Claim Conflict with Our Personal Observations?

Our own observations provide the most reliable source of information about the world. It is therefore only reasonable to be suspicious of any claim

What Americans Believe

According to a recent *Time*/CNN poll:

- Eighty-two percent of Americans believe in the healing power of personal prayer.
- Seventy-three percent believe that praying for someone else can help cure that person's illness.
- Seventy-seven percent believe that God sometimes intervenes to cure people who have a serious illness.

These poll results, if they reflect reality, say something about some of the claims Americans find credible. One thing that makes these high percentages interesting is that evidence of the effectiveness of prayer in healing is so difficult to document. More on that subject in Chapter 11.

that comes into conflict with what we have observed. Let's say that Parker tells Moore there will be no mail delivery today. But Moore has just a short while ago seen the mail carrier arrive and deliver the mail. Under such circumstances, obviously, neither Moore, nor you, nor anybody else needs instruction in critical thinking to decide not to accept Parker's claim. Anyone would reject it outright, in fact.

But observations are not infallible, and critical thinkers need to keep this in mind. Our observations may not be reliable if we make them when the lighting is poor or the room is noisy; when we are distracted, emotionally upset, or mentally fatigued; when our senses are impaired; or when our measuring instruments are inexact, temperamental, or inaccurate.

In addition, critical thinkers recognize that people vary in their powers of observation. Some people see and hear better than others and for this reason may be better at making observations than those whose vision or hearing is less acute. Some people have special training or experience that makes them better observers. Customs agents and professional counselors, even those who wear glasses or hearing aids, are better able than most of us to detect signs of nervousness or discomfort in people they observe. Laboratory scientists accustomed to noticing subtle changes in the properties of substances they are investigating are doubtless better than you or we at certain sorts of observations, though perhaps not at certain other sorts. In fact, some professional magicians actually prefer an audience of scientists, believing that such a group is particularly easy to fool.

Our beliefs, hopes, fears, and expectations also affect our observations. Tell someone that a house is infested with rats, and he is likely to believe he sees evidence of rats. Inform someone who believes in ghosts that a house is haunted, and she may well believe she sees evidence of ghosts. At séances staged by the Society for Psychical Research to test the observational powers of people under séance conditions, some observers insist that they see numerous phenomena that simply do not exist. Teachers who are

More on What Americans Believe

Long the most religious of the developed countries, the United States may have seen an increase in secularism in recent years. According to the Gallup Poll 2001, Americans' views of the Christian Bible have undergone substantial change in the last twenty years. The percentage who believed that the Bible is "the actual word of God" has decreased steadily since the early 1960s, from 65 percent in 1963 to 27 percent in 2001. Those who believe the Bible should not be taken literally have increased in number from 18 percent in 1963 to 49 percent in 2001. Finally, the percentage of Americans who believe the Bible is a book of fables, legends, history, and moral lessons recorded by humans has risen from 11 percent in 1963 to 20 percent in 2001.

Another survey, the *American Religious Identification Survey,* 2001, published by the Graduate Center of the City University of New York, reports that the number of Americans who claim to have no religion at all has grown from 8.16 percent in 1990 to 14.17 percent in 2001. Numbering 29.5 million, those with no religion are outnumbered only by Roman Catholics (50.9 million) and Baptists (33.8 million).

Figures reported in *Free Inquiry,* Summer 2002

---

told that the students in a particular class are brighter than usual are very likely to believe that the work those students produce is better than average, even when it is not.

In Chapter 5 we cover a fallacy (a fallacy is a mistake in reasoning) called *wishful thinking,* which occurs when we allow hopes and desires to influence our judgment and color our beliefs. The case of Mr. Keller, mentioned at the beginning of this chapter, more than likely involved wishful thinking: Mr. Keller must have been aware that the chances of making $50,000 by giving two strangers $10,000 are so small as to be almost nonexistent; but the thought of having or spending so much money must have blinded him to this obvious fact.

Our personal interests and biases affect our perceptions and the judgments we base on them. We overlook many of the mean and selfish actions of the people we like or love—and when we are infatuated with someone, everything that person does seems wonderful. By contrast, people we detest can hardly do anything that we don't perceive as mean and selfish. If we desperately wish for the success of a project, we are apt to see more evidence for that success than is actually present. On the other hand, if we wish for a project to fail, we are apt to exaggerate flaws that we see in it or imagine flaws that are not there at all. If a job, chore, or decision is one that we wish to avoid, we tend to draw worst-case implications from it and thus come up with reasons for not doing it. However, if we are predisposed to want to do the job or make the decision, we are more likely to focus on whatever positive consequences it might have.

"Eyewitness" Accounts

Associated Press/Rob Ostermaier

A Fairfax, Virginia, police officer signals cars after a man was shot to death October 11, 2002, at a gas station near Fredericksburg, Va. Police were looking for a white van.

When the series of sniper shootings occurred in the Washington, D.C., area during the fall of 2002, there were eyewitness reports that the perpetrators were two white men traveling in a white enclosed truck. In fact, two men who fit that description were detained and investigated by authorities. But when those doing the shootings were caught, they turned out to be two black men in a blue Chevrolet Caprice.

This is further evidence that we must be careful when deciding how much credence to put in eyewitness testimony, especially when the stakes are high. An appalling number of people have been convicted of heinous crimes by eyewitness accounts, only later to be freed from prison by conclusive DNA evidence.

Finally, the reliability of our observations is no better than the reliability of our memories, except in those cases where we have the means at our disposal to record our observations. And memory, as most of us know, can be deceptive. Critical thinkers are always alert to the possibility that what they remember having observed may not be what they did observe.

But even though firsthand observations are not infallible, they are still the best source of information we have. Any factual report that conflicts with our own direct observations is subject to serious doubt.

The Lake Wobegon Effect

In radio humorist and author Garrison Keillor's fictitious town of Lake Wobegon, "the women are strong, the men are good-looking, and all the children are above average." Thus the town lends its name to the utterly reliable tendency of people to believe that they are better than average in a host of different ways. A large majority of the population believe that they are more intelligent than average, more fair-minded, less prejudiced, and better automobile drivers.

A huge study was done not long ago by the Higher Education Research Institute at UCLA on high school seniors, with a million respondents to the survey. Seventy percent of them believed they were above average in leadership ability and only 2 percent thought they were below average. In the category of getting along with others, *fully 100 percent* of those seniors believed they were above average. What's more, in this same category 60 percent believed they were in the top 10 percent, and 25 percent believed they were in the top 1 percent!

People are more than willing to believe—it is probably safe to say *anxious* to believe—that they are better in lots of ways than the objective evidence would indicate. This tendency can make us susceptible to all kinds of trouble, from falling victim to con artists to overestimating our abilities in areas that can cost us our fortunes.

—*Adapted from* THOMAS GILOVICH, How We Know What Isn't So

■ Does the Claim Conflict with Our Background Information?

Factual claims must always be evaluated against our **background information**—that immense body of justified beliefs that consists of facts we learn from our own direct observations and facts we learn from others. Such information is "background" because we may not be able to specify where we learned it, unlike something we know because we witnessed it this morning. Much of our background information is well confirmed by a variety of sources. Factual claims that conflict with this store of information are usually quite properly dismissed, even if we cannot disprove them through direct observation. We immediately reject the claim "Palm trees grow in abundance near the North Pole," even though we are not in a position to confirm or disprove the statement by direct observation.

Indeed, this is an example of how we usually treat claims when we first encounter them: We begin by assigning them a certain *initial plausibility*, a rough assessment of how credible a claim seems to us. This assessment depends on how consistent the claim is with our background information—how well it "fits" with that information. If it fits very well, we give the claim a high degree of initial plausibility; we lean toward accepting it. If, however, the claim conflicts with our background information, we give it low initial plausibility and lean toward rejecting it unless very strong evidence can be produced on its behalf. The claim "More guitars were sold in the United States last year than saxophones" fits very well with the background information most of us share, and we would hardly require detailed evidence before accepting it. However, the claim "Charlie's eighty-seven-year-old grandmother swam across Lake Michigan in the middle of winter" cannot command much initial plausibility because of the obvious way it conflicts with our background information about eighty-seven-year-old people, about Lake Michigan, about swimming in cold water, and so on. In fact, short of observing the swim ourselves, it isn't clear just what *could* persuade us to accept such a claim. And even *then* we should consider the likelihood that we're being tricked or fooled by an illusion.

Not every suspicious claim is as outrageous as the one about Charlie's nautical grandmother. Some may not have enough initial plausibility to be immediately acceptable—we may even lean toward rejecting them—but they may still be worth investigating. For example, let's say someone shows up at your door and presents you with what he says is a check for a hundred dollars from an anonymous benefactor. Before you rush right out and spend the hundred, you should, as a critical thinker, consider the possibility that somebody is playing a joke on you. However, even though it's more likely that there is a practical joke going on (practical jokes of this sort are probably more common than actual hundred-dollar surprises, after all), it's worth investigating the possibility that the gift is legitimate. A trip to your bank certainly seems in order. As in most other cases, this is a situation where nothing can take the place of investigating further and gathering more information. Then, depending on how the investigation turns out, one can hope for a reasonable

There are three types of men in the world. One type learns from books. One type learns from observation. And one type just has to urinate on the electric fence.

—DR. LAURA SCHLESSINGER (reported by Larry Englemann)

The authority of experience.

Do Your Ears Stick Straight Out?
Do You Have a Short Neck?

According to Bill Cordingley, an expert in psychographicology—that's face-reading, in case you didn't know (and we certainly didn't)—a person's facial features reveal "the whole rainbow collection" of a person's needs and abilities. Mr. Cordingley *(In Your Face: What Facial Features Reveal About People You Know and Love)* doesn't mean merely that you can infer moods from smiles and frowns. No, he means that your basic personality traits are readable from facial structures you were born with.

Do your ears stick out? You have a need to perform in public. The more they stick out, the greater the need. Is your neck short and thin? You are stubborn and dominate conversations. Large lips? You love attention. The length of your chin, location of your eyebrows, size of your ears, length of your neck, are such reliable predictors of personality that an expert can tell by looking at two people whether their relationship will succeed.

Former President Carter apparently loves attention. President Bush is an introvert (thin lips) and a control freak (small eyelids—Hey! At least they cover his eyes.)

We leave it to you to determine how credible this is. Cordingley is the former mayor of San Anselmo, California. Does that fact make this more credible?

conclusion to the matter. Unfortunately, there are no neat formulas that can resolve conflicts between what you already believe and new information. Your job as a critical thinker is to trust your background information when considering claims that conflict with that information—that is, claims with low initial plausibility—but at the same time to keep an open mind and realize that further information may cause you to give up a claim you had thought was true. It's a difficult balance, but it's worth getting right.

Finally, keep in mind that you are handicapped in evaluating a factual report on a subject in which you have no background information. This means that the broader your background information, the more likely you are to be able to evaluate any given report effectively. Without some rudimentary knowledge of economics, for example, we are in no position to evaluate claims about the dangers of a large federal deficit; unless we know what cholesterol does in the body, we cannot appreciate the benefits of low-cholesterol foods. The single most effective means of increasing your ability as a critical thinker, regardless of the subject, is to increase what you know: Read widely, converse freely, and develop an inquiring attitude! There is simply no substitute for broad, general knowledge.

The President's Own Credibility

Bush on Iraq: Is He Stretching the Facts?

WASHINGTON — President Bush, speaking to the nation this month [October 2002] about the need to challenge Saddam Hussein, warned that Iraq has a growing fleet of unmanned aircraft that could be used "for missions targeting the United States."

Last month, asked if there was new and conclusive evidence of Saddam's nuclear weapons capabilities, Bush cited a report by the International Atomic Energy Agency saying the Iraqis were "six months away from developing a weapon."

And last week, the president said objections by a labor union to having customs officials wear radiation detectors has the potential to delay the policy "for a long period of time."

All three assertions were powerful arguments for the actions Bush sought. And all three statements were dubious, if not wrong. Further information revealed that the aircraft lack the range to reach the United States; there was no such report by the IAEA; and the customs dispute over the detectors was resolved long ago.

By Dana Milbank
WASHINGTON POST

Associated Press/Dan Loh

President Bush speaks in Downingtown, Pa. In making his case for war with Iraq, Bush has made statements that some may find misleading. But his spokesman says his statements are "well documented and supported by the facts."

From the *Sacramento Bee*, Oct. 23, 2002. *Washington Post* News Service.

Assessing the Credibility of Sources

The guiding principle in evaluating claims requires that they come from credible sources. We've considered the credibility of claims as to their content. Now let's consider the credibility of sources.

A person may lack credibility in various ways, but most of them relate either to (1) his or her knowledge or to (2) his or her truthfulness, objectivity, or accuracy. If we have reason to think a source may be biased or self-serving or opinionated or conflicted about a subject, or for other reasons may not be giving us a full and complete and impartial report, then we have a reason for concern under criterion (2). We also should have concerns under criterion (2) if we have suspicions about a source's memory or ability or op-

The "Authority" of Experience

. . . a farmer never laid an egg, but he knows more about the process than hens do.

—HENRY DARCY CURWEN

No one knows more about this mountain than Harry. And it don't dare blow up on him. This goddamned mountain won't blow.

—HARRY TRUMAN, *83-year-old owner of a lodge near Mt. St. Helens in Washington, commenting on geologists' predictions that the volcano would erupt. (A few days later it did erupt, killing Harry Truman.)*

Through their "observations," the hen and Harry know well enough what it's like to lay an egg or live near a volcano, but these observations are obviously not enough to qualify them as reliable sources about the biological and geological processes involved in egg laying and volcanic eruptions.

portunity to have made accurate observations. We shall get to questions as to a source's knowledge in a bit.

Now we can have concerns about a person's accuracy, objectivity, or truthfulness without suspecting him or her of evil intent. Generally people don't lie to one another; given an absence of a specific reason to think otherwise, it is reasonable to assume a source speaks honestly and in good conscience.

Of course, assuming that a source is honest does not require us to assume he or she cannot be mistaken. Even individuals operating in good conscience may still say things that are inaccurate, incomplete, or misleading.

For example, the reports people give one another are very frequently subject to innocent **sharpening** and **leveling**—exaggerating what the speaker thinks is the main point and dropping out or de-emphasizing details that seem peripheral. The result can be a distortion of the story.

For example, let's say you've heard reports from time to time from your college roommate about one of his or her high school friends. Based on what you have heard, you may picture this friend as very smart or exceptionally clever or unusually good-looking. This is because your roommate will probably emphasize those features he or she considers trademarks of the person. Typically, if you finally meet the friend, you may find yourself somewhat surprised by how unexceptional he or she is. The same sharpening and leveling phenomenon accounts for why we are often disappointed when we see a movie after we've heard rave reports from friends about it. By placing so much emphasis on what they liked about the movie, they have raised our expectations to a level that is hard to meet.

Why do people exaggerate this way? It's not usually dishonesty. We naturally tend to remember things that make an experience different and worth reporting and tend to report details that make stories interesting. It's always more interesting to our listeners to hear about somebody or

When the Stakes Are High

Generally speaking, it's reasonable to accept a claim from a credible source if it does not conflict with your observations, background information, or other credible claims. But if the stakes are high, it can be reasonable to hesitate.

For example, if your doctor tells you that you have a cancerous tumor on your leg and wants to amputate the leg, you'd have a claim from a credible source that meets the other requirements for credibility. We trust, however, you'd get a second opinion before rolling up your pants leg.

something that is very this or exceptionally that than to hear about someone or something that is perfectly ordinary.

Also, if we are reporting on someone we like or someone we feel indebted to (or are employed by), we may accentuate the positive and understate the negative more or less unconsciously. Of course the reverse holds true if the subject of our report is someone we dislike. This isn't dishonesty; as we mentioned earlier, it's just human nature to see mostly good in people we like or love and to perceive shortcomings in those we don't care for.

If we do have suspicions about a source's objectivity, accuracy, or veracity (truthfulness), the proper response is not to *reject* what he or she says *as false.* That would be overreacting, and is a mistake in reasoning (a fallacy). We shall discuss this in more detail in Chapter 6. The appropriate response to concerns about the objectivity, accuracy, or veracity of a source is to *suspend or reserve judgment* on what the source says.

As to a source's knowledge, just as you cannot tell merely by looking at someone whether he or she is speaking truthfully, objectively, and accurately, you can't judge his or her *knowledge* or *expertise* by looking at mere surface features. A British-sounding scientist may appear more knowledgeable than a scientist who speaks, say, with a Texas drawl, but a person's accent, height, gender, ethnicity, or clothing doesn't have much to do with a person's knowledge. In the municipal park in our town it is difficult to distinguish the people who teach at the university from the people who live in the park, based on physical appearance.

So then how do you judge a person's expertise? Education and experience are often the most important factors, followed by accomplishments, reputation, and position, in no particular order. It is not always easy to evaluate the credentials of an **expert,** and credentials vary considerably from one field to another. Still, there are some useful guidelines worth mentioning.

Education includes but is not strictly limited to formal education—the possession of degrees from established institutions of learning. (Some "doctors" of this and that received their diplomas from mail-order houses that advertise on matchbook covers. The title "doctor" is not automatically a qualification.)

A Fish Story

Randy Fite of Richards, Texas, boated three bass totaling 22 pounds, 2 ounces Wednesday to take the first-day lead at the Big K-mart BASSMaster Top 150 on Lake Champlain near Burlington, Vt. Shawn Penn of Benton, Ky., is second with 20-15. Rick Lillegard of Atkinson, N.H., had the biggest catch, a 6-pound, 4-ounce largemouth.

—USA Today, 9/16/99

We don't think so. Careful reading and a little arithmetic show that this story contains either a typographical error or something even worse, because it can't be true as it stands. Do you see why?

Experience—both the kind and the amount—is an important factor in expertise. Experience is important if it is relevant to the issue at hand, but the mere fact that someone has been on the job for a long time does not automatically make him or her good at it.

Accomplishments are an important indicator of someone's expertise but, once again, only when those accomplishments are directly related to the question at hand. A Nobel Prize winner in physics is not necessarily qualified to speak publicly about toy safety, public school education (even in science), or nuclear proliferation. The last issue may involve physics, it's true, but the political issues are the crucial ones, and they are not taught in physics labs.

A person's reputation always exists only among a particular contingent of people. You may have a strong reputation as a pool player at your local billiards emporium, but that doesn't necessarily put you in the same league as Minnesota Fats. Your friend may consider his friend Mr. Klein the greatest living expert on some particular subject, and he may be right. But you must ask yourself if your friend is in a position to evaluate Mr. Klein's credentials. Most of us have met people who were recommended as experts in some field but who turned out to know little more about that field than we ourselves knew. (Presumably in such cases those doing the recommending knew even less about the subject, or they would not have been so quickly impressed.) By and large, the kind of reputation that counts most is the one a person has among other experts in his or her field of endeavor.

The positions people hold provide an indication of how well *somebody* thinks of them. The director of an important scientific laboratory, the head of an academic department at Harvard, the author of a work consulted by other experts—in each case the position itself is substantial evidence that the individual's opinion on a relevant subject warrants serious attention.

But expertise can be bought. Recall that the last part of our principle cautions us against sources who may be biased on the subject of whatever claim we may be considering. Sometimes a person's position is an indication

Sins of Omission

The Israeli army on Monday released a Palestinian journalist employed by the Reuters news wire, five days after his arrest in the Gaza Strip, the British news agency said. Reuters photographer, Suhaib Jadallah Salem, 22, was detained on May 22 near the southern Gaza Strip border town of Rafah, from which he was planning to cross into Egypt.

—*Agence France-Presse, May 27*

IDF troops arrested a Reuters photographer in the Gaza Strip last Wednesday, for what military sources said were his links to terror activities. Suhaib Jadallah Salem, 22, who was intercepted while traveling to Rafah, at the southern end of the Strip, was in possession of a hand grenade when he was seized, military sources said.

—Ha'aretz, *May 26*

If the detail provided at the end of the second version of the story is true, then the first story creates a huge distortion by leaving it out. In either case, one or the other of these stories is extremely misleading.

From the *New Republic,* June 10, 2002, p. 9

of what his or her opinion, expert or not, is likely to be. The opinion of a lawyer retained by the National Rifle Association, offered at a hearing on firearms and urban violence, should be scrutinized much more carefully (or at least viewed with more skepticism) than that of a witness from an independent firm or agency that has no stake in the outcome of the hearings. It is too easy to lose objectivity where one's interests and concerns are at stake, even if one is *trying* to be objective.

Experts sometimes disagree, especially when the issue is complicated and many different interests are at stake. In these cases, a critical thinker is obliged to suspend judgment about which expert to endorse, unless one expert clearly represents a majority viewpoint among experts in the field or unless one expert can be established as more authoritative or less biased than the others.

Of course, majority opinions sometimes turn out to be incorrect, and even the most authoritative experts occasionally make mistakes. Thus, a claim that you accept because it represents the majority viewpoint or comes from the most authoritative expert may turn out to be thoroughly wrong. Nevertheless, take heart: At the time, you were rationally justified in accepting the majority viewpoint as the most authoritative claim. The reasonable position is one that agrees with the most authoritative opinion but allows for enough open-mindedness to change if the evidence changes.

Finally, we sometimes make the mistake of thinking that whatever qualifies someone as an expert in one field automatically qualifies that

You ain't goin' nowhere . . . son. You ought to go back to driving a truck.

—JIM DENNY, Grand Ole Opry manager, firing Elvis Presley after one performance

Blonds Have Last Laugh

It's unlikely that there was a hair of truth in a report that apparently fell into the category "too good to check."

[Recently] several British newspapers reported that the World Health Organization had found in a study that blonds would become extinct within 200 years, because blondness is a recessive gene that was dying out. The reports were repeated Friday by anchors for the ABC News program *Good Morning America*, and on Saturday by CNN.

There was only one problem, the health organization said in a statement this week: It had never reported that blonds would become extinct, and it had never done a study on the subject.

"WHO has no knowledge of how these news reports originated," the organization, a Geneva-based agency of the United Nations, announced, "but we would like to stress that we have no opinion on the future existence of blonds."

All the news reports, in Britain and the United States, cited a study from the World Health Organization—"a blondeshell study," as the *Daily Star* of London put it. But none reported any scientific details from the study or the names of the scientists who conducted it.

The British accounts were replete with the views of bleached blonds who said hairdressers would never allow blondness to become extinct, and doctors who said that rare genes would pop up to keep natural blonds from becoming an endangered species.

The big networks can be as gullible as everyone else.

From a story by Lawrence Altman in the *New York Times,* October 3, 2002.

person in other areas. Many people seem to think, for example, that business CEOs are automatically experts in economics or public policy. But even if the intelligence and skill required to become an expert in one field could enable someone to become an expert in any field—a doubtful assumption—possessing the ability to become an expert is entirely different from actually being an expert. Thus, claims put forth by experts about subjects outside their fields are not automatically more acceptable than claims put forth by nonexperts.

■ *The News Media*

The most common source of information about current events for many people is their friends and relatives, but for most of us, newspapers, newsmagazines, and the electronic media (radio, TV, and the Internet) are just as important. The Internet aside, newspapers offer the broadest coverage of general news, and radio and TV offer the most severely edited and least detailed (with the exception of extended-coverage programs and Public Broadcasting System news shows). Newsmagazines fall somewhere in the middle, although they can offer extended coverage in feature stories.

News reports, especially those that appear in major metropolitan newspapers (tabloids excepted), national newsmagazines, and television and radio news programs are generally credible sources of information. (The

Democrats have to overcome a new reality: conservative bias in the media.

—*Time* magazine, commenting on obstacles the Democratic Party faces in the 2004 election

© Knight Ridder/Tribune Media Services. Reprinted with permission.

same holds true of the Internet versions of these sources.) These remarks, however, are subject to serious qualification, as we note below.

On the surface, talk radio seems to offer a wealth of information not available in news reports from conventional sources. But much of the information is based on rumor, hearsay, and gossip from heavily biased sources, and it is difficult or impossible to determine which items, if any, are legitimate. A further defect in talk radio as a source of news is that the information is almost always presented from—and colored by—a political perspective. At the same time, many talk radio hosts do scour traditional legitimate news sources for information relevant to their political agenda; and to the extent they document the source, they provide listeners with many interesting and important facts. Internet chat rooms and Web pages (excluding Web pages from major metropolitan newspapers, national newsmagazines, television networks, universities, and other reliable sources) are virtually worthless as sources of news. We'll discuss the Internet in more detail in a moment.

The breadth of coverage from major media news sources is restricted by space, by their audience's interests, and by the concerns of advertisers, pressure groups, and government officials. The accessibility of reliable reports also restricts coverage because governments, corporations, and individuals often simply withhold information.

The location, structure, and headline of a news story in both print and electronic media can be misleading about what is important or essential in the story. The selective presentation of facts is a widely used approach to persuasion in our society, not just by groups with their own agendas but by the supposedly objective news media. There is no guarantee that the media are giving us "the truth, the whole truth, and nothing but the truth." The news they present is subject to shaping by the conscious and unconscious per-

Because amazing feats are entertaining, the media often plays up amazing events for all (or more than) they are worth, distorts many not-so-amazing events to make them appear extraordinary, and sometimes even passes on complete fabrications from unreliable sources.

—THOMAS GILOVICH, *How We Know What Isn't So*, p. 100)

"*Those are the headlines, and we'll be back in a moment to blow them out of proportion.*"

spectives and purposes of publishers, editors, and owners as well as the groups mentioned above. There are several reasons for this state of affairs, and we'll look at some of the most important ones in the sections that follow.

■ Reporting the News

The popular notion of the hardworking investigative reporter who ferrets out facts, tracks down elusive sources, and badgers people for inside information is largely a creation of the moviemakers. No news service can afford to devote more than a small portion of its resources to real investigative reporting. Occasionally, this kind of reporting pays off handsomely, as was the case with Bob Woodward and Carl Bernstein's reporting of the Watergate affair in the early 1970s. The *Washington Post* won awards and sold newspapers at a remarkable rate as a result of that series of articles. But such cases are relatively rare. The great bulk of news is *given* to reporters, not dug up after weeks or days or even hours of investigation. Press conferences and press releases are the standard means of getting news from both government and private industry into the mass media. And because spokespeople in neither government nor industry are especially stupid or self-destructive, they tend to produce news items that they and the people they represent *want* to see in the media. Further, because reporters depend on sources in governmental and private institutions to pass items along, reporters who offend those sources are not likely to have them very long. A remarkable example of this interrelationship occurred over a half century ago. Walter Duranty won a Pulitzer Prize for his reporting on the Soviet Union, including his prediction that Josef Stalin would rise to power. According to Eric Newton, in *Crusaders, Scoundrels, Journalists,* "A year later, in a special report in which

Frank & Ernest reprinted by permission of Newspaper Enterprises Association, Inc.

he purposely lied, he denied the existence of a government-engineered famine that the dictator used to kill 9 million people. He wrote the story to preserve his reputation as a reporter and his access to Soviet officials."

It is important to remember that the news media in this country are private businesses. This situation has both good and bad sides. The good side is that the media are independent of the government, thus making it very difficult for government officials to dictate exactly what gets printed or broadcast. The bad side is that the media, as businesses, have to do whatever it takes to make a profit, even if this affects which items make the headlines and which are left out entirely.

Aside from the sources of news, the media must therefore be careful not to overly offend two other powerful constituencies: their advertisers and their audiences. The threat of canceled advertising is difficult to ignore when the great bulk of a business's revenues comes from advertising. (This is true of newspapers, which receive more money from advertisers than from those of us who purchase the papers, and of the electronic media as well.) The other constituency, the news-reading and news-watching public, has its own unfortunate effects on the quality of the news that is generally available. The most important of these is the oversimplification of the information presented. Too many people would be bored by a competent explanation of the federal budget deficit or the latest crisis in the Middle East to allow the media to offer such accounts often or in much detail without fearing the loss of their audiences. And, in this context at least, it is not important whether American audiences are unwilling to pay attention to com-

plicated issues or whether they are simply unable to understand them. (In other contexts, however, this distinction is highly significant. Between one-third and one-half of adults in the United States would probably be unable to read and understand the page you're now reading.) Whatever the reason, it is clear that complicated issues are lost on a large percentage of American adults.

Notice the level at which television commercials and political advertisements are pitched. These products are made by highly skilled professionals, who are aware that the projection of an "image" for a candidate or a product goes much further than the coverage of facts and issues. A television network that devotes too much of its prime time to complex social issues in a nonsensationalist way will soon be looking for a new vice president for programming.

■ Who Listens to the News?

We come now to another problem with relying on the mass media for a real understanding of events. Much of what goes on in the world, including many of the most important events, is not only complicated but also not

Springfield, Ore., Pearl, Miss., West Paducah, Ky., Littleton, Colo., Jonesboro, Ark., Edinboro, Pa. School killings certainly appear to be on the increase. Actually, it only seems they are. According to data from the Centers for Disease Control and Prevention, school killings are actually down from the levels reached earlier in the decade and are about what they were in the 1970s.

Saran Wrap and Cancer

Did you know that food covered with Saran Wrap and then microwaved causes cancer? That's according to a study done by researchers at the University of California, Davis and reported on the Web. Except that nobody at UC Davis conducted such research. The moral: Yes, Internet information from reliable sources like universities is credible, but anyone can pretend to be a university and say anything he or she feels like. You need to stick to the official Web pages of credible sources.

And go ahead and nuke your food in Saran Wrap.

---

very exciting. If a television station advertised that its late news would offer extended coverage of several South American countries' threats to default on loans from United States banks, a considerable part of its audience would either go to bed early or watch reruns on another channel. And they would do so even though loans to other countries have an enormous impact upon the American economy. (How many Americans could explain the connection between Asia recovering from an economic crisis and the danger of a depressed stock market in the United States?) This story is apt to get only fifteen seconds so as not to shortchange the story of the shoot-out at the local convenience store, accompanied by some exciting film. The point is that sensational, unusual, and easily understood subjects can be counted on to receive more attention than the unexciting, the usual, and the complicated, even if the latter are much more important in the long run.

The same kind of mass preference holds for people as well as issues and events. The number of movie stars and celebrities interviewed on talk shows is wildly disproportionate to the effect these individuals have on most of our lives. But they are entertaining in ways that, say, the chairman of the Federal Reserve Board is not.

Although there is nothing wrong with entertainment and our desire to be entertained, the overindulgence of our desire to be entertained comes at the expense of our need to be informed. In the process, we become passive citizens rather than active participants in society. If we count on the media to indulge us, they will give us what we pay for.

A couple of final notes on the news media: Things have changed a lot since we wrote about this topic in the early editions of this book. Competition for readers and viewers has become fierce, and the tactics the media employ to attract an audience have begun to endanger the very idea of an independent, honest, straightforward press. Television in particular has developed new means of coming up with sensational stories, stories that may not have happened at all but for the presence of the television cameras. Here's an example of a nonstory: A film crew in Orlando, Florida, once appeared on the street in front of a house where a dead woman had been discovered. The reporter went on for some time about how police were "dumbfounded" about the cause of death of the woman inside the house and had

When the decision was made at Time magazine to darken a cover picture of O. J. Simpson, only the lone nonwhite person in the room objected.

—Press critic JEFF COHEN, commenting on how the lack of diversity on media staffs can affect the content of the product

© Dan Piraro. Reprinted with special permission of King Features Syndicate.

"put a clamp" on information about the scene. Although the story got several minutes of high-energy coverage, it turned out the woman was in her late sixties and had simply died in bed from a heart attack.

There are some programs that *look* like investigative programs but aren't. Staged dramas like *The X-Files* usually push a point of view that is more gullible than investigative. The large number of cable channels, some with twenty-four hours of news to fill, often have to scrape pretty hard to find anything newsworthy; and when nothing turns up, they may be forced to make something up—or at least make a story out of something that isn't really worth reporting.

The numerous commentators these days are paid to attract attention, not to provide factual information or insight. Thus we see a lot of "in your face" or "scream" television, where people from different parts of the political spectrum spend a half hour yelling at each other and somehow we're supposed to be better informed afterward. Our advice is to be even more skeptical of what you see on television news now than ever before; and there never was reason for too much confidence.

■ *The Internet*

An important source of information is the Internet—that amalgamation of electronic lines and connections that allows nearly anyone with a computer and a modem to link up with nearly any other similarly equipped person on the planet. Although the Internet offers great benefits, the information it provides must be evaluated with even *more* caution than information from the print media, radio, or television.

There are basically two kinds of information sources on the Internet. The first consists of commercial and institutional sources; the second, of individual and group sites on the World Wide Web. In the first category we include sources like the Lexis-Nexis facility, as well as the online services provided by newsmagazines, large electronic news organizations, and government institutions. The second category includes everything else you'll find on the Web—an amazing assortment of good information, entertainment of widely varying quality, hot tips, advertisements, come-ons, fraudulent offers, outright lies, and even the opportunity to meet the person of your dreams (or your nightmares)! (Rush Limbaugh met his current wife online, and they corresponded via e-mail for a time before meeting in person. On the other hand, in October 1996, a North Carolina man described online to a Pennsylvania woman how, if she would visit him, he would sexually torture her and then kill her. She went, and he did.)

Just as the fact that a claim appears in print or on television doesn't make it true, so it is for claims you run across online. Keep in mind that the information you get from a source is only as good as that source. The Lexis-Nexis information collection is an excellent asset for medium-depth investigation of a topic; it includes information gathered from a wide range of print sources, especially newspapers and magazines, with special collections in areas like the law. But the editorials you turn up there are no more likely to be accurate, fair-minded, or objective than the ones you read in the newspapers—which is where they first appeared anyhow. Remember also that *any* individual or group can put up a Web site. And they can say *anything* they want to on it. You have about as much reason to believe the claims you find on such sites as you would if they came from any other stranger, except you can't look this one in the eye.

FREE MONEY!
If you have an e-mail account, you've probably heard from somebody in Nigeria, South Africa, or Lagos who is trying to get millions of dollars out of the country and wants your help. You will get a very large sum for furnishing your bank account number so they can temporarily put the funds there. Of course the only transfer that occurs is *out* of your account.

It takes only one in a million gullible takers to make this scheme profitable.

Advertising

> Advertising [is] the science of arresting the human intelligence long enough to get money from it.
>
> —*Stephen Leacock*

If there is anything in modern society that truly puts our sense of what is credible to the test, it's advertising. As we hope you'll agree after reading this section, skepticism is almost always the best policy when considering any kind of advertising or promotion.

Ads are used to sell many products other than toasters, television sets, and toilet tissue. They can encourage us to vote for a candidate, agree with a political proposal, take a tour, give up a bad habit, or join the army. They can also be used to make announcements (for instance, about job openings, lectures, concerts, or the recall of defective automobiles) or to create favorable climates of opinion (for example, toward labor unions or offshore oil drilling).

Advertising firms understand our fears and desires at least as well as we understand them ourselves, and they have at their disposal the expertise to exploit them.* Such firms employ trained psychologists and some of the world's most creative artists and use the most sophisticated and well-researched theories about the motivation of human behavior. Maybe most important, they can afford to spend whatever is necessary to get each detail of an advertisement exactly right. (On a per-minute basis, television ads are the most expensively produced pieces that appear on your tube.) A good ad is a work of art, a masterful blend of word and image often composed in accordance with the exacting standards of artistic and scientific genius (some ads, of course, are just plain silly). Can untrained laypeople even hope to evaluate such psychological and artistic masterpieces intelligently?

Fortunately, it is not necessary to understand the deep psychology of an advertisement to evaluate it in the way that's most important to us. When confronted with an ad, we should ask simply: Does this ad give us a good reason to buy this product? And the answer, in general terms, can be simply put: Because the only good reason to buy anything in the first place is to improve our lives, the ad justifies a purchase only if it establishes that we'd be better off with the product than without it (or that we'd be better off with the product than with the money we would trade for it).

However, do we always know when we'll be better off with a product than without it? Do we really want, or need, a bagel splitter or an exercise bike? Do people even recognize "better taste" in a cigarette? Advertisers spend vast sums creating within us new desires and fears—and hence a need to improve our lives by satisfying those desires or eliminating those fears through the purchase of advertised products. They are often successful, and we find ourselves needing something we might not have known existed before. That others can instill in us through word and image, a desire for something we did not previously desire may be a lamentable fact, but it *is* clearly a fact. Still, *we* decide what would make us better off, and *we* decide to part with our money. So it is only with reference to what in *our* view would make life better for us that we properly evaluate advertisements.

There are basically two kinds of ads: those that offer reasons and those that do not. Those that offer reasons for buying the advertised product always promise that certain hopes will be satisfied, certain needs met, or certain fears eliminated. (You'll be more accepted, have a better image, be a better parent, and so on.)

People watching a sexual program are thinking about sex, not soda pop. Violence and sex elicit very strong emotions and can interfere with memory for other things.

—Brad Bushman of Iowa State University, whose research indicated that people tend to forget the names of sponsors of violent or sexual TV shows (reported by Ellen Goodman)

*For an excellent treatment of this and related subjects, we recommend *Age of Propaganda: The Everyday Use and Abuse of Persuasion*, rev. ed., by Anthony R. Pratkanis and Elliot Aronson (New York: W. H. Freeman and Co., 1998).

FTC: No Miracles in Weight Loss

Marketing claims are frequently false or misleading, a study finds.

WASHINGTON — As an overweight child, Lynn McAfee scanned the ads in her mother's magazines looking for hope in the before-and-after photos of people made thin by seemingly miraculous products.

"These smiling, thin people used to be fat like me," said McAfee, 53, an advocate for overweight people. "But all I got out of those boxes and envelopes I bought in my childhood and adolescence was the sinking feeling that becoming thin like those people in the ads was hopeless."

Many people share her frustration. In a first-of-its-kind study, the Federal Trade Commission found that 55 percent of weight-loss ads make claims that are certainly false or misleading.

"There is no miracle pill that will lead to weight loss," Surgeon General Richard Carmona said at a news conference [recently] announcing the results. "Achieving and maintaining a healthy weight requires a lifelong commitment to healthful eating and adequate physical activity." . . .

FTC Chairman Timothy Muris said his agency has increased its efforts against deceptive weight-loss marketing and has prosecuted 93 fraud cases since 1990. He said the FTC will hold a public workshop on the problem in November.

Muris said the agency filed charges this month against Bio Lab, a Canadian company based in Laval, Quebec, that sells its "Quick Slim" weight-loss product in the United States. The FTC said the product, which contains a substance taken from apples, does not live up to promotions such as: "Lose up to 2 pounds daily without diet or exercise."

The company did not immediately return calls seeking comment Tuesday.

In May, the FTC sued the marketers of three electronic exercise belts—the AB Energizer, Ab Tronic and Fast Abs—accusing them of making misleading claims. Television ads showed people strapping the gadgets on their waists in search of "washboard" abs. Court injunctions took most of the ads off the air, and the cases are pending.

The FTC conducted the advertising study with the Partnership for Healthy Weight Management, a coalition that includes scientists, government agencies and legitimate weight-loss companies.

By David Ho
ASSOCIATED PRESS

"Doctor recommended."

This ambiguous ad slogan creates an illusion that many doctors, or doctors in general, recommend the product. However, a recommendation from a single doctor is all it takes to make the statement true.

Those ads that do not rely on reasons fall mainly into three categories: (1) those that bring out pleasurable *feelings* in us (e.g., through humor, glad tidings, pretty images, beautiful music, heartwarming scenes); (2) those that depict the product being used or endorsed by *people* we admire or think of ourselves as being like (sometimes these people are depicted by actors, sometimes not); and (3) those that depict the product being used in *situations* in which we would like to find ourselves. Of course, some ads go all out and incorporate elements from all three categories—and for good measure also state a reason or two why we should buy the advertised product.

Buying a product (which includes joining a group, deciding how to vote, and so forth) on the basis of reasonless ads is, with one minor exception that we'll explain shortly, never justified. Such ads tell you only that the product exists and what it looks like (and sometimes where it is available and how much it costs); if an ad tells you much more than this, then it begins to qualify as an ad that gives reasons for buying the product. Reasonless ads do tell us what the advertisers think of our values and sense of

Celebrity Endorsements We Can Live With

One of America's greatest humorists was once asked to write an endorsement for a certain brand of piano. Because he would not speak on behalf of a product he had not tried, he wrote the following:

> Dear Sirs,
> I guess your pianos are the best I have ever leaned against.
> Yours truly,
> Will Rogers

Opera singer Giovanni Martinelli, when questioned by a reporter about cigarette smoking, replied, "Tobacco, cigarettes, bah! I would not think of it!" The reporter reminded Martinelli that he had appeared in an advertisement for a particular brand of cigarette and had said that those cigarettes did not irritate his throat. "Yes, yes, of course I gave that endorsement," Martinelli said impatiently. "How could they irritate my throat? I have never smoked."

humor (not always a pleasant thing to notice, given that they have us pegged so well), but this information is irrelevant to the question of whether we should buy the product.

Ads that submit reasons for buying the product might have been treated in Part 3 of this book, which is devoted to arguments. However, so little need be said about argumentative ads that we will discuss them here. Such "promise ads," as they have been called, usually tell us more than that a certain product exists—but not much more. The promise, with rare exception, comes with no guarantees and is usually extremely vague (Gilbey's gin promises "more gin taste," Kleenex is "softer").

Such ads are a source of information about what the *sellers* of the product are willing to claim about what the product will do, how well it will do it, how it works, what it contains, how well it compares with similar products, and how much more wonderful your life will be once you've got one. However, to make an informed decision on a purchase, you almost always need to know more than the seller is willing to claim, particularly because no sellers will tell you what's wrong with their products or what's right with those of their competitors.

We are professional grade.

Meaningless but catchy slogan for GMC trucks. (We are professional grade, too.)

Further, the claims of advertisers are notorious for not only being vague but also for being ambiguous, misleading, exaggerated, and sometimes just plain false. Even if a product existed that was so good that an honest, unexaggerated, and fair description of it would justify our buying it without considering competing items (or other reports on the same item), and even if an advertisement for this product consisted of just such a description, we would still not be justified in purchasing the product on the basis of that advertisement alone. For we would be unable to tell, simply by looking at the

"Truth" vs. Cigarette Makers

Philip Morris "uncool" smoking ads may actually encourage teen smokers.

Michael Szymanczyk, chairman of Philip Morris USA, said the other day: "If you're a decent person, you don't want kids to smoke. You don't want them to do something that's going to harm them. In order for this business to exist successfully in this society, we had to accept at least some responsibility for deterring kids from smoking."

Since the sincerity of Philip Morris is forever in ashes, Szymanczyk actually meant: If you want to be seen as a decent business, you must create the illusion that you don't want to harm kids.

As Szymanczyk blathers, another barrage of evidence shows that big tobacco, whose very success depends on hiding cancer under the cowboy hats of macho men and the petite dresses of liberated women and using the imagery of mountain vistas and calm, endless ocean to dissipate fears of premature departure to eternity has—shocking!—come up mysteriously lame in deterring kids from smoking.

Researchers from the American Legacy Foundation, the public health foundation established by the 1998 tobacco settlement with the states, published a study in this month's *American Journal of Public Health* indicating that its youth antismoking campaign is far more effective at creating negative attitudes toward smoking than the efforts of Philip Morris. In some ways, the Philip Morris campaign is creating more-positive attitudes toward smoking.

The study surveyed the attitudes of more than 6,000 minors, ages 12 to 17. The American Legacy "truth" ad campaign included images of young people placing 1,200 body bags at the door of a cigarette company office building and cowboys pulling horses with body bags over the saddle. The Philip Morris "Think, don't smoke" campaign had teenagers from the beach to the basketball court saying why they do not use cigarettes.

Philip Morris had teens talking about cigarettes making you uncool, ugly or unathletic. But the commercials steered very wide from the graveyard. There were no stark, body-bag connections of cigarettes to death.

According to the American Legacy study, a majority of teenagers were aware of both campaigns, with 66 percent aware of "Think, don't smoke" and 75 percent aware of "truth." While the "truth" campaign made youth 163 percent more likely to agree that "taking a stand against smoking is important to me," the Philip Morris campaign made youth less than 10 percent more likely to agree.

While the "truth" campaign resulted in a 35 percent increase in youths agreeing that cigarette companies "deny that cigarettes cause disease," the Philip Morris campaign actually resulted in a 24 percent *decrease* in agreement. While the "truth" campaign nearly doubled the odds that youths would say "cigarette companies lie," the Philip Morris campaign resulted in virtually no change.

While the "truth" campaign resulted in no change in teens wanting to see cigarette companies go out of business, the Philip Morris campaign resulted in a 21 percent *decrease* in teens wanting big tobacco to disappear.

The authors of the American Legacy study concluded that "exposure to 'Think, don't smoke' engendered more favorable feelings toward the tobacco industry than we found among those not exposed to 'Think, don't smoke.'"

The authors said the Philip Morris campaign, with its slick studio productions built on the guaranteed failure of telling adolescents that cigarettes are "uncool" for them at the same time the company sells them with glee and glitter all over the earth, is actually meant "to buy respectability and not to prevent youth smoking."

Philip Morris boasts on its Web page that its "Think, don't smoke" campaign is working and is "measurable."

Of course, Philip Morris has done nothing so measurable as to be worthy of a medical or public health journal.

By Derrick Z. Jackson

From an op-ed piece distributed by the New York Times News Service.

advertisement, that it was uninflated, honest, fair, and not misleading. Our suspicions about advertising in general should undercut our willingness to believe in the honesty of any particular advertisement.

Thus, even advertisements that present reasons for buying an item do not by themselves justify our purchase of the item. Sometimes, of course, an advertisement can provide you with information that can clinch your decision to make a purchase. Sometimes the mere existence, availability, or affordability of a product—all information that an ad can convey—is all you need to make a decision to buy. But if the purchase is justifiable, you must have some reasons apart from those offered in the ad for making it. If, for some reason, you already know that you want or need and can afford a car with an electric motor, then an ad that informs you that a firm has begun marketing such a thing would supply you with the information you need to buy one. If you can already justify purchasing a particular brand of microwave oven but cannot find one anywhere in town, then an advertisement informing you that the local department store stocks them can clinch your decision to make the purchase.

For people on whom good fortune has smiled, those who don't care what kind of whatsit they buy, or those to whom mistaken purchases simply don't matter, all that is important is knowing that a product is available. Most of us, however, need more information than ads provide to make reasoned purchasing decisions. Of course, we all occasionally make purchases solely on the basis of advertisements, and sometimes we don't come to regret them. In such cases, though, the happy result is due as much to good luck as to the ad.

Recap

Claims lack credibility to the extent they conflict with our observations or experience or our background information, or come from sources that lack credibility. The less initial plausibility a claim has, the more extraordinary it seems, and the less it fits with our background information, the more suspicious we should be.

Doubts about sources generally fall into two categories, doubts about the source's knowledge or expertise and doubts about the source's veracity, objectivity, and accuracy. We can form reasonably reliable judgments about a person's knowledge by considering his or her education, experience, accomplishments, reputation, and position. Claims made by experts, those with special knowledge in a subject, are the most reliable, but the claims must pertain to the area of expertise and not conflict with claims made by other experts in the same area.

Major metropolitan newspapers, national newsmagazines, and network news shows are generally credible sources of news, but it is necessary to keep an open mind about what we learn from them. Skepticism is even more appropriate when we obtain information from unknown Internet sources or talk radio.

Advertising assaults us at every turn, attempting to sell us goods, services, beliefs, and attitudes. Because substantial talent and resources are employed in this effort, we need to ask ourselves constantly whether the products in question will really make the differences in our lives that their advertising claims or hints they will make. Advertisers are more concerned with selling you something than with improving your life.

Exercises

Exercise 3-1

1. The text points out that physical conditions around us can affect our observations. List at least four such conditions.

2. Our own mental state can affect our observations as well. Describe at least three of the ways this can happen, as mentioned in the text.

3. According to the text, there are two ways credibility should enter into our evaluation of a claim. What are they?

4. A claim lacks inherent credibility, according to the text, when it conflicts with what?

5. Our most reliable source of information about the world is _____.

6. The reliability of our observations is not better than the reliability of _____.

Exercise 3-2

List as many irrelevant factors as you can think of that people often mistake for signs of a person's truthfulness (for example, the firmness of a handshake).

Exercise 3-3

List as many irrelevant factors as you can think of that people often mistake for signs of expertise on the part of an individual (for example, appearing self-confident).

Exercise 3-4

Expertise doesn't transfer automatically from one field to another: Being an expert in one does not automatically qualify one as an expert (or even as competent) in other areas. Is it the same with dishonesty? Many people think dishonesty does transfer, that being dishonest in one area automatically discredits that person in all areas. For example, when Bill Clinton lied about having sexual encounters with his intern, some said he couldn't be trusted about anything.

If someone is known to have been dishonest about one thing, should we automatically be suspicious of his honesty regarding other things? Discuss.

Exercise 3-5

▲ In your judgment, are any of these claims less credible than others? Discuss your opinions with others in the class to see if any interesting differences in background information emerge.

1. They've taught crows how to play poker.
2. The center of Earth consists of water.
3. Ray Charles is just faking his blindness.
4. The car manufacturers already can build cars that get over 100 miles per gallon; they just won't do it because they're in cahoots with the oil industry.
5. If you force yourself to go for five days and nights without any sleep, you'll be able to get by on less than five hours of sleep a night for the rest of your life.
6. It is possible to read other people's minds through mental telepathy.
7. A diet of mushrooms and pecans supplies all necessary nutrients and will help you lose weight. Scientists don't understand why.
8. Somewhere on the planet is a person who looks exactly like you.
9. The combined wealth of the world's 225 richest people equals the total annual income of the poorest 2.5 billion people, which is nearly half the world's total population.
10. George W. Bush arranged to have the World Trade Center attacked so he could invade Afghanistan. He wants to build an oil pipeline across Afghanistan.
11. Daddy longlegs are the world's most poisonous spider, but their mouths are too small to bite.
12. Static electricity from your body can cause your gas tank to explode if you slide across your seat while fueling and then touch the gas nozzle.
13. Japanese scientists have created a device that measures the tone of a dog's bark to determine what the dog's mood is.

Exercise 3-6

Find a Web page that makes claims of dubious credibility. If your instructor's classroom has the technology, he or she will project the page for class examination.

Exercise 3-7 _____

In groups, decide which is the best answer to each question. Compare your answers with those of other groups and your instructor.

1. "SPACE ALIEN GRAVEYARD FOUND! Scientists who found an extra-terrestrial cemetery in central Africa say the graveyard is at least 500 years old! 'There must be 200 bodies buried there and not a single one of them is human,' Dr. Hugo Schild, the Swiss anthropologist, told reporters." What is the appropriate reaction to this report in the *Weekly World News*?
   a. It's probably true.
   b. It almost certainly is true.
   c. We really need more information to form any judgment at all.
   d. None of these.

2. Is Elvis really dead? Howie thinks not. Reason: He knows three people who claim to have seen Elvis recently. They are certain that it is not a mere Elvis look-alike they have seen. Howie reasons that, since he has absolutely no reason to think the three would lie to him, they must be telling the truth. Elvis must really be alive, he concludes!
   Is Howie's reasoning sound? Explain.

3. VOICE ON TELEPHONE: Mr. Roberts, this is AT&T calling. Have you recently placed several long-distance calls to Lisbon, Portugal?
   MR. ROBERTS: Why, no . . .
   VOICE: This is what we expected. Mr. Roberts, I'm sorry to report that apparently someone has been using your calling card number. However, we are prepared to give you a new number, effective immediately, at no charge to you.
   MR. ROBERTS: Well, fine, I guess . . .
   VOICE: Again let me emphasize that there will be no charge for this service. Now, for authorization, just to make sure that we are calling Mr. Roberts, Mr. Roberts, please state the last four digits of your calling card number, your PIN number, please.
   Question: What should Mr. Roberts, as a critical thinker, do?

4. On Thanksgiving Day 1990, an image said by some to resemble the Virgin Mary was observed in a stained glass window of St. Dominic's Church in Colfax, California. A physicist asked to investigate said the image was caused by sunlight shining through the window and reflecting from a newly installed hanging light fixture. Others said the image was a miracle. Whose explanation is more likely true?
   a. The physicist's
   b. The others'
   c. More information is needed before we can decide which explanation is more likely.

5. It is late at night around the campfire when the campers hear the awful grunting noises in the woods around. They run for their lives! Two campers, after returning the next day, tell others they found huge footprints around

the campfire. They are convinced they were attacked by Bigfoot. Which explanation is more likely true?

a. The campers heard Bigfoot.

b. The campers heard some animal and are pushing the Bigfoot explanation to avoid being thought of as chickens, or are just making the story up for unknown reasons.

c. Given this information, we can't tell which explanation is more likely.

6. Megan's aunt says she saw a flying saucer. "I don't tell people about this," Auntie says, "because they'll think I'm making it up. But this really happened. I saw this strange light, and this, well, it wasn't a saucer, exactly, but it was round and big, and it came down and hovered just over my back fence, and my two dogs began whimpering. And then it just, whoosh! It just vanished."

Megan knows her aunt, and Megan knows she doesn't make up stories.

a. She should believe her aunt saw a flying saucer.

b. She should believe her aunt was making the story up.

c. She should believe that her aunt may well have had some unusual experience, but it was probably not a visitation by extraterrestrial beings.

7. According to Dr. Edith Fiore, author of *The Unquiet Dead,* many of your personal problems are really the miseries of a dead soul who has possessed you sometime during your life. "Many people are possessed by earthbound spirits. These are people who have lived and died, but did not go into the afterworld at death. Instead they stayed on Earth and remained just like they were before death, with the fears, pains, weaknesses and other problems that they had when they were alive." She estimates that about 80 percent of her more than 1,000 patients are suffering from the problems brought on by being possessed by spirits of the dead. To tell if you are among the possessed, she advises that you look for such telltale symptoms as low energy levels, character shifts or mood swings, memory problems, poor concentration, weight gain with no obvious cause, and bouts of depression (especially after hospitalization). Which of these reactions is best?

a. Wow! I bet I'm possessed!

b. Well, if a doctor says it's so, it must be so.

c. If these are signs of being possessed, how come she thinks that only 80 percent of her patients are?

d. Too bad there isn't more information available, so we could form a reasonable judgment.

8. **EOC—ENGINE OVERHAUL IN A CAN**

Developed by skilled automotive scientists after years of research and laboratory and road tests! Simply pour one can of EOC into the oil in your crankcase. EOC contains long-chain molecules and special thermoactive metallic alloys that bond with worn engine parts. NO tools needed! NO need to disassemble engine.

Question: Reading this ad, what should you believe?

9. ANCHORAGE, Alaska (AP)—Roped to her twin sons for safety, Joni Phelps inched her way to the top of Mount McKinley. The National Park

Service says Phelps, 54, apparently is the first blind woman to scale the 20,300-foot peak.

This report is
a. Probably true
b. Probably false
c. Too sketchy; more information is needed before we can judge

Exercise 3-8

Within each group of observers, are some especially credible or especially not so?

▲ 1. Judging the relative performances of the fighters in a heavyweight boxing match.
   a. the father of one of the fighters
   b. a sportswriter for *Sports Illustrated* magazine
   c. the coach of the American Olympic boxing team
   d. the referee of the fight
   e. a professor of physical education

2. You (or your family or your class) are trying to decide whether you should buy an Apple Macintosh computer or an IBM model. You might consult
   a. a friend who owns either a Macintosh or an IBM
   b. a friend who now owns one of the machines but used to own the other
   c. a dealer for either Macintosh or IBM
   d. a computer column in a big-city newspaper
   e. reviews in computer magazines

▲ 3. The Surgical Practices Committee of Grantville Hospital has documented an unusually high number of problems in connection with tonsillectomies performed by a Dr. Choker. The committee is reviewing her surgical practices. Those present during a tonsillectomy are
   a. Dr. Choker
   b. the surgical proctor from the Surgical Practices Committee
   c. an anesthesiologist
   d. a nurse
   e. a technician

4. The mechanical condition of the used car you are thinking of buying.
   a. the used-car salesperson
   b. the former owner (who we assume is different from the salesperson)
   c. the former owner's mechanic
   d. you
   e. a mechanic from an independent garage

5. A demonstration of psychokinesis (the ability to move objects at a distance by nonphysical means).
   a. a newspaper reporter
   b. a psychologist

c. a police detective
d. another psychic
e. a physicist
f. a customs agent
g. a magician

Exercise 3-9

For each of the items below, discuss the credibility and authority of each source relative to the issue in question. Whom would you trust as most reliable on the subject?

▲ 1. Issue: Is Crixivan an effective HIV/AIDS medication?
  a. *Consumer Reports*
  b. Stadtlander Drug Company (the company that makes Crixivan)
  c. the owner of your local health food store
  d. the U.S. Food and Drug Administration
  e. your local pharmacist

▲ 2. Issue: Should possession of handguns be outlawed?
  a. a police chief
  b. a representative of the National Rifle Association
  c. a U.S. senator
  d. the father of a murder victim

▲ 3. Issue: What was the original intent of the Second Amendment to the U.S. Constitution, and does it include permission for every citizen to possess handguns?
  a. a representative of the National Rifle Association
  b. a justice of the U.S. Supreme Court
  c. a Constitutional historian
  d. a U.S. senator
  e. the President of the United States

4. Issue: Is decreasing your intake of dietary fat and cholesterol likely to reduce the level of cholesterol in your blood?
  a. *Time* magazine
  b. *Runner's World* magazine
  c. your physician
  d. the National Institutes of Health
  e. the *New England Journal of Medicine*

5. Issue: When does a human life begin?
  a. a lawyer
  b. a physician
  c. a philosopher
  d. a minister
  e. you

Exercise 3-10

Each of these items consists of a brief biography of a real or imagined person followed by a list of topics. On the basis of the information in the biography, discuss the credibility and authority of the person described on each of the topics listed.

▲ 1. Jeff Hilgert teaches sociology at the University of Illinois and is the director of its Population Studies Center. He is a graduate of Harvard College, where he received a B.A. in 1975, and of Harvard University, which granted him a Ph.D. in economics in 1978. He taught courses in demography as an assistant professor at UCLA until 1982; then he moved to the sociology department of the University of Nebraska, where he was associate professor and then professor. From 1987 through 1989 he served as acting chief of the Population Trends and Structure Section of the United Nations Population Division. He joined the faculty at the University of Illinois in 1989. He has written books on patterns of world urbanization, the effects of cigarette smoking on international mortality, and demographic trends in India. He is president of the Population Association of America.

Topics

   a. The effects of acid rain on humans
   b. The possible beneficial effects of requiring sociology courses for all students at the University of Illinois
   c. The possible effects of nuclear war on global climate patterns
   d. The incidence of poverty among various ethnic groups in the United States
   e. The effects of the melting of glaciers on global sea levels
   f. The change in death rate for various age groups in all Third World countries between 1970 and 1990
   g. The feasibility of a laser-based nuclear defense system
   h. Voter participation among religious sects in India
   i. Whether the winters are worse in Illinois than in Nebraska

  2. Tom Pierce graduated cum laude from Cornell University with a B.S. in biology in 1973. After two years in the Peace Corps, during which he worked on public health projects in Venezuela, he joined Jeffrey Ridenour, a mechanical engineer, and the pair developed a water pump and purification system that is now used in many parts of the world for both regular water supplies and emergency use in disaster-struck areas. Pierce and Ridenour formed a company to manufacture the water systems, and it prospered as they developed smaller versions of the system for private use on boats and motor homes. In 1981, Pierce bought out his partner and expanded research and development in hydraulic systems for forcing oil out of old wells. Under contract with the federal government and several oil firms, Pierce's company was a principal designer and contractor for the Alaskan oil pipeline. He is now a consultant in numerous developing countries as well as chief executive officer and chairman of the board of his own company, and he sits on the boards of directors of several other companies.

Topics

a. The image of the United States in Latin America
b. The long-range effects of the Cuban revolution on South America
c. Fixing a leaky faucet
d. Technology in Third World countries
e. The ecological effects of the Alaskan pipeline
f. Negotiating a contract with the federal government
g. Careers in biology

Exercise 3-11

Watch a debate or panel discussion (e.g., a political debate, a discussion on PBS's *NewsHour with Jim Lehrer*), and decide which participant made the better case for his or her side. When the debate is over, being honest with yourself, list as many irrelevant factors as you can that you think influenced your decision.

Exercise 3-12

From what you know about the nature of each of the following claims and its source, and given your general knowledge, assess whether the claim is one you should accept, reject, or suspend judgment on due to ambiguity, insufficient documentation, vagueness, or subjectivity (e.g., "Tom Cruise is cute"). Compare your judgment with that of your instructor.

▲ 1. "Campbell Soup is hot—and some are getting burned. Just one day after the behemoth of broth reported record profits, Campbell said it would lay off 650 U.S. workers, including 175—or 11% of the work force—at its headquarters in Camden, New Jersey."

Time

2. [The claim to evaluate is the first one in this passage.] Jackie Haskew taught paganism and devil worship in her fourth-grade classroom in Grand Saline, Texas, at least until she was pressured into resigning by parents of her students. (According to syndicated columnist Nat Hentoff, "At the town meeting on her case, a parent said firmly that she did not want her daughter to read anything that dealt with 'death, abuse, divorce, religion, or any other issue.'")

3. "By 1893 there were only between 300 and 1,000 buffaloes remaining in the entire country. A few years later, President Theodore Roosevelt persuaded Congress to establish a number of wildlife preserves in which the remaining buffaloes could live without danger. The numbers have increased since, nearly doubling over the past 10 years to 130,000."

Clifford May, in the New York Times Magazine

4. Lee Harvey Oswald, acting alone, was responsible for the death of President John F. Kennedy.

Conclusion of the Warren Commission on the Assassination of President Kennedy

5. "[N]ewly released documents, including the transcripts of telephone conversations recorded by President Lyndon B. Johnson in November and December 1963, provide for the first time a detailed . . . look at why and how the seven-member Warren [Commission] was put together. Those documents, along with a review of previously released material . . . describe a process designed more to control information than to elicit and expose it."

"The Truth Was Secondary," Washington Post National Weekly Edition

6. "Short-sighted developers are determined to transform Choco [a large region of northwestern Colombia] from an undisturbed natural treasure to a polluted, industrialized growth center."

Solicitation letter from the World Wildlife Fund

7. "Frantic parents tell shocked TV audience: space aliens stole our son."

Weekly World News

▲ 8. "The manufacturer of Sudafed 12-hour capsules issued a nationwide recall of the product Sunday after two people in the state of Washington who had taken the medication died of cyanide poisoning and a third became seriously ill."

Los Angeles Times

9. "In Canada, smoking in public places, trains, planes or even automobiles is now prohibited by law or by convention. The federal government has banned smoking in all its buildings."

Reuters

10. "The list of vanishing commodities in Moscow now includes not only sausage and vodka, long rationed, but also potatoes, eggs, bread, and cigarettes."

National Geographic

11. "Maps, files and compasses were hidden in Monopoly sets and smuggled into World War II German prison camps by MI-5, Britain's counter-intelligence agency, to help British prisoners escape, according to the British manufacturer of the game."

Associated Press

▲ 12. "Cats that live indoors and use a litter box can live four to five years longer."

From an advertisement for Jonny Cat litter

13. "A case reported by Borderland Sciences Research Foundation, Vista, California, tells of a man who had attended many of the meetings where a great variety of 'dead' people came and spoke through the body mechanism of Mark Probert to the group of interested persons on a great variety of subjects with questions and answers from 'both sides.' Then this man who had attended meetings while he was in a body, did what is called 'die.' Presumably he had learned 'while in the body' what he might expect at the change of awareness called death, about which organized religion seems to know little or nothing."

George Robinson, Exploring the Riddle of Reincarnation, *undated, no publisher cited*

14. "Because of cartilage that begins to accumulate after age thirty, by the time . . . [a] man is seventy his nose has grown a half inch wider and an-

other half inch longer, his earlobes have fattened, and his ears themselves have grown a quarter inch longer. Overall, his head's circumference increases a quarter inch every decade, and not because of his brain, which is shrinking. His head is fatter apparently because, unlike most other bones in the body, the skull seems to thicken with age."

John Tierney (a staff writer for Esquire*)*

15. "Gardenias . . . need ample warmth, ample water, and steady feeding. Though hardy to 20°F or even lower, plants fail to grow and bloom well without summer heat."

The Sunset New Western Garden Book *(a best-selling gardening reference in the West)*

16. "Exercise will make you feel fitter, but there's no good evidence that it will make you live longer."

Dr. Jordan Tobin, National Institute on Aging

17. "Your bones are still growing until you're 35."

From a national milk ad by the National Fluid Milk Processor Promotion Board

18. "*E. coli* 0157:H7 has become common enough to be the current major cause of acute kidney failure in children." [*E. coli* is a food-borne toxin originally found in the intestines of cows.]

Robin Cook, a physician-turned-novelist. This claim was made by a fictional expert on food-borne illnesses in the novel Toxin.

19. "A woman employed as a Santa Claus at a Wal-Mart in Kentucky was fired by Wal-Mart when a child pinched her breast and complained to his mother that Santa was a woman. The woman complained to store managers."

Associated Press

▲ 20. "Bill Clinton is the biological son of Jimmy Carter."

A claim alleged to have been made in a suit alleged to have been filed in the U.S. District Court for the Southern District of New York. Reported on John E. Schwenkler's home page.

21. "The KGB forged documents to show the CIA was behind the Kennedy assassination. It also 'leaked' the information that the AIDS virus was manufactured by the CIA."

60 Minutes

Exercise 3-13 _____

Find five advertisements that give no reasons for purchasing the products they are selling. Explain how each ad attempts to make the product seem attractive.

Exercise 3-14 _____

Find five advertisements that give reasons for purchasing the products they are selling. Which of the reasons are promises to the purchaser? Exactly what is being promised? What is the likelihood that the product will fulfill that promise?

Exercise 3-15 _____

Watch Fox News and CNN news programs on the same day. Compare the two on the basis of (1) the news stories covered, (2) the amount of air time given to two or three of the major stories, and (3) any difference in the slant of the presentations of a controversial story. From your reading of the chapter, how would you account for the similarities between the two in both selection and content of the stories?

Writing Exercises _____

1. Although millions of people have seen professional magicians like David Copperfield and Siegfried and Roy perform in person or on television, it's probably a safe assumption that almost nobody believes they accomplish their feats by means of real magical or supernatural powers—that is, that they somehow "defy" the laws of nature. But even though they've never had a personal demonstration, a significant portion of the population believe that certain psychics are able to accomplish apparent miracles by exactly such means. How might you explain this difference in belief? (Our answer is given in the *Logical Accessory* instructor's manual.)

2. In the text you were asked to consider the claim "Charlie's eighty-seven-year-old grandmother swam across Lake Michigan in the middle of winter." Because of the implausibility of such a claim—that is, because it conflicts with our background information—it is reasonable to reject it. Suppose, however, that instead of just telling us about his grandmother, Charlie brings us a photocopy depicting a page of a Chicago newspaper with a photograph of a person in a wetsuit walking up onto a beach. The caption underneath reads, "Eighty-seven-year-old Grandmother Swims Lake Michigan in January!" Based on this piece of evidence, should a critical thinker decide that the original claim is significantly more likely to be true than if it were backed up only by Charlie's word? Defend your answer.

3. Turn to the "Essays for Analysis" in Appendix 1 and assess the credibility of the author in Selection 9, "Will Ozone Blob Devour the Earth?" Based on the blurb about the author, say what you can about his likely expertise and his susceptibility to bias on the subject of the essay.

4. Are our schools doing a bad job educating our kids? Do research in the library or on the Internet to answer this question. Make a list (no more than one page long) of facts that support the claim that our schools are not doing as good a job as they should. Then list facts that support the opposite view (or that rebut the claims of those who say our schools aren't doing a good job). Again, limit yourself to one page. Cite your sources.

   Now think critically about your sources. Are any stronger or weaker than the others? Explain why on a single sheet of paper. Come prepared to read your explanation, along with your list of facts and sources, to the class.

5. Jackson says you should be skeptical of the opinion of someone who stands to profit from your accepting that opinion. Smith disagrees, pointing out that salespeople are apt to know a lot more about products of the type they sell than do most people.

   "Most salespeople are honest, and you can trust them," Smith argues. "Those who aren't don't stay in business long."

   Take about fifteen minutes to defend either Smith or Jackson in a short essay. When everyone is finished, your instructor will collect the essays and read three or more to the class to stimulate a brief discussion. After discussion, can the class come to any agreement about who is correct, Jackson or Smith?

6. Your instructor will survey the class to see how many agree with this claim: The media are biased. Then he or she will ask you to list your reasons for thinking that this claim is true. (If you do not think it is true, list reasons people might have for believing it.) After ten minutes your instructor will collect the lists of reasons and read from several of the lists. Then he or she will give you twenty minutes to defend one of these claims:
   a. The media are biased.
   b. Some of the reasons people have for believing the media are biased are not very good reasons.
   c. It is difficult to say whether the media are biased.

      At the end of the period your instructor may survey the class again to see if anyone's mind has changed, and why.

7. If you haven't done Exercise 6, your instructor will give you twenty minutes to defend an answer to the question, Are the media biased? Put your name on the back of your paper. When everyone is finished, your instructor will collect the papers and redistribute them to the class. In groups of four or five, read the papers that have been given to your group, and decide if any of them are convincing. Do not look at the names of the authors. Your instructor will ask each group to read to the class any essay that the group thinks is convincing.

Chapter 4

Persuasion Through Rhetoric

A couple of years ago, tobacco companies spent millions trying to defeat Proposition 10, an antismoking ballot initiative in California. Calling themselves "The Committee Against Unfair Taxes," they mailed out expensive glossy brochures warning voters of the perils of the initiative. They were especially anxious to inform everyone that

> The sponsors of ill-conceived Proposition 10 are perennial political activist millionaire Rob Reiner and four other Hollywood/Los Angeles millionaire social engineers who believe they know more about raising your children than you do.

This warning was accompanied by a grainy black-and-white photograph of Reiner (whom you may remember from reruns of the popular 1960s TV show *All in the Family*) that brought to mind a police mug shot.

Now, when others want us to do something or want to influence our attitudes or beliefs, they may use an argument. That is, they may offer a reason why we should or shouldn't do or believe or not believe whatever it is. They might also use threats, bribery, or even more extreme measures. But another technique is available, one used much more frequently, and it is illustrated in the statement quoted above. This technique employs the persuasive power of words, or what we have called their rhetorical force or emotive meaning—their power to express and elicit images, feelings, and emotional associations. In the next few chapters we examine some of the most common rhetorical techniques used to affect people's attitudes and opinions and behavior.

In English, **rhetoric** refers to the study of persuasive writing. As we use the term, it denotes a broad category of linguistic techniques people use when their primary objective is to influence beliefs and attitudes and behavior. Is Rob Reiner "a perennial political activist millionaire"? Or is he an "untiring advocate of social reform willing to spend his considerable fortune for just causes"? The different impressions these two descriptions create is largely due to their differing rhetorical meaning. Does Juanita "still owe over $1,000 on her credit card"? Or does Juanita "owe only a little over $1,000 on her credit card"? There's no factual difference between the two questions—only a difference in their rhetorical force. The thing to remember through these next few chapters is that rhetorical force may be psychologically effective, but by itself it establishes nothing. If we allow our attitudes and beliefs to be affected by sheer rhetoric, we fall short as critical thinkers.

Now before we get in trouble with your English teacher, let's make it clear that there is nothing wrong with trying to make your case as persuasive as possible by using well-chosen, rhetorically effective words and phrases. Good writers always do this. But we, as critical thinkers, must be able to distinguish the *argument* (if any) contained in what someone says or writes from the *rhetoric*; we must be able to distinguish the *logical* force of a set of remarks from their *psychological* force. The statement above from the tobacco companies about Rob Reiner, for example, contains *no argument whatsoever*; it just uses inflammatory rhetorical techniques aimed at getting us to vote against the antismoking initiative.

One of the things you will become aware of—as you read these pages and do the exercises and apply what you have learned to what you read and write—is that rhetoric is often mixed right in with argument. The message isn't that you should *deduct* points from an argument if it is presented in rhetorically charged language, and it isn't that you should try to get all the rhetoric out of your own writing. The message is simply that you shouldn't *add* points for rhetoric. You don't make an argument stronger by screaming it at the top of your lungs. Likewise, you don't make it stronger by adding **rhetorical devices.**

Rhetorical Devices and Techniques

Many of these rhetorical bells and whistles have names because they are so common and so well understood. Because they are used primarily to give a statement a positive or negative slant regarding a subject, they are sometimes called **slanters.** We'll describe some of the more widely used specimens.

■ Euphemisms and Dysphemisms

Language usually offers us a choice of words when we want to say something. Until recently, the term "used car" referred to an automobile that wasn't new, but the trend nowadays is to refer to such a car as "pre-owned." The people who sell such cars, of course, hope that the different terminology will keep potential buyers from thinking about *how* "used" the car might be—maybe it's *used up*! The car dealer's replacement term, "pre-owned," is a **euphemism**—a neutral or positive expression instead of one that carries negative associations. Euphemisms play an important role in affecting

The Death Tax

Opponents of the estate tax, a tax imposed on inherited money, have coined a new phrase for it: the "death tax." Here is Grover Norquist, who is the head of Americans for Tax Reform in Washington, D.C., in a press release from that organization:

> Over seventy percent of Americans oppose the Death Tax, and with good reason. It is the worst form of double-taxation, where, after taxing you all your life, the government decides to take even more when you die.

"Death Tax" is a dysphemism, of course. Like many dysphemisms, it is designed to mislead the reader or listener. The estate tax is not a tax on death, but on inherited wealth, imposed on the occasion of a person's death. And the person paying the tax is not the deceased, but the inheritors, who have never paid tax on the money.

(Interestingly, the tax does not apply to the first $625,000 of a person's inheritance; and the tax rate is 37 percent of inherited wealth from $625,000 to $3 million. So the tax does not apply to all the great majority of Americans, who do not inherit that kind of money. It isn't clear that the majority of Americans *realize* that it will not apply to them, however.)

our attitudes. People may be less likely to disapprove of an assassination attempt on a foreign leader, for example, if it is referred to as "neutralization." People fighting against the government of a country can be referred to neutrally as "rebels" or "guerrillas," but a person who wants to build support for them may refer to them by the euphemism "freedom fighters." A government is likely to pay a price for initiating a "revenue enhancement," but voters will be even quicker to respond negatively to a "tax hike." The U.S. Department of Defense performs the same function it did when it was called the Department of War, but the current name makes for much better public relations.

The opposite of a euphemism is a **dysphemism.** Dysphemisms are used to produce a negative effect on a listener's or reader's attitude toward

Your Cheatin' Heart?

Here's today's timely if perhaps tasteless question: What's the difference between a rendezvous and an affair?

A rendezvous is what Bill Cosby told Dan Rather he had with the mother of the young woman who allegedly attempted to blackmail him, claiming he was her father.

A rendezvous, not an affair.

"Rendezvous" sounds so much nicer than "affair." So romantic. So pretentious. So French. . . .

An affair is middle class, guilt-ridden and furtive. A rendezvous is urbane, sophisticated and self-conscious. . . .

A rendezvous is twin martinis in the Ritz Bar.

An affair is a couple of Gallo Chardonnays after work.

A rendezvous is the back of a limousine.

An affair is the back of a minivan.

A rendezvous is Bobby Short at the Carlyle.

An affair is Kenny G, wherever. . . .

Basically, though, they boil down to much the same thing.

An affair is what you have if you can't, or won't, admit you're cheating.

A rendezvous is what you have if you can't, or won't, admit you're having an affair.

From a column by DIANE WHITE

something or to tone down the positive associations it may have. Whereas "freedom fighter" is a euphemism for "guerrilla" or "rebel," "terrorist" is a dysphemism.

Euphemisms and dysphemisms are often used in deceptive ways or ways that at least hint at deception. All of the examples in the preceding paragraphs are examples of such uses. But euphemisms can at times be helpful and constructive. By allowing us to approach a sensitive subject indirectly—or by skirting it entirely—euphemisms can sometimes prevent hostility from bringing rational discussion to a halt. They can also be a matter of good manners: "Passed on" may be much more appropriate than "dead" if the person to whom you're speaking is recently widowed. Hence, our *purpose* for using euphemisms and dysphemisms determines whether or not those uses are legitimate.

It bears mentioning that some facts just are repellent, and for that reason even neutral reports of them sound horrible. "Lizzy killed her father with an ax" reports a horrible fact about Lizzy, but it does so using neutral language. Neutral reports of unpleasant, evil, or repellent facts do not automatically count as dysphemistic rhetoric.

The University of Washington admits that many of its faculty members suffer from what it calls "salary compression." The faculty themselves have a different word for it, as you can imagine.

—Rhetorical comparison. (And a great cartoon.)

Horsey. Reprinted with permission, Seattle Post Intelligencer.

■ Rhetorical Comparisons, Definitions, and Explanations

Definitions, explanations, and comparative claims are discussed in other chapters, but the emphasis in those discussions is on uses other than slanting. We want to remind you here that each of these can do quite a bit to prejudice an issue.

Rhetorical comparisons are used to express or influence attitudes. You might make the point that a person is of small stature by comparing her to an elf, but you could make the same point in a much less flattering way by comparing her to a gnome or a Chihuahua. You could describe the fairness of a person's complexion by comparing it to either new-fallen snow or whale blubber, but you'd better make the latter comparison out of the individual's hearing.

Rhetorical definitions smuggle prejudice of one sort or another into the meaning of a term. Abortion and associated issues are fruitful grounds for persuasive definitions. If we define "abortion" as "the murder of an unborn child," there's not much difficulty in arriving at the conclusion that abortion is morally unacceptable. Similarly, "human being" is sometimes taken to refer to any living organism (embryonic, fetal, or post-birth) that is produced by humans. By using this definition, it's easy to conclude that an

Whether you wish to admit it or not, when you approve, morally, of the bombing of foreign targets by U.S. military, you are approving of acts morally equivalent to the bombing in Oklahoma City.

—A rhetorical comparison by TIMOTHY MCVEIGH, *Oklahoma City bomber (reported by Larry Englemann)*

What Does "Rape" Mean?

In an article for Knight-Ridder Newspapers, Joanne Jacobs of the *San Jose Mercury News* disagreed markedly with the current use of the word "rape" on some university campuses. Notice how language and definition are crucial in the items she reported:

- Swarthmore College's Acquaintance Rape Prevention Workshop training manual asserts that acquaintance rape includes behavior ranging "from crimes legally defined as rape to verbal harassment and inappropriate innuendo."

- A former director of Columbia University's date-rape education program says that "every time you have an act of intercourse there must be explicit consent, and if there's not explicit consent, then it's rape."

- According to Andrea Parrot, a Cornell University psychiatry professor who's written a book on date rape, "Any sexual intercourse without mutual desire is a form of rape." It's as bad to be psychologically pressured into sexual contact by an acquaintance, writes Parrot, as to be "attacked on the streets."

Jacobs believes that defining "rape" in this way trivializes the term and is "a cruel insult to all the women who have been raped" by a man who did more than use "inappropriate innuendos." The definition of "rape" as any sex that goes farther than the woman wants, whether she made that clear or not, is "fantastically broad," she says. She also maintains that this definition makes reports of an epidemic of sexual assault on campuses highly suspect. Were the victims *raped,* she asks, or were they just *nagged*?

Others would say that use of the word "rape" for the behavior described is justified because it draws attention to the seriousness of that behavior.

Note two things: First, the definitions of "rape" given here are *precising definitions,* definitions that serve to specify the applicability of a term. But they are also disguised *rhetorical definitions* designed to influence attitudes about a variety of sexual behaviors. Second, notice that "a mere question of semantics" is sometimes a very important matter indeed.

abortion involves the taking of a human being's life and is thus a homicide. Rhetorical definitions used this way are said to "beg the question," a fault discussed in Chapter 6.

Definitions by example can also slant a discussion if the examples are prejudicially chosen. Imagine a liberal who points to a story about a conservative politician found guilty of corruption and says, *"That's* what I mean by conservative politics." Clearly, if we really want to see all sides of an issue, we must be careful to avoid definitions and examples that slant the discussion.

Sex-education classes are like in-home sales parties for abortion.

—PHYLLIS SCHLAFLY, activist

A rhetorical comparison.

▲ *Stereotypes.*

Rhetorical explanations are the same kind of slanting device, this time clothed as explanations. "He lost the fight because he's lost his nerve." Is this different from saying that he lost because he was too cautious? Maybe, but maybe not. What isn't in doubt is that the explanation is certainly more unflattering when it's put the former way.

We recently saw a good example of a rhetorical explanation in a letter to an editor:

> I am a traditional liberal who keeps asking himself, why has there been such a seismic shift in affirmative action? It used to be affirmative action stood for equal opportunity; now it means preferences and quotas. Why the change? It's because the people behind affirmative action aren't for equal rights anymore; they're for handouts.

This isn't a dispassionate scholarly explanation but a way of expressing an opinion on and trying to evoke anger at affirmative action policies.

■ *Stereotypes*

When a writer or speaker lumps a group of individuals together under one name or description, especially one that begins with the word "the" (the liberal, the Communist, the right-winger, the Jew, the Catholic, and so on), such labeling generally results in stereotyping. A **stereotype** is a thought or image about a group of people based on little or no evidence. Thinking that women are emotional, that men are insensitive, that lesbians are man haters, that southerners are bigoted, that gay men are effeminate—all count as stereotypes. Language that reduces people or things to categories can induce an audience to accept a claim unthinkingly or to make snap judgments concerning groups of individuals about whom they know little.

Some of the slanters we've already talked about can involve stereotypes. For example, if we use the dysphemism "right-wing extremist" to denigrate a political candidate, we are utilizing a negative stereotype. Com-

It was a reasonable proposal. That's why they rejected it.

—Former Senator BOB DOLE, *explaining why the Democrats voted against a Republican proposal*

A rhetorical explanation.

The ventilation fans will be taken care of in a more timely manner because we know that women love to clean.

—General YURI GLAZKOV, *expressing the hope that U.S. astronaut Shannon Lucid would clean the fans when she joined the Russians on their space station.*

Houston? Are you hearing this, Houston?

monly, if we link a candidate with a stereotype we like or venerate, we can create a favorable impression of the individual. "Senator Kerry addressed his opponent with all the civility of a gentleman" employs a favorable stereotype, that of a gentleman, in a persuasive comparison.

Our stereotypes come from a great many sources, many from popular literature, and are often supported by a variety of prejudices and group interests. The Native American tribes of the Great Plains were considered noble people by most whites until just before the mid–nineteenth century. But as white people grew more interested in moving them off their lands and as conflicts between the two escalated, popular literature increasingly described Native Americans as subhuman creatures. This stereotype supported the group interests of whites. Conflicts between nations usually produce derogatory stereotypes of the opposition; it is easier to destroy enemies without pangs of conscience if we think of them as less "human" than ourselves. Stereotyping becomes even easier when there are racial differences to exploit.

■ *Innuendo*

The next batch of slanting devices doesn't depend as much on emotional associations as on the manipulation of other features of language. When we communicate with one another, we automatically have certain expectations and make certain assumptions. (For example, when your instructor says, "Everybody passed the exam," she doesn't mean that everybody *in the world* passed the exam. We assume that the scope of the pronoun extends to include only those who took the exam.) These expectations and assumptions help fill in the gaps in our conversations so that we don't have to explain everything we say in minute detail. Because providing such details would be a tedious and probably impossible chore, these underlying conversational factors are crucial to the success of communication.

Consider this statement:

Ladies and gentlemen, I am proof that there is at least one candidate in this race who does not have a drinking problem.

Photographic Innuendo

In the 2002 U.S. Senate race in Montana, Republican Mike Taylor dropped out of the race at the last minute, blaming a thirty-second TV ad sponsored by the Montana Democratic Party. The Democrats, he asserted, were using a stereotype to make him look gay.

In the ad, clips show Taylor dressed in a leisure suit, rubbing lotion on another man's temples, his white shirt unbuttoned to his chest, gold necklaces dangling. (Back in the 1980s Taylor ran a beauty salon and hair-care school; the clips are from a TV show he hosted.)

Democrats denied that they were implying anything about Taylor's sexuality. The ad focused on charges that Taylor diverted federal student loans when he ran the beauty salon.

Taylor had been marketing himself as a tough Montana rancher. "The ad has destroyed the campaign," said a spokesperson for Taylor.

The city voluntarily assumed the costs of cleaning up the landfill to make it safe for developers.

—Opponents of a local housing development

The opponents neglected to mention that the law *required* the city to assume the costs. This bit of innuendo on the part of the opponents suggested, of course, that the city was in bed with the developers.

Notice that this remark does *not* say that any opponent of the speaker *does* have a drinking problem. In fact, the speaker is even allowing for the fact that other candidates may have no such problem by using the words "at least one candidate." But he has still managed to get the idea across that some opponent in the race *may* have a drinking problem. This is an example of **innuendo,** a form of suggestion.

Another example, maybe our all-time favorite, is this remark:

> I didn't say the meat was tough. I said I didn't see the horse that is usually outside.
>
> —*W. C. Fields*

As you can see, the use of innuendo enables us to insinuate something deprecatory about something or someone without actually saying it. For example, if someone asks you whether Ralph is telling the truth, you may reply, "Yes, this time," which would suggest that maybe Ralph doesn't *usually* tell the truth. Or you might say of someone, "She is competent—in many regards," which would insinuate that in some ways she is *not* competent.

Sometimes we condemn somebody with faint praise—that is, by praising a person a small amount when grander praise might be expected, we hint that praise may not really be due at all. This is a kind of innuendo. Imagine, for example, reading a letter of recommendation that says, "Ms. Flotsam has done good work for us, I suppose." Such a letter does not inspire one to want to hire Ms. Flotsam on the spot. Likewise, "She's proved to be useful so far" and "Surprisingly, she seems very astute" manage to speak more evil than good of Ms. Flotsam. Notice, though, that the literal information contained in these remarks is not negative in the least. Innuendo lies between the lines, so to speak.

Innuendo Again, This Time in the Caption

▲ *Enron's Fastow: Charged. Are Skilling and Lay Next?*
Comment: Notice that the caption doesn't say *that Skilling and Lay have been charged.*
This is innuendo.

■ *Loaded Questions*

If you overheard someone ask, "Have you always loved to gamble?" you would naturally assume that the person being questioned did in fact love to gamble. This assumption is independent of whether the person answered yes or no, for it underlies the question itself. Every question rests on assumptions. Even an innocent question like "What time is it?" depends on the assumptions that the hearer speaks English and has some means of finding out the time, for instance. A **loaded question** is less innocent, however. It rests on one or more *unwarranted* or *unjustified* assumptions. The world's oldest example, "Have you stopped beating your wife?" rests on the assumption that the person asked has in the past beaten his wife. If there is no reason to think that this assumption is true, then the question is a loaded one.

Innocent Until Reported Guilty

In what may be the most obvious indicator of the administration's priorities, Wen Ho Lee and Peter Lee, who reportedly passed the most clamorous secrets, are not even in jail.

—Boston University professor ANTHONY CODEVILLA, in the Washington Times, talking about the alleged passing of nuclear secrets to China

Notice how the weaseler, "reportedly," is used here. We didn't realize that people who have been *reported* to have committed crimes should be put in jail.

The loaded question is technically a form of innuendo, because it permits us to insinuate the assumption that underlies a question without coming right out and stating that assumption.

■ Weaselers

Overall, Dodge trucks are the most powerful.
—Ad for Dodge

Overall? What does this weaseler mean?

Weaselers are linguistic methods of hedging a bet. When inserted into a claim, they help protect it from criticism by watering it down somewhat, weakening it, and giving the claim's author a way out in case the claim is challenged.

There used to be an advertisement for a brand of sugarless gum that claimed, "Three out of four dentists surveyed recommend sugarless gum for their patients who chew gum." This claim contains two weaseling expressions. The first is the word "surveyed." Notice that the ad does not tell us the criteria for choosing the dentists who were surveyed. Were they picked at random, or were only dentists who might not be unfavorably disposed toward gum chewing surveyed? Nothing indicates that the sample of dentists surveyed even remotely represents the general population of dentists. If 99 percent of the dentists in the country disagree with the ad's claim, its authors could still say truthfully that they spoke only about those dentists surveyed, not all dentists.

The second weaseler in the advertisement appears in the last phrase of the claim: "for their patients who chew gum." Notice the ad does not claim that *any* dentist believes sugarless-gum chewing is as good for a patient's teeth as no gum chewing at all. Imagine that the actual question posed to the dentists was something like this: "If a patient of yours insisted on chewing gum, would you prefer that he or she chew sugarless gum or gum with sugar in it?" If dentists had to answer that question, they would almost certainly be in favor of sugarless gum. But this is a far cry from recommending that any person chew any kind of gum at all. The weaselers allow the advertisement to get away with what *sounds* like an unqualified recommendation for sugarless gum, when in fact nothing in the ad supports such a recommendation.

Innuendo with Statistics

Taxpayers with incomes over $200,000 could expect on average to pay about $99,000 in taxes under Mr. Bush's plan.

—Wall Street Journal

Wow! Pity the poor taxpayer who makes over $200,000! Apparently, he or she will pay almost half of that amount in taxes.

But think again: In the words of the *New Republic* (February 3, 2003), "The *Journal's* statistic is about as meaningful as asserting that males over the age of six have had an average of three sexual partners." Bill Gates and many others like him are among those who make over $200,000.

Let's make up a statistic. Let's say that 98 percent of American doctors believe that aspirin is a contributing cause of Reye's syndrome in children and that the other 2 percent are unconvinced. If we then claim that "some doctors are unconvinced that aspirin is related to Reye's syndrome," we cannot be held accountable for having said something false, even though our claim might be misleading to someone who did not know the complete story. The word "some" has allowed us to weasel the point.

Words that sometimes weasel—such as "perhaps," "possibly," "maybe," and "may be," among others—can be used to produce innuendo, to plant a *suggestion* without actually making a claim that a person can be held to. We can suggest that Berriault is a liar without actually saying so (and thus without making a claim that might be hard to defend) by saying that Berriault *may be* a liar. Or we can say that it is *possible* that Berriault is a liar (which is true of all of us, after all). "*Perhaps* Berriault is a liar" works nicely too. All of these are examples of weaselers used to create innuendo.

Not every use of words and phrases like these is a weaseling one, of course. Words that can weasel can also bring very important qualifications to bear on a claim. The very same word that weasels in one context may not weasel at all in another. For example, a detective considering all the possible angles on a crime who has just heard Smith's account of events may say to an associate, "Of course, it is *possible* that Smith is lying." This need not be a case of weaseling. The detective may simply be exercising due care. Other words and phrases that are sometimes used to weasel can also be used legitimately. Qualifying phrases such as "it is arguable that," "it may well be that," and so on have at least as many appropriate uses as weaseling ones. Others, such as "some would say that," are likely to be weaseling more often than not, but even they can serve an honest purpose in the right context. Our warning, then, is to be watchful when qualifying phrases turn up. Is the speaker or writer adding a reasonable qualification, insinuating a bit of innuendo, or preparing a way out? We can only warn; you need to assess the speaker, the context, and the subject to establish the grounds for the right judgment.

Great Western pays up to 12 percent more interest on checking accounts.

—Radio advertisement

Even aside from the "up to" weaseler, this ad can be deceptive about what interest rate it's promising. Unless you listen carefully, you might think Great Western is paying 12 percent on checking accounts. The presence of the word "more" changes all that, of course. If you're getting 3 percent now and Great Western gives you "up to 12 percent more" than that, they'll be giving you about 3⅓ percent—hardly the fortune the ad seems to promise.

■ *Downplayers*

Downplaying is an attempt to make someone or something look less important or significant. Stereotypes, rhetorical comparisons, rhetorical explanations, and innuendo can all be used to downplay something. Consider this statement, for example: "Don't mind what Mr. Pierce says in class; he's a liberal." This attempt to downplay Mr. Pierce and whatever views he expresses in class makes use of a stereotype. We can also downplay by careful insertion of certain words or other devices. Let's amend the preceding example like this: "Don't mind what Mr. Pierce says in class; he's just another liberal." Notice how the phrase "just another" denigrates Mr. Pierce's status still further. Words and other devices that serve this function are known as **downplayers.**

Perhaps the words most often used as downplayers are "mere" and "merely." If Kim tells you that she has a yellow belt in the Tibetan martial art of Pujo and that her sister has a mere green belt, you would quite naturally make the assumption that a yellow belt ranks higher than a green belt. We'd probably say that Kim's use of the word "mere" gives you the *right* to make that assumption. Kim has used the word to downplay the significance of her sister's accomplishment. But notice this: It could still be that Kim's sister's belt signifies the higher rank. If called on the matter, Kim might claim that she said "mere" simply because her sister has been practicing the art for much longer and is, after all, not *that* far ahead. Whether Kim has such an out or not, she has used a downplayer to try to diminish her sister's accomplishment.

The term "so-called" is another standard downplayer. We might say, for example, that the woman who made the diagnosis is a "so-called doctor," which downplays her credentials as a physician. Quotation marks can be used to accomplish the same thing:

> She got her "degree" from a correspondence school.

Use of quotation marks as a downplayer is different from their use to indicate irony, as in this remark:

> John "borrowed" Hank's umbrella, and Hank hasn't seen it since.

The idea in the latter example isn't to downplay John's borrowing the umbrella; it's to indicate that it wasn't really a case of borrowing at all. But the use of quotation marks around the word "degree" and the use of "so-called" in the earlier examples are designed to play down the importance of their subjects. And, like "mere" and "merely," they do it in a fairly unsubtle way.

Many conjunctions—such as "nevertheless," "however," "still," and "but"—can be used to downplay claims that precede them. Such uses are more subtle than the first group of downplayers. Compare the following two versions of what is essentially the same pair of claims:

> (1) The leak at the plant was a terrible tragedy, all right; however, we must remember that such pesticide plants are an integral part of the "green revolution" that has helped to feed millions of people.

Don't Get Carried Away!

Once you're familiar with the ways slanting devices are used to try to influence us, you may be tempted to dismiss a claim or argument *just because it contains strongly slanted language.* But true claims as well as false ones, good reasoning as well as bad, can be couched in such language. Remember that the slanting *itself* gives us no reason to accept a position on an issue; that doesn't mean that there *are* no such reasons. Consider this example, written by someone opposed to using animals for laboratory research:

> It's morally wrong for a person to inflict awful pain on another sensitive creature, one that has done the first no harm. Therefore, the so-called scientists who perform their hideous and sadistic experiments on innocent animals are moral criminals just as were Hitler and his Nazi torturers.

Comment: These are strong words, and you'll find the passage contains more than one of the slanters we've discussed. But before we dismiss the passage as shrill or hysterical, it behooves us as critical thinkers to notice that it contains a piece of reasoning that may shed some light on the issue. A critical thinker who disagrees with the position taken by our speaker should do so because of a problem with the *argument* presented, not because of the strong *language* in which it is presented. (See if you can restate the reasoning in a somewhat less inflammatory way.)

(2) Although it's true that pesticide plants are an integral part of the "green revolution" that has helped to feed millions of people, it was just such a plant that developed a leak and produced a terrible tragedy.

The differences may not be as obvious as those in the cases of "mere" and "so-called," but the two versions give an indication of where their authors' sympathies lie.

The context of a claim can determine whether it downplays or not. Consider the remark "Daschle won by only six votes." The word "only" may or may not downplay Daschle's victory, depending on how thin a six-vote margin is. If ten thousand people voted and Daschle won by six, then the word "only" seems perfectly appropriate: Daschle just won by the skin of his teeth. But if the vote was in a committee of, say, twenty, then six is quite a substantial margin (it would be thirteen votes to seven, if everybody voted—almost two-to-one), and applying the word "only" to the result is clearly a slanting device designed to give Daschle's margin of victory less importance than it deserves.

As mentioned earlier, slanters really can't—and shouldn't—be avoided altogether. They can give our writing flair and interest. What *can* be avoided is being unduly swayed by slanters. Learn to appreciate the effects that subtle and not-so-subtle manipulations of language can have on you. By being

Sarcasm on the Editorial Page

Steve Benson. Reprinted by permission of United Feature Syndicate, Inc.

▲ *Background: This cartoon mocks the 1999 decision of the Kansas Board of Education to eliminate evolution as a principle of biological science. (The decision was later reversed.)*

aware, you decrease your chances of being taken in unwittingly by a clever writer or speaker.

■ *Horse Laugh/Ridicule/Sarcasm*

The kind of rhetorical device we call the **horse laugh** includes the use of ridicule of all kinds. Ridicule is a powerful rhetorical tool—most of us really hate being laughed at. So it's important to remember that somebody who simply gets a laugh at the expense of another person's position has not raised any objection to that position.

One may simply laugh outright at a claim ("Send aid to Russia? Har, har, har!"), laugh at another claim that reminds us of the first ("Support the Equal Rights Amendment? Sure, when the ladies start buying the drinks! Ho, ho, ho!"), tell an unrelated joke, use sarcastic language, or simply laugh at the person who is trying to make the point.

The next time you watch a debate, remember that the person who has the funniest lines and who gets the most laughs may be the person who *seems* to win the debate, but critical thinkers should be able to see the difference between argumentation on one hand and entertainment on the other. (Notice that we didn't say there's anything *wrong* with entertainment;

▲   *"Ari" refers to President George W. Bush's press secretary. Satire is a form of ridicule.*

just like most of you, we wouldn't like to spend *all* of our time watching people be serious, even if they *were* making good arguments.)

■ *Hyperbole*

Hyperbole is extravagant overstatement. A claim that exaggerates for effect is on its way to becoming hyperbole, depending on the strength of its language and the point being made. To describe a hangnail as a serious injury is hyperbole; so is using the word "fascist" to describe parents who insist that their teenager be home by midnight. Not all strong or colorful language is hyperbole, of course. "Oscar Peterson is an unbelievably inventive pianist" is a strong claim, but it is not hyperbolic—it isn't really extravagant. However, "Oscar Peterson is the most inventive musician who ever lived" goes beyond emphasis and crosses over the line into hyperbole. (How could one know that Oscar Peterson is more inventive than, say, Mozart?)

Dysphemisms often involve hyperbole. So do rhetorical comparisons. When we use the dysphemisms "traitorous" or "extremist" to describe the secretary of state's recommendations to Congress, we are indulging in hyperbole. If we say that the secretary of state is less well informed than a beet, that's hyperbole in a rhetorical comparison. In similar ways, rhetorical explanations and definitions can utilize hyperbole.

Hyperbole is also frequently used in ridicule. If it involves exaggeration, a piece of ridicule counts as hyperbole. The example above, saying that the secretary of state is less well informed than a beet, is hyperbole in a rhetorical comparison used to ridicule that official.

A claim can be hyperbolic without containing excessively emotive words or phrases. Neither the hangnail nor the Oscar Peterson examples contain such language; in fact, the word "unbelievably" is probably the most emotive word in the two claims about Peterson, and it occurs in the nonhyperbolic claim. But a claim can also be hyperbole as a result of the use of such language. "Parents who are strict about a curfew are fascists" is an example. If the word "mean" were substituted for "fascists," we might find the claim strong or somewhat exaggerated, but we would not call it hyperbole. It's when the colorfulness of language becomes *excessive*—a matter of judgment—that the claim is likely to turn into hyperbole.

Hyperbole is an obvious slanting device, but it can also have more subtle—perhaps unconscious—effects. Even if you reject the exaggeration, you may be moved in the direction of the basic claim. For example, you may reject the claim that Oscar Peterson is the most inventive musician who ever lived, but you may now believe that Oscar Peterson must certainly be an extraordinary musician—otherwise, why would someone make that exaggerated claim about him? Or suppose someone says, "Charlotte Church has the most fabulous voice of any singer around today." Even if you reject the "fabulous" part of the claim, you may still end up thinking Charlotte Church must have a pretty good voice. But be careful: Without support, you have no more reason to accept the milder claims than the wilder ones. Hyperbole can add a persuasive edge to a claim that it doesn't deserve. A hyperbolic claim is pure persuasion.

A feminazi is a woman to whom the most important thing in life is seeing to it that as many abortions as possible are performed.

—RUSH LIMBAUGH

A rhetorical definition with hyperbole.

■ *Proof Surrogates*

An expression used to suggest that there is evidence or authority for a claim without actually citing such evidence or authority is a **proof surrogate.** Sometimes we can't *prove* the claim we're asserting, but we can hint that there *is* proof available, or at least evidence or authority for the claim, without committing ourselves to what that proof, evidence, or authority is. Using "informed sources say" is a favorite way of making a claim more authoritative. Who are the sources? How do we know they're informed? How does the person making the claim know they're informed? "It's obvious that" sometimes precedes a claim that isn't obvious at all. But we may keep our objections to ourselves in the belief that it's obvious to everybody but us, and we don't want to appear more dense than the next guy. "Studies show" crops up in advertising a lot. Note that this phrase tells us nothing about how many studies are involved, how good they are, who did them, or any other important information.

Here's a good example of a proof surrogate from the *Wall Street Journal:*

There is no other country in the Middle East except Israel that can be considered to have a stable government. . . . Is Saudi Arabia more stable? Egypt? Jordan? Kuwait? Judge for yourself!

—"Facts and Logic About the Middle East"

Proof surrogates often take the form of questions. This strategy can also be analyzed as switching the burden of proof (see Chapter 6).

> We hope politicians on this side of the border are paying close attention to Canada's referendum on Quebec. . . .
>
> Canadians turned out en masse to reject the referendum. There's every reason to believe that voters in the U.S. are just as fed up with the social engineering that lumps people together as groups rather than treating them as individuals.

There *may* be "every reason to believe" that U.S. voters are fed up, but nobody has yet told us what any of those reasons are. Until we hear more evidence, our best bet is to figure that the quotation mainly reflects what the writer at the *Journal* thinks is the proper attitude for U.S. voters. Without a context, such assertions are meaningless.

Remember: Proof surrogates are just that—surrogates. They are not real proof or evidence. Such proof or evidence may exist, but until it has been presented, the claim at issue remains unsupported. At best, proof surrogates suggest sloppy research; at worst, they suggest propaganda.

Recap

Speakers and writers sometimes win acceptance for a claim or influence a person's attitude or behavior without presenting reasons. We call this rhetoric. A primary means of wielding such influence is the use of rhetorical devices—words or phrases that have positive or negative emotional associations, suggest favorable or unfavorable images, or manipulate the assumptions or expectations that always underlie communication. Common rhetorical devices include euphemisms and dysphemisms; rhetorical comparisons, definitions, and explanations; stereotypes; innuendo; loaded questions; weaselers; downplayers; horse laugh, ridicule, and sarcasm; hyperbole; and proof surrogates. Such devices are often used deliberately, but subtle uses can creep into people's speech or writing even when they think they are being objective. Some such phrases, especially euphemisms and weaselers, have both valuable, nonprejudicial uses and slanting ones. Only by speaking, writing, listening, and reading carefully can we use this type of language appropriately and distinguish prejudicial from nonprejudicial uses of these devices.

Exercises

Exercise 4-1

You will want to recognize when someone is using rhetorical slanting devices to influence your attitudes and beliefs. Let's see if you can identify some of the more common devices. Select the *best* answer.

▲ 1. A question on the "North Poll," a poll associated with the Oliver North talk radio program: "Are you outraged that President Clinton may be selling burial plots in Arlington National Cemetery to big DNC donors?" In which category does this quotation most clearly fall?
   a. dysphemism
   b. stereotype
   c. weaseler/innuendo
   d. hyperbole
   e. not a slanter

2. Larry Kudlow, of *Kudlow and Cramer* on CNBC (in an *American Spectator* interview): "[Former Treasury Secretary] Bob Rubin's a smart guy, a nice man, but he hates tax cuts. To listen to Rubin on domestic issues, you could just die. He's a free-spending left-winger." Which category applies best to the last phrase of the quotation?
   a. rhetorical comparison
   b. stereotype
   c. downplayer
   d. loaded question
   e. not a slanter

3. "Making a former corporate CEO the head of the Securities and Exchange Commission is like putting a fox in charge of the henhouse." This is best seen as an example of
   a. rhetorical comparison
   b. rhetorical explanation
   c. innuendo
   d. dysphemism
   e. not a slanter

▲ 4. "Even with no sleep the night before, Margaret was beautiful the day of the ceremony." The phrase "even with no sleep the night before" is best seen as
   a. innuendo
   b. euphemism
   c. hyperbole
   d. weaseler
   e. not a slanter

5. "Right. George Bush 'won' the election in 2000, didn't he?" The use of quotation marks around "won" has the effect of a
   a. weaseler
   b. dysphemism
   c. downplayer
   d. rhetorical explanation
   e. not a slanter

6. "'Democrat' equals 'ideologically homeless ex-communist.'"

   Linda Bowles

▲ 7. The obvious truth is bilingual education has been a failure." In this statement, "the obvious truth" might best be viewed as
   a. proof surrogate
   b. weaseler
   c. innuendo
   d. dysphemism
   e. not a slanter

8. After George W. Bush announced he wanted to turn a substantial portion of the federal government operation over to private companies, Bobby L. Harnage, Sr., president of the American Federation of Government Em-

ployees, said Bush had "declared all-out war on federal employees." Would
you say that the quoted passage is
a. a rhetorical explanation
b. a euphemism
c. a weaseler
d. hyperbole
e. not a slanter

9. "You say you are in love with Oscar but are you sure he's right for you?
Isn't he a little on the mature side?" This statement contains
a. loaded question
b. a euphemism
c. both a and b
d. neither a nor b

▲ 10. "Before any more of my tax dollars go to the military, I'd like answers to
some questions, such as why are we spending billions of dollars on weapons
programs that don't work?" This statement contains an example of
a. a downplayer
b. a dysphemism
c. a proof surrogate
d. a loaded question
e. hyperbole and a loaded question

11. "Can Governor Davis be believed when he says he will fight for the death
penalty? You be the judge." This statement contains
a. dysphemism
b. proof surrogate
c. innuendo
d. hyperbole
e. no slanters

▲ 12. "Which is it George W. Bush lied about, whether he used cocaine, when he
used cocaine, or how much cocaine he used?" This statement contains
a. hyperbole
b. dysphemism
c. loaded question
d. proof surrogate
e. no slanter

13. "Studies confirm what everyone knows: smaller classes make kids better
learners."

Bill Clinton

This statement contains:
a. proof surrogate
b. weaseler
c. hyperbole
d. innuendo
e. no slanter

14. MAN SELLING HIS CAR: "True, there's a little wear and tear, but what are a few dents?" This statement contains what might best be called
    a. loaded question
    b. innuendo
    c. dysphemism
    d. euphemism

▲ 15. MAN THINKING OF BUYING THE CAR IN EXERCISE 14, TO HIS WIFE: "Okay, okay, so it's got a few miles on it. Still, it may be the only Mustang in the whole country for that price." In this item, "few" and "still" could be said to belong to the same category of slanter. (T or F)

16. In Exercise 15, "it may be" is
    a. a weaseler
    b. a proof surrogate
    c. a downplayer
    d. not a slanter

17. Still in Exercise 15, "in the whole country" is an example of
    a. innuendo
    b. hyperbole
    c. a euphemism
    d. none of these

Exercise 4-2

▲ Determine which of the numbered, italicized words and phrases are used as rhetorical devices in the following passage. If the item fits one of the text's categories of rhetorical devices, identify it as such.

The National Rifle Association's campaign *to arm every man, woman, and child in America*[1] received a setback when the President signed the Brady Bill. But the *gun-pushers*[2] know that the bill was only *a small skirmish in a big war*[3] over guns in America. They can give up some of their more *fanatic*[4] positions on such things as *assault weapons*[5] and *cop-killer bullets*[6] and still win on the one that counts: regulation of manufacture and sale of handguns.

Exercise 4-3

▲ Follow the directions for Exercise 4-2.

The *big money guys*[1] who have *smuggled*[2] the Rancho Vecino development onto the November ballot *will stop at nothing to have this town run just exactly as they want.*[3] *It is possible*[4] that Rancho Vecino will cause traffic congestion on the east side of town, and *it's perfectly clear that*[5] the number of houses that will be built will overload the sewer system. *But*[6] a small number of individuals have taken up the fight. *Can the developers be stopped in their desire to wreck our town?*[7]

Exercise 4-4

Follow the directions for Exercise 4-2.

> The U.S. Congress has cut off funds for the superconducting supercollider that the *scientific establishment*[1] wanted to build in Texas. The *alleged*[2] virtues of the supercollider proved no match for the *huge*[3] *cost-overruns*[4] that had piled up *like a mountain alongside a sea of red ink*.[5] Despite original estimates of five to six billion dollars, the latest figure was over eleven billion and *growing faster than weeds*.[6]

Exercise 4-5

Read the passage below and answer the questions that follow it. Your instructor may have further directions.

> Not to contribute to millennial paranoia, but we do live in strange times. The president is being tried over an extramarital affair.
>
> I know, I know. I have heard Republican after Republican solemnly declaim that this is not about sex—it is about the rule of law, about the sanctity of oaths, about the preservation of the nation and about the retention of the very concept of virtue.
>
> I have heard from Republicans that this is about absolute truth vs. relative truth; it is about where we went wrong in the '60s; it is about high crimes and misdemeanors; and it is about the law treating the mightiest in the land with the same impartial hand as the lowliest among us.
>
> I have never listened to so much bilge in my life, and I have covered the Texas Legislature for 30 years. New records for overreaching are set hourly.
>
> *From a column by Molly Ivins* (Fort Worth Star-Telegram)

1. What issue is the author addressing?
2. What position does the author take on that issue?
3. If the author supports this position with an argument, state that argument in your own words.
4. Does the author use rhetorical devices discussed in this chapter? If so, identify the main types of devices used.

Exercise 4-6

Follow the directions for Exercise 4-5, using the same list of questions.

> Schools are not a microcosm of society, any more than an eye is a microcosm of the body. The eye is a specialized organ which does something that no other part of the body does. That is its whole significance. You don't use your eyes to lift packages or steer automobiles. Specialized organs have important things to do in their own specialties. So

schools, which need to stick to their special work as well, should not become social or political gadflies.

Thomas Sowell

Exercise 4-7

Follow the directions for Exercise 4-5, using the same list of questions.

"It is not the job of the state, and it is certainly not the job of the school, to tell parents when to put their children to bed," declared David Hart of the National Association of Head Teachers, responding to David Blunkett's idea that parents and teachers should draw up "contracts" (which you could be fined for breaching) about their children's behaviour, timekeeping, homework and bedtime. Teachers are apparently concerned that their five-to-eight-year-old charges are staying up too late and becoming listless truants the next day.

While I sympathise with Mr. Hart's concern about this neo-Stalinist nannying, I wonder whether it goes far enough. Is it not high time that such concepts as Bathtime, Storytime and Drinks of Water were subject to regulation as well? I for one would value some governmental guidance as to the number of humorous swimming toys (especially Hungry Hippo) allowable per gallon of water. Adopting silly voices while reading Spot's Birthday or Little Rabbit Foo-Foo aloud is something crying out for regulatory guidelines, while the right of children to demand and receive wholly unnecessary glasses of liquid after lights-out needs a Statutory Minimum Allowance.

John Walsh, The Independent

Exercise 4-8

Identify any rhetorical devices you find in the following selections, and classify those that fit the categories described in the text. For each, explain its function in the passage.

▲ 1. I trust you have seen Janet's file and have noticed the "university" she graduated from.

2. The original goal of the Milosovic government in Belgrade was ethnic cleansing in Kosovo.

3. "National Health Care: The compassion of the IRS and the efficiency of the post office, all at Pentagon prices."

From a letter to the editor, Sacramento Bee

▲ 4. Although it has always had a bad name in the United States, socialism is nothing more or less than democracy in the realm of economics.

5. We'll have to work harder to get Representative Burger reelected because of his little run-in with the law.

▲ 6. It's fair to say that, compared with most people his age, Mr. Beechler is pretty much bald.

   7. During World War II, the U.S. government resettled many people of Japanese ancestry in internment camps.

▲ 8. "Overall, I think the gaming industry would be a good thing for our state."
   From a letter to the editor, Plains Weekly Record

   9. Morgan has decided to run for state senator. I'm sorry to hear that he's decided to become a politician.

  10. I'll tell you what capitalism is: Capitalism is Charlie Manson sitting in Folsom Prison for all those murders and still making a bunch of bucks off T-shirts.

▲ 11. Clearly, Antonin Scalia is the most corrupt Supreme Court Justice in the history of the country.

  12. In a February 1 article, writer Susan Beahay says Bush's abortion decision will return abortions to secrecy, risking the mother's life having a back-alley abortion. That's really juicy. The ultra-left pro-abortion crowd sure can add a little levity into a deadly serious subject.

  13. It may well be that many faculty members deserve some sort of pay increase. Nevertheless, it is clearly true that others are already amply compensated.

▲ 14. I wouldn't say he copied her answers. I'd say he adjusted his own answers.

  15. I love some of the bulleting and indenting features of Microsoft Word. I think it would have been a nice feature, however, if they made it possible to turn some of them off when you don't need them.

Exercise 4-9

Identify any rhetorical devices you find in the following passages, and explain their purposes. Note: Some items may contain *no* rhetorical devices.

▲ 1. "If the United States is to meet the technological challenge posed by Japan, Inc., we must rethink the way we do everything from design to manufacture to education to employee relations."
   Harper's

   2. According to UNICEF reports, several thousand Iraqi children died each month because of the U.N. sanctions.

   3. Maybe Professor Lankirshim's research hasn't appeared in the first-class journals as recently as that of some of the other professors in his department; that doesn't necessarily mean his work is going downhill. He's still a fine teacher, if the students I've talked to are to be believed.

▲ 4. "Let's put it this way: People who make contributions to my campaign fund get access. But there's nothing wrong with constituents having access to their representatives, is there?"
   Loosely paraphrased from an interview with a California state senator

5. In the 2000 presidential debates, Al Gore consistently referred to his own tax proposal as a "tax plan" and to George W. Bush's tax proposal as a "tax scheme."

6. "This healthy food provides all the necessary ingredients to help keep your bird in top health."

—Hartz Parrot Diet

▲ 7. [*Note:* Dr. Jack Kevorkian was instrumental in assisting a number of terminally ill people in committing suicide during the 1990s.] "We're opening the door to Pandora's Box if we claim that doctors can decide if it's proper for someone to die. We can't have Kevorkians running wild, dealing death to people."

Larry Bunting, assistant prosecutor, Oakland County, Michigan

8. "LOS ANGELES—Marriott Corp. struck out with patriotic food workers at Dodger Stadium when the concession-holder ordered them to keep working instead of standing respectfully during the National Anthem. . . . Concession stand manager Nick Kavadas . . . immediately objected to a Marriott representative.

   "Marriott subsequently issued a second memo on the policy. It read: 'Stop all activities while the National Anthem is being played.'

   "Mel Clemens, Marriott's general manager at the stadium, said the second memo clarified the first memo."

Associated Press

9. These so-called forfeiture laws are a serious abridgment of a person's constitutional rights. In some states, district attorneys' offices only have to *claim* that a person has committed a drug-related crime to seize the person's assets. So fat-cat DAs can get rich without ever getting around to proving that anybody is guilty of a crime.

▲ 10. "A few years ago, the deficit got so horrendous that even Congress was embarrassed. Faced with this problem, the lawmakers did what they do best. They passed another law."

Abe Mellinkoff, in the San Francisco Chronicle

11. "[U]mpires are baseball's designated grown-ups and, like air-traffic controllers, are paid to handle pressure."

George Will

12. "Last season should have made it clear to the moguls of baseball that something still isn't right with the game—something that transcends residual fan anger from the players' strike. Abundant evidence suggests that baseball still has a long way to go."

Stedman Graham, Inside Sports

▲ 13. "As you know, resolutions [in the California State Assembly] are about as meaningful as getting a Publishers' Clearinghouse letter saying you're a winner."

Greg Lucas, in the San Francisco Chronicle

14. The entire gain in the stock market in the first four months of the year was due to a mere fifty stocks.

15. I believe that no matter what Saddam Hussein does—including agreeing to let the inspectors back in—Bush will go to war against him. And now there's nothing to stop him. Congress has already rolled over.

16. "[Supreme Court Justice Antonin] Scalia's ideology is a bald and naked concept called 'Majoritarianism.' Only the rights of the majority are protected."

    Letter to the editor of the San Luis Obispo Telegram-Tribune

17. "Mimi Rumpp stopped praying for a winning lottery ticket years ago. . . . But after a doctor told her sister Miki last year that she needed a kidney transplant, the family began praying for a donor. . . . Less than a year later, Miki has a new kidney, courtesy of a bank teller in Napa, Calif., to whom she had told her story. The teller was the donor; she was so moved by Miki's plight she had herself tested and discovered she was a perfect match. Coincidence? Luck? Divine intervention? Rumpp is sure: 'It was a miracle.'"

    Newsweek

▲ 18. "We are about to witness an orgy of self-congratulation as the self-appointed environmental experts come out of their yurts, teepees, and grant-maintained academic groves to lecture us over the impending doom of the planet and agree with each other about how it is evil humanity and greedy 'big business' that is responsible for it all."

    Tim Worstall, in New Times

19. "In the 1980s, Central America was awash in violence. Tens of thousands of people fled El Salvador and Guatemala as authoritarian governments seeking to stamp out leftist rebels turned to widespread arrests and death squads."

    USA Today

Exercise 4-10

Discuss the following stereotypes in class. Do they invoke the same kind of images for everyone? Which are negative and which are positive? How do you think they came to be stereotypes? Is there any "truth" behind them?

1. soccer mom
2. Religious Right
3. dumb blonde
4. tax-and-spend liberal
5. homosexual agenda
6. redneck
7. radical feminist
8. contented housewife
9. computer nerd
10. tomboy
11. interior decorator
12. Washington insider
13. Earth mother
14. frat rat
15. Deadhead
16. trailer trash

Exercise 4-11 _____

Your instructor will give you three minutes to write down as many positive and negative stereotypes as you can. Are there more positive stereotypes on your list or more negative ones? Why do you suppose that is?

Exercise 4-12 _____

Write two brief paragraphs describing the same person, event, or situation—that is, both paragraphs should have the same informative content. The first paragraph should be written in a *purely* informative way, using language that is as neutral as possible; the second paragraph should be slanted as much as possible either positively or negatively (your choice).

Exercise 4-13 _____

Explain the difference between a weaseler and a downplayer. Find a clear example of each in a newspaper, magazine, or other source. Next find an example of a phrase that is sometimes used as a weaseler or downplayer but that is used appropriately or neutrally in the context of your example.

Exercise 4-14 _____

Explain how rhetorical definitions, rhetorical comparisons, and rhetorical explanations differ. Find an example of each in a newspaper, magazine, or other source.

Exercise 4-15 _____

Look through an issue of *Time, Newsweek,* or another newsmagazine, and find a photograph that portrays its subject in an especially good or bad light—that is, one that does a nonverbal job of creating slant regarding the subject.

Exercise 4-16 _____

In groups, invent two newspaper or magazine headlines to go with this photograph of Senator Hillary Rodham Clinton. Make one headline create a favorable impression in combination with the photo and make the other create an unfavorable impression.

Writing Exercises ————————————————————————

1. Turn to Selection 9 in Essays for Analysis, "Will Ozone Blob Devour the Earth?" and identify as many rhetorical devices as you can find. (Your instructor may narrow the scope of the assignment to just certain paragraphs.)

2. Over the past decade, reportedly more than 2,000 illegal immigrants have died trying to cross the border into the southwestern United States. Many deaths have resulted from dehydration in the desert in the desert heat and from freezing to death on cold winter nights. A San Diego–based nonprofit humanitarian organization now leaves blankets, clothes, and water at stations throughout the desert and mountain regions for the immigrants. Should the organization do this? Its members say they are providing simple humanitarian aid, but critics accuse them of encouraging illegal activity. Take a stand on the issue and defend your position in writing. Then identify each rhetorical device you used.

3. Until recently, tiny Stratton, Ohio, had an ordinance requiring all door-to-door "canvassers" to obtain a permit from the mayor. Presumably the ordinance was intended to protect the many senior citizens of the town from harm by criminals who might try to gain entry by claiming to be conducting a survey. The ordinance was struck down by the Jehovah's Witnesses, who thought it violated their First Amendment right to free speech. The Supreme Court agreed and struck down the law in 2002. Should it have? Defend your position in a brief essay without using rhetoric. Alternatively, defend your position and use rhetorical devices, but identify each device you use.

Chapter 5

More Rhetorical Devices: Psychological and Related Fallacies

*E*very now and then we hear snatches of the syndicated national talk radio program called *Savage Nation.* The show, which also airs on MSNBC, stars Michael Savage, who has a Ph.D. in epidemiology and nutrition science from the University of California, Berkeley (according to his Web page).

Mr. Savage is perhaps the most vociferously angry of all national talk radio hosts; his opponents refer to his program as "hate radio." And it does seem undeniable that Savage's default position is outrage.

Aristotle warned against making decisions when the emotions are inflamed and cautioned us to be wary of orators who, like Mr. Savage, try to bring us to a boil with rage or stir up some powerful emotion. In this chapter we examine several rhetorical devices a speaker or writer might employ to tempt us to substitute knee-jerk emotional responses for sound judgment and careful thinking.

These devices go beyond the rhetorical coloration we talked about in the last chapter. They can be made to look like reasonable arguments containing premises and conclusions. But they don't really provide legitimate proof of what they supposedly are proving. The "reasons" they proffer for their "conclusions" do not really support those conclusions. A good argument provides a justification for accepting its conclusion. The devices in this chapter tempt us with considerations that are emotionally or psychologically linked to an issue, but they don't really support the claims they are supposed to support. They offer only *pretended* support; you might think of them as pieces of pretend reasoning, or *pseudo*reasoning.

The devices in this chapter thus all count as fallacies (a fallacy is a mistake in reasoning). The rhetorical devices we discussed in the last chapter—euphemisms, innuendo, and so forth—aren't fallacies. Of course, we commit a fallacy if we think a claim has been supported when the "support" is nothing more than rhetorically persuasive language.

People constantly accept fallacies as legitimate arguments; but the reverse mistake can also happen. We must be careful not to dismiss *legitimate* arguments as fallacies just because they *remind* us of a fallacy. Often beginning students in logic have this problem. They read about fallacies like the ones we cover here and then think they see them everywhere. These fallacies are common, but they are not everywhere; and you sometimes must consider a specimen carefully before accepting or rejecting it. The exercises at the end of the chapter will help you learn to do this, because they contain a few reasonable arguments mixed in with the fallacies.

The Rush Limbaugh Fallacy: The "Argument" from Outrage

If you tune in Mike Savage or Rush Limbaugh (who is still better known than Savage), chances are the first thing you will hear (if not a commercial) is a voice quivering with disbelief and indignation, saying something like, "It's your money . . . and the liberals want to do *what* with it? Spend it on *midnight basketball leagues*??!"

Whether the speaker is the host or someone calling in, he or she will be "mad as hell" (to borrow an expression popular in the 1980s). And the speaker wants you to be mad too—wants you to rise up in righteous indignation and *smite down* the basketball league and every other liberal proposal you run into.

Now, in case you think we are just being anti-Rush, anticonservative, or anti-Republican, we'll be getting around to the Democrats in a bit. Democrat talk-show hosts use the same techniques as Rush Limbaugh, although unfortunately for them, they aren't nearly as well known.

Surely, you might say, if someone is wasting your money, don't you have a right to be angry? The answer is "yes": Anger is not a fallacy, and sometimes it is appropriate to be angry. However, when we are angry (especially if we are in a blind rage), it's easy to become illogical in two very predictable ways. *First, we may think we have been given a reason for being angry when in fact we have not.* Is spending money on midnight basketball leagues a waste of money? Well, you need to investigate to see. If a midnight basketball league reduces crime in a neighborhood, then maybe it isn't a waste of money. It is a mistake to think that something is wrong simply because it makes us angry. Presumably we are angry because it is wrong, not the other way around.

Second, we may let the anger we feel as the result of one thing influence our evaluations of an unrelated thing. The next liberal proposal to come along must be examined on its own merits, and we shouldn't let anger over the midnight basketball leagues poison our evaluation of it. If a person, Johnson, has done something to anger us, that wouldn't be a reason for downgrading Johnson's performance on some *other* matter, nor would it be a reason for upgrading our opinion of someone else.

Rush Limbaugh Hyperventilates

Reprinted by permission of Tim Brinton.

What happens when Rush Limbaugh attacks those of us in public life is that people aren't just content to listen. People want to act because they get emotional . . . and the threats to those of us in public life go up dramatically, against us and our families, and it's very disconcerting.

—SENATOR TOM DASCHLE (D-N.D.)

What more do you want to do to destroy this country than what you've already tried? It is unconscionable what this man has done. This stuff gets broadcast around the world, Senator. What do you want your nickname to be? Hanoi Tom? Tokyo Tom? You sit there and pontificate on the fact that we're not winning the war on terrorism when you and your party have done nothing but try to sabotage it.

—RUSH LIMBAUGH, responding the next day

The **"argument" from outrage,** then, consists in inflammatory words (or thoughts) followed by a "conclusion" of some sort. It substitutes anger for reason and judgment in considering an issue. It is a favorite strategy of demagogues. In fact, it is *the* favorite strategy of demagogues. Let's say the issue is whether gay marriages should be legal. Left-of-center demagogues may wax indignantly about "narrow-minded fundamentalist bigots dictating what people can do in their bedrooms"—talk calculated to get us steamed although it really has nothing to do with the issue. On the other side, conservative demagogues may allude to gays' demanding "special rights." No-

Another Angry American

▲ *Recording artists such as Toby Keith sometimes try to boost popularity by riding the patriotism bandwagon.*

body wants someone else to get special rights, and when we hear about somebody "demanding" them, our blood pressure goes up. But wanting a right other people have is not wanting a special right; it's wanting an equal right.

A particularly dangerous type of "argument" from outrage is known as **scapegoating**—blaming a certain group of people—or even a single person (like George W. Bush or Bill Clinton) for all of life's troubles. George Wallace, the former governor of Alabama who ran for president in 1968 on a "states' rights platform" (which then was a code word for white supremacy) said he could get good old Southern boys to do anything by whooping them into a frenzy over Northern civil rights workers.

"Arguments" based on outrage are so common that the fallacy ranks high on our list of the top ten fallacies of all time, which is contained in Appendix 2. It's unfortunate they are so common—history demonstrates constantly that anger is a poor lens through which to view the world. Policies adopted in anger are seldom wise, as any parent will tell you who has laid down the law in a fit of anger.

The idea behind [talk radio] is to keep the base riled up.

—Republican political advisor Brent Lauder, explaining what talk radio is for ["the base" refers to the Republican rank and file]

Scare Tactics by Omission

The June 2002 risk assessment estimates that there would be a 95% probability that lead made airborne by the projected housing development would result in an increase of 3.7 to 4.3 µg/dL in the lead levels in children's blood.

—From a newsletter by a group opposed to a projected housing development

Sounds scary, all that lead going into kids' blood. However, here's what the June 2002 risk assessment really said:

The June 2002 risk assessment estimates that there would be a 95% probability that lead made airborne by the projected housing development would result in a child blood level of 3.7 to 4.3 µg/dL, well below the 10 µg/dL level of concern.

—Office of Environmental Health Hazard Assessment

Moral: Scare tactics can involve both distorting and omitting.

Scare Tactics

George Wallace didn't just try to anger the crowds when he told them what Northern civil rights workers were up to; he tried to *scare* them. When people become angry or afraid, they don't think clearly. They follow blindly. Demagogues like Wallace like to dangle scary scenarios in front of people.

Trying to scare people into doing something or accepting a position is using **scare tactics.** One way this might be done is the George Wallace method—dangling a frightening picture in front of someone. A simpler method might be just to threaten the person, a special case of scare tactics known as **"argument" by force.** Either way, if the idea is to get people to substitute fear for reason and judgment when taking a position on an issue, it is a fallacy. Likewise, it is a fallacy to succumb to such techniques when others use them on us.

Fear can befuddle us as easily as can anger, and the mistakes that happen are similar in both instances. Wallace's listeners may not have noticed (or not cared) that Wallace didn't actually give them *proof* that civil rights workers were doing whatever it was he portrayed them as doing; the portrayal was its own evidence, you might say. When we are befuddled with fear, we may not notice we lack evidence that the scary scenario is real. Imagine someone talking about global warming: The speaker may paint a picture so alarming we don't notice that he or she doesn't provide evidence that global warming is actually happening. Or take gay marriages again. Someone might warn us of presumably dire consequences if gay people are allowed to marry—we'll be opening "Pandora's box"; marriage will become meaningless; homosexuality will become rampant; society will collapse—but he or she may issue these warnings without providing details as to why (or how) the consequences might actually come about. The consequences are so frightening they apparently don't need proof.

Fear of one thing, X, may also affect evaluation of an unrelated thing, Y. You have your eye on a nice house and are considering buying it, and then the real estate agent frightens you by telling you the seller has received other offers and will sell soon. Some people, in this situation, might overestimate what they really can afford to pay.

To avoid translating fear of one thing into an evaluation of some unrelated thing, we need to be clear on what issues our fears are relevant to. Legitimate warnings do not involve irrelevancies and do not qualify as scare tactics. "You should be careful of that snake—it's deadly poisonous" might be a scary thing to say to someone, but we don't make a mistake in reasoning when we say it, and neither does the other person if he or she turns and runs into the house. Suppose, however, that the Michelin tire people show an ad featuring a sweet (and vulnerable) baby in a ring of automobile tires. Showing pictures of car tires around infants will produce disquieting associations in any observer, and it wouldn't be unreasonable to check our tires when we see this ad. But the issue raised by the Michelin people is whether to buy *Michelin* tires, and the fear of injuring or killing a child by driving on unsafe tires does not bear on the question of *which* tires to buy. The Michelin ad isn't a legitimate warning; it's scare tactics.

Other Fallacies Based on Emotions

Other emotions work much like anger and fear as sources of mistakes in reasoning. *Compassion,* for example, is a fine thing to have. There is absolutely nothing wrong with feeling sorry for someone. But when feeling sorry for someone drives us to a position on an unrelated matter, the result is the fallacy known as **"argument" from pity.** We have a job that needs doing; Helen can barely support her starving children and needs work desperately. But does Helen have the skills we need? We may not care if she does; and if we don't, nobody can fault us for hiring her out of compassion. But feeling sorry for Helen may lead us to misjudge her skills or overestimate her abilities, and that is a mistake in reasoning. Her skills are what they are regardless of her need. Or, suppose you need a better grade in this course to get into law school or avoid academic disqualification or whatever. If you think you *deserve* or have *earned* a better grade because you need a better grade, or you try to get your instructor to think you deserve a better grade by trying to make him or her feel sorry for you, that's the "argument" from pity. Or, if you think someone *else* deserves a better grade because of the hardships he or she (or his or her parents) suffered, that's also the "argument" from pity.

Envy and *jealousy* can also confuse our thinking. Compassion, a desirable emotion, may tempt us to emphasize a person's good points; envy and

Knee Operation Judged Useless

Fake surgery worked just as well in cases of osteoarthritis.

Here we are doing all this surgery on people and it's all a sham.
—DR. BARUCH BRODY, Baylor College of Medicine

Wishful thinking—allowing our desires and hopes to color our beliefs and influence our judgment—is common indeed. A powerful illustration of wishful thinking is the placebo effect, where subjects perceive improvement in a medical condition when they receive what they think is a medication but in fact is an inactive substance. Even surgical procedures, apparently, are subject to a placebo effect, judging from a study of a popular and expensive knee operation for arthritis. People who have had this procedure swear by it as significantly reducing pain. But researchers at the Houston Veterans Affairs Medical Center and Baylor College of Medicine discovered that subjects who underwent placebo (fake) surgery said exactly the same thing. Furthermore, when they tested knee functions two years after the surgery the researchers discovered that the operation doesn't improve knee functions at all.

Source: *Sacramento Bee*, July 11, 2002. From *New York Times* News Service.

jealousy tempt us to exaggerate someone's bad points. When we find fault with a person because of envy, we are guilty of the fallacy known as **"argument" from envy.** "Well, he may have a lot of money but he certainly has bad manners" would be an example of this if it is envy that prompts us to criticize him.

Pride, on the other hand, can lead us to exaggerate *our own* accomplishments and abilities and lead to our making other irrelevant judgments as well. It especially makes us vulnerable to **apple polishing.** Moore recently sat on a jury in a criminal case involving alleged prostitution and pandering at a strip club; the defendant's attorney told the members of the jury it would take *"an unusually discerning jury"* to see that the law, despite its wording, wasn't really intended to apply to someone like his client. Ultimately the jury members did find with the defense, but let us hope it wasn't because the attorney flattered their ability to discern things. Allowing praise of oneself to substitute for judgment about the truth of a claim, or trying to get others to do this, as the lawyer did, is the apple polishing fallacy.

Feelings of *guilt* work similarly. "How could you not invite Trixie to your wedding? She would never do that to you and you know she must be very hurt." The remark is intended to make someone feel sorry for Trixie, but even more fundamentally it is supposed to induce a sense of guilt. Eliciting feelings of guilt to get others to do or not do something, or to accept

Positive Outlook Won't Delay Cancer Death, Study Says

NICE, France — New research has dealt a blow to the idea that a positive outlook might improve a patient's chances of surviving cancer, scientists said Saturday.

However, experts said it is still worthwhile for patients to improve their attitude, perhaps by joining a cancer support group, because often it does make them feel better.

The findings were presented Saturday at a meeting of the European Society of Medical Oncology in Nice, France. The researchers reviewed evidence to determine whether psychologist-run support groups kept patients alive.

"There were some studies out there showing that positive-thinking type of support will not only improve your quality of life—which undoubtedly it does, I'm not questioning that— but also will prolong the lives of cancer patients," said Dr. Edzard Ernst, a professor of complementary medicine at the University of Exeter in England who led the study.

"One study from 1989 gets cited over and over and over again, and we knew there were one or two negative studies on this, too, so we decided to see if it was true," he said.

The researchers analyzed 11 studies that included a total of 1,500 patients.

"The data provided no evidence at all to show that these types of approaches prolong life in cancer patients," Ernst said.

Associated Press

More wishful thinking, apparently.

Source: *Sacramento Bee,* October 19, 2002.

the view that they should or should not do it, is popularly known as putting a **guilt trip** on someone, which is to commit a fallacy. Parents sometimes use this tactic with children when they (the parents) won't (or can't) offer a clear explanation of why something should or shouldn't be done. Certainly, if the child knowingly does something wrong, he or she should feel guilty; but whatever has been done isn't wrong *because* he or she feels guilty.

Hopes, desires, and aversions can also lead us astray logically. The fallacy known as **wishful thinking** happens when we accept or urge acceptance (or rejection) of a claim simply because it would be pleasant (or unpleasant) if it were true. Some people, for example, may believe in God simply on the basis of wishful thinking or desire for an afterlife. A smoker may refuse to acknowledge the health hazards of smoking. We've had students who are in denial about the consequences of cutting classes. The wishful thinking fallacy also underlies much of the empty rhetoric of "positive thinking"—rhetoric that claims "you are what you want to be" and other such slogans.

Most people desire to be liked or accepted by some circle of other people and are averse to having the acceptance withdrawn. A *desire for acceptance* can motivate us to accept a claim not because of its merits, but because we will gain someone's approval (or will avoid having approval withdrawn). When we do this, or try to get someone else to do it, the fallacy is the **peer** pressure "argument." Now, obviously nobody ever said anything quite so blatant as "Ralph, this claim is true because we won't like you any more if you don't accept it." Peer pressure is often disguised or unstated, but anyone going through an American high school, where you can lose social standing merely by being seen with someone who isn't "in," knows it is a real force. Kids who feel ostracized sometimes take guns to school.

It doesn't have to be one's associates who exert peer pressure, either. In scientific experiments, people will actually revise what they say they saw if a group of strangers in the same room deny having seen the same thing.

One very common fallacy that is closely related to the peer pressure "argument" involves one's sense of *group identification,* which people experience when they are part of a group—a team, a club, a school, a gang, a state, a nation, the Elks, Wal-Mart, the U.S.A., Mauritius, you name it. Let's define the **group think fallacy** as happening when one substitutes pride of membership in a group for reason and deliberation in arriving at a position on an issue; and let's include the fallacy in our list of the top ten fallacies of all time, because it is exceedingly common. One obvious form of this fallacy involves national pride, or **nationalism**—a powerful and fierce emotion that can lead to blind endorsement of a country's policies and practices. ("My country right or wrong" explicitly discourages critical thinking and encourages blind patriotism.) Nationalism is also invoked to reject, condemn, or silence criticism of one's country as unpatriotic or treasonable (and may or may not involve an element of peer pressure). If a letter writer expresses a criticism of America on the opinion page of your local newspaper on Monday, you can bet that by the end of the week there will be a response dismissing the criticism with the "argument" that if so-and-so doesn't like it here, he or she ought to move to Russia (or Cuba or Afghanistan or Iraq).

Group think does not play cultural or political favorites, either. On the opposite side of the political spectrum are what some people call the "blame America first" folks. The group think ethic of this club includes, most importantly, automatically assuming that whatever is wrong in the world is the result of some U.S. policy. The club has no formal meetings or rules for membership, but flying an American flag would be grounds for derision and instant dismissal.

Group think "reasoning" is certainly not limited to political groups either. It occurs whenever one's affiliations are of utmost psychological importance.

Now, these various emotional fallacies, from the "argument" from outrage to the group think fallacy, all share certain properties. They often (though not always) contain assertions you might call "premises" and other assertions that you might call a "conclusion." But the "premises" don't actually *support* the "conclusion"; rather, they evoke emotions that make us want to accept the conclusion without support. So, although they can wear the clothing of *arguments,* they are really pieces of *persuasion* (Chapter 1). Whenever language is used to arouse emotions, it is wise to consider carefully whether any "conclusions" that come to mind have been supported by evidence.

Rationalizing

Let's say Mr. Smith decides to do something really nice for his wife on her birthday and buys her a new table saw. "This saw wasn't cheap," he tells her. "But you're going to be glad we have it, because it will keep me out in the garage and out of your way when you're working here in the house."

The fallacy in the reasoning in this made-up example is pretty obvious. Mr. Smith is confusing his wife's desires with his own.

When we do this, when we use a false pretext to satisfy our own desires or interests, we're guilty of **rationalizing,** a very common fallacy. It almost made our list of the top ten fallacies of all time (Appendix 2).

Now, there is nothing wrong with satisfying one's desires, at least if they don't harm someone or aren't illegal. But in this book we're talking logic, not morals. Rationalizing involves a confusion in thinking, and to the extent we wish to avoid being confused in our thinking, we should try to avoid rationalizing.

"But," you may be saying, "It is good to do nice things for other people. If you do something that helps them, or that they like, or that benefits the world, what difference does motivation make? If, for whatever reason, the table saw makes Mr. Smith's wife happy, that's what counts."

Now, there is something to be said for this argument, because it is good to make people happy. But whether Mr. Smith's wife is happy or not, there has been a confusion in his thinking, a fallacy. And it is a common fallacy indeed. Obviously most instances of rationalizing are not as blatant as Mr. Smith's, but people frequently deceive themselves as to their true motives.

Rationalizing need not be selfish, either. Let's say a former oilman is elected governor of a state that produces oil. He may act in what at some level he thinks are the best interests of his state—when in fact he is motivated by a desire to help the oil industry. (Incidentally, you can't just assume he would do this.) To the extent he is deceiving himself about his true motivation, he is rationalizing. But this isn't *selfish* rationalizing; his actions don't benefit him personally.

Rationalizing, then, involves an element of self-deception, but otherwise it isn't necessarily devious. However, some people encourage others to rationalize because they themselves stand to benefit in some way. "Hey, Smith," his buddy Jones says to him. "That's a fine idea! Really creative. Your wife will really like a saw. Maybe you could build a boat for her, and you and I could go fishing." Jones may or may not say this innocently: If he does, he too is guilty of rationalizing; if he doesn't, he's just cynical.

Everyone Knows . . .

In Chapter 4 we examined proof surrogates like "Everyone knows . . ." and "It's only common sense that. . . ." Phrases like this are used when a speaker or writer doesn't really have an argument.

An "Argument" from Unpopularity

If they're going to talk about "the facts," here's a fact: All the major religions of the world consider homosexuality wrong.

—JANEY PARSHALL, spokesperson for the Family Research Council

Background: Ms. Parshall is responding to a joint announcement by the American Academy of Pediatrics, the National Education Association, and the American Psychological Association that there is no support among health and mental health professional organizations for the idea that homosexuality is abnormal or mentally unhealthy.

Such phrases often appear in peer pressure "arguments" ("Pardner, in these parts everyone thinks . . ."). They also are used in the group think fallacy ("As any red-blooded American patriot knows, . . ."). There is, however, a third way these phrases can be used. An example would be when Robert Novak says on CNN's *Crossfire*, "Liberals are finally admitting what everyone knows, that airline safety demands compromise." Novak isn't applying or evoking peer pressure or group think; he is offering "proof" that airline safety demands compromise (and bad-mouthing liberals to boot). His proof is the fact that everyone knows it.

When we do this, when we urge someone to accept a claim (or fall prey to someone's doing it to us) simply on the grounds that all or most or some substantial number of people (other than authorities or experts, of course) believe it, we commit the fallacy known as the **"argument" from popularity.** That most people believe something is a fact is not evidence that it is a fact—most people believe in God, for example, but that isn't evidence that God exists. Likewise, if most people didn't believe in God, that wouldn't be evidence that God doesn't exist.

Most people seem to assume that bus driving and similar jobs are somehow less desirable than white-collar jobs. The widespread acceptance of this assumption creates its own momentum—that is, we tend to accept it because everybody else does, and we don't stop to think about whether it actually has anything to recommend it. For a lot of people, a job driving a bus might make for a much happier life than a job as a manager.

In *some* instances, we should point out, what people think actually *determines* what is true. The meanings of most words, for example, are determined by popular usage. In addition, it would not be fallacious to conclude that the word "ain't" is out of place in formal speech because most speakers of English believe that it is out of place in formal speech.

There are other cases where what people think is an *indication* of what is true, even if it cannot *determine* truth. If several Bostonians of your acquaintance think that it is illegal to drink beer in their public parks, then you have some reason for thinking that it's true. And if you are told by several Europeans that it is not gauche to eat with your fork in your left hand

The Great White Van

While searching for the serial sniper who terrorized the Washington, D.C., area in the fall of 2002, police encountered the suspects' blue Chevrolet Caprice several times in road checkpoints and once with the sniper asleep in the car on a side street in Baltimore. However, the vehicle did not arouse police suspicions, apparently because it did not fit the beliefs that were forming in the case. As one senior government official put it, everyone was focused on finding an angry white guy in a white van. A witness had reported seeing a white van fleeing a shooting, and then other witnesses at other shootings began to see a white van, too. (One report even turned out to have been fabricated.) One eyewitness did report seeing a fleeing Caprice, but the Caprice sighting just didn't catch on. By the end of the case the white van theory had gained so much momentum that another person who saw a Caprice didn't file a report until after the suspects were apprehended. He said he assumed the sniper's vehicle was a white van. *Time* magazine even printed a picture of the white van the police were looking for.

In Chapter 3 we pointed out that this incident shows how eyewitness accounts can be untrustworthy. Here, we see the influence of popular belief in why that's true.

in Europe, then it is not fallacious to conclude that European manners allow eating with your fork in your left hand. The situation here is one of credibility, which we discussed in Chapter 3. Natives of Boston in the first case and Europeans in the second case can be expected to know more about the two claims in question, respectively, than others know. In a watered-down sense, they are "experts" on the subjects, at least in ways that many of us are not. In general, when the "everyone" who thinks that X is true includes experts about X, then what they think is indeed a good reason to accept X.

Thus it would be incorrect to automatically label as a fallacy any instance in which a person cites people's beliefs to establish a point. (No "argument" fitting a pattern in this chapter should *unthinkingly* be dismissed.) But it is important to view such references to people's beliefs as red alerts. These are cautionary signals that warn you to look closely for genuine reasons in support of the claim asserted.

Two variations of the "argument" from popularity deserve mention: **"Argument" from common practice** consists in trying to justify or defend an *action* or *practice* (as distinguished from an assertion or claim) on the grounds that it is common. "I shouldn't get a speeding ticket because everyone drives over the limit" would be an example. "Everyone cheats on their taxes, so I don't see why I shouldn't" would be another. Now there is something to watch out for here: When a person defends an action by saying that other people do the same thing, he or she might just be requesting fair play. He or she might just be saying, in effect, "O.K., O.K., I know it's wrong, but nobody else gets punished and it would be unfair to single me out." That person isn't trying to justify the action; he or she is asking for equal treatment.

The other variant of the argument from popularity is the **"argument" from tradition,** a name that is self-explanatory. People do things because that's the way things have always been done, and they believe things because that's what people have always believed. But, logically speaking, you don't prove a claim or prove a practice is legitimate on the basis of tradition; when you try to do so, you are guilty of "argument" from tradition. The fact that it's a tradition among most American children to believe in Santa Claus, for instance, doesn't prove Santa Claus exists; and the fact it's also a tradition for most American parents to deceive their kids about Santa Claus doesn't necessarily mean it is okay for them to do so.

Relativism

We talked about relativism in Chapter 1. **Relativism** is the theory that "truth is relative" or, more precisely, this: The very same claim may be true in one culture and false in another; it all depends on what the people within the culture think. If, for example, one culture thinks flu is caused by a virus and another culture thinks flu is caused by the flu-god, then in the first culture flu is caused by a virus and in the second culture it isn't—according to a relativist. Relativists, obviously, don't think the "argument" from popularity is a fallacy. What is true *is* what people within a culture think is true, according to them. (More sophisticated versions of relativism can be found that don't sound as silly as this example; that doesn't make the basic idea any more true, as we will see in a moment.)

Now, when it comes to claims that are about moral or ethical values, such as "It is good to help others" or "Adultery is immoral," relativism has a degree of plausibility. That's because claims like these are nonfactual (see Chapter 1), and that means that if one person thinks adultery is immoral and another doesn't, we don't require one of them to be mistaken. Likewise, if one *culture* thinks adultery is immoral and the next one doesn't, we let each culture have its way on the subject. But moral relativism—the idea that the selfsame moral judgment (like "Adultery is immoral") can be true in one culture and false in another—is a difficult and controversial subject in moral philosophy, because the statement "Adultery is immoral" seems to imply that the claim holds true for people in other cultures too. We talk about moral relativism in Chapter 12, but it is primarily a subject for a course in ethics.

When it comes to *factual* claims, claims about straightforward factual matters (like "Flu is caused by a virus"), at first glance relativism seems less plausible. A factual claim, if true, seems to be true regardless of any particular culture's acceptance of it. However, relativists have sophisticated ways of defending their position. Is flu caused by a virus, as Culture A holds, or is it caused by the flu-god, as Culture B holds? Well, relativists argue, Culture A has its own standards of what counts as relevant evidence, and it has its own standards of what counts as proof; it has its own standards of what counts as true and false and its own standards of what counts as sound argument, and these standards may be different from the standards of Culture B. It is meaningless, relativists argue, to say that one culture's standards of truth and relevance and evidence and proof and argument are more correct than another's. Therefore, it is meaningless to suppose that Culture A's views on the flu are more correct than Culture B's.

This is a clever argument, but defective. The trouble is that it undermines itself. Whatever standards of correct argumentation it assumes, it "proves" that those standards are no more correct than alternative standards. It "proves," in short, that we shouldn't take it seriously.

Underneath relativism is the assumption that what is true for one person is not necessarily true for another, an assumption known as **subjectivism.** A problem with subjectivism, and any doctrine that assumes it, is that it is impossible to make any sense of it, at least when it comes to factual claims. What could it mean to think that a claim, like "The room has a window in it" or "You have type-A flu virus," could be true for Mr. Wong and not true for Ms. Mbutu? Could the room have a window if Wong believes it does and at the same time not have a window if Mbutu believes otherwise? It doesn't matter what standards of evidence each person has or what each person thinks "true" means; the room has a window or it doesn't, and if Mbutu and Wong have different opinions on the subject, one is mistaken.

The bottom line is that if we believe that a factual claim may be true for one person or culture and false for another person or culture, we make a mistake, commit a fallacy. "That may be true for you, but it is not true for me" must be understood as a short way of saying, "Maybe you accept the claim, but I don't." Alternatively, it might be interpreted as a way of saying "Let's not talk about this anymore." If others use the "true for you, not true for me" remark to insist that it's correct for you to believe a factual claim and equally correct for them to reject the very same claim, they have fallen prey to subjectivism and relativism.

Two Wrongs Make a Right

Let's say you get tired of the people upstairs stomping around late at night, and so, to retaliate, you rent a tow truck and deposit their car in the river. From an emotional standpoint, you're getting even. From a reasoning standpoint, you're committing the fallacy known as **"two wrongs make a right."** It's a fallacy because wrongful behavior on someone else's part doesn't convert wrongful behavior on your part into rightful behavior, any more than illegal behavior on someone else's part converts your illegal activity into legal activity. If an act is wrong, it is wrong. Wrong acts don't cross-pollinate such that one comes out shorn of wrongfulness.

However, there is a well-known and somewhat widely held theory known as *retributivism*, according to which it is acceptable to harm someone in return for a harm he or she has done to you. But we must distinguish legitimate punishment from illegitimate retaliation. A fallacy clearly occurs when we consider a wrong to be justification for *any* retaliatory action, as would be the case if you destroyed your neighbors' car because they made too much noise at night. It is also a fallacy when the second wrong is directed at someone who didn't do the wrong in the first place—a brother or a child of the wrongdoer, for example. And it is a fallacy to defend doing harm to another on the grounds that that individual *would* or *might* do the same to us. This would happen, for example, if we didn't return excess change to a salesclerk on the grounds that "if the situation were reversed," the clerk wouldn't have given us back the money.

On the other hand, it isn't a fallacy to defend an action on the grounds it was necessary to prevent harm from befalling oneself; bopping a mugger to prevent him from hurting you would be an instance. To take another example, near the end of World War II, the United States dropped two atomic bombs on Japanese cities, killing tens of thousands of civilians. Politicians, historians, and others have argued that the bombing was justified because it helped end the war and thus prevented more casualties from the fighting, including the deaths of more Americans. People have long disagreed on whether the argument provides *sufficient* justification for the bombings, but there is no disagreement about its being a real argument and not empty rhetoric.

Red Herring/Smokescreen

When a person brings a topic into a conversation that distracts from the original point, especially if the new topic is introduced in order to distract, the person is said to have introduced a **red herring.** (It is so called because dragging a herring across a trail will cause a dog to leave the original trail and follow the path of the herring.) In the strip-joint jury trial we mentioned earlier, the defendant was charged with pandering; but the prosecuting at-

"Yes, yes, I *know* that, Sidney—*every*body knows *that!* ... But look: Four wrongs *squared*, minus two wrongs to the fourth power, divided by this formula, *do* make a right."

torney introduced evidence that the defendant had also sold liquor to minors. That was a red herring that had nothing to do with pandering.

The difference between red herrings and their close relatives, **smokescreens,** is subtle (and really not a matter of crucial importance). Generally speaking, red herrings distract by pulling one's attention away from one topic and toward another; smokescreens tend to pile issues on or to make them extremely complicated until the original is lost in the (verbal) "smoke." When Bill Clinton had missiles fired at terrorists in Sudan, he was accused of creating a red herring to deflect public scrutiny from the Monica Lewinsky

Red Herrings in Presidential Politicking

Al Gore: I believe there are 1.4 million children in Texas who do not have health insurance; 600,000 of whom . . . were actually eligible for it, but they couldn't sign up for it because of the barriers that they had to surmount.

Jim Lehrer [Moderator]: Let's let the governor respond to that. . . . Are those numbers correct? Are his charges correct?

George W. Bush: If he's trying to allege that I'm a hard-hearted person and I don't care about children, he's absolutely wrong. We spent $4.7 billion a year in the state of Texas for uninsured people and they get health care . . . somehow the allegation that we don't care and we're going to give money for this interest or that interest and not for children in the state of Texas is totally absurd. And let me tell you who the jury is: the people of Texas. There's only been one governor ever elected to back-to-back four-year terms, and that was me.

—From the 2000 presidential debates

business. When George W. Bush talked about Iraq having missiles capable of threatening the United States, about that country's potential of having a nuclear weapon "within six months," and about similar possible Iraqi threats, he was accused of putting up a smokescreen to hide his real reasons for wanting to attack Iraq, which were said to be oil interests and his own personal desire to complete his father's unfinished business.

Let's look at another example, this one made up but fairly typical of what often happens. Let's say a reporter asks Tom Ridge (secretary of the Department of Homeland Security) whether his office has made the country substantially safer from attacks by terrorists. "I'm pleased to say," Ridge answers, "that the United States is the safest country in the world when it comes to terrorist attacks. Certainly nobody can give an absolute, one hundred percent guarantee of safety, but you are certainly safer here than in any other country of the world."

Ridge has steered clear of the original question (whether his agency had made the country safer) and is leading the reporter on a tangent, toward the comparative safety of the United States (the United States may already have been the safest country before the creation of the agency). He has dragged a red herring across the trail, so to speak.

We admit that this measure is popular. But we also urge you to note that there are so many bond issues on this ballot that the whole concept is getting ridiculous.

—A generic red herring (unclassifiable irrelevance) from a California ballot pamphlet

A Red Herring in a Letter to TIME

Time's coverage of the medical marijuana controversy was thoughtful and scrupulously researched. But what argues most persuasively for a ban on marijuana is the extraordinary threat the drug poses for adolescents. Marijuana impairs short-term memory, depletes energy and impedes acquisition of psychosocial skills. Perhaps the most chilling effect is that it retards maturation for young people. A significant number of kids who use lots of pot simply don't grow up. So it is hardly surprising that marijuana is the primary drug for more than half the youngsters in the long-term residential substance-abuse programs that Phoenix House operates throughout the country.

—*Mitchell S. Rosenthal, M.D.*
President, Phoenix House
New York City

The issue is legalization of marijuana for *adults;* the question of what it would do to children, who presumably would be prohibited from its use, is a red herring.

Source: *Time*, November 28, 2002

Imagine the conversation continues this way:

Reporter: "Mr. Ridge, polls say about half of the public think your agency has failed to make them safer. How do you answer your critics?"

Tom Ridge: "We are making progress toward reassuring people, but quite frankly our efforts have been hampered by the tendency of the press to concentrate on the negative side of the issue."

Once again Ridge brings in a red herring to sidestep the issue raised by the reporter.

Whether a distraction or an obfuscation is a plain red herring or a smokescreen is often difficult to tell in real life, and it's better to spend your energy getting a discussion back on track rather than worrying which type you have before you.

Many of the other fallacies we have been discussing in this chapter (and will be discussing in the next chapter) qualify, in some version or other, as red herrings/smokescreens. For example, a defense attorney might talk about a defendant's miserable upbringing to steer a jury's attention away from the charges against the person; doing this would qualify as an argument from pity as well as a smokescreen/red herring. Likewise, a prosecuting attorney may try to get a jury so angry about a crime it doesn't notice the weakness of the evidence pointing to the defendant. This would be an argument from outrage—and a red herring.

To simplify things, your instructor may reserve the red herring/smokescreen categories for irrelevancies that don't qualify as one of the other

Could somebody please show me one hospital built by a dolphin? Could somebody show me one highway built by a dolphin? Could someone show me one automobile invented by a dolphin?

—Rush Limbaugh, responding to the *New York Times'* claim that dolphins' "behavior and enormous brains suggest an intelligence approaching that of human beings."

Good point. Anyone know of a hospital or highway built by Rush Limbaugh or an automobile invented by him?

fallacies mentioned in this or the next chapter. In other words, he or she may tell you that if something qualifies as, say, an argument from outrage, you should call it that rather than a red herring or smokescreen.

Recap

In a recent interview with CNN's Connie Chung (photo below), tennis champion Martina Navratilova asserted that when she left Communist Czechoslovakia for the United States she changed one system that suppresses free opinion for another. Connie Chung told Navratilova to go ahead and think that at home, but asserted that celebrities shouldn't "spill out" such thoughts in public, because "people will write it down and talk about what you said." (Chung thus ineptly confirmed the very point Navratilova was making.)

One can only speculate as to what exactly was going on in Connie Chung's head, if anything. Maybe she was worried that Navratilova's comment would make people think bad things about the United States. Maybe she thinks the tennis star's comment is unpatriotic. Maybe criticism of the United States just upsets her. Whatever her thoughts, the example nicely illustrates what we have been talking about in this chapter. Sometimes, instead of bringing forth considerations relevant to an issue, people give an unrelated "argument." Many of the fallacies we have examined are like Connie Chung's: The unrelated argument involves some kind of emotion, though it may be hard to pin down exactly what it is.

Thus we have

- ▲ "Argument" from outrage
- ▲ Scare tactics
- ▲ "Argument" by force
- ▲ "Argument" from pity
- ▲ "Argument" from envy
- ▲ Apple polishing
- ▲ Guilt trip
- ▲ Wishful thinking
- ▲ Peer pressure "argument"
- ▲ Group think fallacy
- ▲ Nationalism

Other fallacies discussed in this chapter don't invoke emotions directly, but are closely related to emotional appeals. These include

- ▲ Rationalization
- ▲ "Argument" from popularity
- ▲ "Argument" from common practices
- ▲ "Argument" from tradition
- ▲ Relativism
- ▲ Subjectivism
- ▲ Red herring/smokescreen

In all these specimens, there is something one might call a "premise" and something one might call a "conclusion," but the "premise" either fails to support the conclusion or "supports" some tangential claim. In any case a mistake in reasoning has been made, a fallacy has been committed.

Exercises

In the exercises that follow, we ask you to name fallacies, and your instructor may do the same on an exam.

Exercise 5-1

Working in groups, invent a simple, original, and clear illustration of each type of fallacy covered in this chapter. Then, in the class as a whole, select the illustrations that are clearest and most straightforward. Go over these illustrations before doing the remaining exercises in this chapter, and review them before you take a test on this material.

Exercise 5-2

Identify any instances of fallacious reasoning that occur in the following passages, either by naming them or, where you think they do not conform to any of the patterns we have described, by explaining in one or two sentences why the "argument" is irrelevant to the point at issue. (There are a few passages that contain no fallacies. Be sure you don't find one where none exist!)

▲ 1. The tax system in this country is unfair and ridiculous! Just ask anyone!

2. SHE: I think it was exceedingly boorish of you to finish off the last of their expensive truffles like that.
   HE: Bosh. They certainly would have done the same to us, if given the chance.

3. Overheard:

"Hmmmm. Nice day. Think I'll go catch some rays."

"Says here in this magazine that doing that sort of thing is guaranteed to get you a case of skin cancer."

"Yeah, I've heard that, too. I think it's a bunch of baloney, personally. If that were true, you wouldn't be able to do anything—no tubing, skiing, nothing. You wouldn't even be able to just plain lay out in the sun. Ugh!"

▲ 4. I've come before you to ask that you rehire Professor Johnson. I realize that Mr. Johnson does not have a Ph.D., and I am aware that he has yet to publish his first article. But Mr. Johnson is over forty now, and he has a wife and two high-school-aged children to support. It will be very difficult for him to find another teaching job at his age, I'm sure you will agree.

5. JUAN: But, Dad, I like Horace. Why shouldn't I room with him, anyway?

JUAN'S DAD: Because I'll cut off your allowance, that's why!

6. That snake has markings like a coral snake. Coral snakes are deadly poisonous, so you'd better leave it alone!

▲ 7. RALPH: He may have done it, but I don't hold him responsible. I'm a determinist, you know.

SHARON: What's that?

RALPH: A determinist? Someone who doesn't believe in free will. There's no free will.

SHARON: Oh. Well, I disagree.

RALPH: Why's that?

SHARON: Because. Determinism may be real for you, but it certainly isn't for me.

8. The animal rights people shouldn't pick on rodeos. If they'd come out and see the clowns put smiles on kids' faces and see horses buck off the cowboys and hear the crowd go "ooh" and "ahh" at the bull riding, why then they'd change their minds.

9. HE: Tell you what. Let's get some ice cream for a change. Sunrise Creamery has the best—let's go there.

SHE: Not that old dump! What makes you think their ice cream is so good, anyway?

HE: Because it is. Besides, that old guy who owns it never gets any business any more. Every time I go by the place I see him in there all alone, just staring out the window, waiting for a customer. He can't help it that he's in such an awful location. I'm sure he couldn't afford to move.

▲ 10. Student speaker: "Why, student fees have jumped by more than 300 percent in just two years! This is outrageous! The governor is working for a balanced budget, but it'll be on the backs of us students, the people who have the very least to spend! It seems pretty clear that these increased student fees are undermining higher education in this state. Anybody who isn't mad about this just doesn't understand the situation."

11. "Jim, I'm very disappointed you felt it necessary to talk to the media about the problems here in the department. When you join the FBI, you join a family, and you shouldn't want to embarrass your family."

12. HANK: Where I come from, there are lots of outdoor cafés and bars; people think nothing of drinking in public.
    FRANK: That may be true, but this is Salt Lake City, and they disapprove of that sort of thing very seriously around here.

▲ 13. A fictitious Western governor: "Yes, I have indeed accepted $550,000 in campaign contributions from power companies. But as I stand here before you, I can guarantee you that not one dime of that money has affected any decision I've made. I make decisions based on data, not on donors."

14. "If you ask me, you are making a mistake to break up with Rasheed. Have you forgotten how he stood by you when you needed him last year? Is this how you repay him?"

15. "What? You aren't a Cornhuskers fan? Listen, around here *everybody* is for the Huskers! This is Nebraska!"

Exercise 5-3 _____

Answer the following questions and explain your answers.

▲ 1. How relevant is the fact that a brand of toothpaste is advertised to the issue of whether to buy that brand?

   2. How relevant is the fact that a brand of toothpaste *is* best-selling to the issue of whether to buy that brand?

▲ 3. How relevant is the fact that an automobile is a best-seller in its class to the issue of whether to buy that kind of automobile?

   4. How relevant is the fact that a movie is a smash hit to the issue of whether to see it?

   5. How relevant is the fact that a movie is a smash hit to your opinion of it?

   6. How relevant is the fact that your friends like a movie to the issue of whether to see it?

▲ 7. How relevant is the fact that your friends like a movie to your opinion of it?

   8. How relevant is the fact that your friends like a movie to the issue of whether to say that you like it?

   9. How relevant is the fact that movie critics like a movie to the issue of whether to see it?

▲ 10. Is advertising a product as best-selling an example of the peer pressure "argument"?

Exercise 5-4

Which of the following do you believe? Which of the following do you *really* have evidence for? Which of the following do you believe on an "everyone knows" basis? Discuss your answers with other members of your class.

1. Small dogs tend to live longer than large dogs.
2. Coffee has a dehydrating effect.
3. Most people should drink at least eight glasses of water a day.
4. If you are thirsty, it means you are already dehydrated.
5. Rape is not about sex; it's about aggression.
6. Marijuana use leads to addiction to harder drugs.
7. The news media are biased.
8. You get just as much ultraviolet radiation on a cloudy day as on a sunny day.
9. If you don't let yourself get angry every now and then, your anger will build up to the exploding point.
10. Carrots make you see better.
11. Reading in poor light is bad for your eyes.
12. Sitting too close to the TV is bad for your eyes.
13. Warm milk makes you sleepy.
14. Covering your head is the most effective way of staying warm in cold weather.
15. Smoking a cigarette takes seven minutes off your life.

Exercise 5-5

For each of the passages that follow, determine whether fallacies are present and, if so, whether they fit the categories described in this chapter.

▲ 1. Boss to employee: "I'll be happy to tell you why this report needs to be finished by Friday. If it isn't ready by then, you'll be looking for another job. How's that for a reason?"

2. Mother: "I think he has earned an increase in his allowance. He doesn't have any spending money at all, and he's always having to make excuses about not being able to go out with the rest of his friends because of that."

3. Mother to father: "You know, I really believe that our third-grader's friend Joe comes from an impoverished family. He looks to me as though he doesn't get enough to eat. I think I'm going to start inviting him to have dinner at our house once or twice a week."

▲ 4. "Aw, c'mon, Jake, let's go hang out at Dave's. Don't worry about your parents; they'll get over it. You know, the one thing I really like about you is that you don't let your parents tell you what to do."

5. FIRST PERSON: You know, I might not agree with it, but I could understand it if a society decided to look down on a woman who had a child out of wed-lock. But stoning to death? My God, that's barbaric and hideously immoral!
SECOND PERSON: But remember that you come from a background much different from that of the people in that part of Nigeria. It's less immoral in that situation. Besides, in Iran stoning to death has been a common pun-ishment for adultery under the current regime.

6. FRED: I think we should just buy the new truck and call it a business ex-pense so we can write it off on our taxes.
ETHEL: I don't know, Fred. That sounds like cheating to me. We wouldn't really use the truck very much in the business, you know.
FRED: Oh, don't worry about it. This kind of thing is done all the time.

▲ 7. I'm going to use the textbook that's on reserve in the library. I'll have to spend more time on the campus, but it's sure better than shelling out over a hundred bucks for one book.

8. Imagine yourself alone beside your broken-down car at the side of a coun-try road in the middle of the night. Few pass by and no one stops to help. Don't get caught like that—don't get caught without your Polytech cellu-lar phone!

9. One political newcomer to another: "I tell you, Sam, you'd better change those liberal views of yours. The general slant toward conservatism is obvi-ous. You'll be left behind unless you change your mind about some things."

▲ 10. Reporter COKIE ROBERTS: Mr. Cheney, aside from the legal issues that stem from the various United Nations resolutions, isn't there an overriding moral dimension to the suffering of so many Kurdish people in Iraq?
DICK CHENEY: Well, we recognize that's a tragic situation, Cokie, but there are tragic situations occurring all over the world.

Adapted from an interview on National Public Radio's Morning Edition

Exercise 5-6

For each of the passages that follow, determine whether fallacies are present and, if so, whether they fit the categories described in this chapter.

▲ 1. "Grocers are concerned about *sanitation problems* from beverage residue that Proposition 11 could create. Filthy returned cans and bottles—*over 11 billion a year*—don't belong in grocery stores, where our food is stored and sold. . . . Sanitation problems in other states with similar laws have caused increased use of *chemical sprays* in grocery stores to combat rodents and insects. Vote no on 11."

Argument against Proposition 11, California ballot pamphlet

2. Overheard: "I'm not going to vote for Riordan for governor, and you shouldn't either. I lived in L.A. when Riordan was mayor, and the crime was so bad we had to leave. You couldn't walk around alone, for fear of your life, and I mean this was the middle of the day!"

3. STUDENT: I think I deserve a better grade than this on the second question.
PROF: Could be. Why do you think so?
STUDENT: You think my answer's wrong.
PROF: Well, your answer *is* wrong.
STUDENT: Maybe you think so, but I don't. You can't mark me wrong just because my answer doesn't fit your opinion.

▲ 4. C'mon, George, the river's waiting and everyone's going to be there. You want me to tell 'em you're gonna worry on Saturday about a test you don't take 'til Tuesday? What're people going to think?

5. ATTENDANT: I'm sorry, sir, but we don't allow people to top off their gas tanks here in Kansas. There's a state law against it, you know.
RICHARD: What? You've got to be kidding! I've never heard of a place that stopped people from doing that!

6. One roommate to another: "I'm telling you, Ahmed, you shouldn't take Highway 50 this weekend. In this weather, it's going to be icy and dangerous. Somebody slides off that road and gets killed nearly every winter. And you don't even have any chains for your car!"

▲ 7. That, in sum, is my proposal, ladies and gentlemen. You know that I trust and value your judgment, and I am aware I could not find a more astute panel of experts to evaluate my suggestion. Thank you.

8. JARED: In Sweden, atheists and agnostics outnumber believers 2 to 1, and in Germany, less than half the population believes in God. Here in the United States, though, over 80 percent believe in God. I wonder what makes the States so different.
ALICE: You've answered your own question. If I didn't believe in God, I'd feel like I stuck out like a sore thumb.

9. Businessman to partner: "I'm glad Brownell has some competition these days. That means when we take estimates for the new job, we can simply ignore his, no matter what it is. That'll teach him a lesson for not throwing any business our way last year."

▲ 10. One local to another: "I tell you, it's disgusting. These idiot college students come up here and live for four years—and ruin the town—and then vote on issues that affect us long after they've gone. This has got to stop! I say, let only those who have a real stake in the future of this town vote here! Transient kids shouldn't determine what's going to happen to local residents. Most of these kids come from Philadelphia . . . let them vote there."

Exercise 5-7

For each of the passages that follow, determine whether fallacies are present and, if so, whether they fit the categories described in this chapter.

▲ 1. Chair, Department of Rhetoric (to department faculty): "If you think about it, I'm certain you'll agree with me that Mary Smith is the best candidate for department secretary. I urge you to join with me in recommending her to the administration. Concerning another matter, I'm now setting up next semester's schedule, and I hope that I'll be able to give you all the classes you have requested."

2. NELLIE: I really don't see anything special about Sunquist grapefruit. They taste the same as any other grapefruit to me.
   NELLIE'S MOM: Hardly! Don't forget that your Uncle Henry owns Sunquist. If everyone buys his fruit, you may inherit a lot of money some day!

3. Letter to the editor: "It's unfortunate that there are so many short-sighted people who don't want to increase the military budget. There may be little threat from Russia these days, but they aren't the only ones we have to worry about. Most of the countries of the world are jealous of our success and would like to take what we've got for themselves. It'll be our military that keeps them from doing it."

   Miltonville Gazette

▲ 4. *"Don't risk letting a fatal accident rob your family of the home they love—on the average more than 250 Americans die each day because of accidents.* What would happen to your family's home if you were one of them?
   Your home is so much more than just a place to live. It's a community you've chosen carefully . . . a neighborhood . . . a school district . . . the way of life you and your family have come to know. And you'd want your family to continue sharing its familiar comforts, even if suddenly you were no longer there. . . . Now, as a Great Western mortgage customer, you can protect the home you love. . . . Just complete the Enrollment Form enclosed for you."

   Advertisement from Colonial Penn Life Insurance Company

5. "You've made your mark and your scotch says it all."

   Glen Haven Reserve

6. Dear Senator Jenkins,
   I am writing to urge your support for higher salaries for state correctional facility guards. I am a clerical worker at Kingsford Prison, and I know whereof I speak. Guards work long hours, often giving up weekends, at a dangerous job. They cannot afford expensive houses or even nice clothes. Things that other state employees take for granted, like orthodontia for their children and a second car, are not possibilities on their salaries, which, incidentally, have not been raised in five years. Their dedication deserves better.
   Very truly yours, . . .

▲ 7. In *Shelley v. Kraemer,* 334 U.S.1 (1948), the "argument" was put before the Supreme Court that "state courts stand ready to enforce restrictive covenants excluding white persons from the ownership or occupancy of property covered by such agreements," and that therefore "enforcement of covenants excluding colored persons may not be deemed a denial of equal protection of the laws to the colored persons who are thereby affected."

The court decided that "this contention does not bear scrutiny." In fact, the contention seems to be an example of what form of pseudoreasoning?

8. The suggestion was once made to replace the word "manpower" in the course description of Dartmouth's Business Administration 151, Management of Human Resources, because of the sexist connotation of the word. This brought delighted responses in the press. Shouldn't they get rid of "management" in the course title, wrote someone, in favor of "personagement"? Shouldn't "human" in the title give way to "huperson," asked another? Is the course open to "freshpersons"? a third wondered. In fact, is the course open to any person? the same individual queried. "Son" is a masculine word; therefore "person" itself is sexist.

9. There are very good reasons for the death penalty. First, it serves as a deterrent to those who would commit capital offenses. Second, it is just and fair punishment for the crime committed. Third, reliable opinion polls show that over 70 percent of all Americans favor it. If so many people favor it, it has to be right.

▲ 10. Two famous political opponents are said to have had this exchange in the English Parliament:
GLADSTONE: You, sir, will either die upon the gallows or succumb to a horrible social disease!
DISRAELI: That will depend, I suppose, upon whether I embrace your principles or your mistress.

11. Frankly, I think the Salvation Army, the Red Cross, and the Wildlife Fund will put my money to better use than my niece Alison and her husband would. They've wasted most of the money I've given them. So I think I'm going to leave a substantial portion of my estate to those organizations instead of leaving it all to my spendthrift relatives.

12. "The President's prosecution of the war on terror is being handled exactly right. He wasn't elected to do nothing!"

13. Student to teacher: "I've had to miss several classes and some quizzes because of some personal matters back home. I know you have a no-make-up policy, but there was really no way I could avoid having to be out of town; it really was not my fault."

▲ 14. BUD: So here's the deal. I'll arrange to have your car "stolen," and we'll split the proceeds from selling it to a disposer. Then you file a claim with your insurance company and collect from it.
LOU: Gee, this sounds seriously illegal and dangerous.
BUD: Illegal, yeah, but do you think this is the first time an insurance company ever had this happen? Why, they actually expect it—they even budget money for exactly this sort of thing.

15. Kibitzer, discussing the job Lamar Alexander did as secretary of education: "It was absolutely clear to me that Alexander was not going to do any good for American education. He was way too involved in money-making schemes to give any attention to the job *we* were paying him for. Do you know that back before he was appointed, he and his wife invested five thou-

sand dollars in some stock deal, and four years later that stock was worth over eight hundred thousand dollars? Tell me there's nothing fishy about a deal like that!"

16. My opponent, the evolutionist, offers you a different history and a different self-image from the one I suggest. While I believe that you and I are made in the image of God and are only one step out of the Garden of Eden, he believes that you are made in the image of a monkey and are only one step out of the zoo.

▲ 17. "EPA study—Environmental Protection Agency officials have launched a new study to assess the risks of—brace yourself—taking showers. Seems someone in the EPA worries people might be injured by inhaling water vapor while taking a shower. No, we didn't make it up. It's true, according to a publication called EPA Watch.

   "Doubtless, several bureaucrats will take home paychecks for months while the imagined hazard is studied. Unfortunately, there's probably a 50-50 chance someone in Washington will decide showers really are dangerous.

   "Wonder if anyone at the EPA ever has studied the hazards of pouring billions of dollars into overregulation?"

Charleston (W. Va.) Gazette

18. "Boomers beware! The 76 million people born between 1946 and 1964 are beginning to think about retirement. They'd better listen carefully. Douglas Bernheim, an economics professor at Princeton, says current retirees were 'extraordinarily lucky' in that their home values climbed, high inflation took the sting out of their fixed-rate mortgages, and there were big increases in private and public pensions. 'The average baby boomer must triple his or her rate of savings to avoid a precipitous decline of living standards during retirement,' Bernheim said. . . .

   "To be on the safe side, baby boomers should have an aggressive savings plan and not rely on government assurances of cushy retirement years. It is always best to err on the side of caution."

Charleston (W. Va.) Daily Mail

Writing Exercises

1. Find an example of a fallacy in a newspaper editorial or opinion magazine (substitute an example from an advertisement or a letter to the editor only as a last resort and only if your instructor permits it). Identify the issue and what side of the issue the writer supports. Explain why the passage you've chosen does not really support that position—that is, why it involves a fallacy. If the writer's claims do support some other position (possibly on a different, related issue), describe what position they do support.

2. In 1998 the police in Harris County, Texas, responded to a false report about an armed man who was going crazy. They did not find such an individual; but when they entered the home of John Geddes Lawrence, they found him

and another man, Tyron Garner, having sex. Both men were arrested and found guilty of violating a Texas law that criminalizes homosexual sex acts. The men challenged their conviction, and the case went to the United States Supreme Court in March 2003. A district attorney from the county argued, "Texas has the right to set moral standards of its people."

Do you agree or disagree with the district attorney's statement? Defend your answer in a one-page essay written in class. Your instructor will have other members of the class read your essay to see if they can find your basic argument in the midst of any rhetoric you may have used. They also will note any fallacies that you may have employed.

3. Should there be an amendment to the U.S. Constitution prohibiting desecration of the U.S. flag? In a one-page essay, defend a "yes" or "no" answer to the question. Your instructor will have other members of the class read your essay, following the instructions in Exercise 2.

Chapter 6

More Fallacies

W hat is the most common (and seductive) error in reasoning on the planet? You are about to find out. In this chapter we examine the infamous *argumentum ad hominem,* as well as other common fallacies.

To remind you of the overall picture, in Chapter 4 we explored ways the rhetorical content of words and phrases can be used to affect belief and attitude. In Chapter 5 we considered emotional appeals and related fallacies. The fallacies we turn to now, like the devices in the preceding chapters, can tempt us to believe something without giving us a legitimate reason for doing so.

The Ad Hominem Fallacy

The ad hominem fallacy (*argumentum ad hominem*) is the most common of all mistakes in reasoning. The fallacy rests on a confusion between the qualities of the person making a claim and the qualities of the claim itself. ("Claim" is to be understood broadly here, as including beliefs, opinions, positions, arguments, proposals and so forth.)

Parker, let us assume, is an ingenious fellow. It follows that Parker's opinion on some subject, whatever it is, is the opinion of an ingenious person. But it does not follow that Parker's *opinion itself* is ingenious. To think that it does follow would be to confuse the content of Parker's claim with Parker himself. Or let's suppose you are listening to somebody, your teacher perhaps, whom you regard as a bit strange or maybe even weird. Would it follow that the *car* your teacher drives is strange or weird? Obviously not. Likewise, it would not follow that some specific proposal that the teacher has put forth is strange or weird. A proposal made by an oddball is an oddball's proposal, but it does not follow that it is an oddball proposal. We must

181

not confuse the qualities of the person making a claim with the qualities of the claim itself.

Now, we commit the **ad hominem** fallacy when we think that considerations about a person "refute" his or her assertions. *Ad hominem* is Latin for "to the man," indicating it is not really the subject matter that's being addressed, but the person. The most common varieties of the ad hominem fallacy are as follows.

■ *Personal Attack Ad Hominem*

"Johnson has such-and-such a negative feature; therefore, his claim (belief, opinion, theory, proposal, etc.) stands refuted." This is the formula for the **personal attack ad hominem** fallacy. The name "personal attack" is self-explanatory, because attributing a negative feature to Johnson is attacking him personally.

Now there are many negative features that we might attribute to a person: Perhaps Johnson is said to be ignorant or stupid. Maybe he is charged with being self-serving or feathering his own nest. Perhaps he is accused of being a racist or a sexist or a fascist or a cheat or of being cruel or uncaring or soft on communism or prone to kick dogs or what-have-you. The point to remember is that shortcomings in *a person* are not equivalent to shortcomings in that person's ideas, proposals, theories, opinions, claims, or arguments.

Now it is true that there are exceptional circumstances we can imagine in which some feature of a person might logically imply that what that person says is false; but these circumstances tend to be far-fetched. "Johnson's claim is false because he has been paid to lie about the matter" might qualify as an example. "Johnson's claim is false because he has been given a drug that makes him say only false things" would qualify too. But such situations are rare. True, when we have doubts about the credibility of a source, we must be careful before we accept a claim from that source. But the doubts are rarely sufficient grounds for outright rejection of the claim. No matter what claim Johnson might make and no matter what his faults might be, we are rarely justified in rejecting the claim as false simply because he has those faults.

■ *The Inconsistency Ad Hominem*

"Moore's claim is inconsistent with something else Moore has said or done; therefore, his claim (belief, opinion, theory, proposal, etc.) stands refuted." This is the formula for the **inconsistency** ad hominem, and you encounter versions of this fallacy all the time. Suppose a political commentator exclaims (as we heard Rush Limbaugh say about George W. Bush), "The president says now that he believes in global warming, but ladies and gentlemen, when the president was campaigning he scoffed at the idea." Do we have a reason here for thinking something is wrong with the president's current view? Not at all. The fact that people change their minds has no bearing on the truth of what they say either before or after.

They believe the Boy Scouts' position on homosexuality was objectionable, but they gave no heed to people's objections about using state money to fund displays about sodomy in the people's Capitol.

—California Assemblyman Bill Leonard (R-San Bernardino), criticizing the legislature for funding a gay pride display in the state's Capitol

Man! As if sodomy in the people's Capitol isn't bad enough, they have to go and fund displays about it!

Leonard's remark is an example of an inconsistency ad hominem. (It also contains a wild syntactical ambiguity, as noted above.)

Ye Olde Double Standards Argument

Beware the old "double standards" argument. It comes in various disguises. For example, in the furious discussion of whether the United States should go to war with Iraq, it surfaced in various "rebuttals" of the administration's contention that invading Iraq was justified:

- "America has supported far worse dictators than Saddam. Take the shah, for example."
- "If we are so concerned about the spread of nuclear weapons, why don't we do something about Israel?"
- "I just want to know, why don't we apply the same standard to North Korea?"
- "Twenty years ago we were supporting Saddam! You know how he got all those weapons? Most of them he bought from us."
- "Tell me this. We know for a fact that Iran is developing nuclear weapons. We also know Iran has supported terrorism. Why aren't we going to war with Iran?"

The double standard argument is actually the inconsistency ad hominem. As such, it has zero logical force either for or against doing some specific thing. That the U.S. government adhered to a double standard, if true, only proves its policies are inconsistent. It does not prove which policy is the incorrect policy.

Sometimes a person's claim seems inconsistent, not with previous statements but with that person's behavior. For example, Johnson might tell us to be more generous, when we know Johnson himself is as stingy as can be. Well, Johnson may well be a hypocrite, but we would be guilty of the inconsistency ad hominem fallacy if we regarded Johnson's stinginess or hypocrisy as grounds for rejecting what he says. This type of reasoning, where we reject what somebody says because what he or she says seems inconsistent with what he or she does, even has a Latin name: *tu quoque*, meaning "you too." This version of the inconsistency ad hominem often boils down to nothing more than saying "You too" or "You do it, too!" If a smoker urges another smoker to give up the habit, the second smoker commits the inconsistency ad hominem if she says, "Well, you do it too!"

■ Circumstantial Ad Hominem

"Parker's circumstances are such and such; therefore, his claim (belief, opinion, theory, proposal, etc.) stands refuted." This is the formula for the **circumstantial ad hominem.** An example would be "Well, you can forget about what Father Hennesy says about the dangers of abortion because Father Hennesy's a priest, and priests are required to hold such views." The speaker in this example is citing Father Hennesy's circumstances (being a

I get calls from nutso environmentalists who are filled with compassion for every snail darter that is threatened by some dam somewhere. Yet, they have no interest in the 1.5 million fetuses that are aborted every year in the United States. I love to argue with them and challenge their double standard.

—RUSH LIMBAUGH

Often an inconsistency ad hominem will accuse someone of having a double standard. Notice how this example is combined with ridicule.

priest) to "refute" Father Hennesy's opinion. This example isn't a personal attack ad hominem because the speaker may think very highly of priests in general and of Father Hennesy in particular. Clearly, though, a person could intend to issue a personal attack by mentioning circumstances that (in the opinion of the speaker) constituted a defect on the part of the person attacked. For example, consider "You can forget about what Father Hennesy says about the dangers of abortion because he is a priest and priests all have sexual hang-ups." That would qualify as both a circumstantial ad hominem (he's a priest) and a personal attack ad hominem (priests have sexual hang-ups).

■ Poisoning the Well

Poisoning the well can be thought of as an ad hominem in advance. If someone dumps poison down your well, you don't drink from it. Similarly, when A poisons your mind about B by relating unfavorable information about B, you may be inclined to reject what B says to you.

Well-poisoning is easier to arrange than you might think. You might suppose that to poison someone's thinking about Mrs. Jones, you would have to say or at least insinuate something deprecatory or derogatory about her. In fact, recent psycholinguistic research suggests you can poison someone's thinking about Mrs. Jones by doing just the opposite! If we don't know Mrs. Jones, even a sentence that expresses an outright denial of a connection between her and something unsavory is apt to make us form an unfavorable impression of her. Psychological studies indicate that people are more apt to form an unfavorable impression of Mrs. Jones from a sentence like "Mrs. Jones is not an ax murderer" than from a sentence like "Mrs. Jones has a sister."

Moral: Because it might be easy for others to arrange for us to have a negative impression of someone, we must be extra careful not to reject what a person says *just because* we have an unfavorable impression of the individual.

Genetic Fallacy

The **genetic fallacy** occurs when we try to "refute" a claim (or urge others to do so) on the basis of its origin or history. For example, a person might "refute" the idea that God exists on the grounds that belief in God first rose in superstitious times or on the grounds we come to believe in God to comfort ourselves in the face of death. We have heard people declare the U.S. Constitution "invalid" because it was (allegedly) drafted to protect the interests of property owners. These are examples of the genetic fallacy.

If we "refute" a proposal (or urge someone else to reject it) on the grounds it was part of the Republican (or Democratic) party platform, we commit the genetic fallacy. If we "refute" a policy (or try to get others to reject it) on the grounds that a slave-holding state in the nineteenth century originated the policy, that also would qualify. If we "rebut" (or urge others to reject) a ballot initiative on the grounds that the insurance industry or the association of trial lawyers or the American Civil Liberties Union or "Big Tobacco" or "Big Oil" or multinational corporations or the National Education Association or the National Rifle Association or the National Organization for Women proposed it or back it, we commit the fallacy. Knowing that the NRA or the NEA or NOW proposed or backs or endorses a piece of legislation may give one reason (depending on one's politics) to be suspicious of it or to have a careful look at it; but a perceived lack of merit on the part of the organization that proposed or backs or endorses a proposal is not equivalent to a lack of merit in the proposal itself. Knowing the NRA is behind a particular ballot initiative is not the same as knowing about a specific defect in the initiative itself, even if you detest the NRA.

Obviously, the genetic fallacy is similar to the ad hominem arguments we have examined. The following examples might help you see a difference between the two:

Example of (Personal Attack) Ad Hominem

The Democrats never tire of criticizing the president's war plans, but all this criticism is completely unwarranted. These people just can't get over the fact that Al Gore lost the election in 2000.

Example of Genetic Fallacy

All this criticism of the president's war plans is completely unwarranted. It all just comes from the Democrats.

Just remember what "ad hominem" means. It means "to the person." If we reject *Moore's* claim because *Moore* is, let us say, a member of the National Rifle Association, that's a (circumstantial) ad hominem. If we reject a *claim* because it originated with the National Rifle Association, that's the genetic fallacy.

Don't get into nuclear warfare with your classmates or with your instructor over whether a particular item is a genetic fallacy or this or that type of *argumentum ad hominem*. These mistakes all belong to the same family and resemble each other closely.

> Whom are they kidding? Where are NOW's constitutional objections to the billions of dollars (including about $1 million to NOW itself) that women's groups receive under the Violence Against Women Act?
>
> —ARMIN BROTT, issuing an ad hominem response to opposition by the National Organization for Women to a proposal to provide poor fathers with parenting and marital-skills training and classes on money management

Gender-based inconsistency ad hominem

"Positive Ad Hominem Fallacies"

An ad hominem fallacy, then, is committed if we rebut a person on the basis of considerations that, logically, apply to the person rather than to his or her claims. Strictly speaking, if we automatically transfer the positive or favorable attributes of a person to what he or she says, that's a mistake in reasoning, as well. The fact you think Moore is clever does not logically entitle you to conclude that any specific opinion of Moore's is clever. The fact that in your view the NRA represents all that is good and proper does not enable you to infer that any specific proposal from the NRA is good and

Dilbert reprinted by permission of United Features Syndicate, Inc.

▲ *A straw man generator in Dilbert's office.*

proper. Logicians did not always limit the ad hominem fallacy to cases of rebuttal, but that seems to be the usage now, and we shall follow that policy in this book. You should just remember that a parallel mistake in reasoning happens if you confuse the favorable qualities of a person with the qualities of his or her assertion.

Straw Man

Let's say Mrs. Herrington announces it is time to clean out the attic. Mr. Herrington groans and says, "What, again? Do we have to clean it out every day?"

The **straw man** fallacy happens when you "refute" a position or claim by distorting or oversimplifying or misrepresenting it. Mr. Herrington has distorted his wife's position, creating a straw man.

The straw man fallacy is very common. We can easily imagine Mrs. Herrington coming back with "Just because you think we should keep every last piece of junk forever doesn't mean I do." Now Mrs. Herrington is using a straw man herself, assuming that Mr. Herrington's position is only that they shouldn't clean out the attic right now.

The straw man fallacy is so common that it ranks next to the top on our list of the top ten fallacies of all time (see Appendix 2). One person will say he wants to eliminate the words "under God" from the Pledge of Allegiance, and his opponent will act as if he wants to eliminate the entire pledge. A conservative will oppose tightening emission standards for sulfur dioxide, and a liberal will accuse him of wanting to relax the standards. A Democratic congresswoman will say she opposes cutting taxes, and her Republican opponent will accuse her of wanting to raise taxes.

The ad hominem fallacy attempts to "refute" a claim on the basis of considerations that logically apply to its source. The straw man fallacy attempts to "refute" a claim by altering it so that it seems patently false or even ridiculous.

> *I'm a very controversial figure to the animal rights movement. They no doubt view me with some measure of hostility because I am constantly challenging their fundamental premise that animals are superior to human beings.*
>
> —RUSH LIMBAUGH, setting up another straw man for the kill

False Dilemma

Suppose Mrs. Herrington in the example above says to Mr. Herrington, "Look, either we clean out the attic or all this junk will run us out of house and home. Would you prefer that?" Now Mrs. Harrington is offering Mr. Harrington a "choice": either clean out the attic or let the junk run them out of house and home. But the choice Mrs. Harrington offers is limited to just two alternatives, and there are alternatives that deserve consideration, such as doing it later or not acquiring additional junk.

The **false dilemma** fallacy occurs when you limit considerations to only two alternatives although other alternatives may be available. Like the straw man fallacy, it is encountered all the time. You say you don't want to drill for oil in the Alaskan National Wildlife Reserve? Would you prefer letting Saddam Hussein dictate the price of oil?

Or take a look at this example:

CONGRESSMAN Guess we're going to have to cut back expenditures
CLAGHORN: on social programs again this year.

YOU: Why's that?

CLAGHORN: Well, we either do that or live with this high deficit, and that's something we can't allow.

Here, Claghorn maintains that either we live with the high deficit or we cut social programs, and that therefore, because we can't live with the high deficit, we have to cut social programs. But this reasoning works only if cutting social programs is the *only* alternative to a high deficit. Of course, that is not the case (taxes might be raised or military spending cut, for example). Another example:

DANIEL: Theresa and I both endorse this idea of allowing prayer in public schools, don't we, Theresa?

THERESA: I never said any such thing!

DANIEL: Hey, I didn't know you were an atheist!

Here, Daniel's "argument" amounts to this: Either you endorse prayer in public schools, or you are an atheist; therefore, because you do not endorse school prayer, you must be an atheist. But a person does not have to be an atheist in order to feel unfavorable toward prayer in public schools. The alternatives Daniel presents, in other words, could both be false. Theresa might not be an atheist and still not endorse school prayer.

The example Daniel provides us shows how this type of fallacy and the preceding one can work together: A straw man is often used as part of a false dilemma. A person who wants us to accept X may not only ignore other alternatives besides Y but also exaggerate or distort Y. In other words,

Hot Air and Fallacies in the Pledge Debate

Comments—and some hot air—on the Federal Appeals Court ruling that the phrase "one nation under God" makes the Pledge of Allegiance unconstitutional:

The decision is out of step with America.

—George W. Bush

Argument from popularity

This is a case of old men . . . pushing logic to a ridiculous extreme.

—George W. Bush

Personal attack ad hominem

Taking out the phrase "under God" at this time could be seen by Muslims as another example of godless Americanism.

—Richard Bruce in letter to Sacramento Bee, July 4, 2002

Scare tactics

Only 2.5 percent of the world's population is atheist. On the other hand, a majority of the world's population—Christians, Muslims and Jews—worship the god of Abraham.

—Richard Bruce

Red herring

Enough is enough. We need to put an end to judicial tyranny!

—Robert C. Balliet, letter in Sacramento Bee, July 4, 2002

Argument from outrage

The judge obviously does not appreciate the heritage and legacy upon which our founding fathers built this freedom-loving land. Why not move elsewhere? Russia will embrace him.

—Nancy Woodard, letter in Sacramento Bee, July 4, 2002

Nationalism, tradition, and outrage

The decision is just nuts.

—Senate Majority Leader Tom Daschle

Ridicule

Ridiculous. This decision will not sit well with the American people. The Supreme Court and Congress open each session with references to God, and the Declaration of Independence refers to God or the creator four times.

—Presidential spokesperson Ari Fleischer

Argument from tradition

The decision is directly contrary to two centuries of American tradition.

—Attorney General JOHN ASHCROFT

Argument from tradition

An atheist lawyer. I hope his name never comes before this body for any promotion, because he will be remembered.

—Senate President Pro Tem ROBERT BYRD

Scare tactics

Our founding fathers must be spinning in their graves.

—Senator KIT BOND, *R-Mo.*

Argument from outrage

What is next? Will the courts now strip "so help me God" from the pledge taken by new presidents? This is the worst kind of political correctness run amok.

—Senator KIT BOND, *R-Mo.*

Slippery slope, ridicule, outrage, hyperbole

Will our courts, in their zeal to abolish all religious faith from public arenas, outlaw "God Bless America" too? The great strength of the United States is that we are and will continue to be, despite the liberal court's decision, one nation under God.

*—*ROY BLUNT, *R-Mo.*

More of the same

There may have been a more senseless, ridiculous decision issued by a court at some time, but I don't remember it.

—Senator JOE LIEBERMAN, *D-Conn.*

More ridicule

The phrase "under God" defines who we are as a nation. This nation was founded by believers in God. . . . Should we also recall all the currency so these words can be removed? The economy would go into crisis if that were to happen. No one would ever suggest such an insane thing. So why would we be forced to change the words of the P of A? Sure, the words under God were added in 1954, but they are needed. Since taking God out of our public schools and other public forums there has been a rise in crime and overall immorality. What right does the court have to keep us from expressing our beliefs in the places where our tax dollars are spent? We are citizens of this great country, too.

*—*MICHELLE COS, *letter to Sacramento Bee, June 29, 2002*

Scare tactics, begging the question, ridicule, outrage, slippery slope

▲   *Has Mom fallen for a false trilemma?*

this person leaves only *one* "reasonable" alternative because the only other one provided is really a straw man. You can also think of a false dilemma as a false dichotomy.

It might help in understanding false dilemmas to look quickly at a *real* dilemma. Consider: You know that the Smiths must heat their house in the winter. You also know that the only heating options available in their location are gas and electricity. Under these circumstances, if you find out that they do *not* have electric heat, it must indeed be true that they must use gas heat because that's the only alternative remaining. False dilemma occurs only when reasonable alternatives are ignored. In such cases, both X and Y may be false, and some other alternative may be true.

Therefore, before you accept X because some alternative, Y, is false, make certain that X and Y cannot *both* be false. Look especially for some third alternative, some way of rejecting Y without having to accept X. Example:

> MOORE: Look, Parker, you're going to have to make up your mind. Either you decide that you can afford this stereo, or you decide that you're going to do without music for a while.

Parker could reject both of Moore's alternatives (buying this stereo and going without music) because of some obvious third possibilities. One, Parker might find a less expensive stereo. Or, two, he might buy a part of this stereo now—just the CD player, amplifier, and speakers, say—and postpone until later purchase of the rest.

Before moving on, we should point out that there is more than one way to present a pair of alternatives. Aside from the obvious "either X or Y" version we've described so far, we can use the form "if not X, then Y." For instance, in the example at the beginning of the section, Congressman

Claghorn can say, "Either we cut back on expenditures, or we'll have a big deficit," but he can accomplish the same thing by saying, "If we don't cut back on expenditures, then we'll have a big deficit." These two ways of stating the dilemma are equivalent to one another. Claghorn gets the same result: After denying that we can tolerate the high deficit, he concludes that we'll have to cut back expenditures. Again, it's the artificial narrowness of the alternatives—the falsity of the claim that says "if not one, then surely the other"—that makes this a fallacy.

■ *Perfectionist Fallacy*

A particular subspecies of false dilemma and common rhetorical ploy is something we call the **perfectionist fallacy.** It comes up when a plan or policy is under consideration, and it goes like this:

> If policy X will not meet our goals as well as we'd like them met (i.e., "perfectly"), then policy X should be rejected.

This principle downgrades policy X simply because it isn't perfection. It's a version of false dilemma because it says, in effect, "Either the policy is perfect, or else we must reject it."

An excellent example of the perfectionist fallacy comes from the National Football League's experience with the instant replay rule, which allows an off-field official to review videotape of a play to determine whether the on-field official's ruling was correct. To help the replay official, tape from several angles can be viewed, and the play run in slow motion.

One of the most often heard arguments against the use of videotape replays goes like this: "It's a mistake to use replays to make calls because no matter how many cameras you have following the action on the field, you're still going to miss some calls. There's no way to see everything that's going on."

According to this type of reasoning, we should not have police unless they can prevent *every* crime or apprehend *every* criminal. You can probably think of other examples that show perfectionist reasoning to be very unreliable indeed.

■ *Line-Drawing Fallacy*

Another version of the false dilemma is called the line-drawing fallacy. An example comes from the much-publicized Los Angeles case in which four police officers were acquitted of beating a man named Rodney King. After the trial, one of the jurors indicated that an argument like the following finally convinced her and at least one other juror to vote "not guilty":

> Everybody agrees that the first time one of the officers struck King with a nightstick it did not constitute excessive force. Therefore, if we are to conclude that excessive force was indeed used, then sometime during the course of the beat-

ing (during which King was hit about fifty times) there must have been a moment—a particular blow—at which the force *became* excessive. Since there is no point at which we can determine that the use of force changed from warranted to excessive, we are forced to conclude that it did not become excessive at any time during the beating; and so the officers did not use excessive force.

[People] who are voyeurs, if they are not irredeemably sick, . . . feel ashamed at what they are witnessing.

—IRVING KRISTOL, "Pornography, Obscenity, and the Case for Censorship"

False dilemma

These jurors accepted the **line-drawing fallacy,** the fallacy of insisting that a line must be drawn at some precise point when in fact it is not necessary that such a line be drawn.

To see how this works, consider another example: Clearly, it is impossible for a person who is not rich to become rich by our giving her one dollar. But, equally clearly, if we give our lucky person fifty million dollars, one at a time (very quickly, obviously—maybe we have a machine to deal them out), she will be rich. According to the line-drawing argument, however, *if we cannot point to the precise dollar that makes her rich, then she can never get rich, no matter how much money she is given!*

The problem, of course, is that the concepts referred to by "rich" and "excessive force" (and many others) are vague concepts. We can find cases where the concepts clearly apply and cases where they clearly do not apply. But it is not at all clear exactly where the borderlines are.

Many logicians interpret line drawing as a variety of slippery slope (discussed next). The King case might be seen this way: If the first blow struck against King did not amount to excessive violence, then there's nothing in the series of blows to change that fact. So there's no excessive violence at the end of the series, either.

Our own preference is to see the line-drawing fallacy as a version of false dilemma. It presents the following alternatives: Either there is a precise place where we draw the line, or else there is no line to be drawn (no difference) between one end of the scale and the other. Either there is a certain blow at which the force used against King became excessive, or else the force never became excessive.

Again, remember that our categories of fallacy sometimes overlap. When that happens, it doesn't matter as much which way we classify a case as that we see that an error is being made.

Slippery Slope

We've all heard people make claims of this sort: "If we let X happen, the first thing you know Y will be happening." This is one form of the **slippery slope.** Such claims are fallacious when in fact there is no reason to think that X will lead to Y. Sometimes X and Y can be the same kind of thing or can bear some kind of similarity to one another, but that doesn't mean that one will inevitably lead to the other.

Opponents of handgun control sometimes use a slippery slope argument, saying that if laws to register handguns are passed, the next thing we

$8 Billion, Down the Tube!

Eight billion dollars in utility ratepayers' money and 20 years of effort will be squandered if this resolution is defeated.

—Senator FRANK MURKOWSKI, R-Alaska, using a slippery slope fallacy
to argue for going forward with government plans to bury
radioactive waste in Yucca Mountain, Nevada

The fact that we've spent money on it already doesn't make it a good idea.

know there will be laws to make owning any kind of gun illegal. This is fallacious if there is no reason to think that the first kind of law will make the second kind more likely. It's up to the person who offers the slippery slope claim to show *why* the first action will lead to the second.

It is also argued that one should not experiment with certain drugs because experimentation is apt to lead to serious addiction or dependence. In the case of drugs that are known to be addictive, there is no fallacy present—the likelihood of the progression is clear.

The other version of slippery slope occurs when someone claims we must continue a certain course of action simply because we have already begun that course. It was said during the Vietnam War that because the United States had already sent troops to Vietnam, it was necessary to send more troops to support the first ones. Unless there is some reason supplied to show that the first step *must* lead to the others, this is a fallacy. (Notice that it's easy to make a false dilemma out of this case as well; do you see how to do it?)

Sometimes we take the first step in a series, and then we realize that it was a mistake. To insist on taking the remainder when we could admit our mistake and retreat is to fall prey to the slippery slope fallacy. (If you insist on following one bad bet with another one, we'd like to invite you to a friendly poker game.)

The slippery slope fallacy has considerable force because *psychologically* one item does often lead to another even though *logically* it does no such thing. When we think of X, say, we may be led immediately to think of Y. But this certainly does not mean that X itself is necessarily followed by Y. Once again, to think that Y has to follow X is to engage in slippery slope thinking; to do so when there is no particular reason to think Y must follow X is to commit a slippery slope fallacy.

Misplacing the Burden of Proof

Let's say Moore asks Parker, "Say, did you know that if you rub red wine on your head your gray hair will turn dark again?"

Parker, of course, will say, "Baloney."

Let's suppose Moore then says, "Baloney? Hey, how do you know it won't work?"

Moore's question is odd, because the **burden of proof** rests on him, not on Parker. Moore has misplaced the burden of proof on Parker, and this is a mistake, a fallacy.

Misplacing the burden of proof occurs when the burden of proof is placed on the wrong side of an issue. This is a common rhetorical technique, and sometimes you have to be on your toes to spot it. People are frequently tricked into thinking they have to prove their opponent's claim is wrong, when in fact the opponent should be proving that the claim is right. For example, you often heard people trying their darnedest to prove that we shouldn't go to war with Iraq, in a context in which the burden of proof rests on those who think we should go to war.

What reasonable grounds would make us place the burden of proof more on one side of an issue than the other? There are a variety of such grounds, but they fall mainly into three categories. We can express them as a set of rules of thumb:

1. *Initial plausibility.* In Chapter 3, we said that the more a claim coincides with our background information, the greater its initial plausibility. The general rule that most often governs the placement of the burden of proof is simply this: The less initial plausibility a claim has, the greater the burden of proof we place on someone who asserts that claim. This is just good sense, of course. We are quite naturally less skeptical about the claim that Charlie's now-famous eighty-seven-year-old grandmother drove a boat across Lake Michigan than we are about the claim that she swam across Lake Michigan. Unfortunately, this rule is a rule of thumb, not a rule that can be applied precisely. We are unable to assess the specific degree of a claim's plausibility and then determine with precision just exactly how much evidence its advocates need to produce to make us willing to accept the claim. But, as a rule of thumb, the initial plausibility rule can keep us from setting the requirements unreasonably high for some claims and allowing others to slide by unchallenged when they don't deserve to.

2. *Affirmative/negative.* Other things being equal, the burden of proof falls automatically on those supporting the affirmative side of an issue rather than on those supporting the negative side. In other words, we generally want to hear reasons why something *is* the case before we require reasons why it is *not* the case. Consider this conversation:

MOORE: The car won't start.

PARKER: Yeah, I know. It's a problem with the ignition.

MOORE: What makes you think that?

PARKER: Well, why not?

© Dan Piraro. Reprinted with special permission of King Features Syndicate.

▲ *Paleological misplacement of the burden of proof!*

Parker's last remark seems strange because we generally require the affirmative side to assume the burden of proof; it is Parker's job to give reasons for thinking that the problem *is* in the ignition.

This rule applies to cases of existence versus nonexistence, too. Most often, the burden of proof should fall on those who claim something exists rather than on those who claim it doesn't. There are people who believe in ghosts, not because of any evidence that there *are* ghosts, but because nobody has shown there are no such things. (When someone claims that we should believe in such-and-such because nobody has proved that it *isn't* so, we have a subtype of burden cf proof known as **appeal to ignorance.**) This is a burden-of-proof fallacy because it mistakenly places the requirement of proving their position on those who do not believe in ghosts. (Of course, the first rule applies here, too, because ghosts are not part of background knowledge for most of us.)

In general, the affirmative side gets the burden of proof because it tends to be much more difficult—or at least much more inconvenient—to prove the negative side of an issue. Imagine a student who walks up to the ticket window at a football game and asks for a discounted student ticket. "Can you prove you're a student?" he is asked. "No," the student replies, "can you prove I'm not?" Well, it may be possible to prove he's not a student, but it's no easy chore, and it would be unreasonable to require it.

Incidentally, some people say it's *impossible* to "prove a negative." But difficult is not the same as impossible. And some "negatives" are even easy to prove. For example, "There are no elephants in this classroom."

3. *Special circumstances.* Sometimes getting at the truth is not the only thing we want to accomplish, and on such occasions we may purposely

Innocent Until Proved Guilty

We must point out that sometimes there are specific reasons why the burden of proof is placed entirely on one side. The obvious case in point is in criminal court, where it is the prosecution's job to prove guilt. The defense is not required to prove innocence; it must only try to keep the prosecution from succeeding in its attempt to prove guilt. We are, as we say, "innocent until proved guilty." As a matter of fact, it's possible that more trials might come to a correct conclusion (i.e., the guilty get convicted and the innocent acquitted) if the burden of proof were equally shared between prosecution and defense. But we have wisely decided that if we are to make a mistake, we would rather it be one of letting a guilty person go free than one of convicting an innocent person. Rather than being a fallacy, then, this lopsided placement of the burden of proof is how we guarantee a fundamental right: the presumption of innocence.

place the burden of proof on a particular side. Courts of law provide us with the most obvious example. Specific agreements can also move the burden of proof from where it would ordinarily fall. A contract might specify, "It will be presumed that you receive the information by the tenth of each month unless you show otherwise." In such cases, the rule governing the special circumstances should be clear and acceptable to all parties involved.

One important variety of special circumstances occurs when the stakes are especially high. For example, if you're thinking of investing your life savings in a company, you'll want to put a heavy burden of proof on the person who advocates making the investment. However, if the investment is small, one you can afford to lose, you might be willing to lay out the money even though it has not been thoroughly proved that the investment is safe. In short, it is reasonable to place a higher burden of proof on someone who advocates a policy that could be dangerous or costly if he or she is mistaken.

These three rules cover most of the ground in placing the burden of proof properly. Be careful about situations where people put the burden of proof on the side other than where our rules indicate it should fall. Take this example:

PARKER: I think we should invest more money in expanding the interstate highway system.

MOORE: I think that would be a big mistake.

PARKER: How could anybody object to more highways?

With his last remark, Parker has attempted to put the burden of proof on Moore. Such tactics can put one's opponent in a defensive position; Moore

So Much for Presumed Innocence . . .

I would rather have an innocent man executed than a guilty murderer go free.

—caller on Talk Back Live (CNN), May 2, 2000

This not uncommon thought is a bizarre false dilemma, since if the innocent man is executed, the guilty murderer *does* go free.

now has to show why we should *not* spend more on roads rather than Parker having to show why we *should* spend more. This is an inappropriate burden of proof.

You should always be suspicious when an inability to *disprove* a claim is said to show that one is mistaken in doubting the claim or in saying that it's false. It does no such thing, unless the burden was on that person to disprove the claim. Inability to disprove that there is extrasensory perception (ESP) is no reason to think that one is mistaken in doubting that ESP exists. But psychics' repeated failure to prove that ESP exists *does* weaken *their* case because the burden of proof is on them.

Begging the Question

Here's a version of a simple example of begging the question, one that's been around a long time (we'll return to it later):

> Two gold miners roll a boulder away from its resting place and find three huge gold nuggets underneath. One says to the other, "Great! That's one nugget for you and two for me," handing one nugget to his associate.
> "Wait a minute!" says the second miner. "Why do you get two and I get just one?"
> "Because I'm the leader of this operation," says the first.
> "What makes you the leader?" asks miner number two.
> "I've got twice the gold you do," answers miner number one.

This next example is as famous as the first one was silly: Some people say they can prove God exists. When asked how, they reply, "Well, the Scriptures say very clearly that God must exist." Then, when asked why we should believe the Scriptures, they answer, "The Scriptures are divinely inspired by God himself, so they must be true."

The problem with such reasoning is that the claim at issue—whether it's the case that God exists—turns out to be one of the very premises the argument is based on. If we can't trust the Scriptures, then the argument isn't any good, but the reason given for trusting the Scriptures requires the

"I believe the President is telling the truth because he says he's telling the truth."

Gay marriages should not be legal because if there wasn't anything wrong with them they would already be legal, which they aren't.

—From a student essay

If you examine this "reasoning" closely, it says that gay marriages shouldn't be legal because they aren't legal. This is not quite "X is true just because X is true," but it's close. The issue is whether the law should be changed. So giving the existence of the law as a "reason" for its *not* being changed can carry no weight, logically.

existence of God, the very thing we were arguing for in the first place! Examples like this are sometimes called circular reasoning or arguing in a circle because they start from much the same place as they end up.

Rhetorical definitions can beg questions. Consider an example from an earlier chapter: If we define abortion as "the murder of innocent children," then it's obvious that abortion is morally wrong. But of course anyone who doubts that abortion is morally wrong is certainly not going to accept the definition just given. That person will most likely refuse to recognize an embryo or early-stage fetus as a "child" at all and will certainly not accept the word "murder" in the definition.

And this brings us to the real problem in cases of question begging: a misunderstanding of what premises (and definitions) it is reasonable for one's audience to accept. We are guilty of **begging the question** when we ask our audience to accept premises that are as controversial as the conclusion we're arguing for and are controversial on the same grounds. The sort of grounds on which people would disagree about the morality of abortion are much the same as those on which they would disagree about the definition of abortion above. The person making the argument has not "gone back far enough," as it were, to find common ground with the audience whom he or she wishes to convince.

Let's return to our feuding gold miners to illustrate what we're talking about. Clearly, the two disagree about who gets the gold, and, given what being the leader of the operation means, they're going to disagree just as much about that. But what if the first miner says, "Look, I picked this spot, didn't I? And we wouldn't have found anything if we'd worked where you wanted to work." If the second miner agrees, they'll have found a bit of common ground. Maybe—*maybe*—the first miner can then convince the second that this point, on which they agree, is worth considering when it comes to splitting the gold. At least there's a chance of moving the discussion forward when they proceed this way.

In fact, if you are ever to hope for any measure of success in trying to convince somebody of a claim, you should always try to argue for it based on whatever common ground you can find between the two of you. Indeed, the attempt to find common ground from which to start is what underlies the entire enterprise of rational debate.

Recap

The fallacies in this chapter, like those in Chapter 5, may resemble legitimate arguments, but none gives a reason for accepting (or rejecting) a claim. The discussions in this part of the book should help make you sensitive to the difference between relevant considerations and emotional appeals, factual irrelevancies, and other dubious argumentative tactics.

In this chapter we examined:

- ▲ Personal attack ad hominem—thinking a person's defects refute his or her beliefs
- ▲ Circumstantial ad hominem—thinking a person's circumstances refute his or her beliefs
- ▲ Inconsistency ad hominem—thinking a person's inconsistencies refute his or her beliefs
- ▲ Poisoning the well—encouraging others to dismiss what someone will say, by citing the speaker's defects, inconsistencies, circumstances, or other personal attributes
- ▲ Genetic fallacy—thinking that the origin or history of a belief refutes it
- ▲ Straw man—"rebutting" a claim by offering a distorted or exaggerated version of it
- ▲ False dilemma—an erroneous narrowing down of the range of alternatives; saying we have to accept X or Y (and omitting that we might do Z)
- ▲ Perfectionist fallacy—arguing that we either do something completely or not at all
- ▲ Line-drawing fallacy—requiring that a precise line be drawn someplace on a scale or continuum when no such precise line can be

drawn; usually occurs when a vague concept is treated like a precise one

▲ Slippery slope—refusing to take the first step in a progression on unwarranted grounds that doing so will make taking the remaining steps inevitable or insisting erroneously on taking the remainder of the steps simply because the first one was taken

▲ Misplacing burden of proof—requiring the wrong side of an issue to make its case

▲ Begging the question—assuming as true the claim that is at issue and doing this as if you were giving an argument

Exercises

Exercise 6-1

Working in groups, invent a simple, original, and clear illustration of each fallacy covered in this chapter. Then in the class as a whole select the illustrations that are clearest and most straightforward. Go over these illustrations before doing the remaining exercises in this chapter, and review them before you take a test on this material.

Exercise 6-2

Identify any examples of fallacies in the following passages. Tell why you think they are present, and identify which category they belong in, if they fit any category we've described.

▲ 1. Of course Chinese green tea is good for your health. If it weren't, how could it be so beneficial to drink it?

2. Overheard: "No, I'm against this health plan business. None of the proposals are gonna fix everything, you can bet on that."

3. You have a choice: Either you let 'em out to murder and rape again and again, or you put up with a little prison overcrowding. I know what I'd choose.

▲ 4. "The legalization of drugs will not promote their use. The notion of a widespread hysteria sweeping across the nation as every man, woman, and child instantaneously becomes addicted to drugs upon their legalization is, in short, ridiculous."

From a student essay

5. Way I figure is, giving up smoking isn't gonna make me live forever, so why bother?

6. "The trouble with [syndicated columnist Joseph] Sobran's gripe about Clinton increasing the power of the federal government is that Sobran is the one who wants the government to tell women they can't have abortions."

From a newspaper call-in column

▲ 7. Aid to Russia? Gimme a break! Why should we care more about the Russians than about our own people?

8. Bush's tax cut stinks. He's just trying to please big business.

9. I believe Tim is telling the truth about his brother because he just would not lie about that sort of thing.

▲ 10. I think I was treated unfairly. I got a ticket out on McCrae Road. I was doing about sixty miles an hour, and the cop charged me with "traveling at an unsafe speed." I asked him what would have been a *safe* speed on that particular occasion—fifty? forty-five?—and he couldn't tell me. Neither could the judge. I tell you, if you don't know what speeds are unsafe, you shouldn't give tickets for "unsafe speeds."

Exercise 6-3

Classify each of the following cases of ad hominem as personal attack ad hominem, circumstantial ad hominem, inconsistency ad hominem, poisoning the well, or genetic fallacy. Identify the cases, if any, in which it might be difficult or futile to assign the item to any single one of these categories, as well as those cases, if any, where the item doesn't fit comfortably into any of these categories at all.

▲ 1. The proponents of this spend-now–pay-later boondoggle would like you to believe that this measure will cost you only one billion dollars. That's NOT TRUE. In the last general election some of these very same people argued against unneeded rail projects because they would cost taxpayers millions more in interest payments. Now they have changed their minds and are willing to encourage irresponsible borrowing. Connecticut is already awash in red ink. Vote NO.

2. Rush Limbaugh argues that the establishment clause of the First Amendment should not be stretched beyond its intended dimensions by precluding voluntary prayer in public schools. This is a peculiar argument, when you consider that Limbaugh is quite willing to stretch the Second Amendment to include the right to own assault rifles and Saturday night specials.

3. I think you can safely assume that Justice Scalia's opinions on the cases before the Supreme Court this term will be every bit as flaky as his past opinions.

▲ 4. Harvard now takes the position that its investment in urban redevelopment projects will be limited to projects that are environmentally friendly. Before you conclude that that is such a swell idea, stop and think. For a long time Harvard was one of the biggest slumlords in the country.

5. REPUBLICAN: Finally! Finally the governor is getting around to reducing taxes—as he promised. What do you think of his plan?
   DEMOCRAT: Not much. He's just doing it so the Democrats won't get all the credit.

6. Dear Editor—

   I read with amusement the letter by Leslie Burr titled "It's time to get tough." Did anyone else notice a little problem in her views? It seems a little odd that somebody who claims that she "loathes violence" could also say that "criminals should pay with their life." I guess consistency isn't Ms. Burr's greatest concern.

▲ 7. YOU: Look at this. It says here that white males still earn a lot more than minorities and women for doing the same job.
   YOUR FRIEND: Yeah, right. Written by some woman, no doubt.

8. "Steve Thompson of the California Medical Association said document-checking might even take place in emergency rooms. That's because, while undocumented immigrants would be given emergency care, not all cases that come into emergency rooms fall under the federal definition of an emergency.

   "To all those arguments initiative proponents say hogwash. They say the education and health groups opposing the initiative are interested in protecting funding they receive for providing services to the undocumented."

   Article in Sacramento Bee

9. Horace, you're new to this town, and I want to warn you about the local newspaper. It's in cahoots with all them left-wing environmental nutcakes that are wrecking the economy around here. You can't believe a thing you read in it.

10. Are Moore and Parker guilty of the ad hominem fallacy or poisoning the well in their discussion of Michael Savage on p. 152?

▲ 11. Creationism cannot possibly be true. People who believe in a literal interpretation of the Bible just never outgrew the need to believe in Santa Claus.

12. "Americans spend between $28 billion and $61 billion a year in medical costs for treatment of hypertension, heart disease, cancer and other illnesses attributed to consumption of meat, says a report out today from a pro-vegetarian doctor's group.

   "Dr. Neal D. Barnard, lead author of the report in the *Journal of Preventive Medicine,* and colleagues looked at studies comparing the health of vegetarians and meat eaters, then figured the cost of treating illnesses suffered by meat eaters in excess of those suffered by vegetarians. Only studies that controlled for the health effects of smoking, exercise and alcohol consumption were considered.

   "The American Medical Association, in a statement from Dr. M. Roy Schwarz, charged that Barnard's group is an 'animal rights front organization' whose agenda 'definitely taints whatever unsubstantiated findings it may claim.'"

   USA Today

Exercise 6-4

Elegant Country Estate

- Stunning Federal-style brick home with exquisite appointments throughout
- 20 picturesque acres with lake, pasture, and woodland
- 5 bedrooms, 4.5 baths
- 5800 sq. ft. living space, 2400 sq. ft. basement
- Formal living room; banquet dining with butler's pantry; luxurious foyer, gourmet kitchen, morning room
- 3 fireplaces, 12 chandeliers

Maude and Clyde are discussing whether to buy this nice little cottage. Identify as many fallacies as you can in their conversation. Most are from this chapter, but you may see something from Chapters 4 and 5 as well.

CLYDE: Maude, look at this place! This is the house for us! Let's make an offer right now. We can afford it!

MAUDE: Oh, Clyde, be serious. That house is way beyond our means.

CLYDE: Well, I think we can afford it.

MAUDE: Honey, if we can afford it, pigs can fly.

CLYDE: Look, do you want to live in a shack? Besides, I called the real estate agent. She says it's a real steal.

MAUDE: Well, what do you expect her to say? She's looking for a commission.

CLYDE: Sometimes I don't understand you. Last week you were pushing for a really upscale place.

MAUDE: Clyde, we can't make the payments on a place like that. We couldn't even afford to heat it! And what on earth are we going to do with a lake?

CLYDE: Honey, the payments would only be around $5000 a month. How much do you think we could spend?

MAUDE: I'd say $1800.

CLYDE: Okay, how about $2050?

MAUDE: Oh, for heaven's sake! Yes, we could do $2050!

CLYDE: Well, how about $3100?

MAUDE: Oh, Clyde, what is your point?

CLYDE: So $3100 is okay? How about $3200? Stop me when I get to exactly where we can't afford it.

MAUDE: Clyde, I can't say *exactly* where it gets to be too expensive, but $5000 a month is too much.

CLYDE: Well, I think we can afford it.

MAUDE: Why?

CLYDE: Because it's within our means!

MAUDE: Clyde, you're the one who's always saying we have to cut back on our spending!

CLYDE: Yes, but this'll be a great investment!

MAUDE: And what makes you say that?

CLYDE: Because we're bound to make money on it.

MAUDE: Clyde, honey, you are going around in circles.

CLYDE: Well, can you prove we can't afford it?

MAUDE: Once we start spending money like drunken sailors, where will it end? Next we'll have to get a riding mower, then a boat for that lake, a butler for the butler's pantry—where will it end?

CLYDE: Well, we don't have to make up our minds right now. I'll call the agent and tell her we're sleeping on it.

MAUDE: Asleep and dreaming.

Exercise 6-5

In groups, vote on which option best depicts the fallacy found in each passage; then compare results with other groups in the class. *Note:* The fallacies include those found in Chapter 5 and Chapter 6.

▲ 1. The Health Editor for *USA Today* certainly seems to know what she is talking about when she recommends we take vitamins, but I happen to know she works for Tishcon, Inc., a large manufacturer of vitamin supplements.
   a. smokescreen/red herring
   b. subjectivism
   c. argument from popularity
   d. circumstantial ad hominem
   e. no fallacy

2. The president is right. People who are against attacking Iraq are unwilling to face up to the threat of terrorism.
   a. common practice
   b. peer pressure
   c. false dilemma
   d. straw man
   e. begging the question

3. Well, I for one think the position taken by our union is correct, and I'd like to remind you before you make up your mind on the matter that around here we employees have a big say in who gets rehired.
   a. wishful thinking
   b. circumstantial ad hominem
   c. scare tactics
   d. apple polishing
   e. begging the question

▲ 4. On the whole, I think global warming is a farce. After all, most people
       think winters are getting colder, if anything. How could that many people
       be wrong?
       a. argument from outrage
       b. argument from popularity
       c. straw man
       d. no fallacy

   5. MARCO: I think global warming is a farce.
      CLAUDIA: Oh, gad. How can you say such a thing, when there is so much
      evidence behind the theory?
      MARCO: Because. Look. If it isn't a farce, then how come the world is
      colder now than it used to be?
      a. begging the question
      b. subjectivism
      c. red herring
      d. circumstantial ad hominem
      e. no fallacy

   6. Of course you should buy a life insurance policy! Why shouldn't you?
      a. smokescreen/red herring
      b. wishful thinking
      c. scare tactics
      d. peer pressure argument
      e. misplacing burden of proof

▲ 7. My opponent, Mr. London, has charged me with having cheated on my
       income tax. My response is, When are we going to get this campaign out of
       the gutter? Isn't it time we stood up and made it clear that vilification has
       no place in politics?
       a. smokescreen/red herring
       b. wishful thinking
       c. argument from common practice
       d. argument from popularity
       e. circumstantial ad hominem

   8. Either we impeach the man or we send a message to the kids of this
      country that it's all right to lie under oath. Seems like an easy choice
      to me.
      a. smokescreen/red herring
      b. straw man
      c. false dilemma
      d. inconsistency ad hominem
      e. none of these

   9. If cigarettes aren't bad for you, then how come it's so hard on your health
      to smoke?
      a. circumstantial ad hominem
      b. genetic fallacy
      c. slippery slope
      d. begging the question

▲ 10. Global warming? I don't care what the scientists say. Just 'cause it's true for them doesn't make it true for me.
   a. smokescreen/red herring
   b. subjectivism
   c. argument from tradition
   d. argument from common practice

Exercise 6-6

In groups, vote on which option best depicts the fallacy found in each passage, and compare results with other groups. (It is all right with us if you ask anyone who is not participating in the discussions in your group to leave.) *Note:* The fallacies include those found in Chapter 5 and Chapter 6.

▲ 1. So what if the senator accepted a little kickback money—most politicians are corrupt, after all.
   a. argument from envy
   b. argument from tradition
   c. common practice
   d. subjectivism
   e. no fallacy

2. Me? I'm going to vote with the company on this one. After all, I've been with them for fifteen years.
   a. genetic fallacy
   b. group think fallacy
   c. slippery slope
   d. no fallacy

3. Public opinion polls? They're rigged. Just ask anyone.
   a. argument from common practice
   b. guilt trip
   c. begging the question
   d. argument from popularity
   e. no fallacy

▲ 4. Hey! It can't be time for the bars to close. I'm having too much fun.
   a. false dilemma
   b. misplacing burden of proof
   c. wishful thinking
   d. argument from tradition
   e. no fallacy

5. A mural for the municipal building? Excuse me, but why should public money, *our* tax dollars, be used for a totally unnecessary thing like art? There are potholes that need fixing. Traffic signals that need to be put up. There are a *million* things that are more important. It is an *outrage,* spending taxpayers' money on unnecessary frills like art. Give me a break!
   a. inconsistency ad hominem

    b. argument from outrage

    c. slippery slope

    d. perfectionist fallacy

    e. no fallacy

6. Mathematics is more difficult than sociology, and I *really* need an easier term this fall. So I'm going to take a sociology class instead of a math class.
   a. circumstantial ad hominem
   b. argument from pity
   c. false dilemma
   d. begging the question
   e. no fallacy

▲ 7. Parker says Macs are better than PCs, but what would you expect him to say? He's owned Macs for years.
   a. personal attack ad hominem
   b. circumstantial ad hominem
   c. inconsistency ad hominem
   d. perfectionist fallacy
   e. no fallacy

8. The congressman thought the president's behavior was an impeachable offense. But that's nonsense, coming from the congressman. He had an adulterous affair himself, after all.
   a. inconsistency ad hominem
   b. poisoning the well
   c. circumstantial ad hominem
   d. genetic fallacy
   e. no fallacy

9. Your professor wants you to read Moore and Parker? Forget it. Their book is so far to the right it's falling off the shelf.
   a. poisoning the well
   b. inconsistency ad hominem
   c. misplacing burden of proof
   d. argument from tradition
   e. no fallacy

▲ 10. How do I know God exists? Hey, do you know he doesn't?
   a. perfectionist fallacy
   b. inconsistency ad hominem
   c. misplacing burden of proof
   d. slippery slope
   e. begging the question

Exercise 6-7

In groups, vote on which option best depicts the fallacy found in each passage, and compare results with other groups. *Note:* The fallacies include those found in Chapter 5 and Chapter 6.

Getting Really Worked Up over Ideas

Not long ago the editor of *Freethought Today* magazine won a court case upholding the constitutional separation of church and state. Following are a few samples of the mail she received as a result (there was much more), as they were printed in the magazine. We present them to remind you of how worked up people can get over ideas.

> Satan worshipping scum . . .

> If you don't like this country and what it was founded on & for *get the f—— out of it* and go straight to hell.

> F— you, you communist wh—.

> If you think that mathematical precision that governs the universe was established by random events then you truly are that class of IDIOT that cannot be aptly defined.

These remarks illustrate extreme versions of more than one rhetorical device mentioned in this part of the book. They serve as a reminder that some people become defensive and emotional when it comes to their religion. (As Richard Dawkins, professor of Public Understanding of Science at Oxford University, was prompted to remark, "A philosophical opinion about the nature of the universe, which is held by the great majority of America's top scientists and probably the elite intelligentsia generally, is so abhorrent to the American electorate that no candidate for popular election dare affirm it in public.")

—Adapted from Free Inquiry, *Summer 2002*

▲ 1. Laws against teenagers drinking?—They are a total waste of time, frankly. No matter how many laws we pass, there are always going to be some teens who drink.
   a. misplacing burden of proof
   b. perfectionist fallacy
   c. line-drawing fallacy
   d. no fallacy

2. Even though Sidney was old enough to buy a drink at the bar, he had no identification with him and the bartender would not serve him.
   a. perfectionist fallacy
   b. inconsistency ad hominem
   c. misplacing burden of proof
   d. slippery slope
   e. no fallacy

3. Just how much sex has to be in a movie before you call it pornographic? Seems to me the whole concept makes no sense.
   a. perfectionist fallacy

   b. line-drawing fallacy
   c. straw man
   d. slippery slope
   e. no fallacy

▲ 4. Studies confirm what everyone already knows: Smaller classes make students better learners.
   a. argument from common practice
   b. begging the question
   c. misplacing burden of proof
   d. argument from popularity
   e. no fallacy

  5. The trouble with impeaching Bill Clinton is this: Going after every person who occupies the presidency will take up everyone's time and the government will never get anything else done.
   a. inconsistency ad hominem
   b. straw man
   c. group think
   d. argument from envy
   e. red herring

  6. The trouble with impeaching Bill Clinton is this. If we start going after Clinton, next we'll be going after senators, representatives, governors. Pretty soon no elected official will be safe from partisan attack.
   a. inconsistency ad hominem
   b. slippery slope
   c. straw man
   d. false dilemma
   e. misplacing burden of proof

▲ 7. MR. IMHOFF: That does it. I'm cutting down on your peanut butter cookies. Those things blimp me up.
   MRS. IMHOFF: Oh, Imhoff, get real. What about all the ice cream you eat?
   a. circumstantial ad hominem
   b. subjectivism
   c. straw man
   d. slippery slope
   e. inconsistency ad hominem

  8. KEN: I think I'll vote for Andrews. She's the best candidate.
   ROBERT: Why do you say she's best?
   KEN: Because she's my sister-in-law. Didn't you know that?
   a. apple polishing
   b. argument from pity
   c. scare tactics
   d. peer pressure argument
   e. none of the above

  9. MOE: You going to class tomorrow?
   JOE: I s'pose. Why?

MOE: Say, don't you get tired of being a Goody Two-shoes? You must have the most perfect attendance record of anyone who ever went to this school—certainly better than the rest of us; right, guys?
a. poisoning the well
b. argument from pity
c. scare tactics
d. no fallacy
e. none of the above

▲ 10. Morgan, you're down to earth and I trust your judgment. That's why I know I can count on you to back me up at the meeting this afternoon.
a. apple polishing
b. argument from pity
c. scare tactics
d. guilt trip
e. no fallacy

11. "Do you want to sign this petition to the governor?"
"What's it about?"
"We want him to veto that handgun registration bill that's come out of the legislature."
"Oh. No, I don't think I want to sign that."
"Oh, really? So are you telling me you want to get rid of the Second Amendment?"
a. false dilemma
b. personal attack ad hominem
c. genetic fallacy
d. misplacing burden of proof
e. no fallacy

12. Outlaw gambling? Man, that's a strange idea coming from you. Aren't you the one who plays the lottery all the time?
a. inconsistency ad hominem
b. circumstantial ad hominem
c. genetic fallacy
d. scare tactics
e. no fallacy

Exercise 6-8

Most of the following passages contain fallacies from Chapter 5 or Chapter 6. Identify them where they occur and try to place them in one of the categories we have described.

▲ 1. "People in Hegins, Pennsylvania, hold an annual pigeon shoot in order to control the pigeon population and to raise money for the town. This year, the pigeon shoot was disrupted by animal rights activists who tried to release the pigeons from their cages. I can't help but think these animal rights activists are the same people who believe in controlling the human population through the use of abortion. Yet, they recoil at a similar means of controlling pigeons. What rank hypocrisy."

Rush Limbaugh

2. Dear Mr. Swanson: I realize I'm not up for a salary increase yet, but I thought it might make my review a bit more timely if I pointed out to you that I have a copy of all the recent e-mail messages between you and Ms. Flood in the purchasing department.

3. I don't care if Nike has signed up Michael Jordan, Tiger Woods, and even Santa Claus to endorse their shoes. They're a crummy company that makes a crummy product. The proof is the fact that they pay poor women a dollar sixty for a long day's work in their Vietnamese shoe factories. That's not even enough to buy a day's worth of decent meals!

▲ 4. I don't care if Nike has signed up Michael Jordan, Tiger Woods, and even Santa Claus to endorse their shoes. They're a crummy company, and I wouldn't buy their shoes no matter what the circumstance. You don't need any reason beyond the fact that they pay poor women a dollar sixty for a long day's work in their Vietnamese shoe factories. That's not even enough to buy a day's worth of decent meals!

5. JULIA: Even this long after the 2000 presidential election, I still feel sort of unsettled about it. It still feels sort of illegitimate—do you know what I mean?
   JEFF: Look, it's a done deal, and Bush is the president. Get over it!

6. POWELL FAN: Colin Powell says that diplomatic efforts to avoid war with Iraq were serious and genuine, and his word is good enough for me.
   SKEPTIC: And what makes you so sure he's telling it like it is?
   FAN: Because he's the one guy in the administration you can trust.

▲ 7. I know the repair guy in the service center screwed up my computer; he's the only one who's touched it since it was working fine last Monday.

8. If you give the cat your leftover asparagus, next thing you know you'll be feeding him your potatoes, maybe even your roast beef. Where will it all end? Pretty soon that wretched animal will be sitting up here on the table for dinner. He'll be eating us out of house and home.

9. Look, either we refrain from feeding the cat table scraps, or he'll be up here on the table with us. So don't go giving him your asparagus.

▲ 10. We have a simple choice. Saving Social Security is sure as hell a lot more important than giving people a tax cut. So write your representative now and let him or her know how you feel.

11. Let gays join the military? Give me a break. God created Adam and Eve, not Adam and Steve.

12. So my professor told me if he gave me an A for getting an 89.9 on the test, next he'd have to give people an A for getting an 89.8 on the test, and pretty soon he'd have to give everyone in the class an A. How could I argue with that?

▲ 13. Those blasted Democrats! They want to increase government spending on education again. This is the same outfit that gave us $10,000 toilets and government regulations up the wazoo.

14. All this yammering about Clinton having been a liar—spare me! Every president under the sun has lied—Reagan, Kennedy, Nixon, both Bushes, probably even Jimmy Carter.

15. Lauren did a better job than anyone else at the audition, so even if she has no experience, we've decided to give her the part in the play.

▲ 16. TERRY: I failed my test but I gave my prof this nifty argument. I said, "Look, suppose somebody did 0.0001 percent better than I, would that be a big enough difference to give him a higher grade?" And he had to say "no," so then I said, "And if someone did 0.0001 percent better than that second person, would that be a big enough difference?" And he had to say "no" to that, too, so I just kept it up, and he never could point to the place where the difference was big enough to give the other person a higher grade. He finally saw he couldn't justify giving anyone a better grade.
TERRI: Well? What happened?
TERRY: He had to fail the whole class.

17. There is an increasingly large likelihood of a war with Iraq this winter [of 2003]. It is not clear just what Saddam Hussein has in the way of nuclear, chemical, and biological weapons, but we believe he is not without some capability with weapons of those kinds. In fact, we are sure that, if it were within his capabilities at this moment, he would launch a preemptive strike against the United States. All the more reason that the United States strike first.

Loosely adapted from a conversation on NBC's Meet the Press

18. Look, maybe you think it's okay to legalize tribal casinos, but I don't. Letting every last group of people in the country open a casino is a ridiculous idea, bound to cause trouble.

▲ 19. What, you of all people complaining about violence on TV? You, with all the pro football you watch?

20. You have three Fs and a D on your exams, and your quizzes are on the borderline between passing and failing. I'm afraid you don't deserve to pass the course.

Exercise 6-9

Identify any fallacies in the following passages. Tell why you think they are present, and identify which category they belong in, if they fit any of those we've described. Instances of fallacies are all from the types found in Chapter 6.

▲ 1. Suspicious: "I would forget about whatever Moore and Parker have to say about pay for college teachers. After all, they're both professors themselves; what would you *expect* them to say?"

2. It's obvious to me that abortion is wrong—after all, everybody deserves a chance to be born.

3. Overheard: Well, I think that's too much to tip her. It's more than 15 percent. Next time it will be 20 percent, then 25 percent—where will it stop?

▲ 4. CARLOS: Four A.M.? Do we really have to start that early? Couldn't we leave a little later and get more sleep?
JEANNE: C'mon, don't hand me that! I know you! If you want to stay in bed until noon and then drag in there in the middle of the night, then go by yourself! If we want to get there at a reasonable hour, then we have to get going early and not spend the whole day sleeping.

5. I know a lot of people don't find anything wrong with voluntary euthanasia, where a patient is allowed to make a decision to die and that wish is carried out by a doctor or someone else. What will happen, though, is that if we allow voluntary euthanasia, before you know it we'll have the patient's relatives or the doctors making the decision that the patient should be "put out of his misery."

6. "Congress was elected to make laws, not Dr. David Kessler, commissioner of the Food and Drug Administration, who convinced Clinton to have the FDA regulate nicotine. What's next? Will Dr. Kessler have Clinton regulate coffee and Coca-Cola? Will Big Macs be outlawed and overeating prohibited?"

▲ 7. Whenever legislators have the power to raise taxes, they will always find problems that seem to require for their solution doing exactly that. This is an axiom, the proof of which is that the power to tax always generates the perception on the part of those who have that power that there exist various ills the remedy for which can only lie in increased governmental spending and hence higher taxes.

8. Don't tell me I should wear my seat belt, for heaven's sake. I've seen you ride a motorcycle without a helmet!

9. People who own pit bulls show a lack of respect for their friends, their neighbors, and anybody else who might come in contact with their dogs. They don't care if their dogs chew other people up.

▲ 10. When it comes to the issue of race relations, either you're part of the solution or you're part of the problem.

11. What! So now you're telling me we should get a new car? I don't buy that at all. Didn't you claim just last month that there was nothing wrong with the Plymouth?

12. Letter to the editor: "The Supreme Court decision outlawing a moment of silence for prayer in public schools is scandalous. Evidently the American Civil Liberties Union and the other radical groups will not be satisfied until every last man, woman and child in the country is an atheist. I'm fed up."

Tri-County Observer

▲ 13. We should impeach the attorney general. Despite the fact that there have been many allegations of unethical conduct on his part, he has not done anything to demonstrate his innocence.

14. What do you mean, support Amnesty International? They only defend criminals.

15. Overheard: "Hunting immoral? Why should I believe that coming from you? You fish, don't you?"

16. "Will we have an expanding government, or will we balance the budget, cut government waste and eliminate unneeded programs?"

Newt Gingrich, in a Republican National Committee solicitation

17. "Former House Speaker Newt Gingrich, who is amazingly still welcomed on television despite his fall from power in disgrace, groused that Mondale had served on a panel recently that recommended privatizing Social Security. But don't Republicans adore this notion, even though they now call it 'private retirement investments' in deference to the idea's unpopularity?"

Political columnist Marianne Means

Exercise 6-10

Identify any fallacies in the following passages. Tell why you think they are present, and identify which category they belong in, if they fit in any of those we've described.

▲ 1. Despite all the studies and the public outcry, it's still true that nobody has ever actually *seen* cigarette smoking cause a cancer. All the antismoking people can do is talk about statistics; as long as there isn't real proof, I'm not believing it.

2. There is only one way to save this country from the domination by the illegal drug establishment to which Colombia has been subjected, and that's to increase tenfold the funds we spend on drug enforcement and interdiction.

3. I believe that the great flood described in the Bible really happened. The reason is simple: Noah would not have built the ark otherwise.

▲ 4. In 1996 a University of Chicago study gave evidence that letting people carry concealed guns appears to sharply reduce murders, rapes, and other violent crimes. Gun-control backer Josh Sugarman of the Violence Policy Center commented: "Anyone who argues that these laws reduce crime either doesn't understand the nature of crime or has a preset agenda."

5. Letter to the editor: "I strongly object to the proposed sale of alcoholic beverages at County Golf Course. The idea of allowing people to drink wherever and whenever they please is positively disgraceful and can only lead to more alcoholism and all the problems it produces—drunk driving, perverted parties, and who knows what else. I'm sure General Stuart, if he were alive today to see what has become of the land he deeded to the county, would disapprove strenuously."

Tehama County Tribune

6. Letter to the editor: "I'm not against immigrants or immigration, but something has to be done soon. We've got more people already than we

can provide necessary services for, and, at the current rate, we'll have people standing on top of one another by the end of the century. Either we control these immigration policies or there won't be room for any of us to sit down."

Lake County Recorder

▲ 7. Letter to the editor: "So now we find our local crusader-for-all-that-is-right, and I am referring to Councilman Benjamin Bostell, taking up arms against the local adult bookstore. Is this the same Mr. Bostell who owns the biggest liquor store in Chilton County? Well, maybe booze isn't the same as pornography, but they're the same sort of thing. C'mon, Mr. Bostell, aren't you a little like the pot calling the kettle black?"

Chilton County Register

8. Letter to the editor: "Once again the *Courier* displays its taste for slanted journalism. Why do your editors present only one point of view?

"I am referring specifically to the editorial of May 27, regarding the death penalty. So capital punishment makes you squirm a little. What else is new? Would you prefer to have murderers and assassins wandering around scot-free? How about quoting someone who has a different point of view from your own, for a change?"

Athens Courier

9. "Clinton should have been thrown in jail for immoral behavior. Just look at all the women he has had affairs with since he left the presidency."

"Hey, wait a minute. How do you know he has had affairs since he was president?"

"Because if he didn't, then why would he be trying to cover up the fact that he did?"

▲ 10. It's practically a certainty that the government is violating the law in the arms deals with Saudi Arabians. When a reporter asked officials to describe how they were complying with the law, he was told that details about the arms sales were classified.

Exercise 6-11

Where we (Moore and Parker) teach, the city council recently debated relaxing the local noise ordinance. One student (who favored relaxation) appeared before the council and stated: "If 250 people are having fun, one person shouldn't be able to stop them."

We asked our students to state whether they agreed or disagreed with that student and to support their position with an argument. Here are some of the responses.

Divide into groups, and then identify any instances of fallacious reasoning you find in any answers, drawing from the materials in the last two chapters. Compare your results with those of other students, and see what your instructor thinks.

1. I support what the person is saying. If 250 people are having fun, one person shouldn't be able to stop them. Having parties and having a good time are a way of life for Chico State students. The areas around campus have always been this way.

2. A lot of people attend Chico State because of the social aspects. If rules are too tight, the school could lose its appeal. Without the students, local businesses would go under. Students keep the town floating. It's not just bars and liquor stores, but gas stations and grocery stores and apartment houses. This town would be like Orland.

3. If students aren't allowed to party, the college will go out of business.

4. We work hard all week long studying and going to classes. We deserve to let off steam after a hard week.

5. Noise is a fact of life around most college campuses. People should know what they are getting into before they move there. If they don't like it, they should just get earplugs or leave.

6. I agree with what the person is saying. If 250 people want to have fun, what gives one person the right to stop them?

7. I am sure many of the people who complain are the same people who used to be stumbling down Ivy Street twenty years ago doing the same thing that the current students are doing.

8. Two weeks ago I was at a party and it was only about 9:00 P.M. There were only a few people there and it was quiet. And then the police came and told us we had to break it up because a neighbor complained. Well, that neighbor is an elderly lady who would complain if you flushed the toilet. I think it's totally unreasonable.

9. Sometimes the noise level gets a little out of control, but there are other ways to go about addressing this problem. For example, if you are a neighbor and you are having a problem with the noise level, why don't you call the "party house" and let them know, instead of going way too far and calling the police?

10. I'm sure that these "narcs" have nothing else better to do than to harass the "party people."

11. You can't get rid of all the noise around a college campus no matter what you do.

12. The Chico noise ordinance was put there by the duly elected officials of the city and is the law. People do not have the right to break a law that was put in place under proper legal procedures.

13. The country runs according to majority rule. If the overwhelming majority want to party and make noise, under our form of government they should be given the freedom to do so.

14. Students make a contribution to the community, and in return they should be allowed to make noise if they want.

15. Your freedom ends at my property line.

16. A majority rule must recognize individual rights. A majority does not have the freedom to invade another person's home with anything, whether it's noise, litter, smoke, garbage, their bodies, or anything else.

Exercise 6-12

Identify any examples of fallacies in the following passages. Tell why you think these are fallacies, and identify which category they belong in, if they fit any category we've described.

▲ 1. Letter to the editor: "I would like to express my feelings on the recent conflict between county supervisor Blanche Wilder and Murdock County Sheriff Al Peters over the county budget.

   "I have listened to sheriffs' radio broadcasts. Many times there have been dangerous and life-threatening situations when the sheriff's deputies' quickest possible arrival time is 20 to 30 minutes. This is to me very frightening.

   "Now supervisor Wilder wants to cut two officers from the Sheriff's Department. This proposal I find ridiculous. Does she really think that Sheriff Peters can run his department with no officers? How anyone can think that a county as large as Murdock can get by with no police is beyond me. I feel this proposal would be very detrimental to the safety and protection of this county's residents."

2. Letter to the editor: "Andrea Keene's selective morality is once again showing through in her July 15 letter. This time she expresses her abhorrence of abortion. But how we see only what we choose to see! I wonder if any of the anti-abortionists have considered the widespread use of fertility drugs as the moral equivalent of abortion, and, if they have, why they haven't come out against them, too. The use of these drugs frequently results in multiple births, which leads to the death of one of the infants, often after an agonizing struggle for survival. According to the rules of the pro-lifers, isn't this murder?"

North-State Record

3. In one of her columns, Abigail Van Buren printed the letter of "I'd rather be a widow." The letter writer, a divorcée, complained about widows who said they had a hard time coping. Far better, she wrote, to be a widow than to be a divorcée, who are all "rejects" who have been "publicly dumped" and are avoided "like they have leprosy." Abby recognized the pseudo-reasoning for what it was, though she did not call it by our name. What is our name for it?

▲ 4. Overheard: "Should school kids say the Pledge of Allegiance before class? Certainly. Why shouldn't they?"

5. Letter to the editor: "Once again the Park Commission is considering closing North Park Drive for the sake of a few joggers and bicyclists. These so-called fitness enthusiasts would evidently have us give up to them for their own private use every last square inch of Walnut Grove. Then anytime anyone wanted a picnic, he would have to park at the edge of the park

and carry everything in—ice chests, chairs, maybe even grandma. I certainly hope the Commission keeps the entire park open for everyone to use."

6. "Some Christian—and other—groups are protesting against the placing, on federal property near the White House, of a set of plastic figurines representing a devout Jewish family in ancient Judaea. The protestors would of course deny that they are driven by any anti-Semitic motivation. Still, we wonder: Would they raise the same objections (of unconstitutionality, etc.) if the scene depicted a modern, secularized Gentile family?"

National Review

▲ 7. "It's stupid to keep on talking about rich people not paying their fair share of taxes while the budget is so far out of balance. Why, if we raised the tax rates on the wealthy all the way back to where they were in 1980, it would not balance the federal budget."

Radio commentary by Howard Miller

8. From a letter to the editor: "The counties of Michigan clearly need the ability to raise additional sources of revenue, not only to meet the demands of growth but also to maintain existing levels of service. For without these sources those demands will not be met, and it will be impossible to maintain services even at present levels."

9. In February 1992, a representative of the Catholic Church in Puerto Rico gave a radio interview (broadcast on National Public Radio) in which he said that the Church was against the use of condoms. Even though the rate of AIDS infection in Puerto Rico is much higher than on the U.S. mainland, the spokesman said that the Church could not support the use of condoms because they are not absolutely reliable in preventing the spread of the disease. "If you could prove that condoms were absolutely dependable in preventing a person from contracting AIDS, then the Church could support their use."

▲ 10. A 1991 book by a former member of the National Security Council indicated that supporters of Ronald Reagan may have made a deal with the Iranians who had been holding American hostages for months. The Iranians agreed not to release the hostages until after the 1980 election (in which Reagan defeated Jimmy Carter), and, it was alleged, the new administration promised to make weapons available to Iran. Here's one reaction to the announcement of the deal:

"I'm not surprised about Reagan's using trickery to get himself elected president. After all, he was nothing but an old actor, and he was used to using Hollywood trickery to fool people during his first career."

Exercise 6-13

Identify any examples of fallacies in the following passages. Tell why you think they are present, and identify which category they belong in, if they fit any category we've described.

▲   1. The U.S. Congress considered a resolution criticizing the treatment of ethnic minorities in a Near Eastern country. When the minister of the interior was asked for his opinion of the resolution, he replied, "This is purely an internal affair in my country, and politicians in the U.S. should stay out of such affairs. If the truth be known, they should be more concerned with the plight of minority peoples in their own country. Thousands of black and Latino youngsters suffer from malnutrition in the United States. They can criticize us after they've got their own house in order."

2. It doesn't make any sense to speak of tracing an individual human life back past the moment of conception. After all, that's the beginning, and you can't go back past the beginning.

3. MOE: The death penalty is an excellent deterrent for murder.
   JOE: What makes you think so?
   MOE: Because there's no evidence that it's *not* a deterrent.
   JOE: Well, states with capital punishment have higher murder rates than states that don't have it.
   MOE: Yes, but that's only because there are so many legal technicalities standing in the way of executions that convicted people hardly ever get executed. Remove those technicalities, and the rate would be lower in those states.

▲   4. Overheard: "The new sculpture in front of the municipal building by John Murrah is atrocious and unseemly, which is clear to anyone who hasn't forgotten Murrah's mouth in Vietnam right there along with Hayden and Fonda calling for the defeat of America. I say: Drill holes in it so it'll sink and throw it in Walnut Pond."

5. Overheard: "Once we let these uptight guardians of morality have their way and start censoring *Playboy* and *Penthouse*, the next thing you know they'll be dictating everything we can read. We'll be in fine shape when they decide that *Webster's* should be pulled from the shelves."

6. It seems the biggest problem the nuclear industry has to deal with is not a poor safety record, but a lack of education of the public on nuclear power. Thousands of people die each year from pollution generated by coal-fired plants. Yet to date, there has been no death directly caused by radiation at a commercial nuclear power plant in the United States. We have a clear choice: an old, death-dealing source of energy or a safe, clean one. Proven through the test of time, nuclear power is clearly the safest form of energy and the least detrimental to the environment. Yet it is perceived as unsafe and an environmental hazard.

▲   7. A high school teacher once told my class that if a police state ever arose in America, it would happen because we freely handed away our civil rights in exchange for what we perceived would be security from the government. We are looking at just that in connection with the current drug crisis.

   For almost thirty years we've seen increasing tolerance, legally and socially, of drug use. Now we are faced with the very end of America as we know it, if not from the drug problem, then from the proposed solutions to it.

First, it was urine tests. Officials said that the innocent have nothing to fear. Using that logic, why not allow unannounced police searches of our homes for stolen goods? After all, the innocent would have nothing to fear.

Now we're looking at the seizure of boats and other property when even traces of drugs are found. You'd better hope some drug-using guest doesn't drop the wrong thing in your home, car, or boat.

The only alternative to declaring real war on the real enemies—the Asian and South American drug families—is to wait for that knock on the door in the middle of the night.

8. The mayor's argument is that because the developers' fee would reduce the number of building starts, ultimately the city would lose more money than it would gain through the fee. But I can't go along with that. Mayor Tower is a member of the Board of Realtors, and you know what *they* think of the fee.

9. Letter to the editor: "Next week the philosopher Tom Regan will be in town again, peddling his animal rights theory. In case you've forgotten, Regan was here about three years ago arguing against using animals in scientific experimentation. As far as I could see then and can see now, neither Regan nor anyone else has managed to come up with a good reason why animals should not be experimented on. Emotional appeals and horror stories no doubt influence many, but they shouldn't. I've always wondered what Regan would say if his children needed medical treatment that was based on animal experiments."

▲ 10. Not long before Ronald and Nancy Reagan moved out of the White House, former Chief of Staff Don Regan wrote a book in which he depicted a number of revealing inside stories about First Family goings-on. Among them was the disclosure that Nancy Reagan regularly sought the advice of a San Francisco astrologer. In response to the story, the White House spokesman at the time, Marlin Fitzwater, said, "Vindictiveness and revenge are not admirable qualities and are not worthy of comment."

Exercise 6-14

Listen to a talk-radio program (e.g., Rush Limbaugh, Michael Reagan, Michael Savage), and see how many minutes (or seconds) go by before you hear one of the following: ad hominem, straw man, ridicule, argument from outrage, or scare tactics. Report your findings to the class and describe the first item from the above list that you heard.

Exercise 6-15

Watch one of the news/public affairs programs on television (*NewsHour with Jim Lehrer, Nightline, Face the Nation,* and so on), and make a note of any examples of fallacies that occur. Explain in writing why you think the examples contain fallacious reasoning.

Alternatively, watch *Real Time* with Bill Maher. It usually doesn't take long to find a fallacy there.

Exercise 6-16 ────────────────────────────────

The following passages contain fallacies from both this and the preceding chapter. Identify the category in which each item belongs.

▲ 1. "I can safely say that no law, no matter how stiff the consequence is, will completely stop illegal drug use. Outlawing drugs is a waste of time."

From a student essay

2. "If we expand the commuter bus program, where is it going to end? Will we want to have a trolley system? Then a light rail system? Then expand Metrolink to our area? A city this size hardly needs and certainly cannot afford all these amenities."

From a newspaper call-in column

3. YAEKO: The character Dana Scully on *The X-Files* really provides a good role model for young women. She's a medical doctor and an FBI agent, and she's intelligent, professional, and devoted to her work.
   MICHAEL: Those shows about paranormal activities are so unrealistic. Alien abductions, government conspiracies—it's all ridiculous.

4. Overheard: "The reason I don't accept evolution is that ever since Darwin, scientists have been trying to prove that we evolved from some apelike primate ancestor. Well, they still haven't succeeded. Case closed."

▲ 5. Ladies and gentlemen, as you know, I endorsed Council Member Morrissey's bid for reelection based on his outstanding record during his first term. Because you are the movers and shakers in this community, other people place the same high value on your opinions that I do. Jim and I would feel privileged to have your support.

6. It's totally ridiculous to suppose that creationism is true. If creationism were true, then half of what we know through science would be false, which is complete nonsense.

7. KIRSTI: I counted my CDs this weekend, and out of twenty-seven, ten of them were by U2. They are such a good band! I haven't heard anything by Bono for a long time. He has such a terrific voice!
   BEN: Is he bisexual?

8. Was Gerhard a good committee chair? Well, I for one think you have to say he was excellent, especially when you consider all the abuse he put up with. Right from the start people went after him—they didn't even give him a chance to show what he could do. It was really vicious—people making fun of him right to his face. Yes, under the circumstances he has been quite effective.

▲ 9. Medical research that involves animals is completely unnecessary and a waste of money. Just think of the poor creatures! We burn and blind and torture them, and then we kill them. They don't know what is going to happen to them, but they know something is going to happen. They are scared to death. It's really an outrage.

10. Dear Editor—
    If Christians do not participate in government, only sinners will.

    From a letter to the Chico Enterprise Record

11. The HMO people claim that the proposal will raise the cost of doing business in the state to such a degree that insurers will be forced to leave the state and do business elsewhere. What nonsense. Just look at what we get from these HMOs. I know people who were denied decent treatment for cancer because their HMO wouldn't approve it. There are doctors who won't recommend a procedure for their patients because they are afraid the HMO will cancel their contract. And when an HMO does cancel some doctor's contract, the patients have to find a new doctor themselves—*if* they can. Everybody has a horror story. Enough is enough.

12. HOWARD: Dad, I really appreciate your letting me borrow the Chevy. But Melanie's parents just bought her a brand new Mercedes!
    DAD: Some people just refuse to buy American!

▲ 13. [Dole campaign chairman] SCOTT REID: There is a clear pattern of campaign finance abuse by the [Clinton] administration. Indonesian business interests have steered millions into the President's campaign using a gardener as a front, and [Democratic fund-raiser] John Huang, who apparently laundered money at a fund-raiser at a Buddhist temple in California, is suddenly nowhere to be found.
    [White House Senior Adviser] GEORGE STEPHANOPOULOS: I can't let these charges go unrefuted. Dole has received millions from foreign supporters like José Fanjul, and his vice chairman for finance, Simon Fireman, had to pay the largest fine in the history of the country for massive violations of campaign-finance laws.

    On NBC's Meet the Press

14. The proposal to reduce spending for the arts just doesn't make any sense. We spend a paltry $620 million for the NEA [National Endowment for the Arts], while the deficit is closing in on $200 billion. Cutting support for the arts isn't going to eliminate the deficit; that's obvious.

15. Year-round schools? I'm opposed. Once we let them do that, the next thing you know they'll be cutting into our vacation time and asking us to teach in the evenings and on the weekends and who knows where it will end. We teachers have to stand up for our rights.

▲ 16. [NBC's *Meet the Press* host] TIM RUSSERT: Mr. Perot, Bob Dole says that every vote for Ross Perot is a vote for Bill Clinton. True?
    ROSS PEROT: Of course that's not true. They've been programmed from birth to say that if they are a Republican.

17. Even if we outlaw guns we're still going to have crime and murder. So I really don't see much point in it.

    From a student essay

18. Do you think affirmative action programs are still necessary in the country?
    Answers:
    a. Yes, of course. I don't see how you, a woman, can ask that question. It's obvious we have a very long way to go still.
    b. No. Because of affirmative action, my brother lost his job to a minority who had a lot less experience than he did.
    c. Yes. The people who want to end affirmative action are all white males who just want to go back to the good-old-boy system. It's always the same: Look out for number one.
    d. No. The people who want it to continue know a good deal when they see one. You think I'd want to end it if I were a minority?

Exercise 6-17

Explain in a sentence or two how each of the following passages involves a type of fallacy mentioned in either this or the preceding chapter. *Many of these examples are difficult* and should serve to illustrate how fallacies sometimes conform only loosely to the standard patterns.

▲ 1. I believe that the companies that produce passenger airliners should be more strictly supervised by the FAA. I mean, good grief, everybody knows that you can make more money by cutting corners here and there than by spending extra time and effort getting things just right, and you know there have got to be airlines that are doing exactly that.

2. From a letter to a college newspaper editor: "I really appreciated the fact that your editorial writer supports the hike in the student activity fee that has been proposed. Since the writer is a senior and won't even be here next year, he will escape having to pay the fee himself, so of course there's no downside to it as far as he's concerned. I'm against the fee, and I'll be one of those who pay it if it passes. Mine is an opinion that should count."

3. "'There's a certain sameness to the news on the Big Three [ABC, NBC, and CBS] and CNN,' says Moody, . . . who is in charge of Fox News's day-to-day editorial decisions. That's the message, Moody says, that 'America is bad, corporations are bad, animal species should be protected, and every cop is a racist killer. That's where "fair and balanced" [Fox's slogan] comes in. We don't think all corporations are bad, every forest should be saved, every government spending program is good. We're going to be more inquisitive.'"

   From an interview with John Moody, vice president for news editorial at Fox News Network, in Brill's Content *magazine*

▲ 4. During the Reagan and Bush administrations, Democratic members of Congress pointed to the two presidents' economic policies as causing huge deficits that could ultimately ruin the country's economy. President Bush dismissed such charges as "the politics of doom and gloom." "These people will find a dark cloud everywhere," he has said. Was this response fallacious reasoning?

▲   5. "Louis Harris, one of the nation's most influential pollsters, readily admits he is in the polling business to 'have some impact with the movers and shakers of the world.' So poll questions are often worded to obtain answers that help legitimize the liberal Establishment's viewpoints."

Conservative Digest

6. "At a White House meeting in February of 1983 with Washington, D.C., anchormen, Ronald Reagan was asked to comment on 'an apparent continuing perception among a number of black leaders that the White House continues to be, if not hostile, at least not welcome to black viewpoints.' President Reagan replied as follows: 'I'm aware of all that, and it's very disturbing to me, because anyone who knows my life story knows that long before there was a thing called the civil-rights movement, I was busy on that side. As a sports announcer, I didn't have any Willie Mayses or Reggie Jacksons to talk about when I was broadcasting major league baseball. The opening line of the Spalding Baseball Guide said, "Baseball is a game for Caucasian gentlemen." And as a sports announcer I was one of a very small fraternity that used that job to editorialize against that ridiculous blocking of so many fine athletes and so many fine Americans from participating in what was called the great American game.' Reagan then went on to mention that his father refused to allow him to see *Birth of a Nation* because it was based on the Ku Klux Klan and once slept in a car during a blizzard rather than stay at a hotel that barred Jews. Reagan's 'closest teammate and buddy' was a black, he said."

James Nathan Miller, The Atlantic

7. From a letter to the editor of the *Atlantic Monthly:* "In all my reading and experience so far, I have found nothing presented by science and technology that precludes there being a spiritual element to the human being. . . . The bottom line is this: Maybe there are no angels, afterlife, UFOs, or even a God. Certainly their existence has not yet been scientifically proved. But just as certainly, their *nonexistence* remains unproved. Any reasonable person would therefore have to reserve judgment."

8. Stop blaming the developers for the fact that our town is growing! If you want someone to blame, blame the university. It brings the new people here, not the developers. Kids come here from God knows where, and lots of them like what they find and stick around. All the developers do is put roofs over those former students' heads.

▲   9. Two favorite scientists of the Council for Tobacco Research were Carl Seltzer and Theodore Sterling. Seltzer, a biological anthropologist, believes smoking has no role in heart disease and has alleged in print that data in the huge 45-year, 10,000-person Framingham Heart Study—which found otherwise—have been distorted by anti-tobacco researchers. Framingham Director William Castelli scoffs at Seltzer's critique but says it "has had some impact in keeping the debate alive."

   Sterling, a statistician, disputes the validity of population studies linking smoking to illness, arguing that their narrow focus on smoking obscures the more likely cause—occupational exposure to toxic fumes.

For both men, defying conventional wisdom has been rewarding. Seltzer says he has received "well over $1 million" from the Council for research. Sterling got $1.1 million for his Special Projects work in 1977–82, court records show.

From "How Tobacco Firms Keep Health Questions 'Open' Year After Year,"
Alix Freedman and Laurie Cohen. The article originally appeared in the
Wall Street Journal *and was reprinted in the* Sacramento Bee.

10. We have had economic sanctions in effect against China ever since the Tienanmen Square massacre. Clearly they haven't turned the Chinese leadership in Beijing into a bunch of good guys. All they've done, in fact, is cost American business a lot of money. We should get rid of the sanctions and find some other way to make them improve their human rights record.

Writing Exercises

1. Turn to any of the following selections from the Essays for Analysis in Appendix 1 and follow the instructions: 10A,B; 11A,B; 12; 14A,B; 15A,B; 16A,B; 17A,B; or 18A,B. You might also turn to Selection 19 or 20, and follow the second set of instructions of each.

2. When Marshall Gardiner married J'Noel Gardiner, he was aware that his new wife had been born a male and had recently had a sex-change operation. When Marshall Gardiner died, his wife's right to inherit half of her husband's $2.5 million estate was challenged by Marshall Gardiner's son. The son claimed the marriage was invalid under Kansas law, which declares same-sex marriages invalid, but does not address marriages involving transsexuals. The case went to the state supreme court in 2002. How should the court have ruled, in your opinion? Defend your position, using whatever rhetorical devices from Chapters 4, 5, and 6 you want. When everyone is finished, read the essays in groups looking for fallacies and other rhetorical devices. Your instructor may have groups select one essay to read to the class.

3. Should Kansas (or any other state) make same-sex marriages illegal? Take a position and defend it following the instructions for Exercise 2.

Chapter 7

The Anatomy and Varieties of Arguments

In a box back in Chapter 4, we presented this passage:

> It's morally wrong for a person to inflict awful pain on another sensitive creature, one that has done the first no harm. Therefore, the so-called scientists who perform their hideous and sadistic experiments on innocent animals are moral criminals just as were Hitler and his Nazi torturers.

The rhetorical devices used here are pretty obvious, and a person may read this passage and, especially after the comparison with Hitler and Nazi torturers, simply pass the whole thing off: "No, those scientists are not like Hitler—this person is nuts." But doing so would be a mistake, because it ignores the argument contained in the passage. Remember from Chapter 1 that an argument provides reasons (premises) in support of a conclusion. Stripped of its window dressing (mainly, the slanting devices), the argument looks like this:

1. Animals used in some experiments suffer serious pain as a result.
2. Those same animals have caused no harm (i.e., they do not deserve to suffer).
3. It is morally wrong to inflict pain upon a creature that does not deserve it.
4. Therefore, those who use animals in some experiments are committing morally wrong actions.

Now, we've laid this argument out in more detail than is probably necessary (it could be done in two steps instead of four), but we want it to be clear what is going on. If you remember the anatomy of arguments from Chapter 1,

Conclusion Indicators

When the words in the following list are used in arguments, they usually indicate that a premise has just been offered and that a conclusion is about to be presented. (The three dots represent the claim that is the conclusion.)

Thus . . .	Consequently . . .
Therefore . . .	So . . .
Hence . . .	Accordingly . . .
This shows that . . .	This implies that . . .
This suggests that . . .	This proves that . . .

Example:

Stacy drives a Porsche. This suggests that either she is rich or her parents are.

The conclusion is

Either she is rich or her parents are.

The premise is

Stacy drives a Porsche.

you'll notice that the first three claims in the argument are **premises,** and the fourth claim is the argument's **conclusion.**

This argument is actually a pretty good one. As indicated back in Chapter 1 and as we'll discuss in detail later, what this means is that the premises do in fact offer good support for the conclusion. Indeed, if all three of the premises are true, it is not going to be easy to reject the conclusion. So, if one is going to be able to reject the conclusion, it will be necessary to find something wrong with at least one of the premises. (There. We just made another argument, this time the subject was the first argument. Can you identify the premise and the conclusion of the new argument?)

Here are some further examples of arguments with the premises and conclusions identified:

[Premise] Every officer on the force has been certified, and [premise] nobody can be certified without scoring above 70 percent on the firing range. Therefore, [conclusion] every officer on the force must have scored above 70 percent on the firing range.

[Premise] Mr. Conners, the gentleman who lives on the corner, comes down this street on his morning walk every day, rain or shine. So [conclusion] something must have happened to him, because [premise] he has not shown up today.

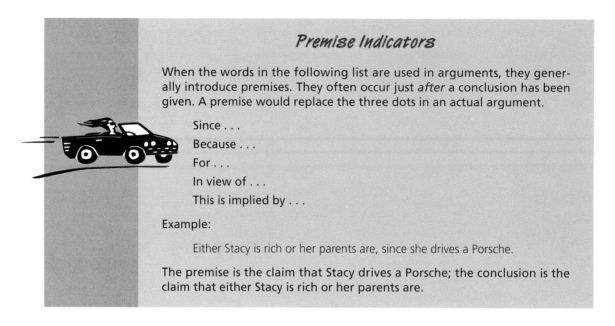

Premise Indicators

When the words in the following list are used in arguments, they generally introduce premises. They often occur just *after* a conclusion has been given. A premise would replace the three dots in an actual argument.

Since . . .

Because . . .

For . . .

In view of . . .

This is implied by . . .

Example:

Either Stacy is rich or her parents are, since she drives a Porsche.

The premise is the claim that Stacy drives a Porsche; the conclusion is the claim that either Stacy is rich or her parents are.

Notice that sometimes the conclusion of one argument can serve as the premise of another:

[Premise] Every student who made 90 percent or better on the midterms has already been assigned a grade of A. [Premise] Since Margaret made 94 percent on her midterms, [conclusion] she already has her A.

[Premise] All those students who have been assigned A's are excused from the final exam. [Premise] Margaret got an A, so [conclusion] she is excused from the final.

The claim that Margaret has a grade of A is the conclusion in the first argument but a premise in the second.

Notice also that arguments can have unstated premises:

[Premise] You can't check books out of the library without an ID card. So [conclusion] Bill won't be able to check any books out.

The unstated premise must be "Bill has no ID card." We'll have more to say about unstated premises later.

Arguments can have unstated conclusions as well:

[Premise] The political party that best reflects mainstream opinion will win the most seats in the next election, and [premise] the Republicans certainly best reflect mainstream opinion.

The unstated conclusion is "The Republicans will win the most seats in the next election."

Don't Confuse Arguments with Explanations!

Remember from Chapter 1, you use an *argument* to show that a claim is true. You use an **explanation** to show what caused something, or what it is, or how it works, or what purpose it serves.

> The reason I believe in God is because the universe couldn't just happen by chance.

> The reason I believe in God is because my parents were devout Christians who took great pains to instill this belief in me.

These two statements have similar wording, but the first one argues God exists, the second one explains the cause of my believing in God.
Arguments and explanations often use exactly the same words, and we have to be careful not to confuse them.

Notice finally that there is a difference between **independent premises** and **dependent premises** for a conclusion.

> [Premise] Raising the speed limit will wear out the highways faster. In addition, [premise] doing so will result in more highway deaths. Therefore, [conclusion] we should not raise the speed limit.

> [Premise] Raising the speed limit will waste gas. [Premise] We don't have any gas to waste. Therefore, [conclusion] we should not raise the speed limit.

The first example gives two independent premises for the conclusion that we should not raise the speed limit (doing so would wear out the highways; doing so would waste lives). The premises are independent of one another because the falsity of one would not cancel the support the other provides for the conclusion.

But the premises in the second example (raising the speed limit will waste gas; we don't have any gas to waste) are dependent on one another. The falsity of either premise would automatically cancel the support the other provides for the conclusion that the speed limit should not be raised.

If you want, you can think of an argument with two independent premises as *two arguments for the same conclusion:*

> Raising the speed limit will wear out the highways faster. In addition, doing so will result in more highway deaths. Therefore, we should not raise the speed limit.

You can view this either as one argument with two independent premises, or as two separate arguments for one conclusion (the conclusion that we should not raise the speed limit). There are subtle theoretical differences between these views, but these differences are of little consequence to the practical task of evaluating arguments.

Exercise 7-1

Indicate which blanks would ordinarily contain premises and which would ordinarily contain conclusions.

▲ 1. __a__ , and __b__ . Therefore, __c__ .

▲ 2. __a__ . So, since __b__ , __c__ .

▲ 3. __a__ , because __b__ .

▲ 4. Since __a__ and __b__ , __c__ .

▲ 5. __a__ . Consequently, __b__ , since __c__ and __d__ .

Exercise 7-2

Identify the premises and conclusions in each of the following arguments.

▲   1. Since all Communists are Marxists, all Marxists are Communists.

2. The Lakers almost didn't beat the Kings. They'll never get past Dallas.

3. If the butler had done it, he could not have locked the screen door. Therefore, since the door was locked, we know that the butler is in the clear.

▲ 4. That cat is used to dogs. Probably she won't be upset if you bring home a new dog for a pet.

5. Hey, he can't be older than his mother's daughter's brother. His mother's daughter has only one brother.

6. Moscone will never make it into the state police. They have a weight limit, and he's over it.

▲ 7. Presbyterians are not fundamentalists, but all born-again Christians are. So no born-again Christians are Presbyterians.

8. I guess he doesn't have a thing to do. Why else would he waste his time watching daytime TV?

9. "There are more injuries in professional football today than there were twenty years ago," he reasoned. "And if there are more injuries, then today's players suffer higher risks. And if they suffer higher risks, then they should be paid more. Consequently, I think today's players should be paid more," he concluded.

▲ 10. Let's see . . . If we've got juice at the distributor, the coil isn't defective, and if the coil isn't defective, then the problem is in the ignition switch. So the problem is in the ignition switch.

Exercise 7-3

Identify the premises and the conclusions in the following arguments.

▲ 1. The darned engine pings every time we use the regular unleaded gasoline, but it doesn't do it with super. I'd bet that there is a difference in the octane ratings between the two in spite of what my mechanic says.

2. Kera, Sherry, and Bobby were all carded at JJ's, and they all look as though they're about thirty. Chances are I'll be carded too.

3. Seventy percent of freshmen at Wharfton College come from wealthy families; therefore, probably about the same percentage of all Wharfton College students come from wealthy families.

▲ 4. When blue jays are breeding, they become very aggressive. Consequently, scrub jays, which are very similar to blue jays, can also be expected to be aggressive when they're breeding.

5. A cut in the capital gains tax will benefit wealthy people. Marietta says her family would be much better off if capital gains taxes were cut, so I'm sure her family is wealthy.

6. According to *Nature*, today's thoroughbred racehorses do not run any faster than their grandparents did. But human Olympic runners are at least 20 percent faster than their counterparts of fifty years ago. Most likely, racehorses have reached their physical limits but humans have not.

▲ 7. It's easier to train dogs than cats. That means they're smarter than cats.

8. "Let me demonstrate the principle by means of logic," the teacher said, holding up a bucket. "If this bucket has a hole in it, then it will leak. But it doesn't leak. Therefore, obviously it doesn't have a hole in it."

9. I know there's a chance this guy might be different, but the last person we hired from Alamo Polytech was a rotten engineer, and we had to fire him. Thus I'm afraid that this new candidate is somebody I just won't take a chance on.

▲ 10. If she were still interested in me, she would have called, but she didn't.

Exercise 7-4

Identify the conclusion in each of the following arguments. For any that have more than one premise, determine whether the premises provide dependent or independent reasons for the conclusion.

▲ 1. North Korea was always a much greater threat to its neighbors than Iraq. After all, North Korea has a million-man army lined up on the border ready to attack. Oh yes! They also have nuclear weapons and have said they'd use them.

2. Jim is going to ride with Mary to the party, and Sandra is going to ride with her too. So Mary won't be driving all by herself to the party.

▲ 3. Michael should go ahead and buy another car. The one he's driving is just about to fall apart, and he just got a new job and he can certainly afford another car now.

4. If Parker goes to Las Vegas, he'll wind up in a casino; and if he winds up in a casino, it's a sure thing he'll spend half the night at a craps table. So you can be sure: If Parker goes to Las Vegas, he'll spend half the night at a craps table.

5. It's going to be rainy tomorrow, and Moore doesn't like to play golf in the rain. It's going to be cold as well, and he *really* doesn't like to play when it's cold. So you can be sure Moore will be someplace other than the golf course tomorrow.

▲ 6. Hey, you're overwatering your lawn. See? There are mushrooms growing around the base of that tree—a sure sign of overwatering. Also, look at all the worms on the ground. They come up when the earth is oversaturated.

7. "Will you drive me to the airport?" she asked. "Why should I do that?" he wanted to know. "Because I'll pay you twice what it takes for gas. Besides, you said you were my friend, didn't you?"

8. If you drive too fast, you're more likely to get a ticket, and the more likely you are to get a ticket, the more likely you are to have your insurance premiums raised. So, if you drive too fast, you are more likely to have your insurance premiums raised.

▲ 9. If you drive too fast, you're more likely to get a ticket. You're also more likely to get into an accident. So you shouldn't drive too fast.

10. The war on terrorism had as its original goal the capture or elimination of Osama bin Laden. Since he's still at large, as of this point the war has not been very successful.

11. DANIEL: Where did that cat go, anyway?
THERESA: I think she ran away. Look, her food hasn't been touched in two days. Neither has her water.

▲ 12. There are several reasons why you should consider installing a solarium. First, you can still get a tax credit. Second, you can reduce your heating bill. Third, if you build it right, you can actually cool your house with it in the summer.

13. From a letter to the editor: "By trying to eliminate Charles Darwin from the curriculum, creationists are doing themselves a great disservice. When read carefully, Darwin's discoveries only support the thesis that species change, not that they evolve into new species. This is a thesis that most creationists can live with. When read carefully, Darwin actually supports the creationist point of view."

14. Editorial comment: "The Supreme Court's ruling that schools may have a moment of silence but not if it's designated for prayer is sound. Nothing stops someone from saying a silent prayer at school or anywhere else. Also, even though a moment of silence will encourage prayer, it will not favor any particular religion over any other. The ruling makes sense."

▲ 15. We must paint the house now! Here are three good reasons: (a) If we don't, then we'll have to paint it next summer; (b) if we have to paint it next summer, we'll have to cancel our trip; and (c) it's too late to cancel the trip.

Exercise 7-5

Identify the conclusion in the following arguments. For any that have more than one premise, determine whether the premises provide dependent or inde-
▲ pendent reasons for the conclusion.

1. All mammals are warm-blooded creatures, and all whales are mammals. Therefore, all whales are warm-blooded creatures.

2. Jones won't plead guilty to a misdemeanor, and if he won't plead guilty, then he will be tried on a felony charge. Therefore, he will be tried on a felony charge.

▲ 3. John is taller than Bill, and Bill is taller than Margaret. Therefore, John is taller than Margaret.

4. Rats that have been raised in enriched environments, where there are a variety of toys and puzzles, have brains that weigh more than the brains of rats raised in more barren environments. Therefore, the brains of humans will weigh more if humans are placed in intellectually stimulating environments.

5. From a letter to the editor: "In James Kilpatrick's July 7 column it was stated that Scientology's 'tenets are at least as plausible as the tenets of Southern Baptists, Roman Catholics . . . and prayer book Episcopalians.' Mr. Kilpatrick seems to think that all religions are basically the same and fraudulent. This is false. If he would compare the beliefs of Christianity with the cults he would find them very different. Also, isn't there quite a big difference between Ron Hubbard, who called himself God, and Jesus Christ, who said 'Love your enemies, bless them that curse you, do good to them that hate you'?"

6. We've interviewed two hundred professional football players, and 60 percent of them favor expanding the season to twenty games. Therefore, 60 percent of all professional football players favor expanding the season to twenty games.

7. Exercise may help chronic male smokers kick the habit, says a study published today. The researchers, based at McDuff University, put thirty young male smokers on a three-month program of vigorous exercise. One year later, only 14 percent of them still smoked, according to the report. An equivalent number of young male smokers who did not go through the exercise program were also checked after a year and it was found that 60 percent still smoked. Smokers in the exercise program began running three miles a day and gradually worked up to eight miles daily. They also spent

five and a half hours each day in modestly vigorous exercise such as soccer, basketball, biking, and swimming.

▲ 8. Letter to the editor: "I was enraged to learn that the mayor now supports the initiative for the Glen Royale subdivision. Only last year he himself proclaimed 'strong opposition' to any further development in the river basin. Besides, Glen Royale will only add to congestion, pollution, and longer lines at the grocery store, not that the grocers will mind."

9. Believe in God? Yes, of course I do. The universe couldn't have arisen by chance, could it? Besides, I read the other day that more and more physicists believe in God, based on what they're finding out about the Big Bang and all that stuff.

▲ 10. From an office memo: "I've got a good person for your opening in Accounting. Jesse Brown is his name, and he's as sharp as they come. Jesse has a solid background in bookkeeping, and he's good with computers. He's also reliable, and he'll project the right image. Best of all, he's a terrific golfer. As you might gather, I know him personally. He'll be contacting you later this week."

Exercise 7-6

Which five of the following statements are probably intended to explain the cause of something, and which five are probably intended to argue that some claim is true?

▲ 1. The reason we've had so much hot weather recently is that the jet stream is unusually far north.

2. The reason Ms. Mossbarger looks so tired is that she hasn't been able to sleep for three nights.

3. The reason it's a bad idea to mow the lawn in your bare feet is that you could be seriously injured.

▲ 4. The reason Ken mows the lawn in his bare feet is that he doesn't realize how dangerous it is.

5. You can be sure that Ryan will marry Beth. After all, he told me he would.

6. If I were you, I'd change before going into town. Those clothes look like you slept in them.

▲ 7. Overeating can cause high blood pressure.

8. Eating so much salt can cause high blood pressure, so you'd better cut back a little.

▲ 9. It's a good bet the Saddam Hussein regime wanted to build nuclear weapons, because the U.N. inspectors found devices for the enrichment of plutonium.

10. The reason Saddam wanted to build nuclear weapons was to give him the power to control neighboring Middle Eastern countries.

Good and Bad, Valid and Invalid, Strong and Weak

When we say that an argument is a **good argument,** we are saying that it gives us grounds for accepting its conclusion. "Good" and "bad" are relative terms: Arguments can be better or worse depending on the degree to which they furnish support for their conclusions.

There is more than one way in which an argument might qualify as good; before we explain them, however, we need to describe some important technical distinctions.

An argument whose premises provide absolutely conclusive support for the conclusion is "valid." In other words,

> A **valid argument** has this characteristic: It is necessary, on the assumption that the premises are true, that the conclusion be true.

This is merely a precise way of saying that the premises of a valid argument, if true, absolutely guarantee a true conclusion. Here's an example:

> [Premise] Every philosopher is a good mechanic, and [premise] Emily is a philosopher. So, [conclusion] Emily is a good mechanic.

Valid? Yes. These premises, if true, guarantee that the conclusion is true.

But notice this: Although the argument about Emily is valid, it so happens the premises aren't true. Not every philosopher is a good mechanic, and Emily is no philosopher; she's Parker's cat. So the argument is not a good one, from the standpoint of offering us justification for accepting the claim that Emily is a good mechanic. However, the argument is valid nonetheless, because the conclusion *must* follow from the premises. Thus, an argument's being valid does not depend on its premises being true. What determines whether an argument is valid is whether the conclusion *absolutely follows* from the premises. (Once again: When we say that a conclusion absolutely follows from the premises, we mean that, *if* the premises were true, the conclusion would then *have to be* true as well.) If the conclusion does absolutely follow from the premises, the argument is valid, whether or not the premises are true.

Now, a valid argument whose premises *are* true is called a "sound" argument:

> A **sound argument** has these two characteristics: It is valid, and its premises are all true.

And here is an example:

> [Premise] Some pesticides are toxic for humans, and [premise] anything that is toxic for humans is unsafe for most humans to consume. Therefore, [conclusion] some pesticides are unsafe for most humans to consume.

Abe Lincoln Knew His Logic
Validity and Soundness in the Lincoln-Douglas Debates

Here's Abraham Lincoln speaking in the fifth Lincoln-Douglas debate:

> I state in syllogistic form the argument:
> Nothing in the Constitution . . . can destroy a right distinctly and expressly affirmed in the Constitution.
>   The right of property in a slave is distinctly and expressly affirmed in the Constitution.
>   Therefore, nothing in the Constitution can destroy the right of property in a slave.

Lincoln goes on to say:

> There is a fault [in the argument], but the fault is not in the reasoning; but the falsehood in fact is a fault of the premises. I believe that the right of property in a slave is *not* distinctly and expressly affirmed in the Constitution.

In other words, the argument is valid, Lincoln says, but unsound, and thus not a good argument.
    Syllogisms, by the way, are covered in Chapter 8.

This is a sound argument: It is valid, because the conclusion absolutely follows from the premises, and its premises (and hence its conclusion) are true.

It should be clear, then, that an argument can be valid without necessarily being a good argument. The argument about Emily isn't good because, even though it's valid, it doesn't justify accepting the conclusion: Its premises are false.

A sound argument, by contrast, normally does justify accepting the conclusion. We say "normally" because cases can arise in which even sound arguments aren't particularly good. Arguments that beg the question, as explained in Chapter 6, fall in this category. Or consider the argument:

> If a person travels in the same direction far enough, he or she will arrive back at the place he or she started from. Therefore, if Kotar travels in the same direction far enough, he will arrive back at the place he started from.

This is a sound argument; but, if Kotar happened to live in the middle of Europe in the tenth century, it seems odd to suppose *he* would have been justified in believing that if he traveled far enough in the same direction, he would arrive back where he started.

Arguments can be useful—they can qualify as good arguments—even though we don't intend them to be valid or sound. Look at this example, in which Moore says to Parker:

Everyday English Definitions

In everyday English, "valid argument," "sound argument," and "strong argument" are used interchangeably and often just mean "good argument." Other terms people use to praise arguments include "cogent," "compelling," and "telling." Further, people often apply "valid," "sound," "cogent," and "compelling" to things other than arguments—claims, theories, and explanations, for example. So before you pounce on someone for using these words incorrectly, consider the context in which you hear them.

In everyday English, "weak argument" includes arguments that are invalid and unsound and has numerous synonyms, including "poor," "faulty," "fallacious," "specious," "unreasonable," "illogical," "stupid," and so forth.

[Premise] Every year as far back as I can remember my roses have developed mildew in the spring. [Conclusion] Therefore, my roses will develop mildew this spring, too.

This argument doesn't qualify as valid (or sound) because it is *possible* that the conclusion is false even assuming the premise is true (it might be an incredibly dry winter). Nevertheless, this isn't a bad argument. In fact, it's really quite good: It may not be absolutely impossible that Moore's roses won't get mildew this spring, but, given the premise, it is very, very likely that they will. Moore is certainly justified in believing his conclusion.

Arguments like this one, which only show that the conclusion is *probably* true, are said to be relatively *strong*. More precisely:

> A **strong argument** has this distinguishing characteristic: It is *unlikely*, on the assumption that the premises are true, that the conclusion is false.

Again, notice that the premises don't actually have to be true for the argument to be strong.

When someone advances an argument like Moore's, an argument that he or she intends only to be a strong argument, it is somewhat inappropriate to discuss whether it is valid. Yes, technically, Moore's argument is invalid, but because Moore is only trying to demonstrate that the conclusion is *likely*, the criticism doesn't amount to much—he never intended that the argument be valid.

Let's summarize these various points:

1. A *good* argument justifies acceptance of the conclusion.
2. A *valid* argument has this defining characteristic: It is necessary, on the assumption that the premises are true, that the conclusion be true.

The "Science of Deduction"

Possibly the most familiar uses of the term "deduction" occur in the Sherlock Holmes stories of Arthur Conan Doyle. The most famous of these is no doubt the second chapter of *A Study in Scarlet,* which is titled "The Science of Deduction." In that chapter Dr. Watson discovers and describes Holmes's methods of observation and inference-drawing. As it turns out, Holmes engages more frequently in *inductive* inference than deductive.

When he identifies a man crossing the street below from their sitting-room window as a retired sergeant of marines, Holmes does so from such clues as an anchor tattoo on the hand and "a military carriage . . . and regulation side whiskers." Thus, he says, "There we have the marine." Notice that these facts about the man do not *absolutely guarantee* the conclusion that the man is a marine, and hence we do not have a valid deductive argument. At best, we have a strong inductive argument. (Just how strong is arguable; but never argue the point with a serious Sherlock Holmes fan.)

3. A valid argument whose premises are all true is called a *sound* argument.

4. A *strong* argument has this defining characteristic: It is unlikely, on the assumption that the premises are true, that the conclusion is false.

5. Normally, sound arguments and strong arguments with true premises are good arguments.

6. The best policy is not to speak of valid arguments as strong or weak; speak of them as sound or unsound. Likewise, the best policy is not to speak of strong or weak arguments as valid or invalid; just refer to them as strong or weak.

Finally, notice that the terms "valid" and "invalid" are absolute terms. Either an argument is valid, or it is not, and that's that. By contrast, "strong" and its opposite, "weak" (like "good" and "bad"), are relative terms. Arguments can be evaluated as stronger or weaker depending on how likely the premises show the conclusion to be.

Deduction and Induction

Philosophers have traditionally distinguished deductive arguments from inductive arguments, though the best way of defining the difference is controversial. One way to view them is to say that **deductive arguments** are those that are either valid or intended by their author to be so. **Inductive**

Deductive and Inductive; General and Specific

A lot of people, including a few authors of articles and books on logic, think it's possible to explain the difference between deductive arguments and inductive arguments by referring to whether they go from general to particular or vice versa. In the words of one such author, "Deductive arguments involve reasoning from general premises to specific conclusions, while inductive arguments involve reasoning from specific instances to general conclusions."

It's true that *many* arguments fit this scheme, but many do not. Here's an argument that is clearly deductive (and valid, in fact):

> Maria scored higher on the logic exam than Ron did, and Ron scored higher than Kenneth. So Maria must have scored higher than Kenneth.

Notice that there are no general claims at all in this argument; each one is about a specific relationship, and so it is false that all deductive arguments proceed from general claims to specific ones.

Now look at this argument:

> I've watched *The Sopranos* television show four or five times recently, and I've really liked it each time. So it's a safe bet that I'll enjoy it again when I watch it tonight.

This argument is inductive (of a type called arguments by analogy, which are described in Chapter 10), but it too contains no general claims at all; it proceeds from particular cases to a further particular case. And so it is also false that all inductive arguments proceed from specific instances to general conclusions.

arguments are neither valid nor intended to be so by their authors; nevertheless, such arguments can be quite strong.

Exercise 7-7

Fill in the blanks where called for, and answer true or false where appropriate.

1. Valid arguments are said to be strong or weak.
2. Valid arguments are always good arguments.
▲ 3. Sound arguments are _____ arguments whose premises are all _____.
4. The premises of a valid argument are never false.
5. If a valid argument has a false conclusion, then not all its premises can be true.
▲ 6. If a strong argument has a false conclusion, then not all its premises can be true.

Speaking of Holmes and Watson

Here's another item that mentions them and has (a little) to do with observation, a topic from Chapter 3.

It seems Sherlock Holmes and Dr. Watson were out on a camping trip. One night, a couple of hours after they'd gone to sleep, Holmes woke up and nudged his faithful friend.

"Watson, look up at the sky and tell me what you see."

Watson replied, "I see hundreds and hundreds of stars."

"What does that tell you?" said Holmes.

Watson pondered for a moment. "Astronomically, it tells me that there are many galaxies and potentially many, many planets similar to ours. Astrologically, I notice Saturn is in Leo. Horologically, I deduce that the time is approximately a quarter past three. Theologically, I can see that God is all powerful and we are small and insignificant. Meteorologically I suspect that we will have a beautiful day tomorrow. What does it tell you, Holmes?"

After a moment, Holmes spoke, "Watson, you idiot. Someone has stolen our tent."

7. A sound argument cannot have a false conclusion.

8. A true conclusion cannot be derived validly from false premises.

▲ 9. "Strong" and "weak" are absolute terms.

10. The following argument is valid: All swans are orange. You're a swan, so you're orange.

11. The following argument is *both* valid *and* sound: Jesse Helms is one of New York's U.S. senators, and he is a Democrat. Therefore, he is a Democrat.

12. A strong argument with true premises has a conclusion that probably is true.

Exercise 7-8

Go back to Exercises 7-2 and 7-3, and determine which arguments are valid. In the answer section at the back of the book, we have answered items 1, 4, 7, and 10 in each set.

Exercise 7-9

Given the premises, discuss whether the conclusion of each argument that follows is (a) true beyond a reasonable doubt, (b) probably true, or (c) possibly true or possibly false. You should expect disagreement on these items, but the closer your answers are to your instructor's, the better.

▲ 1. The sign on the parking meter says "Out of Order," so the meter isn't working.

2. The annual rainfall in California's north valley averages twenty-three inches. So the rainfall next year will be about twenty-three inches.

3. You expect to get forty miles to the gallon in *that?* Why, that old wreck has a monster V8; besides, it's fifty years old and needs an overhaul.

4. In three of the last four presidential races, the winner of the Iowa Republican primary has not captured the Republican nomination. Therefore, the winner of the next Iowa Republican primary will not capture the Republican nomination.

▲ 5. The New York steak, the Maine lobster, and the beef stroganoff at that restaurant are all exceptionally good. You can probably count on all the entrees being excellent.

6. The number of cellular telephones has increased dramatically in each of the past few years. It's a safe bet that there will be even more of them in use this coming year.

7. Since the graduates of Harvard, Yale, Princeton, and other Ivy League schools generally score higher on the Graduate Record Examination than students from Central State, it follows that the Ivy League schools do more toward educating their students than Central State does.

8. Michael Jackson has had more plastic surgery than anybody else in California. You can bet he's had more than anybody in Connecticut!

▲ 9. Although Max bled profusely before he died, there was no blood on the ground where his body was found. Therefore, he was killed somewhere else and brought here after the murder.

10. When liquor was banned in 1920, hospitalizations for alcoholism and related diseases plummeted; in 1933, when Prohibition was repealed, alcohol-related illnesses rose sharply again. Legalization of cocaine, heroin, and marijuana would not curb abuse of those substances.

11. Relax. The kid's been delivering the paper for, how long? Three, four years maybe? And not once has she missed us. The paper will be here, just wait and see. She's just been delayed for some reason.

▲ 12. First, it seems clear that even if there are occasional small dips in the consumption of petroleum, the general trend shows no sign of a real permanent decrease. Second, petroleum reserves are not being discovered as fast as petroleum is currently being consumed. From these two facts we can conclude that reserves will eventually be consumed and that the world will have to do without oil.

Unstated Premises

As we've noted, arguments can have unstated premises. Suppose a husband and wife are dining out, and she says to him: "It's getting very late, so we should call for the check." For this little argument to be valid, this principle must be assumed: *"If it is getting very late, then we should call for the check."* The wife, no doubt, is assuming this principle (among many other things), and even if *she* is not assuming it, *we* must assume it if we are to credit her with a valid argument.

Most real-life arguments are like this argument in that some unstated proposition must be assumed for the argument to be valid. And it is useful to be able to specify what must be assumed for a given argument to be valid. What, for example, must be assumed for the following argument to be valid?

> Moore's dog is a bloodhound, so it has a keen sense of smell.

What must be assumed, of course, is that all bloodhounds have a keen sense of smell. We are also assuming here that we know nothing about the dog in question except that it is a bloodhound. The argument is valid if and only if we assume that all bloodhounds have a keen sense of smell. Further, since it is questionable whether *all* bloodhounds have a keen sense of smell, we see that the validity of this argument depends on a questionable assumption.

However, recall that an argument need not necessarily be valid to be a good argument. Invalid arguments may still be strong; that is, their premises may provide strong support, though less than absolutely conclusive support, for their conclusions. Let's therefore ask, Is there an unstated premise that would make the bloodhound argument a strong one? The answer is that if we plug into the argument the premise that *most* bloodhounds have a keen sense of smell, which is certainly a reasonable premise, the result is quite a strong argument:

> Moore's dog is a bloodhound.
> [Unstated: Most bloodhounds have a keen sense of smell.]
> Therefore, Moore's dog has a keen sense of smell.

If you can spot what must be assumed for an argument to be valid or to be strong, you'll be in a much better position to evaluate it. If a plausible assumption suffices to make an argument valid, then it is a good argument. If a plausible assumption can at best only make an argument strong, then it may still be a good argument, depending on how strong it is and how plausible the stated premises and unstated assumption are.

Let's consider another example:

> He's related to Edward Kennedy, so he's rich.

For the argument to be valid, one must assume that *everyone* related to Kennedy is rich, and that's not a very plausible assumption. For the argument to be strong, one must assume that *most* people related to Kennedy are rich, and although that assumption is a good bit more plausible than the first assumption, it still isn't very plausible. The most plausible relevant assumption is that *many* people related to Kennedy are rich, but with

Unstated Arguments

There aren't any. An argument can contain unstated premises or an unstated conclusion, but the argument *itself* cannot be entirely unstated. A masked bandit who waves you against a wall with a gun is not presenting you with an argument, though his actions give you a good reason for moving; if you do not want to get shot, then you need to construct an argument in your own mind—and quickly, too—with the conclusion "I'd better move" and the premise "If I don't move, I may get my head blown off."

There is no such thing, then, as an unstated argument, though an argument can have unstated premises or an unstated conclusion. Further, it isn't possible for *all* the premises in an argument to be unstated. If the bandit waves his gun and says "Better move over against that wall," he has given you a reason for moving, but he hasn't given you an argument. Maybe you think that waving a gun around *implies* a premise such as "If you don't move, I'll shoot you." But no, implication is a relationship that holds only between claims: Only a claim can imply a claim.

that assumption plugged into the argument, the argument is not particularly strong.

Another example:

It's March, so the apple trees will bloom by the end of the month.

This argument is valid only if it is assumed that the apple trees *always* bloom by the end of March. Maybe that's true where you live, but it's not true where we live. Still, where we live it's true that the apple trees *almost always* bloom by the end of March, and with this assumption plugged into the argument, the argument turns out to be pretty strong.

Identifying Unstated Premises

When someone produces an argument, it isn't productive to ask, What is this person assuming? No doubt the person is assuming many things, such as that you can hear (or read), that you understand English, that what he or she is saying is worth saying, that you aren't comatose or dead, and who knows what else. There are two better questions to ask, and the first is, "Is there a reasonable assumption I could make that would make this argument valid?" And if what you must assume for the argument to be valid isn't plausible, then you ask the second question: "Is there a reasonable assumption I could make that would make this argument strong?"

As may be clear from the preceding section, an argument can be made valid by adding to it a general claim that appropriately connects the stated premises to the conclusion and says, in effect, that if the stated premises are true, then the conclusion is true. For example, take this argument:

> It's raining, so there is a south wind.

This argument is automatically valid if you add to it the general claim "Any time it is raining, there is a south wind." (Any variation of this claim will do equally well, such as "Whenever it is raining, there is a south wind," or "Only in the presence of a south wind is there rain.") But this claim isn't particularly plausible, at least where Moore and Parker live. And thus we should modify the claim by asking the second question, "What must I assume for this argument to be strong?" When we do this, we obtain a more plausible premise: "When it rains, there is usually a south wind." This is a reasonable claim (in our neck of the woods), and including it in the argument makes the argument a strong one.

Here's another example:

> You shouldn't let her pass. After all, this is the second time you caught her cheating.

A general claim that ties the stated premise to the conclusion and yields a valid argument would be "No person caught cheating two times should be permitted to pass." We'd accept this claim and, consequently, we'd also accept the conclusion of the argument, assuming that the stated premise (that this is the second time she was caught cheating) is true.

Here's one final example:

> Yes, Stacy and Harold are on the brink of divorce. They're remodeling their house.

When we try to make the argument valid by connecting the premise to the conclusion with a general claim, we get something like "A couple that remodels their house must be on the brink of divorce." But we can suppose that nobody would seriously subscribe to such a claim. So we attempt to make the argument strong, rather than valid, by modifying the claim to "Most couples remodeling their house are on the brink of divorce." However, this claim is also implausible. The only plausible thing to say is that *some* couples who remodel are on the brink of divorce. But this claim doesn't justify a conclusion that states unqualifiedly that Stacy and Harold are on the brink of divorce.

To summarize: Formulate a general claim that connects the stated premise with the conclusion in such a way as to make the argument valid, as explained above. If that premise isn't plausible, modify or qualify it as necessary to make the argument strong. If the argument just can't be made either valid or strong except by using an implausible premise, then you really shouldn't accept the argument—or use it yourself.

In your own essays, it is good practice always to consider exactly what your readers must accept if they are to agree with your reasoning. You shouldn't require them to accept any claim that is implausible or unreasonable, whether you actually state that claim or not.

Exercise 7-10

For each passage, supply a claim that turns it into a valid argument.

Example

> The fan needs oil. It's squeaking.

Claim That Makes It Valid

> Whenever the fan squeaks, it needs oil.

▲ 1. Jamal is well mannered, so he had a good upbringing.
   2. Bettina is pretty sharp, so she'll get a good grade in this course.
   3. It must have rained lately because there are puddles everywhere.
▲ 4. He'll drive recklessly only if he's upset, and he's not upset.
   5. Let's see . . . they have tons of leftovers, so their party could not have been very successful.
   6. I think we can safely conclude that the battery is still in good condition. The lights are bright.
▲ 7. Either the dog has fleas, or its skin is dry. It's scratching a lot.
   8. Melton was a good senator. He'd make an excellent president.
   9. Gelonek doesn't own a gun. He's sure to be for gun control.
▲ 10. The Carmel poet Robinson Jeffers is one of America's most outstanding poets. His work appears in many Sierra Club publications.

Exercise 7-11

Go back to Exercise 7-10, and supply a claim that will produce a strong (but not valid) argument. In the answer section at the back of the book, we have answered items 1, 4, 7, and 10.

Example

> The fan needs oil. It's squeaking.

Claim That Makes It Strong

> When the fan squeaks, it usually needs oil.

Exercise 7-12 _____

For each passage, supply a claim that turns it into a valid argument.

▲ 1. Prices in that new store around the corner are going to be high, you can bet. All they sell are genuine leather goods.

2. I had a C going into the final exam, but I don't see how I can make less than a B for the course because I managed an A on the final.

3. He's a good guitarist. He studied with Pepe Romero, you know.

▲ 4. That plant is an ornamental fruit tree. It won't ever bear edible fruit.

5. The Federal Reserve Board will make sure that inflation doesn't reach 8 percent again. Its chair is an experienced hand at monetary policy.

6. Murphy doesn't stand a chance of getting elected in this county. His liberal position on most matters is well known.

▲ 7. Jesse Ventura, the former Governor of Minnesota, was a professional wrestler. He couldn't have been a very effective governor.

8. Half the people in the front row believe in God; therefore, half the entire class believes in God.

9. Kennesaw State students are all career-oriented. I say this because every Kennesaw State student I ever met was career-oriented.

▲ 10. Population studies show that smoking causes lung cancer; therefore, if you smoke, you will get lung cancer.

Exercise 7-13 _____

Do Exercise 7-12 again, making the arguments strong but not valid. In the answer section at the back of the book, we have answered items 1, 4, 7, and 10.

Techniques for Understanding Arguments

A good argument provides justification for accepting its conclusion. To do this, its premises must be reasonable—that is, it must be likely that they are true—and they must support the conclusion—that is, the argument must be either valid or strong. So, to evaluate an argument, we must answer these two questions:

1. Are the premises reasonable (i.e., is it likely that they are true)?

2. Do the premises support the conclusion (i.e., is the argument either valid or strong)?

Before we can proceed with the evaluation of an argument, we have to understand it. Many arguments are difficult to understand because they are

Don't Forget Fallacies

In Chapters 5 and 6 we say that people will sometimes make statements in order to establish a claim when in reality their remarks have nothing to do with the claim. For example, "Margaret's qualifications are really quite good; after all, she'd be terribly hurt to think that you didn't think highly of them." Although Margaret's disappointment would no doubt be a reason for *something* (e.g., for keeping your views to yourself), it would not be a reason for altering your opinion of her qualifications. Her feelings are, in fact, thoroughly irrelevant to her qualifications. Extraneous material of all sorts can often be eliminated as argumentative material if you ask yourself simply, "Is this really relevant to the conclusion? Does this matter to what this person is trying to establish?" If you have worked through the exercises in Chapters 5 and 6, you have already had practice in spotting irrelevances. If you haven't done these exercises, it would be useful to go back and do them now.

spoken and thus go by so quickly that we cannot be sure of the conclusion and the premises. Others are difficult to understand because they have a complicated structure. Still others are difficult to understand because they are embedded in nonargumentative material consisting of background information, prejudicial coloring, illustrations, parenthetical remarks, digressions, subsidiary points, and other window dressing. And some arguments are difficult to understand because they are confused or because the reasons they contain are so poor that we are not sure whether to regard them as reasons.

In understanding any argument, the first task is to find the conclusion—the main point or thesis of the passage. The next step is to locate the reasons that have been offered for the conclusion—that is, to find the premises. Next, we look for the reasons, if any, given for these premises. To proceed through these steps, you have to learn both to spot premises and conclusions when they occur in spoken and written passages and to understand the interrelationships among these claims—that is, the structure of the argument.

■ Clarifying an Argument's Structure

Let's begin with how to understand the relationships among the argumentative claims, because this problem is sometimes easiest to solve. If you are dealing with written material that you are free to mark up, one useful technique is to number the premises and conclusions and then use the numbers to lay bare the structure of the argument. Let's start with this argument as an example:

I don't think we should get Carlos his own car. As a matter of fact, he is not responsible, because he doesn't care for his things. And anyway, we don't have enough money for a car, since even now we have trouble making ends meet. Last week you yourself complained about our financial situation, and you never complain without really good reason.

We want to display the structure of this argument clearly. First, circle all premise and conclusion indicators. Thus:

> I don't think we should get Carlos his own car. As a matter of fact, he is not responsible (because) he doesn't care for his things. And anyway, we don't have enough money for a car, (since) even now we have trouble making ends meet. Last week you yourself complained about our financial situation, and you never complain without really good reason.

Next, bracket each premise and conclusion, and number them consecutively as they appear in the argument. So what we now have is this:

> ① [I don't think we should get Carlos his own car.] As a matter of fact, ② [he is not responsible] because ③ [he doesn't care for his things.] And anyway, ④ [we don't have enough money for a car], since ⑤ [even now we have trouble making ends meet.] ⑥ [Last week you yourself complained about our financial situation], and ⑦ [you never complain without really good reason.]

And then we diagram the argument as follows: Using an arrow to mean "therefore" or "is intended as evidence [or as a reason or as a premise] for," the first three claims in the argument can be diagrammed as follows:

Now ⑥ and ⑦ together support ④; that is, they are *dependent* reasons for ④. To show that ⑥ and ⑦ are dependent reasons for ④, we simply draw a line under them, put a plus sign between them, and draw the arrow from the line to ④, like this:

Because ⑤ and ⑥ + ⑦ are *independent* reasons for ④, we can represent the relationship between them and ④ as follows:

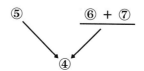

Finally, because ④ and ② are *independent* reasons for ①, the diagram of the entire argument is this:

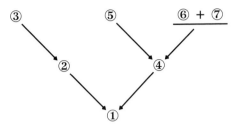

So, the conventions governing this approach to revealing argument structure are very simple: First, circle all premise- and conclusion-indicating words. Then, assuming you can identify the claims that function in the argument (a big assumption, as you will see before long), simply number them consecutively. Then display the structure of the argument using arrows for "therefore" and plus signs over a line to connect together two or more dependent premises.

Some claims, incidentally, may constitute reasons for more than one conclusion. For example:

① [Carlos continues to be irresponsible.] ② [He certainly should not have his own car], and, as far as I am concerned, ③ [he can forget about that trip to Hawaii this winter, too.]

Structure:

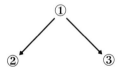

Frequently, too, we evaluate counterarguments to our positions. For example:

① We really should have more African Americans on the faculty. ② That is why the new diversity program ought to be approved. True, ③ it may involve an element of unfairness to whites, but ④ the benefits to society of having more black faculty outweigh the disadvantages.

Notice that claim ③ introduces a consideration that runs counter to the conclusion of the argument, which is stated in ②. We can indicate counterclaims by crossing the "therefore" arrow with lines, thus:

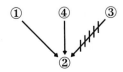

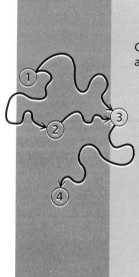

Beer Makes You Smarter

Can you diagram this argument from Tony Graybosch? (The answer is in the answer section at the back of the book.)

A herd of buffalo can only move as fast as the slowest buffalo and when the herd is hunted, it is the slowest and weakest ones at the back that are killed first. This natural selection is good for the herd as a whole, because the general speed and health of the whole group keeps improving by the regular attrition of the weakest members.

In much the same way, the human brain can only operate as fast as the slowest brain cells. Excessive intake of alcohol, we all know, kills brain cells; but naturally it attacks the slowest and weakest brain cells first. In this way, regular consumption of beer eliminates the weaker brain cells, making the brain a faster and more efficient machine. This shows that you always are smarter after a few beers.

This diagram indicates that item ③ has been introduced by the writer as a consideration that runs counter to ②.

Of course, one might adopt other conventions for clarifying argument structure—for example, circling the main conclusion and drawing solid lines under supporting premises and wavy lines under the premises of sub-arguments. The technique we have described is simply one way of doing it; any of several others might work as well for you. However, *no* technique for revealing argument structure will work if you cannot spot the argumentative claims in the midst of a lot of background material.

■ Distinguishing Arguments from Window Dressing

We should point out that it is not always easy to isolate the argument in a speech or a written piece. Often, speakers and writers think that because their main points are more or less clear to them, they will be equally apparent to their listeners or readers. But it doesn't always work that way.

If you are having trouble identifying a conclusion in what you hear or read, it *could* be that the passage is not an argument at all. Make sure that the passage in question is not a report, a description, an explanation, or something else altogether, rather than an argument. The key here is determining whether the speaker or writer is offering reasons intended to convince you of one or more of the claims made in the passage.

The problem could also be that the conclusion is left unstated. Sometimes it helps simply to put the argument aside and ask yourself, "What is this person trying to prove?" In any case, the first and essential step in understanding an argument is to spot the conclusion.

[Premise] Sixty percent of a random sample of two hundred residents of Minneapolis believe in God; [conclusion] therefore, approximately 60 percent of all residents of Minneapolis believe in God.

[Premise] Sixty percent of a random sample of five hundred residents of Minneapolis believe in God; [conclusion] therefore, approximately 60 percent of all residents of Minneapolis believe in God.

The second argument is a better argument because its premise provides a greater justification for believing the conclusion. Weighing the relative strengths of inductive arguments is something we'll consider in Chapters 10 and 11.

Recap

An argument consists of a conclusion and premises. Valid arguments (arguments whose conclusions absolutely follow from the premises) are either sound or unsound, depending on whether or not their premises are all true. On the other hand, conclusions of some arguments are not even intended to follow absolutely from their premises; the premises are supposed to show only that it is likely that the conclusion is true. We speak of such arguments as being to varying degrees strong or weak.

Before you can evaluate an argument, you must understand it; to do this, find the conclusion. Then locate the reasons that have been offered in support of the conclusion—these reasons are stated in the premises. One technique for clarifying the structure of a written argument—if you can identify the claims that function in the argument—is to number the claims consecutively as they are written and then use the numbers to lay out the structure of the argument.

Whatever method helps you to understand an argument, make sure that the premises are reasonable and that they support the conclusion. In Chapters 8–11, we consider more carefully when the premises of an argument do support the conclusion.

Having read this chapter and the chapter recap, you should be able to

- ▲ Explain what the parts of an argument are and identify premises and conclusions in a variety of arguments
- ▲ Discuss the distinctions among good, valid, and strong arguments and relate them to inductive and deductive arguments
- ▲ Explain what unstated premises are, and supply premises in a variety of arguments that have unstated premises making those arguments valid
- ▲ Describe techniques for understanding arguments and for evaluating arguments

Additional Exercises

Exercise 7-14

Diagram the following "arguments," using the method explained in the text.

▲ 1. ①, because ② and ③. [Assume that ② and ③ are dependent.]

2. ① and ②; therefore ③. [Assume that ① and ② are independent.]

3. Since ①, ②; and since ③, ④. And since ② and ④, ⑤. [Assume that ② and ④ are independent.]

▲ 4. ①; therefore ② and ③. But because ② and ③, ④. Consequently, ⑤. Therefore, ⑥. [Assume ② and ③ are independent.]

5. ①, ②, ③; therefore ④. ⑤, in view of ①. And ⑥, since ②. Therefore ⑦. [Assume ①, ②, and ③ are dependent.]

Exercise 7-15

Go back to Exercises 7-2, 7-3, 7-4, and 7-5, and diagram the arguments using the method explained in the text. In the answer section at the back of the book, we have answered items 1, 4, 7, and 10 in Exercises 7-2 and 7-3; items 6, 9, 12, and 15 in Exercise 7-4; and items 1, 4, 8, and 10 in Exercise 7-5.

Exercise 7-16

Diagram the arguments contained in the following passages, using the method explained in the text.

▲ 1. Dear Jim,
Your distributor is the problem. Here's why. There's no current at the spark plugs. And if there's no current at the plugs, then either your alternator is shot or your distributor is defective. But if the problem were in the alternator, then your dash warning light would be on. So, since the light isn't on, the problem must be in the distributor. Hope this helps.
Yours,
Benita Autocraft

2. The federal deficit must be reduced. It has contributed to inflation, and it has hurt American exports.

3. It's high time professional boxing was outlawed. Boxing almost always leads to brain damage, and anything that does that ought to be done away with. Besides, it supports organized crime.

▲ 4. They really ought to build a new airport. It would attract more business to the area, not to mention the fact that the old airport is overcrowded and dangerous.

5. Vote for Jackson? No way. He's too radical, and he's too inexperienced, and his lack of experience would make him a dangerous council member.

Exercise 7-17 _____

Diagram the arguments contained in the following passages, using the method explained in the text. (Your instructor may have different instructions for you to follow.)

▲ 1. Cottage cheese will help you to be slender, youthful, and more beautiful. Enjoy it often.

2. If you want to listen to loud music, do it when we are not at home. It bothers us, and we're your parents.

3. If you want to see the best version of *The Three Musketeers,* try the 1948 version. Lana Turner is luscious; Vincent Price is dastardly; Angela Lansbury is exquisitely regal; and nobody ever has or ever will portray D'Artagnan with the grace, athleticism, or skill of Gene Kelly. Rent it. It's a must.

▲ 4. From a letter to the editor: "The idea of a free press in America today is a joke. A small group of people, the nation's advertisers, control the media more effectively than if they owned it outright. Through fear of an advertising boycott they can dictate everything from programming to news report content. Politicians as well as editors shiver in their boots at the thought of such a boycott. This situation is intolerable and ought to be changed. I suggest we all listen to National Public Radio and public television."

5. Too many seniors, disabled veterans, and families with children are paying far too much of their incomes for housing. Proposition 168 will help clear the way for affordable housing construction for these groups. Proposition 168 reforms the outdated requirement for an election before affordable housing can even be approved. Requiring elections for every publicly assisted housing venture, even when there is no local opposition, is a waste of taxpayers' money. No other state constitution puts such a roadblock in front of efforts to house senior citizens and others in need. Please support Proposition 168.

6. More than forty years after President John F. Kennedy's assassination, it's no easier to accept the idea that a loser like Lee Harvey Oswald committed the crime of the century all by himself with a $12.78 mail-order rifle and a $7.17 scope. Yet even though two-thousand-plus books and films about the episode have been made, there is no credible evidence to contradict the Warren Commission finding that "the shots which killed President Kennedy and wounded Governor Connally were fired by Lee Harvey Oswald" and that "Oswald acted alone."

After all these years, it's time to accept the conclusion. The nation pays a heavy price for chronic doubts and mistrust. Confidence in the government has declined. Participation in the voting process has steadily slid downward. The national appetite for wild theories encourages peddlers to persist. Evil is never easy to accept. In the case of JFK, the sooner we let it go, the better.

▲ 7. "Consumers ought to be concerned about the Federal Trade Commission's dropping a rule that supermarkets must actually have in stock the items they advertise for sale. While a staff analysis suggests costs of the rule outweigh the benefits to consumers, few shoppers want to return to the practices that lured them into stores only to find the advertised products they sought were not there.

"The staff study said the rule causes shoppers to pay $200 million to receive $125 million in benefits. The cost is a low estimate and the benefits a high estimate, according to the study.

"However, even those enormously big figures boil down to a few cents per shopper over a year's time. And the rule does say that when a grocer advertises a sale, the grocer must have sufficient supply of sale items on hand to meet reasonable buyer demand."

The Oregonian

8. "And we thought we'd heard it all. Now the National Rifle Association wants the U.S. Supreme Court to throw out the ban on private ownership of fully automatic machine guns.

"As the nation's cities reel under staggering murder totals, as kids use guns simply to get even after feuds, as children are gunned down by random bullets, the NRA thinks it is everybody's constitutional right to have their own personal machine gun.

"This is not exactly the weapon of choice for deer hunting or for a homeowner seeking protection. It is an ideal weapon for street gangs and drug thugs in their wars with each other and the police.

"To legalize fully automatic machine guns is to increase the mayhem that is turning this nation—particularly its large cities—into a continual war zone. Doesn't the NRA have something better to do?"

Capital Times, Madison, Wisconsin

9. From a letter to the editor: "Recently the California Highway Patrol stopped me at a drunk-drive checkpoint. Now, I don't like drunk drivers any more than anyone else. I certainly see why the police find the checkpoint system effective. But I think our right to move about freely is much more important. If the checkpoint system continues, then next there will be checkpoints for drugs, seat belts, infant car seats, drivers' licenses. We will regret it later if we allow the system to continue."

▲ 10. "Well located, sound real estate is the safest investment in the world. It is not going to disappear, as can the value of dollars put into savings accounts. Neither will real estate values be lost because of inflation. In fact, property values tend to increase at a pace at least equal to the rate of inflation. Most homes have appreciated at a rate greater than the inflation rate (due mainly to strong buyer demand and insufficient supply of newly constructed homes)."

Robert Bruss, The Smart Investor's Guide to Real Estate

11. "The constitutional guarantee of a speedy trial protects citizens from arbitrary government abuse, but it has at least one other benefit, too. It prevents crime.

"A recent Justice Department study found that more than a third of those with serious criminal records—meaning three or more felony convictions—are arrested for new offenses while free on bond awaiting federal court trial. You don't have to be a social scientist to suspect that the longer the delay, the greater the likelihood of further violations. In short, overburdened courts mean much more than justice delayed; they quite literally amount to the infliction of further injustice."

Scripps Howard Newspapers

12. "There is a prevailing school of thought and growing body of opinion that one day historians will point to a certain television show and declare: This is what life was like in small-town America in the mid-20th century.

"We like to think those future historians will be right about *The Andy Griffith Show.* It had everything, and in rerun life still does: humor, wisdom, wholesomeness and good old red-blooded American entertainment.

"That's why we are so puzzled that a group of true-blue fans of the show want to rename some suitable North Carolina town Mayberry. The group believes there ought to be a real-live city in North Carolina called Mayberry, even though everybody already knows that Mayberry was modeled on Mount Airy, sort of, since that is Sheriff Taylor's, uh, Andy Griffith's hometown.

"No, far better that the group let the whole thing drop. For one thing, we don't know of anyplace—hamlet, village, town, or city—that could properly live up to the name of Mayberry. Perhaps the best place for Mayberry to exist is right where it is—untouched, unspoiled and unsullied by the modern world. Ain't that right, Ernest T. Bass?"

Greensboro *(N.C.)* News & Record

▲ 13. As we enter a new century, about 100 million Americans are producing data on the Internet as rapidly as they consume it. Each of these users is tracked by technologies ever more able to collate essential facts about them—age, address, credit rating, marital status, etc.—in electronic form for use in commerce. One Web site, for example, promises, for the meager sum of seven dollars, to scan "over two billion records to create a single comprehensive report on an individual." It is not unreasonable, then, to believe that the combination of capitalism and technology pose a looming threat to what remains of our privacy.

Loosely adapted from Harper's

14. Having your car washed at the carwash may be the best way to go, but there are some possible drawbacks. The International Carwashing Association (ICA) has fought back against charges that automatic carwashes, in recycling wash water, actually dump the salt and dirt from one car onto the next. And that brushes and drag cloths hurt the finish. Perhaps there is some truth to these charges.

The ICA sponsored tests that supposedly demonstrated that the average home car wash is harder on a car than an automatic wash. Maybe. But what's "the average" home car wash? And you can bet that the automatic carwashes in the test were in perfect working order.

There is no way you or I can tell for certain if the filtration system and washing equipment at the automatic carwash are properly maintained. And even if they are, what happens if you follow some mud-caked pickup through the wash? Road dirt might still be caught in the bristles of the brushes or strips of fabric that are dragged over your car.

Here's my recommendation: Wash your own car.

15. Letter to the editor: "The worst disease of the next decade will be AIDS, Acquired Immune Deficiency Syndrome.

"AIDS has made facing surgery scary. In the last ten years several hundred Americans got AIDS from getting contaminated blood in surgery, and it is predicted that within a few years more hundreds of people will receive AIDS blood each year.

"Shouldn't we be tested for AIDS before we give blood? As it is now, no one can feel safe receiving blood. Because of AIDS, people are giving blood to themselves so they will be safe if they have to have blood later. We need a very sensitive test to screen AIDS donors."

North State Record

▲ 16. **Argument in Favor of Measure A**

"Measure A is consistent with the City's General Plan and City policies directing growth to the City's non-agricultural lands. A 'yes' vote on Measure A will affirm the wisdom of well-planned, orderly growth in the City of Chico by approving an amendment to the 1982 Rancho Arroyo Specific Plan. Measure A substantially reduces the amount of housing previously approved for Rancho Arroyo, increases the number of parks and amount of open space, and significantly enlarges and enhances Bidwell Park.

"A 'yes' vote will accomplish the following: • Require the development to dedicate 130.8 acres of land to Bidwell Park • Require the developer to dedicate seven park sites • Create 53 acres of landscaped corridors and greenways • Preserve existing arroyos and protect sensitive plant habitats and other environmental features • Create junior high school and church sites • Plan a series of villages within which, eventually, a total of 2,927 residential dwelling units will be developed • Plan area which will provide onsite job opportunities and retail services. . . ."

County of Butte Sample Ballot

17. **Rebuttal to Argument in Favor of Measure A**

"Villages? Can a project with 3,000 houses and 7,000 new residents really be regarded as a 'village'? The Sacramento developers pushing the Rancho Arroyo project certainly have a way with words. We urge citizens of Chico to ignore their flowery language and vote no on Measure A.

"These out-of-town developers will have you believe that their project protects agricultural land. Hogwash! Chico's Greenline protects valuable farmland. With the Greenline, there is enough land in the Chico area available for development to build 62,000 new homes. . . .

"They claim that their park dedications will reduce use of our overcrowded Bidwell Park. Don't you believe it! They want to attract 7,000 new residents to Chico by using Rancho Arroyo's proximity to Bidwell Park to outsell other local housing projects.

"The developers imply that the Rancho Arroyo project will provide a much needed school site. In fact, the developers intend to sell the site to the school district, which will pay for the site with taxpayers' money.

"Chico doesn't need the Rancho Arroyo project. Vote no on Measure A."

County of Butte Sample Ballot

18. Letter to the editor: "A relative of mine is a lawyer who recently represented a murderer who had already had a life sentence and broke out of prison and murdered someone else. I think this was a waste of the taxpayers' money to try this man again. It won't do any good. I think murderers should be executed.

"We are the most crime-ridden society in the world. Someone is murdered every 27 minutes in the U.S., and there is a rape every ten minutes and an armed robbery every 82 seconds. According to the FBI, there are 870,000 violent crimes a year, and you know the number is increasing.

"Also according to the FBI, only 10 percent of those arrested for the crimes committed are found guilty, and a large percentage are released on probation. These people are released so they can just go out and commit more crimes.

"Why are they released? In the end it is because there aren't enough prisons to house the guilty. The death sentence must be restored. This would create more room in prisons. It would also drastically reduce the number of murders. If a robber knew before he shot someone that if he was caught his own life would be taken, would he do it?

"These people deserve to die. They sacrificed their right to live when they murdered someone, maybe your mother. It's about time we stopped making it easy for criminals to kill people and get away with it."

Cascade News

▲ 19. Letter to the editor: "In regard to your editorial, 'Crime bill wastes billions,' let me set you straight. Your paper opposes mandatory life sentences for criminals convicted of three violent crimes, and you whine about how criminals' rights might be violated. Yet you also want to infringe on a citizen's right to keep and bear arms. You say you oppose life sentences for three-time losers because judges couldn't show any leniency toward the criminals no matter how trivial the crime. What is your definition of trivial, busting an innocent child's skull with a hammer?"

North State Record

▲ 20. Freedom means choice. This is a truth antiporn activists always forget when they argue for censorship. In their fervor to impose their morality, groups like Enough Is Enough cite extreme examples of pornography, such as child porn, suggesting that they are easily available in video stores.

This is not the way it is. Most of this material portrays, not actions such as this, but consensual sex between adults.

The logic used by Enough Is Enough is that if something can somehow hurt someone, it must be banned. They don't apply this logic to more harmful substances, such as alcohol or tobacco. Women and children are more adversely affected by drunken driving and secondhand smoke than by pornography. Few Americans would want to ban alcohol or tobacco even though they kill hundreds of thousands of people each year.

Writing Exercises

1. Write a one-page essay in which you determine whether, and why, it is better (you get to define "better") to look younger than your age, older than your age, or just your age. Then number the premises and conclusions in your essay and diagram it.

2. Should there be a death penalty for first-degree murder? On the top half of a sheet of paper, list considerations supporting the death penalty, and on the bottom half, list considerations opposing it. Take about ten minutes to compile your two lists. (Selection 6 in Appendix 1 may give you some ideas. Your instructor may provide extra time for you to read this selection.)

   After everyone is finished, your instructor will call on people to read their lists. He or she will then give everyone about twenty minutes to write a draft of an essay that addresses the issue, Should there be a death penalty for first-degree murder? Put your name on the back of your paper. After everyone is finished, your instructor will collect the papers and redistribute them to the class. In groups of four or five, read the papers that have been given to your group. Do not look at the names of the authors. Select the best essay in each group. Your instructor will ask each group to read the essay they have selected as best.

   As an alternative, your instructor may have each group rank-order the papers. He or she will have neighboring groups decide which of their top-ranked papers is the best. The instructor will read the papers that have been top-ranked by two (or more) groups, for discussion.

3. Follow the instructions for Exercise 2, but this time address the question, Are free-needle programs a good idea? (Selections 14A and 14B at the end of the book may give you some ideas. Your instructor may provide extra time for you to read those selections.)

4. If you have not done so already, turn to Selection 7, 12, 19, or 20 at the end of the book and follow the first set of instructions.

5. Turn to Selection 8 or 13 at the end of the book and follow the instructions.

6. Turn to Selections 10A,B; 11A,B; 14A,B; 15A,B; 16A,B; 17A,B; or 18A,B at the end of the book and discuss which side has the stronger argument and why.

Chapter 8

Deductive Arguments I: Categorical Logic

M any deductive arguments can be clarified and evaluated using techniques of categorical logic, which is a way of studying inferences that dates back to the time of Aristotle. During more than two thousand years of history, all manner of embellishments were developed and added to the basic theory, especially by monks and other scholars during the medieval period. Although some of these developments are interesting, we don't want to weight you down with lots of fine distinctions that are best appreciated by professional logicians. So we'll be content to set forth the basics of the subject in this chapter.

Categorical logic is logic based on the relations of inclusion and exclusion among classes ("categories"). Like truth-functional logic (the subject of Chapter 9), categorical logic is useful in clarifying and analyzing arguments. But this is only one reason to study the subject: There is no better way to understand the underlying logical structure of our everyday language than learning to put it into the kinds of formal terms we'll introduce in these two chapters. Just exactly what is the difference between the claim (1) "Everybody who is ineligible for physics 1A must take physical science 1" and (2) "No students who are required to take physical science 1 are eligible for physics 1A"? Here's another pair of claims: (3) "Harold won't attend the meeting unless Vanessa decides to go" and (4) "If Vanessa decides to go, then Harold will attend the meeting." You might be surprised how many college students have a very hard time trying to determine whether the claims in each pair say the same thing. In this chapter and the next, we'll show you a foolproof method for determining how to unravel the logical implications of such claims and for seeing how any two such claims relate to one another. [Incidentally, claims (1) and (2) do not say the same thing at all, and neither do (3) and (4).] If you're signing a lease or entering into a contract of any kind, it pays to be able to figure out just what is being said in it and what is not; people who have trouble with claims like those just stated are at risk when there is more at stake.

Studying categorical and truth-functional logic can teach us to become more careful and precise in our own thinking. Getting comfortable with this type of thinking can be helpful in general, but for those who will some-day be applying to law school, medical school, or graduate school, it has the added advantage that many admittance exams for such programs deal with the kinds of reasoning discussed in this chapter.

Let's start by looking at the four basic kinds of claims on which categorical logic is based.

Categorical Claims

A **categorical claim** says something about classes (or "categories") of things. Our interest lies in categorical claims of certain standard forms. A **standard-form categorical claim** is a claim that results from putting names or descriptions of classes into the blanks of the following structures:

A: All _____ are _____.
    (*Example:* All Presbyterians are Christians.)

E: All _____ are _____.
    (*Example:* No Muslims are Christians.)

I: Some _____ are _____.
    (*Example:* Some Christians are Arabs.)

O: Some _____ are not _____.
    (*Example:* Some Muslims are not Sunnis.)

The phrases that go in the blanks are **terms;** the one that goes into the first blank is the **subject term** of the claim, and the one that goes into the second blank is the **predicate term.** Thus, "Christians" is the predicate term of the first example above and the subject term of the third example. In many of the examples and explanations that follow, we'll use the letters *S* and *P* (for "subject" and "predicate") to stand for terms in categorical claims. And we'll talk about the subject and predicate *classes,* which are just the classes that the terms refer to.

But first a caution: Only nouns and noun phrases will work as terms. An adjective alone, such as "red," won't do. "All fire engines are red" does *not* produce a standard-form categorical claim because "red" is not a noun or noun phrase. To see that it is not, try switching the places of the terms: "All red are fire engines." This doesn't make sense, right? But "red vehicles" (or even "red things") will do because "All red vehicles are fire engines" makes sense (even though it's false).

Looking back at the standard-form structures given above, notice that each one has a letter to its left. These are the traditional names of the four types of standard-form categorical claims. The claim "All Presbyterians are

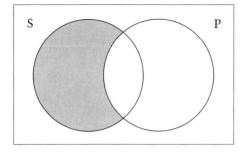

FIGURE 1   A-claim: All S are P.

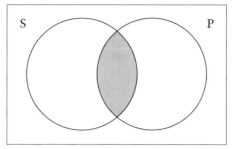

FIGURE 2   E-claim: No S are P.

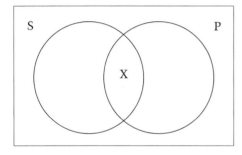

FIGURE 3   I-claim: Some S are P.

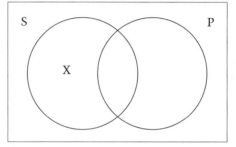

FIGURE 4   O-claim; Some S are not P.

Christians" is an A-claim, and so are "All idolators are heathens," "All people born between 1946 and 1964 are baby boomers," and any other claim of
the form "All S are P." The same is true for the other three letters and the
other three kinds of claims.

■ *Venn Diagrams*

Each of the standard forms has its own graphic illustration in a **Venn diagram,** as shown in Figures 1 through 4. Named after British logician John
Venn, these diagrams exactly represent the four standard-form categorical
claim types. In the diagrams, the circles represent the classes named by the
terms, shaded areas represent areas that are empty, and areas containing Xs
represent areas that are not empty—that contain at least one item. An area
that is blank is one that the claim says nothing about; it may be occupied,
or it may be empty.*

Notice that in the diagram for the A-claim, the area that would contain any members of the S class that were not members of the P class is
shaded—that is, it is empty. Thus, that diagram represents the claim "All S
are P," since there is no S left that isn't P. Similarly, in the diagram for the

---

*There is one exception to this, but we needn't worry about it for a few pages yet.

E-claim, the area where S and P overlap is empty; any S that is also a P has been eliminated. Hence: "No S are P."

For our purposes in this chapter, the word "some" means "at least one." So, the third diagram represents the fact that at least one S is a P, and the X in the area where the two classes overlap shows that at least one thing inhabits this area. Finally, the last diagram shows an X in the area of the S circle that is outside the P circle, representing the existence of at least one S that is not a P.

We'll try to keep technical jargon to a minimum, but here's some terminology we'll need: The two claim types that *include* one class or part of one class within another, the A-claims and I-claims, are **affirmative claims;** the two that *exclude* one class or part of one class from another, the E-claims and O-claims, are **negative claims.**

Although there are only four standard-form claim types, it's remarkable how versatile they are. A large portion of what we want to say can be rewritten, or "translated," into one or another of them. Because this task is sometimes easier said than done, we'd best spend a little while making sure we understand how to do it. And we warn you in advance: A lot of standard form translations are not very pretty—but it's accuracy we seek here, not style.

■ *Translation into Standard Form*

The main idea is to take an ordinary claim and turn it into a standard-form categorical claim that is exactly equivalent. We'll say that two claims are **equivalent claims** if, and only if, they would be true in all and exactly the same circumstances—that is, under no circumstances could one of them be true and the other false. (You can think of such claims as "saying the same thing," more or less.)

Lots of ordinary claims in English are easy to translate into standard form. A claim of the sort "Every X is a Y," for example, more or less automatically turns into the standard-form A-claim "All Xs are Ys." And it's easy to produce the proper term to turn "Minors are not eligible" into the E-claim "No minors are eligible people."

All standard-form claims are in the present tense, but even so we can use them to talk about the past. For example, we can translate "There were creatures weighing more than four tons that lived in North America" as "Some creatures that lived in North America are creatures that weighed more than four tons."

What about a claim like "Only sophomores are eligible candidates"? It's good to have a strategy for attacking such translation problems. First, identify the terms. In this case, the two classes in question are "sophomores" and "eligible candidates." Now, which do we have on our hands, an A-, E-, I-, or O-claim? Generally speaking, nothing but a careful reading can serve to answer this question. So, you'll need to think hard about just what relation between classes is being expressed and then decide how that relation is best turned into a standard form. Fortunately, we can provide some rules of thumb that help in certain frequently encountered problems, in-

▲ *Aristotle was interested in a lot of subjects besides logic—practically everything, in fact. Fortunately for his reputation, these remarks (which he did make!) are not typical of him.*

cluding one that applies to our current example. If you're like most people, you don't have too much trouble seeing that our claim is an A-claim, but *which* A-claim? There are two possibilities:

> All sophomores are eligible candidates

and

> All eligible candidates are sophomores.

If we make the wrong choice, we can change the meaning of the claim significantly. (Notice that "All sophomores are students" is very different from "All students are sophomores.") In the present case, notice that we are saying something about *every* eligible candidate—namely, that he or she must be a sophomore. (*Only* sophomores are eligible—i.e., no one else is eligible.) In an A-claim, the class so restricted is always the subject class. So, this claim should be translated into

> All eligible candidates are sophomores.

In fact, *all claims of the sort "Only Xs are Ys" should be translated as "All Ys are Xs."*

But there are other claims in which the word "only" plays a crucial role and which have to be treated differently. Consider, for example, this claim: "The only people admitted are people over twenty-one." In this case, a restriction is being put on the class of people admitted; we're saying that *nobody else is admitted* except those over twenty-one. Therefore, "people admitted" is the subject class: "All people admitted are people over twenty-one." And, in fact, *all claims of the sort "The only Xs are Ys" should be translated as "All Xs are Ys."*

The two rules of thumb that govern most translations of claims that hinge on the word "only" are these:

The word "only," used by itself, introduces the *predicate* term of an A-claim.

The phrase "the only" introduces the *subject* term of an A-claim.

Note that, in accordance with these rules, we would translate both of these claims

Only matinees are half-price shows

and

Matinees are the only half-price shows

as

All half-price shows are matinees.

The kind of thing a claim directly concerns is not always obvious. For example, if you think for a moment about the claim "I always get nervous when I take logic exams," you'll see that it's a claim about *times*. It's about getting nervous and about logic exams indirectly, of course, but it pertains directly to times or occasions. The proper translation of the example is "All times I take logic exams are times I get nervous." Notice that the word "whenever" is often a clue that you're talking about times or occasions, as well as an indication that you're going to have an A-claim or an E-claim. "Wherever" works the same way for places: "He makes trouble wherever he goes" should be translated as "All places he goes are places he makes trouble."

There are two other sorts of claims that are a bit tricky to translate into standard form. The first is a claim about a single individual, such as "Aristotle is a logician." It's clear that this claim specifies a class, "logicians," and places Aristotle as a member of that class. The problem is that categorical claims are always about *two* classes, and Aristotle isn't a class. (We certainly couldn't talk about *some* of Aristotle being a logician.) What we want to do is treat such claims as if they were about classes with exactly one member—in the present case, Aristotle. One way to do this is to use the term "people

The Most Versatile Word in English

Question:

There's only one word that can be placed successfully in any of the 10 numbered positions in this sentence to produce 10 sentences of different meaning (each sentence has 10 words): (1) *I* (2) *helped* (3) *my* (4) *dog* (5) *carry* (6) *my* (7) *husband's* (8) *slippers* (9) *yesterday* (10). What is that word?

—GLORIA J., Salt Lake City, Utah

Answer:

The word is "only," which makes the following 10 sentences:

1. Only *I* helped my dog carry my husband's slippers yesterday. (Usually the cat helps too, but she was busy with a mouse.)
2. I only *helped* my dog carry my husband's slippers yesterday. (The dog wanted me to carry them all by myself, but I refused.)
3. I helped only *my* dog carry my husband's slippers yesterday. (I was too busy to help my neighbor's dog when he carried them.)
4. I helped my only *dog* carry my husband's slippers yesterday. (I considered getting another dog, but the cat disapproved.)
5. I helped my dog only *carry* my husband's slippers yesterday. (I didn't help the dog eat them; I usually let the cat do that.)
6. I helped my dog carry only *my* husband's slippers yesterday. (My dog and I didn't have time to help my neighbor's husband.)
7. I helped my dog carry my only *husband's* slippers yesterday. (I considered getting another husband, but one is enough.)
8. I helped my dog carry my husband's only *slippers* yesterday. (My husband had two pairs of slippers, but the cat ate one pair.)
9. I helped my dog carry my husband's slippers only *yesterday.* (And now the dog wants help again; I wish he'd ask the cat.)
10. I helped my dog carry my husband's slippers yesterday only. (And believe me, once was enough—the slippers tasted *terrible.*)

—MARILYN VOS SAVANT, author of the "Ask Marilyn" column (Reprinted with permission from Parade and Marilyn vos Savant. Copyright © 1994, 1996.)

who are identical with Aristotle," which of course has only Aristotle as a member. (Everybody is identical with himself or herself, and nobody else is.) The important thing to remember about such claims can be summarized in the following rule of thumb:

> Claims about single individuals should be treated as A-claims or E-claims.

"Aristotle is a logician" can therefore be translated into "All people identical with Aristotle are logicians," an A-claim. Similarly, "Aristotle is not left-handed" becomes the E-claim "No people identical with Aristotle are left-handed people." (Your instructor may prefer to leave the claim in its original form and simply *treat* it as an A-claim or an E-claim. This avoids the awkward "people identical with Aristotle" wording and is certainly okay with us.)

It isn't just people that crop up in individual claims. Often this kind of treatment is called for when we're talking about objects, occasions, places, and other kinds of things. For example, the preferred translation of "St. Louis is on the Mississippi" is "All cities identical with St. Louis are cities on the Mississippi."

Other claims that cause translation difficulty contain what are called *mass nouns*. Consider this example: "Boiled okra is too ugly to eat." This claim is about a *kind of stuff*. The best way to deal with it is to treat it as a claim about *examples* of this kind of stuff. The present example translates into an A-claim about *all* examples of the stuff in question: "All examples of boiled okra are things that are too ugly to eat." An example such as "Most boiled okra is too ugly to eat" translates into the I-claim "Some examples of boiled okra are things that are too ugly to eat."

As we noted, it's not possible to give rules or hints about every kind of problem you might run into when translating claims into standard-form categorical versions. Only practice and discussion can bring you to the point where you can handle this part of the material with confidence. The best thing to do now is to turn to some exercises.

Exercise 8-1

Translate each of the following into a standard-form claim. Make sure that each answer follows the exact form of an A-, E-, I-, or O-claim and that each term you use is a noun or noun phrase that refers to a class of things. Remember that you're trying to produce a claim that's equivalent to the one given; it doesn't matter whether the given claim is actually true.

▲ 1. Every salamander is a lizard.

2. Not every lizard is a salamander.

3. Only reptiles can be lizards.

▲ 4. Snakes are the only members of the suborder Ophidia.

5. The only members of the suborder Ophidia are snakes.

6. None of the burrowing snakes is poisonous.

▲ 7. Anything that's an alligator is a reptile.

8. Anything that qualifies as a frog qualifies as an amphibian.

9. There are frogs wherever there are snakes.

▲ 10. Wherever there are snakes, there are frogs.

11. Whenever the frog population decreases, the snake population decreases.

12. Nobody arrived except the cheerleaders.

▲ 13. Except for vice presidents, nobody got raises.

14. Unless people arrived early, they couldn't get seats.

15. Home movies are often as boring as dirt.

▲ 16. Socrates is a Greek.

17. The bank robber is not Jane's fiancé.

18. If an automobile was built before 1950, it's an antique.

▲ 19. Salt is a meat preservative.

20. Most corn does not make good popcorn.

Exercise 8-2

Follow the instructions given in the preceding exercise.

▲ 1. Students who wrote poor exams didn't get admitted to the program.

2. None of my students is failing.

3. If you live in the dorms, you can't own a car.

▲ 4. There are a few right-handed first basemen.

5. People make faces every time Joan sings.

6. The only tests George fails are the ones he takes.

▲ 7. Nobody passed who didn't make at least 50 percent.

8. You can't be a member unless you're over fifty.

9. Nobody catches on without studying.

▲ 10. I've had days like this before.

11. Roofers aren't millionaires.

12. Not one part of Michael Jackson's face is original equipment.

▲ 13. A few holidays fall on Saturday.

14. Only outlaws own guns.

15. You have nothing to lose but your chains.

▲ 16. Unless you pass this test you won't pass the course.

17. If you cheat, your prof will make you sorry.

18. If you cheat, your friends couldn't care less.

▲ 19. Only when you've paid the fee will they let you enroll.

20. Nobody plays who isn't in full uniform.

■ *The Square of Opposition*

Two categorical claims *correspond* to each other if they have the same subject term and the same predicate term. So, "All Methodists are Christians"

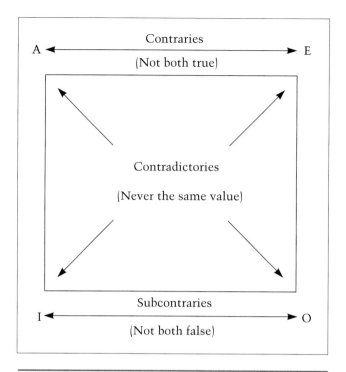

FIGURE 5   The square of opposition.

corresponds to "Some Methodists are Christians": In both claims, "Methodists" is the subject term, and "Christians" is the predicate term. Notice, though, that "Some Christians are not Methodists" does *not* correspond to either of the other two; it has the same terms but in different places.

We can now exhibit the logical relationships between corresponding A-, E-, I-, and O-claims. The **square of opposition,** in Figure 5, does this very concisely. The A- and E-claims, across the top of the square from each other, are **contrary claims**—they can both be false, but they cannot both be true. The I- and O-claims, across the bottom of the square from each other, are **subcontrary claims**—they can both be true, but they cannot both be false. The A- and O-claims and the E- and I-claims, which are at opposite diagonal corners from each other, respectively, are **contradictory claims**—they never have the same truth values.

Notice that these logical relationships are reflected on the Venn diagrams for the claims (see Figures 1 through 4). The diagrams for corresponding A- and O-claims say exactly opposite things about the left-hand area of the diagram, namely, that the area *has* something in it and that it *doesn't*; those for corresponding E- and I-claims do the same about the center area. Clearly, exactly one claim of each pair is true no matter what—either the relevant area is empty, or it isn't.

The diagrams show clearly how both subcontraries can be true: There's no conflict in putting X's in both left and center areas. In fact, it's possible to diagram an A-claim and the corresponding E-claim on the same diagram; we just have to shade out the entire subject class circle. This amounts to saying that *both* an A-claim and its corresponding E-claim can be true *as long as there are no members of the subject class.* We get an analogous result for subcontraries: They can both be false as long as the subject class is empty.* We can easily avoid this result by making an assumption: When making inferences from one contrary (or subcontrary) to another, we'll assume that the classes we're talking about are not entirely empty. On this assumption, the A-claim or the corresponding E-claim (or both) must be false, and the I-claim or the corresponding O-claim (or both) must be true.

If we have the truth value of one categorical claim, we can often deduce the truth values of the other three corresponding claims by using the square of opposition. For instance, if we hear "All aluminum cans are recyclable items," we can immediately infer that its contradictory, "Some aluminum cans are not recyclable items," is false; the corresponding E-claim, "No aluminum cans are recyclable items," is also false because it is the contrary of the original A-claim and cannot be true if the A-claim is true. The corresponding I-claim, "Some aluminum cans are recyclable items," must be true because we just determined that *its* contradictory, the E-claim, is false.

However, we cannot *always* determine the truth values of the remaining three standard-form categorical claims. For example, if we know only that the A-claim is false, all we can infer is the truth value (true) of the corresponding O-claim. Nothing follows about either the E- or the I-claim. Because the A- and the E-claim can both be false, knowing that the A-claim is false does not tell us anything about the E-claim—it can still be either true or false. And if the E-claim remains undetermined, then so must its contradictory, the I-claim.

So here are the limits on what can be inferred from the square of opposition: Beginning with a *true* claim at the top of the square (either A or E), we can infer the truth values of all three of the remaining claims. The same is true if we begin with a *false* claim at the bottom of the square (either I or O): We can still deduce the truth values of the other three. But if we begin with a false claim at the top of the square or a true claim at the bottom, all we can determine is the truth value of the contradictory of the claim in hand.

---

*It is quite possible to interpret categorical claims this way. By allowing both the A- and the E-claims to be true and both the I- and the O-claims to be false, this interpretation reduces the square to contradiction alone. We're going to interpret the claims differently; however, at the level at which we're operating, it seems much more natural to see "All Cs are Ds" as conflicting with "No Cs are Ds."

Exercise 8-3

Translate the following into standard-form claims, and determine the three corresponding standard-form claims. Then, assuming the truth value in parentheses for the given claim, determine the truth values of as many of the other three as you can.

Example

Most snakes are harmless. (True)
Translation (I-claim): Some snakes are harmless creatures. (True)
Corresponding A-claim: All snakes are harmless creatures. (Undetermined)
Corresponding E-claim: No snakes are harmless creatures. (False)
Corresponding O-claim: Some snakes are not harmless creatures. (Undetermined)

▲ 1. Not all anniversaries are happy occasions. (True)
   2. There's no such thing as a completely harmless drug. (True)
   3. There have been such things as just wars. (True)
▲ 4. There are allergies that can kill you. (True)
   5. Woodpeckers sing really well. (False)
   6. Mockingbirds can't sing. (False)
   7. Some herbs are medicinal. (False)
   8. Logic exercises are easy. (False)

Three Categorical Operations

The square of opposition allows us to make inferences from one claim to another, as you were doing in the last exercise. We can think of these inferences as simple valid arguments, because that's exactly what they are. We'll turn next to three operations that can be performed on standard-form categorical claims. They, too, will allow us to make simple valid arguments and, in combination with the square, some not-quite-so-simple valid arguments.

■ Conversion

You find the **converse** of a standard-form claim by switching the positions of the subject and predicate terms. The E- and I-claims, but not the A- and O-claims, contain just the same information as their converses; that is,

> All E- and I-claims, but not A- and O-claims, are equivalent to their converses.

Each member of the following pairs is the converse of the other:

E: No Norwegians are Slavs.
No Slavs are Norwegians.

I: Some state capitals are large cities.
Some large cities are state capitals.

Notice that the claims that are equivalent to their converses are those with symmetrical Venn diagrams.

■ *Obversion*

To discuss the next two operations, we need a couple of auxiliary notions. First, there's the notion of a *universe of discourse.* With rare exceptions, we make claims within contexts that limit the scope of the terms we use. For example, if your instructor walks into class and says, "Everybody passed the last exam," the word "everybody" does not include everybody in the world. Your instructor is not claiming, for example, that your mother and the President of the United States passed the exam. There is an unstated but obvious restriction to a smaller universe of people—in this case, the people in your class who *took* the exam. Now, for every class within a universe of discourse, there is a *complementary class* that contains everything in the universe of discourse that is *not* in the first class. Terms that name complementary classes are complementary terms. So "students" and "non-students" are **complementary terms.** Indeed, putting the prefix "non" in front of a term is often the easiest way to produce its complement. Some terms require different treatment, though. The complement of "people who took the exam" is probably best stated as "people who did not take the exam" because the universe is pretty clearly restricted to people in such a case. (We wouldn't expect, for example, the complement of "people who took the exam" to include *everything* that didn't take the exam, including your shoes and socks!)

Now, we can get on with it: To find the **obverse** of a claim, (a) change it from affirmative to negative, or vice versa (i.e., go horizontally across the square—an A-claim becomes an E-claim; an O-claim becomes an I-claim, and so on); then (b) replace the predicate term with its complementary term.

> All categorical claims of all four types, A, E, I, and O, are equivalent to their obverses.

Here are some examples; each claim is the obverse of the other member of the pair:

A: All Presbyterians are Christians.
No Presbyterians are non-Christians.

E: No fish are mammals.
All fish are nonmammals.

Venn Diagrams for the Three Operations

One way to see which operations work for which types of claim is to put them on Venn diagrams. Here's a regular diagram, which is all we need to explain *conversion:*

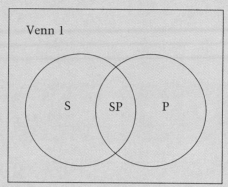

Imagine an I-claim, Some S are P, diagrammed on the above. It would have an X in the central area labeled SP, where S and P overlap. But its converse, Some P are S, would also have an X in that area, since that's where P and S overlap. So, the symmetry of the diagram shows that conversion works for I-claims. The same situation holds for E-claims, except we're shading the central area in both cases rather than placing Xs.

Now, let's imagine an A-claim, All S are P, the diagram for which requires us to shade all the subject term that's not included in the predicate term—i.e., the leftmost area above. But its converse, All P are S, would require that we shade out the *rightmost* area of the diagram, since the subject term is now over there on the right. So the claims with asymmetrical diagrams cannot be validly converted.

We need a somewhat more complicated diagram to explain the other two operations. Let's use a rectangular box to represent the universe of discourse (see text for an explanation) within which our classes and their complements fall. Then we'll label S all the areas that would contain Ss, P the areas where we'd find Ps, S̄ anywhere we would *not* find Ss, and P̄ any-

I: Some citizens are voters.
  Some citizens are not nonvoters.

O: Some contestants are not winners.
  Some contestants are nonwinners.

■ *Contraposition*

You find the **contrapositive** of a categorical claim by (a) switching the places of the subject and predicate terms, just as in conversion, and (b) replacing both terms with complementary terms. Each of the following is the contrapositive of the other member of the pair:

where we would *not* find Ps. Here's the result (make sure you understand what's going on here—it's not all that complicated):

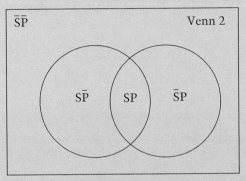

Now let's look at *obversion.* Imagine an A-claim, All S are P, diagrammed on the above. We'd shade out the area labeled S̄P, wouldn't we? (All the subject class that's not part of the predicate class.) Now consider its obverse, "No S are P̄." Since it's an E-claim, we shade where the subject and predicate overlap. And that turns out to be exactly the same area we shaded out for it's obverse! So these two are equivalent: They produce the same diagram. If you check, you'll find you get the same result for each of the other three types of claim, since obversion is valid for all four types.

Finally, we'll see how *contraposition* works out on the diagram. The A-claim, All S are P, once again is made true by shading out the S̄P area of the diagram. But now consider this claim's contrapositive, All P̄ are S̄. Shading out all the subject class that's outside the predicate class produces the same diagram as the original, thus showing that they are equivalent. Try diagramming an O-claim and it's contrapositive, and you'll find yourself putting an X in exactly the same area for each.

But if you diagram an I-claim, Some S are P, putting an X in the central SP area, and then diagram its contrapositive, Some P̄ are S̄, you'll find that the X would have to go entirely outside both circles, since that's the only place P̄ and S̄ overlap! Clearly, this says something different from the original I-claim. You'll find a similarly weird result if you consider an E-claim, since contraposition does *not* work for either I- or E-claims.

A: All Mongolians are Muslims.
    All non-Muslims are non-Mongolians.

O: Some citizens are not voters.
    Some nonvoters are not noncitizens.

All A- and O-claims, but not E- and I-claims, are equivalent to their contrapositives.

The operations of conversion, obversion, and contraposition are important to much of what comes later, so make sure you can do them correctly and that you know which claims are equivalent to the results.

Exercise 8-4

Find the claim described, and determine whether it is equivalent to the claim you began with.

▲ 1. Find the contrapositive of "No Sunnis are Christians."
   2. Find the obverse of "Some Arabs are Christians."
   3. Find the obverse of "All Sunnis are Muslims."
▲ 4. Find the converse of "Some Kurds are not Christians."
   5. Find the converse of "No Hindus are Muslims."
   6. Find the contrapositive of "Some Indians are not Hindus."
▲ 7. Find the converse of "All Shiites are Muslims."
   8. Find the contrapositive of "All Catholics are Christians."
   9. Find the converse of "All Protestants are Christians."
▲ 10. Find the obverse of "No Muslims are Christians."

Exercise 8-5

Follow the directions given in the preceding exercise.

▲ 1. Find the obverse of "Some students who scored well on the exam are students who wrote poor essays."
   2. Find the obverse of "No students who wrote poor essays are students who were admitted to the program."
   3. Find the contrapositive of "Some students who were admitted to the program are not students who scored well on the exam."
▲ 4. Find the contrapositive of "No students who did not score well on the exam are students who were admitted to the program."
   5. Find the contrapositive of "All students who were admitted to the program are students who wrote good essays."
   6. Find the obverse of "No students of mine are unregistered students."
▲ 7. Find the contrapositive of "All people who live in the dorms are people whose automobile ownership is restricted."
   8. Find the contrapositive of "All commuters are people whose automobile ownership is unrestricted."
   9. Find the contrapositive of "Some students with short-term memory problems are students who do poorly in history classes."
▲ 10. Find the obverse of "No first basemen are right-handed people."

Exercise 8-6

For each of the following, find the claim that is described.

Two Common Mistakes

Some of the nuts and bolts on the imported Fords are metric; therefore, some are not.

Some of the nuts and bolts on the imported Fords are not metric; therefore, some are.

It's common for a person to make an inference from "Some As are Bs" to Some As are not Bs," as the first sentence above specifies. But this is a mistake; it's an invalid inference. One might try to defend the mistake by saying, Look, if *all* As are Bs, why not just say so? By not saying so, you're implying that some As are *not* Bs, aren't you? Nope, in fact you're not. One reason we might say "Some As are Bs" is because we only *know* about some As; we may not know whether all of them are Bs.

The same kind of account, naturally, works for going the other way, as in the sentence above.

Example

Find the contrary of the contrapositive of "All Greeks are Europeans." First, find the contrapositive of the original claim. It is "All non-Europeans are non-Greeks." Now, find the contrary of that. Going across the top of the square (from an A-claim to an E-claim), you get "No non-Europeans are non-Greeks."

1. Find the contradictory of the converse of "No clarinets are percussion instruments."

▲ 2. Find the contradictory of the obverse of "Some encyclopedias are definitive works."

3. Find the contrapositive of the subcontrary of "Some English people are Celts."

▲ 4. Find the contrary of the contradictory of "Some sailboats are not sloops."

5. Find the obverse of the converse of "No sharks are freshwater fish."

Exercise 8-7

For each of the numbered claims below, determine which of the lettered claims that follow are equivalent. You may use letters more than once if necessary. (Hint: This is a lot easier to do after all the claims are translated, a fact that indicates at least one advantage to putting claims into standard form.)

1. Some people who have not been tested can give blood.

▲ 2. People who have not been tested cannot give blood.

3. Nobody who has been tested can give blood.

▲ 4. Nobody can give blood except those who have been tested.

   a. Some people who have been tested cannot give blood.

b. Not everybody who can give blood has been tested.

c. Only people who have been tested can give blood.

d. Some people who cannot give blood are people who have been tested.

e. If a person has been tested, then he or she cannot give blood.

Exercise 8-8

Try to make the claims in the following pairs correspond to each other—that is, arrange them so that they have the same subject and the same predicate terms. Use only those operations that produce equivalent claims; for example, don't convert A- or O-claims in the process of trying to make the claims correspond. You can work on either member of the pair or both. (The main reason for practicing on these is to make the problems in the next two exercises easier to do.)

Example

a. Some students are not unemployed people.

b. All employed people are students.

These two claims can be made to correspond by obverting claim (a) and then converting the result (which is legitimate because the claim has been turned into an I-claim before conversion). We wind up with "Some employed people are students," which corresponds to (b).

▲ 1. a. Some Slavs are non-Europeans.
   b. No Slavs are Europeans.

2. a. All Europeans are Westerners.
   b. Some non-Westerners are non-Europeans.

3. a. All Greeks are Europeans.
   b. Some non-Europeans are Greeks.

▲ 4. a. No members of the club are people who took the exam.
   b. Some people who did not take the exam are members of the club.

5. a. All people who are not members of the club are people who took the exam.
   b. Some people who did not take the exam are members of the club.

6. a. Some cheeses are not products high in cholesterol.
   b. No cheeses are products that are not high in cholesterol.

▲ 7. a. All people who arrived late are people who will be allowed to perform.
   b. Some of the people who did not arrive late will not be allowed to perform.

8. a. No nonparticipants are people with name tags.
   b. Some of the people with name tags are participants.

9. a. Some perennials are plants that grow from tubers.
   b. Some plants that do not grow from tubers are perennials.

> ## One More Time
>
> If this is true:
>
> >   Some people who work in the casino are not members of the tribe.
>
> Doesn't this next claim have to be true too?
>
> >   Some members of the tribe are not people who work in the casino.
>
> Nope. Wrong again. The O-claim, however much it might remind us of its converse, says something quite different from it and cannot be inferred from it.

▲ 10. a. Some decks that play digital tape are not devices equipped for radical oversampling.
   b. All devices that are equipped for radical oversampling are decks that will not play digital tape.

Exercise 8-9

Which of the following arguments is valid? (Remember, an argument is valid when the truth of its premises guarantees the truth of its conclusion.)

▲ 1. Whenever the battery is dead, the screen goes blank; that means, of course, that whenever the screen goes blank, the battery is dead.

 2. For a while there, some students were desperate for good grades, which meant some weren't, right?

 3. Some players in the last election weren't members of the Reform Party. Obviously, therefore, some members of the Reform Party weren't players in the last election.

▲ 4. Since some of the students who failed the exam were students who didn't attend the review session, it must be that some students who weren't at the session failed the exam.

 5. None of the people who arrived late were people who got good seats, so none of the good seats were occupied by latecomers.

 6. Everybody who arrived on time was given a box lunch, so the people who did not get a box lunch were those who didn't get there on time.

▲ 7. None of the people who gave blood are people who were tested, so everybody who gave blood must have been untested.

 8. Some of the people who were not tested are people who were allowed to give blood, from which it follows that some of the people who were *not* allowed to give blood must have been people who were tested.

9. Everybody who was in uniform was able to play, so nobody who was out of uniform must have been able to play.

▲ 10. Not everybody in uniform was allowed to play, so some people who were not allowed to play must not have been people in uniform.

Exercise 8-10

For each pair of claims, assume that the first has the truth value given in parentheses. Using the operations of conversion, obversion, and contraposition along with the square of opposition, decide whether the second claim is true, false, or remains undetermined.

Example

a. No aardvarks are nonmammals. (True)

b. Some aardvarks are not mammals.

Claim (a) can be obverted to "All aardvarks are mammals." Because all categorical claims are equivalent to their obverses, the truth of this claim follows from that of (a). Because this claim is the contradictory of claim (b), it follows that claim (b) must be false.

Note: If we had been unable to make the two claims correspond without performing an illegitimate operation (such as converting an A-claim), then the answer is automatically *undetermined.*

▲ 1. a. No mosquitoes are poisonous creatures. (True)
   b. Some poisonous creatures are mosquitoes.

2. a. Some students are not ineligible candidates. (True)
   b. No eligible candidates are students.

▲ 3. a. Some sound arguments are not invalid arguments. (True)
   b. All valid arguments are unsound arguments.

4. a. Some residents are nonvoters. (False)
   b. No voters are residents.

▲ 5. a. Some automobile plants are not productive factories. (True)
   b. All unproductive factories are automobile plants.

Many of the following will have to be rewritten as standard-form categorical claims before they can be answered.

6. a. Most opera singers take voice lessons their whole lives. (True)
   b. Some opera singers do not take voice lessons their whole lives.

7. a. The hero gets killed in some of Gary Brodnax's novels. (False)
   b. The hero does not get killed in some of Gary Brodnax's novels.

8. a. None of the boxes in the last shipment are unopened. (True)
   b. Some of the opened boxes are not boxes in the last shipment.

▲ 9. a. Not everybody who is enrolled in the class will get a grade. (True)
      b. Some people who will not get a grade are enrolled in the class.

10. a. Persimmons are always astringent when they have not been left to ripen. (True)
    b. Some persimmons that have been left to ripen are not astringent.

Categorical Syllogisms

A **syllogism** is a two-premise deductive argument. A **categorical syllogism** (in standard form) is a syllogism whose every claim is a standard-form categorical claim and in which three terms each occur exactly twice in exactly two of the claims. Study the following example:

> All Americans are consumers.
> Some consumers are not Democrats.
> Therefore, some Americans are not Democrats.

Notice how each of the three terms "Americans," "consumers," and "Democrats" occurs exactly twice in exactly two different claims. The *terms of a syllogism* are sometimes given the following labels:

Major term: the term that occurs as a predicate term of the syllogism's conclusion

Minor term: the term that occurs as the subject term of the syllogism's conclusion

Middle term: the term that occurs in both of the premises but not at all in the conclusion

The most frequently used symbols for these three terms are P for major term, S for minor term, and M for middle term. We use these symbols throughout to simplify the discussion.

In a categorical syllogism, each of the premises states a relationship between the middle term and one of the other terms, as shown in Figure 6. If both premises do their jobs correctly—that is, if the proper connections between S and P are established via the middle term, M—then the relationship between S and P stated by the conclusion will have to follow—that is, the argument is valid.

In case you're not clear about the concept of validity, remember: An argument is valid if, and only if, it is not possible for its premises to be true while its conclusion is false. This is just another way of saying that *were* the premises of a valid argument true (whether they are in fact true or false), then

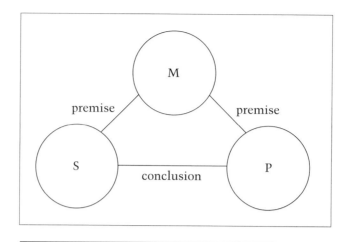

FIGURE 6   Relationship of terms in categorical syllogisms.

the truth of the conclusion would be guaranteed. In a moment we'll begin developing the first of two methods for assessing the validity of syllogisms.

First, though, let's look at some candidates for syllogisms. In fact, only one of the following qualifies as a categorical syllogism. Can you identify which one? What is wrong with the other two?

1. All cats are mammals.
   Not all cats are domestic.
   Therefore, not all mammals are domestic.

2. All valid arguments are good arguments.
   Some valid arguments are boring arguments.
   Therefore, some good arguments are boring arguments.

3. Some people on the committee are not students.
   All people on the committee are local people.
   Therefore, some local people are nonstudents.

We hope it was fairly obvious that the second argument is the only proper syllogism. The first example has a couple of things wrong with it: Neither the second premise nor the conclusion is in standard form—no standard-form categorical claim begins with the word "not"—and the predicate term must be a noun or noun phrase. The second premise can be translated into "Some cats are not domestic creatures" and the conclusion into "Some mammals are not domestic creatures," and the result is a syllogism. The third argument is okay up to the conclusion, which contains a term that does not occur anywhere in the premises: "nonstudents." However, because "nonstudents" is the complement of "students," this argument can be turned into a proper syllogism by obverting the conclusion, producing "Some local people are not students."

© 2003 by Sidney Harris. Reprinted with permission.

Once you're able to recognize syllogisms, it's time to learn how to determine their validity. We'll turn now to our first method, the Venn diagram test.

■ The Venn Diagram Method of Testing for Validity

Diagramming a syllogism requires three overlapping circles, one representing each class named by a term in the argument. To be systematic, in our diagrams we put the minor term on the left, the major term on the right, and the middle term in the middle, but lowered a bit. We will diagram the following syllogism step by step:

No Republicans are collectivists.
All socialists are collectivists.
Therefore, no socialists are Republicans.

In this example, "socialists" is the minor term, "Republicans" is the major term, and "collectivists" is the middle term. See Figure 7 for the three circles required, labeled appropriately.

We fill in this diagram by diagramming the premises of the argument just as we diagrammed the A-, E-, I-, and O-claims earlier. The premises in the above example are diagrammed like this: First: No Republicans are

Paradoxes

The first claim in this box is false.

If the preceding claim is true, then it must be false, right? But if it's false, then what it says is true. So, if it's true, it's false, and if it's false, it's true. Which is it?

This is an example of a liar paradox. Logicians do not agree about what to say about this type of problem. Neither do we. (Although the most frequently said thing about such paradoxes is that they show you can't let a claim refer to itself.)

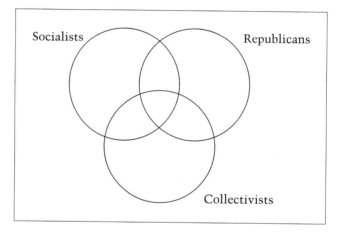

FIGURE 7 Before either premise has been diagrammed.

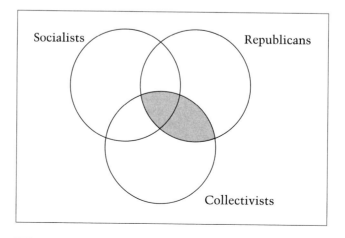

FIGURE 8 One premise diagrammed.

collectivists (Figure 8). Notice that in this figure we have shaded the entire area where the Republican and collectivist circles overlap.

Second: All socialists are collectivists (Figure 9). Because diagramming the premises resulted in the shading of the entire area where the socialist and Republican circles overlap, and because that is exactly what we would do to diagram the syllogism's conclusion, we can conclude that the syllogism is valid. In general, a syllogism is valid if and only if diagramming the premises automatically produces a correct diagram of the conclusion.* (The one exception is discussed later.)

---

*It might be helpful for some students to produce *two* diagrams, one for the premises of the argument and one for the conclusion. The two can then be compared: Any area of the conclusion diagram that is shaded must also be shaded in the premises diagram, and any area of the conclusion diagram that has an X must also have one in the premises diagram. If both of these conditions are met, the argument is valid. (Thanks to Professor Ellery Eells of the University of Wisconsin, Madison, for the suggestion.)

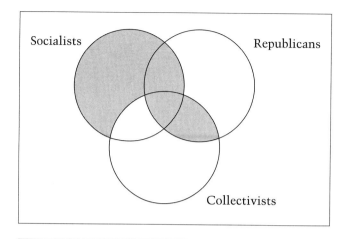

FIGURE 9   Both premises diagrammed.

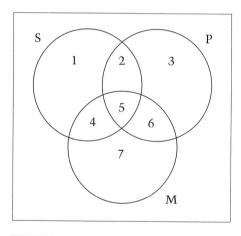

FIGURE 10

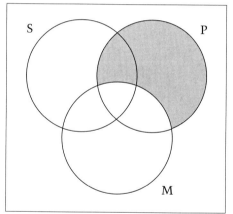

FIGURE 11

When one of the premises of a syllogism is an I- or O-premise, there can be a problem about where to put the required X. The following example presents such a problem (see Figure 10 for the diagram). Note in the diagram that we have numbered the different areas in order to refer to them easily.

Some S are not M.
All P are M.
Some S are not P.

(The horizontal line separates the premises from the conclusion.)

An X in either area 1 or area 2 of Figure 10 makes the claim "Some S are not M" true, because an inhabitant of either area is an S but not an M. How do we determine which area should get the X? In some cases, the de-

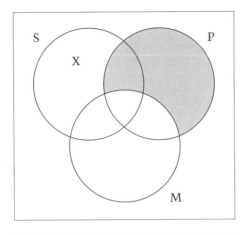

FIGURE 12

cision can be made for us: *When one premise is an A- or E-premise and the other is an I- or O-premise, diagram the A- or E-premise first.* (Always shade before putting in Xs.) Refer to Figure 11 to see what happens with the current example when we follow this rule.

Once the A-claim has been diagrammed, there is no longer a choice about where to put the X—it has to go in area 1. Hence, the completed diagram for this argument looks like Figure 12. And from this diagram we can read the conclusion "Some S are not P," which tells us that the argument is valid.

In some syllogisms, the rule just explained does not help. For example,

All P are M.
Some S are M.
Some S are P.

A syllogism like this one still leaves us in doubt about where to put the X, even after we have diagrammed the A-premise (Figure 13): Should the X go in area 4 or 5? When such a question remains unresolved, here is the rule to follow: *An X that can go in either of two areas goes on the line separating the areas,* as in Figure 14.

In essence, an X on a line indicates that the X belongs in one or the other of the two areas, maybe both, but we don't know which. When the time comes to see whether the diagram yields the conclusion, we look to see whether there is an X *entirely* within the appropriate area. In the current example, we would need an X entirely within the area where S and P overlap; because there is no such X, the argument is invalid. An X *partly* within the appropriate area fails to establish the conclusion.

Please notice this about Venn diagrams: When both premises of a syllogism are A- or E-claims and the conclusion is an I- or O-claim, diagramming the premises cannot possibly yield a diagram of the conclusion (because

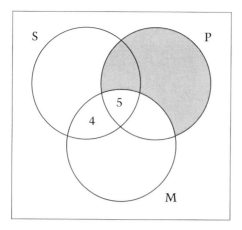

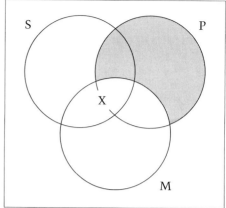

FIGURE 13   **FIGURE 14**

A- and E-claims produce only shading, and I- and O-claims require an X to be read from the diagram). In such a case, remember our assumption that every class we are dealing with has at least one member. This assumption justifies our looking at the diagram and determining whether any circle has all but one of its areas shaded out. *If any circle has only one area remaining unshaded, an X should be put in that area.* This is the case because any member of that class has to be in that remaining area. Sometimes placing the X in this way will enable us to read the conclusion, in which case the argument is valid (on the assumption that the relevant class is not empty); sometimes placing the X will not enable us to read the conclusion, in which case the argument is invalid, with or without any assumptions about the existence of a member within the class.

■ *Categorical Syllogisms with Unstated Premises*

Many "real-life" categorical syllogisms have unstated premises. For example, suppose somebody says:

> You shouldn't give chicken bones to dogs. They could choke on them.

The speaker's argument rests on the unstated premise that you shouldn't give dogs things they could choke on. In other words, the argument, when fully spelled out, is this:

> All chicken bones are things dogs could choke on.
> [No things dogs could choke on are things you should give dogs.]
> Therefore, no chicken bones are things you should give dogs.

The unstated premise appears in brackets.

Our Coach, the Expletive

A few years ago, exasperated with his team's losing, former coach Dick Motta of the Sacramento Kings threatened to trade his star player, Wayman Tisdale, because Motta doubted Tisdale's ability to be a leader.

"You almost have to be a bad guy to be a great leader in this league," Motta said. "Very few guys can do it. There's no way Wayman Tisdale is ever going to be a bad guy."

You can put this argument into standard form, can't you?

Motta also said, "The great ones turn into pieces of [expletive] when they assume leadership. Wayman is a preacher's son. He's not going to be an [expletive]."

Same type of argument, right? Roughly:

All great leaders are pieces of expletive.
Tisdale is not a piece of expletive.
Therefore, Tisdale is not a great leader.

(When he was told what Motta said about him, Tisdale said he thought Motta should know, since Motta was a leader: All great leaders are pieces of expletive; Motta is a great leader; therefore, Motta is a piece of expletive.)

To take another example:

Driving around in an old car is dumb, since it might break down in a dangerous place.

Here the speaker's argument rests on the unstated premise that it's dumb to risk a dangerous breakdown. In other words, when fully spelled out, the argument is this:

All examples of driving around in an old car are examples of risking dangerous breakdown.
[All examples of risking dangerous breakdown are examples of being dumb.]
Therefore, all examples of driving around in an old car are examples of being dumb.

When you hear (or give) an argument that looks like a categorical syllogism that has only one stated premise, usually a second premise has been assumed and not stated. Ordinarily this unstated premise remains unstated because the speaker thinks it is too obvious to bother stating. The unstated premises in the arguments above are good examples: "You shouldn't give dogs things they could choke on," and "It is dumb to risk a dangerous breakdown."

When you encounter (or give) what looks like a categorical syllogism that is missing a premise, ask: Is there a reasonable assumption I could make that would make this argument valid? We covered this question of unstated premises in more detail on pages 241–245, and you might want to look there for more information on the subject.

At the end of this chapter, we have included a few exercises that involve missing premises.

■ *Real-Life Syllogisms*

We'll end this section with a word of advice. Before you use a Venn diagram or the rules method for determining the validity of real-life arguments, it helps to use a letter to abbreviate each category mentioned in the argument. This is mainly just a matter of convenience: It is easier to write down letters than to write down long phrases.

Take the first "real-life" categorical syllogisms given above:

> You shouldn't give chicken bones to dogs because they could choke on them.

The argument spelled out, once again, is this:

> All chicken bones are things dogs could choke on.
> [No things dogs could choke on are things you should give dogs.]
> Therefore, no chicken bones are things you should give dogs.

Abbreviating each of the three categories with a letter, we get

> C = chicken bones; D = things dogs could choke on; and S = things you should give dogs.

Then, the argument is

> All C are D
> [No D are S]
> Therefore, no C are S.

Likewise, the second argument was this:

> Driving around in an old car is dumb, since it might break down in a dangerous place.

When fully spelled out, the argument is

> All examples of driving around in an old car are examples of risking dangerous breakdown.
> [All examples of risking dangerous breakdown are examples of being dumb.]
> Therefore, all examples of driving around in an old car are examples of being dumb.

Abbreviating each of the three categories, we get

> D = examples of driving around in an old car; R = examples of risking danger-
> ous breakdown; S = examples of being dumb.

Then, the argument is

> All D are R
> [All R are S]
> Therefore, all D are S.

A final tip: Take the time to write down your abbreviation key clearly.

Exercise 8-11 _____

Use the diagram method to determine which of the following syllogisms are
valid and which are invalid.

▲ 1. All paperbacks are books that use glue in their spines.
      No books that use glue in their spines are books that are sewn in signatures.
      No books that are sewn in signatures are paperbacks.

   2. All sound arguments are valid arguments.
      Some valid arguments are not interesting arguments.
      Some sound arguments are not interesting arguments.

   3. All topologists are mathematicians.
      Some topologists are not statisticians.
      Some mathematicians are not statisticians.

▲ 4. Every time Louis is tired he's edgy. He's edgy today, so he must be tired
      today.

   5. Every voter is a citizen, but some citizens are not residents. Therefore,
      some voters are not residents.

   6. All the dominant seventh chords are in the mixolydian mode, and no
      mixolydian chords use the major scale. So no chords that use the major
      scale are dominant sevenths.

▲ 7. All halyards are lines that attach to sails. Painters do not attach to sails,
      so they must not be halyards.

   8. Only systems with removable disks can give you unlimited storage capac-
      ity of a practical sort. Standard hard disks never have removable disks, so
      they can't give you practical, unlimited storage capacity.

   9. All citizens are residents. So, since no noncitizens are voters, all voters
      must be residents.

▲ 10. No citizens are nonresidents, and all voters are citizens. So all residents
      must be nonvoters.

Brodie!

"Otterhounds are friendly, are fond of other dogs, bark a lot, and like to chase cats."

"That describes Brodie exactly! He must be an otterhound."

Not so fast, dog lover. The argument seems to be

All otterhounds are friendly, fond of other dogs, and like to chase cats. Brodie is friendly, fond of other dogs, and likes to chase cats. Therefore, Brodie is an otterhound.

This argument has the form

All As are X
All Bs are X.
Therefore, all Bs are As.

If you use techniques described in this chapter, you will see that arguments with this form are invalid. If you just stumbled on this box or your instructor referred you to it, common sense should tell you the same. It's like arguing "All graduates of Harvard are warm-blooded, and Brodie is warm-blooded;" therefore, Brodie is a graduate of Harvard."

Moral: Be on the watch for arguments like the one above. They aren't uncommon. Plus, see the next box, "Additional Common Invalid Argument Forms."

Exercise 8-12

Put the following arguments in standard form (you may have to use the obversion, conversion, or contraposition operations to accomplish this); then determine whether the arguments are valid by means of diagrams.

▲ 1. No blank disks contain any data, although some blank disks are formatted. Therefore, some formatted disks do not contain any data.

2. All ears of corn with white tassels are unripe, but some ears are ripe even though their kernels are not full-sized. Therefore, some ears with full-sized kernels are not ears with white tassels.

3. Prescription drugs should never be taken without a doctor's order. So no over-the-counter drugs are prescription drugs, because all over-the-counter drugs can be taken without a doctor's order.

▲ 4. All tobacco products are damaging to people's health, but some of them are addictive substances. Some addictive substances, therefore, are damaging to people's health.

Additional Common Invalid Argument Forms

Other common invalid argument forms (see the box about Brodie) include these:

> All As are X.
> No As are Y.
> Therefore, Xs are Ys.

> All Xs are Ys; therefore, all Ys are Xs.

> Some Xs are not Ys. Therefore, some Ys are not Xs.

> Some Xs are Ys. Therefore, some Xs are not Ys.

> Some Xs are not Ys. Therefore, some Xs are Ys.

So you don't get lost in all the Xs and Ys, and to help you remember them, we recommend you make up examples of each of these forms and share them with a classmate.

5. A few compact disc players use 24× sampling, so some of them must cost at least a hundred dollars, because you can't buy any machine with 24× sampling for less than a hundred dollars.

6. Everything that Pete won at the carnival must be junk. I know that Pete won everything that Bob won, and all the stuff that Bob won is junk.

▲ 7. Only people who hold stock in the company may vote, so Mr. Hansen must not hold any stock in the company, because I know he was not allowed to vote.

8. No off-road vehicles are allowed in the unimproved portion of the park, but some off-road vehicles are not four-wheel-drive. So some four-wheel-drive vehicles are allowed in the unimproved part of the park.

9. Some of the people affected by the new drainage tax are residents of the county, and many residents of the county are already paying the sewer tax. So it must be that some people paying the sewer tax are affected by the new drainage tax, too.

▲ 10. No argument with false premises is sound, but some of them are valid. So some unsound arguments must be valid.

■ *The Rules Method of Testing for Validity*

The diagram method of testing syllogisms for validity is intuitive, but there is a faster method that makes use of three simple rules. These rules are based on two ideas, the first of which has been mentioned already: affirmative and negative categorical claims. (Remember, the A- and I-claims are affirmative; the E- and O-claims are negative.) The other idea is that of *distribution*.

A-claim: All $\widehat{S}$ are $\widehat{P}$.
E-claim: No $\widehat{S}$ are $\widehat{P}$.
I-claim: Some $\widehat{S}$ are $\widehat{P}$.
O-claim: Some $\widehat{S}$ are not $\widehat{P}$.

FIGURE 15   Distributed terms.

Terms that occur in categorical claims are either distributed or undistributed: Either the claim says something about every member of the class the term names, or it does not.* Three of the standard-form claims distribute one or more of their terms. In Figure 15, the circled letters stand for distributed terms, and the uncircled ones stand for undistributed terms. As the figure shows, the A-claim distributes its subject term, the O-claim distributes its predicate term, the E-claim distributes both, and the I-claim distributes neither.

We can now state the three *rules of the syllogism*. A syllogism is valid if and only if all of these conditions are met:

1. **The number of negative claims in the premises must be the same as the number of negative claims in the conclusion.** (Because the conclusion is always one claim, this implies that no valid syllogism has two negative premises.)

2. **At least one premise must distribute the middle term.**

3. **Any term that is distributed in the conclusion of the syllogism must be distributed in its premises.**

These rules are easy to remember, and with a bit of practice, you can use them to determine quickly whether a syllogism is valid.

Which of the rules is broken in this example?

All pianists are keyboard players.
Some keyboard players are not percussionists.
Some pianists are not percussionists.

---

*The above is a rough-and-ready definition of distribution. If you'd like a more technical version, here's one: A term is *distributed* in a claim if, and only if, on the assumption that the claim is true, the class named by the term can be replaced by *any* subset of that class without producing a false claim. Example: In the claim "All senators are politicians," the term "senators" is distributed because, assuming the claim is true, you can substitute *any* subset of senators (Democratic ones, Republican ones, tall ones, short ones . . .) and the result must also be true. "Politicians" is not distributed: The original claim could be true while "All senators are honest politicians" was false.

Leibniz, Boole and Gödel worked with logic.
I work with logic.
I am Leibniz, Boole and Gödel.

© 2003 by Sidney Harris. Reprinted with permission.

▲   *Nope. 'Fraid it's none of the above. Is it clear what error our logician is making?*

The term "keyboard players" is the middle term, and it is undistributed in both premises. The first premise, an A-claim, does not distribute its predicate term; the second premise, an O-claim, does not distribute its subject term. So this syllogism breaks rule 2.
    Another example:

> No dogs up for adoption at the animal shelter are pedigreed dogs.
> Some pedigreed dogs are expensive dogs.
> _____
> Some dogs up for adoption at the animal shelter are expensive dogs.

This syllogism breaks rule 1 because it has a negative premise but no negative conclusion.
    A last example:

> No mercantilists are large landowners.
> All mercantilists are creditors.
> _____
> No creditors are large landowners.

The minor term, "creditors," is distributed in the conclusion (because it's the subject term of an E-claim) but not in the premises (where it's the predicate term of an A-claim). So this syllogism breaks rule 3.

Recap

In this chapter, we have developed the basics of categorical logic. There are four types of standard-form categorical claims (A: "All S are P"; E: "No S are P"; I: "Some S are P"; and O: "Some S are not P"). These can be represented by Venn diagrams. Translation of ordinary English claims into standard-form categorical ones is made somewhat easier by a few rules of thumb: "Only" introduces a predicate term; "the only" introduces a subject term; "whenever" means we're probably talking about times or occasions; "wherever" means we're probably talking about places. Claims about individuals should be treated as A- or E-claims.

The square of opposition displays the relations of contradiction, contrariety, and subcontrariety among corresponding standard-form claims. Conversion, obversion, and contraposition are three operations that can be performed on categorical claims—depending on the kind of claim on which you perform the operation, the result can be an equivalent claim.

Categorical syllogisms are standardized deductive arguments. We can test them for validity by two methods, one of which relies on Venn diagrams and one of which relies on rules about affirmative/negative claims and about the distribution of terms.

Having read this chapter and the chapter recap, you should be able to

▲ Define categorical logic and explain the four kinds of standard-form categorical claims

▲ Diagram standard-form categorical claims using Venn diagrams and translate claims into equivalent claims

▲ Exhibit the logical relationships among corresponding claims and determine the truth value of those claims using the square of opposition

▲ Make inferences about claims using the operations conversion, obversion, and contraposition

▲ Explain categorical syllogisms and determine their validity using both the diagram method and the rules method

Additional Exercises

Exercise 8-13

In each of the following items, identify whether A, B, or C is the middle term.

▲ 1. All A are B.
   <u>All A are C.</u>
   All B are C.

2. All B are C.
   No C are D.
   No B are D.

3. Some C are not D.
   All C are A.
   Some D are not A.

▲ 4. Some A are not B.
   Some B are C.
   Some C are not A.

5. No C are A.
   Some B are A.
   Some C are not B.

Exercise 8-14

Which terms are distributed in each of the following?

▲ 1. All A are B.
   a. A only
   b. B only
   c. Both A and B
   d. Neither A nor B

2. No A are B.
   a. A only
   b. B only
   c. Both A and B
   d. Neither A nor B

3. Some A are B.
   a. A only
   b. B only
   c. Both A and B
   d. Neither A nor B

▲ 4. Some A are not B.
   a. A only
   b. B only
   c. Both A and B
   d. Neither A nor B

Exercise 8-15

How many negative claims appear in the premises of each of the following arguments? (In other words, how many of the premises are negative?) Your options are 0, 1, or 2.

▲ 1. All A are B.
   <u>All A are C.</u>
   Therefore, all B are C.

2. All B are C.
   <u>No C are D.</u>
   Therefore, no B are D.

3. Some C are not D.
   <u>All C are A.</u>
   Therefore, some D are not A.

▲ 4. Some A are not B.
   <u>Some B are C.</u>
   Therefore, some C are not A.

5. No A are B.
   <u>Some B are not C.</u>
   Some A are C.

Exercise 8-16 _____

Which rules (if any) are broken in each of the following? Select from these options:

   a. Breaks rule 1 only
   b. Breaks rule 2 only
   c. Breaks rule 3 only
   d. Breaks more than one rule
   e. Breaks no rule

▲ 1. All A are B.
   <u>All A are C.</u>
   Therefore, all B are C.

2. All B are C.
   <u>No C are D.</u>
   Therefore, no B are D.

3. Some C are not D.
   <u>All C are A.</u>
   Therefore, some D are A.

▲ 4. Some A are not B.
   <u>Some B are C.</u>
   Therefore, some C are not A.

5. Some A are C.
   <u>Some C are B.</u>
   Therefore, some A are B.

6. Some carbostats are framistans.
   <u>No framistans are arbuckles.</u>
   Some arbuckles are not carbostats.

▲ 7. All framistans are veeblefetzers.
      Some veeblefetzers are carbostats.
      Some framistans are carbostats.

   8. No arbuckles are framistans.
      All arbuckles are carbostats.
      No framistans are carbostats.

   9. All members of the class are registered students.
      Some registered students are not people taking fifteen units.
      Some members of the class are not people taking fifteen units.

▲ 10. All qualified mechanics are people familiar with hydraulics.
      No unschooled people are people familiar with hydraulics.
      No qualified mechanics are unschooled people.

Exercise 8-17 _____

Which rules (if any) are broken in each of the following?

   Note: If an argument breaks a rule, *which* rule is broken depends on how you translate the claims in the argument. For example, the claim "Dogs shouldn't be given chicken bones" could be translated as an *E-claim:* "No dogs are animals that should be given chicken bones." But it also could be translated as an *A-claim:* "All dogs are animals that shouldn't be given chicken bones." If the original claim appeared in an invalid argument, one rule would be broken if you translated it as the E-claim. A different rule would be broken if you translated it as the A-claim.

▲ 1. All tigers are ferocious creatures. Some ferocious creatures are zoo animals. Therefore, some zoo animals are tigers. (For this and the following items, it will help if you abbreviate each category with a letter. For example, let T = tigers, F = ferocious creatures, and Z = zoo animals.)

   2. Some pedestrians are not jaywalkers. Therefore, some jaywalkers are not gardeners, since no gardeners are pedestrians.

   3. Because all shrubs are ornamental plants, it follows that no ornamental plants are cacti, since no cacti qualify as shrubs.

▲ 4. Weightlifters aren't really athletes. Athletics requires the use of motor skills; and few, if any, weightlifters use motor skills.

   5. The trick to finding syllogisms is to think categorically, as well as to focus on the key argument in a passage. For example, some passages contain a good bit of rhetoric, and some passages that do this make it hard to spot syllogisms, with the result that it is hard to spot syllogisms in some passages.

   6. Every broadcast network has seen its share of the television audience decline during the past six years. But not every media outlet that has a decline in television audience share has lost money. So not every broadcast network has lost money.

▲ 7. Many students lift papers off the Internet, and this fact is discouraging to teachers. However, it must be noted that students who do this are only cheating themselves, and anyone who cheats himself or herself loses in the long run. Therefore, lifting papers off the Internet is a losing proposition in the long run.

8. When he was Speaker of the House, Mr. Newt Gingrich could be counted on to advance Republican causes. At the time, nobody who would do that could be accused of being soft on crime, which explains why at the time Gingrich could hardly be accused of being soft on crime.

9. It would be in everyone's interest to amend the Constitution to permit school prayer. And it is obviously in everyone's interest to promote religious freedom. It should be no surprise, then, that amending the Constitution to permit school prayer will promote religious freedom.

▲ 10. If you want to stay out all night dancing, it is fine with me. Just don't cry about it if you don't get good grades. Dancing isn't a total waste of time, but dancing the whole night certainly is. There are only so many hours in a day, and wasting time is bound to affect your grades negatively. So, fine, stay out dancing all night. It's your choice. But you have to expect your grades to suffer.

Exercise 8-18

Refer back to Exercises 8-11 and 8-12 and check the arguments for validity using the rules. We recommend abbreviating each category with a letter.

Once again, remember: If an argument breaks a rule, *which* rule is broken depends on how you translate the claims in the argument. For example, the claim "Dogs shouldn't be given chicken bones" could be translated as an E-claim: "No dogs are animals that should be given chicken bones." But it also could be translated as an A-claim (the obverse of the other version): "All dogs are animals that shouldn't be given chicken bones." If the original claim appeared in an invalid argument, one rule would be broken if you translated it as an E-claim. A different rule would be broken if you translated it as an A-claim.

Answers to 2, 5, 7, and 8 are given in the answer section.

Exercise 8-19

For each of the following items: Abbreviate each category with a letter, then translate the argument into standard form using the abbreviations. Then test the argument for validity using either the diagram method or the rules method.

Note: For many of these items it can be difficult to translate the arguments into standard form.

▲ 1. Some athletes are not baseball players, and some baseball players are not basketball players. Therefore, some athletes are not basketball players.

2. Rats are disease-carrying pests, and as such should be eradicated, because such pests should all be eradicated.

▲ 3. This is not the best of all possible worlds, because the best of all possible worlds would not contain mosquitoes, and *this* world contains plenty of mosquitoes!

4. From time to time, the police have to break up parties here on campus, since some campus parties get out of control, and when a party gets out of control, well, you know what the police have to do.

5. I know that all fundamentalist Christians are evangelicals, and I'm pretty sure that all revivalists are also evangelicals. So, if I'm right, at least some fundamentalist Christians must be revivalists.

▲ 6. "Their new lawn furniture certainly looks cheap to me," she said. "It's made of plastic, and plastic furniture just looks cheap."

7. None of our intramural sports are sports played in the Olympics, and some of the intercollegiate sports are not Olympic sports either. So some of the intercollegiate sports are also intramural sports.

8. The moas were all Dinornithidae, and no moas exist anymore. So there aren't any more Dinornithidae.

▲ 9. Everybody on the district tax roll is a citizen, and all eligible voters are also citizens. So everybody on the district tax roll is an eligible voter.

10. Any piece of software that is in the public domain may be copied without permission or fee. But that cannot be done in the case of software under copyright. So software under copyright must not be in the public domain.

11. None of the countries that have been living under dictatorships for these past few decades are familiar with the social requirements of a strong democracy—things like widespread education and a willingness to abide by majority vote. Consequently, none of these countries will make a successful quick transition to democracy, since countries where the above-mentioned requirements are unfamiliar simply can't make such a transition.

▲ 12. Trust Senator Cobweb to vote with the governor on the new tax legislation. Cobweb is a liberal, and liberals just cannot pass up an opportunity to raise taxes.

13. Investor-held utilities should not be allowed to raise rates, since all public utilities should be allowed to raise rates, and public utilities are not investor-held.

14. Masterpieces are no longer recorded on cassettes. This is because masterpieces belong to the classical repertoire, and classical music is no longer recorded on cassettes.

15. It isn't important to learn chemistry, since it isn't very useful, and there isn't much point in learning something that isn't useful.

16. Stockholders' information about a company's worth must come from the managers of that company, but in a buy-out, the managers of the company are the very ones who are trying to buy the stock from the stockholders. So,

ironically, in a buy-out situation, stockholders must get their information about how much a company is worth from the very people who are trying to buy their stock.

▲ 17. All of the networks devoted considerable attention to reporting poll results during the last election, but many of those poll results were not especially newsworthy. So the networks have to admit that some unnewsworthy items received quite a bit of their attention.

▲ 18. If a person doesn't understand that the earth goes around the sun once a year, then that person can't understand what causes winter and summer. Strange as it may seem, then, there are many American adults who don't know what causes winter and summer, because a survey a year or so ago showed that many such adults don't know that the earth goes around the sun.

19. Congress seems ready to impose trade sanctions on China, and perhaps it should. China's leaders cruelly cling to power. They flout American interests in their actions in Tibet, in their human-rights violations, in their weapons sales, and in their questionable trade practices. Any country with a record like this deserves sanctions.

▲ 20. Since 1973, when the U.S. Supreme Court decided *Miller v. California,* no work can be banned as obscene unless it contains sexual depictions that are "patently offensive" to "contemporary community standards" and unless the work as a whole possesses no "serious literary, artistic, political or scientific value." As loose as this standard may seem when compared with earlier tests of obscenity, the pornographic novels of "Madame Toulouse" (a pseudonym, of course) can still be banned. They would offend the contemporary standards of *any* community, and to claim any literary, artistic, political, or scientific value for them would be a real joke.

21. All creationists are religious, and all fundamentalists are religious, so all creationists are fundamentalists.

22. Every sportscaster is an athlete, and no athlete is a college professor. Therefore, no sportscasters are college professors.

23. Anyone who voted for the Democrats favors expansion of medical services for the needy. So, the people who voted for the Democrats all favor higher taxes, since anyone who wants to expand medical services must favor higher taxes.

24. All cave dwellers lived before the invention of the radio, and no one alive today is a cave dweller. Thus, no person who lived before the invention of the radio is alive today.

25. Conservationists don't vote for Republicans, and all environmentalists are conservationists. Thus, environmentalists don't vote for Republicans. (Hint: Remember, the order of the premises can be reversed.)

26. Since all philosophers are skeptics, it follows that no theologian is a skeptic, since no philosophers are theologians. (Hint: Don't forget to convert, if necessary—and if permissible!)

27. Each philosopher is a skeptic, and no philosopher is a theologian. Therefore, no skeptic is a theologian. (Hint: Remember, you can convert some claims about classes.)

28. Peddlers are salesmen, and confidence men are, too. So, peddlers are confidence men.

29. Should drug addicts be treated as criminals? Well, addicts are all excluded from the class of decent people, yet all criminals belong to that class. Accordingly, no addicts are criminals.

30. Critical thinkers recognize invalid syllogisms: therefore, critical thinkers are logicians, since logicians can spot invalid syllogisms, too.

31. The Mohawk Indians are Algonquin, and so are the Cheyenne. So, the Mohawks are really just Cheyenne.

32. Idiots would support the measure, but no one else would. Whatever else you may think of the school board, you can't say they are idiots. [Therefore . . .]

Exercise 8-20

This exercise is a little different, and you may need to work one or more such items in class in order to get the hang of them. Your job is to try to prove each of the following claims about syllogisms true or false. You may need to produce a general argument—that is, to show that *every* syllogism that does *this* must also do *that*—or you may need to produce a counterexample, that is, an example that proves the claim in question false. The definition of categorical syllogism and the rules of the syllogism are of crucial importance in working these examples.

▲ 1. Every valid syllogism must have at least one A- or E-claim for a premise.

2. Every valid syllogism with an E-claim for a premise must have an E-claim for a conclusion.

3. Every valid syllogism with an E-claim for a conclusion must have an E-claim for a premise.

▲ 4. It's possible for a syllogism to break two of the rules of the syllogism.

5. No syllogism can break three of the rules of the syllogism.

Exercise 8-21

The following is an anonymous statement of opinion that appeared in a newspaper call-in column.

This is in response to the person who called in that we should provide a shelter for the homeless, because I think that is wrong. These people make the downtown area unsafe because they have nothing to lose by robbing, mugging, etc. The young boy killed by the horseshoe pits was attacked by some of these bums, assuming that witnesses really saw people who were homeless, which no doubt they did, since the so-called homeless all wear that old worn-out hippie gear, just like the people they saw. They also lower property values. And don't tell me they are down and out because they can't find work. The work is there if they look for it. They choose for themselves how to live, since if they didn't choose, who did?

A lot of things might be said in criticism of this tirade, but what we want you to notice is the breakdown of logic. The piece contains, in fact, a gross logic error, which we ask you to make the focus of a critical essay. Your audience is the other members of your class; that is, you are writing for an audience of critical thinkers.

Exercise 8-22

Pornography violates women's rights. It carries a demeaning message about a woman's worth and purpose and promotes genuine violence. This is indeed a violation of women's civil rights and justifies the Minneapolis City Council in attempting to ban pornography.

This letter to the editor is, in effect, two syllogisms. The conclusion of the first is that pornography violates women's rights. This conclusion also functions as a premise in the second syllogism, which has as its own conclusion the claim that the Minneapolis City Council is justified in attempting to ban pornography. Both syllogisms have unstated premises. Translate the entire argument into standard-form syllogisms, supplying missing premises, and determine whether the reasoning is valid.

Exercise 8-23

Each of the following arguments contains an unstated premise, which, together with the stated premise, makes the argument in question valid. Your job is to identify this unstated premise, abbreviate each category with a letter, and put the argument in standard form.

▲   1. Ladybugs eat aphids; therefore, they are good to have in your garden.

  2. CEOs have lots of responsibility; therefore, they should be paid a lot.

  3. Anyone who understands how a computer program works knows how important logic is. Therefore, anyone who understands how a computer program works understands how important unambiguous writing is.

▲   4. Self-tapping screws are a boon to the construction industry. They make it possible to screw things without drilling pilot holes.

  5. No baseball player smokes anymore. Baseball players all know that smoking hampers athletic performance.

  6. You really ought to give up jogging. It is harmful to your health.

7. Camping isn't much fun. It requires sleeping on the hard ground and getting lots of bug bites.

8. Having too much coffee makes you sleep poorly. That's why you shouldn't do it.

9. Do you have writer's block? No problem. You can always hire a secretary.

10. "You think those marks were left by a—snake? That's totally crazy. Snakes don't leave footprints."

Writing Exercises

1. Should dogs be used in medical experiments, given that they seem to have the capacity to experience fear and feel pain? Write a short paper defending a negative answer to this question, taking about five minutes to do so. When you have finished, exchange your argument with a friend and rewrite each other's argument as a categorical syllogism or a combination of categorical syllogisms. Remember that people often leave premises unstated.

2. Follow the instructions for Exercise 1, but this time defend the position that it is not wrong to use dogs in medical experiments.

3. Turn to Selection 7 in Appendix 1 and follow the second set of instructions.

4. Turn to Selection 14A, 14B, 15A, 15B, 16A, or 16B and follow the second alternative assignment.

Chapter 9

Deductive Arguments II: Truth-Functional Logic

The earliest development of truth-functional logic took place among the Stoics, who flourished from about the third century B.C.E. until the second century C.E. But it was in the late nineteenth and twentieth centuries that the real power of **truth-functional logic** (known also as *propositional* or *sentential logic*) became apparent.

The "logic of sentences" is one of the bases on which modern symbolic logic rests, and as such it is important in such intellectual areas as set theory and the foundations of mathematics. It is also the model for electrical circuits of the sort that are the basis of digital computing. But truth-functional logic is also a useful tool in the analysis of arguments.

The study of truth-functional logic can benefit you in several ways. For one thing, you'll learn something about the structure of language that you wouldn't learn any other way. For another, you'll get a sense of what it's like to work with a very precise, nonmathematical system of symbols that is nevertheless very accessible to nearly any student willing to invest a modest amount of effort. The model of precision and clarity that such systems provide can serve you well when you communicate with others in ordinary language.

If you're not comfortable working with symbols, the upcoming sections on truth-functional arguments and deductions might look intimidating. But they are not as forbidding as they may appear. We presume that the whole matter of a symbolic system is unfamiliar to you, so we'll start from absolute scratch. Keep in mind, though, that everything builds on what goes before. It's important to master each concept as it's explained and not fall behind. Catching up can be very difficult. If you find yourself having difficulty with a section or a concept, put in some extra effort to master it before moving ahead. It will be worth it in the end.

Truth Tables and the Truth-Functional Symbols

Our "logical vocabulary" will consist of claim variables and truth-functional symbols. Before we consider the real heart of the subject, truth tables and the symbols that represent them, let's first clarify the use of letters of the alphabet to symbolize terms and claims.

■ *Claim Variables*

In Chapter 8, we used uppercase letters to stand for terms in categorical claims. Here we use uppercase letters to stand for claims. Our main interest is now in the way that words such as "not," "and," "or," and so on affect claims and link them together to produce compound claims out of simpler ones. So don't confuse the Ps and Qs, called **claim variables,** that appear in this chapter with the term variables used in Chapter 8.*

■ *Truth Tables*

Let's now consider truth tables and symbols. In truth-functional logic, any given claim, P, is either true or false. The following little table, called a **truth table,** displays both possible truth values for P:

P
—
T
F

*Whichever truth value the claim P might have, its negation or contradictory, which we'll symbolize ~P, will have the other. Here, then, is the truth table for **negation:***

P	~P
T	F
F	T

The left-hand column of this table sets out both possible truth values for P, and the right-hand column sets out the truth values for ~P based on

---

*It is customary to use one kind of symbol, usually lowercase letters or Greek letters, as *claim variables* and plain or italicized uppercase letters for *specific claims.* Although this use has some technical advantages and makes possible a certain theoretical neatness, students often find it confusing. Therefore, we'll use uppercase letters for both variables and specific claims and simply make it clear which way we're using the letters.

P's values. This is a way of defining the negation sign, ~, in front of the P. The symbol means "change the truth value from T to F or from F to T, depending on P's values." Because it's handy to have a name for negations that you can say aloud, we read ~P as "not-P." So, if P were "Parker is at home," then ~P would be "It is not the case that Parker is at home," or, more simply, "Parker is not at home." In a moment we'll define other symbols by means of truth tables, so make sure you understand how this one works.

Because any given claim is either true or false, two claims, P and Q, must both be true, both be false, or have opposite truth values, for a total of four possible combinations. Here are the possibilities in truth-table form:

P	Q
T	T
T	F
F	T
F	F

A **conjunction** is a compound claim made from two simpler claims, called *conjuncts. A conjunction is true if and only if both of the simpler claims that make it up (its conjuncts) are true.* An example of a conjunction is the claim "Parker is at home and Moore is at work." We'll express the conjunction of P and Q by connecting them with an ampersand (&). The truth table for conjunctions looks like this:

P	Q	P & Q
T	T	T
T	F	F
F	T	F
F	F	F

P & Q is true in the first row only, where both P and Q are true. Notice that the "truth conditions" in this row match those required in the italicized statement above.

Here's another way to remember how conjunctions work: If either part of a conjunction is false, the conjunction itself is false. Notice finally that although the word "and" is the closest representative in English to our ampersand symbol, there are other words that are correctly symbolized by the ampersand: "but" and "while," for instance, as well as phrases such as "even though." So, if we let P stand for "Parsons is in class" and Q stand for "Quincy is absent," then we should represent "Parsons is in class even though Quincy is absent" by P & Q. The reason is that the compound claim is true only in one case: where both parts are true. And that's all it takes to require an ampersand to represent the connecting word or phrase.

A **disjunction** is another compound claim made up of two simpler claims, called *disjuncts. A disjunction is false if and only if both of its dis-*

"VERY CREATIVE. VERY IMAGINATIVE. LOGIC... THAT'S WHAT'S MISSING."

juncts are false. Here's an example of a disjunction: "Either Parker is at home or Moore is at work." We'll use the symbol ∨ ("wedge") to represent disjunction when we symbolize claims—as indicated in the example, the closest word in English to this symbol is "or." The truth table for disjunctions is this:

P	Q	P ∨ Q
T	T	T
T	F	T
F	T	T
F	F	F

Notice here that a disjunction is false only in the last row, where both of its disjuncts are false. In all other cases a disjunction is true.

The third kind of compound claim made from two simpler claims is the **conditional claim.** In ordinary English, the most common way of stating conditionals is by means of the words "if . . . then . . . ," as in the example "If Parker is at home, then Moore is at work."

We'll use an arrow to symbolize conditionals: P→Q. The first claim in a conditional, the P in the symbolization, is the **antecedent,** and the second—Q in this case—is the **consequent.** *A conditional claim is false if and only if its antecedent is true and its consequent is false.* The truth table for conditionals looks like this:

P	Q	P→Q
T	T	T
T	F	F
F	T	T
F	F	T

Only in the second row, where the antecedent P is true and the consequent Q is false, does the conditional turn out to be false. In all other cases it is true.

Of the four types of truth-functional claims—negation, conjunction, disjunction, and conditional—the conditional typically gives students the most trouble. Let's have a closer look at it by considering an example that may shed light on how and why conditionals work. Let's say that Moore promises you that, if his paycheck arrives this morning, he'll buy you lunch. So now we can consider the conditional,

If Moore's paycheck arrives this morning, then Moore will buy you lunch.

We can symbolize this using P (for the claim about the paycheck) and L (for the claim about lunch): P→L. Now let's try to see why the truth table above fits this claim.

The easiest way to see this is by asking yourself what it would take for Moore to break his promise. A moment's thought should make this clear: Two things have to happen before we can say that Moore has fibbed to you. The first is that his paycheck must arrive this morning. (After all, he didn't say what he was going to do if his paycheck *didn't* arrive, did he?) Then, it being true that his paycheck arrives, he must then *not* buy you lunch. Together, these two items make it clear that Moore's original promise was false. Notice: Under no other circumstances would we say that Moore broke his promise. And *that* is why the truth table has a conditional false in one and only one case, namely, where the antecedent is true and the consequent is false. Basic information about all four symbols is summarized in Figure 1.

Our truth-functional symbols can work in combination. Consider, for example, the claim "If Paula doesn't go to work, then Quincy will have to work a double shift." We'll represent the two simple claims in the obvious way, as follows:

Negation (~)	Conjunction (&)
Truth table: P ~P T F F T Closest English counterparts: "not," or "it is not the case that"	Truth table: P Q (P & Q) T T T T F F F T F F F F Closest English counterparts: "and," "but," "while"
Disjunction (∨)	Conditional (→)
Truth table: P Q (P ∨ Q) T T T T F T F T T F F F Closest English counterparts: "or," "unless"	Truth table: P Q (P → Q) T T T T F F F T T F F T Closest English counterparts: "if . . . . then," "provided that"

FIGURE 1   The Four Basic Truth-Functional Symbols

P = Paula goes to work.
Q = Quincy has to work a double shift.

And we can symbolize the entire claim like this:

~P→Q

Here is a truth table for this symbolization:

P	Q	~P	~P→Q
T	T	F	T
T	F	F	T
F	T	T	T
F	F	T	F

Notice that the symbolized claim ~P→Q is false in the *last* row of this table. That's because here and only here the antecedent, ~P, is true and its consequent, Q, is false. Notice that we work from the simplest parts to the most complex: The truth value of P in a given row determines the truth value of ~P, and that truth value in turn, along with the one for Q, determines the truth value of ~P→Q.

Consider another combination: "If Paula goes to work, then Quincy and Rogers will get a day off." This claim is symbolized this way:

P→(Q & R)

This symbolization requires parentheses in order to prevent confusion with (P→Q) & R, which symbolizes a different claim and has a different truth table. Our claim is a conditional with a conjunction for a consequent, whereas (P→Q) & R is a conjunction with a conditional as one of the conjuncts. The parentheses are what make this clear.

You need to know a few principles to produce the truth table for the symbolized claim P→(Q & R). First you have to know how to set up all the possible combinations of true and false for the three simple claims P, Q, and R. In claims with only one letter, there were two possibilities, T and F. In claims with two letters, there were four possibilities. *Every time we add another letter, the number of possible combinations of T and F doubles, and so, therefore, does the number of rows in our truth table.* The formula for determining the number of rows in a truth table for a compound claim is $r = 2^n$, where r is the number of rows in the table and n is the number of letters in the symbolization. Because the claim we are interested in has three letters, our truth table will have eight rows, one for each possible combination of T and F for P, Q, and R. Here's how we do it:

P	Q	R
T	T	T
T	T	F
T	F	T
T	F	F
F	T	T
F	T	F
F	F	T
F	F	F

The systematic way to construct such a table is to alternate Ts and Fs in the right-hand column, then alternate *pairs* of Ts and *pairs* of Fs in the next column to the left, then sets of *four* Ts and sets of *four* Fs in the next, and so forth. The leftmost column will always wind up being half Ts and half Fs.

The second thing we have to know is that the truth value of a compound claim in any particular case (i.e., any row of its truth table) depends entirely upon the truth values of its parts; and if these parts are themselves compound, their truth values depend upon those of their parts; and so on, until we get down to letters standing alone. The columns under the letters, which you have just learned to construct, will then tell us what we need to know. Let's build a truth table for P→(Q & R) and see how this works.

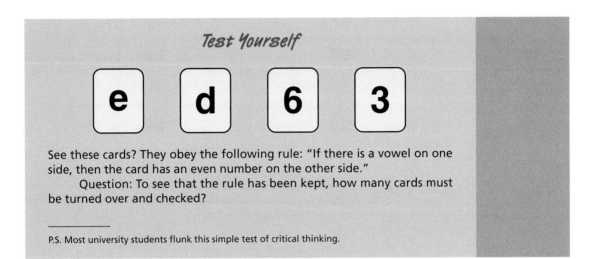

Test Yourself

See these cards? They obey the following rule: "If there is a vowel on one side, then the card has an even number on the other side."

Question: To see that the rule has been kept, how many cards must be turned over and checked?

P.S. Most university students flunk this simple test of critical thinking.

P	Q	R	Q & R	P→(Q & R)
T	T	T	T	T
T	T	F	F	F
T	F	T	F	F
T	F	F	F	F
F	T	T	T	T
F	T	F	F	T
F	F	T	F	T
F	F	F	F	T

The three columns at the left, under P, Q, and R, are our *reference columns,* set up just as we discussed above. They determine what goes on in the rest of the table. From the second and third columns, under the Q and the R, we can fill in the column under Q & R. Notice that this column contains a T only in the first and fifth rows, where both Q and R are true. Next, from the column under the P and the one under Q & R, we can fill in the last column, which is the one for the entire symbolized claim. It contains Fs in only rows two, three, and four, which are the only ones where its antecedent is true and its consequent is false.

What our table gives us is a *truth-functional analysis* of our original claim. Such an analysis displays the compound claim's truth value, based on the truth values of its simpler parts.

If you've followed everything so far without problems, that's great. If you've not yet understood the basic truth table idea, however, as well as the truth tables for the truth-functional symbols, then by all means stop now and go back over this material. You should also understand how to build a truth table for symbolizations consisting of three or more letters. What comes later builds on this foundation, and like any construction project, without a strong foundation the whole thing collapses.

A final note before we turn to tips for symbolizing compound claims: Two claims are **truth-functionally equivalent** if they have exactly the same truth table—that is, if the Ts and Fs in the column under one claim are in the same arrangement as those in the column under the other. Generally speaking, when two claims are equivalent, one can be used in place of another—truth-functionally, they each imply the other.*

■ *Symbolizing Compound Claims*

Most of the things we can do with symbolized claims are pretty straightforward; that is, if you learn the techniques, you can apply them in a relatively clear-cut way. What's less clear-cut is how to symbolize a claim in the first place. We'll cover a few tips for symbolization in this section and then give you a chance to practice with some exercises.

Remember, when you symbolize a claim, you're displaying its truth-functional structure. The idea is to produce a version that will be truth-functionally equivalent to the original informal claim—that is, one that will be true under all the same circumstances as the original and false under all the same circumstances. Let's go through some examples that illustrate some standard symbolization problems.

"If" and "Only If"   In symbolizing truth-functional claims, as in translating categorical claims in Chapter 8, nothing can take the place of a careful reading of what the claim in question says. It always comes down to a matter of exercising careful judgment. Nonetheless, there are some tips we can give you that should make the job a little easier.

Of all the basic truth-functional types of claim, the conditional is probably the most difficult for students to symbolize correctly. There are so many ways to make these claims in ordinary English that it's not easy to keep track. Fortunately, the phrases "if" and "only if" account for a large number of conditionals, so you'll have a head start if you understand their uses. Here are some rules of thumb to remember:

1. The word "if," used alone, introduces the antecedent of a conditional.
2. The phrase "only if" introduces the consequent of a conditional.

---

*The exceptions to this remark are due to *non*-truth-functional nuances that claims sometimes have. Most compound claims of the form P & Q, for example, are interchangeable with what are called their commutations, Q & P. That is, it doesn't matter which conjunct comes first. (Note that this is also true of disjunctions but *not* of conditionals.) But at times "and" has more than a truth-functional meaning, and in such cases which conjunct comes first can make a difference. In "Daniel got on the train and bought his ticket," the word "and" would ordinarily mean "and then." So "Daniel bought his ticket and got on the train" turns out to be a different claim from the previous one; it says that he did the two things in a different order from that stated in the first claim. This temporal-ordering sense of "and" is part of the word's occasional non-truth-functional meaning.

Truth-Functional Trickery

Using what you know about truth-functional logic, can you identify how the sender of this encouraging-looking notice can defend the claim (because it *is* true) even though the receiver is not really going to win one nickel?

You Have Absolutely Won
$1,000,000.00
If you follow the instructions inside
and return the winning number!

Answer: Because there is not going to be any winning number inside (there are usually several *losing* numbers, in case that makes you feel better), the conjunction "You follow the instructions inside and [you] return the winning number" is going to be false, even if you do follow the instructions inside. Therefore, because this conjunction is the antecedent of the whole conditional claim, the conditional claim turns out to be true.

Of course, uncritical readers will take the antecedent to be saying something like "If you follow the instructions inside *by returning the winning number inside* (as if there were a winning number inside). These are the people who may wind up sending their own money to the mailer.

To put it another way: It's not the location of the part in a conditional that tells us whether it is the antecedent or the consequent; it's the logical words that identify it. Consider this example:

> Moore will get wet *if* Parker capsizes the boat.

The "Parker" part of the claim is the antecedent, even though it comes *after* the "Moore" part. It's as though the claim had said,

> If Parker capsizes the boat, Moore will get wet.

We would symbolize this claim as $P \rightarrow M$. Once again, it's the word "if" that tells us what the antecedent is.

> Parker will beat Moore at 9-ball *only if* Moore has a bad day.

This claim is different. In this case, the "Parker" part is the antecedent because "only if" introduces the consequent of a conditional. This is truth-functionally the same as

> If Parker beats Moore at 9-ball (P), then Moore had (or must have had) a bad day (B).

Hell Hath Enlarged Herself

The fearful, and unbelieving, and the abominable, and murderers, and whoremongers, and sorcerers, and idolators, and all liars, shall have their part in the lake which burneth with fire and brimstone.

—Revelation 21:8

This came to us in a brochure from a religious sect offering salvation for the believer. Notice, though, that the passage from the Bible doesn't say that, if you believe, you won't go to hell. It says, if you don't believe, you will go to hell.

Using the letters indicated in parentheses, we'd symbolize this as

$P \rightarrow B$

Don't worry about the grammatical tenses; we'll adjust those in whatever way necessary so that the claims make sense. We can use "if" in front of a conditional's antecedent, or we can use "only if" in front of its consequent; we produce exactly equivalent claims in the two cases. As in the case with "if," it doesn't matter where the "only if" part of the claim occurs. The part of this claim that's about Moore is the consequent, even though it occurs at the beginning of this version:

Only if Moore has a bad day will Parker beat him at 9-ball.

Exercise 9-1

Symbolize the following using the claim variables P and Q. (You can ignore differences in past, present, and future tense.)

▲ 1. If Quincy learns to symbolize, Paula will be amazed.

▲ 2. Paula will teach him if Quincy pays her a big fee.

▲ 3. Paula will teach him only if Quincy pays her a big fee.

▲ 4. Only if Paula helps him will Quincy pass the course.

▲ 5. Quincy will pass if and only if Paula helps him.

Claim 5 in the preceding exercise introduces a new wrinkle, the phrase "if and only if." Remembering our rules of thumb about how "if" and "only if" operate separately, it shouldn't surprise us that "if and only if" makes both antecedent and consequent out of the claim it introduces. We can make P both antecedent and consequent this way:

$(P \rightarrow Q) \ \& \ (Q \rightarrow P)$

Okay, Lew, the deal is, you can use the car tonight only if you wash and wax it this afternoon.

▲ *Comment: We often use "only if" when we mean to state both necessary and sufficient conditions, even though, literally speaking, it produces only the former. If Lew were a critical thinker, he'd check this deal more carefully before getting out the hose and bucket.*

There are other ways to produce conditionals, of course. In one of its senses, the word "provided" (and the phrase "provided that") works like the word "if" in introducing the antecedent of a conditional. "Moore will buy the car, provided the seller throws in a ton of spare parts" is equivalent to the same expression with the word "if" in place of "provided."

Necessary and Sufficient Conditions Conditional claims are sometimes spelled out in terms of necessary and sufficient conditions. Consider this example:

The presence of oxygen is a necessary condition for combustion.

This tells us that we can't have combustion without oxygen, or "If we have combustion (C), then we must have oxygen (O)." Notice that the necessary condition becomes the consequent of a conditional: $C \rightarrow O$.

A sufficient condition *guarantees* whatever it is a sufficient condition for. Being born in the United States is a sufficient condition for U.S. citizenship—that's *all* one needs to be a U.S. citizen. Sufficient conditions are expressed as the antecedents of conditional claims, so we would say, "If John was born in the United States (B), then John is a U.S. citizen (C)": $B \rightarrow C$.

You should also notice the connection between "if" and "only if" on the one hand and necessary and sufficient conditions on the other. The word "if," by itself, introduces a sufficient condition; the phrase "only if" introduces a necessary condition. So the claim "X is a necessary condition for Y" would be symbolized "Y→X."

From time to time, one thing will be both a necessary and a sufficient condition for something else. For example, if Jean's payment of her dues to the National Truth-Functional Logic Society (NTFLS) guaranteed her continued membership (making such payment a sufficient condition) and there was no way for her to continue membership *without* paying her dues (making payment a necessary condition as well), then we could express such a situation as "Jean will remain a member of the NTFLS (M) if and only if she pays her dues (D)": (M→D) & (D→M).

We often play fast and loose with how we state necessary and sufficient conditions. A parent tells his daughter, "You can watch television only if you clean your room." Now, the youngster would ordinarily take cleaning her room as both a necessary and a sufficient condition for being allowed to watch television, and probably that's what a parent would intend by those words. But notice that the parent actually stated only a necessary condition; technically, he would not be going back on what he said if room cleaning turned out not to be sufficient for television privileges. Of course he'd better be prepared for more than a logic lesson from his daughter in such a case, and most of us would be on her side in the dispute. But, literally, it's the necessary condition that the phrase "only if" introduces, not the sufficient condition.

"Unless" Consider the claim "Paula will foreclose unless Quincy pays up." Asked to symbolize this, we might come up with ~Q→P because the original claim is equivalent to "If Quincy doesn't pay up, then Paula will foreclose." But there's an even simpler way to do it. Ask yourself, what is the truth table for ~Q→P? If you've gained familiarity with the basic truth tables by this time, you realize that it's the same as the table for P ∨ Q. And, as a matter of fact, you can treat the word "unless" exactly like the word "or" and symbolize it with a "∨."

Playoff Logic

Following is an excerpt from an Ira Miller piece in the sports pages of the *San Francisco Chronicle* in which he sets up the Oakland Raiders' possibilities leading up to the 2002 Superbowl. When you're setting out lots of possible combinations of events, and drawing conclusions about what happens in which case, you are nearly bound to use some truth-functional logic. (You could arrange these things in terms of categorical logic, but the results would look very bizarre.)

Just for fun, symbolize the three sentences we've put in bold type. Use the following symbol dictionary. (Symbolizations are given in the answer section in the back of the book.)

B = The Raiders get a bye. (A bye is the ability to skip a game.)

C = The Raiders win one more game.

J = Raiders beat the Jets.

L = Jets (New York) lose to Buffalo next Sunday.

D = Raiders beat Denver.

F = Raiders finish 11–5.

A = The AFC East winner is 11–5.

N = New England wins the AFC East.

W = The Dolphins lose this week.

This has been such a wacky, unpredictable NFL season that something like this almost figured: Suddenly, the New York Jets are a much more important opponent for the Raiders than the Denver Broncos.

In fact, the Raiders' game on Sunday at Denver might have no bearing at all on whether they get a first-round bye in the playoffs.

Here's why: **(1) For the Raiders to earn that elusive bye** (forget home-field advantage; as a practical matter, it belongs to Pittsburgh), **they must win at least one more game, and it must be against the Jets, unless New York loses at home to Buffalo next Sunday.**

Because the Jets-Bills game will be over when the Raiders begin their game at Denver, they will know where they stand.

Basically, this is the deal for the Raiders. **(2) If they win both of their remaining games—at Denver, home vs. the Jets—and finish 12–4, they get a bye. But if they finish 11–5, and the AFC East winner also is 11–5, which is likely, they get a bye only if New England wins the AFC East.**

For the record, **(3) there is one combination that could clinch a bye for the Raiders this week: if the Dolphins and Jets lose and the Raiders beat Denver.**

At the moment, the Patriots are 10–5 and the Jets and Miami both are 9–5. The Raiders are the only opponent with a winning record remaining for any of those three teams. New England has a bye, then plays at 1–13 Carolina. The Jets play 2–12 Buffalo at home before coming to Oakland. Fast-fading Miami has two home games remaining, against 7–7 Atlanta and then Buffalo.

—San Francisco Chronicle, *December 25, 2001*

"Either . . ."   Sometimes we need to know exactly where a disjunction begins; it's the job of the word "either" to show us. Compare the claims

Either P and Q or R

and

P and either Q or R.

These two claims say different things and have different truth tables, but the only difference between them is the location of the word "either"; without that word, the claim would be completely ambiguous. "Either" tells us that the disjunction begins with P in the first claim and Q in the second claim. So we would symbolize the first (P & Q) ∨ R and the second P & (Q ∨ R).

The word "if" does much the same job for conditionals as "either" does for disjunctions. Notice the difference between

P and if Q then R

and

If P and Q then R.

"If" tells us that the antecedent begins with Q in the first example and with P in the second. Hence, the second must have P & Q for the antecedent of its symbolization.

In general, the trick to symbolizing a claim correctly is to pay careful attention to exactly what the claim says—and this often means asking yourself just exactly what would make this claim false (or true). Then try to come up with a symbolization that says the same thing—that is false (or true) in exactly the same circumstances. There's no substitute for practice, so here's an exercise to work on.

Exercise 9-2

When we symbolize a claim, we're displaying its truth-functional structure. Show that you can figure out the structures of the following claims by symbolizing them. Use these letters for the first ten items:

P = Parsons objects.
Q = Quincy turns (or will turn) the radio down.
R = Rachel closes (or will close) her door.

Use the symbols ~, &, ∨, and →. We suggest that, at least at first, you make symbolization a two-stage process: First, replace simple parts of claims with letters; then, replace logical words with logical symbols, and add parentheses as required. We'll do an example in two stages to show you what we mean.

Example

If Parsons objects, then Quincy will turn the radio down but Rachel will not close her door.

Stage 1: If P, then Q but ~R.
Stage 2: P→(Q & ~R)

▲   1. If Parsons objects then Quincy will turn the radio down, and Rachel will close her door.

▲   2. If Parsons objects, then Quincy will turn the radio down and Rachel will close her door.

3. If Parsons objects and Quincy turns the radio down then Rachel will close her door.

4. Parsons objects and if Quincy turns the radio down Rachel will close her door.

▲   5. If Parsons objects then if Quincy turns the radio down Rachel will close her door.

6. If Parsons objects Quincy turns the radio down, and if Rachel closes her door Quincy turns the radio down.

7. Quincy turns the radio down if either Parsons objects or Rachel closes her door.

8. Either Parsons objects or, if Quincy turns the radio down, then Rachel will close her door.

9. If either Parsons objects or Quincy turns the radio down then Rachel will close her door.

10. If Parsons objects then either Quincy will turn the radio down or Rachel will close her door.

For the next ten items, use the following letters:

C = My car runs well.
S = I will sell my car.
F = I will have my car fixed.

▲ 11. If my car doesn't run well, then I will sell it.

▲ 12. It's not true that if my car runs well, then I will sell it.

13. I will sell my car only if it doesn't run well.

14. I won't sell my car unless it doesn't run well.

15. I will have my car fixed unless it runs well.

▲ 16. I will sell my car, but only if it doesn't run well.

17. Provided my car runs well, I won't sell it.

18. My car's running well is a sufficient condition for my not having it fixed.

19. My car's not running well is a necessary condition for my having it fixed.

▲ 20. I will neither have my car fixed nor sell it.

Exercise 9-3 _____

Construct truth tables for the symbolizations you produced for Exercise 9-2. Determine whether any of them are truth-functionally equivalent to any others. (Answers to items 1, 5, and 12 are provided in the answer section at the end of the book.)

Truth-Functional Arguments

Categorical syllogisms (discussed in Chapter 8) have a total of 256 forms. A truth-functional argument, by contrast, can take any of an infinite number of forms. Nevertheless, we have methods for testing for validity that are flexible enough to encompass every truth-functional argument. In the remainder of this chapter, we'll look at three of them: the truth-table method, the short truth-table method, and the method of deduction.

Before doing anything else, though, let's quickly review the concept of validity. An argument is *valid*, you'll recall, if and only if the truth of the premises guarantees the truth of the conclusion—that is, if the premises were true, the conclusion could not then be false. (In logic, remember, it doesn't matter whether the premises are *actually* true.)

The *truth-table test for validity* requires familiarity with the truth tables for the four truth-functional symbols, so go back and check yourself on those if you think you may not understand them clearly. Here's how the method works: We present all of the possible circumstances for an argument by building a truth table for it; then we simply look to see if there are any circumstances in which the premises are all true and the conclusion false. If there are such circumstances—one row of the truth table is all that's required—then the argument is invalid.

Let's look at a simple example. Let P and Q represent any two claims. Now look at the following symbolized argument:

P→Q
~P
Therefore, ~Q

We can construct a truth table for this argument by including a column for each premise and one for the conclusion:

P	Q	~P	P→Q	~Q
T	T	F	T	F
T	F	F	F	T
F	T	T	T	F
F	F	T	T	T

A Real-Life Example

Sometimes it takes effort to see the truth-functional logic in a passage. To illustrate, consider the letter to Marilyn vos Savant in the column printed below:

> If homosexuality is a product of nature and not chosen, are fewer homosexuals being born every year? As homosexuality becomes more acceptable, it seems logical that passing genetic material to offspring should be declining.
>
> —Wes Alexander, Lilburn, Ga.

We don't know how many gay people there were in the past any more than we know how many gay people there are now. Estimates of today's gay population range from 1% to 10%, but we don't know whether the number is up, down or stable.

The genetic situation is also unknown. For example, genes can be passed on by people who are themselves unaffected by them. And gay people could be born that way for reasons other than strictly genetic ones: Conditions during pregnancy play an important role in the development of the unborn baby.

Even the social situation is unclear. Gay people may want to have families as much (or as little) as straight people do. After all, when heterosexual couples make love, they usually aren't trying to conceive. A desire for a heterosexual relationship is definitely not the same thing as a desire to become a parent.

Anyway, population is determined by women—heterosexual *and* homosexual. A gay woman is just as capable of having a baby and, if she wants to be a mother, that's what she'll do. And like a straight woman, she has a choice of straight and gay partners. The main difference is that she's less likely to live with him afterward. Either way, the population statistics are unaffected.

Can you see that the letter has this form?

H = Homosexuality is becoming more acceptable
P = Homosexuality is a product of nature and not chosen
F = Fewer homosexuals are born every year

H
H $\rightarrow$ (P $\rightarrow$ F)
~F
/ $\therefore$ ~P

Be sure to notice which premises Ask Marilyn disputes (and why).

The first two columns are reference columns; they list truth values for the letters that appear in the argument. The third and fourth columns appear under the two premises of the argument, and the fifth column is for the conclusion. Note that in the third row of the table, both premises are true and the conclusion is false. This tells us that it is possible for the premises of this argument to be true while the conclusion is false; thus, the argument is invalid. Because it doesn't matter what claims P and Q might stand for, the same is true for *every* argument of this pattern. Here's an example of such an argument:

> If the Saints beat the Forty-Niners, then the Giants will make the playoffs. But the Saints won't beat the Forty-Niners. So the Giants won't make the playoffs.

Using S for "The Saints beat (or will beat) the Forty-Niners," and G for "The Giants make (or will make) the playoffs," we can symbolize the argument like this:

$$S \rightarrow G$$
$$\underline{\sim S}$$
$$\sim G$$

The first premise is a conditional, and the other premise is the negation of the antecedent of that conditional. The conclusion is the negation of the conditional's consequent. It has exactly the same structure as the argument for which we just did the truth table; accordingly, it too is invalid.

Let's do another simple one:

> We're going to have large masses of arctic air (A) flowing into the Midwest unless the jet stream (J) moves south. Unfortunately, there's no chance of the jet stream's moving south. So you can bet there'll be arctic air flowing into the Midwest.

Symbolization gives us

$$A \vee J$$
$$\underline{\sim J}$$
$$A$$

Here's a truth table for the argument:

1	2	3	4
A	J	A ∨ J	~J
T	T	T	F
T	F	T	T
F	T	T	F
F	F	F	T

Note that the first premise is represented in column 3 of the table, the second premise in column 4, and the conclusion in one of the reference columns, column 1. Now, let's recall what we're up to. We want to know whether this argument is valid—that is to say, is it possible for the premises to be true and the conclusion false? If there is such a possibility, it will turn up in the truth table because, remember, the truth table represents every possible situation with respect to the claims A and J. We find that the premises are both true in only one row, the second, and when we check the conclusion, A, we find it is true in that row. Thus, there is *no* row in which the premises are true and the conclusion false. So the argument is valid.

Here's an example of a rather more complicated argument:

> If Scarlet is guilty of the crime, then Ms. White must have left the back door unlocked and the colonel must have retired before ten o'clock. However, either Ms. White did not leave the back door unlocked, or the colonel did not retire before ten. Therefore, Scarlet is not guilty of the crime.

Let's assign some letters to the simple claims so that we can show this argument's pattern.

> S = Scarlet is guilty of the crime.
> W = Ms. White left the back door unlocked.
> C = The colonel retired before ten o'clock.

Now we symbolize the argument to display this pattern:

$$S \rightarrow (W \,\&\, C)$$
$$\underline{\sim W \vee \sim C}$$
$$\sim S$$

Let's think our way through this argument. As you read, refer back to the symbolized version above. Notice that the first premise is a conditional, with "Scarlet is guilty of the crime" as antecedent and a conjunction as consequent. In order for that conjunction to be true, both "Ms. White left the back door unlocked" and "The colonel retired before ten o'clock" have to be true, as you'll recall from the truth table for conjunctions. Now look at the second premise. It is a disjunction that tells us *either* Ms. White did not leave the back door unlocked *or* the colonel did not retire before ten. But if either or both of those disjuncts are true, at least one of the claims in our earlier conjunction is false. So it cannot be that *both* parts of the conjunction are true. This means the conjunction symbolized by W & C must be false. And so the consequent of the first premise is false. How can the entire premise be true, in that case? The only way is for the antecedent to be false as well. And that means that the conclusion, "Scarlet is not guilty of the crime," must be true.

All of this reasoning (and considerably more that we don't require) is implicit in the following truth table for the argument:

If Saddam doesn't back down, the U.S. should launch an air strike. He backed down. Therefore, we shouldn't launch a strike.

—WILLIAM COHEN, Secretary of Defense

Whoops, a mistake in logic.

Rule 1

Hitler was clearly intent on taking over the world. Mr. Buchanan thinks we should have stood by silently watching, and so, no, I would not say he belongs in the Republican Party.

The remark by Senator John McCain of Arizona is actually Rule 1 (page 331) with an unstated premise, in effect:

[If Mr. Buchanan thinks we should have stood by silently watching, then Mr. Buchanan does not belong in the Republican Party.]
    Mr. Buchanan thinks we should have stood by silently watching.
    Therefore, Mr. Buchanan does not belong in the Republican Party.

1	2	3	4	5	6	7	8	9
S	W	C	~W	~C	W & C	S → (W & C)	~W ∨ ~C	~S
T	T	T	F	F	T	T	F	F
T	T	F	F	T	F	F	T	F
T	F	T	T	F	F	F	T	F
T	F	F	T	T	F	F	T	F
F	T	T	F	F	T	T	F	T
F	T	F	F	T	F	T	T	T
F	F	T	T	F	F	T	T	T
F	F	F	T	T	F	T	T	T

We've numbered the columns at the top to make reference somewhat easier. The first three are our reference columns, columns 7 and 8 are for the premises of the argument, and column 9 is for the argument's conclusion. The remainder—4, 5, and 6—are for parts of some of the other symbolized claims; they could be left out if we desired, but they make filling in columns 7 and 8 a bit easier.

Once the table is filled in, evaluating the argument is easy. Just look to see whether there is any row in which the premises are true and the conclusion is false. One such row is enough to demonstrate the invalidity of the argument.

In the present case, we find that both premises are true only in the last three rows of the table. And in those rows, the conclusion is also true. So there is no set of circumstances—no row of the table—in which both premises are true and the conclusion is false. Therefore, the argument is valid.

Although filling out a complete truth table always produces the correct answer regarding a truth-functional argument's validity, it can be quite a tedious chore—in fact, life is much too short to spend much of it filling in truth tables. Fortunately, there are shorter and more manageable ways of finding such an answer. The easiest systematic way to determine the validity or invalidity of truth-functional arguments is the *short truth-table method*. Here's the idea behind it: Because, if an argument is invalid, there has to be at least one row in the argument's truth table where the premises are true and the conclusion is false, we'll look directly for such a row. Consider this symbolized argument:

P→Q
~Q→R
~P→R

We begin by looking at the conclusion. Because it's a conditional, it can be made false only one way, by making its antecedent true and its consequent false. So we do that, by making P false and R false.

Can we now make both premises true? Yes, as it turns out, by making Q true. This case,

P	Q	R
F	T	F

makes both premises true and the conclusion false and thus proves the argument invalid. What we've done is produce the relevant row of the truth table without bothering to produce all the rest. Had the argument been valid, we would not have been able to produce such a row.

Here's how the method works with a valid argument. Consider this example:

(P ∨ Q)→R
S→Q
S→R

The only way to make the conclusion false is to make S true and R false. So we do that:

P	Q	R	S
		F	T

Now, with S true, the second premise requires that we make Q true. So we do that next:

P	Q	R	S
	T	F	T

But now, there is no way at all to make the first premise true, because P ∨ Q is going to be true (because Q is true) and R is already false. Because there is no other way to make the conclusion false and the second premise true and because this way fails to make the first premise true, we can conclude that the argument is *valid.*

In some cases, there may be more than one way to make the conclusion false. Here's a symbolized example:

P & (Q ∨ R)
R → S
P → T
―――――――
S & T

Because the conclusion is a conjunction, it is false if either or both of its conjuncts are false, which means we could begin by making S true and T false, S false and T true, or both S and T false. This is trouble we'd like to avoid if possible, so let's see if there's someplace else we can begin making our assignment. (Remember: The idea is to try to assign true and false to the letters so as to make the premises true and the conclusion false. If we can do it, the argument is invalid.)

In this example, to make the first premise true, we *must* assign true to the letter P. Why? Because the premise is a conjunction and both of its parts must be true for the whole thing to be true. That's what we're looking for: places where we are *forced* to make an assignment of true or false to one or more letters. Then we make those assignments and see where they lead us. In this case, once we've made P true, we see that, to make the third premise true, we are forced to make T true (because a true antecedent and a false consequent would make the premise false, and we're trying to make our premises true).

After making T true, we see that, to make the conclusion false, S must be false. So we make that assignment. At this point we're nearly done, needing only assignments for Q and R.

P	Q	R	S	T
T			F	T

Are there any other assignments that we're forced to make? Yes: We must make R false to make the second premise true. Once we've done that, we see that Q must be true to preserve the truth of the first premise. And that completes the assignment:

P	Q	R	S	T
T	T	F	F	T

This is one row in the truth table for this argument—the only row, as it turned out—in which all the premises are true and the conclusion is false; thus, it is the row that proves the argument invalid.

Some Common Truth-Functional Argument Patterns

Some truth-functional patterns are so built into our thinking process that they almost operate at a subverbal level. But, rather than trust our subverbal skills, whatever those might be, let's identify three common patterns that are perfectly valid—their conclusions follow with certainty from their premises—and three invalid imposters—each of the imposters bears a resemblance to one of the good guys. We'll set them up in pairs:

Valid Argument Forms	**Invalid Argument Forms**
In this case, the premises guarantee the conclusion.	Here, the premises can be true while the conclusion is false.

Valid Argument Forms

1. Modus ponens (or affirming the antecedent)

    If P, then Q

    P
    ___

    Q

2. Modus tollens (or denying the consequent)

    If P, then Q

    Not-Q
    _____

    Not-P

3. Chain argument

    If P, then Q

    If Q, then R
    _____

    If P, then R

Invalid Argument Forms

1-A. Affirming the consequent

    If P, then Q

    Q
    ___

    P

2-A. Denying the antecedent

    If P, then Q

    Not-P
    _____

    Not-Q

3-A. Undistributed middle (truth-functional version)

    If P, then Q

    If R, then Q
    _____

    If P, then R

In the preceding example, there was a premise that forced us to begin with a particular assignment to a letter. Sometimes neither the conclusion nor any of the premises forces an assignment on us. In that case we must use trial and error: Begin with one assignment that makes the conclusion false (or some premise true) and see if it will work. If not, try another assignment. If all fail, then the argument is valid.

Often, several rows of a truth table will make the premises true and the conclusion false; any one of them is all it takes to prove invalidity. Don't get the mistaken idea that, just because the premises are all true in one row and so is the conclusion, the conclusion follows from the premises—that is, that the argument must be valid. To be valid, the conclusion must be true in *every* row in which all the premises are true.

To review: Try to assign Ts and Fs to the letters in the symbolization so that all premises come out true and the conclusion comes out false. There may be more than one way to do it; any of them will do to prove the argument invalid. If it is impossible to make the premises and conclusion come out this way, the argument is valid.

Exercise 9-4

Construct full truth tables or use the short truth-table method to determine which of the following arguments are valid.

▲ 1. P ∨ ~Q
$$\frac{\text{~Q}}{\text{~P}}$$

2. P→Q
$$\frac{\text{~Q}}{\text{~P}}$$

3. ~ (P ∨ Q)
$$\frac{\text{R}\rightarrow\text{P}}{\text{~R}}$$

▲ 4. P→(Q→R)
$$\frac{\text{~ (P}\rightarrow\text{Q)}}{\text{R}}$$

5. P ∨ (Q→R)
$$\frac{\text{Q \& ~R}}{\text{~P}}$$

6. (P→Q) ∨ (R→Q)
$$\frac{\text{P \& (~P}\rightarrow\text{~R)}}{\text{Q}}$$

▲ 7. (P & R)→Q
$$\frac{\text{~Q}}{\text{~P}}$$

8. P & (~Q→~P)
$$\frac{\text{R}\rightarrow\text{~Q}}{\text{~R}}$$

9. L ∨ ~J
$$\frac{\text{R}\rightarrow\text{J}}{\text{L}\rightarrow\text{~R}}$$

10. ~F ∨ (G & H)
$$\frac{\text{P}\rightarrow\text{F}}{\text{~H}\rightarrow\text{~P}}$$

Exercise 9-5

Use either the long or short truth-table method to determine which of the following arguments are valid.

▲ 1. K→(L & G)
M→(J & K)
$$\frac{\text{B \& M}}{\text{B \& G}}$$

▲ 2. L ∨ (W → S)
   P ∨ ~S
   ~L → W
   ─────────
   P

▲ 3. M & P
   R → ~P
   F ∨ R
   G → M
   ─────────
   G & F

▲ 4. (D & G) → H
   M & (H → P)
   M → G
   ─────────
   D & P

▲ 5. R → S
   (S & B) → T
   T → E
   ─────────
   (R ∨ B) → E

Deductions

The next method we'll look at is less useful for proving an argument *invalid* than the truth-table methods, but it has some advantages in proving that an argument is valid. The method is that of **deduction.**

When we use this method, we actually deduce (or "derive") the conclusion from the premises by means of a series of basic truth-functionally valid argument patterns. This is a lot like "thinking through" the argument, taking one step at a time to see how, once we've assumed the truth of the premises, we eventually arrive at the conclusion. We'll consider some extended examples showing how the method works as we explain the first few basic argument patterns. We'll refer to these patterns as truth-functional rules because they govern what steps we're allowed to take in getting from the premises to the conclusion. (Your instructor may ask that you simply learn some or all of the basic valid argument patterns. It's a good idea to be able to identify these patterns whether you go on to construct deductions from them or not.)

■ *Group I Rules: Elementary Valid Argument Patterns*

This first group of rules should be learned before you go on to the Group II rules. Study them until you can work Exercise 9-6 with confidence.

***Rule 1: Modus ponens (MP), also known as* affirming the antecedent**   Any argument of the pattern

$$P \rightarrow Q$$
$$\underline{P}$$
$$Q$$

is valid. If you have a conditional among the premises and if the antecedent of that conditional occurs as another premise, then by **modus ponens** the consequent of the conditional follows from those two premises. The claims involved do not have to be simple letters standing alone—it would have made no difference if, in place of P, we had had something more complicated, such as (P ∨ R), as long as that compound claim appeared everywhere that P appears in the pattern above. For example:

1. (P ∨ R)→Q   Premise
2. P ∨ R   Premise
3. Q   From the premises, by modus ponens

The idea, once again, is that if you have *any conditional whatsoever* on a line of your deduction and if you have the antecedent of that conditional on some other line, you can write down the consequent of the conditional on your new line.

If the consequent of the conditional is the conclusion of the argument, then the deduction is finished—the conclusion has been established. If it is not the conclusion of the argument you're working on, the consequent of the conditional can be listed just as if it were another premise to use in deducing the conclusion you're after. An example:

1. P→R
2. R→S
3. P   Therefore, S

We've numbered the three premises of the argument and set its conclusion off to the side. (Hereafter we'll use a slash and three dots [/∴] in place of "therefore" to indicate the conclusion.) Now, notice that line 1 is a conditional, and line 3 is its antecedent. Modus ponens allows us to write down the consequent of line 1 as a new line in our deduction:

4. R   1, 3, MP

At the right we've noted the abbreviation for the rule we used and the lines the rule required. These notes are called the *annotation* for the deduction. We can now make use of this new line in the deduction to get the conclusion we were originally after, namely, S.

5. S   2, 4, MP

Again, we used modus ponens, this time on lines 2 and 4. The same explanation as that for deriving line 4 from lines 1 and 3 applies here.

THE BORN LOSER reprinted by permission of Newspaper Enterprise Association, Inc.

Notice that the modus ponens rule and all other Group I rules can be used only on whole lines. This means that you can't find the items you need for MP as *parts* of a line, as in the following:

$$P \lor (Q \to R)$$
$$\underline{Q \qquad\qquad}$$
$$P \lor R \qquad\qquad \text{(erroneous!)}$$

This is *not* a legitimate use of MP. We do have a conditional as *part* of the first line, and the second line is indeed the antecedent of that conditional. But the rule cannot be applied to parts of lines. The conditional required by rule MP must take up the entire line, as in the following:

$$P \to (Q \lor R)$$
$$\underline{P \qquad\qquad}$$
$$Q \lor R$$

Rule 2: Modus tollens (MT), also known as denying the consequent   The **modus tollens** pattern is this:

$$P \to Q$$
$$\underline{\sim Q}$$
$$\sim P$$

If you have a conditional claim as one premise and if one of your other premises is the negation of the consequent of that conditional, you can write down the negation of the conditional's antecedent as a new line in your deduction. Here's a deduction that uses both of the first two rules:

1. $(P \& Q) \to R$
2. S
3. $S \to {\sim}R$ $/ \therefore {\sim}(P \& Q)$
4. ${\sim}R$ 2, 3, MP
5. ${\sim}(P \& Q)$ 1, 4, MT

In this deduction we derived line 4 from lines 2 and 3 by modus ponens, and then 4 and 1 gave us line 5, which is what we were after, by modus tollens. The fact that the antecedent of line 1 is itself a compound claim, $(P \& Q)$, is not important; our line 5 is the antecedent of the conditional with a negation sign in front of it, and that's all that counts.

Rule 3: Chain argument (CA)

$$\frac{\begin{array}{l} P \to Q \\ Q \to R \end{array}}{P \to R}$$

The **chain argument** rule allows you to derive a conditional from two you already have, provided the antecedent of one of your conditionals is the same as the consequent of the other.

Rule 4: Disjunctive argument (DA)

$$\frac{\begin{array}{l} P \vee Q \\ {\sim}P \end{array}}{Q} \qquad \frac{\begin{array}{l} P \vee Q \\ {\sim}Q \end{array}}{P}$$

From a disjunction and the negation of one disjunct, the other disjunct may be derived.

Rule 5: Simplification (SIM)   This one is obvious, but we need it for obvious reasons:

$$\frac{P \& Q}{P} \qquad \frac{P \& Q}{Q}$$

If the conjunction is true, then of course the conjuncts must all be true. You can pull out one conjunct from any conjunction and make it the new line in your deduction.

If the Dollar Falls . . .

The valid argument patterns are in fact fairly common. Here's one in an article in *Time* as to why a weakening dollar is a threat to the stock market:

> Why should we care? . . . If the dollar continues to drop, investors may be tempted to move their cash to currencies on the upswing. That would drive the U.S. market lower. . . . Because foreigners hold almost 40% of U.S. Treasury securities, any pullout would risk a spike in interest rates that would ultimately slaughter the . . . market.

The chain argument here is reasonably obvious. In effect: If the dollar falls, then investors move their cash to currencies on the upswing. If investors move their cash to currencies on the upswing, then the U.S. market goes lower. If the U.S. market goes lower, then interest rates on U.S. Treasury securities rise. If interest rates on U.S. Treasury securities rise, then the . . . market dies. [Therefore, if the dollar falls, then the . . . market dies.]

Rule 6: Conjunction (CONJ)

$$\frac{\begin{array}{l} P \\ Q \end{array}}{P \,\&\, Q}$$

This rule allows you to put any two lines of a deduction together in the form of a conjunction.

Rule 7: Addition (ADD)

$$\frac{P}{P \lor Q} \qquad \frac{Q}{P \lor Q}$$

Clearly, no matter what claims P and Q might be, if P is true then *either* P or Q must be true. The truth of one disjunct is all it takes to make the whole disjunction true.

Rule 8: Constructive dilemma (CD)

$$\frac{\begin{array}{l} P \to Q \\ R \to S \\ P \lor R \end{array}}{Q \lor S}$$

The disjunction of the antecedents of any two conditionals allows the derivation of the disjunction of their consequents.

Rule 9: Destructive dilemma (DD)

$$P \rightarrow Q$$
$$R \rightarrow S$$
$$\underline{\sim Q \vee \sim S}$$
$$\sim P \vee \sim R$$

The disjunction of the negations of the consequents of two conditionals allows the derivation of the disjunction of the negations of their antecedents. (Refer to the pattern above as you read this, and it will make a lot more sense.)

Exercise 9-6

For each of the following groups of symbolized claims, identify which of the Group I rules was used to derive the last line.

▲ 1. $P \rightarrow (Q \& R)$
   $(Q \& R) \rightarrow (S \vee T)$
   $P \rightarrow (S \vee T)$

▲ 2. $(P \& S) \vee (T \rightarrow R)$
   $\sim(P \& S)$
   $T \rightarrow R$

▲ 3. $P \vee (Q \& R)$
   $(Q \& R) \rightarrow S$
   $P \rightarrow T$
   $S \vee T$

▲ 4. $(P \vee R) \rightarrow Q$
   $\sim Q$
   $\sim(P \vee R)$

▲ 5. $(Q \rightarrow T) \rightarrow S$
   $\sim S \vee \sim P$
   $R \rightarrow P$
   $\sim(Q \rightarrow T) \vee \sim R$

Exercise 9-7

Construct deductions for each of the following using the Group I rules. Each can be done in just a step or two (except number 10, which takes more).

▲ 1. 1. $R \rightarrow P$
      2. $Q \rightarrow R$ $/ \therefore Q \rightarrow P$

   2. 1. $P \rightarrow S$
      2. $P \vee Q$
      3. $Q \rightarrow R$ $/ \therefore S \vee R$

God and Evil

An age-old argument that God is either not all powerful or not all good goes like this:

> If God is all powerful, then he would be able to abolish evil.
> If God is all good, then he would not allow evil to be.
> Either God is not able to abolish evil, or God allows evil to be.
> Therefore, either God is not all powerful, or God is not all good.

This argument, you can see, is just an instance of Rule 9 (see page 336).

3. 1. R & S
    2. S→P   /∴P

▲ 4. 1. P→Q
    2. ~P→S
    3. ~Q   /∴S

5. 1. (P ∨ Q)→R
    2. Q   /∴R

6. 1. ~P
    2. ~(R & S) ∨ Q
    3. ~P→~Q   /∴~(R & S)

▲ 7. 1. ~S
    2. (P & Q)→R
    3. R→S   /∴~(P & Q)

8. 1. P→~(Q & T)
    2. S→(Q & T)
    3. P   /∴~S

9. 1. (P ∨ T)→S
    2. R→P
    3. R ∨ Q
    4. Q→T   /∴S

▲ 10. 1. (T ∨ M)→~Q
    2. (P→Q) & (R→S)
    3. T   /∴~P

■ *Group II Rules: Truth-Functional Equivalences*

A claim or part of a claim may be replaced by any claim or part of a claim to which it is equivalent by one of the following equivalence rules. Don't despair if this sounds complicated. The way such replacement works will become clear after a few examples.

There are a few differences between these rules about equivalences and the Group I rules. First, these rules allow us to go two ways instead of one—from either claim to its equivalent. Second, these rules allow us to re-place part of a claim with an equivalent part, rather than having to deal with entire lines of a deduction all at once. In the examples that follow the first few rules, watch for both of these differences.

We'll use a double-headed arrow (↔) to indicate the equivalence of two claims.

Rule 10: Double negation (DN)

P ↔ ~~P

This rule allows you to add or remove two negation signs in front of any claim, whether simple or compound. For example, this rule allows the derivation of either of the following from the other

P → (Q ∨ R)   P → ~~(Q ∨ R)

because the rule guarantees that (Q ∨ R) and its double negation, ~~(Q ∨ R), are equivalent. This in turn guarantees that P → (Q ∨ R) and P → ~~(Q ∨ R) are equivalent, and hence that each implies the other.

Here's an example of DN at work:

1. P ∨ ~(Q → R)
2. (Q → R)   / ∴ P
3. ~~(Q → R)   2, DN
4. P   1, 3, DA

Rule 11: Commutation (COM)

(P & Q) ↔ (Q & P)
(P ∨ Q) ↔ (Q ∨ P)

This rule simply allows any conjunction or disjunction to be "turned around," so that the conjuncts or disjuncts occur in reverse order. Here's an example:

P → (Q ∨ R)   P → (R ∨ Q)

Notice that commutation is used on *part* of the claim—just the consequent.

Rule 12: Implication (IMPL)   This rule allows us to change a conditional into a disjunction and vice versa.

(P → Q) ↔ (~P ∨ Q)

Notice that the antecedent always becomes the negated disjunct, or vice versa, depending on which way you're going. Another example:

(P ∨ Q) → R ↔ ~(P ∨ Q) ∨ R

Rule 13: Contraposition (CONTR) This rule may remind you of the categorical operation of contraposition (see Chapter 8)—this rule is its truth-functional version.

$(P \rightarrow Q) \leftrightarrow (\sim Q \rightarrow \sim P)$

This rule allows us to exchange the places of a conditional's antecedent and consequent, but only by putting on or taking off a negation sign in front of each. Here's another example:

$(P \& Q) \rightarrow (P \vee Q) \leftrightarrow \sim(P \vee Q) \rightarrow \sim(P \& Q)$

Sometimes you want to perform contraposition on a symbolization that doesn't fit either side of the equivalence because it has a negation sign in front of either the antecedent or the consequent but not both. You can do what you want in such cases, but it takes two steps, one applying double negation and one applying contraposition. Here's an example:

$(P \vee Q) \rightarrow \sim R$	
$\sim\sim(P \vee Q) \rightarrow \sim R$	Double negation
$R \rightarrow \sim(P \vee Q)$	Contraposition

Your instructor may allow you to combine these steps (and refer to both DN and CONTR in your annotation).

Rule 14: DeMorgan's Laws (DEM)

$\sim(P \& Q) \leftrightarrow (\sim P \vee \sim Q)$
$\sim(P \vee Q) \leftrightarrow (\sim P \& \sim Q)$

Notice that when the negation sign is "moved inside" the parentheses, the "&" changes into a "$\vee$," or vice versa. It's important not to confuse the use of the negation sign in DeMorgan's Laws with that of the minus sign in algebra. Notice that when you take $\sim(P \vee Q)$ and "move the negation sign in," you do *not* get $(\sim P \vee \sim Q)$. The wedge must be changed to an ampersand or vice versa whenever DEM is used. You can think of $\sim(P \vee Q)$ and $(\sim P \& \sim Q)$ as saying "neither P nor Q," and you can think of $\sim(P \& Q)$ and $(\sim P \vee \sim Q)$ as saying "not both P and Q."

Rule 15: Exportation (EXP)

$[P \rightarrow (Q \rightarrow R)] \leftrightarrow [(P \& Q) \rightarrow R]$

Square brackets are used exactly as parentheses are. In English, the exportation rule says that "If P, then if Q, then R" is equivalent to "If both P and Q, then R." (The commas are optional in both claims.) If you look back to Exercise 9-2, items 3 and 5 (page 321), you'll notice that, according to the exportation rule, each of these can replace the other.

Rule 16: Association (ASSOC)

[P & (Q & R)] ↔ [(P & Q) & R]
[P ∨ (Q ∨ R)] ↔ [(P ∨ Q) ∨ R]

Association simply tells us that, when we have three items joined together with wedges or with ampersands, it doesn't matter which ones we group together. If we have a long disjunction with more than two disjuncts, it still requires only one of them to be true for the entire disjunction to be true; if it's a conjunction, then all the conjuncts have to be true, no matter how many of them there are, in order for the entire conjunction to be true. Your instructor may allow you to drop parentheses in such symbolizations, but if you're developing these rules as a formal system, he or she may not.

Rule 17: Distribution (DIST)   This rule allows us to "spread a conjunct across a disjunction" or to "spread a disjunct across a conjunction." In the first example below, look at the left-hand side of the equivalence. The P, which is conjoined with a disjunction, is picked up and dropped (distributed) across the disjunction by being conjoined with each part. (This is easier to understand if you see it done on a chalkboard than by trying to figure it out from the page in front of you.) The two versions of the rule, like those of DEM, allow us to do exactly with the wedge what we're allowed to do with the ampersand.

[P & (Q ∨ R)] ↔ [(P & Q) ∨ (P & R)]
[P ∨ (Q & R)] ↔ [(P ∨ Q) & (P ∨ R)]

Rule 18: Tautology (TAUT)

(P ∨ P) ↔ P
(P & P) ↔ P

This rule allows a few obvious steps; they are sometimes necessary to "clean up" a deduction.

Figure 2 shows some deductions that use rules from both Group I and Group II. Look at them carefully, covering up the lines with a piece of paper and uncovering them one at a time as you progress. This gives you a chance to figure out what you might do before you see the answer. In any case, make sure you understand how each line was achieved before going on. If necessary, look up the rule used to make sure you understand it.

The first example is long, but fairly simple. Length is not always proportional to difficulty.

1. P→(Q→R)
2. (T→P) & (S→Q)
3. T & S   /∴ R
4. T→P   2, SIM

Group I

1. Modus ponens (MP) $P \rightarrow Q$ $\underline{P}$ Q	2. Modus tollens (MT) $P \rightarrow Q$ $\underline{\sim Q}$ $\sim P$	3. Chain argument (CA) $P \rightarrow Q$ $\underline{Q \rightarrow R}$ $P \rightarrow R$
4. Disjunctive argument (DA) $P \lor Q$ $P \lor Q$ $\underline{\sim P}$ $\underline{\sim Q}$ Q P	5. Simplification (SIM) $\underline{P \& Q}$ $\underline{P \& Q}$ P Q	6. Conjunction (CONJ) P $\underline{Q}$ $P \& Q$
7. Addition (ADD) $\underline{P}$ $\underline{Q}$ $P \lor Q$ $P \lor Q$	8. Constructive dilemma (CD) $P \rightarrow Q$ $R \rightarrow S$ $\underline{P \lor R}$ $Q \lor S$	9. Destructive dilemma (DD) $P \rightarrow Q$ $R \rightarrow S$ $\underline{\sim Q \lor \sim S}$ $\sim P \lor \sim R$

Group II

10. Double negation (DN) $P \longleftrightarrow \sim\sim P$	11. Commutation (COM) $(P \& Q) \longleftrightarrow (Q \& P)$ $(P \lor Q) \longleftrightarrow (Q \lor P)$	12. Implication (IMPL) $(P \rightarrow Q) \longleftrightarrow (\sim P \lor Q)$
13. Contraposition (CONTR) $(P \rightarrow Q) \longleftrightarrow (\sim Q \rightarrow \sim P)$	14. DeMorgan's Laws (DEM) $\sim(P \lor Q) \longleftrightarrow (\sim P \& \sim Q)$ $\sim(P \& Q) \longleftrightarrow (\sim P \lor \sim Q)$	15. Exportation (EXPORT) $[P \rightarrow (Q \rightarrow R)] \longleftrightarrow [(P \& Q) \rightarrow R]$
16. Association (ASSOC) $[P \& (Q \& R)] \longleftrightarrow [(P \& Q) \& R]$ $[P \lor (Q \lor R)] \longleftrightarrow [(P \lor Q) \lor R]$	17. Distribution (DIST) $[P \& (Q \lor R)] \longleftrightarrow [(P \& Q) \lor (P \& R)]$ $[P \lor (Q \& R)] \longleftrightarrow [(P \lor Q) \& (P \lor R)]$	18. Tautology (TAUT) $(P \lor P) \longleftrightarrow P$ $(P \& P) \longleftrightarrow P$

FIGURE 2   Truth-Functional Rules for Deductions

5. $S \rightarrow Q$	2, SIM
6. T	3, SIM
7. S	3, SIM
8. P	4, 6, MP
9. Q	5, 7, MP
10. P & Q	8, 9, CONJ
11. (P & Q) $\rightarrow$ R	1, EXP
12. R	10, 11, MP

It's often difficult to tell how to proceed when you first look at a deduction problem. One strategy is to work backward. Look at what you want to get, look at what you have, and see what you would need in order to get what you want. Then determine where you would get *that*, and so on. We'll explain in terms of the following problem.

1. P→(Q & R)
2. S→~Q
3. S   /∴ ~P

4. ~Q   2, 3, MP
5. ~Q ∨ ~R   4, ADD
6. ~(Q & R)   5, DEM
7. ~P   1, 6, MT

We began by wanting ~P as our conclusion. If we're familiar with modus tollens, it's clear from line 1 that we can get ~P if we can get the negation of line 1's consequent, which would be ~(Q & R). That in turn is the same as ~Q ∨ ~R, which we can get if we can get either ~Q or ~R. So now we're looking for someplace in the first three premises where we can get ~Q. That's easy: from lines 2 and 3, by modus ponens. A little practice and you'll be surprised how easy these strategies are to work, at least *most* of the time!

Exercise 9-8

The annotations that explain how each line was derived have been left off the following deductions. For each line, supply the rule used and the numbers of any earlier lines the rule requires.

▲ 1.   1. P→Q   (Premise)
        2. R→S   (Premise)
        3. Q→~S   (Premise) /∴ P→~R
        4. P→~S
        5. ~S→~R
        6. P→~R

   2.   1. ~P   (Premise)
        2. (Q→R) & (R→Q)   (Premise)
        3. R ∨ P   (Premise) /∴ Q
        4. R
        5. R→Q
        6. Q

   3.   1. P→Q   (Premise)
        2. R→(~S ∨ T)   (Premise)
        3. ~P→R   (Premise) /∴ (~Q & S)→T

    4. ~Q→~P
    5. ~Q→R
    6. ~Q→(~S ∨ T)
    7. ~Q→(S→T)
    8. (~Q & S)→T

▲ 4. 1. (P & Q)→T   (Premise)
      2. P   (Premise)
      3. ~Q→~P   (Premise) /∴ T
      4. P→Q
      5. Q
      6. P & Q
      7. T

  5. 1. ~(S ∨ R)   (Premise)
      2. P→S   (Premise)
      3. T→(P ∨ R)   (Premise) /∴~T
      4. ~S & ~R
      5. ~S
      6. ~P
      7. ~R
      8. ~P & ~R
      9. ~(P ∨ R)
     10. ~T

Exercise 9-9

Derive the indicated conclusions from the premises supplied.

▲ 1. 1. P & Q
      2. P→R   /∴R

▲ 2. 1. R→S
      2. ~P ∨ R   /∴P→S

  3. 1. P ∨ Q
      2. R & ~Q   /∴P

▲ 4. 1. ~P ∨ (~Q ∨ R)
      2. P   /∴Q→R

  5. 1. T ∨ P
      2. P→S   /∴~T→S

  6. 1. Q ∨ ~S
      2. Q→P   /∴S→P

  7. 1. ~S ∨ ~R
      2. P→(S & R)   /∴~P

▲ 8. 1. ~Q & (~S & ~T)
      2. P→(Q ∨ S)   /∴~P

9. 1. P ∨ (S & R)
    2. T → (~P & ~R)   / ∴ ~T
10. 1. (S & P) → R
    2. S   / ∴ P → R

Exercise 9-10

Derive the indicated conclusions from the premises supplied.

▲ 1. 1. P → R
      2. R → Q   / ∴ ~P ∨ Q

   2. 1. ~P ∨ S
      2. ~T → ~S   / ∴ P → T

   3. 1. F → R
      2. L → S
      3. ~C
      4. (R & S) → C   / ∴ ~F ∨ ~L

▲ 4. 1. P ∨ (Q & R)
      2. (P ∨ Q) → S   / ∴ S

   5. 1. (S & R) → P
      2. (R → P) → W
      3. S   / ∴ W

   6. 1. ~L → (~P → M)
      2. ~(P ∨ L)   / ∴ M

▲ 7. 1. (M ∨ R) & P
      2. ~S → ~P
      3. S → ~M   / ∴ R

   8. 1. Q → L
      2. P → M
      3. R ∨ P
      4. R → (Q & S)   / ∴ ~M → L

   9. 1. Q → S
      2. P → (S & L)
      3. ~P → Q
      4. S → R   / ∴ R & S

▲ 10. 1. P ∨ (R & Q)
      2. R → ~P
      3. Q → T   / ∴ R → T

■ Conditional Proof

Conditional proof (CP) is both a rule and a strategy for constructing a deduction. It is based on the following idea: Let's say we want to produce a deduction for a conditional claim, P→Q. If we produce such a deduction, what have we proved? We've proved the equivalent of "If P were true, then Q would be true." One way to do this is simply to *assume* that P is true (that is, to add it as an additional premise) and then to prove that, on that assumption, Q has to be true. If we can do that—prove Q after assuming P—then we'll have proved that if P then Q, or P→Q. Let's look at an example of how to do this; then we'll explain it again.

Here is the way we'll use CP as a new rule: Simply write down the antecedent of whatever conditional we want to prove, drawing a circle around the number of that step in the deduction; in the annotation, write "CP Premise" for that step. Here's what it looks like:

```
1. P ∨ (Q→R)    Premise
2. Q            Premise      /∴ ~P→R
③. ~P           CP Premise
```

Then, after we've proved what we want—the consequent of the conditional—in the next step, we write the full conditional down. Then we draw a line in the margin to the left of the deduction from the premise with the circled number to the number of the line we deduced from it. (See below for an example.) In the annotation for the last line in the process, list *all the steps from the circled number to the one with the conditional's consequent*, and give CP as the rule. Drawing the line that connects our earlier CP premise with the step we derived from it indicates we've stopped making the assumption that the premise, which is now the antecedent of our conditional in our last step, is true. This is known as *discharging the premise*. Here's how the whole thing looks:

```
1. P ∨ (Q→R)        Premise
2. Q                Premise    /∴ ~P→R
③. ~P               CP Premise
4. Q→R         1, 3, DA
5. R           2, 4, MP
6. ~P→R        3–5, CP
```

Here's the promised second explanation. Look at the example. Think of the conclusion as saying that, given the two original premises, *if* we had ~P, we could get R. One way to find out if this is so is to *give ourselves* ~P and then see if we can get R. In step 3, we do exactly that: We give ourselves ~P. Now, by circling the number, we indicate that *this is a premise we've given ourselves* (our "CP premise") and therefore that it's one we'll have to get rid of before we're done. (We can't be allowed to invent, use, and keep just any old premises we like—we could prove *anything* if we could do that.) But once we've given ourselves ~P, getting R turns out to be easy! Steps 4 and 5 are pretty obvious, aren't they? (If not, you need more practice with the other rules.) In steps 3 through 5 what we've actually proved is that *if* we had ~P, then we could get R. So we're justified in writing down step 6 because that's exactly what step 6 says: If ~P, then R.

Once we've got our conditional, ~P→R, we're no longer dependent on the CP premise, so we draw our line in the left margin from the last step that depended on the CP premise back to the premise itself. We *discharge* the premise.

Here are some very important restrictions on the CP rule:

1. CP can be used only to produce a conditional claim: After we discharge a CP premise, the very next step must be a conditional with the preceding step as consequent and the CP premise as antecedent. [Remember that lots of claims are equivalent to conditional claims. For example, to get (~P ∨ Q), just prove (P→Q), and then use IMPL.]

2. If more than one use is made of CP at a time—that is, if more than one CP premise is brought in—they must be discharged in exactly the reverse order from that in which they were assumed. This means that the lines that run from different CP premises must not cross each other. See examples below.

3. Once a CP premise has been discharged, no steps derived from it—those steps encompassed by the line drawn in the left margin—may be used in the deduction. (They depend on the CP premise, you see, and it's been discharged.)

4. All CP premises must be discharged.

This sounds a lot more complicated than it actually is. Refer back to these restrictions on CP as you go through the examples, and they will make a good deal more sense.

Here's an example of CP in which two additional premises are assumed and discharged in reverse order.

```
    1. P→[Q ∨ (R & S)]     Premise
    2. (~Q→S)→T            Premise          /∴ P→T
 ┌ (3.) P                  CP Premise
 │  4. Q ∨ (R & S)         1, 3, MP
 │ ┌ (5.) ~Q              CP Premise
 │ │  6. R & S             4, 5, DA
 │ └  7. S                 6, SIM
 │  8. ~Q→S                5–7, CP
 └  9. T                   2, 8, MP
   10. P→T                 3–9, CP
```

Notice that the additional premise added at step 5 is discharged when step 8 is completed, and the premise at step 3 is discharged when step 10 is completed. Once again: Whenever you discharge a premise, you must make that premise the antecedent of the next step in your deduction. (You might try the preceding deduction without using CP; doing so will help you appreciate having the rule, however hard to learn it may seem at the moment. Using CP makes many deductions shorter, easier, or both.

Here are three more examples of the correct use of CP:

```
1. (R→~P)→S          Premise
2. S→(T ∨ Q)         Premise        /∴ ~(R & P)→(T ∨ Q)
③. ~(R & P)          CP Premise
4. ~R ∨ ~P           3, DEM
5. R→~P              4, IMPL
6. S                 1, 5, MP
7. (T ∨ Q)           2, 6, MP
8. ~(R & P)→(T ∨ Q)  3–7, CP
```

In this case, one use of CP follows another:

```
1. (P ∨ Q)→R                    Premise
2. (S ∨ T)→U                    Premise     /∴ (~R→~P) & (~U→~T)
③. ~R                           CP Premise
4. ~(P ∨ Q)                     1, 3, MT
5. ~P & ~Q                      4, DEM
6. ~P                           5, SIM
7. ~R→~P                        3–6, CP
⑧. ~U                           CP Premise
9. ~(S ∨ T)                     2, 8, MT
10. ~S & ~T                     9, DEM
11. ~T                          10, SIM
12. ~U→~T                       8–11, CP
13. (~R→~P) & (~U→~T)           7, 12, CONJ
```

In this case, one use of CP occurs "inside" another:

```
1. R→(S & Q)              Premise
2. P→M                    Premise
3. S→(Q→~M)               Premise
4. (J ∨ T)→B              Premise        /∴ R→(J→(B & ~P))
⑤. R                      CP Premise
⑥. J                      CP Premise
7. J ∨ T                  6, ADD
8. B                      4, 7, MP
9. (S & Q)                1, 5, MP
10. (S & Q)→~M            3, EXP
11. ~M                    9, 10, MP
12. ~P                    2, 11, MT
13. B & ~P                8, 12, CONJ
14. J→(B & ~P)            6–13, CP
15. R→(J→(B & ~P))        5–14, CP
```

Before ending this section on deductions, we should point out that our system of truth-functional logic has a couple of properties that are of great theoretical interest: It is both sound and complete. To say that a logic system is sound (in the sense most important to us here) is to say that *every deduction that can be constructed using the rules of the system constitutes a valid argument.* Another way to say this is that no deduction or string of deductions allows us to begin with true sentences and wind up with false ones.

To say that our system is complete is to say that *for every truth-functionally valid argument that there is (or even could be), there is a deduction in our system of rules that allows us to deduce the conclusion of that argument from its premises.* That is, if conclusion C really does follow validly from premises P and Q, then we know for certain that it is possible to construct a deduction beginning with just P and Q and ending with C.

We could have produced a system that is both sound and complete and that had many fewer rules than our system has. However, in such systems, deductions tend to be very difficult to construct. Although our system is burdened with a fairly large number of rules, once you learn them, producing proofs is not too difficult. So, in a way, every system of logic is a trade-off of a sort. You can make the system small and elegant but difficult to use, or you can make it larger and less elegant but more efficient in actual use. (The smaller systems are more efficient for some purposes, but those purposes are quite different from ours in this book.)

Recap

This chapter has been concerned with the truth-functional structures of claims and arguments. These structures result from the logical connections between claims, connections that are represented in English by words such as "and," "or," "if . . . then," and so forth.

Truth-functional connections are explained by truth tables, which display the conditions under which a claim is true and those under which it is false. Truth tables are used to define truth-functional symbols, which stand for logical connections between claims in symbolizations. To symbolize a claim is simply to reduce it to its truth-functional structure.

To evaluate truth-functional arguments—which are just arguments that depend on truth-functional connections for their validity—we learned first about the full truth-table method. This method requires that we set out a truth table for an argument and then check to see whether there are circumstances in which the premises are true and the conclusion false. In the short truth-table method, we look directly for such rows: If we find one, the argument is invalid; if we don't, it is valid.

We can improve our efficiency in determining the validity of many truth-functional arguments by learning to recognize certain elementary valid argument patterns. These patterns can be linked up to produce deductions, which are sequences of valid truth-functional inferences.

Exercise 9-11

Display the truth-functional structure of the following claims by symbolizing them. Use the letters indicated.

> D = We do something to reduce the deficit.
> B = The balance of payments gets worse.
> C = There is (or will be) a financial crisis.

▲ 1. The balance of payments will not get worse if we do something to reduce the deficit.

2. There will be no financial crisis unless the balance of payments gets worse.

3. Either the balance of payments will get worse or, if no action is taken on the deficit, there will be a financial crisis.

▲ 4. The balance of payments will get worse only if we don't do something to reduce the deficit.

5. Action cannot be taken on the deficit if there's a financial crisis.

6. I can tell you about whether we'll do something to reduce the deficit and whether our balance of payments will get worse: Neither one will happen.

▲ 7. In order for there to be a financial crisis, the balance of payments will have to get worse and there will have to be no action taken to reduce the deficit.

8. We can avoid a financial crisis only by taking action on the deficit and keeping the balance of payments from getting worse.

9. The *only* thing that can prevent a financial crisis is our doing something to reduce the deficit.

Exercise 9-12

For each of the numbered claims below, there is exactly one lettered claim that is equivalent. Identify the equivalent claim for each item. (Some lettered claims are equivalent to more than one numbered claim, so it will be necessary to use some letters more than once.)

▲ 1. Oil prices will drop if the OPEC countries increase their production.

2. Oil prices will drop only if the OPEC countries increase their production.

3. Neither will oil prices drop nor will the OPEC countries increase their production.

▲ 4. Oil prices cannot drop unless the OPEC countries increase their production.

5. The only thing that can prevent oil prices dropping is the OPEC countries' increasing their production.

6. A drop in oil prices is necessary for the OPEC countries to increase their production.

▲ 7. All it takes for the OPEC countries to increase their production is a drop in oil prices.

8. The OPEC countries will not increase their production while oil prices drop; each possibility excludes the other.

a. It's not the case that oil prices will drop, and it's not the case that the OPEC countries will increase their production.

b. If OPEC countries increase their production, then oil prices will drop.

c. Only if OPEC countries increase their production will oil prices drop.

d. Either the OPEC countries will not increase their production, or oil prices will not drop.

e. If the OPEC countries do not increase production, then oil prices will drop.

Exercise 9-13

Construct deductions for each of the following. (Try these first without using conditional proof.)

▲ 1. 1. P
    2. Q & R
    3. (Q & P)→S   /∴S

2. 1. (P ∨ Q) & R
    2. (R & P)→S
    3. (Q & R)→S   /∴S

3. 1. P→(Q→~R)
    2. (~R→S) ∨ T
    3. ~T & P   /∴Q→S

▲ 4. 1. P ∨ Q
    2. (Q ∨ U)→(P→T)
    3. ~P
    4. (~P ∨ R)→(Q→S)   /∴T ∨ S

5. 1. (P→Q) & R
    2. ~S
    3. S ∨ (Q→S)   /∴P→T

6. 1. P→(Q & R)
    2. R→(Q→S)   /∴P→S

▲ 7. 1. P→Q   /∴P→(Q ∨ R)
   8. 1. ~P ∨ ~Q
       2. (Q→S)→R   /∴P→R
   9. 1. S
       2. P→(Q & R)
       3. Q→~S   /∴~P
▲ 10. 1. (S→Q)→~R
       2. (P→Q)→R   /∴~Q

Exercise 9-14 _____

Use the rule of conditional proof to construct deductions for each of the
following.

▲ 1. 1. P→Q
       2. P→R   /∴P→(Q & R)
   2. 1. P→Q
       2. R→Q   /∴(P ∨ R)→Q
   3. 1. P→(Q→R)   /∴(P→Q)→(P→R)
▲ 4. 1. P→(Q ∨ R)
       2. T→(S & ~R)   /∴(P & T)→Q
   5. 1. ~P→(~Q→~R)
       2. ~(R & ~P)→~S   /∴S→Q
   6. 1. P→(Q→R)
       2. (T→S) & (R→T)   /∴P→(Q→S)
▲ 7. 1. P ∨ (Q & R)
       2. T→~(P ∨ U)
       3. S→(Q→~R)   /∴~S ∨ ~T
   8. 1. (P ∨ Q)→R
       2. (P→S)→T   /∴R ∨ T
   9. 1. P→~Q
       2. ~R→(S & Q)   /∴P→R
▲ 10. 1. (P & Q) ∨ R
       2. ~R ∨ Q   /∴P→Q

Exercise 9-15 _____

Display the truth-functional form of the following arguments by symbolizing
them; then use the truth-table method, the short truth-table method, or the
method of deduction to prove them valid or invalid. Use the letters provided.
(We've used underscores in the example and in the first two problems to help
you connect the letters with the proper claims.)

Example

If <u>M</u>aria does not go to the movies, then she will <u>h</u>elp Bob with his logic homework. Bob will <u>f</u>ail the course unless Maria <u>h</u>elps him with his logic homework. Therefore, if <u>M</u>aria goes to the movies, Bob will <u>f</u>ail the course. (M, H, F)

Symbolization

1. ~M→H   (Premise)
2. ~H→F   (Premise)   /∴ M→F

Truth Table

M	H	F	~M	~H	~M→H	~H→F	M→F
T	T	T	F	F	T	T	T
T	T	F	F	F	T	T	F

We need to go only as far as the second row of the table, since both premises come out true and the conclusion comes out false in that row.

▲ 1. If it's <u>c</u>old, Dale's motorcycle won't <u>s</u>tart. If Dale is not <u>l</u>ate for work, then his motorcycle must have <u>s</u>tarted. Therefore, if it's <u>c</u>old, Dale is <u>l</u>ate for work. (C, S, L)

2. If profits depend on <u>u</u>nsound environmental practices, then either the <u>q</u>uality of the environment will deteriorate, or profits will <u>d</u>rop. <u>J</u>obs will be plentiful only if profits do not drop. So, either jobs will not be plentiful, or the quality of the environment will deteriorate. (U, Q, D, J)

3. The new <u>r</u>oad will not be built unless the planning commission <u>a</u>pproves the funds. But the planning commission's approval of the funds will come only if the environmental impact report is positive, and it can't be positive if the road will ruin <u>M</u>ill Creek. So, unless they find a way for the road not to ruin Mill Creek, it won't be built. (R, A, E, M)

▲ 4. The <u>m</u>essage will not be understood unless the <u>c</u>ode is broken. The <u>k</u>iller will not be caught if the message is not understood. Either the code will be broken, or <u>H</u>olmes's plan will fail. But Holmes's plan will not fail if he is given enough <u>t</u>ime. Therefore, if Holmes is given enough time, the killer will be caught. (M, C, K, H, T)

5. If the senator <u>v</u>otes against this bill, then he is <u>o</u>pposed to penalties against tax evaders. Also, if the senator is a <u>t</u>ax evader himself, then he is opposed to penalties against tax evaders. Therefore, if the senator votes against this bill, he is a tax evader himself. (V, O, T)

6. If you had gone to class, taken good notes, and studied the text, you'd have done well on the exam. And, if you'd done well on the exam, you'd have passed the course. Since you did not pass the course and you did go to class, you must not have taken good notes and not studied the text.

▲ 7. Either John will go to class, or he'll miss the review session. If John misses the review session, he'll foul up the exam. If he goes to class, however, he'll miss his ride home for the weekend. So John's either going to miss his ride home or foul up the exam.

8. If the government's position on fighting crime is correct, then if more people are locked up, then the crime rate should drop. But the crime rate has not dropped despite the fact that we've been locking up record numbers of people. It follows that the government's position on fighting crime is not correct.

9. The creation story in the Book of Genesis is compatible with the theory of evolution, but only if the creation story is not taken literally. If, as most scientists think, there is plenty of evidence for the theory of evolution, the Genesis story cannot be true if it is not compatible with evolution theory. Therefore, if the Genesis story is taken literally, it cannot be true.

▲ 10. The creation story in the Book of Genesis is compatible with the theory of evolution, but only if the creation story is not taken literally. If there is plenty of evidence for the theory of evolution, which there is, the Genesis story cannot be true if it is not compatible with evolution theory. Therefore, if the Genesis story is taken literally, it cannot be true.

11. If there was no murder committed, then the victim must have been killed by the horse. But the victim could have been killed by the horse only if he, the victim, was trying to injure the horse before the race; and, in that case, there certainly was a crime committed. So, if there was no murder, there was still a crime committed.

12. Holmes cannot catch the train unless he gets to Charing Cross Station by noon; and if he misses the train, Watson will be in danger. Because Moriarty has thugs watching the station, Holmes can get there by noon only if he goes in disguise. So, unless Holmes goes in disguise, Watson will be in danger.

▲ 13. It's not fair to smoke around nonsmokers if secondhand cigarette smoke really is harmful. If secondhand smoke were not harmful, the American Lung Association would not be telling us that it is. But they are telling us that it's harmful. That's enough to conclude that it's not fair to smoke around nonsmokers.

14. If Jane does any of the following, she's got an eating disorder: If she goes on eating binges for no apparent reason, if she looks forward to times when she can eat alone, or if she eats sensibly in front of others and makes up for it when she's alone. Jane does in fact go on eating binges for no apparent reason. So it's clear that she has an eating disorder.

15. The number of business majors increased markedly during the past decade; and if you see that happening, you know that younger people have developed a greater interest in money. Such an interest, unfortunately, means that greed has become a significant motivating force in our society; and

if greed has become such a force, charity will have become insignificant. We can predict that charity will not be seen as a significant feature of this past decade.

Exercise 9-16

Use the box on page 329 to determine which of the following are valid arguments.

1. If Bobo is smart, then he can do tricks. However, Bobo is not smart. So he cannot do tricks.

2. If God is always on America's side, then America wouldn't have lost any wars. America has lost wars. Therefore, God is not always on America's side.

3. If your theory is correct, then light passing Jupiter will be bent. Light passing Jupiter is bent. Therefore, your theory is correct.

4. Moore eats carrots and broccoli for lunch, and if he does that, he probably is very hungry by dinner time. Conclusion: Moore is very hungry by dinner time.

5. If you value your feet, you won't mow the lawn in your bare feet. Therefore, since you do mow the lawn in your bare feet, we can conclude that you don't value your feet.

6. If Bobo is smart, then he can do tricks; and he can do tricks. Therefore, he is smart.

7. If Charles walked through the rose garden, then he would have mud on his shoes. We can deduce, therefore, that he did walk through the rose garden, because he has mud on his shoes.

8. If it rained earlier, then the sidewalks would still be wet. We can deduce, therefore, that it did rain earlier, because the sidewalks are still wet.

9. If you are pregnant, then you are a woman. We can deduce, therefore, that you are pregnant, because you are a woman.

10. If this stuff is on the final, I will get an A in the class because I really understand it! Further, the teacher told me that this stuff will be on the final, so I know it will be there. Therefore, I know I will get an A in the class.

11. If side A has an even number, then side B has an odd number, but side A does not have an even number. Therefore, side B does not have an odd number.

12. If side A has an even number, then side B has an odd number, and side B does have an odd number. Therefore, side A has an even number.

13. If the theory is correct, then we would have observed squigglyitis in the specimen. However, we know the theory is not correct. Therefore, we did not observe squigglyitis in the specimen.

14. If the theory is correct, then we would have observed dilation in the specimen. Therefore, since we did not observe dilation in the specimen, we know the theory is not correct.

15. If we observe dilation in the specimen, then we know the theory is correct. We observed dilation—so the theory is correct.

16. If the comet approached within 1 billion miles of the earth, there would have been numerous sightings of it. There weren't numerous sightings. So it did not approach within 1 billion miles.

17. If Baffin Island is larger than Sumatra, then two of the five largest islands in the world are in the Arctic Ocean. And Baffin Island, as it turns out, is about 2 percent larger than Sumatra. Therefore, the Arctic Ocean contains two of the world's largest islands.

18. If the danger of range fires is greater this year than last, then state and federal officials will hire a greater number of firefighters to cope with the danger. Since more firefighters are already being hired this year than were hired all last year, we can be sure that the danger of fires has increased this year.

19. If Jack Davis robbed the Central Pacific Express in 1870, then the authorities imprisoned the right person. But the authorities did not imprison the right person. Therefore, it must have not been Jack Davis who robbed the Central Pacific Express in 1870.

20. If the recent tax cuts had been self-financing, then there would have been no substantial increase in the federal deficit. But they turned out not to be self-financing. Therefore, there will be a substantial increase in the federal deficit.

21. The public did not react favorably to the majority of policies recommended by President Ronald Reagan during his second term. But if his electoral landslide in 1984 had been a mandate for more conservative policies, the public would have reacted favorably to most of those he recommended after the election. Therefore, the 1984 vote was not considered a mandate for more conservative policies.

22. Alexander will finish his book by tomorrow afternoon only if he is an accomplished speed reader. Fortunately for him, he is quite accomplished at speed reading. Therefore, he will get his book finished by tomorrow afternoon.

23. If higher education were living up to its responsibilities, the five best-selling magazines on American campuses would not be *Cosmopolitan, People, Playboy, Glamour,* and *Vogue.* But those are exactly the magazines that sell best in the nation's college bookstores. Higher education, we can conclude, is failing in at least some of its responsibilities.

24. Broc Glover was considered sure to win if he had no bad luck in the early part of the race. But we've learned that he has had the bad luck to be involved in a crash right after the start, so we're expecting another driver to be the winner.

25. If Boris is really a spy for the KGB, then he has been lying through his teeth about his business in this country. But we can expose his true occupation if he's been lying like that. So, I'm confident that if we can expose his true occupation, we can show that he's really a KGB spy.

26. The alternator is not working properly if the ammeter shows a negative reading. The current reading of the ammeter is negative. So, the alternator is not working properly.

27. Fewer than 2 percent of the employees of New York City's Transit Authority are accountable to management. If such a small number of employees are accountable to the management of the organization, no improvement in the system's efficiency can be expected in the near future. So, we cannot expect any such improvements any time soon.

28. If Charles did not pay his taxes, then he did not receive a refund. Thus, he did not pay his taxes, since he did not receive a refund.

29. If they wanted to go to the party, then they would have called by now. But they haven't, so they didn't.

30. "You'll get an A in the class," she predicted.
    "What makes you say that?" he asked.
    "Because," she said, "if you get an A, then you're smart, and you *are* smart."

31. If Florin arrived home by eight, she received the call from her attorney. But she did not get home by eight, so she must have missed her attorney's call.

32. The acid rain problem will be solved, but only if the administration stops talking and starts acting. So far, however, all we've had from the president is words. Words are cheap. Action is what counts. The problem will not be remedied, at least not while this administration is in office.

Writing Exercises

1. a. In a one-page essay, evaluate the soundness of the argument in the box on page 337. Alternatively, in a one-page essay evaluate the soundness of the argument in the letter to Ask Marilyn (page 323). Write your name on the back of your paper.
   b. When everyone is finished, your instructor will collect the papers and redistribute them to the class. In groups of four or five, read the papers that have been given to your group and select the best one. The instructor will select one group's top-rated paper to read to the class for discussion.

2. Take about fifteen minutes to write an essay responding to the paper the instructor has read to the class in Exercise 1. When everyone is finished, the members of each group will read each other's responses and select the best one to share with the class.

Chapter 10

Inductive Arguments

*I*n July 2002, George Kapidian, of New Braunfels, Texas, lost his new home in a devastating flood. After four days of violent thunderstorms around San Antonio, the Guadalupe River rose twenty feet—flooding Mr. Kapidian's house and destroying everything in it. Seven people died in the raging floodwaters.

Seventy-two-year-old Mr. Kapidian should have known better. He was aware that the area was subject to flooding; in fact, his house was built on the foundation of a house washed away by a flood just three years earlier. But when he bought the house he reasoned, "What are the odds it would happen again?"

People often think this way, and the old saying "lightning never strikes twice in the same place" would seem to confirm their reasoning—if it were true. But it isn't true, and this kind of thinking can lead to some very bad conclusions indeed. Mr. Kapidian's reasoning would lead us to go swimming at a beach where someone was recently mauled by a shark because, after all, "What are the odds it would happen again?" Well, in truth, the odds are much greater that it will happen again at *this* beach than at a beach where there has been no recent shark attack. And in the original case, the Guadalupe River's having flooded in that spot on earlier occasions made it *more* likely that it would flood there again, not less likely.

When we extend what we have already observed to things or situations we have not observed, we are reasoning inductively; we are producing inductive arguments. This chapter is all about making and evaluating such arguments. Mr. Kapidian's problem was that his inductive reasoning was improper. His part of Texas had been subject to floods in past years; if he had properly extended that information to future years, he would not have been surprised to see flooding again.

We can sketch a general formula that practically all inductive arguments fit, some maybe a little more obviously than others. Here is the formula followed by a bit of explanation:

Lightning never strikes the same place twice.

Do you believe this? Try the idea out on someone who works at a fire lookout on a mountaintop in thunderstorms.

> Premise: X has properties a, b, and c.
> Premise: Y has properties a, b, and c.
> Premise: X has further property p.
> Conclusion: Y also has property p.

X is something we know some things about—namely, that it has properties a, b, c, and p. Y is something we know less about—only that it has properties a, b, and c. We don't *know* whether it has property p, but we make an inference that it does, because X has that property and Y is like X in several other ways. We are counting on the fact that if Y is like X in *some* ways we know about (a, b, and c), this makes it to some degree likely that Y is like X in some *further* way that we don't know about (i.e., in having p). Indeed, this last sentence sums up the most fundamental principle of inductive arguments. We could put it another way by saying that a number of similarities make *further* similarities likely. You can probably find other ways to formulate this general principle yourself. In what follows, we'll use X and Y to indicate which items fit which part of the formula above.

Moore knows several people with Harley-Davidson motorcycles, and they all leak oil. He wouldn't be surprised to learn, therefore, that the Harley that Parker just bought leaks oil. Why does he think so? Because he draws an inference from the Harley-Davidsons he knows about (Xs, more than one of them in this case) to the one Parker just bought (Y), reckoning that what's true of the ones he knows is likely to be true of the one he just learned about.

Your neighbor's new dog barks at you when you take out the trash on Monday, Tuesday, and Wednesday; you expect the dog to bark at you when you take out the trash on Thursday. You are arguing inductively from what happened when you took out the trash on Monday, Tuesday, and Wednesday (Xs) to what will happen when you take out the trash on Thursday (Y).

The peaches you bought at the A&P Market looked okay but were mushy; you don't buy any more of that batch because you think the other peaches in the batch will be mushy as well. Here again you are making an inductive inference—you are extending what you know about the peaches you bought (Xs) to the rest of the peaches in that particular batch (Ys).

It is common to describe inductive arguments as arguments in which the premises, if true, are claimed to support some degree of *probability* for the conclusion's truth, but not its *certainty*. In none of the arguments above do the facts we know guarantee the truth of the conclusion we draw, but they do make it *more likely* that that conclusion is true. We'll adopt the following as a working characterization of inductive arguments:

> An **inductive argument** is an argument the premises of which are intended to provide some degree of *probability* for the truth of the conclusion.*

---

*A somewhat more precise way to put this: An inductive argument is one for which it is claimed that the conclusion has a greater likelihood of being true if its premises are all true than if its premises are not all true.

Remember that this distinguishes such arguments from deductive ones, since the link between the premises and conclusion of the latter is one of certainty, not probability: If the premises are true in a valid deductive argument, the conclusion cannot fail to be true. In a good inductive argument, the conclusion can still fail to be true, but it is less likely to be so.

Inductive arguments give us a way of extending our belief from things we know about to things unknown. Throughout our discussion, we'll use the term **sample** to refer to an item or items we believe something about, and we'll use the term **target** (or **target class** or **target population**) to refer to an item or group of items to which we wish to extend our belief. The feature we know about in the sample and we extend to the target object (or target class) is the **feature** (or **property**) **in question.** Referring back to our formula at the beginning of this section, X is our sample, Y is our target, and p is our property in question.

Applying the same terminology to the peach example, our *sample* is the peaches we bought, the *target class* is the entire batch of peaches, and the *property in question* is being mushy. Are you catching on to the terminology?

In the dog example, our sample is the three previous occurrences of taking out the trash (we already believe the dog barked on those occasions), the target is the next occurrence of taking out the trash (we're reasoning that the dog will bark on this further occasion), and the property in question is the dog's barking when this further deed is done.

And in the motorcycle example, the sample consists of the Harleys that Moore knew about previously, the target is Parker's recently acquired bike, and the property in question is that of leaking oil. Make sure you understand this terminology or most of the rest of the chapter is going to be hard to follow. It isn't as complicated as it might sound at first.

Analogical Arguments and Generalizations

Remember that the target in an argument can be either an entire class (or "population")* of objects, events, occasions, and the like, or it can be just a single such item. Notice also that sometimes the members of the sample are *drawn from* the target (when it is an entire class), and sometimes they are not. For instance, in the example about peaches, the members of the sample (the peaches you bought at the A&P) are also members of the target class (the entire batch of A&P peaches). In the barking-dog example, the members of the sample (Monday's, Tuesday's, and Wednesday's trash haulings) are *not* drawn from the target (Thursday's trash hauling). It is common to divide inductive arguments into categories, and the differences we've just been describing characterize the two main such categories: **inductive generalizations** and **analogical arguments** (or **arguments by analogy**). Ordinarily, arguments by analogy have one thing or event for a target while generalizations always have

---

*The terms "class" and "population" are used interchangeably in this discussion.

Analogical Arguments and Generalizations

Let's take a closer look at the relationship between analogical arguments and generalizations. Consider an analogical argument such as the following:

> We observe that, in a sample of ten American cities of over 750,000 population, every one of them has a public rail transportation system. We also observe that San Francisco is like the cities in the sample in lots of ways (geography, tax base, etc.) We then draw the conclusion that San Francisco probably also has a public rail transportation system.

This is purely inductive argument. Now let's look at another way of reaching the same conclusion from the same premises.

> (A) We make the first observation we made previously, that every one of a sample of ten American cities of over 750,000 population has a public rail system. Since our sample is representative of American cities over three-quarters of a million in general, we go on to generalize that, probably, *all* American cities over 750,000 population have rail transportation systems.
> (B) We then construct another argument, from the premise that all American cities over 750,000 population probably have rail transportations systems to the conclusion that San Francisco, an American city, probably has a rail transportation system.

In this argument scenario, there are two separate arguments: argument A is inductive, and argument B is deductive. Notice that both argument scenarios produce the conclusion that San Francisco's rail system is probable. It should be obvious why in the first scenario; in the second, part A, being inductive, only establishes its conclusion with probability. Therefore, since no deductive argument can provide more certainty for its conclusion than is present for its premises, the final conclusion of B is also merely probable.

So, although analogical argument proceeds by a more direct route to its conclusion, there is always a side trip possible via a generalization to an entire class followed by a return trip to the conclusion by means of a deductive argument.

a class of things or events. In all cases, generalizations have their samples drawn from the target class, while this is never true of arguments by analogy.* (So, the example above about peaches is a generalization; the other examples are arguments by analogy.) Fortunately, although they differ in these ways, the two kinds of arguments both follow the same principles and are evaluated on similar criteria. Thus we will be able to treat them similarly and

---

*Since we do not have a target class in an argument by analogy, the "sample" cannot be a sample of such a class. So we'll just say that the "sample" in such arguments is an item or a group of items from which we are attempting to draw an inference about the target object. This preserves the parallel with generalizations.

will point out any differences that are necessary as we go along. So far, once again, the principal difference is that, in a generalization, the sample is drawn from the target class; in an analogical argument, the sample and the target are distinct—one is not a part of the other. (As mentioned, generalizations always have a class as a target and analogical arguments usually have a single object or just a few objects, but there are exceptions to this way of distinguishing between them.) For an interesting aspect of the relationship between analogical arguments and generalizations, see the box opposite.

Principles of Constructing and Evaluating Inductive Arguments

The most important principle of good inductive reasoning is that the item or items from which we are arguing—our sample—must be similar to the item or items we are arguing to—our target. Let's put that in boldface:

The more like one another our sample and target are, the stronger our argument; the less alike they are, the weaker the argument.

The more Parker's Harley-Davidson is like the others Moore knows (e.g., similar in model, in age, in maintenance history), the more likely it is to leak oil like the others. The more Thursday's trash hauling is like the previous three days' (e.g., done at the same time of day, with the same amount of noise, and with nobody having shot the dog in the meantime), the more likely it is to provoke the neighbor's dog to bark as it did on the earlier occasions.

Exactly the same principle holds when we are generalizing to an entire class of things (as distinct from arguing by analogy from one item to another). When we generalize from the peaches we bought to the *entire* batch of peaches they came from, we don't want our sample peaches to be atypical for that batch of peaches—we want them to be as much like the rest of the batch as possible. When speaking of the similarity between sample and target class of an inductive generalization, the word that is most often used is "representativeness." This word, though awkward to say, is important. We can characterize it and its opposite, "bias," in the following way: The more an inductive argument's sample is similar to its target in all relevant respects, the more **representative** the sample is said to be. (We'll give another definition below, one that applies especially to formal generalizations, but which is essentially the same as this one.) A sample that is significantly different from the target in one or more relevant respects is said to be a **biased sample.**

One special circumstance arises when we don't know whether the target in an analogical argument has a particular relevant feature. For example, what if we don't know whether Parker's Harley is a flathead or a knucklehead or has some other type of engine configuration? In that case, Parker's cycle is more likely to resemble some members of the sample if the latter

Typical Sample Cases: Deduction in Disguise

Let's say you want to know whether gold is denser than lead. To find out, you take exactly one cubic centimeter of gold and the same volume of lead, and you place them on the two sides of a balance scale. The weighing shows that the cubic centimeter of gold is heavier than the lead (in fact, the weight of the lead is about 59 percent of that of the gold). You then conclude that, as a matter of fact, gold in general is more dense than lead—that is, that *every instance* of gold is more dense than *every instance* of lead.

Are you committing the fallacy of hasty generalization by basing your conclusion on only one case? No, of course not. The reason is that you can safely assume that every instance of gold is similar in density to every other instance of gold and that instances of lead are likewise similar with regard to properties like density, because gold and lead are elements.

Reasoning such as this is often made out to be a peculiar kind of inductive argument, one where a little sample goes a long way, as it were. In fact, though, arguments like this are really *deductive* arguments in disguise. The assumption that makes the argument inductively strong (namely, that in the matter of density, gold is all alike and so is lead) also makes the argument deductively valid! The argument turns out to be this:

> General assumption: All gold is alike with regard to density, and so is
> all lead.
> Result of weighing: This gold is denser than this lead.
> Conclusion: Gold is denser than lead.

contains a variety of engine types. In such cases, *variety* in the sample strengthens the argument. So, if we don't know whether our target has some property, we're better off if some of our sample have it and some don't. (This is no different from how we treat generalizations: Our generalization about peaches is stronger if the sample contains peaches from the top, bottom, and middle of the container.) Of course, if we *do* know that the target has a certain property, then the more of the sample that have it, the better.

The second principle governing inductive arguments is about the number of items in the sample, and we may as well put that in boldface too:

**In general, the larger the sample the better—that is, the stronger the
argument.**

The more peaches in our sample, the more confidence we are entitled to feel in extending our observations about those peaches to the rest of the peaches in the batch. The more times an area has flooded in the past, the more insurance we should buy. The more times the dog has barked at us, the more we should expect it to bark at us tomorrow. The more leaky Harleys we know about, the more we ought to consider buying Parker a bag of kitty litter for his garage floor.

The two principles work together, but the first is more important. Everything else being equal, the most important thing is for the items in the sample to be similar to the target—to the items we are arguing to. We can't argue very well even from a huge sample if the items in that sample are different in some significant way from those we are arguing to. On the other hand, a sample consisting of just a single item can warrant a generalization to an entire class if that item is sufficiently similar to the other items in the class. If one copy of a book has a misprint on page 327, then it is extremely likely that every other copy of the book from that printing has a similar misprint. (If we know that all the items in a target class are exactly similar, then we get a different kind of situation. See the box titled "Typical Sample Cases: Deduction in Disguise.")

However, if the items in the class of things we are concerned about are not known to be homogeneous or are known to be nonhomogeneous, then we cannot get away with a small sample. For example, to draw conclusions about the political views of, say, Canadian truck drivers, we'd need a sample large enough to represent the diverse array of personalities, socioeconomic levels, educational background, life experiences, and so forth, that are found in such a large and nonhomogeneous population. How this is accomplished will be explained in the following section.

Formal and Informal Inductive Arguments

Before we go on, we need to separate two "levels" of inductive argumentation. The first level, and the one that concerns most of us most of the time, is that of everyday, informal arguments. We encounter both generalizations and arguments by analogy at this level, and, if you pay close attention, you'll discover that you are making inferences like these practically all the time. They are integral to our rational process, with many such inferences occurring at a subargumentative level—that is, we don't reach the point of putting our thinking into words (which, you might recall, is a necessary condition of making an argument).

The second level is that of what we might call "scientific" or "formal" argumentation. At this level we find generalizations in the form of public opinion polls, ratings for television and radio stations, and research surveys of all sorts. We also find scientific studies undertaken to discover what causes what, but that's the subject of the next chapter.

There are a couple of obvious differences between formal and informal arguments. First, most formal arguments turn out to be generalizations. You could certainly produce a formal, scientifically controlled argument by analogy, but you just won't happen upon one in your everyday life very often. Second, formal arguments require an attention to detail—including reliance on some moderately sophisticated mathematics—that is usually impossible to achieve and generally not necessary in their informal counterparts. But, as we're going to see, however different they may look, arguments at both levels depend entirely on exactly the same general principles. A long and complicated statistical study of the population decline of spotted owls and a quick inference about whether to return to a restaurant you tried last week are measured against the very same general criteria.

So here's our plan: We're going to make an excursus into formal arguments, explaining how several concepts operate at that level as we trace an example from beginning to end. It may get a little technical along the way, but it isn't really complicated. After the formal treatment, we'll show how the principles at that level have corresponding versions at the informal level and how we apply them to our everyday reasoning.

Example of a Formal Generalization: A Political Poll

Imagine that you graduate from college one of these days. (We're serious.) You take a position at a survey research firm that does political polling. Your first job is to manage a poll commissioned by a potential candidate for the presidency—say, Jeb Bush, who figures that it's his time to be president, his father and older brother having already had their turn. Bush wants to know whether the Republican voters of New Hampshire would vote for him rather than, say, John McCain in the presidential primary in that state. Your job is to produce a reasonably accurate assessment of public opinion on the issue at the lowest reasonable cost.

Of course, the most accurate way to settle the issue would be to ask all the Republican voters of New Hampshire. But that would be excessively expensive. So you will have to make do with asking a sample of those voters and extending the knowledge gained thereby to a conclusion about the whole class. How well that knowledge will extend to the whole class depends on how good a job you do. The two most important decisions you have to make are these: How will you select the sample so that it is representative of the entire target class of voters? How large will the sample need to be in order to adequately represent that class?

Clearly, it would be a mistake to let the Republicans on the faculty at the University of New Hampshire be the sample, even if doing so would produce a sample of the right size. It would be a mistake because those faculty at UNH are very likely *not* to represent the entire target class. Indeed, they are too different in too many relevant ways from many, many members of the target class. Which brings us to the question, Just what does it mean for a sample to be representative of a target? In a nutshell, we can say that *a sample is representative of a target class if, and only if, it possesses all the relevant features of the target class in proportions similar to those of the target class.* We should add that a feature is said to be relevant when it is likely that its presence or absence would affect the presence or absence of the property in question. The following should make this clear.

Let's ask ourselves whether there are features of the Republican UNH faculty that are not features of the entire target class. Sure there are: For example, there is a much higher percentage of advanced degrees among the former than among the latter. Now we ask ourselves, Is the possession of advanced degrees relevant to the property in question (which, you'll recall, is the likelihood of voting for Jeb Bush rather than John McCain)? It is very likely that the level of one's education affects the way a person votes, which means that this feature is relevant. Notice that this doesn't mean that *every* highly educated voter must vote differently from every less educated voter, only that the former *tend* to vote in certain ways somewhat more frequently than the latter.

So, by possessing relevant properties that make them different from the target class, the UNH faculty members can't form the sample you want for your survey—they would give you a very biased sample indeed.

Notice also that we don't have to know with absolute certainty that higher levels of education have significant effects on the way a person votes; we only need a strong suspicion. The general rule is, If we're not sure whether a feature is relevant, the safe path is to presume that it *is* relevant. It's a matter of being safe rather than sorry.

You now have a problem: How are you going to find a group that will share relevant features with the entire target class? Setting out to construct such a group (which, as we'll see in the next chapter, actually has to be done sometimes) is a real chore. For starters, we don't usually know what all the relevant features are—Does the age of a Republican in New Hampshire help determine how he or she votes? Probably. Does being left-handed? Probably not. What about being a commuter versus working at home? Who knows? And, even if we knew what all the relevant features are, it would still be a big job getting our sample to exhibit them in just the proportions in which they occur in the target class.

Fortunately for you and the job at hand, there is a method that solves the problem we've been describing. You've probably already figured out what it is: a randomly selected sample. A **random selection process** gives every member of the target class an equal chance of becoming a member of the sample. If we randomly select our sample from the target population, we should, within certain limits, get a sample that possesses all the characteristics of the target class in proportions more or less similar to those found in the target class. We should point out that, as a practical matter, coming up with actual procedures for getting a random selection process can be a difficult problem; statisticians often have to settle for something that works only *pretty* well because it's affordable. Another word on that in a few moments.

> Some people use statistics like a drunk uses a lamp post, more for support than illumination.

Random Variation: Problems and Solutions

A lot of people misunderstand what random selection is all about in cases like the ones we're discussing. They often think that randomness is the goal of the process. And indeed it is *a* goal, but it is merely the means by which we try to achieve the *real* goal, which is representativeness. Fortunately, randomness does a pretty good job of helping us reach this goal; unfortunately, it doesn't do a perfect job. In fact, any given randomly selected sample from a target class might misrepresent that target to some extent or other. That is, there's nothing in a sample, even if it is perfectly random, that *guarantees* that it will be representative.

What does this mean for our need to come up with an answer to the question "How many New Hampshire Republican voters would prefer to vote for Bush over McCain?" It means we have to build some "wiggle room" into the answer. Fortunately, there are some perfectly reasonable ways to do that.

For one thing, nobody would expect you and your firm to be able to determine *exactly, to the precise number,* how many voters would prefer Bush. If you can get reasonably close, you'll have done a good job. So, you want to be able to tell the Bush people, "approximately such-and-such percent" prefer one candidate to the other, or "about so-and-so percent" prefer him. Giving yourself this wiggle room is known as providing an "error margin." An **error margin** is a range of percentage points within which an answer is claimed to fall (rather than a precise percentage point or even fraction of a percentage point). The idea is the same here as in the situation where a potential automobile buyer asks a salesman about a car's gas mileage: The salesman should only have to say *approximately* what mileage can be expected, not *exactly* how many miles per gallon one will get. Of course, if the error margin is too big, the survey is useless. If you have to tell the Bush campaign that somewhere between 10 and 90 percent of New Hampshire Republican voters prefer Bush, you haven't given them much for their money, have you? We'll return to this in a moment.

The second way to make the argument provided by the survey a reasonable one is to allow that there is *some* chance that it might be wrong. Indeed, it would be a rare argument where a person is extending knowledge about a known thing to claims about an unknown thing and where the person would be willing to bet his or her life on the outcome. We don't usually bet the farm unless we *know* we're right. What you want to do here is identify a "confidence level" for your argument. A **confidence level** is a measure of the argument's strength; the higher the confidence level, the more likely the argument's conclusion is to be true. So, we can say that it is "very likely" that such and such a percentage of voters prefer Bush, or we can actually quantify mathematically what percentage of likelihood the conclusion has.

Given that the sample of a survey is randomly selected from the target class, the error margin of the generalization's conclusion and the argument's confidence level are determined entirely by the size of the sample. Here's the principle:

> **As the sample size of a generalization gets larger, either the error margin becomes more narrow, or the confidence level of the argument becomes greater, or a combination of both.**

Notice that this principle relates three variables—sample size, confidence level, and error margin—and shows how they are interdependent. It's very important that you understand how these three factors relate to one another.

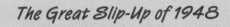

The Great Slip-Up of 1948

Because of a strike, the *Chicago Daily Tribune* had to go to press earlier than usual the night of the 1948 presidential election. So they relied on some early returns, some "expert" opinion, and public opinion polls to decide on the famous "Dewey Defeats Truman" headline. But the polls were not sufficiently accurate, as Truman edged Dewey in a narrow upset victory.

Okay. We've determined that your sample of voters in the New Hampshire primary needs to be randomly selected—we're going to leave it to you to determine how to do that, but we'll warn you to be careful: You don't want to make the mistake the *Chicago Daily Tribune* made. (See the box above.)

Sample Size

By the end of the last section, we had come back around to the topic of **sample size,** and now it's time to deal with it. When it comes to surveys of the sort we're talking about, the question "How many individuals does the sample need to contain?" has a neat, short answer: somewhere between 1,000 and 1,500. You may find it interesting that, no matter what the survey is about, it usually requires between 1,000 and 1,500 in the sample regardless of whether the target class is Republican primary voters in New Hampshire, voters of all parties in New England, citizens of the United States, or human beings on the entire planet. In short, as long as we're not talking about very small target classes (which we can safely ignore, since generalizations like

TABLE 10-1   Approximate Error Margins for Random Samples of Various Sizes

Confidence level of 95 percent in all cases.

Sample Size	*Error Margin (%)*	*Corresponding Range (Percentage Points)*
10	±30	60
25	±22	44
50	±14	28
100	±10	20
250	±6	12
500	±4	8
1,000	±3	6
1,500	±2	4

The error margin decreases rapidly as the sample size begins to increase, but this decrease slows markedly as the sample gets larger. It is usually pointless to increase the sample beyond 1,500 unless there are special requirements of precision or confidence level.

(We presume, both here and in the text, that the target population is large—that is, 10,000 or larger. When the target is small, there is a correction factor that can be applied to determine the appropriate error margin. But most reported polls have large enough targets that we need not concern ourselves with the calculation methods for correcting the error margin here.)

these don't apply to them), the size of the target is not relevant to the size of the sample required. The reason for this will become clear in what follows.

We are not going to discuss the mathematics that lie behind the facts we're describing here, but they are among the most basic mathematics in this field—you can trust them. They guarantee the details that you'll find in Table 10-1, which you should look at now. Notice that the column on the left represents a series of increasing sample sizes. In the second column, we find the error margins that correspond to the various sample sizes expressed as "plus or minus *x* percentage points." This expression is the means by which statisticians express the amount of "wiggle room" an argument requires for its conclusion. The third column just shows the range of percentage points that the error margin produces.

The two most important things to notice about Table 10-1 are, first, that the error margin narrows very quickly as the size of the sample increases from 10 to 25 and then to 50, but as we go down the columns the narrowing of the error margin slows down. So, by the time we get to a sample size of 500, with an error margin of plus or minus 4 percent, we have to *double* the sample to 1,000 in order to decrease the error margin by one percentage point, to plus or minus 3 percent. It takes another 500 added to the sample to get it down one more percentage point. (These error margins are approximate; they've been rounded off for convenience's sake.) In order to get the error margin down to, say, 1 percent or less, we have to vastly increase the size of the sample, and for most practical purposes, the gain in a more precise conclusion (one with a narrow error margin) is outweighed by the difficulty and expense of having to add so many new members to the sample. And this explains the claim made above that we very seldom go beyond 1,000 to 1,500 members of a sample—because the trouble of increasing a sample beyond that is hardly worth the modest decrease produced in the error margin.

Now, throughout the preceding paragraphs, we've been making a presumption about confidence levels. We've been presuming that the people who hired you, the Bush campaign, would be satisfied with a 95 percent confidence level for the survey. We chose this figure because it is the standard of the polling industry—if a confidence level is not mentioned in a professionally done poll, the assumption is that it is 95 percent. This means simply that one can be 95 percent sure that the conclusion of the argument does indeed follow from the premises—in other words, that we can have considerable confidence in the conclusion.* If we have a smaller sample, we will, of course, have a lower confidence level or a larger error margin. On the other hand, if it is important to have a confidence level higher than 95 percent, then we need a larger sample, if the error margin is not to be affected.

In general, professional surveying and polling organizations have settled on the 95 percent figure for a standard level of confidence and an error margin of around 3 percentage points. And thus the figure of 1,000 or a little over is the standard sample size for most professionally done scientific polls.

This brings us to the second thing—or, actually, things—you should remember about Table 10.1. You should make a mental note of the three or four lines near the bottom of the table so that, if you run across a survey with a suspiciously small sample, you'll know what kind of error margins it would be reasonable—or unreasonable—to claim for it. We recently saw in a golf magazine that approximately 200 golfers had been surveyed about something or other and that less than half of them, 45 percent, in fact, had agreed with the poll question. Does this mean that less than half of all golfers can be expected to agree with the poll question? Not at all. If it has the usual confidence level, a sample of 200 must have an error margin of around plus or minus 8 percent, which means it may be that as many as 53 percent—a majority—actually agree with the poll question.

Informal Inductive Arguments

Most of the inductive arguments we have to deal with in everyday life—both generalizations and arguments by analogy—are most definitely *not* of the formal, scientific variety we've been talking about in the last few pages. The most important difference between formal or scientific inductive arguments

---

*If you held our feet to the fire about this, we'd have to say that the *real* meaning of a 95 percent confidence level is that an average of 95 out of every 100 random samples from this target class will have the property in question within the range of the error margin. But this translates into practical terms in the way we've put it in the text.

on one hand and informal, everyday arguments on the other is that the latter do not make use of randomly selected samples. In everyday inductive reasoning, we make use of whatever is at hand—the Harleys Moore is familiar with, the peaches you bought at the A&P this morning, and the like. For this reason, it should be obvious, we cannot make the kinds of calculations that make formal arguments precise.

■ *Informal Error Margins: "Cautious" Conclusions*

But many of the concepts and principles we've described in connection with formal arguments do have their place in their everyday counterparts—they just take a more easygoing form. We probably wouldn't say to a colleague, "Based on my experience in freshman courses, I'd conclude that the grade point for the freshman class is 2.1036." And we're not likely to say to a fishing partner, "Ed, I've looked at the Fish and Game Department statistics, and your chance of catching a fish in that lake that weighs more than three pounds is 12 percent, plus or minus 3 percent." Ed would probably throw us out of the boat. We'd actually say something like, "I'd conclude that the freshman grade point is somewhere around a two-point," or "You're not likely to catch a three-pounder in that lake." These are informal ways of providing error margins for our inferences. Increasing the informal "error margins" in everyday arguments makes the conclusion more cautious.

■ *Informal Confidence Levels: Hedging Our Bets*

In informal arguments we also set informal confidence levels (as distinct from informal error margins) by adjusting the cautiousness of the conclusion. For example, it's more cautious to conclude "there's a good chance" the Yankees will win the pennant than to conclude "You can be pretty sure" the Yankees will win the pennant. "There's a good chance" and "You can be pretty sure" are informal ways of expressing our "confidence level." Obviously, the informal "confidence level" or "error margin" for our conclusion should be appropriate for the premises of our argument. If the Yankees have won only 40 games, we must be more cautious in concluding they will win the pennant than if they won over a hundred.

■ *Summing Up: Evaluating Inductive Generalizations and Analogical Arguments*

When all is said and done, the issue of whether we have a strong inductive generalization (or analogical argument) or a weak one, or something in between, can be reduced to three key questions:

1. **Are the sample and the target sufficiently alike?** If it's a generalization we have before us, this can be stated in terms of representativeness: Does the sample adequately represent the target class? If

we're looking at an argument by analogy, the question is simply, Do members of the sample adequately resemble the target item in relevant respects?

If sample and target are alike with regard to features that are relevant to the property in question, then the argument will be strong; it will be weaker to the extent they are unlike regarding these features. Notice that a significant difference with respect to only one relevant feature is all it takes to ruin an otherwise good inductive argument. If, in our A&P example, the peaches in the sample had all come from the bottom layer of the container, this would make a big difference in the strength of the argument. If the argument is a formal generalization, the best we can do is to make sure the sample was drawn from the target class by a random process.

2. **Is the sample of sufficient size?**

In a formal argument, we've seen that the size of the sample determines the calculated confidence level and error margin. In informal arguments, the general rule is, The bigger the sample, the better (given the same degree of caution in the conclusion). It is possible, as we've indicated, to get a pretty good argument from a small sample—as small as one single item, in fact. But this can occur only when we know that the sample item or items are very, very similar to the target item or to all the members of the target class. It's also helpful in such cases if we can stand a fairly large error margin. To illustrate, consider this example: Let's say that Moore owns a five-year-old Toyota Camry, and he recommends a similar car to Parker when the latter decides he needs to buy one. Parker finds one that is the same year and model as Moore's; it has approximately the same number of miles on it; it has been driven and maintained in similar fashion, and the various fittings—transmission, air conditioning, etc.—are also similar. Parker asks Moore what kind of gas mileage his car gets, and Moore says he averages 30 miles to the gallon. Now, on the assumption that Parker's driving habits are similar to Moore's,* it would be a pretty safe bet that Parker will get *fairly similar* mileage. If Parker were to conclude that he'll get, say, between 25 and 35 miles per gallon, we'd say he had a very strong argument. It's a high probability that his mileage will fall somewhere within this substantial range. Were he to conclude that he'll probably average between 28 and 32 mpg, he'd still have a pretty good case, but considerably less strong than when he adopted the wider error margin—the more cautious conclusion.

These considerations about Moore's gas mileage point to the third key question about inductive generalizations and analogical arguments:

3. **Is the cautiousness of the conclusion appropriate for the premises?**

---

*They aren't.

Cautious conclusions are easier to support than incautious conclusions. If, to revert to an earlier example, the Yankees have won 50 percent of their games so far, we need to be more cautious before we bet on their winning their next game than we would if they've won 66 percent. In short, we shouldn't go out on a limb with our conclusion, unless we have a nice, sturdy trunk (premise) to support us. To say that the conclusion of an inductive generalization or analogical argument is not sufficiently cautious is just a way of saying that it is not a strong argument.

Fallacies

Given what we've been discussing in the previous sections, learning about inductive argument fallacies will amount to little more than learning a little terminology. There are two main categories of fallacy. The first of these is biased sample (sometimes called "biased generalization" when it occurs in that type of argument and "biased analogy" when it appears in an analogical argument). A **biased sample** occurs when the sample is not sufficiently similar to the target; if the argument is a generalization, this means the sample is not *representative* of the target. Let's say that, for whatever reason, Moore knows quite a number of people who own Harleys. But all of those machines are older models, built when Harley-Davidson was owned by an organization different from the one that now owns it, and built to a design different from the more recently manufactured machine that Parker is buying. In this case, even if Moore's sample is sufficiently large (even if it contained hundreds and hundreds of such motorcycles!), the argument would be a very poor one: The sample is simply biased—it is too different from the target in ways that are highly relevant to the property of leaking oil.*

If we jump to a conclusion from a sample that is too small, we are guilty of the fallacy of **hasty conclusion** (or, if the argument is a generalization, **hasty generalization**). If Moore knew of one Harley-Davidson motorcycle, and from the fact that it leaked oil he concluded that Parker's Harley (never mind the entire class of Harleys!) would leak oil, he'd be guilty of hasty conclusion (or, in the case of the entire class, of hasty generalization). On the other hand, if he knew of fifteen or twenty people who owned Harleys of a considerable variety of types, he'd have a much stronger argument, and we wouldn't accuse him of committing any such fallacy.

Within the category of the hasty conclusion/generalization fallacy, a couple of special subspecies bear mention. The first and best known of these is the **fallacy of anecdotal evidence.** The fallacy has this name because it always consists in taking a story about one case (or, sometimes, more than one, but always a very small number) and drawing an unwarranted conclusion from it. If a student is considering taking a philosophy course at our

---

*Something like this is actually true, if we understand recent Harley-Davidson history correctly.

Who Do You Trust?

- When it comes to deciding which kind of car to buy, which do you trust more—the reports of a few friends or the results of a survey based on a large sample?
- When it comes to deciding whether an over-the-counter cold remedy (e.g., vitamin C) works, which do you trust more—a large clinical study or the reports of a few friends?

Many people trust the reports of friends over more reliable statistical information. We hope you aren't among 'em. (According to R. E. Nisbett and L. Ross, *Human Inference: Strategies and Shortcomings of Human Social Judgment* [Englewood Cliffs, N.J.: Prentice Hall, 1980], people tend to be insensitive to sample size when evaluating some product, being swayed more by the judgments of a few friends than by the results of a survey based on a large sample.)

university, and in conversation with another student she hears, "You'd better be careful, because philosophy courses are *hard*, I'm seriously telling you. I know what I'm talking about too, because I took a course from Dr. Moore last term and it was *way hard!*" If, on the basis of this conversation, our student decides that all philosophy courses are difficult and therefore she won't take one, she's guilty of the fallacy of anecdotal evidence. She's jumped to a conclusion from the story (the anecdote) her friend told her and simply doesn't have enough evidence to reach that conclusion.

Before completing this section, we should add that most samples that are too small are also biased. Indeed, small size is one typical source of biased samples, and this is more true as one finds more variety in the target class. However, when a sample is too small, we'll apply the hasty conclusion label to it, and we'll reserve the biased sample label for arguments whose samples are biased, however large they might be.

Polls: Problems and Pitfalls

One of the most frequently encountered uses of inductive arguments is in polls, especially public opinion polls (and most especially in election years). We explained many of the concepts that are important in conducting and reporting polls a bit earlier in the chapter, but it's time now to look at a couple of the problems that crop up in this important use of inductive argumentation.

We should emphasize first that a properly conducted and accurately reported poll can be a very reliable source of information. But we hasten to add that a lot of polls that you hear or read about are *not* properly done and

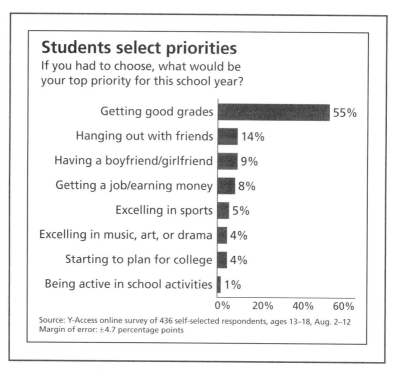

▲   *Comment: A good example of a worthless poll . . .*

often the people who report on the results cannot tell the difference between a good poll and a bad one. We can't go into every possible way a poll can fail, but in what follows we'll take notice of two of the most common ones.

■ *Self-Selected Samples*

Recall that a generalization from a sample is only as good as the representativeness of the sample. Therefore, keep this in mind: *No poll should be trusted if the members of the sample are there by their own choice.* When a television station asks its viewers to call in to express an opinion on some subject, the results tell us very, very little about what the entire population thinks about that subject. There are all kinds of differences possible—indeed, likely—between the people who call in and the population in general. The same goes for polls conducted by mail-in responses. One of the most massive polls ever processed—and one of the most heavily flawed, we should add—was done in 1993. The political organization of H. Ross Perot, a very wealthy businessman who ran for president as a member of the Reform Party, paid for a poll that was conducted by means of the magazine *TV Guide.* People were asked to answer questions posed in the magazine, then tear out or reproduce the pages and send them in for processing. There were other things wrong with this poll, and we'll get to some of them in a minute, but you've already heard all you need to know to discount any results that it produced. In such polls, the sample consists only of people who have strong enough feelings on the issues to respond and who have the time to go to the trouble of doing it. Such a situation almost guarantees that the sample will have views that are significantly different from those of the target population as a whole.

Another example, just for fun: A few years ago the late Abigail Van Buren ("Dear Abby") asked her female readers to write in answering the question "Which do you like more, tender cuddling or 'the act' (sex)?" More of her responders preferred cuddling, it turned out, and when she published this fact it provoked another columnist, Mike Royko, to ask his male readers which *they* liked better, tender cuddling or bowling. Royko's responders preferred bowling. Although both surveys were good fun, neither of them could be taken to reflect accurately the views of either columnist's readership, let alone society in general.

It should go without saying that person-on-the-street interviews (which have become extremely popular in our neck of the woods) should be utterly discounted as indications of popular opinion. They include small samples, almost always biased because, among other reasons, the interviews are usually conducted at a single location and include only people willing to stick their faces in front of a camera. You should read these interviews as fun, not as a reflection of the views of the general public.

■ *Slanted Questions*

A major source of unreliability in polling practices is the wording of the questions that are asked. It is possible to ask nearly any question of importance in many different ways. Consider this pair of questions:

▲ Do you think the school board should agree to teachers' demands for higher pay?

▲ Do you think it is reasonable for local public school teachers to seek pay raises?

These questions ask essentially the same thing, but you would be smart to expect more negative answers to the first version than to the second. The context in which a question is asked can be important too. Imagine a question asking about approval of pay raises for public school teachers, but imagine it coming after one or the other of the following questions:

▲ Are you aware that teachers in this district have not had a salary increase for the past six years?

▲ Are you aware that the school district is facing a budget shortfall for the coming fiscal year?

We'd expect the approval of raises would fare better when asked after the first of these questions than after the second.

Studies indicate that more brunettes than blondes or redheads have high-paying corporate jobs.

—From a letter in the *San Francisco Chronicle*

Is this evidence of discrimination against blondes and redheads, as the writer of the letter thought?

Nope; there are more brunettes to begin with. We'd be suspicious if *fewer* brunettes had high-paying corporate jobs.

We might add that the inclusion of slanted questions is not always accidental. Oftentimes a group or an organization will want to produce results that are slanted in their direction, and so they will include questions that are designed to do exactly that. This is an exercise in deception, of course, but unfortunately it is more widespread than we'd wish.

Have a look at the box ("Ask Us No [Loaded] Questions . . . ") and you'll see how one large, very expensive poll can contain most of the errors we've been discussing.

Playing by the Numbers

What if your instructor were to flip a coin ten times, and it came up heads seven times out of that ten? Would this make you think your teacher is a wizard or a sleight-of-hand artist? Of course not. There's nothing unusual in the coin coming up heads seven times out of ten, despite the fact that we all know the chance of heads in a fair coin flip is 50-50 (or 1 in 2, or, as the statisticians put it, 0.5). But what if your instructor were to get heads 70 percent of the time after flipping the coin a *hundred* times? This would be much more unexpected, and if he or she were to flip the coin a *thousand* times and get 70 percent heads—well, the whole class should leave right now for Las Vegas, where your instructor will make you all rich.

Why is 70 percent heads so unsurprising in the first case and so nearly miraculous in the last? The answer lies in what we call the **law of large numbers,** which says,

> **The larger the number of chance-determined repetitious events considered, the closer the alternatives will approach predictable ratios.**

This is not as complicated as it sounds; metaphorically, it just says that large numbers "behave" better than small numbers. Here's the idea: Because a single fair coin flip (i.e., no weighted coin, no prestidigitation) has a 50 percent chance of coming up heads, we say that the **predictable ratio** of heads to tails is 50 percent. The law of large numbers says that the more flips you include, the closer to 50 percent the heads-to-tails ratio will get.

The reason smaller numbers don't fit the percentages as well as bigger ones is that any given flip or short series of flips can produce nearly any kind of result. There's nothing unusual about several heads in a row or several tails—in fact, if you flip a thousand times, you'll probably get several "streaks" of heads and tails (and a sore thumb, too). Such streaks will balance each other out in a series of a thousand flips, but even a short streak can skew a small series of flips. The idea is that when we deal with small numbers, every number counts for a very large number of percentage points. Just two extra cases of heads can produce a 70:30 ratio in ten flips, a ratio that would be astounding in a large number of flips.

Ask Us No [Loaded] Questions; We'll Tell You No Lies

In the spring of 1993, H. Ross Perot did a nationwide survey that received a lot of publicity. But a survey is only as good as the questions it asks, and loaded questions can produce a biased result. *Time* and CNN hired the Yankelovich Partners survey research firm to ask a split random sample of Americans two versions of the questions; the first was Perot's original version, the second was a rewritten version produced by the Yankelovich firm. Here is what happened for three of the topics covered.

Question 1
PEROT VERSION: "Do you believe that for every dollar of tax increase there should be two dollars in spending cuts with the savings earmarked for deficit and debt reduction?"

YANKELOVICH VERSION: "Would you favor or oppose a proposal to cut spending by two dollars for every dollar in new taxes, with the savings earmarked for deficit reduction, even if that meant cuts in domestic programs like Medicare and education?"

RESULTS: Perot version: 67 percent yes; 18 percent no
Yankelovich version: 33 percent in favor; 61 percent opposed

Question 2
PEROT VERSION: "Should the President have the Line Item Veto to eliminate waste?"

YANKELOVICH VERSION: "Should the President have the Line Item Veto, or not?"

RESULTS: Perot version: 71 percent in favor; 16 percent opposed
Yankelovich version: 57 percent in favor; 21 percent opposed

Question 3
PEROT VERSION: "Should laws be passed to eliminate all possibilities of special interests giving huge sums of money to candidates?"

YANKELOVICH VERSION: "Should laws be passed to prohibit interest groups from contributing to campaigns, or do groups have a right to contribute to the candidate they support?"

RESULTS: Perot version: 80 percent yes; 17 percent no
Yankelovich version: 40 percent for prohibition; 55 percent for right to contribute

The law of large numbers operates in many circumstances. It is the reason we need a minimum sample size even when our method of choosing a sample is entirely random. To infer a generalization with any confidence, we need a sample of a certain size before we can trust the numbers

Welcome to Saint Simpson's

There are many ways that statistical reports can be misleading. The following illustrates one of the stranger ways, known to statisticians as "Simpson's Paradox":

Let's say you need a fairly complicated but still routine operation, and you have to pick one of the two local hospitals, Mercy or Saint Simpson's, for the surgery. You decide to pick the safer of the two, based on their records for patient survival during surgery. You get the numbers: Mercy has 2,100 surgery patients in a year, of which 63 die—a 3 percent death rate. Saint Simpson's has 800 surgery patients, of whom 16 die—a 2 percent death rate. You decide it's safer to have your operation done at Saint Simpson's.

The fact is, you could actually be *more* likely to die at Saint Simpson's than at Mercy Hospital, despite the former hospital's lower death rate for surgery patients. But you would have no way of knowing this without learning some more information. In particular, you need to know how the total figures break down into smaller, highly significant categories.

Consider the categories of high-risk patients (older patients, victims of trauma) and low-risk patients (e.g., those who arrive in good condition for elective surgery). Saint Simpson's may have the better looking overall record, not because it performs better, but because *Saint Simpson's gets a higher proportion of low-risk patients than does Mercy.* Let's say Mercy had a death rate of 3.8 percent among 1,500 high-risk patients, whereas Saint Simpson's, with 200 high-risk patients, had a death rate of 4 percent. Mercy and Saint Simpson's each had 600 low-risk patients, with a 1 percent death rate at Mercy and a 1.3 percent rate at Saint Simpson's.

So, as it turns out, it's a safer bet going to Mercy Hospital whether you're high risk or low, even though Saint Simpson's has the lower overall death rate.

The moral of the story is to be cautious about accepting the interpretation that is attached to a set of figures, especially if they lump together several categories of the thing being studied.

—Adapted from a story in the Washington Post, *which quoted extensively from* DAVID S. MOORE, *professor of statistics at Purdue University*

to "behave" as they should. Smaller samples increase the likelihood of random sampling error.

The law of large numbers also keeps knowledgeable gamblers and gambling establishments in business. They know that if they make a bet that gives them even a modest advantage in terms of a predictable ratio, then all they have to do is make the bet often enough (or, more frequently, have some chump make the opposing bet against them often enough) and they will come out winners.

Let's consider an example. A person who plays roulette in an American casino gives away an advantage to the house of a little over 5 percent. The odds of winning are 1 in 38 (because there are slots for thirty-six numbers plus a zero and a double-zero), but when the player wins, the house pays

Reverse Gambler's Fallacy

If you want to get tails, your chances are greater if you flip after there's been a run of heads, right? Well, no—that's the gambler's fallacy.

And if you *don't* want to get tails, you are better off if you flip *before* there's been a run of heads—Whoops! That's the gambler's fallacy in reverse.

off only at the rate of 1 in 36 (as if there were no zeros). Now, this advantage to the house doesn't mean you might not walk up to a table and bet on your birthday and win four times in a row. But the law of large numbers says that if you pull up a chair and play long enough, eventually the house will win it all back—and the rent money, too.

A final note while we're speaking about gambling. There is a famous error known as the **gambler's fallacy,** and it is as seductive as it is simple. Let's say that you're flipping a coin, and it comes up heads four times in a row. You succumb to the gambler's fallacy if you think that the odds of its coming up heads the next time are anything except 50 percent. It's true that the odds of a coin coming up heads five times in a row are small—only a little over 3 in 100—but once it has come up heads four times in a row, the odds are still 50-50 that it will come up heads the next time. Past performance may give you a clue about a horse race, but not about a coin flip (or any other event with a predictable ratio).

Recap

Inductive arguments include inductive generalizations, analogical arguments, and a third kind we'll discuss in the next chapter.

Inductive generalizations are made when we infer that an entire class or population of things (a target class or population) has a property or feature (the property or feature in question) because a sample of those things has that property or feature. Analogical arguments are made when we infer that some specific thing or things have a property or feature because some other similar things have that property or feature. We can evaluate generalizations and analogical arguments by considering three questions:

1. Are the sample and the target sufficiently alike? If we're considering a generalization, this is the same as asking if the sample is representative of the target class. If it's an analogical argument we have before us, it's simply a matter of whether the sample and target item are sufficiently similar in relevant respects.

2. Is the sample of sufficient size?

3. Is the cautiousness of the conclusion appropriate for the premises?

Which Is More Likely?

Which is more likely to have a week in which over 60 percent of new babies are boys—a large hospital or small hospital? If you thought it was a large hospital, you need to reread the section on the law of large numbers. (The example is from R. E. Nisbett and L. Ross, *Human Inference: Strategies and Shortcomings of Human Social Judgment.* Englewood Cliffs, N.J.:Prentice Hall, 1980.)

Several fallacies occur regularly when people use inductive arguments like the ones described in this chapter. We discussed biased sample (when the sample is not sufficiently representative), hasty conclusion or hasty generalization (when the sample is not big enough to warrant confidence in the conclusion), and anecdotal evidence (a version of hasty conclusion in which a single story or anecdote—or at most a very small number of such stories—is taken to support a conclusion that is actually unwarranted; this is often used to "refute" a larger body of statistical evidence).

Finally, we learned that the law of large numbers governs statistically predictable outcomes of events—the longer you play against the house, the more odds will favor it in precisely predictable ways. (Stay away from bets in the center of the craps table.)

Exercises

Exercise 10-1

In groups (or individually, if the instructor prefers), decide whether each of the following comments is (a) an analogical argument or (b) an analogy that isn't in an argument.

▲ 1. These shrubs look rather similar to privet. I bet they keep their leaves in the winter, just like privet.

   2. Working in this office is just about exactly as much fun as driving in Florida without air conditioning.

   3. The last version of Word was filled with bugs; that's why I know the new version will be the same way.

▲ 4. If you ask me, Dr. Walker has as much personality as a piece of paving concrete.

   5. I hate math, and soon as I saw all those formulas and junk, I knew I'd hate this symbolic logic stuff too.

   6. Your new boyfriend has been married five times?! Say—do you seriously think it'll work out better the next time?

▲ 7. Too much sun will make your face all wrinkly; I suppose it would have that effect on your hands, too.

8. A CEO who had sex with an intern would be fired on the spot. A senior officer who slept with a recruit would be court-martialed. That's why Bill Clinton should not have been acquitted by the Senate.

9. I go, "Mom, don't buy me any more Levi's"; and she's like, "What?" So I told her, "They're for people your age," and she's like, "Well, I'll just give them to your sister"; and so I told her you're just like me and you won't want them either.

▲ 10. Here, you can use your screwdriver just like a chisel, if you want. Just give it a good whack with this hammer.

11. In elections during economic hard times, the party out of the White House has always made major gains in the Congress. That's why we can expect the president's party to suffer losses in the next election.

12. "Those thinkers in whom all stars move in cyclic orbits are not the most profound. Whoever looks into himself as into vast space and carries galaxies in himself, also knows how irregular all galaxies are; they lead into the chaos and labyrinth of existence."

Friedrich Nietzsche

▲ 13. "Religion . . . is the *opium* of the people. To abolish religion as the *illusory* happiness of the people is to demand their *real* happiness."

Karl Marx

14. "Publishing is to thinking as the maternity ward is to the first kiss."

Friedrich Von Schlegel

15. She's not particularly good at tennis, so I doubt she'd be good at racquetball.

▲ 16. "A book is like a mirror. If an ape looks in, a saint won't look out."

Ludwig Wittgenstein

17. As odd as it sounds, historically the stock market goes up when there is bad news on unemployment, and the latest statistics show the unemployment rate is skyrocketing. This could be a good time to buy stocks.

18. I never met anyone from North Carolina who didn't have an accent thicker than molasses.

▲ 19. Yamaha makes great motorcycles, so I'll bet their pianos are pretty good, too.

20. The last time we played Brazil in soccer we got run through the washer and hung out to dry.

Exercise 10-2 ——————————————

In groups (or individually if the instructor prefers), decide whether each of the following is (a) an analogical argument or (b) an inductive generalization.

▲ 1. Paulette, Georgette, Babette, and Brigitte are all Miami University English majors, and they are all atheists. Therefore, probably all the English majors at Miami University are atheists.

2. Paulette, Georgette, Babette, and Brigitte are all Miami University English majors, and they are all atheists. Therefore Janette, who is a Miami University English major, probably is also an atheist.

3. Gustavo likes all the business courses he has taken at Harvard to date. Therefore, he'll probably like all the business courses he takes at Harvard.

▲ 4. Gustavo likes all the business courses he has taken at Harvard to date. Therefore, he'll probably like the next business course he takes at Harvard.

5. Gustavo likes all the business courses he has taken at Harvard to date. Therefore, his brother, Sergio, who also attends Harvard, will probably like all the business courses he takes there, too.

6. Forty percent of Moore's 8:00 A.M. class are atheists. Therefore, 40 percent of all Moore's students are atheists.

▲ 7. Forty percent of Parker's 8:00 A.M. class are atheists. Therefore, probably 40 percent of Parker's 9:00 A.M. class are atheists.

8. Forty percent of Moore's 8:00 A.M. class are atheists. Therefore, probably 40 percent of all the students at Moore's university are atheists.

9. Bill Clinton lied to the American public about his relationship with Monica Lewinsky; therefore, he probably lied to the American public about Iraq, too.

▲ 10. Bill Clinton lied to the American public about his relationship with Monica Lewinsky; therefore, he probably lied to the American public about most things.

Exercise 10-3

In groups (or individually, if the instructor prefers), decide whether each of the following is (a) an analogical argument or (b) an inductive generalization.

▲ 1. With seven out of the last nine El Niños, we saw below-average rainfall across the northern United States and southern Canada. Therefore, chances are we'll see the same with the next El Niño.

2. I've been to at least twenty Disney movies in my lifetime, and not one of them has been especially violent. I guess the Disney people just don't make violent movies.

3. Most of my professors wear glasses; it's a good bet most professors everywhere do the same.

▲ 4. Seems like the Christmas decorations go up a little earlier each year. I bet next year they're up by Halloween.

5. The conservatives I've met dislike Colin Powell. Based on that, I'd say most conservatives feel the same way.

6. FIRST PROF: I can tell after just one test exactly what a student's final grade will be.

SECOND PROF: How many tests do you give, just a couple?
FIRST PROF: As a matter of fact, I give a test every week of the semester.

▲ 7. MRS. BRUDER: Bruder! Bruder! Guess what! The music department is selling two of their grand pianos!
MR. BRUDER: Well, let's check it out, but remember the last pianos they
sold were way overpriced. Probably it'll be the same this time.

8. A 60 percent approval rating? Those polls are rigged by the liberal media.
Most of the people I know say the man ought to be impeached.

9. The New England Patriots have won six of their last seven home games.
They're playing at home on Sunday. Don't bet against them.

▲ 10. You're going out with a Georgette? Well, don't expect much because I've
known three Georgettes and all of them have been stuck up, spoiled yuck-
heads. I'll bet this one will be a yuckhead, too.

Exercise 10-4

Go through the preceding exercise and for each item identify (1) the sample,
(2) the target or target class (population), and (3) the feature (property) in question.

Exercise 10-5

In groups (or individually, if the instructor prefers), determine in which of the following inductive generalizations or analogical arguments the sample, the target
or target class, or feature (property) in question is excessively vague or ambiguous.

▲ 1. The tests in this class are going to be hard, judging from the first midterm.

2. Judging from my experience, technical people are exceedingly difficult to
communicate with sometimes.

3. The transmissions in 1999 Chrysler minivans tend to fail prematurely, if
my Voyager is any indication.

▲ 4. Women can tolerate more stress than can men. My husband even freaks
when the newspaper is a little late.

5. Movies are much too graphic these days. Just go to one—you'll see what
I mean.

6. Violent scenes in the current batch of movies carry a message that degrades
women. Just go see any of the movies playing downtown right now.

▲ 7. You'll need to get better clothing than that if you're going to Iowa. In my
experience, the weather there sucks much of the year.

8. Entertainment is much too expensive these days, judging from the cost
of movies.

9. Art majors sure are weird! I roomed with one, once. Man.

▲ 10. Artsy people bug me. They're all like, aaargh, you know what I mean?

11. The French just don't like Americans. Why, in Paris, they won't give you the time of day.

12. All the research suggests introverts are likely to be well-versed in computer skills.

▲ 13. Suspicious people tend to be quite unhappy, judging from what I've seen.

Exercise 10-6

How well does the sample in each of the following arguments represent the target or target class? In groups (or individually), choose from the following options:

a. Very well

b. Pretty well

c. Not very well

d. Very poorly

▲ 1. The coffee in that pot is lousy—I just had a cup.

2. The coffee at that restaurant is lousy—I just had a cup.

3. The food in that restaurant is lousy—I just ate there.

▲ 4. This student doesn't write very well, judging from how poorly she wrote on the first paper.

5. Terrence didn't treat her very well on their first date; I can't imagine he will treat her better next time.

6. I expect I'll have trouble sleeping after this huge feast. Whenever I go to bed on a full stomach, I never sleep very well.

▲ 7. Lupe's sister and mother both have high blood pressure; chances are Lupe does too.

8. Every movie Courtney Cox Arquette has been in has been great! I bet her next movie will be great, too!

▲ 9. Women don't play trombones; leastwise, I never heard of one who did.

10. I'm sure Blue Cross will cover that procedure because I have Blue Cross and they covered it for me.

11. Cocker spaniels are nice dogs, but they eat like pigs. When I was a kid, we had this little cocker that ate more than a big collie we had.

▲ 12. Cadillacs are nice cars, but they are gas hogs. When I was a kid, we owned this Cadillac that got around ten miles per gallon.

13. Cockatoos are great birds, but they squawk. A lot. Had one when I was a kid, and even the neighbors complained, it was so raucous.

14. The parties at the university always turn into drunken brawls. Why, just last week the police had to go out and break up a huge party at Fifth and Ivy.

Exercise 10-7

Arrange the alternative conclusions of the following arguments in order of increasing cautiousness and then compare your rankings with those of three or four classmates. Some options are pretty close to being tied. Don't get in feuds with classmates or the teacher over close calls.

▲ 1. Not once this century has this city gone Republican in a presidential election. Therefore . . .
   a. I wouldn't count on it happening this time.
   b. it won't happen this time.
   c. in all likelihood it won't happen this time.
   d. there's no chance whatsoever it will happen this time.
   e. it would be surprising if it happened this time.
   f. I'll be a donkey's uncle if it happens this time.

2. Byron doesn't know how to play poker, so . . .
   a. he sure as heck doesn't know how to play blackjack.
   b. it's doubtful he knows how to play blackjack.
   c. there's a possibility he doesn't know how to play blackjack.
   d. don't bet on him knowing how to play blackjack.
   e. you're nuts if you think he knows how to play blackjack.

3. Every time I've used the Beltway, the traffic has been heavy, so I figure that . . .
   a. the traffic is almost always heavy on the Beltway.
   b. frequently the traffic on the Beltway is heavy.
   c. as a rule, the traffic on the Beltway is heavy.
   d. the traffic on the Beltway can be heavy at times.
   e. the traffic on the Beltway is invariably heavy.
   f. typically, the traffic on the Beltway is heavy.
   g. the traffic on the Beltway is likely to be heavy most of the time.

Exercise 10-8

In which of the following arguments is the conclusion insufficiently cautious or overly cautious or neither, given the premises? After you have decided, compare your results with those of three or four classmates.

▲ 1. We spent a day on the Farallon Islands last June, and was it ever foggy and cold! So dress warmly when you go there this June. Based on our experience, it is 100 percent certain to be foggy and cold.

2. We've visited the Farallon Islands on five different days, two during the summer and one each during fall, winter, and spring. It's been foggy and cold every time we've been there. So dress warmly when you go there. Based on our experience, there is an excellent chance it will be foggy and cold whenever you go.

3. We've visited the Farallon Islands on five different days, all in June. It's been foggy and cold every time we've been there. So dress warmly when you go there in June. Based on our experience, it could well be foggy and cold.

▲ 4. We've visited the Farallon Islands on five different days, all in June. It's been foggy and cold every time we've been there. So dress warmly when you go there in June. Based on our experience, there is a small chance it will be foggy and cold.

5. We've visited the Farallon Islands on five different days, all in January. It's been foggy and cold every time we've been there. So dress warmly when you go there in June. Based on our experience, it almost certainly will be foggy and cold.

Exercise 10-9

For the past four years, Clifford has gone on a hundred-mile bicycle ride on the Fourth of July. He has always become too exhausted to finish the entire hundred miles. He decides to try the ride once again, but thinks, "Well, I probably won't finish it this year, either, but no point in not trying."

▲ 1. Does the conclusion as expressed seem appropriately cautious given the size of the sample and its similarity to the target?

2. Suppose the past rides were done in a variety of different weather conditions. With this change, would the argument be stronger than the original, weaker, or about the same?

3. Suppose that Clifford is going to ride the same bike this year that he's ridden in all the previous rides. Does this make the argument stronger than the original version, weaker, or neither?

▲ 4. Suppose the past rides were all done on the same bike, but that bike is not the bike Clifford will ride this year. Does this make the argument stronger than the original version, weaker, or neither?

5. Suppose Clifford hasn't yet decided what kind of bike to ride in this year's ride. Now, is the argument made stronger or weaker or neither if the past rides were done on a variety of different kinds of bike (e.g., road bikes, mountain bikes, racing bikes, etc.)?

6. Suppose the past rides were all done on flat ground and this year's ride will also be on flat ground. Does this make the argument stronger than the original version, weaker, or neither?

7. Suppose the past rides were all done on flat ground and this year's ride will be done in hilly territory. Does this make the argument stronger than the original version, weaker, or neither?

▲ 8. Suppose Clifford doesn't know what kind of territory this year's ride will cover. Is his argument made stronger or weaker or neither if the past rides were done on a variety of kinds of territory?

9. Suppose the past rides were all done in hilly territory and this year's ride will be done on flat ground. Does this make the argument stronger than the original version, weaker, or neither?

▲ 10. In answering the preceding item, did you take into consideration information you have about bike riding in different kinds of terrain or did you consider only the criteria for evaluating inductive arguments in general?

Exercise 10-10 _____

During three earlier years Kirk has tried to grow artichokes in his backyard garden, and each time his crop has been ruined by mildew. Billie prods him to try one more time, and he agrees to do so, though he secretly thinks: "This is probably a waste of time. Mildew is likely to ruin this crop, too." (Note: To say that the new information warrants a more cautious conclusion is the same as saying the new information makes the argument weaker. Similarly for "less cautious" and "stronger.")

▲ 1. Suppose this year Kirk plants the artichokes in a new location. Does this information warrant a more cautious conclusion or less cautious conclusion than the original?

2. Suppose on the past three occasions Kirk planted his artichokes at different times of the growing season. Does this information warrant a more cautious conclusion or less cautious conclusion than the original?

3. Suppose this year Billie plants marigolds near the artichokes. Does this information warrant a more cautious conclusion or less cautious conclusion than the original?

▲ 4. Suppose the past three years were unusually cool. Does this information warrant a more cautious conclusion or less cautious conclusion than the original?

5. Suppose only two of the three earlier crops were ruined by mildew. Does this information warrant a more cautious conclusion or less cautious conclusion than the original?

6. Suppose one of the earlier crops grew during a dry year, one during a wet year, and one during an average year. Does this information warrant a more cautious conclusion or less cautious conclusion than the original?

▲ 7. Suppose this year, unlike the preceding three, there is a solar eclipse. Does this information warrant a more cautious conclusion or less cautious conclusion than the original?

8. Suppose this year Kirk fertilizes with lawn clippings for the first time. Does this information warrant a more cautious conclusion or less cautious conclusion than the original?

9. Suppose this year Billie and Kirk acquire a large dog. Does this information warrant a more cautious conclusion or less cautious conclusion than the original?

▲ 10. Suppose this year Kirk installs a drip irrigation system. Does this information warrant a more cautious conclusion or less cautious conclusion than the original?

Exercise 10-11

Every student I've met from Ohio State believes in God. My conclusion is that most of the students from Ohio State believe in God.

▲ 1. Suppose (as is the case) that Ohio State has no admission requirements pertaining to religious beliefs. Suppose further the students in the sample were all interviewed as they left a local church after Sunday services. Does this supposition warrant a more cautious conclusion or less cautious conclusion than the original?

2. Suppose all those interviewed were first-year students. Does this supposition warrant a more cautious conclusion or less cautious conclusion than the original?

3. Suppose all students interviewed were on the Ohio State football team. Does this supposition warrant a more cautious conclusion or less cautious conclusion than the original?

▲ 4. Suppose the speaker selected all the students interviewed by picking every tenth name on an alphabetical list of students' names. Does this supposition warrant a more cautious conclusion or less cautious conclusion than the original?

5. Suppose the students interviewed all responded to a questionnaire published in the campus newspaper titled "Survey of Student Religious Beliefs." Does this supposition warrant a more cautious conclusion or less cautious conclusion than the original?

6. Suppose the students interviewed were selected at random from the record office's list of registered automobile owners. Does this supposition warrant a more cautious conclusion or less cautious conclusion than the original?

Exercise 10-12

Read the passage below, and answer the questions that follow.

> In the Georgia State University History Department, students are invited to submit written evaluations of their instructors to the department's personnel committee, which uses those evaluations to help determine whether history instructors should be recommended for retention and promotion. In his three history classes, Professor Ludlum has a total of one hundred students. Six students turned in written evaluations of Professor Ludlum; four of these evaluations were negative, and two were positive. Professor Hitchcock, who sits on the History Department Personnel Committee, argued against recommending Ludlum for promotion. "If a majority of the students who bothered to evaluate Ludlum find him lacking," he stated, "then it's clear to me that a majority of all his students find him lacking."

▲ 1. What is the sample in Professor Hitchcock's reasoning?

2. What is the target or target class?

3. What is the property in question?

▲ 4. Is this an analogical argument or a generalization?

5. Are there possibly important differences between the sample and the target (or target class) that should reduce our confidence in Professor Hitchcock's conclusion?

▲ 6. Is the sample random?

7. How about the size of Professor Hitchcock's sample? Is it large enough to help ensure that the sample and target or target classes won't be too dissimilar?

▲ 8. Based on the analysis of Professor Hitchcock's reasoning that you have just completed in the foregoing questions, how strong is his reasoning?

Exercise 10-13

George's class is doing a survey of student opinion on social life and drinking. One of the questions is "Do you believe that students who are members of fraternities and sororities drink more alcohol than students who are not members of such organizations?" Let's say that George would wager a large sum on a bet only if he had a 95 percent chance of winning. Answer the following questions for him. (Use Table 10-1 on page 368 to help you with this exercise.)

▲ 1. Exactly 60 percent of a random sample of 250 students at George's college believe that students in fraternities and sororities drink more than others. Should George bet that exactly 60 percent of the student population hold the same belief?

2. Should he bet that *at least* 60 percent hold that belief?

3. Should he bet that *no more than* 66 percent hold the belief?

▲ 4. According to the text, the largest number that he can safely bet hold the belief mentioned is _____ percent.

5. Let's change the story a bit. Let's make the sample size 100 and once again say exactly 60 percent of it hold the belief in question. Should George bet that at least 55 percent of the total student population hold the belief?

6. Should he bet that at least 50 percent of the total hold it?

▲ 7. It is safe for George to bet that no more than _____ percent of the total population share the belief in question. What's the smallest number that can be placed in the blank?

Exercise 10-14

Critically analyze these arguments by analogy. Do this by specifying what things are compared in the analogy and then identifying important relevant dissimilarities between these things. If any of the arguments seem pretty strong— that is, if the dissimilarities are not relevant or particularly important—say so.

Example

> Earth and Venus are twin planets; both have atmospheres that protect them from the harsh environment of outer space, both are about the same size and exert similar gravitational force, both have seasons, both have days and nights, and there are other similarities. Earth has life; therefore, Venus may well have life, too.

Analysis

1. The terms of the analogy are Earth and Venus.
2. Important relevant dissimilarities are that Venus's atmosphere is composed of 96 percent carbon dioxide, whereas Earth's contains about 0.03 percent. Also, the surface temperature of Venus is higher than the boiling point of lead.

▲ 1. Tiger Woods began playing golf at the age of two, and look what he's accomplished. I'm going to start golf lessons for my two-year-old, too, by golly.

2. A household that doesn't balance its budget is just asking for trouble. It's the same with the federal government. Balance the federal budget, or watch out.

3. Saccharin has been determined to cause cancer in rats; in similar doses, it's likely to do so in humans as well.

▲ 4. Senator Clinton has been an excellent senator for New York; therefore, she would make an excellent president.

5. Ross Perot really knows how to run a business; therefore, he'd have made an excellent president.

6. There's no way I can actually feel other people's pain or experience their fear or anxiety. Nevertheless, I can be quite sure other people experience these things because they behave as I do when I experience them. Therefore, I can be equally sure animals experience these things when they behave in a similar fashion.

▲ 7. Bush has an 80 percent approval rating in Georgia. It's safe to assume he is equally popular in Massachusetts.

8. Suppose you had never seen a clock before and one day you found one lying on the ground somewhere. You'd be certain it had been made by an intelligent being who was up to the task of creating such a thing. Now think of the Earth. It displays a far more complex design than a clock. It is safe to conclude that it, too, just like a clock, is the product of an intelligent creator, and one up to such a momentous task as creating a world. This is a very good reason to think that God exists.

9. Trying to appease Adolf Hitler was totally counterproductive; trying to appease Saddam Hussein would likely have had similar results.

▲ 10. The Barneses are traveling to Europe for a year and decide to find a student to house-sit for them. They settle on Warren because he is neat and tidy in his appearance. "If he takes such good care of his personal appearance," Mrs. Burns thinks, "he's likely to take good care of our house, too."

11. Abortion consists in killing a living person. If abortion is wrong, therefore so is capital punishment.

Exercise 10-15 _____

Identify any fallacies that are present in the following passages:

▲ 1. From a letter to the editor: "I read with great interest the May 23 article on the study of atheists in federal prisons, according to which most of these atheists identify themselves as socialists, communists, or anarchists. That most atheists, along with all their other shortcomings, turn out to be political wackos surprises yours truly not one bit."

2. I ordered a packet of seeds from Hansen Seed Company last year, and only half of them germinated. I'll bet you get only half the plants you're expecting from the order you just sent them.

3. My cousin has a Dodge truck that he drives around on the ranch and back and forth to town as well. It now has 120,000 miles on it without any major overhaul. I've started to believe the commercials: Dodge does build tough trucks!

▲ 4. Drug abuse among professional athletes is a serious and widespread problem. Three players from a single team admitted last week that they had used cocaine.

5. Most Americans favor a national lottery to reduce the federal debt. In a poll taken in Las Vegas, more than 80 percent said they favored such a lottery.

6. I think collies are one of the easiest breeds of dog to train. I had one when I was young, and she was really easy to teach. Of course, if you need more evidence, there's Lassie.

7. From a letter to the editor: "Last week, members of the Animal Liberation Front stole several hours of videotapes of experiments done to animals at the University of Pennsylvania's Head Injury Clinical Research Center. According to reliable reports, one of the tapes shows baboons having their brains damaged by a piston device that smashed into their skulls with incredible force. The anesthetic given the baboons was allegedly insufficient to prevent serious pain. Given that this is what animal research is all about, Secretary of Health and Human Services Margaret Heckler acted quite properly in halting federal funding for the project. Federal funding for animal research ought to be halted, it seems to me, in the light of these atrocities."

▲ 8. Overheard: "You're not going to take a course from Harris, are you? I know at least three people who say he's terrible. All three flunked his course, as a matter of fact."

9. A majority of Ohio citizens consider the problem of air pollution critical. According to a survey taken in Cleveland, more than half the respondents identified air pollution as the most pressing of seven environmental issues and as having either "great" or "very great" importance.

▲ 10. Comment: What are we getting for the billions we spend on these new weapons? Absolutely nothing. Just look at the Apache helicopter. I read where they can't fly one of them for more than ten hours without grounding it for major repairs.

11. The IRS isn't interested in going after the big corporations, just middle-class taxpayers like you and me. I was audited last year, and I know several people who have been audited. You ever hear of ExxonMobil getting nailed?

12. I am totally outraged by the ethics of our elected representatives. Just look around: Bill Clinton accepted illegal campaign donations; Newt Gingrich was fined $300,000 for misusing money. I tell you, they are all just a bunch of crooks.

▲ 13. When he was president, Ronald Reagan occasionally would wave a copy of the "Help Wanted" section of some newspaper to demonstrate that U.S. unemployment wasn't really so bad. What was the fallacy in this "argument"?

Exercise 10-16

1. In the text, it is said that a person playing roulette has a disadvantage of about 5 percent compared with the house. Explain what this means.

2. Choose two or three other gambling bets for which you can calculate or look up the odds against the bettor. (Stick with games of pure chance; those that can involve a bit of skill, such as twenty-one—also known as blackjack—are much more difficult to deal with.) Determine which bet gives the house the smallest advantage.

3. Choose a state lottery and determine the odds against winning for a given ticket. Can you determine what kind of advantage the "house" has in such a gamble?

Exercise 10-17

1. Explain in your own words what the law of large numbers says.

2. "Over the long term, human births will approximate 50 percent males and 50 percent females." Is this a reasonable application of the law of large numbers? What facts can you think of that are relevant to determining whether this is a reasonable application of the law?

3. A person is betting on flips of a coin. He always bets on heads. After he loses three in a row, you hear him say, "Okay, I'm due for one now!" as he raises his bet. Write a brief note in which you give him some friendly advice.

Writing Exercises

1. Which of the following general claims do you accept? Select one that you accept and write a one-page essay presenting evidence (giving arguments) for the claim. When you are finished, write down on a separate piece of paper a number between 1 and 10 that indicates how strong you think your overall "proof" of the general claim is, with 10 = very strong and 1 = very weak. Take about two minutes to complete your essay. Write your name on the back of your paper.

    General claims:

    You get what you pay for.
    Nice guys finish last.
    Everything else being equal, men prefer blondes.
    Women are more gentle and nurturing than men.
    Politicians are untrustworthy.
    Government intrudes into our private lives/business affairs too much.
    Too many welfare recipients don't deserve assistance.
    College teachers are liberals.
    Jocks are dumb.
    The superwealthy pay less in taxes than they should.

2. When everyone is finished, the instructor will collect the papers and redistribute them to the class. In groups of four or five, read the papers and assign a number from 1 to 10 to each one (10 = very strong; 1 = very weak). When all groups are finished, return the papers to their authors. When you get your paper back, compare the number you assigned to your work with the number the group assigned to it. The instructor may ask volunteers to defend their own judgment of their work against the judgment of the group. Do you think there is as much evidence for the claim you selected as you did when you argued for it initially?

Chapter 11

Causal Arguments

A few months ago Uncle Pete (an actual uncle of one of your authors) came down with pneumonia. His condition was life-threatening. Aunt Clara prayed for Pete—and, we are happy to report, Uncle Pete recovered. Did Aunt Clara's prayers help? Aunt Clara is certain they did. ("Can you prove they didn't?" she asks.) Pete's doctor is keeping his thoughts to himself.

We don't know whether Aunt Clara's prayers helped Uncle Pete—and, in fact, neither does Aunt Clara. We know only that Aunt Clara prayed and then Uncle Pete recovered. Knowing that, however, is not the same as knowing that the prayers had anything at all to do with the recovery.

Thinking that X being followed by Y means that X caused Y is a common mistake in reasoning. It's known as the **post hoc fallacy** and is one of the most common mistakes people make.

"Post hoc fallacy" is short for the Latin phrase *post hoc, ergo propter hoc,* which translates as "After that, therefore because of that," or to put it a bit more loosely, "This thing happened right after that other thing, so this thing was caused by that other thing." Whenever we reason this way, we make that mistake. Uncle Pete's cure happened when, or just after, Aunt Clara prayed; therefore the praying caused the cure—this is post hoc reasoning.

That this kind of reasoning is a mistake should be obvious. Moore's birthday happened just before Parker's, but Moore's birthday didn't cause Parker's.

On the other hand, suppose Howard topples over with a heart attack immediately after sprinting up two flights of stairs. Doesn't the fact that the heart attack came on the heels of the sprint suggest the one thing caused the other? The answer is no. After all, the heart attack also came on the heels of Howard's sweating heavily, and it would be peculiar to suppose on these grounds that his sweating caused his heart attack. Something else obviously needs to be considered before we can say that the sprinting caused the heart attack. What other things? To answer this, we need to look at the principles involved when we reason our way to a statement about cause and effect.

Informal Causal Reasoning

A **causal claim** (or **cause-and-effect claim**) states or suggests the presence (or absence) of causation. An example is worth several words, so here are examples of causal claims:

- ▲ Aunt Clara's praying helped Uncle Pete recover from pneumonia.
- ▲ At high concentrations, some enzymes damage healthy tissue.
- ▲ Pressing the two surfaces together produced a permanent bond.
- ▲ Dnase significantly improves lung function.
- ▲ Vitamin C does not cure colds.
- ▲ Fertilizing with Scott's didn't do anything for my lawn.
- ▲ Sprinting caused Howard to have a heart attack.

A statement to the effect that X causes or caused Y can, of course, be offered as a hypothesis rather than as a claim: A **hypothesis** is a supposition offered as a starting point for further investigation. When you hypothesize, you aren't actually claiming something, you are conjecturing (which is a rather fancy way to say you're making a guess). However, the reasoning that supports a claim and the reasoning that tests a hypothesis is the same, so, for practical purposes we can treat causal claims and **causal hypotheses** as the same kind of thing.

Two Basic Patterns of Causal Reasoning

When we assert (or hypothesize) that one thing causes or caused another, our reasoning often will rely on either (or some combination) of two distinct patterns of inference: relevant-difference reasoning and common-thread reasoning.

■ Relevant-Difference Reasoning

If your prized azaleas aren't doing well this year, and you want to know why, you will ask yourself, What else is different this year? If a single relevant difference stands out about this year's azaleas (in addition to their not doing well), you naturally suspect it had something to do with the poor bloom. For example, if this year was hotter and drier than previous years, you would

The Spotted Owl

After logging in old-growth forests was halted in the Pacific Northwest to save the spotted owl, the general economy in that area went into a dramatic tailspin. To conclude from this fact that the measures taken to save the spotted owl *caused* the economic decline is to be guilty of *post hoc* reasoning. To have caused the economic decline, these measures would have to have been the *only relevant thing to happen before the decline*. (In fact, the most severe economic recession since the Great Depression also happened at the same time, throughout the American economy. Maybe the attempt to save the spotted owl caused that, too.)

naturally suspect that the hot, dry weather was involved causally with the poor bloom.

In other words, if an effect is present in one situation and it isn't present in other, similar, situations, we try to find the **relevant difference** between the situation in which the effect is present and the situations in which it isn't. If we find a relevant difference, we suspect that it caused (or was a part of the cause of) the effect we are interested in.

If, for example, the cancer rate in one neighborhood is higher than in others, you want to know what's different about that neighborhood. Is there a former toxic dump nearby? Are chemical solutions present for some reason? If so, we will think that the dump or solutions have something to do with the high cancer rate. If you wake up one morning with a splitting headache and you remember having done something different the night before (reading for several hours in poor light, for example), you will suspect that this variation in behavior caused the headache.

It isn't just bad effects we are interested in. If somebody who hasn't been especially friendly starts going out of his way to be nice, you might wonder what caused the change. Again, you consider what's different about this situation that would explain why Trent, or whoever it is, suddenly wants to be friendly. Did you just get promoted and are now Trent's supervisor? You might well suspect that fact had something to do with the change.

In each instance, an effect has occurred that didn't occur in other, similar, situations. *If there is some important relevant difference between this situation, in which the effect occurred, and the other situations, in which it didn't, we suspect that difference played a causal role in the effect we are concerned about.* The fact that Howard had a heart attack after sprinting up two flights of stairs doesn't give us a real reason for suspecting the sprinting caused the heart attack. If we have legitimate grounds for that suspicion, it is that the sprinting is the only difference between this occasion, on which Howard had a heart attack, and other occasions, on which he did not. Likewise, the fact that Pete recovered after Aunt Clara prayed does not in itself show that her praying played a part in his recovery. You need to know that her praying was the only relevant difference between the time when he was sick and the time when he was well.

■ *Common-Thread Reasoning*

Now, a different pattern of reasoning is illustrated by considering what happens when you have *multiple* occurrences of some effect. For example, what happens when a group of people all come down with the same mysterious health symptoms such as those that accompany SARS? The first thing health officials do is look for a **common thread** among the victims. Did they all breathe the same air? Were they all present in the same location? Is there some person they all contacted? Common threads are immediately suspicious as possible causes of the illness.

Or let's suppose some years your azaleas bloom prolifically and other years they don't. You will look for a common thread among the years they bloom well (or among the years they don't). For example, if, whenever they bloomed well you always pruned in the fall rather than the winter, you'd suspect time of pruning might have something to do with quality of bloom.

Or let's suppose every Saturday evening during the summer the mosquitoes in the backyard are present in hideous numbers but on other evenings they aren't. What is it about Saturday evening? you should wonder. What do Saturday evenings have in common that might be relevant to mosquitoes? For example, do you mow the grass every Saturday in the early evening? If so, you should suspect that is the explanation.

Common-thread reasoning and relevant-difference reasoning are similar, so let's summarize them abstractly:

RELEVANT-DIFFERENCE REASONING:

If an effect is present in one situation but not present in other similar situations, any relevant differences between the situation where the effect is present and the other situations where it isn't are suspicious as the cause (or part of it) of the effect in question.

COMMON-THREAD REASONING

If an effect is present in multiple situations, any common thread that unites those situations in which the effect is present are suspicious as the cause (or part of it) of the effect in question.

Common-thread reasoning is best suited as a method for forming hypotheses to be tested in some other way, usually involving relevant-difference reasoning. That's because the multiple occurrences of the effect could always have resulted from different causes. In other words, coincidence may have

"Teen Smoking Surge Reported After Joe Camel's 1988 Debut"

So said the Associated Press headline that accompanied this graphic. The article, however, stated that the Centers for Disease Control *attributed* the surge in part to kid-friendly cartoon advertising like the Joe Camel ads. It also noted that then-president Clinton had blamed the increase on the Joe Camel ad campaign.

It is true that the CDC studies showed that the rate for beginning smokers had been *declining* steadily for over a decade before it started increasing again in 1988. It is also true that *something* caused the rate to start increasing again, beginning in 1988. And finally, it is true that R. J. Reynolds introduced *Joe Camel* in its advertising for Camel cigarettes in 1988.

But *this* information, while suspicious, does not show that the increase was *caused*, even in part, *by* the ad campaign. For example, reportedly there was a notable increase in the appearance of cigarette smoking in movies beginning about then, too. As it stands, this is *post hoc, ergo propter hoc.*

Rise in youth smoking
The incidence of youths becoming regular smokers jumped 50 percent between 1988 and 1996. A look at the rate of youths who started smoking daily, age 12–17:

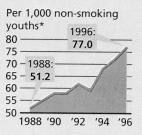

Per 1,000 non-smoking youths*

*Considered non-smokers at start of year
Source: Centers for Disease Control and Prevention

played a role—maybe Sally got sick for one reason, Tony got sick for an entirely unrelated reason, Cisco for some third reason, and so forth. Also, multiple occurrences of an effect are likely to have many other things in common besides whatever it was that caused them—for example, Tony and Sally and Cisco might all be friends, might have traveled together in the same car, and so on.

Relevant-difference reasoning, on the other hand, can be reasonably conclusive—at least in experimental conditions. If a piece of litmus paper turns red when you dip it in a liquid, it is safe to think the liquid caused the alteration. That's because the litmus paper prior to dipping and the litmus paper during dipping were exactly the same, except for the liquid. This may sound like post hoc reasoning, but it isn't: The reason for saying the liquid caused the litmus paper to turn red isn't just that it turned red after touching the liquid. The reason is that touching the liquid was the only relevant difference that could explain the paper's turning red.

Even apart from carefully controlled experimental conditions, relevant-difference reasoning yields conclusions that are certain by everyday standards of rigor. If the car won't start immediately after you installed a new electrical

According to a report in the Journal of the American Medical Association, *infants who are breastfed have higher IQs later in life.*

—On the other hand, maybe parents with higher IQs are more aware of the health advantages of breastfeeding and as a result are more likely to breastfeed their children.

device, and you are sure the before-installation and after-installation conditions are the same except for the device and the failure to start, you are right to think the device or its installation is the problem.

Common Mistakes in Informal Causal Reasoning

When we use relevant-difference or common-thread reasoning, we can make mistakes. Those that follow are pretty common. The strength of the reasoning depends on the extent to which we can eliminate these mistakes. Post hoc reasoning essentially involves a failure to eliminate the following mistakes.

▲ *We can overlook alternative common threads or differences.* Let's say the azaleas don't bloom well this spring. Well, we failed to fertilize the azaleas last winter; but we must be careful not to focus just on that and, as a result, overlook other relevant differences— for example, the fact that we also failed to prune them last winter.

▲ *We can focus on irrelevant differences or common threads.* What if, this year, we read two new books on ornamental horticulture? This is a difference between this year and previous years all right, but it isn't a *relevant* difference if we don't actually *do* anything different to our azaleas. Just *reading* about them plays no role in whether or not they bloom. What do we mean when we say a difference or thread is relevant? We mean it is not unreasonable to suppose the difference might have caused the feature in question. Obviously, the more you know about a subject, the better able you are to say whether a given difference is relevant. If you knew nothing about anything, then you would have no idea what is relevant and what isn't.

▲ *We can overlook the possibility that causation is the reverse of what has been asserted.* Did his fall cause the rope to break? Or did the rope's breaking cause his fall? Does your back ache because of your poor posture? Or is your poor posture the result of your aching back?

▲ *We can overlook the possibility that the stated cause and the stated effect are both effects of some third, underlying cause.* For example: Did Sarah become nauseated because of the hot peppers she ate? Maybe, but if Sarah is pregnant, maybe *that* caused her to want hot peppers and also caused her to be nauseated.

▲ *We can fail to consider the possibility of coincidence.* This is especially likely in common-thread reasoning. For example, if your throat, head, and chest all hurt, it is reasonable to look for a common thread. But possibly the pains result from separate and unrelated

See What Happens If You Watch the Tube?

Recent research indicates that people who watch several hours of TV each day, as compared with those who don't watch very much, express more racially prejudiced attitudes, perceive women as having more limited abilities and interests than men, overestimate the prevalence of violence in society, and underestimate the number of old people. Does this suggest to you that these attitudes and misconceptions are the *result* of watching a lot of TV, that TV *causes* people to have these attitudes and misconceptions? If so, that's fine. But remembering not to overlook possible underlying common causes should lead you to contemplate another idea. It's possible that what accounts for these attitudes and misconceptions isn't too much TV, but rather *ignorance,* and that ignorance is also what makes some people happy spending hours in front of the tube. An underlying common cause?

causes: In other words, possibly there is no relevant common thread. Coincidences happen all the time, and they are frequently taken as cause and effect when they are nothing of the sort.

If we fail to take these possibilities into consideration, we are apt to be left thinking A caused B merely because A happened before B. In other words, if we don't consider these possibilities, we are likely to commit the post hoc fallacy.

Exercise 11-1

Which of the following are causal claims (hypotheses)?

▲ 1. Webber's being healthy is probably what made the Kings so much better this year.

2. Dress more warmly! It's windy out there.

3. Gilbert's disposition has deteriorated since he and his wife separated; it isn't just coincidence.

▲ 4. When Senator Lott praised Strom Thurman, that forced some conservatives to call for his resignation.

5. Getting that new trumpet player certainly improved the brass section.

6. When men wear a swimsuit, they have difficulty doing math problems.

▲ 7. Despite all the injuries, the Dolphins manage to keep winning. It must have something to do with their positive attitude.

8. Too little sleep will slow down your reaction time.

9. Women are worse drivers than men.

Carrots Can Kill!

A few orange and green ruminations about life, death and vegetables

from the Miner Institute, Chazy, New York

Nearly all sick people have eaten carrots. Obviously, the effects are cumulative.

An estimated 99.9 percent of all people who die from cancer have eaten carrots.

Another 99.9 percent of people involved in auto accidents ate carrots within 60 days before the accident.

Some 93.1 percent of juvenile delinquents come from homes where carrots are served frequently.

Among the people born in 1839 who later dined on carrots, there has been a 100 percent mortality rate.

All carrot eaters born between 1900 and 1910 have wrinkled skin, brittle bones, few teeth, and failing eyesight . . . if the perils of carrot consumption have not already caused their deaths.

—Priorities *magazine, Summer 1990*

What errors in causal reasoning can you find in these "arguments"?

▲ 10. Randomized clinical trials produce unbiased data on the benefits of drugs.

11. The batteries in that flashlight are completely dead.

12. The reason that flashlight won't work is that the batteries are completely dead.

▲ 13. Believe me, the batteries in that flashlight are dead. Try it, and you'll see it doesn't work.

14. Sandra's true reason for running for treasurer is she likes all the attention.

15. Aunt Clara thinks her prayers cured Uncle Pete.

▲ 16. The risk of having a heart attack is 33 percent higher in the winter than in the summer, even in Los Angeles.

Exercise 11-2

What is the cause and what is the effect in each of the following?

▲ 1. The cat won't eat, so Mrs. Quibblebuck searches her mind for a reason. "Now, could it be," she thinks, "that I haven't heard mice scratching around in the attic lately?" "That's the explanation," she concludes.

2. Each time one of the burglaries occurred, observers noticed a red Mustang in the vicinity. The police, of course, suspect the occupants as being responsible.

The Great 9/11 Mystery

How could all these facts be mere coincidence?

- The day was 9/11 (9 plus 1 plus 1 equals 11).
- American Airlines Flight 11 was the first to hit the World Trade Center.
- 92 people were on board (9 plus 1 plus 1 equals 11).
- September 11 is the 254th day of the year (2 plus 5 plus 4 equals 11).
- "New York City" has 11 letters in it.
- "The Pentagon" has 11 letters in it.
- "Saudi Arabia" (where most of the 9/11 terrorists were from) has 11 letters in it.
- "Afghanistan" has 11 letters in it.
- And get this: Within eleven months of September 11, 2001, eleven men, all connected to bioterror and germ warfare, died in strange and violent circumstances: One suffocated; another was stabbed, another hit by a car; another shot dead by a fake pizza delivery boy; one was killed in an airplane crash; one died from a stroke while being mugged; and the rest met similar ends.

Could this possibly be coincidence? What are the odds against all these things happening and being connected by the number 11?

Well if you think these events must somehow be causally interconnected, you'd have a lot of company. But it wouldn't include mathematicians—or us. Why not? In a world where so many things happen, strange and

3. Violette is a strong Cowboys fan. Because of her work schedule, however, she has been able to watch their games only twice this season, and they lost both times. She resolves not to watch any more. "It's bad luck," she decides.

▲ 4. Giving the little guy more water could have prevented him from getting dehydrated, said Ms. Delacruz.

5. OAXACA, Mexico (AP)—Considered by many to be Mexico's culinary capital, this city took on McDonald's and won, keeping the hamburger giant out of its colonial plaza by passing around tamales in protest.

6. Eating fish or seafood at least once a week lowers the risk of developing dementia, researchers have found.

▲ 7. It has long puzzled researchers why people cannot detect their own bad breath. One theory is that people get used to the odor.

8. Researchers based at McDuff University put thirty young male smokers on a three-month program of vigorous exercise. One year later, only 14 percent of them still smoked. An equivalent number of young male smokers who

Touchstone Pictures/Frank Masi

In the movie *Signs*, Mel Gibson is worried about the dark secrets of the universe.

seemingly improbable coincidences are bound to happen every second of every day.

Not convinced? Ask each of your classmates to think of as many events or things connected with the 9/11 attack that involve the number 11. Give each person a week to work on this. We'll bet the collected list of "suspicious" coincidences is very long.

As for the men connected with bioterrorism and germ warfare, you might be interested to know that the American Society for Microbiology alone has 41,000 members, and the total number of people "connected" in some way or other with bioterrorism and germ warfare would be indefinitely larger than that. We'd bet our royalties that in the eleven months following September 11, a lot more than just eleven people connected with bioterrorism and germ warfare died mysteriously and/or violently.

Lisa Belkin of the *New York Times Magazine* wrote an article on this subject (August 18, 2002), from which we learned about the coincidences mentioned above.

Incidentally, "Moore/Parker" has eleven letters.

---

did not go through the exercise program were also checked after a year, and it was found that 60 percent still smoked. The experiment is regarded as supporting the theory that exercise helps chronic male smokers kick the habit.

9. The stronger the muscles, the greater the load they take off the joint, thus limiting damage to the cartilage, which explains why leg exercise helps prevent osteoarthritis.

▲ 10. Many judges in Oregon will not process shoplifting, trespassing, and small-claims charges. This saves the state a lot of money in court expenses.

Exercise 11-3

Identify each reasoning pattern as (a) relevant-difference reasoning or (b) common-thread reasoning.

▲ 1. Pat never had trouble playing that passage before. I wonder what the problem is. It must have something to do with the piano she just bought.

Coincidence

One of your authors has a son whose birthday is exactly six months from that of his mother. Is this relationship of birthdays a coincidence or evidence of some "deeper" mother-son connection? Well, regardless of what the fortune-tellers might say about it, it's coincidence, of course. It would be silly to wonder whether these dates are a "sign" of some other connection between the two people. But coincidences are taken as such signs all the time. People who trade in conspiracy theories and prophecy depend on our willingness to place unwarranted weight on coincidence.

But think for a minute about what makes a coincidence. It is nothing but an event or, more usually, a combination of events, that we take to be unlikely—events that "beat the odds." For example, the odds of the birthdays described above being the case for any two randomly chosen people are very small (1 in 365). But notice: For any two people, *whatever* the relation between their birthdays, the odds of their being related in exactly that way are also very small! (In fact, ignoring leap years, the odds are 1 in 365 against it.) Indeed, no matter what combination of events happens, the odds of exactly that combination of events happening is likely to be very small. So, when we think about it that way, most *everything* is a coincidence! Which amounts to saying that nothing is more coincidental than most other things.

Of course, we are able to identify and characterize some combinations of events, and we don't readily do that for others. Having birthdays exactly six months apart is easier to characterize than having them, say, 80 days apart. For no good reason, we find it easier to attribute significance to the former than to the latter. But, as far as odds are concerned, one is just as unlikely as the other.

The point here is just that, given the extraordinary number of things that happen every day, it would be amazing if some combination of them, unlikely as it may be, didn't occasionally get our attention. In the words of mathematician John Allen Paulos, "In reality, the most astonishingly incredible coincidence imaginable would be the complete absence of all coincidences."

2. Sometimes the fishing is pretty good around here; sometimes it isn't. When I try to pin down why, it seems like the only variable is the wind. Strange as it sounds, the fish just don't bite when it's windy.

3. The price of gas has gone up more than 40 cents a gallon in the past three or four weeks. It all started when they had that fire down there in that refinery in Texas. Must have really depleted supplies somehow.

▲ 4. You know, it has occurred to me, whenever we have great roses like this, it's always after a long period of cloudy weather. I'll bet they don't like direct sun.

5. All of a sudden he's all "Let's go to Beano's for a change." Right. Am I supposed to think it's just coincidence his old girlfriend started working there?

6. You really want to know what gets into me and makes me be so angry? It's you! You and your stupid habit of never lifting up the toilet seat.

▲ 7. Why in heck am I so tired today? Must be all the studying I did last night. Thinking takes a lot of energy.

8. The computer isn't working again. Every time it happens the dang kids have been playing with it. Why can't they just use the computers they have down at school?

9. What makes your dog run away from time to time? I bet it has to do with that garbage you feed him. You want him to stay home? Feed him a better brand of dog food.

▲ 10. I'll tell you what caused all these cases of kids taking guns to school and shooting people. Every single one of those kids liked to play violent videogames, that's what caused it.

11. Gag! What did you do to this coffee, anyway—put Ajax in it?

12. Can you beat that? I set this battery on the garage floor last night and this morning it was dead. I guess the old saying about cement draining a battery is still true.

▲ 13. Clinton was impeached. Then his standing went up in the opinion polls. Just goes to show: No publicity is bad publicity.

14. Why did the dog yelp? Are you serious? You'd yelp if someone stomped on your foot, too.

15. Dennis certainly seems more at peace with himself these days. I guess starting up psychotherapy was good for him.

▲ 16. Every time we have people over, the next morning the bird is all squawky and grumpy. The only thing I can figure is it must not get enough sleep when people are over in the evening.

17. The lawn mower started up just fine last week, and now it won't start at all. Could the fact I let it stand out in the rain have something to do with it?

18. Every time Moore plays soccer, his foot starts to hurt. It also hurts right after he goes jogging. But when he bicycles, or uses an elliptical trainer, he doesn't have a problem. He decides to avoid exercise that involves pounding his feet on the ground.

▲ 19. You know, all of a sudden she's been, like, cool toward me? A bit, you know, icy? I don't think she liked it when I told her I was going to start playing poker with you guys again.

20. Your Chevy Suburban is hard to start. My Chevy Suburban starts right up. You always use Chevron gas; I use Texaco. You'd better switch to Texaco.

Exercise 11-4 _____

Identify the numbered items as A, B, or C, where

A = possible coincidence

B = possible reversed causation

C = possibly a case in which the stated cause and stated effect both result
from some third thing, an underlying cause

▲ 1. Whenever I mow the lawn I end up sneezing a lot more than usual. Must
be gas fumes from the mower.

2. Maybe the reason he's sick is all the aspirin he's taking.

3. The only thing that could possibly account for Clark and his two brothers
all having winning lottery tickets is that all three had been blessed by the
Reverend Dim Dome just the day before. I'm signing up for the Reverend's
brotherhood.

▲ 4. What else could cause the leaves to turn yellow in the fall? It's got to be
the cold weather!

5. Perhaps Jason is nearsighted because he reads so many books.

6. First, Rodrigo gets a large inheritance. Then Charles meets the girl of his
dreams. And Amanda gets the job she was hoping for. What did they all
have in common? They all thought positively. It can work for you, too.

▲ 7. It's common knowledge that osteoarthritis of the knee causes weakness in
the quadriceps.

8. Ever since the country lost its moral direction, the crime rate has gone
through the ceiling. What more proof do you need that the cause of skyrocketing crime is the breakdown in traditional family values?

9. Wow! Is Johnson hot or what? After that rocky start, he has struck out the
last nine batters to face him. That's what happens when ol' Randy gets his
confidence up.

▲ 10. Research demonstrates that people who eat fish are smarter. I'm going to
increase my intake.

11. What a night! All those dogs barking made the coyotes yap, and nobody
could get any sleep.

12. Isn't it amazing how when the leaves drop off in the winter, it makes the
branches brittle?

▲ 13. What explains all the violence in society today? TV. Just look at all the
violence they show these days.

14. On Monday, Mr. O'Toole came down with a cold. That afternoon, Mrs.
O'Toole caught it. Later that evening, their daughter caught it, too.

15. Retail sales are down this year. That's because unemployment is so high.

16. Yes, they're saying electric blankets aren't really a health threat, but I know
better. A friend had cancer and know what? He slept with an electric blanket.

17. At finals time, the bearded man on the front campus offers prayers in return for food. Donald is thinking, "Sure. Why not?—can't hurt anything." He approaches the bearded man with a tidbit. Later: The bearded man prays. Donald passes his finals. To skeptical friends: "Hey, you never know. I'll take all the help I can get."

18. It is an unusually warm evening, and the birds are singing with exceptional vigor. "Hot weather does make a bird sing," Uncle Irv observes.

▲ 19. Why did Uncle Ted live such a long time? A good attitude, that's why.

20. Studies demonstrate that people who are insecure about their relationships with their partners have a notable lack of ability to empathize with others. That's why we recommend that partners receive empathy training before they get married.

21. Lack of self-confidence can be difficult to explain, but common sense suggests that stuttering is among the causes, judging from how often the two things go together.

▲ 22. When I went to Munich last summer I went to this movie, and who was there? This guy I went to school with and hadn't seen in 15 years! No way that could be coincidence!

23. It's odd. I've seen a huge number of snails this year, and the roses have mildew. Don't know which caused which, but one of them obviously caused the other.

24. Her boyfriend is in a bad mood, you say? I'll bet it's because she's trying just a bit too hard to leash him. Probably gets on his nerves.

▲ 25. Many people note that top executives wear expensive clothes and drive nice cars. They do the same, thinking these things must be a key to success.

26. ". . . and let's not underestimate the importance of that home field advantage, guys."

   "Right, Dan. Six of the last seven teams that had the home field advantage went on to win the Super Bowl."

27. On your trip across the country, you note that the traffic is awful at the first intersection you come to in Jersey. "They certainly didn't do anyone a favor by putting a traffic light at this place," you reflect. "Look at all the congestion it caused."

Causation in Populations

Many causal claims do not apply in any straightforward way to individuals but rather apply to populations.* The claim "Drinking causes cancer of the mouth," for example, should not be interpreted as meaning that drinking

*For our analysis of causal factors in populations, we are indebted to Ronald N. Giere, *Understanding Scientific Reasoning*, 3d ed. (Fort Worth: Holt, Rinehart, and Winston, 1991).

Let's say increased TV watching does correlate with increased incidence of violent behavior. Maybe watching TV leads to the violent behavior—but, equally as likely, maybe both are a result of an underlying psychological cause (whatever causes kids to act violently also causes them to like to watch violent behavior). Or maybe kids who act violently become more interested in looking at violence (in the same way that signing up for a sport may heighten one's interest in watching the sport on TV).

Correlations, like the one reported in this graphic, do not establish causation. Thinking they do is a type of post hoc reasoning.

Television's effect on kids

Adolescents who watch more television are more prone to violent behavior later in life, a new study shows.

Percent of kids who acted aggressively at ages 16 or 20, based on number of hours of TV watched daily at age 14:

	Male	Female
One hour or less	8.9	2.3
One to three hours	32.5	11.8
Three or more hours	45.2	12.7

Note: The study was conducted over a 17-year interval with 707 individuals.
Source: *Science* Magazine

Associated Press

will cause mouth cancer for any given individual, or even that drinking will cause mouth cancer for the majority of individuals. The claim is that drinking is a *causal factor* for mouth cancer—that is, there would be more cases of mouth cancer if everyone drank than if no one did. And so it is with other claims about causation in populations: To say that X causes Y in population P is to say that there would be more cases of Y in population P if every member of P were exposed to X than if no member of P were exposed to X.

The evidence on which such claims may be soundly based comes principally from three kinds of investigations or "arguments" All are applications of relevant-difference reasoning, which we explained in the previous section.

The basic idea behind these investigations is this. Suppose someone wants to know if port wine applied to bald men's heads promotes hair growth. How would you find out? You would round up a group of bald men and divide them into two subgroups. On the heads of the men in one subgroup. you'd apply port wine. If the men in both subgroups are the same in all respects except for the port wine applied to the heads of the men in the one subgroup (a big "if"), then if the men in that subgroup sprout hair, you conclude that the port wine caused the hair growth, because it, the port wine, is the only relevant difference between the two groups.

With that as background, let's look at these types of investigations just a bit more closely.

■ *Controlled Cause-to-Effect Experiments*

In **controlled cause-to-effect experiments,** a random sample of a target population is itself randomly divided into two groups: (1) an experimental group, all of whose members are exposed to a suspected causal factor, C (e.g., exposure of the skin to nicotine), and (2) a control group, whose members are all treated exactly as the members of the experimental group, except that they are not exposed to C. Both groups are then compared with respect to frequency of some effect, E (e.g., skin cancer). If the difference, *d*, in the frequency of E in the two groups is sufficiently large, then C may justifiably be said to cause E in the population.

This probably sounds complicated, but the principles involved are matters of common sense. You have two groups that are essentially alike, except that the members of one group are exposed to the suspected causal agent. If the effect is then found to be sufficiently more frequent in that group, you conclude that the suspected causal agent does indeed cause the effect in question.

Familiarizing yourself with these concepts and abbreviations will help you understand cause-to-effect experiments:

experimental group—the sample of the target population whose members are all exposed to the suspected causal agent

control group—the sample of the target population whose members are treated exactly as the members of the experimental group are except that they are not exposed to the suspected causal agent

C—the suspected causal agent

E—the effect whose cause is being investigated

d—the difference in the frequency of this effect in the experimental group and in the control group

Let us suppose that the frequency of the effect in the experimental group is found to be greater than in the control group. *How much greater* must the frequency of the effect in the experimental group be for us to say that the suspected causal agent actually is a causal factor? That is, how great must *d* be for us to believe that C is really a causal factor for E? After all, even if nicotine does *not* cause skin cancer, the frequency of skin cancer found in the experimental group *might* exceed the frequency found in the control group because of some chance occurrence.

Suppose that there are one hundred individuals in our experimental group and the same number in our control group, and suppose that *d* was greater than 13 percentage points—that is, suppose the frequency of skin cancer in the experimental group exceeded the frequency in the control group by more than 13 percentage points. Could that result be due merely to chance? Yes, but there is a 95 percent probability that it was *not* due to chance. If the frequency of skin cancer in the experimental group were to

Hey, Couch Potato—Read This!

You sometimes hear people say they avoid exercise because they know of someone who had a heart attack while working out. We trust you now recognize such reasoning as *post hoc:* To show that Uncle Ned's heart attack was caused by lifting weights, or whatever, you have to do more than show that the one thing happened and then the other thing happened.

There is a further problem here. To wonder whether strenuous exercise causes heart attacks is to wonder whether such exercise is a **causal factor** for heart attacks. And you can't really answer such questions by appealing to specific incidents, as we explain in the text. To think that you can show that something is (or isn't) a causal factor in a population by referring to isolated, specific events is to appeal illegitimately to anecdotal evidence, as we discuss at the end of this chapter.

(For the record: According to research published in the *New England Journal of Medicine* at the end of 1993, males who undertake heavy physical exertion increase their risk of a heart attack by around 100 times *if* they are habitually sedentary—out of shape. For those who do exercise regularly, strenuous exertion doesn't increase their risk of heart attack.)

exceed the frequency in the control group by more than 13 percentage points (given one hundred members in each group), then this finding would be *statistically significant at the .05 level,* which simply means that we could say with a 95 percent degree of confidence that nicotine is a cause of skin cancer. If we were content to speak with less confidence, or if our samples were larger, then the difference in the frequency of skin cancer between the experimental group and control group would not have to be as great to qualify as statistically significant.

Thus, saying that the difference in frequency of the effect between the experimental and control groups is **statistically significant** at some level (e.g., .05) simply means that it would be unreasonable to attribute this difference in frequency to chance. Just how unreasonable it would be depends on what level is cited. If no level is cited—as in reports of controlled experiments that stipulate only that the findings are "significant"—it is customary to assume that the results were significant at the .05 level, which simply means that the result could have arisen by chance in about five cases out of one hundred. Recall from Chapter 10 that we referred to a 95 percent confidence level, which is another way of saying we have the kind of probability found in a .05 significance level.

Media reports of controlled experiments usually state or clearly imply whether the difference in frequency of the effect found in the experimental and control groups is significant. However, if, as occasionally happens, there is a question about whether the results are statistically significant (i.e., are unlikely to have arisen by chance), it is important not to assume uncritically or automatically either of the following:

If you look at the new cases of death from AIDS, the fastest growing category could be ladies over the age of 70. If last year one woman over 70 died from AIDS and this year two do, you get a 100 percent increase in AIDS deaths for that category.

—JOHN ALLEN PAULOS

Percentage increases from a small baseline can be misleading.

E-Commerce: The Buyer Wins

Will the Net pull prices ever lower?

Theoretically, Internet retailing is supposed to be a positive sum game that allows consumers to win via easy access to price and product information and retailers to gain by obtaining detailed information on buyers' needs, tastes, and price sensitivity. If a new study on life insurance costs by Jeffrey R. Brown of Harvard University and Austan Goolsbee of the University of Chicago is any indication, however, it's consumers who really come out ahead.

The two economists noticed that the average prices of term life insurance, which had been edging down in response to falling mortality rates during the 1990s, suddenly fell sharply in 1996 and 1997. Meanwhile, a growing number of Web sites had begun offering price quotes from a variety of companies to potential buyers of term life insurance.

To see whether there was a connection, the authors matched highly detailed data from 1992 to 1997 on household Internet usage in the 50 states with similar data on the cost of term life insurance policies purchased by households over the same period. (The first set of data came from Forrester Research, the second from LIMRA International, a life insurance trade group.)

The results are striking. Among households of different age groups in different states, the study shows a strong correlation between rising Net usage and falling costs of term insurance. By contrast, costs of whole life insurance, for which online quotes were unavailable, hardly changed. The authors estimate that each 10% increase in the share of individuals in a group (say, those age 30–34 in Utah, 40–44 in Texas) using the Net reduced the group's average term insurance costs by 3% to 5%. By 1997, annual nationwide savings to consumers totaled $115 million to $215 million.

Thus, the Internet—by enabling people to comparison shop without having to first contact a variety of insurance agents—apparently drove insurance prices down not only for Web surfers but for their neighbors as well. And that was in a period in which Net usage was relatively low by today's standards.

For buyers, of course, this picture spells good news, suggesting that markets will become increasingly competitive as more and more people log on and search for low prices. Sellers both online and offline, however, are likely to find their profit margins under chronic pressure—forcing them to engage in continual cost-cutting. "That's a lesson the stock market finally appears to be learning," says Goolsbee.

By Gene Koretz

Source: Business Week, January 8, 2001.

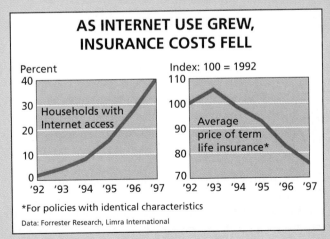

AS INTERNET USE GREW, INSURANCE COSTS FELL

Percent

Households with Internet access

Index: 100 = 1992

Average price of term life insurance*

'92 '93 '94 '95 '96 '97   '92 '93 '94 '95 '96 '97

*For policies with identical characteristics

Data: Forrester Research, Limra International

Caution! These graphs show only correlation, not cause and effect!

TABLE 11-1   Approximate Statistically Significant *d*'s at .05 Level

Number in Experimental Group (with Similarly Sized Control Group)	Approximate Figure That d Must Exceed to Be Statistically Significant (in Percentage Points)
10	40
25	27
50	19
100	13
250	8
500	6
1,000	4
1,500	3

1. *That the sample is large enough to guarantee significance.* A large sample is no guarantee that the difference (*d*) in the frequency of the effect in the experimental group and in the control group is statistically significant. (However, the larger the sample, the smaller *d*—expressed as a difference in percentage points—need be to count as significant.) People are sometimes overly impressed by the mere size of a study.

2. *That the difference in frequency is great enough to guarantee significance.* The fact that there seems to be a pronounced difference in the frequency of the effect in the experimental group and in the control group is no guarantee that the difference is statistically significant. If the sample size is small enough, it may not be. If there are fifty rats in an experimental group and fifty more in a control group, then even if the frequency of skin cancer found in the experimental group exceeds the frequency of skin cancer found in the control group by as much as 18 percentage points, this finding would not be statistically significant (at the .05 level). If each group contained a thousand rats, a difference in frequency of 3 points would not qualify as significant. (And remember that a 3-point difference can be referred to as a "whopping" 50 percent difference if it is the difference between 6 points and 3 points.) Unless you have some knowledge of statistics, it is probably best not to assume that findings are statistically significant unless it is clearly stated or implied that they are.

Nevertheless, it may be helpful to you to have some rough idea of when a difference in frequency of effect in the experimental and control groups may be said to be statistically significant at the .05 level. Table 11-1 provides some examples.

Suppose there are ten individuals each in the randomly selected experimental and control groups. To be statistically significant at the .05 level, the difference between experimental and control group in frequency of the

Cigarettes, Cancer, and the Genetic-Factors Argument

For years the tobacco industry challenged the data showing that smoking causes lung cancer. The industry argued that certain unknown genetic factors (1) predispose some people to get cancer and also (2) predispose those same people to smoke.

The tobacco people are just doing what critical thinkers should do, looking for a common underlying cause for two phenomena that seem on the face of things to be related directly by cause and effect. However, they have never found any such common cause, and the claim that there *might be* one is not equivalent to the claim that there *is* one. Further, even if the industry theory is correct, that wouldn't diminish the other risks involved in smoking.

We should point out, however, that scientists have recently discovered that a chemical found in cigarette smoke damages a gene, known as p53, that acts to suppress the runaway growth of cells that lead to tumors. Researchers regard the discovery as finding the exact mechanism of causation of cancer by cigarette smoke; the discovery seems to put the whole issue of whether cigarettes cause cancer beyond much doubt.

effect must exceed 40 percentage points. If there are twenty-five people in each group, then d must exceed 27 points to be statistically significant, and so forth.

Even if it is clear in a controlled experiment that d is significant, there are a few more considerations to keep in mind when reviewing reports of experimental findings. First, the results of controlled experiments are often extended analogically from the target population (e.g., rats) to another population (e.g., humans). Such analogical extensions should be evaluated in accordance with the criteria for analogical arguments discussed in the previous chapter. In particular, before accepting such extensions of the findings, you should consider carefully whether there are important relevant differences between the target population in the experiment and the population to which the results of the experiment are analogically extended.

Second, it is important in controlled experiments that the sample from which the experimental and control groups are formed be representative of the target population, and thus it is essential that the sample be taken at random. Further, because the experimental and control groups should be as similar as possible, it is important that the assignment of subjects to these groups also be a random process. In reputable scientific experiments it is safe to assume that randomization has been so employed, but one must be suspicious of informal "experiments" in which no mention of randomization is made.

And while we're talking about reputable sources, remember that any outfit can call itself the "Cambridge Institute for Psychological Studies" and publish its reports in its own "journal." Organizations with prestigious-sounding place names (Princeton, Berkeley, Palo Alto, Bethesda, and so on), proper names (Fulbright, Columbia, and so on), or concepts (institute, academy, research, advanced studies) *could* consist of little more than a couple of university dropouts with a dubious theory and an axe to grind.

◼ *Nonexperimental Cause-to-Effect Studies*

A **nonexperimental cause-to-effect study** (or argument) is another type of study designed to test whether something is a causal factor for a given effect. In this type of study, members of a target population (say, humans) who have not yet shown evidence of the suspected effect E (e.g., cancer of the colon) are divided into two groups that are alike in all respects except one. The difference is that members of one group, the experimental group, have all been exposed to the suspected cause C (fatty diets, for example), whereas the members of the other group, the control group, have not. Such studies differ from controlled experiments in that the members of the experimental group are not exposed to the suspected causal agent *by the investigators*— clearly, one limit of experimental studies is that investigators can't purposely expose human subjects to potentially dangerous agents. Eventually, just as in the controlled experiment, experimental and control groups are both compared with respect to the frequency of E. If the frequency in the experimental group exceeds the frequency in the control group by a statistically significant margin, we may conclude that C is the cause of E in the target population.

In reports of nonexperimental cause-to-effect studies, as in reports of controlled experiments, if it is not stated or clearly implied that the findings are significant, do not assume that they are merely because either (1) the samples are large or (2) the difference in the frequency of the effect in absolute terms or percentages is striking.

Likewise, (3) if a causal relationship found to hold in the target population on the basis of such a study is extended analogically to other populations, you should evaluate this analogical extension very carefully, especially with respect to any relevant differences between the target population and the analogical population.

And, finally, (4) note the following important difference between controlled experiments and nonexperimental cause-to-effect studies: In a *controlled* experiment, the subjects are assigned to experimental and control groups by a random process, after which the experimental subjects are exposed to C. This randomization ensures that experimental and control groups will be alike except for the suspected causal agent that the experimental group is then exposed to. But in the *nonexperimental* study, the experimental group (which is still so called even though no experiment is performed) is composed of randomly selected individuals who have already been exposed to the suspected causal agent or who say they were. And the individuals who have already been exposed to C (or who say they were) may differ from the rest of the target population in some respect in addition to having been exposed to C. For example, there is a positive correlation between having a fatty diet and drinking alcoholic beverages. Thus, an experimental group

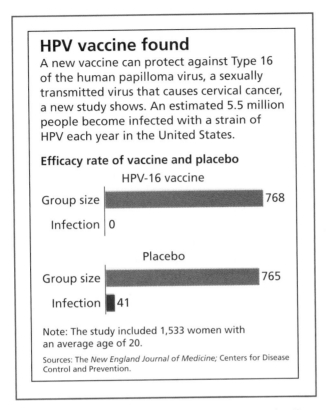

HPV vaccine found

A new vaccine can protect against Type 16 of the human papilloma virus, a sexually transmitted virus that causes cervical cancer, a new study shows. An estimated 5.5 million people become infected with a strain of HPV each year in the United States.

Efficacy rate of vaccine and placebo

HPV-16 vaccine

Group size | 768
Infection | 0

Placebo

Group size | 765
Infection | 41

Note: The study included 1,533 women with an average age of 20.

Sources: The *New England Journal of Medicine*; Centers for Disease Control and Prevention.

▲ Graphics like this, while leaving out important details, help readers understand cause-and-effect investigations.

composed by random means from those in the general population who have fatty diets would include more than its fair share of drinkers. Consequently, the high rate of colon cancer observed in this experimental group might be due in part to the effects of drinking.

It is important, then, that the process by which the individuals in the general population "self-select" themselves (regarding their exposure to C) not be biased in any way related to the effect. In good studies, any factors that might bias the experimental group are controlled by one means or another. Often, for example, the control group is not randomly selected, but rather is selected to match the experimental group for any other relevant factors. Thus, in a study that seeks to relate fatty diets to cancer of the colon, an experimenter will make certain that the same percentage of drinkers is found in the control group as in the experimental group.

Nonexperimental studies of the variety explained here and in the next section are *inherently* weaker than controlled experiments as arguments for causal claims. Because we do not have complete knowledge of what factors are causally related to what other factors, it is impossible to say for certain that all possibly relevant variables in such studies have been

Learning Australian, Lesson 42:
That's Australian for "Another Round!"

Long-term heavy drinkers are not damaging their mental faculties, contrary to previous suggestions, despite consuming the equivalent of eight pints of beer a day, researchers say.

Drinkers scored just as well on psychological and intellectual tests and showed no more signs of brain shrinkage than non-drinkers or those who consumed alcohol only in moderation.

A number of studies have suggested that heavy long-term alcohol use can cause brain damage, including a type of alcohol-induced dementia.

However, researchers from Australia say they can find no evidence that persistent lifelong consumption of alcohol is related to cognitive functioning in men in their 70s.

The researchers, from the Australian National University, Canberra, and the Center for Education and Research on Aging in Sydney, looked at 209 men with an average age of 73 who were veterans of World War II. They were given 18 tests to measure intellectual function, as well as scans for brain atrophy.

The daily alcohol consumption of the men [averaged two and one-half pints of beer and went as high as eight pints].

—CHRIS MIHILL, *Scripps Howard News Service*

We'd look for confirmation of these results before we'd recommend regular trips to the pub. Notice that this study ignores *other* problems caused by alcohol.

controlled. It is good policy to imagine what characteristics those who have been exposed to the suspected causal agent might have and contemplate whether any of these factors may be related to the effect. If you can think of any relevant variables that have not been controlled, you should have doubts about any causal claim that is made on the basis of such studies.

■ *Nonexperimental Effect-to-Cause Studies*

A **nonexperimental effect-to-cause study** is a third type of study designed to test whether something is a causal factor for a given effect. In this type of study, the "experimental group," whose members already display the *effect* being investigated, E (e.g., cancer of the mouth), is compared with a control group none of whose members have E, and the frequency of the suspected cause, C (e.g., using chewing tobacco), is measured. If the frequency of C in the experimental group significantly exceeds its frequency in the control group, then C may be said to cause E in the target population.

Cautionary remarks from the discussion of nonexperimental cause-to-effect studies apply equally to nonexperimental effect-to-cause studies. That is, if it isn't clear that the findings are significant, don't assume that they are merely because (1) the samples seem large or (2) the difference in the frequency expressed in absolute terms seems striking; and (3) evaluate carefully analogical extensions of the results to other populations.

Notice further that (4) the subjects in the experimental group may differ in some important way (in addition to showing the effect) from the rest of the target population. Thus, for instance, former smokers are more likely than others to use chewing tobacco and are also more likely to get mouth cancer. If you sample randomly from a group of victims of mouth cancer, therefore, you are likely to produce more ex-smokers in your sample than occur in the general population. The result is that you are likely to discover more chewing-tobacco users in the sample, even if chewing tobacco plays no role whatsoever in causing cancer of the mouth. Any factor that might bias the experimental group in such studies should be controlled. If, in evaluating such a study, you can think of any factor that has not been controlled, you can regard the study as having failed to demonstrate causation.

Notice, finally, that (5) effect-to-cause studies show only the probable frequency of the cause, not the effect, and thus provide no grounds for estimating the percentage of the target population that would be affected if everyone in it were exposed to the cause.

Appeal to Anecdotal Evidence

In Chapter 10, we discussed the mistake of trying to reach a conclusion by citing one or two examples—i.e., by appealing to anecdotes. There is another type of appeal to anecdotal evidence that relates to causal arguments, as distinct from generalizations: trying to either prove or disprove that X is a causal factor for Y by citing an example or two of an X that did (or didn't) cause a Y. Arguing that red wine prevents colds because we know someone who drinks red wine and never catches cold would be an instance of this type of weak argument. It would be similarly weak to argue that red wine doesn't prevent colds because we've never observed that effect in us. Those who submit that smoking doesn't cause cancer because they know a smoker who lived to ninety-eight and died of old age are guilty of the causal type of anecdotal argument.

To establish that X is a causal factor for Y, we have to show that there would be more cases of Y if everyone did X than if no one did, and you can't really show this—or demonstrate that X isn't a causal factor for Y—by citing an example or two. Isolated incidents can reasonably serve to raise suspicions about causal factors, but that is all.

One of the most common mistakes made in everyday reasoning is to appeal to anecdotal evidence in one or the other of the ways discussed here and in Chapter 10. And now that you know what is involved in saying that something is a causal factor, you know why you can't establish this sort of causal claim by appealing to an anecdote.

Weird Science

Here are some of the criticisms made by research authorities who, in the June 5 *Journal of the American Medical Association,* reviewed the quality of current medical research literature.

- Studies often fail to put results in the context of previous studies.
- Authors of studies do not adequately address legitimate scientific criticism after publication, and journal editors do not require them to do so.
- Study collaborators frequently disagree about their study's conclusions, and the published reports often fail to represent the full range of authors' opinions.
- Guidelines showing how results ought to be reported are often ignored or badly applied.

As indicated, these criticisms were part of an assessment of current literature in medical research. It was found that research reported in company-funded studies was often inaccurate, incomplete, or otherwise misleading in ways that favored the drug companies.

Source: *Business Week,* June 24, 2002.

Exercise 11-5

Identify each of the following as (a) a claim about causation between specific occurrences, (b) a claim about causal factors in populations, or (c) neither of these.

▲ 1. The hibiscus died while we were away. There must have been a frost.

2. Carlos isn't as fast as he used to be; that's what old age will do.

3. Kent's college education helped him get a high-paying job.

▲ 4. The most frequently stolen utility vehicle is a 2000 Jeep Wrangler.

5. Vitamin C prevents colds.

6. The man who put this town on the map was Dr. Jaime Diaz.

▲ 7. The high reading on the thermometer resulted from two causes: This thermometer was located lower to the ground than at other stations, and its shelter was too small, so the ventilation was inadequate.

8. Oily smoke in the exhaust was caused by worn rings.

9. The initial tests indicate that caffeine has toxic effects in humans.

▲ 10. Neonatal sepsis is usually fatal among newborns.

11. WIN 51,711 halted development of paralysis in mice that had recently been infected with polio-2.

12. A stuck hatch cover on *Spacelab* blocked a French ultraviolet camera from conducting a sky survey of celestial objects.

13. An experimental drug has shown broad antiviral effects on a large number of the picornaviruses against which it has been tested.

▲ 14. Investigation revealed the problem was a short-circuited power supply.

15. Arteriovenous malformations—distortions of the capillaries connecting an arteriole and a small vein in the brain—can bleed, causing severe headaches, seizures, and even death.

16. Because of all the guns that its citizens own, the United States has never been invaded.

▲ 17. According to two reports in the *New England Journal of Medicine,* oil from fish can prevent heart disease.

18. The most important cause in the growing problem of illiteracy is television.

19. "Raymond the Wolf passed away in his sleep one night from natural causes; his heart stopped beating when the three men who slipped into his bedroom stuck knives in it."

> Jimmy Breslin, The Gang That Couldn't Shoot Straight

▲ 20. The dramatic increases in atmospheric CO_2, produced by the burning of fossil fuels, are warming the planet and will eventually alter the climate.

Exercise 11-6

Go to Church and Live Longer

According to Bill Scanlon, a reporter for the Scripps Howard News Service, researchers from the University of Colorado, the University of Texas, and Florida State University determined that twenty-year-olds who attend church at least once a week for a lifetime live on the average seven years longer than twenty-year-olds who never attend. The data came from a 1987 National Health Interview Survey that asked 28,000 people their income, age, church-attendance patterns, and other questions. The research focused on 2,000 of those surveyed who subsequently died between 1987 and 1995.

Propose two different causal hypotheses to explain these findings.

Hypothesis One:

What data would you need to have greater confidence in this hypothesis?

Hypothesis Two:

What data would you need to have greater confidence in this hypothesis?

Other Hypotheses:

What data would you need to have greater confidence in these hypotheses?

Auberry Drive Cancer Toll at 7

A new state review will try to see if more than coincidence is at work.

Another two related cancers on Auberry Drive have surfaced, bringing to seven the number of similarly stricken children and adults within one square mile just north of Elk Grove.

The news reinforces some residents' suspicions that they are witnessing a disease cluster perpetuated by some unknown toxic exposure.

"It's no surprise to me," said Dee Lewis, a nearby resident who has been investigating possible environmental links for the past three years.

At The Bee's request, state officials on Wednesday agreed to re-evaluate the area's cancer incidence to determine whether the concentration is more than a coincidence.

Rosemary Cress, an epidemiologist with the state-run California Cancer Registry, has concluded from earlier statistical analyses that the area's incidence of childhood leukemia is not outside the range of chance. She does not expect further analyses to change the conclusion.

By Chris Bowman

The incidence of cancer will be higher on some streets in a community than on others simply due to chance. But as a matter of psychology, it may be difficult to accept that fact if you live on one of the streets that has a high incidence of cancer.

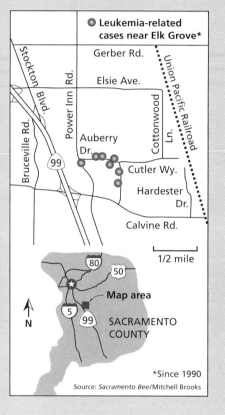

Exercise 11-7

There is no single event, activity, decision, law, judgment, in this period of time that I call the "three strikes" era—other than "three strikes"—that could explain the tremendous acceleration in the drop in crime.

Dan Lungren, former California Attorney General, who helped draft California's Three Strikes law

Under this law, conviction for a third felony carried with it a mandatory sentence of twenty-five years to life. Although the crime rate in California had been falling before the law took effect in 1994, it reportedly fell even faster after the law was enacted, and California's crime rate dropped to levels not seen since the 1960s.

Provide two reasonable alternative hypotheses to explain the acceleration of the drop in the crime rate in California. What data would you need to be convinced that Lungren's hypothesis is the best?

Exercise 11-8

Suppose a university teacher wants to know whether or not requiring attendance improves student learning. How could she find out? In groups (or individually if the instructor prefers), describe an experiment that an instructor might actually use. Groups may then compare proposals to see who has the best idea.

Exercise 11-9

For each of the following investigations:

a. Identify the causal hypothesis at issue.

b. Identify what kind of investigation it is.

c. Describe the control and experimental groups.

d. State the difference in effect (or cause) between control and experimental groups.

e. Identify any problems in either the investigation or in the report of it, including but not necessarily limited to uncontrolled variables.

f. State the conclusion you think is warranted by the report.

▲ 1. Scientists have learned that people who drink wine weekly or monthly are less likely to develop dementia, including Alzheimer's disease. (Daily wine drinking, however, seems to produce no protective effect.) The lead researcher was Dr. Thomas Truelsen, of the Institute of Preventive Medicine at Kommunehospitalet in Copenhagen. The researchers identified the drinking patterns of 1,709 people in Copenhagen in the 1970s and then assessed them for dementia in the 1990s, when they were aged 65 or older. When they were assessed two decades later, 83 of the participants developed dementia. People who drank beer regularly were at an increased risk of developing dementia.

Adapted from BBC News *(online), November 12, 2002*

2. Learning music can help children do better at math. Gordon Shaw of the University of California, Irvine, and Frances Rauscher at the University of Wisconsin compared three groups of second-graders: 26 received piano instruction plus practice with a math videogame, 29 received extra English lessons plus the game, and 28 got no special lessons. After four months the piano kids scored 15 to 41 percent higher on a test of ratios and fractions than the other participants.

Adapted from Sharon Begley, Newsweek, *July 24, 2000*

3. The Carolina Abecedarian Project [A-B-C-D, get it?] selected participants from families thought to be at risk for producing mildly retarded children. These families were all on welfare, and most were headed by a single mother, who had scored well below average on a standardized IQ test (obtaining IQs of 70 to 85). The project began when the participating children were 6 to 12 weeks old, and continued for the next 5 years. Half of the

participants were randomly assigned to take part in a special day-care program designed to promote intellectual development. The program ran from 7:15 to 5:15 for 5 days a week for 50 weeks each year until the child entered school. The other children received the same dietary supplements, social services, and pediatric care, but did not attend day-care. Over the next 21 years, the two groups were given IQ tests and tests of academic achievement. The day-care program participants began to outperform their counterparts on IQ tests starting at 18 months and maintained this IQ advantage through age 21. They also outperformed the others in all areas of academic achievement from the third year of school onward.

Adapted from Developmental Psychology, *6th ed., David R. Schaffer*

▲ 4. Research at the University of Pennsylvania and the Children's Hospital of Philadelphia indicates that children who sleep in a dimly lighted room until age two may be up to five times more likely to develop myopia (nearsightedness) when they grow up.

The researchers asked the parents of children who had been patients at the researchers' eye clinic to recall the lighting conditions in the children's bedroom from birth to age two.

Of a total of 172 children who slept in darkness, 10 percent were nearsighted. Of a total of 232 who slept with a night light, 34 percent were nearsighted. Of a total of 75 who slept with a lamp on, 55 percent were nearsighted.

The lead ophthalmologist, Dr. Graham E. Quinn, said that, "just as the body needs to rest, this suggests that the eyes need a period of darkness."

Adapted from an AP report by Joseph B. Verrengia

5. You want to find out if the coffee grounds that remain suspended as sediment in French press, espresso, and Turkish and Greek coffee can cause headaches.

You divide fifty volunteers into two groups and feed both groups a pudding at the same time every day. However, one group mixes eight grams of finely pulverized used coffee grounds into the pudding before eating it (that's equivalent to the sediment in about one and a half liters of Turkish coffee). Within three weeks, you find that 50 percent of the group that has eaten grounds have had headaches; only 27 percent of the other group have experienced a headache. You conclude that coffee grounds may indeed cause headaches and try to get a grant for further studies. (This is a fictitious experiment.)

6. Do you enjoy spicy Indian and Asian curries? That bright yellow-orange color is due to curcumin, an ingredient in the spice turmeric. An experiment conducted by Bandaru S. Reddy of the American Health Foundation in Valhalla, New York, and reported in *Cancer Research*, suggests that curcumin might suppress the development of colon cancer

Places where turmeric is widely used have a low incidence of colon cancer, so the research team decided to investigate. They administered a powerful colon carcinogen to sixty-six rats, and then added curcumin at the rate of 2,000 parts per million to the diet of thirty of them. At the end of a year, 81 percent of the rats eating regular rat food had developed cancerous tumors, compared with only 47 percent of those that dined on the curcumin-enhanced diet. In addition, 38 percent of the tumors in rats eating regular food were invasive, and that was almost twice the rate in rodents eating curcumin-treated chow.

Adapted from Science News

▲ 7. Does jogging keep you healthy? Two independent researchers interested in whether exercise prevents colds interviewed twenty volunteers about the frequency with which they caught colds. The volunteers, none of whom exercised regularly, were then divided into two groups of ten, and one group participated in a six-month regimen of jogging three miles every other day. At the end of the six months the frequency of colds among the joggers was compared both with that of the nonjoggers (group A) and with that of the joggers prior to the experiment (group B). It was found that, compared with the nonjoggers, the joggers had 25 percent fewer colds. The record of colds among the joggers also declined in comparison with their own record prior to the exercise program.

8. "In the fifty-seven-month study, whose participants were all male physicians, 104 of those who took aspirin had heart attacks, as compared with 189 heart attacks in those who took only a sugar pill. This means ordinary aspirin reduced the heart attack risk for healthy men by 47 percent. At least seven long-term studies of more than 11,000 heart attack victims have shown that one-half or one aspirin per day can reduce the risk of a second attack by up to 20 percent."

Adapted from the Los Angeles Times

9. "Although cigarette ads sometimes suggest that smoking is 'macho,' new studies indicate that smoking can increase the risk of impotence. In a study of 116 men with impotence caused by vascular problems, done at the University of Pretoria, South Africa, 108 were smokers. Two independent studies, one done by the Centre d'Etudes et de Recherches di l'Impuissance in Paris, and reported in the British medical journal *Lancet*, and the other done by Queen's University and Kingston General Hospital in Ontario, found that almost two-thirds of impotent men smoked.

"To test whether smoking has an immediate effect on sexual response, a group of researchers from Southern Illinois and Florida State universities fitted 42 male smokers with a device that measures the speed of arousal. The men were divided into three groups, one group given high-nicotine cigarettes, one group cigarettes low in nicotine, and one group mints. After smoking one cigarette or eating a mint, each man was placed in a private room and shown a two-minute erotic film while his sexual response was monitored. Then he waited ten minutes, smoked two more cigarettes or ate another mint, and watched a different erotic film, again being monitored.

"The results: Men who smoked high-nicotine cigarettes had slower arousal than those who smoked low-nicotine cigarettes or ate mints."

Adapted from Reader's Digest

10. "A study published in the July 27 *Journal of the American Medical Association* indicates that taking androgen (a male sex hormone) in high doses

for four weeks can have important effects on the high density lipoproteins (HDLs) in the blood, which are believed to protect against the clogging of vessels that supply the heart. Ben F. Hurley, an exercise physiologist from the University of Maryland in College Park who conducted the study at Washington University, monitored the levels of HDL in the blood of sixteen healthy, well-conditioned men in their early thirties who were taking androgens as part of their training program with heavy weights. Prior to use of the hormone, all had normal levels of HDLs. After four weeks of self-prescribed and self-administered use of these steroids the levels dropped by about 60 percent.

"Hurley is cautious in interpreting the data. 'You can't say that low HDL levels mean that a specified person is going to have a heart attack at an earlier age. All you can say is that it increases their risk for heart disease.'"

D. Franklin, Science News

11. "New studies reported in the *Journal of the American Medical Association* indicate that vasectomy is safe. A group headed by Frank Massey of UCLA paired 10,500 vasectomized men with a like number of men who had not had the operation. The average follow-up time was 7.9 years, and 2,300 pairs were followed for more than a decade. The researchers reported that, aside from inflammation in the testes, the incidence of diseases for vasectomized men was similar to that in their paired controls.

"A second study done under federal sponsorship at the Battelle Human Affairs Research Centers in Seattle compared heart disease in 1,400 vasectomized men and 3,600 men who had not had the operation. Over an average follow-up time of fifteen years, the incidence of heart diseases was the same among men in both groups."

Edward Edelson, New York Daily News; *reprinted in* Reader's Digest

12. "A new study shows that the incidence of cancer tumors in rats exposed to high doses of X-rays dropped dramatically when the food intake of the rats was cut by more than half. Dr. Ludwik Gross of the Veterans Administration Medical Center noted that this study is the first to demonstrate that radiation-induced tumors can be prevented by restricting diet.

"The experimenters exposed a strain of laboratory rats to a dose of X-rays that produced tumors in 100 percent of the rats allowed to eat their fill—about five or six pellets of rat food a day.

"When the same dose of X-rays was given to rats limited to two pellets of food a day, only nine of 29 females and one of 15 males developed tumors, the researchers reported.

"The weight of the rats on the reduced diet fell by about one-half, but they remained healthy and outlived their counterparts who died of cancer, Gross said. He noted that the restricted diet also reduced the occurrence of benign tumors. There is no evidence that restriction of food intake will slow the growth of tumors that have already formed in animals, he said."

Paul Raeburn, Sacramento Bee

13. "Encephalitis, or sleeping sickness, has declined greatly in California during the past thirty years because more people are staying inside during

prime mosquito-biting hours—7 P.M. to 10 P.M., researchers said. Paul M. Gahlinger of San Jose State University and William C. Reeves of the School of Public Health at UC Berkeley conducted the study. 'People who watch television on warm summer evenings with their air conditioners on are less likely to be exposed during the peak biting period of mosquitoes that carry encephalitis,' Reeves said.

"The researchers found that those counties in California's Central Valley with the highest television ownership had the lowest encephalitis rates for census years. Of 379 Kern County residents interviewed by telephone, 79 percent said they used their air conditioners every evening and 63 percent said they watched television four or more evenings a week during the summer.

"The percentage of residents who spend more time indoors now because of air conditioning than in 1950 more than doubled, from 26 percent to 54 percent, the researchers said."

Associated Press, Enterprise-Record (Chico, California)

▲ 14. "A study released last week indicated that Type A individuals, who are characteristically impatient, competitive, insecure and short-tempered, can halve their chances of having a heart attack by changing their behavior with the help of psychological counseling.

"In 1978, scientists at Mt. Zion Hospital and Medical Center in San Francisco and Stanford University School of Education began their study of 862 predominantly male heart attack victims. Of this number, 592 received group counseling to ease their Type A behavior and improve their self-esteem. After three years, only 7 percent had another heart attack, compared with 13 percent of a matched group of 270 subjects who received only cardiological advice. Among 328 men who continued with the counseling for the full three years, 79 percent reduced their Type A behavior. About half of the comparison group was similarly able to slow down and cope better with stress.

"This is the first evidence 'that a modification program aimed at Type A behavior actually helps to reduce coronary disease,' says Redford Williams of Duke University, an investigator of Type A behavior."

Science News

Exercise 11-10

Here's a news report on the costs of drug abuse that appeared during the administration of George H. W. Bush. See if you can find any flaws in the reasoning by which the figures were reached.

J. Michael Walsh, an officer of the National Institute on Drug Abuse, has testified that the "cost of drug abuse to U.S. industry" was nearly $50 billion a year, according to "conservative estimates." President Bush has rounded this figure upward to "anywhere from $60 billion to $100 billion." This figure would seem to be a difficult one to determine. Here's how Walsh arrived at it. After a survey of 3,700 households, a NIDA

The Wrong Initials Can Shorten Your Life

Researchers at the University of California, San Diego, looking at twenty-seven years of California death certificates, found that men with "indisputably positive" initials like JOY and WOW and ACE and GOD and WIN and VIP lived 4.48 years longer than a control group of men with neutral initials and ambiguous initials, like DAM and WET and RAY and SUN, that had both positive and negative interpretation. Further, men with "plainly negative" initials like ASS or DUD died on average 2.8 years earlier than did the men in the control group.

As an exercise, propose an explanation for these findings that isn't defective in terms of the criteria discussed in this chapter. Explain how you would test the explanation.

contractor analyzed the data and found that the household income of adults who had *ever* smoked marijuana daily for a month [or at least twenty out of thirty days] was 28 percent less than the income of those who hadn't. The analysts called this difference "reduced productivity due to daily marijuana use." They calculated the total "loss," when extrapolated to the general population, at $26 billion. Adding the estimated costs of drug-related crimes, accidents, and medical care produced a grand total of $47 billion for "costs to society of drug abuse."

Doubtful Causal Claims/Hypotheses

Lets say you have trouble sleeping at night and go to the doctor to find out what's wrong. He runs several tests on you and then informs you that, at long last, he has an explanation.

"The cause of your not sleeping," he announces, "is insomnia."

Here you wouldn't need to know what your doctor's reasoning was. His explanation is *inherently* defective. Some causal claims and hypotheses are like that. They suffer from inherent difficulties, problems that lie within the claim itself.

Here are some common defects of this nature.

1. **Circularity:** A causal claim might be circular. This happens when the "cause" merely restates the effect. When your doctor says you can't sleep because of insomnia, his claim is circular. Insomnia just *is* not sleeping. "She has trouble writing because she has writer's block" would be another example of circular causal explanation. Having trouble writing and having writer's block amount to the same thing. A circular causal claim cannot teach you anything new about causes and effects (although it might teach you a new *name* for an effect).

Keeping Secrets

▲   *If somebody told you that cows stand up and talk to each other, you'd say he or she was nuts, right? After all, nobody has ever seen cows doing that. But suppose the person were to respond by telling you that cows stand up only when nobody can see them? Or suppose a psychic were to explain that his or her psychic abilities disappear when threatened or contaminated by the skepticism of an investigating scientist.*

What's wrong with those explanations? Read the rest of the chapter to find out.

2. **Nontestability:** "AIDS is God's punishment for evil." The problem with this assertion, quite obviously, lies in the fact you cannot test whether it is true. There is no way to detect when the hand of God is present and when it isn't. "You can't sleep because you are possessed" would be another example of a nontestable claim.

Now, one does have to bear in mind that being *difficult* to test is *not* necessarily a defect in a causal hypothesis. Scientific hypotheses can be enormously difficult to test. The problem is with claims for which a test is *unimaginable*, like the examples above.

3. **Excessive vagueness:** "The reason Nasim has trouble in his relationships? Bad karma." The trouble with this claim is that "bad karma" is too vague: What does "bad karma" include, and what does it exclude? A similar problem of vagueness would arise if someone tried to attribute Nasim's difficulty to his having "impure thoughts." ("Trouble in his relationships" is pretty vague as well, but if we know Nasim and the speaker we might have an idea what the phrase refers to.)   These two claims about Nasim are also untestable; indeed, in general, if a claim is too vague, we won't know exactly how to go about testing it. But not all untestable claims are too vague: "An invisible gremlin inside your car's engine explains why it misses occasionally" is untestable, but not especially vague.

4. A claim might also involve **unnecessary assumptions.** The assertion above about an invisible gremlin causing the engine to miss does that, since there are *other* explanations for the car's missing that do *not* assume gremlins. Similarly, you sometimes hear déjà vu explained as being caused by memories from previous lives. But that claim, as well as being untestable, requires assuming past lives, which isn't necessary since déjà vu experiences can be induced by electric stimulation of areas within the brain.

5. A similar problem is that a causal claim or hypothesis might **conflict with well-established theory.** For example, an explanation of a person's height as due to something the person did in a past life conflicts with genetic theory. Now, that an assertion conflicts with a well-established theory is not necessarily a *fatal* flaw, because theoretical advances often involve rejecting ideas that once counted as well-established. The theory of continental drift, for example, conflicted with other geological principles. Likewise, the theories of Einstein and other twentieth-century physicists conflicted with Newtonian physics. But evidence was required to establish the new ideas. If a proffered explanation conflicts with well-established theory, we have good reasons to look for an alternative account and to demand powerful evidence of its truth. Also, the fact that there are many examples in which well-established theory is later rejected is itself no evidence at all of the truth of any specific causal claim.

Exercise 11-11

Use letter grades (A, B, C, etc., with pluses and minuses if you wish) to evaluate each of the following explanations as to its degree of testability, its freedom from vagueness, and its noncircularity. ("A, A, A," for example, would indicate an explanation that was clearly testable, contained little or no vagueness, and was not at all circular.)

Indians Flock to See Statues "Drink" Milk

NEW DELHI, India — Millions rushed to Hindu temples across India on Thursday after reports of a miracle—the statues of one of their gods were drinking milk.

The faithful—bearing milk in everything from earthen and steel pots to tumblers and jugs—converged on temples that had reproductions of the elephant-headed Lord Ganesha.

"It is a miracle," said A. K. Tiwari, a priest at a temple in Southern New Delhi.

Scientists dismissed that explanation, saying the offered milk trickled down the granite or marble idols in a thin film that was not easily visible.

Crowds thronged temples in dozens of cities, including New Delhi, Bombay, Calcutta and Madras.

Police and paramilitary soldiers were called out to guard temples across India. In the northern town of Jamshedpur, police waved bamboo canes to control a crowd of 500 that tried to storm a temple.

Tiwari, the priest, said the excitement began early Thursday when a devotee dreamed that the deity wanted milk. When the man, who wasn't identified, held a spoonful of milk near the statue's trunk, the milk disappeared.

Word spread quickly, and people began lining up at temples as early as 4 a.m.

The faithful dismissed suggestions of a hoax.

"It cannot be a hoax. Where would all the milk being offered go? It is such a small idol, it can't take in so much," said Parmesh Soti, a business executive who stood in line. "The gods have come down to Earth to solve our problems."

So widespread were the reports of a miracle that the federal Department of Science and Technology was asked to investigate.

Their scientists offered milk mixed with colored pigment to an idol in a New Delhi temple. Although it disappeared from the spoon, it soon coated the idol. The scientists credited the "miracle" to surface tension, saying molecules of milk were pulled from the spoon by the texture of the statues.

Associated Press, in *The Sacramento Bee*

▲ 1. What causes your engine to miss whenever you accelerate? Perhaps a spark plug is firing sporadically.

2. Antonio certainly had a run of hard luck—but then I warned him not to throw out that chain letter he got.

3. Divine intervention can cure cancer.

▲ 4. Having someone pray for you can cure cancer.

5. Having your mother pray for you brings good luck.

6. Good luck is the result of being very fortunate.

▲ 7. Why did Claudia catch a cold? She's just prone to that sort of thing, obviously.

8. Agassi won the match simply because he wanted it more than Sampras.

9. Tuck can play high notes so well because he has an excellent command of the upper register.

10. Parker's trip to Spain last year is the reason his Spanish is not quite as bad as it used to be.

Exercise 11-12 _____

Use the criteria discussed in the text to evaluate the following causal explanations. The criteria, once again, are

  Testability
  Noncircularity
  Freedom from excessive vagueness
  Freedom from conflict with well-established theory

▲ 1. The reason he has blue eyes is that he acquired them in a previous incarnation.

  2. The Pacers did much better in the second half of the game. That's because they gained momentum.

  3. "Men are biologically weaker than women and that's why they don't live as long, a leading expert declares."

Weekly World News

▲ 4. PARKER: Gad, there are a lot of lawyers in this country today.
     MOORE: That's because there are so many lawsuits that there's a huge demand for lawyers.

  5. Why did he come down with the flu? He's just prone to that sort of thing, I guess.

  6. If God had meant for people to fly, he'd have given them wings.

▲ 7. Alcoholics find it so difficult to give up drinking because they have become physiologically and psychologically addicted to it.

  8. Alcoholics find it so difficult to give up drinking because they have no willpower.

  9. The reason the latest episode of *The Rings* was a box-office hit is that Roger Ebert, one of the most influential movie critics, gave it a good review.

▲ 10. According to some psychologists, we catch colds because we want to. Most of the time we are not aware of this desire, which may, therefore, be said to be subconscious. Viruses are present when we have a cold, but unless we desire to catch cold, the viruses do not affect us.

 11. Why does she sleep so late? I guess she's just one of those people who have a hard time waking up in the morning.

 12. I wonder what made me choose Budweiser. Maybe I've been subjected to subliminal advertising.

▲ 13. The area along this part of the coast is especially subject to mudslides because of the type of soil that's found on the slopes and because there is not enough mature vegetation to provide stability with root systems.

 14. How did he win the lottery twice? I'd have to say he's psychic.

 15. "Parapsychologist Susan Blackmore failed to find evidence of ESP in numerous experiments over more than a decade. *Fate* magazine's consulting editor D. Scott Rogo explained her negative results as follows: 'In the course of my conversations with Blackmore I have come to suspect that

When Is Affirmative Action Not Affirmative Action?

A letter-writer to the "Ask Marilyn" column in *Parade* magazine mentions a case worth your attention:

The writer, Christopher McLaughlin, of Orange Park, Florida, reported a company that opened a factory generating 70 white-collar jobs and 385 blue-collar jobs.

For the 70 white-collar positions, 200 males and 200 females applied; 15 percent of the males were hired, and 20 percent of the females were hired.

For the 385 blue-collar jobs, 400 males applied and 100 females applied; 75 percent of the males were hired, and 85 percent of the females were hired.

In short, the percentage of female applicants hired was greater than the percentage of male applicants hired in both categories. A victory for women, yes?

Nope—at least not according to the Equal Employment Opportunity Commission. The EEOC calculated that a female applying for a job at the factory had a 59 percent chance of being denied employment, whereas a male applicant had only a 45 percent chance of not being hired. So the company actually discriminated against women, the EEOC said.

Here's why: When you consider *total* applications, out of 600 male applicants, 330 were hired, and out of 300 female applicants, 125 were hired. That means that 55 percent of the male applicants—but only 41 percent of the female applicants—were hired.

(This case should remind you of the Saint Simpson's example in Chapter 10—see the box on page 378.)

she resists—at a deeply unconscious level—the idea that psychic phenomena exist. . . . *If* Blackmore is ever fortunate enough to witness a poltergeist or see some other striking display of psychic phenomena, I am willing to bet that her experimental results will be more positive.'"

The Skeptical Inquirer

16. According to a report by Scotty Paul, writing in *Weekly World News*, thousands of tourists who defied an ancient curse and took home souvenir chunks of lava rock from Hawaii's Volcanoes National Park have felt the bitter wrath of a vengeful volcano goddess. The curse explains why a Michigan man tumbled to his death down a stairway and why a Canadian tourist died in a head-on auto accident, as well as why a Massachusetts widow lost her life savings in the stock market. The three suffered their misfortunes after taking lava rocks from the volcano and defying the curse of Madame Pele, the volcano goddess.

▲ 17. Why is there so much violence these days? Rap music, that's why. That and the fact that there's so much violence on TV and in the movies.

18. The reason I got into so much trouble as a kid is that my father became a heavy drinker.

19. In their book *The Gulf Breeze Sightings: The Most Astounding Multiple Sightings of UFOs in U.S. History,* authors Ed and Frances Walters provide photographs of UFOs. According to the *Pensacola News Journal,* the man who now lives in the house the Walters family occupied at the time they photographed the UFOs discovered a flying saucer hidden under some insulation in the attic of the garage. With this model, news photographers were able to create photos of "UFOs" that looked like the Walterses' photos. Walters denies the model was his, but the paper in which it was wrapped contains part of a house plan Walters drew up (Walters is a building contractor). Walters says that model was planted by someone who wished to discredit him.

A Gulf Breeze youngster named Tom Smith also has come forth, admitting that he and Walters's son helped produce the fake saucer photos. He has five UFO photos of his own to substantiate his claim. Walters says Smith's photos are genuine. Which of the two explanations given for the original UFO photos is more plausible, and why? (This report is from *The Skeptical Inquirer.*)

▲ 20. According to mathematician and science writer Martin Gardner, in Shivpuri, a village near Poona, India, there is a large stone ball weighing about 140 pounds in front of a mausoleum containing the remains of Kamarali Darvesh, a Muslim saint. It is possible for five men to stand around the ball and touch it with a forefinger at spots on or below the ball's equator. After they recite in unison "Kamarali Darvesh," an attendant gives a signal, and the ball slowly rises. The explanation of this phenomenon, according to devout Muslims, is that it is a miracle of Allah's.

Causal Explanations and Arguments

Recently we complained to a friend about the paint on a Chevy truck one of us owns.

"What's wrong with it?" he asked.

"It's peeling all over the place," we said.

"Why's it doing that?" he asked.

"That's what we would like to know," we said.

Our friend wasn't confused about this conversation. He understood clearly what it was we wanted to know: We wanted an *explanation* of why the paint on the truck was peeling. We wanted to know what caused it to peel.

Although our friend wasn't confused, causal explanations—explanations of the causes of something—are often confused with arguments, mainly because they involve some of the same words and phrases. But arguing and explaining are different enterprises. They respond to different questions. If someone questions *whether* the paint is peeling, he needs an argument that it either is or isn't. The question of what caused it to peel, on the other hand, assumes that it is peeling and just asks why.

People shouldn't get causal explanations confused with arguments because, with the exception of one complication we shall get to in a second, it is easy to tell them apart. A causal explanation is a causal claim or hypothesis, and as such it asserts or implies cause and effect. A few examples make this clear. The items on the left all assert cause and effect. Those on the right do not—they are arguments.

CAUSAL CLAIM/EXPLANATION	ARGUMENT
The reason Republicans are fiscally conservative is that big business gives a lot of money to fiscal conservatives.	The reason you should vote for Republicans is that they are fiscally conservative.
Dennis didn't live up to expectations; that's why they fired him.	I heard that Dennis didn't live up to expectations; that's why I am pretty sure they will fire him.
The water in the eastern Pacific was 5 degrees above normal in 2002, making 2002 a wetter year across the southern tier of states.	The water in the eastern Pacific was 5 degrees above normal in to assume 2002 was a wetter year across the southern tier of states.

So, the moral is obvious. Don't be fooled by superficial similarities. Causal explanations can look superficially like arguments, but they assert cause and effect. Arguments try to prove that something is the case.

The only complication is that, unfortunately, *sometimes causal explanations are used as premises in arguments.* For example, you might explain *what caused* the carburetor on a lawn mower to be gummed up to prove (argue) *that* the carburetor is gummed up. A Ford ad might explain *what causes* Ford trucks to be best-sellers in order to prove (argue) *that* they are great trucks. Your doctor might explain what caused you to get scarlet fever in order to prove (argue) that what you have is indeed scarlet fever.

Consider the following additional example:

You won't be able to see the comet from here tonight. The reason is that we're too close to the lights of the city to see anything that faint in the sky.

The speaker explains what it is that *will cause* you to not be able to see the comet from here; but it is very likely that he or she wants to provide you with evidence that you won't be able to see the comet from here. In that case, he or she is using this causal explanation to *argue* that you won't be able to see the comet from here tonight.

The moral is simple: Causal explanations can function as premises of arguments.

■ *Explanations and Excuses*

Let's suppose you are late to class—maybe you had a flat tire or your battery was dead. You don't want the professor to mark you down, so you explain why you are late. This is a causal explanation used as an argument; you are arguing that, since your lateness was caused by a flat tire (or whatever), *there-*

Study Backs an Old Idea about Crime

Target petty offenses to prevent serious ones, the theory goes.

Maybe you know about the Broken Windows theory of law enforcement. According to that theory, graffiti, petty vandalism, and other misdemeanor crimes in a neighborhood lead to more serious crime problems; and curtailing blight in a neighborhood prevents it from descending into a world of burglary, drugs, car theft, and violent crimes. Mayor Rudolph Giuliani vigorously subscribed to this theory during his term in office, during which New York City became a cleaner, safer place.

However, the theory had apparently not had much statistical support until an analysis of crime rates for all California counties from 1989 to 2000, by the California Institute for County Government, disclosed that the rate of arrests for serious property crimes indeed fell (overall by 3.6 percent) as the ratio for misdemeanor arrests increased. In some counties, where efforts were made to streamline prosecution for petty crimes, the decline in serious crimes was quite dramatic.

This makes sense, because some people who commit petty crimes also commit more serious crimes, and locking them up automatically reduces the number of serious crimes. But the findings (or at least the newspaper reports of them) leave room for skepticism as to the Broken Windows idea. One, the reduction in the rate of serious crimes might be the cause rather than the effect of an increase in petty crime arrest. The study was limited to property crimes like burglary and auto theft, which may well have declined anyway in this period of relative prosperity, thus increasing enforcement opportunities for misdemeanors. Two, apparently the crime rates were the averages for the counties, which gives rural counties disproportionate weight in the final results. In rural counties, even a small increase or decrease in absolute numbers can produce dramatic percentage shifts.

The most you can say is that the findings at least don't contradict the Broken Windows theory.

Source: *Sacramento Bee*, August 12, 2002.

fore you should be excused for being late. When we try to justify or defend or excuse something we (or someone else) did, we sometimes explain its causes. Imagine David is caught stealing a loaf of bread; he then explains to the authorities that he needed the bread to feed his children. David is attempting to argue that his behavior should be excused, and his explanation of what caused him to steal the bread provides grounds for his argument.

Not every attempt to explain behavior is an attempt to excuse or defend the behavior. After the September 11 suicide attacks on the World Trade Center, speakers on some university campuses offered explanations of the causes of the attacks. In some cases they were accused of trying to excuse the attacks, and perhaps some of them were trying to do that. But others weren't trying to excuse the attacks; they were simply trying to explain their causes. The distinction was lost on some people; attempts to explain the causes of the attacks were frequently condemned as traitorous. (In fact, one of our own colleagues who tried to express his views about the geopolitical causes of the attacks made the national news and was promptly invited by Rush Limbaugh to move to Afghanistan.)

If you assume without thinking about it that anyone who tries to explain the causes of bad behavior is trying to excuse it, you commit a fallacy. (Remember from Chapter 2 that a fallacy is a mistake in reasoning.) This fallacy, which we shall call **confusing explanations and excuses,** is common. For example, someone may try to explain why many Germans adopted the vicious anti-Semitic views of the Nazi party during the 1930s. The speaker may point out that the German economy was in a mess, that the country still suffered from terms imposed on it at the end of World War I, and, furthermore, that people often look for scapegoats on whom they can blame their troubles. These remarks may help us understand why the German people were easily led into anti-Semitism, but an uncritical listener may believe that the speaker is trying to muster sympathy for those people and their actions. As a matter of fact, just such an explanation was given by Phillipp Jenniger in the German parliament on the fiftieth anniversary of the *Kristallnacht,* a night when Nazis attacked Jewish neighborhoods and shops. Many of Jenniger's fellow Germans misunderstood his account, thinking it was designed to make excuses for the Nazis' actions. But it does not follow that a person proposing explanations has any sympathy at all for the views or actions being explained.

To sum things up, sometimes we explain the causes of something that sounds bad in order to excuse it; but sometimes such explanations are simply used to explain. Causal explanations can be entirely neutral with regard to approval or disapproval of an action. If we fail to see the difference between explaining behavior and excusing it, we may stifle attempts to explain it. The speaker may worry that we will think he or she is defending the bad behavior. This would be unfortunate because, lacking an explanation for such behavior, we may be hampered in our attempts to discourage it.

Recap

When we reason about cause and effect in informal, real-life situations, we often (but not always) want to know what caused some effect that interests us. It is never legitimate to conclude that one thing caused another just because the two things occurred around the same time—this is the *post hoc* fallacy. Instead, we must use either relevant-difference reasoning or common-thread reasoning. However, we must be careful not to focus on irrelevant differences or common threads, and we must not overlook alternative relevant common threads and differences. In addition, we should consider the possibilities that causation is the reverse of what we suspect and that the effect and suspected cause are both the effects of some third, underlying cause. Further, we must consider whether the effect and suspected cause are connected merely through coincidence.

We establish that something is a causal factor in a population on the basis of cause-to-effect experimentation or on the basis of cause-to-effect or effect-to-cause studies. We cannot establish causation on the basis of an example or two that looks like cause and effect.

Causal claims can suffer from inherent defects, most notably circularity, nontestability, excessive vagueness, dependence on unnecessary assumptions, or conflict with well-established theory.

Despite superficial similarities in phrasing, causal claims and arguments respond to different questions. A complication is that causal claims can be used as premises of arguments. Sometimes arguments that use causal claims in this way attempt to excuse something, but it is a mistake to think that they all do.

Additional Exercises

Exercise 11-13

In twenty-five words or less, argue *that* a lot of kids cheat in high school. Then, again in twenty-five words or less, explain *why* a lot of kids cheat in high school. You may be asked by your instructor to read your response. The idea here is to make sure you understand the difference between an argument and an explanation.

Exercise 11-14

In this exercise, five items are best seen as arguments and five are best seen as explanations. Sort the items into the proper categories.

1. I got sick because I didn't get enough rest over the weekend.
2. You shouldn't buy that computer because it doesn't have very much memory.
3. The president should resign because foreign leaders are going to lose faith in his leadership skills.
4. Pine trees are called "evergreens" because they don't lose their leaves in the fall.
5. You are making a mistake to wear that outfit because it makes you look too old.
6. Stephanie won't wear outfits like that because she thinks they are tacky.
7. Used couches are so plentiful here because the students give them away at the end of each school year.

8. If I were you, I wouldn't open up a furniture store in this town, because students give good furniture away at the end of each school year.

9. Freestone peaches are better than clingstone, because they are easier to eat and have a crisper texture.

10. The sky appears blue because of the way light reflects off the atmosphere.

Exercise 11-15

In this exercise there are ten items. Four are arguments and seven are explanations. That adds up to eleven, and that's because one explanation is used to justify or excuse behavior. Determine which items fall into which categories.

▲ 1. Collins is absent again today because she's ill.

2. I told you Collins was ill. Just look at her color.

3. Did Bobbie have a good time last night? Are you kidding? She had a *great* time! She stayed up all night, she had such a good time.

▲ 4. The reason Collins was ill is that she ate and drank more than she should have.

5. For a while there, Jazzercise was really popular. But you don't hear much about it anymore. That's probably because rock makes you feel more like moving than jazz. Jazz just makes you feel like moving into another room.

6. How come light beer has fewer calories than real beer? It has less alcohol, that's why.

▲ 7. The senator's popularity goes up and down, up and down. That's because sometimes he says things that make sense, and other times he says things that are totally outrageous.

8. VIKKI: Say, remember the California Raisins? Whatever happened to them, anyway?
   NIKKI: They faded. I guess people got tired of them or something.

9. Programs of preferential treatment may seem unfair to white males, but a certain amount of unfairness to individuals can be tolerated for the sake of the common good. Therefore, such programs are ethically justified.

▲ 10. I'm telling him, "Let's book!" and he's like, "Relax, dude," and since it's his wheel, what can I do? That's why we're late, man.

Exercise 11-16

Which of the following are arguments and which are explanations? Do any of the arguments use explanations to justify or excuse behavior?

▲ 1. What a winter! And to think it's all just because there's a bunch of cold water off the California coast.

2. Hmmmm. I'm pretty sure you have the flu. You can tell because if you had a cold you wouldn't have aches and a fever. Aches and fever are a sure sign you have the flu.

3. You know, it occurs to me the reason the band sounded so bad is the new director. They haven't had time to get used to him or something.

▲ 4. For a while there it seemed like every male under thirty was shaving his head. It was probably the Michael Jordan influence.

▲ 5. Believe it or not, couples who regard each other as relatively equal are more likely to suffer from high blood pressure than are couples who perceive one or the other as dominant. This is an excellent reason for marrying someone you think is beneath (or above) you.

6. Believe it or not, couples who regard each other as relatively equal are more likely to suffer from high blood pressure than are couples who perceive one or the other as dominant. The reason seems to be that couples who see their partners as relatively equal tend to argue more often and more vigorously, pushing up blood pressure.

Exercise 11-17

Which of the following explanations is used to argue that a certain behavior is or was justified or should be excused, and which isn't?

▲ 1. You don't believe me when I say sometimes you can see Pluto with the naked eye? Just think of how the solar system works. The planets all orbit around the sun, and at a certain point, Pluto's orbit gets very close to Earth's orbit.

▲ 2. Yes, yes, I know Susan never goes out, but you can hardly blame her. She thinks it is important to study, and she may be right.

▲ 3. Harold didn't return the book when it was due, but he couldn't help it. Somebody broke into his car and stole his backpack, and the book was in it.

4. Maria really does have perfect pitch. See, her parents were both musicians, and perfect pitch is hereditary. She inherited it from them.

5. It takes time and effort to write decent essays. If you doubt it, just let me tell you what all is involved. You have to pick a topic. You have to think about the audience. You have to organize things and compose an outline. And you have to write about 50 million drafts.

6. Believe it or not, in Germany, it is illegal to name a child "Osama bin Laden." Maybe you think no government has the right to tell someone they can't name their kid what they want; but the way the Germans look at it, the government should protect children, and children with names like "Osama bin Laden" could be picked on at school.

7. "I'm really sorry our dogs made so much noise last night. They thought someone was trying to break into the house, and it took a long time for us to quiet them down."

8. Well, ordinarily I wouldn't be happy if I saw someone cutting down trees in his yard, but those trees are diseased and constantly losing branches. He's doing the right thing to take them out.

9. Sanctions against South Africa are usually cited as bringing down apartheid, but the real explanation was the diplomatic isolation and the threat from guerrillas in neighboring countries.

10. Why in heck would anyone plant zucchini when they could be growing something good like tomatoes or eggplant? The only reason I can see is she must be nuts.

Exercise 11-18

Several of these items seem clearly to be arguments. Among these arguments are three explanations used to argue something. Of these, one is used to justify someone's behavior. Identify which of the items fall into these three categories, and then state what the remaining three items are.

1. If you ask a person to pick a number between 12 and 5, he'll probably say 7. That's because his brain automatically subtracts 5 from 12.

2. Ask a person to name a vegetable, and the first thing to come to mind is a carrot. Don't believe me? Ask someone.

3. To be acid-free, take Pepcid AC.

4. Why, just look at all the dog hair on this keyboard. Where do you let your dog sleep, anyway? No wonder your computer isn't working right.

5. On September 19, Mike Tyson will make the holy trek to Las Vegas, to ask the Nevada Athletic Commission to reinstate his boxing license, and chances are they will. Why would they do that for a man who bit another boxer, attacked a sixty-two-year-old man, and kicked another older man in the groin? The answer can be stated in a sentence: Vegas likes big money.

6. Feet hurt? Really hurt? Chances are you have weak or fallen arches.

7. "Let me explain to you why you know the public opinion polls are all rigged. Everybody I know hates Clinton, and if you listen to talk radio you can tell that most people agree with us."

Adapted from a newspaper call-in column

8. Linda takes care of two youngsters, puts in a forty-hour week at her job, is troop leader for the local chapter of the Girl Scouts, serves half a day each week as a teacher's aide, and is almost finished with a book on time management. So it shouldn't surprise you that she hasn't been able to come out to the coast for a visit lately.

9. I believe God exists because my parents brought me up that way.

10. I believe God exists because there obviously had to be a first cause for everything, and that first cause was God.

Exercise 11-19

For each of the following passages, determine whether the writer or speaker is most likely presenting an argument or an explanation, or both. Determine also whether any argument present attempts to justify an action or practice. If more information about the context is needed to answer, or if your answer assumes some unusual context for the passage, explain.

▲ 1. I can't understand how the garage got this cluttered. Must be we just never throw anything away.

   2. When the sun reaches its equinox, it is directly over the equator, a fact that explains why the days are exactly the same length every place on the globe at that time.

   3. Awww, don't get on her, Mom. The reason she didn't rake the leaves is that her stomach began hurting. She *had* to go lie down.

▲ 4. Watch out! Parker's giving a test today. I saw him carrying a big manila folder into the classroom.

   5. Moore gave a test on Friday because he wanted to surprise everyone.

   6. They get so many fires in southern California in the fall because that's when you get the Santa Ana winds, which blow in from the desert and make everything hot and extremely dry.

   7. LATE SLEEPER: Those idiot garbage collectors. Why do they have to make so much noise so early in the morning, anyway?
      EARLY RISER: Give 'em a break. They gotta make noise when they work, and they gotta start work early to get done on time.

▲ 8. HOSTESS: You know, I really think you should shave. The company's coming in less than an hour and you look like a hairy pig.
      HOST: Hey, I told you! I can't shave because my razor is broken. What am I supposed to do, shave with my pocket knife?

   9. So you actually believe he knows the names of the people in the audience because God tells him? Here's how he really does it. He's got a wireless radio receiver in his ear, okay? And the people fill out prayer cards before the show. Then his wife collects them and radios the details to him during the show.

▲ 10. Let me explain to you why that was a great movie. The acting was good, the story was interesting, the photography was a knockout, and the ending was a killer.

   11. Harold must be rich. Just look at the car he's driving.

▲ 12. I agree that the water contamination around here is getting worse and worse. It's because we've allowed people to install septic systems whenever and wherever they want.

   13. "I could never endure to shake hands with Mr. Slope. A cold, clammy perspiration always exudes from him, the small drops are ever to be seen standing on his brow, and his friendly grasp is unpleasant."
      Anthony Trollope, Barchester Towers

14. I know you think I wasted my vote by voting for Ralph Nader in the 2000 election, but that isn't so. I didn't approve of either of the main candidates, and if I'd voted for one of them and he had won, then my vote would have counted as an endorsement of him and his views. I don't want to add my endorsement to what I think are bad policies.

▲ 15. It could be that you're unable to sleep because of all that coffee you drink in the evening.

16. The coffee I drink in the evening can't be why I'm not sleeping, because I drink only decaffeinated coffee.

17. Of *course* the real estate industry depends on tax benefits. Just look at how hard the real estate lobby fought to preserve those benefits.

▲ 18. Although the computer case is double-insulated, the pins in the cable connections are not insulated at all. In fact, they are connected directly to the logic board inside the machine. So, if you are carrying a charge of static electricity and you touch those connector pins, you can fry the logic circuits of your computer.

19. The orange is sour because it didn't have a chance to ripen properly.

20. "In 1970 Chrysler abandoned reverse-thread lug bolts on the left-hand side of its cars and trucks. One of those engineers must have realized, after about fifty years of close observation, that sure enough, none of the wheels were falling off the competition's cars, which had your ordinary, right-hand wheel fastenings."

John Jerome, Truck

▲ 21. "Economically, women are substantially worse off than men. They do not receive any pay for the work that is done in the home. As members of the labor force their wages are significantly lower than those paid to men, even when they are engaged in similar work and have similar educational backgrounds."

Richard Wasserstrom, "On Racism and Sexism," in Today's Moral Problems

22. Some people think that Iowa has too much influence in determining who becomes a leading presidential candidate, but I don't agree. Having early caucuses in a relatively small state makes it possible for a person to begin a run at the presidency without spending millions and millions of dollars. And it may as well be Iowa as any other state.

23. "The handsome physical setting of Los Angeles is more threatened than the settings of most of the world's major cities. All the region's residents are affected by the ever-present threat of earthquakes, foul-smelling and chemical-laden tap water, and the potential for water shortages by the year 2000."

Charles Lockwood and Christopher Leinberger, "Los Angeles Comes of Age," in the Atlantic Monthly

24. She appears hard-hearted, but in reality she is not. She maintains strict discipline because she believes that if her children learn self-discipline, in the long run they will lead happier, more productive lives.

▲ 25. Letter to the editor: "Many people are under the impression that the Humane Society is an animal rights organization. Let me correct this misconception. Our purpose is that of an animal welfare organization. Animal rights groups have some excellent ideas but they are too broad of scope for us. We focus our attention on the care and welfare of the animals in our community. It is our primary purpose to provide humane care for homeless and owner-surrendered animals. However, we are against mistreatment of animals and we assist the State Humane Officer who investigates cases of animal abuse."

▲ 26. "American Airlines, United Airlines and TWA confirmed [before September 11, 2001] to the *Los Angeles Times* that some of their crews bypass passenger metal detectors. They said their current security procedures are effective and argued that switching to metal detectors would be costly and inefficient. Furthermore, they said, their security policies have been approved by the Federal Aviation Administration."

Los Angeles Times

Exercise 11-20

Which of the following items is an explanation used as an argument? Remember, trying to justify behavior is giving an argument.

▲ 1. The reason the door keeps banging is that the windows are open on the south side of the house and there is a strong south breeze.

2. Moore always starts his class exactly on the hour, because he has a lot to cover, and he can't afford to waste time.

3. Moore always starts his class exactly on the hour. Don't believe me? Just ask anyone.

▲ 4. The Rotary Club provides free dinners on Thanksgiving, the reason being that they want to help the poor.

5. So you think the lawn mower won't start because it's too old? Here's what's really going on. We let gas sit in the carburetor all winter and it gums up the works—that's why it won't start. It has nothing to do with its being too old.

6. "If slavery had been put to a vote, slavery never would have been overturned. We don't accept that you can put civil rights to a vote in a sexist and racist society."

Heather Bergman, member of the Coalition to Defend Affirmative Action by Any Means Necessary

▲ 7. Rejecting a last-minute bid by Democrats to give President Clinton's lawyers a few days for review, House Republican leaders on Thursday arranged to make public today 445 pages of Independent Counsel Kenneth Starr's report on possible impeachable offenses by the president. It may seem as if the Republicans are seeking to embarrass the president as much as they can, but in fact they called for the release because they feel the public deserves to know as soon as possible why Starr believes he has credible evidence that Clinton committed several crimes.

8. The share of public-school budgets devoted to regular education plummeted from 80 percent in 1967 to less than 59 percent in 1996. The rest goes to students with special needs. This at least partially accounts for declines in test scores among America's average students.

9. Just as Mary Shelley's Frankenstein monster fascinated us, computer viruses have come to do the same because they are much like a man-made life-form. Because of this fascination, you'll see more and more viruses turn up in hoaxes, urban legends, television shows, and movies.

Adapted from Scientific American

▲ 10. It is becoming more and more clear that child molestation is passed down from one generation to another. What isn't clear is whether the tendency is actually an inherited trait or whether a young child's being molested by a parent psychologically conditions the child to be a molester when he becomes an adult. Either way, it looks as though a child with a molester for a parent is not the same as the rest of us; he has something to overcome that we don't, and it may be impossible for us to understand what a difficult problem he may have in avoiding the forbidden behavior.

Exercise 11-21 _____

According to statistics from the Department of Transportation (reported by John Lang of the Scripps Howard News Service), in 1996 men accounted for a little more than 50 percent of licensed drivers and women for a little more than 49 percent. But according to these 1996 statistics, 41,010 male drivers were involved in fatal crashes, as opposed to only 14,145 female drivers. (This works out to about 74 percent male.) Of the drivers killed in those crashes, it was 17,822 male and 6,632 female. (This works out to about 73 percent male.)

   Also, according to DOT statistics, in 1996, 86 percent of the people killed while pedaling a bicycle were men.

   Which of the following conclusions can you derive from this information? (Assume all the conclusions pertain to 1996.)

1. Men were more likely than women to be involved in a fatal automobile crash.
2. If an automobile crash involved a fatality, it was more likely to have been a man than a woman.
3. Your brother was more likely to have been involved in a fatal automobile crash than your sister.
4. If your sibling was involved in a fatal automobile crash, it's more likely a brother than a sister.
5. Male drivers were more likely than females drivers to be killed in an automobile crash.
6. Men were less skilled than women at avoiding fatal automobile crashes.
7. Men were less able than women to avoid fatal automobile crashes.

8. Men were more likely than women to be killed while pedaling a bicycle.

9. If someone you know was killed while pedaling a bicycle, chances are the victim was a male.

10. If someone you know was killed in a traffic accident, it is more likely it was a bicycle accident than an auto accident.

Exercise 11-22

Men are involved in far more fatal automobile crashes than are women. List as many plausible explanations for this as you can.

Exercise 11-23

Let's say you randomly divide 700 men in the early stages of prostate cancer into two groups. The men in one group have their prostate removed surgically; those in the other group are simply watched to let the disease take its course. Researchers did this to 700 Scandinavian men and reported the results in the *New England Journal of Medicine* in fall 2002. As it turns out, 16 of those who underwent surgery died from prostate cancer, as compared with 31 of those who did not undergo surgery. On the face of it, these figures suggest your chances of not dying from prostate cancer are better if you have surgery. But put on your thinking caps and answer the following questions.

1. Suppose that despite these findings, there was no statistically significant difference in how long the men in each group lived. What would that suggest?

2. The follow-up comparison lasted six years. Suppose that after ten years the death rates from prostate cancer were the same for the two groups. What would that suggest?

3. Suppose Scandinavian men are not screened for prostate cancer as aggressively as American men and tend to be older when they get their first diagnosis.

4. Suppose Scandinavian men are screened more aggressively for prostate cancer than are American men and tend to be younger when they get their first diagnosis.

Here, as elsewhere, you need to know the whole picture to make a judgment. How old were the men to begin with? If they were relatively young men, how long did the study last? Was there a difference in how long the men in the two groups lived? (Note that prostate removal has risks and sometimes produces important negative side effects.)

Writing Exercises

1. How might one determine experimentally whether or not gender-based or ethnicity-based discrimination occurs in hiring? (Differences in unemploy-

Are Women Less Competitive?

Studies uncover a striking pattern

Although women have made huge strides in catching up with men in the workplace, a gender gap persists both in wages and levels of advancement. Commonly cited explanations for this gap range from charges of sex discrimination to claims that women are more sensitive than men to work-family conflicts and thus less inclined to make sacrifices for their careers.

Now, however, two new studies by economists Uri Gneezy of the University of Chicago and Aldo Rustichini of the University of Minnesota suggest that another factor may be at work: a deeply ingrained difference in the way men and women react to competition that manifests itself even at an early age.

The first study focused on short races run by some 140 9- and 10-year-old boys and girls in a physical education class. At that age, there was no significant difference between the average speeds of boys and girls when each child ran the course alone. But when pairs of children with similar initial speeds ran the race again, things changed. Boys' speeds increased appreciably when running against either a boy or a girl, but more so when paired with a girl. Girls showed no increase when running against a boy and even ran a bit more slowly when paired with a girl.

The second study, by Gneezy, Rustichini, and Muriel Niederle of Stanford University, involved several hundred students at an elite Israeli technical university. Groups of six students were paid to solve simple maze problems on a computer. In some groups, subjects were paid 50¢ for each problem they solved during the experiment. In others, only the person solving the most problems got rewarded—but at the rate of $3 for each maze solved.

Regardless of the sexual makeup of the groups, men and women, on average, did equally well when students were paid for their own performance. But when only the top student was paid, average male performance rose sharply—by about 50%—while female performance remained the same.

The authors conclude that females tend to be far less responsive to competition than males—a tendency with important implications for women and business. It may hurt women in highly competitive labor markets, for example, and hamper efficient job placement—especially for positions in which competitiveness is not a useful trait.

That's something companies with highly competitive atmospheres may need to consider, says Rustichini. If they don't, the results could be "both a subtle bias against women and, in many cases, foregone worker productivity."

By Gene Koretz

Let's Race: Boys' speeds went up.

Source: *Business Week*, December 9, 2002.

ment rates can be powerful evidence, but it is not experimental.) See if you can design an experiment, and describe it in a brief essay.

2. Are women less competitive than men? In a brief essay, (a) explain what you think the investigations in the box show, if anything; or (b) set forth alternative explanations for the results; or (c) describe what implications you think these investigations have.

Prayer Heals!

A study by Duke University researchers found that angioplasty patients with acute heart ailments who were prayed for by seven religious groups did 50 to 100 percent better while in the hospital than patients who received no prayers.

The researchers divided 150 patients at the Durham Veterans Affairs medical center into five 30-person groups. All groups received traditional medical treatment.

The control group received only the traditional medical treatment. Three groups received nontraditional therapies: stress relaxation, guided imagery, or touch therapy. The names of the remaining group were sent to seven different prayer groups, which prayed for all 30 patients by name.

Neither the patients nor the hospital staff knew the treatment assignment of the control group or the group whose members were prayed for.

The researchers, Dr. Mitch Krucoff, a Duke cardiologist, and nurse Suzanne Crater, developed statistical measures using EEG monitoring heart rate, blood pressure, and clinical outcomes to examine the effectiveness of the four different treatments.

In addition to the improvements recorded in those who received prayers, the patients who received the other nontraditional treatments did 30 to 50 percent better than the control group.

"If people are seeking to put themselves into the right relationship with God, which is what prayer is, then you would expect it to have some beneficial effects," said Dr. Geoffrey Wainwright, a professor at the Duke University School of Divinity.

—Adapted from an *Associated Press* report written by Scott Mooneyham

3. Carefully read the story about prayer in the box and note what conclusions you think would most likely be drawn from it by most readers. Then determine how well those conclusions are actually supported by the evidence as described in the account. Identify any problems of vagueness or any other lack of clarity, missing information, and so forth.

4. Which of the following causal hypotheses do you accept? Select one that you accept and write a one-page essay presenting evidence (giving arguments) for the claim. When you are finished, write down on a separate piece of paper a number between 1 and 10 that indicates how strong you think your overall "proof" of the claim is, with 10 = very strong and 1 = very weak. Take about ten minutes to complete your essay. Write your name on the back of your paper.

Causal hypotheses:

Pot leads to heroin/crack addiction.
The death penalty is a deterrent to murder.
The death penalty is not a deterrent to murder.
Chocolate causes acne.
Welfare makes people lazy.
Spare the rod and spoil the child.
People think better after a few beers.
Pornography contributes to violence against women.
If you want to get someone to like you, play hard to get.
Vitamin C prevents colds.
Rap music (TV, the movies) contributes to the crime problem.

When everyone is finished, the instructor will collect the papers and redistribute them to the class. In groups of four or five, read the papers and assign a number from 1 to 10 to each one (10 = very strong; 1 = very weak). When all groups are finished, return the papers to their authors. When you get your paper back, compare the number you assigned to your work with the number the group assigned to it. The instructor may ask volunteers to defend their own judgment of their work against the judgment of the group. Do you think there is as much evidence for the claim you selected as you did when you argued for it initially?

Chapter 12

Moral, Legal, and Aesthetic Reasoning

with Nina Rosenstand and Anita Silvers

In early winter 2003, our friend Elliott Howard was offered a position as the manager of a country club near Winston-Salem, North Carolina.* For Elliott, this was the chance of a lifetime. Unfortunately, there was a problem. Elliott's father, who lived near Elliott in California, was old and had non-Hodgkins lymphoma. It was vital that Elliott live close by to care for him.

This presented a dilemma. On the one hand, Elliott's father depended on Elliott. On the other, Elliott knew he would not get another chance like this. He was getting on in years himself, and the job offer had come to him only through an unusual chain of events, one that would not likely ever be repeated.

Elliott attempted to find a way to accept the position and still provide for his father. Unfortunately, this turned out not to be possible. In the end, Elliott was faced with this hard choice: He could take the job, or he could take care of his father, but he could not do both.

Finally, reluctantly, Elliott turned down the job offer.

From time to time we all face tough moral decisions like this. Sometimes we must weigh our own interests against our responsibilities to people we love. Sometimes we face moral dilemmas of a different sort. When he was governor of Texas, George W. Bush had to decide whether to bestow clemency on Karla Faye Tucker, a woman whose life had undergone a transformation in prison and whose execution many people believed to be unwarranted. A mother may have to determine whether her daughter's birthday takes precedence over a professional responsibility. It can be excruciating to decide whether a dear pet's time has finally come.

When we think abstractly, we might have the idea that morality is a mere matter of personal opinion and that moral issues offer little room for reasoning. But when we are confronted with a real moral dilemma, we see

---

*We have altered the circumstances slightly and changed our friend's name to protect his identity.

instantly that this view is not correct. We reason about moral issues. We consider options, weigh consequences, think about what is right and wrong and why. Some considerations are more important than others, and some arguments are better than others. If Elliott decided what to do by flipping a coin, we would think there was something wrong with him.

In the first part of this chapter, we want to look at the basic concepts and principles that are always involved in moral reasoning.

Factual Issues and Nonfactual Issues

Think back to Chapter 1 again, or reread pages 11–13; better yet, do both. There we called your attention to the distinction between factual issues and claims and nonfactual issues and claims. You take a position on a factual issue by making a factual claim. And there are two ways of telling whether an issue is factual. If, when two people take opposite positions on an issue, it is necessary that one of them be mistaken, the issue is factual. Alternatively, if methods or criteria can be described for settling an issue, it is factual.

Thus, the claims in the left column are factual claims (that is, a person making them is taking a position on a factual issue). The claims in the right column are not.

CLAIMS

Factual Claims	Nonfactual Claims
1. Gina is twenty-seven.	1. Gina is mature for her age.
2. Karl Rove is the president's campaign adviser.	2. Karl Rove has excellent people skills.
3. The third *Rings* movie lasts over three hours.	3. The third *Rings* movie is tedious and boring.
4. Senator Kennedy did not graduate from Harvard.	4. Senator Kennedy is generous and kind.
5. Anyone under eighteen is still legally a child.	5. Anyone under eighteen is too inexperienced to make wise decisions.

Now, at this point we must subdivide the category of nonfactual issues/claims. First, there are nonfactual issues/claims that express values and those that do not. You may well have heard the expression "value judgment": That's a widely used name for a claim that expresses a value, and we shall use that term as well.*

---

*Philosophers often use the term "prescriptive claims" interchangeably with value judgment, and your instructor may wish to do so as well.

NONFACTUAL CLAIMS

Nonfactual Claims That Are Value Judgments	*Nonfactual Claims That Are Not Value Judgments*
1. Gina shouldn't ask people for money.	1. Gina has unusually dainty features.
2. Karl Rove ought to spend more time with his family.	2. Karl Rove looks a lot like Santa Claus.
3. The third *Rings* movie is better than the first.	3. The third *Rings* movie is scarier than the first.
4. Senator Kennedy is an excellent cellist.	4. Senator Kennedy is a tough-minded advocate of civil liberties.
5. Children should be taught to appreciate music.	5. Children are avid learners.

And we need to make one further distinction, this one between value judgments that assign moral values to something and value judgments that do not. For example, if you assert that one should not permit dogs to suffer needlessly, you are assigning a (negative) moral value to dogs' being allowed to suffer needlessly. If you assert that dogs make wonderful companions for old people, you are assigning a (positive) nonmoral value to dogs. Thus:

VALUE JUDGMENTS

Moral Value Judgments	*Nonmoral Value Judgments*
1. Gina is a really good person.	1. Gina has beautiful eyes.
2. Karl Rove frequently deceives other people.	2. Karl Rove dresses very well.
3. The third *Rings* movie is a very bad influence on people's behavior.	3. The third *Rings* movie has some of the best special effects of any movie made this year.
4. It was wrong of Senator Kennedy to have withheld information.	4. It was terrible timing for Senator Kennedy to have been involved in the trouble during the election.
5. Sam's son was responsible for the bullying incident at school.	5. Children are responsible for decisions made about many families' vacations.

Now we can bring into clear focus what moral reasoning is all about. *Moral reasoning is distinguished from other kinds of reasoning by the fact that the conclusions it tries to reach are moral value judgments.*

Typically, moral value judgments employ such words as "good," "bad," "right," "wrong," "ought," "should," "proper," and "justified." But one must also bear in mind that, although these words often signal a moral evaluation, they do not always do so. For example, if you tell somebody she ought to keep her promises, that's a moral value judgment. If you tell her she ought to keep her knees bent when skiing, you are assigning a positive value to her keeping her knees bent when skiing, but not a moral value.

Before continuing, do these exercises to help clarify these various important distinctions.

Exercise 12-1

Determine which of the following are factual claims and which are nonfactual claims.

▲ 1. Lizards make fine pets.
2. You can get a clothes rack at True Value for less than $15.00.
3. The last haircut I got at Supercuts was just totally awful.
▲ 4. It was a great year for regional politics.
5. Key officials of the Department of Defense are producing their own unverified intelligence reports about an arms buildup.
6. Texas leads the nation in accidental deaths caused by police chases.
▲ 7. Napoleon Bonaparte was the greatest military leader of modern times.
8. Racial segregation is immoral anytime anywhere.
9. President Bush is deploying a "missile defense" that hasn't been adequately tested.
▲ 10. Air consists mainly of nitrogen and oxygen.

Exercise 12-2

Determine which of the following nonfactual claims are value judgments and which are not.

▲ 1. T-shirts made by Fruit of the Loom are soft and luxurious.
2. Rumsfeld is not nearly as detailed in reports to the press as is Powell.
3. The Pentagon was not nearly as supportive of a war as it should have been.
▲ 4. Tens of billions of dollars have been wasted on worthless public transportation schemes.
5. Atlanta is sultry in the summer.
6. Religious school teachers are stricter than their nonreligious counterparts.
▲ 7. Six Flags has the scariest rides in the state.
8. The politician with the best sense of humor? That would have to have been Bob Dole.
9. Eugene is not nearly as happy as his wife, Polly.
10. Polly is more selfish than she should be.

Exercise 12-3 _____

Determine which of the following value judgments are moral value judgments and which are not.

▲   1. Marina's car puts out horrible smoke; for the sake of us all, she should get it tuned up.

    2. After the surgery, Nicky's eyesight improved considerably.

    3. Ms. Beeson ought not to have embezzled money from the bank.

▲   4. Violence is always wrong.

    5. Matthew ought to wear that sweater more often; it looks great on him.

    6. Sandy, you are one of the laziest people I know!

▲   7. My computer software is really good; it even corrects my grammar.

    8. "Little Lisa has been very good tonight," said the babysitter.

    9. Judge Ramesh is quite well-informed.

▲ 10. Judge Ramesh's decision gave each party exactly what it deserved.

   11. The editor couldn't use my illustrations; she said they were not particularly interesting.

   12. Wow. That was a tasty meal!

   13. The last set of essays was much better than the first set.

   14. Do unto others as you would have them do unto you.

   15. People who live in glass houses shouldn't throw stones.

   16. You really shouldn't make so much noise when the people upstairs are trying to sleep.

   17. It is unfair the way Professor Smith asks questions no normal person can answer.

   18. "Allegro" means fast, but not that fast!

   19. Being in touch with God gives your life meaning and value.

   20. Thou shalt not kill.

■ *Getting an "Ought" from an "Is": Factual Premises and Moral Value Judgments*

It has been said that no claim about plain fact—that is, no factual claim—can imply a claim that attributes a moral value or a moral obligation. In other words, we cannot legitimately infer what *ought (morally) to be* the case from a claim about what *is* the case. Using the terminology from the previous section, we can put it this way: No value judgment can follow from a set of purely factual premises. For example, consider the following argument:

   1a. Mr. Jones is the father of a young child. Therefore, Mr. Jones ought to contribute to his child's support.

We hear such claims often in the context of everyday life, and we are often persuaded by such conclusions. But we should beware, for the conclusion doesn't follow from the premise; it is, in other words, a **non sequitur.** Jumping to such conclusions is a common phenomenon that philosophers generally refer to as the **naturalistic fallacy.** Let us take a closer look. The premise of this argument states a (nonmoral) fact, but the conclusion is an "ought" claim that prescribes a moral obligation for Mr. Jones. As the philosopher David Hume pointed out, there is nothing in a purely factual statement that implies a moral duty—the concept of duty is added to the statement by our sense of right and wrong.

However, this seems to lead to an unacceptable situation, for shouldn't we, on the basis of facts and statistics, be able to make policies about social and moral issues? If we know that child abuse is likely to create a new generation of adults who will also practice child abuse, shouldn't we then have the right to state that "child abuse should be prevented by all means" without being accused of committing a non sequitur? The answer is that of course facts and statistics can be used to formulate social and moral policies, but not the facts and statistics *by themselves:* We must add a premise to make the argument valid. In this case the premise is the moral statement that "child abuse is wrong." Adding this statement may seem trivial, and you may object that this evaluation is already given with the word "abuse," but the fact is that our moral sense, whether it is based on feelings or reasoning or both, is what allows us to proceed from a set of facts to a conclusion about what ought to be done. Let us return to Mr. Jones:

1b. Premise: Mr. Jones is the father of a young child.

Premise: Parents ought to contribute to the support of their young children.

Conclusion: Therefore, Mr. Jones ought to contribute to his child's support.

The result is now a valid deductive argument. Now, in a real-life dispute about whether Mr. Jones has an obligation toward his child, an argument like example 1b is not likely to convince anyone that Mr. Jones has this obligation if he or she doesn't agree with the premise that parents should take care of their children. But the fact that the argument is valid does help clarify matters. It may be an undisputed fact that Jones is a parent, but it may not be as undisputed that parents ought to take care of their young children under all circumstances. In other words, if you disagree with the conclusion, you need not start disagreeing with the factual claim of the first premise, but you can enter into a discussion about the truth of the second premise. It is important to be able to enter into such discussions apart from the case of Mr. Jones, for otherwise it would be too easy for us to let our thinking be prejudiced by the facts presented in the first premise.

Let us consider a few additional examples:

2a. Sophie promised to pay Jennifer five dollars today. So Sophie ought to pay Jennifer five dollars today.

It is perfectly natural to want to see this as a convincing argument as it stands. But, in keeping with our strategy, we require that an "ought" claim be explicitly stated in the premises. In this case, what would you say is the missing premise? The most obvious way of putting it is "One ought to keep one's promises." If we add the required premise we get this:

> 2b. Sophie promised to pay Jennifer five dollars today. One ought to keep one's promises. So Sophie ought to pay Jennifer five dollars today.

If it seems to you that the second premise is not really necessary in order for the conclusion to follow, that is probably because you accept that making a promise involves a moral duty to keep the promise. In that case, the second premise would just be true by definition, but even so, there is no harm in stating it. And suppose it is *not* true by definition? Not all implicit moral evaluations are; and if they are not, then adding the premise is absolutely necessary in order for the argument to be valid. Consider this example:

> 3a. It is natural for women to have children. Therefore, women ought to have children, and those who try to avoid having children are unnatural.

The trouble with this argument is the concept of "natural," a very ambiguous word. There are several hidden assumptions behind this argument, and one of these—which we'll call the missing premise here—is the following: "What is natural is morally good, and going against nature is evil." We might also say, "What is natural reveals God's intentions with his creation, and if one goes against God's intentions, one is evil (or sinful)." This also explains the concept of "unnatural," for although "natural" in the premise looks completely descriptive, "unnatural" in the conclusion has a judgmental slant. If we add the missing premise, the argument will look like this:

> 3b. It is natural for women to have children. What is natural is morally good, and going against nature is evil. Therefore, women ought to have children, and those who try to avoid having children are evil.

Now the argument is valid, but you may not want to agree with the conclusion. If so, although in some sense it is, overall, "natural" for females to have offspring, you may want to dispute the premise that what is natural is always morally good. After all, many "natural" things, such as leaving severely disabled individuals to fend for themselves or scratching oneself in public, are considered unacceptable in our culture.

Whether an "ought" claim can ever follow directly from an "is" claim has been a controversial issue among philosophers; some point out that a statement such as "An open flame can burn you, so stay away from the campfire" makes an unproblematic transition from an "is" to an "ought"; other philosophers point out that there is a hidden premise that is taken for granted: that getting burned is bad because it hurts.

Exercise 12-4 _____

In each of the following passages, a moral principle must be added as an extra premise to make the argument valid. Supply the missing principle.

Example

> Mrs. Montez's new refrigerator was delivered yesterday, and it stopped working altogether. She has followed the directions carefully but still can't make it work. The people she bought it from should either come out and make it work or replace it with another one.

Principle

> People should make certain that the things they sell are in proper working order.

1. After borrowing Morey's car, Leo had an accident and crumpled a fender. So, Leo ought to pay whatever expenses were involved in getting Morey's car fixed.

▲ 2. When Sarah bought the lawn mower from Jean, she promised to pay another fifty dollars on the first of the month. Since it is now the first, Sarah should pay Jean the money.

3. Kevin worked on his sister's car all weekend. The least she could do is let him borrow the car for his job interview next Thursday.

4. Harold is obligated to supply ten cords of firewood to the lodge by the beginning of October, since he signed a contract guaranteeing delivery of the wood by that date.

▲ 5. Since it was revealed yesterday on the 11:00 news that Mayor Ahearn has been taking bribes, he should step down any day now.

6. As a political candidate, Ms. Havenhurst promised to put an end to crime in the inner city. Now that she is in office, we'd like to see some results.

▲ 7. Since he has committed his third felony, he should automatically go to prison for twenty-five years.

▲ 8. Laura's priest has advised Laura and her husband not to sign up for the in vitro fertilization program at the hospital because such treatments are unnatural.

9. Ali has been working overtime a lot lately, so he should receive a bonus.

10. It is true there are more voters in the northern part of the state. But that shouldn't allow the north to dictate to the south.

■ *Consistency and Fairness*

A common mistake made in moral reasoning is inconsistency—treating cases that are similar as if they weren't that way. For example, suppose Moore announces on the first day of class that the final in the class will be optional "except," he says, pointing at some person at random, "for the young woman there in the third row. For you," he says, "the final is mandatory."

The problem is that Moore is treating a student who is similar to the rest of the class as if she were different. And the student will want to know, of course, why she has been singled out: "What's so different about me?" she will wonder (as she drops the course).

A similar sort of problem occurs if Moore gives two students the same grade despite the fact that one student did far better in the course than the other. Treating dissimilar cases as if they were similar isn't inconsistent; it just involves a failure to make relevant discriminations.

As you have probably foreseen by now, when we treat people inconsistently, the result is often *unfair*. It is unfair to the woman in the third row to require her alone to take the final; alternatively, it would be unfair to the rest of the class to make the final mandatory for them and to make it optional for the woman in the third row. From the perspective of critical thinking, what is so troubling about unfairness is that it is illogical. It is like saying "All Xs are Y" and then adding, "Some Xs are not Y."

This doesn't mean, of course, that if you have been treating people unfairly, you should *continue* to treat them unfairly on the grounds that it would be inconsistent for you to change your policy. There is nothing illogical in saying, in effect, "I treated the Xs I encountered to date wrongly, but now I will not treat the Xs I encounter wrongly." People sometimes adhere to a bad policy simply on the grounds that it would be inconsistent of them to change, but there is no basis in logic for this idea.

Not all cases of inconsistency result in unfairness. For example, let's imagine that Harlan approved of the war in Iraq and opposed the war in Vietnam but is unable to point out any relevant differences between the two cases. This isn't a matter of unfairness, exactly; it's just inconsistency on Harlan's part.

When are two or more cases sufficiently similar to warrant our calling somebody inconsistent who treats them as if they were different? The answer is that there is no hard-and-fast rule. The point to keep in mind, however, is that the burden of proof is on the person who appears inconsistent to show that he or she is not treating similar cases dissimilarly. If, when challenged, Harlan cannot *tell* us what's different about Vietnam and Iraq that justifies his difference in attitude between the two, then we are justified in regarding him as inconsistent.

Imagine that Carol is a salesperson who treats black customers and white customers differently: She is, let us imagine, much more polite to customers of her own racial group (we needn't worry about which group that is). Can Carol explain to us what is so different about black and white customers that would justify her treating them differently? If not, we are justified in regarding her practices as inconsistent.

Suppose, however, that Carol thinks that skin color itself is a difference between blacks and whites relevant to how people should be treated, and she charges us with failing to make relevant discriminations. Here it would be easy for us to point out to Carol that skin color is an immutable characteristic of birth like height or eye color; does Carol adjust her civility to people depending on those characteristics?

It isn't difficult to perceive the inconsistency on the part of a salesperson who is more polite to customers of one group; but other cases are far tougher, and many are such that reasonable people will disagree about their proper assessment. Is a person inconsistent who approves of abortion but not capital punishment? Is a person inconsistent who, on the one hand, believes that the states should be free to reduce spending on welfare but, on the other, does not think that the states should be able to eliminate ceilings on punitive damages in tort cases? No harm is done in asking "What's the difference?" and because much headway can be made in a discussion by doing so, it seems wise to ask.

In Chapter 6, we talked about the inconsistency ad hominem, a fallacy we commit when we think we rebut the content of what someone says by pointing out inconsistency on his or her part. Now, let's say Ramesh tells us it is wrong to hunt, and then we find out Ramesh likes to fish. And let's say that, when we press Ramesh, he cannot think of any relevant moral difference between the two activities. Then he is being inconsistent. But that does not mean that it is right to hunt; nor does it mean that it is wrong to fish. An inconsistency ad hominem occurs if we say something like "Ramesh, you are mistaken when you say it is wrong to hunt, because you yourself fish." It is not an inconsistency ad hominem to say "Ramesh, you are being inconsistent. You must change your position either on the hunting or on the fishing."

Similarly, let's suppose Professor Moore gives Howard an A and gives James a C, but cannot think of any differences between their performance in his course. It would be the inconsistency ad hominem if we said "Moore, James does not deserve a C, because you gave Howard an A." Likewise, it would be the inconsistency ad hominem if we said "Moore, Howard does not deserve an A, because you gave James a C." But it is *not* illogical for us to say "Moore, you are being inconsistent. You have misgraded one of these students."

Exercise 12-5

Answer the question or respond to the statement that concludes each item.

▲ 1. Tory thinks that women should have the same rights as men. However, he also thinks that although a man should have the right to marry a woman, a woman should not have the right to marry a woman. Is Tory being consistent in his views?

▲ 2. At Shelley's university, the minimum GPA requirement for admission is relaxed for 6 percent of the incoming students. About half of those allowed to attend the university under one of its special admissions programs are affirmative action students—women and members of minorities. The other half are athletes, children of alumni, and talented art students. Shelley is opposed to special admissions programs for affirmative action students. She is not opposed to special admissions programs for art students, athletes, and children of alumni. Is Shelley consistent?

▲ 3. Marin does not approve of abortion because the Bible says explicitly, "Thou shalt not kill." "'Thou shalt not kill' means thou shalt not kill," he says. Marin does, however, approve of capital punishment. Is Marin consistent?

4. Koko believes that adults should have the unrestricted right to read whatever material they want to read, but she does not believe that her seventeen-year-old daughter Gina should have the unrestricted right to read whatever she wants to read. Is Koko consistent?

5. Jack maintains that the purpose of marriage is procreation. On these grounds he opposes same-sex marriages. "Gays can't create children," he explains. However, he does not oppose marriages between heterosexual partners who cannot have children due to age or medical reasons. "It's not the same," he says. Is Jack being consistent?

6. Alisha thinks the idea of outlawing cigarettes is ridiculous. "Give me a break," she says. "If you want to screw up your health with cigarettes, that's your own business." However, Alisha does not approve the legalization of marijuana. "Hel-loh-o," she says. "Marijuana is a *drug,* and the last thing we need is more druggies." Is Alisha being consistent?

7. California's Proposition 209 amends the California state constitution to prohibit "discrimination or preferential treatment" in state hiring based on race, gender, or ethnicity. Opponents say that Proposition 209 singles out women and members of racial and ethnic minorities for unequal treatment. Their argument is that Proposition 209 makes it impossible for members of these groups to obtain redress for past discrimination through preferential treatment, whereas members of other groups who may have suffered past discrimination (gays, for example, or members of religious groups) are not similarly restricted from seeking redress. Evaluate this argument.

▲ 8. Harold prides himself on being a liberal. He is delighted when a federal court issues a preliminary ruling that California's Proposition 209 (see previous item) is unconstitutional. "It makes no difference that a majority of California voters approved the measure," Harold argues. "If it is unconstitutional, then it is unconstitutional." However, California voters also recently passed an initiative that permits physicians to prescribe marijuana, and Harold is livid when the U.S. Attorney General says that the federal government will ignore the California statute and will use federal law to prosecute any physician who prescribes marijuana. Is Harold consistent?

9. Graybosch is of the opinion that we should not perform medical experiments on people against their will, but he has no problem with medical experiments being done on dogs. His wife disagrees. She sees no relevant difference between the two cases.

"What, no difference between people and dogs?" Graybosch asks.

"There are differences, but no differences that are relevant to the issue," Graybosch's wife responds. "Dogs feel pain and experience fear just as much as people."

Is Graybosch's wife correct?

10. Mr. Bork is startled when a friend tells him he should contribute to the welfare of others' children as much as to his own.

"Why on earth should I do that?" Mr. Bork asks his friend.

"Because," his friend responds, "there is no relevant difference between the two cases. The fact that your children are yours does not mean that there is something different about them that gives them a greater entitlement to happiness than anyone else's children."

How should Mr. Bork respond?

11. The university wants to raise the requirements for tenure. Professor Peterson, who doesn't have tenure, says that doing so is unfair to her. She argues that those who received tenure before she did weren't required to meet such exacting standards; therefore, neither should she. Is she correct?

12. Reverend Heintz has no objection to same-sex marriages, but is opposed to polygamous marriages. Is there a relevant difference between the two cases, or is Reverend Heintz being inconsistent?

■ *Major Perspectives in Moral Reasoning*

Moral reasoning usually takes place within one or more frameworks or perspectives. Here we present some of the perspectives that have been especially influential in Western moral thought.

Relativism   A popular view of ethics, especially perhaps among undergraduates taking a first course in philosophy, is **moral relativism,** which we'll define as the idea that what is right and wrong depends on and is determined by one's group or culture.

A mistake sometimes made in moral reasoning is to confuse the following two claims:

1. What is *believed* to be right and wrong may differ from group to group, society to society, or culture to culture.
2. What *is* right and wrong may differ from group to group, society to society, or culture to culture.

The second claim, but not the first, is moral relativism. Please read the two claims carefully. They are so similar it takes a moment to see they are actually quite different. But they are different. The first claim is incontestable; the second claim is controversial and problematic. It may well have been the majority belief in ancient Greece that there was nothing wrong with slavery. But that does not mean that at that time there was nothing wrong with slavery.

Another popular moral perspective is an extreme form of relativism known as **subjectivism,** according to which what is right and wrong is just a matter of subjective opinion. The theory often loses its appeal when some of its implications are noticed. For example, what if Parker thinks that it is right for a person to do X, and Moore thinks that it is wrong for a person to do X? Is it right or wrong for a person to do X? Subjectivism seems to place its adherents in a logically untenable situation.

Subjectivists sometimes think they avoid problems by saying something like "Well, it's right for Parker to do X, but wrong for Moore to do it." However, let's suppose that X in this case is torturing a pet. Would we wish to say, "Well, because Parker doesn't think there is anything wrong with torturing a pet, then there isn't anything wrong with *his* doing it"?

Relativism and subjectivism were popular among cultural anthropologists in the 1920s and still have popular appeal. But neither has won widespread acceptance among moral philosophers.

Utilitarianism   The perspective we call **utilitarianism** rests on the idea that if an individual can feel pleasure and pain, then he or she deserves moral consideration. (This broad criterion has led many utilitarians to include animals in their considerations.) The theory is based on one principle, the **principle of utility:** Maximize happiness and minimize unhappiness. This means that the utilitarian is concerned with the *consequences* of actions and decisions. If an act will produce more happiness than will the alternatives, the act is the right one to do; if it will produce less happiness, it would be morally wrong to do it in place of one of its alternatives.

Many of us use a pro-and-con list of consequences as a guideline when considering what course of action to take. Let's assume you have to decide whether to go home for Thanksgiving or to spend the long weekend writing a term paper that is due. Your family is expecting you, and they will be disappointed if you don't come home. However, their disappointment will be lessened by their knowing that your studies are important to you. On the other hand, if you stay to finish the paper, you will miss seeing your friends at home. But if you go home, you will not finish the paper on time, and your final grade may be adversely affected. As a utilitarian, you try to weigh the consequences of the alternatives on everyone's happiness. You also have to factor in how *certain* the outcomes of each alternative are with respect to happiness, assigning relatively more weight to relatively more certain positive outcomes. Because you can generally be more certain of the effect of an act on your own happiness and on the happiness of others you know well, it is often morally proper to favor that act that best promotes your own or their happiness. Of course, you must not use this as an excuse to be entirely self-serving: Your own happiness isn't more important morally than another's. The best course of action morally is not always the one that best promotes your own happiness.

In sum, utilitarians weigh the consequences of the alternatives, pro and con, and then choose the alternative that maximizes happiness. One of the original and most profound intellects behind utilitarianism, Jeremy

Bentham (1748–1832), even went so far as to devise a *hedonistic calculus*—a method of assigning actual numerical values to pleasures and pains based on their intensity, certainty, duration, and so forth. Other utilitarians think that some pleasures are of a higher quality (e.g., reading Shakespeare is of a higher quality than watching Daffy Duck). Although there are other important issues in utilitarianism, the basic idea involves weighing the consequences of possible actions in terms of happiness. Utilitarianism has considerable popular appeal, and real-life moral reasoning is often utilitarian to a considerable extent.

Nevertheless, some aspects of the theory are problematic. Typically, when we deliberate whether or not to do something, we don't always take into consideration just the effect of the action on happiness. For example, other people have *rights* that we sometimes take into account. We would not make someone in our family a slave even if the happiness produced for the family by doing so outweighed the unhappiness it created for the slave. We also consider our *duties* and *obligations*. We think it is our duty to return a loan to someone even if we are still short on cash and the other person doesn't need the money and doesn't even remember having loaned it to us. If we make a date and then want to break it because we've met the love of our life, we think twice about standing up our original date, even if we believe that our overall happiness will far outweigh the temporary unhappiness of our date. To many, the moral obligation of a promise cannot be ignored for the sake of the overall happiness that might result from breaking it.

In estimating the moral worth of what people do, utilitarianism also seems to discount people's *intentions*. Suppose a mugger attacks somebody just as a huge flower pot falls from a balcony above. The mugger happens to push the individual the instant before the flower pot lands on the exact spot where the victim had been standing. The mugger has saved the victim's life, as it turns out. But would we say that the mugger did a morally good deed just because his action had a happy result? According to utilitarianism we would, assuming the net result of the action was more happiness than would otherwise have been the case. So utilitarianism doesn't seem to be the complete story in moral reasoning.

Duty Theory/Deontologism   Immanuel Kant (1724–1804), who witnessed the beginning phases of the utilitarian philosophy, found that philosophy deficient because of its neglect, among other things, of moral duty. Kant's theory is a version of what is called **duty theory,** or **deontologism.**

Kant acknowledged that our lives are full of imperatives based on our own situations and our objectives. If we want to advance at work, then it is imperative that we keep our promises; if we are concerned about our friends' happiness, then it is imperative that we not talk about them behind their backs. But this type of **hypothetical imperative,** which tells us we ought to do (or ought not to do) something in order to achieve such and such a result, is not a *moral* imperative, Kant argued. Keeping a promise so we'll get a solid reputation is neither morally praiseworthy nor morally blameworthy, he said. For our act to be *morally* praiseworthy, it must be done, not

Acts and Rules

Let's say you are thinking of cheating on a test. It isn't inconceivable that the sum total of happiness in the world would be increased by this single *act* of cheating. But it also isn't inconceivable that if the *principle* involved were adopted widely, the sum total of happiness would be decreased.

This raises an interesting question: When calculating happiness outcomes, should we contemplate happiness outcomes of the particular *act* in question? Or should we contemplate happiness outcomes of adoption of the *principle* involved in the act?

Clearly there are difficult issues here. Accordingly, some philosophers have made a distinction between "act utilitarianism," which evaluates the moral worth of an act on the happiness it would produce, and "rule utilitarianism," which evaluates the moral worth of an act on the happiness that would be produced by adoption of the principle it exemplifies. (A possible middle ground might be to attempt to factor in, as a part of the happiness outcomes of a particular act, the likelihood that doing it will contribute to a general adoption of the principle involved. This is often what we do when we ask, "But what if everyone did this?")

for the sake of some objective, but simply because *it is right.* Our action of keeping our promise is morally praiseworthy, he said, only if we do it simply because it is right to keep our promises. A moral imperative is unconditional or **categorical;** it prescribes an action, not for the sake of some result, but simply because that action is our moral duty.

It follows from this philosophy that when it comes to evaluating an action morally, what counts is not the result or consequences of the action, as utilitarianism maintains, but the intention from which it is done. And the morally best intention, indeed in Kant's opinion the *only* truly morally praiseworthy intention, is that according to which you do something just because it is your moral duty.

But what makes something our moral duty? Some deontologists ground duty in human nature; others ground it in reason; in Western culture, of course, many believe moral duty is set by God. How can we tell what our duty is? Some believe our duty is to be found by consulting conscience; others believe that it is just self-evident or is clear to moral intuition. Those who maintain that human moral duties are established by God usually derive their specific understanding of these duties through interpretations of religious texts such as the Bible, though there is disagreement over what the correct interpretation is and even over who should do the interpreting.

Kant answered the question, How can we tell what our moral duty is? as follows: Suppose you are considering some course of action—say, whether to borrow some money you need very badly. But suppose that you know you can't pay back the loan. Is it morally permissible for you to borrow money under such circumstances? Kant said to do this: First, find the *maxim* (prin-

Inmate Who Got New Heart While Still in Prison Dies

A California prison inmate believed to be the first in the nation to receive a heart transplant while incarcerated has died, officials said Tuesday.

Department of Corrections spokesman Russ Heimerich said the inmate, whose identity has been withheld, died late Monday at Stanford University Medical Center.

Heimerich said the exact cause of death was still undetermined, "but it looks like his body was rejecting the heart" he received in an expensive and controversial taxpayer-financed operation in January.

Officials estimated the surgery and subsequent care—including the $12,500 a day it cost to keep him in the Stanford facility after he was admitted Nov. 23—have cost more than $1.25 million. Heimerich said that figure does not include transportation, medication or providing round-the-clock security while the inmate was in the hospital.

"It could easily reach $2 million when it's all added in," Heimerich said.

The prisoner was a 32-year-old two-time felon serving a 14-year sentence for robbing a Los Angeles convenience store in 1996. He was eligible for parole in October 2008.

He became the center of a national controversy after The Bee disclosed the surgery, which also took place at Stanford.

The operation raised questions about whether there should be limits on the kinds of medical care to which prison inmates are entitled.

At the time of the transplant, prison officials said they were required under numerous court orders, including a 1976 U.S. Supreme Court decision, to provide necessary health care to all inmates.

The decision to provide the inmate, who had longtime heart problems caused by a viral infection, with a new heart was made by a medical panel at Stanford. The surgery was performed on a day when at least 500 other Californians were waiting for similar operations.

But medical professionals and organ transplant centers said they can make decisions about who gets organs and who doesn't based only on medical protocols and not social factors.

While the first of its kind, the transplant is not likely to be the last. As California's prison population ages, authorities are concerned the cost of inmate health care will soar far above last fiscal year's $663 million.

Compounding the problem, Heimerich said, is that many inmate patients don't follow doctor's orders. He said the heart recipient apparently did not follow all of the medical recommendations, although it wasn't clear his failure to do so played a role in his death.

"We can treat them," Heimerich said, "but we can't baby-sit them."

By Steve Wiegand
BEE STAFF WRITER

Comment: Such cases involve legal reasoning (see next section of this chapter) as well as moral reasoning. The position taken here by medical professionals is duty theory; they are explicitly ruling out utilitarian considerations in deciding whom to give transplants to.

Source: The *Sacramento Bee*, December 17, 2002.

ciple of action) involved in what you want to do. In the case in question, the maxim is "Every time I'm in need of money, I'll go to my friends and promise I'll pay it back, even if I know I can't." Next, ask yourself, "Could I want this maxim to be a *universal* law or rule, one that everyone should follow?" This process of *universalization* is the feature that lets you judge whether something would work as a moral law, according to Kant. Could you make it a universal law that it is okay for everybody to lie about paying back loans? Hardly: If everyone adopted this principle, then there would be no such thing as loan making. In short, the universalization of your principle undermines the very principle that is universalized. If everyone adopted the principle, then nobody could possibly follow it. The universalization of your principle is illogical, so it is your duty to pay back loans.

As you can see, the results of acting according to Kant's theory can be radically different from the results of acting according to utilitarianism. Utilitarianism would condone borrowing money with no intention of repaying it, assuming that doing so would produce more happiness than would be produced by not doing so. But Kant's theory would not condone it.

Kant also noted that if you were to borrow a friend's money with no intention of repaying it, you would be treating your friend merely as a means to an end. If you examine cases like this, in which you use other people as mere tools for your own objectives, then, Kant said, you will find in each case a transgression of moral duty, a principle of action that cannot be universalized. Thus, he warned us, it is our moral duty never to treat someone else *merely* as a tool, as means to an end. Of course Kant did not mean that Moore cannot ask Parker for help on some project; doing so would not be a case of Moore's using Parker *merely* as a tool.

Kant's theory of the moral necessity of never treating other people as mere tools can be modified to support the ideas that people have rights and that treatment of others must always involve fair play. Regardless of whether you subscribe to Kant's version of duty theory, the chances are that your own moral deliberations are more than just strictly utilitarian and may well involve considerations of what you take to be other moral requirements, including your duties and the rights of others.

Divine Command Theory   Those who believe moral duty is set by an authority of some sort subscribe to "command" duty theory. (Some authorities view Kant's duty theory as a version of command duty theory, with the authority for Kant being *reason*.) In our culture, probably the most popular version of command duty theory is that according to which our moral duty is set by God; it thus is known as **divine command theory.** As noted above, those who hold this view generally derive their understanding of God's commandments by interpretation of religious texts like the Bible. How, for example, is "Thou shalt not kill" to be understood? Should it be understood as prohibiting capital punishment, or killing in self-defense, or killing in wartime? Does it cover abortion? There is obviously room for disagreement here; likewise, there is room for disagreement over who is to do the interpreting, whether it is to be oneself, or one's minister, or the head of one's church, or whoever.

A philosophical difficulty in divine command theory lies in the question, Is God *justified* in decreeing an act to be right, or is his decree simply arbitrary? On the one hand, we don't want to say that God hands down rules arbitrarily; yet the first alternative seems to imply that there is some stan-

I'm not guilty of murder. I'm guilty of obeying the laws of the Creator.

—Benjamin Matthew Williams, who committed suicide while awaiting sentencing for having murdered a gay couple

dard of rightness and wrongness *above* or *apart* from God, in terms of which God finds the justification for his edicts.

Virtue Ethics   Utilitarianism and duty theory, as well as most other classical approaches to moral theory, focus on the question of what to do. For that reason they are commonly referred to as ethics of conduct. However, another approach was predominant in classical Greek thinking and has regained popularity in recent years. That approach, known as **virtue ethics,** focuses not on *what to do,* but on *how to be;* in other words, the moral issue is not one of single actions or types of actions but of developing a *good character.*

The ancient Greeks believed it was supremely important for a person to achieve psychological and physical balance; and to do that, the person needed to develop a consistently good character. A person out of balance will not be able to assess a situation properly and will tend to overreact or not react strongly enough; moreover, such a person will not know his or her proper limits. People who recognize their own qualifications and limitations and who are capable of reacting to the right degree, at the right time, toward the right person, and for the right reason are virtuous persons. They understand the value of the idea of moderation: not too much and not too little, but in each case a response that is just right.

Aristotle (384–322 B.C.E.) regarded virtue as a trait, like wisdom, justice, or courage, that we acquire when we use our capacity to reason to moderate our impulses and appetites. The largest part of Aristotle's major ethical writing, the *Nicomachean Ethics,* is devoted to analysis of specific moral virtues as means between extremes (for example, courage is the mean between fearing everything and fearing nothing). He also emphasized that virtue is a matter of habit; it is a trait, a way of living.

Virtue ethics is not an abstruse ethical theory. Many of us (fortunately) wish to be (or become) persons of good character. And as a practical matter, when we are deliberating a course of action, our approach often is to consider what someone whose character we admire would do in the circumstances.

Still, it is possible that virtue theory alone cannot answer all moral questions. Each of us may face moral dilemmas of such a nature that it simply isn't clear what course of action is required by someone of good character.

■ *Moral Deliberation*

Before you began this chapter, you may have assumed that moral discussion is merely an exchange of personal opinion or feeling, one that reserves no place for reason or critical thinking. But moral discussion usually assumes some sort of perspective like those we have mentioned here. Actually, in real life, moral reasoning is often a mixture of perspectives, a blend of utilitarian considerations weighted somewhat toward one's own happiness, modified by ideas about duties, rights, and obligations, and mixed often with a thought, perhaps guilty, about what the ideally virtuous person (a parent, a teacher) would do in similar circumstances. It also sometimes

Why Moral Problems Seem Unresolvable

Differences of opinion over ethical issues sometimes seem irreconcilable. Yet this fact often strikes thoughtful people as amazing, because ethical opponents often share a great deal of common ground. For example, pro-life and pro-choice adherents agree on the sanctity of human life. So why in the world can't they resolve their differences? Likewise, those who favor affirmative action and those who agree that racism and sexism still exist and are wrong and need to be eradicated—why on earth can't they resolve their differences?

The answer, in some cases, comes down to a difference in moral perspective. Take affirmative action. Those who favor affirmative action often operate within a utilitarian perspective: They assume that whether a policy should be adopted depends on whether adopting the policy will produce more happiness than will not adopting it. From this perspective, if policies of affirmative action produce more happiness over the long run, then they should be adopted—end of discussion. But those who oppose affirmative action (on grounds other than blatant racism) do so because they believe deontologism trumps utilitarianism. From the deontologist perspective, even if affirmative action policies would produce more happiness in the long run, if they involve even temporarily using some people as a means to that objective, then they are wrong—end of discussion.

In other disputes the root difference lies elsewhere. Pro-life and pro-choice adherents often both are deontologists and agree, for example, that in the absence of a powerful justification, it is wrong to take a human life. They may disagree, however, either as to what counts as a human life or as to what counts as a powerful justification. This difference, then, comes down to a difference in basic definitions—which fact, incidentally, illustrates how silly it can be to dismiss a discussion as "mere semantics."

involves mistakes—factual and nonfactual claims may be confused, inconsistencies may occur, inductive arguments may be weak or deductive arguments may be invalid, fallacious reasoning may be present, and so forth.

We can make headway in our own thinking about moral issues by trying to get clear on what perspective, if any, we are assuming. For example, suppose we are thinking about the death penalty. Our first thought might be that society is much better off if murderers are executed. Are we then assuming a utilitarian perspective? Asking ourselves this question might lead us to consider whether there are *limits* on what we would do for the common good—for example, would we be willing to risk sacrificing an innocent person? It might also lead us to consider how we might *establish* whether society is better off if murderers are executed—if we are utilitarians, then ultimately we will have to establish this if our reasoning is to be compelling.

Or suppose we have seen a friend cheating on an exam. Should we report it to the teacher? Whatever our inclination, it may be wise to consider our perspective. Are we viewing things from a utilitarian perspective? That is, are we assuming that it would promote the most happiness overall to report our friend? Or do we simply believe that it is our duty to report him or her, come what may? Would a virtuous person report the person? Each of these questions will tend to focus our attention on a particular set of considerations—those that are the most relevant to our way of thinking.

It may occur to you to wonder at this point if there is any reason for choosing among perspectives. The answer to this question is yes: Adherents of these positions, philosophers such as those we mentioned, offer grounding or support for their perspectives in theories about human nature, the natural universe, the nature of morality, and other things. In other words, they have *arguments* to support their views. If you are interested, we recommend a course in ethics.

Exercise 12-6

1. Roy needs to sell his car, but he doesn't have money to spend on repairs. He plans to sell the vehicle to a private party without mentioning that the rear brakes are worn. Evaluate Roy's plan of action from a Kantian perspective—that is, can the maxim of Roy's plan be universalized?

2. Defend affirmative action from a utilitarian perspective.

3. Criticize affirmative action from a Kantian perspective. (Hint: Consider Kant's theory that people must never be treated as means only.)

4. Criticize or defend medical experimentation on animals from a utilitarian perspective.

5. Criticize or defend medical experimentation on animals from a divine command perspective.

6. A company has the policy of not promoting women to be vice presidents. What might be said about this policy from the perspective of virtue ethics?

7. What might be said about the policy mentioned in item 6 from the perspective of utilitarianism?

8. Evaluate bisexuality (in humans) from a divine command perspective.

9. In your opinion, would the virtuous person, the person of the best moral character, condemn, approve, or be indifferent to bisexuality?

10. "We can't condemn the founding fathers for owning slaves; people didn't think there was anything wrong with it at the time." Comment on this remark from the standpoint of duty theory.

11. "Let's have some fun and see how your parrot looks without any feathers." (The example is from philosopher Joseph Grcic.) Which of the following perspectives seems best equipped to condemn this suggestion?
    a. utilitarianism
    b. Kantian duty theory
    c. divine command theory
    d. virtue ethics
    e. relativism

12. "Might makes right." Could a utilitarian accept this? Could a virtue ethicist? Could Kant? Could a relativist? Could someone who subscribes to divine command theory?

Exercise 12-7

> This is Darwin's natural selection at its very best. The highest bidder gets youth and beauty.

These are the words of fashion photographer Ron Harris, who auctioned the ova of fashion models via the Internet. The model got the full bid price, and the Web site took a commission of an additional 20 percent. The bid price included no medical costs, though it listed specialists who were willing to perform the procedure. Harris, who created the video "The 20 Minute Workout," said the egg auction gave people the chance of reproducing beautiful children who would have an advantage in society. Critics, however, were numerous. "It screams of unethical behavior," one said. "It is acceptable for an infertile couple to choose an egg donor and compensate her for her time, inconvenience and discomfort," he said. "But this is something else entirely. Among other things, what happens to the child if he or she turns out to be unattractive?"

Discuss the (moral) pros and cons of this issue for five or ten minutes in groups. Then take a written stand on the question "Should human eggs be auctioned to the highest bidder?" When you are finished, discuss which moral perspective seems to be the one in which you are operating.

Exercise 12-8

> Jesus wants to be our best friend. . . . If we let him direct our lives, he will give us the desires of our heart . . . we must yield to God and let him direct our future plans.

Selection 12 in Appendix 1 concerns a student who was denied permission to deliver a valedictorian speech at his public high school graduation that contained this statement.

The principal held that religious points of view cannot be presented at an event sponsored by a publicly funded agency. If the student had given his speech, the principal said, it would be as if the school had condoned the strong religious message the words contained.

The student, on the other hand, said he should be free to express his opinions and beliefs in a graduation speech. He said:

> I would not object if a Buddhist student gave a graduation speech and thanked Buddha, or a graduation speaker thanked Satan and urged people to follow Satan's example. Our country was founded on godly principles. God is on all of our money. It's pretty absurd that you can't even mention God at a graduation.

Should the student have been denied permission to include the statement we've quoted in his graduation speech? Think about the issue for a few minutes and then discuss it in groups of four or five for another five or ten minutes. After that, take a position on the issue and, on a piece of paper, defend the position with the best argument you can think of. Groups should then decide which moral perspective or combination of moral perspectives your defense seems to adopt.

Legal Reasoning

When we think about arguments and disputes, the first image to come to most minds is probably that of an attorney arguing a case in a court of law. Although it's true that lawyers require a solid understanding of factual matters related to their cases and of psychological considerations as well, especially where juries are involved, it is still safe to say that a lawyer's stock-in-trade is argument. Lawyers are successful—in large part—to the extent that they can produce evidence in support of the conclusion that most benefits their client—in other words, their success depends on how well they can put premises and conclusions together into convincing arguments.

■ *Legal Reasoning and Moral Reasoning Compared*

There are some obvious similarities between moral and legal claims. For example, they are both often prescriptive—they tell us what we should do. Both play a role in guiding our conduct, but legal prescriptions carry the weight of society behind them in a way that moral prescriptions do not: We can be punished if we fail to follow the former. In terms of specific actions, it's obvious that the class of illegal activities and the class of immoral activities greatly overlap. Indeed, a society whose moral and legal codes were greatly at odds would be very difficult to understand.

In our own society, we use the term "morals offenses" for a certain class of crimes (usually related to sexual practices), but in reality, most of the crimes we list in our penal codes are also offenses against morality: murder, robbery, theft, rape, and so on. There are exceptions both ways. Lying is almost always considered immoral, but it is illegal only in certain circumstances—under oath or in a contract, for example. On the other hand, there are many laws that have little or nothing to do with morality: laws that govern whether two people are married, laws that determine how far back from the street you must build your house, laws that require us to drive on the right side of the road, and so on. (It may be morally wrong to endanger others by driving on the wrong side of the road, but it is of no moral consequence whether the *correct* side is the right or the left. Hence, the actual content of the law is morally neutral.)

■ *Two Types of Legal Studies: Justifying Laws and Interpreting Laws*

Two principal kinds of questions are asked in legal studies. (Such studies, incidentally, are known variously as "jurisprudence" and "philosophy of law.") The first kind of question asks what the law *should be* and what procedures for making law should be adopted; the second kind asks what the law *is* and how it should be applied. Typically, philosophers are more interested in the former type of question and practicing attorneys in the latter.

By and large, reasoning about the first kind of question differs very little from moral reasoning as described in the first part of this chapter. The difference is simply that the focus is on justifying *laws* rather than on justifying moral statements or principles. We are often most interested in the justification of laws when those laws forbid us from doing something we might otherwise want to do or when they require us to do something we might otherwise want not to do. Justifications are simply arguments that try to establish the goodness, value, or acceptability of something. Here, they are used to try to answer the question, What should the law be?

Consider whether a law that forbids doing X should be enacted by your state legislature.* Typically, there are four main grounds on which a supporter of a law can base his or her justification. The first is simply that doing X is immoral. The claim that the law should make illegal anything that is immoral is the basis of the position known as **legal moralism.** One might use such a basis for justifying laws forbidding murder, assault, or unorthodox sexual practices. For a legal moralist, the kinds of arguments designed to show that an action is immoral are directly relevant to the question of whether the action should be illegal.

The next ground on which a law can be justified is probably the one that most people think of first. It is very closely associated with John Stuart Mill (1806–1873) and is known as the **harm principle:** The only legitimate basis for forbidding X is that doing X causes harm to others. Notice that the harm principle states not just that harm to others is a good ground for forbidding an activity, but that it is the *only* ground. (In terms of the way we formulated such claims in Chapter 9, on truth-functional logic, the principle would be stated, "It is legitimate to forbid doing X *if and only if* doing X causes harm to others.") A person who defends this principle and who wants to enact a law forbidding X will present evidence that doing X does indeed cause harm to others. Her arguments could resemble any of the types covered in earlier chapters.

A third ground on which our hypothetical law might be based is legal paternalism. **Legal paternalism** is the view that laws can be justified if they prevent a person from doing harm to him- or herself; that is, they forbid or make it impossible to do X, *for a person's own good.* Examples include laws

---

*The example here is of a criminal law—part of a penal code designed to require and forbid certain behaviors and to punish offenders. The situation is a little different in civil law, a main goal of which is to shift the burden of a wrongful harm (a "tort") from the person on whom it falls to another, more suitable person—usually the one who caused the harm.

that require that seat belts be worn while riding in automobiles and that helmets be worn while riding on motorcycles.

The last of the usual bases for justifying criminal laws is that some behavior is generally found offensive. The **offense principle** says that a law forbidding X can be justifiable if X causes great offense to others. Laws forbidding burning of the flag are often justified on this ground.

The second question mentioned earlier—What *is* the law and how should it be applied?—may be more straightforward than the first question, but it can still be very complicated. We needn't go into great detail here about why this is the case, but an example will provide an indication. Back in Chapter 2 we discussed vague concepts, and we found that it is impossible to rid our talk entirely of vagueness. Here's an example from the law. Let's suppose that a city ordinance forbids vehicles on the paths in the city park. Clearly, a person violates the law if he or she drives a truck or a car down the paths. But what about a motorbike? A bicycle? A go-cart? A child's pedal car? Just what counts as a vehicle and what does not? This is the kind of issue that must often be decided in court because—not surprisingly—the governing body writing the law could not foresee all the possible items that might, in somebody's mind, count as a vehicle.

The process of narrowing down when a law applies and when it does not, then, is another kind of reasoning problem that occurs in connection with the law.

■ *The Role of Precedent in Legal Reasoning*

Generally speaking, legal reasoning is like other reasoning insofar as it makes use of the same kinds of arguments: They are deductive or inductive; if the former, they can be valid or invalid; if the latter, they can range from strong to weak. The difference between legal and other types of reasoning is mainly in the subject matter to which the argumentative techniques are applied, a taste of which we have sampled in the preceding paragraphs.

There is one kind of argument that occupies a special place in reasoning about legal matters, however: the **appeal to precedent.** This is the practice in the law of using a case that has already been decided as an authoritative guide in deciding a new case that is similar. The appeal to precedent is a variety of argument by analogy in which the current case is said to be sufficiently like the previous case to warrant deciding it in the same way. The general principle of treating like cases alike, discussed in the previous section on moral reasoning, applies here for exactly the same reasons. It would be illogical—and most would say, in some cases at least, unfair or immoral—to treat similar cases in different ways. The Latin name for the principle of appeal to precedent is ***stare decisis*** ("don't change settled decisions," more or less). Using terminology that was applied in the Chapter 10 discussion of analogical argument, we'd say that the earlier, already settled case is the sample, the current case is the target, and the property in question is the way in which the first case was decided. If the target case is sufficiently like the sample case, then, according to the principle of *stare decisis*, it should be decided the same way. Arguments in such situations, naturally, tend to focus on whether the current case really is like the precedent in enough relevant respects. Aside from the fact that such disputes sometimes have more significant consequences for the parties involved, they are not importantly different from those over analogies in other subject matters.

Exercise 12-9

For each of the following kinds of laws, pick at least one of the four grounds for justification discussed in the text—legal moralism, the harm principle, legal paternalism, and the offense principle—and construct an argument designed to justify the law. You may not agree either with the law or with the argument; the exercise is to see if you can connect the law to the (allegedly) justifying principle. For many laws, there is more than one kind of justification possible, so there can be more than one good answer for many of these.

▲ 1. Laws against shoplifting
▲ 2. Laws against forgery
   3. Laws against suicide
▲ 4. Laws against spitting on the sidewalk
   5. Laws against driving under the influence of drugs or alcohol
▲ 6. Laws against adultery
   7. Laws against marriage between two people of the same sex
   8. Laws that require people to have licenses before they practice medicine
   9. Laws that require drivers of cars to have driver's licenses
▲ 10. Laws against desecrating a corpse
  11. Laws against trespassing
  12. Laws against torturing your pet (even though it may be legal to kill your pet, if it is done humanely)

Exercise 12-10

This exercise is for class discussion or a short writing assignment. In the text, "Vehicles are prohibited on the paths in the park" was used as an example of a law that might require clarification. Decide whether the law should be interpreted to forbid motorcycles, bicycles, children's pedal cars, and battery-powered remote-control cars. On what grounds are you deciding each of these cases?

Exercise 12-11

The U.S. Supreme Court came to a decision not long ago about the proper application of the word "use." Briefly, the case in point was about a man named

John Angus Smith, who traded a handgun for cocaine. The law under which Smith was charged provided for a much more severe penalty—known as an enhanced penalty—if a gun was used in a drug-related crime than if no gun was involved. (In this case, the enhanced penalty was a mandatory thirty-year sentence; the "unenhanced" penalty was five years.) Justice Antonin Scalia argued that Smith's penalty should not be enhanced because he did not use the gun in the way the writers of the law had in mind; he did not use it *as a gun.* Justice Sandra Day O'Connor argued that the law only requires the *use* of a gun, not any particular *kind* of use. If you were a judge, would you vote with Scalia or with O'Connor? Construct an argument in support of your position. (The decision of the court is given in the answer section at the back of the book.)

Aesthetic Reasoning

Like moral and legal thinking, aesthetic thinking relies on a conceptual framework that integrates fact and value. Judgments about beauty and art—even judgments about whether something is a work of art or just an everyday object—appeal to principles that identify sources of aesthetic or artistic value. So when you make such a judgment, you are invoking aesthetic concepts, even if you have not made them explicit to yourself or to others.

■ Eight Aesthetic Principles

Here are some of the aesthetic principles that most commonly support or influence artistic creation and critical judgment about art. The first three identify value in art with an object's ability to fulfill certain cultural or social functions.

1. *Objects are aesthetically valuable if they are meaningful or teach us truths.* For example, Aristotle says that tragic plays teach us general truths about the human condition in a dramatic way that cannot be matched by real-life experience. Many people believe art shows us truths that are usually hidden from us by the practical concerns of daily life.

2. *Objects are aesthetically valuable if they have the capacity to convey values or beliefs that are central to the cultures or traditions in which they originate or that are important to the artists who made them.* For example, John Milton's poem *Paradise Lost* expresses the seventeenth-century Puritan view of the relationship between human beings and God.

3. *Objects are aesthetically valuable if they have the capacity to help bring about social or political change.* For instance, Abraham Lincoln commented that Harriet Beecher Stowe's *Uncle Tom's Cabin* contributed to the antislavery movement, which resulted in the Civil War.

Another group of principles identifies aesthetic value with objects' capacities to produce certain subjective—that is, psychological—states in persons who experience or appreciate them. Here are some of the most common or influential principles of the second group:

4. *Objects are aesthetically valuable if they have the capacity to produce pleasure in those who experience or appreciate them.* For instance, the nineteenth-century German philosopher Friedrich Nietzsche identifies one kind of aesthetic value with the capacity to create a feeling of ecstatic bonding in audiences.

5. *Objects are aesthetically valuable if they have the capacity to produce certain emotions we value, at least when the emotion is brought about by art rather than life.* In the *Poetics*, Aristotle observes that we welcome the feelings of fear created in us by frightening dramas, whereas in everyday life fear is an experience we would rather avoid. The psychoanalyst Sigmund Freud offers another version of this principle: While we enjoy art, we permit ourselves to have feelings so subversive that we have to repress them to function in everyday life.

6. *Objects are aesthetically valuable if they have the capacity to produce special nonemotional experiences, such as a feeling of autonomy or the willing suspension of disbelief.* This principle is the proposal of the nineteenth-century English poet Samuel Taylor Coleridge. One of art's values, he believes, is its ability to stimulate our power to exercise our imaginations and consequently to free ourselves from thinking that is too narrowly practical.

Notice that principles 4 through 6 resemble the first three in that they identify aesthetic value with the capacity to fulfill a function. According to these last three, the specified function is to create some kind of subjective or inner state in audiences; according to the first three, however, art's function is to achieve such objective outcomes as conveying information or knowledge or preserving or changing culture or society. But there are yet other influential aesthetic principles that do not characterize art in terms of capacities for performing functions. According to one commonly held principle, art objects attain aesthetic value by virtue of their possessing a certain special aesthetic property or certain special formal configurations:

7. *Objects are aesthetically valuable if they possess a special aesthetic property or exhibit a special aesthetic form.* Sometimes this aesthetic property is called "beauty," and sometimes it is given another name. For instance, the early-twentieth-century art critic Clive Bell insists that good art is valuable for its own sake, not because it fulfills any function. To know whether a work is good aesthetically, he urges, one need only look at it or listen to it to see or hear whether it has "significant form." "Significant form" is valuable for itself, not for any function it performs.

Finally, one familiar principle insists that no reasons can be given to support judgments about art. Properly speaking, those who adhere to this principle think that to approve or disapprove of art is to express an unreasoned preference rather than to render judgment. This principle may be stated as follows:

8. *No reasoned argument can conclude that objects are aesthetically valuable or valueless.* This principle is expressed in the Latin saying *"De gustibus non est disputandum,"* or "Tastes can't be disputed."

The principles summarized here by no means exhaust the important views about aesthetic value, nor are they complete expositions of the views they represent. Historically, views about the nature of art have proven relatively fluid, for they must be responsive to the dynamics of technological and cultural change. Moreover, even though the number of familiar conceptions of aesthetic value is limited, there are many alternative ways of stating these that combine the thoughts behind them in somewhat different ways.

Consequently, to attempt to label each principle with a name invites confusion. For example, let's consider whether any of the principles might be designated *formalism*, which is an important school or style of art. Although the seventh principle explicitly ascribes aesthetic value to a work's form as opposed to its function, the formal properties of artworks also figure as valuable, although only as means to more valuable ends, in certain formulations of the first six principles. For instance, some scholars, critics, and artists think certain formal patterns in works of art can evoke corresponding emotions, social patterns, or pleasures in audiences—for example, slow music full of minor chords is commonly said to make people feel sad.

You should understand that all of the principles presented here merely serve as a basic framework within which you can explore critical thinking about art. If you are interested in the arts, you will very likely want to develop a more complex and sophisticated conceptual framework to enrich your thinking about this subject.

The story is told of the American tourist in Paris who told Pablo Picasso that he didn't like modern paintings because they weren't realistic. Picasso made no immediate reply. A few minutes later the tourist showed him a snapshot of his house.

"My goodness," said Picasso, "is it really as small as that?"

—JACOB BRAUDE

■ *Using Aesthetic Principles to Judge Aesthetic Value*

The first thing to notice about the aesthetic principles we've just discussed is that some are compatible with each other. Thus, a reasonable thinker can appeal to more than one in reaching a verdict about the aesthetic value of an object. For instance, a consistent thinker can use both the first and the fifth principle in evaluating a tragic drama. Aristotle does just this in his *Poetics*. He tells us that tragedies are good art when they both convey general truths about the human condition and help their audiences purge themselves of the pity and fear they feel when they face the truth about human limitations. A play that presents a general truth without eliciting the proper catharsis (release of emotion) in the audience, or a play that provokes tragic emotions unaccompanied by recognition of a general truth, is not as valuable as a play that does both.

However, some of these principles cannot be used together consistently to judge aesthetic value. These bear the same relationship to each other as do contrary claims (recall the square of opposition in Chapter 8). They cannot both be true, although both might be false. For instance, the principle that art is valuable in itself, by virtue of its form or formal configuration (not because it serves some function), and the principle that art is valuable because it serves a social or political function cannot be used consistently together. You might have noticed also that the eighth principle contradicts the others; that is, the first seven principles all specify kinds of reasons for guiding and supporting our appreciation of art, but the last principle denies that there can be any such good reasons.

Finally, it is important to understand that the same principle can generate both positive and negative evaluations, depending on whether the work in question meets or fails to meet the standard expressed in the principle. For example, the fourth principle, which we might call aesthetic hedonism, generates positive evaluations of works that produce pleasure but negative evaluations of works that leave their audiences in pain or displeased.

Exercise 12-12

Suppose that the two statements in each of the following pairs both appear in a review of the same work of art. Identify which of the eight aesthetic principles each statement in the pair appeals to. Then state whether the principles are compatible (that is, they do not contradict each other) and thus form the basis for a consistent critical review or whether they contradict each other and cannot both be used in a consistent review.

▲ 1. a. Last weekend's performance of the Wagnerian operatic cycle was superb; the music surged through the audience, forging a joyous communal bond.
   b. Smith's forceful singing and acting in the role of Siegfried left no doubt why Wagner's vision of heroic morality was attractive to his Teutonic contemporaries.

2. a. Leni Riefenstahl's film *Triumph of the Will* proved to be effective art because it convinced its audiences that the Nazi party would improve the German way of life.
   b. Despite its overtly racist message, *Triumph of the Will* is great art, for films should be judged on the basis of their visual coherence and not in terms of their moral impact.

3. a. All lovers of art should condemn Jackson Pollock's meaningless abstract expressionist splatter paintings.
   b. These paintings create neither sadness nor joy; those who view them feel nothing, neither love nor hate nor any of the other passions that great art evokes.

▲ 4. a. Laurence Olivier's film production of *Hamlet* has merit because he allows us to experience the impact of the incestuous love that a son can feel for his mother.

▲   *Pablo Picasso,* Guernica

   b. Nevertheless, Olivier's *Hamlet* is flawed because it introduces a dimension inconceivable to an Elizabethan playwright.

5. a. There is no point arguing about or giving reasons for verdicts about art, because each person's tastes or responses are so personal.
   b. Those who condemn sexually explicit performance art do not recognize that art is valuable to the extent it permits us to feel liberated and free of convention.

■ *Evaluating Aesthetic Criticism: Relevance and Truth*

Is any evaluation of a work of art as good as any other in creating a critical treatment of that work? The answer is no, for two reasons: (1) the principles of art one adopts function as a conceptual framework that distinguishes relevant from irrelevant reasons; (2) even a relevant reason is useless if it is not true of the work to which it is applied.

   Let's consider the first reason. What would convince you of the value of a work if you accepted principles 4 through 6—all of which maintain that aesthetic value resides in the subjective responses art evokes in its audiences? In this case, you are likely to be drawn to see Picasso's *Guernica* if you are told that it has the power to make its viewers experience the horrors of war; but you would not be attracted by learning, instead, that *Guernica* explores the relationship of two- and three-dimensional spatial concepts. Suppose you reject principles 1 through 3, which conceive of aesthetic value in terms of the work's capacity to perform an objective, cognitive, moral, social, or political function. The fact that Picasso was a Communist will strike you as irrelevant to appreciating *Guernica* unless you accept one or more of the first three principles.

   To illustrate the second reason, look at the nearby reproduction of *Guernica*. Suppose a critic writes, "By giving his figures fishlike appearances and showing them serenely floating through a watery environment,

Picasso makes us feel that humans will survive under any conditions." But no figures in *Guernica* look anything like fish; moreover, they are surrounded by fire, not water, and they are twisted with anguish rather than serene. So, this critic's reasons are no good. Because they are not true of the work, they cannot guide us in perceiving features that enhance our appreciation. A similar problem occurs if reasons are implausible. For instance, an interpretation of *Guernica* as a depiction of the Last Supper is implausible, because we cannot recognize the usual signs of this theme, the twelve disciples and Christ at a table (or at least at a meal), in the far fewer figures of the painting.

Exercise 12-13

State whether each of the reasons below is relevant according to any one of the aesthetic principles. If the reason is relevant, identify the principle that makes it so. If no principle makes the reason relevant, state that it is irrelevant.

▲ 1. Raphael's carefully balanced pyramidal compositions give his paintings of the Madonna such beautiful form that they have aesthetic value for Christian and atheist alike.

2. By grouping his figures so that they compose a triangle or pyramid, Raphael directs the viewer's eye upward to heaven and thereby teaches us about the close connection between motherhood and God.

3. The melody from the chorus "For unto Us a Child Is Born" in Handel's *Messiah* was originally composed by Handel for an erotic love song. Consequently, it evokes erotic responses which distract and detract from the devotional feeling audiences are supposed to experience when they hear *Messiah* performed.

▲ 4. Vincent van Gogh tells us that he uses clashing reds and greens in *The Night Café* to help us see his vision of "the terrible passions of humanity"; it is the intensity with which he conveys his views of the ugliness of human life that makes his work so illuminating.

5. The critics who ignored van Gogh's painting during his lifetime were seriously mistaken; by damaging his self-esteem, they drove him to suicide.

6. Moreover, these critics misjudged the aesthetic value of his art, as evidenced by the fact that his paintings now sell for as much as $80 million.

▲ 7. By showing a naked woman picnicking with fully clothed men in *Déjeuner sur l'herbe*, Édouard Manet treats women as objects and impedes their efforts to throw off patriarchal domination.

Exercise 12-14

June the chimp has been very sad and lonely, so the zoo director gives her some paper, paints, and brushes to keep her busy. Look at the photograph of the chim-

▲ *June the chimpanzee*

▲ *June's painting*

panzee painting and at the photograph of the chimpanzee's painting. Does June's work have aesthetic value? Use each of the eight principles to formulate one reason for or against attributing aesthetic value to June's work. You should end up with eight reasons, one appealing to each principle.

■ Why Reason Aesthetically?

The various aesthetic principles we've introduced are among those most commonly found, either explicitly or implicitly, in discussions about art. Moreover, they have influenced both the creation of art and the selection of art for both private and public enjoyment. But where do these principles come from? There is much debate about this; to understand it, we can draw on notions about definition (introduced in Chapter 2) as well as the discussion of generalizations (Chapter 10).

Some people think that aesthetic principles are simply elaborate definitions of our concepts of art or aesthetic value. Let's explain this point. We use definitions to identify things; for example, by definition we look for three sides and three angles to identify a geometric figure as a triangle. Similarly, we can say that aesthetic principles are definitions; that is, these principles provide an aesthetic vocabulary to direct us in recognizing an object's aesthetic value.

If aesthetic principles are true by definition, then learning to judge art is learning the language of art. But because artists strive for originality, we are constantly faced with talking about innovative objects to which the critic's familiar vocabulary does not quite do justice. This aspect of art challenges even the most sophisticated critic to continually extend the aesthetic vocabulary.

Others think that aesthetic principles are generalizations that summarize what is true of objects treated as valuable art. Here, the argument is by analogy from a sample class to a target population. Thus, someone might hold that all or most of the tragic plays we know that are aesthetically valuable have had something important to say about the human condition; for this reason, we can expect this to be true of any member of the class of tragic plays we have not yet evaluated. Or, also by inductive analogy, musical compositions that are valued so highly that they continue to be performed throughout the centuries all make us feel some specific emotion, such as joy or sadness; so we can predict that a newly composed piece will be similarly highly valued if it also evokes a strong, clear emotion. Of course, such arguments are weakened to the extent that the target object differs from the objects in the sample class. Because there is a drive for originality in art, newly created works may diverge so sharply from previous samples that arguments by analogy sometimes prove too weak.

It is sometimes suggested that these two accounts of the source of aesthetic principles really reinforce each other: Our definitions reflect to some extent our past experience of the properties or capacities typical of valuable art, and our past experience is constrained to some extent by our definitions. But if art changes, of what use are principles, whether analytic or inductive, in guiding us to make aesthetic judgments and—even more difficult—in fostering agreement about these judgments?

At the very least, these principles have an emotive force that guides us in perceiving art. You will remember that emotive force (discussed briefly in Chapter 2) is a dimension of language that permits the words we use to do something more than convey information. In discussion about art, the words that constitute reasons can have an emotive force directing our attention to particular aspects of a work. If the critic can describe these aspects accurately and persuasively, it is thought, the audience will focus on these aspects and experience a favorable (or unfavorable) response similar to the critic's. If a critic's reasons are too vague or are not true of the work to which they are applied, they are unlikely to bring the audience into agreement with the critic.

The principles of art, then, serve as guides for identifying appropriate categories of favorable or unfavorable response, but the reasons falling into these categories are what bring about agreement. They are useful both in developing our own appreciation of a work of art and in persuading others. The reasons must be accurately and informatively descriptive of the objects to which they are applied. The reasons enable us (1) to select a particular way of viewing, listening, reading, or otherwise perceiving the object and (2) to recommend, guide, or prescribe that the object be viewed, heard, or read in this way.

So, aesthetic reasons contain descriptions that prompt ways of perceiving aspects of an object. These prescribed ways of seeing evoke favorable (or unfavorable) responses or experiences. For instance, suppose a critic states that van Gogh's brush strokes in *Starry Night* are dynamic and his colors intense. This positive critical reason prescribes that people focus on these features when they look at the painting. The expectation is that persons whose vision is swept into the movement of van Gogh's painted sky and pierced by the presence of his painted stars will, by virtue of focusing on these formal properties, enjoy a positive response to the painting.

To learn to give reasons and form assessments about art, practice applying these principles as you look, listen, or read. Consider what aspects of a painting, musical performance, poem, or other work each principle directs you to contemplate. It is also important to expand your aesthetic vocabulary so that you have words to describe what you see, hear, or otherwise sense in a work. As you do so, you will be developing your own aesthetic expertise. And, because your reasons will be structured by aesthetic principles others also accept, you will find that rational reflection on art tends to expand both the scope and volume of your agreement with others about aesthetic judgments.

Recap

Reasoning about morality is distinguished from reasoning about matters of fact only in that the former always involves claims that express moral values. Such claims are moral value judgments. Claims about matters of fact, by contrast, are factual because they describe what is believed to be the case. Certain words—especially "ought," "should," "right," and "wrong"—are used in value judgments in a moral sense, although these terms can also be used in a nonmoral sense.

A conclusion containing a value judgment can't be reached solely from premises that are purely factual; in other words, you can't get an "ought" from an "is." To be valid, an argument that has a value-expressing conclusion—that is, a moral argument—must also have a value-expressing premise. Although such premises often remain unstated, we need to state them deliberately to avoid assuming that the value somehow can be derived from a factual premise. Then, in case we disagree with the moral conclusion but can't dispute the factual premise, we can point to the value-expressing premise as the point with which we disagree.

People are sometimes inconsistent in their moral views: They treat similar cases as if the cases were different even when they cannot tell us what is importantly different about them. When two or more cases that are being treated differently seem similar, the burden of proof is on the person who is treating them differently to explain what is different about them.

Moral reasoning is usually conducted within a perspective or framework. In this chapter we considered relativism, subjectivism, utilitarianism, Kantian duty theory, divine command theory, and virtue ethics. Often, different perspectives converge to produce similar solutions to a moral issue;

on other occasions they diverge. Keeping in mind our own perspective can help focus our own moral deliberations on relevant considerations.

There are similarities between the subjects of moral reasoning and legal reasoning, especially in that they both are often prescriptive. Legal studies are devoted to problems like that of justifying laws that prescribe conduct; we looked at legal moralism, the harm principle, legal paternalism, and the offense principle as grounds for such justification. Problems in determining just when and where a law applies often require making vague claims specific.

Precedent is a kind of analogical argument by means of which current cases are settled in accordance with guidelines set by cases decided previously. Whether a precedent governs in a given case is decided on grounds similar to those of any other analogical argument.

To reason aesthetically is to make judgments within a conceptual framework that integrates facts and values. Aesthetic value is often identified as the capacity to fulfill a function, such as to create pleasure or promote social change. Alternatively, aesthetic value is defined in terms of a special aesthetic property or form found in works of art. Still another view treats aesthetic judgments as expressions of tastes. Reasoned argument about aesthetic value helps us to see, hear, or otherwise perceive art in changed or expanded ways and to enhance our appreciation of art. A critic who gives reasons in support of an aesthetic verdict forges agreement by getting others to share perceptions of the work. The greater the extent to which we share such aesthetic perceptions, the more we can reach agreement about aesthetic value.

Additional Exercises

Exercise 12-15

State whether the following reasons are (a) helpful in focusing perception to elicit a favorable response, (b) helpful in focusing perception to elicit an unfavorable response, (c) too vague to focus perception, (d) false or implausible and therefore unable to focus perception, or (e) irrelevant to focusing perception. The information you need is contained in the reasons, so try to visualize or imagine what the work is like from what is said. All of these are paraphrases of testimony given at a hearing in 1985 about a proposal to remove *Tilted Arc*, an immense abstract sculpture, from a plaza in front of a federal office building.

▲ 1. Richard Serra's *Tilted Arc* is a curved slab of welded steel 12 feet high, 120 feet long, weighing over 73 tons, covered completely with a natural oxide coating. The sculpture arcs through the plaza. By coming to terms with its harshly intrusive disruption of space, we can learn much about how the nature of the spaces we inhabit affects our social relations.

▲ *Richard Serra,* Tilted Arc

2. Richard Serra is one of our leading artists, and his work commands very high prices. The government has a responsibility to the financial community. It is bad business to destroy this work because you would be destroying property.

3. *Tilted Arc*'s very tilt and rust remind us that the gleaming and heartless steel and glass structures of the state apparatus can one day pass away. It therefore creates an unconscious sense of freedom and hope.

▲ 4. *Tilted Arc* looks like a discarded piece of crooked or bent metal; there's no more meaning in having it in the middle of the plaza than in putting an old bicycle that got run over by a car there.

5. *Tilted Arc* launches through space in a thrilling and powerful acutely arched curve.

6. *Tilted Arc* is big and rusty.

▲ 7. Because of its size, thrusting shape, and implacably uniform rusting surface, *Tilted Arc* makes us feel hopeless, trapped, and sad. This sculpture would be interesting if we could visit it when we had time to explore these feelings, but it is too depressing to face every day on our way to work.

8. Serra's erotically realistic, precise rendering of the female figure in *Tilted Arc* exhibits how appealingly he can portray the soft circularity of a woman's breast.

9. *Tilted Arc* is sort of red; it probably isn't blue.

▲ *Artemisia Gentileschi,* Judith

Exercise 12-16

The artist Artemisia Gentileschi was very successful in her own time. Success came despite the trauma of her early life, when she figured as the victim in a notorious rape trial. But after she died, her work fell into obscurity; it was neither shown in major museums nor written about in art-history books. Recently, feminist scholars have revived interest in her work by connecting the style and/or theme of such paintings as her *Judith* with her rape and with feelings or issues of importance to women. But other scholars have pointed out that both her subject matter and her treatment of it are conventionally found as well in the work of male painters of the Caravaggist school, with which she is identified. Based on this information, and using one or more of the aesthetic principles described in this chapter, write an essay arguing either that the painting *Judith* has aesthetic value worthy of our attention or that it should continue to be ignored.

Writing Exercises

1. In the movie *Priest*, the father of a young girl admits to the local priest—
   in the confessional—that he has molested his daughter. However, the man
   lacks remorse and gives every indication that he will continue to abuse the
   girl. For the priest to inform the girl's mother or the authorities would be
   for him to violate the sanctity of the confessional, but to not inform anyone
   would subject the girl to further abuse. What should the priest do? Take
   about fifteen minutes to do the following:
   a. List the probable consequences of the courses of action available to the
      priest.
   b. List any duties or rights or other considerations that bear on the issue.
   When fifteen minutes are up, share your ideas with the class.
       Now take about twenty minutes to write an essay in which you do the
   following:
   a. State the issue.
   b. Take a stand on the issue.
   c. Defend your stand.
   d. Rebut counterarguments to your position.
       When you are finished, write down on a separate piece of paper a num-
   ber between 1 and 10 that indicates how strong you think your argument
   is (10 = very strong; 1 = very weak). Write your name on the back of your
   paper.
       When everyone is finished, the instructor will collect the papers and
   redistribute them to the class. In groups of four or five, read the papers and
   assign a number from 1 to 10 to each one (10 = very strong; 1 = very weak).
   When all groups are finished, return the papers to their authors. When you
   get your paper back, compare the number you assigned to your work with
   the number the group assigned it. The instructor may ask volunteers to defend their own judgment of their work against the judgment of the group.
   Do you think there is as much evidence for your position as you did at the
   beginning of the period?
2. Follow the same procedure as above to address one of the following issues:
   a. A friend cheats in the classes he has with you. You know he'd just laugh
      if you voiced any concern. Should you mention it to your instructor?
   b. You see a friend stealing something valuable. Even though you tell your
      friend that you don't approve, she keeps the item. What should you do?
   c. Your best friend's fiancé has just propositioned you for sex. Should you
      tell your friend?
   d. Your parents think you should major in marketing or some other practi-
      cal field. You want to major in literature. Your parents pay the bills. What
      should you do?

Essays for Analysis (and a Few Other Items)

THREE STRIKES AND THE WHOLE ENCHILADA

In this first selection, we've taken a real-life case of some importance and identified how various sections of the book bear on the issue and various aspects of the controversy that surround it. As we said at the beginning, this material is not designed to operate just in the classroom.

As you no doubt know, several states have "three strikes" laws, which call for life terms for a criminal convicted of any felony—if the criminal already has two prior felony convictions that resulted from serious or violent crime.

Have such laws helped to reduce crime in the states that have them? This is a factual question (Chapter 1), a question of causation (Chapter 11). How might the issue be resolved?

In California, Frank Zimring, a University of California, Berkeley, law professor, analyzed the records of 3,500 criminal defendants in Los Angeles, San Diego, and San Francisco before and after California's law was enacted. Zimring found no evidence the law was a deterrent to crime. For our purposes we do not need to go into the details of the study.

People Against Crime, an organization that favors tougher penalties for criminals, denounced the study as "so much more left-wing propaganda coming out of a notoriously liberal university."

This charge is an ad hominem fallacy (Chapter 6). But is it nevertheless a reasonable challenge to Zimring's credibility that warrants not outright rejection of the study but suspension of judgment about its findings (Chapter 3)? The answer is no. Stripped of its rhetoric (Chapter 4), the charge is only that the author of the study is a professor at Berkeley; and that charge gives no reason to suspect bias on his part.

Other criticisms of the study were reported in the news. A spokesperson for the California secretary of state said, "When you see the crime rate going down 38 percent since three strikes you can't say it doesn't work."

This remark is an example of the fallacy "post hoc, ergo propter hoc," discussed in Chapter 11. In fact, that's being charitable. According to Zimring's research, the crime rate had been declining at the same rate before the law was passed.

The same spokesperson also criticized the Zimring study for ignoring the number of parolees leaving the state (to avoid getting a third strike, presumably). This is a red herring (Chapter 5). If the decline in the crime rate was unaffected after the law passed, as the Zimring study reportedly learned, then the law had no effect regardless of what parolees did or did not do.

The spokesperson also said, "Clearly when people are committing 20 to 25 crimes a year, the year they are taken off the street, that's 20 to 25 crimes that aren't going to happen." This too is a red herring (Chapter 5): If the decline in the crime rate remained the same before and after the "three strikes" law, then that's the end of the story. The criticism assumes criminals will continue to commit crimes at the same rate if there is no mandatory life sentence for a third felony. It therefore also begs the question (Chapter 6)—it assumes the law works in order to prove the law works. You will also have noticed the proof surrogate "clearly" (Chapter 4) in the criticism.

One might, of course, maintain that without the law the crime rate would have *stopped* declining, which would mean that the law had an effect after all. But the burden of proof (Chapter 6) is not on Zimring to *disprove* the possibility that the crime rate would have stopped declining if the law had not been passed.

A critic might also say that Zimring's study was conducted too soon after the law for the effects of the law to show up. This is another red herring (Chapter 5). It is not a weakness in the study that it failed to find an effect that might show up at a *later* time.

Selection 2

CONTROLLING IRRATIONAL FEARS AFTER 9/11

We present this selection as an example of a fairly well-reasoned argumentative essay. There is more here than arguments—there's some window dressing and you'll probably find some slanters here and there as well. You should go through the selection and identify the issues, the positions taken on those issues, and the arguments offered in support of those arguments. Are any arguments from opposing points of view considered? What is your final assessment of the essay?

The terrorist attacks of September 11, 2001, produced a response among American officials, the media, and the public that is probably matched only by the attack on Pearl Harbor in 1941. Since it is the very nature of terrorism not only to cause immediate damage but also to strike fear in the hearts of the population under attack, one might say that the terrorists were extraordinarily successful, not just as a result of their own efforts but also in consequence of the American reaction. In this essay, I shall argue that this reaction was irrational to a great extent and that to that extent Americans unwittingly cooperated with the terrorists in achieving a major goal: spreading fear and thus disrupting lives. In other words, we could have reacted more rationally and as a result produced less disruption in the lives of our citizens.

There are several reasons why one might say that a huge reaction to the 9/11 attacks was justified. The first is simply the large number of lives that were lost. In the absence of a shooting war, that 2,800 Americans should die from the same cause strikes us as extraordinary indeed. But does the sheer size of the loss of life warrant the reaction we saw? Clearly sheer numbers do not always impress us. It is unlikely, for example, that many Americans remember that, earlier in 2001, an earthquake in Gujarat, India, killed approximately 20,000 people. One might explain the difference in reaction by saying that we naturally respond more strongly to the deaths of Americans closer to home than to those of others halfway around the world. But then consider the fact that, every *month* during 2001 more Americans were killed in automobile crashes than were killed on 9/11 (and it has continued every month since as well). Since the victims of car accidents come from every geographical area and every social stratum, one can say that those deaths are even "closer to home" than the deaths that occurred in New York, Washington, and Pennsylvania. It may be harder to identify with an earthquake victim in Asia than with a 9/11 victim, but this cannot be said for the victims of fatal automobile accidents.

One might say that it was the *malice* of the perpetrators that makes the 9/11 deaths so noteworthy, but surely there is plenty of malice present in the 15,000 homicides that occur every year in the United States. And while we have passed strict laws favoring prosecution of murderers, we do not see the huge and expensive shift in priorities that has followed the 9/11 attacks.

It seems clear, at least, that sheer numbers cannot explain the response to 9/11. If more reasons were needed, we might consider that the *actual total* of the number of 9/11 deaths seemed of little consequence in post-attack reports. Immediately after the attacks, the estimated death toll was about 6,500. Several weeks later it was clear that fewer than half that many had actually died, but was there a great sigh of relief when it was learned that over 3,000 people who were believed to have died were still alive? Not at all. In fact, well after it was confirmed that no more than 3,000 people had died, Secretary of Defense Donald Rumsfeld still talked about "over 5,000" deaths on 9/11. So the actual number seems to be of less consequence than one might have believed.

We should remember that fear and outrage at the attacks are only the beginning of the country's response to 9/11. We now have a new cabinet-level Department of Homeland Security; billions have been spent on beefing up

Note: This essay borrows very heavily from "A Skeptical Look at September 11th", an article in the *Skeptical Inquirer* of September/October 2002 by Clark R. Chapman and Alan W. Harris. Rather than clutter the essay with numerous references, we simply refer the reader to the original, longer piece.

security and in tracking terrorists and potential terrorists; billions more have been spent supporting airlines whose revenues took a nosedive after the attacks; the Congress was pulled away from other important business; the National Guard was called out to patrol the nation's airports; air travelers have been subjected to time-consuming and expensive security measures; you can probably think of a half-dozen other items to add to this list.

It is probable that a great lot of this trouble and expense is unwarranted. We think that random searches of luggage of elderly ladies getting on airplanes in Laramie, Wyoming, for example, is more effective as a way of annoying elderly ladies than of stopping terrorism.

If we had been able to treat the terrorist attacks of 9/11 in a way similar to how we treat the carnage on the nation's highways—by implementing practices and requirements that are directly related to results (as in the case of speed limits, safety belts, and the like, which took decades to accomplish in the cause of auto safety)— rather than by throwing the nation into a near panic and using the resulting fears to justify expensive but not necessarily effective or even relevant measures.

But we focused on 9/11 because of its terrorist nature and because of the spectacular film that was shown over and over on television, imprinting forever the horrific images of the airliner's collision with the World Trade Center and the subsequent collapse of the two towers. The media's instant obsession with the case is understandable, even if it is out of proportion to the actual damage, as awful as it was, when we compare the actual loss to the loss from automobile accidents.

Finally, our point is that marginal or even completely ineffective expenditures and disruptive practices have taken our time, attention, and national treasure away from other matters with more promise of making the country a better place. We seem to have all begun to think of ourselves as terrorist targets, but, in fact, reason tells us we are in much greater danger from our friends and neighbors behind the wheels of their cars.

The remainder of the essays in this section are here for analysis and evaluation. Your instructor will probably have specific directions if he or she assigns them, but at a minimum, they offer an opportunity to identify issues, separate arguments from other elements, identify premises and conclusions, evaluate the likely truth of the premises and the strength of the arguments, look for unstated assumptions or omitted premises, and lots of other stuff besides. We offer sample directions for many of the pieces.

Selection 3

EXCERPTS FROM FEDERAL COURT RULING ON THE PLEDGE OF ALLEGIANCE

The following are excerpts from the ruling yesterday by a three-judge federal appeals court panel in San Francisco that reciting the Pledge of

Allegiance in public schools is unconstitutional because it includes the phrase "one nation under God." The vote was 2 to 1. Judge Alfred T. Goodwin wrote the majority opinion, in which Judge Stephen Reinhardt joined. Judge Ferdinand F. Fernandez wrote a dissent.

From the Opinion by Judge Goodwin

In the context of the pledge, the statement that the United States is a nation "under God" is an endorsement of religion. It is a profession of a religious belief, namely, a belief in monotheism. The recitation that ours is a nation "under God" is not a mere acknowledgment that many Americans believe in a deity. Nor is it merely descriptive of the undeniable historical significance of religion in the founding of the republic. Rather, the phrase "one nation under God" in the context of the pledge is normative. To recite the pledge is not to describe the United States; instead, it is to swear allegiance to the values for which the flag stands: unity, indivisibility, liberty, justice, and—since 1954— monotheism. The text of the official pledge, codified in federal law, impermissibly takes a position with respect to the purely religious question of the existence and identity of god. A profession that we are a nation "under God" is identical, for Establishment Clause purposes, to a profession that we are a nation "under Jesus," a nation "under Vishnu," a nation "under Zeus," or a nation "under no god," because none of these professions can be neutral with respect to religion. "The government must pursue a course of complete neutrality toward religion." Furthermore, the school district's practice of teacher-led recitation of the pledge aims to inculcate in students a respect for the ideals set forth in the pledge, and thus amounts to state endorsements of these ideals. Although students cannot be forced to participate in recitation of the pledge, the school district is nonetheless conveying a message of state endorsement of a religious belief when it requires public school teachers to recite, and lead the recitation, of the current form of the pledge. . . . 1

The pledge, as currently codified, is an impermissible government endorsement of religion because it sends a message to unbelievers "that they are outsiders, not full members of the political community, and an accompanying message to adherents that they are insiders, favored members of the political community." 2

From the Dissent by Judge Fernandez

We are asked to hold that inclusion of the phrase "under God" in this nation's Pledge of Allegiance violates the religion clauses of the Constitution of the United States. We should do no such thing. We should, instead, recognize that those clauses were not designed to drive religious expression out of public thought; they were written to avoid discrimination. We can run through the litany of tests and concepts which have floated to the surface from time to time. Were we to do so, the one that appeals most to me, the one I think to be correct, is the concept that what the religion clauses of the First Amendment require is neutrality; that those clauses are, in effect, an early kind of equal 3 protection provision and assure that government will neither discriminate for nor discriminate against religion or religions. But, legal world abstractions and ruminations aside, when all is said and done, the danger that "under God" in our Pledge of Allegiance will tend to bring about a theocracy or suppress somebody's belief is so miniscule as to be de minimis. The danger that phrase presents to our First Amendment freedoms is picayune at most.

Selection 4

GAYS' IMPACT ON MARRIAGE UNDERESTIMATED

Jeff Jacoby

1   It was a year ago last month that the Vermont law authorizing same-sex civil unions—a marriage by another name—took effect, and the *New York Times* marked the anniversary with a story July 25. "Quiet Anniversary for Civil Unions," the double headline announced. "Ceremonies for Gay Couples Have Blended Into Vermont Life." It was an upbeat report, and its message was clear: Civil unions are working just fine.

2   The story noted in passing that most Vermonters oppose the law. Presumably, they have reasons for not wanting legal recognition conferred on homosexual couples, but the *Times* had not room to mention them. It did have room, though, to dismiss those reasons—whatever they might be—as meritless: "The sky has not fallen," Gov. Howard Dean said, "and the institution of marriage has not collapsed. None of the dire predictions have come true. . . . There was a big rhubarb, a lot of fear-mongering, and now people realize there was nothing to be afraid of."

3   In the *Wall Street Journal* two days later, much the same point was made by Jonathan Rauch, the esteemed Washington journalist and vice president of the Independent Gay Forum. Opponents of same-sex marriage, he wrote, worry "that unyoking marriage from its traditional male-female definition will destroy or severely weaken it. But this is an empirical proposition, and there is reason to doubt it. Opponents of same-sex marriage have done a poor job of explaining why the health of heterosexual marriage depends on the exclusion of a small number of homosexuals."

4   The assertion that same-sex marriage will not damage traditional family life is rarely challenged, as Rep. Barney Frank, D-Mass., said during the 1996 congressional debate over the Defense of Marriage Act.

5   "I have asked and I have asked and I have asked, and I guess I will die . . . unanswered," Frank taunted. "How does the fact that I love another man and live in a committed relationship with him threaten your marriage? Are your relations with your spouses of such fragility that the fact that I have a committed, loving relationship with another man jeopardizes them?"

6   When another congressman replied that legitimizing gay unions "threatens the institution of marriage," Frank said, "That argument ought to be made by someone in an institution because it has no logical basis whatsoever."

But Frank's sarcasm, Rauch's doubts and Dean's reassurances notwith- 7
standing, the threat posed by the same-sex unions to traditional marriage
and family life is all too real. Marriage is harmed by anything that diminishes its privileged status. It is weakened by anything that erodes the social
sanctions that Judeo-Christian culture developed over the centuries for
channeling men's naturally unruly sexuality into a monogamous, lasting,
and domestic relationship with one woman. For proof, just look around.

Over the past 40 years, marriage has suffered one blow after another. 8
The sexual revolution and the pill made it much easier for men to enjoy
women sexually without having to marry them. Legalized abortion reduced
pressure on men to marry women they impregnated and reduced the need
for women to wait for lasting love. The widespread acceptance of unmarried cohabitation—which used to be disdained as "shacking up"—diminished marrage further. Why get married if intimate companionship can be
had without public vows and ceremony?

The rise of the welfare state with its subsidies for single mothers sub- 9
verted marriage by sending the unmistakable message that husbands were
no longer essential for family life. And the rapid spread of no-fault divorce
detached marriage from any assumption of permanence. Where couples
were once expected to stay married "for as long as you both shall live"—
and therefore to put effort into making their marriage work—the expectation today is that they will remain together only "for as long as you both
shall love."

If we now redefine marriage so it includes the union of two men or two 10
women, we will be taking this bad situation and making it even worse.

No doubt the acceptance of same-sex marriage would remove whatever 11
stigma homosexuality still bears, a goal many people would welcome. But
it would do so at a severe cost to the most basic institution of our society.
For all the assaults marriage has taken, its fundamental purpose endures: to
uphold and encourage the union of a man and a woman, the framework that
is the healthiest and safest for the rearing of children. If marriage stops
meaning even that, it will stop meaning anything at all.

Selection 5

BUSH'S ENVIRONMENTAL RECORD

Bob Herbert

Bob Herbert is a *New York Times* columnist.

Do you remember the character "Pig-Pen" in the "Peanuts" cartoons? He 1
was always covered with dirt and grime. He was cute, but he was a walking
sludge heap, filthy and proud of it. He once told Charlie Brown, "I have affixed to me the dirt and dust of countless ages. Who am I to disturb history?"

For me, Pig-Pen's attitude embodies President Bush's approach to the   2
environment.

We've been trashing, soiling, even destroying the wonders of nature for   3
countless ages. Why stop now? Who is Bush to step in and curb this venerable orgy of pollution, this grand tradition of fouling our own nest?

Oh, the skies may once have been clear and the waters sparkling and   4
clean. But you can't have that and progress too. Can you?

Last week we learned that the Bush administration plans to cut fund-   5
ing for the cleanup of 33 toxic waste sites in 18 states. As the *New York
Times'* Katharine Seelye reported, this means "that work is likely to grind
to a halt on some of the most seriously polluted sites in the country."

The cuts were ordered because the Superfund toxic waste cleanup is run-   6
ning out of money. Rather than showing the leadership necessary to replenish
the fund, the president plans to reduce its payouts by cleaning up fewer sites.

Pig-Pen would have been proud.   7

This is not a minor matter. The sites targeted by the Superfund pro-   8
gram are horribly polluted, in many cases with cancer-causing substances.
Millions of Americans live within a few miles of these sites.

The Superfund decision is the kind of environmental move we've come   9
to expect from the Bush administration. Mother Nature has been known to
tremble at the sound of the president's approaching footsteps. He's an environmental disaster zone.

In February, a top enforcement official at the Environmental Protection   10
Agency, Eric Schaeffer, quit because of Bush administration policies that he
said undermined the agency's efforts to crack down on industrial polluters.

Schaeffer said he felt he was "fighting a White House that seems deter-   11
mined to weaken the rules we are trying to enforce."

That, of course, is exactly what this White House is doing.   12

Within weeks of Schaeffer's resignation came official word that the ad-   13
ministration was relaxing the air quality regulations that applied to older
coal-fired power plants, a step backward that delighted the administration's
industrial pals.

During this same period the president broke his campaign promise to   14
regulate the industrial emissions of carbon dioxide, a move that, among
other things, would have helped in the fight to slow the increase in global
warming. Bush has also turned his back on the Kyoto Protocol, which
would require industrial nations to reduce their emissions of carbon dioxide and other greenhouse gases.

The president was even disdainful of his own administration's report   15
on global warming, which acknowledged that the U.S. would experience
far-reaching and, in some cases, devastating environmental consequences
as a result of the climate change.

The president's views on global warming seem aligned with those of   16
the muddle-headed conservative groups in Texas that have been forcing
rewrites in textbooks to fit their political and spiritual agendas. In one
environmental science textbook, the following was added: "In the past,
Earth has been much warmer than it is now, and fossils of sea creatures

show us that the sea level was much higher than it is today. So does it really matter if the world gets warmer?"

Sen. Joseph Lieberman, not exactly a left-winger on the environment or anything else, gave a speech in California in February in which he assailed the president's lack of leadership on global warming and other environmental issues. He characterized the president's energy policy as "mired in crude oil" and said Bush had been "AWOL in the war against environmental pollution."   17

Several states, fed up with Bush's capitulation to industry on these matters, have moved on their own to protect the environment and develop more progressive energy policies.   18

Simply stated, the president has behaved irresponsibly toward the environment and shows no sign of changing his ways.   19

You could laugh at Pig-Pen. He was just a comic strip character. But Bush is no joke. His trashing of the environment is a deadly serious matter.   20

Selection 6

DEATH PENALTY HAS NO PLACE IN U.S.

Cynthia Tucker

Many Americans will applaud the decision of a Jasper, Texas, jury to condemn John William King to die. They will argue that the death penalty is exactly what King deserves for chaining James Byrd Jr. to the back of a pickup truck and dragging him until his body was torn apart—his head and right arm here, his torso there.   1

If there is to be capital punishment in this country, isn't this just the sort of case that demands it? King is the epitome of cold-blooded evil, a man who bragged about his noxious racism and attempted to win converts to his views. He believed he would be a hero after Byrd's death. He has proved himself capable of the sort of stomach-churning cruelty that most of us would like to believe is outside the realm of human behavior.   2

Besides, there is the matter of balancing the books. King is a white man who (with the help of accomplices, apparently) killed a black man. For centuries, the criminal justice system saw black lives as so slight, so insignificant, that those who took a black life rarely got the death penalty. Isn't it a matter of fairness, of equity, of progress, that King should be put to death?   3

No. Even though King is evil. Even though he is utterly without remorse. Even though he is clearly guilty. (After the prosecution mounted a case for five days, King's lawyers mounted a defense of only one hour. The jury of 11 whites and one black then deliberated only two and half hours to determine King's guilt.)   4

This is no brief for King, who would probably chain me to the back of a pickup truck as quickly as he did Byrd. This is a plea for America, which is strong enough, just enough and merciful enough to have put aside, by now, the thirst for vengeance.   5

The question is not, Does John William King deserve the death penalty?   6
The question is, Does America deserve the death penalty?

Capital punishment serves no good purpose. It does not deter crime. If it   7
did, this country would be blessedly crime-free. It does not apply equally to all.
King notwithstanding, the denizens of death row are disproportionately blacks
and Latinos who have killed whites. It remains true that the lives of blacks and
Latinos count for less, that their killers are less likely to be sentenced to die.

Death row also counts among its inmates a high quotient of those who   8
are poor, dumb and marginalized. Those criminals blessed with education,
status and connections can usually escape capital punishment:

Last Tuesday, William Lumpkin, an attorney in Augusta, Ga., was   9
found guilty of capital murder in the death of real estate agent Stan White,
who owned the title to Lumpkin's home and was about to evict him. Lumpkin beat White to death with a sandbag and dumped the body in the Savan-
nah River. But Lumpkin descends from Georgia gentry; one ancestor was a
state Supreme Court justice. He was sentenced to life in prison.

Worse than those inequities, capital punishment is sometimes visited   10
upon the innocent. Lawrence C. Marshall, a law professor at Northwestern
University, is director of the National Conference on Wrongful Convictions
and the Death Penalty. Since 1972, he says, 78 innocent people have been
released from death row.

It does not strain the imagination to think that maybe, just maybe, the   11
system did not catch all of its errors and some of those who were wrongly convicted have already been sent to their deaths. How many? There is no way to
know, but even one is too many. The execution of even one innocent man puts
us law-abiding citizens uncomfortably close to the level of a John William King.

Selection 7

Instructions: Identify the main issue in this essay and the author's position
on this issue. Then state in your own words three arguments given by the
author in support of his position.

As an additional exercise, show how at least two of these arguments
can be treated as categorical syllogisms (Chapter 8), as truth-functional
arguments (Chapter 9), or as common deductive argument patterns.

HETERO BY CHOICE?

A radio commentary by Richard Parker

For a while there, everybody who could get near a microphone was claiming   1
that only he or she and his or her group, party, faction, religion, or militia
stood for real American family values.

Now, it was seldom made clear just what those values were supposed   2
to be. I have a notion that if [my son] Alex and I were to go out and knock

over a few gas stations and convenience stores, the mere fact that we did it together would make it count as somebody's family values.

For some, the phrase "family values" never amounted to more than a 3 euphemism for gay-bashing. I remember a [few] years ago, during the loudest squawking about values, when a reporter asked Dan Quayle whether he believed that a gay person's homosexuality was a matter of his or her psychological makeup or whether it was a matter of choice. He answered that he believed it was mainly a matter of choice. Two weeks later, Barbara Bush was quoted as saying that sexual orientation is mainly a matter of choice. Since then, it's turned up frequently.

It seems to me that people who make such a remark are either being re- 4 markably cynical (if they don't really believe it themselves) or remarkably fatuous (if they do believe it).

If it were *true* that a person's sexual preference were a matter of choice, 5 then it must have happened that each of us, somewhere back along the way, *decided* what our sexual preference would be. Now, if we'd made such decisions, you'd think that somebody would remember doing it, but nobody does.

In my case, I just woke up one morning when I was a kid and discov- 6 ered that girls were important to me in a way that boys were not. I certainly didn't sit down and *decide* that it was girls who were going to make me anxious, excited, terror-struck, panicky, and inclined to act like an idiot.

Now, if the people who claim to hold the "choice" view were right, it 7 must mean that gay people have always chosen—they've *decided*—to have the sexual orientation they have. Can you imagine a person, back in the '50s, say, who would *choose* to have to put up with all the stuff gay people had to put up with back then? It's bad enough now, but only the mad or the criminally uninformed would have *chosen* such a life back then.

(Actually, it seems clear to me that the whole idea of a preference rules out 8 the notion of choice. I choose to eat chocolate rather than vanilla, but I don't choose to *prefer* chocolate to vanilla. One simply discovers what one prefers.)

If it's clear that people don't consciously choose their sexual prefer- 9 ences, why would anybody make such claims? I can think of a cynical reason: It only makes sense to condemn someone for something they choose, not for things they can't do anything about.

Is it just a coincidence that people who claim we choose our sexual pref- 10 erences are often the same people who demonize homosexuals? No, of course not. In fact, their cart comes before their horse: They are damned sure going to condemn gay people, and so, since you can only condemn someone for voluntary actions, it *must* be that one's sexuality is a voluntary choice. Bingo! Consistent logic. Mean, vicious, and mistaken. But consistent.

Selection 8

In a brief essay, argue for whether Bonnie and Clyde should receive the same or different punishment.

BONNIE AND CLYDE

Bonnie and Clyde are both driving on roads near a mountain community in   1
northern California. Both are driving recklessly—much faster than the
posted speed limit. Each of them has a passenger in the car.

At a sharp and very dangerous curve, Bonnie loses control of her car and   2
crashes into some nearby trees; only moments later, on another dangerous
section of road, Clyde's car goes into a skid, leaving the road and rolling
over several times down an embankment.

As a result of their accidents, Bonnie and Clyde are bruised and shaken,   3
but not seriously hurt. However, both of their passengers are hurt badly and
require medical attention. Passersby call an ambulance from the next town,
and soon it arrives, taking the injured passengers to the only medical facility in the area.

A neurosurgeon who is on duty examines both passengers when they   4
arrive at the medical center. She determines that both have suffered serious
head injuries and require immediate cranial surgery if they are to survive.
However, she is the only person available who is competent to perform the
surgery, and she cannot operate on both patients at once. Not knowing what
to do, she tries to find someone to call for advice. But she can reach nobody.
So she flips a coin.

As a result of the coin flip, the surgeon operates on Bonnie's passenger   5
and leaves Clyde's passenger in the care of two medical technicians. The
latter do the best they can, but Clyde's passenger dies. Because of the attention of the physician, Bonnie's passenger survives and, in time, makes a
complete recovery.

Selection 9

Identify rhetorical devices, including slanters and fallacious reasoning, and
evaluate the credibility and potential bias of the author.

WILL OZONE BLOB DEVOUR THE EARTH?

Edward C. Krug

At the time this essay was first published, Edward C. Krug was director
of environmental projects for CFACT, Committee For A Constructive
Tomorrow, a group based in Washington, D.C., that promotes free-market
solutions to current environmental concerns.

The great and good protectors of the environment are hailing the latest bit   1
of science-by-press-release.

Simultaneously staged with the announcement that the National Aero-   2
nautics and Space Administration (NASA) and the National Oceanic and

Atmospheric Administration have begun studying the possible development of an ozone hole over the Northern Hemisphere, the [first] Bush administration, Greenpeace, Sen. Al Gore (D-Tenn.), and others proclaimed that this "discovery" justifies the banning of all human-made substances that contain chlorine and bromine. Chlorine and bromine, by the way, are common elements also found in seawater, in plants, and in you and me.

3   I've been asked why I, a scientist, would dare question all this concern for the environment. The answer: Real science operates quite differently from what passes for environmental "science" these days.

4   Take physics, for example. Cold-fusion scientists were vilified by the scientific community a few years ago for their violation of scientific ethics when they went to the media with their supposed findings before their articles had been peer-reviewed.

5   The release of such unproven scientific findings is materially and sociologically devastating. It is highly misleading and causes the misallocation of limited social resources. It also undermines the very process of science and the public's faith in the objectivity of science.

6   The science of physics, which stones to death false prophets and therefore produces new miracles every day, operates in a way exactly the opposite of the new environmental "science," which stones to death the real prophets and therefore produces new scandals every day.

7   The Feb. 3 NASA press release on the new ozone hole over the Northern Hemisphere is being widely lauded. But there has been no peer review of the data. Indeed, peer review cannot even begin until after the end of March, because all the data have not yet been collected.

8   Even so, the details of the press release give pause. It states that according to ozone hole theory, we could have been losing 1 percent to 2 percent of our ozone per day in mid-January. But this appears not to have happened. Furthermore, it notes that we have information on only about half of the theoretically relevant chemical processes in the atmosphere, and only about 30 percent of the theoretically relevant range of altitude.

9   Why couldn't NASA obey the time-honored ethics of science and wait until sometime after March to make its findings known?

10   For the same reason that biologist Paul Ehrlich in 1968 dared anybody to prove wrong his theory that the world would run out of food in 1977. And for the same reason that the Environmental Protection Agency Administrator William Reilly announced last April that unpublished NASA satellite data would show ozone depletion had speeded up to twice its previous rate. By the time 1977 rolled around and people were still eating food, and by the time NASA finally published its satellite data showing that ozone is increasing (not decreasing), nobody cared anymore.

11   Environmentalists also abuse the scientific method. In real science, one must prove the positive, not the negative. If a hypothesis cannot explain all the data, it gets thrown out. In the new environmental "science," by contrast, utterances of doom are automatically assumed to be correct. And it is assumed that society must act on them, unless someone can prove the environmental fantasy has absolutely no possibility of ever coming true.

Since it is logically impossible to prove the negative, environmentalists   12
have a sweet racket going: We must subordinate ourselves to their will lest
their horrible predictions come to pass.

These are also the rules for the "space invaders" game. In this game,   13
the leader says we face dreadful danger from hostile invaders from space.
Since it is impossible to prove there is no alien life in the universe, we are
forced to make the leader emperor of Earth so that he or she can save us all
from this potential horror.

One might wish that NASA played "space invaders" instead of issuing   14
"ozone-blob-that-may-swallow-Earth" press releases into the real world.

Selections 10A and 10B

Evaluate the arguments on both sides. Who has the stronger arguments,
and why? Make certain your response does not rest too heavily on rhetorical devices. As an alternative assignment, determine which author relies
more heavily on rhetorical devices to persuade the audience.

EQUAL TREATMENT IS REAL ISSUE—NOT MARRIAGE

USA Today

Our view: The fact is that marriage is already a messy entanglement of
church and state.

With shouting about "gay marriage" headed for a new decibel level . . .   1
chances for an amicable resolution seem bleak.

Traditionalists see the issue in private, religious terms, and with legis-   2
lators in many states mobilizing around their cause, they're in no mood to
compromise. They say marriage, by common definition, involves a man
and a woman. And for most people, it's just that. In polls, two-thirds of the
public supports the status quo.

But looking through the lenses of history and law, as judges must, mar-   3
riage is far from a private religious matter. So much so that short of a constitutional amendment, compromise is inevitable.

Not only does the state issue marriage licenses and authorize its offi-   4
cers to perform a civil version of the rite, it gives married couples privileged
treatment under law.

For example, when one spouse dies the house and other property held   5
jointly transfer easily for the other's continued use and enjoyment. The survivor gets a share of continuing Social Security and other benefits. Joint
health and property insurance continues automatically.

If there's no will, the law protects the bereaved's right to inherit. There's   6
no question of who gets to make funeral arrangements and provide for
the corpse.

It's the normal order of things, even for households that may have ex-   7
isted for only the briefest time, or for couples who may be long estranged
though not divorced.

But some couples next door—even devoted couples of 20 or 30 years'   8
standing—don't have those rights and can't get them because of their sex.

Support for marriage is justified as important to community stability,   9
and it undoubtedly is. But when it translates into economic and legal discrimination against couples who may be similarly committed to each other,
that should be disturbing.

The U.S. Constitution says every person is entitled to equal protection   10
under law. Some state constitutions go farther, specifically prohibiting sexual discrimination. . . .

Ironically, people who oppose gay marriages on religious grounds would   11
have their way but for the fact that marriage has evolved as a messy entanglement of church and state. To millions, marriage is a sacrament, and the
notion that the state would license or regulate a sacrament ought to be an
outrage. Imagine the uproar if a state legislature tried to license baptisms or
communions, and wrote into law who could be baptized or who could receive bread and wine. Or worse yet gave tax breaks to those who followed
those practices.

Short of getting out of the marriage business altogether, which isn't   12
likely to happen, the state must figure a way to avoid discrimination. The
hundreds of employers now extending workplace benefits to unmarried but
committed couples and the handful of municipalities offering informal "domestic partner" status may be pointing in the right direction.

The need is not necessarily to redefine marriage but to assure equal   13
treatment under the law.

GAY MARRIAGE "UNNATURAL"

The Rev. Louis P. Sheldon

*The Rev. Louis P. Sheldon is chairman of the Traditional Values Coali-
tion, a California-based organization of some 32,000 churches.*

Opposing view: Opinion polls show that nearly 80% of Americans
don't accept "homosexual marriage."

In everything which has been written and said about . . . homosexual   1
marriage . . . , the most fundamental but important point has been overlooked. Marriage is both culturally and physiologically compatible but
so-called homosexual marriage is neither culturally nor physiologically
possible.   2

Homosexuality is not generational. The family tree that starts with a
homosexual union never grows beyond a sapling. Without the cooperation
of a third party, the homosexual marriage is a dead-end street. In cyber language, the marriage is not programmed properly and there are hardware
problems as well.

. . . Across America, "rights" are being created and bestowed routinely   3
by judges indifferent to the wishes and values of their communities. This
new wave of judicial tyranny confers special rights upon whichever group
can cry the shrillest claim of victimhood.

At the core of the effort of homosexuals to legitimize their behavior is   4
the debate over whether or not homosexuality is some genetic or inherited
trait or whether it is a chosen behavior. The activists argue that they are a
minority and homosexuality is an immutable characteristic.

But no school of medicine, medical journal or professional organization   5
such as the American Psychological Association or the American Psychi-
atric Association has ever recognized the claim that homosexuality is genetic, hormonal or biological.

While homosexuals are few in number, activists claim they represent   6
about 10% of the population. More reliable estimates suggest about 1% of
Americans are homosexual. They also are the wealthiest, most educated
and most traveled demographic group measured today. Per capita income
for the average homosexual is nearly twice that for the average American.
They are the most advantaged group in America.

Homosexuality is a behavior-based life-style. No other group of Ameri-   7
cans have ever claimed special rights and privileges based solely on their
choice of sexual behavior, and the 1986 Supreme Court decision of Bowers
vs. Hardwick said sodomy is not a constitutionally protected right.

When the state enacts a new policy, it must be reflected in its public   8
school curriculum. Textbook committees and boards of education will ensure that all of that flows into the classroom. American families do not
want the "normalcy" of homosexual marriage taught to their children.

Churches may not be forced to perform homosexual weddings but indi-   9
vidual churches that resist may be subjected to civil suit for sexual discrimination. Resistance may be used as a basis for denying them access to federal,
state or local government programs. In the Archdiocese of New York, Catholic
churches were singled out by the city and denied reimbursement given to
every other church for providing emergency shelter to the city's homeless.
The reason cited was Catholic opposition to homosexual "rights" ordinances.

Whatever the pronouncements of the . . . nation's highest court, Ameri-   10
cans know that "homosexual marriage" is an oxymoron. Calling a homosexual relationship a marriage won't make it so. There is no use of rhetoric that
can sanitize it beyond what it is: unnatural and against our country's most
basic standards. Every reputable public opinion poll demonstrates that nearly
8 of every 10 Americans don't accept the pretense of "homosexual marriage."

Selections 11A and 11B

Evaluate the arguments on both sides. Who has the stronger arguments,
and why? Make certain your response does not rest too heavily on rhetorical devices.

Alternative assignment: Determine which author relies more heavily on rhetorical devices to persuade his or her audience.

LATEST RULING IS GOOD SCOUT MODEL

USA Today

Groups that take in public funds can't shut out minorities.

The Boy Scouts still don't get it. They want all the perquisites and support due a welcoming, public-service organization—while retaining a '50s-style right to discriminate against minorities.   1

Fortunately, most similar organizations understand by now that it's wrong to invite the public in, use taxpayer-financed facilities and then claim the right to act like a small and sectarian private club.   2

The Scouts, virtually alone, still want it both ways. And they're vowing to go to the U.S. Supreme Court in defense of their hypocrisy.   3

The national Scout bureaucracy professes shock at the New Jersey Supreme Court's decision this week in favor of a highly honored Scout who was expelled by the organization for being gay.   4

The reason ought to be obvious: The Scouts solicit and get generous public support in troop sponsorship, free space for their activities and cash contributions. And they attempt, nonselectively, to bring in almost every pre-teen in pants. Therefore they are subject to New Jersey's law barring discrimination on the basis of race, religion or sexual orientation.   5

Their claim to "freedom of association" smacks of what the Jaycees argued in the Supreme Court in 1984 in defense of barring women—and lost. It's what Rotary clubs argued in 1987—and lost. It's what businessmen's clubs argued in 1988—and lost.   6

The argument is embarrassingly similar to that used by the white-supremacist defenders of racial segregation as a way of life in the 1950s and '60s.   7

Civil rights watchdogs say the Scouts are about the last national organization to make such a claim. Others, including Jaycees and Rotary, say they are stronger for having abandoned their former discriminatory practices.   8

Unfortunately, the Scouts' arguments have worked in states with weaker anti-discrimination laws or more pliable courts. California's Supreme Court even found a tortuous distinction between the Boys Clubs, covered by that state's law, and the Scouts, who aren't. The ayatollahs of California Scouting were permitted to kick out otherwise model Scouts who made the mistake of being honest about their religious uncertainties or sexual makeup.   9

The reaction has been a backlash: In California, Illinois, Maine and elsewhere, some governments, local charities and corporations are unwilling to fund and sponsor Scout programs because the Scouts insist on discriminating. And more lawsuits are soaking up Scout funds.   10

The Scouts' lofty objectives and 80-year record of success with nearly   11
100 million boys are commendable. The Scout Law soundly directs a Scout
to "seek to understand others," to "treat others as he wants to be treated."

The bosses of Scouting—and their apologists—ought to reread that good   12
advice.

DECISION ASSAULTS FREEDOM

Larry P. Arnn

Larry P. Arnn is president of the Claremont Institute in California.

Opposing view: The organization has the right to select its associates.

Since 1911, Boy Scouts have been required to swear: "On my honor I will   1
do my best to do my duty to God and country and to obey the Scout Law;
To help other people at all times; To keep myself physically strong, mentally awake, and morally straight." This oath has shaped the character and
habits of generations of young men.

In New Jersey, the Boy Scout oath and the principles behind it are effec-   2
tively voided. The state's Supreme Court ruled that the Scouts must re-
admit James Dale, a scoutmaster who revealed his homosexuality in 1990.

The court said that the Scouts' decision to expel Dale was "based on   3
little more than prejudice" and was "not justified by the need to preserve
the organization's expressive rights."

Nonsense. The court's decision [tramples] on the rights of privacy, free   4
association, and religious liberty to promote homosexuality as a civil right.
The implications go far beyond the Boy Scouts.

Under American civil rights law, once a group becomes a protected   5
minority, a myriad of constitutional forces is called to its defense. If homosexuality is granted this special status, it will affect every family in the land.

Furthermore, the court said that the Scouts are a "place of public ac-   6
commodation" because they have a broad-based membership and work
closely with other public service groups. But the Boy Scouts do not hang a
sign that says all may enter. As the name implies, they are looking first of
all for boys. More specifically, boys who will promise to follow the creed
that makes scouting what it is. We can change that only by destroying
scouting.

The rights of the Boy Scouts, not James Dale, are being violated. Each   7
of us has under the Constitution the right to associate with whom we
please. That means necessarily the right not to associate, too. If that right
cannot be protected here, it cannot be protected anywhere. Indeed, churches
will be next.

This case likely will go to the U.S. Supreme Court. But maybe the next   8
president will sign a Defense of the Boy Scouts Act. We must hope for that,
and also for a good decision from the U.S. Supreme Court.

Selection 12

Determine whether this essay contains an argument and, if it does, what it is.

Alternative assignment: Identify rhetorical devices, including slanters and fallacious reasoning.

IS GOD PART OF INTEGRITY?

Editorial from *Enterprise Record*, Chico, California

What Oroville High School was trying to do last Friday night, said Superin-   1
tendent Barry Kayrell, was "maintain the integrity of the ceremony."

The ceremony was graduation for approximately 200 graduates.   2

The way to maintain "integrity," as it turned out, was to ban the words   3
"God" and "Jesus Christ."

The result was a perfect example of out-of-control interpretation of the   4
separation of church and state.

The high school's action in the name of "integrity" needlessly disrupted   5
the entire proceeding as almost the entire graduating class streamed out of
their seats in support of Chris Niemeyer, an exemplary student who had
been selected co-valedictorian but was barred from speaking because he
wanted to acknowledge his belief in God and Jesus Christ.

The speech, said Kayrell, "was more of a testimonial."   6

It was preaching, added OHS principal Larry Payne.   7

"I truly believe in the separation (of church and state)," explained Kayrell.   8

It was a complicated story that led to last Friday night.   9

Niemeyer and fellow senior Ferin Cole had prepared their speeches   10
ahead of time and presented them to school officials. Cole, who plans to attend Moody Bible College, had been asked to deliver the invocation.

Both mentioned God and Jesus Christ. Both were told that was   11
unacceptable.

Both filed a last-minute action in federal court, challenging the school's   12
censorship. At a hearing Friday, just hours before graduation, a judge refused to overrule the school on such short notice. The suit, however, continues, and the judge acknowledged it will involve sorting out complex constitutional questions.

Defeated in court, Niemeyer and Cole met with school officials to see   13
what could be salvaged.

Both agreed to remove references to the deity, but Cole wanted to men-   14
tion why, in an invocation—by definition a prayer—he was not allowed to
refer to God. That was nixed, and Cole simply bowed out.

Niemeyer was supposed to deliver his revised draft to Payne by 5 P.M.   15

He missed the deadline, but brought the draft with him to the ceremony.   16

When it was his turn to speak, Niemeyer came forward, but Payne   17
instead skipped over the program listing for the valedictory address and

announced a song. The two debated the question on stage as the audience and graduates-to-be looked on.

Finally turned away, Niemeyer left the stage, tears of frustration on his cheeks, and his classmates ran to his side in a dramatic show of support.   18

You might say they were inspired by integrity.   19

The object of the First Amendment to the U.S. Constitution is to bar government-enforced religion. It was not designed to obliterate belief in God.   20

To stretch that command to denying a student the right to acknowl-edge what has spurred him on to the honor he has won is a bitter perversion.   21

That would apply whether the student was Islamic, Buddhist or any be-lief—atheist included. There is room, it would seem, for diversity in vale-dictory speeches, too.   22

Not at Oroville High School. There God and integrity don't mix.   23

It's spectacles like that played out last Friday night that have prompted Congress to consider a constitutional amendment aimed at curbing such misguided excess.   24

Earlier in the week it drew a majority vote in House, but fell short of the two-thirds margin needed.   25

Maybe such actions as witnessed locally can push it over the top.   26

Selection 13

Identify the conclusion of the article. How many arguments are given for it?

SHORTEN FEDERAL JAIL TIME

From the *Washington Post*

Don Edwards

> U.S. Rep. Don Edwards, D-Calif., was chairman of the House Judiciary Subcommittee on Civil and Constitutional Rights.

A debate is raging in the federal judiciary about whether the number of fed-eral judges should be increased. This question should not be decided by Congress until it straightens out the mess it has made of the courts' jurisdiction.   1

Beginning in 1981, with the highly publicized "War on Crime," Con-gress has swamped the federal courts with thousands of criminal cases, many unsuited for the federal judiciary. The result is that civil cases are ne-glected and our federal prisons are bulging with prisoners, many of whom should not be there at all.   2

In its effort to be tough on drug offenders and other criminals, Congress enacted mandatory minimum laws that require judges to impose a certain sentence on the offender regardless of the person's background and the   3 nature of the crime. The statutes now number over 100, with many more proposed. Not only must we not enact new ones, but Congress should repeal the ones now on the books for the following reasons:

- The federal prison population has tripled since 1980 and is projected to quintuple by the year 2000. Three-quarters of new arrivals are drug offenders; by 1996, these offenders will comprise two-thirds of the prison population. Ninety-one percent of mandatory minimum sentences are imposed on nonviolent first-time offenders.

- To accommodate this influx of new prisoners, we opened 29 new federal prisons between 1979 and 1990. They cost $1.7 billion to maintain, and each new cell cost $50,000. With the continued precipitous increase in population, by 2002 we will be spending $2.8 billion annually.

  If we had instead sentenced these inmates to an alternative to prison, such as probation or a halfway house, we could have saved $384 million in 1992 and, projections show, an astounding $438 million in 1993. The Bureau of Prisons says that the recidivism rate for offenders sentenced to alternatives is extremely low—only 14 percent.

- Because of mandatory sentencing laws, federal district judges have no discretion in sentencing. They must send offenders to prison for long terms. Between 1984, when mandatory minimums were first enacted, and 1991, the sentence length for a nonviolent offender increased an average of 48 months. Ironically, mandatory sentencing laws often mean that nonviolent offenders receive longer sentences than violent offenders.

- We are filling our prisons with nonviolent, first-time offenders while dangerous and violent criminals often avoid mandatory minimum sentences.

- Morale in the federal judiciary is often low. The Judicial Conference, representing all federal judges, has come out against mandatory minimum sentences, and two senior judges recently announced publicly that they would no longer preside over drug trials because of these inequities. . . .

We must reverse the congressional policies that have created this mind-less mandatory minimum sentencing, which overcrowds our federal prisons. At $30,000 per year to maintain these inmates, we not only cannot afford to implement these senseless policies, but we are creating a criminal training ground for first-time nonviolent offenders who might otherwise be turned into law-abiding citizens.   4

Selections 14A and 14B

Evaluate the arguments on both sides. Who has the stronger arguments, and why?

Alternative assignment: Identify rhetorical devices and determine which author relies more heavily on such devices.

Second alternative assignment: In each of the two essays, find four arguments that can be treated as categorical syllogisms. Set up a key, letting a letter stand for a relevant category. Be sure you identify the category in plain English. Then circle all and only the distributed terms. Then state whether each syllogism is valid, identifying rules broken by any syllogisms that are not.

CLEAN NEEDLES BENEFIT SOCIETY

USA Today

Our view: Needle exchanges prove effective as AIDS counterattack. They warrant wider use and federal backing.

Nothing gets knees jerking and fingers wagging like free needle-exchange programs. But strong evidence is emerging that they're working. 1

The 37 cities trying needle exchanges are accumulating impressive data that they are an effective tool against spread of an epidemic now in its 13th year. 2

- In Hartford, Conn., demand for needles has quadrupled expectations—32,000 in nine months. And free needles hit a targeted population: 55% of used needles show traces of AIDS virus.
- In San Francisco, almost half the addicts opt for clean needles.
- In New Haven, new HIV infections are down 33% for addicts in exchanges.

Promising evidence. And what of fears that needle exchanges increase addiction? The National Commission on AIDS found no evidence. Neither do new studies in the *Journal of the American Medical Association.* 3

Logic and research tell us no one's saying, "Hey, they're giving away free, clean hypodermic needles! I think I'll become a drug addict!" 4

Get real. Needle exchange is a soundly based counterattack against an epidemic. As the federal Centers for Disease Control puts it, "Removing contaminated syringes from circulation is analogous to removing mosquitoes." 5

Addicts know shared needles are HIV transmitters. Evidence shows drug users will seek out clean needles to cut chances of almost certain death from AIDS. 6

Needle exchanges neither cure addiction nor cave in to the drug scourge. They're a sound, effective line of defense in a population at high risk. (Some 28% of AIDS cases are IV drug users.) And AIDS treatment costs taxpayers far more than the price of a few needles. 7

It's time for policymakers to disperse the fog of rhetoric, hyperbole and scare tactics and widen the program to attract more of the nation's 1.2 million IV drug users. 8

We're a pragmatic society. We like things that work. Needle exchanges   9
have proven their benefit. They should be encouraged and expanded.

PROGRAMS DON'T MAKE SENSE

Peter B. Gemma Jr.

Opposing view: It's just plain stupid for government to sponsor dangerous, illegal behavior.

If the Clinton administration initiated a program that offered free tires to   1
drivers who habitually and dangerously broke speed limits—to help them
avoid fatal accidents from blowouts—taxpayers would be furious. Spending
government money to distribute free needles to junkies, in an attempt to
help them avoid HIV infections, is an equally volatile and stupid policy.

It's wrong to attempt to ease one crisis by reinforcing another.   2

It's wrong to tolerate a contradictory policy that spends people's hard-   3
earned money to facilitate deviant behavior.

And it's wrong to try to save drug abusers from HIV infection by per-   4
petuating their pain and suffering.

Taxpayers expect higher health-care standards from President Clinton's   5
public-policy "experts."

Inconclusive data on experimental needle-distribution programs is no   6
excuse to weaken federal substance-abuse laws. No government bureaucrat
can refute the fact that fresh, free needles make it easier to inject illegal
drugs because their use results in less pain and scarring.

Underwriting dangerous, criminal behavior is illogical: If you subsidize   7
something, you'll get more of it. In a Hartford, Conn., needle-distribution
program, for example, drug addicts are demanding taxpayer-funded needles
at four times the expected rate. Although there may not yet be evidence of
increased substance abuse, there is obviously no incentive in such schemes
to help drug-addiction victims get cured.

Inconsistency and incompetence will undermine the public's confi-   8
dence in government health-care initiatives regarding drug abuse and the
AIDS epidemic. The Clinton administration proposal of giving away needles hurts far more people than [it is] intended to help.

Selections 15A and 15B

Evaluate the arguments on both sides. Who has the stronger arguments,
and why?

Alternative assignment: Identify rhetorical devices and determine
which author relies more heavily on them.

Second alternative assignment: In the first essay, find as many arguments as you can that can be treated as categorical syllogisms. Set up a key,

letting a letter stand for a relevant category. Be sure you identify the category in plain English. Then circle all and only the distributed terms. Then state whether each syllogism is valid, identifying rules broken by any syllogisms that are not.

MAKE FAST FOOD SMOKE-FREE

USA Today

Our view: The only thing smoking in fast-food restaurants should be the speed of the service.

Starting in June, if you go to Arby's, you may get more than a break from burgers. You could get a break from tobacco smoke, too. 1

The roast-beef-sandwich chain on Tuesday moved to the head of a stampede by fast-food restaurants to limit smoking. 2

Last year, McDonald's began experimenting with 40 smokeless restaurants. Wendy's and other fast-food chains also have restaurants that bar smoking. 3

But Arby's is the first major chain to heed a call from an 18-member state attorneys general task force for a comprehensive smoking ban in fast-food restaurants. It will bar smoking in all its 257 corporate-owned restaurants and urge its 500 franchisees to do the same in their 2,000 restaurants. 4

Other restaurants, and not just the fast-food places, should fall in line. 5

The reason is simple: Smoke in restaurants is twice as bad as in a smoker's home or most other workplaces, a recent report to the *Journal of the American Medical Association* found. 6

Fast-food restaurants have an even greater need to clear the air. A quarter of their customers and 40% of their workers are under 18. 7

Secondhand smoke is a class A carcinogen. It is blamed for killing an estimated 44,000 people a year. And its toxins especially threaten youngsters' health. 8

The Environmental Protection Agency estimates that secondhand smoke causes up to 1 million asthma attacks and 300,000 respiratory infections that lead to 15,000 hospitalizations among children each year. 9

All restaurants should protect their workers and customers. If they won't, then local and state governments should do so by banning smoking in them, as Los Angeles has. 10

A person's right to a quick cigarette ends when it threatens the health of innocent bystanders, and even more so when many of them are youngsters. 11

They deserve a real break—a meal in a smoke-free environment that doesn't threaten their health. 12

DON'T OVERREACT TO SMOKE

Brennan M. Dawson

Opposing view: With non-smoking sections available, and visits brief, what's the problem?

If the attorneys general from a handful of states—those charged with up- 1
holding the law—were to hold a forum in Washington, you might expect
them to be tackling what polls say is the No. 1 public issue: crime.

Not these folks. They're worried someone might be smoking in the smok- 2
ing section of a fast-food restaurant. And, there might be children in the non-
smoking section. Thus, they say, fast-food chains should ban all smoking.

Some would argue that this raises serious questions about priorities. 3
But it may be worth debating, since this is supposed to be about protecting
children. Everyone is (and should be) concerned with children's health and
well-being.

But what are we protecting them from—the potential that a whiff of 4
smoke may drift from the smoking section to the non-smoking section during the average 20-minute visit for a quick burger?

Anyone knowledgeable would tell you that none of the available stud- 5
ies can reasonably be interpreted to suggest that incidental exposure of a
child to smoking in public places such as restaurants is a problem. After
all, with the almost universal availability of non-smoking sections, parents
have the option of keeping their kids out of the smoking section.

A recent study published in the *American Journal of Public Health* re- 6
ported that the separate smoking sections in restaurants do a good job of
minimizing exposure to tobacco smoke. According to the figures cited, customers would have to spend about 800 consecutive hours in the restaurants
to be exposed to the nicotine equivalent of one cigarette.

That would represent about 2,400 fast-food meals. Under those condi- 7
tions, most parents would worry about something other than smoking.

Selections 16A and 16B

Evaluate the arguments on both sides. Who has the stronger arguments,
and why?

Alternative assignment: Identify rhetorical devices and determine
which author relies more heavily on them.

Second alternative assignment: In each of the two essays, find as many
arguments as you can that can be treated as categorical syllogisms. Set up a
key, letting a letter stand for a relevant category. Be sure you identify the
category in plain English. Then circle all and only the distributed terms.
Then state whether each syllogism is valid, identifying rules broken by any
syllogisms that are not.

BUYING NOTES MAKES SENSE AT LOST-IN-CROWD CAMPUSES

USA Today

Our view: Monster universities and phantom professors have only
themselves to blame for note-selling.

Higher education got a message last week from a jury in Gainesville, Fla.: 　1
Its customers, the students across the nation, deserve better service.

The jury found entrepreneurs are free to sell notes from college profes- 　2
sors' lectures. And Ken Brickman is an example of good, old free enterprise,
even if his services encourage students to skip class.

Brickman is a businessman who pays students to take notes in classes 　3
at the University of Florida. From a storefront a block off campus, he re-
sells the notes to other students with a markup.

Professors and deans bemoan Brickman's lack of morals. They even use 　4
the word "cheating." They'd be more credible if their complaints—and the
university's legal resources—were directed equally at Brickman's competi-
tor in the note-selling business a few blocks away.

The difference: The competition pays professors for their notes; Brick- 　5
man pays students. Morals are absent, it seems, only when professors aren't
getting their cut.

The deeper issue is why Brickman has found a lucrative market. It's 　6
easy to say that uninspired students would rather read someone else's notes
than spend time in class, but that's not the point.

Why are students uninspired? Why are they required to learn in auditorium- 　7
size classes where personal attention is non-existent, taking attendance impossible, and students can "cut" an entire semester with no one noticing?

Why are students increasingly subjected to teaching assistants—gradu- 　8
ate students who know little more than they—who control classes while
professors are off writing articles for esoteric journals that not even their
peers will read?

Why are there not more professors—every former student can remem- 　9
ber one—who transmit knowledge of and enthusiasm for a subject with a
fluency and flair that make students eager to show up? No one would prefer to stay away and buy that professor's notes.

The debate over professorial priorities—students vs. research—is old. 　10
But so long as students come in second, they'll have good reasons to go to
Ken Brickman for their notes.

BUYING OR SELLING NOTES IS WRONG

Opposing view: Note-buyers may think they're winners, but they lose
out on what learning is all about.*

It's tough being a college student. Tuition costs and fees are skyrocketing. 　1
Classes are too large. Many professors rarely even see their students, let
alone know their names or recognize their faces. The pressure for grades is
intense. Competition for a job after graduation is keen.

---

*The author of the companion piece to the *USA Today* editorial on this subject would not
give us permission to reproduce her essay in a critical thinking text, so we wrote this item
ourselves.

But that's no excuse for buying the notes to a teacher's course. What 2
goes around comes around. Students who buy someone else's notes are only
cheating themselves—by not engaging in the learning process to the fullest
extent. They aren't learning how to take notes. Or how to listen. Or how to
put what someone is saying into their own words.

What happens if the notes are inaccurate? Will a commercial note-taker 3
guarantee the notes? Would you want to take a test using someone else's
notes?

Besides, what the professor says is her own property. It is the result of 4
hard work on her part. A professor's lectures are often her principal means
of livelihood. Nobody but the professor herself has the right to sell her property. Buying the notes to her lectures without her permission is just like
selling a book that she wrote and keeping the money for yourself.

And buying the notes from someone who is selling them without the 5
teacher's permission is the same as receiving stolen goods.

And that's assuming that there will be anyone out there to buy the 6
notes in the first place. After all, most students will want to take notes for
themselves, because they know that is their only guarantee of accuracy.
People who think they can get rich selling the notes to someone's lectures
should take a course in critical thinking.

The pressure for good grades doesn't justify buying or selling the notes 7
to a professor's lectures without her permission. If you can't go to class,
you shouldn't even be in college in the first place. Why come to school if
you don't want to learn?

Selections 17A and 17B

Evaluate the arguments on both sides. Who has the stronger arguments,
and why?

Alternative assignment: Identify rhetorical devices and determine
which author relies more heavily on them.

NEXT, COMPREHENSIVE REFORM OF GUN LAWS

USA Today

Our view: Waiting periods and weapon bans are welcome controls, but
they're just the start of what's needed.

The gun lobby got sucker-punched by the U.S. Senate last weekend. It 1
couldn't happen to a more deserving bunch.

For seven years, gun advocates have thwarted the supersensible Brady 2
bill, which calls for a national waiting period on handgun purchases.
Through a mix of political intimidation, political contributions and perverse constitutional reasoning, gun lobbyists were able to convince Congress to ignore the nine out of 10 Americans who support that idea.

But suddenly, after two days of filibuster, the Senate abruptly adopted   3
the Brady bill. The House has already acted, so all that remains is to do
some slight tinkering in a House-Senate conference, and then it's off to the
White House for President Clinton's signature.

That's not the end of welcome gun control news, though. As part of the   4
anti-crime bill adopted last week, the Senate agreed to ban the manufacture
and sale of 19 types of assault-style semiautomatic weapons. Although
these weapons constitute fewer than 1% of all guns in private hands, they
figure in nearly 10% of all crime. The bill also bans some types of ammunition and restricts gun sales to, and ownership by, juveniles.

These ideas are worthy, but they can't do the whole job. Waiting periods   5
and background checks keep criminals from buying guns from legal dealers.
Banning certain types of anti-personnel weapons and ammunition will keep
those guns and bullets from growing more common and commonly lethal.

Yet the wash of guns and gun violence demands much, much more. The   6
judicial ability to process firearm-related crimes with certainty and speed is
part of the solution. But even more so is the adoption of laws that permit
gun licensing, gun registration and firearm training and education.

After years of denying the popular mood, Congress appears ready to   7
honor it. That merits applause. But its new laws are just a start. Without
truly comprehensive controls, the nationwide slick of gun carnage is bound
to continue its bloody, inexorable creep.

GUN LAWS ARE NO ANSWER

Alan M. Gottlieb

Opposing view: Disarming the law-abiding populace won't stop crime.
Restore gun owners' rights.

Every time another gun control law is passed, violent crime goes up, not   1
down, and the gun-ban crowd starts to yelp for more anti-gun laws.

So it's no surprise that the gun-banners are already snapping at the heels   2
of our Bill of Rights.

They turn a blind eye to the fact that California, with a 15-day waiting   3
period, experienced a 19% increase in violent crime and a 20% increase in
homicide between 1987 and 1991. And that a 1989 ban on "assault weapons"
in that state has also resulted in increased violent crime.

In Illinois, after a 30-day waiting period was installed, that state experi-   4
enced a 31% increase in violent crime and a 36% increase in the homicide rate.

And, a handgun ban in Washington, D.C., has made it the murder capi-   5
tal of the world!

The results are in. Gun control makes the streets safe for violent crimi-   6
nals. It disarms their victims—you and me. The people's right to protect
themselves should be restored, not restricted.

Case in point: Bonnie Elmasri of Wisconsin, who was being stalked by   7
her estranged husband despite a court restraining order, was killed along

with her two children while she waited for the handgun she purchased under that state's gun-waiting-period law.

Bonnie and her children are dead because of gun control laws, as are thousands of other victims each year.   8

Anybody who believes that disarming the law-abiding populace will help reduce crime has rocks in the head.   9

The next time a violent criminal attacks you, you can roll up your copy of USA TODAY and defend yourself with it. It may be all you'll have left for self-protection.   10

Selections 18A and 18B

Evaluate the arguments on both sides. Who has the stronger arguments, and why?

Alternative assignment: Identify rhetorical devices and determine which author relies more heavily on them.

HOW CAN SCHOOL PRAYER POSSIBLY HURT? HERE'S HOW

USA Today

Our view: Mississippi case shows how people's rights can be trampled by so-called "voluntary prayer."

What harm is there in voluntary prayer in school?   1

That's the question . . . House Speaker Newt Gingrich and others pose in their crusade to restore prayer to the classroom. They argue that a constitutional amendment to "protect" so-called voluntary school prayer could improve morals and at worst do no harm.   2

Well, a mother's lawsuit filed Monday against Pontotoc County, Miss., schools says otherwise. It shows government-sponsored voluntary prayer in school threatens religious liberty.   3

All the mother, Lisa Herdahl, wants is that her six children get their religious instruction at home and at their Pentecostal church, not at school.   4

But their school hasn't made that easy. Prayers by students go out over the public address system every day. And a Bible study class is taught at every grade.   5

School officials argue that since no one is ordered to recite a prayer or attend the class, everything is voluntary.   6

But to Herdahl's 7-year-old son, it doesn't seem that way. She says he was nicknamed "football head" by other students after a teacher told him to wear headphones so he wouldn't have to listen to the "voluntary" prayers.   7

And she says her 11-year-old son was branded a "devil worshiper" after a teacher told students he could leave a Bible class because he didn't believe in God.   8

Indeed, Herdahl's children have suffered exactly the kind of coercion to conform that the Supreme Court found intolerable when it banned state-written prayers in 1962 and outlawed Alabama's moment of silence for meditation or voluntary prayer in 1985.   9

As the court noted in those cases, when government—including schools—strays from neutrality in religious matters, it pits one religion against another. And youngsters especially can feel pressured to submit to a majority's views.   10

That's why a constitutional amendment to protect "voluntary prayer" in school is so dangerous.   11

Students don't need an amendment to pray in school now. They have that right. And they can share their religious beliefs. They've formed more than 12,000 Bible clubs nationwide that meet in schools now, only not during class time.   12

For the Herdahls, who refused to conform to others' beliefs, state-sponsored voluntary prayer and religious studies have made school a nightmare.   13

For the nation, a constitutional amendment endorsing such ugly activities could make religious freedom a joke.   14

WE NEED MORE PRAYER

Armstrong Williams

Armstrong Williams is a Washington, D.C.–based business executive, talk-show host, and author of The Conscience of a Black Conservative.

Opposing view: The tyranny of the minority was never envisioned by the nation's Founding Fathers.

The furor aroused by . . . Newt Gingrich's remarks about renewing school prayer illustrates how deep cultural divisions in American society really are.   1

A few moments of prayer in schools seems a small thing—harmless enough, almost to the point of insignificance. Yet it has provoked an impassioned firestorm of debate about the dangers of imposing viewpoints and the potential for emotionally distressing non-religious children.   2

The Constitution's framers were wary of a "tyranny of the majority," and so they imposed restraints on the legislature. They never foresaw, nor would they have believed, the tyranny of the minority made possible through an activist judiciary changing legal precedents by reinterpreting the Constitution.   3

The American ideal of tolerance has been betrayed by its use in directly attacking the deeply held convictions of millions of Americans.   4

The fact that this country was once unashamedly Christian did not mean that it was necessarily intolerant of other views—at least not nearly so intolerant of them as our rigid secular orthodoxy is toward all religious expression. Through the agency of the courts, a few disgruntled malcontents have managed to impose their secular/humanist minority views on the majority.   5

But it has not always been so.   6

The confidence with which some maintain that school prayer is mani-   7
festly unconstitutional belies an ignorance of our nation's history. America
was founded by religious men and women who brought their religious beliefs and expressions with them into public life.

It was in 1962 that an activist Supreme Court ruled that denomination-   8
ally neutral school prayer was judged to violate the establishment clause of
the First Amendment. Since then, the "wall of separation" between church
and state has rapidly become a prison wall for religious practice.

The drive to protect the delicate sensibilities of American children from   9
the ravages of prayer is particularly ironic when our public schools have become condom clearinghouses that teach explicit sex.

The real heart of the school prayer issue is the role of religion in our   10
public life.

Selection 19

Summarize the author's argument or arguments, if any.

Alternative assignment: Identify rhetorical devices, if any.

PLANET OF THE WHITE GUYS

Barbara Ehrenreich

On the planet inhabited by the anti–affirmative action activists, the only   1
form of discrimination left is the kind that operates against white males.
There, in the name of redressing ancient wrongs, white men are routinely
shoved aside to make room for less qualified women and minorities. These
favored ones have no problems at all—except for that niggling worry that
their colleagues see them as underqualified "affirmative-action babies."
Maybe there was once an evil called racism in this charmed place—30 or
300 years ago, that is—but it's been replaced by affirmative action.

Now I agree that discrimination is an ugly thing no matter who's at the   2
receiving end, and that it may be worth reviewing affirmative action . . . to
see whether it's been fairly applied. People should not be made to suffer for
the wicked things perpetrated by their ancestors or by those who merely
looked like them. Competent white men should be hired over less competent women and minorities; otherwise, sooner or later, the trains won't run
on time and the planes will fall down from the sky.

But it would be a shame to sidestep the undeniable persistence of   3
racism in the workplace and just about everywhere else. Consider the recent lesson from Rutgers University. Here we have a perfectly nice liberal
fellow, a college president with a record of responsiveness to minority concerns. He opens his mouth to talk about minority test scores, and then—
like a Tourette's syndrome victim in the grip of a seizure—he comes out

with the words "genetic hereditary background." Translated from the academese: minorities are dumb, and they're dumb because they're born that way.

Can we be honest here? I've been around white folks most of my life— from left-wingers to right-wingers, from crude-mouthed louts to prissy-minded élitists—and I've heard enough to know that *The Bell Curve* is just a long-winded version of what an awful lot of white people actually believe. Take a look, for example, at a survey reported by the National Opinion Research Center in 1991, which found a majority of whites asserting that minorities are lazier, more violence-prone and less intelligent than whites. Even among the politically correct, the standard praise word for a minority person is "articulate," as if to say, "Isn't it amazing how well he can speak!" 4

Prejudice of the quiet, subliminal kind doesn't flow from the same place as hate. All you have to do to be infected is look around: at the top of the power hierarchy—filling more than 90% of top corporate-leadership slots and a grossly disproportionate share of managerial and professional positions—you see white men. Meanwhile, you tend to find minorities clustered in the kind of menial roles—busing dishes, unloading trucks—that our parents warned were waiting for us too if we didn't get our homework done. 5

So what is the brain to make of this data? It does what brains are designed to do: it simplifies and serves up the quickie generalizations that are meant to guide us through a complex world. Thus when we see a black colleague, who may be an engineer or a judge, the brain, in its innocence, announces helpfully, "Janitor-type approaching, wearing a suit." 6

Maybe it's easier for a woman to acknowledge this because subliminal prejudice hurts women too. Studies have shown, for example, that people are more likely to find an article convincing if it is signed by "Bob Someone" instead of, say, "Barbara Someone." It's just the brain's little habit of parceling reality into tidy equations, such as female = probable fluffhead. The truth is that each of us carries around an image of competence in our mind, and its face is neither female nor black. Hence our readiness to believe, whenever we hear of a white male losing out to a minority or a woman, that the white guy was actually more qualified. In Jesse Helms' winning 1990 campaign commercial, a white man crumples up a rejection letter, while the voice-over reminds him that he was "the best-qualified." But was he? Is he always? And why don't we ever hear a white guy worry out loud that his colleagues suspect he got the job—as white men have for centuries—in part because he's male and white? 7

It's a measure of the ambient racism that we find it so hard to believe that affirmative action may actually be doing something right: ensuring that the best guy gets the job, regardless of that guy's race or sex. Eventually, when the occupational hierarchy is so thoroughly integrated that it no longer makes sense for our subconscious minds to invest the notion of competence with a particular skin color or type of genitalia, affirmative action can indeed be cast aside like training wheels. 8

Meanwhile, aggrieved white men can console themselves with the gains their wives have made. Numerically speaking, white women are the 9 biggest beneficiaries of affirmative action, and because white women tend to marry white men, it follows that white men are, numerically speaking, among the top beneficiaries too. On this planet, Bob Dole and Pat Buchanan may not have been able to figure that out yet, but most white guys, I like to think, are plenty smart enough.

Selection 20

Summarize the author's argument or arguments, if any.
   Alternative assignment: Identify rhetorical devices, if any.

DO WOMEN REALLY NEED AFFIRMATIVE ACTION?

Joanne Jacobs

Joanne Jacobs is a columnist for the *San Jose Mercury News.*

I was an affirmative action hire, back in 1978. As a result, I became the first 1
woman on my newspaper's editorial board. Was I qualified? I certainly didn't have much experience. My boss took a chance on me. I like to think it paid off for him, as well as for me.

At any rate, I made the editorial board safe for women, who now make 2
up half the editorial pages staff.

Defenders of affirmative action have a new strategy: Consider the 3
ladies. Recently, a liberal coalition kicked off a campaign to cast affirmative action as a gender issue rather than a racial issue. Affirmative action is essential for women's advancement, argued feminist leaders. "Women will not quietly accept a rollback of our rights that have been part of the American scene for over a generation," said Katherine Spillar of the Feminist Majority.

It's a smart, though doomed, strategy. 4

The smart part is that women (if not feminists) are a majority of the 5
population. If women were persuaded that affirmative action guarantees not only our benefits but also our "rights," public thinking would shift dramatically.

Poll data suggest that the majority of Americans accept preferences for 6
women in jobs and college admissions; most are deeply hostile to preferences for minority group members.

The doomed part is that most white women don't think they'd lose 7
their rights or their jobs if affirmative action ceased to exist; some are concerned about the job prospects of their white male husbands.

Affirmative action used to mean reaching out to people excluded from 8
opportunities by overt discrimination or old-boy networks, giving everyone a chance to compete fairly. Now it means judging members of previously excluded groups by different, generally lower, standards than others.

Are women less able to compete [today] because of past discrimination? 9
Certainly not women under the age of 45.

In the past, affirmative action helped women—especially educated, 10
middle-class women—enter occupations previously reserved for men. Some
argue that affirmative action primarily benefited white women who were
most prepared to take advantage of opportunities, and most likely to have
connections to powerful men.

Certainly, it appears that affirmative action has done most for those 11
who needed it least. It has opened doors; it has not built ladders.

While women haven't achieved parity in income or in top executive 12
ranks or in all careers, affirmative action can provide little help in achieving those goals in the future.

In education, affirmative action already is passé. Women don't need to 13
be judged by different standards to get into college, and they're not. Females
earn higher high school grades than males, and are more likely to attend
college.

Women are underrepresented in the mathematical sciences and in engi- 14
neering studies. An affirmative action program that admitted calculus-
deficient women to physics class would do them no favor. The solution is
not changing the standards; it's persuading more girls to take more math in
high school.

Women do worry about workplace equity, but here too affirmative ac- 15
tion isn't likely to play much of a role. The much-discussed "glass ceiling"
for women in business is related to two factors: family and calculus.

Women are more likely than men to put family needs ahead of career 16
advancement. These choices have consequences. Women who work more
at home don't get to be CEO.

Selection 21

WANT BETTER SCHOOLS? SUPPORT TEACHERS

Dean Simpson Jr.

Dean Simpson Jr. is a journalist and English teacher at Mount Whitney
High School in Visalia.

In the education business we have a saying—you get what you get. That 1
means we teachers deal with everyone who comes through the schoolhouse
door and do our best to send them out at the end of their term as educated,
productive citizens. We don't always succeed, but overall I think we do a
pretty good job, particularly considering the circumstances.

So it really aggravated me when my local newspaper, the *Visalia Times-* 2
Delta, ran an editorial the essence of which was that teachers need to quit
complaining about things we can't control, "a diverse population, poorly

prepared students, television, parents who don't want to be involved and per-pupil spending," and focus on what we can control, "more rigorous course offerings, return to heavy concentration of reading and writing, longer school sessions and updated curriculum." Unfortunately, such admonitions seem to be on the lips of every legislator and a good many citizens in California today.

The implication is that teachers are a bunch of crybabies for complaining about conditions. Has it ever occurred to anyone that education is the only industry that cannot control the raw materials with which it works? No other industry is accused of whining if it advocates and lobbies for better conditions within its business. 3

If US Steel receives a shipment of substandard pig iron from a supplier, no one accuses it of whining when it sends it back. If Ford Motor Co. is sent a batch of defective tires, it's not expected to refabricate them. It sends them back. In education—we get what we get. 4

Imagine the newspaper has 10 openings to fill and must hire the first 10 people who walk through the door. A blind man running the press? Non-English speakers writing and editing copy? Editors who have no concept of personnel management? Advertising sales people with no communication skills? And you'd better be able to show an increase in profit and decline in the number of errors in the copy. After all, standards must be raised and met. Oh, and there's no money for additional training. 5

Now let's take a look at some of the things teachers are told we need to do to improve education, which will apparently overcome all the problems in the schools today. The first need supposedly is for more rigorous course offerings. 6

From kindergarten up, the curriculum is more rigorous than it has ever been. When I attended kindergarten in 1951, children learned "socialization," finger-painting and maybe their ABCs. Now they learn to read and write and do addition and subtraction up to 10s and sometimes beyond. Kindergartners also learn geometric shapes and handle thought problems using, among other things, simple graphs. 7

All the math I needed for admission to college in 1964 was geometry. Now students must have Algebra II and trigonometry to have much of a chance of college acceptance. 8

Advanced Placement classes—high school classes where students may earn college credit after completing the course and passing a rigorous test—didn't exist when I graduated. Today, virtually every high school in California offers these. 9

As far as the need for a heavy concentration on reading and writing, my students not only read a lot, I also expect them to interpret what they read. No fill-in-the-bubbles tests. They dig into what they've read, understand it, apply it to their lives and write about what they learn from the literature and its application in the world. None of these skills, unfortunately, show up on standardized tests. 10

Freshman enter my journalism class as experienced writers having written much more, and with greater depth of thought, than was expected of my generation until we entered college. 11

The next element that supposedly would make "for better trained, better educated students" is "longer school sessions." I'm not sure whether that refers to a longer school day or longer school year, but it doesn't really matter. If it's a longer school day, we must do away with all school sports, as they do in other countries. Those who wish can play club sports—for a fee. Other electives and extracurricular activities? Gone. We can't cherry pick here. We either do it as it's done in Europe and Japan, or we don't.   12

Is a longer school year the goal? Why? I know the Japanese have a longer school year, and they score better on standardized tests, too. But if those schools are so far superior to ours, why are their students, and students from nearly every country in the world, clamoring to get into our universities? And why are our high school graduates able to compete successfully with those foreign students at our universities?   13

A graduate from Redwood, one of our Central Valley high schools, received a Nobel Prize last year. If foreign schools are so great, why is the United States, with its allegedly awful schools, so far ahead in technology, industrial production and every other category worth mentioning?   14

As far as updating curriculum, can anyone update his home or business without money? Textbooks cost money. Lab materials, technology, and continuing education and training for teachers cost money. Show me the money or shut up and get out of the way.   15

What we need in education is support. If we set standards—and we've actually been fighting for high standards for a long time—parents and community members must help us enforce them, not just by paying lip service, but by feeling real pain when necessary. Standards are inherently hard to reach and not everyone will reach them, at least not easily. We in education have been doing our part for years. Remember CLAS and CAP tests? Educators didn't dump them in the trash can. Politicians and the public did.   16

Instead of accusing us of being whiners, see us as advocates for improvement of all conditions that contribute to educational success for all children. Help us by addressing the social problems we can't control and we'll continue to improve the conditions we can.   17

---

*This article appeared in the *Sacramento Bee*, October 30,1999.

The Top Ten Fallacies of All Time

irst, we should say that we are not exactly reporting on a scientific study of reasoning errors. Far from it. We're simply listing for you what we think, based on our experience, are the top ten fallacies that crop up in spoken and written discourse. While we're disclaiming, we should point out that some of the items on our list are not really *errors* of reasoning, but *failures to reason at all.* We've included a remark or two for each, but we remind you that all of them are discussed at greater length in the text.

Ad Hominem/Genetic Fallacy   If it operated no place but in politics, ad hominem would still be among the most ubiquitous examples of flawed thinking. While the fact that a claim came from a prominent Democrat or Republican might be reason to suspect that it is political spin, we constantly see people refuse even to consider—and sometimes even to hear—claims from a political perspective not their own.

But what goes in politics also goes in other arenas. Adversaries in practically every field of endeavor put less, or no, stock in their opponents' claims. We've seen this happen in disputes between physical scientists and behavioral scientists, between factions of a single academic department and between members of the same family. If they're the enemy, they cannot be right.

Wishful Thinking   It may surprise some that this category makes the list, but in observing people's behavior, we've come to think that they frequently put a lot more trust in what they *want* to be true than in what they have sufficient evidence for. Advertising, our beliefs about famous people, and many other aspects of life play upon this tendency. How frequently do we see people buy products that claim they'll make them younger-looking or more attractive only to see no results? And then those same people will buy the next product that comes along, possibly with the same ingredients but with a bigger advertising budget or better-looking models. We can't explain this in any other way except the triumph of wishful thinking.

Neither of your authors is sure about capital punishment. But we'd seriously consider an exception for people who invent scams based on wishful thinking and milk elderly people out of money they've spent their lives saving up.

"Argument" from Popularity   Operating in conjunction with wishful thinking, appeal to the authority of one's peers and predecessors accounts for a great proportion of the belief in the supernatural aspects of the world's religions down through history. This alone is sufficient to place it prominently on the list.

If a person thought that none of her peers believed in miracles, and she had never witnessed anything firsthand herself that she considered miraculous, what are the chances that she would believe in miracles in general? We believe it is very slight. But most people (well over half of Americans) do believe in miracles, and we're pretty sure such a belief rests at least in substantial part on the fact that most of one's peers share the belief.

Hasty Conclusion   The human mind is quick to notice connections between and among things, events, people, and such. Often, its ability to latch on to similarities is a matter of survival. ("A snake like that bit Oog the other day, and he died. I'm leaving this one alone.") But connections that we make "naturally" need to be sifted through our reasoning faculty. ("When I dreamed that it would rain on Easter, it did. Last night I dreamed the stock market would go up . . . I think I'll add to the portfolio.") We must learn to distinguish psychological connections from logical ones and real causes from coincidence.

Some related fallacies fall under this headline: One is the mistake of making generalizations from evidence that is merely anecdotal. We are very quick to conclude that our own experience or that of our friends is sufficient to arrive at a conclusion about the entire world. But the fact that your friend had no trouble quitting smoking is no reason to think that others won't.

A second variation is drawing conclusions (or generalizing) based on a bad analogy. The fact that two (or more) things are alike in *some* way is not all by itself a reason to think they're alike in some other way. Analogous items have to be very carefully scrutinized for relevant similarities and dissimilarities before we can draw conclusions about one from the other.

A final variety of the hasty conclusion fallacy, and a sad one, is the way the public (following the media) jump to conclusions about a person's guilt when in fact the person has merely been mentioned as a possible suspect. Examples range from the famous Sam Shepard case (on which the TV series and movie titled *The Fugitive* were based) to the case of Richard Jewell, who was actually a hero in the bomb incident at the 1996 Atlanta Olympics—he spotted the fateful knapsack and warned people away from what turned out to be a pipe bomb, but later his name was leaked as a suspect.

Although he was soon cleared by the authorities, the stigma of suspicion stuck to him and he was hounded for months at ball games and other events. His hopes for a career in law enforcement went down the drain. This, and similar stories, are the result of a public that makes up its mind on the basis of mere slivers of information.

"Argument" from Outrage   This category makes the list because of the increase in polarizing political discourse. The mere mention of Bill Clinton (on one hand) or George Bush (on the other) has the power to raise the blood pressure of large numbers of citizens. Animosity and bile have taken over from the presentation and evaluation of evidence, with the result that national political discussion has been on a downward trend for over a generation. We attribute at least part of the decline to the electronic media. Radio and television, although they had the potential to raise the level of public discourse, have had the opposite effect. ("Television is a medium," said Ernie Kovacs, "because it is neither rare nor well done.") Many television shows where politics are "discussed" have turned into shouting matches in which several morons compete for the loudest volume and sharpest insults. They may look vaguely like debates, but they are nothing of the sort. "Talk radio" is even worse, with made-up statistics, false allegations, the spread of rumor, and general name-calling taking the place of argument and evidence. (We've given this fallacy the nickname "the Rush Limbaugh fallacy," based on the fact that he has mastered it so thoroughly.) Most of these shows are designed solely to raise the heart rate of people who already believe the nonsense the host is talking about (no matter which side of the political spectrum he or she claims to represent).

It's worth pointing out that this "argument" almost always involves straw men (see the following entry) and very often involves scapegoating as well. The easiest way to deal with an issue is to identify an enemy, distort his real position, and blame everything on him amidst a flurry of name-calling.

Straw Man   One principal weapon of the indignation-mongers just described is the distortion and exaggeration of the views of others. This tactic is widespread in the talk media and in mainstream politics as well. In fact, the latter seem to be taking their cue from the former in recent years. (We recall the advice given by Newt Gingrich, former Speaker of the House of Representatives, to Republican candidates in 1994. When discussing Democratic candidates and proposals, he said, you should use terms such as —, —, and "traitor" as frequently as possible.) Nobody, it seems, trusts the American people to make political choices based on straightforward information. And, judging from the number of Americans who make no effort to obtain such information, there may be some basis for such an attitude.

Post Hoc   We usually explain post hoc reasoning in terms of simple-minded examples of cause and effect. But the real bamboozling takes place at more complicated levels. For example, while Jimmy Carter was president, interest rates went through the roof; while George Bush (the first such) was president, the economy went into a recession. Both presidents were blamed for these occurrences, but were the policies of either the cause of these effects? Probably not. At least most people did not stop to ask whether it was anything the president did or refused to do that might have caused them. It is our tendency to seize on easy, simple answers that helps give this fallacy, and most of the others, such a high profile.

Red Herring/Smokescreen   One might think of this as the "attention span" fallacy, since it seems to work best on listeners and readers who are unable to stay focused on an issue when they are tempted by distractions. These days, when politicians and other public figures dodge hard questions, they are generally allowed to get away with it. How long has it been since you've heard a reporter say, "But sir, you didn't answer the question!" Probably even longer than it's been since you heard a politician give a direct answer to a hard question. We've all heard it said that the media and the people they interview are interdependent—that is, they each depend on the other and have fundamentally the same interests (at the highest government and corporate levels), and that accounts for the easy treatment officials often get from interviewers. But we think it's also true that reporters themselves are prone to fall for the fallacy in question here. They are too likely to move on to the next question on their list when the public would be better served by a pointed follow-up question on an issue that has just been ducked. We think this is a serious failure and damaging to the public good.

Group Think Fallacy   This fallacy has undergone a major revival since the terrorist attacks of September 11, 2001. An understandable emotional shock followed those attacks, but even before the shock began to wear off, the motto for the country became "United We Stand." There's nothing unreasonable about a united stand, of course, but this motto soon began to stand for an intolerance of any position that ran counter to the official view of the United States government (as voiced by the White House). The prevailing view in many quarters, including many places in the government itself, was that all "good" Americans agreed with this official position. Considerable force was brought to bear in enforcing this group thinking, including describing critics as "giving aid and comfort" and so forth.

Of course, it isn't just national policy that can cause this phenomenon to occur. At what may be the other end of the spectrum, notice how the referees in a basketball game call way too many fouls on your team and not nearly enough on the other team. Our loyalty to our group can affect our judgments in ways that range from the amusing to the dangerous.

Scare Tactics   Reasoning that uses scare tactics is like rain in Seattle: It happens all the time. The scare tactics fallacy ranks up there with the "argument" from outrage as a staple of talk radio hosts and other demagogues, which means it is easy to find examples. What's wrong with raising fuel-efficiency standards? Well, how would you like to drive around in an elfmobile?—Because that's what you're going to be driving if the car companies are forced to make cars more efficient. What's wrong with the proposed big tax cut? Nothing—if you'd just as soon eliminate Social Security and Medicare.

The fallacy of scare tactics is an appeal to fear, and fear can often provide a perfectly reasonable motivation: Evidence that some frightening thing will result from X is certainly a good reason for avoiding X. But mere *mention* of the frightening thing is not, nor is a vivid portrayal or depiction of the ominous event. You don't prove we should overthrow a dictator merely by describing what would happen if he set off a nuclear bomb in New York City—or by showing gruesome photos of what he did to someone else. When the danger depicted is not shown to be relevant, the thinking is fallacious.

The use of scare tactics is not uncommon in advertising, either. But getting consumers to worry about having dandruff or halitosis or odors in their carpets is no argument for purchasing some particular brand of shampoo or mouthwash or carpet cleaner. People are often motivated to buy a house or car because they fear someone else will make an offer ahead of them and they'll lose the opportunity. This is unreasonable when it results in a purchase they would not otherwise have made. We wonder how many people live in houses they decided they liked just because they were afraid they'd lose them if they didn't make their offer *right now*.

The Scrapbook of Unusual Issues

*T*his section includes a number of items from here and there, mainly from newspapers, that struck us as amusing, out-of-the-ordinary, or just plain weird. All of them involve issues that can make for interesting discussions or subjects for writing assignments. (There's no reason critical thinking subjects have to be dull.) Like the essays in Appendix 1, these present opportunities to exercise your critical thinking skills.

Strange Stuff: News Out of the Ordinary

Golf-ball retriever's jail term tees off public

London — For 10 years John Collinson made a modest living diving for lost golf balls and selling them for 20 cents each. But after police caught him on a midnight expedition, a court frowned on his unusual job and jailed him for six months for theft.

The sentence provoked a public outcry, made the 36-year-old father of two a cause célèbre and prompted lawmakers, celebrities and the British media to campaign for this release.

Police arrested Collinson in August at Whetstone Golf Course in Leicester, central England. He and a colleague had fished 1,158 balls from Lily Pond, the bane of hundreds of golfers playing the difficult par 3 fifth hole.

Collinson, who made roughly $21,500 a year collecting balls at courses throughout the country, claimed in his defense they didn't belong to anyone. He told a jury at Leicester Crown Court last week he even filled out tax returns on his earnings. But the judge jailed him, saying, "It is obvious you show no remorse and no intention of quitting."

Gavin Dunnett, managing director of UK Lakeballs—a firm that buys balls from hundreds of divers and sells more than a million worldwide every year via the Internet—said Thursday that recovered balls are not stolen, but are abandoned property. He said more than 2,000 people had signed his online petition to "Free the Whetstone One."

Doodle crosses the line

MOUNT LEBANON, Pa. — A school suspended an 11-year-old girl for drawing two teachers with arrows through their heads, saying the stick figures were more death threat than doodle.

Becca Johnson, an honor-roll sixth grader at Mellon Middle School, drew the picture on the back of a vocabulary test on which she had gotten a D.

"That's my way of saying I'm angry," Becca said, adding she meant no harm to the teachers.

The stick figures, on a crudely drawn gallows with arrows in their heads, had the names of Becca's teacher and a substitute teacher written underneath. Another teacher spotted the doodle in the girl's binder Tuesday and reported it, prompting the three-day suspension.

Becca's parents, Philip and Barbara Johnson, denied the school's contention the drawings were "terrorist threats."

"She had done poorly on a test that was handed back to her. We've always told her that you can't take your feelings out on your teacher, so write about it or draw it, as a catharsis," Barbara Johnson said.

Brief lapse in judgment?

Poway — A high school vice principal was placed on leave Wednesday after she allegedly lifted girls' skirts—in front of male students and adults—to make sure they weren't wearing thong underwear at a dance.

"Everyone saw everything," said Kim Teal, whose 15-year-old daughter attended the dance but was not checked. "It was a big peep show."

Rita Wilson was placed on administrative leave Wednesday while the school district investigates the allegations concerning the dance last Friday for students at Rancho Bernardo High School.

Girls said they were told to line up outside the gym before entering the dance so Wilson could check their underwear. Those wearing thongs were turned away.

Cities Give Psychics License to Entertain, Not Exorcise

Believers in the occult and thrill-seekers are the bread and butter of psychic parlors.

For a few dollars, psychics foretell a client's future by gazing into a crystal ball or reading a palm.

Such transactions are legal entertainment, said Detective Jan Cater of the Sacramento County Sheriff's Department.

What is illegal, she said, is when clients are led to believe that they are cursed and must pay large sums of money to purge the evil.

"So-called psychics terrify people into believing that awful things will happen to them . . . . unless they pay to have the curses removed, Cater said.

Sacramento and other cities require that psychics and others in the "seer" trades undergo background checks before receiving a license issued by the police department.

But other communities, including unincorporated sections of Sacramento County, have no requirements—a situation that, according to Cater, can lead to big-time swindles.

"Psychic fraud isn't a major problem in the unincorporated areas (of the county), but it has the potential to be," Cater said.

In the city of Sacramento, the safeguards in place for several years are working, officials said. . . .

In San Francisco and other cities without license laws, psychic swindles are on the rise, authorities said.

One of Cater's biggest cases involved Ingrid, a Sonoma County woman who said she paid more than $265,000 over five years to "psychics" Louie Stevens and Nancy Uwnawich. The couple convinced Ingrid, 66, that she had a curse that had to be removed by them—or she would suffer, Ingrid said.

After wiping out her savings, Ingrid told the pair she had no more cash—except for $2,000 in her son's college trust fund.

"What's more important—your health or your son?" Uwnawich shouted, according to Ingrid.

Ingrid said she gave them the $2,000. . . .

An arrest warrant has been issued for Uwnawich, who remains at large.

By Edgar Sanchez
BEE STAFF WRITER

Monogamy? Don't Cheat Yourself

DEAR ANN: Why do you come down so hard on married men who cheat? I don't understand why you believe monogamy is the only "right" way when it comes to marriage. If monogamy works so well, how come so many men cheat and so many women want divorces?

Many other cultures openly accept polygamy. Even the Bible says a man can have more than one wife. And in many countries, adultery is accepted and shrugged off as nothing to get excited about. The wife gets the financial support and social position, and the mistress gets the sex. This sounds like a fair exchange to me.

The truth is, there is not a woman on the planet who can fully satisfy the needs of a powerful male. When women quit asking men to make them the center of their universe and learn to accept a strong man's desire for more than one female companion, families will live together longer, more happily and with less rancor.

–Truth Hurts In Texas

DEAR AUSTIN: Men in biblical times had multiple wives in order to ensure an abundance of offspring because so many women died in childbirth. It was not a license to cheat.

I am sure your approach appeals to a lot of men, but I doubt that the women in your life would go for it. And without their cooperation, you're not going to get very far. Of course, if you would like to live in one of those foreign countries that condones adultery, be my guest. Most women would be delighted to see you go. A few might even be willing to buy you a ticket.

Ann Landers (By permission of Esther P. Lederer and Creators Syndicate, Inc.)

Writer to 'Dear Abby' Faces Child-Porn Counts

MILWAUKEE — A man who wrote to "Dear Abby" for advice on how to handle his fantasies about having sex with young girls was charged Wednesday with possessing child pornography after the columnist turned him in, authorities say.

Paul Weiser, 28, was charged with three counts of possession of child pornography. He was ordered released on $10,000 bail on condition he avoid computers and contact with anyone under 18.

Police said 40 pornographic photographs of children were found in his computer after his arrest Monday.

"He acknowledges he needs help and denies ever acting upon any desires," said prosecutor Paul Tiffin.

"Dear Abby" columnist Jeanne Phillips, daughter of column founder Pauline Phillips, called police after receiving the letter.

Jeanne Phillips, who shares her mother's pen name, Abigail Van Buren, told the Milwaukee Journal Sentinel she agonized over whether to report Weiser, since the column's credibility is based on anonymity given those seeking advice.

"I lost sleep. Didn't sleep for days, because I really believe this man wrote to me genuinely seeking help," she said. "I was torn, because my readers do turn to me for help, yet there was the priority of the safety of those young girls."

She was said to be traveling on Wednesday and did not return a call seeking comment.

Weiser could get up to five years in prison and a $10,000 fine on each count, if convicted.

He did not speak during Wednesday's court hearing except to answer "yes" when asked if he understood the proceedings.

A preliminary hearing was set for March 25.

Associated Press

Strange Stuff: News Out of the Ordinary

'Whites' Send a Message

GREELEY, Colo — Unable to persuade a local school to change a mascot name that offends them, some students at the University of Northern Colorado have named their intramural basketball team "The Fighting Whites."

The students are upset with Eaton High School for using an American Indian caricature on the team logo. The team is called the Reds.

The university team said it chose a white man as its mascot to try to raise awareness of stereotypes that some cultures endure.

"The message is 'Let's do something that will let people see the other side of what it's like to be a mascot,'" said Solomon Little Owl, a team member and director of Native American Student Services at the university.

The team, made up of American Indians, Latinos and Anglos, wears jerseys that say, "Every thang's going to be all white."

"It's not meant to be vicious. It is meant to be humorous," said Ray White, a Mohawk American Indian on the team.

No 'ATHEISTS' Allowed

ST. PETERSBURG, Fla — Steven Miles has tooled around Gainesville, Fla., for 16 years with a license plate that says "ATHEIST."

To Miles, it is a form of self-expression, one he is happy to spend a few extra dollars every year to keep.

But to the state of Florida now, the tag is "obscene or objectionable," according to a letter Miles received last month from the Department of Highway Safety and Motor Vehicles.

"The plate must be canceled," the department ordered.

Miles said he is incensed and that giving up his tag is out of the question.

The vice president of Atheists of Florida said he intends to fight and is in touch with the American Civil Liberties Union.

Dummy is no woman

RENTON, Wash. — A woman with a full-size mannequin buckled into the front seat of her minivan tried to sneak into the car pool lane on Interstate 405 and became involved in a chain of crashes that backed up traffic for hours Tuesday morning, according to the Washington State Patrol.

Susan Aeschliman-Hill, 59, of Kent, Wash., acknowledged using the dummy—complete with a woman's sweater, makeup and wig—to drive illegally in the car pool lane. "When traffic gets real bad, I sometimes use (the mannequin)," she said.

State Patrol Trooper Monica Hunter said the woman likely will be cited for driving in the high-occupancy vehicle lane and making an unsafe lane change.

"Her passenger wasn't breathing, and that's one of our requirements," Hunter said.

By Rick Nelson, compiled from the Associated Press

Christian School Expels Nude Dancer's Daughter

A Sacramento private school has expelled a kindergarten student three weeks before the end of the school year, saying her mother's work as a nude dancer clashes with its Christian philosophy.

Christina Silvas, 24, of Rancho Cordova said one reason she took the job as a dancer at Gold Club Centerfolds off Highway 50 was so she could afford the $400 monthly tuition at Capital Christian School.

"If you choose to do the wrong thing willfully, then God's word instructs me as to what my responsibility is," said Rick Cole, head pastor of Capital Christian Center. "I need to be faithful to my calling."

Silvas, who previously worked at the church as a Sunday school teacher, has a different view.

"I thought the church was supposed to accept everybody," she said. "That's my understanding of what God is about. My daughter is the one who goes to school there, not me, and they're turning her away."

Silvas, who is single, said school administrators telephoned her last week and told her that in response to persistent rumors, a parent went to the strip club's Web site, downloaded pictures of Silvas and brought them to school staff. Silvas told them she does dance at the all-nude bar, which doesn't serve alcohol, and subsequently met with Cole.

As head pastor of Capital Christian, which at 4,000 members is one of the largest Assemblies of God churches in the country, Cole had final say on the matter.

Silvas said Cole told her not only will her 5-year-old daughter be expelled, but as long as she keeps her job at the club, neither she nor her daughter can attend church there.

Silvas said her daughter enjoys the school's daily worship sessions and was looking forward to the end-of-the-year pool party and graduation ceremony. She hasn't told the girl why she won't be able to go to school next week.

"I'm still trying to figure out what I'm going to tell her," she said.

Silvas said that when she enrolled her daughter, she signed a statement that stipulated she would abide by the school's Christian commitment/philosophy and that "weekly church attendance is necessary for continued enrollment at CCS."

The statement also reads, in part, "CCS is interested in maintaining a partnership with our parents regarding the standards and criteria of a Christian learning structure that involves the entire family."

By working as an exotic dancer and refusing to change her vocation, Silvas violated the school's philosophy, Cole said.

"She was not in compliance with what she agreed to," he said. "There's a commitment form that parents sign. They agree to our Christian commitment and standard of living."

Seeking a resolution, Silvas and Cole met privately Tuesday. "I talked to her for over an hour, letting her know how much God cares about her and loves her and that he has a much better plan for her than what she is doing with her life right now," Cole said.

He promised that if Silvas quit her job, he and the church would forgive her, help her and even waive tuition for the month of June.

"My understanding was she would come back to me after thinking about it and let me know her decision," Cole said. "It's very stunning to me to see it on the news. We've been thrust into this media craze, but not by our desire."

Silvas said the decision is not simple and couldn't be made that quickly.

"I have other expenses besides my daughter's tuition," she said. "I am also concerned with providing a comfortable life for my daughter."

Silvas said she feels she is being demonized.

"I'm trying to let them know we're not evil because of our work, and we do have an awareness of God. My job is legal, it's not illegal."

Though Silvas is considering legal action, experts in education law say that may not be a good option.

"In a private school, it's hard to make a breach of contract action because they have the right to kick a kid out at any time for any reason," said Michael Sorgen, an education lawyer in San Francisco.

Even though the school has the legal right to expel the child, that doesn't make it right, Sorgen said.

By Erika Chavez
BEE STAFF WRITER

Ill Boy's New Hurdle: School Drug Rules

Life just got more complicated for an 8-year-old boy and his mother who's had great success battling his mental disorders with a doctor-approved marijuana therapy.

The youngster's medical condition has improved so dramatically that he can now attend public school, but school officials won't permit a school nurse to administer his cannabis capsules and won't let him take the pills himself on campus, the child's mother said Monday.

"Other kids get their medication," she complained. But the drug her son needs daily at 1 p.m. must be delivered by her personally, off the school grounds, she said.

"It makes him feel he's not normal, that he's being treated differently. He wonders why he's being targeted. He just wants to be normal," she said.

She hopes to persuade school officials to change their minds and allow the capsules to be given on campus.

The woman, whose name is being withheld by The Bee to protect the boy's identity, has been treating her son with medical cannabis for the past year, at home and at the private school he had been attending.

But in April, they moved from Rocklin to El Dorado County. She presented her son's new school with the required permission slip for students who need medication at school, a form she and the boy's doctor signed.

The day before the boy was to report to his new school, however, a message left on the family answering machine informed the mother that her son's recommended medication could not be administered on campus.

So, she says, she's been forced to drive a round trip of 26 miles each noontime to remove him from the school grounds, give him his capsules, and return him to class.

Vicki L. Barber, superintendent of the El Dorado County Office of Education, said she could not comment on any specific case. But she said state law permits schools to dispense drugs only when they are formally "prescribed" by a physician. The boy's doctor made a "recommendation," and there is a difference, Barber added, between a "prescription" and a doctor's "recommendation."

Proposition 215, the Compassionate Use Act of 1996, permits the use of medical marijuana and the possession of the drug by the patient or a primary caregiver.

A school district cannot be legally defined as a "primary caregiver," Barber said.

Because the district has a zero-tolerance policy, students are not permitted to have in their possession or to self-administer drugs of any kind, she said.

The boy's mother said the school district is relying on a technicality to dodge responsibility.

"There is no 'prescription' for 'oral cannabis,'" the capsule she prepares in a five-hour cooking and drying process at home, she said.

But it is a part of her son's "individual education plan" or IEP, the state's "prescription" for helping children with special needs, the mother said.

School district officials would not comment on the boy's specific IEP.

By Wayne Wilson
BEE STAFF WRITER

A letter to the editor of the local paper by one of your authors.

5 Million Teen-agers Perfect

You recently ran an Associated Press story that first appeared in the New York Times about a Port Orchard, Wash., Eagle Scout who was given a week to abandon his atheism or be kicked out of Boy Scouts of America. Boy Scout law requires scouts to be trustworthy, loyal, helpful, friendly, courteous, kind, obedient, cheerful, thrifty, brave, clean and reverent. An atheist, of course, finds it difficult to be reverent in the intended sense of the term.

Unfortunately, you ended the article with a paraphrase from Gregg Shields, a Boy Scout spokesman: "Mr. Shields said for the Boy Scouts to insist on anything less would be unfair to the 5 million members."

But there were two more sentences in the New York Times article and you missed the best point of the piece by leaving them out. Shields said: "It would be a disservice to all the other members to allow someone to selectively obey or ignore our rules."

The Times piece then added, "As for the other 11 points of the Scout Law, Mr. Shields could not say whether anyone had been ejected for being untrustworthy, disloyal, unhelpful, unfriendly, discourteous, unkind, disobedient, cheerless, unthrifty, cowardly or sloppy."

I for one believe it is a credit to scouting that none of those 5 million teen-age boys has ever been found guilty of any of those other 11 defects.

—Richard Parker, Chico

Strange Stuff: News Out of the Ordinary

As if the goings-on of government weren't scary enough, the state of North Carolina is calling in a Ghostbusters-like group to do a paranormal screening of the old state Capitol.

Although there have been no sightings of a huge marshmallow man, staffers at the Capitol say they have heard floorboards creak with invisible footsteps, keys jangle, and doors squeak open and shut.

"To be honest with you, I've always made it a rule to be out of the building at quitting time," said Raymond Beck, the Capitol historian. "I've had enough of those strange vibes here that I don't like sticking around after it gets dark."

Researchers from the Ghost Research Foundation will give the 162-year-old Raleigh landmark where the governor has his offices a "spectral inspection," using infrared cameras, electromagnetic field detectors and audio recorders.

The Capitol has gone through a number of night watchmen who begged off the duty after one night of creepy noises, Beck said.

But Owen J. Jackson, 84, managed the job for 12 years before retiring in 1990. He apparently never was slimed, and strains of gospel hymns, the thump-thump-thumps that followed him down stairs and the angry slams of doors never bothered him, Jackson said.

"You get used to something like that," Jackson said. "I think there's a couple (of) million dollars buried somewhere there, and they're just trying to tell us where it was."

It's hormonal

A Harvard research team has found that married men, whether fathers or not, have significantly less testosterone than single men, regardless of age or overall health status.

Anthropology graduate student Peter Gray and his colleagues also found that men who spent lots of time with their wives had even lower testosterone levels than other married men. The team tested 58 men in the Boston area, according to an article in a recent Harvard University Gazette.

Gray suggests that married men don't produce as much testosterone after they shift from dating to parenting roles.

"Married men, particularly fathers, are less likely to engage in dominance interaction and aggression," he said. "That means less risk taking, less conflicts or other behaviors that could lead, for example, to death by homicide.

And couldn't it be Mother Nature's way of keeping men from straying? After all, other studies have found that men with high levels of testosterone were more likely to commit adultery than those with less of the stuff.

"That's an interesting question," Gray said. "That's a topic that will be harder to get at."

Is skydiving dog cruelty?

Animal lovers are howling over a skydiving dog.

The parachuting dachshund, known as Brutus the Skydiving Dog, is set to perform at this weekend's Air and Space show at Vandenberg Air Force Base.

"What we feel is this is is cruelty to animals," said Shirley Cram, shelter director and treasurer for the Volunteers for Inter-valley Animals. "It's exploiting the dog. It certainly isn't fun for that dog to jump out of the plane."

Brutus' skydiving partner disagrees.

"He gets all excited when I'm getting my gear ready," said Ron Sirull of Delray Beach, Fla., contending Brutus enjoys his aerial activities. He added, "He's totally up for it."

Sirull said his dog's veterinarian and the Arizona Humane Society have signed off the activity being safe for man's best friend.

Brutus is tucked into a special pouch affixed to his owner's chest for the jump and dons custom-made goggles for what Sirull calls his "fleafall." While Sirull has 1,000 jumps, Brutus has logged 100.

"That's equal to 700 jumps in dog years," his owner joked.

—Compiled by Caroline Liss from the Associated Press and Washington Post. In SAC BEE, 1 Nov. 2002.

10 Commandments Monument Must Go

MONTGOMERY, Ala. — A Ten Commandments monument in the rotunda of Alabama's judicial building violates the Constitution's ban on government promotion of religion, a federal judge ruled Monday.

U.S. District Judge Myron Thompson gave Alabama Chief Justice Roy Moore, who had the 5,300-pound granite monument installed in the state building, 30 days to remove it.

Thompson said previous court rulings have allowed displays on government property if they have a secular purpose and do not foster "excessive government entanglement with religion." He said the Ten Commandments monument fails this test.

"His fundamental, if not sole, purpose in displaying the monument was non-secular; and the monument's primary effect advances religion," Thompson said.

Moore testified during the trial that the commandments are the moral foundation of American law. He said the monument acknowledges God, but does not force anyone to follow the justice's conservative Christian religious beliefs.

A lawsuit seeking removal of the monument argued that it promoted the judge's faith in violation of the Constitution.

Moore installed the monument after the building closed on the night of July 31, 2001, without telling any other justices.

"This monument was snuck in during the middle of the night, and they can sneak it out just as easily. It's a gross violation of the rights of the citizens of Alabama," said Morris Dees, lead counsel and co-founder of the Southern Poverty Law Center. He urged Moore to remove the monument immediately.

The monument, which features the King James Bible version of the Ten Commandments sitting on top of a granite block, is one of the first things visitors see upon entering the building.

"Its sloping top and the religious air of the tablets unequivocally call to mind an open Bible resting on a podium," Thompson said.

In his ruling, Thompson said he found the monument to be more than just a display of the Ten Commandments.

"The court is impressed that the monument and its immediate surroundings are, in essence, a consecrated place, a religious sanctuary, within the walls of the courthouse," Thompson wrote.

An appeal is expected. Neither Moore nor his lead attorney, Stephen Melchior, had any immediate comment. An assistant to Melchior said they were reserving comment until they had read the opinion.

Moore fought to display a wooden plaque of the commandments on his courtroom wall in Etowah County before he won election as chief justice in November 2000.

"The basic issue is whether we will still be able to acknowledge God under the First Amendment, or whether we will not be able to acknowledge God," Moore testified.

The Rev. Barry Lynn, executive director of Americans United for Separation of Church and State, call the ruling a setback for "Moore's religious crusade."

"It's high time Moore learned that the source of U.S. law is the Constitution and not the Bible," Lynn said.

One of Moore's supporters, Alabama Christian Coalition President John Giles, said he believes there may be a backlash against the ruling in Alabama.

"I am afraid the judge's order putting a 30-day limit on removal of the monument will lead to an uprising of citizens protesting removal of that monument," Giles said.

Associated Press

Teen Claims Bias after School Bars Her Bid to Be Prom King

SACRAMENTO — An 18-year-old Encina High School senior is not being allowed to run for prom king because . . . she is a female.

Kristine Lester said Thursday that school officials told her that her name would not be on the ballot because only males can run for king and only females can run for queen.

"I just wanted to do something different," said Lester, who is openly gay and plans to wear a tuxedo at Saturday's prom. "I'm not the feminine type to run for queen."

Lester said she believes the school is discriminating against her because of her sexual orientation.

Officials with the San Juan Unified School District back Encina Principal Myrtle Berry's decision and denied that there was any discrimination involved.

"We don't discriminate," said district spokeswoman Deidra Powell.

—**Sandy Louey**

Stop Barring Men, Feminist Professor Told

Boston College gives her an ultimatum to allow males in class—or quit teaching

BOSTON — A radical feminist professor at Boston College has been given an ultimatum from the school: admit men to her classes or stop teaching.

Theologian Mary Daly lets only women take her courses.

Daly, whose seven major books, including "Outercourse," have made her a pioneer in feminist circles, has said she won't back down. Opening her classes to men would compromise her belief that women tend to defer to a man whenever one is in the room, she said.

Daly took a leave of absence from the Jesuit college this semester rather than bow to demands that she admit senior Duane Naquin into her class in feminist ethics.

Naquin, who claimed discrimination, has the backing of the Center for Individual Rights, a conservative law firm in Washington whose lawsuit ended affirmative action at the University of Texas. The firm sent a letter to BC in the fall threatening legal action if Daly did not relent.

Daly has argued Naquin did not have the prerequisite of another feminist studies course.

College officials said a second male student also complained of discrimination.

Daly has rejected a retirement offer from the college, the Boston Globe reported.

Daly, who is 70, taught only men when she first arrived at the Newton campus in 1966. The college of arts and science did not begin admitting women until 1970. In the early '70s, she said, she observed problems in her coed classes.

"Even if there were only one or two men with 20 women, the young women would be constantly on an overt or a subliminal level giving their attention to the men because they've been socialized to nurse men," she said.

Boston College officials said Daly's ground rules violate federal civil rights laws and school policy.

If a male professor tried to bar women from his classes, college spokesman Jack Dunn said, "we'd be run out of town."

Daly, who abandoned her Roman Catholic faith in the early 1970s, describes herself as a radical feminist, which she interprets as "going to the roots" of societal problems.

Her books include "The Church and the Second Sex," "Gyn/Ecology: The Metaethics of Radical Feminism" and "Outercourse," a theological autobiography.

She said she views the controversy as an attack on academic freedom and an assault on feminism by "an extreme right-wing organization" trying to "assert white male supremacy."

Daly's students are rallying around her, and 14 of them wrote a letter to college administrators.

"I think there comes a point where women need to claim their own space," said Kate Heekin, a Greenwich, Conn., senior. "If that needs to be a classroom, so be it."

By Robin Estrin
Associated Press

In Short . . .

She is Barbie's oldest friend, happily married and visibly pregnant with her second child—and some parents think she's a little too real for children. The pregnant version of Midge, which pops out a curled-up baby when her belly is opened, has been pulled from Wal-Mart shelves across the nation following complaints from customers, a company spokeswoman said.

Associated Press; from *The Sacramento Bee*, December 26, 2002.

Glossary

ad hominem Attempting to rebut a source's argument or claim or position, etc., on the basis of considerations that logically apply to the source rather than to the argument or claim or position.

affirmative claim A claim that includes one class or part of one class within another: A- and I-claims.

affirming the antecedent *See* modus ponens.

affirming the consequent An argument consisting of a conditional claim as one premise, a claim that affirms the consequent of the conditional as a second premise, and a claim that affirms the antecedent of the conditional as the conclusion.

ambiguous claim A claim that could be interpreted in more than one way and whose meaning is not made clear by the context.

analogical argument An argument in which something that is said to hold true of a sample of a certain class is claimed also to hold true of another member of the class.

analogy A comparison of two or more objects, events, or other phenomena.

analytic claim A claim that is true or false by virtue of the meanings of the words that compose it. *Contrast with* synthetic claim.

analytical definition A definition that specifies (1) the type of thing the defined term applies to and (2) the difference between that thing and other things of the same type.

antecedent *See* conditional claim.

appeal to anecdotal evidence, fallacy of A form of hasty generalization presented in the form of an anecdote or story. Also the fallacy of trying to prove (or disprove) that x is a causal factor for y by citing an example or two of an x that did (or didn't) cause y.

appeal to ignorance The view that an absence of evidence *against* a claim counts as evidence *for* that claim.

appeal to indignation *See* outrage, "argument" from.

appeal to pity *See* pity, "argument" from.

appeal to precedent The claim (in law) that a current case is sufficiently similar to a previous case that it should be settled in the same way.

apple polishing A pattern of fallacious reasoning in which flattery is disguised as a reason for accepting a claim.

argument An attempt to support a claim or an assertion by providing a reason or reasons for accepting it. The claim that is supported is called the conclusion of the argument, and the claim or claims that provide the support are called the premises.

argument from analogy *See* analogical argument.

argument pattern The structure of an argument. This structure is independent of the argument's content. Several arguments can have the same pattern (e.g., modus ponens) yet be about quite different subjects. Variables are used to stand for classes or claims in the display of an argument's pattern.

background information The body of justified beliefs that consists of facts we learn from our own direct observations and facts we learn from others.

bandwagon *See* peer pressure.

begging the question *See* question-begging argument.

biased generalization, fallacy of A generalization about an entire class based on a biased sample.

biased sample A sample that is not representative.

burden of proof, misplacing A form of fallacious reasoning in which the burden of proving a point is placed on the wrong side. One version occurs when a lack of evidence on one side is taken as evidence for the other side, in cases where the burden of proving the point rests on the latter side.

categorical claim   Any standard-form categorical claim or any claim that means the same as some standard-form categorical claim. *See* standard-form categorical claim.

categorical imperative   Kant's term for an absolute moral rule that is justified because of its logic: If you can wish that your maxim were a universal law, your maxim qualifies as a categorical imperative.

categorical logic   A system of logic based on the relations of inclusion and exclusion among classes ("categories"). This branch of logic specifies the logical relationships among claims that can be expressed in the forms "All Xs are Ys," "No Xs are Ys," "Some Xs are Ys," and "Some Xs are not Ys." Developed by Aristotle in the fourth century B.C.E., categorical logic is also known as Aristotelian or traditional logic.

categorical syllogism   A two-premise deductive argument in which every claim is categorical and each of three terms appears in two of the claims—for example, all soldiers are martinets and no martinets are diplomats, so no soldiers are diplomats.

causal claim   A statement that says or implies that one thing caused or causes another.

causal factor   A causal factor for some specific effect is something that contributes to the effect. More precisely, in a given population, a thing is a causal factor for some specified effect if there would be more occurrences of the effect if every member of the population were exposed to the thing than if none were exposed to the thing. To say that C is a causal factor for E in population P, then, is to say that there would be more cases of E in population P if every member of P were exposed to C than if no member of P were exposed to C.

causal hypothesis   A statement put forth to explain the cause or effect of something, when the cause or effect has not been conclusively established.

cause-and-effect claim   *See* causal claim.

chain argument   An argument consisting of three conditional claims, in which the antecedents of one premise and the conclusion are the same, the consequents of the other premise and the conclusion are the same, and the consequent of the first premise and the antecedent of the second premise are the same.

circularity   The property of a "causal" claim where the "cause" merely restates the effect.

circumstantial ad hominem   Attempting to discredit a person's claim by referring to the person's circumstances.

claim variable   A letter that stands for a claim.

common practice, "argument" from   Attempts to justify or defend an action or a practice on the grounds that it is common—that "everybody," or at least lots of people, do the same thing.

common thread   In common-thread reasoning, multiple occurrences of a feature are said to be united by a single relevant common thread.

complementary term   A term is complementary to another term if and only if it refers to everything that the first term does not refer to.

composition, fallacy of   To think that what holds true of a group of things taken individually necessarily holds true of the same things taken collectively.

conclusion   In an argument, the claim that is argued for.

conclusion indicator   A word or phrase (e.g., "therefore") that ordinarily indicates the presence of the conclusion of an argument.

conditional claim   A claim that state-of-affairs A cannot hold without state-of-affairs B holding as well—e.g., "If A then B." The A-part of the claim is called the *antecedent*; the B-part is called the *consequent*.

conditional proof   A deduction for a conditional claim "If P then Q" that proceeds by assuming that P is true and then proving that, on that assumption, Q must also be true.

confidence level   *See* statistical significance.

conflicting claims   Two claims that cannot both be correct.

confusing explanations and excuses, fallacy of   Mistaking an explanation of something for an attempt to excuse it.

conjunction   A compound claim made from two simpler claims. A conjunction is true if and only if both of the simpler claims that compose it are true.

consequent   *See* conditional claim.

contradictory claims   Two claims that are exact opposites—that is, they could not both be true at the same time and could not both be false at the same time.

contrapositive   The claim that results from switching the places of the subject and predicate terms in a claim and replacing both terms with complementary terms.

contrary claims   Two claims that could not both be true at the same time but could both be false at the same time.

control group   *See* controlled cause-to-effect experiment.

controlled cause-to-effect experiment   An experiment designed to test whether something is a causal factor for a given effect. Basically, in such an experiment two groups are essentially alike, except that the members of one group, the *experimental group*, are exposed to the suspected causal factor, and the members of the other group, the *control group*, are not. If the effect is then found to occur with significantly more frequency in the experimental group, the suspected causal agent is considered a causal factor for the effect.

converse   The converse of a categorical claim is the claim that results from switching the places of the subject and predicate terms.

deduction (proof)   A numbered sequence of truth-functional symbolizations, each member of which validly follows from earlier members by one of the truth-functional rules.

deductive argument   An argument that is either valid or intended by its author to be so.

definition by example   Defining a term by pointing to, naming, or describing one or more examples of something to which the term applies.

definition by synonym   Defining a term by giving a word or phrase that means the same thing.

denying the antecedent   An argument consisting of a conditional claim as one premise, a claim that denies the antecedent of the conditional as a second premise, and a claim that denies the consequent of the conditional as the conclusion.

denying the consequent   *See* modus tollens.

deontologism   *See* duty theory.

dependent premises   Premises that depend on one another as support for their conclusion. If the assumption that a premise is false cancels the support another provides for a conclusion, the premises are dependent.

descriptive claim   A claim that states facts or alleged facts. Descriptive claims tell how things are, or how they were, or how they might be. *Contrast with* prescriptive claims.

difference in question   In relevant-difference reasoning, one item is said to have a feature ("the feature in question") that other similar items lack, and there is said to be only one other relevant difference ("the difference in question") between the item that has the feature in question and the other items that don't have the feature in question.

disjunction   A compound claim made up of two simpler claims. A disjunction is false only if both of the simpler claims that make it up are false.

divine command theory   The view that our moral duty (what's right and wrong) is dictated by God.

division, fallacy of   To think that what holds true of a group of things taken collectively necessarily holds true of the same things taken individually.

downplayer   An expression used to play down or diminish the importance of a claim.

duty theory   The view that a person should perform an action because it is his or her moral duty to perform it, not because of any consequences that might follow from it.

dysphemism   A word or phrase used to produce a negative effect on a reader's or listener's attitude about something or to tone down the positive associations the thing may have.

emotive force   The feelings, attitudes, or emotions a word or an expression expresses or elicits.

envy, "argument" from   Trying to induce acceptance of a claim by arousing feelings of envy.

equivalent claims   Two claims are equivalent if and only if they would be true in all and exactly the same circumstances.

error margin   A range of possibilities; specifically, a range of percentage points within which the conclusion of a statistical inductive generalization falls, usually given as "plus or minus" a certain number of points.

euphemism   An agreeable or inoffensive expression that is substituted for an expression that may offend the hearer or suggest something unpleasant.

experimental group   *See* controlled cause-to-effect experiment.

expert   A person who, through training, education, or experience, has special knowledge or ability in a subject.

explanation   A claim or set of claims intended to make another claim, object, event, or state of affairs intelligible.

explanatory comparison   A comparison that is used to explain.

explanatory definition   A definition used to explain, illustrate, or disclose important aspects of a difficult concept.

extension   The set of things to which a term applies.

factual claim   A claim about a factual issue.

factual issue/question   An issue that there are generally accepted methods for settling.

fallacy   An argument in which the reasons advanced for a claim fail to warrant acceptance of that claim.

false dilemma   This pattern of fallacious reasoning: "X is true because either X is true or Y is true, and Y isn't," said when X and Y could both be false.

feature in question   *See* difference in question.

force, "argument" by   Using a threat rather than legitimate argument to "support" some "conclusion."

functional explanation   An explanation of an object or occurrence in terms of its function or purpose.

gambler's fallacy   The belief that recent past events in a series can influence the outcome of the next event in the series. This reasoning is fallacious when the events have a predictable ratio of results, as is the case in flipping a coin, where the predictable ratio of heads to tails is 50–50.

generalization   An argument offered in support of a general claim.

genetic fallacy   Rejecting a claim on the basis of its origin or history.

good argument   An argument that provides grounds for accepting its conclusion.

grouping ambiguity   A kind of semantic ambiguity in which it is unclear whether a claim refers to a group of things taken individually or collectively.

group think fallacy   Fallacy that occurs when someone lets identification with a group take the place of reason and deliberation when arriving at a position on an issue.

guilt trip   Trying to get someone to accept a claim by making him or her feel guilty for not accepting it.

harm principle   The claim that the only way to justify a restriction on a person's freedom is to show that the restriction prevents harm to other people.

hasty conclusion, fallacy of   A fallacy of inductive arguments that occurs when conclusions are drawn from a sample that is too small.

hasty generalization, fallacy of   A generalization based on a sample too small to be representative.

horse laugh   A pattern of fallacious reasoning in which ridicule is disguised as a reason for rejecting a claim.

hyperbole   Extravagant overstatement.

hypothesis   A supposition offered as a starting point for further investigation.

hypothetical imperative   Kant's term for a command that is binding only if one is interested in a certain result; an "if-then" situation.

implied target class   In an analogical argument, the class to which the sample items and the target item belong.

inconsistency (form of ad hominem)   A pattern of fallacious reasoning of the sort, "I reject your claim because you act inconsistently with it yourself," or "You can't make that claim now because you have in the past rejected it."

independent premises   Premises that do not depend on one another as support for the conclusion. If the assumption that a premise is false does not cancel the support another premise provides for a conclusion, the premises are independent.

indirect proof   Proof of a claim by demonstrating that its negation is false, absurd, or self-contradictory.

inductive analogical argument   *See* analogical argument.

inductive argument   An invalid argument whose premises are intended to provide some support, but less than conclusive support, for the conclusion.

inductive generalization   *See* generalization.

innuendo   An insinuation of something deprecatory.

intension   The set of characteristics a thing must have for a term correctly to apply to it.

invalid argument   An argument whose conclusion does not necessarily follow from the premises.

issue   A point that is or might be disputed, debated, or wondered about. Issues are often introduced by the word "whether," as in the example "whether this train goes to Chattanooga."

law of large numbers   A rule stating that the larger the number of chance-determined, repetitious events considered, the closer the alternatives will approach predictable ratios. Example: The more times you flip a coin, the closer the results will approach 50 percent heads and 50 percent tails.

legal moralism   The theory that, if an activity is immoral, it should also be illegal.

legal paternalism   The theory that a restriction on a person's freedom can sometimes be justified by showing that it is for that person's own benefit.

leveling   The process of omitting or de-emphasizing features in a description of a person, thing, or event, when the person doing the describing does not think those features are important. The result can be a distorted description.

line-drawing fallacy   The fallacy of insisting that a line must be drawn at some precise point when in fact it is not necessary that such a line be drawn.

loaded question   A question that rests on one or more unwarranted or unjustified assumptions.

logic   The branch of philosophy concerned with whether the reasons presented for a claim, if those reasons were true, would justify accepting the claim.

matter of fact   *See* factual matter.

matter of pure opinion   *See* nonfactual issue.

mean   A type of average. The arithmetic mean of a group of numbers is the number that results when their sum is divided by the number of members in the group.

median   A type of average. In a group of numbers, as many numbers of the group are larger than the median as are smaller.

mode   A type of average. In a group of numbers, the mode is the number occurring most frequently.

modus ponens   An argument consisting of a conditional claim as one premise, a claim that affirms the antecedent of the conditional as a second premise, and a claim that affirms the consequent of the conditional as the conclusion.

modus tollens   An argument consisting of a conditional claim as one premise, a claim that denies the consequent of the conditional as a second premise, and a claim that denies the antecedent of the conditional as the conclusion.

moral relativism   The view that what is morally right and wrong depends on and is determined by one's group or culture.

nationalism   A powerful and often fierce emotional attachment to one's country that can lead a person to blind endorsement of any policy or practice of that country. ("My country, right or wrong!") It is a subdivision of the group think fallacy.

naturalistic fallacy   The assumption that one can conclude directly from a fact (what "is") what a rule or a policy should be (an "ought") without a value-premise.

negation   The contradictory of a particular claim.

negative claim   A claim that excludes one class or part of one class from another: E- and O-claims.

nonexperimental cause-to-effect study   A study designed to test whether something is a causal factor for a given effect. Such studies are similar to controlled cause-to-effect experiments, except that the members of the experimental group are not exposed to the suspected causal agent by the investigators; instead, exposure has resulted from the actions or circumstances of the individuals themselves.

nonexperimental effect-to-cause study   A study designed to test whether something is a causal factor for a given effect. Such studies are similar to non-experimental cause-to-effect studies, except that the members of the experimental group display *the effect*, as compared with a control group whose members do not display the effect. Finding that the suspected cause is significantly more frequent in the experimental group is reason for saying that the suspected causal agent is a causal factor in the population involved.

nonfactual claim   A claim about a nonfactual issue.

nonfactual issue/question   An issue concerning which neither of two disagreeing parties is required to be mistaken or concerning which there is no generally accepted method of resolving.

non sequitur   The fallacy of irrelevant conclusion; an inference that does not follow from the premises.

objective claim   A claim about a factual matter. Objective claims are true or false regardless of our personal experiences, tastes, biases, and so on.

obverse   The obverse of a categorical claim is that claim that is directly across from it in the square of opposition, with the predicate term changed to its complementary term.

offense principle   The claim that an action or activity can justifiably be made illegal if it is sufficiently offensive.

opinion   A claim that somebody believes to be true.

outrage, "argument" from   An attempt to persuade others by provoking anger in them, usually by inflammatory words, followed by a "conclusion" of some sort.

peer pressure "argument"   A fallacious pattern of reasoning in which you are in effect threatened with rejection by your friends, relatives, etc., if you don't accept a certain claim.

perfectionist fallacy   Concluding that a policy or proposal is bad simply because it does not accomplish its goal to perfection.

personal attack ad hominem   A pattern of fallacious reasoning in which we refuse to accept another's argument because there is something about the person we don't like or of which we disapprove. A form of ad hominem.

pity, "argument" from   Supporting a claim by arousing pity rather than offering legitimate argument.

poisoning the well   Attempting to discredit in advance what a person might claim by relating unfavorable information about the person.

popularity, "argument" from   Accepting or urging others to accept a claim simply because all or most or some substantial number of people believe it—to do this is to commit a fallacy.

***post hoc, ergo propter hoc*, fallacy of**   Reasoning that X caused Y simply because Y occurred after X.

precising definition   A definition that limits the applicability of a term whose usual meaning is too vague for the use in question.

predicate term   The noun or noun phrase that refers to the second class mentioned in a standard-form categorical claim.

predictable ratio   The ratio that results of a series of events can be expected to have given the antecedent conditions of the series. Examples: The predictable ratio of a fair coin flip is 50 percent heads and 50 percent tails; the predictable ratio of sevens coming up when a pair of dice is rolled is 1 in 6, or just under 17 percent.

premise   The claim or claims in an argument that provide the reasons for believing the conclusion.

premise indicator   A word or phrase (e.g., "because") that ordinarily indicates the presence of the premise of an argument.

prescriptive claim   A claim that states how things ought to be. Prescriptive claims impute values to actions, things, or situations. *Contrast with* descriptive claims.

principle of utility   The basic principle of utilitarianism, to create as much overall happiness and/or to limit unhappiness for as many as possible.

proof surrogate   An expression used to suggest that there is evidence or authority for a claim without actually saying that there is.

property in question   In inductive generalizations and analogical arguments, the members of a sample are said to have a property. This property is the "property in question."

pseudoreason   A consideration offered in support of a position that is not relevant to the truth or falsity of the issue in question.

question-begging argument   An argument whose conclusion restates a point made in the premises or clearly assumed by the premises. Although such an argument is technically valid, anyone who doubts the conclusion of a question-begging argument would have to doubt the premises, too.

random sample   *See* random selection process.

random selection process   Method of drawing a sample from a target population so that each member of the target population has an equal chance of being selected.

rationalizing   Using a false pretext in order to satisfy our desires or interests.

red herring   *See* smokescreen.

reductio ad absurdum   An attempt to show that a claim is false by demonstrating that it has false or absurd logical consequences; literally, "reducing to an absurdity."

refutation via hasty generalization   When we ask someone to reject a general claim on the basis of an example or two that run counter to the claim, we commit a fallacy known as "refutation via hasty generalization."

relativism   The view that two different cultures can be correct in their differing opinions on the same factual issue.

relevant/relevance A consideration is relevant to an issue if it is not unreasonable to suppose that its truth has some bearing on the truth or falsity of the issue. *See also* relevant difference

relevant difference A relevant difference is one that is not unreasonable to suppose caused the feature in question.

representative sample A sample that possesses all relevant features of a target population and possesses them in proportions that are similar to those of the target population.

rhetoric In our usage, "rhetoric" is language used primarily to persuade or influence beliefs or attitudes rather than to prove logically.

rhetorical comparison A comparison used to express or influence attitudes or affect behavior; such comparisons often make use of images with positive or negative emotional associations.

rhetorical definition A definition used to convey or evoke an attitude about the defined term and its denotation; such definitions often make use of images with positive or negative emotional associations.

rhetorical devices Rhetorical devices are used to influence beliefs or attitudes through the associations, connotations, and implications of words, sentences, or more extended passages. Rhetorical devices include slanters and fallacies. While rhetorical devices may be used to enhance the persuasive force of arguments, they do not add to the logical force of arguments.

rhetorical explanation An explanation intended to influence attitudes or affect behavior; such explanations often make use of images with positive or negative emotional associations.

rhetorical force *See* emotive force.

sample That part of a class referred to in the premises of a generalizing argument.

sample size One of the variables that can affect the size of the error margin or the confidence level of certain inductive arguments.

scapegoating Placing the blame for some bad effect on a person or group of people who are not really responsible for it, but who provide an easy target for animosity.

scare tactics Trying to scare someone into accepting or rejecting a claim. A common form includes merely describing a frightening scenario rather than offering evidence that some activity will cause it.

self-contradictory claim A claim that is analytically false.

semantically ambiguous claim An ambiguous claim whose ambiguity is due to the ambiguity of a word or phrase in the claim.

sharpening The process of exaggerating those points that a speaker thinks are important when he or she is describing some person, thing, or event. The result can be a distorted description.

slanter A linguistic device used to affect opinions, attitudes, or behavior without argumentation. Slanters rely heavily on the suggestive power of words and phrases to convey and evoke favorable and unfavorable images.

slippery slope A form of fallacious reasoning in which it is assumed that some event must inevitably follow from some other, but in which no argument is made for the inevitability.

smokescreen An irrelevant topic or consideration introduced into a discussion to divert attention from the original issue.

social utility A focus on what is good for society (usually in terms of overall happiness) when deciding on a course of action. *See also* principle of utility.

sound argument A valid argument whose premises are true.

spin A type of rhetorical device, often in the form of a red herring or complicated euphemism to disguise a politician's statement or action that might otherwise be perceived in an unfavorable light.

square of opposition A table of the logical relationships between two categorical claims that have the same subject and predicate terms.

standard-form categorical claim Any claim that results from putting words or phrases that name classes in the blanks of one of the following structures: "All _____ are _____"; "No _____ are _____"; "Some _____ are _____"; and "Some _____ are not _____."

stare decisis "Letting the decision stand." Going by precedent.

statistical significance To say that some finding is statistically significant at a given confidence level—say, .05—is essentially to say that the finding could have arisen by chance in only about five cases out of one hundred.

stereotype An oversimplified generalization about the members of a class.

stipulative definition A definition used to introduce an unfamiliar term or to assign a new meaning to a familiar term.

straw man A type of fallacious reasoning in which someone ignores an opponent's actual position and presents in its place a distorted, exaggerated, or misrepresented version of that position.

strong argument An argument that has this characteristic: On the assumption that the premises are true, the conclusion is unlikely to be false.

subcontrary claims Two claims that can both be true at the same time but cannot both be false at the same time.

subject term The noun or noun phrase that refers to the first class mentioned in a standard-form categorical claim.

subjective claim A claim expressing a person's subjective state of mind. *See* nonfactual claim.

subjectivism The assumption that what is true for one person is not necessarily true for another.

subjectivist fallacy   This pattern of fallacious reasoning: "Well, X may be true for you, but it isn't true for me," said with the intent of dismissing or rejecting X.

syllogism   A deductive argument with two premises.

syntactically ambiguous claim   An ambiguous claim whose ambiguity is due to the structure of the claim.

synthetic claim   A claim whose truth value cannot be determined simply by understanding the claim—an observation of some sort is also required. *Contrast with* analytic claim.

target   In the conclusion of an inductive generalization, the members of an entire class of things is said to have a property or feature. This class is the "target" or "target class." In the conclusion of an analogical argument, one or more individual things is said to have a property or feature. The thing or things is the "target" or "target item."

target class   The population, or class, referred to in the conclusion of a generalizing argument.

target item   *See* target.

term   A word or an expression that refers to or denotes something.

tradition, "argument" from   "Arguing" that a claim is true on the grounds that it is traditional to believe it is true.

truth-functional equivalence   Two claims are truth-functionally equivalent if and only if they have exactly the same truth table.

truth-functional logic   A system of logic that specifies the logical relationships among truth-functional claims—claims whose truth values depend solely upon the truth values of their simplest component parts. In particular, truth-functional logic deals with the logical functions of the terms "not," "and," "or," "if . . . then," and so on.

truth table   A table that lists all possible combinations of truth values for the claim variables in a symbolized claim or argument and then specifies the truth value of the claim or claims for each of those possible combinations.

two wrongs make a right   This pattern of fallacious reasoning: "It's acceptable for A to do X to B because B would do X to A," said where A's doing X to B is not necessary to prevent B's doing X to A.

utilitarianism   The moral position unified around the basic idea that we should promote happiness as much as possible and weigh actions or derivative principles in terms of their utility in achieving this goal.

vague claim   A claim that lacks sufficient precision to convey the information appropriate to its use.

valid argument   An argument that has this characteristic: On the assumption that the premises are true, it is impossible for the conclusion to be false.

Venn diagram   A graphic means of representing a categorical claim or categorical syllogism by assigning classes to overlapping circles. Invented by English mathematician John Venn (1834–1923).

virtue ethics   The moral position unified around the basic idea that each of us should try to perfect a virtuous character that we exhibit in all actions.

weaseler   An expression used to protect a claim from criticism by weakening it.

wishful thinking   Accepting a claim because you want it to be true, or rejecting it because you don't want it to be true.

Answers, Suggestions, and Tips for Exercises Marked with a Triangle

Chapter 1: Thinking Critically: It Matters

Exercise 1-1

1. An argument is an attempt to support a claim or an assertion by providing a reason or reasons for accepting it.

4. Yes

7. No

10. It can be implied.

13. No

16. False

19. True

20. True

Exercise 1-3

1. Argument

4. No argument

7. Argument

10. Argument

Exercise 1-4

1. No argument

4. Argument

7. Argument

10. No argument

13. No argument (Warren says that there are reasons for her conclusion, but she doesn't tell us what they are.)

16. No argument

19. No argument

Exercise 1-5

1. a. The other three claims in the paragraph are offered as *reasons* for the claim that Hank ought not to take the math course.

3. d. The other two claims are given as reasons for believing claim d; they are premises of an argument and claim d is the conclusion.

4. c. The remainder of the passage provides examples of what is claimed in c. The claims in which these examples appear function as premises for c.

13. b. There is a lot of information in this passage, but answer (b) is certainly the *main* issue of the selection. The easiest way to see this is to notice that almost all of the claims made in the passage support this one. We'd put answer (c) in second place.

14. c. Answers (a) and (b) don't capture the futility of the prison policy expressed in the passage; answer (d) goes beyond what is expressed in the passage.

15. c. Answer (b) is in the running, but it is clear from the passage that the author isn't interested so much in whether the permit holders are individuals or corporations as in whether they are big-time operators or small-time operators.

16. b

20. c

Exercise 1-6

1. Whether police brutality happens very often

4. Whether there exists a world that is essentially independent of our minds

7. Whether a person who buys a computer should take some lessons

10. Whether Native Americans, as true conservationists, have something to teach readers about our relationship to the earth. There are other points made in the passage, but they are subsidiary to this one.

Exercise 1-7

1. There are two issues: whether they're going on Standard Time the next weekend and whether they'll need to set the clocks forward or back. Both speakers address both issues.

4. The issue is whether complaints about American intervention abroad are good or bad. Both speakers address this issue.

Exercise 1-8

1. Suburbanite misses Urbanite's point. Urbanite addresses the effects of the requirement; Suburbanite addresses the issue of whether he and his neighbors can afford to comply with it.

3. On the surface, it may seem that both Hands address the issue of whether a person such as One Hand can feel safe in her own home. But it's clear that One Hand's real issue is whether the large number of handguns makes one unsafe in one's own home. Other Hand ignores this issue completely.

5. The issue for both parties is whether Fed-Up will be happier if he retires to Arkansas.

Exercise 1-9

The distinction used is between factual claims and nonfactual claims.

1. Factual

4. Nonfactual

7. Factual

10. Factual

Exercise 1-10

1. Factual

4. Factual

7. Factual. Although this is not possible to resolve as a practical matter, we can at least imagine circumstances under which we could test for life on any given planet.

10. Nonfactual

13. Nonfactual. This one is very controversial, of course, but if there were an agreed-upon method of settling the issue, it would have long since ceased to be so controversial. This is an obvious case where the claim's being nonfactual does not in the least prevent its being very important.

Exercise 1-11

The factual claims are 2, 4, 6, 8.

Exercise 1-12

1. The house, a two-story colonial-style building, burned to the ground.

4. The Coors Brewing Company began offering gay partners the same benefits as spouses in 1995.

7. McCovey was able to finish the assignment on time.

10. The Reverend Jesse Jackson's visit to Pakistan has been delayed due to lack of support from the Bush administration.

13. For two years the rapidly expanding global computer matrix had concerned Gates.

Exercise 1-13

1. No nonfactual elements present

4. No nonfactual elements present

7. The second sentence is arguably nonfactual, since the phrase "on the face of it" is something about which disagreements might be difficult to settle. On the other hand, the phrase "might be" is such a strong weaseler (see Chapter 4), one might say that this sentence commits to very little at all.

10. One might change the second sentence to read, "Those who felt character was important went on to elect one who some think has the feeblest intellect of any president in history."

Exercise 1-14

1. b
4. b
7. b
10. a

Exercise 1-15

1. Argument
4. Argument
7. Explanation
10. Explanation

Chapter 2: Critical Thinking and Clear Writing

Exercise 2-2

7, 6, 4, 1, 3, 2, 5

Exercise 2-3

1. "Piano" is defined analytically.
4. "Red planet" is defined by synonym. (This one is tricky because it looks like a definition by example. But there is only *one* red planet, so the phrase refers to exactly the same object as the word "Mars.")
8. "Chiaroscuro" is defined by synonym.
11. "Significant other" is defined by example—several of them.

Exercise 2-4

1. The Raider tackle blocked the Giants linebacker.
4. How Therapy Can Help Victims of Torture
7. Susan's nose resembles Hillary Clinton's.
10. 6 Coyotes That Maul Girl Are Killed by Police
13. Second sentence: More than one disease can be carried and passed along to humans by a single tick.
16. We give to life good things.
19. Dunkelbrau—for those who crave the best-tasting real German beer
22. Jordan could write additional profound essays.
25. When she lay down to nap, she was disturbed by a noisy cow.
28. When Queen Elizabeth appeared before her troops, they all shouted "harrah."
32. AT&T, for as long as your business lasts.
33. This class might have had a member of the opposite sex for a teacher.
34. Married 10 times before, woman gets 9 years in prison for killing her husband.

Exercise 2-5

1. As a group
4. As a group
7. It's more likely that the claim refers to the Giants as a group, but it's possible that it refers to the play of individuals.
10. As individuals
12. Probably as individuals
15. Ambiguous. If the claim means that people are living longer than they used to, the reference is to people as individuals. If the claim means that the human race is getting older, then the reference is to people as a group. If the claim expresses the truism that to live is to age, then the reference is to people as individuals.

Exercise 2-6

In order of decreasing vagueness:

1. (d), (e), (b), (c), (f), and (a). Compare (e) and (b). If Eli and Sarah made plans for the future, then they certainly discussed it. But just discussing it is more vague—they could do that with or without making plans.
4. (c), (d), (e), (a), (b)

Exercise 2-7

1. a
4. b
7. a. But it's close.
10. b
15. a

Exercise 2-8

1. Too vague. Sure, you can't say exactly how much longer you want it cooked, but you can provide guidelines; for example, "Cook it until it isn't pink."
4. Not too vague.
7. Not too vague.
10. If this is the first time you are making frosting or if you are an inexperienced cook, this phrase is too vague.

Exercise 2-10

"Feeding" simply means "fertilizing" and is not too vague. "Frequently" is too vague. "No more than half" is not too vague. "Label-recommended amounts" is not too vague. "New year's growth begins" and "each bloom period ends" are pretty vague for a novice gardener, but because pinpoint timing apparently isn't crucial, the vagueness here is acceptable. "Similar" is too vague for a novice gardener. "Immediately after bloom" suggests that precise timing is important here, and we find the phrase too vague, at least for novices. "When the nights begin cooling off" is too vague even if precision in timing isn't terribly important.

Exercise 2-11

1. Twenty percent more than what? (You might wonder what "real dairy butter" is, but it's probably safe to assume that it's just plain old butter.)
4. This is not too bad, but the word "desert" covers a lot of territory—not all deserts are like the Sahara.
7. The comparison is okay, but don't jump to the conclusion that today's seniors are better students. Maybe the teachers are easier graders.
10. In the absence of absolute figures, this claim does not provide any information about how good attendance was (or about how brilliant the season was).

Exercise 2-12

1. Superior? In what way? More realistic character portrayal? Better expression of emotion? Probably the claim means only "I like Paltrow more than I like Blanchett."
4. Fine, but don't infer that they both grade the same. Maybe Smith gives 10 percent A's and 10 percent F's, 20 percent B's and 20 percent D's, and 40 percent C's, whereas Jones gives everyone a C. Who do you think is the more discriminating grader, given this breakdown?
7. Well, first of all, what is "long-distance"? Second, and more important, how is endurance measured? People do debate such issues, but the best way to begin a debate on this point would be by spelling out what you mean by "requires more endurance."
10. This is like a comparison of apples and oranges. How can the popularity of a movie be compared with the popularity of a song?

Exercise 2-13

1. The price-earnings ratio is a traditional (and reasonable) measure of a stock, and the figure is precise enough. Whether this is good enough reason to worry about the stock market is another matter; such a conclusion may not be supported by the price-earnings figure.
4. "Attend church regularly" is a bit vague; a person who goes to church each and every Christmas and Easter is a regular, although infrequent, attender. We don't find "majority" too vague in this usage.
7. "Contained more insights" is much too vague. The student needs to know more specifically what was the matter with his or her paper, or at least what was better about the roommate's paper.
10. These two sorts of things are much too different to be compared in this way. If you're starving, the chicken looks better; if you need to get from here to there, it's the Volkswagen. (This is the kind of question Moore likes to ask people. Nobody can figure out why.)

Exercise 2-15

1. Students should choose majors with considerable care.
4. If a nurse can find nothing wrong with you in a preliminary examination, a physician will be recommended to you. However, in this city physicians wish to protect themselves by having you sign a waiver.

7. Soldiers should be prepared to sacrifice their lives for their comrades.

10. Petitioners over sixty should have completed form E-7. ["He(she)" is awkward and retains sexist implications.]

13. Language is nature's greatest gift to humanity.

16. The proof must be acceptable to the rational individual.

17. The country's founders believed in the equality of all.

20. Athletes who want to play for the National Football League should have a good work ethic.

24. Most U.S. senators are men.

27. Mr. Macleod doesn't know it, but Ms. Macleod is a feminist.

30. To be a good politician, you have to be a good salesperson.

Exercise 2-16

In case you couldn't figure it out, the friend is a woman.

Chapter 3: Credibility

Exercise 3-5

Something like number 9 is probably true, given the huge, almost unimaginable difference in wealth between the richest and the poorest people on the planet, but we have no idea what the actual numbers are. We've seen number 12 going around the Web, but we don't know whether there's anything to it and we're not interested in conducting the appropriate experiments. We think the rest of these don't have much of a chance (although there are conspiracy theorists who seem to believe number 10.)

Exercise 3-8

1. In terms of expertise, we'd list (d), (c), and (b) first. Given what we've got to go on, we wouldn't assign expert status to either (a) or (e). We'd list all entries as likely to be fairly unbiased except for (a), which we would expect to be very biased.

3. Expertise: (b) first, then (a), then (c) and (d) about equal, and (e) last. We'd figure that (b) is most likely to be unbiased, with (c), (d), and (e) close behind; Choker would be a distant last on this scale. Her bad showing on the bias scale more than makes up for her high showing on the expertise scale.

Exercise 3-9

1. The most credible choices are either the FDA or *Consumer Reports,* both of which investigate health claims of the sort in question with reasonable objectivity. The company that makes the product is the least credible source because it is the most likely to be biased. The owner of the health food store would not qualify in the area of drugs, even if she is very knowledgeable regarding nutrition. Your local pharmacist can reasonably be regarded as credible, but he or she may not have access to as much information as the FDA or *CR*.

2. It would probably be a mistake to consider any of the individuals on this list more expert than the others, although different kinds and different levels of bias are fairly predictable on the parts of the victim's father, the NRA representative, and possibly the police chief. The senator might be expected to have access to more data that are relevant to the issue, but that would not in itself make his or her credibility much greater than that of the others. The problem here is that we are dealing with a value judgment that depends very heavily upon an individual's point of view rather than his or her expertise. What is important to this question is less the credibility of the person who gives us an answer than the strength of the supporting argument, if any, that he or she provides.

3. Although problem 2 hinges on a value judgment, this one calls for an interpretation of the original intent of a constitutional amendment. Here our choices would be either the Supreme Court justice or the constitutional historian, with a slight preference for the latter because Supreme Court justices are concerned more with constitutional issues as they have been interpreted by other courts than with original intent. The NRA representative is paid to speak for a certain point of view and would be the least credible in our view. The senator and the U.S. President would fall somewhere in between: Both might reasonably be expected to be knowledgeable about constitutional issues, but much less so than our first two choices.

Exercise 3-10

1. Professor Hilgert would possess the greatest degree of credibility and authority on (d), (f), and (h), and, compared with someone who had not lived in both places, on (i).

Exercise 3-12

1. We'd accept this as probably true—but probably only *approximately* true. It's difficult to be precise about such matters; Campbell will most likely lay off *about* 650 workers, including *about* 175 at its headquarters.

8. We'd accept this as likely.

12. No doubt cats that live indoors do tend to live longer than cats that are subject to the perils of outdoor life. If statistics on how much longer indoor cats live on the average were available, we'd expect the manufacturer to know them. But we suspect that such statistics would be difficult to establish (and probably not worth the effort), and we therefore have little confidence in the statistic cited here.

20. We'd reject this.

Chapter 4: Persuasion Through Rhetoric

Exercise 4-1

1. c
4. e
7. a
10. c
12. c
15. T

Exercise 4-2

(1) hyperbole (in Chapter 6 we'll call this "straw man"), (2) dysphemism, (3) not a rhetorical device, (4) dysphemism, (5) not a rhetorical device, (6) dysphemism

Exercise 4-3

(1) dysphemism, (2) dysphemism, (3) hyperbole, (4) weaseler, (5) proof surrogate, (6) not a downplayer in this context, (7) loaded question

Exercise 4-8

1. The quotation marks downplay the quality of the school.
4. Persuasive definition
6. No rhetorical device present
8. "Gaming" is a euphemism for "gambling."

11. "Clearly" is a proof surrogate; the final phrase is hyperbole.
14. "Adjusted" is a euphemism.

Exercise 4-9

1. "Japan, Inc." is a dysphemism.
4. "Getting access" is a euphemism, and, in this context, so is "constituents." We'll bet it isn't just *any* old constituent who gets the same kind of "access" as big campaign contributors.
7. The last sentence is hyperbolic.
10. (We really like this one.) "Even," in the first sentence, is innuendo, insinuating that members of Congress are more difficult to embarrass than others. The remainder is another case of innuendo with a dash of downplaying. Although it's a first-class example, it's different from the usual ones. Mellinkoff makes you think that Congress *merely* passes a law in response to the situation. But stop and think for a moment: Aside from the odd impeachment trial or congressional hearing, *all that Congress can do is pass laws!* So Mellinkoff's charge really should not be seen as belittling Congress at all.
13. "As you know" is a variety of proof surrogate. The remainder is a persuasive comparison.
18. Lots of them here! To begin, "orgy" is a dysphemism; "self-appointed" is a downplayer. The references to yurts and teepees is ridicule, and "grant-maintained" is a downplayer. The rest of it employs a heavy dose of sarcasm.

Chapter 5: More Rhetorical Devices

Exercise 5-2

1. "Argument" from popularity

4. "Argument" from pity

7. Subjectivism

10. "Argument" from outrage. There is also an example of straw man in the last sentence—we'll meet straw man in Chapter 6.

13. Rationalism (almost certainly)

Exercise 5-3

1. Not very.

3. Relevant. A popular automobile may have continued support from its maker, and this can be advantageous to the owner of such a car.

7. Not relevant. Notice, though, that our likes and dislikes seem to be influenced by the opinions of others, whether we want them to be or not.

10. It can be. Advertising a product as best-selling may create a feeling on the part of consumers that they will be out of step with the rest of society if they don't purchase the advertised product. (But within limits almost any product can be *said* to be popular or a best-seller, so the fact that such a claim is made is no reason for one to feel out of step by not purchasing the product.) Usually, however, the "best-seller" tag is intended to make us think that the product must be good because so many people cannot be wrong. In other words, such ads in effect are appeals to the "authority of the masses" or the "wisdom of society." However, unless you have some reason to believe (a) that the claims made in the ad about *unusual* popularity are *true,* and either (b) that the buyers of the product have themselves bought the product for some reason that applies in your case as well or (c) that you could indirectly benefit from the popularity of the item (popular cars, for instance, hold their resale value), then to buy a product on the basis of such advertising would be falling prey to a fallacy.

Exercise 5-5

1. Scare tactics

4. Apple polishing, with a touch of peer pressure

7. No fallacy

10. Smokescreen/red herring

Exercise 5-6

1. No fallacy

4. Peer pressure

7. Apple polishing

10. "Argument" from outrage

Exercise 5-7

1. Scare tactics

4. *Issue:* Whether this is fallacious depends largely on one's assessment of how likely he or she is to be among the 250 who die from accidents on a given day. In any case, it is not an argument for buying *this company's* accident insurance.

7. Two wrongs make a right

10. Ridicule

14. Common practice

17. We find ridicule in the first paragraph, a red herring in the second, and a loaded question in the third.

Chapter 6: More Fallacies

Exercise 6-2

1. Begging the question

4. Straw man

7. Straw man

10. Line-drawing fallacy (false dilemma)

Exercise 6-3

1. Inconsistency ad hominem

4. Inconsistency ad hominem

7. Circumstantial ad hominem

11. Personal attack ad hominem

Exercise 6-5

1. d
4. b
7. a
10. b

Exercise 6-6

1. c
4. c
7. b
10. c

Exercise 6-7

1. b
4. b
7. e
10. a

Exercise 6-8

1. Straw man, smokescreen/red herring
4. No fallacy. Notice that the passage is designed to attack the company, not the company's product. The wages it pays are relevant to the point at issue.
7. No fallacy
10. False dilemma
13. Genetic fallacy
16. Line-drawing fallacy (false dilemma)
19. Inconsistency ad hominem

Exercise 6-9

1. Circumstantial ad hominem
4. Straw man (Jeanne responds as if Carlos wanted to sleep until noon). Can also be analyzed as false dilemma ("Either we get up right now, at 4:00 A.M., or we sleep until noon.")
7. This begs the question. The conclusion merely restates the premise.
10. False dilemma
13. Misplaced burden of proof

Exercise 6-10

1. This is an example of burden of proof. Yes, it is indeed slightly different from the varieties explained in the text, and here's what's going on. The speaker is requiring proof of a sort that *cannot be obtained*—actually *seeing* smoke cause a cancer. So he or she is guilty of one type of "inappropriate burden of proof."
4. This is false dilemma because Sugarman's alternatives are certainly not the only ones. Notice that he is giving *no argument* against the Chicago study; he is simply using the false dilemma to deny the study's conclusion.
7. Inconsistency ad hominem
10. This is a case of misplaced burden of proof. The speaker maintains that the government is violating the law. The burden of proof therefore falls on the speaker to justify his or her opinion. Instead of doing that, he or she acts as if the fact that officials haven't disproved the claim is proof that the claim is true.

Exercise 6-12

1. Assuming that the sheriff's department has more than two officers, the speaker is misrepresenting her opponent's position. Straw man.
4. Misplaced burden of proof
7. Perfectionist fallacy (false dilemma)
10. This is an ad hominem. It rides the border between personal attack and the circumstantial variety.

Exercise 6-13

1. Ad hominem: inconsistency. You hear this kind of thing a lot.
4. Ad hominem: personal attack
7. Slippery slope
10. Ad hominem: personal attack

Exercise 6-16

1. Perfectionist fallacy (false dilemma)
5. Apple polishing
9. "Argument" from pity and "argument" from outrage
13. Two wrongs; a case can easily be made for common practice as well.
16. Straw man, technically, but this is also ridicule.

Exercise 6-17

1. This is an example of misplaced burden of proof. The fact that the airplane builders *might* be cutting corners is not evidence that they are *in fact* cutting corners. The speaker's contention that the manufacturers may be tempted to cut corners may be good grounds for scrutinizing their operations, but it's not good grounds for the conclusion that they really are cutting corners.

4. Yes—this is clearly fallacious. Bush's remark is irrelevant to the Democrats' claim. This case does not fit one of our regular categories. It's a smokescreen.

5. The quoted remark from Harris is not relevant to the conclusion drawn in this passage. This passage doesn't fit neatly into any of our categories, although ad hominem would not be a bad choice. Notice a possible ambiguity that may come into play: "Having an impact" might mean simply that Harris wants his work to be noticed by "movers and shakers"—or it could mean that he wishes to sway people toward a certain political view. It's likely that he intended his remark the first way, but it's being taken in the second way in this passage.

9. This is a borderline circumstantial ad hominem. It certainly does not follow that Seltzer and Sterling are making false claims from the fact that they are being paid by an interested party. But remember the cautions from Chapter 3: Expertise can be bought, and we should be very cautious about accepting claims made by experts who are paid by someone who has a vested interest in the outcome of a controversy.

Chapter 7: The Anatomy and Varieties of Arguments

Exercise 7-1

1. a. Premise
   b. Premise
   c. Conclusion

2. a. Premise
   b. Premise
   c. Conclusion

3. a. Conclusion
   b. Premise

4. a. Premise
   b. Premise
   c. Conclusion

5. a. Premise
   b. Conclusion
   c. Premise
   d. Premise

Exercise 7-2

1. Premise: All Communists are Marxists.
   Conclusion: All Marxists are Communists.

4. Premise: That cat is used to dogs.
   Conclusion: Probably she won't be upset if you bring home a new dog for a pet.

7. Premise: Presbyterians are not fundamentalists.
   Premise: All born-again Christians are fundamentalists.
   Conclusion: No born-again Christians are Presbyterians.

10. Premise: If we've got juice at the distributor, the coil isn't defective.
    Premise: If the coil isn't defective, then the problem is in the ignition switch.
    [Unstated premise: We've got juice at the distributor.]
    Conclusion: The problem is in the ignition switch.

Exercise 7-3

1. Conclusion: There is a difference in the octane ratings between the two grades of gasoline.

4. Conclusion: Scrub jays can be expected to be aggressive when they're breeding.

7. Conclusion: Dogs are smarter than cats.

10. Unstated conclusion: She is not still interested in me.

Exercise 7-4

1. Independent
3. Independent
6. Independent
9. Independent
12. Independent
15. Dependent

Exercise 7-5

1. Dependent

4. Dependent; one premise is unstated: What's true for rats is probably true for humans.

8. Independent

10. Independent

Exercise 7-6

1. To explain

4. To explain

7. To explain

9. To argue

Exercise 7-7

3. Valid; true

6. False

9. False

Exercise 7-8

(Refer to Exercise 7-2)

1. Invalid

4. Invalid

7. Valid

10. Valid

(Refer to Exercise 7-3)

1. Invalid

4. Invalid

7. The unstated premise "Being more easily trained is a *sure* sign of greater intelligence" would make the argument valid. An unstated premise such as "Being more easily trained is a *good* sign of greater intelligence" would make the argument fairly strong but invalid.

10. Valid

Exercise 7-9

1. Probably true; had "out of order" been written in pencil on the meter, we'd have a different opinion, since most of the meters in our town have those words scrawled on them.

5. Probably true; a restaurant that does a good job on these three different kinds of entrees will probably do a good job on the rest.

9. Probably true; it's possible that the killer cleaned up very thoroughly, but it's more likely that the body was brought from somewhere else.

12. True beyond a reasonable doubt; it may be that consumption will drop in the future or that discovery of new reserves will increase, but as long as consumption at some level continues, *eventually* all the oil will be used up.

Exercise 7-10

1. Assumed premise: All well-mannered people had a good upbringing.

4. Assumed conclusion: He will not drive recklessly.

7. Assumed premise: All dogs that scratch a lot have fleas or dry skin.

10. Assumed premise: Every poet whose work appears in many Sierra Club publications is one of America's outstanding poets.

Exercise 7-11

1. Assumed premise: Most people who are well-mannered had a good upbringing.

4. Assumed conclusion: He will drive safely.

7. Assumed premise: Most dogs that scratch a lot have fleas or dry skin. (Or: When this dog scratches a lot, he usually has either fleas or dry skin.)

10. Assumed premise: Most poets whose work appears in many Sierra Club publications are among America's outstanding poets.

Exercise 7-12

1. Assumed premise: All stores that sell only genuine leather goods have high prices.

4. Assumed premise: No ornamental fruit trees bear edible fruit.

7. Assumed premise: No former professional wrestler could be a very effective governor.

10. Assumed premise: If population studies show that smoking causes lung cancer, then all smokers will get lung cancer.

Exercise 7-13

1. Assumed premise: Most stores that sell only genuine leather goods have high prices.

4. Assumed premise: Few ornamental fruit trees bear edible fruit.

7. Assumed premise: Few people who were professional wrestlers could be very effective governors.

10. Assumed premise: If population studies show that smoking causes lung cancer, then most smokers will get lung cancer.

Exercise 7-14

1.

4.

Exercise 7-15

(See Exercise 7-2)

1. ① All Communists are Marxists.
   ② All Marxists are Communists.

4. ① That cat is used to dogs.
   ② She won't be upset if you bring home a new dog for a pet.

5.

7. ① Presbyterians are not fundamentalists.
   ② All born-again Christians are fundamentalists.

③ No born-again Christians are Presbyterians.

10. ① If we've got juice at the distributor, the coil isn't defective.
    ② If the coil isn't defective, then the problem is in the ignition switch.
    ③ The problem is in the ignition switch.

(See Exercise 7-3)

1. ① The engine pings every time we use the regular unleaded gasoline.
   ② The engine doesn't ping when we use super.
   ③ There is a difference in octane ratings between the two.

4. ① When blue jays are breeding, they become very aggressive.
   ② Scrub jays are very similar to blue jays.
   ③ Scrub jays can be expected to be aggressive when breeding.

7. ① It's easier to train dogs than cats.
   ② Dogs are smarter than cats.

10. ① If she were still interested in me, she would have called.
    ② She didn't call.
    ③ [Unstated] She's not still interested in me.

(See Exercise 7-4)

6. ①You're overwatering your lawn.
   ②There are mushrooms growing around the base of the tree.
   ③Mushrooms are a sure sign of overwatering.
   ④There are worms on the ground.
   ⑤Worms come up when the earth is oversaturated.

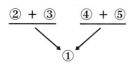

9. ①If you drive too fast, you're more likely to get a ticket.
   ②(If you drive too fast,) you're also more likely to get into an accident.
   ③You shouldn't drive too fast.

12. ①You should consider installing a solarium.
    ②Installing a solarium can get you a tax credit.
    ③Installing a solarium can reduce your heating bill.
    ④Installing a solarium correctly can help you cool your house in the summer.

15. ① We must paint the house now.
    ②If we don't, we'll have to paint it next summer.
    ③If we have to paint it next summer, we'll have to cancel our trip.
    ④It's too late to cancel our trip.

(See Exercise 7-5)

1. ①All mammals are warm-blooded creatures.
   ②All whales are mammals.
   ③All whales are warm-blooded creatures.

4. ①Rats that have been raised . . . have brains that weigh more. . . .
   ②The brains of humans will weigh more if humans are placed in intellectually stimulating environments.

8. ①The mayor now supports the initiative for the Glen Royale subdivision.
   ②Last year the mayor proclaimed strong opposition to further development in the river basin.
   ③Glen Royale will add to congestion.
   ④Glen Royale will add to pollution.
   ⑤Glen Royale will make the lines longer at the grocery.
   ⑥[Unstated] The mayor should not support the Glen Royale subdivision.

10. ①Jesse Brown is a good person for your opening in Accounting.
    ②He's as sharp as they come.
    ③He has a solid background in bookkeeping.
    ④He's good with computers.
    ⑤He's reliable.
    ⑥He'll project the right image.
    ⑦He's a terrific golfer.
    ⑧I know him personally.

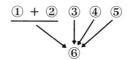

Exercise 7-16

1. ①Your distributor is the problem.
   ②There's no current at the spark plugs.
   ③If there's no current at the plugs, then either your alternator is shot or your distributor is defective.
   ④[Unstated] Either your alternator is shot or your distributor is defective.
   ⑤If the problem were in the alternator, then your dash warning light would be on.

⑥The light isn't on.

4. ①They really ought to build a new airport.
②It [a new airport] would attract more business to the area.
③The old airport is overcrowded and dangerous.

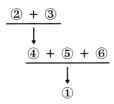

Note: Claim number ③ could be divided into two separate claims, one about overcrowding and one about danger. This would be important if the overcrowding were clearly offered as a reason for the danger.

Exercise 7-17

1. ①Cottage cheese will help you to be slender.
②Cottage cheese will help you to be youthful.
③Cottage cheese will help you to be more beautiful.
④Enjoy cottage cheese often.

4. ①The idea of a free press in America is a joke.
②The nation's advertisers control the media.
③Advertisers, through fear of boycott, can dictate programming.
④Politicians and editors shiver at the thought of a boycott.
⑤The situation is intolerable.
⑥I suggest we all listen to NPR and public television.

Note: The writer may see claim ① as the final conclusion and claim ⑤ as his comment upon it. Claim ⑥ is probably a comment on the results of the argument, although it too could be listed as a further conclusion.

7. ①Consumers ought to be concerned about the FTC's dropping the rule requiring markets to stock advertised items.
②Shoppers don't like being lured to stores and not finding advertised products.
③The rule costs at least $200 million and produces no more than $125 million in benefits.
④The figures boil down to a few cents per shopper over time.
⑤The rule requires advertised sale items to be on hand in reasonable numbers.

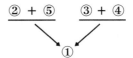

10. ①Well-located, sound real estate is the safest investment in the world.
②Real estate is not going to disappear as can dollars in savings accounts.
③Real estate values are not lost because of inflation.
④Property values tend to increase at a pace at least equal to the rate of inflation.
⑤Most homes have appreciated at a rate greater than the inflation rate. . . .

13. ①About 100 million Americans are producing data on the Internet. . . . .
②Each user is tracked so private information is available in electronic form.
③One Web site . . . promises, for seven dollars, to scan . . . etc.
④The combination of capitalism and technology poses a threat to our privacy.

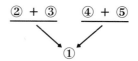

⑤What is your definition of trivial, busting an innocent child's skull with a hammer?

16. ①Measure A is consistent with the City's General Plan and city policies. . . .
②A "yes" vote will affirm the wisdom of well-planned, orderly growth. . . .
③Measure A substantially reduces the amount of housing previously approved for Rancho Arroyo.
④Measure A increases the number of parks and amount of open space.
⑤Measure A significantly enlarges and enhances Bidwell Park.
⑥Approval of Measure A will require dedication of 130.8 acres to Bidwell Park.
⑦Approval of Measure A will require the developer to dedicate seven park sites.
⑧Approval of Measure A will create 53 acres of landscaped corridors and greenways.
⑨Approval of Measure A will preserve existing arroyos and protect sensitive plant habitats. . . .
⑩Approval of Measure A will create junior high school and church sites.
⑪Approval of Measure A will plan villages with 2,927 dwellings.
⑫Approval of Measure A will provide onsite job opportunities and retail services.
⑬[Unstated conclusion:] You should vote for Measure A.

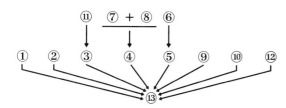

20. ①Freedom means choice.
②This is a truth antiporn activists always forget when they argue for censorship.
③In their fervor to impose their morality, groups like Enough Is Enough cite extreme examples of pornography, such as child porn, suggesting that they are available in video stores.
④This is not the way it is.
⑤Most of this material portrays, not actions such as this, but consensual sex between adults.
⑥The logic used by Enough Is Enough is that if something can somehow hurt someone, it must be banned.
⑦They don't apply this logic to more harmful substances, such as alcohol or tobacco.
⑧Women and children are more adversely affected by drunken driving and secondhand smoke than by pornography.
⑨Few Americans would want to ban alcohol or tobacco even though they kill hundreds of thousands of people each year.
⑩[Unstated conclusion] Enough Is Enough is inconsistent.
⑪[Unstated conclusion] Enough Is Enough's antiporn position is incorrect.

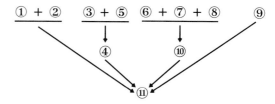

19. ①In regard to your editorial, "Crime bill wastes billions," let me set you straight. [Your position is mistaken.]
②Your paper opposes mandatory life sentences for criminals convicted of three violent crimes, and you whine about how criminals' rights might be violated.
③Yet you also want to infringe on a citizen's rights to keep and bear arms.
④You say you oppose life sentences for three-time losers because judges couldn't show any leniency toward the criminals no matter how trivial the crime.

Box: Beer Makes You Smarter

1 = A herd of buffalo can only move as fast as the slowest buffalo.
2 = When the herd is hunted, it is the slowest and weakest ones at the back that are killed first.
3 = This natural selection is good for the herd as a whole.
4 = The general speed and health of the whole group keeps improving by the regular attrition of the weakest members.
5 = The human brain can only operate as fast as the slowest brain cells.

6 = Excessive intake of alcohol kills brain cells.
7 = Alcohol attacks the slowest and weakest brain cells first.
8 = Regular consumption of beer eliminates the weaker brain cells.
9 = The brain is made a faster and more efficient machine.
10 = You always are smarter after a few beers.

Note: 1–4 are offered as an analogy, not as an argument that 10 is true.

Chapter 8: Deductive Arguments I: Categorical Logic

Exercise 8-1

1. All salamanders are lizards.

4. All members of the suborder Ophidia are snakes.

7. All alligators are reptiles.

10. All places there are snakes are places there are frogs.

13. All people who got raises are vice presidents.

16. All people identical with Socrates are Greeks.

19. All examples of salt are things that preserve meat.

Exercise 8-2

1. No students who wrote poor exams are students who were admitted to the program.

4. Some first basemen are right-handed people.

7. All passers are people who made at least 50 percent.

10. Some prior days are days like this day.

13. Some holidays are holidays that fall on Saturday.

16. All people who pass the course are people who pass this test. Or: No people who fail this test are people who pass the course.

19. All times they will let you enroll are times when you've paid the fee.

Exercise 8-3

1. Translation: Some anniversaries are not happy occasions. (True)
Corresponding A-claim: All anniversaries are happy occasions. (False)
Corresponding E-claim: No anniversaries are happy occasions. (Undetermined)

Corresponding I-claim: Some anniversaries are happy occasions. (Undetermined)

4. Translation: Some allergies are things that can kill you. (True)
Corresponding A-claim: All allergies are things that can kill you. (Undetermined)
E-claim: No allergies are things that can kill you. (False)
Corresponding O-claim: Some allergies are not things that can kill you. (Undetermined)

Exercise 8-4

1. No non-Christians are non-Sunnis. (Not equivalent)

4. Some Christians are not Kurds. (Not equivalent)

7. All Muslims are Shiites. (Not equivalent)

10. All Muslims are non-Christians. (Equivalent)

Exercise 8-5

1. Some students who scored well on the exam are not students who didn't write poor essays. (Equivalent)

4. No students who were not admitted to the program are students who scored well on the exam. (Not equivalent)

7. All people whose automobile ownership is not restricted are people who don't live in the dorms. (Equivalent)

10. All first basemen are people who aren't right-handed. (Equivalent)

Exercise 8-6

2. All encyclopedias are nondefinitive works.

4. No sailboats are sloops.

Exercise 8-7

Translations of lettered claims:

a. Some people who have been tested are not people who can give blood.

b. Some people who can give blood are not people who have been tested.

c. All people who can give blood are people who have been tested.

d. Some people who have been tested are not people who can give blood.

e. No people who have been tested are people who can give blood.

2. Equivalent to (c).

4. Equivalent to (c).

Exercise 8-8

1. Obvert (a) to get "Some Slavs are not Europeans."

4. Obvert the conversion of (b) to get "Some members of the club are not people who took the exam."

7. Contrapose (a) to get "All people who will not be allowed to perform are people who did not arrive late." Convert (b) to get "Some people who will not be allowed to perform are people who did not arrive late."

10. Convert the obverse of (b) to get "No decks that will play digital tape are devices that are equipped for radical oversampling."

Exercise 8-9

1. Invalid (this would require the conversion of an A-claim).

4. Valid (this requires the conversion of an I-claim, which is valid).

7. Valid (the premise is the obverse of the conclusion).

10. Invalid (this conclusion does not follow, although its subcontrary, "Some people not allowed to play are people in uniform," does follow).

Exercise 8-10

1. The converse of (a) is the contradictory of (b), so (b) is false.

3. The contrapositive of (a) is a true O-claim that corresponds to (b); and that means that (b), its contradictory, is false.

5. Contrapose (a) to get "Some unproductive factories are not plants not for automobiles." Then obvert (b) to get "No unproductive factories are plants not for automobiles." Because (a) is true, (b) is undetermined.

9. The translation of (a) is "Some people enrolled in the class are not people who will get a grade." The obverse of the converse of (b) is "Some people enrolled in the class are not people who will get a grade." Wow! They're identical! So (b), too, is true.

Exercise 9-11

1. Valid:
   All P are G.
   No G are S.
   No S are P.

4. Invalid:
   All T are E.
   All T-T are E.   (T = times Louis is tired, etc.)
   All T-T are T.   (T-T = times identical with today)

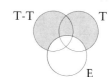

7. Valid:
   All H are S.
   No P are S.
   No P are H.

10. Invalid:
    All C are R.
    All V are C.
    No R are V.

    (Note: There is more than one way to turn this into standard form. Instead of turning nonresidents into residents, you can do the opposite.)

Exercise 8-12

1. No blank disks are disks that contain data.
   Some blank disks are formatted disks.
   Some formatted disks are not disks that contain data.
   Valid:

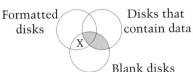

   Formatted disks — Disks that contain data — Blank disks

4. All tobacco products are substances damaging to people's health.
   Some tobacco products are addictive substances.
   Some addictive substances are substances damaging to people's health.

   Valid:

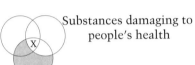

   Addictive substances — Substances damaging to people's health — Tobacco products

7. All people who may vote are stockholders in the company.
   Mr. Hansen is not a person who may vote.
   Mr. Hansen is not a stockholder in the company.

   Invalid:

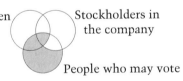

   Mr. Hansen — Stockholders in the company — People who may vote

Note: Remember that claims with individuals as subject terms are treated as A- or E-claims.

10. After converting, then obverting the conclusion:
    No arguments with false premises are sound arguments.
    Some arguments with false premises are valid arguments.
    Some valid arguments are not sound arguments.

    Valid:

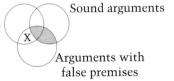

    Valid arguments — Sound arguments — Arguments with false premises

Exercise 8-13

1. A
4. B

Exercise 8-14

1. a
4. b

Exercise 8-15

1. 0
4. 1

Exercise 8-16

1. c
4. c
7. b
10. e

Exercise 8-17

1. All T are F.
   Some F are Z.
   Some Z are T.

   Invalid; breaks rule 2

4. There are two versions of this item, depending on whether you take the first premise to say *no* weightlifters use motor skills or only some don't. We'll do it both ways:

   All A are M.
   No W are M.
   No W are A.

   Valid

   All A are M.
   Some W are not M.
   No W are A.

   Invalid; breaks rule 3

7. Using I = people who lift papers from the Internet
   C = people who are cheating themselves
   L = people who lose in the long run

   All I are C.
   All C are L.
   All I are C.

   Valid

10. D = people who dance the whole night
W = people who waste time
G = people whose grades will suffer

All D are W.
All W are G.
All D are G.

Valid

Exercise 8-18 _____

(Refer to Exercise 8-11 for these first four items.)

2. (Given in standard form in the text)
Invalid: breaks rule 2

5. All voters are citizens.
Some citizens are not residents.
Some voters are not residents.
Invalid: breaks rule 2

7. All halyards are lines that attach to sails.
No painters are lines that attach to sails.
No painters are halyards.
Valid

8. All systems that can give unlimited storage . . . are systems with removable disks.
No standard hard disks are systems with removable disks.
No standard hard disks are systems that can give unlimited storage. . . .
Valid

(Refer to Exercise 8-12 for the next four items.)

2. After obverting both premises, we get:
No ears with white tassels are ripe ears.
Some ripe ears are not ears with full-sized kernels.
Some ears with full-sized kernels are not ears with white tassels.
Invalid: breaks rule 1

5. After obverting the second premise:
Some compact disc players are machines with 24x sampling.
All machines with 24x sampling are machines that cost at least $100.
Some compact disc players are machines that cost at least $100.
Valid

7. All people who may vote are people with stock.
No [people identical with Mr. Hansen] are people who may vote.
No [people identical with Mr. Hansen] are people with stock.
Invalid: breaks rule 3 (major term)

8. No off-road vehicles are vehicles allowed in the unimproved portion of the park.
Some off-road vehicles are not four-wheel-drive vehicles.

Some four-wheel-drive vehicles are allowed in the unimproved portion of the park.
Invalid: breaks rule 1

Exercise 8-19 _____

1. A = athletes; B = baseball players;
C = basketball players
Some A are not B.
Some B are not C.
Some A are not C.
Invalid: breaks rule 1

3. T = worlds identical to this one; B = the best of all possible worlds; M = mosquito-containing worlds
No B are M.
All T are M.
No T are B.
Valid

6. P = plastic furniture; C = cheap furniture;
L = their new lawn furniture
All L are P.
All P are C.
All L are C.
Valid

9. D = people on the district tax roll; C = citizens;
E = eligible voters
All D are C.
All E are C.
All D are E.
Invalid: breaks rule 2

12. C = people identical to Cobweb; L = liberals;
T = officials who like to raise taxes
All C are L.
All L are T.
All C are T.
Valid

17. P = poll results; U = unnewsworthy items;
I = items receiving considerable attention from the networks
All P are I.
Some P are U.
Some I are U.
Valid

18. E = people who understand that the Earth goes around the Sun; W = people who understand what causes winter and summer; A = American adults
All W are E.
Some A are not E.
Some A are not W.
Valid

20. N = the pornographic novels of "Madame Toulouse"; W = works with sexual depictions patently offensive to community standards and with no serious literary, artistic, political, or

scientific value; O = works that can be banned as obscene since 1973

All O are W.
All N are W.
All N are O.
Invalid: breaks rule 2

Exercise 8-20

1. True. A syllogism with neither an A- nor an E-premise would have (I) two I-premises, which would violate rule 2; or (II) two O-premises, which would violate rule 1; or (III) an I-premise and an O-premise. Alternative (III) would require a negative conclusion by rule 1, and a negative conclusion would require premises that distribute at least two terms, the middle term and (by rule 3) at least one other. Because an I-premise and an O-premise collectively distribute only one term, alternative (III) won't work either.

4. True. An AIE syllogism whose middle term is the subject of the A-premise breaks exactly two rules. If the middle term is the predicate of the A-premise, this syllogism breaks three rules.

Exercise 8-23

1. L = ladybugs; A = aphid-eaters; G = good things to have in your garden

   All L are A.
   [All A are G.]
   All L are G.

   Valid

4. S = self-tapping screws; B = boons to the construction industry; P = things that allow screws to be inserted without pilot holes

   All S are P.
   [All P are B.]
   All S are B.

   Valid

Chapter 9: Deductive Arguments II: Truth-Functional Logic

Exercise 9-1

1. Q → P
2. Q → P
3. P → Q
4. Q → P
5. (P → Q) & (Q → P)

Exercise 9-2

1. (P → Q) & R
2. P → (Q & R)

   Notice that the only difference between (1) and (2) is the location of the comma. But the symbolizations have two different truth tables, so moving the comma actually changes the meaning of the claim. And we'll bet you thought that commas were there only to tell you when to breathe when you read aloud.

5. P → (Q → R). Compare (5) with (3).

11. ~C → S

12. ~(C → S)

16. S → ~C. Ordinarily, the word *but* indicates a conjunction, but in this case it is present only for

emphasis—*only if* is the crucial truth-functional phrase.

20. ~(F ∨ S) or (~F & ~S). Notice that when you "move the negation sign in," you have to change the wedge to an ampersand (or vice versa). Don't treat the negation sign as you would treat a minus sign in algebra class, or you'll wind up in trouble.

Box: Sports Logic Symbolizations

(1) B → (C & (J ∨ L))

(2) ((D & J) → B) & ((F & A) → (B → N))

(3) ((W & L) & D) → B

Exercise 9-3

1.

P	Q	R	(P → Q)	(P → Q) & R
T	T	T	T	T
T	T	F	T	F
T	F	T	F	F
T	F	F	F	F
F	T	T	T	T
F	T	F	T	F
F	F	T	T	T
F	F	F	T	F

5.

P	Q	R	Q → R	P → (Q → R)
T	T	T	T	T
T	T	F	F	F
T	F	T	T	T
T	F	F	T	T
F	T	T	T	T
F	T	F	F	T
F	F	T	T	T
F	F	F	T	T

12.

C	S	C → S	~(C → S)
T	T	T	F
T	F	F	T
F	T	T	F
F	F	T	F

Exercise 9-4

1. Invalid:

		(Premise)	(Premise)	(Conclusion)
P	Q	~Q	P ∨ ~Q	~P
T	T	F	T	F
T	F	T	T	F
F	T	F	F	T
F	F	T	T	T

(Row 2)

4. Invalid:

			(Conclusion)	(Premise)		(Premise)
P	Q	R	(P → Q)	~(P → Q)	(Q → R)	P→(Q→R)
T	T	T	T	F	T	T
T	T	F	T	F	F	F
T	F	T	F	T	T	T
T	F	F	F	T	T	T
F	T	T	T	F	T	T
F	T	F	T	F	F	T
F	F	T	T	F	T	T
F	F	F	T	F	T	T

(Row 4)

7. Invalid:

			(Premise)		(Premise)	(Conclusion)
P	Q	R	~Q	P & R	(P & R) → Q	~P
T	T	T	F	T	T	F
T	T	F	F	F	T	F
T	F	T	T	T	F	F
T	F	F	T	F	T	F
F	T	T	F	F	T	T
F	T	F	F	F	T	T
F	F	T	T	F	T	T
F	F	F	T	F	T	T

(Row 4)

Exercise 9-5

We've used the short truth-table method to demonstrate invalidity.

1. Valid. There is no row in the argument's table that makes the premises all T and the conclusion F.

2. Invalid. There are two rows that make the premises T and the conclusion F. (Such rows are sometimes called "counterexamples" to the argument.) Here they are:

L	W	S	P
T	F	F	F
T	T	F	F

(Remember: You need to come up with only *one* of these rows to prove the argument invalid.)

3. Invalid. There are two rows that make the premises T and the conclusion F:

M	P	R	F	G
T	T	F	F	T
T	T	F	T	F

4. Invalid. There are three rows that make the premises true and the conclusion F:

D	G	H	P	M
F	T	T	T	T
F	T	F	T	T
F	T	F	F	T

5. Invalid. There are two rows that make the premises T and the conclusion F:

R	S	B	T	E
T	T	F	F	F
F	F	T	F	F

Exercise 9-6

1. Chain argument

2. Disjunctive argument

3. Constructive dilemma

4. Modus tollens

5. Destructive dilemma

Exercise 9-7

1. 1. R → P   (Premise)
   2. Q → R   (Premise) /∴ Q → P
   3. Q → P   1, 2, CA

4. 1. P → Q   (Premise)
   2. ~P → S   (Premise)
   3. ~Q   (premise) / ∴ S
   4. ~P   1, 3, MT
   5. S   2, 4, MP

7. 1. ~S   (Premise)
   2. (P & Q) → R   (Premise)
   3. R → S   (Premise) / ∴ ~(P & Q)
   4. ~R   1, 3, MT
   5. ~(P & Q)   4, 2, MT

10. 1. (T ∨ M) → ~Q   (Premise)
    2. (P → Q) & (R → S)   (Premise)
    3. T   (Premise) /∴ ~P
    4. T ∨ M   3,   ADD
    5. ~Q   1, 4, MP
    6. P → Q   2,   SIM
    7. ~P   5, 6, MT

Exercise 9-8

1. 4. 1,3, CA
   5. 2,   CONTR
   6. 4,5, CA

4. 4. 3,   CONTR
   5. 2,4, MP
   6. 2,5, CONJ
   7. 1,6, MP

Exercise 9-9

There is usually more than one way to do these.

1. 1. P & Q   (Premise)
   2. P → R   (Premise) /∴ R
   3. P   1,   SIM
   4. R   2, 3, MP

2. 1. R → S   (Premise)
   2. ~P ∨ R   (Premise) /∴ P → S
   3. P → R   2,   IMPL
   4. P → S   1, 3, CA

4. 1. ~P ∨ (~Q ∨ R)   (Premise)
   2. P   (Premise) /∴ Q → R
   3. P → (~Q ∨ R)   1,   IMPL
   4. ~Q ∨ R   2, 3, MP
   5. Q → R   4,   IMPL

8. 1. ~Q & (~S & ~T)   (Premise)
   2. P → (Q ∨ S)   (Premise) /∴ ~P
   3. (~Q & ~S) & ~T   1,   ASSOC
   4. ~Q & ~S   3,   SIM
   5. ~(Q ∨ S)   4,   DEM
   6. ~P   2,5, MT

Exercise 9-10

1. 1. P → R   (Premise)
   2. R → Q   (Premise) /∴ ~P ∨ Q
   3. P → Q   1, 2, CA
   4. ~P ∨ Q   3,   IMPL

4. 1. P ∨ (Q & R)   (Premise)
   2. (P ∨ Q) → S   (Premise) /∴ S
   3. (P ∨ Q) & (P ∨ R) 1,   DIST
   4. P ∨ Q   3,   SIM
   5. S   2,4, MP

7. 1. (M ∨ R) & P   (Premise)
   2. ~S → ~P   (Premise)
   3. S → ~M   (Premise) /∴ R
   4. P → S   2,   CONTR
   5. P   1,   SIM
   6. S   4,5, MP
   7. ~M   3,6, MP
   8. M ∨ R   1,   SIM
   9. R   7,8, DA

10. 1. P ∨ (R & Q)   (Premise)
    2. R → ~P   (Premise)
    3. Q → T   (Premise) /∴ R → T
    4. (P ∨ R) & (P ∨ Q)   1,   DIST
    5. P ∨ Q   4,   SIM
    6. ~P → Q   5,   DN/IMPL
    7. R → Q   2,6, CA
    8. R → T   3,7, CA

Exercise 9-11

1. D → ~B

4. B → ~D

7. C → (B & ~D)

Exercise 9-12

1. Equivalent to (b)

4. Equivalent to (c)

7. Equivalent to (c)

E-22 Answers, Suggestions, and Tips for Exercises Marked with a Triangle

Exercise 9-13

1. 1. P   (Premise)
   2. Q & R   (Premise)
   3. (Q & P) → S   (Premise) /∴ S
   4. Q   2,   SIM
   5. Q & P   1,4,   CONJ
   6. S   3,5,   MP

4. 1. P ∨ Q   (Premise)
   2. (Q ∨ U) → (P → T)   (Premise)
   3. ~P   (Premise)
   4. (~P ∨ R) → (Q → S)   (Premise) /∴ T ∨ S
   5. Q   1,3,   DA
   6. Q ∨ U   5,   ADD
   7. P → T   2,6,   MP
   8. ~P ∨ R   3,   ADD
   9. Q → S   4,8,   MP
   10. T ∨ S   1,7,9, CD

7. 1. P → Q   (Premise) /∴ P → (Q ∨ R)
   2. ~P ∨ Q   1,   IMPL
   3. (~P ∨ Q) ∨ R   2,   ADD
   4. ~P ∨ (Q ∨ R)   3,   ASSOC
   5. P → (Q ∨ R)   4,   IMPL

10. 1. (S → Q) → ~R   (Premise)
   2. (P → Q) → R   (Premise) /∴ ~Q
   3. ~R → ~(P → Q)   2,   CONTR
   4. (S → Q) → ~(P → Q)   1,3,   CA
   5. ~(S → Q) ∨ ~(P → Q)   4,   IMPL
   6. ~(~S ∨ Q) ∨ ~(~P ∨ Q)   5,   IMPL (twice)
   7. (S & ~Q) ∨ (P & ~Q)   6,   DEM/DN (twice)
   8. ((S & ~Q) ∨ P) & ((S & ~Q) ∨ ~Q)   7,   DIST
   9. (S & ~Q) ∨ ~Q   8,   SIM
   10. ~Q ∨ (S & ~Q)   9,   COM
   11. (~Q ∨ S) & (~Q ∨ ~Q)   10,   DIST
   12. ~Q ∨ ~Q   11,   SIM
   13. ~Q   12,   TAUT

Exercise 9-14

1. 1. P → Q   (Premise)
   2. P → R   (Premise) /∴ P → (Q & R)
   3. P   CP Premise
   4. Q   1,3, MP
   5. R   2,3, MP
   6. Q & R   4,5, CONJ
   7. P → (Q & R)   3–6, CP

4. 1. P → (Q ∨ R)   (Premise)
   2. T → (S & ~R)   (Premise) /∴ (P & T) → Q
   3. P & T   CP Premise
   4. P   3,   SIM
   5. T   3,   SIM
   6. Q ∨ R   1,4,   MP
   7. S & ~R   2,5,   MP
   8. ~R   7,   SIM
   9. Q   6,8,   DA
   10. (P & T) → Q   3–9,   CP

7. 1. P ∨ (Q & R)   (Premise)
   2. T → ~(P ∨ U)   (Premise)
   3. S → (Q → ~R)   (Premise) /∴ ~S ∨ ~T
   4. S   CP Premise
   5. Q → ~R   3,4,   MP
   6. ~Q ∨ ~R   5,   IMPL
   7. ~(Q & R)   6,   DEM
   8. P   1,7,   DA
   9. P ∨ U   8,   ADD
   10. ~~(P ∨ U)   9,   DN
   11. ~T   2,10, MT
   12. S → ~T   4–11, CP
   13. ~S ∨ ~T   12,   IMPL

10. 1. (P & Q) ∨ R   (Premise)
   2. ~R ∨ Q   (Premise) /∴ P → Q
   3. P   CP Premise
   4. ~Q   CP Premise
   5. ~R   2,4,   DA
   6. P & Q   1,5,   DA
   7. Q   6,   SIM
   8. ~Q → Q   4–7, CP
   9. Q ∨ Q   8,   IMPL
   10. Q   9,   TAUT
   11. P → Q   3–10, CP

Exercise 9-15

1. C → ~S
  ~L → S
  ――――――
  C → L
  Valid

4. ~M ∨ C
  ~M → ~K
  C ∨ H
  T → ~ H
  ――――――
  T → K
  Invalid

7. C ∨ S
  S → E
  C → R
  ――――――
  R ∨ E
  Valid

10. C → ~L
  (E → (~C → ~T)) & E
  ――――――
  L → ~T
  Valid

13. S → ~F
  ~S → ~T
  T
  ――――――
  ~F
  Valid

Chapter 10: Inductive Arguments

Exercise 10-1

1. a
4. b
7. a
10. b
13. b
16. b
19. a

Exercise 10-2

1. b
4. a
7. a
10. b

Exercise 10-3

1. a
4. a
7. a
10. a

Exercise 10-5

1. "Tests in this class" is perfectly clear, "hard" (the property in question) is a little bit vague.

4. "Tolerant of stress" (the property in question) is vague.

7. Weather in Iowa is pretty vague, and so is "weather that sucks."

10. Everything in this one is way too vague.

13. "Suspicious people" is pretty vague, and "quite unhappy" is too.

Exercise 10-6

1. a
4. a
7. b

9. d
12. c

Exercise 10-7

1. In order of increasing cautiousness, we'd say d, f, b, e, a

Exercise 10-8

1. Insufficiently cautious
4. Overly cautious

Exercise 10-9

1. This argument is fairly strong; the conclusion exhibits a reasonable amount of caution.

4. This makes the argument weaker; a relevant difference between the sample and the target.

8. This makes the argument stronger. Since we don't know the terrain of the target ride, a variety within the sample strengthens the argument.

10. If you considered the fact that bicycling is generally easier on flat land than on hilly land, you went beyond the criteria for inductive arguments in general. This is "specialized" knowledge and is part of a different argument: "This year's ride will be on flat land, while previous rides were in hilly terrain. Bicycling is easier on flat land. Therefore, this year's ride will be easier."

Exercise 10-10

1. More cautious
4. More cautious
7. Neither more nor less cautious; the fact of the eclipse is not significant to the property in question.
10. More cautious

Exercise 10-11

1. More cautious
4. Less cautious

Exercise 10-12 _____

1. The six students who turned in written evaluations

4. Generalization

6. No

8. It's not very strong. The sample is small, given that it's not random, and it's very likely to be unrepresentative: The students who bothered to write have relatively strong feelings about Ludlum one way or the other, and there is no reason to think that the spread of their opinions reflects the spread among Ludlum's students in general.

Exercise 10-13 _____

1. No. If he bet that exactly 60 percent held the belief, he would be allowing for no error margin whatsoever. If he makes that bet, take him up on it.

4. 66 percent.

7. With a sample of 100, he can safely bet that no more than 70 percent share that belief, because the error margin is now ± 10 percent.

Exercise 10-14 _____

1. The terms of the analogy are Tiger Woods and the speaker's child. The analogy is strong only if the speaker's child has other features in common with Tiger Woods, such as great powers of concentration, great physical ability, and a proper temperament for golf. It's unlikely that early lessons are as important as these traits.

4. The terms of the analogy are Senator Clinton's performance as a senator and her potential performance as president. There is a great deal more exposure as president and a greater number of constituencies to serve. Still, the similarities are significant. This conclusion gets more support from its premise than many in this exercise.

7. The terms of the analogy are Bush's ratings in Georgia and his ratings in Massachusetts. Since the political scene in these two states is very different, and since they have voted very differently in recent elections, this is not a very good argument.

10. The terms of the analogy are Warren's personal appearance and his care of the Barnes's house. These terms are sufficiently dissimilar—especially as regards their motivations—as to make for a weak argument. It *may* be that Warren's personal grooming will carry over to taking care of the house, but we suspect that many neat-appearing people live in messy abodes.

Exercise 10-15 _____

1. Biased generalization

4. Hasty generalization; quite likely biased, too

8. Hasty generalization; biased, too

10. Biased generalization

13. Refutation by hasty generalization

Chapter 11: Causal Arguments

Exercise 11-1 _____

1. Causal claim

4. Causal claim

7. Causal claim, although a very vague one

10. Not a causal claim

13. Not a causal claim

16. Not a causal claim

Exercise 11-2 _____

1. Effect: cat is not eating; cause: cat is eating mice

4. Effect: the little guy's not dehydrating; cause: giving him water

7. Effect: that people cannot detect their own bad breath; cause: becoming used to the odor

10. Effect: a savings to the state in court expenses; cause: judges' failure to process shoplifting, trespassing, and small-claims charges

Exercise 11-3 _____

1. a

4. b

7. a

10. b

13. a

16. b

19. a

Exercise 11-4

1. C; mowing the grass results in both fumes and grass dust.

4. C; shorter days contribute to both.

7. C; getting older can result in both conditions.

10. B; maybe smarter people eat more fish.

13. B; if there is more violence, there is likely to be more on TV.

16. A

19. A; C is also possible, since good health may have contributed both to Uncle Ted's attitude and to his longevity.

22. A; yes it could.

25. B; top executives can easily afford expensive clothes and nice cars.

Exercise 11-5

1. a

4. c

7. a

10. b, but the claim is vague

14. a

17. b

20. a

Exercise 11-9

1. There are are three causal hypotheses mentioned: One is that drinking wine weekly or monthly may cause dementia; a second is that drinking wine daily probably does *not* prevent dementia; and the third is that regular beer drinking is probably a cause of dementia.

The study is cause-to-effect, but the study is largely nonexperimental because of the self-selection of the experimental group(s)—i.e., the drinkers—and the control group—the nondrinkers. Nothing is mentioned of the nature or size of either group. The description of the study is quite vague. Although the source of the study appears to be a legitimate authority, the account given here would lead us to want more details of the study before we'd give more than a very tentative acceptance of the results.

4. Causal claim: Sleeping in a room with a light until age two is a cause of nearsightedness in later years. The study is nonexperimental, cause to effect. No differences between the experimental groups (children who slept with lights on) and the control group (children who slept in darkness). The differences in effect were 24 percent between night light and darkness, and 45 percent between a lamp and darkness. From what is reported, no problems can be identified.

Although the study is fairly small, the results indicate it is likely that there is a causal connection between the described cause and effect—a *d* of about 11 percent would be necessary in an experimental study; the higher numbers here help compensate for the nonexperimental nature of the study.

7. Causal claim: Exercise prevents colds. The study is a controlled cause-to-effect experiment, with one experimental group and two control groups. The first control group consists of the ten non-exercising volunteers; the second consists of the experimental group prior to the jogging program.

The experimental group had 25 percent fewer colds than the first control group and some non-indicated percent fewer than the second control group. We don't know enough about the groups and how they were chosen to tell if there are significant differences. Given the small size of the groups, a *d* of 40 percent is necessary to have statistical significance. The 25 percent figure is substantial and may indicate a causal connection, but it isn't enough to convince us to take up jogging.

Exercise 11-11

Our evaluations:

1. A, A, A

4. F, C, B

7. A, B, F

Exercise 11-12

1. This explanation is full of problems. It is untestable; its relevance is questionable (we couldn't have predicted blue eyes from a previous incarnation

unless we knew more about the incarnation, but the explanation is too vague to enable us to do that); it contains unnecessary assumptions; and it conflicts with well-established theory about how we get our eye color.

4. Reasonable explanation

7. Poor explanation; circular

10. Poor explanation; untestable (given the fact that subconscious desires are allowed); excessively vague; questionable relevance; conflicts with well-established theory

13. Reasonable explanation

17. Before we accepted this explanation of violence we'd want to consider alternatives: poverty, hopelessness, and discrimination, to name a few. Rap music and TV/movie violence may be reflections of violence rather than causes.

20. This poor explanation is untestable, unreliable, and extremely vague. The nondevout might add that it requires unnecessary assumptions.

Exercise 11-15

1. Explanation

4. Explanation

7. Explanation

10. Explanation

Exercise 11-16

1. Explanation

4. Explanation

5. Explanation used to excuse

Exercise 11-17

Of the first three items, numbers 2 and 3 use explanations to help justify an action (or omission, in the case of 3). Item 1 does not justify any behavior.

Exercise 11-19

1. This is just an explanation of how the garage got this cluttered.

4. This is an argument that Parker is giving a test today.

8. Host gives an explanation of why he can't shave. The explanation is intended to serve as a justification.

10. The speaker is explaining why this is a great movie, and in doing so is arguing that it is a great movie.

12. This is an explanation of why the contamination is getting worse and worse, and it might be used to argue that we should change what we allow people to do.

15. This is an explanation of why you are unable to sleep.

18. The speaker explains how touching the pins can fry the logic circuits and in doing so argues that if you touch the pins, you can expect this to be the result.

21. The author is explaining why women are worse off than men economically and in doing so is arguing that women are worse off. Don't be discouraged by the subtlety of real-life specimens.

25. The writer is explaining what the purpose of the Humane Society is and by doing so is arguing that the Humane Society is not an animal rights organization.

26. This is an explanation used as a justification.

Exercise 11-20

1. There is no argument present.

4. This is an explanation that *might* be used to justify the Rotary's behavior, but given the behavior in question, there is certainly no need to justify it.

7. Explanation used in an argument to justify making the 445 pages of report public.

10. This is an explanation that could easily *and mistakenly* be taken as an attempt to justify a molester's behavior.

Chapter 12: Moral, Legal, and Aesthetic Reasoning

Exercise 12-1

1. Nonfactual
4. Nonfactual
7. Nonfactual
10. Factual

Exercise 12-2

1. Not a value judgment, although it surely hints at one.
4. Value judgment
7. Not a value judgment in the ordinary sense, but since rides are often evaluated by degree of scariness, this may imply such a judgment.

Exercise 12-3

1. Not a moral value judgment
4. Moral value judgment
7. Not a moral value judgment
10. Moral value judgment

Exercise 12-4

2. People ought to keep their promises.
5. A mayor who takes bribes should resign.
7. Anyone who commits a third felony should automatically go to prison for twenty-five years.
8. Whatever is unnatural is wrong and should be avoided.

Exercise 12-5

1. Tory is being consistent in that what he is proposing for *both* sexes is that members of both should have the right to marry members of the *other* sex.
2. To avoid inconsistency, Shelley must be able to identify characteristics of art students, athletes, and children of alumni—for whom she believes special admissions programs are acceptable—and show that, unless they are also in one of the listed categories, these characteristics are not true of affirmative action students: women and members of minorities. Furthermore, the characteristics she identifies must be relevant to the issue of whether an individual should be admitted into the university. It may well be possible to identify the characteristics called for. (Remember that consistency is a necessary condition for a correct position, but it is not a sufficient one.)

3. Marin could be consistent only if he could show that the process of abortion involves killing and capital punishment does not. Because this is impossible—capital punishment clearly does involve killing—he is inconsistent. However, Marin's inconsistency is the result of his blanket claim that *all* killing is wrong. He could make a consistent case if he were to maintain only that the killing of *innocent* people is wrong, and that abortion involves killing innocent people but capital punishment does not. (Each of these last claims would require strong arguments.)

8. To avoid inconsistency, Harold would have to identify a relevant difference between the discrimination law and the marijuana law. In fact, there is one fairly obvious one to which he can appeal: The former has been declared contrary to the state constitution; the latter has not been alleged to be contrary to any constitution. So Harold may object to the failure to implement the latter on grounds that it, unlike the discrimination law, is unconstitutional. (It is a separate matter, of course, whether he can build a strong argument in the case of the marijuana law.)

Exercise 12-9

1. The harm principle: Shoplifting harms those from whom one steals.
2. The harm principle: Forgery tends to harm others.
4. We think the offense principle is the most relevant, because the practice in question is found highly offensive by most people (at least we believe—and hope—so). But one might also include the harm principle, because spitting in public can spread disease-causing organisms.
6. Legal moralism, because many people find adultery immoral; and, to a lesser extent, both the harm principle and legal paternalism, because adultery can increase the spread of sexually transmitted diseases.
10. The offense principle.

Exercise 12-12

1. a. Principle 4
   b. Principle 2
   Compatible

4. a. Principle 5
   b. Principle 2
   Compatible

Exercise 12-13

1. Relevant on Principle 7

4. Relevant on Principle 1

7. Relevant on Principle 3

Exercise 12-14

Principle 1: June's picture does not teach us anything, for no chimp can distinguish between truth and falsity; it is a curiosity rather than a work of art.

Principle 2: By looking at June's very symbolic paintings, we are compelled to accept her vision of a world in which discourse is by sight rather than by sound.

Principle 3: Perhaps the most far-reaching impact of June's art is its revelation of the horrors of encaging chimps; surely beings who can reach these heights of sublimely abstract expression should not see the world through iron bars.

Principle 4: Dear Zookeeper: Please encourage June to keep painting, as the vibrant colors and intense brushstrokes of her canvases fill all of us with delight.

Principle 5: I never thought I would wish to feel like a monkey, but June's art made me appreciate how chimps enjoy perceiving us humans as chumps.

Principle 6: This is not art, for no monkey's product can convey the highest, most valuable, human states of mind.

Principle 7: Whether by the hand of monkey or man, that the canvases attributed to June show lovely shapes and colors is indisputable.

Principle 8: What is art is simply what pleases a person's taste, and June obviously finds painting tasty, as she tends to eat the paint.

Exercise 12-15

1. a

4. b

7. b

Credits

Index

About the Authors

Both Moore and Parker have taught philosophy at California State University, Chico, for more years than they care to count. Aside from courses in logic and critical thinking, Moore also tries to teach epistemology and analytic philosophy. He is also chair of the department and once was selected as the university's Outstanding Professor. Parker's other teaching duties include courses in the history of modern philosophy and philosophy of law; he has chaired the academic senate and once upon a time was dean of undergraduate education.

Moore majored in music at Antioch College; his Ph.D. is from the University of Cincinnati. For a time he held the position of the world's most serious amateur volleyball player. He and Marianne currently share their house with three large dogs. Moore has never sold an automobile.

Parker's undergraduate career was committed at the University of Arkansas; his doctorate is from the University of Washington. Single now, he has been married to three wonderful women (not at the same time), is a serious amateur flamenco guitarist, drives two British automobiles and a motorcycle (also not at the same time), and goes to Spain when he gets the chance.

Moore and Parker have been steadfast friends through it all.

Brooke Noel Moore and Richard Parker, not necessarily in the order pictured above.